2019 ROCK DYNAMICS SUMMIT

PROCEEDINGS OF THE 2019 ROCK DYNAMICS SUMMIT (RDS 2019), OKINAWA, JAPAN, 7–11 MAY 2019

2019 Rock Dynamics Summit

Edited by

Ömer Aydan
University of the Ryukyus, Okinawa, Japan

Takashi Ito
University of the Ryukyus, Okinawa, Japan

Takafumi Seiki
Utsunomiya University, Japan

Katsumi Kamemura
Fukada Geological Research Institute, Tokyo, Japan

Naoki Iwata
Chuden Engineering Consultants Co., Ltd., Hiroshima, Japan

CRC Press
Taylor & Francis Group
Boca Raton London New York

CRC Press is an imprint of the
Taylor & Francis Group, an **informa** business

A BALKEMA BOOK

Published by:
CRC Press/Balkema
P.O. Box 447, 2300 AK Leiden, The Netherlands
e-mail: Pub.NL@taylorandfrancis.com
www.crcpress.com – www.taylorandfrancis.com

First issued in paperback 2023

ISBN: 978-1-03-257087-7 (pbk)
ISBN: 978-0-367-34783-3 (hbk)
ISBN: 978-0-429-32793-3 (ebk)

DOI: https://doi.org/10.1201/9780429327933

Typeset by Integra Software Services Pvt. Ltd., Pondicherry, India

Table of contents

T2: Estimation Procedures and Numerical Techniques of Strong Motions Associated with the Earthquakes

T3: Dynamic Response and Stability of Rock Foundations, Underground Excavations in Rock, Rock Slopes, and Stone Masonry Historical Structures

T4: Induced Seismicity

T5: Dynamic Simulation of Loading and Excavation

T6: Blasting and machinery induced vibrations

T7: Rockburst, Outburst, Impacts

T8: Nondestructive Testing Using Shock Waves

T9: Case Histories of Failure Phenomenon in Rock Engineering

2019 Rock Dynamics Summit– Aydan et al. (eds)
© *2019 Taylor & Francis Group, London, ISBN 978-0-367-34783-3*

Preface

The 2019 Rock Dynamics Summit in Okinawa (abbreviated as 2019RDS) as a specialized conference of the International Society for Rock Mechanics and Rock Engineering (ISRM) is held in Bankoku Shinryokan on Busena Point in Nago, one of the most scenic spots in Okinawa (Japan), surrounded by the beautiful emerald ocean and clear blue sky and hosted 2000 Kyushu-Okinawa G-8 Summit.

The 2019 Rock Dynamic Summit in Okinawa is jointly organized by the Rock Dynamics Committee of Japan Society of Civil Engineers (JSCE-RDC) and Japan Society for Rock Mechanics (JSRM) and supported by the International Society for Rock Mechanics and Rock Engineering (ISRM) and Turkish National Society for Rock Mechanics (TNSRM).

Rock dynamics involves diverse research scopes ranging from earthquake engineering, blasting, impacts, failure of rock engineering structures as well as the occurrence and prediction of earthquakes, induced seismicity, rockbursts and non-destructive testing and explorations. Rock dynamics has wide applications in civil and infrastructural, resources and energy, geological and environmental engineering, geothermal energy, and earthquake hazard management. This topic has become one of the important area of research and engineering activities in Rock Mechanics and Rock Engineering. The 2019 Rock Dynamic Summit in Okinawa is organized to address scientific research as well as engineering applications.

There are 8 keynotes related to each theme of the summit by the eminent speakers involved in various aspects of Rock Dynamics. The abstract submission was open from April 18, 2018 till September 12, 2018. During this duration, 163 abstract were submitted and they were, at least, reviewed by two reviewers from the reviewing committee of the 2019RDS consisting of members from the organizing committee and international scientific and advisory committee of the 2019RDS. Finally, we have accepted 128 full-papers, which were also peer-reviewed by the reviewing committee. The 2019 Rock Dynamics Summit in Okinawa is hoped to be the representative forum of the science and engineering of Rock Dynamics and to be a milestone for the advance of Rock Mechanics and Rock Engineering.

2019 Rock Dynamics Summit– Aydan et al. (eds)
© 2019 Taylor & Francis Group, London, ISBN 978-0-367-34783-3

Introduction

Rock dynamics has become one of the most important topics in the field of rock mechanics and rock engineering. The spectrum of rock dynamics is very wide and it includes, the failure of rocks, rock masses and rock engineering structures such as rockbursting, spalling, popping, collapse, toppling, sliding, blasting, non-destructive testing, geophysical explorations, and, impacts.

The 2019RDS is held at the Bankoku Shinryokan in Okinawa between May 7 and 11, 2019 and it is devoted to topics on the dynamic behavior of rock and rock masses and scientific and engineering applications. The themes include:

T1 Laboratory tests on Dynamic Responses of Rocks and Rock Masses; Fracturing of Rocks and Associated Strong Motions

T2 Estimation Procedures and Numerical Techniques of Strong Motions Associated with the Rupture of Earth's Crust and Some Strong Motion

T3 Dynamic Response and Stability of Rock Foundations, Underground Excavations in Rock, Rock Slopes Dynamic Responses and Stability of Stone Masonry Historical Structures and Monuments

T4 Induced Seismicity

T5 Dynamic Simulation of Loading and Excavation

T6 Blasting and machinery induced vibrations

T7 Rockburst, Outburst, Impacts

T8 Nondestructive Testing Using Shock Waves

T9 Case Histories of Failure Phenomenon in Rock Engineering

This e-book include 128 full-papers related one or several themes listed above an they represent the state of art and current achievements on Rock dynamics. It is hoped that the 2019RDs would be a mile-stone in advancing the knowledge in this field and leading to the new techniques for experiments, analytical and numerical modelling as well as monitoring in dynamics of rocks and rock engineering structures.

Acknowledgements

The 2019 Rock Dynamics Summit in Okinawa could not have been successful without the great efforts and tremendous help and efforts of the members of the organizing committee, the steering committee and supporting members, which are gratefully acknowledged. The generous financial support offered by Okinawa Convention Bureau (OCVB), Japan Society of Civil Engineers (JSCE), Kajima Foundation, Shimizu Construction Company as Platinum sponsors, JUTEN, ESA-RAAX, Nishimatsu Construction Co., Okabe Maintenance Co. & Green Earth NPO as gold sponsors, TAISEI Construction Co., MIDAS, NIIMO, Choda Co. - Kiso Jiban Consultant Co. as silver sponsors. The exhibitors KAJIMA Co., Okinawa Construction Consultants Association, and SHIMADZU, Crustal Engineering and Disaster Prevention Research Association besides sponsors and supporters were of great financial help to hold the 2019RDS. We gratefully acknowledge our sponsors, supporters and exhibitors. We sincerely acknowledge our reviewing committee consisting of members from the organizing and steering committees and international scientific and advisory committee for peer review of the abstracts as well as full-papers. The great help and tremendous efforts and understanding of Ms. A. Nago and her staff at Bankoku Shinryokan-MICE, Ms. A Kaneshiro of OCVB-MICE with the spirit of "Chimu Gukuru", Ms. M. Morita and K. Tamaki of OTS and Prof. Yaga of the University of the Ryukyus are sincerely acknowledged. Furthermore, the Busena Terrace Hotel offered a special rate for the participants of the 2019RDS. Last but not the least, none of the conferences and symposia like the 2019RDS could not be successful without the attendance of participants and presenters. We thank you all for making this unique event successful and we wish you all prosperous and good health for years to come.

Prof. Dr. Ömer AYDAN
Chairman

Prof. Dr. Takashi KYOYA
Co-Chairman

Emeritus Prof. Dr. Toshikazu KAWAMOTO
Honorary Chairman

2019 Rock Dynamics Summit– Aydan et al. (eds)
© *2019 Taylor & Francis Group, London, ISBN 978-0-367-34783-3*

Committees

ORGANIZING COMMITTEE

Ömer Aydan, University of the Ryukyus, Chairman
Takashi Kyoya, Tohoku University, Co-Chairman
Takafumi Seiki, Utsunomiya University, Secretariat

Members:
Tomoyuki Aoki (Taisei Corporation)
Takashi Ito (University of the Ryukyus)
Takatoshi Ito (Tohoku University)
Naoki Iwata (Chuden Engineering Consultants Co., Ltd.)
Katsumi Kamemura (Fukada Geological Research Institute)
Yuzo Obara (Kumamoto University)
Yuzo Ohnishi (Kyoto University, Former Vice-President at large of ISRM)
Yoshimi Ohta (Japan Nuclear Safety Regulation)
Tetsuji Okada (Central Research Institute of Electric Power Industry)
Tetsuo Okuno (Shimizu Corporation)
Masahiko Osada (Saitama University)
Yasunori Otsuka (Japan Society of Engineering Geology)
Koichi Shin (Central Research Institute of Electric Power Industry)
Naohiko Tokashiki (University of the Ryukyus)
Reşat Ulusay (Hacettepe University)
Atsushi Yokoo (Kajima Corporation)
Jun Yoshida (Suncoh Consultants Co., Ltd.)

INTERNATIONAL ADVISORY & SCIENTIFIC COMMITTEE

Leandro Alejano (Spain)
Aydin Bilgin (Turkey)
Tore Borvik (Norway)
Ezio Cadoni (Switzerland)
Evgeny Dolmatov (Russia)
Xia-Ting Feng (China)
Melih Geniş (Turkey)
Giovanni Graselli (Canada)
Seokwon Jeon (Korea)
Sang Ho Cho (Korea)
Petr Konicek (Czech Republic)
Suseno Kramadibrata (Indonesia)
Stanislav Lenart (Slovenia)
Jose Muralha (Portugal)
D.J.M. Ngan-Tillard (The Netherlands)
Eda Quadros (ISRM President)
Yuri Petrov (Russia)
Antonio Samaniego (Peru)

Wulf Schubert (Austria)
Thomas R. Stacey (South Africa)
Mostafa Sharifzadeh (Australia)
Norikazu Shimizu (Japan)
Giovanni Barla (Italy)
Antonio Bobet (USA)
Ismet Canbulat (Australia)
Feng Dai (China)
Sevket Durucan (UK)
Sergio Fontura (VP, ISRM)
R.K. Goel (India)
Yossef Hatzor (Israel)
Erik Johanson (Finland)
Tohid Kazerani (Switzerland)
Heinz Konietzky (Germany)
Louis Lamas (ISRM Secretary)
Charlie C. Li (Norway)
Pawel Nawrocki (UAE)
Uğur Özbay (USA)
Frederic Pellet (France)
Laura J. Pyrak-Nolte (USA)
Serkan Saydam (Australia)
Kourosh Shahriar (Iran)
Alexandros Sofianos (Greece)
Yingxin Zhou (Singapore)
Yujing Jiang (Japan)

STEERING COMMITTEE

Ömer Aydan, (University of the Ryukyus)
Takashi Kyoya, (Tohoku University)
Takafumi Seiki, (Utsunomiya University)
Seiji Ebisu (Murasakikensetsu Co., Ltd.)
Takaaki Ikeda (Nagaoka University of Technology)
Takashi Ito (University of the Ryukyus)
Takatoshi Ito (Tohoku University)
Naoki Iwata (Chuden Engineering Consultants Co., Ltd.)
Katsumi Kamemura (Fukada Geological Research Institute)
Shinjuro Komata (Nippon Koei Co., Ltd.)
Tomofumi Koyama (Kansai University)
Atsushi Kusaka (Public Works Research Institute)
Guichen Ma (Oyo Corpration)
Satoshi Mori (Newjec Inc.)
Kenichi Nakaoka (Obayashi Corporation)
Tsuyoshi Nishimura (Tottori University)
Yasunori Ohtsuka (Engineering Geology Society of Japan)
Yoshimi Ohta (Nuclear Regulation Authority, Japan)
Tetsuji Okada (Central Research Institute of Electric Power Industry)
Toshinori Sato (Japan Atomic Energy Agency)
Yuya Suda (University of the Ryukyus)

Naohiko Tokashiki (University of the Ryukyus)
Jun Tomiyama (University of the Ryukyus)
Masaaki Wani (Chubu Electric Power Co., Inc.)
Jun Yoshida (Suncoh Consultants Co., Ltd.)

Scientific Sub-Committee:
Ö. Aydan (UR), T. Ito (UR), T. Ito (TU)), Koyama, T. Seiki, Y. Ohnishi, T. Kyoya, T. Nishimura, G. C. Ma, T. Ikeda

Organizing Sub-Group:
T. Tokashiki, J. Tomiyama, Y. Suda, N. Iwata, M. Wani, T. Okada, T. Sato, S. Mori, K. Nakaoka, S. Ebisu, Yoshida, S. Komata, Y. Ohta

General Affairs Sub-Group:
T. Kyoya, T. Seiki, N. Iwata, K. Kamemura

Supporting Members:
A. Nago (Bankoku Shinryokan-MICE), A. Kaneshiro (OCVB-MICE), K. Tamaki (OTS), Morita (OTS), Prof. M. Yaga (UR)

2019 Rock Dynamics Summit– Aydan et al. (eds)
© 2019 Taylor & Francis Group, London, ISBN 978-0-367-34783-3

Venue

The summit is held in **Bankoku Shinryokan** on Busena Point, one of the most scenic spots in Okinawa (Japan), surrounded by the beautiful emerald ocean and clear blue sky and hosted 2000 Kyushu-Okinawa G-8 Summit.

2019 Rock Dynamics Summit– Aydan et al. (eds)
© 2019 Taylor & Francis Group, London, ISBN 978-0-367-34783-3

Sponsors

PLATINUM

Shimizu Corporation

GOLD

JUTEN (Japan Cavity Filling Association), ESA-RAAX (Earth Scanning Research Association & RAAX), Okabe Maintenance- Green Earth NPO, Nishimatsu Const. Co.

SILVER

TAISEI Corporation, MIDAS, NIIMO, CHODAI-KISO JIBAN, OYO Co.

SUPPORTERS

Okinawa Convention Bureau, Japan Society of Civil Engineers, Kajima Foundation, Okinawa Prefectural Gov., Society for the Promotion of Construction Engineering

Keynotes

2019 Rock Dynamics Summit– Aydan et al. (eds)
© *2019 Taylor & Francis Group, London, ISBN 978-0-367-34783-3*

Application of high performance computing for earthquake hazard and disaster estimation

M. Hori

Earthquake Research Institute, The University of Tokyo, Tokyo, Japan
Department of Mathematical Science and Advanced Technology, JAMSTEC, Yokohama, Japan

ABSTRACT: Large-scale numerical simulation is being enabled by utilizing high performance computing. A socially important target is the estimation of earthquake hazard and disaster. Integrated Earthquake Simulation is being developed to this end. This simulation combines the ground motion simulation and the urban are seismic response simulation. A parallel finite element method has been developed so that an analysis model of a few ten trillion degree-of-freedoms can be solved. An analysis model is automatically constructed for an urban area using available data resources.

1 INTRODUCTION

Systems have been developed for the estimation of earthquake hazard and disaster; as shown in Figure 1, the methodology of the estimation is being shifted from the empirical to the simulation (Climellaro et al. 2014, Sahin et al. 2016, Lu & Guan, 2017, Hori 2018). They share the following two elements: 1) an empirical attenuation equation for ground motion; and 2) fragility curves for structure damage. These elements provide a unique solution for the estimation of earthquake hazard and disaster in an urban area of a few kilometers in which more than ten thousand structures are located.

There are new technologies that could be an alternative of the two elements. Numerical analysis of earthquake wave propagation is used for the estimation of ground motion distribution (Bao et al. 1996). Many numerical analysis methods are available for structural seismic responses analysis.

To apply these alternatives to the estimation of earthquake hazard and disaster, we need to enhance them with high performic computing (HPC) capability. For instance, required is a finite element method (FEM) that can solve an analysis model of a few billion of degree-of-freedom (DOF) for a target crust structure or a target ground structure (Quinay et al. 2013, Agata et al. 2016). Also, required is an automated model construction of a suitable analysis model for structures the number of which is of the order of 100,000 or 1,000,000.

The author and his colleagues have been developing a system, called *Integrated Earthquake Simulation* (IES), for the estimation of earthquake hazard and disaster enhanced with HPC. IES integrates numerical analysis methods, together with modules for the automated model construction. The representative analysis methods which are implemented in the current IES are those for the ground

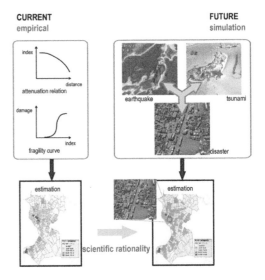

Figure 1 Shift of methodology for the estimation of earthquake hazard and disaster.

motion simulation and the urban area seismic response simulation.

This paper summarizes recent achievements of developing IES, which are made by using a large-scale parallel computer such as K computer in Japan (Miyazaki et al. 2012). The contents of the present paper are organized as follows: First, in Section 2, we briefly explain the key features of IES. In Section 3, we discuss a parallel FEM to which a conjugate gradient (CG) method is implemented to analyze a large-scale analysis model (Saad, 2003). Examples of using IES for the earthquake hazard and disaster assessment are presented in Section 4. The target is Tokyo Metropolis.

2 KEY FEATURES OF IES

For the estimation of earthquake hazard, a large-scale numerical computation is needed to analyze a three-dimensional model by using HPC. A model of more than 1,000,000,000,000 DOF can be analyzed by using a parallel FEM which is implemented in IES. The number of time steps is of the order of 10,000, since the time increment, and the time duration are 0.01 s and 100 s, respectively. We must mention that numerical analysis of this class of a model is a challenge in the field of HPC

A program of the automated model construction is implemented in IES. In general, automated model construction is regarded as data conversion, in the sense that data stored in several data resources are processed to data which correspond to requested analysis models. There are no single data resource for material and mechanical properties for all structures. We thus must guess them in constructing analysis models for seismic response analysis.

2.1 Parallel FEM Implemented In Ies

FEM is suitable to ground motion simulation, since a major concern of the simulation is the identification of sites at which larger ground motion is concentrated due to the topographical effects (Ichimura et al. 2007, Ichimura et al. 2009, Ichimura et al. 2014, Ichimura et al. 2014b, Ichimura et al. 2015). An analysis model of high fidelity is needed, and DOF of the model is inevitably large. Therefore, parallel FEM is a unique solution for the ground motion simulation; parallel FEM is used for the structural seismic response analysis (Miyamura et al. 2015).

The major portion of the numerical computation of parallel FEM is used in solving a matrix equation for unknown displacement,

$$
\left(\frac{4}{dt^2} M + \frac{2}{dt} C^n + K^n \right) \delta u^n = F^n - Q^{n-1}
$$
$$
+ C^n v^n + M \left(a^n + \frac{4}{dt} v^{n-1} \right)
\tag{1}
$$

where u, v, a, f and q are displacement, velocity, acceleration, external force and residual force vectors, respectively, M, C and K are mass, damping and stiffness matrices, dt is the time increment, and superscript n stands for the n-th time step. A simple Ramberg-Osgood model and Masing rule are used for k, and Rayleigh damping is used for C (Idriss et al. 1978, Masing 1926).

Parallel FEM has the solver based on the CG method that is tuned for K computer; the strong scalability is almost ideal, and the peak performance is 10% to 15% of K computer. The following two are the major tunings: 1) the geometric multi-grid which uses coarse and fine solutions (a coarse solution of

less DOF and serves an initial solution for a fine solution of full DOF); and 2) the mixed precision arithmetic which uses single and double precision for the coarse and fine solution, respectively.

These two tunings contribute high scalability of parallel FEM, as presented; as the number of CPU cores increase, the CPU time decreases linearly. Other tunings, such as an element-by-element method, compressed row storage and higher order predictor are made for the improvement of the scalability (Fujita et al. 2016).

2.2 Automated Model Construction Of Ies

The automated model construction consists of interpreting data stored in data resources and converting data to data of an analysis model (Ohtani et al. 2014, Hori et al. 2015, Jayasinghe et al. 2015). A program of taking these two steps is developed for IES. While the key simulations are the ground motion simulation and the urban area seismic response simulation, IES is implementing the crustal-scale seismic wave propagation simulation and the tsunami inundation simulation, as well as social-response-against-disaster simulation such as mass evacuation simulation, traffic simulation and economic activity simulation. Hence, the program of the automated model construction ought to be flexible so that it can handle models of various numerical analysis methods.

Data resources currently available are commercial Geographical Information System (GIS) or a set of inventories operated by local government. The commercial GIS has data of external configuration of structures, and the inventories have data used for specific purposes such as the registration of real estate, the structure type and construction year.

Interpreting data is made by understanding the meta-data of the data resource. The meta-data means the structure and property of attributes of the data resource; there are cases where an attribute consists of a few attributes, which makes the meta-data complicated. Location information is a key attribute. The automated model construction program uses this attribute and links an item in different data resource to one common structure.

The difficulty of converting the interpreted data to data of an analysis model depends on the complexity of the model. For a simpler analysis model, the interpreted data are converted to a fewer model parameters. The simplest model for the structural seismic response analysis is a linear one-degree-of-freedom system with two model parameters, a mass and a stiffness.

The accuracy of the model parameters determines the quality of the model that is automatically constructed. To increase the accuracy, a natural frequency, a key characteristic of a structure, is used. In IES, the natural frequency of the analysis model is compared with an empirical relation between the natural frequency and the structure height.

We do not expect that complete information is available in automatically constructing an analysis model. To account for the limitation of available data, IES can construct 10,000 or more analysis models for a structure, suitably varying a set of model parameters. It is another challenge of HPC to construct and analyze numerous models for one target problem considering the uncertainty of the model; the number of analysis models reaches 10,000,000,0000 if IES analyzed 1,000,000 structures using 10,000 analysis models for each.

3 CG METHOD OF PARALLEL FEM

A key feature of parallel FEM is a solver, which solves a matrix equation; for simplicity, we convert Eq. (1) to the following matrix equation of a simpler form:

$$Ax = y, \qquad (2)$$

where x is an unknown vector, and A are y given matrix and vector, respectively, and the dimension of A is N (which coincides with DOF). The CG method employed in the parallel FEM is categorized as an iterative solver, which seeks a solution from a given initial solution in an iterative manner. Most of FEM uses a direct solver such as the Gauss Zeidel method which can provide the exact solution of Eq. (2) within the accuracy of numerical computation. The CG method does not provide the exact solution. The solution of the CG method is close to the exact solution, and the difference between them is practically negligible.

A major advantage of employing the CG method as a solver is that the number of computing processes in solving Eq. (1) by the CG method increases linearly to N, while that by the direct method increases linearly to N^2. That is, the CPU time of the CG method is linear to N, while the CPU time of the direct method is linear to N^2; the CPU time becomes 10 or 100 times larger for the CG method and the direct method, respectively, if N becomes 10 times larger. Therefore, the use of the CG method as a solver of FEM is inevitable for an analysis model of N which is the order of 1,000,000 or larger.

The increase of the CPU time becomes larger if the iteration number of the CG method increases. In general, as the condition number of the matrix A (the ratio of the maximum and minimum eigen/singular-values) increases, the iteration number increases for A of the same dimension N. If Eq. (2) is multiplied by A^{-1}, the inverse matrix of A, the resulting matrix equation has the matrix of the condition number 1. Thus, the CG method seeks to find a matrix which is close to A^{-1}, which is called a pre-conditioning matrix.

As for a continuum at quasi-static state, A coincides with K, a stiffness matrix. Locally, a nodal displacement is separated to a rigid body motion (translation and rotation) and the remaining displacement which accompanies strain. Since an analysis domain is decomposed into a set of small domains in parallel computation, a smart way to decrease the iteration number of the CG method is to compute rigid body motion for each small domain. This domain decomposition technique is effective in reducing the iteration number or making fast computation of FEM.

On the interface between two neighboring small domains, rigid body motion is discontinuous. This discontinuity produces traction on the interface. It is expected that the produced traction is close to the exact traction on the interface; for each small domain, the traction computed by the discontinuity of the rigid body motion traction become traction boundary conditions, and the CG method solves the remaining displacement locally (without considering interactions effects from neighboring small domains).

As for a continuum at dynamic state, the domain decomposition technique is applicable in reducing the iteration number. However, the rigid body motion that changes temporally generates acceleration, unlike the quasi-static state at which the rigid body motion generates no force terms. The domain decomposition technique might not be effective for the dynamic state.

At dynamic state, the matrix A of Eq. (2) is

$$A = \frac{4}{dt^2} M + \frac{2}{dt} C^n + K^n \qquad (3)$$

As a smaller time increment of dt is used, the mass matrix M becomes dormant in A. If a ramped mass matrix is used, M is diagonal, and it is easy to find a suitable pre-conditioning matrix for this A. Parallel FEM often uses smaller dt, which is effective in reducing the CPU time for non-linear problems.

An alternative method of reducing the iteration number of the CG method is to use an initial solution which is close to the exact solution. As mentioned, parallel FEM uses a coarse solution of less DOF to set the initial solution. The use of an analysis model of reduced fidelity to find a good initial solution is a unique approach in utilizing the CG method as a solver of FEM.

As for a structure at dynamic state, a displacement function is given as the sum of distinct dynamic modes. Every instance, modes which are dominant change because each mode has its own frequency. A coarse solution is thus required to include dominant modes in it.

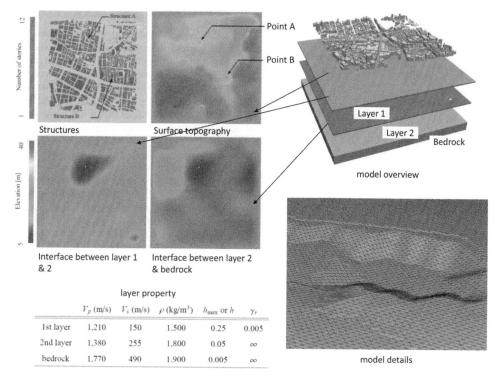

Figure 2. Analysis model of surface layers.

Table for layer property:

layer property	V_p (m/s)	V_s (m/s)	ρ (kg/m³)	h_{max} or h	γ_r
1st layer	1,210	150	1,500	0.25	0.005
2nd layer	1,380	255	1,800	0.05	∞
bedrock	1,770	490	1,900	0.005	∞

4 EXAMPLE OF IES

4.1 *Ground motion simulation*

An analysis model of surface layers is presented in Figure 2; the domain is 1,250 x 1,250 m, and consists of three layers including bedrock. The number of nodes, elements, and degree-of-freedom are 340,876,783, 252,737,051, and 1,022,630,349, respectively, so that the spatial and temporal resolution are up to 10 m and 10 Hz, respectively.

Figure 3 presents the distribution of SI, which is defined as

$$SI = \frac{1}{2.4} \int_{0.1}^{2.5} S_v(T)\,dT, \quad (4)$$

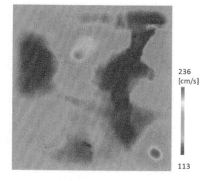

Figure 3. Distribution of SI.

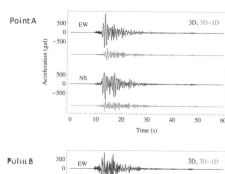
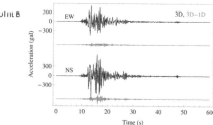

Figure 4. Comparison of 3D-analysis and 1D-analysis of ground motion amplification process.

with S_v being the velocity response spectra; SI is the average of the velocity response taken over 0.1 sec and 2.4 sec (Housner, 1955). Ground motion of JR Takatori is used as input seismic wave. As is seen, the distribution of SI is far from being uniform, which is due to the topographical effects of the three layers of complicated configuration. There are two

6

0.02
[rad.]

0.00

Figure 5 Distribution of MSDA.

sites at which SI is concentrated, and the value of SI exceeds 200 kine.

The results of parallel FEM are compared with the conventional analysis that uses a one-dimensional (1D) stratified model. In Figure 4, the waveform of acceleration in the EW and NS directions is presented; Points A and B indicated in Figure 4 are used. The waveform at Point B seems similar to each other; the largest difference is around 30 gal. However, the difference in the waveform at Point A is substantial, since the largest difference reaches 100 gal. Point A is located near the valley of the underground structure, and greater topographical effect leads to ground motion concentration.

4.2 Urban Area Seismic Response Simulation

There are 4,066 residential buildings in the area presented in Figure 2. It is assumed that the structure type of all the buildings are reinforced concrete for simplicity. A non-linear multi-degree-of-freedom system is made; the number of the mass is the floor number and a bi-linear spring is used. The stiffness for the linear relation is determined from an empirical equation between the first natural frequency and the building height.

The distribution of the maximum story drift angle (MSDA) is presented in Figure 5. As is seen, the distribution of MSDA is different from the distribution of SI shown in Figure 3; for instance, MSDA is relatively smaller at the two spots where SI is greater. Building models which have greater MSDA are located at sites where SI is not large. The discrepancy between the ground motion distribution (SI) and the structural response distribution (MSDA) is due to the difference in the dominant frequency of the ground motion and the natural frequency of the structure. As for the earthquake hazard estimation, the distribution of MSDA is more important.

Like the preceding subsection, we examine the necessity of making ground motion simulation of IES, for the estimation of structural seismic response. Identical analysis models of residential buildings are used, and input ground motion is either the one computed by using the ground motion simulation or the conventional 1D analysis. The results are presented in Figure 6; SI and MSDA, together

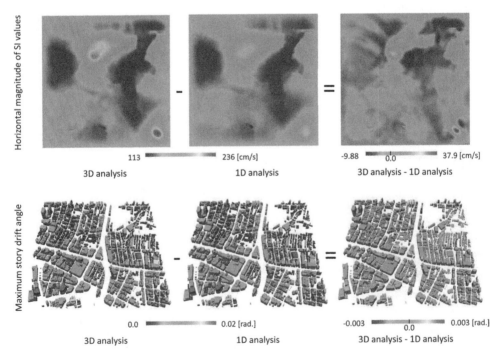

Horizontal magnitude of SI values

113 ▬▬ 236 [cm/s]   -9.88 ▬▬ 0.0 ▬▬ 37.9 [cm/s]

3D analysis   1D analysis   3D analysis - 1D analysis

Maximum story drift angle

0.0 ▬▬ 0.02 [rad.]   -0.003 ▬▬ 0.0 ▬▬ 0.003 [rad.]

3D analysis   1D analysis   3D analysis - 1D analysis

Figure 6. Comparison of structural seismic response due to different in-site ground motion synthesized by 3D-analysis (parallel FEM) and conventional 1D-analysiss.

7

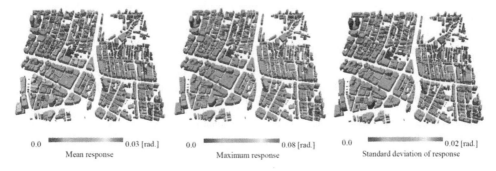

0.0 ▬▬▬▬ 0.03 [rad.]   0.0 ▬▬▬▬ 0.08 [rad.]   0.0 ▬▬▬▬ 0.02 [rad.]

Mean response   Maximum response   Standard deviation of response

Figure 7. Mean, maximum and standard deviation of maximum drift angle using 10,000 models for each residential building.

with the difference, are plotted. The difference for MSDA reaches 0.003, which might be crucial for the disaster estimation.

There is larger uncertainty in determining the strength of an analysis model for the residential buildings. We thus apply capacity computing of generating 10,000 models for each residential building, assuming a normal distribution of the strength; the mean is determined by using an empirical relation between the stiffness and the strength, and the standard deviation is assumed to be 10% of the mean. A typical distribution of MSDA for 10,000 analysis models is shown in Figure 7; the mean, the maximum and the standard deviation of MSDA are plotted. In general, A wider distribution of MSDA is observed for greater input ground motion, due to the non-linearity of the analysis models.

Effects of model uncertainty on the seismic response can be evaluated by the standard deviation of MSDA. For instance, providing the standard deviation of 10% of the mean for the structural strength results in the standard deviation of around 0.01 for MSDA. MSDA of this scale is crucial for the earthquake disaster estimation because structures whose the residual value of MSDA is greater than 0.01 could be regarded as a structure damaged by ground motion in Japan.

4.3 Combined Simulation Of Ground Motion And Urban Area Seismic Response Simulation

Using K computer, IES combines the ground motion simulation and the urban area seismic response simulation for a domain of 10,250 x 9,250 m. The ground motion simulation is regarded as capability computing since it needs the whole 705,024 compute cores of K computer, with time-to-solution being less than 12 hours. Used is a ground structure model similar to the one in Figure 2; DOF is 133,000,000,000 and the number of time steps is 6,600.

An example of the combined simulation is presented in Figure 8; there are 32,800 analysis models for residential buildings. The input ground

0 ▬▬▬▬ 250
SI [kine]

0. ▬▬▬▬ 0.02
MSDA [rad]

Figure 8 Combined simulation of ground motion amplification and structural seismic response analysis for 10 x 10 km area; gray color on ground is SI, and green-to-red color of building is MSDA.

motion is computed for an assumed earthquake scenario of Tokyo Metropolis Earthquake. In IES, the earthquake hazard and disaster are quantified in terms of SI and MSDA, respectively, and the distribution of these two indices is the result of the combined simulation.

New findings are net made in the combined simulation, because IES simply combines ground motion simulation to well-established structural seismic response analysis. However, applying the combined simulation to a large area, identified are sites of large SI or greater MSDA. These sites are worth being examined for more detailed investigation, in order to apply suitable disaster mitigation measures.

5 CONCLUDING REMARKS

In this paper, we present recent achievement of IES, which is enhanced with HPC. This system is used to make an estimation of earthquake hazard and disaster for Tokyo Metropolis for a given earthquake scenario. Similar estimation of IES can be made for any urban area if suitable data resources are available.

We are implementing in IES numerical analysis methods of social-response-against-disaster simulation. It is important to add an estimation of earthquake disaster response to the earthquake hazard and disaster estimation, since the estimation of social response contributes strengthening of the resilience of an earthquake.

REFERENCES

Agata, R., Ichimura, T., Hirahara, K., Hyodo, M., Hori, T. & Hori, M. 2016. Robust and portable capacity computing method for many finite element analyses of a high-fidelity crustal structure model aimed for coseismic slip estimation, *Computers & Geosciences*, 94: 121-130, 10.1016/j.cageo. 2016.05.015.

Bao, H., Bielak, J., Ghattas, O., O'Hallaron, D.R., Kallivokas, L.F., Shewchuk, J.R. & Xu, J. 1996. Earthquake ground motion modeling on parallel computers, *the 1996 ACM\ IEEE conference on Supercomputing*, Pittsburgh, USA, 10.1145/369028.369053.

Cimellaro, G.P., Renschler, C. & Bruneau, M. 2014. *Introduction to Resilience-Based Design, Computational Methods, Seismic Protection, Hybrid Testing and Resilience in Earthquake Engineering*, Springer International Publishing AG, New York, USA.

Fujita, K., Yamaguchi, T., Ichimura, T., Hori, M. & Maddegedara, L. 2016. Acceleration of Element-by-Element Kernel in Unstructured Implicit Low-order Finite-element Earthquake Simulation using OpenACC on Pascal GPUs, *Third Workshop on Accelerator Programming Using Directives (WACCPD, WACCPD' 16)*, 1-12, 10.1109/WACCPD. 2016.6.

Hori, M., Wijerathne, L., Ichimura, T. & Tanaka, S. 2015. Meta-Modeling for Constructing Model Consistent with Continuum Mechanics, *Journal of Japan Society of Civil Engineers*, A2, 71 (1),10.2208/journalofjsce.2.1_269.

Hori, M. 2018. *Introduction to computational earthquake engineering*, 3rd edition, Singapore, World Science.

Housner, G.W. 1952. Spectrum intensity of strong motion earthquakes, Proceedings of the Symposium on Earthquakes and Blast Effects on Structures, *Earthquake Engineering Research Institute*, California, 20-36.

Ichimura, T. Hori, M. & Kuwamoto, H. 2007. Earthquake Motion Simulation with Multi-Scale Finite Element Analysis on Hybrid Grid, *Bulletin of the Seismological Society of America*, 97: 1133-1143,10.1785/0120060175.

Ichimura, T., Hori, M. & Bielak, J. 2009. A Hybrid multiresolution meshing technique for finite element three-dimensional earthquake ground motion modeling in basins including topography, *Geophysical Journal International*, 177: 1221-1232, 10.1111/j.1365-246X.2009.04154.

Ichimura, T., Fujita, K., Hori, M., Sakanoue, T. & Hamanaka. R. 2014a. Three-dimensional nonlinear seismic ground response analysis of local site effects for estimating seismic behavior of buried pipelines, *Journal of Pressure Vessel Technology, American Society of Mechanical Engineers*, 136, 10.1111/j.1365-246X.2009.04154.x.

Ichimura, T., Fujita, K., Tanaka, S., Hori, M., Lalith, M., Shizawa, Y. & Kobayashi, H. 2014b. Physics-based urban earthquake simulation enhanced by 10.7 BlnDOF x 30 K time-step unstructured FE non-linear seismic wave simulation, *the International Conference on High Performance Computing, Networking, Storage and Analysis (SC'14)*" IEEE Computer Society Press, 15-26, 10.1109/SC.2014.7.

Ichimura, T., Fujita, K., Quinay, P.E.B., Maddegedara, L., Hori, M., Tanaka, S., Shizawa, Y., Kobayashi, H. & Minami, K. 2015. Implicit Nonlinear Wave Simulation with 1.08T DOF and 0.270T Unstructured Finite Elements to Enhance Comprehensive Earthquake Simulation, *the International Conference for High Performance Computing, Networking, Storage and Analysis (SC' 15)*, IEEE Computer Society Press, 10.1145/2807591.2807674.

Idriss, I.M., Singh, R.D. & Dobry. R. 1978. Nonlinear behavior of soft clays during cyclic loading, *Journal of the Geotechnical Engineering*, 104: 1427-1447.

Jayasinghe, J.A.S.C., Tanaka, S., Wijerathne, L., Hori, M. & Ichimura, T. 2015. Automated Construction of Consistent Lumped Mass Model for Road Network, *Journal of Japan Society of Civil Engineers*, 71(4): I_547-I_556, 10.2208/jscejseee.71.I_547.

Lu, X. & Guan, H. 2017. *Earthquake Disaster Simulation of Civil Infrastructures, From Tall Buildings to Urban Areas*, Springer, New York, USA.

Masing, G. 1926. Eigenspannungen und verfestigung beim messing (in German), *the 2nd International Congress of Applied Mechanics*, 332-335.

Miyamura, T., Akiba, H. & Hori, M. 2015. Large-scale seismic response analysis of super-high-rise steel building considering soil-structure interaction using K computer, *High Rise*, 4(1): 75-83, 10.1142/ S1793431116400145.

Miyazaki, H., Kusano, Y., Shinjou, N., Shoji, F., Yokokawa, M. & Watanabe, T. 2012. Overview of the K computer system, *FUJITSU Sci. Tech, J.*,48(3): 302-309.

Ohtani, H., Chen, J. & Hori, M. 2014. Automatic combination of the 3D shapes and the attributes of buildings in different GIS data, *Japan Society of Civil Engineers*, 70(2): I_631-I_639, 10.2208/jscejam.70. I631.

Saad, Y., 2003. *Iterative methods for sparse linear systems*, 2nd edition, New York, SIAM.

Sahin, A., Sisman, R., Askan, A. & Hori, M. 2016. Development of integrated earthquake simulation system for Istanbul, *Earth, Planets and Space*, 68: 115, 10.1186/s40623-016-0497-y.

Quinay, P.E.B., Ichimura, T., Hori, M., Nishida, A. & Yoshimura. S. 2013. Seismic Structural Response Estimates of a High-Fidelity Fault-Structure System Model Using Multiscale Analysis with Parallel Simulation of Seismic Wave Propagation, *Bulletin of the Seismological Society of America*, 103: 2094-2110, 10.1785/0120120216.

2019 Rock Dynamics Summit– Aydan et al. (eds)
© 2019 Taylor & Francis Group, London, ISBN 978-0-367-34783-3

Monitoring, assessment and mitigation of rock burst and gas outburst induced seismicity in longwall top coal caving mining

S. Durucan, W. Cao, W. Cai, J-Q, Shi, A. Korre & G. Si
Department of Earth Science and Engineering, Royal School of Mines, Imperial College London, London, UK

S.Jamnikar & J.Roser
Coal Mine Velenje, Velenje, Slovenia

ABSTRACT: Underground coal extractions lead to continuous stress and pressure redistribution around mine openings which, in some cases, may lead to coal and gas outburst and rock bursts. This paper presents seismic monitoring research, which aimed at characterising the dynamic behaviour of the coal seam in response to gateroad development and longwall top coal caving mining at Coal Mine Velenje. The early campaigns involved time-lapse active seismic tomography and microseismic monitoring at the two selected longwall panels. Based on the findings of this early work, a more comprehensive microseismic monitoring programme was set up. Seismic monitoring data were processed and used for model development and rock burst/gas outburst risk assessment purposes, which also led to the development of measures to mitigate against seismic hazards.

1 INTRODUCTION

Underground coal extractions lead to continuous stress and pressure redistribution around mine openings. Rock bursts are generally triggered in conditions where a slight change in stress equilibrium can lead to an instantaneous release of a large amount of stored strain energy ejecting rock and/or coal particles in the mine opening. Rock bursts have been extensively studied in hard rock mines, and less so in coal mines (Ortlepp and Stacey, 1994; Wang and Park, 2001; Haramy and McDonnell, 1988, Zhao and Jiang, 2009, Calleja and Nemcik, 2016).

Numerous experimental investigations have been carried out to elucidate the burst process (Kidybinski, 1981; Huang et al., 2001; He et al., 2010; Ohta and Aydan, 2010). Zubelewicz and Mroz (1983) applied a dynamic failure approach to numerically study the process of rock bursts. A finite element method was proposed by Sharan (2007) to predict the occurrence of rock bursts in underground openings. Numerical models simulating fracture initiation and propagation were used in the modelling of pillar rock bursts (Wang et al., 2006) and rock bursts induced by dynamic disturbance (Zhu et al., 2010).

Although high gas content coal seams may not store much elastic strain energy, the presence of high-pressure gas may provide sufficient energy to break coal and eject a considerable amount of coal-gas mixture into mine openings, referred to as gas outbursts or uncontrolled gas emissions. Since the first documented coal and gas outburst occurred in the Issac Colliery in France (1843),

as many as 30,000 outbursts have occurred in the world coal mining industry (Lama and Bodziony, 1998). Using in situ borehole pressure measurements, Diaz Aguado and Gonzalez Nicieza (2007) have concluded that the gas outburst occurrences in the Asturias coal basin of Spain were controlled by the mining stresses and gas pressure build up in virgin areas. As the understanding of the structural conditions and mechanisms leading to gas outbursts improve, more effective preventative measures are being developed and implemented. In recent years, researchers at CSIRO in Australia carried out a series of projects investigating coal and gas outbursts, developed a coupled geomechanical-flow model and applied to field conditions at a number of Australian coal mines (Wold and Choi, 1999; Choi and Wold, 2002; Wold et al., 2008). Although less often, the European coal industry still experiences gas outburst events, which lead to loss of production, and even loss of life.

This paper presents seismic monitoring research carried out at Coal Mine Velenje since 2010, which aimed at characterising the dynamic behaviour of the coal seam in response to gateroad development and longwall top coal caving (LTCC) mining. The early campaigns involved time-lapse active seismic tomography and microseismic monitoring at the two selected LTCC panels which were in production during 2011. Based on the findings of this early work, a more comprehensive microseismic monitoring programme was set up for the period 2016-19, which included nine LTCC panels operating at -80 and -95m production levels. The monitoring data were processed and used for model development and rock burst/gas outburst risk assessment purposes.

2 SEISMIC MONITORING AT COAL MINE VELENJE

2.1 *Field Site*

Located in Slovenia, Coal Mine Velenje currently produces around 3.4 million tonnes of lignite per annum from a lens-shaped deposit, which is up to 165 m thick at the centre and pinches out towards the margins (Figure 1). Depth of the seam varies from 200 to 500 m. The Velenje lignite is highly heterogeneous, with varying abundance in xylite, detrite, and mineral matter. Therefore, it possesses different rock mechanical and reservoir properties with respect to gas content, sorption characteristics, permeability. The seam gas at Coal Mine Velenje is predominantly CO_2, with variable proportions of CH_4 also being present. In terms of geomechanical strength, detrite is easily fractured once it loses moisture, and it is weaker compared to the wooden structured xylite component, which may be twice as strong as detrite. It has also been established that, depending on the ratio of xylite to detrite, the P-wave velocities in xylite-rich zones maybe up to 1.5 times that of detrite-rich areas.

Rock bursts and coal/gas outbursts have affected coal production at Coal Mine Velenje since the early days of mine production. While rock burst potential is mostly a concern during development in large coal blocks at the mine, excessive gas emissions during LTCC face advance are occasionally experienced due to the relatively high gas content of the Velenje lignite (up to 8 m^3/tonne mixed gas at a pressure of up to 4.0 MPa) and large volumes coal handled during top coal caving.

The mining method used at Coal Mine Velenje is a combination of multi-level mining and longwall top coal caving (LTCC), developed over the decades as the most effective method due to extreme seam thickness, depth and prevailing geotechnical conditions (Figure 2). From the top to the bottom, the entire coal deposit is divided into a series of mining levels ranging from 10 to 20 m thick, mined in time-sequence with at least six months between the mining of each underlying longwall panel. At each level, the lower part of the panel, which is 3 to 4

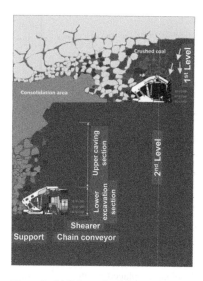

Figure 2. (a) Schematic of the multi-level longwall top coal caving mining method implemented at the ultra-thick coal deposit of Coal Mine Velenje (after Jeromel et al., 2010).

metres high, is cut by a shearer under the hydraulic supports while the upper section is allowed to cave and be recovered in front of the supports.

2.2 *Seismic Monitoring Campaigns in 2011*

Long-term observations at Coal Mine Velenje has suggested that transition from the "hard" xylite to "soft" detrite, or the existence of large pockets of weak porous detrite-rich lignite surrounded by strong heterogeneous xylite-rich lignite, or vice versa, often lead to coal and gas outbursts after de-stressing of the coal seam around the longwall face (Zavšek et al. 1997). Therefore, the designs for the first seismic monitoring campaign focused at LTCC extraction at two longwall panels K.-50/C and K.-130/A and aimed at coupling time-lapse seismic tomography with real time microseismic monitoring to investigate the role of lithological heterogeneity and stress/pressure on gas emissions. Here, only the data obtained and interpretation of the results from panel K.-50/C will be briefly discussed. For further details of the monitoring systems used, please refer to Si et al. (2015). Figure 3 illustrates the 100m long time-lapse seismic tomography zone, the configuration of sensors used for microseismic monitoring and ventilation measurements at this LTCC panel.

The time-lapse seismic tomography was performed by K-UTEC AG Salt Technologies over a 100 m × 141 m area ahead of the advancing longwall face (Figure 3). A total of 40 receiver and 40 source boreholes were placed along the intake and return gateroads with a regular spacing of around 2.5 m. Each receiver was equipped with 3 receiver components, i.e., X-, Y- and Z-components, monitoring the P-wave velocities parallel to the gateroad, perpendicular to the gateroad and in the vertical direction,

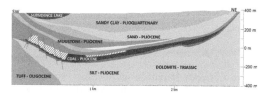

Figure 1. Schematic SW–NE trending geological cross-section of the Velenje lignite deposit (after Markič and Sachsenhofer, 2010).

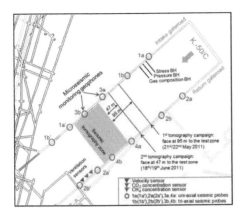

Figure 3. The time-lapse seismic tomography zone and the configuration of sensors used for microseismic monitoring and ventilation measurements at LTCC panel K.-50/C.

respectively. Explosives placed in boreholes were used to generate P-waves, and P-wave velocities were computed and interpolated to produce velocity tomograms for every receiver component.

Figure 4 presents the P-wave velocity tomograms of the 1st and 2nd campaigns for the X component when the longwall face was 95 m and 47 m away from the tomography area respectively. During the first campaign, a relatively high velocity zone was detected diagonally across the centre of the tomography zone. This heterogeneous zone was considered to be xylite-dominated coal with a relatively high strength, as compared to the lower strength detritic coal. The second tomogram, taken when the LTCC face advanced further closer to the tomography zone, indicates even higher velocities over the heterogeneous zone, and a much lower seismic velocities over the weaker coal, which is now closer to the advancing face, indicating intensified but preferential fracturing of the weaker zones due to abutment stresses. In view of these observations, it was also possible to hypothesise that gas pressure may also increase beyond the high strength xylite-dominated zone.

A 32-channel flameproof automated seismic observation system (SOS) developed by the Laboratory of Mining Geophysics of Central Mining Institute (GIG) in Poland was used for the real time microseismic monitoring which ran in parallel with the active seismic tomography. In total, eight geophones (four uniaxial and four triaxial) were installed to record microseismicity induced by coal extraction from 27 April to 30 August 2011. The geophones initially located at 1a-2a (1b-2b) were later moved to 1a'-2a' (1b'-2b') as the face-line reached their initial locations (Figure 3). Microseismic data recorded were collected automatically in these sensors and transferred to surface data loggers via fibreoptic cables, and then processed to obtain source parameters and event locations.

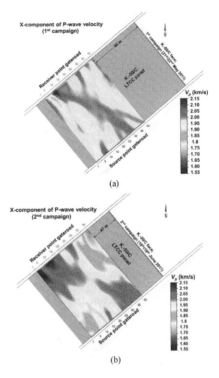

Figure 4. P-wave velocity tomograms in X-component for the (a) 1st tomography campaign and (b) 2nd tomography campaign at K.-50/C LTCC panel.

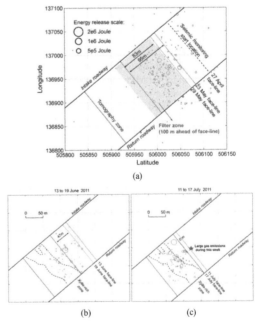

Figure 5. Weekly records of microseismic events within 100 m ahead of the face-line at K.-50/C LTCC panel, (a) at the time of first time-lapse tomography campaign, (b) at the time of second first time-lapse tomography campaign and (c) evolution of microseismic events leading to an episode of excessive gas emission on 15 July 2011.

13

Figure 5 shows some examples of the spatial distribution of weekly microseismic events over the monitoring period during which the coal face advanced over 130 m. It was observed that seismic events predominantly occurred within 70 m of the advancing coal face.

As the face-line reached the high strength xylite-dominated zone in week 11-17 July 2011, a decrease in microseismic event counts was observed. This was accompanied by a significant increase in the average released seismic energy, coinciding with 80% increase in gas emission within two hours, which interrupted coal production at the face on 15 July 2011 (Figure 5c).

These early monitoring campaigns confirmed the experienced based hypotheses on the role of lithological heterogeneities on excessive gas emissions and gas outbursts hazards at the mine. As a follow up project, a three-year long comprehensive microseismic monitoring campaign aimed at investigating potential rock burst and gas outburst hazards was set up in 2016.

2.3 Microseismic Monitoring Campaigns During 2016-2019

The new microseismic monitoring campaigns earmarked eight LTCC panels operating at -80 m and -95 m production levels from 2016 onwards. Longwall panels K.-80/B, K.-80/C, K.-80/D, K.-80/E, and CD2 were scheduled to mine during 2016-18 with longwall panels K.-95/A, K.-95/E, K.-95/D and CD3G coming into production during 2018-19 as the -80 m level is being mined-out. It was also recognised that a relatively high stress concentration on areas around the central coal pillar could increase the risk of seismic hazards in this region considerably, particularly during the development of longwall headings. Therefore, the number and position of seismic sensors were determined on the basis that the whole area, including the operating longwall panels, should be surrounded by the sensors.

The seismic monitoring system installed consisted of triaxial and uniaxial seismic sensors. The triaxial seismic sensors remained in the main/permanent roadways throughout the whole period of the project, whereas the uniaxial sensors, mostly located at the gateroads, were moved progressively as the face-line advanced or when a panel is mined-out and a new longwall panel started production. Similarly, the seismic sensors were gradually moved to active panels at -95 m level as mining at -80 m level is completed (Figure 6).

Figure 7 illustrates examples of spatial distribution of weekly recorded microseismic events at two different periods at Coal Mine Velenje. These figures demonstrate the correlation between mining activity and induced microseismicity. Most of the time, induced microseismicity clustered in the vicinity and ahead of the advancing longwall faces (Figure 7a).

Field observations and intensified seismic activity experienced during the development of longwall headings for LTCC panel K.-95/D immediately below

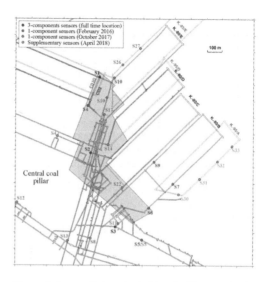

Figure 6. Longwall layout at -80 and -95 m production levels at Coal Mine Velenje during 2016-2019. The grey zone indicates the approximate shape of the central coal pillar protecting the main infrastructure in the production district.

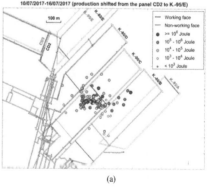

(a)

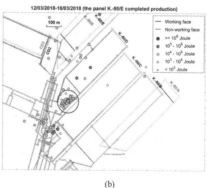

(b)

Figure 7. Spatial distribution of weekly recorded microseismic events around the longwall panels during (a) 10-16 July 2017 and (b) 12-18 March 2018.

the already abandoned panel K.-80/D was also confirmed by the clustering of events recorded during week 12-18 March 2018 (Figure 7b). Further analysis of these data will be presented in Section 3.2.

3 MICROSEISMICITY MODELLING AND RISK ASSESSMENT

3.1 Microseismicity Modelling

A microseismicity modelling approach applicable to longwall coal mining, was developed based on the shear slip mechanism of mining-induced microseismicity (Cao et al., 2018). This approach makes use of the built-in discrete fracture network (DFN) capability in the commercial software FLAC^{3D} to form stochastically distributed hypocentres for generating synthetic seismic events in a 3D continuum model. The fracture intensity and range of the fracture sizes are constrained by recorded number of microseismicity and the range of recorded energy release magnitude, respectively. The frequency distribution of fracture sizes is described by the widely used power law relationship (Bonnet et al., 2001):

$$n(l) = \alpha \cdot l^{-a} \tag{1}$$

where $n(l)$ is the number of fractures with size l per unit volume, α is the density term, and a is the power law scaling exponent.

After each longwall face advance, deterministic global elastoplastic stress analysis is first performed, followed by evaluation for fracture slip at all discrete fractures. The Mohr-Coulomb slip condition is used to evaluate if a fracture has slipped or not depending on the prevailing stress states:

$$\tau_{sf} = \mu_{sf}\sigma + c_f \tag{2}$$

where τ_{sf} is the resistance to slip, σ is the normal stress on the fracture surface, and μ_{sf} and c_f are the static friction coefficient and the cohesion along the fracture, respectively.

The released energy from a microseismic event is estimated by considering both the fracture size R and stress drop $\Delta\sigma$ along the fracture surface (Salamon, 1993):

$$E_k = \frac{4(1-v)\Delta\tau^2 R^3}{3G(1-v/2)} \tag{3}$$

where v and G are the Poisson's ratio and shear modulus of the host rock, respectively.

The microseismic modelling approach developed was applied to simulate microseismic events induced by the progressive advance of the coal face towards the xylite-rich zone as described in Section 2.2 above (Cao et al., 2017). A 3D model was constructed to represent the LTCC panel K. -50/C, the floor coal and the overlying strata (Figure 8(a)).

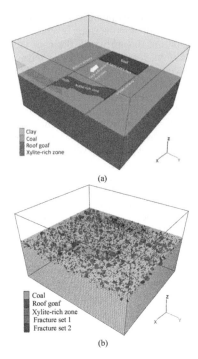

(a)

(b)

Figure 8. 3D model geometry and geological implementation of the xylite-rich zone at LTCC panel K.-50/C: (a) 3D model, and (b) distribution of DFNs located at the panel level [-330 m, -350 m].

The geometry of the xylite-rich zone inferred from the active seismic tomography measurements was digitised from Figure 4 and implemented into the model, assuming full penetration within the coal panel (Figure 8).

The influence of both strength of xylite and fracture attributes within the xylite-rich zone on microseismic characteristics were investigated. First, while fractures in the xylite-rich zone remained the same as those in the coal seam, four modelling scenarios were considered by varying both the compressive and tensile strength of the xylite-rich zone($\sigma_c h$ and $\sigma_t h$) from 1.0, 1.5, 2.0 to 2.5 times those of the coal seam ($\sigma_c c$ and $\sigma_t c$). Then, while maintaining the strength of the xylite-rich zone the same as that of the coal seam, another four modelling scenarios were considered by adopting two groups of discrete fractures characterised by different intensity terms (a_c and a_h) and power law scaling exponents (α_c and α_h) for the coal seam and the xylite-rich zone, respectively: (a) $a_h = 2.7$, $\alpha_h = \alpha_c$; (b) $a_h = 2.4$, $\alpha_h = 0.9\alpha_c$; (c) $a_h = 2.1$, $\alpha_h = 0.8\alpha_c$, and (d) $a_h = 1.8$, $\alpha_h = 0.7\alpha_c$ (Figure 8(b)).

The model run covered a period of twelve weeks, starting on 23th May to 28th August 2011, with an interruption of a two-week holiday period starting from 16th July. A total of twelve longwall extraction steps were then modelled, each extraction step representing one production week during which 12 m of coal was extracted. Multiple runs were carried out for each modelling scenario to account for the stochastic nature of DFN.

The modelled microseismicity were investigated in terms of frequency-magnitude distribution and logarithmic energy of microseismicity. The seismicity frequency-magnitude distribution, i.e., relative abundance of small seismic events with respect to large ones, was characterized by the Gutenberg-Richter relationship (Richter, 1958):

$$\log_{10} N(> M) = \log a - bM \qquad (4)$$

where N is the cumulative number of seismic events with a magnitude greater than M, and a and b are constants. The parameter b gives a measure of the scaling of seismicity magnitude distribution, with a higher value of b indicating a higher proportion of small seismic events in the population.

Figure 9 presents the variation of fitted b values from frequency-magnitude distribution of recorded and simulated microseismicity over the monitoring period. When the rock strength of elements and fracture attributes within the xylite-rich zone remain the same as those of coal, b values of microseismicity are quite consistent over the whole period. When the rock strength of elements is higher or the power law scaling exponent within the xylite-rich zone is lower, there is a notable reduction in b values several weeks before reaching the xylite-rich zone. An average b value as low as around 0.7 can be fitted when approaching the xylite-rich zone for modelling scenarios $\sigma_{ch} = 2.5\sigma_{cc}$, $\sigma_{th} = 2.5\sigma_{tc}$ and $a_h = 1.8$, $a_h = 0.7\alpha_c$.

Figure 10 shows examples of the frequency-magnitude distribution of recorded and simulated microseismicity during a two-week period after reaching the xylite-rich zone. The frequency-magnitude distribution for different modelling scenarios approximately follows a power law relationship, but the fitted curve becomes less steep with increasing rock strength of elements or decreasing power law scaling exponent within the xylite-rich zone, i.e. the b value decreases. When the rock strength of elements within the xylite-rich zone is 2.0-2.5 times of that of coal or the power law scaling exponent for the xylite-rich zone is 0.7 times of that for coal, the fitted b values match well with that of recorded microseismicity.

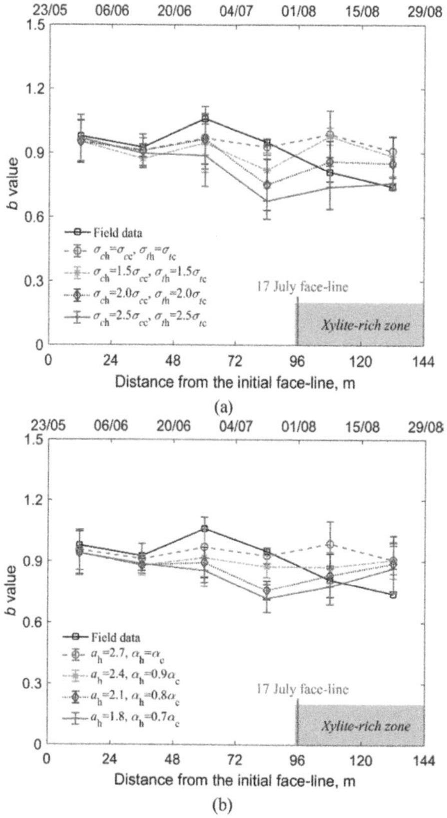

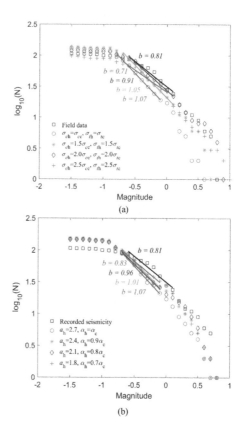

Figure 9. The bi-weekly b values for frequency-magnitude distribution of the recorded and simulated microseismicity over the twelve-week monitoring period: the effect of (a) rock strength, and (b) fracture attributes.

Figure 10. Examples of b values obtained through regression fitting for the recorded and simulated microseismic events during 1 to 14 August 2011: the effect of (a) rock strength, and (b) fracture attributes.

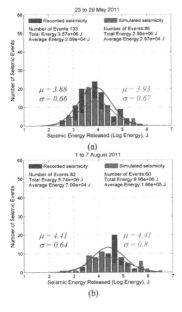

Figure 11. Example of weekly histograms of the logarithmic event energy for the recorded and simulated seismicity for the scenario $\sigma_{ch} = 2\sigma_{cc}$, $\sigma_{th} = 2\sigma_{tc}$: (a) during 23 to 29 May 2011, and (b) during 1 to 7 August 2011.

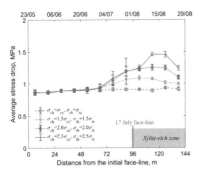

Figure 12. Simulated average stress drops over the twelve-week monitoring period.

Figure 11 presents examples of weekly histograms of the logarithmic event energy for the field data and modelling scenario $\sigma_{ch} = 2\sigma_{cc}$, $\sigma_{th} = 2\sigma_{tc}$ before and after reaching the xylite-rich zone, respectively. It can be observed that after reaching the xylite-rich zone, blue bars (the recorded microseismicity) have a notable shift to the right (higher energy levels), with a decrease in total number of microseismic events. The rightward shift can be modelled by increasing the rock strength of elements within the xylite-rich zone. In addition, the number of simulated microseismic events decreases due to a lower chance of slippage of fractures within the xylite-rich zone.

Figure 12 presents the variation of simulated average resultant stress drops in each week during face advance towards the xylite-rich zone. It can be

seen that for the scenario with high xylite strength, the resultant average stress drop remains quite consistent during the first few production weeks, begins to increase up to three weeks before the xylite-rich zone is encountered and fluctuates till the end of the monitoring period.

3.2 Risk Assessment

As briefly mentioned in Section 2.3, Figure 7b, and confirmed through numerical modelling, longwall gateroad developments in zones of high stress concentration in a new production level (such as moving from level -80 m to -95 m level) often face high risk of rock bursts. Figure 13 presents one such stress analysis carried out in FLAC^{3D} to establish the peak stress zone ahead of gateroad development for longwall panel K.-95/D, which was being driven below the mined-out panel K.-80/D (Figures 13 and 14)

Numerical modelling has clearly shown that driving a heading through the stress abutment zone within the central pillar poses a rock burst hazard. Figure 14 presents the spatial distribution of seismic events recorded during the period when

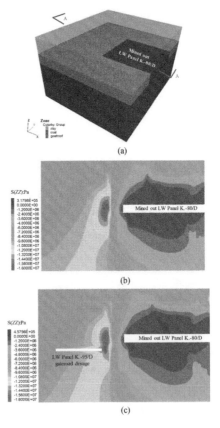

Figure 13. (a) 3D model geometry and geological representation of the area around longwall panel K.-80/D and stress distribution (b) before and (c) after the longwall K.-95/D gateroad is developed.

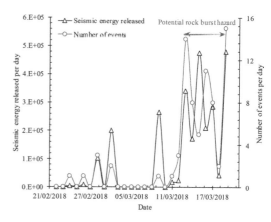

Figure 16. Risk assessment based on the sequential evolution of seismic events and energy released.

Figure 14. Longwall layout at -80 and -95 m production levels, production schedule, microseismic sensors installed and the spatial distribution of seismic events occurred in the period 18/02/2018-18/03/2018 at Coal Mine Velenje

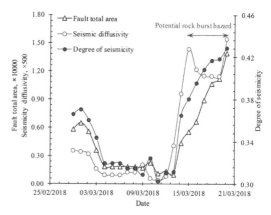

Figure 17. Temporal risk assessment based on the sequential evolution of seismic indicators.

Figure 15. Risk assessment based on the density distribution of seismic events recorded in the period 18/02/2018 to 18/03/2018.

longwall panel K.-95/D gateroad was being driven. As marked with a rectangle, a spatial clustering of the events around the region where the gateroad was located can be easily identified. Figure 15 presents the results of a risk assessment study performed using the spatially smoothed seismicity model (Frankel. 1995; Cai et al., 2015) based on the number of seismic events per m² in the study area. This aimed at identifying potential rock burst zones during the period 18/02/2018 to 18/03/2018.

Figure 16 presents risk analysis based on temporal evolution of seismic events and energy released per day due mainly to driving the gateroad heading for longwall panel K.-95/D. The figure illustrates that there is a dramatic increase in the number of

seismic events and the energy towards the end of the period considered, suggesting a potential rock burst risk. Further analysis of the same set of recorded data utilising some commonly used micro seismicity forecasting indices (Cai et al., 2018), such as fault total area (a measure of seismic energy level and event count), seismic diffusivity (spatial and temporal interval between seismic events) and degree of seismicity (based on seismic event counts and magnitudes) also confirms a similar and dramatic increase in these indices and seismic hazard risk (Figure 17). Such analysis can be conducted on a daily basis and help plan safety measures.

4 MEASURES TAKEN TO MITIGATE AGAINST SEISMIC HAZARDS

Since a rock burst is triggered by excessive stress concentration, the prevention methods have been mostly focused on stress relief or redistribution. The creation of boreholes or slots is believed to provide free space for stressed rock/coal to deform, and thus the accumulated strain energy can be reduced. In

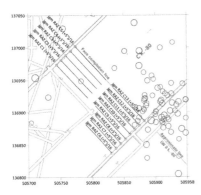

Figure 18. Large diameter stress relief boreholes and recorded seismic events over the barrier pillar of panel C k. -80 as longwall face B k. -80 retreated towards the face completion line.

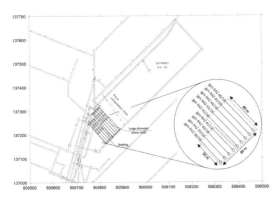

Figure 19. Large diameter borehole slotting experimental site in the barrier pillar of panel D k. -95.

German coal mines, drilling large diameter boreholes (∅~150 mm) is mostly used for this purpose. Another common method is to employ mechanical and/or hydraulic slotting in the stress build up zones. The application of high-pressure water jet slotting along the length of long in-seam pre-drainage boreholes in outburst prone coal seams in China was recently reported as another borehole stimulation technique by Lu et al. (2009; 2011).

Destress blasting, whereby stress concentrations are expected to be reduced by locally fragmenting rock/coal is also used in some coalfields in the Czech Republic, Poland and Canada (Tang, 2000).

On the other hand, outburst prevention focuses on reducing gas pressure and gas content prior to mining. To achieve this, gas drainage can serve as the most effective method. However, drainage performance is only effective in high permeability coal formations. Well stimulation techniques, such as high-pressure water injection has been applied to prevent gas outbursts in some coalfields.

As indicated by the analysis of seismic monitoring data and numerical modelling results discussed earlier, the central coal pillar left to protect the main

access roads in the production district (Figure 6) and the longwall panel barrier pillars result in relatively high stress concentration over this structure. In order to protect gateroad developments and longwall salvage operations from high levels of seismicity and risk of rock bursts, Coal Mine Velenje started drilling groups of 160 mm diameter and 50 to 80 m long boreholes in and around the barrier pillars of longwall developments. Figure 18 illustrates one such large diameter borehole zone implemented in the barrier pillar of longwall panel K.-80/C as it was being developed. Microseismicity recorded over a period of two months while the neighbouring panel K.-80/B was approaching this zone is also included in this figure. When compared with high levels of seismicity and events experienced in other but similar regions in the production district, it was noted that the density of recorded seismic events in this zone was relatively low. This practice was found to be very successful and has been extended to longwall panel developments in both -80 and -95 production levels since.

More recently, the high-pressure hydraulic borehole slotting system developed by the Central Mining Institute (GIG) was used to create 10 slots each along four 50 m long large diameter boreholes in the barrier pillar of longwall panel K.-95/D (Figure 19). This LTCC face started production in February 2019 and will be retreating towards the experimental site during the first half of 2019. Microseismic monitoring will continue and recorded data will be evaluated to assess the performance of the slotting trials in the future.

5 CONCLUSIONS

Active seismic tomography carried out at Coal Mine Velenje confirmed the role of lithological, mechanical strength and gas pressure heterogeneities in coal and gas outbursts experienced during LTCC mining. However, this technique is costly and can also interfere with production as a routine means of monitoring. On the other hand, real-time microseismic monitoring, coupled with numerical modelling can be readily used in forecasting potential rock bursts hazards in LTCC mining. Long-term research at Coal Mine Velenje has shown that large diameter stress relief boreholes can help reduce seismic events considerably.

ACKNOWLEDGEMENTS

This research was carried out as part of a European Commission Research Fund for Coal and Steel (RFCS) funded projects "Development of Novel Technologies for Predicting and Combating Gas Outbursts and Uncontrolled Emissions in Thick Seam Coal Mining - CoGasOUT" and "Monitoring, Assessment, Prevention and Mitigation of Rock

Burst and Gas Outburst Hazards in Coal Mines -MapROC", Grant Nos: RFCR-CT-2010-00002 and RFCR-CT-2015-00005 respectively. The authors would like to express their gratitude to the research partners at K-UTEC AG Salt Technologies of Germany, who implemented the time-lapse seismic tomography underground and colleagues from the Central Mining Institute (GIG), of Poland who carried out the passive microseismic monitoring system and borehole slotting experiment at the mine.

REFERENCES

Bonnet, E., Bour, O., Odling, N. E., Davy, P., Main, I., Cowie, P., Berkowitz, B. (2001). Scaling of fracture systems in geological media. *Reviews of geophysics*, 39(3), 347–383.

Cai, W., Dou, L., Gong, S., Li, Z., Yuan, S. (2015). Quantitative analysis of seismic velocity tomography in rock burst hazard assessment. *Natural Hazards*, 75(3), 2453–2465.

Cai, W., Dou, L., Zhang, M., Cao, W., Shi, J. Q., Feng, L. (2018). A fuzzy comprehensive evaluation methodology for rock burst forecasting using microseismic monitoring. *Tunn. Undergr. Sp. Technol*, 80, 232–245.

Calleja, J., Nemick, J. (2016). Coalburst causes and mechanisms. Proc. 2016 Coal Operator's Conference, 10-12 Feb 2016 Wollongong, Australia. 310-320.

Cao, W., Shi, J. Q., Durucan, S., Si, G., Korre, A. (2017). Modelling the influence of heterogeneity on microseismic characteristics in longwall coal mining. Proc. 51st US Rock Mechanics/Geomechanics Symposium. American Rock Mechanics Association.

Cao, W., Shi, J. Q., Si, G., Durucan, S., Korre, A. (2018). Numerical modelling of microseismicity associated with longwall coal mining. *Int, J, Coal Geol.*, 193, 30–45.

Choi, S.K., Wold, M.B. (2002). Numerical modelling of outburst mechanisms and the role of mixed gas desorption.CSIRO Petroleum ACARP Project No.C9023, December, Australian Coal Association Research Program. 37p.

Diaz Aguado, M.B., Gonzalez Nicieza, C. (2007). Control and prevention of gas outbursts in coal mines, Riosa–Olloniego coalfield, Spain. *Int, J, Coal Geol.*, 69, 253–266.

Frankel A (1995) Mapping seismic hazard in the central and eastern United States. *Seismol Res Lett.* 66:8–21. doi:10.1785/gssrl.66.4.8.

Haramy, K.Y., McDonnell, J.P. (1988). Causes and control of coal mine bumps.Report of Investigation 9225, 1988 Spokane, WA. United States Bureau of Mines, 40p.

He, M.C., Miao, J.L., Feng, J.L. (2010. Rock burst process of limestone and its acoustic emission characteristics under true-triaxial unloading conditions. *Int. J. Rock Mech. Min. Sci,* 47(2), 286–298.

Huang, R.Q., Wang, X.N., Chan, L.S. (2001). Triaxial unloading test of rocks and its implication for rock burst. *Bulletin of Engineering Geology and the Environment*, 60(1), 37–41.

Jeromel, G., Medved, M., Likar, J. (2010). An analysis of the geomechanical processes in coal mining using the Velenje mining method. *Acta Geotechnica Slovenica*, vol. 7, no. 1, pp.31–45.

Kidybiński, A. (1981). Bursting liability indices of coal. *Int. J. Rock Mech. Min. Sci.*, 18, 295–304.

Lama, R.D., Bodziony, J. (1998). Management of outburst in underground coal mines. *Int. J. Coal Geol.* 35, 83–115.

Lu, T., Yu, H., Zhou, T., Mao, J., Guo, B. (2009). Improvement of methane drainage in high gassy coal seam using waterjet technique. *Int. J. Coal Geol.* 79, 40–48.

Lu, T., Zhao Z. and Hu H. (2011). Improving the gateroad development rate and reducing outburst occurrences using the waterjet technique in high gas content outburst-prone soft coal seam. *Int. J. Rock Mech. Min. Sci.*, pp. 1271–1282.

Markič, M. and Sachsenhofer, R.F. (2010). The Velenje lignite - its petrology and genesis. Geological Survey of Slovenia, Ljubljana. ISBN 978-961-6498-20-3.

Ohta, Y., Aydan, Ö. (2010). The dynamic responses of geo-materials during fracturing and slippage. *Rock Mech. Rock Eng.*, 43(6), 727–740.

Ortlepp, W.D., Stacey, T.R. (1994). Rockburst mechanisms in tunnels and shafts. *Tunn. Undergr. Sp. Technol*, 9, 59–65.

Richter, C.F. (1958). Elementary Seismology. WH. Freeman and Company, San Francisco.

Salamon, M.D.G. (1993). Keynote address: Some applications of geomechanical modelling in rockburst and related research. Proc. 3rd Int. Symp. Rockbursts and Seismicity in Mines, 297-309. Rotterdam: AA Balkema.

Sharan, S. (2007). A finite element perturbation method for the prediction of rockburst. *Computers & Structures* 85, 1304–1309.

Si, G., Durucan, S., Jamnikar, S., Lazar, J., Abraham, K., Korre, A., Shi, J.Q., Zavšek, S., Mutke, G., Lurka, A. (2015). Seismic monitoring and analysis of excessive gas emissions in heterogeneous coal seams. *Int. J. Coal Geol.*, 149, 41–54.

Tang, B. (2000). Rockburst control using destress blasting. PhD thesis, McGill University.

Wang, J.-A., Park, H.D. (2001). Comprehensive prediction of rockburst based on analysis of strain energy in rocks. *Tunn. Undergr. Sp. Technol.* 16, 49–57.

Wang, S., Lam, K., Au, S., Tang, C., Zhu, W., Yang, T. (2006). Analytical and numerical study on the pillar rockbursts mechanism. *Rock Mech. Rock Eng.*, 39, 445–467.

Wold, M.B., Choi, S.K. (1999).Outburst mechanisms: coupled fluid flow geomechanical modelling of mine development. CSIRO Petroleum ACARP Project No.C6024, December, Australian Coal Association Research Program. 58p.

Wold, M.B., Connell, L.D., Choi, S.K. (2008). The role of spatial variability in coal seam parameters on gas outburst behaviour during coal mining. *Int. J. Coal Geol.* 75, 1–14.

Zavšek, S., Tamše, M., Koćevar, M., Markić, M., Likar, J. Žigman, F. (1997). Nenadni vdori premoga in plina v jamske prostore. – RLV, IGGG, IRGO - Zakljućno gradivo za 17. sejo projektnega sveta, 40 str (in Slovene).

Zhao, Y.X., Jiang, Y.D. (2009). Investigation on the mechanism of coal mine bumps and relating microscopic experiments. Proc. 3rd CANUS Rock Mech. Symp., May 2009 Toronto, Canada. Paper 4158, 10p.

Zhu, W., Li, Z., Zhu, L., Tang, C. (2010). Numerical simulation on rockburst of underground opening triggered by dynamic disturbance. *Tunn. Undergr. Sp. Technol.* 25, 587–599.

Zubelewicz, A., Mroz, Z. (1983). Numerical simulation of rock burst processes treated as problems of dynamic instability. *Rock Mech. Rock Eng.*, 16(4), 253–274.

2019 Rock Dynamics Summit– Aydan et al. (eds)
© *2019 Taylor & Francis Group, London, ISBN 978-0-367-34783-3*

An overview of research into understanding coal burst

Ismet Canbulat, Chengguo Zhang & John Watson
School of Minerals and Energy Resources Engineering, University of New South Wales, Sydney, Australia

ABSTRACT: Coal burst is a high safety and production risk in underground coal mines. It involves a sudden release of strain energy stored within the coal or rock mass due to the disturbance of an unstable state of equilibrium. To improve our understanding of coal burst, it is important to quantify the energy transition involved in this dynamic failure process, since energy sources that cause coal burst are highly complex. This paper provides an overview of progress on understanding coal burst mechanisms, particularly in evaluating two of the energy components associated with coal burst, strain energy and gas expansion energy, using analytical and numerical modelling. Energy release, particularly research on cleating and jointing, is also assessed for a clear understanding of the energy transition process. The energy requirements of the ground support system for coal burst control are then discussed. The paper reports our progress on a multi-element coal burst research in Australia mostly funded by ACARP.

1 INTRODUCTION

Coal burst is a form of violent and dynamic failure of coal or rock, with high velocity ejection of materials from an abutment or longwall face in underground mines, and is a significant threat to mining safety and productivity. The occurrence of a coal burst is a highly complex process as a range of mining and geological factors are involved that are challenging to quantify. Extensive studies have been conducted to assess the contributing factors, the causal mechanisms and the control of coal burst (Dou and He, 2001; Holland, 1955; Peng, 2008; Zhang et al., 2017, Zhao et al., 2010). Although this dynamic rock or coal failure has been investigated previously from different perspectives, understanding the circumstances in which coal bursts occur and the likelihood of their occurrence in coal mines still needs improvement.

A key element limiting the understanding of coal burst mechanisms is the uncertainty in assessing and quantifying the energy sources, accumulation and release. There are a range of energy components involved and they have different roles and contributions. Recent research in Australia into understanding coal burst (Canbulat et al., 2018) highlighted the following four main research topics:

1. Coal burst mechanisms (both direct and indirect)

   a. Energy sources
   b. Energy storage (geological units and material properties)
   c. Energy release and damage mechanisms

2. Triggers (or causative factors)

   a. What causes or initiates failure?
   b. How is the energy release manifested?

3. Prediction

   a. Monitoring
   b. Identification of "coal burst risk domains" at a mine site

4. Controls

   a. Preventative measures for elimination of risk (i.e., mine design criteria)
   b. Mitigation of risk (i.e., support design)

The key considerations and factors associated with these four identified areas in current Australian coal burst research are summarized in the following sections. This paper does not examine coal burst mechanisms in detail, rather it summarizes the coal burst research and discusses how the associated critical factors should be addressed. It provides a basis for further research to quantify the effects of energy involved in coal burst to help establish effective coal burst control strategies.

2 COAL BURST OCCURENCES

2.1 Terminology

A wide range of terminologies have been used to describe the rock and coal dynamic and violent failure. A pressure or coal bump is a dynamic release of energy within the rock (or coal) mass in a coal mine, often due to intact rock failure or failure and/or displacement along a geological structure, that generates an audible signal, ground vibration and potential for displacement of existing loose or fractured material into mine openings. A pressure bump is also sometimes referred to as a bounce. A pressure or coal burst is a pressure bump that actually causes consequent dynamic rock or coal failure in the vicinity of a mine opening, resulting in high velocity expulsion of this broken and failed material into the mine opening (Galvin, 2016). The energy levels, and hence velocities, involved in pressure or coal burst can cause significant damage to, or destruction of, conventional installed ground support elements such as bolts and mesh.

In Australian mining terminology, an outburst is also a dynamic energy release leading to some form of rock failure. However, the source of energy is primarily associated with *in situ* gas pressure, sometimes also accompanied by stress-related energy. Gas pressure can be a source of energy in some coal bursts, but the dominant energy source in coal or pressure bursts is stress. Brauner (1994) also describes the characteristics of an outburst that may help to distinguish it from a coal burst as below:

−   The process of an outburst is slower and less violent than a coal burst.
−   There is no or relatively little seismic reaction in an outburst.
−   The ejected coal is broken into smaller particles in an outburst.
−   Occurrence of an outburst is predominately associated with geological disturbances and is usually in the coal face or its close vicinity.
−   High gas content in the seam is required for an outburst (i.e. more than 10 m³ of methane per tonne of solid coal).

The contribution of gas expansion energy is discussed further in Section 4.

2.2 *Coal Burst Hazard Risk Factors*

International experience on coal burst has been widely reported in Canada, South Africa, the United States, the former Soviet Union, China, India, France, Germany, Poland and Czech Republic. Over the years, a significant amount of research has been conducted into understanding, prevention and monitoring of coal and rock bursts. These studies indicate that dynamic failures occur under the effects of complex environments of geology, stress and mining conditions, and there is no one set of defining characteristics that is responsible. The relative roles of these different contributing factors can change from one site to another.

Vardar et al. (2018a) summarised the different types of mining and geological and geotechnical factors associated with coal burst. Although over twenty contributing factors have been identified in the literature, nine factors were identified as the main contributors of coal bursts in that study: depth of cover, topography, thick competent unit in the roof, significant geological structures, past seismic activity, coal cleating and jointing, abutment stresses, multi-seam mining and gas content. Based on these factors, a semi-quantitative coal burst risk classification system was developed by Vardar et al. (2018a) to assist mine operators to identify coal

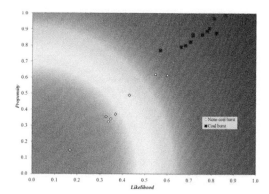

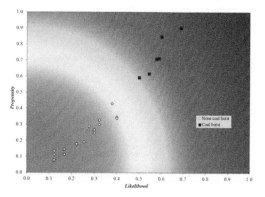

Figure 1. BurstRisk – Risk classification for longwall (top) and development roadway (bottom) (Vardar et al., 2018a). Note: circular points indicate none-coal burst cases while solid square points include coal burst cases.

burst risk level in underground coal mines. The database used in the back-analysis of the classification system consists of 41 cases from Australia, China and the United States. The back-analysis conducted for the longwall and development sections is shown in Figure 1, where low risk level is shown in green, medium risk in yellow and high risk in red.

3 CONSIDERATIONS IN ENERGY STORAGE AND RELEASE

Four critical conditions have identified for coal burst occurrence (Galvin, 2016), which are presented in Figure 2, together with the major factors contributing to each condition.

These four conditions have to be satisfied simultaneously to trigger a dynamic coal burst. When a system is in a state of equilibrium, energy has to be expended to induce further deformation. Conversely, a system that is in a state of unstable equilibrium requires only a very small change to transform it from a stable to an unstable state and to release surplus energy which then overwhelms and accelerates the deformation process (Galvin, 2016).

In an underground mining environment, a range of energy components are available due to *in situ*

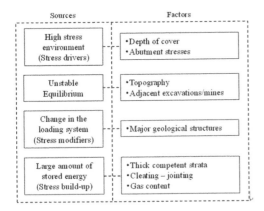

Figure 2. Critical conditions for coal burst occurrence (modified from Galvin, 2016).

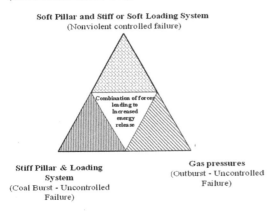

Soft Pillar and Stiff or Soft Loading System
(Nonviolent controlled failure)

Combination of forces leading to Increased energy release

Stiff Pillar & Loading System
(Coal Burst - Uncontrolled Failure)

Gas pressures
(Outburst - Uncontrolled Failure)

Figure 3. Different loading mechanisms contributing to coal burst energy release (Zhang et al., 2017).

status and the impact of mining. Potential energy is the stored energy of the position possessed by coal including gravitational potential energy and elastic potential energy. Gravitational potential energy is the result of coal's vertical position. Strain energy is the energy stored in coal due to deformation, and the external mining work done on the coal causing it to distort from its unstressed state is usually transformed into strain energy. Kinetic energy is the energy possessed by coal in motion.

Strain energy is the main energy source in most coal bursts, due to the work done by the boundary loading and body forces. Energy is dissipated in several forms, including the plastic work within the rock or coal mass, kinetic energy of the ejected blocks, and the frictional work in fracturing development. The unstable release of potential energy from the rock mass around excavations, mainly in the form of kinetic energy, causes coal burst. The loading mechanisms can affect the energy storage and release process. The energy can be released in either a controlled or uncontrolled manner depending on the loading mechanisms and the system stiffness. A stiff loading system is mainly due to *in situ* and/

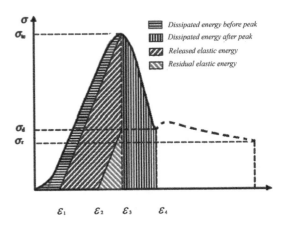

Figure 4. Analytical calculation of dissipated energy and released energy (Xie et al., 2009).

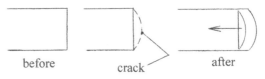

before   crack   after

Figure 5. Illustration of ejected coal block in an opening.

or mining induced stresses that result in uncontrolled coal bursts (Figure 3). Evaluation of other loading mechanisms, such as gas pressures, as a potential contributor to the magnitude of energy release is discussed in the following section.

Figure 4. demonstrates the calculation of the dissipated, released and residual energies. Energy dissipation is a key characteristic of the deformation and failure of the coal mass. The energy evolution declines the strength of the coal, which may result in total failure. In this context, the dissipated energy is directly related to the damage as well as the mitigating strength of coal.

4 ENERGY ASSESSMENT

4.1 *Strain Energy*

It is postulated that, in broad terms, the strain energy of the rock and coal is converted to the kinetic energy of the ejected material. The following analysis further assumes that the strain energy is just that of the rock or coal ejected when the burst takes place, and the kinetic energy of that material after the burst equals all of that strain energy minus the work that has been done to create a single crack to detach it from the surrounding rock or coal as shown in Figure 5.

It is implied, therefore, that:

a. there is no transfer of strain energy from the surrounding rock or coal to the ejected material;

b. there is no fragmentation of the ejected material, as that would require additional work to be done;

c. there is, in thermodynamic terms, 100% efficiency of conversion of strain energy to kinetic energy.

These conditions are more likely to exist in hard rock than in coal. In mathematical terms,

$$K = S - W \qquad (1)$$

where K is the kinetic energy after the burst, S is the strain energy before the burst, and W is the work done to create a crack.
We have:

$$K = 0.5mc^2 \qquad (2)$$

where m is the mass of ejected rock and c is the velocity.

$$S = \int_V U \, dV \qquad (3)$$

where U is the strain energy density and V is the volume of ejected material, and

$$W = 2A\gamma_S \qquad (4)$$

where A is the area of crack and γ_s is the specific surface energy of the material. Specific surface energy is the work that has to be done to create a unit area of new surface in the material, as defined in the Griffith theory (1920) of fracture. Strain energy density is given by

$$U = [\sigma_1\varepsilon_1 + \sigma_2\varepsilon_2 + \sigma_3\varepsilon_3]/2 \qquad (5)$$

where E = Young's modulus of elasticity, σ_1 etc are principal stresses and ε_1 etc are corresponding principal strains.

As an exploratory analysis, the case is considered of a tunnel of circular cross section in a rock mass in which there is hydrostatic pre-mining stress (Figure 6). Let the radius of the tunnel be a, and suppose that a burst takes place over an angle α (radians) of its perimeter to a depth $(b - a)$ into the rock.

The principal stresses in a cylindrical polar coordinate system (r, θ, z) before the burst are σ_r, σ_θ and σ_z, and corresponding principal strains are ε_r, ε_θ and ε_z. We have

$$\sigma_r = \sigma_0(1 - a^2/r^2)$$
$$\sigma_\theta = \sigma_0(1 + a^2/r^2) \qquad (6)$$
$$\sigma_z = \sigma_0$$

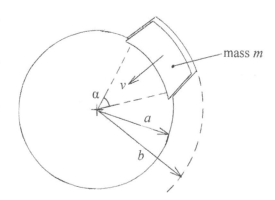

Figure 6. A tunnel of circular cross section in a rock mass.

where σ_0 = hydrostatic stress at infinity, and according to Hooke's Law

$$\varepsilon_r = \sigma_0[(1-2v) - (1+v)a^2/r^2]/E$$
$$\varepsilon_\theta = \sigma_0[(1-2v) + (1+v)a^2/r^2]/E \qquad (7)$$
$$\varepsilon_z = \sigma_0(1-2v)/E$$

where v = Poisson's ratio.
Taking 1m run of tunnel,

$$S = a\int_a^b \sigma_0^2[3(1-2v) + 2(1+v)a^4/r^4]/2Er \, dr \qquad (8)$$
$$= a\sigma_0^2[3(1-2v)b^2 - (1-8v)a^2 - 2(1+v)a^4/b^2]/4E$$

Also,

$$m = \alpha\rho(b^2 - a^2)/2 \qquad (9)$$

where ρ = density of ejected material. Neglecting the term W in equation (1), the following numerical examples are considered (Table 1).

Table 1. Calculated ejection velocity for two cases.

	Hard rock case	Coal case
Depth (m)	2000	300
E (GPa)	50	3
v	0.2	0.3
ρ (kg/m³)	2700	1400
a (m)	2	2
b (m)	2.5	2.5
α (degree)	45	45
σ₀ (MPa)	54	8.1
S (kJ)	85.9	27.7
m (kg)	2385	1236
c (m/s)	8.5	6.7

24

4.2 Friction work consumed in cleating and jointing

The quantification of the friction work consumed in cleating and jointing is important as it determines how much energy is left in the system that could contribute to coal burst. The effect of the jointing pattern was studied by Tahmasebinia et al. (2018a) to determine the effect of different joint densities on the trajectory of the failure which can be a combination of the shear and tension failure. Figure 7 illustrates the modelled joint arrangements either as a single direction or as crossed joints.

A non-linear regression method (normalised cases) was used to consider the effect of the joint densities in a damage model which can represent the level of energy dissipation or work done in jointing fracturing (Tahmasebinia et al., 2018a). Equation (10) was proposed to determine the amount of the fracture energy due to the different joint density.

$$E\,fracture = \left| \left(\frac{J1}{\varepsilon_{uf}} \right) \times (1-D) \times \ln(K) - P1 \right| \tag{10}$$

where $J1$ and $P1$ are constant values, $39<J1<45$ and $1389<P1<1405$; D is the damage value; K is the number of joints per square metre and ε_{uf} is the ultimate strain value.

In addition, the energy release characteristics of large coal mass samples under uniaxial compression were studied (Vardar et al., 2018b). It was hypothesised that the intensity and rapidity of a failure decreases as cleat density increases. To examine this, the UDEC Trigon method was used to model large coal mass samples. Input properties for the models were obtained through a comprehensive literature survey and calibration procedure. The input properties were validated by conducting coal pillar compression simulations and comparing the obtained pillar strength values to the UNSW pillar strength formulae (Salamon et al., 1996).

A non-linear relationship between the total energy release and the number of cleats per sample width was observed, as shown in Figure 8. In a broad sense, higher energy release values were observed as cleat density reduced. In other words, as the coal brightness decreased, the released energies increased, and intensity of the failure increased.

For future numerical modelling work, it is necessary to trace the propagation of the crack through the rock on an incremental basis, taking account of the stress concentrating effect of the crack root. At each increment it is necessary to compute stress intensity factors and compare them with fracture toughness as determined by standard laboratory tests to predict how the crack extends through the rock (Ewalds et al., 1989). Analysis by reference to tensile strength alone is unlikely to yield satisfactory results. Finally, the need for dynamic analysis must be considered.

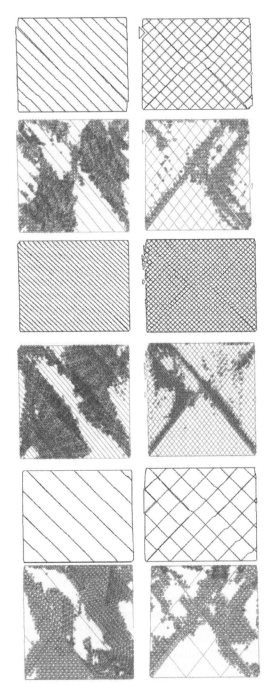

Figure 7. Failure modes due to different jointing densities, where purple indicates tensile failure, red indicates at yield surface, and green indicates yielded in past (Tahmasebinia et al., 2018a).

In the laboratory environment, dimensional similitude is important, which means ensuring the same balance between factors influencing physical behaviour at a reduced scale as at full scale in the field. For example, in fluid mechanics, setting

25

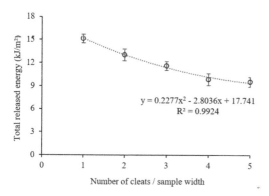

Figure 8. Relationship between released energy and number of cleats across the width of the coal mass sample (Vardar et al., 2018b).

Reynold's Number (Re) to the same value in the laboratory as in the field assures the same balance between viscous and dynamic forces (forces due to fluid particle acceleration). A characteristic feature of indices such as Re is that they are dimensionless. If the factors which govern coal burst are unconfined compressive strength σ_c, fracture toughness K_{1c} and a characteristic dimension such as mining height h, then the numerical value of $K_{1c}/\sigma_c h^{0.5}$ should be the same in the laboratory as in the field. This implies that if the laboratory model is at a scale 1:50, the ratio K_{1c}/σ_c of the material making up the model should be about 0.14 times that of rock in the field.

4.3 Gas expansion energy

Free gas may exist in cleats, and in pore spaces within intact fragments of coal. It is assumed that the quantity of gas which escapes from the pore spaces to cleats during a burst is negligible, and that the gas contained in cleats is a closed thermodynamic system which undergoes adiabatic expansion, defined as expansion which takes place quickly enough for the amount of heat entering or leaving the system to be negligible. Supposing a worst case that initial gas pressure p_1 is uniform and equal to the pre-mining pressure p_0, the work done by the system on its surroundings is then:

$$W = (p_1 V_1 - p_2 V_2)/(\gamma - 1) \tag{11}$$

where p_2 is atmospheric pressure, V_1 and V_2 are the initial and final volumes of gas, and the ratio of specific heats γ equals 1.304 for methane or carbon dioxide. The initial cleat volume V_1 is calculated from cleat porosity which is generally less than 0.005, and V_2 is given by

$$p_2 V_2{}^\gamma = p_1 V_1{}^\gamma \tag{12}$$

Note that P_1 and P_2 are absolute pressures, and a closed system is one in which there is no transfer of matter to or from the surroundings during the process of expansion. The surroundings on which work is done are the fragments of coal between the cleats, so if frictional and other losses are negligible then the kinetic energy of the fragments after the burst equals W.

The analysis for zero gas pressure assumed that there is no fragmentation of the ejected material. In that case, elastic waves can propagate freely within the material, undergoing reflections at its surface. On detachment of the ejected material from the surrounding stable mass of rock or coal, strain energy becomes the energy of wave propagation, and there are considerable spatial variations of particle velocity. However, wave amplitudes are progressively reduced by internal damping and ultimately the ejected rock or coal mass moves as a rigid body. If rotational components of velocity can be neglected and energy loss due to damping is small, all particles eventually move with practically the same velocity and it can be proposed that:

$$0.5mc^2 = W \tag{13}$$

where m is mass of ejected material and c is velocity.

For a mass of coal containing gas under pressure, fragments are not constrained to move with the same velocity and it can only be concluded on the basis of the preceding analysis that:

$$\sum_j 0.5 m_j c_j{}^2 = W \tag{14}$$

where m_j and c_j are the mass and velocity of fragment j.

It should be noted that a coal seam may contain significant quantities of pulverised coal, especially near geological structures such as dykes and faults. That material, by virtue of its high specific surface area, can hold large amounts of adsorbed gas which is considered to be capable of more rapid desorption than gas adsorbed to larger fragments of coal. If desorption can occur sufficiently rapidly, then such adsorbed gas may cause a coal burst. However, it is out of the scope of this paper to discuss the gas desorption process and its contribution to coal burst.

5 GROUND SUPPORT REQUIREMENTS FOR COAL BURST CONTROLS

Current coal burst control techniques can be classified into two groups: preventative controls and mitigating controls. Preventative controls to avoid occurrence of coal bursts are usually implemented before starting underground mining by optimising the mine design, while mitigating controls are applied as risk mitigation measures to minimise the risks of coal bursts during mining. Coal burst risks still exist even when preventative controls are implemented (Wei et al., 2018). Due to the unpredictability of coal burst, ground support is usually the final and most common line of protection to ensure safety in high risk zones (Cai, 2013). Ground support has a pivotal role in a dynamic environment, which has been well recognised by the mining industry (Cai, 2013; Jiang et al., 2014; Mikula and Brown, 2018).

Studies on yielding support in underground mines have been conducted in many countries, and Wei et al. (2018) presented a comprehensive review of current coal burst control techniques. Wei et al. (2018) proposed a coal burst management framework that includes three stages: identification of risk, development of a management plan, and management of coal burst. Current knowledge gaps were identified and discussed. To implement a successful control technique, systematic methods should be developed to recognise and address all the complex contributing factors that may be found in a specific mine. Establishment of control strategies must be based on a monitoring program and risk classification technique.

Numerical and analytical analyses were conducted by Tahmasebinia et al. (2018b) to study the structural behaviour of fully grouted cable bolts under impulsive loading, which was represented by a dropped ball. As illustrated in Figure 9, the momentum energy from the dropped mass would initially be transferred to the concrete surfaces and the transmitted energy due to the impulsive loading will reach the cable bolt. Figure 9 demonstrates the failure process of the cable bolt under impact loading starting with the initial deformed shape followed by the brittle shear failure.

In order to provide a better understanding of the effect of the velocity on the induced impact loading, the influence of the different velocities of the dropped ball was also tested (Tahmasebinia et al., 2018c). Figure 10 demonstrates the induced impact loading due to the different applied velocity. As can be seen, the applied velocity can significantly influence the reaction of the impact loading. Also, it can affect the initial inertia forces in the impact loading at the first stage of transmitting the momentum energy from the dropped ball to the concrete surfaces.

Rock and cable bolts can be combined with surface support elements to form an integrated system support in coal burst prone conditions. The surface elements, such as shotcrete, mesh and strap, also help to increase the support performance. Nevertheless, the bolt-mesh linkage and the retaining elements themselves are often the weakest part of the system (Wei et al., 2018). In addition, the support components are subject to loading by the holding rock mass and interact with other elements. Because these elements are always subject to combinations of loading and failure mechanisms and their complex interactions, theoretical determination of the capacity of a support system is highly unlikely to

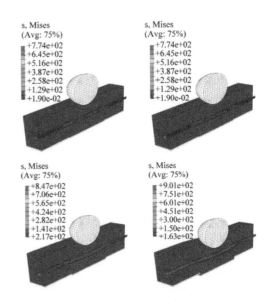

Figure 9. Shear ductile failure in the cable bolt under impact loading at four loading stages (Tahmasebinia et al., 2018b).

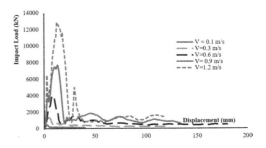

Figure 10. .Impact loading versus mid-span deflection under different velocities (Tahmasebinia et al., 2018c).

be successful at present, particularly in a dynamic loading environment (Stacey, 2012). Therefore, a systematic approach is needed to effectively assess ground support system capability and requirements.

6 CONCLUSIONS

Coal burst occurs under the effects of complex environments of geology, stress and mining conditions, and there is no one set of defining characteristics that is responsible. Coal bursts are relatively rare in the Australian coal industry, however they seem to be increasing, particularly in deeper mines. The understanding of coal burst in Australia is primarily qualitative and based on overseas experience that may or may not be applicable to Australian conditions. Based on research to date, the four most important areas of research are coal burst mechanisms, triggers, forecasting and controls.

To understand and manage coal burst it is important to apply analytical and computational methods to determine the energy magnitude required to cause a coal burst. Quantification of the energy elements and the energy release process are two critical aspects in understanding coal burst mechanisms. This paper presents a preliminary assessment of the energy components involved in coal burst and discusses the strain energy, frictional work and potential gas expansion energy, and the associated kinetic energy, using analytical and/or numerical modelling methods. Further studies using computational methods should be conducted to explicitly represent the coal mass, the cleating and gas desorption to quantify the energy components and the energy levels in greater detail and accuracy. Seismic energy has been identified as another source of energy that contributes to coal burst. Further studies using analytical and numerical modelling should also be conducted to understand the magnitude of seismic energy in and around excavations. In addition, the contribution of gas expansion energy to coal burst needs to be quantified considering desorption and free gas expansion.

To understand the energy requirements for ground support in coal burst control, dynamic numerical modelling and laboratory testing need to be conducted to quantify the energy build-up and transfer within the support rock and coal mass. Importantly, a systematic approach is required to understand the roles and interactions between different support elements as well as the host rock and coal mass.

ACKNOWLEDGEMENTS

The authors would like to acknowledge the support and the funding of the Australian Coal Association Research Program (ACARP) to conduct the research summarised in this paper.

REFERENCES

Brauner, G. 1994. *Rockbursts in coal mines and their prevention.* Rotterdam: A.A. Balkema.

Cai, M. 2013. Principles of rock support in burst-prone ground, *Tunnelling and Underground Space Technology*, 36, 46–56.

Canbulat, I., Hebblewhite, B., Zhang, C., Watson, J., Thomas, R., Shen, B., Gale, W. and Luo, X. 2018. *Review of Australian and International Coal Burst Experience and Control Technologies.* Australian Coal Research Association Program. Draft Report. Project C25004. August 2018.

Dou, L. and He, X. 2001. *Prevention theory and technology of rockburst.* Xuzhou: China University of Mining and Technology Press.

Ewalds, H.L., Wanhill, R.J.H. and Arnold, W. 1989. *Fracture mechanics.*

Galvin, J.M. 2016. *Ground engineering and management in coal mining.* Springer.

Griffith, A.A. 1920. The phenomena of rupture and flow in solids. *Philosophical Transactions*, Series A, 221, 163–198.

Holland, C.T. 1955. Rockburst or bumps in coal mines, *Colliery Engineering*, 145–153.

Itasca, 2014. Universal Distinct Element Code (UDEC) Version 6.0. Itasca Consulting Group Inc., Minneapolis.

Jiang, Y., Pan, Y., Jiang, F., Dou, L. and Ju, Y. 2014. State of the art review on mechanism and prevention of coal bumps in China, *Journal of China Coal Society*, 39(2), 205–213.

Mikula, P. and Brown, B. 2018. The need for additional dynamic testing methods for ground support elements, in *Proceedings of the 3rd International Conference on Rock Dynamics and Applications (RocDyn-3)*, 26-27 June 2018, Trondheim, Norway.

Peng, S.S. 2008. *Coal Mine Ground Control*, Department of Mining Engineering/College of Engineering and Mineral Resources. West Virginia University, USA.

Salamon, M., Galvin, J., Hocking, G. and Anderson, I. 1996. Coal pillar strength from back-calculation. Part of a New South Wales Joint Coal Board Research Project-Strata Control for Coal Mine Design. Sydney, Australia: University of New South Wales, School of Mines.

Stacey, T.R. 2012. A philosophical view on the testing of rock support for rockburst conditions, *Journal of the Southern African Institute of Mining and Metallurgy*, 112, 703–710.

Tahmasebinia, F., Zhang, C., Canbulat, I., Vardar, O. and Saydam, S. 2018a. Computing the damage and fracture energy in a coal mass based on joint density, *International Journal of Mining Science and Technology*, 28, 813–817.

Tahmasebinia, F., Zhang, C., Canbulat, I., Vardar, O. and Saydam, S. 2018b. Numerical and analytical simulation of the structural behaviour of fully grouted cable bolts under impulsive loading, *International Journal of Mining Science and Technology*, 28, 807–811.

2019 Rock Dynamics Summit– Aydan et al. (eds)
© 2019 Taylor & Francis Group, London, ISBN 978-0-367-34783-3

Fracture tests on rocks under different loading rates: Progressive fracture mechanism and rate dependence of fracturing profiles

F. Dai

State Key Laboratory of Hydraulics and Mountain River Engineering, College of Water Resources and
Hydropower, Sichuan University, P.R. China

ABSTRACT: Myriads of testing methods for measuring fracture toughness of rocks under different load-ing rates have been proposed in the literature. However, the reason for the measured discrepancy of fracture toughness using different methods has not been thoroughly revealed. In this Keynote, we report our recent numerical and theoretical investigations in this regard. First, progressive fracture mechanisms of typical fracture testing specimens are systematically investigated by numerical modelling. The Fracture Process Zone (FPZ) neglected in Linear Elastic Fracture Mechanics (LEFM) is simulated. The FPZ lengths of the typical specimens and their effects on the K_{Ic} measurements are compared. Second, theoretical derivations are conducted to estimate and compare FPZ lengths of these testing specimens for conventional rocks, and their effects on the fracture toughness measurements are assessed using the effective crack model. The FPZs are revealed to be partly responsible for the measuring discrepancy among these methods. Third, our study reveals that, in dynamic fracture tests using split Hopkinson pressure bar (SHPB), even if the dynamic force balance can be achieved by careful pulse shaping, the positions of the critical fracturing profiles remain highly dependent on the loading rates. The loading-rate-dependent cracking profiles of chevron-notched specimens cannot be ignored in order to accurately determine the dynamic rock fracture toughness.

1 INTRODUCTION

Natural rocks often contain a large number of cracks with sizes ranging from a micro scale to possibly even a few kilometers. On one hand, the sizes, den-sities and orientations of cracks have influences on mechanical behavior of rocks. On the other hand, the failure/breakage of rocks is often a result of crack initiation and propagation. Thus, cracks play an important role in rock failure processes; it is impor-tant to study crack propagation in rocks and espe-cially the resistance of rocks against crack/fracture growth. To this end, rock fracture mechanics have been widely applied to many diverse areas associ-ated with brittle fracture/cracking of rocks.

In rock fracture mechanics, the most important parameter is the mode I (i.e., tensile) fracture tough-ness K_{Ic}, also known as the critical mode I stress intensity factor (SIF), which represents the resist-ance of rocks against opening-mode crack propaga-tion. Since the 1960s, diverse testing methods with different specimen geometries and loading config-urations have been developed for K_{Ic} measurements of rocks; some of them are shown in Table 1. Among them, the first four methods (i.e., CB, SR, CCNBD and SCB) have been suggested by the International Society for Rock Mechanics (ISRM). Many stud-ies indicated that different testing methods usually yield inconsistent fracture toughness results. For example, a variation of ~20-30% was often observed between the toughness values measured using the

SR and CB specimens for the same rock. Moreover, K_{Ic} measured using the ISRM-suggested CCNBD method is significantly lower than that using the SR and CB methods. In addition, the toughness result of the SCB specimen is usually very conservative compared with those of other specimens.

The measuring discrepancy of K_{Ic} for these ISRM-suggested methods has received much atten-tion from rock fracture mechanics community. The potential explanations offered for this discrepancy include rock anisotropy, size effects, the storage conditions of specimens, inaccuracies of dimen-sionless SIFs of specimens, etc. However, it has not been thoroughly clarified why the ISRM-suggested CCNBD and SCB methods usually produce consid-erably conservative fracture toughness compared to the CB and SR methods, and the progressive frac-ture mechanisms of these typical specimens have not been fully understood.

Some studies have noticed the importance of fracture process zone (FPZ) on fracture behavior of rock-like materials (Bazant and Kazemi 1990). Due to the FPZ, the classical Linear Elastic Fracture Mechanics (LEFM) theory is not strictly applicable to rock fracture issues. Only when the FPZ is com-parably small, the LEFM is approximately appli-cable and then the measured value of the fracture toughness may approximate to the true value of it. However, in conventional fracture tests, the fracture toughness is just calculated using the critical load (e.g., the maximum load) and the critical crack length

Table 1 Some typical methods for K_{Ic} measurements of rocks.

Testing method	Loading type
Chevron bend (CB) (Ouchterlony 1988; Wei et al. 2016a)	Three-point bending
Chevron-notched short rod (SR) (Ouchterlony 1988)	Direct tension
Cracked chevron notched Brazilian disc (CCNBD) (Fowell 1995; Xu et al. 2016a)	Brazilian-type compression
Semi-circular bend (SCB) (Kuruppu et al. 2014; Wei et al. 2016b,2017a)	Three-point bending
Cracked chevron notched semi-circular bend (Wei et al. 2015;2016c)	Three-point bending
Cracked straight through Brazilian disc (Wei et al. 2018a)	Brazilian-type compression
Straight notched disc bend (Aliha and Bahmani 2017)	Three-point bending
Edge-cracked ring (Mirsa-yar 2014)	Indirect tension
Edge-cracked triangular specimen (Aliha et al. 2013)	Three-point bending
Chevron-notched rectangular beam bend (Jenkins et al. 1987)	Three-point bending
Chevron notched short rod bend (Wei et al. 2018b)	Three-point bending

ac, generally determined prior to tests from the theory of LEFM. Such a critical crack length ac underestimates the true value of the effective crack length when unstable fracture occurs, and thus, leads to a certain error in the fracture toughness measurement, Until now, FPZs in the ISRM-suggested specimens and their effects on the K_{Ic} measurements have never been evaluated and compared. Thus, the FPZs and their effects are estimated and compared in this study by numerical modelling and theoretical analysis.

On the other hand, the sample configurations used for quasi-static rock fracture tests have been extended to dynamic loading conditions for characterizing the dynamic mode I fracture toughness of rocks using the Split Hopkinson Pressure Bar (SHPB) testing system (Zhang and Zhao 2014; Zhang et al. 2000; Dai et al. 2010a; Dai et al. 2011; Wang et al. 2010; Du et al. 2017; Xu et al. 2018). In particular, the straight through notched semi-circular bend (NSCB) method using the SHPB testing system has been suggested by ISRM to measure the dynamic mode I fracture toughness of rocks (Zhou et al. 2012). Compared to the specimens with straight through notched cracks, the chevron notched

specimens require no tedious fabrication of a sharp crack, and thus they are widely used in rock fracture toughness measurements. As the latest ISRM-suggested chevron notched specimen for static fracture tests (Fowell 1995), the CCNBD method can be conveniently applied to the SHPB loading system in a diametrical compression frame, compared with the other two ISRM-suggested chevron-notched specimens (SR and CB) (Ouchterlony 1988).

Under high-rate loading conditions, however, the progressive fracturing of CCNBD specimens has never been assessed before. Most importantly, in the existing dynamic fracture tests using the SHPB system, it is generally believed that, as long as a dynamic force equilibrium of the sample can be achieved by careful pulse shaping, the test sample will be in a quasi-static state and a quasi-static data reduction can be used. Indeed, for straight through notched specimens, e.g. SCB, this practice has been verified via the discrete element method (DEM) simulations (Xu et al. 2016b). However, the physical fracturing (especially the profile of the critical fracture front) of CCNBD specimens under high loading rates, can be distinct from that under static loads. Whether that concept can be directly adopted has not been investigated. Therefore, sufficient attention should be paid to the realistic progressive fracture mechanism of the CCNBD rock specimen subjected to dynamic loads.

2 TYPICAL FRACTURE TOUGHNESS TESING METHODS

2.1 The CB method

The CB sample geometry is schematically shown in Figure 1, where the related nomenclatures are also given. This sample configuration has a chevron notch and is loaded by three-point bending. The dimensions suggested by ISRM for the standard CB specimen are tabulated in Table 2. The mode I fracture toughness K_{Ic}^* can be calculated by the following formula (Ouchterlony 1988):

$$K_{Ic} = \frac{P_{max}}{D^{1.5}} Y_c \qquad (1)$$

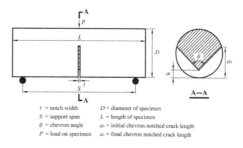

t = notch width	D = diameter of specimen
S = support span	L = length of specimen
θ = chevron angle	a_0 = initial chevron notched crack length
P = load on specimen	a_1 = final chevron notched crack length

Figure 1 The CB sample geometry.

Table 2 ISRM-suggested value or range for the geometric parameters of the standard CB specimen.

Parameter	Suggested value or range
D	> 10 × grain size
L	$4D$
a_0	$0.15D$
θ	90°
t	$\leqq 0.03D$ or 1mm
S	$3.33D$

where Y_c is the critical/minimum dimensionless SIF. For the standard CB specimen, the relation between the dimensionless SIF (Y) and the normalized crack length ($\alpha = a/R$) can be written as (Wei et al. 2016a):

$$Y = \begin{cases} 20.363 - 40.422\alpha + 36.531\alpha^2, & \alpha < 0.56 \\ 15.852 - 22.491\alpha + 19.048\alpha^2, & \alpha \geq 0.56 \end{cases} \quad (2)$$

2.2 The SR method

A schematic of the SR sample geometry and the related nomenclatures are given in Figure 2. Table 3 present the standard geometric dimensions of the SR specimen. During testing, the SR specimen is loaded by applying a direct tensile load perpendicular to the initial notched plane. The mode I fracture toughness is also calculated using Eq. (1). For the standard SR specimen, the variation of Y versus α can be expressed as (Wei et al. 2016a):

$$Y = \begin{cases} 97.244 - 96.447\alpha + 31.347\alpha^2, & \alpha < 1.56 \\ 87.799 - 81.096\alpha + 25.402\alpha^2, & \alpha \geq 1.56 \end{cases} \quad (3)$$

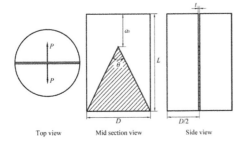

Figure 2. The SR sample geometry.

Table 3. ISRM-suggested value or range for the geometric parameters of the standard SR specimen.

Parameter	Suggested value or range
D	> 10 × grain size
L	$1.45D$
a_0	$0.48D$
θ	54.6°
t	$\leqq 0.03D$ or 1mm

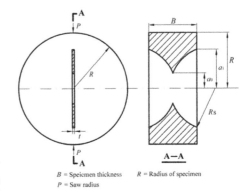

B = Speicmen thickness R = Radius of specimen
P = Saw radius

Figure 3. The CCNBD sample geometry.

Table 4. ISRM-suggested value or range for the geometric parameters of the standard CCNBD specimen.

Parameter	Suggested value or range
D	75.0 mm
B	30.0 mm ($\alpha_B = B/R = 0.8$)
a_0	9.89 mm ($\alpha_0 = a_0/R = 0.2637$)
a_1	24.37 mm ($\alpha_0 = a_1/R = 0.65$)
Rs	26.0 mm ($\alpha_0 = Rs/R = 0.6933$)
t	$\leqq 1.5$ mm

2.3 The CCNBD method

Figure 3 illustrates the CCNBD specimen geometry; Table 4 describes the standard dimensions for the CCNBD specimen. The CCNBD specimen has two chevron notches, which are perpendicular to those of CB and SR if all of them are prepared from the same rock core. During the test, the CCNBD specimen is subjected to a Brazilian-type indirect tensile load. K_{Ic} of the CCNBD specimen is calculated using the following formula (Fowell 1995):

$$K_{Ic} = \frac{P_{max}}{B\sqrt{D}} Y_c \quad (4)$$

For the standard sample geometry, the relation between Y and α is determined as (Wei et al. 2016a):

$$Y = \begin{cases} 3.427 - 10.129\alpha + 10.353\alpha^2, & \alpha < 0.5 \\ 2.850 - 7.505\alpha + 7.420\alpha^2, & \alpha \geq 0.5 \end{cases} \quad (5)$$

2.4 The SCB method

Figure 4 presents a schematic of the SCB specimen; Table 5 lists the recommended configurations for the SCB specimen, which has a straight-through notch and is loaded by three-point bending. The mode I fracture toughness can be calculated using the following formula (Kuruppu et al. 2014):

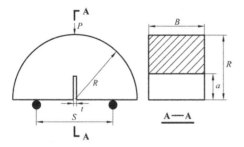

Figure 4. The SCB sample geometry.

Table 5. ISRM-suggested value or range for the geometric parameters of the SCB specimens.

Parameter	Suggested value or range
R	Larger of 5 × grain size or 38 mm
B	Larger of $0.4D$ or 30 mm
a	$0.4 \leqq a/R \leqq 0.6$
S	$0.5 \leqq S/2R \leqq 0.8$

$$K_{Ic} = \frac{P_{\max} \sqrt{\pi a}}{2RB} Y \qquad (6)$$

According to Wei et al. (2016a), Y of the SCB specimens can be determined from the following formula:

$$Y = \left(0.4122 + 5.06355\frac{S}{D}\right) + \left(-16.65 + 3.319\frac{S}{D}\right)\alpha$$
$$+ \left(52.939 + 76.910\frac{S}{D}\right)\alpha^2 + \left(-67.027 - 257.726\frac{S}{D}\right)\alpha^3 \qquad (7)$$
$$+ \left(29.247 + 252.8\frac{S}{D}\right)\alpha^4$$

3 NUMERICAL MODELLING OF THE TESTING METHODS

The finite element code RFPA3D (Rock Failure Process Analysis-3D) (Tang and Kaiser 1998; Liang et al. 2012) is employed for the modelling. The validity of the RFPA3D code for modeling crack growth in rock has been verified through diverse investigations on the fracture behavior of rock specimens. In the modelling of the K_{Ic} testing specimens, the input mechanical parameters are determined by simulating Brazilian tests and uniaxial compression tests of Changtai granite. Figure 5 gives four generated numerical models for the ISRM-suggested specimens. The CB, SR and CCNBD models have the same dimensions as the standard specimen geometries. As for the SCB model, $\alpha = 0.5R$ and three different support spans ($S/D = 0.8$, 0.5 and 0.3) are considered. For

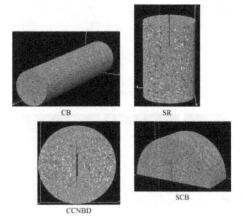

Figure 5. Four typical numerical models.

Figure 6. The evolution of the minimum principal stress in the notched plane of the CB model.

all the models, $D = 75$ mm. During the modelling, the loads are controlled by displacements and are applied with a rate of 0.002 mm per step.

Figure 6 shows the evolution of the minimum principal stress in the notched plane of the CB mode. It can be observed that as the load is applied on the specimen, a severe stress concentration occurs at the tip of the chevron notched ligament. When the loading force reaches around 40% of its peak value, the main crack is initiated from the chevron ligament tip. Then the main crack continues to propagate along the ligament toward the load point.

Special attention should be paid to the peak force stage, because the corresponding loading force and crack length are crucial for the fracture toughness determination. It can be found the critical crack front is a little curved due to three-dimensional effect. As a straight through crack front is usually assumed in the chevron notched samples, the simulated curved crack front at the onset of unstable crack growth is approximately treated as a straight crack front, on the basis of constant crack area. Then the simulated critical crack length (a_c) is estimated as $0.759R$. Note that, according to LEFM, ac should be $0.56R$ for the CB specimen (Wei et al. 2016a). The difference between the numerically simulated a_c value and the LEFM-based a_c value can be interpreted as follows.

Figure 7. The evolution of the minimum principal stress in the notched plane of the SR model.

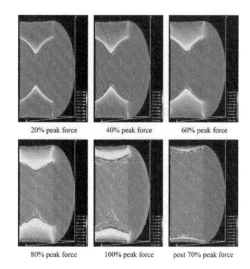

Figure 8. The evolution of the minimum principal stress in the notched plane of the CCNBD model.

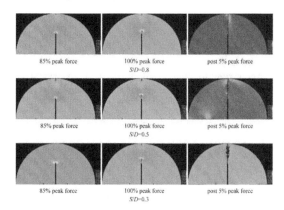

Figure 9. The evolutions of the minimum principal stresses in the central cross-sections of the SCB models.

In the LEFM, the stress at a crack tip should be infinite due to the singularity of $r^{-1/2}$ (where r is the distance from the crack tip). In fact, for rock materials, the region with "excessive stresses/strains/energies" will generate microcracking and even a length of macrocrack; the combination of the microcracking region and the unanticipated macrocrack can be regarded as the fracture process zone (FPZ). Thus, the difference between the simulated and the LEFM-based a_c values can be regarded as the length of FPZ. For a given specimen, the shorter the FPZ, the closer the fracture of the specimen is to the LEFM theory. Thus, the FPZ length is determined as $0.199R$ for the numerical CB test.

Figure 7 shows the evolution of the numerically simulated minimum principal stress in the notched plane of the SR specimen. The main crack is initiated from the chevron ligament tip when the loading force is between ~40-60% of its peak value. After that, the main crack propagates along the ligament. From Figure 7, the critical effective crack length can be determined as $1.739R$. Since the critical crack length determined in the LEFM theory is $1.56R$, the FPZ length can be estimated as $0.179R$ for the numerical SR test.

Figure 8 presents the stress evolution in the notched plane of the CCNBD specimen. As the CCNBD model is under mode I loading, high stress concentration appears at tips of the chevron ligaments. The main crack start to propagate after the loading force reaches to 40-60% of its peak value. Most importantly, it seems that the crack front at the peak force stage is already beyond the root of the chevron ligament. The numerically simulated critical effective crack length is determined as $0.722R$. Note that the critical crack length determined from the LEFM theory is only $0.5R$, thus the numerically simulated FPZ length can be estimated as $0.222R$.

Figure 9 gives the stress distributions in the center sections of the SCB models. For any of the three support spans, there is always a length of FPZ. The numerically determined FPZ lengths are $0.076R$, $0.089R$ and $0.121R$ for $S/D = 0.8$, 0.5 and 0.3, respectively, suggesting that the FPZ is longer for a smaller support span. In addition, the smaller the support span, the wider the FPZ in the SCB model.

In the effective crack model proposed by Nallathambi and Karihaloo (1986) it is considered that a more representative or size-independent/geometry-independent fracture toughness value can be determined using the effective crack length, which has taken the FPZ into account. The concept of effective crack has also been supported/employed by many studies and documents (Jenq and Shah 1985; Xu and Reinhardt 1999), and therefore, it is used in this study to estimate the effects of FPZs on the fracture toughness determinations. According to the K_{Ic} calculation formulas, the influence of FPZ on the K_{Ic} measurement can be evaluated by

Table 6 Estimates of the influences of the numerically simulated FPZs on the K_{Ic} measurements

Test	α_c	$Y(\alpha_c)$	$\alpha\alpha_{ec}$	$Y(\alpha_{ec})$	Effect of FPZ
CB	0.56	9.167	0.759	9.755	6.0%
SR	1.56	23.06	1.739	23.59	2.3%
CCNBD	0.50	0.946	0.722	1.300	27%
SCB (a/R = 0.5, S/D = 0.8)	0.5	6.534	0.576	7.917	23%
SCB (a/R = 0.5, S/D = 0.5)	0.5	3.612	0.589	4.664	29%
SCB (a/R = 0.5, S/D = 0.3)	0.5	1.672	0.621	2.631	43%

$$K_a/K_c = \begin{cases} Y(\alpha_c)/Y(\alpha_{ec}) & \text{(CB, SR, CCNBD)} \\ \left[Y(\alpha)\sqrt{\alpha}\right]/\left[Y(\alpha_{ec})\sqrt{\alpha_{ec}}\right] & \text{(SCB)} \end{cases} \quad 8$$

where K_a is the apparent fracture toughness determined by ignoring the FPZ. K_c denotes the fracture toughness obtained using the critical effective crack length a_{ec}, and it can approximate the inherent fracture toughness accordiong to the effective crack model. $Y(\alpha_{ec})$ denotes the dimensionless SIF at the normalized critical effective crack length $\alpha_{ec} = a_{ec}/R$ and $Y(\alpha_c)$ represents the dimensionless SIF at the conventional normalized critical crack length $\alpha_c = a_c/R$. Table 3 summarizes the effects of FPZs on the K_{Ic} measurements. It can be found that the K_{Ic} measurements using the CCNBD and SCB specimens are influenced much more seriously by FPZs than those using CB and SR specimens. Moreover, with a smaller support span, the effect of FPZ is more significant for the SCB test.

4 THEORETICAL ESTIMATIONS OF FRACTURE PROCESS ZONES

The maximum normal stress criterion suggested by Schmidt and Lutz (1979) is applied herein to estimate the length of FPZ. It is assumed in this criterion that the FPZ is caused by excessive tensile stresses near the crack tip. Thus, the FPZ length along the direction of crack propagation can be estimated as the distance from the crack tip to the point where the tensile stress is equal to the tensile strength (σ_t) of rock.

To estimate the FPZ lengths of the ISRM-suggested standard specimens at their critical stages (only $\alpha = 0.5$ and $S/D = 0.8$ are considered for the SCB specimen), tensile stresses in their critical residual ligaments are needed and thus determined from finite element analysis. To qualitatively estimate the FPZ lengths, some mechanical parameters of these two models are also required. Thereupon, $K_a = k$ MPa·m$^{0.5}$ and $\sigma_t = mk$ MPa are assumed in the models. This assumption is acceptable, because for a given specimen for determining K_{Ic}, the relation between its apparent fracture toughness K_a and the

tensile strength σ_t of the rock can be mathematically depicted as

$$\sigma_t/\sigma_0 = m \times \left(K_a/K_0\right) \quad (9)$$

where $\sigma_0 = 1$ MPa, $K_0 = 1$ MPa·m$^{0.5}$, and m is not only dependent on properties of the rock but also on the fracture toughness testing method.

As the critical dimensionless SIFs of the CB, SR, CCNBD and SCB specimens are 9.167, 23.06, 0.946 and 6.534, respectively, the critical loads for the intact CB, SR, CCNBD and SCB specimens are determined as 484.84k N, 1219.63k N, 4727.42k N and 772.32k N, respectively, where k is the value of apparent fracture toughness corresponding to the unit of MPa·m$^{0.5}$. After applying appropriate loads and boundary conditions on finite element models, the normal stresses σ_N in the critical residual ligaments of the specimens can be determined, as shown in Figure 10 and fitted in Eq. (10).

$$\left(\sigma_N/k\right) = \begin{cases} -2.7553 - 3.5096 \times \ln\left(X/l_{rc} - 0.0082\right) & \text{(CB)} \\ -3.0704 - 3.0415 \times \ln\left(X/l_{rc} - 0.0167\right) & \text{(SR)} \\ 1.2142 - 3.9150 \times \ln\left(X/l_{rc} - 0.0269\right) & \text{(CCNBD)} \\ -5.3015 - 5.8585 \times \ln\left(X/l_{rc} - 0.0182\right) & \text{(SCB)} \end{cases} \quad (10)$$

where the fitted expression is suitable only for $\sigma_{yy}/k > 3$. By substituting $\sigma_t = mk$ MPa into Eq. (10), the ratio of the FPZ length (l_{FPZ}) to the critical residual ligament length (l_{rc}) for the specimens can be approximatively determined as follows:

$$l_{FPZ}/l_{rc} = \begin{cases} e^\wedge\left(-0.7851m - 0.2849\right) + 0.0082 & \text{(CB)} \\ e^\wedge\left(-1.0095m - 0.3288\right) + 0.0167 & \text{(SR)} \\ e^\wedge\left(0.3101m - 0.2554\right) + 0.0269 & \text{(CCNBD)} \\ e^\wedge\left(-0.9049m - 0.1707\right) + 0.0182 & \text{(SCB)} \end{cases} \quad (11)$$

By substituting l_{rc} of these specimens into Eq. (11), the FPZ lengths of the specimens are estimated as

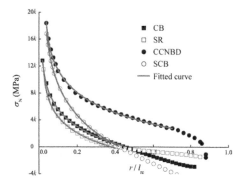

Figure 10. The stress component σ_{yy} in the critical residual ligaments of the SR, CB, CCNBD and SCB specimens.

$$l_{FPZ} = \begin{cases} \left[e^\wedge(-0.7851m-0.2849)+0.0082\right]\times 1.44R & \text{(CB)} \\[6pt] \left[e^\wedge(-1.0095m-0.3288)+0.0167\right]\times 1.34R & \text{(SR)} \\[6pt] \left[e^\wedge(0.3101m-0.2554)+0.0269\right]\times 0.5R & \text{(CCNBD)} \\[6pt] \left[e^\wedge(-0.9049m-0.1707)+0.0182\right]\times 0.5R & \text{(SCB)} \end{cases}$$
(12)

According to the effective crack model, the critical effective crack lengths of the specimens can be written as

$$a_{ec} = \begin{cases} \left\{\left[e^\wedge(-0.7851m-0.2849)+0.0082\right]\times 1.44+0.56\right\}R & \text{(CB)} \\[6pt] \left\{\left[e^\wedge(-1.0095m-0.3288)+0.0167\right]\times 1.34+1.56\right\}R & \text{(SR)} \\[6pt] \left\{\left[e^\wedge(0.3101m-0.2554)+0.0269\right]\times 0.5+0.5\right\}R & \text{(CCNBD)} \\[6pt] \left\{\left[e^\wedge(-0.9049m-0.1707)+0.0182\right]\times 0.5+0.5\right\}R & \text{(SCB)} \end{cases}$$
(13)

Then the normalized values of the critical effective crack lengths are as follows.

$$\alpha_{ec} = \begin{cases} \left[e^\wedge(-0.7851m-0.2849)+0.0082\right]\times 1.44+0.56 & \text{(CB)} \\[6pt] \left[e^\wedge(-1.0095m-0.3288)+0.0167\right]\times 1.34+1.56 & \text{(SR)} \\[6pt] \left[e^\wedge(0.3101m-0.2554)+0.0269\right]\times 0.5+0.5 & \text{(CCNBD)} \\[6pt] \left[e^\wedge(-0.9049m-0.1707)+0.0182\right]\times 0.5+0.5 & \text{(SCB)} \end{cases}$$
(14)

Thus, the dimensionless stress intensity factors (Y) corresponding to the critical effective crack lengths can be obtained for the specimens. Subsequently, the effects of FPZs on these fracture toughness tests can be reflected by Eq. (8).

Since the FPZ lengths are related to values of m, the effects of FPZs on these fracture toughness tests are also dependent on m, as shown in Figure 11(a). For all the specimens, when m is rather small (e.g., $m = 2$), the test error induced by ignoring the FPZ is severe; with the increase of m, the test error becomes smaller and smaller. Most importantly, for any fixed value of m, the test error induced by

ignoring the FPZ is more significant in the SCB and CCNBD tests than in the CB and SR tests.

To explicate more clearly the difference between these methods in terms of the influence of l_{FPZ} on the K_{Ic} determination, the following analysis is made. A parameter n is defined as follows:

$$\sigma_t/\sigma_0 = n\times(K_c/K_0) \tag{15}$$

where n represents the ratio of values of tensile strength to inherent fracture toughness for a rock material. The value of n only depends on properties of the rock. According to Eqs. (9) and (15), the following relation can be obtained.

$$n = m\times(K_a/K_c) \tag{16}$$

Since the variations of K_a/K_c versus m for the specimens have been obtained (Figure 11(a)), the curves of K_a/K_c versus n can also be drawn, as given in Figure 11(b). This figure presents that, the CB and SR methods appears to be superior to the CCNBD and SCB methods for quantifying the fracture toughness of rocks, due to the smaller effects nduced by the presence of FPZ.

Because l_{FPZ} is related to m, and the relation between m and n can be derived from Figure 11, the relation between l_{FPZ} and n can also be determined, as plotted in Figure 12. For all the specimens, when n is small, l_{FPZ} cannot be ignored; only when n is relatively large, the FPZ length becomes insignificant compared with the specimen size. Most importantly,

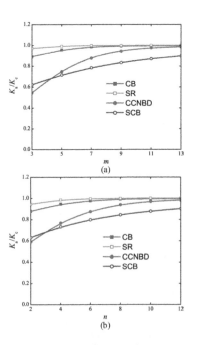

Figure 11. Effects of FPZs on K_{Ic} measurements for the tests with different (a) m and (b) n values.

35

for the same rock material, l_{FPZ}/l_{rc} is larger in the CCNBD and SCB tests than in the CB and SR tests.

5 LOADING-RATE-DEPENDENT RACTURING PROFILES

The CCNBD method for determining quasi-static fracture toughness, along with its measuring assumptions, is extended to SHPB tests to measure dynamic fracture toughness. The CCNBD-SHPB test and its measuring assumptions are schematically depicted in Figure 13. The idea of the dynamic CCNBD method in SHPB tests assumes that the CCNBD specimen is in a quasi-static state as long as the dynamic forces at the specimen-bar interfaces are balanced. Then dynamic mode I fracture toughness can be calculated via the quasi-static method (Equations 5-6). In this way, the measuring principle of the dynamic CCNBD-SHPB tests directly adopts that of the ISRM-suggested CCNBD method for the static mode I fracture toughness measurements (Fowell 1995). In the tests, it postulates that the primary cracks initiate from the chevron-notched tip and grow symmetrically towards both loading ends with a straight through cracking front (Figure 13c). When the loading force reaches its peak value P_{max}, i.e. the critical stage to calculate the fracture toughness, the dimensionless SIF reaches its minimum Y_c, which corresponds to the critical fracture profile.

Due to the challenges of directly observing the internal fracturing of the CCNBD specimen in the laboratory tests especially under high loading rates,

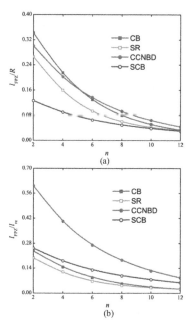

Figure 12. Theoretical estimates of the FPZ lengths of the ISRM-suggested specimens: (a) the ratio of l_{FPZ} to R; (b) the ratio of l_{FPZ} to l_{rc}.

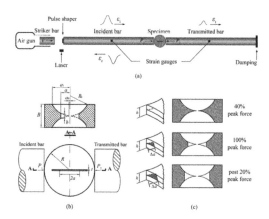

Figure 13. Schematics of (a) the SHPB testing system, (b) the CCNBD sample geometries, and (c) the conventional measuring assumption of the CCNBD method.

DEM is employed to simulate the dynamic CCNBD-SHPB tests for it provide a convenient access to physic details of the fracture tests. The DEM model of the CCNBD-SHPB testing system is established as shown in Figure 14. A wide range of loading rates covering 27~140 GPa·m$^{0.5}$/s can be achieved via changing the velocity of the striker bar through the implementation of the numerical pulse shaper.

The dynamic equilibrium of nine loading cases is assessed in Figure 15 along with the test without pulse shaping. Dynamic forces on both sides of the specimen match well with each other under the loading rates lower than 71 GPa·m$^{0.5}$/s. This equilibrium state is also demonstrated by the slight fluctuation of force equilibrium coefficient μ around the zero level throughout the loading period. When the loading rate reaches 83 GPa·m$^{0.5}$/s, μ fluctuates to a larger extent as the loading rate increases. Under a rather high loading rate of 140 GPa·m$^{0.5}$/s, the incident force and the transmitted force can never match, similar to the results of the test without pulse shaper.

The critical fracture profiles of CCNBD specimens under a range of loading rates from 27 GPa·m$^{0.5}$/s to 71 GPa·m$^{0.5}$/s, where dynamic force equilibrium is satisfied, are exhibited in Figure 16. The three-dimensional fracture profiles are demonstrated via

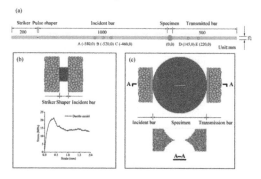

Figure 14. Three-dimensional DEM model of the CCNBD-SHPB testing system with a pulse shaper.

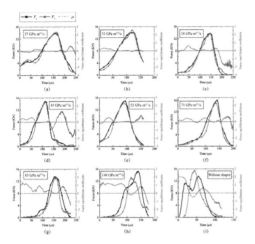

Figure 15. Dynamic force check on both loading sides of the CCNBD specimen in SHPB testing under different loading rates.

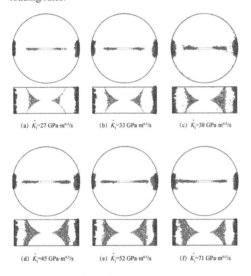

(a) $\dot{K}_I$=27 GPa·m$^{0.5}$/s   (b) $\dot{K}_I$=33 GPa·m$^{0.5}$/s   (c) $\dot{K}_I$=38 GPa·m$^{0.5}$/s

(d) $\dot{K}_I$=45 GPa·m$^{0.5}$/s   (e) $\dot{K}_I$=52 GPa·m$^{0.5}$/s   (f) $\dot{K}_I$=71 GPa·m$^{0.5}$/s

Figure 16. Loading rate dependence of the critical fracture profile of the CCNBD specimen in the view of the direction perpendicular to the disc surface and the direction perpendicular to the diametric cut plane through the notch tip.

the AE distribution in the view of the direction perpendicular to the disc surface and the diametrically cut plane through the chevron notch tip. It can be seen that the critical fracture propagates farther away from the initial crack with the increasing loading rate. More specifically, the critical fracture is still confined within the chevron ligament for a lower loading rate range (e.g. 27 to 38 GPa·m$^{0.5}$/s), while it surpasses the root of the chevron notched ligament under a higher loading rate regime (e.g. 45 to 71 GPa·m$^{0.5}$/s).

According to the measuring principle of the CCNBD method, the critical dimensionless SIF (Y_c) plays an essential role in accurately measuring the mode I fracture toughness values. However, the Y_c value is numerically calibrated based on a certain hypothesis highly related to the progressive fracturing process

of the CCNBD specimen especially the critical fracture profile. Previous SHPB tests on chevron notched specimens simply adopt the measuring principle in the static counterpart that assumes a constant critical fracture profile. Our numerical assessment reveals that the actual fracturing of the CCNBD specimen in SHPB tests is significantly loading rate dependent. This scenario can influence severely the measuring accuracy of the dynamic mode I fracture toughness of rocks. An unexamined extension of the quasi-static assumptions to SHPB testing is problematic for measuring the dynamic mode I fracture toughness using chevron notched specimens involving the CCNBD specimen.

6 CONCLUSIONS

The progressive fracture mechanisms of typical fracture testing specimens (i.e., the ISRM-suggested specimens) are systematically revealed by numerical modelling. The numerical results show that unstable crack propagation always occurs at the crack length which is larger than the critical crack length determined from the LEFM theory. Thus, the fracture process zone neglected in LEFM is modeled. The CCNBD and SCB tests are found to be affected more significantly by FPZs than the CB and SR tests.

To further check the numerical results, theoretical investigations are conduced to estimate and compare the FPZs and their effects on those four tests for conventional rock materials. The results indicate that the CCNBD and SCB tests are indeed easier to significantly underestimate the fracture toughness due to the presence of FPZ than the CB and SR tests. Only when the rock has a relatively large ratio of values of tensile strength to fracture toughness, K_{Ic} results of the specimens are close to each other.

The dynamic fracturing of chevron-notched specimens in SHPB testing is numerically investigated via assessing the acoustic emission evolution. As the loading rate increases, the critical fracture profile features significant rate dependence even if the dynamic equilibrium can be achieved by careful pulse shaping. None of the critical fracture under varied loading rates matches what is calibrated based on the straight through crack assumption. The rate-dependent critical fracture profile cannot be ignored in the dynamic mode I fracture toughness determination using chevron-notched specimens.

ACKNOWLEDGEMENTS

The authors are grateful for the financial support from the National Program on Key Basic Research Project (no. 2015CB057903), the National Natural Science Foundation of China (no. 51779164).

REFERENCES

Aliha, M.R.M., Hosseinpour, G.R. & Ayatollahi, M.R. 2013. Application of cracked triangular specimen subjected to three-point bending for investigating fracture behavior of rock materials. *Rock Mechanics and Rock Engineering* 46(5): 1023–1034.

Bazant, Z.P. & Kazemi, M.T. 1990. Determination of fracture energy, process zone length and brittleness number from size effect, with application to rock and concrete. *International Journal of Fracture* 44: 111-131.

Dai, F., Chen, R., Iqbal, M.J. & Xia, K. 2010. Dynamic cracked chevron notched Brazilian disc method for measuring rock fracture parameters. *International Journal of Rock Mechanics and Mining Sciences* 47:606–613.

Dai, F., Xia, K., Zheng, H. & Wang, Y.X. 2011. Determination of dynamic rock mode-I fracture parameters us-ing cracked chevron notched semi-circular bend specimen. *Engineering Fracture Mechanics* 78(15):2633–2644.

Du, H.B., Dai, F., Xu, N.W., Zhao, T. & Xu, Y. 2017. Numerical investigation on the dynamic progressive fracture mechanism of cracked chevron notched semi-circular bend specimens in split Hopkinson pressure bar tests. *Engineering Fracture Mechanics* 184: 202–217

Fowell, R.J. 1995. ISRM commission on testing methods. Suggested method for determining mode I fracture toughness using cracked chevron notched Brazilian disc (CCNBD) specimens. *International Journal of Rock Mechanics and Mining Sciences & Geomechanics Abstract* 32(1): 57–64.

Jenkins, M.G., Kobayashi, A.S., White, K.W. & Bradt, R.C. 1987. A 3-D finite element analysis of a chevron-notched, three-point bend fracture specimen for ceramic materials. *International Journal of Fracture* 34: 281–295.

Jenq, Y.S. & Shah, S.P. 1985. Two-parameter fracture model for concrete. *ASCE Journal of Engineering Mechanics* 111(10): 1227–1241.

Kuruppu, M.D., Obara, Y., Ayatollahi, M.R., Chong, K.P. & Funatsu, T. 2014. ISRM-Suggested method for determining the mode I static fracture toughness using semi-circular bend specimen. *Rock Mechanics and Rock Engineering* 47(1): 267–274.

Liang, Z.Z., Xing, H., Wang, S.Y., Williams, D.J. & Tang, C.A. 2012. A three dimensional numerical investigation of fracture of rock specimen containing a pre-existing surface flaw. *Computers and Geotechnics* 45: 19–33.

Mirsayar, M.M. 2014. A new mixed mode fracture test specimen covering positive and negative values of T-stress. *Engineering Solid Mechanics* 2(2): 67–72.

Nallathambi, P. & Karihaloo, B.L. 1986. Determination of the specimen size independent fracture toughness of plain concrete. *Magazine of Concrete Research* 38(135): 67–76.

Ouchterlony, F. 1988. ISRM commission on testing methods. Suggested methods for determining fracture toughness of rock. *International Journal of Rock Mechanics and Mining Sciences & Geomechanics Abstracts* 25: 71–96.

Schmidt, R.G. & Lutz, T.J. 1979. K_{Ic} and J_{Ic} of Westerly granite—effects of thickness and in-plane dimensions// In: Frehman SW, editor. *Fracture mechanics applied to brittle materials.* ASTM STP 678, 166-182.

Tang, C.A. & Kaiser, P.K. 1998. Numerical simulation of cumulative damage and seismic energy release during brittle rock failure—part I: fundamentals. *International Journal of Rock Mechanics and Mining Sciences* 35: 113–121.

Wang, Q.Z., Zhang, S. & Xie, H.P. 2010. Rock dynamic frac-ture toughness tested with holed-cracked flattened Brazilian discs diametrically impacted by SHPB and its size effect. *Experimental Mechanics* 50:877–885.

Wei, M.D., Dai, F., Xu, N.W., Xu Y. & Xia K. 2015. Three-dimensional numerical evaluation of the progressive fracture mechanism of cracked chevron notched semi-circular bend rock specimens. Engineering Fracture Mechanics 134: 286–303.

Wei, M.D., Dai, F., Xu, N.W. & Zhao, T. 2016a. Stress intensity factors and fracture process zones of ISRM-suggested chevron notched specimens for mode I fracture toughness testing of rocks. *Engineering Fracture Mechanics* 168: 174–189.

Wei, M.D., Dai, F., Xu, N.W., Zhao, T. & Xia, K. 2016b. Experimental and numerical study on the fracture process zone and fracture toughness determination for ISRM-suggested semi-circular bend rock specimen. *Engineering Fracture Mechanics.*, 154: 43–56.

Wei, M.D., Dai, F., Xu, N.W., Liu, J.F. & Xu, Y. (2016c) Experimental and numerical study on the cracked chevron notched semi-circular bend method for characterizing the mode I fracture toughness of rocks. *Rock Mechanics and Rock Engineering* 49(5): 1595–1609.

Wei, M.D., Dai, F., Xu, N.W., Liu, Y. & Zhao, T. 2017a. Fracture prediction of rocks under mode I and mode II loading using the generalized maximum tangential strain criterion. *Engineering Fracture Mechanics* 186: 21–38.

Wei, M.D., Dai, F., Xu, N.W., Zhao, T. & Liu, Y. 2017b. Experimental and theoretical assessment of semi-circular bend specimens with chevron and straight-through notches for mode I fracture toughness testing of rocks. *International Journal of Rock Mechanics and Mining Sciences* 99: 28–38.

Wei, M.D., Dai, F., Zhou, J.W., Liu, Y. & Luo J. 2018a. A further improved maximum tangential stress criterion for assessing mode I fracture of rocks considering non-singular stress terms of the Williams expansion. *Rock Mechanics and Rock Engineering* DOI: 10.1007/s00603-018-1524-z.

Wei, M.D., Dai, F., Xu, N.W., Liu, Y. & Zhao, T. 2018b. A novel chevron notched short rod bend method for measuring the mode I fracture toughness of rocks. *Engineering Fracture Mechanics* 190: 1-15.

Xu, S. & Reinhardt, H.W. 1999. Determination of double-K criterion for crack propagation in quasi-brittle materials, Part I: experimental investigation of crack propagation. *International Journal of Fracture* 98: 111-149.

Xu, Y., Dai, F., Zhao, T., Xu, N.W. & Liu, Y. 2016a. Fracture toughness determination of cracked chevron notched Brazilian disc rock specimen via Griffith energy criterion incorporating realistic fracture profiles. *Rock Mechanics and Rock Engineering* 49(8): 3083-3093.

Xu, Y., Dai, F., Xu, N.W. & Zhao, T. 2016b. Numerical investi-gation of dynamic rock fracture toughness determi-nation using a semi-circular bend specimen in split Hopkinson pressure bar testing. *Rock Mechanics and Rock Engineering* 49(3):731-745.

Xu, Y. & Dai, F. (2018) Dynamic response and failure mechanism of brittle rocks under combined compression-shear loading experiments. *Rock Mechanics and Rock Engineering* 51(3): 747-764.

Zhang, Q.B. & Zhao, J. 2014. A review of dynamic experi-mental techniques and mechanical behaviour of rock materials. *Rock Mechanics and Rock Engineering* 47:1411-1478.

Zhang, Z.X., Kou, S.Q., Jiang, L.G. & Lindqvist, P.A. 2000 Ef-fects of loading rate on rock fracture: fracture char-acteristics and energy partitioning. *International Journal of Rock Mechanics and Mining Sciences* 37(5): 745–762.

Zhou, Y.X., Xia, K., Li, X.B., Li, H.B., Ma, G.W., Zhao, J., Zhou, Z.L. & Dai, F. 2012. Suggested methods for determin-ing the dynamic strength parameters and mode-I fracture toughness of rock materials. *International Journal of Rock Mechanics and Mining Sciences* 49:105–112.

2019 Rock Dynamics Summit– Aydan et al. (eds)
© 2019 Taylor & Francis Group, London, ISBN 978-0-367-34783-3

Spatial variation in stress in seismogenic zones in South African gold mines

H. Ogasawara, A. Ishida & K. Sugimura
Ritsumeikan University, Kusatsu, Japan

Y. Yabe, S. Abe & T. Ito
Tohoku University, Sendai, Japan

A. Funato
Fukada Geological Institute, Tokyo, Japan

H. Kato
3D Geoscience, Tokyo, Japan

B. Liebenberg
Moab Khotsong mine, Viljoenskroon, South Africa

G. Hofmann & L. Scheepers
Anglogold Ashanti, Western Levels, South Africa

R. J. Durrheim
University of the Witwatersrand, Johannesburg, South Africa

The SATREPS & DSeis teams
Science and Technology Research Partnership for Sustainable Development (SATREPS)
Drilling into Seismogenic zones of M2.0-5.5 earthquakes in South Africa (DSeis)
Japan, South Africa, Switzerland, US, Germany, India, Israel, and Australia

ABSTRACT: We review our endeavor to probe spatial variation in stress in seismogenic zones in deep hard-rock gold mines in South Africa. Highly stressed ground, especially in remnants or pillars, often increases the risks posed by normal-faulting earthquakes to mining operations. Some of these seismogenic zones were identified by AE monitoring, allowing us to explore them with drill holes shorter than several tens of meters in length. A M5.5 strike-slip earthquake, not a usual mining-induced earthquake, took place in 2014 on an unknown nearly-vertical geological structure, with the upper fringe of the aftershock zone several hundreds of meters below the mining horizons. The aftershock zone was elucidated by in-mine dense geophone network and surface strong motion network. This allowed us to drill two holes of 817 and 700 m length from 2.9 km depth. Integration of in-situ stress measurements by both conventional and new methods allows us to constrain the stress field accurately.

1 INTRODUCTION

Understanding and controlling the behavior of highly-stressed heterogeneous brittle material is a significant challenge. Complicated and difficult conditions are often encountered in deep mines, especially during excavations in highly-stressed rock masses penetrated by faults and dykes, or in the seismogenic zones of natural earthquakes (e.g. Ortlepp, 1997; Brown, 2012; Mori & Ellsworth, 2013). Tau Tona gold mine in South Africa, which Fairhurst (2017) describes as the world's deepest mine in the 20th century, was closed in 2017, and other deep South Africa gold mines will close sooner or later.

With increases in depth or stress during mining, accurate quantitative stress information is critical. This keynote paper reviews our endeavor to obtain this information in deep gold mines in South Africa.

2 IN-SITU STRESS MEASUREMENTS IN SOUTH AFRICA

As reviewed by McGarr & Gay (1978), some important techniquesfor in-situ stress measurement have been developed in South Africa, such as the doorstopper or CSIR triaxial cell. Stacey & Wesseloo (1998), Wesseloo & Stacey (2006), and Handley

(2013) reviewed the data base of hundreds of in situ stress measurements in South Africa. During 1998-2002, the DeepMine and FutureMine research programmes gained knowledge and developed technology to support ultra-deep gold mining in South Africa (e.g. Durrheim 2007). Their agenda included studies of the feasibility of in-situ stress measurement in rock masses prone to earthquakes and drilling-induced damage. The programmes measured stress at a depth of 3351.6m using 10 CSIR, door stopper and CSIRO cells, but only achieved one reliable result with vertical stress of 91±11 MPa. Difficulties were mainly caused by inflexibility in the overcoring procedure (especially drilling direction) in highly stressed ground, which is very prone to core discing or borehole breakout. A SATREPS project "Observational studies to mitigate seismic risks in mines (2010-2015)" downsized the Compact Conical-ended Borehole Overcoring (CCBO) technique (Sugawara & Obara, 1999) from N-size to B-size. It was under the auspices of Japan Science and Technology Agency and Japan International Cooperation Agency (JST and JICA, respectively). CCBO proved to be more suitable for adverse condition and difficult access from surface typical of deep level South African gold mines. Several sequential BX CCBO overcorings could be completed in the same time that it took the CSIRO HI technique to complete a single overcoring (Ogasawara et al. 2012, 2014a). Since then, the BX CCBO technique has been used at the deepest mining level in gold mines, in crush pillars in some platinum mines, and in a copper mine in South Africa. Ogasawara et al. (2014a,c) also discussed the effects of anisotropy and inelasticity on overcoring in highly stressed ground.

With the increase in in-situ stress information, Hofmann et al (2012, 2013) numerically modelled stress with Map3D, a commercial Boundary Element Method software (Wiles 2017), successfully constraining stress changes and strengths (a static peak friction angle: 25 degrees) on the faults of a $M_L 2.1$ earthquake at Mponeng mine (Figs. 1 and 2) and seven $M_w 2.9$-4.0 earthquakes at Tau Tona mine, respectively.

As shown by a thick line in Figure 2, Yabe et al. (2013) drilled into the seismogenic fault of the $M_L 2.1$ ($M_w 2.2$) earthquake that was elucidated by microfracturing monitoring by the Japan German Underground Acoustic Emission Research in South Africa (JAGUARS; Nakatani et al. 2008; Yabe et al. 2009; Naoi et al. 2011). The JAGUARS network could detect events as small as M_w -5 and could locate the events within an accuracy of a few decimeters, being much more sensitive and accurate (for micro events inside the JAGUARS network) than the existing in-mine geophone monitoring system (e.g. Mendecki 1997). Although the most critical sections of the drilled core was disced, seriously damaged, or lost, Yabe et al. (2013) could roughly constrain the upper limit of stress through the integrated analysis of core discing and borehole breakouts (gray bars in Fig. 3), which was consistent with the above-mentioned stress

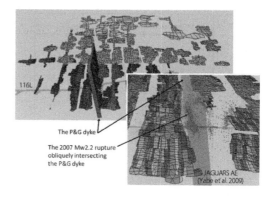

Figure 1. Configuration of thin tabular stope dipping about 20 degree to SSE, the Pink and Green dyke, tunnels on 116 level (3.4 km depth), and aftershocks of the 2007 $M_w 2.2$ event. AE data after JAGUARS (Nakatani et al. 2008; Yabe et al. 2009). Boundary element model after Hofmann et al. (2012).

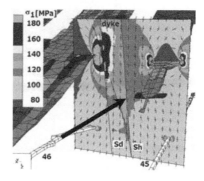

Figure 2. Numerically modelled stress by Map3D on a grid plane normal to the strike of the Pink and Green dyke (Hofmann et al. 2012). Crosscut horizontal tunnels (No. 45 and 46 on 116 L) and inclined thin tabular stope with monthly mining advance typically ~10m/month are also shown. Note that stress is elevated in the dyke and drastically changes at the $M_w 2.2$ fault intersection. Sd and Sh: locations of strainmeters, which recorded strain change consistent with the modelled stress (Ogasawara et al. 2014b).

modelling by Hofmann et al. (2012). Yabe et al. (2015) found that foreshock microfracturing activity accelerated toward the $M_w 2.2$ mainshock, with locations consistent with the fault ride numerically modelled by Hofmann et al. (2012). However, the accuracy of the stress estimate was not as good as measurements constrained with additional data using a new technique that is described below.

3 DIAMETRICAL CORE DEFORMATION ANALYSIS (DCDA) IN-SITU STRESS MEASUREMENTS IN SOUTH AFRICA

Drilled core expands elastically as stress in the rock mass is relieved. In cases where the stress field is anisotropic, the section normal to the core axis is

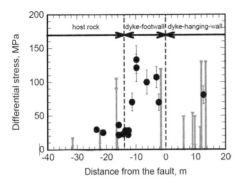

Figure 3. Gray bars: upper limits of differential stress constrained by an integrated analysis of borehole breakout and core discing (Yabe et al. 2013). Black circles with error bars: differential stress measured by DCDA (Abe et al. 2017) with cores from the sections with no core discing.

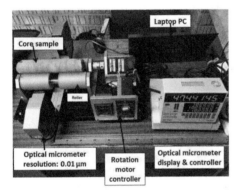

Figure 4. DCDA measurement system (photo after Ishida et al. 2018) used to measure NQ core from Hole A drilled around the M5.5 aftershock zone at Moab Khotsong mine.

elliptical. Funato & Ito (2013) showed the theoretical relationship between the differential stress, S_{max} - S_{min}, in the plane normal to the core axis and the difference in the largest and smallest diameters of the drilled core, d_{max} and d_{min}, respectively: S_{max} - S_{min} = 2G (d_{max} - d_{min})/ $d_0 \sim$ 2G (d_{max} - d_{min})/d_{min}, where G = rigidity, and d_0 = an inner diameter of a drilling bit. Advantages of this method include: (1) only drilled core is needed, (2) no special drilling procedure is required, (3) the method is non-destructive, and (4) the measurement takes much less time than other method to estimate stress using core. Figure 4 shows a measuring system that Funato and Ito (2017) developed. LED light is projected on a core that is rotated on rollers; the width of the core shadow is measured with an optical micrometer with a 10 nm resolution at 2 degree intervals in roll angle of the core. Ito et al. (2013) successfully integrated the DCDA method with hydraulic fracturing stress measurements in the central region of the Kumano forearc basin, Japan, at a water depth of 2054 m, using a 1.6 km riser hole drilled by the International Ocean Drilling Program (IODP) Expedition 319. In the laboratory,

Funato and Ito (2017) drilled core from cubes of five kinds of material while differential stresses of 1, 2.5, 5, 7.5, 10 and 15 MPa were applied. They confirmed that the differences between the applied and DCDA-estimated differential stresses were ± 0.85 MPa.

Abe et al. (2017) applied DCDA to the core (quartzite and gabbro dyke) drilled from the seismogenic fault of the M_w2.2 earthquake at Mponeng mine. Integrated analysis of core discing and borehole breakout enabled them to constrain the stress much better than Yabe et al. (2013) did (Fig. 3).

4 APPLICATION OF DCDA METHOD TO THE CORE FROM A 817M HOLE DRILLIED INTO THE SEISMOGENIC ZONE OF A M5.5 EARTHQUAKE

In deep gold mines in South Africa, vertical stress is the maximum principal stress. Stress in remnants increases with the increase in the mined area of thin tabular reef. Consequently, the typical mechanism of mining-induced earthquakes in South Africa is normal faulting. However, an atypical M_L5.5 strike-slip earthquake occurred in 2014 near Orkney, South Africa. An unknown nearly-vertical geological structure hosted the M_L5.5 event with the upper fringe of the aftershock zone several hundreds of meters below the mining horizons (Figs. 5 and 6). The aftershock zone was elucidated by a surface strong motion network and a dense in-mine geophone network (Manzunzu et al. 2017; Ogasawara et al. 2017). The International Continental Scientific Drilling Program approved our project "Drilling into Seismogenic zones in deep South African gold mines (DSeis)". DSeis drilled two holes of 817 m and 700 m length, respectively, from a chamber 2.9 km below earth's surface, with a ~90m branch hole from the second hole. The holes intersected metamorphosed Archean rock formations (quartzite, siltstone/shale, and basaltic lava of the West Rand Group, Witwatersrand Supergroup), as well as younger intrusives. The drilling direction was carefully chosen to minimize core discing or borehole breakout. Using the S-wave velocity and density logged in the hole, Ishida (2018) carried out DCDA stress measurements on the core from the 817m hole (Figure 7). The dashed line represents numerically- modelled differential stress, assuming that the stress is the same as that measured by the BX CCBO method on 98 level (3.0 km depth) and zero on surface. The DCDA results near the borehole collar were consistent with the BX CCBO results on 98 level. It was notable that higher differential rock stress was measured near the interface between siltstone and intrusive where seepage of brine with pressure >10 MPa was observed at about 420m depth (Rusley et al. 2018). Higher rock stress was also found at the depth (~700m) corresponding to the upper fringe of the M5.5 aftershock zone. Summary and prospect.

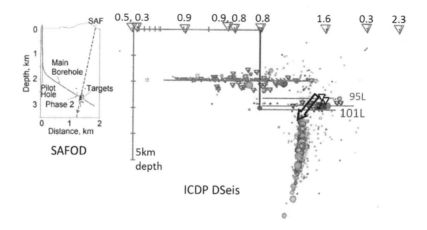

Figure 5. Comparison of scientific drilling projects. Left: San Andreas Fault Observatory at Depth (SAFOD; modified after Hickman et al., 2007. Right: DSeis (Ogasawara et al. 2017). Spheres: aftershocks of the 2014 Orkney $M5.5$ earthquake and earthquakes induced on mining horizons (data courtesy of Anglogold Ashanti ltd). Surface larger triangles: Council for Geoscience National Seismograph Network strong motion meters with PPA, G. Smaller triangles: in-mine network of accelerometers and 4.5Hz geophones. From 95 level (2.9km from surface), we drill holes with a rig used for underground routine drilling of geological exploration.

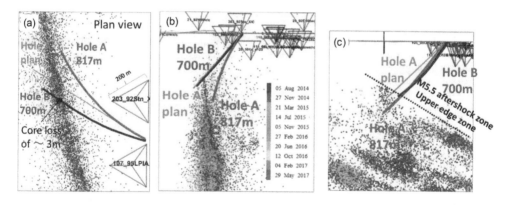

Figure 6. Configuration of the aftershocks of the M5.5 earthquake, Holes A and B, in-mine seismometers, and haulage horizontal tunnels on 95 level. (a): plan view. (b) and (c): vertical sectional view in parallel to and normal to the planar aftershock distribution, respectively. Each seismometer is centered on each triangle with 200m sides. Hole A was deflected too much to intersect the aftershock zone, while, Hole B intersected a fracture zone with ~3m core loss.

In-situ stress measurements in the highly-stressed and geologically-complicated rock mass and adverse conditions encountered in deep South African gold mines were difficult and not cost- or effort-effective until we implemented the BX CCBO overcoring technique. Here we demonstrate that DCDA can supplement CCBO, especially when a stress profile of a long borehole is desired.

5 SUMMARY AND PROSPECT

In-situ stress measurements in the highly-stressed and geologically-complicated rock mass and adverse conditions encountered in deep South African gold mines were difficult and not cost- or effort-effective until we implemented the BX CCBO overcoring technique. Here we demonstrate that DCDA can supplement CCBO, especially when a stress profile of a long borehole is desired.

ACKNOWLEDGEMENTS

We sincerely thank Anglogold Ashanti Ltd. and Harmony Gold Ltd., Open House Management Solutions (Pty) Ltd., Seismogen CC, and Groundwork Consulting (Pty) Ltd. for their kind facilitation of the project. Our work was financially supported by the International Continental Scientific Drilling Program

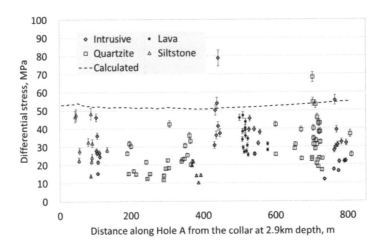

Figure 7. Differential stress in a plane normal to the Hole A axis determined by DCDA (modified after Ishida et al. 2018). Dashed line: calculated differential stress by assuming the stress measured by BX CCBO technique on 98 level (Ogasawara et al. 2012; at 3.0 km depth), no stress on surface, and linear increase with depth.

(ICDP), Japan Science and Technology Agency - Japan International Cooperation Agency SATREPS program, Japan Science Promotion Society (JSPS) Grant-in-Aid, Core-to-Core Program, JSPS - South African National Research Foundation bilateral program, the Ministry of Education, Culture, Sports, Science and Technology (MEXT) of Japan, under its Earthquake and Volcano Hazards Observation and Research Program, Ritsumeikan University, and Tohoku University.

REFERENCES

Abe, S., Yabe, Y., Ito, T., Nakatani, M., Hofmann, G. & Ogasawara, H. 2017. Estimate of the stress state of earthquake source region in a South African deep gold mining by diametrical core deformation analysis (DCDA). *Proc. 14th Domestic symp. in Rock* Mechanics, *Kobe*, January *2017*. Talk No.107.

Brown, E.T. 2012. Progress and challenges in some areas of deep mining. In Y. Potvin (ed.), *Proc. Deep Mining* 2012: 1–24. Perth: Australian Centre for Geomechanics (ACG) https://papers.acg.uwa.edu.au/p/1201_01_brown/

Durrheim, R.J. 2007. The DeepMine and FutureMine research programmes – knowledge and technology for deep gold mining in South Africa. In Y. Potvin et al. (ed.), *Challenges in Deep and High Stress Mining*: 133–141. Perth: ACG.

Fairhurst, C. 2017. Some challenges of deep mining. *Engineering* 3: 527–537.

Funato, A., Ito, T. 2013. Diametrical Core Deformation Analysis (DCDA) – Measuring Device and Laboratory Verification. Takatoshi Ito (ed.), *Proc. 6th intern. symp. on In-situ Rock Stress, Sendai, 20-22 August 2013*: 771–778.

Funato, A. & Ito, T. 2017. A new method of diametrical core deformation analysis for in-situ stress measurements. *Int. J. Rock Mech. Min. Sci.* 91: 112–118. doi: 10.1016/j.ijrmms.2016.11.002

Handley, M.F. 2013. Pre-mining stress model for subsurface excavations in southern Africa. *J. South Afr. Inst. Min. Metallurgy (SAIMM)* 133: 499–471.

Hickman, S., Zoback, M., Ellsworth, W., Boness, N., Malin, P., Roecker, S. & Thurber, C. 2007. Structure and properties of the San Andreas Fault in central California recent results from the SAFOD experiment. *Scientific Drilling* Special Issue (1): 29–32.

Hofmann, G.F., Ogasawara, H., Katsura, T. & Roberts, D. 2012. An attempt to constrain the stress and strength of a dyke that accommodated a ML2.1 seismic event. *Proc. 2nd Southern Hemisphere Intern. Rock Mech. symp.*: 436–50. Johannesburg: SAIMM.

Hofmann, G., Scheepers, L. & Ogasawara, H. 2013. Loading conditions of geological faults in deep level tabular mines. In T. Ito (ed.), *Proc. 6th intern. symp. on In-Situ Rock Stress, Sendai, 20–22 August 2013*: 558–580.

Ishida, A., Sugimura, K., Liebenberg, B., Rickenbacher, M., Mngadi, S., Kato, H., Abe, S., Yabe, Y., Noda, T., Ogasawara, H., Funato, A., Ito, A., Nakatani, M., Ward, A., Durrheim, R., Yasutomi, T. & the ICDP DSeis team. 2018. Stress measurement using cores of drilling into seismogenic zone of M2.0-M5.5 earthquakes in South African gold mine (ICDP DSeis project). *Abstract MIS08-P05 presented at Japan Geophysical Union 2018, Chiba*.

Ito, T., Funato, A., Lin, W., Doan, M.L., Boutt, D.F., Kano, Y., Ito, H., Saffer, D., Mcneill, L.C., Byrne, T. & Moe, K.T. 2013. Determination of stress state in deep subsea formation by combination of hydraulic fracturing in situ test and core analysis: A case study in the IODP Expedition 319. *J. Geophys. Res.* 118: 1203–1215. doi:10.1002/jgrb.50086

McGarr, A. & Gay, N.C. 1978. State of stress in the earth's crust. *Annual Review of Earth and Planetary Sciences* 6: 405–436.

Manzunzu, B., Midzi, V., Mangongolo, A. & Essrich, F. 2017. The aftershock sequence of the 5 August 2014 Orkney earthquake (ML 5.5), South Africa. *J. Seismol.* doi: 10.1007/s10950-017-9667-z.

Mendecki, A. (ed.). 1997. *Seismic Monitoring in Mines*. London: Chapman & Hall.

Mori, J. & Ellsworth, W. 2013. Active faults and earthquakes. In ICDP (ed.) *Unravelling the workings of planet earth (ICDP Science Plan for 2014-2019)*. https://www.icdp-online.org/fileadmin/icdp/media/Science_Conference/ICDP_SciencePlan2014.pdf

Nakatani, M., Yabe, Y., Philipp, J., Morema, G., Stanchits, S., Dresen, G. & JAGUARS-Group. 2008. Acoustic emission measurements in a deep gold mine in South Africa—Project overview and some typical waveforms. *Seismol. Res. Lett.* 79(2): 311.

Naoi, M., Nakatani, M., Yabe, Y., Kwiatek, G., Igarashi, T. & Plenkers, K. 2011. Twenty thousand aftershocks of a very small (M2) earthquake and their relation to the mainshock rupture and geological structures. *Bull. Seismol. Soc. Amer.* 101: 2399–2407, doi: 10.1785/0120100346.

Ogasawara, H., Kato, H., Hofmann, G. & de Bruin, P. 2012. Trial of the BX conical-ended borehole overcoring stress measurement technique. *J. SAIMM* 102(8): 749–754.

Ogasawara, H., Kato, H., Hofmann, G., Roberts, D., Piper, P., Clements, T., Ward, A.K., Yabe Y., Yilmaz, H. & Durrheim, R.J. 2014a. BX CCBO in-situ stress measurements at earthquake prone areas in South African gold mines– a summary of mini-workshop on 13 Feb 2014. *Proc. ARMA 2014, Minneapolis*: CD-ROM, paper number 14–7438.

Ogasawara, H., Katsura, T.,Hofmann, G., Yabe, Y., Nakatani, M., Naoi, M., Ishii, H., Roberts, D., Nakao, S., Okubo, M., Wienand, J., Lenegan, P. & Ward, A.K. 2014b. In-situ monitoring of rock mass response to mining in South African gold mines using the Ishii strainmeters. *Proc. 6th South African Rock Eng. symp. SARES 2014*: 21-34. Johannesburg: SAIMM.

Ogasawara, H., Nakatani, M., Durrheim, R.J., Naoi, M., Yabe, Y., Moriya, H., Hofmann, G.F., Stander, C., Roberts, D.P., de Bruin, P., Oelofse, J., Kato, H., Cichowicz, A., Birch, D., Ngobeni, D., Milev, A., Kgarume, T., Satoh, T., Horiuchi, S., Kawakata, H., Murakami, O., Yoshimitsu, N., Ward, A.K., Wienand, J., Lenegan, P., Yilmaz, H., Mngadi, S., Piper, P.S., Clements, T.N., Nakao, S., Okubo, M., Ishii, H. & Visser, A.V. 2014c. Observational studies of the rock mass response to mining in highly stressed gold mines in South Africa. In M. Hudyma & Y. Potvin (eds), *Proc. 7th intern. congr. on Deep and High Stress Mining, Sudbury, 16-18 September 2014*:123–137. Perth: ACG, https://papers.acg.uwa.edu.au/p/1410_06_Ogasawara/

Ogasawara, H., Durrheim, R.J., Yabe, Y., Ito, T., van Aswegen, G., Grobbelaar, M., Funato, A., Ishida, A., Ogasawara, H. Jnr, Mngadi, S., Manzi, M.S.D., Martin, Z., Ward,

A., Moyer, P., Boettcher, M., Dight, P., Ellsworth, W., Liebenberg, B., Wechsler, N., Onstott, T., Berset, N. & the DSeis Team. 2017. Drilling into seismogenic zones of M2.0-M5.5 earthquakes from deep South African gold mines (DSeis): establishment of research sites. *Proc. Afrirock symp. 2017 (ISRM), Capetown, 30 September - 6 October 2017*: 237–248.

Ortlepp, W.D. 1997. *Rock Fracture and Rockbursts, an illustrative study*. Johannesburg: SAIMM, Monograph Series M9. pp.98.

Rusley, C., Onstott, T.C., Liang, R. Higgins,J.A., Slater, N.W., Ogasawara, H., Cason, E.D., B Sherwood Lollar, B., Wiersberg,T., Zimmer, M., van Heerden, E., Kieft,T.L., Freese, B., Liebenberg, B. & Esterhuizen, V.H. 2018. Exploring the limits of life in a South African deep subsurface brine. *Abstract B23E-2552 presented at 2018 Fall meeting, AGU, Washington DC, 11 December 2018.*

Stacey, T.R. & Wesseloo, J. 1998. Evaluation and upgrading of records of stress measurement data in the mining industry. *Final Project Report Gap511: 1–31.* Johannesburg: Safety in Mines Research Advisory Committee.

Sugawara, K. & Obara, Y. 1999. Draft ISRM suggested method for in situ stress measurement using the compact conical-ended borehole overcoring (CCBO) technique. *Intern. J. Rock Mech. Min. Sci.* 36: 307–322.

Wiles, T.D. 2017. *Map3d User's Manual*. Map3D International Ltd, viewed 29 November 2018, https://www.map3d.com/ftp/Map3D_Help.pdf.

Yabe, Y., Philipp, J., Nakatani, M., Morema, G., Naoi, M., Kawakata, H., Igarashi, T., Dresen, G., Ogasawara, H. & Jaguars-Group. 2009. Observation of numerous aftershocks of an Mw1.9 earthquake with an AE network installed in a deep gold mine in South Africa. *Earth Planets Space* 61: E49–E52.

Yabe, Y., Nakatani, M., Naoi, M., Philipp, J., Janssen, C., Watanabe, Katsura, T., Kawakata, H., Georg, D. & Ogasawara, H. 2015. Nucleation process of an M2 earthquake in a deep gold mine in South Africa inferred from on-fault foreshock activity. *J. Geophys. Res.* 120 (8): 5574–4495, doi: 10.1002/2014JB011680.

Yabe, Y., Nakatani, M., Naoi, M., Iida, T., Satoh, T., Durrheim, R., Hofmann, G., Roberts, D., Yilmaz, H., Morema, G. & Ogasawara, H. 2013. Estimation of the stress state around the fault source of a Mw 2.2 earthquake in a deep gold mine in South Africa based on borehole breakout and core discing. In Takatoshi Ito (ed.), *Proc. 6th Intern. symp. on In-Situ Rock Stress, Sendai, 20–22 August 2013*: 604–613.

2019 Rock Dynamics Summit– Aydan et al. (eds)
© 2019 Taylor & Francis Group, London, ISBN 978-0-367-34783-3

Assessment of a complex large slope failure at Kışlaköy open pit mine, Turkey

R. Ulusay
Department of Geological Engineering, Hacettepe University, Ankara, Turkey

Ö. Aydan
Department of Civil Engineering, University of the Ryukyus, Okinawa, Japan

A. Ersen
Ersen Associates, Istanbul, Turkey

ABSTRACT: A very complex slope failure took place at an open-pit coal mine in 1984 in the eastern Turkey. The earlier investigations showed that a basal clay layer at the bottom of the coal seam was the main cause of the failure. The failure took place while the coal seam was uncovered. In this study, a series of back analyses utilizing the information of geotechnical investigations before and after the failure and static and dynamic limiting equilibrium techniques and finite element analyses were performed, and their outcomes are presented and discussed. It is shown that the failure process was complex and involved the buckling of the lignite seam, sliding along the basal clay layer and shearing failure of overburden layers. The drainage conditions also played an important role in the initiation of the failure. Once the failure as initiated, the failure developed rapidly.

1 INTRODUCTION

The failure of slopes during mining operations results in not only loss of the mining machinery but also human lives which make the mining operations become very costly. Therefore, a good design is necessary before the initiation of mining operations since any change of design later on will result in high cost and time loss.

A very complex failure took place at the north-eastern slope of the Kışlaköy open-pit mine of Afşin-Elbistan Lignite Mining Complex in the Eastern Turkey in 1984. The pit-floor heaved up as a result of the buckling failure of the lignite seam and a combined form of shear and sliding failure of mining benches occurred. The investigations showed that weak clay layers existed in the lignite seam of about 20 m thick and one of these clay layers played an important role in the failure (Ulusay et al. 1986).

In this paper, the authors re-consider the 1984 failure from another point of view. Previous analyses for the assessment of stability of these slopes were based on combined failure surfaces partly passing through the slope forming materials and partly along the basal clay layer (Ulusay et al. 1986). Examining the photographs of the failure, the heaved lignite seam shows a set of buckles. Based on this issue, the failure was re-examined by considering also the possibilities of buckling failure and compressive failure of the coal seam which constitutes the pit floor. For this purpose, a series of back analyses utilizing the

information of geotechnical investigations before and after the failure, and static and dynamic limiting equilibrium techniques and finite element method were performed. The outcomes of these analyses are presented and discussed.

2 GEOLOGY AND HYDROGEOLOGY

2.1 Geology

The Afşin-Elbistan basin is a closed basin and is formed as a result of uplifting of Toros (Taurus) Mountain range at the end of Alps orogenesis. The basin is uplifted and is subjected to a thrust aligned in the direction of NW-SE. There are normal faults, whose strikes are aligned NW-SE and dip to south with an inclination of 70-90°. The orientation of these faults are also indicative of ongoing NW-SE thrust type stress regime (Ketin 1984).

The basal rock unit is limestone of Permo-Carboniferous age (Fig. 1). Both the later units and lignite deposits were formed during the Tertiary period. The deposits above the basal limestone and their characteristics from the bottom to top are briefly described below.

The unit underlaying lignite consists of greenish gray highly plastic clay with some organic content.

Lignite, with a thickness ranging between 10 to 80 m in the pit, is approximately 15-20 m in the location of failure. There are several clayey layers within the lignite seam.

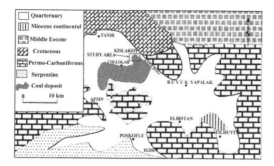

Figure 1. Geology of Afşin-Elbistan basin and location of the Kışlaköy pit (Ural and Yüksel 2004)

Gytja is characterized by a calcereous sandy clayey formation together with remnants of fossils with thickness ranging from a few metres to 50 m and overlies the coal seam. The thickness of this very porous unit is about 0.5-2 m in the failure site.

Blue clay, belonging to Quaternary period, is calcereous unit. It is about 10-15 m thick, dips towards the south with an inclination of 2-4° and has slaking characteristics.

Marl has a grenish colour and a thickness of 0.5-3 m in the failure site.

Loam, which belongs to Quaternary period, is a clayey, sandy and gravelly deposit with a thickness of about 15-40 m. It is a water-bearing porous unit.

2.2 Hydrogeology

Loam and gytja deposits are the main water-bearing units. In particular, the porosity and hydraulic conductivity of the gytja is approximately 50 % and about 1×10^{-7} m/s, respectively. It is also an artesian in the region. An extensive drainage network has been developed to drain the above these two deposits.

3 DESCRIPTION OF THE FAILURE

The main failure occurred on July 1, 1984 at the North-West slope of the Kışlaköy open-pit mine (Polat and Yüksel 1984). The failure involved a region, which was 650 m long and 250 m wide. The length of the region was later extended to 1000 m with subsequent failures. The final horizontal movement of the failed slope was more than 50 m. The lignite seam of the pit-floor was heaved up and it had a set of buckles whose strikes were almost perpendicular to direction of the movement and parallel to the axis of benches (Fig. 2). Fig. 3 shows the buckled lignite seam in the Kışlaköy open pit.

The first indications of the failure were observed on June 20, 1984 and tension cracks developed behind the slope crest. On June 21, the engineers of the open-pit mine installed gauges to monitor the horizontal and vertical movements of these tension

Figure 2. A view of failed benches in the Kışlaköy open pit

Figure 3. A view of buckled lignite seam in the Kışlaköy open pit mine.

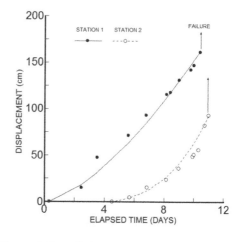

Figure 4. Time-displacement response of monitoring stations at Kışlaköy open pit (arranged from Polat and Yüksel 1984)

cracks. The main tension crack occurred at a distance of 40 m behind the crest of the upper most bench. Fig. 4 shows the vertical movements and associated velocity and acceleration responses as a

46

function of elapsed time at Station 1 (the dynamite depot about 30 m from the slope crest) and Station 2 (a drainage well next to the crest of the upper most bench).

4 GEOMECHANICAL PROPERTIES

Following the failure and concerns about the stability problems on the western slope of the pit near, which the electric power station is situated, an extensive program to investigate geomechanical characteristics of formations for stability analyses was initiated by Ulusay et al. (1986). The shear strength characteristics of the slope forming units are given in Table 1. Their shear responses obtained from direct shear tests are compared with each other in Fig. 5. As seen from the figure, the weak clay has the lowest rigidity and strength while the lignite has the highest rigidity and the strength. On the other hand, blue clay, gytja, marl and loam have similar characteristics.

Table 1. Material properties of the slope forming units (Ulusay et al. 1986; Aydan et al. 1996)

Unit	c_p (kPa)	φ_p (°)	c_r (kPa)	φ_r (°)
Loam	54.0	25.4	38.5	22.9
Marl	51.8	26.8	1.0	25.9
Blue clay	93.0	13.9	78.5	8.9
Gytja	13.5	37.8	9.6	33.5
Weak clay	23.7	2.2	15.5	1.7
Lignite	161.0	33.3	17.9	27.5

c_p, c_r: *Peak and residual cohesion*

φ_p, φ_r: *Peak and residual internal friction angle*

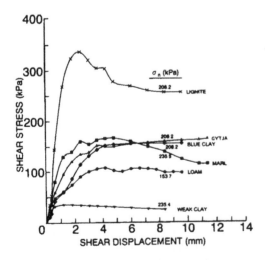

Figure 5. Shear behaviour of the slope forming units.

5 LIMITING EQUILIBRIUM ANALYSES

5.1 *Analyses of the North-West Slope by Conventional Limiting Equilibrium Methods*

First a series of limiting equilibrium analyses were performed by using the Janbu's method (Janbu 1954). The failure surface consisted of two parts: a circular part through loam, marl, blue clay and gytja, and a planar part along the gytja-lignite interface (Fig. 6). All analyses were performed by assuming that the slope was dry in view of the drainage system and in-situ observations. Table 2 gives calculated safety factors for three cross sections shown in Fig. 6 by considering the peak and residual shear strength parameters. The analyses showed that the slope should be stable for these failure surface configurations for both peak and residual values of the strength parameters. The observations on the failure also confirm this conclusion.

A second series of analyses were performed by using the Sarma's method (Sarma 1979). The failure surface passes through a weak clay layer within the lignite seam as shown in Fig. 6. If the peak values are used, the calculated safety factors were again large and the failure was not possible. If the residual values of the strength parameters were used the safety factor was close to 1 or less than that as given

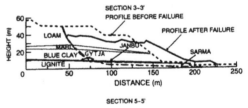

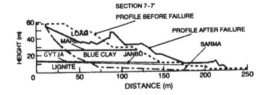

Figure 6. Cross sections used for stability analysis and slope geometries before and after the failure.

Table 2. Safety factors (Janbu's method)

Section	3-3'	5-5'	7-7'
Peak	2.41	2.87	2.18
Residual	1.85	2.26	1.76

Table 3. Safety factors (Sarma's method)

Section	3-3'	5-5'	7-7'
$c_r = 11.5$ kPa	1.070	1.060	0.980
$c_r = 0.0$ kPa	1.007	1.002	-

in Table 3. If the residual cohesion of the weak clay was assumed to be nil, the possibility of failure increased. In other words, the failure of the slope was not possible unless residual values were effective and these analyses did not provide any answer to the buckled configuration of the pit-floor. In the next sub-section, an alternative method to analyse the buckling failure is presented.

5.2 A limiting equilibrium method for a complex shearing, sliding and buckling failure

Ulusay et al. (1995) proposed a limiting equilibrium analysis method for a failure mechanism shown in Fig. 7. It is assumed that failure takes through shearing of intact layers at the back of the slope and sliding along a bedding plane. For the pit floor layer, there may be two possible modes: MODE 1: compressive failure, and MODE 2: buckling failure, MODE 3: combined compressive and buckling failure.

For the sliding and shearing part, the force system acting on a typical block may be modelled as shown in Fig. 8a. Note that a lateral force is also assumed to act in order to consider the lateral stresses. The

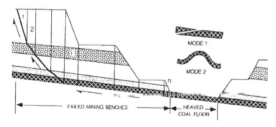

Figure 7. Failure mechanism proposed by Ulusay et al. (1995).

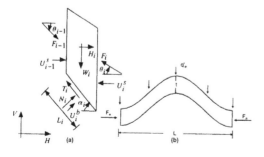

Figure 8. Mechanical models for stability assessment.

equilibrium equations for the chosen coordinate system can be written as:

$$\sum F_s^i = -T_i + W_i \sin\alpha_i + H_i \cos\alpha_i + F_{i-1}\cos(\alpha_i - \theta_{i-1}) -$$
$$F_i \cos(\alpha_i - \theta_i) + (U_{i-1}^s - U_i^s)\cos\alpha_i = 0 \quad (1)$$

$$\sum F_n^i = N_i + U_i^b - W_i\cos\alpha_i + H_i\sin\alpha_i + F_{i-1}\sin(\alpha_i - \theta_{i-1}) -$$
$$F_i \sin(\alpha_i - \theta_i) + (U_{i-1}^s - U_i^s)\sin\alpha_i = 0 \quad (2)$$

Assuming that the rock obeys to the Mohr-Coulomb yield criterion and the ratio of the horizontal force to the weight of the slice is given in the following forms:

$$T_i = \frac{c_i L_i + N_i \tan\varphi_i}{SF}, \quad H_i = \lambda W_i. \quad (3)$$

One easily obtains an equation for inter-slice force F_i, which can be solved *step by step* to obtain the force F_n together with the condition of $F_0 = 0$.

The resistance of the pit floor against compressive failure would be similar to the thrust type faulting. Therefore, no equation is given herein. As for the buckling failure of the coal seam, the following non-homogenous differential equation holds (Fig. 8b):

$$\frac{d^2 u}{dx^2} + \frac{F_n}{EI}u = \frac{q_{o'}}{2EI}x(L-x) \quad (4)$$

where E is elastic modulus, I is second areal inertia moment, u is displacement, and $q_{o'}$ is effective distributed load. Solution of the above equation is given below.

$$u = A\cos kx + B\sin kx + \frac{q_{o'}}{4EIk^2}(Lx - x^2 + \frac{2}{k^2}),$$
$$k^2 = \frac{F_n}{EI} \quad (5)$$

If $u = 0$ and $du/dx = 0$ at the ends of the layer, the integration constants A and B are obtained as follows:

$$A = -\frac{q_{o'}}{F_n k^2}, \quad B = -\frac{q_{o'}L}{F_n k}. \quad (6)$$

Assuming that $du/dx = 0$ at $x = L/2$, the critical buckling load is obtained as:

$$F_n = EI\left(\frac{8.99}{L}\right)^2 \quad (7)$$

Introducing $I = bt^3/12$ and $F_n = \sigma_o bt$, the critical axial stress for buckling is obtained as follows

$$\sigma_o^{cr} = 6.735E\left(\frac{t}{L}\right)^2 \quad (8)$$

48

where t is layer thickness, and L is span. An application of the above approach is shown in Fig. 9. Fig 9a was obtained from force $F_n = \sigma_o bt$ by considering the combined shearing and sliding failure for $SF = 1$ by varying lateral stress coefficient λ.

Fig. 9b was obtained from buckling analysis by assuming that E/σ_c as 65 (continuous line) and 26 (broken line). Since it is more likely that the peak strength values hold, the slope may become unstable and pit floor fails in compression provided that the lateral stress coefficient is 0.13 for a uniaxial strength of 596 kPa. As for buckling failure, the lateral stress coefficient failure should be greater than 0.1 and less than 0.13 in view of the actual range of L/t at the time of failure. The uncovered span of the lignite seam was 113 m at the time of failure. If the thin *Gytja* formation just above the lignite seam near the toe of the slope is neglected, the effective span is about 153 m. For a 5 m thick lignite seam, the value of L/t for compressive failure is found to be 21.95 from Fig. 9b. If the effect of the Gytja formation on L/t ratio is taken into account, the L/t is 22.6. On the other hand, if its effect is neglected, L/t is 30.6. Considering the above numbers, it has been contemplated that the lignite seam would likely be buckling rather than failing in compression. If the thickness of the seam involved in failure is thinner than 5 m, the possibility of failure by buckling increases more rapidly as compared with that by compression.

6 FINITE ELEMENT ANALYSES

The discrete finite element method proposed by Aydan-Mamaghani (Aydan et al. 1996(a,b); Mamaghani et al. 1994) was chosen to simulate the failure process of the slope . This method is based entirely on finite element method and can simulate very large deformation of jointed media. Material properties used in the analyses are given in Table 4. Fig. 10 shows the finite element mesh and boundary conditions.

First, a series of elastic analyses was carried out by varying lateral stress coefficient κ to see the magnitude of the axial stress of lignite seam at the location adjacent to the sliding benches. Fig. 11 shows the relation between lateral stress coefficient κ and the axial stress in lignite seam. As seen from the figure, lateral stress coefficient must be greater than 0.58 and less than 0.78 to cause the buckling of the seam. Otherwise, the lignite seam must fail in compression which is contradictory against field evidents. The results further indicates that if the lateral stress coefficient is less than 0.58, the axial stress in the seam may be tensile. By setting the lateral stress coefficient κ as 0.7, an elasto-plastic analysis was carried out. Fig. 12 shows the deformed configuration of the open-pit for each respective pseudo time step. As seen from this figure, the sliding of the benches on the left-hand side and buckling of the lignite seam at pit-floor are well simulated.

Table 4. Geomechanical parameters used in DFEM analysis.

Unit	λ (MPa)	μ (MPa)	c (kPa)	φ (°)
Loam, Marl, Blue clay	408	5.4	60	25
Lignite	710	9.4	161	33
Base layer	670	38	161	33
Weak clay	3.2	1.1	23	6
Fracture plane	3.2	1.1	23	20

λ, μ:Lame coefficients; c: Cohesion; ϕ : Internal friction angle

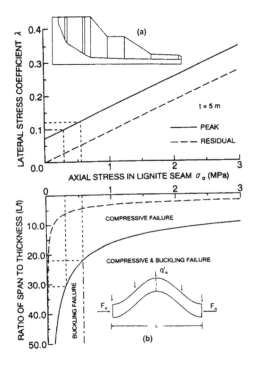

Figure 9. Computed stability chart for the lignite seam.

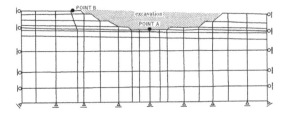

Figure 10. Finite element mesh used in analyses.

49

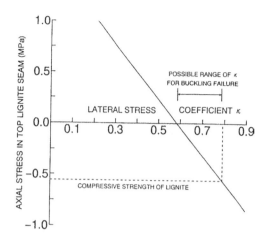

Figure 11. Computed stability chart for buckling failure.

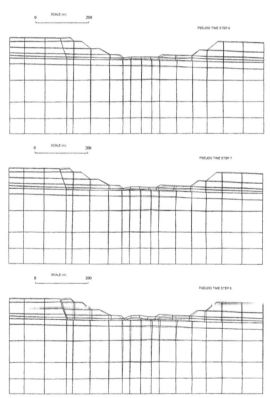

Figure 12. Deformed configurations at various pseudo time steps.

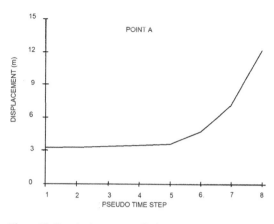

Figure 13 Pseudo time step vs displacement for point A.

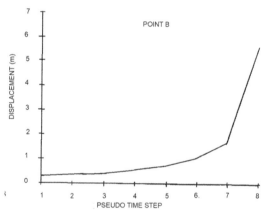

Figure 14 Pseudo time step vs displacement for point B.

response of point A which is located at the center of the pit-floor corresponds to the heaving of the floor. The heaving of the floor proceeds at a constant rate up to pseudo time step 6 and thereafter increases with an increasing rate and the floor buckles.

The displacement response of point B which is selected at the rear top of the sliding benches corresponds to the horizontal displacement of the sliding body. This response is very similar to the measured response shown in Fig. 4.

7 DYNAMIC LIMITING EQUILIBRIUM METHOD

The method used for estimating post-failure motions of the failed body is based on the earlier proposals by Aydan et al. (2006, 2008), Aydan and Ulusay (2002) and Tokashiki and Aydan (2011). Let us consider a landslide body consisting of N number of blocks sliding on a slip surface as shown in Fig. 15. If inter-slice forces are chosen to be nil as assumed in the simple sliding model (Fellenius-type), one may write the following equation of motion for the sliding body

It should be noted if the analysis becomes non-convergent in finite element analysis, this may be taken as the indication of failure of the structure, and each iteration step can be regarded as pseudo time step. With this concept in mind, the displacement responses of open-pit at selected point shown in Fig. 12 are plotted in Figs. 13 and 14. The displacement

50

$$\sum_{i=1}^{n}(S_i - T_i) = \bar{m}\frac{d^2s}{dt^2} \qquad (9)$$

where $\bar{m}, s, t, n, S_i$ and T_i are total mass, travel distance, time, number of slices, shear force and shear resistance, respectively. Shear force and shear resistance may be given in the following forms together with Bingham-type yield criterion:

$$S_i = W_i(1+\frac{a_H}{g})\sin\alpha_i;$$

$$T_i = c_i A_i + (N_i - U_i)\tan\varphi_i + \eta W_i \left(\frac{ds_i}{dt}\right)^b \qquad (10)$$

where $W_i, A_i, N_i, U_i, \alpha_i, a_V, a_H, c_i, \varphi_i, \eta$ and b are weight, basal area, normal force, uplift pore water force, basal inclination, vertical and horizontal earthquake acceleration, cohesion, friction angle of slice i, Bingham type viscosity and empirical coefficient, respectively. If normal force and pore water uplift force related to the weight of each block as given below

$$N_i = W_i(1+\frac{a_V}{g})\cos\alpha_i \ \ U_i = r_u W_i \qquad (11)$$

One can easily derive the following differential equation with the use of Eqs. 9-11:

$$\frac{d^2s}{dt^2} + \eta\left(\frac{ds}{dt}\right)^b - B(t) = 0 \qquad (12)$$

where

$$B(t) = \frac{g}{\bar{m}}\left(\sum_{i=1}^{n} m_i \left((\sin\alpha_i(1+\frac{a_H}{g})\right)\right.$$

$$\left.(\cos\alpha_i(1+\frac{a_V}{g})-r_u)\tan\varphi_i\right) + \frac{c_i A_i}{g}\right) \qquad (13)$$

In the derivation of Eq. (12), the viscous resistance of shear plane of each block is related to the overall viscous resistance in the following form:

$$\eta\bar{m}g\left(\frac{ds}{dt}\right)^b = \sum_{i=1}^{n}\eta m_i g\left(\frac{ds_i}{dt}\right)^b \qquad (14)$$

Eq. (12) can be solved for the following initial conditions together with the definition of the geometry of basal slip plane.

At time $t = t_o$ \qquad (15)

$s = s_o$ and $v = v_o$

There may be different forms of constitutive laws for the slip surface (i.e. Aydan et al. 2006, 2008; Aydan

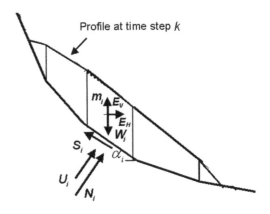

Figure 15 Mechanical model for estimating post-failure motions (Aydan 2016).

and Ulusay, 2002). The simplest model would be elastic-brittle plastic to implement. If this model is adopted, the cohesion will exist at the start of motion and it will disappear thereafter. Therefore, cohesion component introduced in Eq. (13) may be taken as nil as soon as the motion starts. Thereafter, the shear resistance will consist of mainly frictional component together with some viscous resistance.

The method explained above was applied by Tokashiki and Aydan (2011) to Kita-Uebaru landslide in Okinawa (Japan) involving bedding plane and fault plane. This method utilizes Bingham type visco-plastic yield criterion. Although the assumed geometry of the open-pit mine is slightly different from the actual one, it was applied to the failure in the Kışlaköy open-pit mine. Fig. 16 shows the displacement response during failure. The actual displacement of the failed body was about 34 m. The estimated displacement of the failed body is about 33 m. The material properties are shown in Figs. 16 and 17, which are based on those given in Table 4. Fig. 17 shows the deformation configuration of the failed body in space with time. Despite some

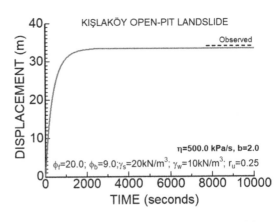

Figure 16. Displacement response of the mass center of the failed body.

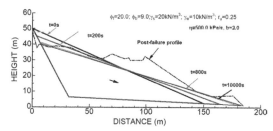

Figure. 17. Deformed configurations of the failed body.

difference between the assumed and actual geometries of the failed body, the estimations are very close to the actual ones.

8 CONCLUSIONS

From the back analyses of the failed slope, the following conclusions may be drawn.

The existence of a weak clay layer within the lignite seam with a very low shear modulus and strength played an important role and caused very high compressive stresses in the lignite seam.

The stresses are further increased by thinning the overlaying layers because of the excavation.

The failure of slope is not possible unless the residual values are utilised which is very unrealistic in limiting equilibrium analyses. If the effect of lateral force is taken into account, the failure becomes possible, provided that the lateral force coefficient is greater than 0.1 in view of the actual span of the lignite seam at the time of failure. The lateral force coefficient is likely to be greater than 0.1, if the stress regime of the region is taken into account.

Finite element computations showed that the buckling failure of the pit-floor was possible when the lateral in-situ stress coefficient was between 0.58 and 0.78. The computed responses were very similar to those measured.

Dynamic limiting equilibirum appraoch was also applied to estimate the post-failure motions of the open-pit failure and the estimations were very close to the actual observations.

In the same open-pit, there were failures at the east slope in 1988 when the depth of excavation was almost the same as the one reported herein. A series of in-situ measurements on the stress state of the region is needed for a better design of future open-pits at the Afşin-Elbistan Mining Complex.

REFERENCES

Aydan, Ö., 2016. Some considerations on a large landslide at the left bank of the Aratozawa Dam caused by the 2008 Iwate-Miyagi intraplate earthquake, Rock Mechanics and Rock Engineering, Special Issue on Deep-seated Landslides. 49(6):2525–2539,

Aydan, Ö., Ulusay, R., 2002. A back analysis of the failure of a highway embankment at Bakacak during the 1999Düzce-Bolu earthquake. Env. Geol. 42: 621–631.

Aydan, Ö., Ulusay, R., Kumsar, H., Ersen, A. 1996a. Buckling failure at an open-pit coal mine. EUROCK96, 641–648.

Aydan, Ö., Mamaghani, I.H.P., Kawamoto, T. 1996b. Application of dicrete finite element method (DFEM) to rock engineering structures. NARMS'96, Montreal.

Aydan, Ö., Daido, M., Ito, T., Tano, H. and Kawamoto, T. 2006. Prediction of post-failure motions of rock slopes induced by earthquakes. 4th Asian Rock Mechanics Symposium, Singapore, Paper No. A0356 (on CD).

Aydan, Ö., Ulusay R., Atak, V.O. 2008. Evaluation of ground deformations induced by the 1999 Kocaeli earthquake (Turkey) at selected sites on shorelines. Environmental Geology, Springer Verlag, 54, 165–182.

Janbu, N. 1954. Application of composite slip circles for stability analysis. European Conf. on Stability of Earth Slopes, Stockholm, 3:43–49.

Ketin, I., 1984. A general view to geology of Turkey. ITU Vakfı Yayını, No.32, İstanbul (in Turkish).

Mamaghani, I.H.P, Baba, S., Aydan, Ö., Shimizu, Y. 1994. Dicrete finite element method for blocky systems. IACMAG, 843–850.

Polat, H. Yüksel, M.F. 1984. Geology and cuases of the landslide area. Report of Geology Division of TKI-AEL, 55 (in Turkish, unpublished).

Sarma, S.K. 1979. Stability analysis of embankments and slopes. J. Geotech. Div., ASCE, 105(12):1511–1524.

Tokashiki, N. Aydan, Ö. 2011. Kita-Uebaru natural rock slope failure and its back analysis. Environmental Earth Sciences, 62 (1): 25–31.

Ulusay, R., Koçak, S., Selçuk, S., Ider, M.H. 1986. Stability of the West slope of Kışlaköy (Elbistan-K.Maraş) open pit. Proc. of the 1st Rock Mechanics Symposium, Turksih National Soceiety for Rock Mechanics, Ankara, 467–487 (in Turkish).

Ulusay, R., Aydan, Ö., Karaca, M. and Ersen, A. 1995. Buckling failure of an open-pit mine and its back analyses. Proc. of the 8th Int. Rock Mech. Symp., Tokyo, 451–454.

Ural, S. Yüksel, F. 2004. Geotechnical characterization of lignite-bearing horizons in the Afsin–Elbistan lignite basin, SE Turkey. Engineering Geology, 75 (2): 129–146.

2019 Rock Dynamics Summit– Aydan et al. (eds)
© *2019 Taylor & Francis Group, London, ISBN 978-0-367-34783-3*

Seismic response and stability of rock tunnels–its history and problems today

K. Kamemura
Fukada Geological Institute, Tokyo Japan

ABSTRACT: In Japan, a lot of underground structures has been constructed and they are in service in spite of complex and poor geological conditions. On the other hand, Japan is famous for the big earthquakes resulting a huge national loss, thus the aseismic design to prevent the severe damage of structures is very important. Though many buildings and structures are designed considering earthquakes, the aseismic design of structures constructed in deep underground has not been well discussed up to now. In this paper, the historical background of aseismic design is reviewed and typical seismic damages of rock tunnels are presented. And the present situation of static and dynamic design for underground structures ranging from the cut and cover tunnels in the shallow underground to the rock tunnels in deep rock formation is presented. Then problems to establish a practical method of aseismic design for the rock tunnels is discussed.

1 HISTORY OF ASEISMIC DESIGN OF UNDER-GROUND STRUCTURES

Underground structures are utilized for various purpose and playing an important role as a social infrastructure. They are classified into the following three types according to the construction method, namely cut and cover tunnel, shield tunnel and drill and blast (rock) tunnel (including caverns). Typical shape and scale of these structures are shown in Figure 1.

As Japan is famous for big earthquakes resulting a huge national loss, thus the aseismic design is very important for every kind of structures. Today we have a strict regulation concerning to the aseismic design of buildings and a various kind of civil structures. It has to have a long way to result in this situation and even now we are repeating "Lesson and Learn".

According to the Japanese standard for tunneling of Japanese Society of Civil Engineering (JSCE), and other aseismic design related documents, the history of aseismic design for buildings and underground structures can be summarized as follows.

Aseismic design in Japan started in 1923 taking the opportunity of the Great Kanto Earthquake. It was the very first huge earthquake that had hit the modernized metropolitan area with magnitude of 7.9. Up to 100,000 people died and about 300,000 houses were collapsed or burnt out. The enormous damage occurred in the lifeline and the life of the citizens was seriously.

As a result of serious damages by Kanto Earthquake, aseismic design regulation for buildings with a horizontal seismic intensity of 0.1 was set in 1924. This is the world's first aseismic design standard. Table 1 shows the major earthquakes fol-

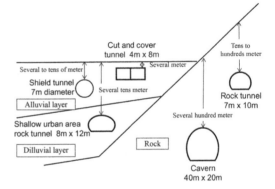

Figure 1. Various kinds of underground structures

lowing to the Kanto Earthquake and corresponding situation in the area of design for buildings and tunnels. As shown in Table 1, aseismic design of buildings have been improved in each time of the serious seismic damage occurred. However, nothing had been done on the tunnels until 1995 Hyogo-ken Nanbu Earthquake, in spite of the serious damages on the tunnels.

1.1 Kanto Earthquake And Damage Of Tunnels

In the Kanto Earthquake, 93 tunnels out of the 149 tunnels (including under construction) of national railway were damaged and repaired. Those tunnels were within the range of about 120 km from the epicenter.

The most severe damage can be seen in seven tunnels out of 11 tunnels located between Odawara and Manazuru of Atami line (present Tokaido Line). They were locating near to epicenter with the distance of 20-25km.

Table 1. Major earthquakes and aseismic design

Earthquakes and aseismic design

1923.9.1	Great Kanto M=7.9

[Buildings] In 1924, Urban Building Law to be the world's first earthquake resistance regulation (**seismic intensity 0.1**) was enact. [Tunnels] non

1930.11.26	Kitaizu M=7.3
1943.9.10	Tottori M=7.2
1944.12.7	Toh-nankai M=7.9
1945.1.13	Mikawa M=6.8
1946.12.21	Nankai M=8.0

Because of World War II, damages were not made public and nothing was feedback to aseismic design

1948.6.28	Fukui M=7.1

[Buildings] The seismic intensity method was examined and the necessity of dynamic analysis were discussed, and in **1950** the Building Standard Law (**seismic intensity 0.2**, idea of long-term and short-term stability) was enacted.[Tunnels] non

1964.6.16 Niigata M=7.5

1968.7.9 Tokachi-oki M=7.9

1978.6.12 Miyagiken-oki M=7.4

[Buildings] In **1981**, the Building Standards Law was revised. In new aseismic design method reflecting the concept of **dynamic design method**, seismic force is set according to the vibration characteristics of buildings, and earthquake resistance is calculated.[Tunnels] non

1983.5.26 Nihonkai-chuubu M=7.7

1993.7.12 Hokkaido-nanseioki M=7.8

1995.1.17 Hyougo-ken Nanbu(Great Hanshin Awaji) M=7.3

[Tunnels] In **1996**, Tunnel Standards states that further investigation is in need after clarifying the importance of aseismic design due to the damage of the tunnel structure by the Hyogo ken Nanbu Earthquake.

[Tunnels] In **1998**, Aseismic design of **cut & cover tunnel** was published.

[Buildings] In **2000**, the Building Standards Law was revised again and the **performance design method** was introduced.

[Tunnels] In **2006,** the tunnel standard specified aseismic design for **shield tunnels** as well as cut & cover tunnels, and aseismic design has been in practice.

2011.3.11 Great East Japan M=9.0

The largest earthquake observed around Japan

2016.4.14/16 Kumamoto M=6.5 and 7.3

In the latest 2016 tunnel standard for rock tunnel, only a basic idea of earthquake resistance is stated and detailed discussions are not described.

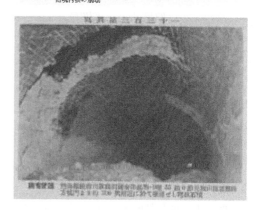

Photo 1. Damage of rock tunnel by Kanto Earthquake in 1923 (from National Diet Library Digital Collections, http://dl.ndl.go.jp/info:ndljp/pid/1175815)

54

Landslides and slope failures caused collapse and burying near the portals, and cracks and cross sectional deformation occurred even in locations away from the entrance. Photo 1 shows examples of tunnel damage.

In 1927, the Ministry of National Railway published a report on the survey results of every kind of facilities related to the railway, and that concluded about the seismic damage of tunnel as follows;

The most of damage of railway tunnels is;

- crack of arch
- crack of side wall
- break of arch
- break of entrance

Most of them can be seen near the entrance and not so many in the central part of tunnel. However, if the central part is in the condition of;

- soft soil/rock
- thin overburden
- fault part or changing hard to/from soft part

there are some serious damages such as crack or collapse.

In addition, those with significant damage near the entrance are necessarily causing slope failure of upper part of the entrance.

1.2 *Hyogo-Ken Nanbu Earthquake and aseismic design of underground structures*

The Hyogo-ken Nanbu Earthquake in 1995, with magnitude of 7.3, was the earthquake causing a largest national loss after World War II until The Great East Japan Earthquake in 2011. More than 6,000 lives were lost, more than 110,000 houses were collapsed or burnt out, and enormous damage also occurred in the lifeline.

Taking these seismic damages seriously, discussions concerning to the seismic resistance of underground structures finally started. The first description on the earthquake resistance of underground structures can be found in the standard of 1996 year edition. It says "Based on experience of the damage of the underground structures by the 1995 Hyogo-ken Nanbu (or Great Hanshin-Awaji) Earthquake, it is necessary to clarify the importance of aseismic design and further investigation should be done."

Photo 2 shows a typical damage of subway station, which was constructed by cut and cover tunneling method in Kobe city. Center pillars were buckled and ceiling subsided. This subsidence was reached to the ground surface.

In 1998, "Aseismic design of Cut and Cover Tunnel" was published by JSCE, and specific concept and method for the examination of aseismic design were presented. Next, in the standard for shield tunneling published in 2006, the aseismic design of shield tunnels in soils was discussed in detail.

For aseismic design for shallow depth underground structures such as cut and cover tunnel and shield tunnel, seismic deformation method, seismic

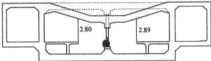

Photo 2. Damage of subway station by Great Hanshin-Awaji Earthquake in 1995 (Kobe Rapid Transit Railway Co. "Record of Disaster Recovery of Tozai Line Daikai Station". Sato Kogyo Co. 1997.1. Great Hanshin-Awaji Earthquake Disaster Materials Collection, Kobe University Library,http://www.lib.kobe-u.ac.jp/directory/eqb/book/11-276/index.html)

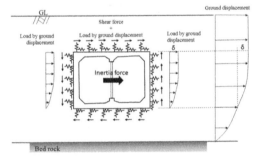

a) Cut and cover tunnel

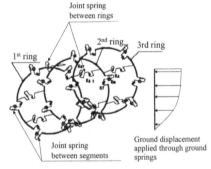

b) Shield tunnel

Figure 2. Analysis model of seismic deformation method

intensity method and dynamic analysis are established and aseismic design of these structures has become common in practical design. Example of seismic deformation method for cut and cover tunnel and shield tunnel are illustrated in Figure 2.

Photo 3. Damage of pilot tunnel of Tanna tunnel by Kitaizu Earthquake in 1930 (Kobe University Library, News Paper Archive: http://www.lib.kobe-u.ac.jp/das/jsp/ja/ ContentViewM.jsp?METAID=00102374&TYPE=PRINT_ FILE&POS=8)

However regarding the tunneling in rock, the latest 2016 Standard for mountain tunneling have little touches on earthquake resistance. Although basic knowledge on seismic damage of rock tunnels has been stated, it does not mention specific concept and how to design.

This is probably because the knowledge needed to examine the earthquake resistance of the rock tunnel is not sufficient yet.

2 SEISMIC DAMAGE OF ROCK TUNNEL

In this section, a typical seismic damage of rock tunnel is reviewed and study result concerning to the damage of rock tunnel is shown.

2.1 Kitaizu Earthquake In 1930(Magnitude 7.3)

The Kitaizu Earthquake occurred due to the activity of Kitaizu fault system, and was a near-field earthquake. Survey after the earthquake revealed many faults and the largest Tanna fault among them was reported to be about 35 km in length and moved 2.4 m in vertical and 2.7 m to the north. This fault intersects the Tanna tunnel under construction for the new line of the Tokaido Line (current Gotemba Line), and as Photo 3 shows, the cross section of pilot tunnel is completely blocked due to a large shear displacement. This is a typical tunnel damage due to fault displacement

2.2 Notohanto-oki Earthquake in 1993 (Magnitude 6.6)

In the Notohanto-oki Earthquake, Kinoura tunnel was heavily damaged as shown in Photo 4. Tunnel crown fell down together with the upper loosened rock and its repair required several months. This tunnels was within the range of about 120 km from the epicenter.

Photo 4. Damage of Kinoura tunnel by Notohanto-oki Earthquake in 1993 (provided by MLIT Hokuriku Regional Development Bureau)

2.3 Niigata-ken Chuetsu-oki Earthquake in 2004 (Magnitude 6.8)

The vicinity of epicenter of the Chuetsu-oki Earthquake is a mountainous area, and many rock tunnels have been constructed for various uses. As the result, many rock tunnels were damaged which had never been experienced after the Kanto Earthquake. Photo 5 shows a damage of road tunnel, where the concrete lining of tunnel crown was fallen down.

In this case, as the result of detailed investigation of tunnel damage, it was concluded that a large repairs was not necessary and restoration with installation of steel rib and shotcrete could be carried out without closing the road. As seen in this photo, the rock behind tunnel lining did not collapsed and was in a stable state although the concrete lining collapsed.

2.4 Kumamoto Earthquake in 2016(Magnitude 7.0)

The damage of Tawarayama tunnel near the earthquake source fault is shown in Photo 6. This road tunnel is located in the distance of about 0.5 - 1.5 km from the fault, and a large scale of lining concrete collapsed from tunnel crown.

Photo 5. Damage of Wanatsu tunnel by Chuetsu Earthquake in 2004 (provided by MLIT Hokuriku Regional Development Bureau)

Photo 6. Damage of Tawarayama tunnel by Kumamoto Earthquake in 2016 (Ministry of Land, Infrastructure, Transport and Tourism. Report on the seismic damages of road structures, http://www.mlit.go.jp/common/001136053.pdf

2.5 Cause of Seismic Damage

Through these tunnel damages due to recent large earthquakes, especially many tunnel damages of railroads, roads and waterways in the Chuetsu Earthquake in 2004, tunnel engineers have realized that the vaguely belief "rock tunnels have sufficient earthquake resistant" is not right anymore. And many researches related to the mechanism of seismic damage in rock tunnels have started after the Chuetsu Earthquake.

Yashiro et al. 2009 examined the structural damage of rock tunnel caused by the earthquake and discussed the mechanism of the damage. Number of damaged tunnel in the afflicted area of investigated earthquake is listed in Table 2, where the damage of rock tunnels is classified into 4 levels. Yashiro also pointed out that the damage of the rock tunnels can be classified into the following three types.

① Damage in the area of portal or shallow overburden section
② Damage in the area of weak rock/fracture zone
③ Damage due to fault displacement

Table 2. Number of damaged tunnels in major earthquakes

Earthquake(Max. Intensity)	Level of tunnel damage*				
	L	M	S	None	Total
Kanto, 1923	25	12	56	55	148
M 7.9 (7)	(17)	(8)	(38)	(37)	(100)
Izu, 1978	2	4	3	22	31
M 7.0 (5)	(6)	(13)	(10)	(71)	(100)
Hyogo, 1995	12	0	18	80	110
M 7.3 (7)	(11)	(0)	(16)	(73)	(100)
Chuetsu, 2004	11	14	24	89	138
M 6.8 (7)	(8)	(10)	(17)	(65)	(100)
Chuetsu-oki, 2007	4	1	1	14	20
M 6.8 (6S)	(20)	(5)	(5)	(70)	(100)

* Level of damage
  L: Large=Large scale repair is required.
  M: Medium=Repair is required.
  S: Small=Repair is not required

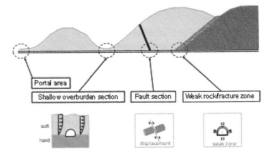

Figure 3. Patterns of seismic damage of rock tunnel (based on Yashiro et al. 2009)

Images of these damage pattern is shown in Figure 3.

It should be noted that the damage types shown here is almost same as the damage patterns concluded in the damage survey report of the Kanto Earthquake in the previous section 1.1.

Investigated result showed the major pattern of "Large" level damaged tunnel shown in Table 2 is "Damage in the area of weak rock zone". Therefore, many researchers had conducted numerical analysis (static and dynamic) and experimental studies using static loading devices in order to clarify the mechanism of seismic damage in rock tunnels.

As a result, followings were pointed out. Damage is likely if a void between lining concrete and rock exists near the tunnel crown. Installation of invert section can improve the rigidity of the entire tunnel structure, and also makes it possible to suppress the deformation of lining and uplift of the bottom.

However, it was very difficult to explain the destructive concentration of stress near the tunnel crown by numerical analysis and experimental studies under the static loading condition of forced displacement.

On the other hand, the results of a few examination in the dynamic conditions indicated the effect of dynamic ground motions could not be explained by the static conditions. Therefore, it seems necessary for further investigations, whether examination result considering dynamic effect shows same result by the static examination related to the damage mechanism and reinforcement effect, or not.

3 SEISMIC DAMAGE MECHANISM OF ROCK TUNNEL

3.1 Seismic damage and tunnel condition

In 2005, Tunnel Engineering Committee of JSCE, started up "Niigata-ken Chuetsu Earthquake Special Subcommittee" and compiled the whole situation of the damaged tunnels, such as roadway tunnels, railway tunnels, and conduits. Number of surveyed tunnel was up to 138, and the damage level of each tunnel was classified into 4 levels as shown in Table 2.

Damage level

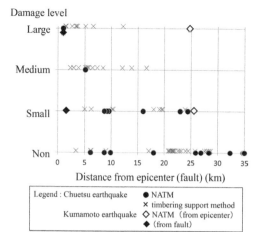

Legend : Chuetsu earthquake ● NATM
× timbering support method
Kumamoto earthquake ◇ NATM （from epicenter）
◆ （from fault）

Figure 4 Relationship between damage level and distance from epicenter (fault)

Figure 4 shows the relationship between the distance from the epicenter and the damage level of each tunnel, where tunnel construction method is also shown. In the tunnel by timbering support method (conventional tunneling method), large and medium damage has occurred even if it is 10 km or more away from the epicenter. On the other hand, tunnels by NATM are heavily damaged only within the distance of 5 km. A big difference in earthquake resistance between tunnels by the conventional method and that by NATM tunnels is obvious.

From the three damage types shown in 2.5 and the results of Figure 4, the relationship between the seismic damage of the rock tunnel and the distance from the epicenter or fault can be conceptually summarized as shown in Figure 5 and Table 3.

If the tunnel is located close to or crossing an earthquake fault, or there are geological structures such as faults, fracture zones and inhomogeneity, the potential of large seismic damage increases. Especially with a structural weakness such as a void between lining and surrounding rock, the possibility of damage becomes higher.

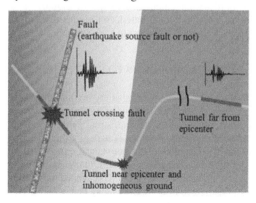

Figure 5. An illustration of the relationship among location and geological condition of a tunnel and seismic damage.

Table 3. Tunnel conditions and seismic damage

Location and effect of earthquake	Geological condition	Structural weakness	Seismic damage
Crossing fault Fault displacement	Fault	-	Large
Near fault Large seismic motion	Inhomogeneous	Yes	Large
		No	Large ~medium
	Homogeneous	Yes	Medium ~ small
		No	Small
Far from fault Small seismic motion	Inhomogeneous	Yes	Medium ~ small
		No	Small
	Homogeneous	Yes	Small
		No	None

3.2 *Mechanism of seismic damage*

Considering these results of examination together with the damage examples shown in Section 2, the mechanism of the seismic damage of rock tunnel can be summarized as illustrated in Figure 5

Pattern A: Seismic motions increase the strain of the rock surrounding tunnel and tunnel, and the lining which cannot withstand that strain increase is damaged. Damages of this pattern have been observed as crack in concrete lining and deformation of roadbed in many tunnels. In some cases, large collapse of concrete lining has occurred (Photo 5, 6).

This strain increase is greatly affected by the rigidity of rock around tunnel, the depth of overburden and the presence of the slope on the ground surface. In order to perform the original function of tunnel lining against this strain increase, it is necessary for the tunnel to be integrated with the surrounding rocks. In this point of view, the void existing between the lining and rock has a large influence. It is very important to investigate voids and fill by grouting.

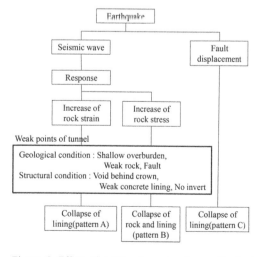

Figure 6. Effect of earthquake and patterns of seismic damage of tunnel

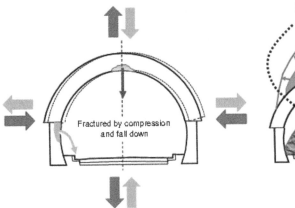

Figure 7. Seismic damage of tunnel: Pattern A

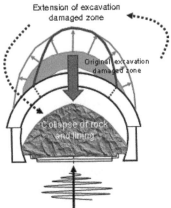

Figure 8. Seismic damage of tunnel: Pattern B

For this pattern, lining with high yield strength that can accommodate the seismic motion of the surrounding rock will be effective. Also a cushioning material between tunnel lining and surrounding rock may be able to reduce the strain increase of tunnel. Grouting to surrounding rock may increase the rigidity of surrounding rock and can reduce the strain increase of surrounding rock and tunnel.

Pattern B: Strain of the rock surrounding tunnel is increased by the seismic motion, and the rock is damaged. Increase of damaged zone of rock causes the increase of rock pressure acting on the tunnel lining, and lining will break. This type of damage has been observed in some tunnels (Photo 4).

In case of a large damaged zone (plastic zone) is already resulted around the tunnel during tunnel excavation, this damaged zone is subjected to strain increase (stress increase) due to the earthquake as an additional external load and is further extended. Additional load due to this enlargement of damaged zone acts on the lining and in case of the lining cannot bear, the lining breaks and then falls down together with the damaged rocks. Especially when there is the void between lining and surrounding rock, the rock easily yields and the lining resistibility against rock pressure decreases, so that the lining is highly likely to collapse.

For this pattern, the integration of surrounding rock and tunnel lining is effective in order not to enlarge the damaged zone around tunnel. This may be possible by means of grouting to the void between lining and surrounding rock and by the placement of rock-bolts. On the other hand, grouting to the damaged zone may improve rock strength and reduce the enlargement of loosened zone due to the earthquake.

Pattern C: Permanent displacement will occur in the fault zone crossing the tunnel due to seismic motion and will fracture the lining. This type of damage was seen in the Tanna tunnel by Kitaizu Earthquake in 1930 (Photo 3).

Many faults in various scales can be usually seen during tunnel excavation. If such faults cause per-

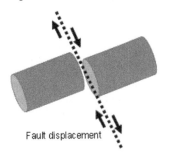

Figure 9. Seismic damage of tunnel: Pattern C

manent shear displacement due to the earthquake, they act as forced displacement against the tunnel, so that the structural damage of the tunnel is inevitable. However, it is very difficult to know which fault will move and how much displacement will occur, beforehand.

For this pattern, a buffer zone using deformable material around the tunnel is proposed. It is expected that this kind buffer zone can reduce the direct effect of permanent displacement of the fault due to the earthquake on the tunnel structure. However, such measures are not effective as long as the fault displacement due to the earthquake cannot be specifically estimated, so it seems difficult to apply this countermeasure to rock tunnels.

4 FURTHER PROBLEMS TO BE STUDIED

From the survey and relating study on tunnel damage caused by the recent earthquakes, as shown in Section 3, rough image on the mechanism of seismic damage has been clarified. Namely, if it is possible to know the positional relationship between the tunnel to be examined and the earthquake fault in the vicinity, and the geological condition around the tunnel, necessary and effective countermeasures may be decided.

However, it is not so easy to design the practical countermeasure for existing tunnel and newly

Table 4. Features in the design of cut and cover tunnel

cut and cover tunnel

geological condition	–	can be estimated by shallow boring
	–	mechanical properties of soil can be evaluated by laboratory test
underground water	–	can be clarified by survey
quality of structure	–	reinforced concrete
	–	easy to control the quality because of open air construction
surrounding ground	–	ground around structure is back filled so its quality can be controlled
seismic design	–	static and dynamic analysis are applied

Table 5. Features in the design of shield tunnel

shield tunnel

geological condition	Same as cut & cover tunnel	
underground water	Same as cut & cover tunnel	
quality of structure	–	reinforced concrete
	–	quality is clear because of plant product
surrounding ground	–	ground around structure is little affected by excavation
seismic design	Same as cut & cover tunnel	

Table 6. Difficulties in the design of rock tunnel and cavern

Rock tunnel and cavern

geological condition	–	only rough condition can be estimated by limited information in advance
	–	can be clarified through the data obtained during excavation
	–	mechanical properties of rock mass is difficult to evaluate because of discontinuities
underground water	–	difficult to clarify by survey
	–	unexpected water inflow will cause difficulties
quality of structure	–	plain concrete
	–	difficult to control the quality because of in-situ condition
surrounding ground	–	EDZ(excavation disturbed zone) is created
	–	character of EDZ is hard to evaluate
	–	ground water condition may change and affect the construction
seismic design	–	usually not carried out

constructed tunnel. Because, we still have many problems to be solved as follows.

In Tables 4, 5 and 6, existing design methods of underground structures are summarized from the viewpoint of performance evaluation. Comparing these tables, it is clear that the situation of rock tunnel in the deep underground is very different from that of tunnels in the shallow soil ground. Many uncertainties and unknowns are still remaining in the evaluation of performance of rock tunnels and caverns, so the aseismic design of these deep rock structures is not easy to generalize.

For the structures constructed in the soft ground such as cut & cover tunnel and shield tunnel, ground conditions can be clarified by shallow boring and laboratory tests, and of course many of laboratory test are prescribed by standard. The groundwater condition also can be clarified by survey. And these structures are made by reinforced concrete, which is constructed in open air condition or is manufactured in plant, so the quality of structure is easy to control. Also, there is no need to worry about the existence of excavation damaged zone which may affect the seismic behavior of structure and surrounding ground. That is, it is easy to clearly define the static stress condition which is the initial condition of the aseismic analysis, and seismic design condition such as boundary condition and material parameter of ground.

On the other hand, as shown in Table 6, many uncertainties and unknowns are still remaining in the evaluation of performance of rock tunnels and caverns. First of all, detailed geological condition around tunnel is not so clear in the design stage because the survey in rock tunnel are usually limited to few borings and field investigation. And even in the completion of construction, we only know the geological condition around the tunnel through the tunnel face observation. So in order to make the aseismic design of deep rock structures practical, following problems should be examined, besides discussing about the method to know more detailed geological condition.

Seismic motion or fault displacement as external load is unknown:

In order to carry out the aseismic analysis, seismic motion or fault displacement should be set up as a loading condition. However there has been no well-established method to evaluate the seismic motion that may possibly occur in tunnel and its surrounding rocks under the deep rock condition. Moreover, it is impossible to judge whether the fault that crosses the tunnel would move or not by the earthquake and how much displacement would occur in case of relative slip.

Stress condition (or safety factor) of tunnel in use is unknown:

As the initial condition of aseismic analysis, the current stress and strength condition (namely safety factor) are important. However it is difficult to have a detailed information about those conditions of the existing tunnel and surrounding rock. Because, it has been common that the performance of the tunnel at the completion of construction has not been evaluated by means of comparing the construction result with that of requested in the design, so far.

There are remaining a lot of uncertainties and unknowns such as the existence of the void between lining concrete and rock, unevenness of excavated surface, actual thickness of lining concrete, location of reinforcing steel bar and so on. Moreover, it is very rare to discuss the difference of geological condition between that of assumed in the design stage and that of clarified in the excavation, and its influence on the tunnel stability at the time of completion. So it is almost impossible to evaluate the safety of the tunnel in use.

Function of concrete lining:

Another important and difficult problem is that the mechanical function of tunnel lining is not clear. Usually tunnel lining is constructed using plain concrete, and there is no design method for the plane concrete lining.

How to evaluate and analyze the mechanical function and the performance of plain (not reinforced) concrete lining of the existing rock tunnel in the aseismic design, may be a problem to be solved first in the discussion on the aseismic design of rock tunnel.

5 CONCLUSIONS

As mentioned in Section 1, since the Hyogo-ken Nanbu Earthquake, research on the seismic resistance of rock tunnels has been undertaken by many research institutions and companies, and the obtained results have been presented at related societies. However, standard as the final target of these researches has not been established up to now.

So what is important to establish the standard? Here, I would like to show a former research result which is describing how to proceed the study on earthquake resistance evaluation of tunnels.

Okamoto, one of the most famous earthquake engineers, et al. 1963 said in their paper titled "On the seismic force acting on structures under the ground" about seismic damage of tunnels as follows;

"Although the tunnel lining is often damaged by the earthquake, there is no empirical formula or theoretical expression to evaluate the behavior at the time of the earthquake. It is a fact that the aseismic design is made only by experiences.

Therefore, establishment of the theory of seismic stress calculation is desired, but there is no other way than to establish a method of analysis step by step considering various environmental conditions. And the accumulation of such theories and analysis, related to the surveyed seismic damage, will eventually give a sufficient engineering solution to this difficult problem."

These are exactly right pointing out, and the fact that this was already made in 1963 is surprising.

The first thing to do may be to reflect on that practical actions have not been taken for a long time.

As discussed in 4.2, it is necessary to solve many problems in order to evaluate the earthquake resistance of rock tunnel and this will have a long time. On the other hand, it is also necessary to take appropriate measures as soon as possible to prevent the damage caused by the structural or geological weak-points of the newly planned tunnel.

On this point of view, although it is not the standard yet, guidelines concerning to the countermeasures has been prepared based on the accumulated discussion on the seismic resistance of rock tunnels up to now. For example, in March 2017, Road Bureau of Ministry of Land, Infrastructure, Transport and Tourism issued the following notice. It is titled as "Points to be noted about aseismic countermeasures for road tunnels on the basis of the Kumamoto earthquake in 2016" and is stating about seismic countermeasures in road tunnel as follows;

1) Special condition of the tunnel considered to be possibly damaged by the earthquake

1-1: Sections where long-term interruption of construction is enforced due to sudden massive groundwater inflow or equivalent thereto

1-2: Sections where construction has been suspended for a long time due to the large scale collapse of the face

1-3: Sections using large-scale countermeasures against for the instability of the rock

1-4: Sections where the geological condition has suddenly changed and two or more ranks of support patterns have been changed (Excluding the connected section with the entrance section)

1-5: Sections where the mechanical property of the rock dynamically changes in longitudinal and/or cross sectional direction

1-6: Sections subjected to unsymmetrical large earth pressure

1-7: Sections where the earth covering is extremely small

1-8: Sections where rock classification is evaluated as D2 or worse (including faults and fracture zones)

2) Supporting in the section with special condition

2-1: Make the tunnel into a ring structure with an invert and make the structure more mechanically stable

2-2: Apply sufficient amount of supporting such as shotcrete, steel rib and rock bolts

2-3: Even if the arch concrete is damaged due to the earthquake, to avoid the fall of large scale concrete blocks, reinforcing with the single steel bar and other is applied

We are still on the way to establish the effective seismic design method for existing and planned tunnels. Many numerical analysis and laboratory model experiments are carried out to search the effective and possible measures to prevent the seismic damage.

However, we must pay careful attention on the evaluation of the results of such studies, because in the numerical analysis and model test, a limited number of analysis and testing conditions is assumed. From those results, we can understand the qualitative effects of aseismic measures, however many problems that should be clarified from the viewpoint of quantitative evaluation are still remaining. In fact, some results of previous research show the inconsistent results between static analysis and dynamic analysis. And some other case show the difference of seismic behavior between model experiment and numerical analysis. Namely, more and more studies and discussions are needed to establish a practical methodology both in the aseismic retrofitting of existing tunnels and in the aseismic design of new tunnels.

In order to evaluate the performance of rock tunnel and to take appropriate aseismic measures, it is most important that stakeholders join together and start discussions. At the same time, we must accumulate analytical and experimental studies from various perspectives on the behavior of tunnels during earthquake. Of course, in those studies, it is very important to focus on the discontinuity of rock and the rock dynamics which have not been well discussed so far.

REFERENCES

Ministry of National Railway. 1927. Report on the seismic damages of railway in 1923. (in Japanese)

Yashiro, K., Kojima, Y, Fukazawa, N. & Asakura, T. 2009. Seismic Damage Mechanism of Mountain Tunnels in Poor Geological Condition. *JSCE paper collection C*, Vol.65 *No.4*. Tokyo: JSCE (in Japanese)

Tunnel Engineering Committee of Japanese Society of Civil Engineers. 2005. Report of Niigata-ken Chuetsu Earthquake Special Subcommittee. Tokyo: JSCE (in Japanese)

Okamoto, S., Kato, K., & Hakuno, M. 1963. On the Seismic Force Acting on Structures under the Ground. *Trans. of JSCE*, No.92. *April*: JSCE (in Japanese)

2019 Rock Dynamics Summit– Aydan et al. (eds)
© *2019 Taylor & Francis Group, London, ISBN 978-0-367-34783-3*

Dynamic analysis and investigation of vibrations induced by train from the subway station

K. Shahriar, M. Mahmoodi & A. Torkashvand
Department of Mining and Metallurgy Engineering, Amirkabir University of Technology, Tehran, Iran

ABSTRACT: As the train passes through the metro station, dynamic load is applied to the station floor, tunnel and the surrounding soil. This dynamic loading can cause adversely impacts on adjacent buildings or disturb the occupants of nearby buildings. This issue becomes more important in cases where the station has crossed a city's historical region. The aim of this study is to investigate the effect of train transit from the inside of the metro station on the ground surface points through dynamic analysis. Parameters such as train passage speed, train weight, distance to the station center, geometrical position of station, characteristics of soil layers and groundwater level are the most important factors affecting the vibration of the soil around the metro station. This study has been carried out by finite difference method and implemented through FLAC3D software modeling. The train wheels applied load was modeled with respect to the speed and weight of the wagons using the Fourier transform, firstly. Finally, the level of ground surface vibration is presented in terms of displacement, velocity and acceleration maps. Numerical modeling results show that when the train passes at 80 km/h from the studied station and the weight of each wagon is 50 tons, the maximum vertical displacement, the maximum vertical component of the velocity, and the maximum vertical component of the acceleration at the ground surface is $6.55*10^{-4}$ m, $4.8*10^{-6}$ m/s, and $1.8*10^{-3}$ m/s^2, respectively.

1 INTRODUCTION

The loads that size, direction and their impact point change during the time are called dynamic loads and are expressed by F(T=P). Static loads are special case of dynamic loads, defined by constant function. In other words, the variable accelerations in a dynamic system provide forces proportional with their size, which apply the loading set with varying intensities on the structure. The average load intensity can be obtained from the differential equation and integration at each section. In general, the dynamic loads are divided into two periodic and non-periodic groups (Clough et al., 2003).

A period or frequency can be considered for periodic loads. The sea waves-induced loads, the tides, the loads caused by vibration of power generating systems or the vibrating loads due to non-regulating part of the system are examples of periodic loads. It is not possible to define a certain period of time for non-periodic loads. An example of such loads is explosion-induced load on a structure or seismic load from the earth during an earthquake. Impact loads are a type of dynamic loads.

Therefore, loads which are suddenly applied to the structure are considered the impact dynamic loads. Impact loads create a vibration in elastic bodies so that body is balanced in terms of physical with continuing the vibration. In this case, the structural damping occurs (Clough et al., 2003). In dynamic science of structures, non-periodic loads can be analyzed by Fourier series transformations in series of periodic loads and sine and cosine diagrams. The most important application of the dynamic rules and principles of natural mandatory loading (severe whirlwind and storms, earthquake and impact) is in systems designed for vibration engineering, the operation of motors, the breaking of the sound wall caused by the vibrating waves of high speed jet engines, structural vibrations in the pillar shear force, earthquake, movement of the foundation, and so on (Clough et al., 2003). The choice of the urban tunnels route depends on factors such as traffic, economic, and environmental issues. Therefore, the feasibility of constructing and operating of urban tunnels should be investigated in two critical stages of drilling (during construction) and train passage (during operation). Vibration from about 2–200 Hz is caused by trains moving on the ground surface or in tunnels.

This dynamic loading can cause adversely impacts on adjacent buildings or disturb the occupants of nearby buildings (Sheng et al., 2006). The general concept of the train-induced vibration propagation and its effect on the building is shown in Fig. 1. By moving the train wheels on the rails, vibration energy is applied to the structure through the existing support system. The value of transmission energy to the structure is significantly dependent on factors such as the wheel and rail roughness, the resonant frequency of the rail suspension system and the support system of the tunnel route. Similar to all mechanical systems, these systems also have a resonance that increases the vibration at certain frequencies

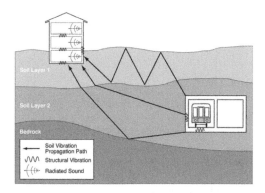

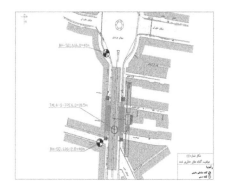

Figure 1. The general concept of the train-induced vibration propagation and its effect on the building General view of test bench.

Figure 2. Overview of the study area and E6 station.

called natural frequency. The vibration of the transitional structure (train) causes the vibration to the adjacent ground, which results the development of seismic waves from different soil and rock regions to the adjacent buildings foundation (Hanson et al., 2006). This seismic motion acts periodic and reverse in the form of a cycle. The range of vibration in this cycle is generally highly variable and includes values less than one Hz to more than 10 Hz (Wang et al., 2001). Parameters such as train passage speed, train weight, distance to the station center, geometrical position of station, characteristics of soil layers and groundwater level are the most important factors affecting the vibration of the soil around the metro station. On the other hand, the importance of this issue has increased with the advancement of rail transportation and the use of high-speed passenger trains and heavier freight trains in recent years (Yuanqiang et al., 2008, Galvin et al., 2007). In this study, the effect of the train transit from the metro station is investigated on the surrounding soil and the level of surface vibration.

2 PROJECT DESCRIPTION

Tehran metro line 6 starts from Dolatabad area in the Toska forest park and arrives to 17th Shahrivar Street with passing of Azadegan highway, Kianshahr region, and Besat highway. This line continues

along the 17th Shahrivar street towards the north and finally reaches to the Kan town. The total length of this line is approximately 30 km and includes of 28 stations. The southern and middle-northern sections of Tehran metro line 6 were excavated by using mechanized (TBM) and New Austrian tunneling method (NATM), respectively. The mechanical properties of the soil layers in the studied station (E6) are listed in Table 1, and overall view of the station and the location of the boreholes are presented in Fig. 2.

3 DYNAMIC LOADING MODEL OF THE TRAIN TRANSIT

A periodic dynamic load is applied into the tunnel floor with train transit through the tunnel. In order to evaluate the vibration, the calculation of the main frequencies of the applied force function in terms of time and the evaluation of the role of each of these frequencies is necessary. For this purpose, Fourier transform can be applied. Fourier transform, transmits a function from time domain to frequency domain without removing any information. Although, a small amount of information is removed in real transformations due to the interruption of the series expansion from one place to the next. However, the error can be reduced to the desired level by using sufficient sentences in extension (Costa, 2012). It should be noted that the participation of high frequencies in the received dynamic

Table 1 Mechanical properties of the soil layers.

Layer No	Description	Depth (m)	$\gamma(KM/m3)$	C(KPa)	ϕ (deg)	E(MPa)	v	K_0
L1-1	(SW-SM) or (SC)	2-7	19	30	25	30	0.35	0.58
L1-2	(SC) or (SM)	7-16.5	20	30	30	55	0.35	0.50
L2	(CL) sat	16.5-23	19	30	25	25	0.4	0.58
L3	(SC-SM) or (SC) sat	23-34	18.5	30	28	60	0.35	0.38
L4	(CL) sat	34-48	20	50	20	50	0.4	0.66
L5	(SC-SM) or (SC) sat	>38	19.5	80	25	75	0.35	0.58

response is less than low frequencies. This is logical due to the low frequency of the system. Therefore, fewer sentences can be used at lower speeds (Ntotsios, 2017). A periodic function with a period of time T can be represented by a Fourier series as follows (Dean et al., 2002):

$$f(t) = a_0 + \sum_{n=1}^{\infty} [a_n \cos\left(\frac{n\pi t}{T}\right) + b_n \sin\left(\frac{n\pi t}{T}\right)] \quad (1)$$

Where a_0, a_n, and b_n are called Fourier series coefficients and are obtained as follows:

$$\begin{cases} a_0 = \frac{1}{T} \int_0^T f(t)\,dt \quad (2) \\ a_n = \frac{2}{T} \int_0^T f(t)\cos\left(\frac{n\pi t}{T}\right)dt \\ b_n = \frac{2}{T} \int_0^T f(t)\sin\left(\frac{n\pi t}{T}\right)dt \end{cases}$$

Therefore, in order to use of Fourier transform, it is necessary to specify the shape of the applied force function in terms of time, firstly. In the following, the force periodic function in terms of time is obtained based on the wagons length, the distance between the two wagons, the distance between the axles of the two adjacent wheels, the speed of the train, and the weight of each axle, as shown in Fig. 3.

In this case, the length of the wagon is 29 m, the distance between the two wagons is 0.6 m, and the distance between the two adjacent wheels is 2 m. Also, the train speed and the train weight are 80 km/h and 50 tons, respectively.

Dynamic analysis is performed for this state and the results are presented in the following.

The dynamic load due to the train transit from inside of the station is considered as a sum of a few sinusoidal pulses with the main frequencies derived from the Fourier series. The applied single-wave amplitude to the software is obtained as $A_n = \sqrt{a_n^2 + b_n^2}$ for each of the coefficients a_n and b_n. Furthermore, the phase difference in waves obtained from the Fourier series of loading is also considered as $\varphi = \tan^{-1}\left(\frac{b_n}{a_n}\right)$. After dynamical analysis of the model, under the effect of each

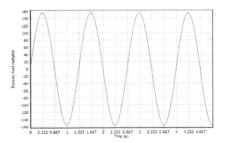

Figure 4. Loading curve (n=1, V=80, W=50).

sinusoidal waves with the calculated amplitude and phase difference, the dynamic response at the related points is obtained by collecting the applied wave displacements. The loading curve in the weight of 50 tons and the speed of 80 km/h for the first sentence (n=1) of the Fourier series is shown in Fig 4.

4 DYNAMIC ANALYSIS RESULTS

In this section, the results of the dynamic analysis of the finite difference model are presented at three points A, B, and C (Fig. 5), where the A point is at middle of the model and top of the tunnel axis in ground surface, B point is shifted 20 m to the right of tunnel axis and finally the C point exactly is on tunnel crown in middle of the numerical model. Output results include displacement, velocity and acceleration diagrams are shown in Figs. 6 – 13. The vertical component of displacement graph in point A, B and C indicates the maximum value, $1.65*10^{-4}$ m, $2.25*10^{-4}$ m and $6.55*10^{-4}$ m, respectively. As a result, vertical displacement at point B demonstrates large amount rather point A, this means the far field effect in the ground surface occurs. The vertical component of velocity in ground surface at point A and B imply $4.8*10^{-6}$ m/s and $1.0*10^{-6}$ m/s, respectively, in addition, the vertical velocity on tunnel crown is $3.5*10^{-6}$ m/s. The vertical component of acceleration plot in point A, B and C shows the maximum value, $1.4*10^{-3}$ m/s², $1.8*10^{-3}$ m/s², $1.4*10^{-3}$ m/s², respectively.

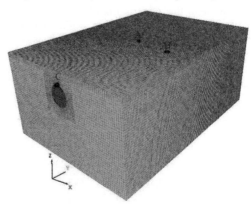

Figure 5. 3D numerical model and monitoring points.

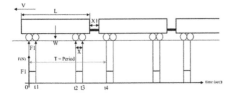

Figure 3. Display the variations of the applied force by the wagons wheel in terms of time.

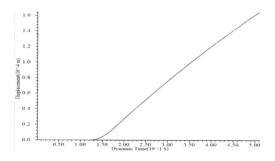

Figure 6. Vertical component of displacement at the speed of 80 km/h (point A).

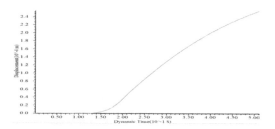

Figure 7. Vertical component of displacement at the speed of 80 km/h (point B).

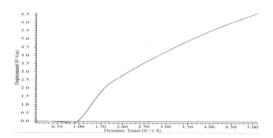

Figure 8. Vertical component of displacement at the speed of 80 km/h (point C).

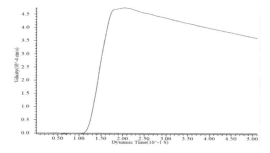

Figure 9. Vertical component of velocity at the speed of 80 km/h (point A).

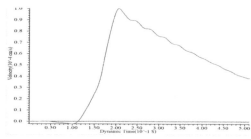

Figure 10. Vertical component of velocity at the speed of 80 km/h (point B).

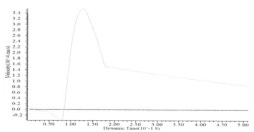

Figure 11. Vertical component of velocity at the speed of 80 km/h (point C).

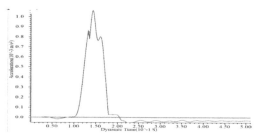

Figure 12. Vertical component of acceleration at the speed of 80 km/h (point A).

5 CONCLUSIONS AND SUGGESTIONS

The results show that when the train passes at 80 km/h from the studied station and the weight of each wagon is 50 tons, the maximum vertical displacement, the maximum vertical component of the velocity, and the maximum vertical component of the acceleration is $6.65*10^{-4}$ m, $4.8*10^{-6}$ m/s, and $1.8*10^{-3}$ m/s^2, respectively. The construction of the station under these conditions and desired route is possible in terms of vibration if these values of displacement, velocity, and acceleration are not problematic for the surface area above the tunnel in terms of structures and machines safety, as well as the comfort of residents. Otherwise, the location of the station should be changed or, as much as possible, the vibration level should be reduced.

In order to reduce vibration levels, different methods can be applied in three stages of generation, propagation, and wave reception. In this case, the wave is placed in the track. Actually, using these techniques apply a damper to the railroad pavement to reduce the amplitude of the wave. In the

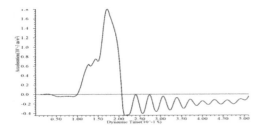

Figure 13. Vertical component of acceleration at the speed of 80 km/h (point B).

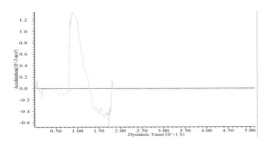

Figure 14. Vertical component of acceleration at the speed of 80 km/h (point C).

present dynamic analysis, the critical state is considered. These dampers reduce surface vibration and increase the level of certainty. A wave propagation barrier around the tunnel can be created at a specified distance, depth and thickness, in cases where the vibration caused by train transit inside the station is not within acceptable limits.

REFERENCES

Clough R. W. Penzien J. 2003. Dynamics of Structures, 3rdEdition, Computers & Structures, Inc., Berkeley.

Costa P.A., Calçada R, Cardoso A.S. 2012. Track–ground vibrations induced by railway traffic: In-situ measurements and of a 2.5 D FEM-BEM model. Soil Dynamics and Earthquake Engineering. 32(1):111-128.

Dean, G. Duffy. 2002. "Advanced engineering mathematics", CRC Press.

Final report of dynamic analysis and train-induced vibrations on adjacent historical buildings along the Isfahan Metro line 2, consultant of Isfahan University of Technology.

Galvin, P. and Dominguez, J. 2007. "Analysis of ground motion due to moving surface loads induced by high-speed trains", Engineering Analysis with Boundary Elements, 31, pp. 931–941.

Hanson, Carl E. David A. Towers & Lance D. Meister. 2006. Transit noise and vibration impact assessment. No. FTA-VA-90-1003-12 06.

Ntotsios E, Thompson D, Hussein M. 2017. The effect of track load correlation on ground-borne vibration from railways. Journal ofSound and Vibration. 402:142-163.

Sheng X, Jones CJ, Thompson D.J. 2006. Prediction of ground vibration from trains using the wavenumber finite and boundary element methods. *Journal of Sound and Vibration.*

Wang, J. N., & Munfakh, G. A., 2001. Seismic design of tunnels. WIT Transactions on the Built Environment, 57.

Yuanqiang, C., Honglei, S.and Changjie, X. 2008. "Response of railway track system on poroelastic halfspace soil medium subjected to a moving train load", International Journal of Solids and Structures, Article in Press.

T1: Laboratory tests on Dynamic Responses of Rocks and Rock Masses; Fracturing of Rocks and Associated Strong Motions

2019 Rock Dynamics Summit– Aydan et al. (eds)
© *2019 Taylor & Francis Group, London, ISBN 978-0-367-34783-3*

Determination of dynamic elastic properties from the frequency of natural vibration by using impact acoustics

T. Yamabe, M. Osada & C. Adachi
Saitama University, Saitama, Japan

ABSTRACT: The dynamic elastic properties of rocks can be determined in a non-destructive manner by using dynamic resonance methods. As mentioned in the industrial standards like JIS A 1127 or ASTM E 1876, the dynamic elastic properties of a material can be computed if the geometry, mass, and mechanical resonant frequencies of a suitable (rectangular or cylindrical geometry) test specimen of that material can be measured. The dynamic Young's modulus is determined using the resonant frequency in either the flexural or longitudinal mode of vibration. The dynamic shear modulus, or modulus of rigidity, is found using torsional resonant vibrations. In this study, prismatic columnar specimen is used to investigate the elastic properties of various types of rocks by measuring the frequencies of natural vibration. The measuring system consists of an impact acoustic generator and a receiver. The impulse tool is used to provide a single elastic strike to the test specimens to induce the fundamental resonant frequency and the receiver senses the resulting mechanical vibrations. Specimen supports, impulse location, and signal pick-up points are selected to induce specific modes of the transient vibrations. The results obtained by the above-mentioned systems are compared with those obtained by the ultrasonic methods with significantly acceptable coincidence between the two methods. Then, some basic experiments are carried out to clarify the influences of defects on the amplitude and resonant frequency of impact acoustics. Even though it is difficult to evaluate the equivalent elastic properties of a cracked specimen by ultrasonic methods, the simple impact acoustic methods are applicable. Numerical analyses are conducted to determine the relationships between the natural frequency and elastic properties.

1 BACKGROUND AND OBJECTIVES

Although hammering tests of acoustic sounds are generally applied to rock masses by referring to rock mass classification, most of these results are qualitative. A new testing apparatus for performing a hammering test was developed and applied to rock blocks that may fall to railroad track (Kawagoe et al. 2011). A laboratory method of quantitatively classifying rock mass by using acoustic pressure was proposed (Inamori et al. 1999). Such methods are often proposed in the field of concrete engineering. The application study of JIS (2017) demonstrated that the influence of crack depth on the resonant frequency of a longitudinal wave to longitudinal orientation is almost negligible and flexural resonances can be easily determined (Uomoto & Ito 1996). In addition, a rotational hammering test method was proposed by inserting a rotating rod to in concrete, and that characteristics were investigated through a numerical analysis and some experiments (Sonoda et al. 2008). In the study, the frequency components greater than 10 kHz cannot be observed in the results of a mortar specimen cured in water for 28 days.

In this study, three simple experimental methods based on JIS A 1127 are developed, and the dynamic elastic properties of a prismatic columnar rock specimen are determined via longitudinal, flexural, and torsional vibration. The results are compared with the values determined by a measuring devise of dynamic elastic moduli (DYoung made by Marui corp.), and then, the effectiveness of the proposed method is validated. Furthermore, the number of cracks (notches) is gradually increased in the same specimen, and the variations in dominant frequency are discussed.

2 VIBRATION OF HOMOGENEOUS PRISMATIC COLUMN

JIS A 1127 describes a method of observing longitudinal, flexural, and torsional vibration, Furthermore, the standard also shows a method of determining the dynamic Young's modulus E_d and dynamic shear modulus G_d. However, this standard only shows the results of the test. Thus, we make the complement of the description of JIS A 1127, and the measurement principal of JIS can be described as followings. The cross-sectional area, length and density of the homogeneous prismatic column are given by A, L, and ρ, respectively.

2.1 *Longitudinal vibration*

As the longitudinal orientation of the prismatic column is set to the x axis, the displacement and normal stress to x orientation are given by u and σ_{xx}, respectively. The equilibrium to the orientation is given as follows.

$$\rho \frac{\partial^2 u}{\partial t^2} = \frac{\partial \sigma_{xx}}{\partial x} \tag{1}$$

The relationship of strains with displacement and that of stress with strain are given by

$$\varepsilon_{xx} = \frac{\partial u}{\partial x}, \sigma_{xx} = E_d \varepsilon_{xx} \tag{2}$$

ε_{xx} is the strain to the longitudinal orientation. The wave equation (4) is given by substituting these eqs in eq.(1).

$$\rho \frac{\partial^2 u}{\partial t^2} = \frac{\partial}{\partial x} \left(E_d \varepsilon_{xx} \right) = E_d \frac{\partial^2 u}{\partial x^2} \tag{3}$$

$$\frac{\partial^2 u}{\partial t^2} = c^2 \frac{\partial^2 u}{\partial x^2}, c = \sqrt{\frac{E_d}{\rho}} \tag{4}$$

The axial stress becomes zero when both the ends of the prismatic column are free to displace. Then, the relationships between the dominant frequency of longitudinal vibration f_1 and the dynamic Young's modulus E_d can be obtained as follows.

$$f_1 = \frac{1}{2L} \sqrt{\frac{E_d}{\rho}} \Rightarrow E_d = 4L^2 f_1^2 \rho = \frac{4 L m f_1^2}{A} \tag{5}$$

2.2 Flexural vibration

The governing equation of flexural vibration is given by using vertical displacement v as follows.

$$\rho A \frac{\partial^2 v}{\partial t^2} + E_d I \frac{\partial^4 v}{\partial x^4} = 0 \tag{6}$$

Here, I denote the geometrical moment of inertia. Solution $v(x,t)$ can be obtained by applying unknown function $\phi(x)$ and number of vibration ω to eq. (6)

$$v(x,t) = \phi(x)e^{i\omega t} \tag{7}$$

$v(x,t)$ is substituted to eq. (6), and $\frac{\rho A}{E_d I} \omega^2$ is replaced with $\frac{\lambda}{L}$,

$$\frac{d^4 \phi}{dx^4} - \left(\frac{\lambda}{L} \right)^4 \phi = 0 \tag{8}$$

The bending moment and shear force become zero in both ends of the freely supported beam, as given by eq.(8),

$$\frac{d^2 \phi}{dx^2} = 0, \frac{d^3 \phi}{dx^3} = 0, (x = 0, L) \tag{9}$$

The frequency equation becomes $\cos \lambda_n \cosh \lambda_n = 1$ from eq. (9), and it is given by the first-order characteristic oscillation mode as follows.

$$\cos \lambda_1 \cosh \lambda_1 = 1 \Rightarrow \lambda_1 = 4.73004 \tag{10}$$

This equation become a joint if the basic function $\phi_n(x) = 0$, and the first-order specific mode is given by the following equation.

$$\phi_1(x) = 0 \Rightarrow x = 0.22415, 0.77584 \tag{11}$$

$\lambda_1 = 4.73004, I = \frac{bt^3}{12}$ are substituted to natural frequency $\omega_1 = 2\pi f_1$,

$$f_1 = \frac{\lambda_1^2}{2\pi L^2} \sqrt{\frac{E_d bt^3}{12\rho A}} = \frac{1.0279}{L^2} \sqrt{\frac{E_d bt^3}{\rho A}} \tag{12}$$

The equation is solved for E_d by considering ρ=m/AL

$$E_d = 0.94645 \frac{L^3}{bt^3} m f_1^2 \tag{13}$$

In JIS(2017) and ASTM(2009), as the modification for variations in the Poisson's ratio, the correction equation of E_d is multiplied by the correction coefficient T.

2.3 Torsional vibration

Following equation is given when a general material possessing uniform sectional parts are simple sheared under Bernoulli-Euler hypothesis.

$$T = G_d J \frac{d\varphi}{dx} \tag{14}$$

T, $G_d J$ and φ are torsional moment, torsional stiffness and rotation angle around the center point of the material, respectively. Then, the following equation is given for square section of the material.

$$\frac{G_d J}{\rho I_p} = \frac{2G}{\rho} \left\{ 1 - \frac{192}{\pi^5} \sum_{n=1}^{\infty} \frac{1}{(2n-1)^5} \tanh \frac{(2n-1)\pi}{2} \right\} \tag{15}$$

$n=3$ is enough for the equation because the series of eq.(15) are rapidly converged Komatsu (1983). G_d can be obtained as follows.

$$f_1 = \frac{1}{2L} \sqrt{\frac{G_d J}{\rho I_p}} \Rightarrow G_d = 4.74172 \frac{m L f_1^2}{A} \tag{16}$$

3 SIMPLE EXPERIMENTAL METHOD AND SPECIMENS

Simple experimental methods are developed in this study. In the method, the external force of a driving circuit is applied to the test specimen through a freely falling glass bead, and a digital voice recorder of a pickup circuit is used to record the sound of impact.

3.1 Longitudinal vibration

Figure 1 shows the experimental setup for measuring longitudinal vibration. Rubber pads are installed at the boundary between a stable desk and devices to isolate vibration. The specimen is horizontally set on marble supports at 22.4% of L from the ends of specimen. The diameter and height of the marble supports are 50 mm, and a stiffer rubber pad having a width of 10 mm is installed at the boundary between the specimen and supports. When a glass bead falls, as shown in the right side of fig. 1, along the arc shape guide, the bead impacts on the center of the right end of the specimen. Then, the bead rebounds several times, and finally stops. The impact sounds are recorded by the microphone of a digital voice recorder. The system has enough sensitivity of a pitch extent (20 to 21,000 Hz) to measure the rebound sound.

The recorded source of left channel is only active not to detect unnecessary information. The initial impact sounds are extracted, and a spectrum analysis is conducted by Audacity. Although similar analysis can be conducted using very expensive software (Mathematica), in this study, the dominant frequency f_1 can be determined with sufficient accuracy by Audacity. The analysis results obtained using two software are shown in figs. 2 and 3, respectively. These figures show the results of the spectrum analysis for the same wave shape, and the results show good agreement with each other.

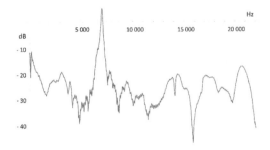

Figure 3. Dominant frequency by Mathematica

3.2 Flexural vibration

Fig. 4 shows the experimental setup for measuring flexural vibration. Although the devices are same as those used in the test to measure longitudinal vibration, the external force is applied on the upper surface of the specimen through a freely falling glass bead. In addition, the method of spectrum analysis is the same as that of longitudinal vibration test, and the dynamic Young's modulus E_d is obtained by eq. (13) after determining the dominant frequency.

3.3 Torsional vibration

Fig. 5 shows the experimental setup for measuring for torsional vibration. In this test, a stiffer steel foundation (SS400) is set on the stable desk, and a cruciform support is set at the boundary between the foundation and specimen. The cross-sectional shape of the support is an isosceles triangle, and the condition of the support is shown in fig. 6. For agreement of testing condition, set $m \to m/2$ and $L \to L/2$, and these are substituted in eq.(16).

$$G_d = \frac{4.74172}{4}\frac{mLf_1^2}{A} = 1.185\frac{mLf_1^2}{A} \qquad (17)$$

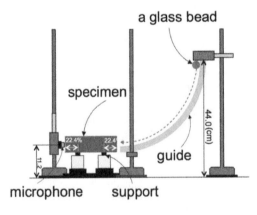

Figure 1. Setup of testing device for measuring longitudinal vibration

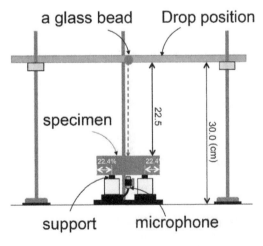

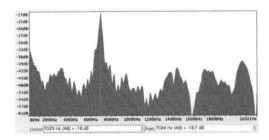

Figure 2. Dominant frequency by Audacity

Figure 4. Setup of testing device for measuring flexural vibration

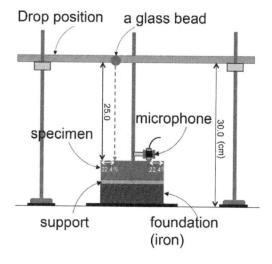

Figure 5. Setup of testing device for measuring torsional vibration

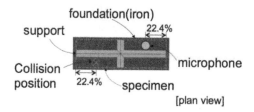

Figure 6. Plan view of supports for torsional vibration measurement

3.4 Specimens

Table 1 lists the size and density of each rock specimens used in this study. These values are the average of values obtained through three separate measurement. The wide variations of elastic parameters can be obtained because the density of rocks. The surface of the prismatic column is made smooth by using a grinding machine, and the errors due to surface roughness can be homogeneously minimized.

Table 1. Dry specimens, cross-sectional area A (mm²), length L (mm), density ρ (g/cm³)

#	Rocks	A	L	ρ
1	Shira-hamasandstone	1906.7	144.56	2.301
2	Oya tuff	2022.7	145.03	1.491
3	Makabe granite	2027.1	145.97	2.634
4	Inada granite	1942.0	145.19	2.626

#	Rocks	A	L	ρ
5	Ashi-yawelded tuff	2015.6	145.16	2.166
6	Kuzuu dolostone	2025.4	145.03	2.712
7	Travertine	2021.1	145.12	2.679
8	Savonnieres limestone	2036.7	145.03	1.618
9	Honkomatsuandesite	2021.4	144.88	2.616
10	Fukken granite	2018.7	144.93	2.605
11	Tage tuff	1996.3	144.89	1.782
12	Kima-chisandstone	2010.2	144.96	2.041

4 COMPARISON OF DYNAMIC ELASTIC MODULUS

4.1 Comparison between longitudinal and flexural vibrations

Figure 7 shows a comparison of the dynamic elastic modulus E_d obtained by each method. The numbers in the figure denote the specimens. The simple methods of measuring longitudinal and flexural vibrations show quite similar results, except # 7.

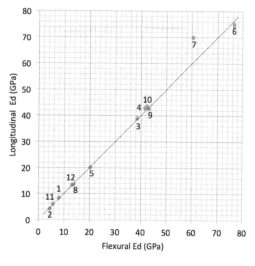

Figure 7. Comparison of E_d between longitudinal and flexural vibrations

4.2 Comparison of torsional vibration

The dynamic shear modulus G_d obtained under torsional vibration is discussed. Figure 8 shows the testing device for measuring dynamic elastic properties for comparison with those obtained by the proposed method. This device is abided by JIS A 1127, and it can indicate the dominant frequency of resonance mode, as shown in fig.9. Figure 10 compares G_d by the method of figs. 5 and 8. Although there are few variations in greater area of G_d, almost similar results are determined in the whole area. Even though a simple method is used to measure torsional vibration, the results are fairly accurate.

Figure 8. The devise of measuring dynamic Young's moduli (DYoung made by Marui corp.)

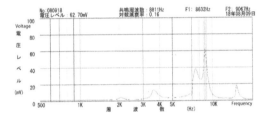

Figure 9. Example of results of # 9 by DYoung

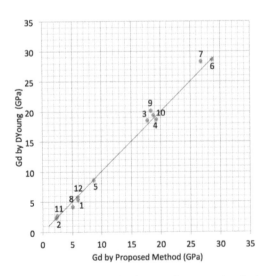

Figure 10. Comparison of G_d between the proposed method and DYoung

5 DYNAMIC BEHAVIOR OF PRISMATIC COLUMN WITH CRACK

Cracks are installed in dry specimens used in chapter 4, and difference of mechanical responses are discussed by using the simple proposed method.

5.1 Preparing specimen and cases

For a specimen with length L, 2 mm crack is made at $L/3$ from the end of specimen. When a crack is made at one third of the specimen height t, this testing case is denoted by '1/3' in the figures. The proposed tests of longitudinal, flexural, and torsional vibration are conducted, and then, the crack is progressed to the same orientation by $2t/3$. The case is called '2/3', and the series of proposed tests are conducted. Furthermore, an additional crack having a depth of $2t/3$ is installed at $2L/3$ from the end of specimen, and the case is called '3/4'. Thus, the specimen in the '3/4' case consists of two symmetric $2t/3$ cracks as shown in fig. 13.

5.2 Dominant frequency reduction due to cracks

Figures 12 and 13 show the variations in the dominant frequency due to cracks. '0' is shown in figs for the specimen with no cracks. The values of the dominant frequency become smaller as both the crack depth and number of cracks increase. The dominant frequency f_1 decreases because the bending stiffness decreases owing to cracks. Furthermore, the decrease in E_d is proportional to the squaring of f_1, as shown in Eq.(13). However, the mass m of specimens decreases slightly owing to cracks, and E_d cannot be determined in the same condition because the structure is clearly different from the continuous specimen. Thus, f_1 is adopted to evaluate the variations in mechanical properties due to cracks.

Figure 14 shows a comparison of the dominant frequency f_1 between cracked and continuous specimens. The number of cracks and specimens are shown in the figure, and the lines denote rough tendencies. Although the data for # 10 deviates from the line in the '1/3' case, the results are almost

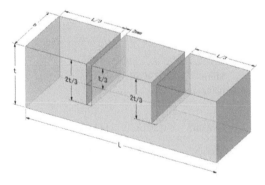

Figure 11. Prismatic columnar specimen possessing cracks

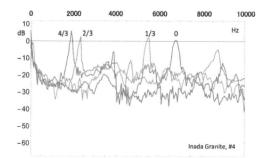

Figure 12. Variations of dominant frequency due to cracks by flexural vibration (example of # 4)

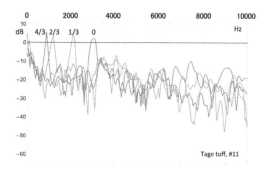

Figure 13. Variations of dominant frequency due to cracks by flexural vibration (example of # 11)

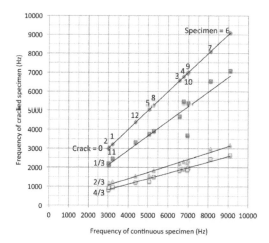

Figure 14. Comparison of dominant frequency between cracked and continuous specimens by flexural vibration

similar because f_1 decreases as the number of cracks increases. Furthermore, although the spectrum shown in e.g. fig. 14 do not show smooth responses because there are several noises, f_1 could be determined without any difficulty. The existence of cracks may be determined by the difference in f_1, as shown in fig. 14.

6 CONCLUSION

The elastic moduli of several rocks were determined using the proposed methods of measuring the longitudinal, flexural and torsional vibrations of a prismatic columnar specimen. The proposed method was verified through a comparison of the dynamic elastic moduli obtained by the methods and conventional method (measuring device for determining DYoung). By using the method, the reductions in stiffnesses due to cracks can be accurately confirmed.

ACKNOWLEDGEMENT

We would like to acknowledge to Mr. Hatakeyama, Technical staff of Saitama University for preparation of rectangular rock samples. This work was supported by JSPS KAKENHI Grant Numbers JP26249068 and JP26630218. This research would not be complete without the supports from them.

REFERENCES

ASTM (2009). *Standard Test Method for Dynamic Young's Modulus, Shear Modulus, and Poisson's Ratio by Impulse Excitation of Vibration, Designation E 1876-09.* 1–15.
Inamori, M., Y. Mitarai, S. Funahiki, M. Sezaki, T. Harada, & H. Yokota (1999). Response sound pressure pulse inclination method as a quantitative index for rock mass classification. *J. JSCE 638/III-9*, 335–351.
JIS (2017). *JIS A 1127, Methods of test for dynamic modulus of elasticity, rigidity and Poisson's ratio of concrete by resonance vibration.* 2288–2292.
Kawagoe, K., T. Ishihara, T. Urakoshi, & T. Ohta (2011). Evaluation methods of rock lump stability on rock slope. *RTRI REPORT 25*(7), 31–36.
Komatsu, S. (1983). *Structural Analysis III.* Maruzen, 93–100.
Sonoda, Y., A. Nanayama, A. Miyoshi, & N. Yoshida (2008). A fundamental study on diagnostic characteristics of the rotary hammering test. *J.JSCE Series A 65*(2), 514–522.
Uomoto, T. & Y. Ito (1996). Non-destructive testing method of concrete using impact acoustics. *Concrete Research and Technology 7*(1), 143–152.

2019 Rock Dynamics Summit– Aydan et al. (eds)
© 2019 Taylor & Francis Group, London, ISBN 978-0-367-34783-3

Analysis of rock fracture toughness (mode I) by dynamics simulation code

Changheon Song, Daeji Kim, Sang-Min Lee, Joo-Young Oh, Jong-Hyoung Kim, Jin-Young Park, Si-Geun Choi & Jung-Woo Cho
Korea Institute of Industrial Technology, Daegu, South Korea

ABSTRACT: To understand the tensile fracturing mechanism of rock, the fracture toughness of rock should be defined. We performed 3-point bending tests for 4 types of rocks. Then, the LS-Dyna code was adopted to simulate the 3-point bending tests and the tensile fracturing procedures of rock. The sample rock was modeled using the HJC (Holmquist-Johnson-Cook) rock material model, and the key parameters were selected by sensitivity analysis and then optimized to precisely describe the mechanism. Finally, the suggested material model and the simulation results were validated with the 3-point bending experiment results. The proposed 3 key parameters (e.g. *PL*, *k*1, and *evol*) can successfully describe the tensile fracturing process of rock model.

1 INTRODUCTION

The application of rock blasting method in urban area is gradually regulated because the noises and vibration induce inconvenience to civilians. To minimize the disadvantages of the blasting method, some methods producing less vibration (e.g. smooth blasting, drilling-splitting, water jetting method, etc.) has been used in urban site. However, these methods are suffering from the low working efficiency and poor excavation rate when encountering the hard rock.

The study is developing a new cutting-splitting method which is for rapid excavation rate for hard rocks. The procedure is that cutting rock at a certain depth by a rock saw, and then splitting the rock into two blocks by inserting a chisel into the crevice. The key mechanism of the method is tensile fracture propagation on the basement of rock blocks. To understand the fracture mechanism of rock, the fracture toughness (mode I) of each rock should be determined. 3-point bending tests were performed for four strength rock classes (i.e., low, moderate, medium and hard strength). From the results, fracture toughness (mode I) values of the samples were obtained.

Then, the LS-Dyna code was used to simulate the tensile fracturing procedures during 3-point bending tests. HJC (Holmquist-Johnson-Cook) material model (Holmquist et al., 1993) was adopted. The model has many input parameters which are very hard to be defined with limited numbers of laboratory experiments. We performed sensitivity analysis to filter out the important parameters before the main simulations. After selecting key parameters, a series of 3 point bending simulations were carried out. The simulation data were compared to the experimental results by the optimization method. After the procedure, we suggested the optimum values for the four classes of rocks. This can help solving tensile fracturing process in rock splitting method to the future study.

2 METHODOLOGY

2.1 *3-point bending test*

The 3-point bending test devices were designed according to the suggested method (ISRM, 1988). These devices consist of a cylindrical rock specimen which is supported by two fixed rollers near the end of its length (Figure 1). Also, the rock specimen is set-up horizontally through a loading roller located on the center of the rock specimen.

2.2 *Dynamic code (Ls-dyna)*

The fracture toughness test of 3-point bending is a quasi-static test. In order to solve the propagation process of tensile crack, it is necessary to calculate velocity, strain rate and stress as a function of time and position. We used LS-DYNA code (Livermore, 2014) to simulate the 3-point bending test of rock specimen.

Figure 1. The setup on the 3-point bending tests.

2.3 HJC material model

The LS-Dyna code was used to simulate the tensile fracturing procedures during 3-point bending tests. HJC (Holmquist-Johnson-Cook) material model was adopted. The model was originally presented by Holmquist et al. (1993) with the purpose of developing a concrete model for impact computations where the material experiences large strains, high strain rates and high pressures (Polanco et al., 2008).

2.4 Optimization method

We adopted the evolutionary algorithm (EA) proposed by Jeong et al. (2005) for enhancing efficiency. They produced offspring by mixing genes of different chromosomes (an intermediate tendency recombination) and sometimes by randomly changing (mutation) some genes of the chromosomes. Then they deterministically selected the best chromosomes for the next generation. For efficient design optimization, a commercial software, PIAnO (Process Integration, Automation and Optimization) (PIDOTECH Inc., 2015) was used to integrate and optimize the simulation results to experimental data. A summary of the optimization procedure is shown in Figure 2.

2.5 Design of experiment (DOE) of 3-point bending tests

To simulate the fracture propagation of rock with HJC material model, we should define each value of 23 parameters for Ls-dyna code. Though 10 parameters can be obtained from static lab-scale test.

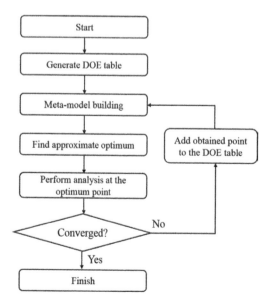

Figure 2. Computational procedure of progressive meta-model based optimization.

We have investigated the range of the parameters adopted in the previous studies (Holmquist et al., 1993; Polanco et al., 2008; Bu Changgen et al., 2009; Fang et al., 2014; Li and Shi, 2016).

Table 1 lists the lower and upper boundary values. The DOE of 13 parameters were made by three level L_{27} (3^{13}) orthogonal array. An orthogonal array of the DOE is listed in Table 2 based on the defined design level. The effect of the chosen parameters on the tensile fracture propagation is then evaluated by statistical analysis using the software by Minitab Inc. (2015).

2.6 Dynamic simulations

The 3-point bending simulation was performed according to the level combinations shown in Table 2. The finite element model used in the 3-point bending simulations is shown in Figure 3. The dimension of the chevron bend specimen follows the suggested method (ISRM, 1988; KSRM, 2008), with the length and V-notch of the cylindrical rock specimen 150.4 mm and 1 mm, respectively. Then a series of simulations were carried out (Figure 4) and the results were obtained (Figure 5). The effect of each parameter on the tensile failure of the rock was evaluated using analysis of variance (ANOVA).

2.7 Sensitivity analysis

For the sensitivity analysis, screening method was utilized to filter out some important design

Table 1. Upper and lower level of 13 parameters in HJC material model.

Design variables	Lower(1)	Middle(2)	Upper(3)
Normalized cohesive strength (A)	0.3	0.65	1
Normalized pressure hardening (B)	1.23	1.865	2.5
Strain rate coefficient (C)	0.0045	0.0071	0.0097
Pressure hardening exponent (N)	0.025	0.4575	0.89
Plastic strain before fracture (EFMIN)	0.004	0.007	0.01
Normalized maximum strength (SFMAX)	7	13.5	20
Locking pressure (PL)	0.8	1.65	2.5
Locking volumetric strain (UL)	0.012	0.196	0.38
Damage constant (D1)	0.01	0.028	0.046
Pressure constant (k1)	3.1	44.05	85
Pressure constant (k2)	-171	-73	25
Pressure constant (k3)	8.4	297.2	550
Volumetric erosion (evol)	0.001	0.0055	0.01

Table 2. DOE of 13 parameters for the simulations.

Exp. No	A	B	C	N	EFMIN	SFMAX	PL	UL	D1	k1	k2	k3	evol
1	1	1	1	1	1	1	1	1	1	1	1	1	1
2	1	1	1	1	2	2	2	2	2	2	2	2	2
3	1	1	1	1	3	3	3	3	3	3	3	3	3
4	1	2	2	2	1	1	1	2	2	2	3	3	3
5	1	2	2	2	2	2	2	3	3	3	1	1	1
6	1	2	2	2	3	3	3	1	1	1	2	2	2
7	1	3	3	3	1	1	1	3	3	3	2	2	2
8	1	3	3	3	2	2	2	1	1	1	3	3	3
9	1	3	3	3	3	3	3	2	2	2	1	1	1
10	2	1	2	3	1	2	3	1	2	3	1	2	3
11	2	1	2	3	2	3	1	2	3	1	2	3	1
12	2	1	2	3	3	1	2	3	1	2	3	1	2
13	2	2	3	1	1	2	3	2	3	1	3	1	2
14	2	2	3	1	2	3	1	3	1	2	1	2	3
15	2	2	3	1	3	1	2	1	2	3	2	3	1
16	2	3	1	2	1	2	3	3	1	2	2	3	1
17	2	3	1	2	2	3	1	1	2	3	3	1	2
18	2	3	1	2	3	1	2	2	3	1	1	2	3
19	3	1	3	2	1	3	2	1	3	2	1	3	2
20	3	1	3	2	2	1	3	2	1	3	2	1	3
21	3	1	3	2	3	2	1	3	2	1	3	2	1
22	3	2	1	3	1	3	2	2	1	3	3	2	1
23	3	2	1	3	2	1	3	3	2	1	1	3	2
24	3	2	1	3	3	2	1	1	3	2	2	1	3
25	3	3	2	1	1	3	2	3	2	1	2	1	3
26	3	3	2	1	2	1	3	1	3	2	3	2	1
27	3	3	2	1	3	2	1	2	1	3	1	3	2

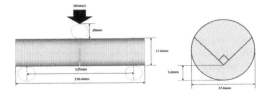

Figure 3. Numerical model of chevron bend specimen.

Figure 4. Simulation result of 3-point bending test.

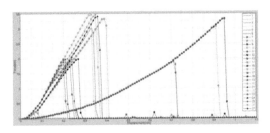

Figure 5. Simulation results: force-displacement curves of 27 cases.

parameters influencing the tensile failure in the simulation. The input design variables are 13 parameters of HJC material model and three output parameters are set to the slope of the force-penetration curve (k), the maximum force (F_{max}) and penetration (d_{max}) obtained from the 3-point bending tests. Table 2 lists the combinations of design parameters used in the orthogonal array for the 27 repetitions of the simulation. The contribution ratios of the design parameters extracted via screening method are listed in Table 3. The design factors PL, $k1$, and $evol$ are significantly associated with the effects of the tensile fracture (95% significance level), thus influencing the tensile fracture, whereas parameters A, B, C, N, $EFMIN$, $SFMAX$, UL, $D1$, $k2$ and $k3$ are not. The pressure constants $k2$ and $k3$ in a tensile fracture has been reported not to contribution to the rock fracture effects (Holmquist et al., 1993).

The three parameters (PL, $k1$, and $evol$) were finally selected with a contribution rate of over 4%, and these were used to optimize the fitting of rock fracture toughness via the Kriging meta-model (Kleijnen JPC, 2009) and evolutionary algorithm.

Table 3. Contribution ratio of configuration variable for tensile fracture of HJC rock model.

Parameters	Contributions		
	F_{max}	d_{max}	k
A	1.2%	0.9%	0.6%
B	0.7%	3.6%	2.7%
C	0.4%	0.8%	0.1%
N	0.4%	0.9%	0.0%
EFMIN	0.5%	0.8%	0.1%
SFMAX	0.6%	0.8%	0.2%
PL	0.5%	4.7%	3.7%
UL	0.3%	1.2%	0.2%
D1	0.4%	0.7%	0.2%
K1	0.5%	43.0%	54.2%
K2	0.3%	0.6%	0.2%
K3	0.4%	0.6%	0.2%
evol	93.7%	41.4%	37.7%
Total	100.0%	100.0%	100.0%

79

2.8 Optimization procedure

The orthogonal array of the DOE and the Kriging meta-model were introduced to fit the selected three parameters to precisely describe the 3-point bending test. For that optimization process, a meta-model with evolutionary algorithm (Deb et al., 2009) was generated by the results of a 3-point bending simulations. An optimization problem for design of rock fracture toughness is formulated considering design requirements and design variables explained in equation (1).

- Find: *PL, kl* and *evol*

- To minimize error (*O1*): $\dfrac{\left| F_{max}^{exp} - F_{max}^{sim} \right|}{F_{max}^{exp}} \leq 1\%$ (1)

- Constraint (*G1*): $\dfrac{\left| d_{max}^{exp} - d_{max}^{sim} \right|}{d_{max}^{exp}} \leq 1\%$

Table 4 shows the results of the optimal levels of the HJC parameters derived for high strength rock. As shown in Table 4, the required design specifications and constraints were satisfied after 19th iterations.

In the case of medium strength rock, the optimal solution could be observed through 18th iterations (Table 5). Furthermore, in the cases of moderate and low strength rocks, the requirements were satisfied in 11th (Table 6), and in 17th iterations (Table 7), respectively.

Table 4. Convergence results of parameter for fracture toughness of high strength rock.

Opt. No.	PL	k1	evol	G1	O1
1	2.4884	83.39	0.0097	5.1%	234.2%
2	2.4978	3.47	0.0005	6.2%	60.9%
3	2.4903	3.74	0.0006	6.2%	59.2%
4	1.9726	4.10	0.0007	3.4%	52.4%
5	1.47719	4.14	0,0007	1 4%	48 9%
6	0.8318	4.14	0.0008	5.4%	43.9%
7	0.8314	5.40	0.0008	6.2%	42.7%
8	0.8309	7.82	0.0008	17.7%	40.2%
9	0.8453	4.64	0.0009	8.2%	35.4%
10	0.8765	5.07	0.0010	5.4%	32.0%
11	0.9803	5.09	0.0011	11.0%	22.9%
12	1.4796	8.00	0.0012	6.2%	19.9%
13	1.7118	8.02	0.0012	3.4%	15.3%
14	1.7881	8.20	0.0013	3.4%	13.9%
15	1.7994	8.04	0.0013	3.4%	14.6%
16	1.7890	8.17	0.0014	0.6%	11.8%
17	1.7534	8.32	0.00151	1.4%	7.3%
18	1.7245	8.40	0.00165	1.4%	6.3%
19	1.7181	8.50	0.00167	0.6%	0.4%

Table 5. Convergence results of parameter for fracture toughness of medium strength rock.

Opt. No.	PL	k1	evol	G1	O1
1	1.6118	4.51	0.00065	62.3%	40.8%
2	1.5557	4.63	0.00069	60.6%	35.0%
3	1.6213	0.66	0.00104	47.3%	3.1%
4	1.0958	1.73	0.00140	10.6%	18.4%
5	0.9745	1.70	0.00128	8.9%	21.6%
6	1.0425	3.64	0.00139	35.6%	21.3%
7	1.5552	0.66	0.00106	47.3%	2.9%
8	1.5555	0.50	0.00105	43.9%	2.9%
9	1.9498	1.17	0.00734	92.9%	253.7%
10	1.5557	0.65	0.00104	47.3%	3.3%
11	1.1995	1.17	0.00107	3.6%	0.9%
12	1.1904	1.25	0.00122	1.4%	13.4%
13	1.1995	1.15	0.00100	55.6%	10.5%
14	1.1025	1.17	0.00120	3.0%	10.1%
15	1.1025	1.17	0.00116	1.4%	9.1%
16	0.907	1.17	0.00117	1.4%	9.2%
17	0.819	1.16	0.00118	0.3%	0.8%
18	0.8320	1.16	0.00118	0.3%	0.7%

Table 6. Convergence results of parameter for fracture toughness of medium rock.

Opt. No.	PL	k1	evol	G1	O1
1	0.9018	3.28	0.00061	49.0%	16.8%
2	0.9387	3.62	0.00061	51.0%	16.2%
3	0.8191	3.82	0.00063	51.0%	12.7%
4	0.8331	8.44	0.00068	67.4%	10.5%
5	1.1008	5.09	0.00071	57.2%	9.4%
6	0.8191	3.42	0 00071	44 9%	4.9%
7	0.8501	4.70	0.00073	53.1%	3.9%
8	0.8488	3.57	0.000760	42.9%	5.3%
9	0.8400	1.18	0.000760	55.1%	3.0%
10	0.8322	1.22	0.000740	6.1%	0.4%
11	0.8215	1.20	0.000745	0.2%	0.7%

The errors between the experiments and analyses data converged to less than 1% for both parameters of F_{max} and d_{max}. Based on the optimization results, the selection of HJC parameters and each level of those were proposed for simulating the tensile fracture characteristics of rock.

Figure 6 shows the numerical simulation results of the HJC model parameters derived through the optimization process for four rock classes in comparison with the experimental results. Figure 6 (a) to (d) show the simulations of tensile fracture toughness

Table 7. Convergence results of parameter for fracture toughness of soft rock.

Opt. No.	PL	k1	evol	G1	O1
1	2.4295	10.16	0.00055	61.9%	1.3%
2	1.2734	44.07	0.00056	76.6%	1.3%
3	1.3400	19.28	0.00055	70.7%	2.1%
4	2.1474	9.72	0.00056	59.0%	6.3%
5	1.0997	9.66	0.00055	59.0%	4.2%
6	0.9365	9.61	0.00055	59.0%	4.7%
7	0.9364	9.56	0.00055	59.0%	4.7%
8	0.9364	3.01	0.00055	29.6%	5.5%
9	0.8930	2.61	0.00056	20.8%	14.3%
10	0.8795	2.35	0.00056	17.9%	10.6%
11	0.8758	1.34	0.00056	11.4%	7.1%
12	0.8750	1.34	0.00056	8.5%	5.8%
13	0.8758	1.33	0.00050	2.6%	0.8%
14	0.8736	1.00	0.00050	56.0%	10.9%
15	0.8736	1.33	0.00052	5.6%	3.0%
16	0.8755	1.32	0.00051	5.6%	0.9%
17	0.8756	1.33	0.00050	0.3%	0.9%

for hard rock, medium hard rock, medium rock, and soft rock, respectively.

3 CONCLUSIONS

In order to analyze the tensile fracture mechanism of rocks, the 3-point bending tests and numerical simulations were performed. The tensile fracture characteristics of rocks were analyzed by the HJC material model and the Kriging meta-model and the evolution algorithm were adopted to fit and optimize the HJC parameters.

The 3 key parameters of the HJC material model, that significantly influence to the tensile fracture of rocks, were identified by the sensitivity analysis. Then, the levels of each parameter were proposed by processing optimization technique. Based on this result, the proposed three parameters (e.g. PL, k1, and evol) can successfully describe the tensile fracturing process of rock model.

The results of this study can be used as the fundamental research data for producing rock blocks using rock cutting and splitting methods. Future study will be required to analyze the tensile crack characteristics of various rock classes considering the real-scale rock mass.

REFERENCES

Bu C, Qu Y, Cheng Z, and Liu B. (2009), Numerical simulation of impact on pneumatic DTH hammer percussive drilling. Journal of Earth Science, 20(5), pp. 868–878.

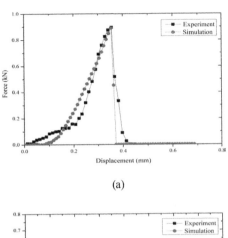

(a)

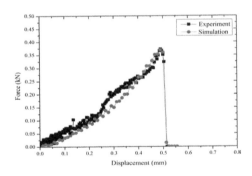

(b)

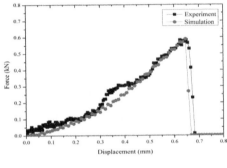

(c)

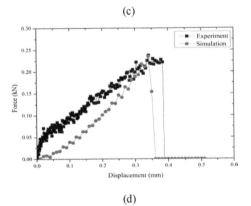

(d)

Figure 6. Optimization results; (a) high, (b) medium, (c) moderate, and (d) low strength rocks.

Deb K., Gupta S, Daum D, Branke J, Mall AK, and Padmanabhan D. (2009), Reliability-based optimization using evolutionary algorithms. IEEE Transactions on 13(5), pp. 1054–1074.

Fang Q, Kong X, Wu H, and Gong Z. (2014), Determination of Holmquist-Cook constitutive model parameters of rock, Engineering Mechanics, 31(3), pp. 197–204.

Holmquist TJ, Johnson GR, Cook WH. (1993), A computational constitutive model for concrete subjected to large strains, high strain rates and high pressures. Proceedings of the 14th international symposium on Ballistics. Quebec, Canada, pp. 591–600.

ISRM suggested method, (1988), Suggested Methods for Determining the Fracture Toughness of Rock, Int J Rock .Mech. Min. Sci. & Geomech. Abstract. Vol. 25, No. 2, pp. 71–96.

Jeong MJ, Dennis BH, Yoshimura S, (2005) Multidimensional clustering interpretation and its application to optimization of coolant passage of a turbine blade, Journal of Mechanical Design, 127(2), pp. 215–221.

Kleijnen JPC. (2009), Kriging metamodeling in simulation: a review. European Journal of Operational Research, 192(3), pp. 707–716.

Korean Society for Rock Mechanics, (2008), Standard test method for level 1 fracture toughness of rock using chevron bend specimens. Korean Society for Rock Mechanics, 18(6), pp. 503–506.

Livermore Software Technology Corporation, (2014), LS-Dyna Keyword User's Manual.

Li HY, and Shi GY. (2016), A dynamic material model for rock materials under conditions of high confining pressures and high strain rates, International Journal of Impact Engineering, 89, pp. 38–48.

Minitab Inc., (2015), Minitab Start guide.

PIDOTECH Inc., (2015), PIAnO User's Manual.

Polanco-Loria M, Hopperstad OS, Borvik T, and Berstad T. (2008), Numerical predictions of ballistic limits for concrete slabs using a modified version of the HJC concrete model, International Journal of Impact Engineering, 35(5), pp. 290–303.

2019 Rock Dynamics Summit– Aydan et al. (eds)
© *2019 Taylor & Francis Group, London, ISBN 978-0-367-34783-3*

Evaluation of elastic region of the surrounding bedrock using the cyclic unconfined compression test of rocks

K. Murakami, K. Yokota & A. Iwamori
Kansai Electric Power Co.,Ltd, Fukui, Japan

T. Okada & A. Sekiguchi
Central Research Institute of Electric Power Industry, Chiba, Japan

ABSTRACT: The seismic response analysis is typically carried out to evaluate the seismic resistance of the civil engineering structure including the surrounding bedrock related to nuclear power facilities. The bedrock is conservatively assumed to be a linear material in the seismic response analysis. It is confirmed that the stress ratios (= stress/failure strength) occurring in the bedrock are within 0.5 from result of the seismic response analysis. Therefore, the multi-stage cyclic compression test of the rock was performed in order to confirm whether the bedrock is generally elastic against the stress ratio. As a result, it was confirmed that the stress-strain relation of the rock specimens were almost linear until the loading stage just before the failure and the elastic wave velocities of the rocks were not generally nearly decreased. Therefore, they has been shown it is reasonable evaluation method to assume the bedrock as a linear material.

1 INTRODUCTION

In Japan, the seismic response analysis is typically carried out to evaluate the seismic resistance of civil engineering structure including the surrounding bedrock related to nuclear power facilities. Stress-strain relation of bedrock is conservatively assumed to be a liner elastic material in the seismic response analysis and Young's modulus are calculated from elastic wave velocity, because the stress ratio (= stress/failure strength = reciprocal of local safety factor) is basically used as an evaluation index in the seismic response analysis.

On the other hand, a condition of bedrock is not clear while the stress is increasing in case that bedrock is received by cyclic loading like seismic wave. Stress-strain relation of bedrock is assumed to keep elastic condition on small stress range, but become a plastic state little by little as the stress increases. Regarding the surrounding bedrock of civil engineering structure in Takahama Nuclear Power Plant (operated by Kansai Electric Power Co., Ltd), it is confirmed that the stress ratios occurring in the bedrock is smaller than about 0.5 from the result of seismic response analysis. The bedrock consists of rock and discontinuity and it is assumed that the rocks bear external shear force like seismic wave. If the rock keeps elastic condition while stress ratios are within 0.5, it is appropriated that the bedrock is assumed to be a liner material in the seismic response analysis. Therefore, elastic region of the surrounding bedrock shall be indirectly evaluated by tests which uses the rocks as specimens in this paper.

Thus, the multi-stage cyclic compression test (hereinafter called "compression test") and measurement

of the ultrasonic wave velocity of rock (hereinafter called "measurement of ultrasonic") were carried out in order to confirm that the rock keeps elastic condition at stress ratios within 0.5, after the rock is loaded by cyclic compression force.

2 OUTLINE OF TEST

Compression test of rocks were carried out referencing Japanese Geotechnical Society standard (JGS2561: Method for multi-stage cyclic undrained triaxial test on rocks). The test was carried out under the condition that confining pressure was zero (which means the uniaxial compression test) to conservatively evaluate elastic region of the rocks.

Measurement of ultrasonic was carried out according to Japanese Geotechnical Society standard (JGS2110: Method for laboratory measurement of ultrasonic wave velocity of rock by pulse test). The frequency of ultrasonic element of test machine was set to 200Hz in P wave and 100Hz in S wave, referencing the unconfined compression strength test results that had been carried out before than this time.

2.1 Specimens of test

The rock types used as test specimens are the rhyolite, andesite and rhyolite tuff, which are main rock types in the site of Takahama Nuclear Power Plant. However, the distribution range of rhyolite tuff is slight compared to that of rhyolite and andesite. It is assumed that the unconfined compression strength of rhyolite and andesite is about 100MPa and that of rhyolite tuff is about 40MPa from the test results carried out before than this time.

The rock specimens using the boring core samples are shaped 5cm in diameter and 10cm in height according to JGS2561. The moisture contents of the rock specimens are 0.91% in rhyolite, 1.25% in andesite and 1.28% in rhyolite tuff on average. Strain gauges are put in an axial direction and in a circumferential direction on both sides position at the center of the specimen height. The length of strain gauges at axial direction is 20mm or 60mm and at circumferential direction is 20mm or 30mm according to the grain size included in the specimens.

2.2 Method of test

The procedure of the test is shown in the Figure 1.

At first, to confirm the initial status of the rock specimen, the ultrasonic wave velocity of the rock specimen was measured before the STEP1 of compression test. In STEP1, compression test

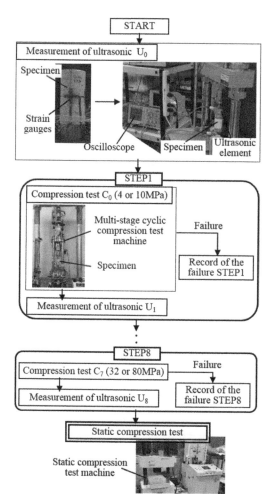

Figure 1. Procedure of test

and measurement of ultrasonic were conducted. Compression test was conducted with 10% load of the unconfined compression strength (10MPa : rhyolite and andesite, 4MPa : rhyolite tuff). Then, the ultrasonic wave velocity of the rock is measured. The cyclic loading was increased 10% for every STEP (the cyclic loading in STEP2 is 20%, the cyclic loading in STEP3 is 30%). 10 cycles was given in one step of compression test. If the rock specimen doesn't reach to the failure by the end of STEP8, the static compression test was conducted to confirm the compression strength. However, in case that the rock specimen reaches to the failure before the end of STEP8, cyclic loading at the time of the failure was recorded as the compression strength. The frequency of cyclic loading of compression test was basically 1Hz according to JGS2561. However, compression test was also implemented with 2Hz frequency, because it is confirmed that one of the basic ground motions in Takahama site responds to around 2Hz of the seismic frequency band. Stress-strain relation in every cyclic compression load was recorded and equivalent Young's modulus was calculated from 10th wave at every STEP of compression test.

3 THE RESULT OF TEST

The initial status including elastic wave velocity of the rock specimens before compression test is shown in Table 1. Initial density (ρ_0), P wave velocity (hereinafter called V_p) and S wave velocity (hereinafter called V_s) from measurement of the ultrasonic wave velocity (U_0), dynamic Young's modulus (E_d) calculated from V_p and V_s, equivalent Young's modulus (E_{eq}) obtained by STEP1 of compression test (C_0), the step of the failure, failure strength, failure strain and frequency of the cyclic loading are shown in this table. In case that the rock specimen reaches to the failure by static compression test, "static" is written at the column of "step of the failure" in Table 1. Rhyolite, andesite and rhyolite tuff are written Rh, An and Rh(T) in Table 1.

3.1 The result of measurement of ultrasonic wave velocity

The results of V_p and V_s obtained from measurement of ultrasonic are shown in Figures 2 and 3, respectively. The result of V_p and V_s of every STEP is normalized by initial ultrasonic wave velocity shown in Table 1 (hereinafter called "normalized V_p, V_s"). Normalized V_p, V_s are plotted on the vertical axis in these Figures. The values of the uniaxial stress is normalized by the failure strength are plotted on the horizontal axis. The uniaxial stress is recorded in 10th cycles of every STEP.

When the value of axial stress/failure strength was 0.5, normalized V_p decreased about 0% to 5%

Table 1. Initial status before compression test and result of test

	V_p (m/s)	V_s (m/s)	E_d (MPa)	E_{eq} (MPa)	ρ_0 (g/cm³)	Step of failure	Failure strength (MPa)	Failure strain	Loading Frequency
Rh1	3,811	2,481	33,654	20,354	2.415	static	124.99	0.554	1
Rh2	3,914	2,539	35,681	20,791	2.435	static	122.08	0.499	1
Rh3	3,266	2,172	24,999	14,130	2.401	7	69.54	0.459	1
Rh4	3,373	2,187	25,876	15,477	2.379	8	79.79	0.483	1
Rh5	3,454	2,219	27,178	15,178	2.402	5	50.05	0.321	1
Rh6	3,215	2,058	23,138	14,862	2.369	5	46.75	0.335	1
Rh7	3,833	2,511	34,188	19,799	2.412	static	104.36	0.458	2
Rh8	3,377	2,190	26,033	16,043	2.388	static	88.10	0.503	2
Rh9	3,498	2,216	27,599	13,271	2.411	5	38.83	0.406	2
An1	4,900	2,850	51,432	51,554	2.545	static	110.79	0.234	1
An2	4,859	2,702	47,472	41,033	2.548	static	83.34	0.232	1
An3	3,067	1,960	21,298	12,341	2.400	6	58.02	0.344	1
An4	4,534	2,634	42,826	36,426	2.479	static	92.27	0.236	2
An5	3,748	2,359	32,272	23,951	2.474	8	77.66	0.331	2
Rh(T)1	3,062	2,026	21,078	12,514	2.311	static	37.71	0.316	1
Rh(T)2	2,919	2,050	19,711	11,934	2.315	static	61.44	0.571	1
Rh(T)3	2,789	1,926	18,109	9,905	2.338	static	81.31	0.778	2
Rh(T)4	2,772	2,035	17,844	10,506	2.352	static	58.30	0.592	2

※About Rh(T)4, it's carried out 11steps and static test

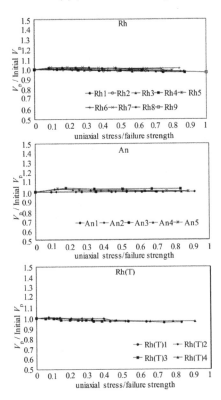

Figure 2. Result of V_p

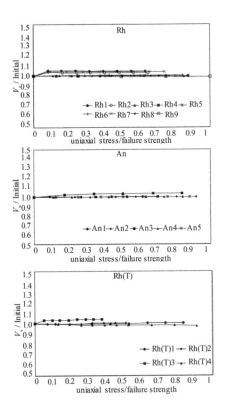

Figure 3. Result of V_s

and normalized V_s basically did not change from initial value. Furthermore, both normalized V_p and normalized V_s decreased within 5% just before the failure. It was confirmed that there were almost no differences of the test results by rock types and by cyclic wave frequency. On the other hand, there were some specimens where the elastic velocity increase from initial value. These facts are considered in *3.3 the result of stress-strain relation*.

The average values for each 0.1 step of axial stress/failure strength from all test results are shown in Table 2. The sample standard deviations from Figures 2 and 3 were 0.143 for normalized V_p and 0.045 for normalized V_s. From above, it was confirmed that though the elastic wave velocity was falling as the cyclic compression load increases, the falling rate of the elastic wave velocity was included within the margin of variability.

3.2 *The result of compression test*

Equivalent Young's modulus from the result of compression test is shown in Figure 4. Calculating methods of Young's modulus are written below for each case. In case of calculating from the cyclic loading test, Young's modulus are calculated from the slope of line that connects the peak load coordinate and the coordinate where shear stress is 0, which is average value between the start coordinate before the compression load and the end coordinate after the compression load, to consider residual strain. In case of calculating from the static compression test, Young's modulus is calculated from the slope of line that connects point of the coordinate before the compression load and the coordinate of the failure. Equivalent Young's modulus from test results are divided by initial equivalent Young's modulus shown in Table 1, and the value of equivalent Young's modulus is normalized such that initial value is 1.0. Normalized equivalent Young's modulus is plotted on the vertical axis in this Figure. The value of the uniaxial stress divided by the failure strength is plotted on the horizontal axis. The uniaxial stress is recorded in 10th cycles of every STEP.

Equivalent Young's modulus of rhyolite slightly increased as the cyclic compression load increases. And that of andesite barely decreased. On the other hand, that of rhyolite tuff decreased at first, but conversely increased from the middle as cyclic load increases. After all, it was confirmed that the range of fluctuation is within 20% when the

value of axial stress/failure strength is 0.5, and the fluctuation barely changes just before the failure. It was also confirmed that there were almost no differences of test results by rock types and by cyclic compression wave frequency. However, the elastic wave velocity of An3 increased from initial value. These facts are considered in *3.3 the result of stress-strain relation*.

The average values for each 0.1 step of uniaxial stress/failure strength from all test results are shown in Table 3. The sample standard deviation from Figure 4 was 0.153. From above, it was confirmed that though the elastic wave velocity was falling as the cyclic compression load increases, the falling rate of the elastic wave velocity was included within the margin of variability.

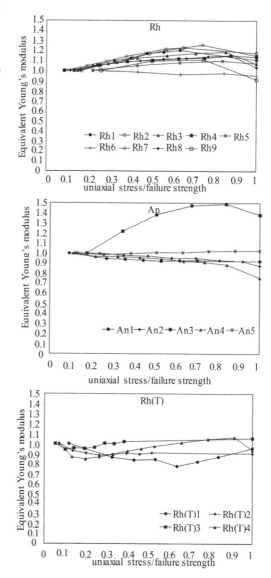

Figure 4. Equivalent Young's modulus

Table 2. Average of V_p and V_s

Stress ratios									
0.1	0.2	0.3	0.4	0.5	0.6	0.7	0.8	0.9	1.0
V_p 1.00	1.00	0.95	0.91	1.00	0.99	0.94	0.96	0.98	–
V_s 1.02	1.01	1.01	1.01	1.01	1.01	0.98	0.97	0.99	–

Table 3. Average of equivalent Young's modulus

Stress ratios									
0.1	0.2	0.3	0.4	0.5	0.6	0.7	0.8	0.9	1.0
Equivalent Young's modulus									
0.98	0.99	1.01	1.03	1.05	1.06	1.07	1.08	1.03	1.01

3.3 The result of stress-strain relation

In this chapter, the facts that normalized V_p, normalized V_s and equivalent Young's modulus from the result of stress-strain relation increased as the cyclic load increases are considered.

Stress-strain relation of Rh2, in which the falling rate of normalized V_s and equivalent Young's modulus are relatively large compared to other specimens, is shown in Figure 5. Stress-strain relation of Rh2 has the feature of convex curve below. As one of the reasons why stress-strain curve is convex below, it is considered that rock specimens had invisible weakened parts or potential cracks (hereinafter called "potential cracks"). So it is assumed that potential cracks are compressing as the cyclic load increases, and the stiffness of the rock specimens increase. It is assumed that the equivalent Young's modulus increased as test STEP progresses due to the same reason.

Stress-strain relation curve at 10th cyclic wave load from STEP1 to STEP8 are shown in Figure 6. Though the slopes are almost same in the range of small uniaxial stress from STEP1 to STEP8, the slope in the range of small uniaxial stress is gentler than that of large axial stress for each STEP. From above, it is assumed that equivalent Young's modulus increases in case that the uniaxial stress exceeds a certain value because potential cracks disappear due to compression at the axial stress. The slope of all STEPs in the range of small uniaxial stress is almost same, so it is considered that the damage of the rock specimens is extremely small.

The stress-strain relation curves of Rh1, Rh3, Rh4, Rh5 and Rh6 are similar to that of Rh2. Especially, the stress-strain relation curve of Rh1 is

almost the same as that of Rh2. However, the stress-strain curves of Rh3, Rh4, Rh5 and Rh6 (hereinafter called "Rh3 to Rh6") are slightly more bent under than those of Rh2. It is assumed that the rock specimens of Rh3 to Rh6 have more potential cracks than that of Rh2, judging from the fact that the initial values of E_d and E_{eq} of Rh3 to Rh6 are smaller than those of Rh2. As a reference, the stress-strain relation of Rh3 is shown in Figure 7.

The results of An1 and An3 shall be considered. About An1, the degree of fluctuation of equivalent Young's modulus is very small. On the other hand, the degree of fluctuation of equivalent Young's modulus of An3 is very large. Stress-strain relation of An1 and An3 are shown in Figures 8 and 9. It is considered that An1 has almost no potential cracks because the shape of stress-strain relation is almost straight line, and furthermore, the falling rate of equivalent Young's modulus does not exceed over about 5% from the initial load to the failure load. On the other hand, stress-strain curve of An3 is convex below like that of Rh2. So it is assumed that potential cracks are distributed in the rock specimen of An3, different from An1. Therefore, it is consid-

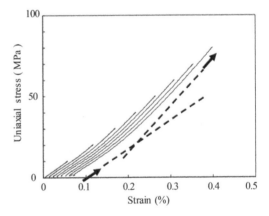

Figure 6. 10th wave (STEP1-8) stress-strain relation

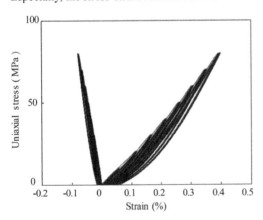

Figure 5. Stress-strain relation of Rh2

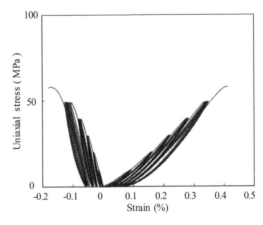

Figure 7. Stress-strain relation of Rh3

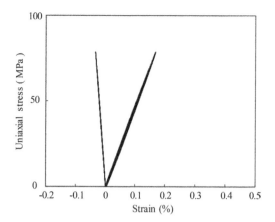

Figure 8. Result of An1 stress-strain relation

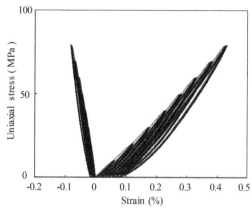

Figure 10. Result of Rh7 stress-strain relation

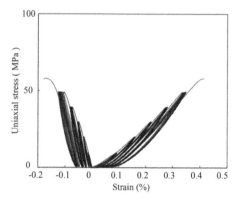

Figure 9. Result of An3 stress-strain relation

ered that the difference of fluctuation of equivalent Young's modulus between An1 and An3 results from the amount of the potential cracks in the rocks. This reasoning is consistent with the fact that the initial values of E_d, E_{eq} and rock density of An3 shown in Table 1 were definitely small compared with that of An1.

In order to confirm above reasoning, the results of the specimens where the initial values of E_d, E_{eq} and rock density are similar, are compared. Stress-strain relation of Rh2 and Rh7 are shown in Figures 5 and 10. It is confirmed that the shapes of stress-strain relation of Rh2 is similar with that of Rh7 which has similar initial value of E_d, E_{eq} and rock density of Rh2, although the frequency of cyclic load of Rh1 is different from that of Rh7.

4 CONCLUSION

The elastic wave velocities of the rocks were hardly falling in spite of the cyclic loads increasing, it is assumed that the rocks generally keep elastic just before the failure. On the other hand, the range of fluctuation of equivalent Young's modulus was within 20% for every STEP in cyclic loading test, although equivalent Young's modulus of the rock specimens change as the cyclic load increases due to the potential cracks. It was assumed that the damage of the rock specimens was very slight and the rocks kept generally elastic, because the rigidity of the rock specimens such as E_d or E_{eq} did not fall from initial measurement in spite of increasing of cyclic load.

Therefore, it was confirmed that the bedrock keeps generally elastic in the stress status just before the failure over the stress ratio 0.5, and it was a reasonable condition in the evaluation method to assume the bedrock as a linear elastic material in the site of Takahama site. Moreover, it was confirmed that the seismic resistance of civil engineering structure including the surrounding bedrock could be conservatively evaluated by using dynamic Young's modulus calculated from elastic wave velocity.

REFERENCES

Japanese Standards and Explanations of Laboratory Tests of Geomaterials (Japanese Geotechnical Society)
Japanese Geotechnical Society Standards Laboratory Testing Standards of Geomaterials (Japanese Geotechnical Society)

2019 Rock Dynamics Summit– Aydan et al. (eds)
© *2019 Taylor & Francis Group, London, ISBN 978-0-367-34783-3*

Experimental study on dynamic fracturing behavior under blasting loading in PMMA

S. Jeon, S. Choi & S. Lee
Department of Energy Systems Engineering, Seoul National University, Seoul, Korea

B. Jeon & H. Jeong
Research Institute of Energy and Resources, Seoul National University, Seoul, Korea

ABSTRACT: This study experimentally investigated dynamic fracturing behavior of brittle material under blasting loading. In the test, a detonator charged with DDNP (Diazodinitrophenol) and a PMMA ((Poly methyl methacrylate) specimen, a transparent and homogeneous material, were used. A high speed camera was employed to observe the cracking behavior, and it successfully provided obvious evidences to understand the crack propagation of brittle material under blasting loading. Two kinds of crack, namely shock-wave driven and gas-driven crack, were observed during the test. We qualitatively identified the initiation and propagation processes of the cracks. The results show that the gas-driven fracture in an ear-like shape was more predominant than shock-wave-driven fractures. Experimental results (i.e., crack length and the number of cracks) reasonably matched with the analytical solution based on fracture mechanics.

1 INTRODUCTION

Borehole fracturing has been widely used in tunneling, mining, gas production and geothermal projects. The fracturing is classified into two types; the 'static' and 'dynamic' fracturing according to the loading speed and method. Blasting and hydraulic fracturing are dynamic fracturing methods induce fractures through the detonation of explosives and the stimulation with air or water, respectively. It is important to predict the extent and orientation of cracks to achieve the purpose of a project, such as control of the fragment size and estimation of the damaged zone at excavation surfaces.

Many experimental studies have been carried out to investigate the crack propagation characteristics induced by dynamic loading. Kutter & Fairhurst (1971) categorized the fracture mechanisms in blasting into stress-wave-generated and gas-expansion-generated types. The respective pressures of the stress-wave and gas-expansion were simulated by underwater spark discharge and pressurized oil, respectively. The plexiglass and rock specimens were used for the test. The study identified the characteristics of crack propagation during the test, and they noted that the gas-generated fractures play an important role in the fragmentation of blasting, and stress-wave-generated fractures play as precondition of gas-generated fractures. Daehnke & Knasmillner (1996) investigated the propagation of dynamic fractures in PMMA (Poly methyl methacrylate) specimen which was loaded with Lead Azide. They found that the stress wave rapidly outpaces the slower dynamic fractures and the majority

of fracturing occurred due to pressurization by detonation gas. Yang et al. (2012) investigated the effect of the pre-exist notches around blasthole on the characteristics of crack propagation. They used PMMA specimen having two blastholes, and created artificial notches with two kinds of modes around the blasthole. The blast loading was simulated with multi-spark discharge. They observed the changes in dynamic stress intensity factor, dynamic energy release rate, and crack velocity during the test, and analyzed the effect of the notches on the test results.

As can be seen in the previous studies, it is very important to understand the mechanism of fracturing and its patterns (i.e., order of crack generation, crack length, crack shape, and the number of crack) according to various conditions (i.e., explosion position, specimen shape, properties of explosive, stemming, etc.).

Due to the facts that the dynamic fracturing by blast loading occurs in a very short period of time and the characteristics of initiation and propagation of cracks is complex, experimental approach has limitations to evaluate the cracking behavior in brittle materials.

In this study, two cases of experiments for dynamic fracturing were carried out to investigate the characteristics of crack propagation. PMMA (Poly methyl methacrylate) and DDNP (Diazodinitrophenol) were used for the specimen and the dynamic loading source, respectively. During the tests, the crack initiation and its propagation process were observed by a high speed camera, and the process was carefully analyzed. This study identified the characteristics of blast induced cracks (i.e., mechanism, shape, number, direction, extent, and initiation order, etc.), and categorized the cracks into several types according to the its direction and mechanism. Also, the results were compared with analytical approach based on the fracture mechanics theories. The results will be helpful to understand the characteristics of crack propagation under dynamic loading in brittle materials.

2 EXPERIMENTAL SETUP

2.1 *Specimen*

The PMMA, a transparent and homogeneous material, was used in the experiment. Two cylindrical specimens having a drilled blast-hole at the center were prepared. The size and dimension of the specimen are illustrated in Figure 1a, and the mechanical properties of the PMMA are listed in Table 1.

2.2 *Charging*

Two kinds of custom-manufactured detonator of 6.2 mm in diameter, EA-4 and New TLD, were used for the blast loading source. The detonators were charged with DDNP, and the amount of charges of the EA-4 and NEW-LTD were 300 and 140 mg, respectively. The one of the specimens was loaded by the EA-4 (Test A), and the other was loaded by the NEW-LTD (Test B). The properties of DDNP explosive are listed in Table 2. The holes were stemmed with sand after loading the detonators.

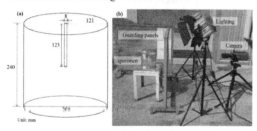

Figure 1. Dimension of the specimen (a) and the testing set-up (b).

Table 1. Mechanical properties of PMMA in a quasi-static condition.

Item	Value
Density (g/cm³)	1.18
Young's Modulus (MPa)	3000
Poisson's ratio	0.4
Compressive strength (MPa)	151
Tensile strength (MPa)	61
Mode-I Fracture toughness (MPa√m)	1.5

Table 2. Material properties and JWL parameters of DDNP (Zong et al. 2011).

Item	Value
Density (g/cm³)	1.00
Velocity of detonation (m/s)	4500
CJ pressure (GPa)	5.06
Internal energy (GPa)	8.5
A* (GPa)	524.2
B* (GPa)	0.769
R1*	4.2
R2*	1.0
ω *	0.3

*A, B, R1, R2 and ω are coefficients of the JWL-EOS

2.3 *Observation*

Under the blasting load, the detonation energy is suddenly released, and the corresponding crack behavior occurred in a very short time. The high speed camera (Phantom-v710) allowed observation of the crack initiation and propagation induced by explosive loading. The blasting experiments were recorded by the high speed camera located 2 m away from the specimen, and the resolution was set by 864 × 768 pixels with 10,000 frame/sec (Figure 1b).

3 EXPERIMENTAL RESULTS

In Test A, the specimen was split into several pieces as presented in Figure 2. It concluded that the 300 mg charge of DDNP was excessive to trace crack lengths and orientations. In Test B, on the other hand, all the cracks were caught inside of the specimen without separating the specimen into pieces. Thus, further analysis was made only for the case of Test B.

3.1 *Identification of crack propagation*

When the detonator was initiated in the circular hole, radial cracks initiated and propagated around the hole. The crack propagation process in the PMMA specimen is presented in Figure 3. Initially, a

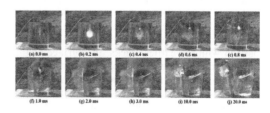

Figure 2. Crack propagation of PMMA specimen caused by an explosion of detonator charged with 300 mg of DDNP (Test A).

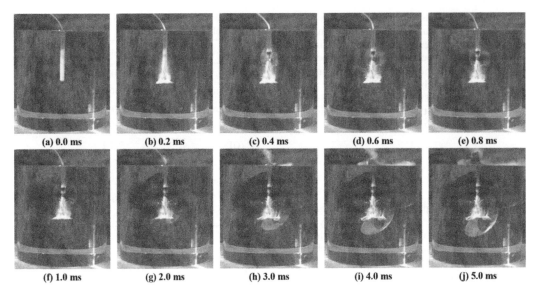

| (a) 0.0 ms | (b) 0.2 ms | (c) 0.4 ms | (d) 0.6 ms | (e) 0.8 ms |

| (f) 1.0 ms | (g) 2.0 ms | (h) 3.0 ms | (i) 4.0 ms | (j) 5.0 ms |

Figure 3. Crack propagation of PMMA specimen caused by an explosion of detonator charged with 140 mg of DDNP (Test B).

conical-shaped fracture zone was developed and propagated at the bottom of the blasthole (Figure 3b), and the radius of the blasthole instantaneously increased due to the expansion of the explosives. The radius of the conical-shaped fracture increased to a certain length (significantly short than radial cracks) because the pressure rapidly decreased with the expansion of the fractures. Thereafter, the conical shaped fracture did not grow anymore, and prominent radial cracks developed at the blasthole wall vicinity.

The radial cracks initiated in a semi-circular shape (Figure 3c), then they gradually propagated in a bean or ear shape; the radial cracks are called "ear-shaped cracks" in the later part of this paper. Initially, the direction of crack propagation was mostly straight, toward the radial direction. But at the later stage, cracks propagated in the downward direction. The phenomenon might be caused by the thermodynamic effect of expanding gas volume and decreasing gas pressure.

Based on observation by the high-speed camera, it was found that the propagation of the ear-shaped cracks is significantly affected by gas expansion. The gas-driven fractures were categorized into two types with time sequence. The earlier gas-driven cracks propagated owing to the detonation gas and expansion of the blasthole (Figure 3c), but the later gas-driven cracks propagated owing to expansion of the detonation gas (Figures 3d–3j). In comparison with the earlier cracks, the surface of the latter cracks had noticeable charcoal color which indicated the gas with full chemical reaction was involved in the cracking process at this stage.

Based on these results, the process of blast induced fracturing was divided into two phases: (i) the cone-shaped fracture zone and small ear-shaped cracks that were initiated around the blasthole

by the shock-wave (or stress-wave), which were induced by the reaction of explosives, and (ii) the prominent ear-shaped cracks that were generated around the blasthole wall due to gas the expansion. The fractures caused by shock-waves were limited to a local area, which was around the blasthole, but ear-shaped cracks propagated into radial directions. Figure 4 presents the overall observation from the test where gas-driven fractures had more important role for the fragmentation of the specimen.

Various experimental and numerical studies (Fourney et al. 1979, Worsey et al. 1981, McHugh

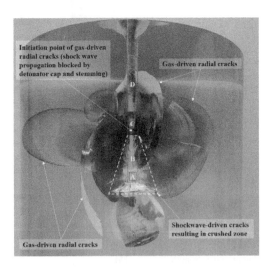

Figure 4. Photograph of the shockwave-driven and the gas-driven cracks in PMMA specimen (Test B). A, B, C and D refer to the part of base charge (DDNP), detonator tube containing a delay element, isolation cap, and stemming (sand), respectively.

1983, Schatz et al. 1987) also concluded that the gas-driven fracture propagation is the dominant mechanism for the extension of long fractures, and it is possible that gas-expansion increases the fracture length by a factor of 10-100 in comparison to crack extension due to stress-wave loading alone. It is known that the amount of energy released by stress-wave accounts for a small fraction, typically of the extent of up 20%, of the total energy released by the explosive (Kutter & Fairhurst 1971).

3.2 *Types of crack propagation*

As mentioned above, a number of ear-shaped cracks were initiated around the blasthole wall by the shock-wave, but only a few of these cracks continuously propagated due to gas expansion. Crack growth depends on the stress intensity factor of the crack which is the function of the pressure on the blasthole wall and the crack length. In this case, the stress intensity factor decreased with the crack length and increased with the pressure. Because the stress intensity factor decreased with time, the crack propagation was ceased at a certain length. Figure 5 shows a photograph of PMMA specimen after explosion of the detonator. Six ear-shaped cracks initiated and propagated. The crack length was about 68 mm on average. The ear-shaped cracks were symmetrically developed with respect to the blasthole axis, and the included angles between adjacent cracks were approximately 60 degrees.

Consequently, ear-shaped gas driven cracks initiated at the top of the borehole first; however, they were categorized into three types, according to the propagation direction (Figure 6). Type-A cracks only propagate in the radial direction, and its length

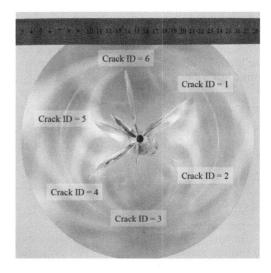

Figure 5. Photograph of the six cracks induced by blast loading.

gradually increases in the radial direction. Of the gas-driven cracks, Type-A cracks were much longer than other types of cracks in the radial direction. Type-B cracks propagate in not only the radial direction but also the vertical direction. Type-A and -B cracks consistently grow, and their shapes were a symmetrical circle or ellipse. While Type-C cracks irregularly grow, and their shape was also non-symmetric. The propagation direction of Type-C cracks was curved downward, and its propagation ceased at the bottom of the blasthole. Type-C cracks were much shorter than other types of cracks in the radial direction.

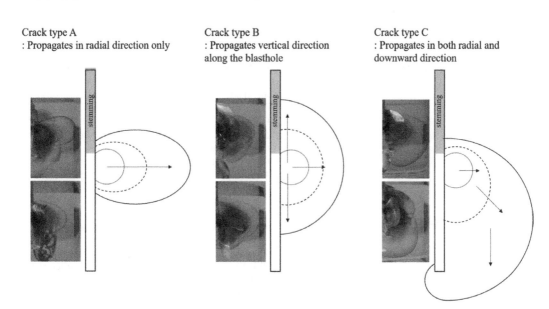

Crack type A
: Propagates in radial direction only

Crack type B
: Propagates vertical direction along the blasthole

Crack type C
: Propagates in both radial and downward direction

Figure 6. Schematic drawing of three types of ear-shaped gas-driven crack.

3.3 Comparing with analytic solution

If the radial cracks around a blast-hole are uniformly distributed, the mode I stress intensity factor KI of the blast-hole under an internal pressure P is determined by (Whittaker et al. 1992):

$$K_I = \frac{2PRk(n)}{\sqrt{\pi a}} \qquad (1)$$

$$k(n) = \sin\left(\frac{\pi}{n}\right)\sqrt{\frac{\pi}{\left(\left(\frac{2\pi}{n}\right) + \sin\left(\frac{2\pi}{n}\right)\right)}}, \quad a > R \qquad (2)$$

where a = the average length of radial cracks; and n = the number of the radial cracks.

The blasting pressure with high explosives was defined by the Jones-Wilkins Lee (JWL) equation of state. The JWL equation defines the pressure as a function of the relative volume (V) and internal energy (E) of explosives.

$$P = A\left(1 - \frac{w}{R_1 V}\right)e^{-R_1 V} + B\left(1 - \frac{w}{R_2 V}\right)e^{-R_2 V} + \frac{wE}{V} \qquad (3)$$

where A, B, R1, R2, w = parameters for the JWL-EOS, and its value are listed in Table 2.

The relative volume of the explosive is increased after blasting, and the pressure decreases according to the increase in the relative volume. Figure 7 shows the pressure change curve of the DDNP with the relative volume of the DDNP. The relative volume is defined as the ratio of the current volume to the initial volume, and the initial pressure was calculated as 10 GPa at 140 cm³. The relative volume of the DDNP, when the empty space in the detonator was filled with the DDNP was approximately 1465 cm³. At that moment, the pressure that acts on the blasthole was calculated as 158 MPa, and then the maximum crack length was calculated as 92.8 mm where six radial cracks were propagated (Figure 8). Considering that the number of cracks were five and seven, the

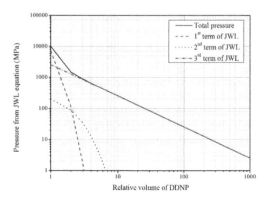

Figure 8. Stress intensity factor versus average length of the radial cracks at different internal pressures when a total number of radial cracks is 6 and a diameter of blast-hole is 0.008m.

maximum crack length was 111.1 mm and 79.6mm, respectively. Although the maximum length was influenced by the number of cracks, the results were reasonable when compared with experimental observations. On the other hand, in the case of the EA-4 detonator, the initial volume was 300 cm³, and the corresponding maximum crack length was estimated as 360.5 mm when six wing cracks were propagated. Because this value exceeded the radius of the specimen (125 mm), the specimen was shattered into several pieces. The crack length, which was calculated by analytical solution, had good agreement with those of experiments.

The blasthole pressure not only initiates the crack, but also propagates the crack to a certain length (Jiang 1996). Therefore, it is important to estimate the optimized blasthole pressure for efficient blasting. The critical pressure (blasthole pressure) is only dependent on the crack length, the number of cracks, and the fracture toughness of the material in Eq. 1 - 2. On the other hand, a few experimental studies have investigated the relationship between the critical pressure and the tensile strength of the material. Persson et al. (1994) estimated the blasthole pressure to be of the order of 2 to 4 times higher than the tensile strength. Dunn & Cocker (1995) and Brent (1995) found that the blasthole pressure must exceed the tensile strength of the material. In this study, the critical pressure was approximately 4 times greater than the tensile strength of the PMMA.

4 CONCLUSIONS

The dynamic fracturing test was performed for PMMA specimens to investigate the characteristic of crack propagation in brittle materials. The blasting loading was simulated by the detonator charged with DDNP, and the high speed camera was employed as the means of observation. It provided sequential

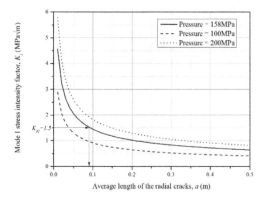

Figure 7. The pressure change of DDNP with different relative volume. Each term means the pressure calculated by the three parts of JWL equation, respectively.

observations of crack initiation and propagation under blast loading. The crack propagation behavior was analyzed through experimental observations. The results show that the gas-driven fractures were more important than shock-wave-driven fracture for fragmentation, and the extent of crack propagation was significantly affected by the amount of charge. The crack extension length and propagation orientation were observed from the experiment, and the results matched well with analytical solution based on the fracture mechanics.

However, the current study is limited to single blast-hole in a cylindrical specimen. The fragmentations in dynamic fracturing were induced by propagation and connection of cracks, and the crack propagation could be influenced by the boundary condition of the specimen. Thus, further study is required to investigate the characteristics of crack propagation in multiple blast-holes, and the influence of the boundary condition. The results of this study can provide useful information for further study. The tests do not consider the effects of in-situ stresses on crack propagation. It is well-known that the characteristics of crack propagation and orientation are significantly affected by the principal stresses; thus, more numerical simulation and testing must be conducted to investigate the influences of the principal stresses.

ACKNOWLEDGEMENT

This research was supported by the National Strategic Project-Carbon Upcycling of the National Research Foundation of Korea(NRF) funded by the Ministry of Science and ICT(MSIT), the Ministry of Environment(ME) and the Ministry of Trade, Industry and Energy(MOTIE). (NRF-2017M3D8A2085654), and was also funded by the Korea Agency for Infrastructure Technology Advancement under the Ministry of Land, Infrastructure and Transport in Korea (Project No.: 19SCIP-B105148-05).

REFERENCES

Brent, G.F. 1995. The design of pre-split blasts, *Proc. EXPLO `95 Conference*: 299–305.

Daehnke, A. Rossmanith, H.P. & Knasmillner, R.E. 1996. Blast-induced dynamic fracture propagation. *Rock fragmentation by Blasting*, Mohanty (ed.), Rotterdam: 13–18.

Dunn, P. & Cocker, A. 1995. The design of pre-split blasts. *Proc. EXPLO `95 Conference*: 307–314.

Fourney, W.L. Holloway, D.C. & Barker, D.B. 1979. Fracture initiation from the packer area. *University of Maryland Research Report prepared for the National Science Foundation.*

Hu, R. Zhu, Z. Xie, J. & Xiao, D. 2015. Numerical study on crack propagation by using softening model under blasting. *Advanced in Materials Science and Engineering*, Vol. 2015 (ID: 158580).

Jiang, J.J. 1996. Study of pre-split blasting using fracture mechanics. *5th Int. Symposium on Rock Fragmentation by Blasting*, Mohanty (ed.), Balkema, Rotterdam: 201–206.

Kutter, H.K. & Fairhurst, C. 1971. On the fracture process in blasting. *International Journal of Rock Mechanics and Mining Sciences & Geomechanics Abstracts* 8(3): 181–202.

McHugh, S. 1983. Crack extension caused by internal gas pressure compared with extension caused by tensile stress. *International Journal of Fracture* 21(3):163–176.

Persson, P. Holmberg, R. & Lee, J. 1994. *Rock blasting and Explosives Engineering*, CRC, Florida.

Schatz, J.F. Zeigler, B.J. Hanson, J. & Christianson, M. 1987. Multiple radial fracturing from a wellbore-experimental and theoretical results. *The 28th U.S. Symposium on Rock Mechanics*, Tucson, Arizona: 821–829.

Whittaker, B.N. Singh, R.N. & Sun, G. 1992. *Rock Fracture Mechanics, Principals, Design and Applications.* Elsevier, The Netherlands.

Worsey, P.N. Farmer, I.W. & Matheson, G.D. 1981. The mechanics of pre-spiltting in discontinuous rock. The *22nd U.S. Symposium on Rock Mechanics*, Cambridge, Massachusetts: 205–210.

Yang, R. Wang, Y. Xue, H. & Wang, M. 2012. Dynamic Behavior Analysis of Perforated Crack Propagation in Two-Hole Blasting. *Procedia Earth and Planetary Science* 5: 254–261.

Yang, X. Zeng, X. Pu, C. & Xiao, D. 2018. Effect of the preexisting fissures with different fillings in PMMA on blasting-induced crack propagation. *Advanced in Materials Science and Engineering*, Vol 2018 (ID: 737882).

Zong, Q. Yan, L.P. & Wang, H.B. 2011. Numerical simulation analysis on explosion stress field of different charge construction. *Advanced Materials Research* 250:2612–2616.

2019 Rock Dynamics Summit– Aydan et al. (eds)
© 2019 Taylor & Francis Group, London, ISBN 978-0-367-34783-3

Study on dynamic shear strength and deformation characteristics of rock discontinuity

J. Yoshida & T. Sasaki
SUNCOH Consultant Co. Ltd., Tokyo, Japan

R. Yoshinaka
Saitama University, Saitama, Japan

ABSTRACT: The authors have developed a new dynamic direct shear test machine, for the purpose of investigating the response to the earthquake motion of rock discontinuities. We conducted a large number of dynamic direct shear tests for the rock discontinuities. Test specimens are Limestone joint, Sandstone joint and Mudstone bedding plane of natural discontinuity made by boring-core, and artificial discontinuities made of mortar. By these test results, we examined dynamic shear strength and dynamic shear deformability of rock discontinuity.

The authors defined dynamic peak shear strength $\tau_{p(d)}$ by the results of Multi-stage amplitude dynamic direct shear tests. Then, we investigated comparison of dynamic shear strength and static shear strength, and dependence on frequency for dynamic shear strength. It is clear that dynamic shear strength exceeds static shear strength for relatively rough planes, and dynamic peak shear strength $\tau_{p(d)}$ does not depend on the frequency in the range from 0.1Hz to 3.0Hz

Also we defined the dynamic diagonal shear stiffness $K_{sd(d)}$ and attenuation h. Furthermore, we examine the stress dependence and frequency dependence of these dynamic deformability parameters. We defined skelton curves and modelled it by hyperbolic function. It is Clear that both the dynamic diagonal shear stiffness $K_{sd(d)}$ and attenuation h, have a dependance about normal stress σn and shear stress amplitude.

1 GENERAL INTRODUCTION

Many rock discontinuities are distributed in hard rock such as bedding planes or joint planes, and influence the strength and deformability of the rock mass. In recent years, dynamic analysis method for rock foundations such as for very important facilities and on large rock slopes was required. With regard to this dynamic analytical method for discontinuous rock mass, the problems due to the conventional elastic analysis methods being insufficient are pointed out, but a new analytical method for rock foundation and rock structure has been proposed in recent years (Iwata et al., 2012; Yoshinaka et al., 2012).

However, in analysis and design using these analytical methods, dynamic strength and deformability of the rock discontinuity become the input parameters for analysis of the problem. Not much data has been accumulated on dynamic strength and deformability of rock discontinuities in the past, unlike that for the static parameters from laboratory core tests or in-situ tests. Furthermore, exclusive test machines are not generally available. Particularly, regarding dynamic cyclic load test machines for investigating the rock discontinuities that are assumed to occur in an earthquake, there are only a few research papers. But, among these research papers, dynamic testing under the condition of assumed real seismic motions is almost non-existent.

The authors developed the new dynamic direct shear test machine which Figure 1 showed for the purpose of investigating the dynamic properties of rock discontinuity at the time of the earthquake (Yoshida et al., 2014). In this study, we examined the dynamic shear strength of rock discontinuity and dynamic shear properties than the test result using this test machine.

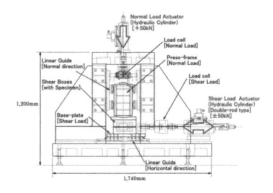

Figure 1. Dynamic direct shear test machine for rock discontinuity

2 DYNAMIC SHEAR STRENGTH OF ROCK DISCONTINUITY

2.1 *Definition of dynamic shear strength*

Figure 2 shows the result of Multi-stage amplitude dynamic direct shear test for limestone joints. The limestone joints in the tests were rough plane (JRC = 10~16 for observation) and were interlocking well (rank B). The frequency of the shear stress wave was 1.0 Hz, the shear stress amplitude arrived at the targeted strength (static shear strength, τs) after loading with 10 stages and 5 cycles of waves. Figure 2(b) shows the time history waveform of shear displacement, and Figure 2(c) shows the shear hysteresis.

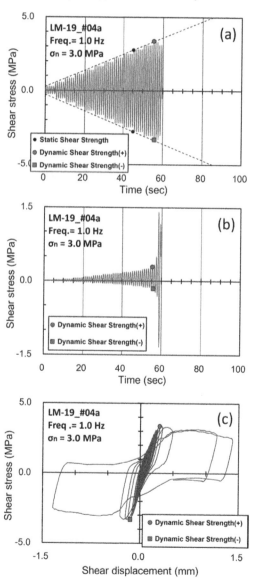

Figure 2. Result of Multi-stage amplitude dynamic direct shear test (Limestone joint, JRC=10~16)

Multi-stage amplitude dynamic direct shear test is performed by loading with M stages of N cycles of sine waves in succession, under a constant normal stress. After loading 10 stages with 5 cycles of sine waves, we planned that shear stress amplitude would reach the targeted strength. We set the target strength as the static shear strength, τ_s of the same discontinuity. So, set the shear stress amplitude of the first stage to S=0.1τ_s. Thereafter, the shear stress amplitude was increased stepwise: 0.2τ_s, 0.3τ_s, 0.4τ_s, 0.5τ_s, ... In reality, the number of loading stages may exceed 10, because shear stress loading is continued until dynamic shear failure.

Figure 3 shows the time history waveform of the stress ratio R_s, which calculated about a shear stress amplitude wave both positive side and negative side. Stress ratio R_s is the ratio of peak value τ_{pi} and target shear stress amplitude τ_o every loading wave pattern 1 cycle, and it is calculated by the next expression.

$$R_s = \tau_{pi} / \tau_o$$

According to Figures 2 and 3, the shear stress gradually than 12th-stage of 3rd-wave decreases in an positive side, and the stress ratio, Rs is less than 0.95 in 12th-stage of 4th-wave. In negative side, the stress ratio, Rs is less than 0.95 in 12th-stage of 4th-wave equally. Similarly, the shear displacement suddenly increases in 12th-stade of 4th-wave in both sides of the positive and negative sides.

We define the dynamic peak shear strength, $\tau_{p(d)}$ as follows.

When shear stress decreases continually after shear stress is less than 95% of target shear stress amplitude τ_o, or when shear displacement exceeds 1.0mm, we define shear failure.

Dynamic peak shear strength $\tau_{p(d)}$ is defined the maximum value of the shear stress amplitude before dynamic shear failure. In addition, dynamic peak shear strength $\tau_{p(d)}$ is defines in both sides of positive and negative.

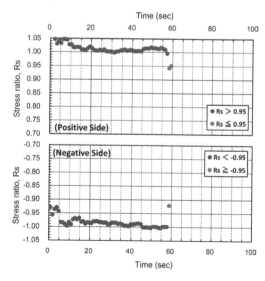

Figure 3. Stress ratio of positive and Negative sides

Figure 4 shows the failure envelope curve of the Multi-stage amplitude dynamic direct shear test result for a mortar tension crack(JRC = 8-12, interlocking = rank A). The mark of ○ and □ in Figure 4 are dynamic peak shear strength $\tau_{p(d)}$ by an positive side and negative side. Solid line is Mohr-Coulomb's criteria which modelled these results, and the dashed line is an envelope curve of the static shear strength in a similar discontinuity. According to this, the shear strength of rock discontinuity understands that a dynamic strength exceeds static strength.

Figure 5 shows the test result of the case which assumed the frequency of the loading wave pattern 0.1Hz, 1.0Hz and 3.0Hz. According to this, the loading frequency dependence in dynamic peak shear strength $\tau_{p(d)}$ is not recognized.

Figure 6 shows the result that modelled by Barton's criteria. Basic friction is φ_b=31.6° is provided from ultimate shear strength by static shear tests. According to this, in JRC, dynamic strength (JRC = 24.1) exceeds static strength (JRC = 18.7).

Figure 7 shows the failure envelope curve which made from the results of Multi-stage amplitude dynamic direct shear tests for Limestone joints and Mud-stone bedding planes. This failure envelope curves are modeled in Mohr-Coulomb's criteria. Limestone joint is relatively rough (JRC=10~16 by observation), and the interlocking is slightly good (rank B).Mud-stone bedding plane is relatively flat (JRC=4~8 by observation), and the interlocking is slightly bad (rank C). Differences between dynamic strength and static strength are big with the relatively surface rough Limestone joints, but, as for the difference, are not almost admitted in the relatively flat Mud-stone bedding planes.

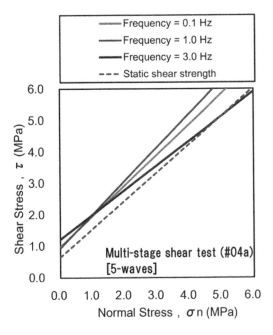

Figure 5. Frequency-dependent of dynamic strength (Mortar tension crack, JRC=8~12)

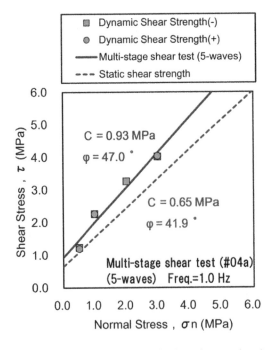

Figure 4. Failure envelope curve for dynamic strength and static strength (Mortar tension crack, JRC=8~12)

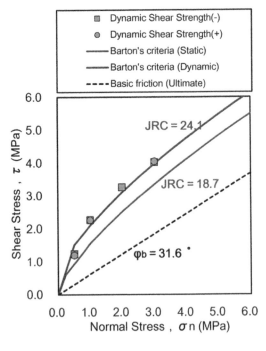

Figure 6. Barton's criteria (Mortar tension crack, JRC=8~12)

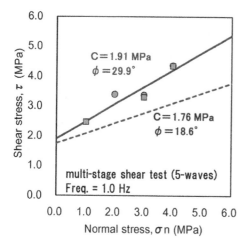

(a) Limestone joints (JRC=10~16, rank B)

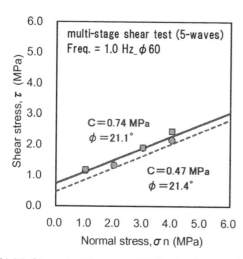

(b) Mudstone bedding plane (JRC=4~8, rank C)

Figure 7. Dynamic shear strength of many type of rocks

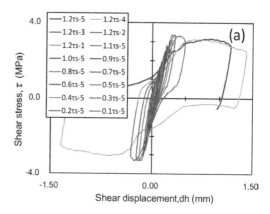

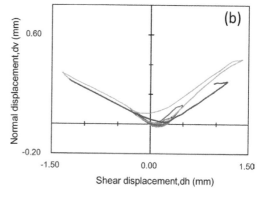

Figure 8. State of the deformation every loading stage (Limestone joint, JRC=10~16)

3 DYNAMIC SHEAR DEFORMABILITY OF ROCK DISCONTINUITY

Like shear strength, Figure 8 shows the results of the Multi-stage amplitude dynamic direct shear tests for Limestone joints. Figure 8 showed the fifth wave shear hysteresis and dilation curve of each loading stage individually. According to this, loading almost presents a rectilinear figure to the sixth phase (0.6τs) from the first, but presents fusiform hysteresis afterwards. Large displacement and a drop of the shear stress are accepted to the 12th-stage (1.2τs) afterwards at both sides of the positive and negative. Such tendency was similar to the dynamic heteromorphic characteristic in the core test specimen.

We define dynamic shear parameters to investigate dynamic shear properties. Figure 9 shows a definition of dynamic diagonal shear stiffness, $K_{sd(d)}$.

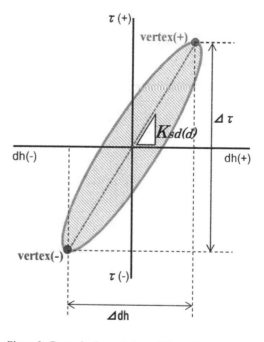

Figure 9. Dynamic diagonal shear stiffness, $K_{sd(d)}$

We define dynamic diagonal shear stiffness $K_{sd(d)}$, it with a straight line to link positive vertex and negative vertex of the shear in a Figure 9.

Dynamic diagonal shear stiffness, $K_{sd(d)}$ is a shear stiffness to define about individual shear hysteresis, and is defined next expression.

$$K_{sd(d)} = \frac{\Delta\tau}{\Delta dh} \quad \text{(MPa/mm)} \quad (1)$$

where $\Delta\tau$: shear stress amplitude both sides (MPa)

Δdh: shear displacement amplitude both sides (mm)

We define attenuation h by the next expression, as decrement energy in area ΔW surrounded by shear hysteresis.

$$h = \frac{1}{2\pi}\frac{\Delta W}{W} \times 100\,(\%) \quad (2)$$

Equivalent elastic energy W is defined by the next expression.

$$W = \frac{1}{4}\Delta\tau \cdot \Delta dh \quad (3)$$

Figure 10 shows the example which calculated a dynamic shear parameter. We define dynamic diagonal shear stiffness, $K_{sd(d)}$ with the incline of the straight line to link an equilateral positive vertex of the shear hysteresis and the negative vertex to show in Figure 9. In addition, attenuation, h is provided by calculating the area of the loop of the closed shear hysteresis.

Figure 11 plots it about a change for shear displacement both amplitude Δdh of dynamic diagonal shear stiffness, $K_{sd(d)}$ and attenuation, h provided from Multi-stage amplitude dynamic direct shear tests result for Limestone joints. According to this, it is identified as both parameters to depend about the shear displacement amplitude namely the shear stress amplitude.

According to Figure 11(a), dynamic diagonal shear stiffness, $K_{sd(d)}$ tends to decrease with increase in shear displacement both amplitude Δdh (i.e., shear stress both amplitude $\Delta\tau$). Furthermore, a tendency to increase with increase in normal stress has it, and this tendency is common in the static shear.

Figure 11(b) plots it about a change for shear displacement both amplitude Δdh of attenuation, h provided than the same test result. Although a tendency to increase with increase in shear displacement both amplitude Δdh is seen, according to this, attenuation, h presents some decrease halfway. These causes include the thing with the reversely sigmoid model shear hysteresis by increase in shear displacement (i.e., the shear stress amplitude) as having been seen in a change of the shear hysteresis that mentioned above.

Figure 12 shows the shear hysteresis of the Multi-stage amplitude dynamic direct shear tests for mortar tension crack. Figure 12(a) is the figures which plotted positive vertex and negative top in the 5th-wave shear hysteresis every stage of the shear stress amplitude. The curve expressed by the set of

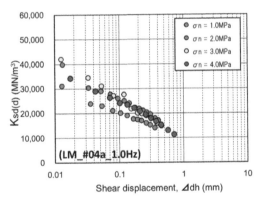

(a) Dynamic diagonal shear stiffness, Ksd(d)

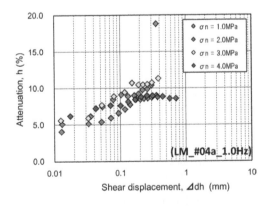

Figure 11. Dependence for the shear displacement amplitude of the dynamic shear parameter

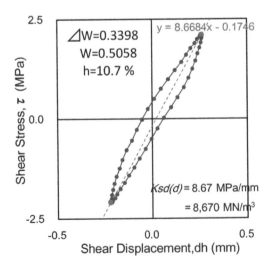

Figure 10. Example of dynamic shear parameters

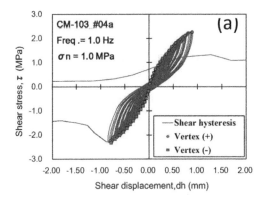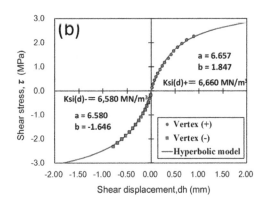

Figure 12. Shear hysteresis and vertex of skelton curve

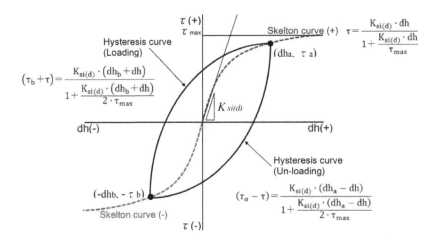

Figure 13. Hyperbola model

these points is called a frame curve (Skelton curve) generally.

Figure 12(b) shows the positive vertex and negative vertex, and was similar by a hyperbola model in each. The hyperbola model models the skelton curve indicating shear stress, τ - shear displacement, dh relations by hyperbolic function.

Figure 13 shows hyperbola model for rock joints. In Figure 13, the skelton curve is modelled by a hyperbola model, and the history curve is displayed by Masing rule.

4 CONCLUSION

In this study, we examined a dynamic strength and deformability of rock discontinuity, than a large number of dynamic direct shear test results for a rock discontinuities and mortar artificial discontinuities.

The authors defined dynamic peak shear strength $\tau p(d)$ and defined dynamic diagonal shear stiffness, $K_{sd(d)}$ and attenuation, h. Also we examine the characteristic about these dynamic parameters and suggest the modeling by the hyperbola model.

REFERENCES

Yoshida, J., Yoshinaka, R. & Sasaki, T. 2014. Study on Dynamic Properties of Rock Discontinuity using Dynamic Direct Shear Test Machine. ARMS8, 8th Asian Rock Mechanics Symposium, 14–16 October 2014, Sapporo Japan

Iwata, N., Sasaki,T., Sasaki, K., Yoshinaka, R., 2012, Static and dynamic response analysis of rock mass considering joint distribution and its applicability, Proc. 12th ISRM Congress, 233–236.

Yoshinaka,R., Iwata,N. and Sasaki,T., 2012, Applicability of Multiple Yield Model to earthquake response analysis for foundation rock of large-scale structure, Jour. of JSCE, 451–465.

2019 Rock Dynamics Summit– Aydan et al. (eds)
© 2019 Taylor & Francis Group, London, ISBN 978-0-367-34783-3

Laboratory observations of fracture plane reactivation induced by pore pressure in Kimachi sandstone

D. Asahina, M. Sato & M. Takahashi

Geological Survey of Japan, AIST, Ibaraki, Japan

ABSTRACT: Pore-fluid pressure is known to play a key role in activating of geological faults at a wide range of scale. Elevated fluid pressure within fractures or pores changes mechanical conditions, resulting in fluid-induced seismicity. This paper presents an experimental technique to reactivate fracture plane induced by elevated pore pressure. Fracture plane of Kimachi sandstone was reactivated through a series of incremental sequence of pore pressure under stress conditions controlled by true triaxial test apparatus. Of particular interest is the relationship between pore pressure evolution and mechanical response within the fractured specimen during reactivation. Stress, displacement, and permeability exhibited instantaneous changes at an elevated pore pressure. Hysteresis effect and anisotropic measurements of displacement indicate fracture reactivation of the specimen.

1 INTRODUCTION

Pore-fluid pressure is known to play a key role in activating of geological faults at a wide range of scale. One of the main factors causing the fault reactivation is increases in pore fluid pressure which generally decreases in effective stress and shear strength. Pore pressure-induced dynamic strains could change mechanical conditions of a critically stressed fault, resulting in seismic waves which sometimes trigger large earthquake. Understanding the hydromechanical interactions associated with fault reactivation in the shallow crust is important for many geoengineering applications, such as evaluating micro-seismicity impacts on impounding reservoirs, extracting shale gas through fracture stimulation, and enhancing geothermal systems (Valko & Economides 1995).

Various field observations have been conducted to investigate fault behavior during reactivation induced by elevated pore pressure. Guglielmi et al. (2015) has been directly measured spatial displacements of the fault during reactivation induced by pore pressure injection. Their in-situ observation showed the movement of the existing faults (i.e., opening and sliding) and its correlation with seismic events. In field experiments at a granite quarry, Cornet et al. (2003) measured displacement-injection pressure hysteresis via hydraulic jacking tests to study the effects of threshold injection pressure on fracture aperture.

Analog laboratory studies of the fault reactivations have also been conducted. Direct shear loading tests have been performed to represent fault reactivation of the fault gouge layer or evaluate

its potential (Cuss & Harrington 2016; Scuderi & Collettini 2016). The cyclic effect of pore pressure on permeability has been experimentally studied. Pore pressure oscillation has been introduced to activate fracture planes and simulate seismic events (Elkhoury et al. 2011, Candela et al. 2015). Ying et al. (2009) have conducted the experiments to study the effect of oscillating pore pressure which cause a seismicity for a longer period than a step increase pore pressure. Huo & Benson (2015) used CT images to measure fracture aperture distributions and to investigate permeability evolution during confining pressure cycles. Their results clearly indicated that the contact area of fracture surfaces changes before and after being subjected to high effective stress.

This paper summarizes recent work of the authors, in which the reactivation of fracture plane induced by elevated pore pressure is investigated through a laboratory test (Asahina et al. 2019). Fracture plane of Kimachi sandstone was reactivated through a series of incremental sequence of pore pressure under stress conditions controlled by true triaxial test apparatus. Kimachi sandstone was used to measure hydromechanical responses (i.e., stress, displacement, and permeability), all of which exhibited instantaneous changes at an elevated pore pressure. In this paper, attention is given to understanding the relationship between pore pressure evolution and mechanical response within the fractured sample. Hysteresis effect and anisotropic measurements of displacement indicate fracture reactivation of the specimen. Finally, plans are described for a better understanding of seismic event during elevated pore pressure-induced reactivation.

2 EXPERIMENTAL TECHNIQEUS

2.1 *Specimen description and test equipments*

Kimachi sandstone, obtained from Shimane Prefecture in western Japan, was selected for use in investigating the effect of elevated pore pressure on fracture plane behavior under triaxial compression. Kimachi sandstone is a sedimentary rock having grains that are composed of rock fragments (andesite, pyroxene, amphibole, plagioclase, potassium feldspar, and quartz grains), with an average grain size of 0.5–1.0 mm, and a grain density of 2.23 g/cm^3. Test sample was cut into a rectangular parallelepiped (70 × 35 × 35 mm^3). The longer axis was set perpendicular to the bedding plane. The porosity and permeability of the sample is 21.3 % and 1.6×10^{-18} m^2, respectively. The test sample was dried in room condition for more than one month.

Figure 1 shows an overview of the experimental setup conducted by Asahina et al. (2019). The TTT apparatus consists of a pressure vessel within a biaxial load frame in conjunction with permeability measurement devices. This TTT apparatus has been successfully used for investigating the mechanical and hydraulic properties of rock samples such as the permeability anisotropy of deformed rock (Li et al. 2002, Takahashi 2007, Takahashi et al. 2012, Panaghi et al. 2018), the permeability evolution during progressive failure (Sato et al. 2018).

Two primary advantages of the TTT apparatus, that uses rectangular prismatic specimens, compared with the conventional triaxial test can be summarized as follows (Sato et al. 2018, Asahina et al. 2019). First, the orientation of fracture plane can be controlled so that fracture plane are parallel to the intermediate principal stress axis. Second, the permeability parallel to fracture plane during progressive deformation can be measured.

2.2 *True triaxial compression*

A fracture plane in a dry, intact sample, was first generated by the true triaxial compression test. We employed so-call Mogi-type TTT apparatus, in which three principal compressive stresses can be independently applied: σ_Y and σ_Z are applied by two rigid pistons, whereas oil pressure in the vessel is used to control σ_X. The X, Y, and Z directions correspond to the minimum, intermediate, and maximum principal directions, respectively. The two loading pistons slowly applied constant stress conditions of $\sigma_X = 0$, $\sigma_Y = 11.5$, and $\sigma_Z = 11.5$ MPa. Note that compression stress is positive. σ_Z was then increased under displacement controlled with a constant loading rate until failure. The direction of the fracture plane is parallel to the Y direction and inclined in the Z direction, as shown in Figure 2.

2.3 *Critical stress setting*

To ensure a nearly critical stress setting of the fracture plane, the fractured sample was reloaded in the Z direction. A critically stressed fault is that the residual shear stresses acting on the fault is close to the frictional strength of the fault; therefore, once the shear stress increases, the fault starts to slide. After shear failure under true triaxial compression, the confining pressure, P_c, was increased to 16.5 MPa. The pore pressure was increased to 5 MPa, and σ_Z was then increased under displacement controlled until the stress responses of the sample showed ductile behavior. After that, the vertical piston was held in its position, and the confining pressure was kept constant. At the end of this loading stage, the critically stressed fracture plane was achieved; therefore the reactivation of the fracture plane will be likely to occur by applying a slight increase in pore pressure.

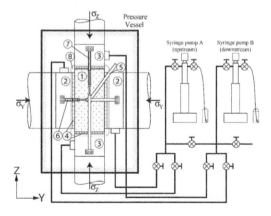

Figure 1. Schematic diagram of the true triaxial apparatus and its assembly (Adapted from Sato et al. 2018). 1 is the specimen, 2 and 3 are the end-loading plugs for the Y and Z directions, respectively; 4 indicates the porous metal plates for water dispersion; 5, 6, and 7 are the LDTs for the X, Y, and Z directions, respectively; and 8 is the silicon sealant.

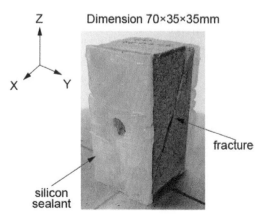

Figure 2. Fracture pattern obtained after a true triaxial compression test.

2.4 *Cyclic pore pressure*

We imposed step-wise incremental pore pressure of the fractured sample through the up and downstream Syringe pump A shown in Figure 1. We have controlled pore pressure as the variable input during the pore pressure test, and other control units (i.e., confining pressure and vertical piston) were kept constant. A series of flow pump tests was performed by injecting distilled de-aired water at a constant rate from an upstream reservoir into the specimen using ISCO syringe pump, and measuring the differential pressure between the one sides to the other side surfaces in the Y direction of the specimen. During the pore pressure test, the flow line in the Y direction was only used, whereas that in the Z direction was closed.

The specimen was connected to the permeability measurement units via porous metals, end pieces, and pipes (Figure 1). Each porous metal with 66 holes (1 mm in diameter) was designed to distribute the pore water to the specimen surface. In this experiment, we used Darcy's law to calculate the bulk permeability of the specimen, including the contributions of matrix, pores, and fractures.

Therefore, two basic aspects can be considered to cause changes in the measured permeability: the volume of pore space or micro-fracture changes, and fracture aperture changes.

3 EXPERIMENTAL RESULTS

Figure 2 shows photographs of the specimen used after the pore pressure test. Figures 3 and 4 show the main measurements during incremental pore pressure cycling sequence. Figure 3b shows the differential stresses, $\sigma_z - \sigma_x$. Large stress changes were observed when approaching the higher pore pressure (the gray region in the figure), whereas during lower pore pressure, the stresses were not strongly influenced. The stress remained nearly constant after peak pore pressure. Figure 3c shows the displacements in the X directions. Although the displacement sensitively responded at all pore pressure levels, large displacements in the X direction were also observed when approaching higher pore pressures. Figure 3d shows the permeabilities of the specimen in the Y direction for each pore pressure. The fracture planes were generated in nearly parallel to the Y direction.

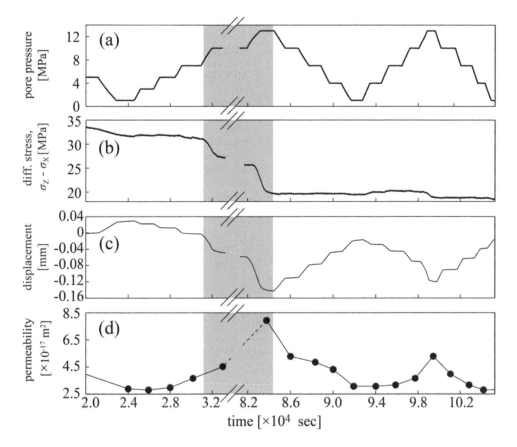

Figure 3. Responses recorded during the pore pressure cycles: (a) pore pressure, (b) differential stress, (c) displacement in X direction, and (d) permeability in the Y-direction. Gray areas indicate periods in which notable changes were observed during measurements.

Here too, changes in the permeability were significant at higher pore pressures. The permeability at the peak pore pressure was about 3 times as large as that at the lowest pore pressure.

Figures 4 shows the displacements in the X and Y directions versus pore pressure fluctuations performed in 2 cycles. As noted, the displacement magnitude increased (i.e., dilated) under increasing pore pressure for each cycle. Figure 5 shows the differences of the displacements between the presented pore pressure in X-axis and its next incremental step. Figure 5a and 5b show differences of the displacements, δu, when pore pressure increases and decreases, respectively. For clarity, we show the absolute values for these figures. It is worth noting that the displacement-pore pressure curves show two types of hysteresis: irreversibility and path dependence.

The irreversible displacement was observed in the fractured specimen about 0.04 mm at the second visit of P_p=1 MPa, as indicated by red arrow in Figure 4. After the specimen deformed in X direction greatly in the 1st cycle, the displacement is irreversible and does not return to the initial state even when the pore pressure decreases. We can interpret this irreversibility as follows. Since the orientation of the pre-generated fracture was nearly perpendicular to the XZ-plane, if opening or sliding occurs on the fracture surface, the displacement in the X direction, rather that the Y direction, is expected to respond. The test results show that the irreversible displacement was clearly observed in the X direction and not in the Y direction, as shown in Figure 4. This suggests that anisotropic displacement in the X and Y directions possibly serves as a means to identify fracture behavior of specimens. It is highly probable that an increase in pore pressure induced migration and overriding of particles along the fractured surface, causing irreversible displacement.

The other type of hysteresis is path dependences. As shown in Figure 4, the displacement nonlinearly decreases (indicating dilation) with increase of pore pressure, while it linearly increased (indicating compaction) with decrease of pore pressure. These differences can be more clearly observed in Figure 5a and b. It is obvious that $|\delta u|$ increases with increase of pore pressure (Figure 5a), while $|\delta u|$ is almost constant when pore pressure decreases (Figure 5b). This type of hysteresis appeared both in the X and Y directions at least to some extents. If this path dependence was associated with the major inclined fracture, the anisotropic effect should be appeared in the same manner as described above. Instead, it is likely related with behavior of pore spaces or micro-fractures.

4 DISCUSSION AND CONCLUSION

This paper presents laboratory experiments on Kimachi sandstone sample to observe mechanical behaviors associated with fracture plane reactivation due to increasing pore pressure. We used a TTT apparatus with rectangular prismatic specimen because the fracture direction of specimen can be controlled as well as the fluid flow parallel to the fracture plane can be measured. The pore pressure was incrementally increased/decreased for two cycles to reactivate the fracture planes. As expected, the elevated pore

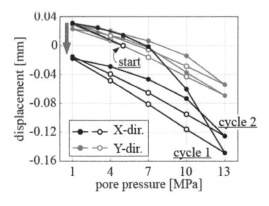

Figure 4. Displacement in X and Y direction vs. pore pressure. Solid and open symbols represent increasing and decreasing pore pressure sequences, respectively. The red arrow indicates an irreversible displacement in the X direction (Asahina et al. 2019).

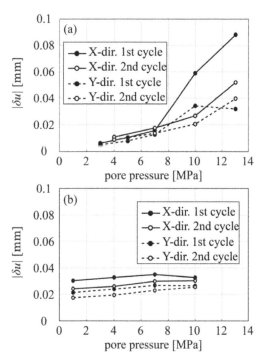

Figure 5. Difference of displacement, δu, between the presented pore pressure (X-axis) and its next incremental step when pore pressure: (a) increases, and (b) decreases.

pressure reduced the effective stress and induced instability of the pre-generated fracture planes.

The hydromechanical couplings occur through the movements of the fracture planes and pore-fluid interactions. According to Wang (2000), this type of hydromechanical coupling could be considered as a fluid-to-solid coupling that occurs when a change in fluid pressure produces a change in mechanical properties. Although we could observe the change in mechanical properties (i.e., fracture surface configuration) and associated influence on the permeability evolution, the detail relationship between pore pressure and a change in mechanical properties, such as stress-dependent permeability and fracture aperture distribution, has not been observed in this paper.

We have observed two types of hysteresis, i.e., irreversibility and path dependence, in the displacement of the fractured specimen. The irreversible displacement was clearly observed in the X direction and not in the Y direction. Such displacement anisotropy was associated with the fracture direction; therefore, the fracture reactivation was able to identify. In general, however, it is not easy to identify the movements (opening, closing, or sliding) of the fracture planes based on the presented measurement devices. The displacement sensor developed by Guglielmi et al. (2015) can directly measure the relative displacement of each fault compartment. Detail measurements of the fracture plane movements under the controlled loading conditions of the TTT instrument lead to a better understanding hydromechanical coupled behavior during fault reactivation.

Although quasi-static loading has been considered in this paper, this step is necessary for the next objective to understand the mechanisms of dynamic displacements induced by pore pressure during seismic events.

REFERENCES

Asahina, D., Pan, P., Sato, M., Takeda, M. and Takahashi, M., 2019, Hydraulic and mechanical responses of porous sandstone during pore pressure-induced reactivation of fracture planes: an experimental study, *Rock Mechanics and Rock Engineering*, https://doi.org/10.1007/s00603-018-1706-8

Candela, T., Brodsky, E. E., Marone, C. and Elsworth, D., 2015, Flow rate dictates permeability enhancement during fluid pressure oscillations in laboratory experiments, *Journal of Geophysical Research: Solid Earth*, 120, 2037–2055.

Cornet, F. H., Li, L., Hulin, J. P., Ippolito, I. and Kurowski, P., 2003, The hydromechanical behaviour of a fracture: an in situ experimental case study, *International Journal of Rock Mechanics and Mining Sciences*, 40, 7–8, 1257–1270.

Cuss, R. J. and Harrington, J. F., 2016, An experimental study of the potential for fault reactivation during changes in gas and pore-water pressure, *International Journal of Greenhouse Gas Control*, 53, 41–55.

Elkhoury, J. E., Niemeijer, A., Brodsky, E. E. and Marone, C., 2011, Laboratory observations of permeability enhancement by fluid pressure oscillation of in situ fractured rock, *Journal of Geophysical Research*, 116, B02311.

Guglielmi, Y., Cappa, F., Avouac, J.-P., Henry, P. and Elsworth, D., 2015, Seismicity triggered by fluid injection–induced aseismic slip, *Science*, 348, 6240, 1224–1226.

Huo, D. and Benson, S. M., 2015, An experimental investigation of stress-dependent permeability and permeability hysteresis behavior in rock fractures. Fluid Dynamics in Complex Fractured-Porous Systems, John Wiley & Sons, Inc: 99–114.

Li, X., Wu, Z., Takahashi, M. and Yasuhara, K., 2002, Permeability anisotropy of Shirahama sandstone under true triaxial stresses, *J. Geotech Eng*, 708, III-59, 1–11.

Panaghi, K., Golshani, A., Sato, M., Takemura, T. and Takahashi, M., 2018, Crack tensor-based evaluation of inada granite behavior due to damage under true-triaxial testing condition, *international Journal of Rock Mechanics & Mining Sciences*, 106, 30–40.

Sato, M., Takemura, T. and Takahashi, M., 2018, Development of the permeability anisotropy of submarine sedimentary rocks under true triaxial stresses, *International Journal of Rock Mechanics and Mining Sciences*, 108, 118–127.

Scuderi, M. M. and Collettini, C., 2016, The role of fluid pressure in induced vs. triggered seismicity: insights from rock deformation experiments on carbonates, *Scientific reports*, 6, 24852.

Takahashi, M., 2007, Permeability and Deformation Characteristics of Shirahama Sandstone under a General Stress State, *Arch. Min. Sci*, 52, 3, 355–369.

Takahashi, M., Kato, M., Takahashi, N., Fujii, Y., Park, H. and Takemura, T., 2012, 3 Dimensional Microscopical Pore Distribution of Kimachi Sandstone and Its Permeability and Specific Storage Change by Hydrostatic Stress and Deviatoric Stress (in Japanese with English abstract), *Jour.JapanSoc.Eng.Geol*, 53, 1, 31–42.

Valko, P. and Economides, M. J., 1995, Hydraulic fracture mechanics, Wiley, New York.

Wang, H. F., 2000, Theory of Linear Poroelasticity: With Applications to Geomechanics and Hydrogeology, Princeton Univ Pr.

Ying, W.-I., Benson, P. M. and Young, R. P., 2009, Laboratory simulation of fluid-driven seismic sequences in shallow crustal conditions, *Geophysical Research Letters*, 36, 20,

2019 Rock Dynamics Summit– Aydan et al. (eds)
© 2019 Taylor & Francis Group, London, ISBN 978-0-367-34783-3

The behaviour of Oya tuff pillars under static and shock loading

T.K.M. Dintwe
Kyushu University, Fukuoka, Japan

T. Seiki
Utsunomiya University, Utsunomiya, Japan

Ö. Aydan & N. Tokashiki
University of the Ryukyus, Okinawa, Japan

ABSTRACT: Earthquakes have caused some damage to abandoned quarries in Oya region, Japan and the behaviour of pillars of Oya tuff quarries is therefore of paramount importance for the region. This study reports investigation results of Oya tuff pillars under static and dynamic conditions. Also, the backfilling of underground quarries is considered as a countermeasure of collapse in the long term and during earthquakes. For this purpose, an experimental laboratory program was initiated to investigate the static and dynamic response of Oya tuff pillars. The shock tests demonstrate that the overall stiffness of samples subjected to shock loading is higher than that of samples under static loading. With respect to backfilling, the results show that, the overall stiffness of backfilled samples is higher than that of unfilled and in some cases, the yielding strength of backfilled samples displays strain hardening behaviour, while unfilled samples show strain softening behaviour. Different degrees of backfilling, indicate that partial backfilling is not effective to prevent ground settlement.

1 INTRODUCTION

Underground structures have been known to be seismically insulated or rather could sustain seismic loads with little damage. However, over recent years the investigations regarding seismic response reveal that underground spaces such tunnels and underground shallow mines could experience significant damages (Wang, et al., 2001; Genis & Aydan, 2008; Aydan, et al., 2010). Unlike in deep mines, where stress-induced failures can occur in a violent manner invoking seismic events, in shallow mines, the type of dynamic loading that could pose a threat to shallow mines is the occurrence of the earthquake. In Japan, there are several cases in which underground spaces have been damaged by the often occurrence of earthquakes. After the 2011 Great Earthquake of East Japan with the magnitude of 9 (Aydan, 2014) embarked on the expedition visiting the affected areas focusing on the various damaged geo-engineering structures, and with special interest to abandoned underground mines and quarries. Aydan, (2014) documented and summarized almost all the accessible damaged underground areas. The records show that the earthquake caused sinkholes in several locations Iwaki in the Fukushima Prefecture, eleven locations in Kurihara, seven locations in Osaki, eleven locations in Higashi Matsushima, three locations in Kurogawa in the Miyagi Prefecture (Aydan & Tano, 2012a). It was later mentioned

by the Ministry of Economy, Trade and Industry (METI) that the actual number of the sinkholes is more than 316. Still from the list, it is interesting to note that one of the collapsed mines is the semi-underground mine located in Oya town in Tochigi prefecture and this collapse is not far from the area of the underground room and pillar quarry that is understudy. There is still an apprehension of similar cases reoccurring due to the influence of dynamic loading. A recent study by Seiki, et al., 2016 on seismic responses of long-wall type quarry 2km away from the current study area, has shown that increase in wave velocities results in high strain levels on the roofs walls. With all these past studies revealing the problems associated with dynamic loading in various underground spaces, a dynamic response investigation of pillars serves to be necessary and an essential study that could give fundamental understandings for underground mine stability.

Laboratory tests are a good start to investigate rock material response under dynamic loading. However, dynamic tests were often viewed as challenging due the loading rate needed and data recording. Specialized equipment is often a requirement to be able to carry out such test. Most studies on dynamic properties of material were demonstrated using the split Hopkinson pressure bar (SHPB) (Niu, et al., 2015). With SHPB the dynamic properties of the solid material could be captured and enables to estimate the work done on the material. Over the years the dynamic response of rock material under dynamic loading has been studied (Okubo, et al., 1990; Hashiba, et al., 2006; Chen, et al., 2005; Li, et al., 2008; Aydan, et al., 2010). Specifically, for shock or impact loading tests, not sufficient studies have been done, however it is appropriate to mention the work by Li, et al., 2007; Niu, et al., 2015; Aziznejad, et al., 2018. The three independent studies highlight the importance of the shock test in rock engineering. To simulate ground motions, a shaking table test could be used on room and pillar models and assess the response as demonstrated by Aydan, et al., 2010.

The subsidence incidents that took place in Japan dictate that regional support is the ultimate solution and the probable support for the abandoned mine would be backfilling. Backfilling in dynamic states hinder stopes and pillar to vibrate in critical resonant frequencies (Goldbarch, 1991), hence reducing risk of collapse. The other component of interest is the pillar size, thus the width to height (w/h) ratio. The optimum pillar size is desirable to have high production (high extraction ratios). High extraction ratios results in slender pillars or few pillars in the layout. The determination of the pillar size does not consider production only, but also the stability of the pillar, which plays important role in the safety success of the mine. That is very slender pillars could cause stability problems and high w/h ratios would reduce production.

Therefore, this study attempts to understand the effect of shock loading in rocks particularly Oya tuff through a series of tests in static and shock loading and observing the rock response. Finally, backfilling is introduced in the experiment and investigated as countermeasure for pillar failure.

2 METHODOLOGY

2.1 Sample preparation

There are specimen dimension standards for performing Uniaxial Compression strength (UCS) tests, typical standard size in width to height ratio is 0.5 with a 100mm height, and this was adopted for the experimental tests. However, in this study the test is pillar oriented laboratory testing, so the pillars are to be in bounded between two cylindrical blocks of (Ryukyu lime stone) that act as the roof and floor of quarry. It is ideal to use sample rock type for roof and floor as it is the case in the actual field. In some instances, the top blocks may also fail when the rock type is same as the pillar during the test or in the field where pillar are observed to have punched into the foundations (Ryder & Jager, 2002). Nevertheless, the pillar behaviour could be assessed despite the rock type of top and bottom blocks. Samples with square cross section were preferred since the pillars in the quarry are mostly square pillars. The samples were saw cut out from large blocks prepared by the mining company to the specified dimensions with w/h ratios, 0.3, 0.5 and 1 respectively.

2.2 Experimental setup and instrumentation

A cylindrical acrylic vessel with internal diameter of 100 mm, a height of 250 mm and 10 mm thickness was used for investigating the support effect of sand granular backfilling on Oya tuff pillar samples. The standard size of pillar sample was 48-51mm in width and 98 to 100 mm in height. The Ryukyu limestone block have dimensions of 100 mm height and 98 mm diameter, see (Fig. 1) for the conceptual model. The experiments were performed under servo controlled machine with capacity of 2000 kN and loading rates ranging between 0.7 to 1.0 MPa/s. The experimental setup allows for multi-parameter monitoring. The load acting on the pillars was continuously monitored together with the displacement, acceleration, and acoustic emission (AE). Before and after each compression test p and s wave velocities were measured.

In dynamic test, shock load was applied on sample to obtain the shock strength of the rock still using the conventional compression test machine. The machine is not actually designed for dynamic loading; however, the loading was done instantly such that it could represent shock loading. With the limitations of the

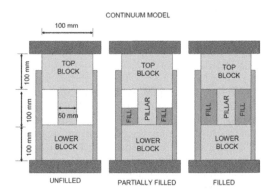

Figure 1. Conceptual setup of the pillar-oriented test

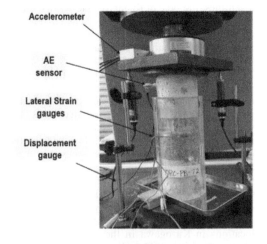

Figure 2. Experimental setup of the pillar-oriented test

107

machine in mind some of the results in this method are to be compared with those of a specialized machine designed for shock loading in future publications. Even though the earthquake loading is complex cyclic loading and could result in different damage mechanism of the rock, the shock test gives a somehow representation of the vertical component of the earthquake wave. The loading rate estimation is from load-time graph and could be calculated as follows:

$$r = \frac{\Delta l}{\Delta t} \equiv \frac{l_p}{t_p} \qquad (1)$$

where l_p is the maximum load (kN) or the peak load and t_p is the time (s) taken from the start of the test until reaching the peak load.

For backfilling, sand is used as the backfill material and similar procedure as the static and dynamic test is used with the difference being that the sample is surrounded by sand. Two types of backfill test were carried out as previously mentioned; the first sample was partially filled to about 70% of the pillar height and load applied until failure and still continuously monitoring. The second part of the backfill experiment is to entirely fill the pillar sample with granular sand material. The tangential strain of the backfill material used in the compression test was monitored on the outer walls of the acrylic cell.

3 EXPERIMENTAL RESULTS

3.1 Static tests (Uniaxial)

Figure 3 shows the load-time graph of the pillar samples. The graph shows all the multi parameters measures recorded during the static test experiment, except the acceleration. The relationship among the parameters surely is commensurate. The increase and decrease in the load is clearly detected by the AE. The load was constantly increasing until the rock failure, where a sudden vertical increase is noticed. The trend on the load and displacement is reflected in the cumulative AE, and the time of events actually coincides.

Figure 4 depicts the typical stress-strain curve of Oya pillar sample obtained from the experiment. The young modulus is estimated from this curve, and the average strength (UCS) of Oya tuff is 8.6MPa. The three phases of the stress strain curve are outlined, yield, peak and residual. Table 1 summarizes physical and mechanical parameters of the intact Oya tuff.

3.2 Shock test (Uniaxial)

The loading rate of the shock test is 187kN/s (75 MPa/s). Similar to the static experiment, the shock test failure was detected by AE and the corresponding displacement was observed see (Fig. 5). Upon assessment of the loading history, all the micro loads acting on the sample while attempting to apply the shock load are recorded. This type of testing will require several

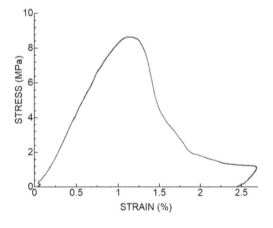

Figure 4. Stress-strain curve for static test

Table 1. Physico-mechanical properties of Oya Tuff (static)

Density	Shear modulus (GPa)	Bulk modulus (GPa)	Compresive strength (MPa)	Cohesion (MPa)	Tensile strength (MPa)
2400	0.91	1.28	8.6	2.1	1.08

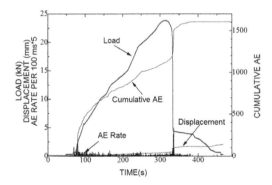

Figure 3. Load displacement curve and AE counts for static test

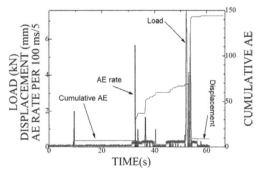

Figure 5. Load-displacement curve and AE counts responses for shock test

samples to get the test right and personnel experience is vital. The records before about 50 seconds are missed shock loadings; this highlights the difficulty of using the conventional machine for dynamic shock testing. Despite the limitations, attempts were made and successful tests were recorded. Figure 6 shows the stress-strain curve of the shock test. In the figure, it is clear that the shock curve peak strength is higher than that of the samples tested under static loading. It is also observed that stiffness of the samples is higher in shock loading. The acceleration has varied response signatures according to position of the accelerometer and the loading rate. Figure 7(a-b) show the acceleration history of the static test and shock tests. The acceleration results presents only the vertical component of the system. This is simply due to the fact that the loading is applied in the vertical direction. Generally, the accelerations from the top platen positions are much higher than the bottom platen where noise is usually recorded, hence only the top platen acceleration presented, and obviously the shock tests result in higher acceleration peaks than static tests. Having much higher peaks in the top platen accelerometers is a result of the mobile part gaining more acceleration after the sample failure than the stationary bottom part (Aydan, 2004).

3.3 Effects of w/h ratio

The aspect ratio is well-known to have effects on the pillar strength, and substantial research has covered this area. Past experimental work on Oya tuff has been done to estimate static mechanical properties regarding the aspect ratio. However, dynamic properties of various aspect ratios are still yet to be determined. The procedure for both static and shock UCS tests was repeated for additional aspect ratios namely, 0.3 and 1. Figures 8 and 9 show the aspect ratio results of the static and shock tests respectively. The results are startling; it is well known that the pillar strength increase with increasing w/h ratio, however the results are showing the opposite of this fact. The strength of the pillar shows to be increasing with decrease in w/h ratio. This relationship trend is observed in both the static and shock tests results.

3.4 Backfilling

3.4.1 Static test

As stated before that the degree of backfilling is of interest, Figure 10 presents load-displacement vs time together with other parameters of the partially

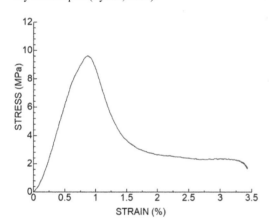

Figure 6. Stress-strain curve for shock test

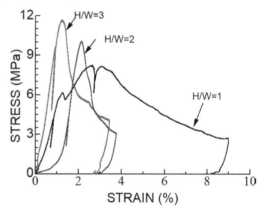

Figure 8. Stress-strain curve for various pillar geometry under static loading

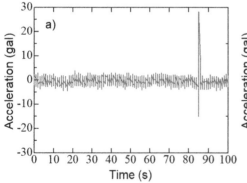

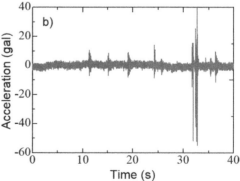

Figure 7. Acceleration response of samples a) top platen (static) b) top platen (shock)

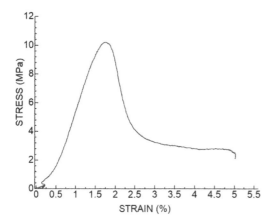

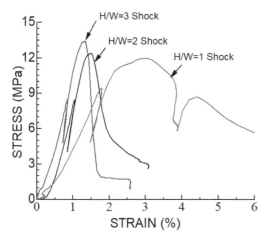

Figure 9. Stress-strain curve for various pillar geometry under shock loading

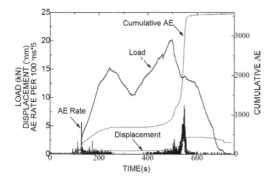

Figure 10. Load-displacement curve, AE counts for partially filled sample

filled pillar sample (PSB). The load curve indicates that the sample had failed under loading and then loaded again (cyclic loading). The pillar samples load bearing capacity has increased after failure and this surely due to the confinement offered by the sand backfill. Just like in the non-backfill sample (NSB) test, the AE rate increases as the load

increases and decrease when the samples undergo unloading. Furthermore, permanent straining is observed after loading and unloading cycle even though it is just one cycle. The full backfill pillar sample (FSB) results are shown in Figure 11; the sample was loaded in three cycles. It is interesting to note that the drop in the loading does not mean that the sample had failed, rather the loading was terminated as the strength was progressively increasing in strain hardening behaviour and could cause bursting of the acrylic cell. Also the displacement shows an increase in the first 200 secs, thereafter it remains relatively constant until experiment termination. Like the previous test, the AE rate is directly proportional to the load, whereas the cumulative AE still shows permanent straining and in this case it was sharply increasing and Kaiser-effect (K.E) is noted from the measured responses. The results are in correspondence with those observed by Aydan (2013). The comparison of the three conditions of the continuum pillar sample; NSB, PSB, FSB are all compared in Figure 12. It is clearly seen that the brittle failure manner is pronounced in the NSB sample, while the PSB sample undergoes the strain softening failure manner,

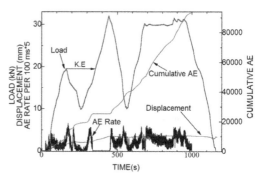

Figure 11. Load-displacement curve, AE counts response for full backfill sample

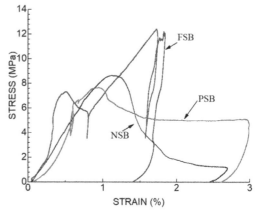

Figure 12. Stress-strain curve for various backfilling degrees under static loading

and it shows a much higher residual strength than the NSB sample. In the case of the FSB, the bearing capacity increased to about 1.5 times that of the unfilled pillar sample. The backfill cases reveal that after yielding or even before that, the backfill material that takes over the load with the rock, and sustain further cyclic loading and display strain-hardening. Thus, the backfill material alters the overall stress-strain response (Aydan, et al., 2013).

3.4.2 Shock test
The three cases of NSB, PSB and FSB compression test are repeated under shock loading. The results are presented in the stress-strain curves in Figure 13. No-backfill curve display low stiffness as compared to other graphs. Stiffness show to be increasing with the addition of backfill and highest stiffness is achieved with full backfilling. The peak strength of the PSB appear to be the highest in this case, however, the FSB samples did not show clear fracturing as the PSB samples.

3.5 Failure characteristics of samples
The pillar samples were then assessed after the experiment and have a closer look at the failure planes. Figure 14 (a-c) shows the photos of the continuum pillar samples after the UCS test, the NSB sample shows well-defined tensional crack (mode I), and the PSB sample has failed more like under triaxial conditions due to confinement provided by the backfill. Clear shear fractures are observed on the partially filled sample forming under the maximum principal stress σ_1. The movement on the shear fractures is parallel to the fracture surface (mode II & III). Since the stress state in the PSB acts as in triaxial ($\sigma_1 > \sigma_2 \geq \sigma_3$), the fracture planes form a conjugate pair (conjugate shear fractures). Under a true triaxial test with well controlled conditions or confinement,

a   b   c

Figure 14. Failed samples under different backfill degrees a) unfilled b) partial-backfill c) full-backfill

the angle between the conjugate fractures may be bisected by the maximum principal stress σ_1. The FSB sample is still intact, no visible fractures with naked eye, and this require more magnified assessment of the sample by using high resolution camera, and minute size extensional fractures were observed on the surface of the pillar sample.

4 CONCLUSION

An experimental study investigating the behaviour of Oya tuff pillars under static and shock loading was carried out. The shock loading has shown to be possible with conventional compression machine despite its limitations. The static and shock loading test results have been compared and analysed. It was found out that the overall stiffness of the pillar sample subjected to shock loading is higher than those of the samples under static loading. The strength of samples tends to increase with the decrease in w/h ratio which is normally not the case, rather the strength has to be increasing with the ratio. The authors are very aware that the aspect ratio results are somewhat different from the previous studies. Therefore, further work is to be conducted to understand what could have caused the outcome.

The experiments demonstrate that the stiffness of backfilled samples is higher than those of unfilled samples. Backfilling increases the bearing capacity, and results in peak and residual strength of backfilled samples being much higher than the unfilled samples. In some cases, backfilled samples display strain hardening behaviour while the unfilled display strain softening behaviour. Different degrees of backfilling indicate that partial backfilling is not effective to prevent settlement as compared to full backfill.

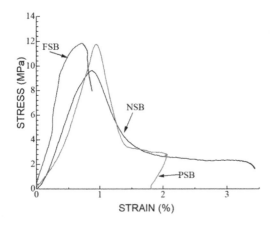

Figure 13. Stress-strain curve for various pillar geometry under shock loading

REFERENCES

Aydan, Ö., 2004. An Experimental Study on Dynamic Responses of Geo-Materials during Fracturing. *Journal of The School of Marine Science and Technology*, 1(2), pp. 1–7.

Aydan, Ö., 2014. Crustal stress changes and characteristics of damageto geo-engineering structures induced by the Great East Japan Earthquake of 2011. *Bull Eng Geol Environ (2015) 74:1057–1070*, p. 1064.

Aydan, Ö. et al., 2010. Response and stability of underground structures in rock mass during earthquakes. *Rock Mech Rock Eng*, pp. 857–875.

Aydan, Ö. & Tano, H., 2012a. The damage to abandoned mines and quarries by the Great East Japan Earthquake on March 11, 2011. Tokyo, s.n., pp. 981–992.

Aydan, Ö., Tokashiki, N. & Tano, H., 2013. An experimental study on te supporting effect of backfilling on an abandoned room and pillar mines, quarries and karstic caves, and its in-situ verification. s.l., American Rock Mechanics Association.

Aziznejad, S., Esmaieli, K., Hadjigeorgiou, J. & Labrie, D., 2018. Responces of jointed rock masses subjected to impact loading. *Journal of Rock Mechanics and Geotechnical Engineering*, pp. 624–634.

Chen, F., MA, C.-d. & XU, J.-c., 2005. Dynamic responce and failure beahaviour of rock under static-dynamic loading. *J. Cent. South Uni. Technol*, 12(3), pp. 354–358.

Genis, M. & Aydan, Ö., 2008. *Assessment of Dynamic Response and Stability of an Abandoned Room and Pillar underground Lignite Mine*. Goa, s.n., pp. 3899–3906.

Goldbarch, O. D., 1991. *Ground motion studies in a backfilled stope at West Driefontein*, Johannesburg: s.n.

Hashiba, K., Okubo, S. & Fukui, K., 2006. A new testing method for investigating the loading rate dependency of peak strength and residual rock strength. *International Journal of Rock Mechanics & Mining Sciences*, pp. 894–904.

Li, X. et al., 2007. Innovative tesiting technique of rock subjected to coupled static and dynamic loads. *International Journal of Rock Mechanics & Mining Sciences*, pp. 739–748.

Li, X. et al., 2008. Innovative testing technique of rock subjected to coupled static and dynamic loads. *International Journal of Rock Mechanics & Mining Sciences*, pp. 739–748.

Niu, Y. et al., 2015. Experimental study on Shock Mechanical Properties of Red Sandstone under Preloaded 3D Static Stress. *Journal of Engineering Science and Technology Review*, pp. 205–2011.

Okubo, S., Nishimatsu, Y. & He, C., 1990. Technical Note: Loading Rate Dependence of Class II Rock Behaviour in Uniaxial and Triaxial Compression Tests-An Application of a Proposed New Control Method. *Int.J.Rock Mech Min.Sci.& Geomech.Abstr.*, 27(6).

Ryder, J. A. & Jager, A. J., 2002. Mechanical Properties of Rock. In: *Rock Mechanics For Tabular Hard Rock Mines*. Johannesburg: The Safety in Mines Research Advisory Committe(SIMRAC), pp. 1–60.

Seiki, T. et al., 2016. Seismic response of numerical analysis and field measurement in Oya tuff qurry. *Eurock*, p. 8p.

Wang, W. L. et al., 2001. Assessment of damage in mountain tunnels due to the Taiwan Chi-Chi Earthquake. *Tunnelling and Underground Space Technology*, Volume 16, pp. 133–150

2019 Rock Dynamics Summit– Aydan et al. (eds)
© 2019 Taylor & Francis Group, London, ISBN 978-0-367-34783-3

Experimental investigation on the fatigue mechanical properties of intermittent jointed rock under random cyclic uniaxial compression

Y. Liu & F. Dai
State Key Laboratory of Hydraulics and Mountain River Engineering, College of Water Resource and Hydropower, Sichuan University, Chengdu, Sichuan, China

ABSTRACT: This study experimentally investigates the fatigue mechanical response of synthetic intermittent jointed rocks subject to random cyclic compression with different loading amplitudes and durations. Our results report the influence of the two loading parameters on the mechanical behavior of synthetic jointed rocks, involving the strength and deformation characteristics, the fatigue damage evolution and the progressive failure behavior. Both the fatigue strength and the deformation modulus of jointed samples decrease with increasing loading duration, while they exhibit opposite variation with increasing loading amplitude. The irreversible plastic strain exists in each hysteresis loop observed in stress-strain curves of jointed samples, which develops in a two-stage manner with increasing cycle number. Under higher loading amplitude or duration, the synthetic jointed rock is characterized by higher cumulative damage. Five crack coalescence patterns are observed in the present study based on the interaction between two adjacent joints, and the progressive failure process of tested jointed sample is captured via a high resolution industrial camera.

1 INTRODCTION

Rock engineering structures are likely to be subjected to seismic loading during their service lives. Structural damage induced by earthquakes has been reported in numerous recent seismic sequences, such as Nepal (2015), Tohoku (Japan, 2011), Christchurch (New Zealand, 2010) and Wenchuan (China, 2008) (Wang & Zhang 2013). Since the earthquake loading is random and cycle in essence, it is thus crucial to characterize the mechanical behavior of rocks under random cyclic loading, especially for the intermittent jointed rocks that widely exist in various engineering structures.

Considerable study has concentrated on the mechanical properties of rock materials subjected to regular cyclic loading via laboratory experiments. The strength and deformation characteristics are the primary results in these cyclic loading tests. Early researches revealed the hysteresis of the cyclic stress-strain curves and the accumulation of the deformation of intact rocks (Burdine 1963; Attewell & Farmer 1973). Scholars further reported that both the fatigue strength and deformation modulus of intact rocks decrease with increasing cycle number (Fuenkajorn & Phueakphum 2010; Ma et al. 2013); they also concluded that the fatigue mechanical properties of intact rocks were significantly affected by the cyclic loading parameters. Moreover, the fatigue failure mechanism of intact rocks has been revealed via a series of regular cyclic loading tests (Cerfontaine & Collin 2017; Liu et al. 2018). In contrast to the intact rocks, the researches on intermittent jointed rocks mostly focused on their static mechanical properties (Singh et al. 2002; Prudencio et al. 2007). Since jointed rocks are highly sensitive to cyclic loading and their mechanical behavior quite differs from that of intact rocks, the fatigue response of jointed rocks to regular cyclic loading was also investigated (Brown & Hudson 1974; Prost 1988). Li et al. (2001) indicated that the preexisting intermittent joints significantly affect the fatigue strength and deformation behavior of rock materials under regular cyclic triangular loading. Liu et al. (2017, 2018) further investigated the influence of joint geometric configurations and cyclic loading parameters on the fatigue response of intermittent jointed rocks subject to regular cyclic sinusoidal loading.

Compared with efforts devoted to studying the mechanical responses of rocks to regular cyclic loading, investigations on random cyclic loading were mostly limited to metallic and reinforced concrete materials (Amadio et al. 2003; Li et al. 2014; Anes 2017). However, according to the available literature, the mechanical response of jointed rocks to random cyclic loading has never been evaluated, and the influence of the loading parameters (i.e., amplitude and duration) on the fatigue mechanical properties of rocks remains far from being understood. The Wenchuan earthquake with a magnitude of Ms 8.0 struck the Sichuan province of China in 2008, which induced severe losses of life and property due to the high seismic intensity (Wang & Zhao 2015; Han et al. 2016). In this study, based on the Wenchuan seismic wave, the random cyclic uniaxial compression tests on synthetic intermittent jointed rocks are conducted to characterize the fatigue mechanical behavior of jointed rocks, and to reveal the influence of random cyclic loading parameters (i.e., amplitude and duration) on these mechanical properties.

2 EXPERIMENTAL METHODOLOGY

2.1 Specimen preparation and test facility

In our tests, the synthetic rock-like materials are used to simulate jointed rocks, which consists of high-strength cement, fine sand, silicon powder, water and water reducing agent, and their mass ratio are 1: 1.2: 0.15: 0.35: 0.015, respectively. The specified jointed mold (length × thickness × height = 200 mm × 100 mm × 100 mm) is designed, and the jointed rock samples with same dimensions are casted by pouring the mixed synthetic materials into the mold, as shown in Fig. 1. In our tests, to independently study the effect of random cyclic loading parameters, all the synthetic jointed rocks have the same geometrical parameters, involving the dip angle $\theta = 45°$, the joint length $a = 15$ mm, the rock bridge length $b = 20$ mm, the spacing $d = 24$ mm, the persistence $k = 0.318$ and the intensity $\rho = 6.75 \times 10^{-3}$ mm^{-1}. Herein, referring the definition by Dershowitz and Einstein (1988), the intensity is the total joint area per unit volume, and the persistence is the ratio of the sum of individual joint surface areas to the surface of a coplanar reference plane. An MTS-793 rock and concrete test machine is employed to perform the random cyclic compression tests in this study. Moreover, to reduce the friction effects, Teflon films coated with petrolatum are glued on the contact surfaces between the loading plates and two sample ends.

2.2 Test procedure and schemes

The Wenchuan seismic wave with a period of 10s is simplistically applied in our random cyclic compression tests. As shown in Fig. 2, this seismic wave

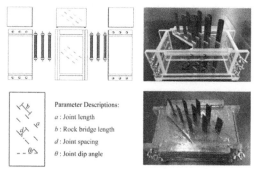

Figure 1. Sketch of the mold for fabricating jointed rocks and the sample preparation

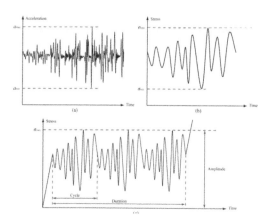

Figure 2. Schematic of the Wenchuan seismic wave and the simplified random cyclic loading

is simplified as the random sinusoidal wave with different amplitudes and frequencies, in which the characteristic parameters of random cyclic loading are defined as follows: σ_{max} and σ_{min} are the cyclic amplitude and the minimum stress, respectively, and the duration is the total period of random cyclic loading acting on tested samples.

Before the random cyclic compression test, monotonic uniaxial compression tests are first conducted on three jointed rock samples. In this study, the measured average uniaxial compression strength of jointed samples is 32.76 MPa, and the average Young's modulus is 8.13 GPa. Twenty-one random cyclic uniaxial compression tests are subsequently conducted to study the influence of random cyclic loading parameters, including different amplitude levels A of 0.60, 0.70, 0.80 and 0.90, and different loading duration T of 20, 40, 60 and 80s; herein, the amplitude level A is defined as the ratio of the cyclic amplitude to the average static uniaxial compressive strength of jointed samples.

3 EXPERIMENTAL RESULTS OF RANDOM CYCLIC UNIAXIAL COMPRESSION TESTS

3.1 Influence on the strength and deformation characteristics

Figures 3 and 4 depict the representative stress-strain curves of intermittent jointed rock samples under random cyclic compression with various loading parameters. These stress-strain curves can be divided into three stages. Initially, the axial stress slowly increases to the initial value of random cyclic loading, and then proceeds to the random cyclic loading and unloading stage; thereafter, the axial stress continuously increases to the fatigue strength and then steeply decreases.

The influence of random cyclic loading on the fatigue strength and deformation characteristics of jointed rock samples is shown in Figs. 5 and 6. The fatigue

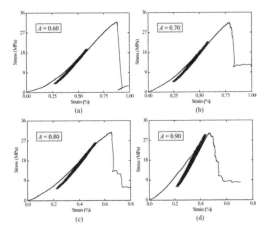

Figure 3. Representative stress-strain curves of the jointed rock samples under random cyclic compression with different loading amplitude levels

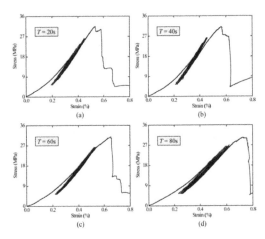

Figure 4. Representative stress-strain curves of the jointed rock samples under random cyclic loading with different durations

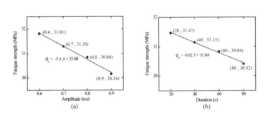

Figure 5. Influence of random cyclic loading parameters on the fatigue strength of the test jointed samples

strength of jointed sample features a linear decrease with increasing amplitude level and loading duration, while the fatigue deformation modulus of tested sample exhibits opposite variation with the two loading parameters. Similar to the regular cyclic loading tests,

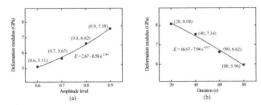

Figure 6. Influence of random cyclic loading parameters on the Young's modulus of the test jointed samples

the hysteresis loops can be observed in stage II of stress-strain curves of jointed samples under random cyclic loading, and two parts of fatigue strain exist in each hysteresis loop, namely, the elastic strain and the plastic strain. Figure 7 depicts the development of the irreversible plastic strain of jointed samples under random cyclic compression. It can be observed that all these curves develop in a similar two-phase manner with increasing cycles under various random cyclic loading. In phase I, the irreversible strain increases quickly, and then the strain remains almost steady in phase II until the fatigue failure occurs.

3.2 Influence on the fatigue damage evolution

The fatigue damage is one of the most crucial mechanical properties of materials in cyclic loading and unloading tests. An appropriate damage variable with a distinct physical significance is required to quantitatively describe the accumulated damage evolution of materials under cyclic loading. Referring to the previous studies (Xiao et al. 2010; Liu et al. 2017), the residual strain method is reasonable to definite the fatigue damage variable for its clear physical meaning and relevant description on the degradation behavior of rock materials. In our study, this definition is employed to describe the fatigue damage of jointed rock samples under random cyclic compression, as expressed in Eq. (1):

$$D = \sum_{j=1}^{N} |\varepsilon_i|_j \left/ \sum_{j=1}^{N_t} |\varepsilon_i|_j \right. \tag{1}$$

where D and ε_i are the fatigue damage variable and the irreversible strain of jointed samples, respectively, N and N_t are the current and total cycle

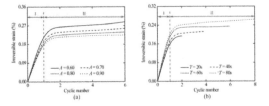

Figure 7. Development of the irreversible strain of the tested jointed samples under different random cyclic loadings

number, respectively. The evolution curves of the fatigue damage variable of tested jointed samples under various random cyclic loading conditions are shown in Fig. 8. There is a relatively large damage cumulated in the first cycle probably due to the closure of preexisting cracks and pores of tested samples, and then the damage variable stably increases with increasing cycle number until the fatigue failure occurs. Although the damage evolution law of jointed samples is similar under different random cyclic compression, the magnitude of damage variable is highly dependent on random cyclic loading parameters. Under higher amplitude lever or higher loading duration, the tested jointed sample is characterized by higher fatigue damage.

The influence of the loading amplitude and duration on the fatigue damage variable of jointed samples can be illustrated by the fracturing mechanism of tested samples during the random cyclic loading process. As the loading amplitude increases, the average stress applied on the tested sample increases, and thus the cracks are easier to initiate and propagate, resulting in higher fatigue damage in jointed samples. Similarly, with increasing loading duration, the random cyclic load acts on the tested samples for a longer time. As a result, the cracks have sufficient incubation time to propagate and coalesce, leading to a higher damage variable cumulated in the random cyclic loading process.

3.3 Influence on the progressive fatigue failure behavior

The representative failure scenarios of the tested jointed samples under various random cyclic loading conditions are depicted in Fig. 9. It can be observed that the tensile wing cracks are the dominating cracks in all recovered jointed samples. Developing along a certain angle with the prefabricated intermittent joints towards to the maximum compression direction (Sagong & Bobet 2002), these tensile cracks trigger an inevitable splitting failure through the entire samples. Under higher amplitude level and loading duration, the surface cracks of tested jointed samples coalesce more thoroughly.

Referring to the classification of crack coalescence types in previous studies (Wong & Einstein 2009), five crack coalescence patterns are classified

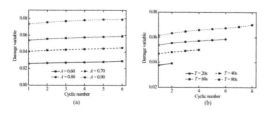

Figure 8. Evolution of the damage variable of the jointed rock samples under different random cyclic loading conditions

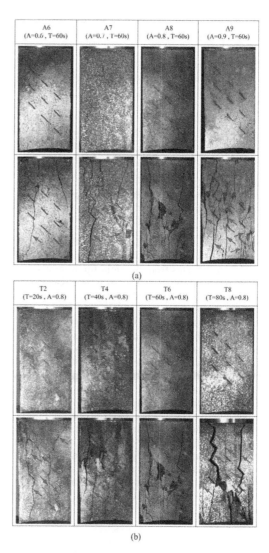

(a)

(b)

Figure 9. Representative failure modes of the jointed rock samples under different random cyclic loading conditions: (a) different amplitude levels, and (b) different durations

in the present study, and the sketch is shown in Fig. 10. For Pattern I, the two adjacent intermittent joints tips at the same side are linked up by a T-I crack dominantly of tensile nature, emerging almost a straight line parallel to the axial loading direction. Note that the T-I crack in Fig. 10a cannot be considered as a wing crack since its propagation direction is opposite to that of the conventional wing crack. In Pattern II, the two adjacent joints are connected by a tensile wing crack T-II. This T-II crack perhaps initiates from the left tip of joint 1 or the right tip of joint 2 and then propagates towards to the face of the adjacent joint at a certain distance from the joint tip, or it possibly emerges from the face of joint 1 or joint 2 at a certain distance from the tips and then develops towards to the left tip of joint 1 or the right

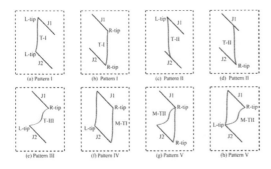

Figure 10. Characteristic graphs of the crack coalescence categories observed in the present study

tip of joint 2. Pattern III denotes that the two adjacent joint tips coalesce at the opposite side by a T-III crack that initiates from the right tip of joint 1 and then propagates to the left tip of joint 2. Generally, T-III crack is a "S" shaped curve, and there may be occasional short shear segments in the coalescence direction. Pattern IV and V are the mixture coalescence types; herein, M-TI crack denotes that the two adjacent joints tips are linked up by two T-I cracks at each side, and pattern V denotes that the crack initiates from one joint tip and then propagates towards to the two tips of adjacent joint. Taking the sample T2, A8, A7 and T8 as examples, Fig. 11 shows the representative coalescence patterns observed in the present random cyclic compression tests.

Employing a JAI SP-5000M high resolution industrial camera, the typical cracking processes of jointed samples under random cyclic compression are captured. Taking the tested sample A8 as an example, Fig. 12 depicts the typical progressive

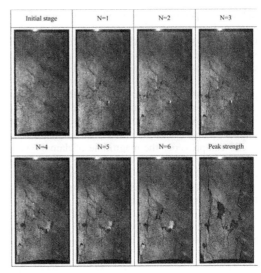

Figure 12. Progressive failure behavior of the jointed rock sample A8 under random cyclic uniaxial compression (N denotes the cycle number)

failure process at eight stages, namely, the initial stage, the 1st cycle, the 2nd cycle, the 3rd cycle, the 4th cycle, the 5th cycle, the 6th cycle and the fatigue failure stage. Form stage I (i.e., initial stage) to stage II (i.e., the 1st cycle), there are small cracks appear on the front surface of the sample A8; these existing cracks further propagate and form the observable surface cracks as the cycle number increases from the 2nd (i.e., stage III) to the 4th (i.e., stage V), accompanied with some new cracks emerging on the sample surface. Through stage VI (i.e., the 5th cycle) to stage VII (i.e., the 6th cycle), these surface cracks further widen and lengthen, eventually triggering the inevitable splitting failure in stage VIII (i.e., the fatigue failure stage). Note that these surface cracks always initiate from the joint tips, and then develop at a certain angle with the intermittent joints towards to the maximum compression direction.

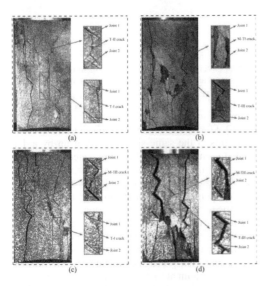

Figure 11. Representative crack coalescence patterns observed in the present study: (a) sample T2, (b) sample A8, (c) sample A7, and (d) (a) sample T8

4 SUMMARY AND CONCLUSIONS

Random cyclic loading is likely to be encountered in various geotechnical earthquake engineering structures. However, the study on the mechanical behavior of rocks under random cyclic compression is quite limited, especially for the intermittent jointed rocks that widely exist in these engineering structures. In the present study, the random cyclic compression tests with different loading parameters are conducted on the synthetic intermittent jointed rocks, regarding four loading amplitudes and four loading durations. Our experimental results reveal the mechanical response of synthetic jointed rocks to the random cyclic loading, involving the fatigue strength and deformation characteristics, the damage evolution and the fatigue failure behavior. The following main conclusions can be drawn:

(1) Similar to the regular cyclic loading tests, the hysteresis loops can be observed in the stress-strain curves of jointed samples under random cyclic loading, and the irreversible plastic strain exist in each hysteresis loop; this irreversible strain develops in a two-phase manner with increasing cycles. Both the fatigue strength and deformation modulus decrease with increasing loading duration, while they oppositely vary with increasing loading amplitude.

(2) The fatigue damage variable of jointed rock samples under random cyclic compression is determined based on the irreversible strain. There is a relatively large damage cumulated in the first cycle, and then the damage variable stably increases with increasing cycle number until the fatigue failure occurs. Under higher loading amplitude or duration, the synthetic jointed rock is characterized by higher cumulative damage.

(3) Five crack coalescence patterns are observed in the present study based on the interaction between two adjacent joints, and the fatigue progressive failure process of jointed sample is captured via a high resolution industrial camera. Under random cyclic compression, the surface cracks always initiate from the joint tips, and then propagate at a certain angle with the intermittent joints towards to the maximum compression direction.

REFERENCES

Amadio, C., Fragiacomo, M., & Rajgelj, S. 2003. The effects of repeated earthquake ground motions on the non-linear response of SDOF systems. *Earthquake Engineering and Structural Dynamics* 32: 291–308.

Anes, V., Caxias, J., Freitas, M. & Reis, L. 2017. Fatigue damage assessment under random and variable amplitude multiaxial loading conditions in structural steels. *International Journal of Fatigue* 100: 591–601.

Attewell, P.B, & Farmer, I.W. 1973. Fatigue behaviour of rock. *International Journal of Rock Mechanics and Mining Sciences* 10: 1–9.

Brown, E.T. & Hudson, J.A. 1974. Fatigue failure characteristics of some models of jointed rock. *Earthquake Engineering and Structural Dynamics* 2: 379–386.

Burdine, N.T. 1963. Rock failure under dynamic loading conditions. *Society of Petroleum Engineers Journal* 3: 1–8.

Cerfontaine, B. & Collin, F. 2017. Cyclic and fatigue behaviour of rock materials: review, interpretation and research perspectives. *Rock Mechanics and Rock Engineering* DOI: 10.1007/s00603-017-1337-5.

Dershowitz, W.S. & Einstein, H.H. 1988. Characterizing rock joint geometry with joint system models. *Rock Mechanics and Rock Engineering* 21(1): 21–51

Fuenkajorn, K. & Phueakphum, D. 2010. Effects of cyclic loading on mechanical properties of Maha Sarakham salt. *Engineering Geology* 112: 43–52.

Han, B., Zdravkovic, L., Kontoe, S. & Taborda, D.M.G. 2016. Numerical investigation of the response of the Yele rockfill dam during the 2008 Wenchuan earthquake. *Soil Dynamics and Earthquake Engineering* 88: 124–142.

Ma, J.L., Liu, X.Y., Wang, M.Y., Xu, H.F., Hua, R.P., Fan, P.X., Jiang, S.R., Wang, G.A. & Yi, Q.K. 2013. Experimental investigation of the mechanical properties of rock salt under triaxial cyclic loading. *International Journal of Rock Mechanics and Mining Sciences* 62: 34–41.

Li, N., Chen, W., Zhang, P. & Swoboda, G. 2001. The mechanical properties and a fatigue-damage model for jointed rock masses subjected to dynamic cyclical loading. *International Journal of Rock Mechanics and Mining Sciences* 38: 1071–1079.

Li, Y., Song, R.Q. & Van De Lindt, J.W. 2014. Collapse fragility of steel structures subjected to earthquake mainshock–aftershock sequences. *Journal of Structural Engineering* 140(12): 04014095.

Liu, Y., Dai, F., Zhao, T. & Xu, N.W. 2017. Numerical investigation of the fatigue properties of intermittent jointed rock models subjected to cyclic uniaxial compression. *Rock Mechanics and Rock Engineering* 50: 89–112.

Liu, Y., Dai, F., Fan, P.X., Xu, N.W. & Dong, L. 2017. Experimental investigation of the influence of joint geometric configurations on the mechanical properties of intermittent jointed rock models under cyclic uniaxial compression. *Rock Mechanics and Rock Engineering* 50: 1453–1471.

Liu, Y., Dai, F., Dong, L. Xu, N.W. & Feng, P. 2018. Experimental investigation on the fatigue mechanical properties of intermittently jointed rock models under cyclic uniaxial compression with different loading parameters. *Rock Mechanics and Rock Engineering* 51: 47–68.

Liu, Y., Dai, F., Xu, N.W., Zhao, T. & Feng, P. 2018. Experimental and numerical investigation on the tensile fatigue properties of rocks using the cyclic flattened Brazilian disc method. *Soil Dynamics and Earthquake Engineering* 105: 68–82.

Prost, C.L. 1988. Jointing at rock contacts in cyclic loading. Int J Rock Mech Min Sci Geomech Abstr 25(5): 263–272.

Prudencio, M. & Van Sint Jan, M. 2007. Strength and failure modes of rock mass models with non-persistent joints. *International Journal of Rock Mechanics and Mining Sciences* 44: 890–902.

Sagong, M. & Bobet, A. 2002. Coalescence of multiple faws in a rockmodel material in uniaxial compression. *International Journal of Rock Mechanics and Mining Sciences* 39:229–241.

Singh, M., Rao, K. & Ramamurthy, T. 2002. Strength and deformational behaviour of a jointed rock mass. *Rock Mechanics and Rock Engineering* 35(1): 45–64.

Wang, Z.J. & Zhao, B.M. 2015. Correlations between structural damage and ground motion parameters during the Ms8.0 Wenchuan Earthquake. *Soil Dynamics and Earthquake Engineering* 72: 129–137.

Wang, Z.Z. & Zhang, Z. 2013. Seismic damage classification and risk assessment of mountain tunnels with a validation for the 2008 Wenchuan earthquake. *Soil Dynamics and Earthquake Engineering* 45: 45–55.

Wong, L.N.Y. & Einstein, H.H. 2009. Crack coalescence in molded gypsum and Carrara marble: part 1. Macroscopic observations and interpretation. *Rock Mechanics and Rock Engineering* 42(3):475–511.

Xiao, J.Q., Ding, D.X., Jiang, F.L. & Xu, G. 2010. Fatigue damage variable and evolution of rock subjected to cyclic loading. *International Journal of Rock Mechanics and Mining Sciences* 47(3): 461–468.

2019 Rock Dynamics Summit– Aydan et al. (eds)
© *2019 Taylor & Francis Group, London, ISBN 978-0-367-34783-3*

Experimental study of scale effect in rock discontinuities on stick-slip behavior

R. Kiyota & N. Iwata
Chuden Engineering Consultants, Hiroshima, Japan

Ömer Aydan & N. Tokashiki
University of the Ryukyus, Okinawa, Japan

ABSTRACT: The stick-slip phenomenon is used to explain a mechanism of earthquake recurrence. A number of stick-slip experiments have been performed to clarify the mechanism of recurring slip instabilities and slip weakening. Although the amplitude of sliding of most experiments is quite smaller than actual earthquakes, and the observed acceleration is larger. The authors have developed an experimental setup, in which blocks move on a conveyor belt and is restrained by the spring, and conducted stick-slip experiments. However, the amplitude of slippage and acceleration observed in these experiments were quite smaller than actual earthquakes. Therefore, a large-scale experimental device was improved to be able to experiment with a larger rock blocks, and the experiment which changed the size and the type of the rock blocks was conducted. In this study, the results of large-scale experimental device were compared with the results of previous experiments, and the scale effect of rock discontinuities on the stick-slip phenomenon was investigated.

1   INTRODUCTION

The stick-slip is a phenomenon that interfaces is repeated sticking (accumulation of stress) and slip (release of stress). In the field of rock engineering, it is very important to explain the periodic occurrence of earthquakes, and seismic moment and displacement accompanying the stress drop at the fault plane, as well as creep behavior of unstable zones of slope movement and large underground cavities. Brace & Byerlee (1966) conducted some laboratory experiments using rocks to explain the mechanism of occurrence of earthquakes, and proposed that the stick-slip phenomenon is associated with this mechanism. However, most were using the compression testing equipment in the past studies, amount of slippage was very small with 1μm-1mm and the peak accelerations during slipping were very large with 10^2-10^5 m/s^2 (Ohnaka 2003). These results are quite different from case of medium/large earthquakes, slip amount of 10cm-1m, peak acceleration of 1-10 m/s^2.

The authors have developed an experimental setup (50cm long), in which blocks move on a conveyor belt and is restrained by the spring, and conducted stick-slip experiments (Ohta & Aydan 2010, Iwata et al. 2016). This experimental setup is able to simulate conditions in actual earthquakes better than previous stick-slip experimental devices. During experiments, the velocity of base block, stiffness of springs and normal load acting on block interface were varied to study their effect on the periodicity and stick-slip response. On the other hand, since the area of discontinuous surface is quite different in the actual earthquake fault and the specimen in the laboratory experiment, for the estimation and evaluation of ground motion and displacement in earthquake faults, it is necessary to consider the scale effect of them. However, there are studies on the scale effect of the shear strength and deformation characteristics of rock discontinuities (Yoshinaka et al. 2006), but there are few studies on the scale effect of the stick-slip behavior. Therefore, in order to confirm the scale effect of the contact surface, a larger scale experimental device was improved to be able to experiment with a larger rock blocks, and the experiment which changed the size of four types of rock blocks was conducted. In this paper, the results obtained from this experiment will be described, which examines the influence of the scale effect of the rock discontinuities on the stick-slip behavior and its factors.

2   OUTLINE OF THE EXPERIMENT

2.1   *Materials*

The rock types of the block used in this experiment are gabbro, granite, andesite and diorite rocks. The stick-slip experiment is carried out on the base block made of each rock type with the upper block of each rock type made so that the contact area is 100 cm^2, 200 cm^2 and 300 cm^2. Where, the contact area indicates the area of bottom of the upper block, it refers to the apparent contact area.

Figure 1 shows the contact surface of each rock type used in this experiment. The contact surface between upper and base block of gabbro and granite are man-made surfaces, and andesite and diorite are natural schistosity surfaces.

Table 1 shows the classification of the characteristics of the contact surfaces of the block of each rock type used in this experiment, by the roughness of discontinuous of the hard rock surface shown in JGS 3811-2011 (JGS 2013).

2.2 Stick-slip experiment

Figure 2 shows a stick-slip experimental device. The experimental equipment consists of a rubber conveyor belt and a fixed frame, and the conveyor belt's moving speed can be changed freely. The base block is on the conveyor belt, and the upper block is fixed to the fixed frame through the spring. When the conveyor belt is operated, the upper and base blocks are moved in the direction where the spring is stretched together, but when it exceeds a certain displacement, a slip is caused by the restoring force of the spring connected to the upper block. The repetition of this behavior is a stick slip phenomenon.

In the experiment, in order to measure the force acting on the upper block due to the stick-slip, the load cell was installed between the spring and the fixed frame, and the accelerometer was installed on the upper block to measure the horizontal acceleration of the conveyor belt movement direction. The horizontal displacement of the upper and base blocks during the experiments are measured as the distance between the fixed frame by the contact type displacementmeter attached to the frame. The measurement sampling interval was 5ms, and the displacement, load and acceleration were recorded on the computer using a dynamic strain amplifier. The experimental conditions were based on the case given in Table 2, and the velocity of the conveyor belt and the normal load were changed. Where, the normal stress shown in Table 2 refers to the apparent normal stress, which is obtained by dividing the normal load (the weight of the upper block and the loaded weights) by the apparent contact area. The normal stresses are adjusted by the the loaded weight. The spring used is an elastic spring with a stiffness of 1.0 N/mm.

3 EXPERIMENTAL RESULTS

Figure 3 shows the time histories of spring force in each rock type and each block size. Figure 4 shows the cumulative slip amount of the upper block (base block displacement minus upper block displacement). The cumulative slip amount is zero when the upper and base blocks are moving on sticking, and is added when slippage occurs. Therefore, it is repeated that the spring force decreases with the

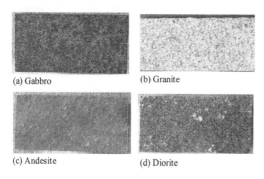

(a) Gabbro   (b) Granite

(c) Andesite   (d) Diorite

Figure 1. Contact surface of rock blocks.

Table 1. Classification of the characteristics of the contact surfaces by the roughness of hard rock discontinuous surface shown in JGS 3811-2011 (JGS 2013). Parentheses indicate the results of classifying the contact surfaces of rock blocks used in this experiment.

Large scale (1~2m) \ Small scale (10cm)	Rough : r	Slightly rough : m	Smooth: s
Stepwise : s	r_{sr}	r_{sm}	r_{ss}
Wavy : w	r_{wr} (Diorite)	r_{wm} (Andesite)	r_{ws}
Planar : p	r_{pr}	r_{pm} (Granite)	r_{ps} (Gabbro)

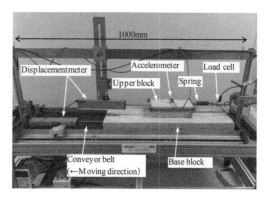

Figure 2. Stick-slip experimental setup.

Table 2. Stick-slip experimental conditions.

Rock types	Contact area (cm²)	Normal stress σ_n (kPa)	Velocity of conveyor velt (mm/s)
Gabbro	100, 200	1.5, 2.0	0.5, 1.0, 1.5, 3.0
Granite	100, 200, 300	1.5	(for all rock types)
Andesite	100, 200, 300	1.5, 2.0	
Diorite	100, 200, 300	2.0	

increase of the cumulative slippage when the spring force reaches the peak load after only the spring force is increased in a state where the cumulative slippage is constant.

As shown in Figures 3, 4, the magnitude of the spring force and the slippage are different depending on the upper block size, i.e. contact area. In addition, the stick-slip behavior varies depending on the rock type, the recurrence time (the time from the end of the slip to the start of the next slip, stress accumulation time) becomes longer as the contact surface becomes rougher, and the change of the spring force before and after the slip (hereinafter referred to as the force drop) and the slippage tend to be large. Thus, it is considered that the state of asperity, such as the roughness of the contact surface shown in Table 1, influences the stick-slip behavior. Figures 3, 4 show the case where the moving speed of the conveyor belt is 1.0 mm/s, in this paper, the results of this case will be discussed.

4 DISCUSSION ON EXPERIMENTAL RESULTS

4.1 *Relation between friction coefficient and contact area*

In order to confirm the effect of the difference of contact area on stick-slip behavior, the friction coefficient (the ratio of the spring force and the normal load) is compared. Figure 5 shows the friction coefficient at the time of the peak load (spring force just before slipping), i.e. the static friction coefficient, for each contact area. As shown in Figures 5(a), (b), in the case of gabbro and granite where the contact surfaces are smooth or the entire surface is planar, the magnitude of the static friction coefficient is almost unchanged depending on the contact area. On the other hand, as shown in Figures 5(c), (d), in the case of andesite and diorite where the contact surfaces are rough or entire surface is wavy, the static friction coefficient sharply decreases in the range of $100cm^2$ to $200cm^2$ where the contact area is small, in the

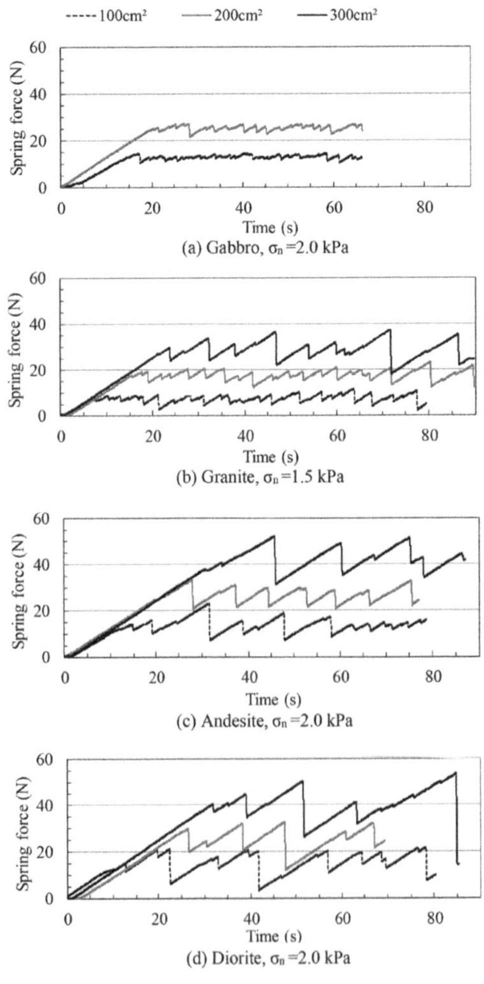

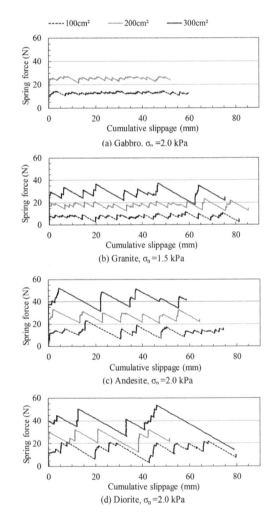

Figure 3. Time histories of spring force.

Figure 4. Cumulative slip amount of upper block.

range of 200cm² to 300cm², the change in static friction coefficient is small. Yoshinaka et al.(2006) confirmed that the shear resistance angle of the smooth surface of the saw-cut granite corresponds to the static friction angle and is no affected by the shear area, and that the peak shear strength of the fracture surface of granite which is man-made has a significant decrease due to the increase of the shear area in the small range of the shear area which is generally less than 1,000cm². The results of experiment in this study are almost consistent with the characteristics of the scale effect on the shear strength.

Figure 6 shows a conceptual diagram of the contact part of the discontinuous surface (Scholz, C. H. 2010). In the actual discontinuous surface, the contact area is only the area of the true contact area A_r of several asperities in the apparent contact area A, and

the frictional force of the discontinuous surface is the sum of the shear strength of the true contact area A_r.

Figure 7 shows the contact part of the upper and base blocks of the gabbro and diorite. In the case of gabbro shown in Figure 7(a), both the upper and base blocks have a smooth surface and the entire surface is planar. Because variations in the shape of the contact surface is small, even if the apparent contact area A increases, the ratio of the true contact area A_r almost unchanged. Therefore, even if the apparent contact area A_r increases, the friction coefficient does not change. In the case of gabbro shown in Figures 7(b), (c), both the upper and base blocks have a rough surface and the entire surface is wavy with a long wavelength. In the case of contact area 100cm² shown in Figure 7(b), variations in shape of the contact surface is large, and it becomes engaged condition when sticking, so that the ratio of the true contact area A_r becomes larger than the smooth surface. Therefore, the frictional force, i.e. the friction coefficient tends to be large. In the case of contact area 300cm² shown in Figure 7(c), due to the influence of the surface roughness and the wavelength of the entire surface shape, the upper and base block surfaces become hardly to contact. Thus, when the variations in the large and small wavelength of the contact surface shape is large, the ratio of a true contact area A_r tends to small when the apparent contact area A increases, and the friction coefficient decreases. However, when the apparent contact area A becomes further large and the shape wavelength of the entire surface contains a certain number of wavelengths within the apparent contact surface, even if the apparent contact area A increases, it is presumed that the ratio of the true contact area A_r hardly changes.

4.2 Relations of slippage and velocity and acceleration

Figures 8, 9 show the relation between slippage, which is relative displacement during sliding, and maximum velocity/acceleration for stick slip events of each rock

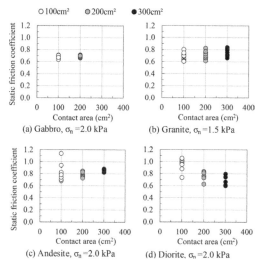

Figure 5. Relation between static friction coefficient and contact area.

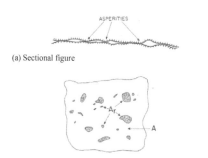

(a) Sectional figure

(b) Plane figure. A indicates the apparent contact area, and stippled area A_r indicates the true contact area where the asperities are in contact.

Figure 6. Conceptual diagram of the contact part of the discontinuous surface (Scholz, C. H. 2010).

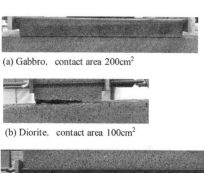

(a) Gabbro, contact area 200cm²

(b) Diorite, contact area 100cm²

(c) Diorite, contact area 300cm²

Figure 7. Contact condition of upper and base block.

type. As result, it is seen positive correlation between the slippage and the maximum velocity/acceleration of each rock type. Moreover, it is assumed that the maximum velocity and maximum acceleration have a positive correlation. These relations are consistent with the biaxial experimental results by Ohnaka (2003) as well as stick-slip experiments reported by Ohta & Aydan (2010). In this experiment, because we use an elastic spring, the force drop is proportional to the slippage as shown in Figure 4. When slippage is substituted with force drop in Figure 8, the maximum velocity is proportional to the force drop. This relation is consistent with the result that is provided from past earthquake records (Kanamori & Anderson 1975). As for the difference of the contact area, a correlation between the slippage and the maximum velocity is not clearly observed. In the relation between slippage and maximum acceleration shown in Figure 9, as the contact surface becomes coarser, the variation of the relationship and the difference of the maximum acceleration due to the magnitude of the contact area become larger. The frictional force at the time of slipping can be described by the relationship between the friction coefficient of the contact surface and the normal load, and it can also be described from the relationship between the block weight and the acceleration from the motion equation. From the above, it is inferred that the maximum acceleration at the time of slipping is related to the characteristic due to the variation of the friction coefficient and the difference in the contact area shown in Figure 5.

4.3 Relation between force drop and slippage

Since the seismic moment is proportional to the amount of force drop on the fault surface (Kanamori & Anderson 1975, Molnar 1975), the characteristics

of the force drop obtained in this experiment is confirmed. Figure 10 shows the relation between the ratio of force drop (ratio of force drop and frictional force) and the slippage for stick-slip events. The ratio of force drop is proportional to the amount of slippage in all conditions. It is also confirmed by the relation between the shear stress drop rate and the slip displacement amount in the frictional slip indicated by Ohnaka (2003) and from the past experimental results by Kiyota et al. (2018). In the case of Figure 10(a), (c), (d) with the same normal stress, the inclination of the linearity is almost the same for the same contact area. In addition, since the

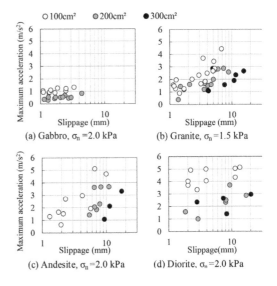

Figure 9. Relation between maximum acceleration and slippage for stick-slip events.

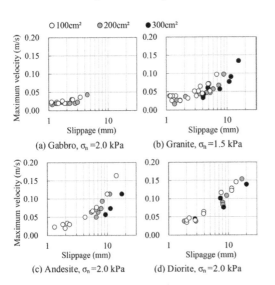

Figure 8. Relation between maximum velocity and slippage for stick-slip events.

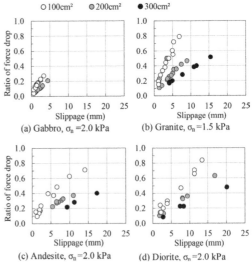

Figure 10. Relation between ratio of force drop and amount of slippage.

magnitude of the inclination of the linearity for each contact area within the same rock type is roughly the inverse ratio of the contact area in any rock type, the relationship between the ratio of force drop per unit area and the slip amount is uniform. Therefore, the relation between the ratio of force drop and the slippage is hardly influenced by the scale effect, and it is thought that it depends on the magnitude of the normal stress and the stiffness of the elastic spring.

5 CONCLUSIONS

In this study, in order to confirm the scale effect in the stick-slip behavior of rock discontinuities, a stick-slip experiment using rock blocks of different contact areas was conducted for four rock types. The findings obtained from this study are summarized as follows:

1. In the case where the contact surface is smooth or the entire surface is planar, the influence of the scale effect is small. On the other hand, In the case where the contact surface is rough or the entire surface is wavy, the friction coefficient significantly decreases with increase of the contact area in the range where contact area is small, and as the contact area becomes larger, the variation in the friction coefficient becomes smaller.

2. The maximum velocity and the maximum acceleration of the block at the time of sliding have a positive correlation with the amount of slippage. However, the influence of the scale effect on the maximum velocity is small regardless of the roughness of the contact surface. As for the maximum acceleration, as the contact surface is rough or the entire surface becomes wavy, the scale effect is more significant as well as the friction coefficient.

3. Although the amount of slippage and the force drop are in linear relation, the scale effect is small regardless of the roughness of the contact surface, and the inclination of the linear is determined by the normal stress and the stiffness of the spring.

As described above, in order to estimate the characteristics of the stick-slip phenomenon on the discontinuous rock surface, when the shape of the discontinuous rock surface is uniformly smooth and the variation is small, it is thought that it can be inferred by the experiment with the contact area of several 100cm^2 as in this experiment. On the other hand, in the case where shape is complicated and there are many variations, it is desirable to evaluate the experimental results to some extent increase the contact area. However, since this study is a qualitative evaluation, we would like to evaluate the geometric shape and pattern of the contact surfaces quantitatively, and experiment and evaluate them with consideration. In the future, these findings are used for parameter setting of displacement and stress drop of fault plane in fault rupture simulations.

REFERENCES

Brace, W. F. & Byerlee, J. D. 1966. Stick-slip as a mechanism for earthquakes. *Science*, 153: 990–992.
Iwata, N., Takahashi, Y., Adachi, K., Aydan, Ö. & Tokashiki, N. 2016. Stick-slip behavior of rock discontinuities and its implications in the estimation of strong motions during earthquakes. *Proc. of the ISRM 9th Asian Rock Mechanics Symposium (ARMS9), Bali.*
JGS (The Japanese Geotechnical Society). 2013. Method for engineering clasification of rock mass, JGS3811–2011. *Japanese Standards and Explanations of Geotechnical and Geoenvironmental Investigation Methods*: 60–76.
Kanamori, H. & Anderson D.L. 1975. Theoretical basis of some empirical relations in seismology. *Bull. Seism. Soc. Am.*, 65(5): 1073–1095.
Kiyota, R., Iwata, N., Takahashi, Y., Adachi, K. & Aydan, Ö. 2018. Stick-slip behavior of rock discontinuities by difference in rock types. *Proc. of the 3rd International Confrence on Rock Dynamics and Applications (RocDyn-3), Trondheim*: 563–568.
Molnar, P. 1975. Earthquake recurrence intervals and plate tectonics. *Bull. Seism. Soc. Am.*, 69(1): 115–133.
Ohta, Y. & Aydan, Ö. 2010. The dynamic responses of geomaterials during fracturing and slippage. *Rock Mechanics and Rock Engineering*, Vol.43, No.6: 727–740.
Ohnaka, M. 2003. A constitutive scaling law and a unified comprehension for frictional slip failure, shear fracture of intact rock, and earthquake rupture. *J. Geophys. Res.*, 108(B2): 6-1-21.
Scholz, C. H. 2010. *The mechanics of earthquakes and faulting. 2nd edition*, Kokon shoin: 51–97.
Yoshinka, R., Yoshida, J., Sasaki, T. & Sasaki, K. 2006. Evaluation of mechanical design parameters of rock discontinuities considering scale effect. *Journal of Japan Society of Civil Engineers, Ser. C*, Vol.62, No.2: 457–470 (in Japanese).

2019 Rock Dynamics Summit– Aydan et al. (eds)
© 2019 Taylor & Francis Group, London, ISBN 978-0-367-34783-3

Evaluation of the dynamic shear strength of rocks under confining pressure

G. Min, S. Oh, S. Park & S. Cho
Chonbuk National University, Jeon-ju, Republic of Korea

Y. You & L. Park
Agency of Defense Development, Dea-jeon, Republic of Korea

ABSTRACT: The shear strength of rocks is one of the significant parameters for designing geo-structures such as rock slope, tunnels, and underground space. This parameter has been widely applied for simulating rock fracture problems, which adopt a frictional failure criterion. In order to apply a shear failure model for simulating a dynamic fracture problem, it is essential to obtain the dynamic shear strength with increasing confining pressure. This paper presents the dynamic tri-axial punch shear test for evaluating the dynamic shear strength of rocks under confining pressure. In this method, a modified Hoek tri-axial pressure cell, which is applicable to split Hopkinson pressure bar (SHPB) system, was utilized to acquire confining pressure dependent dynamic shear strength. The investigation of dynamic shear strength was carried out with increment of constant confining pressure and loading-rate. It was found that dynamic shear strength increases with increasing confining pressure increment and considerably increases with increasing loading-rate. The value of cohesion also increased with loading-rate increment, while the internal friction angle remained constant. Finally, based on the experimental results, loading-rate dependent frictional dynamic failure criterion of rock was suggested.

1 INTRODUCTION

The shear strength is an important material parameter for rocks. This parameter has been widely applied to simulate rock fracture problems, which apply a frictional failure criterion.

There are several suggested methods for quantifying the shear strength of rocks and concrete, such as direct shear-box test and block punch index (BPI) test under static-loading (Schrier van der 1988; Sulukcu & Ulusay 2001).

However, rock-material may be subjected to dynamic loadings in various rock engineerings such as rock drilling, blasting, and projectile penetration. Therefore, it is necessary to come up with a method for determining the dynamic shear strength of the rock.

Thus, several methods have been developed to perform the dynamic shear tests for geo-material, which is subjected to dynamic loadings. A dynamic punch test method utilizing a split Hopkinson pressure bar (SHPB) system was suggested to quantify the dynamic shear strength of rocks (Huang et al. 2011). However, this method was incomplete as the dynamic shear strength is obtained with the condition of the confining pressure.

In this study, the split Hopkinson pressure bar (SHPB) system is utilized to determine the dynamic shear strength of rocks. A Hoek tri-axial pressure cell is designed to apply confining pressure to the rock samples during dynamic punch shear tests.

The shear strength of rock samples was presented as a function of confining pressure and loading-rate to suggest a loading-rate dependent frictional failure criterion. The loading-rate dependent Mohr-Coulomb (RDMC) model, which is based on the dynamic punch shear tests regarding confinement, is also developed and implemented into LS-DYNA.

2 EXPERIMENT

2.1 Description of the experimental approaches

An SHPB system, which consists of an impact bar, an incident bar, and a transmitted bar, is employed to apply the dynamic loadings to the rock sample for dynamic tri-axial punch shear tests as shown in Figure 1. Inside a modified Hoek tri-axial pressure cell, the rock sample is located between two bars: a 20mm diameter incident bar and a 50 mm diameter transmitted bar. Semi-conduct strain-gauges are attached to the bars to measure strain signal histories: incident strain (ε_I), reflected strain (ε_R) and transmitted strain (ε_T).

The pulse-shaping technique using thin copper disc is applied to achieve the dynamic force balance that acts on the sample (Chen & Song 2011; Cho et al. 2014).

When a gas gun fires the impact bar to collide with the incident bar, an elastic compressive wave is generated and propagated toward the sample. At the boundary between the incident bar and the specimen, the incident wave will be divided into two waves: a reflected wave (tensile) and the transmitted wave (compressive).

Using the strain signals (ε_I, ε_R, and ε_T) measured by the strain-gauges on the bars, the applied forces P_1 and P_2 on both sides of a sample can be computed as:

$$P_1 = EA_I(\varepsilon_I + \varepsilon_R) \tag{1}$$

$$P_2 = EA_T\varepsilon_T \tag{2}$$

where E is Young's modulus of bars, A_I is a cross-sectional area of the incident bar, A_T is a cross-sectional area of the transmitted bar.

It is significant to satisfy the dynamic force equilibrium condition (i.e. $P_1 = P_2$), to accomplish a dynamic stress equilibrium state of the sample before its failure. Regarding the dynamic stress equilibrium state, the dynamic punch shear strength of rock sample can be calculated using the following equation:

$$\tau = \frac{P}{\pi DB} \tag{3}$$

where τ is the dynamic punch shear strength, P (= $P_1 = P_2$), is the applied force; D is the diameter of the incident bar; B is the thickness of the rock sample.

The modified Hoek tri-axial pressure cell is designed to apply a constant confining pressure to the rock sample during the dynamic tri-axial punch shear test as shown in Figure 2. The constant confining pressure was applied to the rock sample through the flexible membrane. The incident bar applies the dynamic loadings as a punch hammer, and the rear assistance bar is attached to the transmitted bar. The inner diameter of the rear assistance bar is slightly larger than the diameter of the incident bar to induce a proper shear deformation.

2.2 Preparation of rock-specimens

Table 1 presents the physical and mechanical properties of rock samples. Figures 3(a-c) presents that disc-shaped rock samples, which were processed

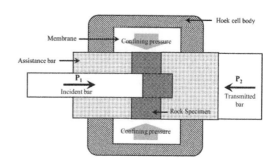

Figure 2. Schematic diagrams of modified tri-axial Hoek cell for dynamic punch shear test.

Table 1. physical and mechanical properties of rock samples.

	Granite gneiss	Gneiss A	Gneiss B
Density (kg/m^3)	2590	2700	2610
Porosity (%)	0.26	0.19	1.53
P-wave velocity (m/s)	3863	5168	4628
S-wave velocity (m/s)	2272	3233	2722
Young's modulus (GPa)	41.30	68.50	64.58
Poisson's ratio	0.23	0.18	0.16
Uni-axial compressive strength (MPa)	231.65	152.2	153.37
Tensile strength (MPa)	11.00	10.8	12.06
Cohesion (MPa)	28.10	25.65	23.66
Internal friction angle ($Degree$)	59.20	53.10	49.12

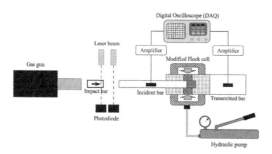

Figure 1. Schematic diagram of the modified SHPB system for dynamic tri-axial punch shear test.

Figure 3. Disc-shaped rock samples before (a), (b), (c) and after (d), (e), (f) the dynamic tri-axial punch shear test.

to place them inside the Hoek tri-axial pressure cell. The diameter and thickness of the samples are 50 mm and 20 mm, respectively.

2.3 Test conditions and results

The launching velocity of the impact bar was controlled to achieve a wide range of loading rate, which represents from 500 GPa/s to 1500 GPa/s. Also, the confining pressure was increased up to a constant value of 0, 5, and 10 MPa for each rock sample. Figures 3(d-f) shows the rock samples after the dynamic triaxial punch shear test.

Dynamic punch shear strength indicates the peak value among the punching shear stresses calculated by using Equation 3. Figures 4-6 plot the dynamic punch shear strength of the rocks. The dynamic punch shear strength increases with increasing confining pressure and significantly increases with loading-rate.

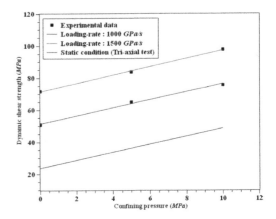

Figure 6. Dynamic shear strength of the Gneiss B as a function of confining pressure under different loading-rates.

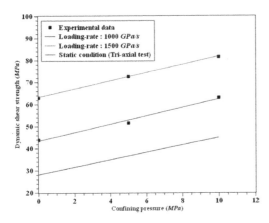

Figure 4. Dynamic shear strength of the Granite gneiss as a function of confining pressure under different loading-rates.

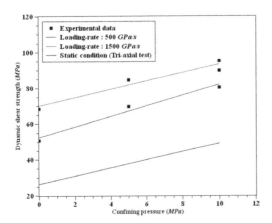

Figure 5. Dynamic shear strength of the Gneiss A as a function of confining pressure under different loading-rates.

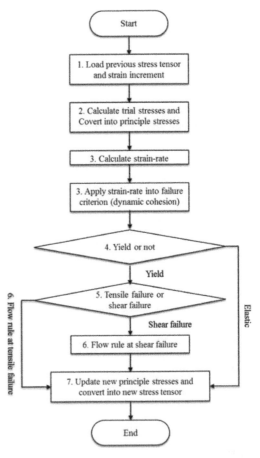

Figure 7. The flowchart of the RDMC model code using the UMAT in LS-DYNA.

2.4 Loading-rate dependent shear failure model

The obtained dynamic shear strength data were presented as a function of confining pressure and different loading-rates based on Mohr-Coulomb failure criterion.

127

It is noted that the cohesion increased with loading-rate increment, while the internal friction angle remained constant as shown in Figures 4-6. Therefore, the loading-rate dependent Mohr-Coulomb (RDMC) criterion can be presented as:

$$\tau_d = c_s(1 + A\ln\frac{\dot{\tau}}{\dot{\tau}_0}) + \sigma_n\tan\phi \qquad (4)$$

Where τ_d is dynamic shear strength, $\dot{\tau}$ is loading-rate, $\dot{\tau}_0$ is reference loading-rate, σ_n is confining pressure, c_s is cohesion, ϕ is internal friction angle and, A is coefficient of loading-rate dependent. In this study, A was obtained 1.1234 (granite gneiss), 0.3914 (gneiss A) and 0.4730 (gneiss B), respectively.

The loading-rate dependent shear failure criteria could be applied as the failure criterion of the numerical simulations considering dynamic rock fracturing and developed for a user-defined material (UMAT) model code of LS-DYNA. Figure 7 shows the flowchart of the RDMC code, which revises stresses by following non-associated flow rule.

3 VALIDATION OF THE SUGGESTION MODEL

The numerical simulation using the finite element (FE) model were performed to validate loading-rate dependent shear failure model, which is applied by UMAT of LS-DYNA. In this study, dynamic tri-axial punch shear test was simulated with different loading-rates.

3.1 Configuration and boundary condition

The FE models were generated using the LS-Prepost, which applied in multi-physical software LS-DYNA. Figure 8 presents the layout of the finite element model for analyzing the dynamic punch tests.

The bars were modeled as the elastic body using the MAT_ELASTIC model in LS-DYNA. Elastic incident waves are applied from 1000 GPa/s to 3000 GPa/s in LS-DYNA simulation. Also, constant 5MPa confining pressure was applied to the rock sample.

The loading-rate dependent shear failure material model was adopted as the constitutive model of the rock sample using UMAT in LS-DYNA. Input parameters of the rock-specimen are listed in Table 2.

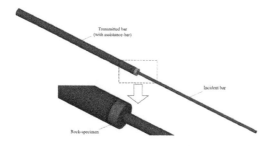

Figure 8. Finite element layout for analyzing dynamic tri-axial punch shear test of rock-specimen.

Table 2. Input parameters of the rock sample.

Material property	Unit	Value
Density (ρ)	kg/m^3	2590
Young's modulus (E)	GPa	41.30
Poisson's ratio (v)		0.23
Uniaxial compressive strength (σ_c)	MPa	231.65
Tensile strength (σ_t)	MPa	11
Cohesion (c)	MPa	28.1
Internal friction angle ($\varnothing$)	$Degree$	59.2
Material constant (A)		1.1234

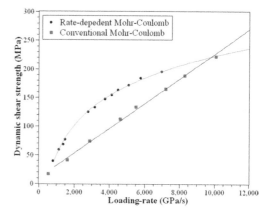

Figure 9. Non-Linear Rdmc and Linear Mc Behavior of the Dynamic Shear Strength With Different Loading-Rate.

3.2 The result of dynamic tri-axial punch test simulation

Figure 9 shows the dynamic shear strength regarding the different dynamic loadings as the rock specimen applies the RDMC and conventional Mohr-Coulomb (MC) model, respectively.

It is noted that the dynamic shear strength presents non-linear behavior with different loading-rates as the RDMC model applied, while conventional MC model shows linear.

4 CONCLUSION

In this study, dynamic tri-axial punch shear tests were performed to quantify the dynamic shear strength of various rock-materials regarding constant confinement.

A dynamic loading was applied by the split Hopkinson pressure bar (SHPB) system and a modified Hoek tri-axial pressure cell was specially designed to apply confining pressure into rock samples during dynamic punch shear test.

The dynamic shear strength of various rock samples was arranged as the function of the confining pressure and loading-rate. It shows that a cohesion increased with loading-rate increment, while an internal friction angle remained constant. Based on these concepts, we suggested a loading-rate dependent shear failure criterion, which was developed to apply into numerical simulations of dynamic rock fracturing.

As the developed code validation, the numerical simulations of the dynamic tri-axial punch shear test were also performed using the RDMC and conventional MC model, respectively. The simulation results presented that the RDMC model can well simulate the non-linear behaviors of loading-rate dependency on the dynamic shear strength.

REFERENCES

Chen, W. 1995. Dynamic failure behavior of ceramics under multiaxial compression. PhD Thesis. California Institute of Technology.

Chen, W.W. & Song, B. 2011. Split Hopkinson (Kolsky) bar: Design, testing and applications. Mechanical Engineering Series.

Cho, S.H. et al. 2014. Determination of dynamic fracture toughness of rock using straight notched disk bending (SNDB) Specimen. International Symposium 8th Asian Rock Mechanics Symposium.

Huang, S. et al. 2011. A dynamic punch method to quantify the dynamic shear strength of brittle solids. Dynamic Behavior of Materials 1:157–163.

Johnson, G.R. & Homlquist, T.J. 1994. An improved computational constitutive model for brittle materials. In: AIP Conference Proceedings: 981–984.

Schrier, van der. 1988. The block punch index test. Bulletin of the International Association of Engineering Geology 38:121–126.

Sulukcu, S. & Ulusay, R. 2001. Evaluation of the block punch index test with particular reference to the size effect, failure mechanism and its effectiveness in predicting rock strength. International Journal of Rock Mechanics & Mining Sciences 38:1091–1111.

2019 Rock Dynamics Summit– Aydan et al. (eds)
© 2019 Taylor & Francis Group, London, ISBN 978-0-367-34783-3

A new constitutive equation for a solid material

E. Nakaza

Department of Civil Engineering, University of the Ryukyus, Nishihara, Okinawa, Japan

ABSTRACT: The present author (Nakaza), introduced a new elastic theory in 2005. According to this theory, the internal stress of an isotropic elastic material consists of an elastic stress governed by Hooke's law and an internal pressure governed by the state equation. In conventional theory, the internal stress consists of the two elastic stresses corresponding to the two elastic moduli and thus the force causes the Poisson effect cannot be explained, even if we observe the occurrence of the lateral strains under a uniaxial loading. According to Nakaza's elastic theory, Hooke's law has only one elastic modulus and in this sense, the new theory is consistent with the assertion of Navier who used only one elastic constant to derive his fundamental elastic theory. The new elastic theory may change the stress evaluation from the existing theory.

1 INTRODUCTION

Even if we observe deformations of a material (i.e., a distribution of strain), we cannot immediately judge whether the distribution corresponds to the distribution of the internal stress. A strain distribution is converted into a stress distribution based on Hooke's law, which is characterized as having two elastic moduli. For example, when we observe the deformation of the material under longitudinal (vertical) uniaxial loading, not only longitudinal strain, but also lateral strains are generally observed in the material. In such a case, non-zero internal stress exists only in the longitudinal direction in the material, i.e., the stresses in the lateral directions are evaluated to be zero.

In this case, the distortions in the transverse directions are caused by the Poisson effect. By relying on the traditional elastic theory, the stress that causes these cannot be explained. Then, by further loading the material, the material shows cracks or splitting in the longitudinal direction, leading to failure. At this time, the fracture surface is in the longitudinal direction. Why did the material crack run vertically? The conventional theory cannot answer this simple question because two elastic moduli are used in the traditional theory

In the present paper, a new elastic theory is introduced, and its characteristics are discussed. According to the new elastic theory, the simple questions stated above are clearly resolved. Furthermore, it is shown that the distribution of internal stresses received from outer loading directly corresponds to the measured deformations (strain distribution). Finally, a new failure criterion is suggested.

2 OUTLINES OF THE NEW ELASTICITY THEORY

Here, we will derive the new elastic theory of Nakaza (2005) as a development of the conventional isotropic elastic theory. First, the linear relationship between stress and strain in the conventional elasticity theory is as follows (Fung, 1994):

$$\sigma_{ij} = \lambda \varepsilon_{kk} \delta_{ij} + 2\mu \varepsilon_{ij} + \beta dT \delta_{ij} \tag{1}$$

where σ_{ij} is the stress tensor, ε_{ij} is the strain tensor, λ and μ are the moduli of elasticity, dT is the temperature change, β is the coefficient of linear thermal expansion, and δ_{ij} is the Kronecker delta.

Equation (1) shows the constitutive equation (general Hooke's law) introduced in the traditional elastic theory, considering the thermal expansion. Nakaza (2005, 2009, 2010) recognized the existence of pressure in the internal stress of the elastic body. The relationship among pressure change, material density change and temperature change is generally given as follows, according to the thermodynamic equation of state:

$$p = R\rho T \left(\frac{d\rho}{\rho} + \frac{dT}{T} \right) \tag{2}$$

where p is the inner pressure change, ρ is the density, $d\rho$ is the density change, T is the temperature, dT is the temperature change, and R is the material constant.

Therefore, by introducing Equation (2) to Equation (1), we have the following relationship:

$$\sigma_{ij} = -p\delta_{ij} + 2E\varepsilon_{ij} \tag{3}$$

where, E is the modulus characterizing the elasticity, which may be a function of a state defined by inner pressure and temperature, and is uni-modular for an elastic isotropic material. From Equation (1) and (3), the relationship $E = \mu$ is clearly identified.

Note that when we consider the plasticity and the viscosity of a material, we may introduce the plasticity strain, the rate of strain and viscosity coefficient to the fundamental relationship, i.e., Eq. (3).

Equation (3) is the linear relationship between the stress and strain defined by Nakaza (2005). The first term on the right-hand side of Equation (3) represents the change in the internal pressure governed by the state equation shown by Equation (2) and the second term represents the elastic stress governed by Hooke's law. In this way, the elastic modulus is defined as only one constant for Hooke's law of an isotropic elastic material, which conforms to Navier's theorem claiming that there should be only one elastic modulus (Timoshenko, 1988).

The internal pressure change shown in Equation (3) causes isotropic expansion or contraction of a material. Therefore, we move the term of internal pressure to the left-hand side in order to obtain the following equation:

$$\sigma_{ij} + p\delta_{ij} = 2E\varepsilon_{ij}. \tag{4}$$

Alternatively, this equation can be written as follows:

$$\tau_{ij} = 2E\varepsilon_{ij} \tag{5}$$

where τ_{ij} is termed *elastic stress*.

The physical meaning of Equation (4) is interpreted such that the left-hand side represents the stresses as the action that causes deformations of an elastic material, and the right-hand side represents the appearance showing the resistance of the material by elastic springs counter to the actions of the stresses. Equation (3) will be further transformed into another type of equation through the discussion in the following sections.

3 PHYSICAL INTERPRETATION OF THE POISSON EFFECT

Although conventional elastic theory recognizes the occurrence of strain due to the Poisson effect, it cannot explain what stress causes it. On the other hand, the theory of Nakaza shown in Equation (3) is explained below. Here, the case of uniaxial compression of an elastic bar in the vertical direction is assumed. At this time, the relationship between stress and strain in the longitudinal direction is given as follows:

$$\sigma_3 = -p + 2E\varepsilon_3 \tag{6}$$

where the suffix 3 indicates the direction of the bar axis.

For the lateral directions, the following relations are given:

$$0 = -p + 2E\varepsilon_1 \tag{7}$$

And

$$0 = -p + 2E\varepsilon_2. \tag{8}$$

From these, we have

$$p = 2E\varepsilon_1 \tag{9}$$

and

$$p = 2E\varepsilon_2. \tag{10}$$

In these equations, $\left(\varepsilon_1,\ \varepsilon_2,\ \varepsilon_3\right)$ and $\left(\sigma_1,\ \sigma_2,\ \sigma_3\right)$ are the principal strain and stress.

From Equations (9) and (10), it is concluded that the Poisson effect is an isotropic deformation caused by the internal pressure change. According to Hooke's law, the change in internal pressure causes isotropic deformation of the material.

In Equation (6), the external load is supported by a change in internal pressure and elastic stress. Since the deformation caused by the pressure change is isotropic, the following relationship can be given:

$$p = 2E\varepsilon_p. \tag{11}$$

where ε_p is the isotropic strain caused by the internal pressure change. This can be measured as the strain due to the Poisson effect being observed. From this equation, we have the pressure change in a material.

Substituting Equation (11) in to Equation (6), we have:

$$\sigma_3 = 2E\left(\varepsilon_3 - \varepsilon_p\right). \tag{12}$$

The stress, indicated by the left-hand side of Equation (12), due to the outer loading on a material is supported by the elastic stress formed by the strain observed after subtraction of the isotropic strain.

The strain $\left(\varepsilon_3 - \varepsilon_p\right)$ can be written in ε'_3, which is termed the *effective strain* corresponding to the stresses purely caused by outer loads.

In traditional theory, for a uniaxial test of an elastic bar, Poisson's ratio has been defined as

$$\varepsilon_1 = \varepsilon_2 = -\nu\varepsilon_3 \tag{13}$$

where ν is the Poisson's ratio.

In Equation (13), the Poisson effect is not properly defined as satisfying the isotropic condition for an isotropic elastic material. Since in the direction of axial loading, any strain caused by the Poisson effect has not explicitly been introduced.

From Equations (3) and (12), considering a constant temperature condition, we have:

$$\varepsilon_p = -\theta\,\varepsilon_{kk} \tag{14}$$

$$\theta = \nu / \left(1 - 2\nu\right). \tag{15}$$

131

Instead of using Poisson's definition for Poisson's ratio, as in Equation (13), we here introduce the *modified Poisson's ratio*, θ, properly defined as satisfying the isotropic condition of the Poisson effect.

From Equation (11) and (14), for the state change for a constant temperature, obtaining the relative volume change ε_{kk}, we have:

$$p = 2E\left(-\theta\,\varepsilon_{kk}\right). \tag{16}$$

4 EFFECTS OF TEMPERATURE

For temperature changes, under no outer loading on a material, the new theory explains that temperature change causes an inner pressure change as shown by Equation (2), so that the pressure change causes strain as shown in Equation (3). From the constitutive equation (Equation (3) and the state Equation (2)), we have

$$p = 2E\varepsilon_{ij} \tag{17}$$

which shows the thermal expansions of an elastic material due to temperature changes. According to Hooke's law, these thermal expansions are dynamically caused by inner pressure changes. When we introduce the inner pressure changes due to the temperature changes, Equation (17) shows that the modulus between the pressure change and the strain can be given by the elasticity, E.

5 FAILURE CRITERIA FOR ISOTROPIC MATERIALS

There are many types of failure criteria, and the criteria of Tresca and von Mises are well known as basic failure criteria for isotropic materials. These are mathematically related to invariants of the stress and/or strain tensor (Chen & Saleeb, 1982).

In the traditional theory, the stress tensor is split into two parts, the pure hydrostatic pressure and the pure shear stress, as follows:

$$\sigma_{ij} = -\bar{p}\delta_{ij} + 2G\left(\varepsilon_{ij} - 1/3\varepsilon_{kk}\delta_{ij}\right). \tag{18}$$

where G is the shear modulus, $\bar{p}$ is the mean stress (pure hydrostatic stress) and the second term is the deviatoric stress (stress deviator) tensor.

Denoting the deviatoric stress tensor by σ'_{ij}, we have:

$$\sigma_{ij} = -\bar{p}\delta_{ij} + \sigma'_{ij} \tag{19}$$

or

$$\sigma_{ij} = -\bar{p}\delta_{ij} + 2G\varepsilon'_{ij}. \tag{20}$$

Here ε'_{ij} is the deviatoric strain tensor and is given as:

$$\varepsilon'_{ij} = \left(\varepsilon_{ij} - 1/3\varepsilon_{kk}\delta_{ij}\right). \tag{21}$$

We can now compare Equation (20) with Equation (3) in order to determine the differences between these equations.

Tresca's failure criteria are related to the maximum shear stress at a point in a material as follows:

$$max\left[\frac{1}{2}\left(\sigma_1 - \sigma_2\right),\ \frac{1}{2}\left(\sigma_2 - \sigma_3\right),\ \frac{1}{2}\left(\sigma_3 - \sigma_1\right)\right] = k \tag{22}$$

where σ_1, σ_2, and σ_3 are the principal stresses, and k is the failure (yield) strength.

On the other hand, von Mises criteria are related to the second invariant of deviatoric stress tensor as follows:

$$J_2 - k^2 = 0 \tag{23}$$

where J_2 is the second invariant of the deviatoric stress tensor.

There are also Drucker-Prager failure criteria relating the two invariants to the above fundamental failure criterions:

$$\alpha I_1 + \sqrt{J_2} - k = 0 \tag{24}$$

where α is a material constant, and I_1 is the first invariant of the stress tensor.

The failure criteria introduced here, and many other failure criteria are thus associated with the invariants of the stress tensor or strain tensor. This is due to the mathematical property whereby the invariants do not depend on how coordinate axes are set. However, there is no obvious physical reason to associate these invariants with failure criteria.

The failure criteria discussed above indicate that at any point in an isotropic elastic body, when the stress or strain reaches a certain level, cracking or splitting occurs. There is a problem, however, with the setting of the failure surface. For example, in the case of uniaxial pure compression for which the loading state is very simple, the fracture surface cannot be set even if we see that the stress state reaches a certain failure criterion. Therefore, according to the conventional theory, dual criteria (Wu, 1974) are recommended based on both the stress tensor for checking the failure point and the strain tensor for setting the fracture surface.

In contrast to these conventional failure criteria, Nakaza's new theory presents a physically clear and simple method.

First, Equation (3) gives the following equation for the work done by the stress and the inner pressure:

$$\int\left(\sigma_{ij} + p\delta_{ij}\right)d\varepsilon_{ij} = E\varepsilon_{ij}^2. \tag{25}$$

The right-hand side of Equation (25) shows the *strain energy* as the work done by the internal pressure p and the stress σ_{ij}.

As shown in Equation (4), it is physically explained that the isotropic elastic material is deformed under the influence of the stresses due to the actions of both the external force and the internal pressure change. The material withstands the actions of the stresses with the reaction of the elastic stress produced by a mechanical function of an elastic spring as an elastic body. That is, as a failure criterion, it is sufficient to investigate the magnitude of the elastic stress and/or the strain energy represented by the right-hand side of Equation (4) and the strain energy term of Equation (25).

Instead, in term of the traditional theory of elasticity, the strain energy is given as follows:

$$\int \sigma_{ij} d\varepsilon_{ij} = \frac{1}{2} K\varepsilon_{kk}^2 + G\varepsilon'^2_{ij} \qquad (26)$$

or

$$\int \sigma_{ij} d\varepsilon_{ij} = \frac{1}{2} \lambda\varepsilon_{kk}^2 + \mu\varepsilon_{ij}^2 \qquad (27)$$

where K and G are the bulk and the shear modulus, respectively, ε_{kk} is the volumetric strain, and ε'_{ij} is the deviatoric tensor of the strain tensor. Comparing Equations (26) and (27) with Equation (25), we can completely explain the physical differences between these equations.

According to Equation (25), for example, if we consider the strain energy, the failure criterion is given as follows:

$$E\varepsilon_{ij}^2 = k^2 \qquad (28)$$

where k is the failure (yield) strength, which may generally be a function of state defined in terms of the inner pressure and temperature.

The failure criteria shown in Equation (28) are similar to the von Mises failure criteria shown in the Equation (23), though the von Mises failure criteria focus on the deviatoric stress tensor of the stress tensor. In the new theory, the criteria, for example, is physically based on the strain energy that shows the material deformation enduring both stresses due to external force and internal pressure. Of course, creating a failure criterion based on the maximum principal stress or the maximum shear stress of the elastic stress is straightforward, shown as follows:

$$E\varepsilon_1 = k \qquad (29)$$

or

$$max\left[\frac{1}{2} E(\varepsilon_1 - \varepsilon_2), \frac{1}{2} E(\varepsilon_2 - \varepsilon_3), \frac{1}{2} E(\varepsilon_3 - \varepsilon_1) \right] = k \qquad (30)$$

In the new elastic theory, there is no necessity to introduce the deviatoric stress or strain tensor as used in the conventional elastic theory. Furthermore, in the new failure criteria, in Equation (28), (29), and (30),

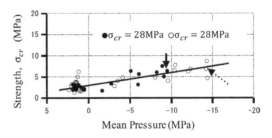

Figure 1. Strengths σ_{cr} of two types of concrete specimens under combined loading of pure compression, pure tension, pure shear, compression and shear, and tension and shear. The solid line indicates Equation (29) for which k is a function of pressure and temperature changes. The two arrows indicate the pure compression strengths for two types of concrete specimens. All experimental data are from Okajima (1970).

the distributions of the stress and the strain directly correspond to each other such that the new theory has the advantage that the failure surface based on the stress completely coincides with the failure surface based on the strain.

Therefore, according to the new theory, there is no need to include a dual failure criteria system. The problem that arose in the case of uniaxial compression discussed earlier is also completely resolved by the new theory.

Figure 1 shows the strengths of two types of concrete specimens under combined loading of pure compression, pure tension, pure shear, compression and shear, and tensile and shear. One of the concrete specimens has normal strength (compression strength, $\sigma_{cr} = 28$ MPa) and the other has high strength (compression strength, $\sigma_{cr} = 43$ MPa). Following Equation (5), the elastic stress is evaluated. In Figure 1, both concrete specimens are shown in the graph.

The result based on the new theory, represented by the solid line, creates a failure criterion in wide range from pure tension to pure compression. The x-axis of Figure 1, although indicated as the pure hydrostatic (mean) pressure $\bar{p}$, can be easily rewritten as the pressure p, as follows:

$$p = \frac{3v}{1+v} \bar{p}. \qquad (31)$$

Let us note that the failure criterion, k, may not be a constant or a linear function, but rather must be a nonlinear function of pressure and temperature.

6 CONCLUSION

Based on the principles of physics, Nakaza presented a new elastic theory and described the differences between the new elastic theory and the conventional elastic theory. In the present paper, based on the new theory, the physical mechanism of the Poisson effect and the dynamics of the thermal expansion of an elastic body were explained. The theory of failure criteria for isotropic elastic materials was then explained. The failure criteria based on the elastic stress was applied to two types of concrete specimens that have different compressive strengths, and the effectiveness of the new theory was demonstrated.

ACKNOWLEDGEMENTS

Throughout this research, the author received a great deal of advice from Professors Emeritus of the University of the Ryukyus Seikoh Tsukayama, Tetsuo Yamakawa and Shigeo Iraha, and former Professor Tasuo Okajima of the Nagoya Institute of Technology and the University of the Ryukyus, who also provided useful data on the experiments of concrete specimens under complex loading. The present paper was proofread in part by Dr. Carolyn Schaab.

REFERENCES

Chen W. F. and Saleeb A.F. (1982): Constitutive equations for engineering material, Vol.1, A Willey Publication, 580p.
Fung Y.C. 1994. A first course in continuum mechanics, Third Edition, Prentice Hall, 311p.
Nakaza E. 2005. Theory of deformation and motion of materials, Boarder Ink, 425p., in Japanese.
Nakaza E. 2009. A study on failure criteria of brittle concrete materials, Japanese concrete association, Proceedings of annual concrete engineering conference, Vol. 31, No.1, pp.481–486.
Nakaza E. 2010. New theory of elasticity, Boarder Ink, 97p., in Japanese.
Okajima T. 1970. Failure experiments on concrete specimens under complex loading (compression and shearing, tension and shearing), Journal of Japanese Architectural association, No.199, pp.1–8.
Timoshenko S.P. 1988. History of strength of material, Dover Publication, 451p.
Wu H.C. 1974). Dual failure criterion for plain concrete, Journal of the Engineering Mechanics Division, ASCE, Vol.100, No.EM6, pp.1167–1181.

2019 Rock Dynamics Summit– Aydan et al. (eds)
© 2019 Taylor & Francis Group, London, ISBN 978-0-367-34783-3

Measuring seismic properties of fine sediments in an off-Earth environment

M.A. Dello-Iacovo & S. Saydam
UNSW Sydney, Sydney, Australia

R.C. Anderson
Jet Propulsion Laboratory, Los Angeles, USA

ABSTRACT: This study expands upon an existing experiment to measure the seismic properties of off-Earth regolith simulants at varying atmospheric pressures. Reducing atmospheric pressure appeared to reduce the measured seismic velocity and increase the attenuation, and possible explanations for this are explored. The Australian Lunar Regolith Simulant (ALRS-1) was measured to have a P-wave velocity of 98.6 m/s using a piezoelectric transducer of with a source frequency of 54 kHz, and the Mars Mojave Simulant (MMS) dust was measured to have a P-wave velocity of 83.1 m/s, 61.3 m/s and 72.6 m/s using three different source frequencies (24 kHz, 54 kHz and 150 kHz respectively).

1 INTRODUCTION

Determining the structure and geomechanics of planetary bodies (e.g. the Moon, Mars and asteroids) is crucial for understanding the history of rocky bodies and for planning exploration, mining, colonization, and (for small planetary bodies, i.e. asteroids and comets) deflection. For example, SPBs may react differently to deflection measures based on their structural type (Gibbings & Vasile 2011). Understanding the subsurface of off-Earth bodies will assist in understanding the presence and distribution of potentially valuable resources, and knowledge of the rock strength will dictate what mining method should be used, if feasible at all.

Seismic exploration is a geophysical method that has seen widespread use in terrestrial exploration in the last century. It is proposed that seismic sources and receivers could be emplaced on planetary bodies to determine their subsurface properties, including structure, porosity and geomechanical strength. However, to date this technique has seen relatively little use off-Earth, and little testing has been performed to determine whether the feasibility and data quality of seismic is adversely affected by various planetary environments (Pike et al 1996), for example differing gravitational strength, atmospheric thickness, electrostatics and large temperature fluctuations.

In this study we will expand upon a novel method of determining the P-wave velocity of fine grained, low compaction regolith using active source piezoelectric transducers to measure it in low atmospheric pressure (Dello-Iacovo et al 2017). This work is amongst the first of its nature.

We will measure the P-wave velocity of two off-Earth regolith simulants (ALRS-1 and MMS dust), seek to understand the impact of atmospheric pressure on seismic data collection, and to begin to constrain the degree of wave attenuation in these conditions. This is critical information for the planning of any space exploration mission, but will also be useful for planning Earth-based seismic surveys and experiments, and will advance the understanding of how fine-grained, weakly compacted sediments act when exposed to seismic waves.

2 SEISMOLOGY AND GEOMECHANICS

Seismic experiments involve the use of sound (or mechanical) waves which travel though some medium and are recorded by a receiver. The source can either be passive (e.g. earthquakes or meteorite impacts) or active (e.g. explosive or vibrating source). The travel time can be used to determine the seismic velocity of the material or subsurface, and with enough receiver and source locations a 2D or 3D model of the subsurface can be developed.

There have been few seismic missions performed in off-Earth environments. Several missions were undertaken on the Moon during manned missions in the 1970's (Knapmeyer & Weber 2015) and provided useful data on the lunar subsurface. The Viking 1 and 2 Mars landers, which arrived on Mars in 1976, were both equipped with seismic receivers, though did not receive any confirmed seismic events (Goins & Lazarewicz 1979). Current and future seismic missions to Mars include InSight and Mars 2020. Relatively little research on the seismic properties of uncompacted soils is available in the literature.

Seismic wave attenuation has been a pervasive problem in seismic studies of unconsolidated material (Purnell 1986), imposing upper and lower bounds on the measurement distance and the source/

receiving power respectively. Uneven pore space distribution and friction between grains are the primary causes of this phenomenon (Sherlock & Evans 2001). A low effective pressure will mean that the energy lost to friction will be high.

The seismic velocity of unconsolidated sediment is primarily influenced by sediment frame rigidity (Sherlock & Evans 2001), which is strongly controlled by grain shape and sorting. Density, porosity, chemical composition, grain size and permeability appear to have little to no effect (Talwani et al 1973; Bell & Shirley 1980).

It is known that different frequencies can travel at different velocities (known as velocity dispersion) however this phenomenon is poorly understood (Sherlock & Evans 2001). Liu & Nur (1996) have suggested that the first arrival, or first break, is typically the lowest frequency.

Vibration of loose sand can have significant impacts on grain packing and therefore velocity. One example described in Sherlock & Evans (2001) suggests that velocity of loose sand can change from 1600 m/s to 1730 m/s after vibration.

3 METHODOLOGY

The apparatus we use consists of a pair of interchangeable piezoelectric transducers (source and receiver) of various frequencies, which are connected to a Pundit PL-200 recording system. This system was initially outlined in Dello-Iacovo et al (2017). The primary seismic properties of interest in this experiment are P-wave velocity and attenuation. Tests performed at Earth's atmospheric pressure were performed in a steel box, while tests performed at a range of reduced atmospheric pressures were performed in a vacuum chamber.

3.1 Recording system

The Pundit PL-200 is connected to two interchangeable piezoelectric transducers, one of which acts as a source, the other as a receiver. The recording system allows for automatic or manual determination of seismic wave delay time based on waveform analysis. Combined with a manual measurement of the distance between the source and receiver, the seismic velocity can be determined. The recordable delay time ranges from $0.1 - 7930$ μs, and has a resolution of 0.1 μs (for < 793 μs) and 1 μs (for > 793 μs). All tests were performed with 200 V. The sources used in this experiment were all P-wave, with 24 kHz (P-24), 54 kHz (P-54) and 150 kHz (P-150).

3.2 Earth atmosphere tests

A 200 x 300 x 300 mm steel box with 1.2 mm thick steel walls is partially filled with a regolith simulant. A schematic of the basic set up is shown in

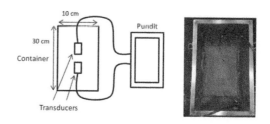

Figure 1. A schematic of the basic recording system set up for Earth atmosphere tests.

Figure 1. The source and receiver are half buried in the regolith, and the distance between them is measured from their closest surfaces. The initial placement of regolith and the source/receiver should be such that they are as close to the centre of the box as possible. Given that the metal of the box and the air have significantly higher seismic velocity than the soil being measured, this ensures that the first wave arrival will be one transmitted completely through the soil, not via any other medium. The source and receiver are then completely buried to a depth of around five cm.

The effect of soil compaction on velocity measurement is of great interest, however unfortunately, there is no available repeatable method of achieving a desired level of compaction. In this experiment, the regolith was either left uncompacted, or was compacted using a flat tool. There is no way to measure the level of compaction without influencing the compaction itself, and so some differences between measurements are to be expected. Care was taken to achieve a consistent degree of compaction between experiments.

3.3 Reduced atmosphere tests

The vacuum chamber and pump (Figure 2) allows for measurements to be taken using the Pundit system as above, but at different atmospheric pressures.

The pressure is measured in inches of mercury (in.Hg) relative to Earth sea level atmospheric pressure. A reading of 0 in.Hg indicates room pressure. A true reading of -30 in.Hg is analogous to vacuum. A reading of around -29.7 in.Hg is approximate to Mars atmosphere (1 % Earth atmosphere). The chamber and pump are capable of achieving pressures of around -29 in.Hg (~3 % Earth atmosphere), which is sufficient for our measurement purposes.

The source and receiver are buried in the regolith using the methodology outlined above. They are connected to the recording system, which sits outside the chamber, via the feedthrough ports. The velocity is also calculated in the way outlined above as the atmosphere in the chamber is pumped down from 100% Earth atmosphere to ~3% Earth atmosphere. Delay time measurements are taken at intervals as the atmosphere is pumped down. The chamber is held at a constant pressure prior to

Figure 2. Chamber and pump set up without Pundit recording system.

taking a measurement to ensure the pressure was reached in the pore spaces.

3.4 *Waveform analysis*

The part of the waveform we are most interested in is the first arrival of the wave. For an ideal measurement, the first arrival will be the first time at which the amplitude increases. However, there will be always be some noise present, and so the first arrival should be the point at which a consistent signal is first visible above the noise.

3.5 *Regolith simulants*

Two regolith simulants were used for these experiments, which have been designed based on known and predicted geochemical and geological properties of the Moon and Mars regolith.

The Australian Lunar Regolith Simulant 1 (ALRS-1) was developed by Garnock & Bernold (2012). The P-wave seismic velocity of the *in-situ* Lunar regolith was first compared with ALRS-1 by Dello-Iacovo et al (2017), however the geomechanical properties of the simulant such as grain shape have not been examined. Lunar regolith is known for its highly angular grains (Heiken et al 1991), and grain shape and sorting are known to be important factors in determining the velocity of a regolith (Sherlock & Evans 2001). It remains to be seen whether the ALRS-1 is representative of *in-situ* Lunar regolith in this way. ~100% of the ALRS-1 grains are smaller than 2 mm, and ~50% are smaller than 0.08 mm.

The Mojave Mars Simulant (MMS) was developed by Peters et al (2008). The samples were created by mechanically crushing boulders, which was intended to recreate the weathering processes on Mars. Four distinct MMS samples are available with distinct grain sizes; dust, fine, medium and

coarse. The MMS dust is used in these experiments. ~100% of the dust grains are smaller than 0.1 mm, and ~50% are smaller than 0.02 mm.

4 RESULTS

The results of the seismic experiments on the ALRS-1 are shown in Figure 3. These measurements were made with the P-54 kHz source, in regular atmospheric conditions and with no compaction. The mean velocity measurement is 98.6 m/s.

The results of the experiments on the MMS dust are shown in Figures 4-8. Of these, Figures 4-6 show the velocity of individual measurements in regular atmospheric conditions with the P-24, P-54 and P-150 kHz sources respectively. The mean velocity measurements are 83.1 m/s, 61.3 m/s and 72.6 m/s respectively.

Figures 7-8 show tests using the vacuum chamber using the P-54 kHz source). Each figure represents an individual run, where a run consists of pumping down the atmospheric pressure from Earth atmosphere to near-Mars atmosphere.

Figure 7 further shows the test run of P-wave velocity of air at different atmospheric pressures. Figure 8 shows five runs in total; the P-wave velocity of the MMS dust with a source-receiver separation distance of 0.6 cm with no compaction (1 run),

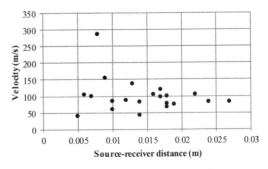

Figure 3. Source-receiver distance vs velocity of ALRS-1 using P-54. Of 23 measurements, 2 had no signal. Mean velocity of 98.6 m/s.

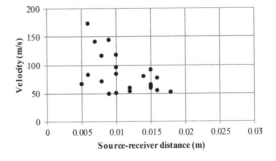

Figure 4. Source-receiver distance vs velocity of MMS dust using P-24. Of 22 measurements, all had signal. Mean velocity of 83.1 m/s.

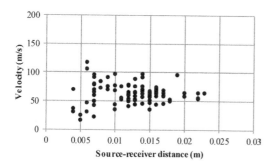

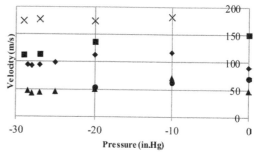

Figure 5. Source-receiver distance vs velocity of MMS dust using P-54. Of 93 measurements, all had signal. Mean velocity of 61.3 m/s.

Figure 8. Chamber pressure vs velocity of MMS dust using P-54 for five different runs. Triangle is 0.6 cm separation with no compaction, circle is 0.8 cm separation with no compaction, the diamond and square are 0.8 cm separation with compaction, and the cross is at 1.5 cm separation with compaction.

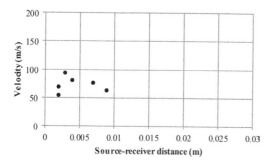

Figure 6. Source-receiver distance vs velocity of MMS dust using P-150. Of 9 measurements, 3 had no signal. Mean velocity of 72.6 m/s.

5 DISCUSSION

The P-wave velocity of the ALRS-1 (98.6 m/s) is comparable to the two available *in-situ* measurements of Lunar regolith from Lunar missions (104 m/s & 114 m/s; Kovach & Watkins 1973). This gives some confidence in the accuracy of the geomechanical and seismic properties of ALRS-1 (Dello-Iacovo et al 2017), however as discussed above, the grain properties of the regolith simulant should be further examined to ensure this result is not a coincidence.

The MMS dust has similar P-wave velocities to the ALRS-1, however no *in-situ* measurements of Mars regolith P-wave velocity are available for comparison. Once *in-situ* measurements are made in the future, this will provide an interesting point of reference.

Previous studies have stated that grain size does not appear to influence velocity of sediments, however the ALRS-1 and MMS dust both have velocities that are uncommonly low compared to terrestrial sediments. Given that the most common factor between the simulants appears to be their small grain size, this leads us to consider the possibility of such a relationship.

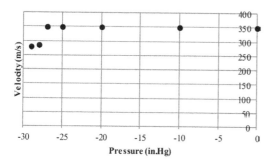

Figure 7. Chamber pressure vs velocity of air using P-54.

0.8 cm with no compaction (1 run) and heavy compaction (2 runs), and 1.5 cm with heavy compaction (1 run).

Some of measurements were unable to produce a signal that could be interpreted above the noise level. No P-wave signal was measured from the ALRS-1 at source-receiver separation distances of greater than 2.7 cm, and no signal was measured from the MMS dust at distances of greater than 2.3 cm. The maximum separation distance at which a signal could be resolved at near-Mars atmosphere was 0.6 cm without compaction, and 1.5 cm with compaction.

From Figures 3-6 we can see that the scatter in the individual measurements is reduced at greater source-receiver separation distances, with the highest and lowest velocities approaching the mean velocity. The nature of the experiment is such that it is non-reproducible. The configuration of the regolith grains are different each time, and thus some variation between measurements over small distances should be expected. At larger distances, e.g. over several metres or more, the small variations in the regolith grain properties should average out, resulting in a more consistent seismic velocity measurement centred on the mean.

The P-wave velocity of the MMS dust differs between different frequencies, which may be explained by velocity dispersion. However, given that the mean velocities of the 24 kHz, 54 kHz and 150 kHz were 83.1 m/s, 61.3 m/s and 72.6 m/s respectively, it would seem that the trend is not just increasing or decreasing with increasing frequency, which seems unusual, and no explanation for this can currently be devised.

As an additional calibration test, the P-wave velocity of air was measured in the vacuum chamber as the atmospheric pressure was pumped down, however this yielded an interesting result. At 0 in.Hg (Earth surface atmospheric pressure), the measured air velocity was 348 m/s, similar to the known mean velocity of 343 m/s. This remained constant as the pressure was pumped down, until -28 in.Hg (6.7% of Earth surface atmospheric pressure), where it rapidly dropped to 283.3 m/s.

It is assumed that the P-wave velocity of air would trend towards zero as the atmospheric pressure approaches -30 in.Hg (approaching true vacuum). It was also assumed that this relationship would be somewhat linear, however this doesn't appear to be the case.

From Figure 8, the P-wave velocity of the MMS dust appears to decrease slightly in general as Mars atmosphere (-29.67 in.Hg) is approached, although there are some exceptions to this. The data are limited, although there are some tentative conclusions we can draw at this time.

If the velocity being measured were purely that of the MMS dust, we shouldn't expect any difference in P-wave velocity as we reduce the atmospheric pressure. Three explanations are provided which may explain this phenomenon, although more research is required to determine which of them, if any, are taking place.

Since the velocity of air is several times higher than that of the regoliths being measured, it is possible that the seismic wave may be partially or completely travelling via the air itself between the grains, increasing the mean velocity. A diagram of how this may take place is depicted in Figure 9.

The coupling, or contact between the source/receiver and the medium being investigated, is known to be a pervasive problem for seismic experiments (Jefferson et al 1998). In particular, this has been observed in geophone receivers that are placed in loose surface regolith. Due to the highly fine grained and loose nature of the regolith simulants being examined, it is possible that the air was providing additional contact between the source/receiver and the regolith. As the atmospheric pressure is lowered, the degree of coupling provided by the air is reduced. This may reduce the apparent seismic velocity, though it seems unlikely to have a major impact, given that the coupling zone would be <1 mm, and the source-receiver separation distances being measured are on the order of 1 cm. If this phenomenon were a major contributor to the observed velocity trend, we should expect its effect to be weaker with increased source-receiver separation. Currently, there is not enough data to determine whether this is the case.

As discussed previously, small vibrations of loose regolith may perturb the grains, thus compacting

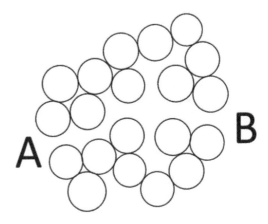

Figure 9. A proposed model for how the measured velocity of sediment might be higher than its true velocity. For a grain velocity lower than air velocity, the fastest point from A to B would be via the air pathway produced by permeability. Given enough of these scenarios in a sample, the measured velocity of the sample could approach that of air.

them slightly and having a non-trivial effect on the velocity. It may be the case that the air provides some weak pore pressure which, once removed, causes the grains to shift and collapse. The flow of air through the pore spaces as it is pumped out of the vacuum chamber may also create some grain movement, thus having a similar effect.

The degree of seismic wave attenuation in the ALRS-1 and MMS dust has not yet been quantified. We can however observe that it appears quite high. It is suspected that this high attenuation is due to the particularly fine grained and uncompacted nature of the regoliths being examined.

Approaching Mars atmosphere also appeared to have a notable effect on signal attenuation. Quantifying the influence of grain size, compaction, source power and source frequency on signal attenuation in future experiments will be enlightening. If coupling is affected by atmospheric pressure, this may explain some of the attenuation.

It is unclear whether the primary issue with resolving the signal is that the signal has completely attenuated, or that the signal to noise ratio is simply too low. It may be worthwhile examining whether generating multiple waveforms for the same measurement and summing their amplitude them can resolve some of the issues with signal strength. Noise filtering processes may also aid in resolving the signal.

From the limited data, it appears that the effect of compaction is to significantly increase the velocity. However, as we have seen from the more extensive measurements outside of the vacuum chamber, the degree of scatter in individual measurements is substantial, and we should gather further data before concluding this too strongly.

6 CONCLUSIONS

In this study, we have extended the methodology of Dello-Iacovo (et al 2017) to measure the seismic properties of regoliths at varying atmospheric pressures. The reduced atmospheric pressure appears to reduce the seismic velocity of the regolith measured. Three possible explanations for this phenomenon are:

- The seismic wave travels partially via the air in pore spaces
- The coupling between the source/receiver and the sediment is affected by the lack of air
- The removal of the air affects the grain packing, and therefore the velocity

The attenuation of the seismic waves also appears to be greatly increased as vacuum is approached.

The MMS dust has a P-wave seismic velocity of 83.1 m/s, 61.3 m/s and 72.6 m/s when measured with 24 kHz, 54 kHz and 150 kHz sources respectively. These results are similar to the fine grained ALRS-1 (98.6 m/s), which was in turn similar to the measured P-wave velocity on the Moon (104 and 114 m/s), prompting a reexamination of the theory that grain size of a sediment does not affect its seismic velocity.

This research will assist future studies of regoliths in off-Earth conditions, including current and future missions to other planetary bodies.

REFERENCES

Bell, D.W. and Shirley, D.J. 1980, Temperature variation of the acoustical properties of laboratory sediments, *Journal of the Acoustical Society of America*. 68: 227–231.

Dello-Iacovo, M., Anderson, R.C. and Saydam, S. 2017. A novel method of measuring seismic velocity in off-Earth conditions: Implications for future research. *48th Lunar and Planetary Science Conference*, The Woodlands, Texas.

Heiken, G. H., Vaniman, D. T., and French, B. M., eds. 1991. Lunar sourcebook, Cambridge University Press, Cambridge, UK.

Garnock, B. and Bernold, L. 2012. Experimental study of hollow-core beams made with waterless concrete. *13th ASCE Aerospace Division Conference on Engineering, Science, Construction, and Operations in Challenging Environments*, Pasadena, California, United States, 15-18 April.

Gibbings, A. and Vasile, M. 2011. A smart cloud approach to asteroid deflection. *62nd Int. Astron. Congress*, IAC-11.A3.4.

Goins, N.R. and Lazarewicz, A.R. 1979. Martian seismicity. *Geophysical Research Letters* 6: 368-370.

Jefferson, R.D., Steeples, D.W., Black, R.A. and Carr, T. 1998. Effects of soil-moisture content on shallow-seismic data. Geophysics. 63: 1357–1362.

Knapmeyer, M., and R. C. Weber 2015, Seismicity and interior structure of the Moon, in Extraterrestrial Seismology, edited by V. Tong and R. Garcia, Cambridge Univ. Press, Cambridge, U. K.

Kovach, R.L. and Watkins, J.S. 1973. Apollo 17 seismic profiling: Probing the Lunar crust. *Science*. 180: 1063–1064.

Liu, X., and Nur, A. 1996, A new experimental method for studying velocity dispersion in rocks (abs.), *Society of Exploration Geophysicists 66th International Meeting*, Expanded Abstracts. 2: 1683–1686.

Peters, G.H. et al 2008. Mojave Mars Simulant – Characterization of a new geologic Mars analog. *Icarus* 197: 470–479.

Pike, W.T., Martin, R.D., Kaiser, W.J. and Banerdt, W.B. 1996. Development of microseismometers for space applications, *Ann. Geophys*. 14: C828.

Sherlock, D.H. and Evans, B.J. 2001, The development of seismic reflection sandbox modelling, *AAPH Bulletin* 85: 1645–1659.

Talwani, P., Nur, A. and Kovak, R.L. 1973, Compressional and shear wave velocities in granular materials to 2.5 kilobars, *Journal of Geophysical Research* 78: 6899–6909.

Walker, J.D., Huebner, W.F., Chocron, S., Gray, W., & Boice, D., 2009, Active seismology of asteroids through impact and/or blast loading, White Paper for Primitive Bodies Decadal Survey.

2019 Rock Dynamics Summit– Aydan et al. (eds)
© 2019 Taylor & Francis Group, London, ISBN 978-0-367-34783-3

Effect of different parameters on post-peak response of rocks

A. Taheri
The University of Adelaide, Australia

H. Munoz
Golder Associates, Adelaide, Australia

ABSTRACT: The mechanism which controls violent/non-violent rock failure is studied through an extensive experimental study through uniaxial and triaxial compression loading. The tests were undertaken under quasi-static and post-peak cyclic loading. Axial displacement control system can be employed to measure the post-peak response of the rocks only if the rock demonstrates class I behaviour. Lateral strain control loading system, however, is applicable to measure true post-peak rock behaviour when they exhibit class II behaviour. It was observed that post-peak behaviour of rocks is an intrinsic property and independent of loading condition. In the triaxial compressive testing when the ratio between confining pressure to the unconfined compressive strength increases, rock post-peak behaviour shifts from class II behaviour to class I behaviour. Rock samples with a higher aspect ratio were observed to behave more brittle. The effect of scale on the post-peak response of rock, however, is found to be insignificant.

1 INTRODUCTION

Rocks in the deep underground are subjected to high in-situ stresses. Therefore, rock behaviour becomes complicated in response to extraction activities of deep earth resources. Rock behaviour and failure in the deep underground is not well understood, leading to a flaw in engineering designs. Inadvertent violent rock failure at large scales translated into injuries to personnel, loss of lives, production delay, damage to equipment and mine infrastructure, mine closure and lose of the entire mineral reserves exploitation are significant consequences of the lack of proper knowledge in rock failure behaviour. Rock violent failure is a major threat to the future exploitation of deep mining resources all around the world (Akdag et al. 2018).

A Proper characterisation of the total process of rock deformation in pre-peak and post-peak regimes, associated with the decrease of rock load-carrying capacity is critical to many civil engineering applications, mining development projects and mineral exploration operations. Rock brittle fracture damage, detrimental to the stability of surface and underground rock excavations, is amongst major practical applications (Munoz et al. 2016a). In this manner, pioneering studies on the complete stress-strain behavior of rocks undergoing quasi-static compression classify rocks into two categories: class I, characterized by a negative post-peak slope, where fracture propagation is controllable, and class II, showing positive post-peak slope, where fracture propagation is uncontrollable (Hudson et al. 1971, Fairhurst & Hudson 1999).

In this study, an extensive experimental study is carried out to investigate the post-peak response of rocks having a wide range of strength properties under uniaxial and triaxial compressive loading condition. Effects of various parameters on post-peak behaviour of rocks including the sample aspect ratio, sample size (scale) and loading type (i.e. monotonic and cyclic loading) were investigated.

2 MEASUREMENT OF COMPLETE STRESS-STRAIN RESPONSE

2.1 *Post-peak behaviour of rocks*

Rock behaviour under axial loading is generally studied in the laboratory using load-controlled or displacement-controlled compressive loading systems. Load-controlled method, however, can only measure pre-peak stress-strain relations. In most of the cases, axial-displacement (or axial-strain) rate is controlled during compression using a servo-controlled compressive machine to measure stress-strain behaviour of rocks (Bieniawski & Bernede 1979, Kumar et al. 2010, Taheri & Tani 2008, Taheri et al. 2016). Axial-displacement feedback, however, is insufficient to accurately measure the post-peak regime for class II rocks (Munoz et al. 2016b). The reason is axial strain in the compressive tests no longer increases monotonically from the moment the rock starts behaving as class II. As a result, a critical response on the rock stress-strain curve takes place leading to eruptive rock failure. Consequently, after the peak point, the stress-strain behaviour demonstrates a rapid strength reduction at a constant axial displacement or in an uncontrolled manner. On the other hand, during lateral-strain controlled method, the lateral displacement increases monotonically before and after the peak stress even if when the specimen shows a snapback behaviour. Therefore, the lateral strain rate could be controlled during a compressive test to measure the post-peak behaviour of brittle rocks.

2.2 Sample preparation

Ten different rocks having Unconfined Compressive Strength (UCS) values ranging from 0.7 to 215 MPa were prepared for this study. A series of uniaxial and triaxial compressive tests under quasi-static monotonic loading and cyclic loading conditions were carried out on different rock samples including a brown coal (Victoria coal), five limestones (Tuffeau, Savonniere, Massangis, Chassagne and Rocheron), two sandstones (Hawkesbury and Gosford) and two granites (Alvand and Harcourt).

The samples were cut to 42 mm in diameter and each 100 mm in height. The diameter of each specimen was at least 20 times larger than the largest grain in the samples (Bieniawski and Bernede, 1979). The aspect ratio was maintained to be 2.4 which complies with the ISRM recommendations.

The study also aimed to investigate the influence of aspect ratio on post-peak stress-strain characteristics of rocks. To do so, a number of the Hawkesbury sandstone (UCS=34 MPa) samples were prepared to have aspect ratios (i.e. the length to diameter ratio) equal to 1.5, 2.4 and 3.5 and diameter of 42 mm. In addition, to investigate the influence of sample scale on the post-peak response, Hawkesbury sandstone specimens with an aspect ratio of 2.4 and diameters of 19 mm, 30 mm, 42 mm and 63 mm, are prepared.

2.3 Experimental set-up

A closed-loop servo controlled testing machine with a loading capacity of 1000 kN was used to apply an axial load by an inbuilt computer system as a function of load or displacement. Axial deformation of the specimens was measured externally by a pair of axial linear variable displacement transducers (LVDTs) mounted at both left and right sides of the specimen. Axial strains were calculated from LVDTs readings.

For uniaxial compressive testing, the rock specimens were instrumented by a direct-contact lateral extensometer (i.e. lateral ring-shape extensometer) which was wrapped around the cylindrical sample. The lateral displacement was measured by the lateral extensometer which was fed back into the computer. The computer program then adjusts the load on the specimen in order to keep the lateral displacement at a constant rate of 0.16 mm/min to apply monotonic loading.

For the triaxial compression loading tests, a conventional Hoek cell was upgraded. Firstly, the membrane was instrumented by a four-strain gauge arrangement. This arrangement was located in the mid-high of the membrane so to measure the average lateral strain along the perimeter at mid-length of the rock specimens. The gauges are wired to a Wheatstone bridge system that connects to a computer system via a feedback loop. This feedback loop is used to control the axial loading. This feedback loop is used to control the axial loading. Lateral strains, ε_L and lateral strain rate, were obtained from the change in diameter of the rocks samples (and therefore membrane). Similar to the uniaxial compressive loading the loading rate of 0.16 mm/min was applied for monotonic loading.

Cyclic loading tests were performed in a damage-controlled fashion; i.e. in each test rock sample is sheared under compressive load under a constant lateral loading rate of 0.2×10^{-4}/s until the lateral strain from the beginning of loading in each cycle is reached to a prescribed value. Axial loading then was reversed under the same loading rate until the axial load is almost equal to 1 MPa. Cyclic loading continues until the sample demonestrates a complete failure. The total lateral strain value in each cycle was chosen considering the strength and the brittleness of each rock type.

3 POST -PEAK BEHAVIOUR OF ROCKS

3.1 Stress-strain relations

The complete stress-strain curve either in uniaxial and triaxial compression conditions becomes fundamental to describe the rock failure behaviour. Figure 1 shows the typical stress-strain results under uniaxial loading condition. The results show that, in general, post-peak behaviour of different rocks follows class II or a combination of class I and class II. Even Victoria coal which is a very soft rock partially demonstrates a class II behaviour. This rock is the only rock that exhibits a residual strength in uniaxial loading after a class I-class II softening behaviour. As it can be seen in the figure, the post-peak regime is demonstrated by progressively dropping and minor recoveries, if any, of the load-carrying capacity accompanied with crack propagation process and localisation.

3.2 Effect of confining pressure

Figure 2 shows the stress-strain curves for Chassagne limestone (UCS= 123 MPa) and Hawkesbury sandstone (UCS= 34 MPa). As it may be seen in this figure, depending on the amount of confining pressure, the sample may demonstrate a combination of class I and class II behaviour or solely class II behaviour at different extent when confining pressure is low. In general, rock failure behaviour becomes more ductile with an increase in confining pressure. This trend significantly depends on the ratio of uniaxial

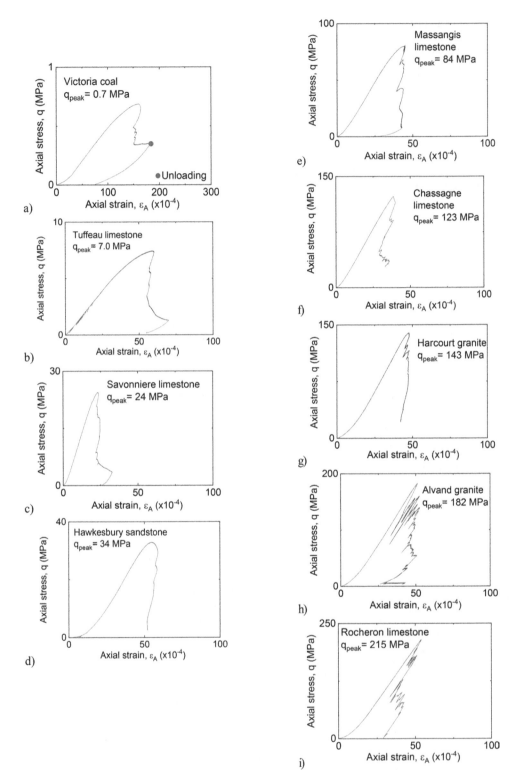

Figure 1. Complete stress-strain relations for different rocks

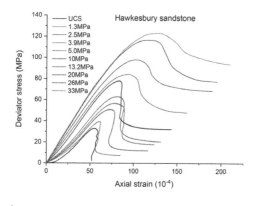

a)

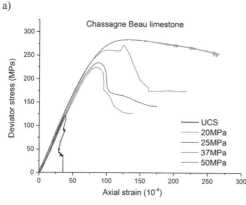

b)

Figure 2. Stress-strain relations at different confining pressures for a) Hawkesbury sandstone and b) Chassagne limestone

compressive strength (UCS) of rock to the amount of confining pressure. Figure 2a clearly shows this trend. In triaxial compressive tests undertaken on Hawkesbury sandstone sample, post-peak behaviour is class II when confining pressure is equal to 1.3 and 2.5 MPa. However, when confining pressure increases to 10 MPa, failure behaviour becomes mainly class I. When confining pressure is 13.2 MPa and higher, rock behaviour becomes entirely ductile.

3.3 Effect of cyclic loading

To investigate the effect of cyclic loading on overall post-peak behavior of samples, in Figure 3 results of three cyclic loading tests on Gosford sandstone (UCS=30 MPa) Tuffea limestone (UCS=7 MPa) and Massangis limestone (UCS=80 MPa) samples are compares with results of monotonic tests, as normalized axial stress- axial strain curves. As it may be seen in this figure, the overall post-peak behaviour is almost similar for each rock in monotonic and cyclic post-peak loading. Therefore, post-peak cyclic loading does not cause any extra damage and,

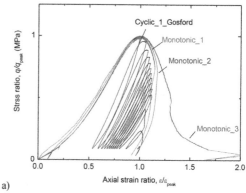

a)

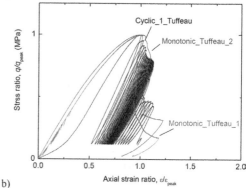

b)

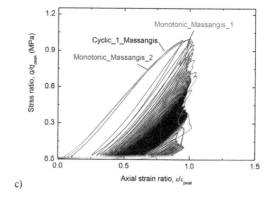

c)

Figure 3. Axial stress-strain relations during post-peak monotonic tests and cyclic loading tests for (a) Gosford sandstone; (b) Tuffea limestone and (c) Massangis limestone.

therefore, doesn't demonstrate a significant effect on the post-failure behaviour of rocks.

As explained earlier, Figure 3 shows that failure behaviour of different rocks during either monotonic or cyclic loading is a combination of class I and class II behaviour. For the case of Gosford sandstone and Tuffea limestone, in general, a progressive drop of stress is followed associated with progressive strength degradation. Whereas, for the case of Massangis limestone sudden drops and recoveries of the load-carrying capacity at different extent characterised post-peak stress-strain curves.

3.4 Effect of sample aspect ratio

Figure 4 shows the normalised stress-axial strain curves for Hawkesbury specimens with length to diameter, L/D, of 1.5, 2.4 and 3.5. The stress-axial strain curves were normalised by dividing the stresses and strains by the peak stress and the axial strain at peak stress of their respective L/D. The results show that the post-peak stress-strain characteristics are dependent upon sample height when other testing parameters are the same. This is mainly due to the degree of confinement by frictional stress on the specimens which influences the characteristics of failure and, therefore, the post-peak stress-strain curves.

Figure 4 shows that rock brittleness increases with an increase in the specimen aspect ratio if brittleness is defined by the slope of the stress-axial strain post peak curve. Post-peak stress-strain results showed that specimens with L/D= 1.5 followed a predominant class I behaviour throughout from the beginning of shearing until complete failure. On the other hand, specimens with L/D= 2.4 and 3.5 showed a combination of class I and class II behaviour. Therefore, the extent of external confinement in the samples, due to specimen-platens frictional stresses, plays a significant role on the failure pattern and therefore on the stress-strain behaviour.

The pre-peak curves in Figure 4 for different aspect ratios almost fall into a unique curve irrespective of the specimen height. In pre-peak regime, it is expected to obtain a single normalised stress-strain curve, and the peak stress to be independent of the specimen length, if specimen-platens frictional stress is entirely eliminated by a useful friction-reduction layer inserted between the specimens and the loading platens.

3.5 Effect of sample size

Figure 5, presents the complete stress-strain relations of uniaxial compressive tests of Hawkesbury sandstone samples having different diameters. The stress-strain curves are normalised to investigate the dependency of scale effects on the pre-peak and post-peak behaviour. The normalised stress-strain curves were obtained by dividing the stresses and strains by peak stress and strain at failure. It can be seen that, in the pre-peak regime, the stress-strain patterns for most of the rock samples with are similar. In the post-peak regime, the stress-strain characteristics for samples with different diameters both showed a combination of class I and class II behaviour. The post-peak stress-strain curves almost overlapped with each other, and their minor variations do not show a conclusive trend. This results indicate that the scale effect on the post-peak stress-strain characteristics is negligible.

4 CONCLUSIONS

Post-peak stress-strain relations of different rocks types having UCS values ranging from 0.7 to 215 MPa were investigated under quasi-static and cyclic uniaxial and triaxial compression loading. Post-peak behaviour of different rocks under quasi-static uniaxial compressive loading is either a combination of class I-II or class II behaviour. Rocks in triaxial loading condition demonstrated similar post-peak trend when confining pressure is low. Predominantly class I behaviour throughout post-peak regime was only observed at relatively high confinement pressure as compared to rock strength. In addition, cyclic post-peak loading did not make any noticeable change of the overall post-peak behaviour of rocks. Therefore, post-peak behaviour was found to be an intrinsic rock behaviour independent of loading condition.

Sample aspect ratio largely influenced the post-peak behaviour of rocks. Rocks demonstrate more

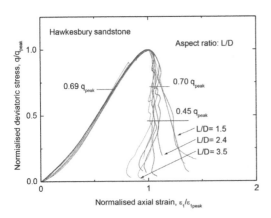

Figure 4. Normalised stress-strain curves of Hawkesbury sandstone samples with different aspect ratios.

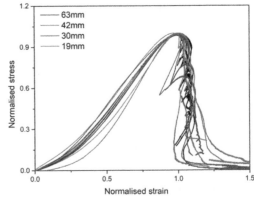

Figure 5. Normalised stress-strain curves of Hawkesbury sandstone samples with different diameters.

brittle behaviour with an increase in aspect ratio. Effect of sample size (i.e. diameter) on post-peak behaviour was found to be insignificant.

REFERENCES

Akdag, S., Karakus, M., Taheri, A., Nguyen, G. & He, M. 2018. Effects of thermal damage on strain burst mechanism for brittle rocks under true-triaxial loading conditions. *Rock Mechanics and Rock Engineering* 51(6): 1657–1682.

Bieniawski, Z. T. & Bernede, M.J. 1979. Suggested methods for determining the uniaxial compressive strength and deformability of rock materials: Part 1. Suggested method for determination of the uniaxial compressive strength of rock materials. *International Journal of Rock Mechanics and Mining Sciences & Geomechanics Abstracts* 16:137.

Fairhurst, C. E. & Hudson, J. A. 1999. Draft ISRM suggested method for the complete stress-stress curve for intact rock in uniaxial compression. *International Journal of Rock Mechanics and Mining Sciences* 36: 279–289.

Kumar, R., Sharma, K. G. & Varadarajan, A. 2010. Post-peak response of some metamorphic rocks of India under high confining pressures. *International Journal of Rock Mechanics and Mining Sciences* 47:1357–1362.

Hudson, J. A., Brown, E. T. & Fairhurst, C. 1971. Optimizing the control of rock failure in servo-controlled laboratory tests. *Rock mechanics* 3, 217–224.

Munoz, H., Taheri, A. & Chanda, E. 2016a. Pre-peak and post-peak rock strain characteristics during uniaxial compression by 3D Digital Image Correlation. *Rock Mechanics and Rock Engineering* 49(7): 2541–2554.

Munoz, H., Taheri, A. & Chanda, E. 2016b. Fracture energy-based brittleness index development and brittleness quantification by pre-peak strength parameters in rock uniaxial compression. *Rock Mechanics and Rock Engineering* 49(12): 4587–4606

Taheri, A. & Tani, K. 2008. Use of down-hole triaxial apparatus to estimate the mechanical properties of heterogeneous mudstone. *International Journal of Rock Mechanics and Mining Sciences* 45:1390–1402.

Taheri A, Yfantidis, N., Olivares, C. L., Connelly, B. J. & Bastian, T.J. 2016. Experimental study on degradation of mechanical properties of sandstone under different cyclic Loadings. *Geotechnical Testing Journal* 39(4): 673–687

2019 Rock Dynamics Summit– Aydan et al. (eds)
© 2019 Taylor & Francis Group, London, ISBN 978-0-367-34783-3

Ultrasonic wave properties of weathered sandstones in Khorat Group and their factors affecting

W. Sukplum & L. Wannakao
Department of Geotechnology, Faculty of Technology, Khon Kaen University, Khon Kaen, Thailand

ABSTRACT: This research intends to use the ultrasonic properties of the rocks to classify weathering grade of sandstones and to correlate their physical and mechanical properties. Sandstone specimens were collected from Nam Phong (NP) and Phra Wihan (PW) Formations, Khorat Group, Northeast Thailand. Three weathering grades were categorized as fresh (I), slightly weathered (II) and moderately weathered (III). Petrofabric of both NP and PW sandstones at each degree of weathering were observed. Physical properties and uniaxial compressive strength tests were also conducted in the laboratory. Both ultrasonic compressional (Vp) and shear (Vs) wave velocities of the rocks were measured. The results of the research indicated that the ultrasonic wave velocities decrease as increasing of degree of weathering and porosity. On the contrary, the wave velocities increase as increasing percentage of quartz, density and uniaxial compressive strength.

1 INTRODUCTION

The strength and other engineering properties of rock materials were usually reduced when the rocks are weathered. In generally, weathering is the process which changes the properties of rock mass due to changing the environment conditions such as physical, chemical and biological. Khorat Group, the non-marine red-beds sandstone, forms mostly in the Khorat Plateau, Northeast Thailand which is in tropical area (Figure 1). Sandstone is a sedimentary rock that composed of sand-sized grains of mineral, rock fragments, or organic material. It also contains the most common cementing materials that are silica and calcium carbonate; accordingly, it is very moisture-sensitive and easily weathered due to reaction with water. Thus, the authors interested in weathered sandstone of the Khorat Group. This Group was named by Ward & Bunnag (1964) who subdivided the group into 9 Formations, namely from older to younger sequences (Triassic to Cretaceous) as, Haui Hin Lat, Nam Phong, Phu Kradung, Phra Wihan, Sao Khua, Phu Phan, Khok Kruat, Mahasarakarm and Phu Thok Formations. The rock samples in this research were taken from Nam Phong and Phra Wihan Formations that Nam Phong Formation deposits in Triassic and Phra Wihan Formation deposits in Cretaceous. The Nam Phong Formation has a type section along the Nam Phong stream and distributes along with the Loei-Phetchabun Fold Belt and the western regime of the Khorat Plateau. The formation consists of resistant red-brown micaceous sandstone, siltstones, mudstones, and conglomerates of fluviatile origin (Chonglakmani 2011). The Phra Wihan Formation is widely distributed, occurring in the Phu Phan Rang and around the western and particularly the southern rims of the Khorat Plateau. Generally,

the rock consists of light buff to gray, fine to coarse-grained quartzitic sandstones and rare siltstones and mudstones with occasional conglomerates (Meesook 2011).

The weathering classification for rock material is differently defined by different researchers, but the most widely used it's in engineering implication that given in Table 1. It has adapted from many

Figure 1. The Khorat Plateau showing distribution of Khorat Group.

researchers (Dearman et al. 1978; ISRM 1981; BS 5930 1999; Hencher et al. 1995; Price 1995; Anon 1995; Santi 2006). Most weathering classification or degree of weathering of rock mass is suggested by International Society for Rock Mechanics (ISRM). Due to the fact that the effect of weathering changes the engineering properties of the rock mass, engineering and geotechnical engineering try to correlate the mechanism or process of weathering with the engineering properties in rock through laboratory tests. Many researchers have been reported the correlations between physical and mechanical properties for different types of rocks. Irfan & Dearman (1978) carried out the relationship between physical properties on the seven stages of weathered granite from east Cornwall, England. Tugrul & Gurpinar (1997) reported the effect of weathering processes on the physical and mechanical properties of basalt from Niksar ring road and Ünye-Akkuş-Niksar highway alignments, Turkey. Tugrul & Zarif (1998) analyzed the influence of weathering grade on mechanical properties of sandstone in Istanbul, Turkey. Basu & Ghosh (2012) categorized the weathering grades and assessed Brazillian tensile strength of quartzitic material in Turamdih region, India and Tating et al. (2013) established the relationship of the reduction of intact rock strength for thick to very-thick bedded sandstone with exposure time in the rock mass for the SST unit of the Crocker formation in Sabah, Malaysia. As mention previously, the weathering grades of rock have been correlated significantly with physical and mechanical properties that completely destroy the rock samples during the test. Therefore, the ultrasonic testing techniques have been applied in the evaluation of rocks properties because they are non-destructive and easy to apply. In rock mechanics, these techniques are regularly used for determination of the dynamic elastic constants of rocks. In addition, these techniques have been used to evaluate the engineering properties of rocks such as to correlate the static and dynamic Young's moduli of rock samples under unconfined and confined stresses (Mavko & Jizba 1994; Chang et al. 2006; Casper et al. 2008; Wanakao et al. 2010; Sukplum et al. 2014; Sukplum 2016). Furthermore, researchers are also applied these techniques to correlation between sonic and seismic wave velocities on sedimentary rock reservoirs (Mantilla 2002; Assefa et al. 2003; Nelson 2010). With knowledge about the P- and S-wave velocities and the density, for instance from well logs or seismics, it should thus in principle be a simple job to obtain the elastic moduli even if they do not have the possibility to perform rock mechanical tests. In reality, this is not quite so simple. Many researchers have been correlated the ultrasonic wave properties with weathering grades for different types of rocks, such as Lednicka & Kalab (2012) for granite, Heidari et al. (2013) for granitic rocks, Boudani et al. (2015) for marble, Momeni et al. (2015) for granitoid rocks and Sajid et al. 2016

Table 1. Weathering grade system for sandstones.

Grade or Degree	Description
I Fresh	Unchanged from original state; no evident micro fracturing; slight discoloration on major discontinuity surfaces
II Slightly weathered	Slight discoloration, slight weakening; weathering penetrates through most discontinuities
III Moderately weathered	Considerably weakened, penetrative discoloration; large pieces cannot be broken by hand
IV Highly weathered	Significantly weaker than the fresh rock; easily breakable by hand
V Completely weathered	Original texture is present; all micro fractures tend to be open; most of strength of fresh rock
VI Residual soil	Soil derived by in situ weathering but retaining none of original texture

for granite. Naturally, the ultrasonic waves velocity depends on the physical and mechanical properties of the rock (such as density, porosity, elastic constants, and structure), Therefore, correlating the changes in these properties with the ultrasonic wave velocity helps to provide classification schemes for evaluating the degree of rock deterioration.

The main objective of the research is to determine weathering grades by using the ultrasonic wave properties of the sandstones. Two sandstone Formations are selected for this study, namely, Nam Phong and Phra Wihan Formations due to their homogeneous. Physical and strength properties are also tested. Their mineralogy and textural of rock are also determined in order to establish relationships to the ultrasonic wave properties.

2 EXPERIMENTS

2.1 Tested rock material and experimental procedures

The sandstones block samples were collected from natural exposures. Their weathering grades were assigned by using the set of four factors (straining, grain boundary, relative strength, and texture). Three weathering grades were found (Table 2). The core specimens were drilled, cut and lapped as specified by the ASTM D4543 standard practice with nominal dimension of 5.45 cm. (NX) in diameter and 10.90-13.63 cm. in length (L/D ratio is 2-2.5). There were 12, 21 and 6 NP sandstone specimens and 5, 22 and 9 PW sandstone specimens, for weathering grade I, II and III, respectively. All specimens were labeled as to their rock Formation (NP and PW). Then, the core specimens were tested the hardness by Schmidt rebound hammer test method following the standard method (ISRM 1981; ASTM C805/C805M) to ensure consistent and reliable values and

Table 2. Weathering classification for the sandstones block sample.

NP	PW	Weathering grade	Schmidt hammer value (R_L)
		I	33-37 (NP)
			39-46 (PW)
		II	28-34 (NP)
			24-37 (PW)
		III	18-30 (NP)
			21-29 (PW)

Table 3. Quantity of sandstone core samples and physical properties.

Formation/ Grade	Quantity	Dry density kg/m³	Porosity (%)
NP/I	12	2423-2605	2.12-5.56
NP/II	21	2341-2598	3.27-6.76
NP/III	6	2314-2346	5.72-7.29
PW/I	5	2445-2493	0.53-1.57
PW/II	22	2237-2384	4.28-6.94
PW/III	9	2141-2384	5.22-7.54

Table 4. Mineral compositions sandstones as determined from optical microscope.

Formation/ Grade	Average minerals (%)					
	Q	WQ	F	WF	R	Other
NP/I	68.0	2.0	20.5	5.5	3.0	1.0
NP/II	62.0	5.0	11.5	13.0	7.5	1.0
NP/III	50.5	9.5	13.5	12.5	10.5	3.5
PW/I	92.5	6.5	-	-	1.0	-
PW/II	84.0	14.5	-	-	1.5	-
PW/III	73.0	25.0	-	-	2.0	-

Remarks: Q=Quartz, QW= Weathered quartz, F=Feldspar, FW= Weathered Feldspar and R= Rock fragment

reproducible correlations for the weathering grades of samples. The 'L' type hammer which is low impact energy rebound hammer was used for this test. The rebound hammer was applied vertically downward at least four different points on each core specimen. The reading obtained from the Schmidt Hammer can provide an approximate indication of the weathering grades of the rock sample by means of rebound hardness number (Aydin & Basu 2005; Basu & Ghosh 2012). Range of the rebound values (R_L) are also summarized in Table 2. The physical properties (dry density and porosity) were obtained from all cylindrical specimens with the aspect ratio (length/diameter) of 2–2.25. Additionally, the representative thin sections of 3 weathering grades were prepared for the petrographic examination.

2.2 Physical properties

Physical properties and mineral composition of the specimens were also determined. Two physical properties, density and porosity were evaluated for the specimens. These properties were conducted by water replacement method following the testing standard of ISRM 1981 and their values are tabulated in Table 3. The mineral compositions of the rock samples were evaluated by polarized microscope methods. The results of the mineral composition are shown in Table 4. Nam Phong Formation can be classified as Arkosic sandstone. Specimens from Phra Wihan Formation are mainly composed of quartz 98-99% with very few rock fragments, for it can be classified as Quartz Arenite sandstone. The mineral compositions at different weathering grades were depicted in Figure 2.

2.3 Experimental procedures

All specimens were tested under dry condition. First, ultrasonic wave velocities were measured by using the GCTS© ULT-100 P&S Ultrasonic

Figure 2.Mineral compositions at different weathering grade (a)-(c) Nam Phong Formation and (d)-(f) Phra Wihan Formtion.

Velocities Measurement System. The zero time and wave form was first calibrated by generating pulses between the transmitter and receiver without a specimen. The specimen is placed between the P and S wave transmitter and receiver, then the ultrasonic pulse is generated. The travel time of P wave (t_p) is the first arrival signal, while the S wave travel time (t_s) is determined from the phase difference of the wave form. The compressional and shear wave velocities (Vp and Vs) are determined from the length of the core specimens (L) and the travel times as in Equation 1 and 2. The representative of Vp and Vs was illustrated in Figure 3.

$$Vp = \left(\frac{L}{t_p} \right) \qquad (1)$$

$$Vs = \left(\frac{L}{t_s} \right) \qquad (2)$$

After that, all specimens were tested under an unconfined compressive strength condition. The test procedure of unconfined compressive strength followed the testing standard of ASTM D7012. Ultrasonic wave velocities (Vp and Vs), and unconfined compressive strength (UCS) are presented in Table 5.

3 RESULTS AND DISCUSSIONS

In order to describe the important relationships between physical properties and weathering grade of the sandstones studied, dry density, porosity, hardness from Schmidt hammer and the main mineral composition have been plotted against weathering grade as shown in Figure 4. As expected, dry density, Schmidt rebound hammer value and quartz mineral decrease with higher weathering grades while porosity increases. Those are similar results on different rocks types from many researchers.

Figure 3. The representative of ultrasonic wave velocities Vp (Left) and Vs (Right).

Table 5. Ultrasonic wave properties and unconfined compressive strength.

Formation/ Grade	UCS (MPa)	Ultrasonic wavevelocities Vp/Vs		
		Vp (m/s)	Vs (m/s)	
NP/I	101-149	3822-4451	2411-2950	1.54
NP/II	89-94	3038-3588	1992-2477	1.49
NP/III	78-84	2172-2968	1495-1990	1.45
PW/I	101-108	4393-5396	2891-3316	1.53
PW/II	38-44	2411-3826	1758-2680	1.47
PW/III	27-30	1159-2239	773-1500	1.47

Ultrasonic wave velocities are plotted relative to weathering grades and wave velocities ratios (Vp/Vs) are also plotted in Figures 5-6, respectively. Figure 5a and 5c illustrate that ultrasonic wave velocities gradually decrease when weathering grades increase. The use of the exponential line is required to optimize the correlation with these properties. Coefficients of determination (R^2) from 2 Formations are best fit-curves. Figure 5b and 5d illustrate the mean value of Vp and Vs which are marked by markers on the rang of standard deviation bars. Ultrasonic wave velocities ratios (Vp/Vs) of Nam Phong Formation (Figure 6a) are 1.54, 1.49, 1.45 for weathering grades (WG) I, II and III, respectively while the ratios of Phra Wihan Formation (Figure 6b) are 1.53, 1.47 and 1.47, respectively. It indicates that if the weathering grades increase, the ratios will slightly decrease.

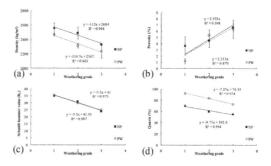

Figure 4. Relationships of weathering grade with (a) dry density, (b) porosity, (c) Schmidt hammer value and (d) mineral composition (quartz).

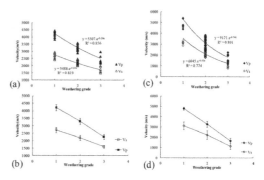

Figure 5. Relationship between weathering grade and ultrasonic wave velocities (a)-(b) Nam Phong Formation and (c)-(d) Phra Wihan Formtion.

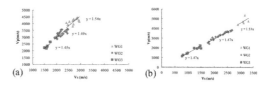

Figure 6. Ultrasonic wave velocities ratio for (a) Nam Phong Formation and (b) Phra Wihan Formtion.

Unconfined compressive strengths relative to weathering grades and Schmidt rebound hammer value are plotted in Figure 7. As seen in Figure 7a, the strength of rock decreases rapidly with higher weathering grades. The correlation between the unconfined compressive strength and Schmidt rebound hardness value respectively for three different grades is presented in Figure 7b. The best fit regression line was demonstrated by exponential correlation with the best coefficient of determination.

Factors affecting of physical properties, hardness, strength and mineral composition of weathered sandstone samples on ultrasonic wave properties were illustrated in Figures 8-9. Figure 8 shows the relationships of wave velocities with percentages of quartz mineral. It is clear that wave velocities are significantly affected by the percentages of quartz mineral. Figure 9 shows the correlation between porosity, density and wave velocities. It can be summarized that if the percentage of porosity increases, wave velocities will significantly decrease. On the other hand, when dry density increases, wave

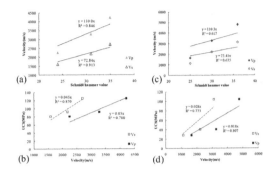

Figure 10. The correlation between wave velocities and Schmidt rebound hammer, unconfined compressive strength (a)-(b) Nam Phong Formation, (c)-(d) Phra Wihan Formation.

velocities also increase. Figure 10 shows the effect of hardness and strength on wave velocities. The best linear correlation clearly shows that an increase of hardness value and unconfined compressive strength of weathered sandstones make ultrasonic wave velocities increased.

4 CONCLUSIONS

The research presented here was undertaken to determine the correlations between weathering grades and ultrasonic wave properties of sandstones from Nam Phong Formation and Phra Wihan Formation. The physical properties, mineral composition, hardness value and strength of samples were also determined to display how ultrasonic wave velocities affected by these properties. The weathering grades of rocks could be determined by ultrasonic velocity and wave velocity ratio. An empirical mathematic correlation between ultrasonic wave velocity and uniaxial compressive strength should be used for certain sandstone that comprises similar mineral compositions.

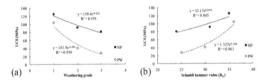

Figure 7. Relationships of unconfined compressive strengths to (a) weathering grades and (b) Schmidt rebound hammer value.

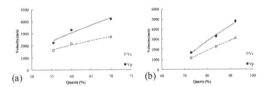

Figure 8. Relationships between percentages of quartz and wave velocities (a) Nam Phong Formation, (b) Phra Wihan Formation.

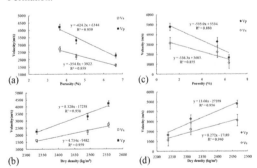

Figure 9. Relationships of wave velocities to physical properties; (a)-(b) Nam Phong Formation, (c)-(d) Phra Wihan Formation.

ACKNOWLEDGEMENTS

The authors would like to express a sincere thanks to the Department of Geotechnology, Faculty of Technology, Khon Kaen University for facilitating the study in the Geomechanics Laboratory. The authors also thanks to the assistance in laboratory provided by Ms. Kedkanok Thaothip, Ms. Ornsiri Thasidum and Mr. Waranchit Chaba.

REFERENCES

Anon. 1995. The description and classification of weathered rock for engineering purposes, Working Party Report. *Quarterly Journal of Engineering Geology* 28: 207–242.

Assefa, S. McCann, C. & Sothcott, J. 2003. Velocities of compressional and shear waves in limestones. *Geophysical Prospecting* 51: 1–13.

ASTM C805/C805M-13a. 2013. Standard Test Method for Rebound Number of Hardened Concrete. West Conshohocken, Pennsylvania: ASTM International.

ASTM D4543-08e1. 2008. Standard Practices for Preparing Rock Core as Cylindrical Test Specimens and Verifying Conformance to Dimensional and Shape Tolerances. West Conshohocken, Pennsylvania: ASTM International.

ASTM D7012-14e1. 2014. Standard Test Methods for Compressive Strength and Elastic Moduli of Intact Rock Core Specimens under Varying States of Stress and Temperatures. West Conshohocken, Pennsylvania: ASTM International.

Aydin, A. & Basu, A. 2005. The Schmidt hammer in rock material characterization. *Engineering Geology* 81(1): 1–14.

Basu, A. & Ghosh, N. 2012. Categorizing weathering grades of quartzitic materials and assessing Brazilian tensile strength with reference to assigned grades. *International Journal of Rock Mechanics and Mining Sciences* 49: 148–155.

BS 5930. 1999. British Standard Code of practice for site investigations. British Standard Institution: London. 1–192.

Boudani, M.E. Wilkie-Chancelliera, N. Martineza, L. Hébertb, R. Rollandc, O. Forstc S. Vergès-Belmind, V. & Serfatya S. 2015. Marble characterization by ultrasonic methods. *Procedia Earth and Planetary Science* 15: 249–256.

Casper, O. Helle, F.C. Ida, L.F. 2008. Static and dynamic Young's moduli of chalk from the North Sea. *GEOPHYSICS* 73 (2): 41–50.

Chang, C. Zoback, D. & Khaksar, A. 2006. Empirical relations between rock strength and physical properties in sedimentary rocks. *Journal of Petroleum Science and Engineering* 51(3-4): 223–237.

Chonglakmani, C. 2011. Triassi. In M.F. Ridd A.J. Barber & M.J. Crow (eds), *The Geology of Thailand*: 137–150. London: The Geological Society.

Dearman, W.R. Baynes, F.J. & Irfan, T.Y. 1978. Engineering grading of weathered granite. *Engineering Geology* 12: 345–374.

Heidari, M. Momeni, A.A. & Naseri, F. 2013. New weathering classifications for granitic rocks based on geomechanical parameters. *Engineering Geology* 166: 65–73.

Hencher, S.R. & McNicholl, D.P. 1995. Engineering in weathered rock. *Quarterly Journal of Engineering Geology and Hydrogeology* 28: 253–266.

Irfan, T.Y. & Dearman, W.R. 1978. Engineering classification and index properties of a weathered granite. *Bulletin of the International Association of Engineering Geology* 17(1): 79–90.

ISRM. 1981. Rock Characterization Testing and Monitoring. In E. Brown (ed). Oxford: Pergamon Press.

Lednicka, M. & Kalab, Z. 2012. Evaluation of granite weathering in the Jeronym mine using non-destructive methods. *Acta Geodynamica et Geromaterialia* 9(2): 211–220.

Mantilla, A.E. 2010. *Predicting petrophysical properties by simultaneous inversion of seismic and reservoir engineering data*. [Doctor of Philosophy Dissertation in Geophysics]. Stanford: Stanford University.

Mavko, G. & Jizba, D. 1994. The relation between seismic P-wave and S-wave velocity dispersion in saturated rocks. *GEOPHYSICS* 59(1): 87–92.

Meesook, A. 2011. Cretaceous. In M.F. Ridd A.J. Barber & M.J. Crow (eds), *The Geology of Thailand*: 169–184.The Geology of Thailand. London: The Geological Society

Momeni, A.A. Khanlari, G.R. Heidari, M. Sepahi, A.A. & Bazvand E. 2015. New engineering geological weathering classifications for granitoid rocks. *Engineering Geology* 185: 43–51.

Nelson, P.H. 2010. *Sonic Velocity and Other Petrophysical Properties of Source Rocks of Cody, Mowry, Shell Creek, and Thermopolis Shales, Bighorn Basin, Wyoming, USA*. Virginia: Geological Survey.

Price, D.G. 1995. Weathering and weathering process. *Quarterly Journal of Engineering Geology and Hydrogeology* 28: 529–252.

Sajid, M. Coggan, J. Arif, M. Andersen, J. & Rollinso, G. 2016. Petrographic features as an effective indicator for the variation in strength of granites. *Engineering Geology* 202: 44–54.

Santi, P.M. 2006. Field methods for characterizing weak rock for engineering. *Environmental & Engineering Geoscience* 12(1): 1–11.

Sukplum, W. Wannakao, L. Chanasuek, P. & Sonlukjai, P. 2014. Static and dynamic elastic properties of the Phu Kradung sandstone at variation applied stress levels. *International Society for Rock Mechanics and Rock Engineering. ISRM International Symposium - 8th Asian Rock Mechanics Symposium, 14-16 October 2014*. Sapporo: Japan.

Sukplum, W. 2016. *Fatigue behavior and ultrasonic wave properties under confining pressures of the Phu Kradung sandstone*. [Doctor of Philosophy Dissertation in Geotechnology]. Khon Kaen: Khon Kaen University.

Tating, F. Hack, R. & Jetten, V. 2013. Engineering aspects and time effects of rapid deterioration of sandstone in the tropical environment of Sabah, Malaysia. *Engineering Geology* 159:20–30.

Tugrul, A. & Zarif, I.H. 1998. The influence of mineralogical textural and chemical characteristics on the durability of selected sandstones in Istanbul, Turkey. *Bulletin of Engineering Geology and the Environment* 57(2): 185–190.

Tugrul, A. & Gurpinar, O. 1997. The effect of chemical weathering on the engineering properties of Eocene basalts in northeastern Turkey. *Environmental and Engineering Geoscience* 3(2): 225–234.

Wannakao, L. Wannakao, P. & Yungme, W. 2010. *The use of ultrasonic in evaluating engineering properties: Phra Wihan, Phu Phan, Nam Phong Sandstones and Permian Limestone. Research Report, No.1*. Khon Kaen: Department of Geotechnology, Faculty of Technology, Khon Kaen University.

Ward D.E. & Bunnag D. 1964. *Stratigraphy of the Mesozoic Khorat Group in Northeast Thailand. Report of Investigation, No.6*. Bangkok: Department of Mineral Resource.

2019 Rock Dynamics Summit– Aydan et al. (eds)
© *2019 Taylor & Francis Group, London, ISBN 978-0-367-34783-3*

Dynamic fracture properties of rocks subjected to static pre-load and hydrostatic confining pressure

Wei Yao
Department of Civil and Mineral Engineering, University of Toronto, Toronto, Ontario, Canada
State Key Laboratory of Hydraulics and Mountain River Engineering, College of Water Resource & Hydropower,
Sichuan University, Chengdu, China

Kaiwen Xia
Department of Civil and Mineral Engineering, University of Toronto, Toronto, Ontario, Canada

Rong Chen
College of Science, National University of Defense Technology, Changsha, Hunan, P.R. China

ABSTRACT: Dynamic fracture failure of rocks subjected to static hydrostatic confining pressure is commonly encountered in deep underground rock engineering. The static fracture behavior of rocks under hydrostatic stress has been well studied in the literature. However, it is desirable to investigate the dynamic fracture failure of rocks under various pre-loads and hydrostatic confining pressures. In this study, a split Hopkinson pressure bar (SHPB) system is modified to measure the dynamic fracture toughness of rocks under three pre-loads and five hydrostatic confining pressures. Pulse shaping technique was used in all dynamic tests to facilitate dynamic force equilibrium in the specimen. Three groups of NSCB specimen are tested under a pre-load of 0, 37 and 74% of the maximum static load and five groups of NSCB specimens are tested under a hydrostatic confining pressure of 0 MPa, 5 MPa, 10 MPa, 15 MPa and 20 MPa. The results show that under a given pre-load or a certain hydrostatic confining pressure, the dynamic fracture toughness of rock increases with the loading rate, resembling the typical rate dependence of materials. Furthermore, the dynamic rock fracture toughness decreases with the static pre-load at a given loading rate, implying that the dynamic load-bearing capacity of the engineering structure is reduced under static pre-load. However, the dynamic fracture toughness increases with the hydrostatic confining pressure at the similar loading rate due to the closure of microcracks in rocks, indicating that dynamic loading capacity of rocks is significantly enhanced under static hydrostatic confining pressure. Empirical equations are proposed to represent the effect of loading rate and pre-load force/hydrostatic confining pressure on the rock dynamic fracture toughness, and the results show that these equations can depict the trend of the experimental data well. The experimental results are of great significance to underground engineering design and assessment.

1 INTRODUCTION

With the further development of the mining industry and the increasing depth of underground engineering, rocks are naturally subjected to dynamic impact (such as blasting and seismicity) under static pre-load or static hydrostatic confining pressure. For example, in underground mining processes, tectonic stress and gravity are typical static pre-loads and hydrostatic confining pressures which act on underground rock engineering structures. In addition, these structures may be exposed to dynamic loads due to production blasting nearby or rock bursts or earthquakes. Under these circumstances, the dynamic response of rock subjected to static pre-load or hydrostatic confining pressure is crucial to the safety of workers, equipment and engineering facilities (Xia and Yao 2015).

There are also many unique phenomena observed in underground engineering activities that cannot be explained by means of existing rock mechanics theories. Especially for rocks at depths of 1000 meters and

more below the ground surface, abnormal phenomena (e.g. rockburst, zonal disintegration, core discing) become more prominent, leading to an enormous challenge to the traditional rock mechanics (Qian 2004; Zhou et al. 2005). Traditional fracture criteria and strength theories are not applicable when the rocks are under the dynamic load superimposed with static pre-load or hydrostatic confining pressure because the measured total rock strength is different from the summation of the rock dynamic strength and the static pre-load or hydrostatic confining pressure (Li et al. 2008a). Therefore, it is necessary to investigate the dynamic properties of rocks subjected to static pre-load and hydrostatic confining pressure. Since 2002, there have been many investigations on the issue of rock strength for rocks subjected the dynamic load superposed with static pre-load (Gong et al. 2010; Li et al. 2008a; Li et al. 2008b; Wu et al. 2015; Zhou et al. 2014) and with hydrostatic confining pressure (Li et al. 2017; Wu et al. 2016).

Despite of recent efforts and results on the effect of pre-load and hydrostatic confining pressure on the dynamic mechanical responses of rocks, the influence of pre-load and hydrostatic confining pressure on the dynamic fracture properties and the crack propagation process remains unknown. It is thus the objective of this paper to fill in this gap. In this paper, a modified split Hopkinson pressure bar (SHPB) test system was used to carry out dynamic tests on the rock specimens under pre-stress conditions and hydrostatic confining conditions. The notched semi-circular bend (NSCB) specimens were selected to investigate the fracture toughness. The NSCB method was used to study of the dynamic fracture initiation toughness and some other fracture parameters by researchers (Chen et al. 2009; Dai et al. 2010; Xia et al. 2013). The merits of the NSCB method include: (a) a short specimen is convenient for the achievement of the dynamic force balance; (b) one only needs to align the two supporting points on the transmitted bar side, thus facilitating specimen alignment. The NSCB method was therefore chosen as the ISRM suggested method for quantification of rock dynamic fracture toughness (Zhou et al. 2012).

This paper is organized as follows. After the introduction, Section 2 introduces the specimen preparation and measurement principles. The results and discussions are detailed in Section 3. Section 4 summarizes the entire paper.

2 METHODOLOGY

2.1 Specimen preparation

Two granites were chosen in this study: Heshuo granite (HG) (is from the Northeast region of Heshuo, Xinjiang province, China) and Laurentian granite (LG) (is from the Laurentian region of Quebec City in Canada). HG was used to conduct the dynamic NSCB tests with static pre-load conditions and LG was used to conduct the NSCB tests with static hydrostatic confining conditions. Figure 1 shows the schematic of a NSCB specimen in the SHPB system.

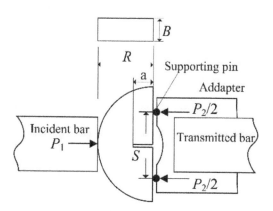

Figure 1. Schematic of NSCB specimen in the SHPB system.

For the dynamic fracture test with the pre-load, the specimen radius R is 24 mm, the thickness B is 18 mm, the length of the notched crack a is 4 mm, and the span of the supporting pins S is 30 mm. For the dynamic fracture test with the hydrostatic confining pressure, the specimen radius R is 25 mm, the thickness B is 25 mm, the length of the notched crack a is 5 mm, and the span of the supporting pins S is 27.5 mm.

The NSCB specimens were cored with a nominal diameter, and then sliced into disc specimens with a nominal thickness. By diametrical cutting, half disc specimens were subsequently made from the full discs. The NSCB specimens in this study were prepared with a 1 mm wide notch firstly and then sharpened at the tip by a diamond wire saw to achieve a tip width of 0.5 mm. The roughness variation for all surfaces of the NSCB specimen is less than 0.04 mm (Zhou et al. 2012).

2.2 The SHPB system with pre-load system

Figure 2. shows the modified system for the dynamic test under static pre-load based on the SHPB test platform. The entire system consists of an axial compressive loading unit and a traditional SHPB test platform. The axial compressive unit is connected to the hydraulic control system through the high-pressure pipeline. The hydraulic confinement is computer controlled so that the pre-stress can be controlled.

At the beginning of the experiment, the rigid mass and the hydraulic cylinder were fixed on the supporting platform (as shown in Figure 2), and the hydraulic cylinder was then pressurized to the desired value. The pressure was transmitted to the bar systems because the rigid mass constrains the leftwards axial motion of the bars through a flange attached to the incident bar (Figure 2). However the motion of the bars associated with impact is rightwards and thus the flange does not affect stress wave propagation. When the axial pressure reached the desired value, the striker bar was launched to apply dynamic loading by the impact on the free end of the incident bar. The incident pulse propagates along the incident bar before it hits the specimen, leading to a reflected stress wave and a transmitted stress wave that are recorded by the strain gauges attached on the incident and transmitted bar surfaces.

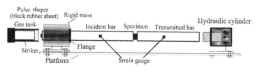

Figure 2. Configuration of the loading system for dynamic tests under pre-load.

2.3 The SHPB system with hydrostatic confining system

The triaxial SHPB system consists of a traditional SHPB system (a striker bar, an incident bar, a transmitted bar and a recording system) and a hydrostatic loading system (Figure 3). The hydrostatic loading system is mainly composed of a pressure vessel that applies lateral confinement pressure to the rock specimens (Cylinder 1) and a pressure chamber that provides axial pressure to the bars and specimens (Cylinder 2). The relative motion between the Cylinder 2 and the rigid mass is restricted by two tie-rods and the leftward axial motion of the bars is constrained by the rigid mass through a flange on the incident bar (Figure 3).

In addition, the NSCB specimen is sandwiched between the incident bar and the transmitted bar. Hence, the axial pressure from Cylinder 2 is applied to the contact areas between the NSCB specimen and two bars (i.e., as shown in Figure 4, σ_1 is the axial pressure from the transmitted bar and σ_2 is the axial pressure from the incident bar.) due to the constrained leftward motion of the incident bar. The remaining parts of the NSCB specimen are surrounded by the hydraulic oil in Cylinder 1, and the oil pressure σ_0 in Cylinder 1 is thus acted on the remaining parts of the NSCB specimen. In addition, the Cylinder 1 is connected to the same hydraulic pressure control system with Cylinder 2. In this case, the pressures in both two cylinders are equal to the oil pressure σ_0 (Wu et al. 2016). According to the force balance of the transmitted bar (Figure 4),

$$\sigma_0(A_b - A_s) + \sigma_1 A_s = \sigma_0 A_b \tag{1}$$

where A_b is the cross-sectional area of the bars, A_s is the contact area between the specimen and the transmitted bar, σ_1 is the stress of the specimen at the transmitted bar end. By solving Equation (1), we can get

$$\sigma_1 = \sigma_0 \tag{2}$$

Similarly, we can obtain

$$\sigma_0 = \sigma_1 = \sigma_2 \tag{3}$$

where σ_2 is the stress of the specimen at the incident bar end (Figure 4). Therefore, the NSCB specimen is under hydrostatic stress state when the pressures in both cylinders are identical (Figure 4).

2.4 Measurement principles

When the desired pre-load or hydrostatic pressure is reached, the impact of striker on the left end of the incident bar generates the incident stress wave ε_i in the incident bar. The stress wave propagation is not affected by the flange since the impact motion of the bars is rightward and the mass of the flange is quite small (Wu et al. 2015; Wu et al. 2016). A reflected stress wave ε_r and a transmitted stress wave ε_t are produced at the interface between the specimen and the bars. These three waves are captured by the strain gauges on the incident and transmitted bar surfaces and recorded by a digital oscilloscope from the Wheatstone bridge after amplification (Figure 5a). As shown in Figure 5a, the solid line represents the original signal of the semi-sinusoidal incident wave and reflected wave and the dash line delineates the original signal of the transmitted wave. It is noteworthy that, with the Wheatstone bridge circuit, only the change of the strain could be measured when the oscilloscope channels for the strain gauges are set to alternating current (AC) coupling in our tests. Namely, the strain caused by the static pre-compression in the transmitted bar is not recorded in the signal from the strain gauges during the dynamic tests. Consequently, the baseline of the transmitted pulse aligns with zero. Furthermore, the pulse shaper placed at the free end of the incident bar (Figures 2 and 3) was applied to achieve the dynamic force balance, which is the prerequisite of valid dynamic NSCB tests (Zhou et al. 2012).

The dynamic force balance condition on two ends of the NSCB specimen (Figure 1) can be written as: $P_1(t) \approx P_2(t)$, where $P_1(t) = A_b E[\varepsilon_i(t) + \varepsilon_r(t)]$ and $P_2(t) = A_b E \varepsilon_t(t)$, A_b and E are the cross-sectional area and Young's modulus of the bars, respectively. Figure 5b shows these two forces in a typical dynamic NSCB test under a hydrostatic pressure of 5 MPa. P_1 and P_2 are approximately identical during the whole dynamic loading period, indicating that the dynamic force balance on both loading ends of the NSCB specimen is achieved. The dynamic force balance has been critically assessed for all dynamic NSCB tests in this study.

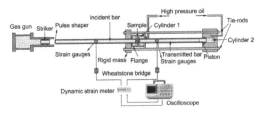

Figure 3. Configuration of the loading system for dynamic tests under hydrostatic confining pressure.

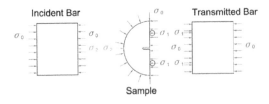

Figure 4. Hydrostatic stress of rock materials in a SHPB test.

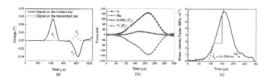

Figure 5. (a) Original signal captured from the strain gauges; (b) dynamic force balance and (c) loading rate determination in a typical NSCB-SHPB test.

The stress intensity factor (SIF) history $K_I(t)$ for mode-I fracture in the NSCB specimen under dynamic force balance can be determined by (Zhou et al. 2012):

$$K_I(t) = \frac{P(t)S}{BR^{3/2}} \cdot Y\left(\frac{a}{R}\right) \qquad (4)$$

where $P(t)$ is the time varying loading force in the specimen, which equals to the transmitted force $P_2(t)$ ($\approx P_1(t)$). $Y(a/R)$ is the dimensionless stress intensity factor and calculated via numerical methods as detailed in references (Chen et al. 2010; Zhou et al. 2012). The dynamic fracture toughness K_{Id} can be calculated with the maximum value of $K_I(t)$. A typical dynamic SIF-time curve in the NSCB specimen is shown in Figure 5c, in which there is an almost linear regime in the curve and the slope of this region (the dash line in Figure 5c) is defined as the loading rate for the NSCB-SHPB tests (Zhou et al. 2012).

3 RESULTS AND DISCUSSIONS

3.1 Failure pattern of the NSCB specimens

Figure 6 shows the typical recovered NSCB specimens after the dynamic fracture tests with the hydrostatic confining pressure and the pre-load. Based on the photos in Figure 6, these NSCB specimens in two types of tests were completely split into two quarter-discs from the notch tip,

3.2 Dynamic fracture toughness under pre-load

Figure 7 show the dynamic testing results of the NSCB specimens under three different pre-load conditions; the corresponding pre-loads in the NSCB specimen are 0%, 37%, and 74% of the critical static load, which is a critical external load on the NSCB specimen when the fracture occurs under static loading.

Figure 7a illustrates the dynamic fracture toughness K_{Id} versus the loading rate under different pre-loads. In this case, the loading force P used in Equation (4) for calculating K_{Id} is the dynamic force P. The tendency of K_{Id} represents the dynamic loading capacity of a specimen under a certain static load which increases with the loading rate. It is observed that under the same loading rate, the initiation fracture toughness decreases with the pre-load. In other words, the specimen's dynamic

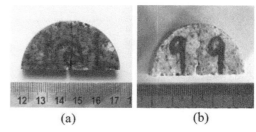

(a)   (b)

Figure 6. (a) a typical recovered NSCB specimen under hydrostatic confining pressure and (b) a typical recovered NSCB specimen under pre-load.

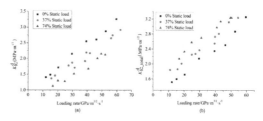

Figure 7. (a) The dynamic fracture toughness and (b) the total fracture toughness versus loading rate.

impact load capacity is reduced by the pre-load and the propagation of crack in the specimen can be initiated by a lower impact load.

Figure 7b shows the total fracture toughness K_{I_total} versus the loading rate. The loading force P used in Equation (4) for calculating K_{I_total} is the sum of the pre-load and the dynamic force, i.e. $P_{total} = P_{dynamic} + P_{pre}$. Consequently, K_{I_total} indicates the total loading capacity under superposed loading conditions. Compared with the results shown in Figure 7a, Figure 7b indicates that under the same dynamic loading rate, the total fracture toughness increases with the pre-load. It can also be seen that K_{I_total} is higher than K_{Id} under the same loading rate. Under the same loading rate, K_{Id} decreases with the increase of the pre-load, which means that pre-load weakens the rock material. However, K_{I_total} increases with the pre-load.

3.3 Empirical relations for dynamic fracture toughness under pre-load

The dynamic fracture toughness of rocks with pre-load can also be described as follows:

$$\frac{K_{Id}}{K_{IC}} = f_{Dp}(\frac{\dot{K}_I}{K_{IC}}, \frac{P_{pre}}{P_0}) = 1 - \frac{P_{pre}}{P_0}$$
$$+ \alpha(\frac{\dot{K}_I}{K_{IC}}) + \beta(\frac{\dot{K}_I}{K_{IC}})\frac{P_{pre}}{P_0} \qquad (5)$$

where P_0 is the critical static load force and P_{pre} is the pre-load force. When the specimen is loaded by the quasi-static condition, the parameter f_{Dp}=1- P_{pre}/P_0. For example, when the pre-load is 74% of the critical static load, the specimen under quasi-static

156

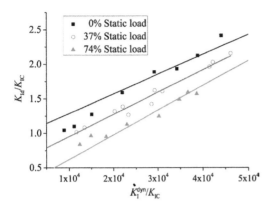

Figure 8. The normalized dynamic fracture toughness versus the normalized loading rate.

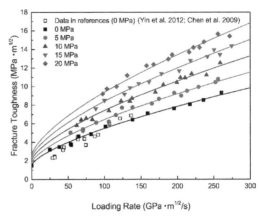

Figure 9. The dynamic fracture toughness versus loading rate for different hydrostatic conditions.

condition will be fractured when the additional quasi-static load is 26% of the critical static load. The experimental data fitting based on Equation (5) results in the values of the parameters $\alpha = 2.86\times10^{-5}$, and $\beta = 9.78\times10^{-6}$. As shown in Figure 8, the function fits well to the data points.

3.4 *Dynamic fracture toughness under hydrostatic confining pressure*

In this study, the NSCB specimens were tested under five different hydrostatic pressures: 0 MPa, 5 MPa, 10 MPa, 15 MPa and 20 MPa. Figure 9 illustrates the dynamic fracture toughness of LG under different loading rates and hydrostatic pressures. The fracture surfaces are subjected to the pressure since the notch itself is also exposed to the oil. Thus the data in Figure 9 give only the effect of hydrostatic pressure on the dynamic fracture toughness. It is obvious that dynamic fracture toughness increases with the loading rate, indicating the rate dependency of fracture toughness that prevails in rocks (Chen et al. 2009; Yao et al. 2017; Yin et al. 2012). Moreover, the dynamic fracture toughness of LG under the similar loading rate increases with the increase of the hydrostatic pressure. This phenomenon is consistent with the results reported in the references, in which the static fracture toughness of rocks increases with the hydrostatic pressures (Schmidt and Huddle 1977; Thiercelin 1987). Also, the increase of dynamic fracture toughness could in part be attributed to the closure of microcracks when the specimen bears the hydrostatic stress.

3.5 *Empirical relations for dynamic fracture toughness under hydrostatic confining pressure*

In addition, an empirical formula is proposed to describe the dynamic fracture toughness K_{Id} under various hydrostatic pressures and loading rates:

$$K_{Id} = \left(1+\alpha\left(\frac{P_{hydro}}{\sigma_t}\right)^{\beta}\right)\left(1+\gamma\left(\frac{\dot{K}_I}{\dot{K}_C}\right)^{n}\right)K_{IC} \qquad (6)$$

where P_{hydro} is the hydrostatic pressure, σ_t =12.8 MPa is the quasi-static tensile strength of LG (Xia et al. 2017), K_{IC} =1.5 MPa·m$^{1/2}$ is the quasi-static fracture toughness of LG (Nasseri and Mohanty 2008), $\dot{K}_I$ is the loading rate, $\dot{K}_C$ is the reference loading rate and equals 0.001 GPa·m$^{1/2}$/s, α, β, γ and n are data fitting constants, which are determined by using Genetic algorithm based on dynamic fracture toughness data in Figure 9. The first and second terms in Equation (6) represent the effect of hydrostatic pressure and loading rate, respectively. The fitting curves in Figure 9 reveal that the empirical formula predicts the data points well, where α=0.4496, β=1.0197, γ=0.0012, and n=0.6723.

In addition, both the data points and the fitting curves indicate that the effect of the hydrostatic pressure on the dynamic fracture toughness of LG is more significant under higher loading rates than that under lower loading rates, i.e., with the same increasing value of the hydrostatic pressure, the increment of the fracture toughness of LG under higher loading rates is larger than that at lower loading rates.

4 CONCLUSIONS

In this study, a modified SHPB test system was established to investigate the dynamic fracture behavior of rocks under static pre-load and static hydrostatic pressure condition. The NSCB method was applied to investigate the initiation fracture toughness. Experimental analyses indicate that the dynamic fracture toughness of rocks under pre-load is different from fracture toughness without pre-load. The test results demonstrate that the dynamic fracture toughness decreases with the pre-load, implying

157

that the dynamic load bearing capacity of engineering structure is relatively reduced under static preload, which is of great significance to underground engineering design and assessment. An empirical equation is used to simulate the effect of loading rate and pre-load force, and the results show that this equation fits well with the experimental data points.

Further, a triaxial SHPB system was utilized to measure the dynamic fracture toughness of rocks under various hydrostatic pressures. The NSCB specimens were tested under five different hydrostatic pressures: 0 MPa, 5 MPa, 10 MPa, 15 MPa and 20 MPa. The results show that dynamic fracture toughness under a certain hydrostatic pressure enhances with the increase of the loading rate. Meanwhile, an empirical formula was proposed to describe the influence of the loading rate and the hydrostatic pressure on dynamic fracture toughness, and successfully reproduced the trend of the experimental results.

REFERENCES

Chen, R., Guo, X., Lu, F., Xia, K. 2010. Research on dynamic fracture behaviors of stanstead granite. *Chin J Rock Mech Eng* 2(23):375–380.

Chen, R., Xia, K., Dai, F., Lu, F., Luo, S.N. 2009. Determination of dynamic fracture parameters using a semi-circular bend technique in split Hopkinson pressure bar testing. *Eng Fract Mech* 76(9):1268–1276.

Dai, F., Chen, R., Xia, K. 2010. A semi-circular bend technique for determining dynamic fracture toughness. *Exp Mech* 50(6):783–791.

Gong, F., Li, X., Liu, X. 2010. Experimental study of dynamic characteristics of sandstone under one-dimensional coupled static and dynamic loads. *Chin J Rock Mech Eng* 29(10):2076–2085.

Li, X. et al. 2017. Failure mechanism and coupled static-dynamic loading theory in deep hard rock mining: A review. *Journal of Rock Mechanics and Geotechnical Engineering* 9(4):767–782.

Li, X., Zhou, Z., Lok, T.-S., Hong, L., Yin, T. 2008a. Innovative testing technique of rock subjected to coupled static and dynamic loads. *Int J Rock Mech Min Sci* 45(5):739–748.

Li, X., Zhou, Z., Ye, Z., Ma, C., Zhao, F., Zuo, Y., Hong, L. 2008b. Study of rock mechanical characteristics under coupled static and dynamic loads. *Chin J Rock Mech Eng* 27(7):1387–1396.

Nasseri, M.H.B., Mohanty, B. 2008. Fracture toughness anisotropy in granitic rocks. *Int J Rock Mech Min Sci* 45(2):167–193.

Qian, Q. 2004. The characteristic scientific phenomena of engineering response to deep rock mass and the implication of deepness. *Journal of East China Institute of Technology* 27(1):1–5.

Schmidt, R.A., Huddle, C.W. 1977. Effect of confining pressure on fracture toughness of Indiana limestone. *Int J Rock Mech Min Sci Geomech Abst* 14(5):289–293.

Thiercelin, M. (1987) Fracture toughness under confining pressure using the modified ting test. Paper presented at the The 28th U.S. Symposium on Rock Mechanics (USRMS), Tucson, Arizona, 1987/1/1

Wu, B., Chen, R., Xia, K. 2015. Dynamic tensile failure of rocks under static pre-tension. *In J Rock Mech Min Sci* 80:12–18.

Wu, B.B., Yao, W., Xia, K.W. 2016. An Experimental Study of Dynamic Tensile Failure of Rocks Subjected to Hydrostatic Confinement. *Rock Mechanics and Rock Engineering* 49(10):3855–3864.

Xia, K., Huang, S., Dai, F. 2013. Evaluation of the frictional effect in dynamic notched semi-circular bend tests. *In J Rock Mech Min Sci* 62(9):148–151.

Xia, K., Yao, W. 2015. Dynamic rock tests using split Hopkinson (Kolsky) bar system – A review. *J Rock Mech Geotech Eng* 7(1):27–59.

Xia, K., Yao, W., Wu, B. 2017. Dynamic rock tensile strengths of Laurentian granite: experimental observation and micromechanical model. *J Rock Mech Geotech Eng* 9(1):116–124.

Yao, W., Xu, Y., Yu, C., Xia, K. 2017. A dynamic punch-through shear method for determining dynamic Mode II fracture toughness of rocks. *Eng Fract Mech* 176:161–177.

Yin, T., Li, X., Xia, K., Huang, S. 2012. Effect of thermal treatment on the dynamic fracture toughness of Laurentian granite. *Rock Mechanics and Rock Engineering* 45(6):1087–1094.

Zhou, H., Xie, H., Zuo, J. 2005. Developments in researches on mechanical behaviors of rocks under the condition of high ground pressure in the depths. *Adv Mech* 35(1):91–99.

Zhou, Y.X. et al. 2012. Suggested methods for determining the dynamic strength parameters and mode-I fracture toughness of rock materials. *In J Rock Mech Min Sci* 49(1):105–112.

Zhou, Z., Li, X., Zou, Y., Jiang, Y., Li, G. 2014. Dynamic Brazilian tests of granite under coupled static and dynamic loads. *Rock Mechanics and Rock Engineering* 47(2):495–505.

2019 Rock Dynamics Summit– Aydan et al. (eds)
© 2019 Taylor & Francis Group, London, ISBN 978-0-367-34783-3

Fatigue life characteristics of limestone in karst tunnel

C.S. Qiao
School of Civil Engineering, Beijing Jiaotong University, Beijing, China

ABSTRACT: Fatigue life prediction of limestone is the basis of long-term stability analysis of surrounding rock in karst tunnel under train cycle loading. Through the low cycle fatigue tests of karst limestone under different upper limit stress, stress amplitude, loading frequency and saturated condition, the basic fatigue characteristics of the limestone are experimented. The fatigue test results show that the fatigue life of rock will be significantly reduced after saturation and increases linearly with the decrease of upper limit stress. There is also a fatigue threshold for the limestone and the fatigue threshold value increases with the loading rate increasing. The stress amplitude plays a key role in the fatigue life of the limestone. The fatigue life of the limestone increases with the loading frequency increasing. The linear relationship between the fatigue life and loading frequency satisfies the double logarithmic coordinate. By introducing a correction coefficient, a fatigue life evaluation model based on load frequency and stress level is established.

1 INTRODUCTION

With the development of high-speed and heavy haul railways, effect of low-cycle fatigue damage of rock wall between karst cave and tunnel caused by the combined action of long-term train vibration cyclic load and the seismic dynamic loading is an important problem which restricts the long-term stability of such tunnels.

The fatigue damage law and fatigue life of limestone are the foundation of long-term stability analysis of surrounding rock of karst tunnel under the train vibration cyclic load and the seismic dynamic loading. However, in the previous research, the fatigue properties and prediction models of metal and concrete materials have been studied a lot, but the fatigue properties of rocks were seldom studied. Burdine (1963) found that the fatigue failure of Berea sandstone occurred within a certain number of cycles when the specimen reached a certain upper limit stress. Haimson & Kim (1972) proved that the fatigue life of the two marbles is closely related to the rock type and stress level. When the stress level is different, the fatigue failure cycles of Georgia marble and Tennessee marble are different. Attewell & Farmer (1973) studied the effect of loading frequency (0.3Hz, 2.5Hz, 10Hz, 20Hz) dolomite fatigue life. Ishizuka & Abe (1990) also proved that the fatigue life of rocks increases with the increase of loading frequency, and the fatigue strength of rocks under wet conditions is about 7% lower than that under dry conditions. Singh (1989) proved that the fatigue life of Australian miscellaneous sandstone increases with the decrease of stress amplitude, and the number of loading cycles increases logarithmically with the decrease of stress

amplitude. Zhen & Hai (1990) show that the deformation caused by sinusoidal wave loading is greater than that of triangular wave, and the longer the cyclic amplitude, the shorter the life of rock. Ge & Jiang (2003) found that the rock fatigue failure has a threshold value, threshold value corresponds to demarcation point of the linear and nonlinear.

As a fundamental research, this paper studies the fatigue characteristics and fatigue life of limestone in karst railway tunnel. Through the low cycle fatigue tests of karst limestone under different upper limit stress, stress amplitude, loading frequency and saturated condition, the basic fatigue characteristics of the limestone are revealed, and the fatigue life characteristics of the limestone are obtained. The fatigue life prediction model of the limestone is established by fitting the test results. The research results in this paper can provide scientific basis for the long-term stability analysis of karst tunnel under the train vibration cyclic load and the seismic dynamic loading and guide tunnel design.

2 TEST INTRODUCTIONS

2.1 Sample preparation and test equipment

The test rock samples were taken from near the entrance of Longlingong karst railway tunnel in Hubei province, China, and all the rock samples were fresh intact limestone block. The specimens were prepared into the standard samples with a diameter 50mm and a height 100mm as shown in Figure 1. The dynamic cyclic fatigue test of the limestone is carried out by the dynamic test system of MTS 810 electro hydraulic servo controlled as shown in Figure 2.

2.2 Deformation acquisition system and testing method

The deformation monitoring and data acquisition system is composed of two pairs of LVDT sensors (supplemented by strain gauges), Focus II high-speed synchronous data acquisition instrument and PC system as shown in Figure 2. The fatigue test is carried out in two stages by means of stress control such as Figure 3. The upper stress limit is σ_{max} and lower stress limit is σ_{min}, the stress amplitude is $\Delta\sigma=\sigma_{max}-\sigma_{min}$.

For convenience, the upper limit stress ratio (S) and lower limit stress ratio ($\overline{S}$) are defined as the ratio of upper and lower stress to rock uniaxial compressive strength (σ_c) respectively. In order to obtain the fatigue characteristics of the limestone, a series of fatigue tests were carried out under different upper limit stress, stress amplitude and loading frequency respectively. Loading waveform selects sine wave, the loading frequency (f)

Figure 1. Limestone samples.

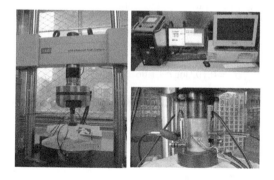

Figure 2. Test equipment and measuring devices.

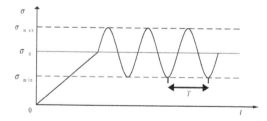

Figure 3. Schematic diagram of two stages loading for cyclic fatigue test.

used in the test is 0.5 ~ 3.0Hz, and the corresponding train operation speed is 40 ~ 250km/h. The vibration frequency of tunnel surrounding rock caused by earthquake is generally low, and this frequency range can cover the seismic response in most cases. The upper limit stress ratio, the lower limit stress ratio and the stress amplitude are set to 0.9, 0.3, $0.6\sigma_c$ respectively. In addition, the fatigue characteristics of the limestone under drying conditions and saturation conditions are tested and compared respectively.

3 TEST RESULTS AND ANALYSES

3.1 Effect of water content on fatigue life of limestone

According to the loading scheme shown in Table 1, the low cycle fatigue tests of limestone were carried out under drying and saturated conditions at 1.0, 2.0, and 3.0Hz loading frequencies. The measured fatigue life (N) and mean fatigue life ($\overline{N}$) are summarized in Table 1.

As shown in Table 1, under the same loading conditions, the fatigue life of saturated rock is significantly lower than that of dry rock. Because of the large discrete of fatigue test results, according to the traditional way of fatigue test data processing, as shown in Figure 4, the average value of each set of fatigue test results is plotted into a double logarithmic coordinate system. The fatigue life of saturated specimen increases with the increase of loading frequency; but the growth rate of fatigue life is obviously slower than that of drying. A dual logarithmic estimation model for saturated fatigue life is obtained by linear fitting (as followed in Equation 1).

$$\log f = -0.9378 + 0.5041\log N \qquad (1)$$

where f = loading frequency; N = fatigue life.

Table 1. Fatigue life statistics of limestone under different saturation state conditions.

f Hz	S	$\Delta\sigma$	Drying Number	N	$\overline{N}$	Saturated Number	N	$\overline{N}$
1	0.9	$0.6\sigma_c$	9	75	188	93	6	69
1	0.9	$0.6\sigma_c$	53	102		79	18	
1	0.9	$0.6\sigma_c$	82	386		90	184	
2	0.9	$0.6\sigma_c$	52	270	382	42	89	325
2	0.9	$0.6\sigma_c$	11	371		59	281	
2	0.9	$0.6\sigma_c$	94	503		80	604	
3	0.9	$0.6\sigma_c$	10	399	839	64	257	581
3	0.9	$0.6\sigma_c$	55	954		91	499	
3	0.9	$0.6\sigma_c$	76	1164		85	986	

The results of uniaxial compressive test before fatigue test show that the compressive strength of the limestone decreases by about 6.4% compared with that of dry limestone, and under the same cyclic stress, the softening effect of water will aggravate the fatigue damage speed of the limestone, reduce the fatigue resistance of rock and shorten the life span. Therefore, the softening of water is the main reason why the fatigue life of saturated limestone is less than that of dry limestone.

3.2 Influence of the upper limit stress on fatigue life

The fatigue test results of the limestone under different upper limit stress are shown in Table 2. The relationship between the fatigue life and stress level fits the exponential function (as Equation 2) and the

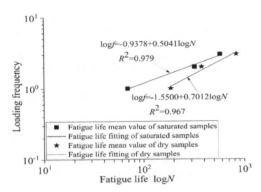

Figure 4. Relationship between fatigue life and loading frequency of drying and saturated specimens.

Table 2. Fatigue life statistics of the limestone under different upper limit stress.

Number	S	$\bar{S}$	$\Delta\sigma$	N
35	0.75	0.15	$0.60\sigma_c$	No destruction(1022022)
19	0.80	0.20	$0.60\sigma_c$	2782
31	0.80	0.20	$0.60\sigma c$	4420
47	0.80	0.20	$0.60\sigma_c$	No destruction(15628)
70	0.85	0.25	$0.60\sigma_c$	752
62	0.85	0.25	$0.60\sigma_c$	1010
17	0.85	0.25	$0.60\sigma_c$	1569
52	0.90	0.30	$0.60\sigma_c$	270
11	0.90	0.30	$0.60\sigma_c$	371
94	0.90	0.30	$0.60\sigma_c$	503
49	0.95	0.35	$0.60\sigma_c$	21
67	0.95	0.35	$0.60\sigma_c$	34
15	0.95	0.35	$0.60\sigma_c$	97

Equation 3 can be obtained by fitting the measured data after linearization.

$$N \cdot e^{15.2439S} = 1.0879 \times 10^{16} \tag{2}$$

$$S = 1.0520 - 0.0656 \log N \quad (R^2 = 0.948) \tag{3}$$

where N = fatigue life; S = upper limit stress ratio.

As shown in Figure 5, the fatigue life of the limestone increases with the decrease of the upper limit stress. Fatigue tests of many metal and concrete materials have confirmed that when the stress amplitude is constant, the lower the upper limit stress level, the longer the fatigue life of the material, but there is a certain critical value of the stress level, when the stress is below this critical value, the fatigue life of the material tends to infinity, which is the same as the Ge & Jiang (2003) proposed by the rock fatigue. According to Ge's point of view (Ge & Jiang 2003), the fatigue threshold corresponds to the boundary stress level between the linear segment and the nonlinear segment of the complete stress and strain relation curve of the rock, and this stress level is called the initiation stress. If the upper limit stress is lower than the initiation stress, even if the cycle is millions, the rock will not have fatigue damage.

The threshold value of different rocks fatigue obtained by Chinese scholars is different, ranging from 60% to 85% of σ_c. The initiation stress (σ_{ci}) and compressive strength of the limestone are not constants, but increase with the increase of loading strain rate ($\dot{\varepsilon}$). The ratio of initiation stress to compressive strength satisfies the relationship shown in Equation 4 within the loading strain rate ($2.44 \times 10^{-6} \sim 1.92 \times 10^{-4}$).

$$\sigma_{ci}/\sigma_c = 0.102 \log \dot{\varepsilon} + 1.06 \tag{4}$$

where σ_{ci} = the initiation stress; $\dot{\varepsilon}$ = strain rate.

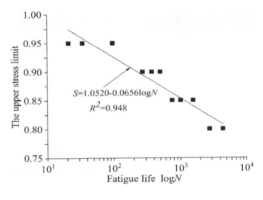

Figure 5. The relationship between the fatigue life and the upper stress limit.

When the loading frequency is 2.0Hz, the average strain rate is 5.749 ×10⁻³ during the whole cyclic process. The fatigue threshold values of the limestone are shown in Table 3, the threshold values of fatigue increase with the increase of loading rate.

3.3 Effect of the stress amplitude on fatigue life

The fatigue life test results of the limestone under different stress amplitudes are shown in Table 4. The loading frequency is 2.0Hz and the upper stress limit ratio is kept at 0.9.

The fatigue life of the limestone decreases with the increase of stress amplitude. Linear fitting of the 9 test data has been done, and the fitting result is shown in Figure 6. Xiao (2009) carried out an experimental study on the relationship between fatigue life and stress amplitude of granite. Compared with the limestone, the stress amplitude has a more significant effect on the fatigue life of the granite. However, they all show that the fatigue life decreases with the increase of stress amplitude.

The regression formula in Figure 6 simply considers that the effect of stress amplitude on fatigue life obviously does not fully reflect the effect of stress level on the fatigue life of rock, and the upper limit stress, stress amplitude and average stress should be considered at the same time. Based on the modification of the fatigue life model of concrete proposed

by Aas-jakobsen (1970), a prediction model of the limestone fatigue life as shown in Equation 4 is proposed.

$$S = 1.04 - 0.00887(1 - R)\log N \qquad (5)$$

Where S = upper limit stress ratio; R = ratio of lower limit stress to upper limit stress.

As shown in Figure 7, Equation 5 can well fit the fatigue test data including Tables 2 and 4.

3.4 Effect of the loading frequency on fatigue life

At present, the relationship between rock fatigue life and loading frequency is not clear, and the relationship between loading frequency and fatigue life obtained by different researchers is even the opposite. Jiang (2003) and Xiao (2009) believe that the loading frequency is actually the effect of loading rate. When loading at high frequency, the loading rate is faster, so the failure rate of rock is faster and the fatigue cycle is shorter. The fatigue test results of sandstone under three different loading strain rates given by Ray et al. (1999) show that the

Table 3. Fatigue threshold value of the limestone under different loading rates.

Loading frequency	Mean strain rate	Threshold value
Hz	×10⁻³	%
0.5	1.500	77.2
1.0	2.908	80.1
1.5	4.509	82.1
2.0	5.749	83.2
2.5	7.248	84.2
3.0	8.601	84.9

Table 4. Statistics of fatigue life of the limestone under different stress amplitudes.

Number	$\Delta\sigma$	S	$\overline{S}$	N
52	$0.6\sigma_c$	0.9	0.3	270
11	$0.6\sigma_c$	0.9	0.3	371
94	$0.6\sigma_c$	0.9	0.3	503
32	$0.5\sigma_c$	0.9	0.4	549
29	$0.5\sigma_c$	0.9	0.4	878
5	$0.5\sigma_c$	0.9	0.4	1228
27	$0.4\sigma_c$	0.9	0.5	735
65	$0.4\sigma_c$	0.9	0.5	3141
60	$0.4\sigma_c$	0.9	0.5	4846

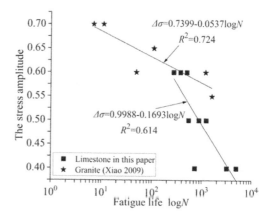

Figure 6. The relationship between fatigue life and stress amplitude.

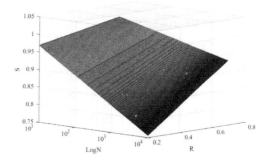

Figure 7. 3D figure of multi-parameter model for fatigue life estimation.

fatigue life of sandstone decreases under high strain rate (high loading frequency). Attewell's (1973) experimental study shows that the fatigue life of dolomite increases with the increase of loading frequency. The results of uniaxial and triaxial fatigue tests carried out by Ishizuka & Abe (1990) show that the relationship between loading frequency and fatigue life of granite satisfies linearity in the double logarithmic coordinate system.

In order to understand the influence of loading frequency on the fatigue life of the limestone, the fatigue test of the limestone was carried out by means of stress control and sine wave loading, respectively, at 0.5, 1.0, 1.5, 2.0, 2.5 and 3.0Hz loading frequencies. The test was divided into 6 groups of 3 specimens each, and the test results are shown in Table 5. As shown in Table 5 and Figure 8, the fatigue life of limestone increases with the increase of loading frequency. The linear fitting of 18 test data in a dual-logarithmic coordinate system shows that the experimental data are discrete and the correlation coefficient is only 0.673. Due to the small number of repetitive test data, it is not possible to meet the minimum sample requirements of fatigue test statistics, so the fatigue life of limestone is analyzed by calculating the average mean value. The average fatigue life is plotted in Figure 8, and under the double logarithmic coordinates, there is a good linear relationship between the average fatigue life

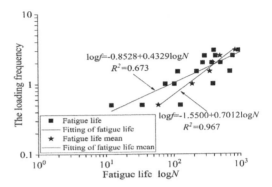

Figure 8. The relationship between fatigue life and loading frequency.

Table 6. Mean strain rate of the limestone under different loading frequencies.

Frequency	Mean strain rate*	Coefficient variation
Hz	s^{-1}	%
0.5	1.500×10^{-3}	5.04
1.0	2.908×10^{-3}	4.08
1.5	4.509×10^{-3}	4.43
2.0	5.749×10^{-3}	2.97
2.5	7.249×10^{-3}	4.41
3.0	8.601×10^{-3}	7.49

*The mean strain amplitude and mean strain rate are the comprehensive average values of the 3 samples.

Table 5. Statistics of fatigue life of the limestone under different loading frequencies.

Number	$\dfrac{f}{Hz}$	S	$\Delta\sigma$	N	$\overline{N}$
40	0.5	0.9	$0.6\sigma_c$	12	
6	0.5	0.9	$0.6\sigma_c$	34	59
13	0.5	0.9	$0.6\sigma_c$	125	
9	1.0	0.9	$0.6\sigma_c$	75	
53	1.0	0.9	$0.6\sigma_c$	102	188
82	1.0	0.9	$0.6\sigma_c$	386	
45	1.5	0.9	$0.6\sigma_c$	115	
57	1.5	0.9	$0.6\sigma_c$	219	344
63	1.5	0.9	$0.6\sigma_c$	697	
52	2.0	0.9	$0.6\sigma_c$	270	
11	2.0	0.9	$0.6\sigma_c$	371	382
94	2.0	0.9	$0.6\sigma_c$	503	
43	2.5	0.9	$0.6\sigma_c$	288	
75	2.5	0.9	$0.6\sigma_c$	422	493
95	2.5	0.9	$0.6\sigma_c$	769	
10	3.0	0.9	$0.6\sigma_c$	399	
55	3.0	0.9	$0.6\sigma_c$	954	839
76	3.0	0.9	$0.6\sigma_c$	1164	

and the loading frequency, and the correlation coefficient reaches 0.967.

$$\log f = -1.5500 + 0.7012 \log \overline{N} \qquad (6)$$

where $\overline{N}$ = mean fatigue life.

The test results of the average strain rate of the limestone at different loading frequencies are shown in Table 6, and the strain rate effect of the limestone is very obvious. In the 0.5~3.0Hz frequency range, the average strain rate of the limestone is at 10^{-3} magnitude, and the dynamic strength (σ_d) of the limestone at this loading rate will be significantly higher than its static strength (σ_c). However, the static strength is used as the reference value when determining the upper and lower limit stress ratios of the fatigue test, which means that the upper limit stress and stress amplitude of the fatigue test are reduced. As mentioned earlier, the lower the stress level, the longer the fatigue life of the rock. Therefore, it is not difficult to explain the test results of fatigue life increasing with the increase of loading frequency.

In order to consider the effect of loading frequency on fatigue life, the stress level needs to be corrected to the true value, for which the stress correction coefficient (M_r) shown in the definition Equation 7 is defined. By using M_r the relationship between the upper stress limit value (S) and the modified upper stress limit ratio (S') can be established.

$$M_r = \frac{\sigma_d}{\sigma_c} = c_1 + c_2 \log f \qquad (7)$$

$$S' = \frac{\sigma_{max}}{\sigma_d} = \frac{\sigma_{max}}{\sigma_c M_r} = \frac{S}{M_r} \qquad (8)$$

where c_1 = material parameter; c_2 = material parameter; M_r = stress correction coefficient; σ_d = dynamic strength of rock; S' = modified upper stress limit ratio.

The Equation 7 was fitted by using the test data of Zhao (2014), and the c_1=1.039 and c_2=0.0802 were obtained. By introducing the stress correction coefficient, a calculation model for predicting the fatigue life of the limestone based on the loading frequency (f) and upper limit stress ratio (S), as shown in Equation 9, can be established.

$$S = M_r \left(A - B \log N \right) = \left(c_1 + c_2 \log f \right) \left(A - B \log N \right) (9)$$

where A = material parameters; B = material parameter; for the limestone, A = 1.0522, B = 0.0746. Substituting these two parameters values into Equation 9, the following equation can be obtained.

$$S = (1.0239 + 0.0802 \log f)(1.0522 - 0.0746 \log N) \quad (10)$$

The two most important factors affecting the fatigue life of the limestone, namely stress level and loading frequency, are considered in the Equation 10.

4 CONCLUSIONS

Through the low cycle fatigue test of the karst limestone under different conditions, the following main conclusions are obtained.

(1) The fatigue life of the limestone increases with the decrease of the upper limit of loading stress, and the relationship satisfies exponential. The fatigue life of the limestone decreases with the increase of stress amplitude and the relationship is linear in half logarithmic space.

(2) A multi-stress parameter fatigue life estimation model of the limestone is established to reflect the influence of the upper stress limit, lower stress limit and mean stress. The fatigue life of the limestone increases with the increase of the loading frequency. In the double logarithmic coordinate system, a good linear relationship is satisfied between the mean fatigue life and the loading frequency.

(3) By introducing a stress correction factor, a fatigue life estimation model of the limestone based on loading frequency and upper limit stress is established.

The fatigue characteristics of rock are related to rock type, water saturation state, loading frequency, stress level and amplitude, and whether the above conclusion is applicable to other rocks has yet to be further tested and verified. At the same time, how to use the rock fatigue life prediction model established in this paper for the long-term stability analysis of karst tunnel under the action of earthquake and train circulation load needs to be further explored.

ACKNOWLEDGEMENTS

This work was financially supported by the China Natural Science Foundation (No. 50978018 and No. 51478031).

REFERENCES

Aas-Jakobson, K. 1970. Fatigue of concrete beams and columns. Bulletin No. 70-1. NTH Institute of Betonkonstruksjoner, Trondheim.
Attewell, P.B. & Farmer, I.W. 1973. Fatigue behaviour of rock. Int J Rock Mech Min Sci 35(10): 1–9.
Burdine, N.T. 1963. Rock failure under dynamic loading conditions. Soc Petr Eng J (03): 1–8.
Ge, X.R., Jiang, Y., Lu, Y.D. & Ren, J.X. 2003. Testing study on fatigue deformation law of rock under cyclic loading. Chinese Journal of Rock Mechanics and Engineering 22(10):1581–1585.
Haimson, B.C. & Kim, C.M. 1972. Mechanical behaviour of rock under cyclic fatigue. In: Cording EJ, editor. Stability of rock slopes Proceedings of the 13th Symposium on Rock Mechanics, New York: ASCE: 845–863.
Ishizuka, Y. & Abe, T. 1990. Fatigue behaviour of granite under cyclic loading. Static and Dynamic Considerations in Rock Engineering, Brummer, Balkema, Rotterdam: 139–147.
Jiang, Y. 2003. Fatigue failure and deformation development law of rock under cyclic load. Shanghai Jiaotong University, Master's Thesis (in Chinese).
Jiang,Y., Ge, X.R. & Ren, J.X. 2004. Deformation rules and acoustic emission characteristics of rocks in process of fatigue failure. Chinese Journal of Rock Mechanics and Engineering 23(11):1810–1814.
Li, S.C. 2008. Deformation and damage law and its nonlinear characteristics of rock under cyclic load. Chongqing: Chongqing University, Ph. D. Thesis (in Chinese).
Ray, S.K., Sarkar, M. & Singh, T.N. 1999. Effect of cyclic loading and strain rate on the mechanical behaviour of sandstone. Int J Rock Mech Min Sci 36(5): 543–549.
Singh, S.K. 1989. Fatigue and strain hardening behaviour of Graywacke from the Flagstaff formation, New South Wales. Engineering Geology 45(26): 171–179.
Xiao, J.Q. 2009. Theoretical and experimental study of rock fatigue characteristics under cyclic loading. Changsha: Central South University, Ph. D. Thesis (in Chinese).
Zhao, K. 2013. Experimental study on fatigue characteristics of Karst areas limestone and engineering application. Beijing: Beijing Jiaotong University, Ph. D. Thesis (in Chinese).
Zhen, Y.T & Hai, H.M. 1990. Technical note: an experimental study and analysis of the behaviour of rock under cyclic loading. Int J Rock Mech Min Sci Geomech Abstr 27(1): 51–56.

2019 Rock Dynamics Summit– Aydan et al. (eds)
© 2019 Taylor & Francis Group, London, ISBN 978-0-367-34783-3

Dynamic shear strength of an artificial rock joint under cyclic and seismic wave loading

T. Okada
Central Research Institute of Electric Power Industry, Chiba, Japan

T. Naya
Dia Consultants Co., Ltd., Saitama, Japan

ABSTRACT: Static strength has traditionally been considered as rock strength in the foundation bedrock for seismic design in Japan. The appropriate evaluation of dynamic strength instead of static strength became necessary for dynamic response analysis following the Great East Japan Earthquake. Therefore, we previously proposed a mathematical model to evaluate the dynamic strength for intact rocks. In this study, we performed laboratory tests and validated the model for discontinuity by using an artificial rock. As a result, the dynamic strength exceeded the static strength in the experimental and calculated results. However, the influence of fatigue and loading rate were less, compared to the cases of intact rocks. Therefore, the dynamic strength of the discontinuity does not greatly differ from static strength.

1 INTRODUCTION

In seismic design for the foundation bedrock of nuclear power plants in Japan, static strength has traditionally been used as rock strength, based on the fact that "dynamic strength ≥ static strength". This relationship has been validated for various rock types (Nishi & Esashi 1982, Yoshinaka et al.1987, Sugiyama et al. 2001). However, those test conditions are limited. Dynamic strength does not have a clear definition. Besides, it is difficult to formulate a single definition of dynamic strength because seismic waveforms are diverse, and the stress waveform inside the ground also varies depending on the location. To resolve these issues, we proposed a mathematical model to evaluate dynamic strength (Okada & Naya 2018). Using the mathematical model, the dynamic strength can be obtained by monotonic loading tests performed at several loading rates in different orders and cyclic loading tests at several different shear stress amplitudes. In the previous research, we performed laboratory tests to validate the model, using intact rocks, but we have never conducted the test using rock joints.

In this study, monotonic loading and cyclic loading direct shear tests were conducted on the artificial rock joint we made from plaster in order to obtain the parameters of the mathematical model. Afterwards, multistep direct shear tests under cyclic and seismic-wave loadings were performed, followed by simulations using the mathematical model. The test results showed that the dynamic strength exceeds the static strength, corroborating previous research. Furthermore, the dynamic strength calculated from the mathematical model of the artificial

rock joint was generally consistent with the dynamic strength obtained from the experimental data.

2 MATHEMATICAL MODEL OUTLINE OF DYNAMIC STRENGTH

2.1 *Effect of fatigue*

When the stress amplitude is fixed, and repeated stress is applied until the failure of specimen, the stress amplitude decreases as the number of cycles increases. This relationship is called stress-cycle (S-N) curve in the field of designing the fatigue strength of metallic materials. The function expressing this relationship is defined as the fatigue function f_1 and is expressed as follows:

$$f_1\left(N_f\right) := \frac{\tau_{f\_N_f}}{\tau_{f\_N=1}} = 1 - a \cdot \log N_f \tag{1}$$

where N_f (≥ 1) is the number of cycles at the time of failure, τ_{f_Nf} is the shear strength at the time of failure following the loading N_f, $\tau_{f\_N=1}$ is the shear strength under monotonic loading (N=1), and a is a parameter defining the slope of the function, which means that the decrease in strength is greater for larger values of a. To simplify the problem, confining pressure is not considered in this function.

2.2 *Effect of loading rate*

It is also known that strength tends to increase as loading rate increases. While the fatigue effect always leads to decrease in the strength as the number of cycles increases, increasing the loading rate tends to increase the strength. In our proposed mathematical model, the relationship between the dynamic and static strengths is determined mainly by these two effects. We define the function f_2 as the function that represents the relationship between the loading rate and shear strength, and define it as follows:

$$f_2(\dot{\varepsilon}) := \tau_f = \alpha + \beta \cdot \log \dot{\varepsilon} \qquad (2)$$

where $\dot{\varepsilon}$ is the axial strain rate (%/min), τ_f is the maximum shear strength (MPa), and α and β are parameters. To simplify the problem, the confining pressure is not considered in this function.

2.3 Integration of fatigue and loading rate effects

Letting τ_f in Eq. (2) be the strength $\tau_{f\_N=1}$ at loading time $N = 1$ in Eq. (1), the following relationship is obtained:

$$\tau_{f\_N_f} = f_1 \cdot f_2 = (\alpha + \beta \cdot \log \dot{\varepsilon})(1 - a \cdot \log N_f) \qquad (3)$$

From this relationship, the relationship between $N_f(\geq 1)$ and τ_{f_Nf} at any strain rate can be obtained. Note that the derivation of Eq. (3) assumes that the relation in Eq. (1) is satisfied, regardless of the loading rate. However, experimental data confirms that this assumption is, to some extent, reasonable (Okada & Ito 2009).

2.4 Application of cumulative damage rule

To express the impact of an arbitrary waveform in the mathematical model, it is necessary to know the impact of the cyclic loading that leads to fracture (hereinafter referred to as the "damage effect"). However, because no such test data was available, we applied the cumulative damage rule used in metal materials design (Otaki 2007). The strength exerted after N wave loading (hereinafter referred to as "residual strength") is assumed to change linearly with respect to its fracture count. Adopting this perspective, the damage function f_3 representing the damage effect due to cyclic loading can be expressed as follows:

$$f_3(N) := \frac{\tau_{d\_N}}{\tau_{f\_N=1}} = 1 - d \cdot (N - 1) \qquad (4)$$

where N is the number of cycles preceding failure, and τ_{d_N} is the strength (residual strength) exerted after N cycles. At the time of failure (when $N=N_f$), Eq. (1) = Eq. (4) obtains. Because $f_1(N_f) = f_3(N_f)$, the following relation is obtained:

$$d = \frac{a \cdot \log N_f}{N_f - 1} = \frac{1 - \dfrac{\tau_{f\_N_f}}{\tau_{f\_N=1}}}{10^{\left(1 - \frac{\tau_{f\_N_f}}{\tau_{f\_N=1}}\right)\Big/a} - 1} \qquad (5)$$

Therefore, the parameter d in Eq. (4) is determined from Eq. (5) when the stress ratio $\dfrac{\tau_{f\_N_f}}{\tau_{f\_N=1}}$ is determined. The following relationship is then obtained from Eqs. (2) and (4):

$$\tau_{f\_N_f} = f_2 \cdot f_3 = (\alpha + \beta \log \dot{\varepsilon})\{1 - d(N - 1)\} \qquad (6)$$

Note that for Eq. (6) to hold in the manner of Eq. (3), the relationship shown in Eq. (4) must hold, regardless of the loading rate. Lacking empirical data in this regard, we make the same assumptions as the cumulative damage rule. This assumption makes it possible to determine the degree of damage even if the stress ratio $\dfrac{\tau_{f\_N_f}}{\tau_{f\_N=1}}$ is determined at different loading rates.

3 VALIDATION OF THE DYNAMIC STRENGTH CALCULATED FROM THE MODEL

3.1 Specimens

An artificial rock discontinuity was used as the test specimen in order to validate the numerical model with fewer variations of geometric and mechanical properties. The discontinuity was made by using a mold that has a regular protruding ridge with an angle of 30 degrees (Figure 1); that is a simulated rocked discontinuity. The discontinuous plane is 200 mm long (shear direction) and 100 mm wide, and the height of the protruding ridge is about 4 mm.

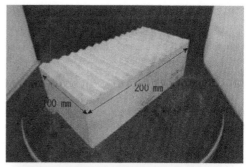

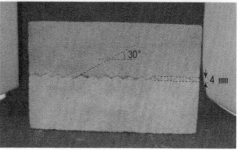

Figure 1. Specimen of discontinuity of artificial rock.

Furthermore, intact specimens with no discontinuity and a discontinuity with an angle of 0 degree (horizontal surface) were made.

The compound ratio of the sand, clay, cement and water of the artificial rock was 67.1:6.9:11.2:14.8 by weight. The curing period of the material was set between 30 to 42 days in order to reduce the variations in the mechanical properties. The wet density was about 1.9 g/cm³ and the unconfined compressive strength was about 14 MPa.

3.2 Test equipment

The direct shear test equipment with two-axis hydraulic servo-type actuators was used for all the tests (Figure 2). A hydraulic cylinder is electrically controlled by a hydraulic servo valve and the load-cell output value is fed back for control. Both the maximum vertical load and shear load are 200 kN. The friction of the vertical and shear load is reduced by a crosshead and linear guide respectively. The structure of the shear box is shown in Figure 3. The pressure plate is not separated from the shear box and is part of the upper box according to the JGS0561 method (Japanese Geotechnical Society 2009). A defect of this shear box is that as loading progress, the gap between the upper and lower (movable) box changes, and, as a result, the thickness of the shear plane varies from the initial value. The initial gap before the vertical loading was fixed at 6 mm in order to ensure that it would be greater than the height of the protruding ridge after the vertical loading.

3.3 Test method

Direct shear tests were carried out in principles corresponding to the standard method for direct shear test on a rock joint (Japanese Geotechnical Society, 2009). A list of purposes and loading conditions is given in Table 1.

Series D30-1 represents the monotonic loading test, and the asssumed static strength test. The displacement rate was 0.05 mm/min, and the vertical

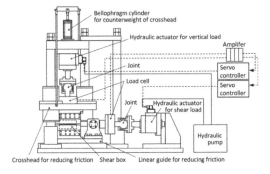

Figure 2. The direct shear test equipment

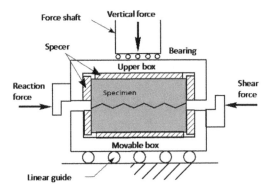

Figure 3. The direct shear box.

stress ranged from 0.2 MPa to 6.0 MPa. Intact specimens with no discontinuity and discontinuities with an angle of 0 degree (horizontal surface) were also used for comparative test. Series D30-2 represents the cyclic loading tests for investigating the fatigue function. The loading frequencies were 0.1 Hz and the waveforms were sine waves. Vertical stress was fixed at 0.2 MPa. Series D30-3 represents the monotonic loading tests for investigating the loading rate function. They were performed at various displacement rates (0.05–15 mm/min). Series D30-4 represents the multistep cyclic loading test for validating the dynamic strength evaluation model. D30-4a and D30-4b represent the sine and seismic wave loading,

Table 1. Information on the direct shear tests

Series	Purpose	Loading method	Loading conditions				Vertical stress (MPa)	Quantity (pcs)
D30-1	Static strength	Monotonic	Displacement rate: 0.05mm/min				0.2 to 6.0	4
D30-2	Fatigue effect	Cyclic	Frequency: 0.1 Hz, Stress amplitude: 1.6 to 2.0 MPa				0.2	6
D30-3	Rate effect	Monotonic	Displacement rate: 0.005 to 15 mm/min				0.2	4
D30-4a	Model validation	Multistep Cyclic	No.	Frequency (Hz)	Waves (cycles)	Number of steps	0.2	3
			D30-4a1	0.1	10	10		
			D30-4a2	0.01	50	10		
			D30-4a3	0.1	5	5		
D30-4b			No.	Seismic wave		Number of steps	0.2	2
			D30-4b1	Time axis 50 times		5		
			D30-4a1	Time axis 20 times		5		

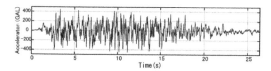

Figure 4. Artificial seismic wave.

respectively. The frequency, number of waves, and number of steps are as indicated in Table 1. In D30-4b, an artificial seismic wave was used, as shown in Figure 4, and the time axis was stretched about 20 or 50 times, owing to the limitations of the shear loading actuator of the test machine.

3.4 Test results

(1) Static Strength
The static strength characteristics obtained from D30-1 are shown in Figure 5 with the results of the intact specimen and the discontinuity with an angle of 0-degree (horizontal surface). The result of the intact and 0-degree discontinuity yielded values of $\phi \fallingdotseq 40°$. The 0 and 30-degree discontinuities yielded vales of $c \fallingdotseq 0$. When 30-degree discontinuity having values of $\phi \fallingdotseq 40°$ slides on its surface, the value of the friction angle theoretically becomes 70 degrees. Although the friction angle of the test of the 30-degree discontinuity is nonlinear, the friction angle is 64 degrees at low confining pressure. The value is slightly lower than 70 degrees, but it is a reasonable value, in consideration of a small breakage of the surface of the discontinuity. The value of the friction angle of the 30-degree discontinuity becomes about 40 degrees at high confining pressure because of the complete destruction of the surface.

(2) Fatigue tests
Examples of the results of the fatigue test on 30D-2 are shown in Figure 6. The shear displacement increases as loading progresses, and then, they sharply increase at 680 seconds once the shear stress can no longer be maintained. All the fatigue tests show the specific point at which shear displacement

rapidly increases. These points were defined as failure points and were counted and recorded as the number of cycles leading up to failure N_f.

Figure 7 shows that the relationship between N_f and the strength normalized with the strength at single wave loading. This relationship corresponds to Eq. (1). The equation approximated using the least square method is shown as a solid line in the figure. The three data that the number of cycles exceeds 1000 times did not show the failure, however the data were assumed as the failure data for approximation using the least square method. Therefore, the actual gradient of the straight line is less steep than that in the figure. The parameter value for Eq. (1) was $a = 0.0288$.

The impact of fatigue on the discontinuity was considerably smaller, compared to the case of intact rocks (Okada & Naya, 2018).

(3) Loading rate function from the tests
The relationship between loading rate and shear strength obtained in D30-3 is shown in Figure 8. Note that the loading rate on the horizontal axis is the shear displacement rate (mm/min). These relationships correspond to Eq. (2). Further, the dispersion of the test data is small. The equation approximated by the least square method is shown as a solid line in the figure. The parameter value for Eq. (2) was $\alpha = 0.441$ and $\beta = 0.00872$. α is the strength at a displacement rate of 1 mm/min. The β/α implied rate of strength gain was

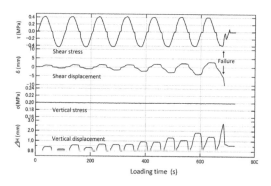

Figure 6. Static strength characteristics.

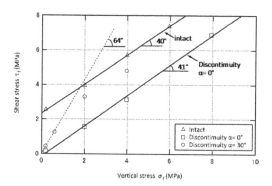

Figure 5. Static strength characteristics.

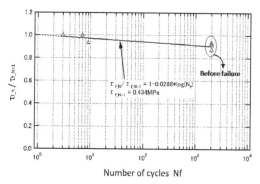

Figure 7. Fatigue test results.

168

0.02. The value was considerably lower than the value for intact rocks (Okada & Naya, 2018).

(4) Multiple step loading tests for validation
Figure 9(a) and Figure 9(b)M show examples of the test results for the sine and seismic wave multistep cyclic loading in the cases of D30-4a3 and D30-4b2. As both loading progresses, the shear displacement gradually increases. Then, the shear displacement rapidly increases when shear stress amplitude can no longer be maintained. In all cases, the specific point at which shear displacement rapidly increases was clear. These points were defined as failure points, and the maximum shear stress provided before the failure points were designated the strength at multi-step cyclic loading, that is, the dynamic strength.

3.5 Discussion

The dynamic strength was obtained from the multistep cyclic test and then compared with the calculation results obtained by the evaluation model. The static strength is assumed to be the strength at the loading rate of 0.05 mm/min as shown in Figure 8.

The static strength should be determined from Figure 5, but the strength characteristics were nonlinear, and the strength was affected by the approximate nonlinear function.

The calculation of the sine wave loading rate, using the mathematical model proceeds in the following manner: letting the loading time be the time required for $N = 1/4$ of a sine wave, the shear displacement at the point of failure can be considered an almost constant value, regardless of the loading rate (Figure 10). The loading rate at failure was calculated from the loading frequency and the failure displacement. Under the present conditions where the shear stress increases at the same frequency, the lower the stress amplitude, the lower the displacement rate. Therefore, when the stress amplitude is $1/n$, the loading rate is also calculated as $1/n$.

The calculation of the seismic wave loading rate using the mathematical model proceeds in the following manner (Figure 11): the point where the seismic wave crosses the X axis (Y=0) is identified, and

the time interval (Tn) is obtained for each half wave. In the same manner as for the sine wave, the shear displacement in relation to the failure rate was calculated. If the seismic wave time axis is stretched by 10 times, the displacement rate will be 1/10. When the stress amplitude is $1/n$, the loading rate is also calculated as $1/n$. In addition, as shown in Figure 11, letting the respective stress histories be the maximum value on both the positive and negative sides, the dynamic strength is calculated separately for each side using the mathematical model, and the side with the maximum absolute value is taken as the dynamic strength.

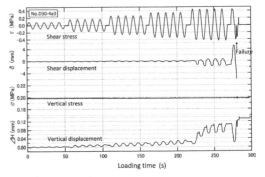

(a) Sine wave results

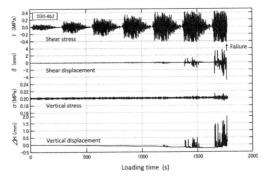

(b) Seismic wave results

Figure 9. Examples of multistep cyclic loading tests.

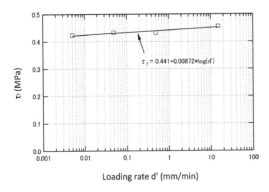

Figure 8. Loading rate function from the tests.

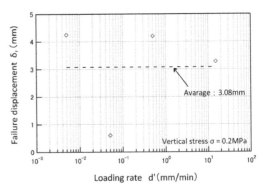

Figure 10. Failure displacement at various loading rate.

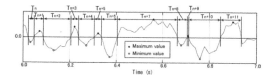

Figure 11. Illustration of frequency and stress history in a seismic wave.

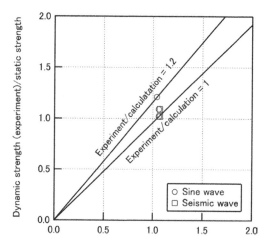

Figure 12. Comparison between experimental and calculated results.

Figure 12 shows the relationship between the dynamic strength obtained from the calculation and the dynamic strength obtained from the experiment normalized with the static strength. On the vertical axis, the dynamic strength (experiment)/static strength ranges from 1.02 to 1.09. On the horizontal axis, the dynamic strength (calculated)/static strength is from 1.02 to 1.07. The dynamic strength is larger than the static strength in the experimental and calculated results. Comparing the experimental and calculated results, the experimental results range from -4–16%. The relative error does not greatly differ from the value for intact rocks (Okada & Naya, 2018). However, the impact of fatigue and loading rate were small; therefore, the influence of variation was relatively large. As a result, the dynamic strength of the discontinuity does not greatly differ from the static strength.

4 CONCLUSION

(1) We had proposed a mathematical model to evaluate the dynamic strength for intact rocks in the previous research. We performed laboratory tests and validated the model for discontinuity by using an artificial rock.

(2) The dynamic strength is larger than the static strength in the experimental and calculated results. However, the influence of fatigue and loading rate were less, compared the case of intact rocks. Therefore, the dynamic strength of the discontinuity does not greatly differ from its static strength.

(3) We plan to conduct laboratory tests in order to get experimental results for cyclic loading of large number of cycles under extreme loading conditions and high loading rate to further clarify the dynamic strength of rock discontinuities.

REFERENCES

Japanese Geotechnical Society. 2009. Method for direct shear test on a rock discontinuity, *Geotechnical materials test methods and commentary*: 912–944 (in Japanese).

Nishi,K. and Esashi, Y.1982. Study on the mechanical properties of mudstone (Part 4), *Central Research Institute of Electric Power Industry Research Report* 382014 (in Japanese).

Okada, T. and Ito, H. 2009. Mathematical modeling of the dynamic strength of soft rock, *Proc. the 38th Symposium on Rock Mechanics* (in Japanese).

Okada, T. and Naya, T. 2018. A mathematical model for evaluating the dynamic shear strength of rock and validation of the model with laboratory test data, *Proc. 3rd Int. Conf. on Rock Dynamics and Applications, Trondheim*: 179–185.

Otaki, H. 2007. Mechanical fatigue strength design method, Nikkan Kogyo Shimbun (in Japanese).

Sugiyama,H, Saotome, A. and Nakamura, Y. 2001. Dynamic strength characteristics of volcanic breccia based on triaxial compression test, *The 56th Annual Scientific Lecture Meeting of the Japan Society of Civil Engineers*: 134–135 (in Japanese).

Yoshinaka, R., Ogino, I., Takada,S. and Kanazawa, K.1987. Strength characteristics of sedimentary soft rock under dynamic cyclic loading, *7th National Symposium on Rock Mechanics in Japan technical papers*: 61–66 (in Japanese).

2019 Rock Dynamics Summit– Aydan et al. (eds)
© 2019 Taylor & Francis Group, London, ISBN 978-0-367-34783-3

Shock test on rounded rock fragments in Suruga Bay sediments and its implications on past mega-earthquakes

I. Sakamoto
Tokai University, Dept. of Marine and Earth Science, Shizuoka, Japan

Ö. Aydan
University of the Ryukyus, Department of Civil Eng., Okinawa, Japan

ABSTRACT: The breakage of boulders/cobbles in conglomeratic deposits are reported to occur along tectonic lines. The authors have been undertaking both in-situ investigations and laboratory experiments to understand the fundamental mechanism and causes of the breakage of the boulders/cobbles. Some sampling and observations are done in the close vicinity of Horai district, which is located along the famous Median Tectonic Line of Japan. In addition, some sampling of broken boulders/cobbles was done in Seno-umi in Suruga Bay, Shizuoka by Bousei-maru investigation ship of Tokai University. A drop-weight equipment was developed to carry out experiments on the boulders/cobbles sampled from Horai district, Kusanagi and Miho district of Shizuoka City, which constitute Nihon-daira formation. The authors report the outcomes of this experimental study in relation to mega earthquakes in the past. Furthermore, some numerical studies on ideal cases were carried out in order to see the stress state causing the breakage of the boulders.

1 INTRODUCTION

It is reported the breakage of boulders/cobbles in conglomeratic deposits occur along tectonic lines (e.g. Futamura 2016). The boulders/cobbles in such environments are found to be broken either in extension, sheared or both (Fig. 1). There is no doubt that these boulders/cobbles are broken due to shock waves induced during the large seismic events. The most important aspect is how to identify the stress state and seismic shock waves causing the breakage of boulders/cobbles. The authors have been undertaking both in-situ investigations and laboratory experiments to understand the fundamental mechanism and causes of the breakage of

Figure 1. Views of breakage of pebbles, cobbles and boulders.

the boulders/cobbles. Some sampling and observations are done in the close vicinity of Horai district, which is located along the famous Median Tectonic Line of Japan. In addition, some sampling of broken boulders/cobbles was done in Seno-umi in Suruga Bay, Shizuoka by Bousei-maru investigation ship of Tokai University.

The authors devised a special drop-weight equipment to carry out experiments on the boulders sampled from Horai district, Kusanagi and Miho district of Shizuoka City, which are constituted Nihon-daira formation, which is also related to the deposits in Seno-umi area in Suruga Bay. The behavior of samples under shock waves induced by the drop of cylindrical steel object from a certain height is observed. During the experiments, shock forces and accelerations were both measured simultaneously.

The maximum velocity during the impact were measured. The second author has developed some empirical relations with maximum ground velocity and earthquake magnitude (Aydan 2012). These relations may be used to infer the magnitude of the earthquakes as a function of the maximum velocity causing the breakage of the boulders/cobbles. This study may lead a new of way inference of magnitude of mega-earthquakes in the past. In addition, some finite element studies have been carried out to investigate the stress state involved in the breakage of the boulders/cobbles under different boundary conditions. The outcomes of these experimental and numerical studies are presented and their implications on the past seismicity of regions are discussed.

2 SHOCK TESTING DEVICE AND PEBBLES

2.1 *Shock Testing Device*

The authors developed a special shock testing device. The device consist of a steel cylinder with a weight of 8300 gf having a diameter of 97 mm, a load cell, an accelerometer (Fig. 2). The plastic pipe container, in which the cylinder was dropped, had an internal diameter of 100 mm and height of 500 mm. The steel cylinder was dropped from certain heights and acceleration and force were measured simultaneously using YOKOGAWA WE7000 data-acquisition system at a sampling rate of 1ms. No digital filtering was imposed on measured force and accelerometer records.

2.2 *Characteristics of Cobbles*

The size of rock fragments having different sizes are called gravel or pebble (2-64 mm), cobble (64-256 mm) or boulders (greater than 264 mm). Therefore, the testing device is limited to gravel/pebble and cobbles with a size less than 100 mm. The samples were collected at Miho-shore along Suruga Bay in which Seno-umi area exist and Kusanagi hill, which is a part of uplifted conglomeratic sea-bed deposits. The first author sampled some pebbles from Seno-umi area in Suruga Bay, Shizuoka by Bousei-maru investigation ship of Tokai University. The pebbles/cobbles in Miho and Kusanagi area are made of sandstone (Fig. 3).

Figure 3. Views of pebbles of Kusanagi.

The authors also visited Horai-cho district in Aichi prefecture, which is located over famous the Median Tectonic Line. In this locality, one can observe deformed and broken boulders and cobbles over a huge area. The deformed and/or broken boulders/cobbles are found in conglomeratic deposits with a fine content consists of mainly sand. Most of boulders/pebbles in this area are made of hardened shale or sandstone (Fig. 4).

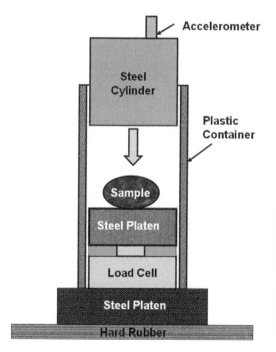

Figure 2. Schematic drawing of shock testing device.

Figure 4. Views of tested pebbles of Horai-cho.

172

3 EXPERIMENTS AND RESULTS

3.1 *Kusanagi Pebbles/Cobbles*

The pebbles/cobbles of Kusanagi were made of sandstone. The maximum resistance depends upon the intrinsic strength of the pebble/cobble as well as its shape and drop height and fracturing state as seen in Figs. 5,6,7 and 8 and noted from Table 1. As noted from photos of the samples shown in Fig. 5, the fracturing state of samples increases as the drop height or maximum nominal velocity computed from the following formula increases:

$$V_{max} = \sqrt{2gH_d} \qquad (1)$$

where g is gravitational acceleration and H_d is drop height.

It is also interesting to note that the maximum resistance achieved before the peak acceleration occurs. Furthermore, the upward acceleration is also high, which causes the rebounding of the sample. This fact was also observed in preliminary tests with regular shapes.

Figure 5. Views of tested pebbles of Kusanagi.

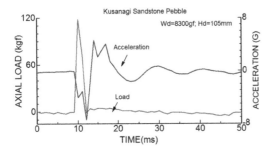

Figure 6. Force and acceleration response of Kusanagi R1 sample during the drop-weight test.

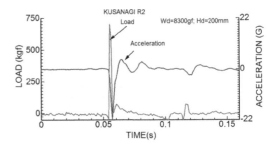

Figure 7. Force and acceleration response of Kusanagi R2 sample during the drop-weight test.

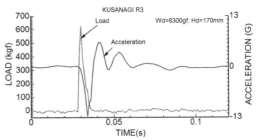

Figure 8. Force and acceleration response of Kusanagi R3 sample during the drop-weight test.

Table 1. Size and measured parameters for Kusanagi pebbles/cobbles.

Kusanagi	D(mm)	F(kgf)	|Amax| (g)	Vmax(cm/s)
R1	55	118.02	7.42	143.53
R2	65	703.04	21.72	199.09
R3	52	626.94	12.76	182.63

3.2 *Miho Pebbles*

The pebbles of Miho were made of sandstone and their resistance were quite similar as they were subjected to the maximum nominal velocity (Table 2) The fracturing state of Miho R1 sample was more destructive than that of the Miho R2 sample. The acceleration responses of the both samples are saturated.

In Miho samples, it is also interesting to note that the maximum resistance achieved before the peak acceleration occurs. Furthermore, the upward acceleration was also high, which causes the rebounding of the sample.

Table 2. Size and measured parameters for Miho pebbles.

Miho	D(mm)	F(kgf)	|Amax| (g)	Vmax(cm/s)
R1	50	103.87	11.42	177.78
R2	50	116.35	12.23	177.78

Figure 9. Views of tested pebbles of Miho after testing.

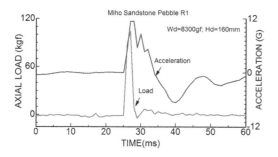

Figure 10. Force and acceleration response of Miho R1 sample during the drop-weight test.

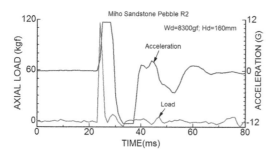

Figure 11. Force and acceleration response of Miho R2 sample during the drop-weight test.

3.3 Horai Pebbles/Cobbles

The pebbles/cobbles of Horai were made of hardened shale and sandstone. The Horai R1 sample, which is hardened shale did not fracture in the first test in which it was dropped from height of 135 mm. However, it was broken when the drop height was increased to 200 mm. As said previously, the maximum resistance depends upon the intrinsic strength of the pebble/cobble as well as its shape and drop height and fracturing state as seen in Figs. 12,13,14 and 15 and noted from Table 3. As noted from photos of the samples shown in Fig. 13, the fracturing state of samples R1 and R2 involved a single rupture surface while Horai R3 sample had several rupture surfaces. As also noted in previous experiments, the acceleration response is not symmetric with respect to time axis.

Figure 12. Views of tested pebbles of Horai after testing.

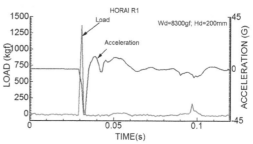

Figure 13. Force and acceleration response of Horai R1 sample during the drop-weight test.

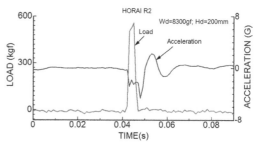

Figure 14. Force and acceleration response of Horai R2 sample during the drop-weight test.

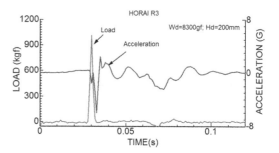

Figure 15. Force and acceleration response of Horai R3 sample during the drop-weight test.

Table 3. Size and measured parameters for Horai pebbles/cobbles.

Horai	D (mm)	F (kgf)	|Amax| (g)	Vmax (cm/s)	Comment
R1-1	50	95.54	12.22	162.75	Not broken
R1-2	50	1367.16	39.88	199.09	Broken
R2	50	558.63	7.65	199.09	Broken
R3	68	1015.21	5.99	199.09	Broken

4 NUMERICAL STUDIES ON ROUNDED ROCK FRAGMENTS EMBEDDED IN SOFTER MATRIX UNDER VARIOUS LOADING CONDITIONS

The site observations indicated that rounded rock fragments were embedded in matrices having softer materials. A series of preliminary numerical analyses were carried out to see the deformation and stress state in the rounded rock fragments and surrounding matrix under static condition assuming that material behaviour was elastic. Material properties used in numerical analyses are given in Table 4 while Fig. 16 shows the displacement and force boundary conditions. The domain was subjected to four force conditions, nameli CASE 1: shearing; CASE 2: normal loading and CASE 3: normal and shearing loading and CASE 4 shearing under non-uniform normal loading. The shear and normal load intensities are given in Table 5. CASE 4 was considered in order to take into account the possibility of non-uniform normal load distribution at the top boundary in reality.

Figs. 17 to 20 shows the maximum shear stress distributions in the analyzed domain. Except the lower boundary due to fixed boundary condition, the numerical analyses are not uniform within the domain and the rock fragments act like stress attraction area within the domain. In other words, the stresses are much higher in the harder rock fragments. This situation would be more amplified as

Table 4. Material properties used in numerical analyses

Material	Elastic Modulus (GPa)	Poisson's ratio
Rock Fragment	10	0.25
Matrix	1	0.30

Table 5. Loading Conditions

CASE No	Shear Traction (MPa)	Normal Traction (MPa)
1	1.0	0.0
2	0.0	1.0
3	1.0	1.0
4	0.0	1.0,5.0 (over cobble)

the stiffness between the inclusions (rock fragments) and matrix become larger. Figs. 21 to 23 show the principal stress vectors under shearing, combined shearing and normal loading and non-uniform normal loading. In this figure it is again noted that the stresses on hard inclusions are higher. The stress

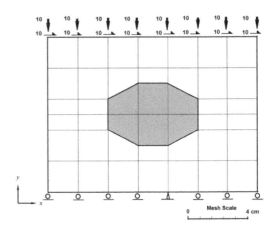

Figure 16. Boundary conditions (CASE 3)

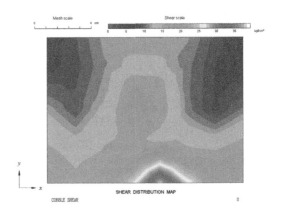

Figure 17. Max. shear stress distribution (CASE 1)

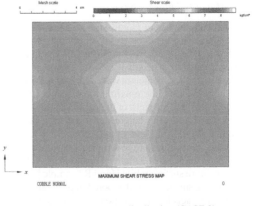

Figure 18. Max. shear stress distribution (CASE 2)

175

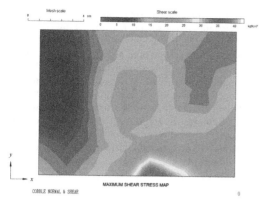

Figure 19. Max. shear stress distribution (CASE 3)

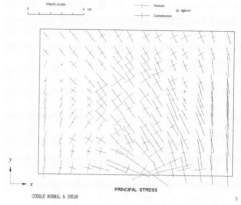

Figure 22. Distribution of principal stresses(CASE 3)

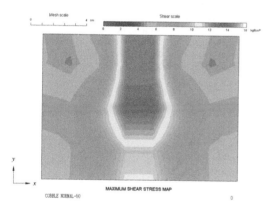

Figure 20. Max. shear stress distribution (CASE 4)

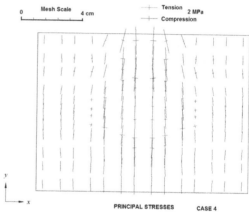

Figure 23. Distribution of principal stresses(CASE 4)

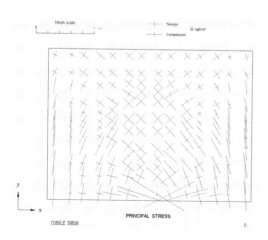

Figure 21. Distribution of principal stresses(CASE 1)

distributions for CASE 1 and CASE 3 indicates that if stress state is enough to rupture the inclusions, tension cracks will occur at an angle of 45 degrees or more with respect to the shearing direction.

Under high normal loads, tensile stresses develop in inclusions perpendicular to applied direction of normal loads as seen in Fig. 23. As the intensity of normal load increases the tensile stresses also increases. In other words, if inclusion ruptures in extension, there will be fractures through the rock fragments parallel to the direction of loading, which may also indicate the maximum stress direction.

5 INFERENCE OF EARTHQUAKE MAGNITUDE

Aydan (2012) proposed some empirical relations between magnitude and seismic parameters of the earthquakes. The equations developed by Aydan may be applied to the source area of the earthquake by assuming that the rupture of the rock fragments occur. Thus, the moment magnitude of the earthquake can be estimated from the following relation for the maximum velocity in the source area as

$$M_w = 1.16 * \ln(6.8 * V_{max}) \tag{2}$$

If the values, which are obtained from the maximum velocity to fracture the rock fragments in experiments on samples from the Suruga Bay deposits, are used in Eq. (2), one would infer the moment magnitude of earthquakes as 7.99, 8.23, 8.26 and 8.37 in Suruga Bay. The moment magnitude of earthquakes in Suruga Bay is known to be ranging between 7.9 and 8.4 since 1400. Thus the estimations from the broken rounded rock fragments are in accordance with the past earthquakes in the region.

6 CONCLUSIONS

The causes of breakage of rounded rock fragments in conglomeratic deposits are experimentally studied through a specially developed experimental device, which is equipped with a load cell and an accelerometer. When the maximum nominal velocity is sufficient to cause the fracture of the rounded rock fragments, it may correspond to the maximum velocity in the source area of the earthquake. With this assumption, it is shown that it is possible to estimate the magnitude of the earthquake from Eq. (2) together with the maximum nominal velocity sufficient enough to cause the rupture of the rounded rock fragments. In additions, the stress state due to imposed deformation conditions imposed during the earthquake is analyzed using numerical methods and it is shown that the fracture of rounded fragments are possible in extension, shearing or both.

REFERENCES

Aydan, Ö. (2012): Ground motions and deformations associated with earthquake faulting and their effects on the safety of engineering structures. Encyclopedia of Sustainability Science and Technology, Springer, R. Meyers (Ed.), 3233–3253.

Futamura, K. 2016. Sinistral brittle shearing in the Fudesute Conglomerate of Lower Miocene Suzuka Group, Kameyama City, Central Japan (in Japanese) (Chikyu Kagaku), 70, 95–108.

2019 Rock Dynamics Summit– Aydan et al. (eds)
© 2019 Taylor & Francis Group, London, ISBN 978-0-367-34783-3

Dynamic-deformation characteristics of granite under uniaxial compressive stress

N. Kamoshida, H. Yamamoto & T. Saito
Iwate University, Morioka, Iwate, Japan

ABSTRACT: Recently, it has become necessary to evaluate the resistance properties of rock structures based on dynamic properties. This is due to the Japanese guidelines for seismic evaluation and the design of civil structures have been reviewed in response to the frequent occurrence of large earthquakes in the country. However, few studies have been performed on the dynamic-deformation and damping characteristics of rocks. To clarify the dynamic-deformation characteristics of granite, we performed cyclic uniaxial tests using "JGS0542-2009 Method". The experimental results are as follows: (1) the measured equivalent Young's modulus increases along with strain in the small- (<10 micro-strain) to medium- (10 to 1,000 micro-strain)-strain range; however, the modulus decreases abruptly at the large (>1,000 micro-strain) strain range where the specimen is broken and (2) the evaluated damping ratio is 0% in the small- to medium-strain range; however, it increases abruptly at the medium- to large-strain range.

1 INTRODUCTION

Previously, rock structures were believed to be resistant to earthquakes; therefore, seismic evaluations were implemented only for critical rock structures. However, recently, the Japanese guidelines for seismic evaluation and the design of civil structures have been reviewed in response to the frequent occurrence of large earthquakes in the country. Hence, seismic evaluation of rock structures has become necessary.

A dynamic-response analysis of the rock structure requires the dynamic-deformation and damping properties of the rock mass. The testing method used to determine the dynamic-deformation properties of soils has been standardized since the 1960s and has been the subject of many reports. However, the test method for rock is not yet standard. Therefore, there have been few reports about the dynamic-deformation and damping properties of intact rock and rock mass that are unclear (Ochi et al. 1993, Ochi et al. 1995, Fukumto et al. 2009).

we discuss the effects of number of cycles and loading rate on the dynamic-deformation properties of a hard rock under uniaxial compressive stress.

2 DYNAMIC-DEFORMATION PROPERTIES

The deformation properties of soil and soft rocks required for numerical analysis of dynamic (e.g., seismic) loads are often called "dynamic-deformation properties." Originally, "dynamic" refers to rapid loading (impact loading) or rapid cyclic loading, but in the field of geotechnical engineering, "dynamic" is used in the sense of cyclic loading by loading/unloading cycles, as opposed to "static," which means monotonic loading. Therefore, the dynamic-deformation characteristics in this case are the equivalent Young's modulus obtained from the cyclic-deviator-stress amplitude, cyclic-axial-strain amplitude, as well as the hysteresis-damping factor obtained from the hysteresis curve of the deviator stress and the axial strain.

The test method for determining the equivalent Young's modulus and hysteretic damping factor is specified as the *Method for cyclic triaxial test to determine deformation properties of geomaterials* (jgs0542-2009), set forth by the Japan Geotechnical Society (2016). This standard specifies methods of testing to determine the deformation properties of geomaterials under isotropic or anisotropic stress when subjected to cyclic loading using a triaxial testing apparatus under drained or undrained conditions. The standard applies to geomaterials such as sandy soils, cohesive soils, gravelly soils, soft rocks, and improved soils.

The procedure for the cyclic triaxial test is shown in Figure 1 (Yoshida 2010). In this test, 11 cycles of loading with axial-strain amplitudes are performed as shown in the upper part of Figure 1a. Figure 1b shows a stress–strain diagram (hysteresis curve) drawn from the result of the 10th cycle. The equivalent Young's modulus, E_{eq}, calculates the slope of the secant that connects the starting point and the unloading point of this hysteresis curve. Conversely, the hysteretic damping factor h is calculated from the following equation:

$$h = \frac{1}{4\pi} \frac{S}{W},$$ (1)

where S is the damping energy in the cycle, and is the area of the hysteresis curve form the axial load difference and the axial displacement, and W is equivalent to the elastic energy in that loading.

In the experiment, cyclic loading starts from a small axial-strain amplitude, and the amplitude is gradually increased. The equivalent Young's modulus, E_{eq}, and the hysteresis-damping factor, h, obtained above are expressed as functions of the single-amplitude-axis strain $(\varepsilon_a)_{SA}$, as shown in Figure 1a.

In soils and soft rocks with high uniformity excluding jointed hard rocks, the results of static-cyclic-loading tests are equivalent to those of dynamic tests (for example, the resonant-column tests of soil and elastic-wave-velocity tests), considering the effect of strain rate and cyclic loading (Japan Geotechnical Society 2009). In particular, the effect of strain rate and cyclic loading is very small at small-strain levels (<10 micro-strain) showing the behavior of an almost-linear elastic body. Moreover, even at the medium-strain level (10 to 1,000 micro-strain) indicating the behavior of the elasto-plastic body, the influence of the strain rate is relatively small compared with the large strain (1,000 micro-strain or more) leading to failure.

Therefore, the results of the static-cyclic-loading test on soils and soft rocks can basically be used as dynamic-deformation characteristics.

However, with regard to hard rocks and rock masses, little is known about the influence of strain rate and cyclic load upon the dynamic-deformation characteristics, and it is unclear whether the results of the static-cyclic-loading test can be applied. Therefore, to clarify the dynamic-deformation characteristics of granite, we perform cyclic uniaxial testing in accordance with the *JGS0542-2009 Method for cyclic triaxial test to determine deformation properties of geomaterials*.

3 MATERIALS AND METHODS

3.1 Rock sample

The rock samples in this study were Himekami granite from Morioka city, Iwate Prefecture. Table 1 shows the properties of the rock. The granite was shaped into a cylinder with a diameter of 30 mm and a length of 75 mm. To avoid the influence of anisotropy upon the experimental result, the specimens cored a rock block in a direction perpendicular to the rift plane. Thereafter, the specimens were dried for more than 48 hours in a dry oven at 110 °C and cooled to room temperature in the desiccator containing silica gel and used for the test.

3.2 Experimental apparatus

The loading device used a hydraulic servo-type universal-testing machine (maximum load capacity: 300 kN) controlled by a computer. The load was measured with a load meter built into the testing machine, and the axis and lateral strain were measured with two cross gauges attached to the center of the side of the specimen. Measurements of the general-purpose foil strain gauges were performed using the opposite side-2-active-gauge 3-wire system.

3.3 Method for cyclic uniaxial testing

The cyclic uniaxial test to determine the deformation properties of intact rock was based on the standards

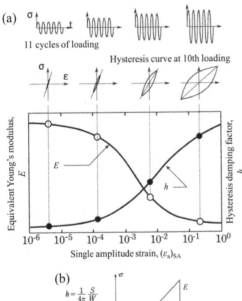

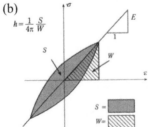

Figure 1. Method for performing the cyclic triaxial test. (a) Loading method and data organization. (b) Calculation of the equivalent Young's modulus and hysteresis-damping factor. (Yoshida 2010)

Table 1. Properties of Himekami granite.

Dry density	(g/cm3)	2.65
Effective porosity	(%)	0.94
Unconfined compression strength	(MPa)	258
Tensile strength	(MPa)	14
P-wave velocity	(km/s)	5.28
S-wave velocity	(km/s)	3.03
Dynamic Young's modulus	(GPa)	61.0
Dynamic Poisson's ratio	(-)	0.254
Dynamic shear modulus	(GPa)	24.33

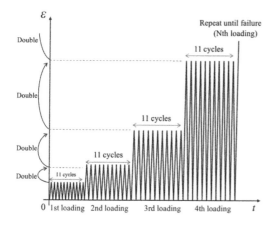

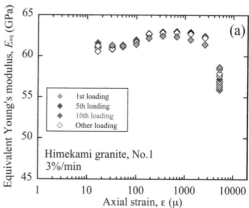

Figure 2. Cyclic-loading method of the cyclic uniaxial test.

of the Geotechnical Engineering Association. In particular, the soil material was changed to a rock specimen, and triaxial compression was changed to uniaxial compression. Figure 2 shows the cyclic loading method of the test. The cyclic-axis displacement between the lower- and the upper-axis strains (approximately 0 to 10 micro-strain) was applied as a sequence of 11 triangular waves. This cyclic loading is called the first loading. Next, a loading equivalent to the first load was performed such that the upper-axis strain would double with the first load (second loading). Thereafter, loading was carried out in the manner of the second loading. This control was repeated as far as loading was possible (nth loading).

To examine the load-rate dependency of the equivalent Young's modulus and the hysteresis-damping factor, cyclic loading was performed at strain rates of 0.3%/min and 3%/min, and a stress rate of 530 MPa/min. Note that the stress rate of 530 MPa/min and the strain rate 3%/min are almost the same as the loading time from the lower-limit strain to the upper-limit strain. Moreover, in the cyclic uniaxial test, it is not possible to unload the lower-limit-axis strain to zero because residual strain occurs every time the number of loadings increases. Therefore, the lower-limit load was set to 300 N for strain-rate control and 500 N for stress-speed control based on the performance of the universal-testing machine. This is because if the lower-limit load is zero, there is a possibility of separating from the upper platen to the specimen.

4 RESULTS AND DISCUSSION

4.1 Effects of the number of cycles upon deformation properties

Figure 3 shows an example of the cyclic uniaxial test results of granite with a loading rate of 3%/min. This figure shows the equivalent Young's modulus, E_{eq}, and hysteresis-damping factor, h, obtained in all 11 cycles under each repetitive loading except the first.

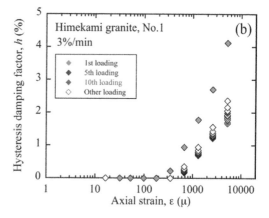

Figure 3. Example of the cyclic-uniaxial-test results. (a) Relation between axial strain and the equivalent Young's modulus. (b) Relation between axial strain and the hysteresis-damping factor.

However, the first loading shows only the values in cycles 2 to 8 due to the failure to obtain the data immediately after the start of the test and the shift to the second loading without running the ninth cycle.

As shown in Figure 3a, the effect of the number of cycles upon the equivalent Young's modulus is small except at the tenth loading just before fracture strength. It is clear from the results for the second to ninth loadings that Young's modulus shows a nearly constant value except at the 1st cycle, but has a tendency to gradually decrease with the increase of the cycle during the tenth loading. Incidentally, although fluctuation is seen in the equivalent Young's modulus under the first loading, there is no correlation between the equivalent Young's modulus and the number of cycles.

Conversely, from Figure 3b, the effect of the number of cycles upon the hysteresis-damping factor is large compared with the equivalent Young's modulus. In particular, the damping factor of the first cycle becomes a large value after the seventh loading, and its value decreases with the increase of the cycle. The decreasing tendency of the value

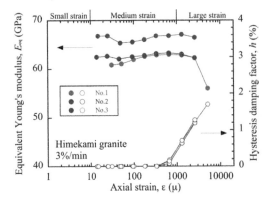

Figure 4. Relation between axial strain and equivalent Young's modulus/hysteresis-damping factor. (Loading rate 3%/min, Number of specimens 3)

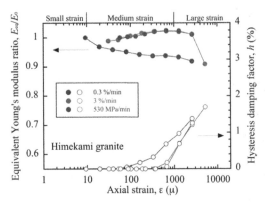

Figure 5. Relation between axial strain and equivalent Young's modulus ratio/hysteresis-damping factor. (Loading rate of 3%/min, 0.3%/min, and 530 MPa/min,)

with increase of the cycle becomes clear as the stage progresses from the seventh to the tenth loadings.

4.2 *Effects of axial-strain level upon deformation properties*

According to the JGS 0542-2009 Method, the equivalent Young's modulus and hysteresis-damping factor are evaluated based on the results of the 5th and 10th cycles; however, in many reports, the dynamic-deformation properties of soils have been discussed using the results for the 10th cycle. Therefore, this study also discusses the equivalent Young's modulus and the hysteresis-damping factor using the results of the 10th cycle.

Figure 4 shows the relation between axial strain and equivalent Young's modulus/hysteresis-damping factor. This figure is the result under the condition of a loading rate of 3%/min. In this study, two out of three specimens broke in the middle of the tenth loading; thus, the tenth-loading data were obtained for only one specimen.

The equivalent Young's modulus increases after decreasing once in the medium-strain region (10 to 1,000 micro-strain), then decreases in the large-strain region (1,000 micro-strain or more), and shows a rapid decrease immediately before the failure. Conversely, the hysteresis-damping factor remains 0% in the latter half of the medium-strain region, showing a behavior that rapidly increases from the medium-strain to large-strain region. Then, the equivalent Young's modulus decreases while drawing a concave down, whereas the hysteresis-damping factor increases almost linearly.

When the variation of each specimen is confirmed, the equivalent Young's modulus varies significantly between specimens; however, the hysteresis-attenuation rate almost never varies.

Thus, it was found that the equivalent Young's modulus and hysteresis-damping factor of the granite under uniaxial compression depend upon the strain level. The increase in the equivalent Young's modulus seen in the middle-strain region is considered to be caused by the closing of preexisting small cracks because the nonlinear of the stage by Price & Farmer (1979) is seen in the middle-strain region of the axial stress–strain curve.

4.3 *Effects of loading rate upon deformation properties*

Figure 5 shows the results of the cyclic uniaxial test conducted at different loading rates. The equivalent Young's modulus ratio E_{eq}/E_0 in the figure is obtained by normalizing the equivalent Young's modulus, E_{eq}, obtained from the cyclic test by the initial secant Young's modulus, E_0. In the figure, three specimens were tested at each loading rate, one of which was shown as a representative result. In addition, for loading rates of 0.3%/min and 530 MPa/min, all specimens were fractured in the middle of the tenth loading, such that the tenth-loading data could not be obtained.

First, we discuss the effect of the loading method upon the dynamic-deformation characteristics. Between the strain rate of 3%/min and the stress rate of 530 MPa/min, there is no difference in the relation between the strain level and the equivalent Young's modulus ratio/hysteresis-attenuation rate. The stress rate of 530 MPa/min and the strain rate of 3%/min are almost the same as the loading time from the lower-limit strain to the upper-limit strain. Therefore, if the loading time is the same, the difference in the loading method is small. However, at a stress rate of 530 MPa/min, it was not possible to obtain the equivalent Young's modulus and the hysteresis-damping factor of the relatively small axial-distortion level of 10 to 50 micro-strain. This is thought to be due to the control performance of the universal-testing machine used in this experiment.

Furthermore, to discuss the effect of loading rate upon the dynamic-deformation properties, we compare the results for strain rates of 3%/min and 0.3%/min. When the strain rate decelerates to one tenth of its value, the equivalent Young's modulus ratio decreases from the small-strain region. In other words, the increased behavior in the medium-strain region at a strain rate of 3%/min is not observed.

In contrast, when the strain rate decelerates to one tenth, the axial-strain level which begins to increase the hysteresis rate decreases to about one tenth (from about 350 micro-strain of strain rate 3%/min to about 40 micro-strain of 0.3%/min).

Thus, the loading-rate dependences of the equivalent Young's modulus and the hysteresis-damping factor of granite were confirmed under uniaxial compression. The difference in the increase or decrease in the equivalent Young's modulus seen in the medium-strain region seems to be due to the loading rate, but its mechanism has yet to be elucidated.

5 CONCLUSIONS

To clarify the dynamic-deformation characteristics of granite, we performed cyclic uniaxial testing using the JGS0542-2009 Method.

The experimental results were as follows:

(1) The measured equivalent Young's modulus increases with strain increasing at small- (<10 micro-strain) to medium- (10 to 1000 micro-strain)-strain values. However, the modulus decreases abruptly at large (>1,000 micro-strain)-strain values, at which the specimen will be broken.

(2) The damping factor is evaluated to be 0% in the small- to medium-strain range, but it increases abruptly in the medium-to large-strain range.

(3) The number of cycles has different effects upon the equivalent Young's modulus and the hysteresis-damping factor; i.e, the effect upon the equivalent Young's modulus is not observed until just before the fracture strength, but that on the hysteresis-damping factor is observed from the middle-strain region before the strength.

(4) The effect of the loading method upon the equivalent Young's modulus and hysteresis-attenuation factor is not observed if the loading times from the lower-limit strain to the upper-limit strain are almost the same. However, if the loading time is reduced to one tenth, a difference in the loading rate is confirmed in the equivalent Young's modulus/history-damping factor.

REFERENCES

Fukumoto, S. Yoshida, N. & Sahara, M. 2009. Dynamic deformation characteristics of sedimentary soft rock. *Journal of japan association for earthquake engineering* 9(1): 46–64.

Japanese Geotechnical Society, 2016. *Japanese geotechnical society standards, Laboratory testing standards of geomaterials*: 2. Tokyo: Japanese Geotechnical Society.

Ochi, K. Kim, Y. S. & Tatsuoka, F. 1993. Study into the deformation characteristics of sedimentary soft rock taking into account the strain level-dipendency and measurement errors. *Doboku Gakkai Ronbunshu* 1993(463): 133–142.

Ochi, K. Tsubouchi, T. & Tatsuoka, F. 1995. Deformation characteristics of sedimentary soft rock examined by the excavation of deep shaft and field tests. *Doboku Gakkai Ronbunshu*. 1993(463): 143–152.

Price, A. M. & Farmer, I. W. 1979. Application of yield models to rock. *International journal of rock mechanics and mining sciences & geomechanics abstracts* 16(2); 157–159.

Yoshida, N. 2010. *Earthquake response analysis of ground*. Tokyo: Kajima institute publishing.

2019 Rock Dynamics Summit– Aydan et al. (eds)
© *2019 Taylor & Francis Group, London, ISBN 978-0-367-34783-3*

Determination of dynamic mode II fracture toughness of rocks using a dynamic punch-through shear method

Wei Yao & Kaiwen Xia
Department of Civil and Mineral Engineering, University of Toronto, Toronto, Ontario, Canada

Ying Xu
State Key Laboratory of Hydraulic Engineering Simulation and Safety, School of Civil Engineering, Tianjin University, Tianjin, China

ABSTRACT: Shearing is the most common mode of failure in rocks under the compressive stress states in geomechanical applications, and the failure of rock bridges between two adjacent discontinuities in rock masses mostly develops as shear fracture. Therefore, mode II fracture (sliding mode) is another important failure mechanism in rock engineering, especially in the macroscopic sense. Although a few methods have been proposed to quantify Mode II fracture toughness of rocks under the static loading condition, no method is available to measure dynamic Mode II fracture toughness of rocks. This paper presents a punch-through shear (PTS) method to measure such material parameter of rocks under different loading rates and confinement pressures. In this method, circular notches are drilled 10 mm deep at both ends of a 54 mm diameter cylinder with 30 mm length to obtain the fracture specimen. The lateral confinement pressure is applied to the cylinder specimen by a 54.74 mm diameter Hoek cell. The dynamic load is exerted on the rock specimen placed in the Hoek cell by using a split Hopkinson pressure bar (SHPB) system, which is modified to guarantee that the specimen assembly remains in contact in the Hoek cell during pressurization by applying a static axial-stress. A specimen holder is designed to allow the punch head to load the specimen directly and, in combination with momentum-trap technique in the SHPB, this specimen holder also enables soft recovery of the rock short rod and rock hollow cylinder produced by punching. Pulse shaping technique is used in all dynamic SHPB tests to facilitate dynamic force equilibrium, under which condition finite element method is employed to obtain equations calculating Mode II fracture toughness under different confinement pressures and loading rates. The application to an isotropic and fine-grained Fangshan marble demonstrates the flexibility and applicability of the proposed PTS method. The dynamic Mode II fracture pattern and modes of the rock are analyzed using the X-ray Micro-CT technique. Five groups of marble specimen under confinements of 0, 5, 10, 15 and 20 MPa are tested with different loading rates. The results show that the failure of the rock specimen is shear dominant and the Mode II fracture toughness increases with the loading rate and the confining pressure.

1 INTRODUCTION

Shear fracturing and mixed mode failure often occur in rock structures, resulting from the complex interaction of tensile and shear fractures (Aliha and Ayatollahi 2011; Chen et al. 2005). Therefore, the studies associated with shear fractures are significant in the assessment of surface and underground stability, earthquakes, hydro-shearing and many other geomechanical applications. Despite the importance of Mode II fracturing in rocks, most of the experimental studies about the rock fracture mechanics have focused on the mode I fracture and experimental methods for determining Mode I fracture toughness for rocks. Compared to the studies about the Mode I fracture toughness determination, fewer studies have been performed on the Mode II fracture toughness determination due to the complexity of the shear fracture toughness determination (Backers et al. 2004; Backers et al. 2005; Backers et al. 2002; Chang et al. 2002; Ko and Kemeny 2006).

The punch through shear (PTS) test is currently popular since the failure in PTS specimen is shown to be dominant in Mode II (Backers and Stephansson 2015). As a result, the International Society for Rock Mechanics and Rock Engineering (ISRM) has recently accepted the PTS method as the suggested method for determining static Mode II fracture toughness of rocks (Backers and Stephansson 2015). Another important merit of the PTS specimen is that it facilitates in the measurement of the Mode II fracture toughness with confinements (Backers et al. 2002), which has wide applications in rock engineering and rock mechanics research due to the natural environment of rocks involved in situ stresses. Also theoretically, many studies have shown that the static Mode II fracture toughness K_{IIC} strongly depends on the confining pressure that is applied to the rock (Backers 2005; Backers et al. 2006; Backers et al. 2004; Backers et al. 2005; Backers et al. 2002; Wu et al. 2017). However, in a large number of underground rock engineering applications, such as deep rock drilling, geothermal excavation and rock blasting, rocks with high in-situ stresses may be dynamically failed. Hence, it is desirable to determinate the dynamic Mode II fracture toughness of rocks under different confining pressures. To our best knowledge, there is only one experimental study in the open literature that has been conducted on the determination of the dynamic Mode II fracture toughness of rocks with confining pressures. Lukić and Forquin (2016) conducted a series of PTS experiments on ultra-high-performance concrete to investigate the confined shear fracture under different strain rates.

As discussed above, in spite of the quantification of the static Mode II fracture toughness of rocks under confining conditions, a method is missing to systematically measure the dynamic Mode II fracture toughness of rocks under different confining pressures. Therefore, this work introduces a modified PTS method to measure the dynamic Mode II fracture toughness of rocks with lateral confinement. A split Hopkinson pressure bar (SHPB) system is used to exert the dynamic load to the rock specimen, with the confining pressure applied to the specimen through a Hoek cell. The dynamic PTS specimen in this study is designed according to our early dynamic PTS specimen without confinement (Yao et al. 2017b). The fracture pattern and modes of rocks under different loading rates and confining pressures are examined by the post-mortem observation through 3D X-ray micro-computed tomography (CT) technique (Desrues et al. 2006).

2 METHODOLOGY

2.1 *Specimen preparation*

Fangshan marble (FM) from Beijing, China is chosen in this study to conduct the proposed dynamic PTS tests with confinement because it has fine grains and homogeneous. Based on the ISRM suggested PTS method for determining the Mode II fracture toughness under static loading (Backers and Stephansson 2015), a valid dynamic PTS specimen geometry has been proposed in our previous study for dynamic tests without confinement (Yao et al. 2017b). Hence, in this study, the same PTS specimen geometry is utilized in the dynamic tests with confinement except for the outer diameter, as shown in Figure 1a. In order to match the inner diameter of the Hoek cell used for applying confinement pressure, the outer diameter of the PTS specimen in this study is changed into 54 mm, which is 4 mm larger than both the static (Backers and Stephansson 2015) and previous dynamic (Yao et al. 2017b) PTS specimen. A little bit increase of outer diameter is acceptable for a valid K_{IIC} measurement in the dynamic PTS tests because it has been proven that the Mode II fracture toughness K_{IIC} derived from the PTS experiment is almost independent of the outer diameter when other dimensions of the PTS specimen are

constant (Backers et al. 2002). Thus the specimen geometry for dynamic PTS tests with confinement in this study is designed as shown in Figure 1a.

To make the PTS specimen, the FM block was first drilled into 54 mm in diameter cores and then cut and grinded into a cylinder with 30 mm in length. Thereafter, two circular notches with a diameter of 25.4 mm, a width of 1.5 mm and a depth of 10 mm were made into two ends surface of the cylindrical specimen. In such way, the length of the intact rock portion is 10 mm (Figure 1b). A small curvature should exist at the bottom of the notches.

2.2 *Dynamic PTS method with confinement*

As a further investigation of the dynamic PTS method without confinement, the split Hopkinson pressure bar (SHPB) system was employed and modified to measure the dynamic Mode II fracture toughness of rocks in this study (Yao et al. 2017b; Zhou et al. 2012). As shown in Figure 2, the modified SHPB system is composed of a traditional SHPB system and a confinement system. The traditional SHPB system consists of a striker, an incident bar, a transmitted bar and a data acquisition system. The confinement system includes a Hoek cell and an axial pre-stress system attached to SHPB.

The schematic of the Hoek cell with the specimen is further shown in Figure 3. Similar to dynamic PTS tests without confinement (Yao et al. 2017b), the punch head in this device is the incident bar (25.4 mm diameter) and the steel rear supporter attached to the transmitted bar has an inner diameter of 30 mm, which is 4.6 mm larger than the diameter of the incident bar to accommodate shear deformation

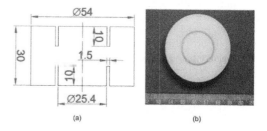

Figure 1. (a) the dimension of the dynamic PTS specimen and (b) a typical virgin PTS specimen.

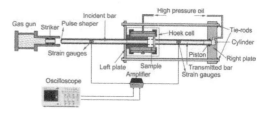

Figure 2. Schematics of a triaxial-SHPB system for dynamic PTS tests with confinement.

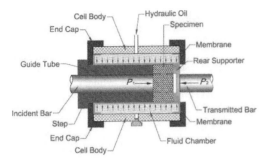

Figure 3. Schematics of the specimen holder and Hoek cell in dynamic PTS tests.

(Figure 3). Base on the geometry of the specimen shown in Figure 3 and Figure 1a, the compressive axial load generates the shear force in the sample through the incident bar and the rear supporter. Meanwhile, the lateral confinement is uniformly acted to the side surface of the specimen by using a Hoek cell.

The specimen, the rear supporter and the transmitted bar move rightwards during the dynamic tests. Thereby, the small contact area between the rear supporter and the membrane of the rear supporter is necessary to minimize the frictional effect during the dynamic loading period. In this study, a guide tube is designed to support the lateral pressure exerted by the remaining part of the Hoek cell, and the incident bar can freely slide in the guide tube and its end is in good contact with the specimen. The outer diameter of the guide tube is 54 mm, the same as that the inner diameter of the membrane and the outer diameter of the rear supporter and the specimen. The inner diameter of the guide tube is 25.6 mm, slightly larger than the incident bar (i.e. punch head) to ensures that the incident bar can move freely and also guarantees that the incident bar is properly aligned with the specimen and the rear supporter.

The axial pre-load on the outer ring part of the specimen (as shown in Figure 2) is indispensable in the dynamic PTS tests with confinement because the axial pre-load on the outer ring part of the specimen reduces the bending force during the dynamic test and avoids the further damage in the specimen (Backers and Stephansson 2015). Therefore, the axial pre-stress system is used to apply axial pre-stress on the sample assembly (Figure 2). This pre-stress system is similar to that used in our early work to measure the static pre-stress effect on the dynamic tension behavior of rocks (Wu et al. 2015). The pressure chamber (Cylinder) at the right end of the transmitted bar provides axial pressure to the bars and thus to the rear supporter, the outer ring part of the dynamic PTS specimen and the guide tube. This pre-stress stops the rightward motion of the transmitted bar and the sample assembly. The leftward axial motion of the sample assembly is constrained by the left plate through a step at the

left end of the guide tube (Figure 2). The relative motion between the right and left plate is restricted by two tie-rods. Therefore, in this testing system, the axial compressive pre-load is applied to the outer ring part of the dynamic PTS specimen and the confining pressure is acted on the side of the PTS specimen.

When the confining pressure and axial pressure are applied to the PTS specimen, an incident elastic compressive wave is generated in the incident bar after impacted by the striker bar. The incident wave propagates rightwards and interacts with the specimen assembly, producing a reflected wave propagating back into the incident bar and a transmitted wave into the transmitted bar. The incident wave ε_i, reflected wave ε_r and transmitted wave ε_t are measured by strain gauges glued on the surfaces of the incident and transmitted bars. A digital oscilloscope is used to record and store the amplified strain signals.

2.3 Dynamic force balance

The dynamic forces P_1 and P_2 on the two ends of the specimen (Figure 3) can be derived from the strains derived from three waves (i.e. the incident wave ε_i, the reflected wave ε_r and the transmitted wave ε_t) in the elastic bars (Kolsky 1953):

$$P_1 = AE(\varepsilon_i + \varepsilon_r) \qquad (1)$$

$$P_2 = AE\varepsilon_t \qquad (2)$$

where A and E are the cross-sectional area and Young's modulus of bars, respectively.

Based on Equations (1) and (2), the dynamic loading forces on two ends of the PTS samples for the typical test with the confinement are calculated and shown in Figure 4. One can see that the dynamic forces on both ends of the specimen with the confinement are almost identical before the peak of these two forces, which is also synchronized with

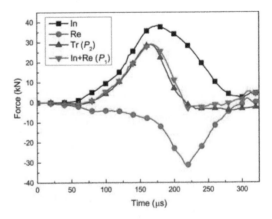

Figure 4. Dynamic force balance check for a typical dynamic PTS test with pulse shaping.

185

the failure point of the PTS specimen (Yao et al. 2017b). This verifies that the force balance is reached during the dynamic loading period (i.e. $P_1 \approx P_2$) and the inertial effects are thus eliminated. It further approves that the design for the dynamic PTS tests with confinement successfully minimizes the friction between the sample and the membrane in this study.

2.4 Deduction of the Mode II facture toughness

According to the suggested method for determining the Mode II fracture toughness under static loading (Backers 2005; Backers and Stephansson 2015), the Mode II fracture toughness of rocks through the PTS tests can be determined as following:

$$K_{IIC} = \alpha \cdot p_L + \beta \cdot p_c \qquad (3)$$

where K_{IIC} is the Mode II fracture toughness (MPa·m$^{1/2}$), p_L is the axial load at failure (MPa), p_c is the applied confining pressure (MPa), α and β are parameters to be obtained from regression.

In addition, it has been proven that the inertial effect in the dynamic test can be ignored when the dynamic force equilibrium condition is satisfied (i.e. $P_1 \approx P_2$) in the dynamic rock specimen (Xia and Yao 2015; Zhang and Zhao 2013b; Zhou et al. 2012). Under such a condition, the dynamic Mode II fracture toughness K_{IIC}^D can be evaluated via the same formula as the static Mode II fracture toughness formula of Equation (3). As a result, the following formula is proposed to determine the K_{IIC}^D from the peak dynamic load P_{max} and the confinement pressure p_c during the dynamic PTS test with confinement:

$$K_{IIC}^D = \eta P_{max} + \lambda p_c \qquad (4)$$

where the unit for K_{IIC}^D is MPa·m$^{1/2}$ and for P_{max} and p_c are MPa. The shape parameters η and λ are determined for the dynamic specimen geometry by using the displacement extrapolation technique (DET) along with a finite element analysis as suggested by the ISRM to obtain the Mode II fracture toughness K_{IIC} under the static PTS test (Backers and Stephansson 2015). In this study, $\eta = 0.04192$ m$^{1/2}$ and $\lambda = -5.6 \times 10^{-3}$ m$^{1/2}$ for the dynamic FM PTS specimen geometry.

With the dynamic force equilibrium across the dynamic PTS specimen, the dynamic SIF can be derived from Equation (4):

$$K_{II}^D(t) = \eta P(t) + \lambda p_c \qquad (5)$$

where K_{II}^D has the unit MPa·m$^{1/2}$, $P(t) = P_1(t) = P_2(t)$ is the axial dynamic loading history (MPa). The confining pressure is constant during the dynamic loading and the value of the parameters η and λ is the same as

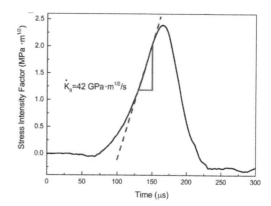

Figure 5. Typical SIF-time curve for determining the loading rate.

those in Equation (4). The maximum value of K_{II}^D is considered as the Mode II facture toughness K_{IIC} of the specimen tested since the peak dynamic loading is synchronized with the failure of the specimen with the achievement of dynamic stress equilibrium (Yao et al. 2017a). The loading rate $\dot{K}_{II}$ is calculated by the time evolution of the SIF in the specimen. Figure 5 gives the SIF-time curve of the PTS specimen during the dynamic loading period. According to the method suggested by the ISRM for determining the dynamic behaviors of rocks (Xia and Yao 2015; Zhang and Zhao 2013b; Zhou et al. 2012), the slope (the dash line in Figure 5) of the almost linear rising region in the SIF-time curve is the loading rate. For each dynamic PTS test, the loading rate was determined using this method.

3 RESULTS AND DISCUSSIONS

3.1 Fracture pattern and modes

In order to study the fracture pattern and modes in the tested specimen, the specimens are softly recovered for valid post-mortem examinations (Song and Chen 2004; Xia and Yao 2015; Zhang and Zhao 2013b). Figure 6 shows the soft-recovered dynamic PTS specimen, which is punched into a short rod and a hollow cylinder. Also, few visible radial cracks are identified on the hollow cylinder.

In addition to the main visible fractures on the short rod and hollow cylinder, the 3D microcracks inside the

Figure 6. The rock short rod and hollow cylinder produced in a typical dynamic PTS test (the unit in the picture is in centimeter).

186

recovered dynamic PTS specimen are examined by using X-ray micro-CT technique (a non-destructive method). In this study, the nanoVoxel2102E system micro-CT by Sanying Precision Instruments was used to scan the recovered PTS specimens at 150 kV and 150 μA. In this study, five groups of FM specimens were tested with the lateral confinements of 0, 5, 10, 15 and 20 MPa. Figure 7 shows the fracture patterns of the specimen under the similar loading rates but different confining pressures. It is clear that the main Mode II fractures are essentially formed in the PTS specimens under all confining pressures, verifying that the dynamic PTS tests are valid for all the confining pressures in this study. Furthermore, the wing cracks occur at the top notch (at the side of the transmitted bar), and the ring fractures ('doughnut fractures') are formed at the bottom notch (at the side of the incident bar). In addition, since the main fracture is always initiated at the top notch, the 'doughnut fracture' sometimes formed at the bottom notch has negligible influence on the stress intensity factor for the main fracture initiation. Moreover, the wing crack first commences at the inner top notch and then propagates outward from the induced shear zone and finally stops. This phenomenon is also observed in the static PTS test and has been verified that the wing crack has a negligible effect on the stress intensity factor in shear (Backers and Stephansson 2015; Backers et al. 2002; Kawakata and Shimada 2000). Therefore, the main fracture formed at the peak force in the dynamic PTS test is a dominated Mode II fracture. Also, the designed PTS specimen and the modified dynamic PTS test with the SHPB are valid and suitable for the dominated Mode II fracture with confinements.

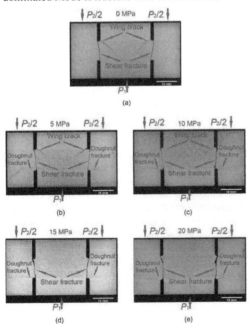

Figure 7. The CT images of specimens under the similar loading rates and different confining pressures.

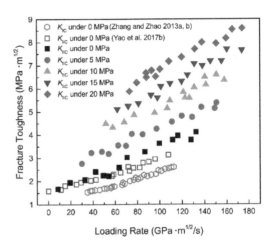

Figure 8. The dynamic fracture toughness versus loading rate.

It is noteworthy in Figure 7 that the wing cracks are vanishing and the 'doughnut fractures' appear as the confining pressure increases. This phenomenon may be caused by the different static stress states at the tips of the notches due to the increasing confining pressures. Another observation is that the gaps in the main fracture increase with the lateral confinement, indicating that more materials in the main fracture are pulverized under a higher confining pressure.

3.2 Dynamic mode II fracture toughness

In this study, dynamic PTS experiments were performed under various loading rates and five different confining pressures. The correlation between the dynamic K_{IIC} of FM and loading rates under confining pressures is illustrated in Figure 8. It can be observed that, under a given lateral confinement, the K_{IIC} increases with the loading rates, showing that the K_{IIC} of FM has strong loading rate dependence. Another important observation is that at a given loading rate, the K_{IIC} increases with the lateral confinement. This trend is also observed about the static K_{IIC} (Backers and Stephansson 2015) of rocks under confining pressures. In addition, Figure 8 gives the K_{IIC} of FM in our early study (Yao et al. 2017b) and the dynamic K_{IC} of FM in the references (Zhang and Zhao 2013a). The K_{IIC} of FM without the confining pressure in this study is very close to that in our early work and the little difference between them is probably caused by the diversity of the FM.

4 CONCLUSIONS

The dynamic Mode II fracture toughness of rocks under different confining pressures was systematically investigated by using a dynamic punch-through shear (PTS) method and a Hoek cell in this work. According to the size of the Hoek cell, the PTS specimen was designed as a cylinder with a diameter of 54 mm and a length of 30 mm as well as a 10 mm depth circular notch at both ends. A split Hopkinson pressure bar (SHPB) system was successfully modified to accommodate Mode II fracture toughness tests with confinements. An isotropic and fine-grained Fangshan marble was used in this study and the dynamic force balance in all dynamic PTS specimen was achieved by using the pulse shaping technique. The rock specimen was punched and broke into a short rod and a hollow cylinder, then it was soft recovered to analyze the dynamic Mode II fracture pattern and modes through the X-ray Micro-CT technique. The dynamic Mode II fracture toughness of rocks is derived from the peak dynamic load and confining pressure via an equation determined by using the finite element method.

The failure pattern inside the recovered PTS specimen was revealed by the CT images and is dependent on the loading rate and confinements. The main Mode II fractures are essentially formed at the peak force in the PTS specimens under all confining pressures. The wing cracks are vanishing and the 'doughnut fractures' appear as the confining pressure increases and the gaps in the main fracture increase with the lateral confinement. Moreover, the main fracture forming at peak force in the dynamic PTS test is dominated Mode II fracture, and the designed PTS specimen and the modified dynamic PTS test with the SHPB are valid and suitable for the dominated Mode II fracture with confinements.

The Mode II fracture toughness K_{IIC} of FM under a certain confining pressure increases with the loading rate, showing clear loading rate dependence. Another important observation is that the K_{IIC} at a given loading rate increases with the lateral confinement. This trend is also observed for the static K_{IIC} of rocks under confining pressures.

REFERENCES

Aliha, M.R.M., Ayatollahi, M.R. 2011. Mixed mode I/II brittle fracture evaluation of marble using SCB specimen. *Procedia Engineering* 10:311–318.
Backers, T. (2005) Fracture toughness determination and micromechanics of rock under mode I and mode II loading. Dissertation, Geoforschungszentrum
Backers, T., Antikainen, J., Rinne, M. Time dependent fracture growth in intact crystalline rock: laboratory procedures and results. In: GeoProc2006—2nd International Conference on coupled THMC processes in geosystems: fundamentals, modelling, experiments, applications, Nanjing, China, 2006.
Backers, T., Dresen, G., Rybacki, E., Stephansson, O. 2004. New data on mode II fracture toughness of rock from the punchthrough shear test. *International Journal of Rock Mechanics and Mining Sciences* 41, Supplement 1:2–7.
Backers, T., Stanchits, S., Dresen, G. 2005. Tensile fracture propagation and acoustic emission activity in sandstone: The effect of loading rate. *International Journal of Rock Mechanics and Mining Sciences* 42(7):1094–1101.
Backers, T., Stephansson, O. (2015) ISRM Suggested Method for the Determination of Mode II Fracture

Toughness. In: Ulusay, R (ed) The ISRM Suggested Methods for Rock Characterization, Testing and Monitoring: 2007-2014. Springer International Publishing, New York City, United States, pp 45–56. doi:10.1007/978-3-319-07713-0_4
Backers, T., Stephansson, O., Rybacki, E. 2002. Rock fracture toughness testing in Mode II—punch-through shear test. *International Journal of Rock Mechanics and Mining Sciences* 39(6):755–769.
Chang, S.-H., Lee, C.-I., Jeon, S. 2002. Measurement of rock fracture toughness under modes I and II and mixed-mode conditions by using disc-type specimens. *Engineering Geology* 66(1–2):79–97.
Chen, F., Cao, P., Rao, Q., Ma, C., Sun, Z. 2005. A mode II fracture analysis of double edge cracked Brazilian disk using the weight function method. *International Journal of Rock Mechanics and Mining Sciences* 3(42):461–465.
Desrues, J., Viggiani, G., Besuelle, P. (2006) Advances in X-ray Tomography for Geomaterials. LSTE USA, Newport Beach, California
Kawakata, H., Shimada, M. 2000. Theoretical approach to dependence of crack growth mechanism on confining pressure. *Earth, Planets and Space* 52(5):315–320.
Ko, T., Kemeny, J. Determination of Mode II stress intensity factor using short beam compression test. In: Rock Mechanics in Underground Construction: ISRM International Symposium 2006: 4th Asian Rock Mechanics Symposium, 8–10 November 2006, Singapore, 2006. World Scientific, p 346
Kolsky, H. (1953) Stress waves in solids. Clarendon, Oxford
Lukić, B., Forquin, P. 2016. Experimental characterization of the punch through shear strength of an ultra-high performance concrete. *International Journal of Impact Engineering* 91:34–45.
Song, B., Chen, W. 2004. Loading and unloading split Hopkinson pressure bar pulse-shaping techniques for dynamic hysteretic loops. *Experimental Mechanics* 44(6):622–627.
Wu, B., Chen, R., Xia, K. 2015. Dynamic tensile failure of rocks under static pre-tension. *International Journal of Rock Mechanics and Mining Sciences* 80:12–18.
Wu, H., Kemeny, J., Wu, S. 2017. Experimental and numerical investigation of the punch-through shear test for mode II fracture toughness determination in rock. *Engineering Fracture Mechanics* 184:59–74.
Xia, K., Yao, W. 2015. Dynamic rock tests using split Hopkinson (Kolsky) bar system-A review. *Journal of Rock Mechanics and Geotechnical Engineering* 7(1):27–59.
Yao, W., He, T., Xia, K. 2017a. Dynamic mechanical behaviors of Fangshan marble. *Journal of Rock Mechanics and Geotechnical Engineering* 9(5):807–817.
Yao, W., Xu, Y., Yu, C., Xia, K. 2017b. A dynamic punch-through shear method for determining dynamic Mode II fracture toughness of rocks. *Engineering Fracture Mechanics* 176:161–177.
Zhang, Q.B., Zhao, J. 2013a. Determination of mechanical properties and full-field strain measurements of rock material under dynamic loads. *International Journal of Rock Mechanics and Mining Sciences* 60:423–439.
Zhang, Q.B., Zhao, J. 2013b. A Review of Dynamic Experimental Techniques and Mechanical Behaviour of Rock Materials. *Rock Mechanics and Rock Engineering* 47(4):1411–1478.
Zhou, Y.X. et al. 2012. Suggested methods for determining the dynamic strength parameters and mode-I fracture toughness of rock materials. *International Journal of Rock Mechanics and Mining Sciences* 49:105–112.

2019 Rock Dynamics Summit– Aydan et al. (eds)
© *2019 Taylor & Francis Group, London, ISBN 978-0-367-34783-3*

The effect of characteristics of back-filling material on the seismic response and stability of castle retaining-walls

Y. Yamashiro, Ö. Aydan, N. Tokashiki, J. Tomiyama & Y. Suda
Department of Civil Engineering, University of the Ryukyus, Nishihara, Okinawa, Japan

ABSTRACT: Recent earthquakes caused severe damage to the retaining walls of various castles in Japan. In this study, the authors describe this instrumentation, some numerical analyses and the results obtained so far. The site investigations clearly showed that river gravels were used as backfilling materials at the collapsed castle retaining walls. In this study, the authors have investigated the effect of the type of backfilling material on the seismic response and stability of the model castle retaining walls using a shaking table in the laboratory. In addition, a dynamic limiting equilibrium approach was used to investigate the responses and stability of the test results on the model castle retaining walls. The experiments and theoretical studies clearly showed that the type of backfilling material has a great effect on the dynamic response and seismic stability of the walls and the results clearly showed that the castle walls utilizing rounded river gravels as backfilling material are quite vulnerable to fail during great earthquakes. The authors would present the outcomes of the this unique experimental and analytical study and discuss their implications in practice.

1 INTRODUCTION

The 2016 Kumamoto earthquake caused huge damage to Kumamoto Castle. Particularly the retaining-walls of the Kumamoto castle was heavily damaged (Aydan et al. 2018). Similar events were also observed on the retaining walls of Sunpu Castle in Shizuoka due to the 2009 Suruga Bay earthquake, Katsuren Castle due to the 2011 Off-Okinawa earthquake, Shirakawa Castle due to 2011 Great East Japan earthquake. The site investigations at the damaged sites clearly indicated that river gravels were used as backfilling materials at the collapsed castle retaining walls (Fig. 1).

The effect of the type of backfilling material on the seismic response and stability of the model castle retaining walls were investigated using a shaking table in the laboratory in the University of the Ryukyus. Three different back-filling materials were used. Furthermore, laboratory shear tests on the backfilling material and friction coefficient between the backfilling material and the retaining wall material were performed. In addition, a dynamic limiting equilibrium approach was used to investigate the responses and stability of the test results on the model castle retaining walls. The shaking table experiments clearly showed that the type of backfilling material has a great effect on the dynamic response and seismic stability of the walls and the results clearly showed that the castle walls utilizing rounded river gravels as backfilling material are quite vulnerable to fail during great earthquakes. The dynamic limiting equilibrium method was able to confirm the experimental results theoretically. The outcomes of the this unique experimental and analytical study are presented and discussed with an emphasis on their implications in practice.

2 SHAKING TABLE AND MODELS

2.1 *Shaking table and instrumentation*

The shaking table used for model tests was produced by AKASHI. Its operation system was recently updated by IMV together with the possibility of applying actual acceleration wave forms from earthquakes. Table 1 gives the specifications of the shaking table and monitoring devices. The size of the shaking table is 1000 x 1000 mm². The maximum acceleration is 600 gals for a model with a weight of 100 kgf. The displacement

Figure 1. Some examples of castle-wall damage in recent earthquakes.

Table 1. The specifications of monitoring sensors and shaking table.

Shaking table and sensors	Specifications	
Shaking Table - AKASHI	Frequency Stroke Acceleration	1-50 Hz 100 mm 600 gals
Accelerometers	Range	10G
Laser Displacement Transducer OMRONKEY-ENCE	Range Range	0-300 mm 0-100 mm

response of models were monitored using laser displacement transducers and the input acceleration of the shaking table and acceleration response of the retaining wall were measured using the two accelerometers.

2.2 Model setup

An acrylic transparent box with 630 mm in length, 300 mm in height and 100 mm width was used as shown in Fig. 2. The wall-thickness was 10 mm so that the box was relative rigid and the frictional resistance of sidewalls was quite low.

The blocks used were made of Ryukyu limestone with a size of 40x40x99.5 mm with the consideration of materials used for the retaining walls of historical castles in Ryukyu Archipelago. Furthermore, the base block was such that the overall wall inclination can be chosen as 70, 83 and 90 degrees. The base block was fixed to two-sided tapes to the base of the acrylic box. In addition, the Ryukyu limestone of the same size was laid over the base as seen in Fig. 2. This was expected to provide a condition similar to the actual conditions observed in many historical castles in Ryukyu Archipelago. The wall height was 240 mm and the ratio of the height to width was 1/6. When the retaining wall inclination is 90 degrees without backfill material, it is expected that the wall would start rocking at an acceleration level of 167 gals.

3 BACKFILL MATERIALS AND THEIR PROPERTIES

3.1 Backfill materials

Three different backfill materials were chosen (Fig. 3). Glass beads were chosen to represent the lowest shear resistant backfill material while the angular fragments of Motobu limestone was selected as the highest shear resistant backfill material. The third backfill material was rounded river gravels having a shear resistant in-between those of the two previous backfill materials.

3.2 Properties of backfill materials

A special shear testing set-up was developed to obtain the shear strength characteristics of backfill materials under low normal stress levels, which are quite relevant to the model tests to be presented in this study. Fig. 4 shows the shear testing device. Fig. 5 shows the shear strength envelopes for three backfill materials. As noted from the figure, shear strength of rounded river gravel is in-between the shear strength envelopes of glass-beads and Motobu limestone gravel. The strength of backfill materials

Glass beads   Rounded river gravel   Motobu limestone gravel

Figure 3. Views of backfill materials.

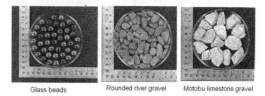

Figure 4. A view of shear testing device under low normal stress level.

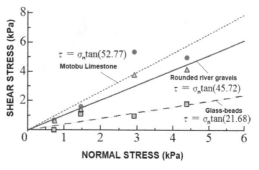

Figure 5. Shear strength envelopes for backfill materials.

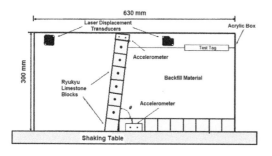

Figure 2. An illustration of model box.

190

is frictional and the friction angle of the glass-beads is about 21.68 degrees.

Another important factor for the stability of the retaining walls of historical castles as well as other similar structures is the frictional resistance between the backfill material and retaining-wall blocks. For this purpose, tilting experiments were carried out. The backfill material contained in a box put upon the Ryukyu limestone platens without any contact and tilted until it slides. This response of the backfill material contained in box was measured using laser-displacement transducers. The inferred friction angles are given in Table 2. The lowest friction angle was obtained in case of glass beads as expected.

4 SHAKING TABLE TESTS ON RETAINING-WALLS WITH GLASS-BEADS BACKFILL

A series of sweep tests were carried before the failure tests. Regarding the glass-beads backfill material, the retaining walls were statically unstable for 90 degrees while they failed during the sweep test on the retaining walls with an inclination of 83 degrees. Therefore, we could show one example for retaining walls for the inclination of 70 degrees (Fig. 6). Its Fourier spectra analysis is shown in Fig. 7. The results indicated no apparent natural frequency was dominant. The situation was quite similar in all experiments. Therefore, more emphasis will be given to the failure experiments.

Although the test on the retaining wall with an inclination of 83 degrees was intended for a sweep test, it resulted in failure. Fig. 8 shows the displacement and base acceleration during the test.

Table 2. Friction angle between Ryukyu limestone and backfill materials.

Parameter	Glass-beads	Rounded river gravel	Motobu limestone fragments
Friction angle	12.5-16.8	25.0-27.5	25.9-27.8

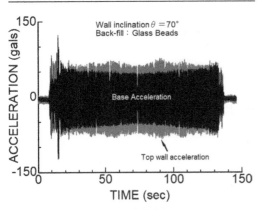

Figure 6. Acceleration records of the shaking table and top of the retaining wall.

Failure tests on the retaining walls with an inclination of 70 degrees were carried out by applying sinusoidal waves with a frequency of 3Hz. The amplitude waves were gradually increased until the failure occurred. Fig. 9 shows an example of failure. The yielding initiated at about 110 gals and the total failure occurred when the input acceleration reached 215 gals. Fig. 10 shows the retaining wall

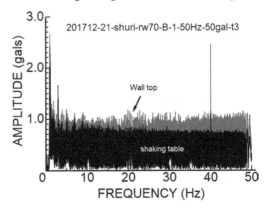

Figure 7. Fourier spectra of acceleration records.

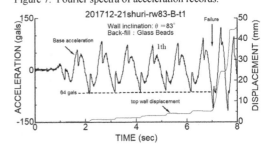

Figure 8. Acceleration and displacement responses on the retaining wall with an inclination of 83 degrees.

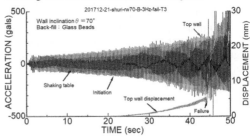

Figure 9. Acceleration and displacement responses on the retaining wall with an inclination of 70 degrees.

(a) Before   (b) After

Figure 10. Views of the model retaining wall with an inclination of 70 degrees before and after the test.

191

before and after the failure test. The retaining wall failed due to toppling (rotation) failure although some relative sliding occurred with the block at the toe of the model retaining wall.

5 SHAKING TABLE TESTS ON RETAINING-WALLS WITH RIVER GRAVEL BACKFILL

A series of sweep tests were carried before the failure tests as explained in the previous section. Regarding the rounded river gravel backfill material, the retaining walls were statically unstable for 90 degrees while the sweep test on the retaining walls with an inclination of 83 and 70 degrees could be carried. We show one example for retaining walls for the inclination of 83 degrees in Fig. 11 and its Fourier spectra analysis in Fig. 12. The results indicated there was no dominant natural frequency for the given range of frequency. The situation was quite similar in all experiments for 83 and 70 degrees retaining wall models.

Failure tests on the retaining walls with inclinations of 83 and 70 degrees were carried out by applying sinusoidal waves with a frequency of 3Hz. The amplitude waves were gradually increased until the failure occurred. Figs. 13 and 14 show acceleration and

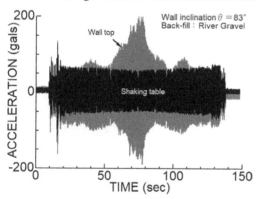

Figure 11. Acceleration records of the shaking table and top of the 83 retaining wall with rounded river gravel backfill.

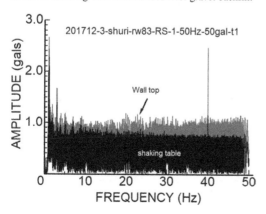

Figure 12. Fourier spectra of acceleration records.

displacement responses of retaining walls with inclinations of 83 and 70 degrees as examples of failure tests. The yielding initiated at about 100 gals and the total failure occurred when the input acceleration reached 210 gals for 83 degrees retaining walls. On the other hand, the yielding initiated at 220 gals and the total failure occurred when the input acceleration was 430 gals for 70 degrees retaining walls as seen in Fig. 15. The retaining wall failed due to toppling (rotation) failure although some relative sliding occurred with the block at the toe of the model retaining wall (Fig. 15).

6 SHAKING TABLE TESTS ON RETAINING-WALLS WITH MOTOBU LIMESTONE GRAVEL BACKFILL

A series of sweep tests were carried before the failure tests as explained in the previous section. Regarding the angular Motobu limestone gravel backfill material, the retaining walls were statically unstable for 90 degrees with a height of 240 mm. However, they were stable when the height was reduced to 160 mm. The sweep test on the retaining walls with an inclination of 90, 83 and 70 degrees wee carried. We show

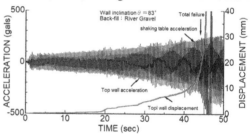

Figure 13. Acceleration and displacement responses on the retaining wall with an inclination of 83 degrees.

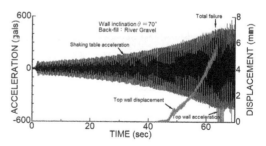

Figure 14. Acceleration and displacement responses on the retaining wall with an inclination of 70 degrees.

(a) Before   (b) After

Figure 15. Views of the model retaining wall with an inclination of 70 degrees before and after the test.

192

one example for retaining walls for the inclination of 70 degrees in Fig. 16 and its Fourier spectra analysis in Fig. 17. Again, the results indicated there was no dominant natural frequency for the given range of frequency. The situation was quite similar in all experiments for 90, 83 and 70 degrees retaining wall models.

Failure tests on the retaining walls with inclinations of 90, 83 and 70 degrees were carried out by applying sinusoidal waves with a frequency of 3Hz. The procedure was the same those in previous experiments. Figs. 18, 19 and 20 show acceleration and displacement responses of retaining walls with

inclinations of 90, 83 and 70 degrees as examples of failure tests. The yielding initiated at about 110 gals and the total failure occurred when the input acceleration reached 260 gals for 90 degrees retaining walls. On the other hand, the yielding initiated at 130 gals and the total failure occurred when the input acceleration was 300 gals for 83 degrees retaining walls as seen in Fig. 19. The retaining walls failed due to toppling (rotation) failure.

The retaining walls with 70 degrees inclination and height of 240 mm did not failed during the entire test up to 400 gals as seen in Figs. 20 and 23. Although some relative sliding occurred with the block at the toe of the model retaining wall when the base acceleration reached to the level of 300 gals (Fig. 23). However, some settlement of the backfill occurred and the retaining wall was pushed in passive sliding mode.

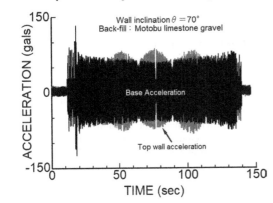

Figure 16. Acceleration records of the shaking table and top of the 70 degrees retaining wall with rounded river gravel backfill.

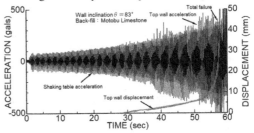

Figure 19. Acceleration and displacement responses on the retaining wall with an inclination of 83 degrees.

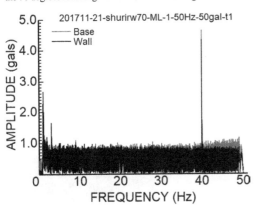

Figure 17. Fourier spectra of acceleration records.

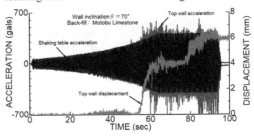

Figure 20. Acceleration and displacement responses on the retaining wall with an inclination of 70 degrees.

(a) Before   (b) After

Figure 21. Views of the model retaining wall with an inclination of 90 degrees before and after the test.

(a) Before   (b) After

Figure 22. Views of the model retaining wall with an inclination of 83 degrees before and after the test.

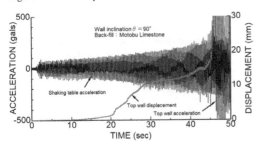

Figure 18. Acceleration and displacement responses on the retaining wall with an inclination of 90 degrees.

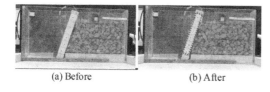

(a) Before   (b) After

Figure 23. Views of the model retaining wall with an inclination of 70 degrees before and after the test.

7 DYNAMIC RESPONSE ANALYSES

Aydan (2017), Aydan et al. (2003) and Tokashiki and Aydan (2007) proposed a dynamic equilibirum method to analyse the acceleration and displacement response as well as the stability of retaining walls. The retaining walls generally fail in three modes; sliding failure, toppling failure and combined sliding and toppling failure as illustrated in Fig. 24. This method applied to the experiments shown in previous sections. Figs. 25 to 27 shows the comparison of computed results with experimental results for selected retaining walls with the consideration of toppling and sliding modes with and without sidewall resistances. Material properties used are also shown in the figures. Although the computed results estimate the failure of retaining walls at lower acceleration levels, it is capable of estimating the dynamic displacement responses during failure.

8 CONCLUSIONS

The recent earthquakes such as the 2016 Kumamoto earthquake damaged the Castles, especially, their retaining-walls. Site investigations showed that

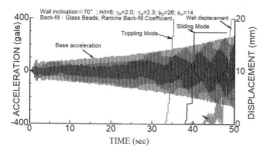

Figure 24. Failure modes of retaining walls.

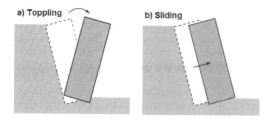

Figure 25. Comparison of computations with experimental data for the 70 degrees retaining wall with glass-beads backfill.

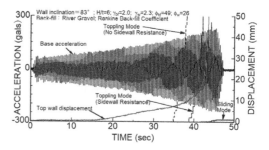

Figure 26. Comparison of computations with experimental data for the 83 degrees retaining wall with river gravel backfill.

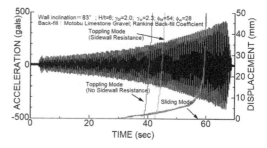

Figure 27. Comparison of computations with experimental data for the 83 degrees retaining wall with limestone gravel backfill.

rounded river gravels were generally used as backfill materials at collapsed castle walls. The effect of three backfilling materials on the seismic stability of the model castle walls was investigated through shaking table model tests. Furthermore, the frictional strength between the backfill material and castle walls is measured. The shaking table experiments showed that the type of backfilling material has a great effect on the seismic stability of the walls and the castle walls utilizing rounded river gravels are quite vulnerable to fail during great earthquakes. Furthermore, the dynamic limiting equilibirum method used for simulating the shaking table tests and it was found that it was possible to evaluate the displacament responses of the retaining walls.

REFERENCES

Aydan, Ö. (2017): Rock Dynamics, CRC Press, ISRM Book Series, No.3, 464 pages.

Aydan, Ö., Ogura, Y., Daido, M. and Tokashiki, N., 2003, A model study on the seismic response and stability of masonary structures through shaking table tests, *Fifth National Conference on Earthquake Engineering*, Istanbul, Turkey, Paper No: AE-041(CD).

Aydan, Ö., Tomiyama, J., Matsubara, H., Tokashiki, N., Iwata, N. (2018). Damage to rock engineering structures induced by the 2016 Kumamoto earthquakes.The 3rd Int. Symp on Rock Dynamics, RocDyn3, Trondheim, 6p, on CD.

Tokashiki, N., Aydan, Ö., Daido, M. and Akagi, T., 2007, Experiments and analyses of seismic stability of masonry retaining walls, *Rock Mechs. Symp., JSCE*, 115–120.

2019 Rock Dynamics Summit– Aydan et al. (eds)
© 2019 Taylor & Francis Group, London, ISBN 978-0-367-34783-3

An experimental study on the formation mechanism of tsunami boulders

K. Shimohira, Ö. Aydan, N. Tokashiki
Department of Civil Engineering, University of the Ryukyus, Okinawa, Japan

K. Watanabe, Y. Yokoyama
School of Marine Science and Technology, Tokai University, Shizuoka, Japan

ABSTRACT: There are many tsunami boulders in the islands of the Ryukyu Archipelago. Among them, tsunami boulders in Okinawa (Kasakanja), Miyako (Higashi-Hennasaki), Shimojiri (Obiwa) and Ishigaki (Ohama) Islands are famous. These tsunami boulders are definitely due to the past mega-tsunamis. An experimental study has been undertaken on the mechanism of tsunami boulders by using a specially developed tsunami generation device. Various single blocks having different densities and cliff models with different geometry and materials are prepared and tested. The experiments shown that tsunami boulders found on terraces are likely to be associated with cliffs having a toe erosion and failing like passive flexural toppling failure. The authors describe these experimental studies and discuss their implications in the tsunami boulder formation as well as the estimation of mega-earthquakes.

1 INTRODUCTION

Tsunami induced boulders are found worldwide and the mechanism involved in the formation of such tsunami boulders is still controversial. There is no doubt that such huge tsunami boulders can only be caused by huge tsunamis resulting from mega-earthquakes. There are many tsunami boulders in the islands of the Ryukyu Archipelago. Among them, tsunami boulders in Okinawa (Kasakanja), Miyako (Higashi-Hennasaki), Shimojiri (Obiwa) and Ishigaki (Ohama) Islands are famous and they are definitely due to the past mega-tsunamis (Aydan and Tokashiki, 2018). The authors have initiated an investigation program to quantify the geometry and position of tsunami boulders and topographical conditions in their close vicinity including the erosion state along the shoreline. However, the formation of tsunami boulders and associated conditions are quite important for both understanding the past mega-earthquakes and future earthquakes in relation to the preparation of disaster mitigation plans.

The authors have developed an experimental device to investigate the formation mechanism of tsunami boulders. The device is capable of inducing both tsunami waves due to thrust faulting and normal faulting. The model cliffs with toe erosion were prepared. The cliffs were either continuum type or blocky type. The continuum type cliffs were prepared using plaster while the blocky cliffs were made of Ryukyu limestone blocks. The experiments were carried under different velocities of faulting and the geometry of the model cliffs. The experiments shown that tsunami boulders found on terraces are likely to be associated with cliffs having a toe erosion and failing like passive flexural toppling failure. The authors will describe the

findings from these experimental studies and discuss their implications in the tsunami boulder formation as well as the estimation of mega-earthquakes.

2 OBSERVATIONS ON TSUNAMI BOULDERS IN RYUKYU ARCHIPELAGO

The present observations were made on selected tsunami boulders in Okinawa (Kasakanja), Miyako (Higashi-Hennasaki), Shimojiri (Obiwa) and Ishigaki (Ohama) Islands (Fig. 1), which are definitely due to the past mega-tsunamis. For this purpose, the authors have been utilizing aerial photogrammetry and laser scanning techniques to map the boulders and their geometrical locations with respect to the topography. Although the tsunami boulders in Ishigaki, Miyako and Shimoji islands were initially believed to be due to the 1771 Meiwa earthquake with an estimated magnitude of 7.4 (Aydan and Tokashiki 2007), the recent studies indicated by Aydan and Tokashiki (2018) that they were much older. Particularly, the tsunami boulder in Shimoji Island is probably the largest in the world. Table 1 gives the size and elevation of the tsunami boulders and their distance to the nearby cliffs in selected locations. In addition, some large boulders of metamorphic origin and sandy tsunami deposits were observed by the authors within Ryukyu limestone layer during an excavation of a large engineering structure in Ishigaki Island (Aydan and Tokashiki 2018). These observations also imply that the events were cyclically occurring in Ryukyu Archipelago.

One can also find many boulders on the tsunami-induced boulders., which we call "Sea Tsunami Boulders" (Fig. 2). However, it is extremely difficult to differentiate from the boulders resulting from cliff

Figure 1. Major islands of the Ryukyu Archipelago and selected huge tsunami boulders in (base map from 11th Regional Coast Guard Headquaters of Japan).

failures into the sea due to toe erosion resulting from the ordinary action of sea waves (Tokashiki and Aydan, 2010). The rockblocks failed due to toe-erosion may look-like tsunami boulders distributed around Ryukyu Islands resulting in mis-interpretations (Fig. 3).

3 MODEL TSUNAMI GENERATION DEVICES

The authors performed two series of experiments using model tsunami generation devices. The first series of experiments were done at Tokai University and the second series of experiments were done in the University of the Ryukyus. Brief summary of these experiments are outlines in this section.

3.1 Experimental facility at Tokai University

The tsunami generation model is made of 2000 mm long, 300 mm wide and 400 mm high acrylic box. Fig. 4 shows illustrates the concept of rising sea wave and receding wave type tsunami modeling. The inclination of the sea bottom was 1/10. The water level change was about 100 mm. Single rectangular prism type blocks made of Ryukyu limestone or porous concrete were tested. Several overhanging configurations were also tested. Furthermore, the performance of boulder type wave-breaks were also tested.

Table 1. Elevation and height of tsunami boulders from selected locations

Location	El.(m)	H(m)	L(m)	W(m)	DtoC(m)
Hennazaki	20	4	6.8	4.5	21
Obiwa	12.5	9.0	16	14	43
Kasakanja	12	3	7.5	5	34
Ohama	8.0	5.9	12	11	96

El.: Elevation; H:Height; L:Length; W: Width; DtoC:Distance to cliff

Figure 2. Sea tsunami boulders in Ishigaki Island.

Figure 3. Toe erosion and subsequent cliff failures.

3.2 Experimental facility at the University of Ryukyus

Following the preliminary attempts at Tokai University, the third author designed a tsunami generation device OA-TGD2000X to study the tsunami waves due to thrust and normal faulting event shown in Fig. 5. The dimensions and characteristics of the

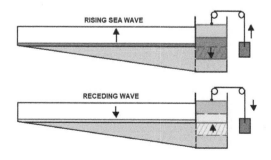

Figure 4. Illustration of the basic concept of tsunami generation device at Tokai University.

Figure 5. A view of the tsunami generation device OA-TGD2000X.

device was quite similar to that used in Tokai university except the wave-induction system. A tank was lowered or raised through pistons with a given velocity to generate rising or receding tsunami waves. The pressure and wave velocity at specified locations were measured using pressure sensors and the amount of tank movement was measured using laser transducers. Fig. 6 shows an example of record during the movement of tanks inducing rising and receding tsunami waves.

4 EXPERIMENTS AT TOKAI UNIVERSITY

4.1 Ryukyu limestone blocks

First Ryukyu limestone blocks with a height of 100 mm and width of 40 mm immersed to a depth of 30mm initially were subjected to different tsunami wave with different wave height. When the final water level variation was less than 40 mm, the blocks slid upward as seen in Fig. 7. However, if the final water level variation was more than 60 mm, the blocks were toppled landward in the direction of tsunami wave propagation as seen in Fig. 8. Fig. 9 shows the experiments at several time intervals. The movements of the blocks occurred mainly during the water level rise stage and there was almost no movement during the receding wave stage.

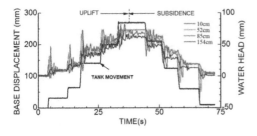

Figure 6. Water head response at specified locations in relation to the tank movement.

Figure 7. A view of Ryukyu limestone blocks before and after the test.

Figure 8. A view of Ryukyu limestone blocks before and after the test.

Figure 9. Views of the test shown in Figure 8 at different time steps.

4.2 Porous concrete blocks

Next porous concrete blocks having dimensions of 150 mm height and 50 mm width were tested as shown in Figs. 10 and 11. As the density of the blocks is lower than that of Ryukyu limestone and the height over width ratio is higher, they easily toppled as seen in the Figures.

4.3 Experiments on boulder type wave-breaks

The final experiment was carried out on the performance of boulder-type wave-breaks. The stone was

Figure 10. A view of porous concrete blocks before and after the test.

Figure 12. Side and top views of experiment on wave-breaks

siliceous sandstone from Abe River in Shizuoka Prefecture belonging the Shimanto River Formation. The height of the wave-break was 50 mm and it was immersed in water up to 25 mm. The final water level rise was about 25 mm after the test. Fig. 12 shows side and top views of the wave-break before and after the experiment. As seen from the figure, the wave-break subsides and spreads due to the tsunami waves. In other words, the boulders are displaced by the generated tsunami waves. Fig. 13 shows the experiment at time intervals. As noted from the figure the boulders are displaced during the overflow process of the tsunami waves. The receding tsunami waves do not cause major movements during the experiment.

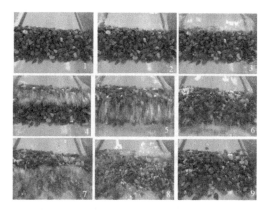

Figure 13. Views of the test shown in Figure 12 at different time steps.

5 EXPERIMENTS AT THE UNIVERSITY RYUKYUS

5.1 Triangular Ryukyu limestone blocks

Rectangular prism blocks were tested and the results were quite similar to those tested in Tokai University. A triangular shaped prismatic blocks shown in Figs. 14 and 15 tested under the same condition. The longest side of the triangular prismatic block shown in Fig. 14 was downward while the longest side of the triangular prismatic block shown in Fig. 15 was upward. While the downward triangular prismatic block was almost non-displaced, the upward triangular prismatic block was considerably displaced. One of the main reasons for such big difference when they are subjected to the tsunami forces is that the tsunami wave apply a surging uplift force on the block. As for the downward triangular prism the surging force increases the normal force on the block. We also put a rectangular prism Ryukyu limestone next to the triangular prismatic block. The displacement of the rectangular block was quite small.

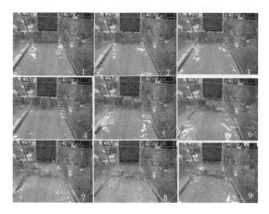

Figure 11. Views of the test shown in Figure 10 at different time steps.

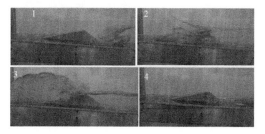

Figure 14. Views of the downward triangular prism block at different time steps

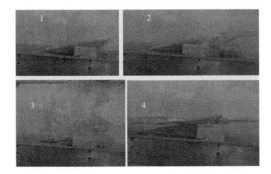

Figure 15. Views of the upward triangular prism block at different time steps

5.2 Plaster blocks

First a rectangular prismatic block made of plaster was subjected to rising tsunami waves as shown in Fig. 16. The overall behaviour is fundamentally similar to those tested in Tokai University. Nevertheless, the block was toppled towards downstream side and displaced horizontally in the direction of receding tsunami waves as seen in Fig. 16.

Next two plaster prismatic blocks laid over the Ryukyu limestone blocks as shown in Fig. 17. The density of the plaster blocks is almost half of that of the Ryukyu limestone blocks. As seen from images 2 and 3 in Fig. 17, the plaster blocks thrown upward and displaced in the direction of the tsunami waves. This experiment clearly demonstrates the importance of the density and overhanging degree of blocks when they are subjected to tsunami waves in nature.

5.3 Breakable overhanging cliffs

The next series of experiments involve the breakable blocks. Finding appropriate material for breakable blocks under the forces induced by tsunami forces by the experimental device was quite cumbersome. Although the materials had a very small density as compared those in nature, it provided an insight view on the mechanism of formation of tsunami boulders,

Figure 17. Top views of the plaster block overhanging over the base Ryukyu limestone blocks at different time steps.

which was the main goal of this study. Figs. 18 and 19 show the images of the models during at different time steps. The surging tsunami wave enters under the overhanging blocks and applies upward forces. As a result, the overhanging block starts to bend upward and they are broken after a certain amount of the displacement. In other words, the failure of the overhanging blocks are quite close to cantilever beams. However, the failure of the overhanging blocks is against the gravity. Once the block is broken, it is dragged by overflowing tsunami waves. This observation is in accordance with the mechanism proposed by Aydan and Tokashiki (2019) for the formation of tsunami boulders. Our experiments clearly indicated that if the inclination of the lower side of the overhanging block ranges between 10-20 degrees, they are quite vulnerable to fail.

6 SOME CONSIDERATIONS ON THE FORMATION OF TSUNAMI BOULDERS AND THEIR MOVEMENTS

6.1 Mechanism of formation of tsunami boulders

Figures 18 and 19 show how tsunami waves breaks at the model overhanging rock cliffs. When the wave hits the cliff or rock block it is reflected and a huge splash

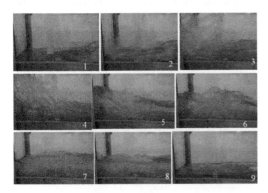

Figure 16. Views of the plaster block test at different time steps.

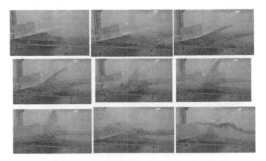

Figure 18. Views of the experiments using a breakable overhanging block at different time steps.

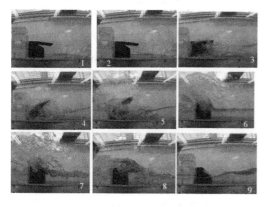

Figure 19. Views of the experiments using a breakable overhanging block at different time steps.

occurs in proportion to sea-wave pressure. As the cliffs of Ryukyu limestone is quite prone to dissolution by sea water and erosion by sea waves, they result in overhanging cliffs. As studied by Tokashiki and Aydan (2010) and Horiuchi et al. (2018), the overhanging cliffs are quite prone to failure by bending under gravitational and seismic forces. However, storm-waves and tsunami waves may also cause the failure of overhanging cliffs. As noted in Figures 18 and 19, the impact force by the storm-waves or tsunamis may particularly cause the bending failure of overhanging cliffs against gravity and throw the broken overhanging blocks onto the cliff terraces. If rock mass is blocky due to the existing discontinuities, they may displace the blocks resulting in active or passive toppling or other failure modes (e.g. Aydan and Tokashiki 2018; Aydan et al. 1989). However, if the rock mass is continuum type, the failure due to huge impact forces induced by tsunamis may result in a failure mode similar to the passive type flexural toppling failure (Aydan and Kawamoto, 1992) as observed previously by Aydan and Amini (2006) as well as by Horiuchi et al. (2018) in shaking table tests on rock slopes. Aydan and Amini (2006) also developed some analytical solutions for the passive flexural failure of cantilever beams under seismic forces. In view of these experiments, observations and analytical studies, the impact force, which is a function of characteristics of tsunami, is a decisive factor to throw the tsunami boulders onto cliff terraces.

6.2 Movement of tsunami boulders

Once the broken overhanging blocks thrown over the terraces, they will be subjected to water pressure, uplift and drag forces. As a result, the broken boulder or boulders may be moved by translation, toppling or both as observed in Fig. 1. If the boulders existing near shorelines may be moved the landward by rising tsunami waves or away from the shoreline by receding tsunami waves as seen in Fig. 2. The amount of the movements would undoubtedly depend upon the resulting tsunami waves as a function of magnitude

of the earthquake or displaced water body in case of slope failure, meteorite impact or volcanic eruption.

7 CONCLUSIONS

The authors reported some experimental studies at the University of the Ryukyus and Tokai University over a period of two years. Various blocks having different densities and cliff models with different geometry and materials are prepared and tested using the tsunami generation devices at both institutes. Some overhanging cliffs consisting of breakable overhanging parts prepared and tested. From this experimental study, the following conclusions may be drawn:

(1) Blocks of different shape and densities can be translated, toppled or both. Most of the movements results from the rising tsunami waves and run-up stage. Nevertheless, some blocks may also be displaced during receding tsunami wave stage. However, if the shape of the boulders such that the blocks may not move at all.

(2) When the surging tsunami wave enters under the overhanging blocks and applies upward impact forces. As a result, the overhanging block bends upward and they may be broken after a certain amount of the displacement. Once the block is broken, it is dragged by overflowing tsunami waves.

(3) Experiments indicated that if the inclination of the lower side of the overhanging block ranges between 10-20 degrees, they are quite vulnerable to fail.

(4) Experiment on the performance of boulder-type wave-breaks indicated that the wave-break subsides and spreads due to the tsunami waves. The boulders are displaced during the overflow process of the tsunami waves. The receding tsunami waves do not cause major movements during the experiment.

REFERENCES

Aydan, Ö. and Kawamoto, T. 1992. The stability of slopes and underground openings against flexural toppling and their stabilisation. Rock Mechanics and Rock Engineering 25(3) (1992), pp.143–165.

Aydan, Ö. and Tokashiki, N. 2007. Some damage observations in Ryukyu Limestone Caves of Ishigaki and Miyako islands and their possible relations to the 1771 Meiwa Earthquake. Journal of The School of Marine Science and Technology, 5(1), pp. 23–39.

Aydan, Ö., Amini, M. G. (2009). An experimental study on rock slopes against flexural toppling failure under dynamic loading and some theoretical considerations for its stability assessment. Journal of Marine Science and Technology, Tokai University, Vol. 7, No. 2, 25–40.

Aydan, Ö. and Tokashiki, N. 2018. Tsunami Boulders and Their Implications on the Mega Earthquake Potential along Ryukyu Archipelago, Japan. Bulletin of Engineering geology and Environments.

Horiuchi, K, Aydan, Ö, Tokashiki, N. (2018). Recent failures of limestone cliffs in Ryukyu Archipelago and their analyses. Proc. of 45th Rock Mechanics Symposium of Japan, JSCE, 131–136.

Tokashiki, N. and Ö. Aydan, 2010. The stability assessment of overhanging Ryukyu limestone cliffs with an emphasis on the evaluation of tensile strength of Rock Mass. Journal of Geotechnical Engineering, JSCE, Vol. 66, No. 2, 397–406.

2019 Rock Dynamics Summit– Aydan et al. (eds)
© 2019 Taylor & Francis Group, London, ISBN 978-0-367-34783-3

Some considerations on the static and dynamic shear testing on rock discontinuities

Ö. Aydan

Department of Civil Engineering, University of the Ryukyus, Okinawa, Japan

ABSTRACT: Although dynamic shearing properties of rock discontinuities are of great importance in many rock engineering applications, it is very rare to see dynamic shear tests on rock discontinuities. The author describes static, cyclic, one-way and two-ways dynamic shearing experiments on various natural rock discontinuities and saw-cut planes and discusses the shear strength characteristics obtained from the different testing techniques and interrelations. The experiments clearly showed that the dynamic shear behavior of discontinuities are very complex and anisotropic. Nevertheless, this study provides some fundamental understanding on the static and dynamic behavior of rock discontinuities and their correlation with each other.

1 INTRODUCTION

Dynamic shearing properties of rock discontinuities are of great importance in many rock engineering applications (Aydan et al. 2011). Nevertheless, dynamic shear tests on rock discontinuities are very rare. In this study, the author describes static, cyclic, one-way and two-ways dynamic shearing experiments on various natural rock discontinuities and saw-cut planes. Natural rock discontinuities involve schistosity planes in quartzite, green-schist, cooling planes in andesite, saw-cut planes of Ryukyu limestone, Motobu limestone, andesite and basalt from Mt. Fuji, dolomite from Kita-Daitojima, granodiorite from Ishigaki and Inada granite and Oya-tuff.

The experiments clearly showed that the dynamic shear behavior of discontinuities are very complex and anisotropic (Aydan et al. 1996). Furthermore, the degradation of roughness of the discontinuity planes results in further complexity in regard with modeling shear behavior under dynamic conditions. Nevertheless, the dynamic response at the first full cycle provides a basis for the overall dynamic shear behavior of discontinuities. and interfaces. Unpolished saw-cut planes of rocks should never be used for evaluating their shear responses under dynamic conditions in the first step (Aydan 2016). The dynamic shear behavior of actual discontinuities should be carried out as the next step.

The experimental results are compared with those from static and one-way dynamic shearing experiments and their implications in practice are discussed. Although more experiments are necessary for the selection of appropriate dynamic loading pattern with the consideration possible loading conditions in-situ, this study provides some fundamental understanding on the static and dynamic behavior of rock discontinuities and their correlation with each other.

2 STATIC SHEAR TESTING TECHNIQUES

2.1 Tilting tests

Tilting tests are known as a laboratory technique for measuring the friction angle in physics and illustrating the concept of friction (Aydan, 2016; Aydan et al. 1995, 2017). This technique is one of the most popular technique due to its simplicity and it is one of the most suitable technique to perform and to obtain the frictional characteristics of rock discontinuities and interfaces in-situ. This technique can be used to determine the apparent friction angle of discontinuities (rough or planar) under low stress levels. It definitely gives the maximum apparent friction angle, which would be one of the most important parameters to determine the shear strength criteria of rock discontinuities as well as various contacts.

2.1.1 Theory of tilting tests

Let us assume that a block is put upon a base block with an inclination α as illustrated in Fig. 1. The

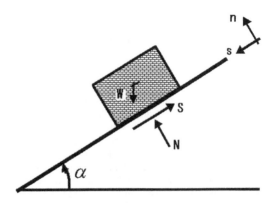

Figure 1. Mechanical model for tilting experiments.

dynamic force equilibrium equations for the block can be easily written as follows:

For s-direction

$$W \sin \alpha - S = m \frac{d^2 s}{dt^2} \quad (1)$$

For n-direction

$$W \cos \alpha - N = m \frac{d^2 n}{dt^2} \quad (2)$$

Let us further assume that the following frictional laws holds at the initiation and during the motion of the block (Aydan et al. 2017) as illustrated in Fig. 2:
At initiation of sliding

$$\frac{S}{N} = \tan \phi_s \quad (3)$$

During motion

$$\frac{S}{N} = \tan \phi_d \quad (4)$$

At the initiation of sliding, the inertia terms are zero so that the following relation is obtained:

$$\tan \alpha = \tan \phi_s \quad (5)$$

The above relation implies that the angle of inclination (rotation) at the initiation of sliding should correspond to the static friction angle of the discontinuity. If the normal inertia term is negligible during the motion and the frictional resistance is reduced to dynamic friction instantenously, one can easily obtain the following relations for the motion of the block

$$\frac{d^2 s}{dt^2} = A \quad (6)$$

where $A = g(\sin \alpha - \cos \alpha \tan \phi_d)$.

The integration of differential equation (6) will yield the following

$$s = A \frac{t^2}{2} + C_1 t + C_2 \quad (7)$$

Since the followings hold at the initiation of sliding

$$s = 0 \text{ and } v = 0 \text{ at } t = T_s \quad (8)$$

Eq. (7) takes the following form

$$s = \frac{A}{2}(t - T_s)^2 \quad (9)$$

Coefficient A can be obtained either from a given displacement s_n at a given time t_n with the condition, that is, $t_n > T_s$

$$A = 2 \frac{s_n}{(t_n - T_s)^2} \quad (10)$$

or from the application of the least square technique to measured displacement response as follows

$$A = 2 \frac{\sum_{i=1}^{n} s_i (t_i - T_s)^2}{\sum_{i=1}^{n} (t_i - T_s)^4} \quad (11)$$

Once constant A is determined, the dynamic friction angle is obtained from the following relation

$$\phi_d = \tan^{-1} \left(\tan \alpha - \frac{1}{\cos \alpha} \frac{A}{g} \right) \quad (12)$$

2.1.2 Tilting test device and setup

An experimental device consists of a tilting device operated manually. During experiments, the displacement of the block and rotation of the base is measured through laser displacement transducers produced by KEYENCE while the acceleration responses parallel and perpendicular to the shear movement are measured by a three component accelerometer (TOKYO SOKKI) attached to the upper block and WE7000 (YOKOGAWA) data acquisation system. The measured displacement and accelerations are recorded onto lap-top computers. The weight of the accelerometer is about 0.96N. Fig. 3 shows the experimental set-up.

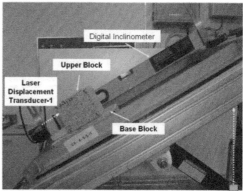

Figure 3. A conceptual illustration of experimental set-up.

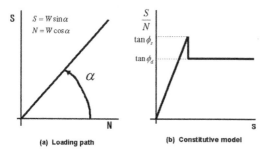

(a) Loading path

(b) Constitutive model

Figure 2. Loading path in tilting experiments and constitutive relation.

202

2.1.3 Tilting tests

Some tilting experiments are describe herein and saw-cut discontinuity planes of Ryukyu limestone samples are chosen as an example. Fig. 4 shows responses measured during a tilting experiment on a saw-cut plane of Ryukyu-limestone. The static and dynamic friction coefficients of the interface were calculated from measured displacement response explained in previous section and they were estimated as 28.8-29.6° and 24.3-29.2° respectively.

2.2 Stick-slip tests

2.2.1 Theory of stick-slip test

In this model, the basal plate is assumed to be moving with a constant velocity v_m and overriding block is assumed to be elastically supported by the surrounding medium as illustrated in Fig. 5 (Ohta & Aydan 2010; Bowden & Leben, 1939; Jaeger & Cook, 1979). The basic concept of modeling assumes that the relative motion between the basal plate and overriding block is divergent. The governing equation of the motion of the overriding block may be written as follow:

During the stick phase, the following holds

$$\dot{x} = v_s, \qquad F_s = k \cdot x \tag{13}$$

where v_s is belt velocity and k is the stiffness of the system. The initation of slip is given as (Fig. 5)

$$F_y = \mu_s N \tag{14}$$

where μ_s is static friction coefficient, and N is normal force. For block shown in Fig. 5, it is equal to block weight W and it is related to the mass m and

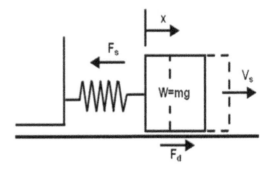

Figure 5. Mechanical modeling of stick-slip phenomenon.

gravitational acceleration g through mg. During slip phase, the force equilibrium yields:

$$-kx + \mu_k W = m \frac{d^2 x}{dt^2} \tag{15}$$

where μ_k is dynamic friction angle. The solution of above equation can be obtained as.

$$x = A_1 \cos \Omega t + A_2 \sin \Omega t + \mu_k \frac{W}{k} \tag{16}$$

If initial conditions $t = t_s$, $x = x_s$ and $\dot{x} = v_s$) are introduced in Eq. (16), the integration constants are obtained as follow.

$$x = \frac{W}{k}(\mu_s - \mu_k) \cos \Omega(t - t_s) + \frac{v_s}{\Omega} \sin \Omega(t - t_s) + \mu_k \frac{W}{k}$$

$$\dot{x} = -\frac{W}{k}(\mu_s - \mu_k)\Omega \sin \Omega(t - t_s) + v_s \cos \Omega(t - t_s)$$

$$\ddot{x} = -\frac{W}{k}(\mu_s - \mu_k)\Omega^2 \cos \Omega(t - t_s) - v_s \Omega \sin \Omega(t - t_s)$$

$$\tag{17}$$

where $\Omega = \sqrt{k/m}$ and $x_s = \mu_s \dfrac{W}{k}$.

At $t = t_l$, velocity becomes equal to belt velocity, which is given as $\dot{x} = v_s$. This yield the slip period as:

$$t_l = \frac{2}{\Omega}\left(\pi - \tan^{-1}\left(\frac{(\mu_s - \mu_k)W\Omega}{k \cdot v_s}\right)\right) + t_s \tag{18}$$

where $x_s = v_s \cdot t_s$. The rise time, which is slip period is given by (Fig. 6)

$$t_r = t_l - t_s \tag{19}$$

Rise time can be specifically obtained from Eqs. 19 and 18 as

$$t_r = \frac{2}{\Omega}\left(\pi - \tan^{-1}\left(\frac{(\mu_s - \mu_k)W\Omega}{k \cdot v_s}\right)\right) \tag{20}$$

If belt velocity is negligible, that is, $v_s \approx 0$, the rise time reduce (t_p) to the following form

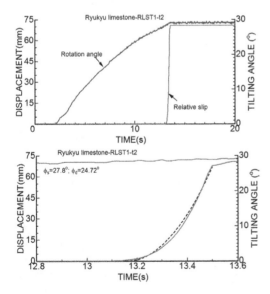

Figure 4. Responses of saw-cut discontinuity planes of Ryukyu limestone samples during a tilting test.

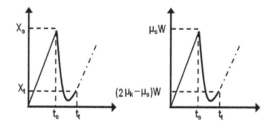

Figure 6. Frictional forces during a stick-slip cycle.

$$t_r = \pi \sqrt{\frac{m}{k}} \qquad (21)$$

The amount of slip is obtained as

$$x_r = \left| x_t - x_s \right| = 2\frac{W}{k}(\mu_s - \mu_k) \qquad (22)$$

The force drop during slip is given by

$$F_d = 2(\mu_s - \mu_k)W \qquad (23)$$

It should be noted that the formulation given above does not consider the damping associated with slip velocity. If the damping resistance is linear, the governing equation (15) will take the following form

$$-kx - \eta\dot{x} + \mu_k W = m\frac{d^2x}{dt^2} \qquad (24)$$

2.2.2 Device of stick-slip tests

Figure 7 shows a view of the experimental device. The experimental device consists of an endless conveyor belt and a fixed frame. The inclination of the conveyor belt can be varied so that tangential and normal forces can be easily imposed on the sample as desired. To study the actual frictional resistance of interfaces of rock blocks, the lower block is stuck to a rubber belt while the upper block is attached to the fixed frame through a spring as illustrated in Figure 5. We conducted the experiment using the rock samples of granite with planes having different surface morphologies. The base blocks were 200-400mm long, 100-150mm wide and 40-100mm thick. The upper block was 100-200mm long, 100mm wide and 150-00 mm high.

When the upper block moves together on the base block with at a constant velocity (stick phase), the spring is stretched at a constant velocity. The shear force increases to some critical value and then a sudden slip occurs with an associated spring force drop. Because the instability sliding of the upper block occurs periodically, the upper block slips violently over the base block. Normal loads can also be easily increased in experiments.

2.2.3 Stick-slip tests

A series of stick-slip experiments are carried out on the saw-cut discontinuities of Ryukyu Limestone. The peak (static) friction angle can be evaluated from the

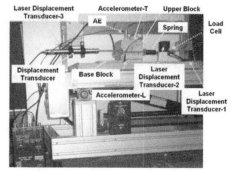

(a) Overall view

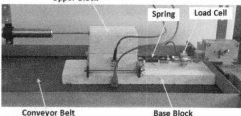

(b) Detailed view

Figure 7. Stick-slip experimental setup.

T/N response while the residual (kinetic) friction angle is obtained from the theoretical relation (Aydan et al. 2018). Fig. 8 show the stick-slip responses of discontinuity planes shown. The residual (kinetic) friction angle of saw-cut discontinuity plane of Ryukyu limestone obtained from stick-slip experiments are very close to those obtained from tilting experiments. Nevertheless, the kinetic or residual friction angle is generally lower than those obtained from the tilting experiments.

2.3 Direct shear tests

Direct shear testing device is commonly used. The loading step may be monotonic under a constant normal loading or multi-stage normal loading. A multi-stage (multi-step) direct shear test on a saw-cut surface of sandy Ryukyu limestone sample, which consist of two blocks with dimensions of 150x75x37.55mm, was varied out (Figs. 9 and 10).

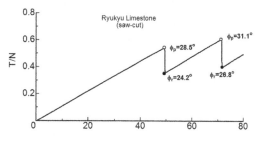

Figure 8. Stick-slip response of saw-cut plane of Ryukyu limestone.

Figure 9. A view of direct shear test on a saw-cut discontinuity plane.

Figure 10. A view of a saw-cut discontinuity plane after a direct shear test.

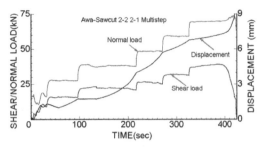

Figure 11. Shear stress-shear load response of the interface of sandy limestone blocks during the multi-stage(step) direct shear experiment.

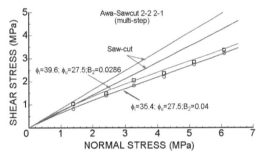

Figure 12. Comparison of shear strength envelope for the interface of sandy limestone blocks with experimental results from the multi-stage(step) direct shear experiment and tilting tests.

The initial normal load was about 17 kN and increases to 30, 40, 50, 60 and 70 kN during the experiment. Fig. 11 shows the shear displacement and shear load responses during the experiment. As noted from the figure, the relative slip occurs between blocks at a constant rate after each increase of normal and shear loads. This experiment is likely to yield shear strength of the interface two blocks under different normal stress levels. Fig. 11 shows the peak and residual levels of shear stress for each level of normal stress increment. Tilting tests were carried out on the same interface and the apparent friction angles ranged between 35.4 and 39.6 degrees. Tilting test results and direct shear tests are plotted in Fig. 12 together with shear strength envelopes using the shear strength failure criterion of Aydan (Aydan 2008; Aydan et al. 1996).

Therefore, the data for determining the parameters of the shear strength criteria for rock discontinuities should utilize both tilting test and direct shear experiment.

As noted from the figure, the friction angles obtained from tilting tests are very close to the initial part of the shear strength envelopes. However, the friction angle becomes smaller as the normal stress level increases. In other words, the friction angle obtained from tilting tests on saw-cut surfaces can not be equivalent to the basic friction angle of planar discontinuities and interfaces of rocks. The basic friction angle of the planar interface of sandy limestone blocks is obtained as 27.5 degrees for the range of given normal stress levels.

3 DYNAMIC SHEAR TESTING TECHNIQUES

3.1 Direct cyclic shear tests

Cyclic shearing tests rock discontinuities and interfaces are performed under cyclic loads with a given period results in relation to machinery vibration and/or earthquakes. Fig. 9 is a shear testing machine capable of cyclic loading (Aydan et al., 1994, 2016).

3.2 Dynamic one-way shear tests

Dynamic one-way shear testing is one of the techniques to investigate the dynamic shear testing of discontinuities (Aydan et al. 1994; Aydan 2018).

Particularly, this procedure is important to evaluate the dynamic behaviour of discontinuities including the linear response. The device is the shear testing machine upgraded by Aydan et al. (2016) with a one way-shear testing capability as shown in Fig. 13.

3.3 *Dynamic two-ways shear tests*

A series of two-ways direct shear tests were carried out saw-cut discontinuity planes of Ryukyu limestone. Fig. 14 shows the normalized shear resistance (T/N) response of the saw-cut planes and acceleration during the two-ways shearing with ±15 mm forced displacement amplitude. As noted the acceleration response is quite different than that anticipated and it has a very irregular response with an amplitude less than 0.42g.

Fig. 15 shows the relative displacement-T/N relation. In the same figure, the trajectory of the first cycle is distinguished. It is quite interesting to note that the shear resistance of the discontinuity plane gradually changes after each cycle of shearing. At the end of the experiment, a thin powder is recognized on the discontinuity plane.

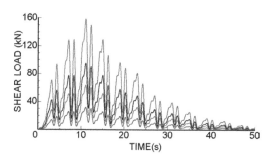

Figure 13. Adopted direct shear loading patterns in the dynamic shear testing machine (OA-DSTM).

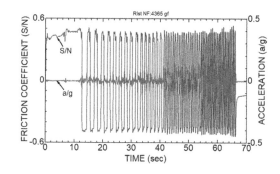

Figure 14. The response of the normalized S/N and acceleration response of a saw-cut plane during two shearing experiment.

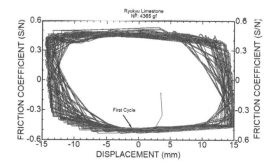

Figure 15. Shear response of a saw-cut discontinuity plane of Ryukyu limestone shown in Figure 14 in the space of relative displacement and T/N.

4 DISCUSSIONS AND CONCLUSIONS

The author performed many shear testing on natural rock discontinuities involve schistosity planes in quartzite, green-schist, cooling planes in andesite, saw-cut planes of Ryukyu limestone, Motobu limestone, andesite and basalt from Mt. Fuji, dolomite from Kita-Daitojima, grano-diorite from Ishigaki and Inada granite and Oya-tuff. The static and dynamic (kinetic) friction angles on these discontinuities are plotted in Fig. 16 in order to compare static and dynamic shear testing results as an example. Most of experimental indicate that the dynamic (kinetic) friction angle of natural and saw-cut discontinuities is about 0.87 times that of the static friction angle.

This theoretical and experimental study on various natural rock discontinuities and saw-cut planes utilizing different experimental techniques

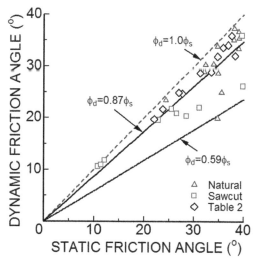

Figure 16. Comparison of static and kinetic friction angles obtained from tilting experiments on various rock discontinuities.

clearly showed that their static and dynamic shear properties would be different. Experimental results indicated that peak (static) friction angle for both discontinuity planes obtained from tilting tests and stick-slip experiments are very close to each other. Furthermore, the residual or dynamic friction angle for rough discontinuity planes and saw-cut discontinuity planes obtained from stick-slip experiments are very close to those obtained from the tilting experiments. Nevertheless, the dynamic or residual friction angle is generally lower than those obtained from the tilting experiments.

REFERENCES

Aydan, Ö. (2008). New directions of rock mechanics and rock engineering: Geomechanics and Geoengineering. 5th Asian Rock Mechanics Symposium (ARMS5), Tehran, 3-21.

Aydan, Ö. (2016). Considerations on Friction Angles of Planar Rock Surfaces with Different Surface Morphologies from Tilting and Direct Shear Tests 6p, Bali, on CD, Paper No.57.

Aydan, Ö. (2018): Rock Reinforcement and Rock Support, CRC Press, ISRM Book Series, No.7, 473 pages.

Aydan, Ö., T. Akagi, H. Okuda, T. Kawamoto (1994). The cyclic shear behaviour of interfaces of rock anchors and its effect on the long term behaviour of rock anchors. *Int. Symp. on New Developments in Rock Mechanics and Rock Engineering*, Shenyang, 15-22.

Aydan, Ö., Shimizu, Y, Kawamoto, T., 1995. A portable system for in-situ characterization of surface morphology and frictional properties of rock discontinuities. *Field Measurements in Geomechanics*. 4th International Symposium 463-470.

Aydan, Ö. Fuse, T. Ito, T. (2015): An experimental study on thermal response of rock discontinuities during cyclic shearing by Infrared (IR) thermography. Proc. 43rd Symposium on Rock Mechanics, JSCE, 123-128.

Aydan, Ö., Y. Shimizu, T. Kawamoto (1996). The anisotropy of surface morphology and shear strength characteristics of rock discontinuities and its evaluation. *NARMS'96*, 1391-1398.

Aydan Ö., Ohta Y., Daido M., Kumsar H. Genis M., Tokashiki N., Ito T. & Amini M. (2011): Chapter 15: Earthquakes as a rock dynamic problem and their effects on rock engineering structures. *Advances in Rock Dynamics and Applications*, Editors Y. Zhou and J. Zhao, CRC Press, Taylor and Francis Group, 341-422.

Aydan, Ö., N. Tokashiki, J. Tomiyama, N. Iwata, K. Adachi, Y. Takahashi (2016). The Development of A Servo-control Testing Machine for Dynamic Shear Testing of Rock Discontinuities and Soft Rocks, EUROCK2016, Ürgüp, 791-796.

Aydan, Ö., N. Tokashiki, N. Iwata, Y. Takahashi, K. Adachi, (2017). Determination of static and dynamic friction angles of rock discontinuities from tilting tests. 14th Domestic Rock Mechanics Symposium of Japan, Kobe, Paper No. 041, 6p (in Japanese).

Bowden, F.P. & Leben, L. (1939): The nature of sliding and the analysis of friction. *Proc. Roy. Soc.*, London, A169, 371-391.

Jaeger, J.C. & Cook, N.W.G. (1979): Fundamentals of Rock Mechanics, 3rd Edition, *Chapman and Hall*, London.

Ohta, Y. and Aydan, Ö. (2010): The dynamic responses of geomaterials during fracturing and slippage. Rock Mechanics and Rock Engineering, Vol.43, No.6, 727-740.

2019 Rock Dynamics Summit– Aydan et al. (eds)
© 2019 Taylor & Francis Group, London, ISBN 978-0-367-34783-3

Inelastic mechanical behavior of granite under spherical indentation loading condition with emphasis on a criteria for damage description

H. Shariati, M. Saadati & P.L. Larsson
Department of Solid Mechanics, KTH Royal Institute of Technology, Stockholm, Sweden

ABSTRACT: The inelastic behavior of Bohus granite is studied based on experimental and numerical results. A number of quasi-static indentation tests are carried out. Several small and large load-drops are noticed in the force-penetration responses during those tests. These load drops are probably due to conical, Hertzian cracks initiated at the surface of the rock specimens. Each of these load drops are corresponding to a material removal event in the region with positive hydrostatic stress (negative pressure) state at the rock surface enclosing the growing contact boundary. Moreover, the *P-h* response after the first large load-drop is highly affected by cracking and fragmentation in this material. Therefore in order to predict the load-drop force levels and the response after the first large load-drop, a pressure-dependent damage criteria seems required.

1 INTRODUCTION

A review of the literature reveals that quasi-brittle materials under quasi-static (Q–S) or dynamic loading condition is being widely investigated (Price & Farmer 1979, Cook et al. 1984, Vermeer & De Borst 1984, Detournay 1986). The cracking behavior in Q-S indentation test on rock specimen is studied in (Cook et al. 1984). Several stages of fragmentation are described by the authors and the formation of a crater, a crushed zone and a region with multiple cracks are investigated. A damage-viscoplastic cap constitutive model is considered by Saksala (Saksala 2010) in order to simulate rock behavior during low-velocity impact. In another study, a numerical tool is developed in order to simulate rock fragmentation during indentation in (Wang et al. 2011). Heterogeneity and isotropic damage is considered by Wang et al in that work. More specifically, the fragmentation response of Bohus granite is investigated in (Saadati et al. 2014, 2018). Pressure dependent plasticity (i.e. Krieg, Swenson and Taylor (KST) model), together with an anisotropic damage model (proposed by Denoual, Forquin and Hild (DFH) model) is taken into consideration (Saadati et al. 2014).

In the study by Shariati et al. (Shariati et al. 2018), a linear Drucker-Prager law with variable dilation angle as for the inelastic constitutive model is employed in order to predict the Q–S force-penetration (*P–h*) response of Bohus granite loaded by a spherical indenter. The constitutive model is calibrated based on quasi oedometric compression tests performed earlier (Saadati et al. 2018). It is shown how to obtain the yield surface and dilation angle. From those experimental results, together with corresponding finite element simulations, it is suggested that a linear Drucker-Prager law, with a variable dilation angle,

can be used to numerically model the inelastic behavior of Bohus granite. In order to avoid the cracking effects in constitutive modelling, the attention is put on the loading up to the large load-drop in the *P-h* response, load capacity of Bohus granite (Weddfelt et al. 2017). Moreover, a high speed camera is utilized to take images from the upper surface of the rock specimen during the Q-S spherical test. It can be seen from those images that each load-drop in *P-h* response during that test corresponds to a material removal occurrence on the specimen surface. Those load drops are probably due to conical, Hertzian cracks initiated at the surface of the rock specimens. This explanation is also supported by the fact that no substantial subsurface is visible in the tested specimens checked by Computed Tomography (CT) in (Shariati et al. 2018).

Here in this work, the stress state close to the contact region under Q-S indentation loading condition is numerically investigated using the constitutive model employed in (Shariati et al. 2018). From finite element simulation results, it is noticed that material removal (load-drop) events in *P-h* response correspond to the region with positive hydrostatic stress (negative pressure) state at the surface enclosing the growing contact boundary. Therefore for potential future study, a pressure-dependent damage description should be employed in order to predict the load-drops in *P-h* response of Bohus granite during Q-S indentation test.

2 EXPERIMENTS

In the study by Shariati et al. (Shariati et al. 2018), the inelastic material behavior is determined using the experimental data produced by quasi-oedometric compression test performed earlier in (Saadati et al. 2018).

During quasi-oedometric compression test, a cylindrical rock specimen which is enclosed in metallic rings, is axially loaded in compression. Both axial and radial stresses increase during the test.

Thereafter, a number of quasi-static indentation tests are performed by Shariati et al. rock blocks of Bohus granite are loaded by a tungsten carbide spherical indenter. The tungsten carbide elastic modulus is ten times larger than the granite elastic modulus. Therefore the indenter is assumed to be rigid in numerical simulations. The size of the rock blocks are chosen large enough to eliminate any boundary effects on stress state close to the contact region. The tests are conducted in displacement control mode with a constant velocity of 5 μm/s to avoid dynamic effects. The force level is produced by the load cell.

The force penetration response (*P-h*) is provided for a number of indentation tests carried out in (Shariati et al. 2018). As it can be seen from Figure 1, the *P-h* response is almost linear. As a result, it can be assumed that strain hardening is small or negligible (Storåkers et al. 1997).

A number of small load-drops prior to the main load drop, at penetration depth of about 0.5 mm, can be observed in Figure 1. These small load-drops are mostly due to ring (Hertzian) cracks initiated immediately outside the contact zone. A high speed camera is utilized by Shariati et al. in order to monitor the surface of specimen during indentation test. It is noticed that each load-drop corresponds to a material removal on the rock block surface. It is also shown that the large load-drop in the *P–h* curve is mainly due to chipping caused by lateral expansion of the inelastically deformed material under the indenter. The tested specimens are thereafter scanned by CT scanner and no visible sub-surface cracks are detected. It should be noted that The physical size of one voxel is 25.4 μm. It can be argued that small subsurface median cracks are

Figure 2. The rock surface images are taken by the high speed camera used in (Shariati et al. 2018), (left figure) prior to the large load-drop and (right figure) after the large load-drop.

possibly formed during the initial stages of indentation loading condition, but they are trapped by the crushed zone. Therefore the effects of these tensile damages are excluded in the employed elastoplastic material model used in numerical simulation. It should also be mentioned that any form of damages rather than the tensile failure is considered as plasticity in (Shariati et al. 2018).

3 CONSTITUTIVE MODELING

In the study by Shariati et al. (Shariati et al. 2018), the emphasis is put on prediction of the stress state distribution in the vicinity of the contact zone in Q-S indentation test. The force-penetration response up to the large load-drop is only considered in the absence of any tensile damage as neither ring cracks nor the possible trapped median cracks have significant effects on the stiffness of the material. The yield surface and dilation angle are obtained from quasi-oedometric test performed in (Saadati et al. 2018). It can be seen from quasi-oedometric test that the yield surface has an almost linear shape. Therefore, a linear Drucker-Prager (D-P) model (Drucker and Prager 1952) with ideal plasticity is assumed. As it is mentioned before, dilation angle plays a significant role in the stiffness of the predicted P-h response, therefore a non-associated flow rule is taken into consideration. The linear Drucker-Prager criterion is written as

$$F = q - p \tan \beta - d = 0 \tag{1}$$

where F is yield function, d the cohesion of the material, β the friction angle and q von Mises equivalent stress. The flow potential is

$$G = q - p \tan \Psi \tag{2}$$

where Ψ is the is the dilation angle, which is shown schematically in Figure 7a in (Shariati et al. 2018). The inelastic flow is assumed to be at angle Ψ with respect to the q-axis in the q-p plane, and $\Psi \neq \beta$. The inelastic strain increment is expressed as

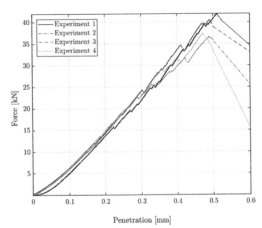

Figure 1. Experimentally determined force-penetration (*P–h*) response for the spherical indentation tests performed in (Shariati et al. 2018).

$$d\varepsilon^i = d\lambda \frac{\partial G}{\partial \sigma} \tag{3}$$

where $d\lambda$ is the inelastic strain rate multiplier, and ε^i the inelastic strain tensor.

The material parameters used in this work are determined in (Shariati et al. 2018) and these parameters are summarized in Table 1. The dilation angle (Ψ), which corresponds to the volume change when the material is subjected to shear strains is

$$\Psi = \tan^{-1}\left(\dfrac{3}{3\dfrac{\dot{\varepsilon}_a^i}{\dot{\varepsilon}_v^i}-1}\right) \qquad (4)$$

where $\dot{\varepsilon}_a^i$ is the inelastic axial strain rate, and $\dot{\varepsilon}_v^i$ the inelastic volumetric strain rate. The dilation angle is introduced as a function of the ratio between the inelastic axial and volumetric strain rates which is determined form quasi-oedometric test results. This ratio is by itself related to the pressure, see Figure 8b in (Shariati et al. 2018). Consequently, the dilation angle is set as a function of pressure in numerical FE simulation.

4 NUMERICAL RESULTS AND DISCUSSION

The numerical simulation is performed based on the constitutive model determined in (Shariati et al. 2018). By means of symmetries, only one quarter of the problem is modeled to reduce the computation time and

8-node linear reduced integration brick elements are used for meshing purposes, see Figure 3. The indenter is modeled as a 3D analytical rigid body, see (Shariati et al. 2018) for detailed FE analysis procedure.

The predicted force-penetration responses by numerical simulations and one of the experimental responses are provided in Figure 11 in (Shariati et al. 2018). The numerical simulation results are in good agreement with the experimental results. It should be mentioned that a series of simulations with D-P model combined with tension cut off are also performed in that study. It is found that the tension cut-off effect is negligible.

As discussed previously, the constitutive behavior is assumed in the absence of any tensile damage in (Shariati et al. 2018), since the force-penetration response up to the first main load-drop is of primary interest in that work. However the P-h response after the large load-drop is highly affected by cracking and fragmentation in the Bohus granite material. Therefore the D-P model in the absence of damage description is not valid after the large load-drop in the P-h response.

During the indentation test, several material removal events on the surface immediately outside the contact surface are detected utilizing high speed camera, Figure 2. The small load-drops are possibly due to Hertzian cracks created by radial stresses, see Figure 4. As it can be seen from this figure, the radial stress level is increased to very high level at very first steps, corresponding to initial small load-drops in the P-h response of the material. The large

Table 1. Material parameters employed in (Shariati et al. 2018).

E (GPa)	52
ν	0.25
ρ (kg/m³)	2630
β (°)	51.7
d (MPa)	153.3
Ψ (°)	Figure 8b in (Shariati et al. 2018)

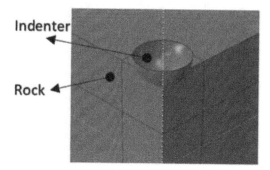

Figure 3. The geometry of the numerical model including the rock specimen and the rigid indenter. By means of symmetries, only one quarter of the problem is modeled. The reader may refer to (Shariati et al. 2018) for detailed description of the numerical modelling.

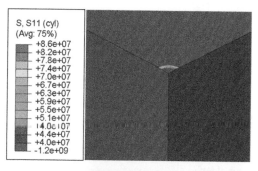

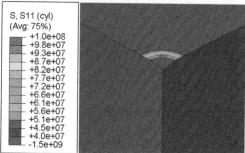

Figure 4. The radial stress fields determined from the finite element simulations of the spherical indentation test corresponding to penetration depths of (top figure) 0.1 mm and (bottom figure) 0.5 mm. The stresses are expressed in Pa.

load-drop is probably facilitated by the first and second principal stresses at the surface of the rock samples, see Figure 12 in (Shariati et al. 2018). These two principal stresses reach their maximum values at the indentation depth of 0.5 mm corresponding to the large load-drop in the *P-h* response. It is also necessary to consider damage description, if those load-drop simulations are of interest.

Figure 5 shows the pressure fields determined from the finite element simulations of the spherical indentation test at different indentation depth levels. It can be seen from the figure that the growing contact boundary is enclosed by a positive hydrostatic stress region (negative pressure). As discussed in previous paragraphs, material removal events, corresponding to load drops, are detected in the same region. Therefore it can be stated that a pressure-dependent damage criteria can be coupled to D-P model in order to account for the load drops in the numerical simulation.

5 CONCLUSION

The stress state in quasi-static indentation test of Bohus granite is numerically investigated using the constitutive model employed and calibrated in

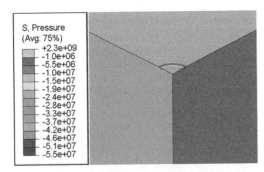

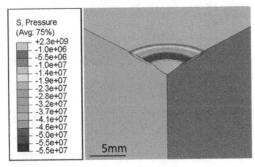

Figure 5. The pressure fields determined from the finite element simulations of the spherical indentation test corresponding to penetration depths of (top figure) 0.1 mm and (bottom figure) 0.5 mm. The stresses are expressed in Pa. Note that a negative pressure corresponds to a positive hydrostatic stress.

(Shariati et al. 2018). A small and narrow region at the surface immediately outside the growing contact boundary with positive hydrostatic stress (negative pressure) state is noticed by numerical simulation of the indentation test, see Figure 5. The radial stress levels in that region are increased to high levels, Figure 4. Such high radial stress levels which are larger than the tensile strength of the material can cause conical, Hertzian cracks on the surface. These cracks are probably the reason behind the load-drops spotted in (Shariati et al. 2018). Therefore, it seems that a pressure-dependent damage description is required to capture those load-drops in numerical simulation.

REFERENCES

Cook NGW, Hood M, Tsai F (1984) Observations of crack growth in hard rock loaded by an indenter. In: International Journal of Rock Mechanics and Mining Sciences & Geomechanics Abstracts. Elsevier, pp 97–107
Detournay E (1986) Elastoplastic model of a deep tunnel for a rock with variable dilatancy. Rock Mech Rock Eng 19:99–108
Drucker DC, Prager W (1952) Soil mechanics and plastic analysis or limit design. Q Appl Math 10:157–165
Price AM, Farmer IW (1979) Application of yield models to rock. In: International Journal of Rock Mechanics and Mining Sciences & Geomechanics Abstracts. Elsevier, pp 157–159
Saadati M, Forquin P, Weddfelt K, et al (2018) On the Mechanical Behavior of Granite Material With Particular Emphasis on the Influence From Pre-Existing Cracks and Defects. J Test Eval 46:1–13
Saadati M, Forquin P, Weddfelt K, et al (2014) Granite rock fragmentation at percussive drilling–experimental and numerical investigation. Int J Numer Anal Methods Geomech 38:828–843
Saksala T (2010) Damage–viscoplastic consistency model with a parabolic cap for rocks with brittle and ductile behavior under low-velocity impact loading. Int J Numer Anal Methods Geomech 34:1362–1386
Shariati H, Saadati M, Bouterf A, et al (2018) On the Inelastic Mechanical Behavior of Granite: Study Based on Quasi-oedometric and Indentation Tests. Rock Mech Rock Eng. https://doi.org/10.1007/s00603-018-1646-3
Storåkers B, Biwa S, Larsson P-L (1997) Similarity analysis of inelastic contact. Int J Solids Struct 34:3061–3083
Vermeer PA, De Borst R (1984) Non-associated plasticity for soils, concrete and rock. HERON, 29 (3), 1984
Wang SY, Sloan SW, Liu HY, Tang CA (2011) Numerical simulation of the rock fragmentation process induced by two drill bits subjected to static and dynamic (impact) loading. Rock Mech Rock Eng 44:317–332
Weddfelt K, Saadati M, Larsson P-L (2017) On the load capacity and fracture mechanism of hard rocks at indentation loading. Int J Rock Mech Min Sci 100:170–176

2019 Rock Dynamics Summit– Aydan et al. (eds)
© 2019 Taylor & Francis Group, London, ISBN 978-0-367-34783-3

A trial to reveal stress recovery at Nojima fault after the 1995 Kobe earthquake by core-based measurement methods

W. Lin, S. Yano & T. Sugimoto
Graduate School of Engineering, Kyoto University, Kyoto, Japan

T. Nishiwaki & A. Lin
Graduate School of Science, Kyoto University, Kyoto, Japan

ABSTRACT: In order to reveal stress recovery at Nojima fault after the 1995 Mw 6.9 Kobe earthquake, Japan, we carried out in-situ stress measurements by using core samples obtained from a vertical scientific drilling penetrated through the Nojima fault which ruptured during the Kobe earthquake. The drilling operations were conducted from 2016 to 2017, therefore the measured stress data using the core samples should be as the time of stress relief i.e. the same as the drilling time, ~22 years passed after the earthquake. We applied two core-based stress measurement methods called Anelastic Strain Recovery (ASR) and Diametrical Core Deformation Analysis (DCDA). Although we are continuously working on core re-orientation and discussions on how interpret the measured stress state, and have not reached conclusions, we believe that a useful data set of the current state of stress around Nojima fault has been obtained by using the two core-based stress measurement methods.

1 INTRODUCTION

Stress and earthquakes are known to be interrelated: stress triggers earthquakes and earthquakes alter the shear and normal stresses on the source faults. On the other hand, the stresses both on the fault plane and in the wall rock around the fault gradually build up in the interseismic period. We have, however, very limited knowledge on the quantitative relationship between stress change and elapsed time after earthquake occurrence. Therefore, in order to reveal stress recovery at Nojima fault after the 1995 Mw 6.9 Kobe earthquake, Japan, we carried out in-situ stress measurements by using core samples obtained from a vertical scientific drilling penetrated through the Nojima fault which ruptured during the Kobe earthquake (also called Hyogo-ken Nanbu earthquake). The drilling operation and ASR measurements were conducted from 2016 to 2017 when ~22 years passed after the earthquake. A recent stress measurement study using borehole breakout analyses suggested that the stress in the deep part of the same borehole as this study looks like to have started to recovery after the Kobe earthquake (Nishiwaki et al., 2018). We have conducted the ASR measurements for 19 core samples and DCDA for 13 core samples, respectively. The core reorientation works and discussion on how to interpret the measured stress state are still on going. Therefore, we only show two representative examples of ASR and DCDA preliminary results following the descriptions on the measurement methods in this paper. We believe that a useful data set of the current state of stress around Nojima

fault has been obtained by using the two core-based stress measurement methods.

2 NOJIMA FAULT DRILLING SITE AND CORE SAMPLES

2.1 Geological background

Our research drilling site is the NFD-1 borehole which penetrated through the Nojima Fault to understand the current stress state at a time of ~22 years elapsed since the 1995 Mw 6.9 Kobe earthquake in Japan (Figure 1). The Nojima Fault is a dextral strike slip fault with minor thrust component. Its strike and down dip are NE-SW and 75 – 80° southeast, respectively. The borehole NFD-1 used in this study was drilled to a maximum depth of ~1000 m in 2017, as a part of the Drilling into Fault Damage Zone (DFDZ) project; has operated at the project site in the past three years from 2016 to 2018. In the depth range above ~230 m of the borehole corresponding to the sedimentary rocks, the rocks are composed of weak consolidated sandstones and conglomerate of the Plio-Pleistocene Osaka-Group and mudstones and sandstones of the Miocene Kobe Group (Figure 2). The basement rocks in the depths below ~230 m consists of Cretaceous granitic rocks. The main fault plane was found at a depth of 529.3 m with a 15 cm thick fault gouge zone and a damage zone of ~60 m thick (those in hanging wall and footwall are ~20 m and ~40 m, respectively), perpendicular to the main fault plane with a dip angle of 72° (Nishiwaki et al., 2018).

2.2 Core samples used for stress measurements

For determining stress state in the vicinity of the Nojima fault, we applied two core-based stress measurements called ASR and DCDA. First, we measure the anelastic strain change of the cores with time (ASR) from stress relief soon to calculate three-dimensional principal in-situ stress orientations and magnitudes. In this study, to ensure the enough amount of ASR, we conducted the measurements using the cores collected within a short time (e.g. 2.5 – 3.5 hours) after stress relief by drilling at an on-site laboratory in the drilling site in Awaji Island, Japan. As tests of DCDA, we measure the core diameters in all (360°) azimuths; and then determine the two horizontal principal stresses orientations and their difference in magnitude. The DCDA tests were conducted after a long time passed from core collecting. Totally, we collected 19 ASR and 13 DCDA core samples for the stress measurements. The lithology of the all samples is granitic rock. Because the stress orientation determined from the

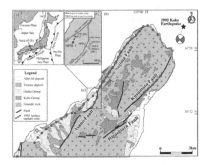

Figure 1. (a) Inset map is the tectonic setting of Japan Islands, (b) the main map is the geological map of the northern part of Awaji Island, its area is shown in (a), and (c) the detailed location of drilling site NFD-1 (Nishiwaki et al., 2018).

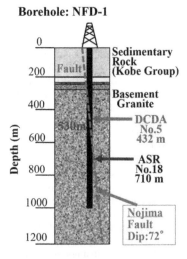

Figure 2. A simplified lithological column showing the distribution of the rock types, location of the fault.

core samples must be restored to the geological coordinates, we are reorienting the cores by correlating the fracture patterns in the drill cores and in BHTV images of the borehole wall. These works are still on going, thus we could not show all the stress measurement results, but only two preliminary results of ASR No.18 and DCDA No.5 cores (Figure 2).

3 METHODS

3.1 Anelastic strain recovery (ASR)

The principal idea behind the ASR technique is that stress-induced strain is released first instantaneously (i.e., as time-independent elastic strain), followed by a more gradual or time-dependent release (Byrne et al., 2009). The ASR method takes advantage of the time-dependent strain. The ASR method for three-dimensional stress determination was proposed in 1990's (Matsuki & Kakeuchi, 1993); and has been frequently applied in deep drilling projects (e.g. Lin et al., 2006).

In this study, we followed the ASR measurement principle proposed by Matsuki & Takeuchi (1993) and the technical procedure established by Lin et al. (2006). Figure 3 shows the layout of strain gaugesstuck on the core surface for the ASR measurements caused by the stress relief and a photo of the ASR sample used in this study.

3.2 Diametrical core deformation analysis (DCDA)

We applied the second core-based method called DCDA to measure the maximum horizontal principal stress azimuth and difference of the maximum and minimum horizontal stresses in magnitude, i.e. $S_{Hmax} - S_{hmin}$ based on a concept mentioned by Ito et al., 2013 and Funato & Ito, 2017. In addition,

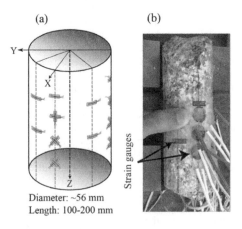

Figure 3. (a) The layout of strain gauges stuck on the core surface for ASR measurement, (b) A photo of the ASR core sample taken from NFD-1.

Optical micrometer

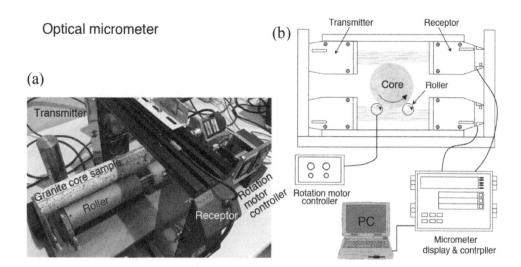

(a)

(b)

Figure 4. (a) A photo of DCDA measurement system of our research group in Graduate School of Engineering, Kyoto University, (b) A schematic diagram of the DCDA system modified from Ito et al., 2013.

Sugimoto et al. (this volume) also described the details on DCDA test procedure applied in this study. DCDA is based on the measurement of azimuthal change of core "diameter" in whole circumference. The basic assumption is that the maximum "diameter" is in the same azimuth as S_{Hmax} because relief of S_{Hmax} may cause the maximum expansion in case of vertical drilling. Therefore, DCDA is a two-dimensional stress measurement, to enable us to know the maximum horizontal stress azimuth by detect the azimuth of the maximum core "diameter" d_{max}. Moreover, the differential stress $S_{Hmax}-S_{hmin}$ can be determined by the following equation:

$$S_{max} - S_{min} = \frac{E}{1+v} \cdot \frac{d_{max} - d_{min}}{d_{min}}$$

$$= 2G \cdot \frac{d_{max} - d_{min}}{d_{min}} \tag{1}$$

where, E, v, G are the Young's modulus, Poisson's ratio and share modulus of the rock core sample, respectively; d_{max} and d_{min} are the maximum and minimum "diameter" of the core sample; and d_0 is the initial core "diameter" and can be assumed to be d_{min} in this calculation. The DCDA measurement system used in this study is as shown in Figure 4.

4 PRELIMINARY RESULTS

4.1 *ASR results*

As a representative example, ASR measurement results of the No.18 core sample are shown in Figure 5. The duration of this measurement was approximately 24 days. During the experiment, the constant thermostatic chamber worked correctly; the temperature was basically controlled within 23±0.1°C. From Figure 5, it is obvious that the anelastic strains in

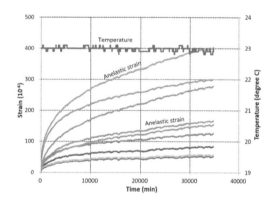

Figure 5. A representative ASR raw data obtained from ASR No.18 core sample retrieved from a depth of ~711 m in NFD-1.

all directions were positive (extensions); and all the curves had a similar variation tendency as time increased. It can be said that ASR in all directions continued for a long period, although the rate of ASR decreased with time. The magnitude of strains in various directions, continuously measured for ~24 days achieved about 60 – 400 microstrains. For the accuracy of the strain measurement system used, these levels of the anelastic strain was sufficiently high and the data could be used for a three-dimensional stress analysis. The preliminary three-dimensional stress regime from the ASR data shown in Figure 5 is of a strike-slip faulting stress regime.

4.2 *DCDA results*

Figure 6 shows a typical DCDA result of the core sample DCDA No.5 retrieved from a depth of ~432 m in NFD-1. Although the diameter data

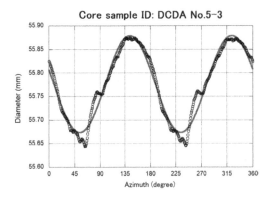

Figure 6. A representative DCDA raw data (open circles) and their regression curve based on the trigonometric function obtained from the 3rd interval of DCDA No.5 core sample retrieved from a depth of ~432 m in NFD-1.

(the open circles) has some irregular trends around the minimum values etc. probably caused by unevenness of the core surface, it clearly showed a regular change pattern of the trigonometric function trend. This change can be attributed to relief of the anisotropic in-situ stress state ($S_{Hmax} > S_{hmin}$) during drilling (coring). Therefore, the in-situ stress information at a time just before the drilling can be obtained from the DCDA data. As a result, the S_{Hmax} azimuth is at ~145° and 325° from the 0° (the reference line of the DCDA core sample).

5 SUMMARY

As a trial to reveal stress recovery at Nojima fault after the 1995 Kobe earthquake, we conducted in-situ stress measurements by two core-based measurement methods called Anelastic Strain Recovery (ASR) and Diametrical Core Deformation Analysis (DCDA). Although we are continuing some additional works needed for determination of stress orientation corresponding to the geologic coordinates and of stress magnitude and have not reached the final conclusions, we believe that a useful data set of the current state of stress around Nojima fault has been obtained by using the two core-based stress measurement methods.

ACKNOWLEDGMENTS

This work was supported by Grants-in-Aid for Scientific Research 16H04065 of the Japan Society for the Promotion of Science (JSPS), and a contract included in the Drilling into Fault Damage Zone (DFDZ) project of the Secretariat of Nuclear Regulation Authority of Japan. We also acknowledge T. Yoshizaki, M. Murakami and the others of OYO Corporation for their kind helps to our core sampling in the drilling site, and T. Ito, K. Omura, T. Yokoyama and A. Funato for their meaningful scientific discussions and helpful suggestions.

REFERENCES

Byrne, T., Lin, W., Tsutsumi, A., Yamamoto, Y., Lewis, J., Kanagawa, K., Kitamura, Y., Yamaguchi, A. & Kimura, G., 2009. Anelastic strain recovery reveals extension across SW Japan subduction zone, Geophys. Res. Lett., 36, L23310.

Funato, A., & Ito, T. 2017. A new method of diametrical core deformation analysis for in-situ stress measurements. International Journal of Rock Mechanics and Mining Sciences, 91, 112–118. http://doi.org/10.1016/j.ijrmms.2016.11.002

Ito, T., Funato, A., Lin, W., Doan, M.-L., Boutt, D. F., Kano, Y., Ito, H., Saffer, D., McNeill, L.C., Byrne, T. & Moe, K.T. 2013. Determination of stress state in deep subsea formation by combination of hydraulic fracturing in situ test and core analysis: A case study in the IODP Expedition 319, J. Geophys. Res. Solid Earth, 118, 1203–1215.

Lin, W., Kwasniewski, M., Imamura, T., & Matsuki, K. 2006. Determination of three-dimensional in-situ stresses from anelastic strain recovery measurement of cores at great depth, Tectonophysics, 426, 221–238.

Matsuki, K. & Takeuchi, K. 1993. Three-dimensional in-situ stress determination by anelastic strain recovery of a rock core, Int. J. Rock Mech. Min. Sci. & Geomech. Abstr., 30(7),1019–1022.

Nishiwaki, T., Lin, A. & Lin, W. 2018. Recovery of stress during the interseismic period around the seismogenic fault of the 1995 Mw 6.9 Kobe earthquake, Japan, Geophysical Research Letters, 45, 12,814–12,820. https://doi.org/10.1029/ 2018GL079317

Sugimoto, T., Lin, W. & Lin, A. 2019. Application of core-based stress measuring method in the vicinity of earthquake source fault: Diametrical Core Deformation Analysis, Proceedings of 2019 Rock Dynamics Summit in OKINAWA (the same volume).

T2: Estimation Procedures and Numerical Techniques of Strong Motions Associated with the Earthquakes

2019 Rock Dynamics Summit– Aydan et al. (eds)
© 2019 Taylor & Francis Group, London, ISBN 978-0-367-34783-3

Methodology to back-analyze the slip-weakening distance of induced seismicity, considering seismic efficiency

A. Sainoki
International Organization for Advanced Science and Technology, Kumamoto University, Kumamoto, Japan

C. Hirohama
Department of Civil and Environmental Engineering, Kumamoto University, Kumamoto, Japan

ABSTRACT: The accurate simulation of induced seismicity is crucial in estimating its risk and evaluating the damage caused by seismic waves. To achieve this, two important mechanical parameters that dictate the severity and magnitude of induced seismicity need to be determined, which are the coefficient of kinetic friction and critical slip-weakening distance. The slip-weakening distance is known to be scale-dependent. The present study proposes a methodology to take into account the scale-dependency by considering seismic efficiency in the dynamic modelling of induced seismicity.

1 INTRODUCTION

1.1 Induced seismicity

Induced seismicity is the dynamic phenomenon of the rockmass at great depths caused by various human activities, such as underground mining, geothermal energy development, the construction of hydraulic power plant, CO_2 sequestration, and so o^{o}n (Ortlepp, 2000, Baisch et al., 2010, Majer et al., 2007, Rutqvist et al., 2007). In underground mines, a severe seismic event can produce devastating damage to mine openings (Ortlepp and Stacey, 1994), whilst for the seismic activity due to geothermal fluid injection, it may raise public concern (Majer et al., 2007). Hence, a better understanding of the mechanism of induced seismicity and developing a methodology to predict its severity are of paramount importance.

1.2 Methodology to simulate induced seismicity

Over the past few decades, various studies have been undertaken, based on numerical analyses, field measurements, and laboratory experiments (Sainoki and Mitri, 2014, Hedley, 1992, Hu, 2011). Laboratory experiments are conducted to elucidate the fundamental mechanism of crack initiation and its propagation that could eventually lead to the failure of rockmass and seismic events (Li et al., 2004, Lockner et al., 1991), while field measurements with seismic monitoring system aim to obtain knowledge on when and where it occurs in relation to orebody extraction, fluid injection, and hydraulic fracturing (Cai et al., 2001, Urbancic and Trifu, 2000, Trifu and Urbancic, 1996). Then, the numerical simulation of induced seismicity can be performed in conjunction with the aforementioned approaches in order to quantitatively estimate the magnitude of a seismic event (Sjöberg et al., 2012) and/or back-analyze the mechanical properties of the causative fault (Hofmann and Scheepers, 2011, Sainoki and Mitri, 2016).

In previous studies, static analysis was predominantly performed to simulate the slip of a causative fault. The analysis allows us to estimate the magnitude of shear displacements on a fault, giving the seismic moment of the simulated event, whereas the energy radiated from the seismic event as the form of seismic waves cannot be simulated with the static analysis. In order to overcome the limitation, dynamic analysis has been recently performed in some studies (Urpi et al., 2016, Sainoki and Mitri, 2014), enabling us to analyze the dynamic behaviour of the fault during seismic event, such as slip rate, particle velocity, and acceleration. Thus, we can quantitatively estimate the amount of seismically radiated energy, giving the seismic source parameters that cannot be obtained from the static analysis.

1.3 Study problem

Although the dynamic analysis has been recently applied to the numerical simulation of induced seismicity, the mechanical properties of faults, especially those of dynamic friction laws, are frequently determined without any verifications in terms of the dynamic behaviour of the faults. That is, even if the magnitude of the simulated seismic event is reasonable, seismically radiated energy and/or the slip rate of the fault may be unrealistic, thus resulting in an inaccurate estimation of the severity of the seismic event in terms of the intensity of seismic waves and released energy. As a matter of fact, the slip-weakening distance D_c, which is one of input parameters of dynamic friction laws and dictates the slip-weakening behaviour of faults during a seismic event, is known to be scale-dependent, meaning that depending on the magnitude of the simulated seismic event, the value needs to be calibrated. Otherwise, the simulated seismic event can yield seismic source parameters far from realistic values. Although several scaling relations have been proposed to estimate D_c based on experimental results and the analysis of seismic wave forms (Ohnaka, 2003), the relation between D_c and the size of seismic events remains to be elucidated. The present study aims to develop a methodology to estiamte the scale-dependency of D_c by considering energy-related seismic source parameters.

2 ENERGY RELATED INDICES

2.1 Energy budget of a seismic event

Figure 1 schematically illustrates the energy budget of a seismic event, where E_G, E_R, and E_F denote fracture energy, seismically radiated energy, and frictional energy loss. As can be seen, the amount of fracture energy is determined by the slip-weakening behaviour. The shear stress acting on the fault decreases from τ_1 (peak stress) to τ_2 (residual stress) over the critical slip-weakening distance D_c. On the other hand, seismically radiated energy is affected by not only the slip-weakening behaviour but also the unloading stiffness of the surrounding rockmass. The sum of the three energy quantities corresponds to the change in total potential energy during the seismic event (Kanamori, 2001).

2.2 Seismic efficiency

Seismic efficiency η is one of seismic source parameters and is defined as the ratio of seismically radiated energy to the total potential energy change as follows.

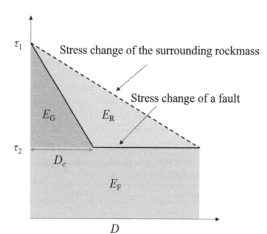

Figure 1. Energy budge of a seismic event.

$$\eta = \frac{E_R}{E_R + E_G + E_F} \tag{1}$$

Previous studies have revealed that seismic efficiency does not exceed 0.06, irrespective of the size of seismic events, including natural earthquakes, induced seismicity, and laboratory-scale slip (McGarr, 1999, McGarr, 1994, Domański and Gibowicz, 2008). A plausible explanation about the scale-independency of η is as follows. As discussed in previous studies, D_c is scale-dependent, varying from a few micrometers for laboratory-scale shear tests to dozens of meters for natural earthquakes with a large magnitude. Importantly, the unloading stiffness of the surrounding rockmass is also dependent upon the size of an earthquakes as estimated below (Scholz, 1998).

$$k = \frac{G \cdot \alpha}{L} \tag{2}$$

where k represents the stiffness of the surrounding rockmass; G, α, and L denote the shear modulus of the rockmass, a constant close to unity, which is affected by the in-situ stress state and the shape of the slipping region, and the length of the fault, respectively. The scale-dependency of both the parameters implies that the proportion of seismically radiated energy to the total potential energy remains the same, despite of the magnitude of earthquakes.

2.3 Postulation about the scale effect of D_c

Importantly, the unloading stiffness of the surrounding rockmass is naturally considered in a numerical simulation because the stiffness of the rockmass obviously varies depending on its dimensions, while D_c is an input parameter that needs to be determined. That is, D_c is unknown. If the assumption that η is predominantly determined by the ratio of the stiffness of the fault and that of the surrounding rockmass is valid, this indicates that D_c can be calibrated under the condition of $\eta \leq 0.06$ for a seismic event with a given magnitude. The present study verifies the postulation by performing numerical analysis, and the obtained scale-relation will be compared to the semi-empirical equation proposed in a previous study.

3 NUMERICAL SIMULATION

3.1 Numerical model description

The present study simulates an induced seismicity, assuming a general case, where fault-slip takes place along an inclined normal fault due to the reduction of the effective normal stress. In the case of underground mining, the reduction of confining stress, i.e. reginal unclamping (Castro et al., 2012), is caused by

$$\mu_s = \tan\phi_s \qquad (4)$$

where τ and σ_n are the shear and normal stresses acting on a fault, respectively; μ_s and μ_d are coefficients of static and kinetic frictions, respectively; d_s and A are a relative shear displacement of the fault and a constant dictating the critical slip-weakening distance; and ϕ_s is a basic friction angle of the fault. Various types of slip-weakening behaviour can be reproduced with this model by adjusting A, but as expected, the exponential type of slip-weakening law cannot precisely determine D_c because the residual stress continuously decreases with the increase in the shear displacement. Therefore, the present study defines D_c as the shear displacement at which the shear stress drops to 1 % of the peak stress.

3.3 Mechanical properties

Table 1 lists the mechanical properties of the rockmass and the fault in the numerical model. The mechanical properties of the rockmass were determined based on those of igneous rock whilst considering the pre-existing fractures in the rockmass in seismically active zones in underground mines and reservoirs. More specifically, Geological Strength Index (GSI) proposed by Hoek and Brown (1997) was assumed to be 60. This is a typical value estimated for the rockmass in deep hard rock mines. According to the GSI, the intact rock properties are decreased with the empirical formulation (Hoek, 2007).

As for the fault, its normal and shear stiffness was approximated assuming the weathering of the fault surface and its degradation due to the shear displacement that the fault experienced in the past. Compared to joints and fractures in a rockmass, fault surfaces are assumed to have undergone intense weathering and deterioration in the past. It should be noted, however, that the normal and shear stiffness of the joint does not exert a large influence on the result of the dynamic analysis because the mechanical behaviour of the fault is predominantly dictated by the slip-weakening behaviour determined with Equation (3) as well as the rigidity of the surrounding rockmass. Regarding the coefficient of static friction μ_s, 0.58 is assumed, which corresponds to that derived from the Coulomb's friction law with a friction angle of 30°. The value falls within the typical range of basic friction angles of various rocks.

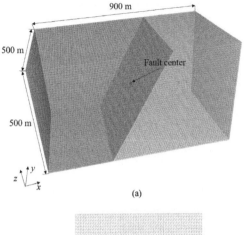

(a)

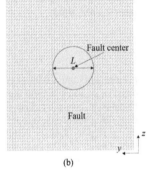

(b)

Figure 2. (a) perspective view, (b) front view of the fault, showing the area where the effective normal stress of the fault is decreased to simulate an induced seismicity.

the extraction of a large amount of orebodies, whilst in geothermal reservoirs, fluid injection increases pore pressure, which decreases the effective stress acting on rockmass discontinuities in the reservoir (Majer et al., 2007).

Figure 2(a) shows the perspective view of the numerical model including the fault that dips at 60° and strikes in the y-direction. The dimensions of the model were determined so that the boundary effect does not affect the behaviour of the fault. Also, the domain was discretized densely to obtain a stable result with respect to the zone size. Figure 2(b) the region where the effective normal stress is reduced to simulate an induced seismicity. The diameter of the region is varied to simulate seismic events with different magnitudes, and for each case D_c is calibrated based on η.

3.2 Friction law

As for the friction law of the fault, a slip-weakening law is employed in this study that is expressed as below.

$$\tau = \sigma_n\left(\mu_d - (\mu_d - \mu_s)\exp\left(\frac{-d_s}{A}\right)\right) \qquad (3)$$

Table 1. Mechanical properties

Rockmass				Fault	
E	ν	γ	kn	ks	μ_s
(GPa)		(kN/m³)	(GPa/m)	(GPa/m)	
10	0.28	25.5	3	1.25	0.58

221

3.4 Analysis procedure

Figure 3 illustrates the analysis procedure taken for this study. As can be seen, the analysis starts with the simulation of the in-situ stress state in a static condition, where the overburden pressure corresponding to the depth of 1000 m is applied to the top boundary of the model, while the horizontal stresses are applied to the lateral boundaries assuming the vertical-to-horizontal stress ratio of 0.7. The bottom boundary is fixed in the direction perpendicular to the boundary.

After the equilibrium in force is reached, the effective normal stress is gradually reduced in the circular region shown in Figure 2 in the static condition until the shear stress acting on the fault reaches its peak strength. When the critical stress state is attained at one of the fault patches in the region, the analysis condition is changed from static to dynamic whilst transforming the boundary condition to viscous boundary condition. Simultaneously, the dynamic friction law expressed with Equation 3 is applied to the interface element. During the dynamic analysis, the stress state of the interface elements in the region is continuously checked, and as soon as the critical condition of the shear stress is satisfied, the dynamic friction law is applied for the other interface elements in the region.

4 RESULT AND DISCUSSIONS

4.1 Model parametric study

Following the analysis procedure shown in Figure 3, a model parametric study is conducted with respect to the diameter of the circular region represented by L. For each analysis, the coefficient kinetic friction is set at 0.42, and the parameter A in Equation 3 is calibrated so that the seismic efficiency of the simulated seismic event becomes 0.06. In this way, the stress drop caused by the slip-weakening behaviour basically remains almost the same amongst the models with different L, whilst the simulated seismic events have different slip-weakening distances, depending on their magnitudes. For the respective models, seismic moment was computed from the simulated result.

Figure 4 shows the result of the model parametric study. It can be seen from the figure that D_c decreases with the decreasing seismic moment, thus suggesting

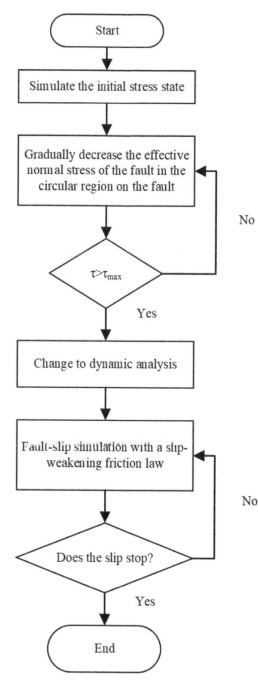

Figure 3. Analysis procedure.

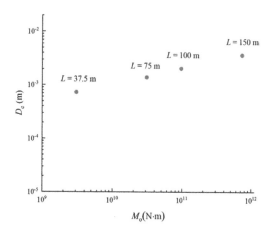

Figure 4. Result of the model parametric study with respect to L in Figure 2.

fully producing the scaling relation of D_c with respect to the size of the seismic events. When $L = 150$ m, the seismic moment of the event and calibrated D_c are 7.3×10^{11} N·m and 3.64 mm, respectively, whilst for the model with $L = 37.5$ m, the seismic moment and D_c reduce to 3.0×10^9 N·m and 0.72 mm, respectively. It should be noted that according to previous studies (Ohnaka, 2003), D_c is also dependent on the apparent stress of seismic events. Therefore, in order to examine only the effect of the size of the slipping region, the coefficient of kinetic friction was kept constant amongst the models.

4.2 Comparison with a semi-empirical equation developed in a previous study

Over the past decades, the scaling law of D_c has remained to be elucidated. Hence, significant efforts have been made to obtain the relation between D_c and seismic source parameters, such as stress drop and seismic moment.

Ohnaka (2000) proposed the following scaling law based on the equation of seismic moment whilst assuming the scaling relation of fault length with the shear displacement induced during an earthquake.

$$M_o = 0.8 \times 10^{19} D_c^3 \qquad (5)$$

Figure 5 shows the comparison of D_c calibrated from the numerical simulation of this study with that obtained from the equation. As can be seen, D_c calibrated with seismic efficiency agrees well with those estimated from Equation 5. It should be noted that Equation 5 uses the same parameters assumed by Ohnaka (2000), so that the magnitude of the stress drop used for Equation 5 is different from that in this numerical simulation. Notwithstanding the fact, it is remarkable that the numerical simulation employed in this study produces almost the same scaling relation.

The result strongly suggests that seismic efficiency can be used to constrain D_c in the numerical simulation of induced seismicity. Furthermore, the employed numerical simulation method may be able to estimate D_c more accurately compared to the scaling relation proposed by Ohnaka (2000) because Equation 5 does not consider the effect of seismically radiated energy in the equation. In reality, D_c exerts a direct influence on the amount of seismically radiated energy, which is taken into account in the proposed method as the form of seismic efficiency. For instance, a small D_c produces more drastic stress change during the slip-weakening behaviour, thus yielding a large η. In the present study, the estimation of D_c was performed based on the condition of $\eta = 0.06$, following the postulation made by McGarr (1994). Therefore, a further study needs to be done in order to gain a comprehensive understanding of the influence of D_c on seismic efficiency, so that a novel scaling relation of D_c would be obtained that considers not only the effect of magnitude, but also the intensity of seismically radiated energy.

5 CONCLUSION

The slip-weakening distance D_c of a fault is known to be scale-dependent. Thus, a difficulty arises in determining the parameter in the numerical simulation of induced seismicity. An inaccurate estimation of D_c as an input parameter for numerical analysis can yield a seismic event with unrealistic seismic source parameters. The present study demonstrated based on the result of dynamic analyses that D_c can be accurately estimated whilst considering its scale dependency when D_c is calibrated in terms of seismic efficiency. The estimated D_c for seismic events with different magnitudes was compared to those calculated from a semi-empirical equation proposed in a previous study. The comparison showed remarkable consistency between the results. Furthermore, the methodology to estimate D_c proposed in this study allows for the effect of seismically radiated energy in its estimation, which is not taken into account in the previous study. This indicates that the proposed method may be able to estimate D_c more accurately compared to the previous study, although a further study needs to be undertaken to examine its applicability.

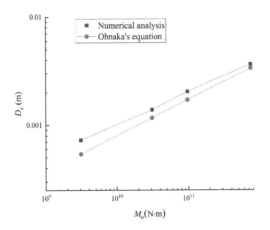

Figure 5. Comparison between D_c obtained from the numerical analysis and Ohnaka's equation.

REFERENCES

Baisch, S., Vörös, R., Rothert, E., Stang, H., Jung, R. & Schellschmidt, R. 2010. A numerical model for fluid injection induced seismicity at Soultz-sous-Forêts. *International Journal of Rock Mechanics and Mining Science*, 47, 405–413.

Cai, M., Kaiser, P. K. & Martin, C. D. 2001. Quantification of rock mass damage in underground excavations from microseismic event monitoring. *International Journal of Rock Mechanics and Mining Science*, 38, 1135–1145.

Castro, L. A. M., Bewick, R. P. & Carter, T. G. 2012. An overview of numerical modelling applied to deep mining. *Innovative Numerical Modelling in Geomechanics*, 393–414.

Domański, B. & Gibowicz, S. J. 2008. Comparison of source parameters estimated in the frequency and time domains for seismic events at the Rudna copper mine, Poland. *Acta Geophysica*, 56, 324–343.

Hedley, D. G. F. 1992. *Rockburst handbook for ontario hard rock mines*, Ontario Mining Association

Hoek, E. 2007. *Practical rock engineering.*

Hoek, E. & Brown, E. T. 1997. Practical estimates of rock mass strength. *International Journal of Rock Mechanics and Mining Science*, 34, 1165–1186.

Hofmann, G. F. & Scheepers, L. J. 2011. Simulating fault slip areas of mining induced seismic tremors using static boundary element numerical modelling. *Mining Technology*, 120, 53–64.

Hu, H. E. 2011. Study of acoustic emission monitoring technology for rockburst. *Rock and Soil Mechanics*, 32, 1262–1268.

Kanamori, H. 2001. Energy budget of earthquakes and seismic efficiency. *International Geophysics*, 76, 293-305.

Li, Y. P., Zeng, J. & Chen, L. 2004. Study on characteristics of Acoustic Emission from Marble with Precut Crack during Failure. *Underground Space*, 24, 290–293.

Lockner, D. A., Byerlee, J. D., Kuksenko, V., Ponomarev, A. & Sidrin, A. 1991. Quasi-static fault growth and shear fracture energy in granite. *Nature*, 350, 39–42.

Majer, E. L., Baria, R., Stark, M., Oates, S., Bommer, J., Smith, B. & Asanuma, H. 2007. Induced seismicity associated with enhanced geothermal systems. *Geothermics*, 36, 185–222.

Mcgarr, A. 1994. Some comparisons between mining-induced and laboratory earthquakes. *PAGEOPH*, 142, 467–489.

Mcgarr, A. 1999. On relating apparent stress to the stress causing earthquake fault slip. *Journal of Geophysical Research*, 104, 3003–3011.

Ohnaka, M. 2000. A physical scaling relation between the size of an earthquake and its nucleation zone size. *Pure and applied geophysics*, 157, 2259–2282.

Ohnaka, M. 2003. A constitutive scaling law and a unified comprehension for frictional slip failure, shear fracture of intact rock, and earthquake rupture. *Journal of Geophysical Research*, 108, ESE 6-1.

Ortlepp, W. D. 2000. Observation of mining-induced faults in an intact rock mass at depth. *International Journal of Rock Mechanics and Mining Science*, 37, 423–426.

Ortlepp, W. D. & Stacey, T. R. 1994. Rockburst Mechanisms in Tunnels and Shafts. *Tunnelling and Underground Space Technology*, 9, 59–65.

Rutqvist, J., Birkholzer, J., Cappa, F. & Tsang, C.-F. 2007. Estimating maximum sustainable injection pressure during geological sequestration of CO2 using coupled fluid flow and geomechanical fault-slip analysis. *Energy conversion and management*, 48, 1798–1807.

Sainoki, A. & Mitri, H. S. 2014. Dynamic behaviour of minig-induced fault slip. *International Journal of Rock Mechanics and Mining Science*, 66c, 19–29.

Sainoki, A. & Mitri, H. S. 2016. Back analysis of fault-slip in burst prone environment. *Journal of Applied Geophysics*, 134, 159–171.

Scholz, C. H. 1998. Earthquakes and friction laws. *Nature*, 391, 37–42.

Sjöberg, J., Perman, F., Quinteiro, C., Malmgren, L., Dahner-lindkvist, C. & Boskovic, M. 2012. Numerical analysis of alternative mining sequences to minimise potential for fault slip rockbursting. *Mining Technology*, 121, 226–235.

Trifu, C. I. & Urbancic, T. I. 1996. Fracture coalescence as a mechanism for earthquake: obsrevations based on mining induced microseismicity. *Tectonophysics*, 261, 193–207.

Urbancic, T. I. & Trifu, C. 2000. Recent advances in seismic monitoring technology at Canadian mines. *Journal of Applied Geophysics*, 45, 225–237.

Urpi, L., Rinaldi, A., Rutqvist, J., Cappa, F. & Spiers, C. 2016. Dynamic simulation of CO2-injection-induced fault rupture with slip-rate dependent friction coefficient. *Geomechanics for Energy and the Environment* 7, 47–65.

2019 Rock Dynamics Summit– Aydan et al. (eds)
© 2019 Taylor & Francis Group, London, ISBN 978-0-367-34783-3

Effects of fault geometry and subsurface structure model on the strong motion and surface rupture induced by the 2014 Kamishiro Fault Nagano Earthquake

N. Iwata & R. Kiyota
Chuden Engineering Consultants Co., Ltd., Hiroshima, Japan

Ö. Aydan
Department of Civil Engineering, University of the Ryukyus, Okinawa, Japan

T. Ito
Institute of Fluid Science, Tohoku University, Sendai, Japan

F. Miura
Graduate School of Science and Engineering, Yamaguchi University, Yamaguchi, Japan

ABSTRACT: The authors used a three-dimensional finite element method (3D-FEM) to examine a series of fault rupture simulations for the 2014 Northern Nagano Earthquake and simultaneously estimate the displacement and strong motions. The computational results confirmed that the maximum responses of ground motions and displacement could be simultaneously evaluated using the appropriate constitutive parameters and fine FEM mesh. However, the duration of the acceleration response and shape of the surface displacement waves were not well simulated. In this study, we examined the influence of the fault bend at a shallow depth and P- and S-wave velocity structure models based on a geological survey. As a result, by taking account of bending of the fault plane and the elastic velocity structure at a shallow depth, it was possible to perform a seismic behaviour analysis using the 3D-FEM approach.

1 INTRODUCTION

Fault rupture simulations have been performed mainly for the reproduction and prediction of strong motions. After the 1999 Chi-Chi and Kocaeli earthquakes damaged many important structures as a result of surface rupture, the displacement and inclination of the ground surface caused by fault rupture have become significant issues in engineering. Many prediction methods for strong motion, displacement and inclination of the surface ground have been suggested. However, most analytical methods do not evaluate displacement and strong motions at the same time. Furthermore, the parameters adopted in these methods (fault length, slip on fault, stress drop, etc.) are empirically determined and the dynamic destruction process is not taken into account. Therefore, it is difficult to evaluate strong motions of magnitude nine, the seismic response around the epicentre and surface rupture with few observed data.

The authors have examined a series of fault rupture simulations to simultaneously estimate the displacement and strong motions at the ground surface using a three-dimensional finite element method (3D-FEM) considering the dynamic destruction process of the fault plane. This analytical method and modelling of faults was proposed by Toki & Sawada (1988) and Mizumoto et al. (2005). The fault plane was assumed

to be discontinuities of bedrock and was modelled by joint elements, and shear failure occurred at the hypocentre and spread to the surrounding areas with increasing shear stress. Iwata et al. (2018) conducted a fault rupture simulation for the 2014 Northern Nagano Earthquake, which was induced by the Kamishiro Fault $(M_w 6.3)$. The computational results confirmed that the maximum responses of the ground motion and displacement could be evaluated using the appropriate constitutive parameters and a fine FEM mesh with a size less than 150 m. However, the duration of the acceleration response and shape of the surface displacement response were not well simulated, partly because the fault plane was assumed to be a straight planar feature and the bedrock was assumed to be homogeneous.

In this study, we examined the influence of a fault bend at a shallow depth and the S-wave velocity structure model based on a geological survey. In addition, we examined the influence of the initial stress distribution on the fault plane.

2 OUTLINE OF ANALYTICAL METHOD

If the equation of motion involves the rupture movement of the fault plane, it is necessary to treat the motion as a nonlinear problem. Therefore, it is appropriate to obtain a solution for the equation of motion, not in the frequency domain but rather in the time domain. The destruction process of dislocation and the dynamic behaviour of the ground are calculated by solving the equation of motion using the stress drop of the dislocation as the external force. The equation of motion at time step n is written as

$$[M]\{\ddot{u}\}_n + [C]\{\dot{u}\}_n + [K]\{u\}_n = \{F(n,s)\} \qquad 1$$

where $[M]$, $[C]$ and $[K]$ are mass, damping and stiffness matrix, respectively. $\{\ddot{u}\}$, $\{\dot{u}\}$, $\{u\}$ are acceleration, velocity and displacement vector, respectively. $\{F(n,s)\}$ is the external force vector calculated from the dynamic stress drop where n and s, respectively, stand for the time step and nodal pairs where the fault rupture takes place. The damping matrix $[C]$ is obtained from the linear combination of $[M]$ and $[K]$, which is called Rayleigh damping. Equation (1) is solved using Newmark's β method, with $\beta = 0.25$, $\gamma = 0.5$ at each time interval. To solve the nonlinear equation of motion, we employed the load transfer method using the initial stiffness method (Toki & Sawada, 1988, Tsuboi & Miura, 1996).

The fault plane was modelled by joint elements shown in Figure 1. The shear springs K_s, K_r and normal spring K_n are connected between nodal points of the solid elements, and sliding occurs according to the Mohr–Coulomb failure criterion. When the calculated shear stress τ is less than the peak stress τ_y, the stress–deformation relationship is linear with the joint stiffness K_s, K_r. Sliding takes place if the shear stress τ reaches the peak stress τ_y, and a stress drop occurs. The shear stress becomes equivalent to the residual strength τ_r, and the stress drop $\Delta\tau_d (= \tau_y - \tau_r)$ is released and spreads to nearby elements. In this way, the released stress drop at the hypocentre is triggered and shear failure spreads to the surrounding areas with increasing shear stress.

3 OVERVIEW OF THE 2014 NORTHERN NAGANO EARTHQUAKE

The 2014 Northern Nagano Earthquake had a moment magnitude of 6.3. It occurred at 22:08 JST on November 22nd, in the northern part of Nagano

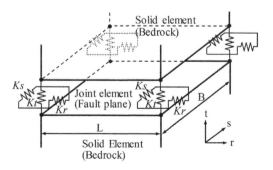

Figure 1. Schematic diagram of joint element.

Prefecture. The slip sense was estimated as thrust faulting with a left–lateral strike slip. There have been many reports by various organizations about the fault geometry and seismic moment. According to the F-net centroid moment tensor solution catalogue (F-net 2014) of the National Research Institute for Earth Science and Disaster Prevention (NIED), the seismic moment M_0 was 2.76×10^{18} N·m, the fault plane was dipping downwards at N16E–50E, the rake angle was 65° and the focal depth was estimated at 5 km. On the other hand, the aftershock activity indicated a non-planar, bending fault plane, which consisted of two planar faults: the shallow plane with a dip angle of 40° connected to a deep plane with a dip angle of 60° at 4 km depth (Panayotopoulos, 2016).

Figure 2 shows the acceleration records at K-NET Hakuba station, which is one of the strong motion stations of the dense strong motion network operated by the NIED. It is located approximately 0.5 km from the surface rupture in the west, on the footwall side of the earthquake fault. The maximum accelerations are 570 Gal horizontally and 278 Gal vertically. The displacement response was calculated using the EPS method proposed by Aydan & Ohta (2011), which is an integration technique for obtaining ground motions in consideration of device operation features, fault rupture duration and arrival time difference of the P- and S-waves. From these results, the residual displacement was obtained as 24 cm horizontally and 10 cm vertically at the Hakuba site.

4 ANALYTICAL CONDITIONS AND PARAMETERS

4.1 Analytical conditions

To investigate the influence of the initial stress distribution in a fault plane, dip angle and velocity structure of the bedrock, we performed a series of fault rupture simulations under various conditions. The validity of the parameters and modelling was evaluated by comparison with the observed surface ground motions recorded at K-NET Hakuba station.

In previous studies, we assumed that the initial shear stress distribution shape was mountain type, in which the hypocentre is highest and the shear stress decreases towards the fault ends and becomes zero, as shown in Figure 3. However, the asperity type, in which the shear stress is concentrated only in protrusions on the fault plane, is used for predicting strong motions. Therefore, we compared the seismic response of asperity type with that of mountain type.

As for the dip angle, according to the aftershock activities and the results of seismic waveform inversion analyses (F-net 2014, JMA 2014 and Panayotopoulos, 2016), we compared the seismic response of the planar model with a dip angle of 50° or 60° and the bending model; the shallow plane with a dip angle of 40° connected the deep plane with a dip angle of 60° at 4 km depth.

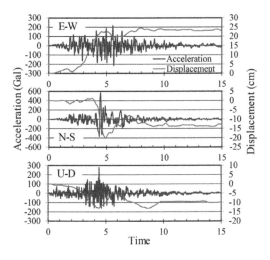

Figure 2. Acceleration records at K-NET Hakuba station and displacement time histories calculated by the EPS method.

As for the velocity of bedrock, in the previous studies we assumed that the bedrock was homogeneous and distributed uniformly with the ground surface. However, the elastic wave speed decreased as it neared ground surface at shallow depths. We examined parametric studies using horizontally layered model by the Japan Seismic Hazard Information Station (J-SHIS 2018): S-wave velocity

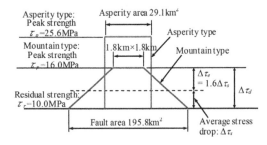

Figure 3. Schematic diagram of initial stress distribution.

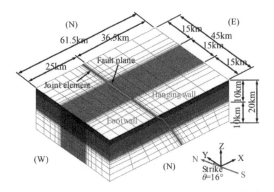

Figure 4. 3D FEM mesh.

was 1.1 km/s from the ground surface to 0.7 km in depth, 2.1 km/s at 0.7–2.6 km, 2.7 km/s at 2.6–2.7 km, 3.1 km/s at 2.7–3.0 km, 3.3 km/s at 3.0–8.0 km and 3.5 km/s below 8.0 km.

Other parameters were set as follows: the shear stiffness of a joint element was 1.0×10^6 kN/m^3, the normal stiffness was 3.0×10^6 kN/m^3 and the damping ratio was 0.03.

4.2 FEM model

Figure 4 shows the FEM model with a bending fault plane. Joint elements were set up from the northern lateral boundary to the southern one, and the strength of joint elements located out of the fault plane was set to have a high value. The distance from the fault edge to the boundaries was more than the fault length (15 km), and viscous dampers were introduced at the lateral and bottom boundaries to absorb scattering wave energy.

4.3 Stress on fault plane

The average static stress drop $\Delta\tau_s$ in the fault plane was 2.3 MPa, calculated from the seismic moment and fault area according to the Recipe for Predicting Strong Ground Motions (Irikura 2006). In the mountain type stress distribution, the peak strength τ_y was determined based on the knowledge that the excess strength $\Delta\tau_e$ is 1.6 times the static stress drop $\Delta\tau_s$ (Andrew, 1976) and $\Delta\tau_d$ was made 6.0 MPa, as shown in Figure 3. In that figure, the shape of the peak stress distribution is a square, and the area of peak strength is calculated so that the volume of the truncated square pyramid described by a solid blue line equals that of the cube described by a dashed line. In the asperity type distribution, the area of asperity was calculated according to the 'Recipe', and $\Delta\tau_d$ was made 15.6 MPa by multiplying $\Delta\tau_s$ by the ratio of the fault area to the asperity area. The residual strength τ_r made 10.0 MPa sufficiently larger than the dynamic stress drop $\Delta\tau_d$. The shear stress at the hypocentre was assumed to be slightly larger than the peak strength τ_y. The rake angle was 60° according to the results of the seismic waveform inversion analysis by the Japan Meteorological Agency (JMA 2014). The behaviour of thrust faulting with a left–lateral strike slip was reproduced by setting the initial shear stress in the rake angle.

5 ANALYSIS RESULTS

5.1 Influence of initial stress distribution

We carried out fault rupture simulations using various initial stress distributions: mountain type and asperity type, and compared the seismic moments and seismic response at the ground surface corresponding to that at K-NET Hakuba station. The parameters except the initial stress distribution were the same, the dip angle of fault plane was 50° and the bedrock was homogeneous.

Figure 5 shows the slip distribution in the fault plane in mountain-type case. The slip spreads from the hypocentre at the rake angle and reaches the ground surface. The maximum slip amount is 1.69 m, and the seismic moment is estimated as 5.93×10^{18} N·m, which is 2.1 times that of the observed value (M = 2.76×10^{18} N·m). Although the shape of asperity-type slip distribution is similar to that of

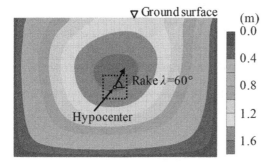

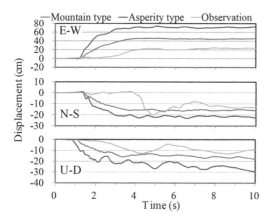

Figure 5. Slip distribution on the fault plane of Mountain type.

Figure 6. Comparison of analysed displacement waves using different initial stress distributions.

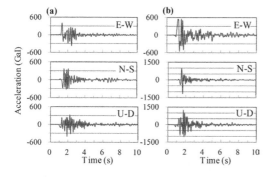

Figure 7. Comparison of acceleration waves using various initial stress distribution: (a) mountain type, (b) asperity type.

mountain type, the maximum slip amount is 3.46 m and the seismic moment is estimated as 9.32×10^{18} N·m, which is 1.6 times that of mountain type.

Figure 6 compares the analysed displacement waves at a distance of 0.5 km from the surface rupture with the observed records. The analysed residual displacements of mountain type are 1.3–2.0 times the observations, and those of asperity type are 1.5–3.5 times the observations.

Figure 7 shows a comparison of the analysed acceleration waves. The amount of stress drop for asperity type is so large that the destruction spreads over a stretch of area. Then, the maximum acceleration and residual displacement of asperity type become larger than those of mountain type and the duration of main shock becomes shorter. The maximum acceleration of asperity type is 1.4 times that of mountain type in the E–W direction and about 3.0 times in the N–S and U–D directions. Comparing the maximum accelerations, those of the mountain type are 1.9 times those of the observations in the E–W direction, 0.7 times in N–S direction and 1.5 times in U–D direction. The directivity effect of the fault rupture could not be reproduced in the acceleration response. It was necessary to conduct parametric studies of the rake angles. When the mesh size becomes less than 100 m, accelerations in the U–D direction become smaller. The vertical response could be improved using a finer FEM mesh.

From these results, even if the amount of stress drop in a fault plane was the same, the seismic response varied greatly according to the difference in initial stress distribution. When a stress drop based on the Recipe is adopted for a fault rupture simulation using the 3D-FEM, in some cases the analysed seismic response becomes considerably larger than expected.

5.2 *Influence of dip angle*

We performed fault rupture simulations using various dipping fault planes: a planar model with a dip angle of 50° or 60° and a bending model, and we examined the influence of the dip angle and shape of the fault plane. All parameters except the geometry of the fault plane were the same, the initial stress distribution was mountain type and the bedrock was homogeneous.

The slip distribution shape of the planar model with a dip angle of 60° was similar to that of the model with a dip angle of 50°. The maximum slip amount became 1.71 m, and the seismic moment was estimated as 5.93×10 N·m, which became slightly larger than the moment at 50°. Figure 8 shows the slip distribution with a bending fault plane. The slip distribution becomes discontinuous at a depth of around 4 km owing to the changing dip angle at that depth. The maximum slip on the shallow fault plane (dip angle of 40°) is 1.17 m and is larger around the hypocentre. The slide amount of the bending model is smaller than that of the planar model. The seismic moment was estimated as 4.45×10 N·m, which is 0.75 times that of the planar model with a dip angle of 50°.

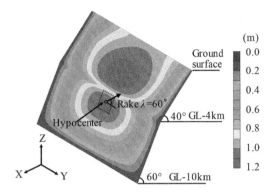

Figure 8. Slip distribution on the bending fault plane.

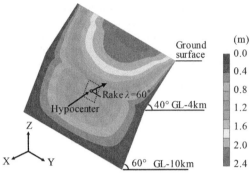

Figure 11. Slip distribution of horizontally layered model.

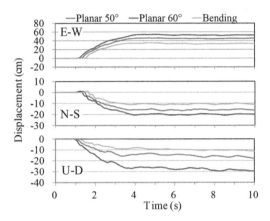

Figure 9. Comparison of analysed displacement waves using different fault geometry.

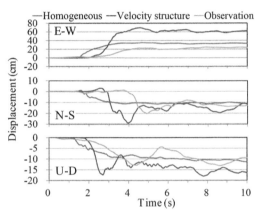

Figure 12. Comparison of analysed displacement waves using different velocity structures.

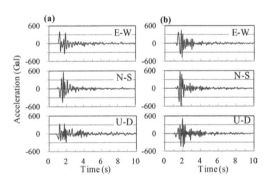

Figure 10. Comparison of acceleration waves using different fault geometry: (a) Planar 60°, (b) Bending 40+60°.

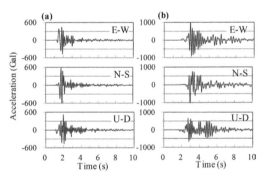

Figure 13. Comparison of acceleration waves using different velocity structures: (a) Homogeneous, (b) Horizontally layer.

Figure 9 compares the analysed displacement waves. The residual displacement of the 60° planar model is about 1.2 times that of the 50° planar model horizontally and 1.6 times vertically. The residual displacement of the bending model is about 0.7 times that of the 50° planar model.

Figure 10 shows the analysed acceleration waves. The waveform of the 60° planar model is similar to

that of the 50° one shown in Figure 7(a). However, in the case of the bending fault, the peak acceleration becomes larger than that of the planar mode, the peak is delayed and the duration is extended. The maximum acceleration of the 60° planar model is at the same level as that of the 50° one in the E–W and U–D directions, and 1.5 times in N–S direction. The acceleration of the bending model is 1.2 times that of the 50° planar model in the E–W and U–D directions, and 2.0 times in N–S direction.

5.3 *Influence of velocity structure of bedrock*

To compare the influence of the velocity structure of bedrock, we conducted fault rupture simulations using homogeneous and horizontally layered models of the elastic wave velocity. In both cases, the initial stress distribution was mountain type and the fault shape was that of the bending model.

Figure 11 shows the slip distribution on the fault plane in the case of the horizontally layered model. The slip distribution becomes discontinuous at 4-km depth, like the homogeneous model shown in Figure 8. The maximum slip amount around the hypocentre is 1.34 m. The slippage become larger towards the ground surface, and the maximum slip amount is 2.36 m. The slippage of the horizontally layered model is larger than that of the homogeneous model. However, the seismic moment of the horizontally layered model $(3.49 \times 10$ N·m) is smaller than that of the homogeneous model because the elastic wave velocity and stiffness of the surface layer are small.

Figures 12 and 13 show a comparison of the analysed displacement and acceleration waves at distances of 0.5 km from the surface rupture, respectively. When an elastic wave velocity structure is adopted, the time at which residual displacement begins to occur and the time of peak acceleration are delayed. The duration of the acceleration wave is extended, resulting in fluctuations. These behaviours resemble the observed records. However, the residual displacement and maximum acceleration become considerably larger.

Figure 14 shows the vertical displacement distribution at the ground surface in the E–W direction (orthogonal to the surface rupture). The gap in the surface rupture is 0.55 m in the homogeneous model and 1.3 m in the horizontally layered model. The distribution shape of the east side of the fault in the homogeneous model is approximately level at less than 1 km from the surface rupture and inclines gently at the outside. On the other hand, the vertical displacement of the east side of the fault in the horizontally layered model decreases sharply as it leaves the surface rupture. The inclination angle of the ground surface in the horizontally layered model

is steeper than that in the homogeneous model. According to an InSAR analysis by the Geospatial Information Authority of Japan (GSI 2014), the incline of the ground surface on the east side of the surface rupture is steeper than that on the west side and the deformation occurs on the side of the surface rupture. Consequently, when an elastic wave velocity structure is adopted, the analysis result can be brought closer to the actual deformation.

6 CONCLUSIONS

In this study, we performed a series of fault rupture simulations using the 3D-FEM for the 2014 Northern Nagano Earthquake and evaluated the influences of the initial stress distribution on the fault plane, dip angle and velocity structure of the bedrock. The findings obtained from this study can be summarized as follows:When the amount of stress drop was defined based on the Recipe, the seismic moment, acceleration response and residual displacement were larger than the observed records.The analysed seismic response using the initial stress distribution of asperity type was larger than that of mountain type because the stress drop was larger and more rapidly released.The slip in the fault plane and displacement response of the bending fault model were smaller than those of the planar model. However, the acceleration response was larger.As the elastic velocity of the surface layer decreases, the seismic response and slippage became larger. However, the seismic moment became smaller because the stiffness of the surface layer was small.

By taking into account the bending of the fault plane and the elastic velocity structure at a shallow depth, it was possible to make the seismic behaviour analysed by the 3D-FEM approach the observation. However, the maximum acceleration and residual displacement were greater than the observations. Therefore, it is necessary to examine a setting method for the parameters and a model to reproduce the actual values.

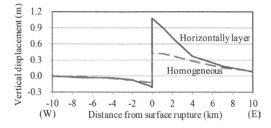

Figure 14. Comparison of vertical displacement distribution at the surface ground in E–W direction using various velocity structures.

REFERENCES

Andrews, D.J. 1976. Rupture velocity of plane strain shear racks, *J. of Geo. Res.*, Vol. 81, No. 32, 5679–5687.

Aydan, O. & Ohta, Y. 2011. The erratic pattern screening (EPS) method for estimation of co-seismic deformation of ground from acceleration records and its application, *Seventh National Conf. on Earth. Eng.*, Turkey.

F-net (Full Range Seismograph Network of Japan), National Research Institute for Earth Science and Disaster Resilience. 2014. Topics: The 2014 North Nagano Earthquake (in Japanese) http://www.hinet.bosai.go.jp/topics/n-nagano141122/?LANG=en.

GSI (Geospatial Information Authority of Japan). 2014. Crustal deformation associated to the Northern Nagano earthquake observed by the "Daichi-2" ALOS-2 satellite (in Japanese). http://www.gsi.go.jp/BOUSAI/h26-nagano-earthquake-index.html.

Irikura, K. 2006. Predicting strong ground motions with "Recipe", *Bull. Earthq. Res. Inst. Univ. Tokyo*, 81: 341–352.

Iwata, N., Kiyota, R., Adachi, K., Takahashi, Y., Aydan, Ö., Miura, F. & Ito, T. 2018. Simulation of strong motions and surface rupture of the 2014 Northern Nagano Earthquake. *Proc. of the 3rd inter. conf. on Rock dynamics and applications (ROCDYN-3), Norway*: 541–546.

JMA (Japan Meteorological Agency). 2014. News release document about the earthquake of North Nagano on November 22, 2014 about 22:08 (in Japanese).

J-SHIS (Japan Seismic Hazard Information Station). 2018. Subsurface structure.

Mizumoto, T., Tsuboi, T. & Miura, F. 2005. Foundamental study on failure rupture process and earthquake motions and near a fault by 3D-FEM. *Journal of Japan Society of Civil Engineers* 780/I-70: 27–40 (in Japanese with English abstract).

Panayotopoulos, Y., Hirata, N., Hashima, A., Iwasaki, T., Sakai, S. & Sato, H. 2016. Seismological evidence of an active footwall shortcut thrust in the Northern Itoigawa–Shizuoka Tectonic Line derived by the aftershock sequence of the 2014 M6.7 Northern Nagano earthquake. *Tectonophysics* 679: 15–28.

Toki, K. & Sawada, S. 1988. Simulation of the fault rupture process and near filed ground motion by the three-dimensional finite element method. *Proc. of 9th World Conf. on Earth. Eng., Japan*, Vol. II: 751–756.

Tsuboi, T. & Miura, F. 1996. Simulation of stick-slip shear failure of rock masses by a nonlinear finite element method. *Journal of Japan Society of Civil Engineers* 537/I-35: 61–76 (in Japanese with English abstract).

2019 Rock Dynamics Summit– Aydan et al. (eds)
© 2019 Taylor & Francis Group, London, ISBN 978-0-367-34783-3

Comparison of stress field change around a fault by dynamic fault rupture simulation using 3D-FEM

N. Iwata & R. Kiyota
Chuden Engineering Consultants Co., Ltd., Hiroshima, Japan

Ö. Aydan
Department of Civil Engineering, University of the Ryukyus, Okinawa, Japan

T. Ito
Institute of Fluid Science, Tohoku University, Sendai, Japan

F. Miura
Graduate School of Science and Engineering, Yamaguchi University, Yamaguchi, Japan

ABSTRACT: When we are able to evaluate earthquake-induced stress changes of the ground around ruptured faults and adjacent faults, it will be possible to improve the prediction accuracy of the magnitude and probability of future earthquakes. Generally, the interaction between active faults is represented by static Coulomb stress changes (ΔCFF) induced by fault rupturing. In most cases, ΔCFF is calculated based on the elasticity theory of dislocation; there are few studies where it is calculated by 3D-FEM. In this study, we conducted fault rupture simulations using 3D-FEM for simple models with a planar fault plane and homogeneous bedrock and examined the influence of fault type and initial stress distribution. As a result, ΔCFF calculated by 3D-FEM became considerably larger than that calculated by the elasticity theory of dislocation. Moreover, even when a fault type and seismic magnitude were the same, the distribution domain and quantity of ΔCFF differed greatly owing to the combination of analytical parameters.

1 INTRODUCTION

When we are able to estimate earthquake-induced stress changes of ground around ruptured faults and adjacent faults, it will be possible to evaluate rupture propagation and improve the prediction accuracy of the magnitude and probability of future earthquakes. Generally, the change of stress before and after an earthquake is evaluated by static Coulomb stress change (ΔCFF) (Stein et al. 1997, Toda et al. 1998, Hashimoto 1996). ΔCFF has been applied to earthquake-forecasting indexes that have been used to assess potential hazards related to earthquake activity. Seismic activities are enhanced by slight ΔCFF increases of 0.1 MPa (Toda et al. 1998). Examination by dynamic ΔCFF, i.e. Coulomb stress change during an earthquake, has been carried out recently, and there is a report that dynamic ΔCFF distribution matches aftershock activities than static ΔCFF distribution well (Kilb et al. 2002, Gomberg et al. 2003). However, in those studies, ΔCFF was calculated from static or dynamic stress changes using the assumed dislocation on the fault plane; the dynamic destruction process was not taken into account.

The present authors have examined a series of fault rupture simulations to estimate displacement and strong motions in the ground surface at the same time using the three-dimensional finite element method (3D-FEM) to consider the dynamic destruction process of the fault plane. This analytical method and modelling was proposed by Toki & Sawada (1988) and Mizumoto et al. (2005). The fault plane is assumed as bedrock discontinuities and is modelled by joint elements, and shear failure occurring at hypocentre spreads to surrounding areas with increasing shear stress. Iwata et al. (2018, 2019) conducted a fault rupture simulation for the 2014 Northern Nagano Earthquake induced by the Kamishiro Fault (M_w 6.3) and found that the quantity and distribution shape of dislocation on the fault plane depended on the initial stress distribution and stiffness of the joint elements. Furthermore, the distribution shape of dislocation calculated by 3D-FEM was markedly different from the rectangle that was used for the strong motion prediction based on 'Recipe' by Irikura (2006).

In this study, we calculated static ΔCFF for simple fault rupture models using two methods: (i) elasticity theory of dislocation (Okada 1992) in a homogeneous half-space and (ii) fault rupture simulation by 3D-FEM. We then compared the influence of analytical method and fault type. Dynamic ΔCFF is also important for evaluating the influence on neighbouring faults; however, this was not considered in the present study.

2 OUTLINE OF ANALYTICAL METHOD

2.1 Elasticity theory of dislocation

In the elasticity theory of dislocation, stress and deformation are calculated mathematically in an elastic half-space with uniform isotropic elastic properties (following Okada (1992)). The fault sliding is given uniformly in rectangular sources. The difference in fault types is expressed by changing the input sliding direction. We used software Coulomb 3.3 (Toda et al. 2011).

2.2 Fault rupture simulation by 3D-FEM

If the equation of motion involves the rupture movement of the fault plane, it is necessary to treat the motion as a nonlinear problem. Therefore, it is appropriate to obtain a solution for the equation of motion not in the frequency domain but in the time domain. The destruction process of dislocation and the dynamic behaviour of ground are calculated by solving the equation of motion using the stress drop of dislocation as an external force. The equation of motion at time step n is written as;

$$[M]\{\ddot{u}\}_n + [C]\{\dot{u}\}_n + [K]\{u\}_n = \{F(n,s)\} \qquad (1)$$

where $[M]$ is mass matrix, $[C]$ is damping matrix, $[K]$ is stiffness matrix, $\{\ddot{u}\}$ is acceleration vector, $\{\dot{u}\}$ is velocity, $\{u\}$ is displacement and $\{F(n,s)\}$ is the external force vector calculated from the dynamic stress drop, where n and respectively, stand for the time step and nodal pairs where fault rupture takes place. The damping matrix $[C]$ is obtained from the linear combination of $[M]$ and $[K]$, which is termed Rayleigh damping. Equation (1) is solved using the Newmark's β method, $\beta = 0.25$, $\gamma = 0.5$, at each time interval. To solve the nonlinear equation of motion, we employed the load transfer method utilising the initial stiffness method (Toki & Sawada 1988, Tsuboi & Miura 1996).

The fault plane is modelled by joint elements, as shown in Figure 1. The shear spring K_s, K_r and normal spring K_n are connected between nodal points of solid elements and a sliding occurs according to the Mohr−Coulomb failure criterion. When the calculated

shear stress τ is less than the peak stress τ_y the stress-deformation relation is linear with joint stiffness K_s, K_r.

Sliding takes place if the shear stress τ reaches the peak stress τ_y and stress drop occurs. The shear stress becomes equivalent to residual strength τ_r and stress drop $\Delta\tau_d$ (= τ_y - τ_r) is released and spread to nearby elements. In this way, the released stress drop in the hypocentre is triggered and the shear failure spreads to surrounding areas with increasing shear stress.

2.3 ΔCFF

ΔCFF caused by main shock rupture effectively explains the aftershock distributions for earthquakes triggered by stress changes of more than 0.1 MPa. For a given fault plane and slip vector, stress change can be quantified as

$$\Delta CFF = \Delta\tau + \mu'\Delta\sigma \qquad (2)$$

where $\Delta\tau$ is the shear stress change in the slip direction on the potential fault, $\Delta\sigma$ is the normal stress change (positive for compression) and μ' is the effective friction coefficient that is often used with the assumed value, e.g. 0.4 (King et al. 1994, Stein et al. 1997). If $\Delta CFF > 0$, slip potential is enhanced; if $\Delta CFF < 0$, it is inhibited. ΔCFF for 3D-FEM is calculated from stress change at the end of fault rupture.

3 ANALYTICAL CONDITIONS AND PARAMETERS

3.1 Fault parameters

The bedrock is a uniform isotropic medium with an elastic velocity of $Vp = 6.1$ km/s and $Vs = 3.5$ km/s. The fault plane is 15 km in length and 10 km in depth from surface ground and hypocentre is located at 5km depth. We examined two fault types: (i) a strike slip fault with left-lateral slip and (ii) a trust fault. The strike slip fault plane dips at 90° with a rake angle of $\lambda = 180°$. The trust fault plane dips at 50° and we assumed two rake angle cases: $\lambda = 90°$ (trust fault) and $\lambda = 120°$ (trust fault with left-lateral slip). Table 1 presents the parameters for the strike slip fault, which were calculated from the fault area based on Recipe for predicting strong motion (Irikura 2006). To compare the influence of fault type, the parameters used for the trust fault were the same as those for the strike slip fault.

3.2 Stress conditions for 3D-FEM

It is necessary to set fault plane strength and initial stress distribution because the amount of dislocation on the fault plane is not an input parameter. To compare the influence of the shape of the initial stress distribution, we examined two stress distribution types, as shown in Figure 2: the mountain type and asperity type. Shear stress in the asperity type was con-

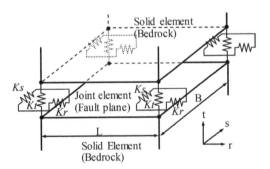

Figure 1. Schematic diagram of joint element.

Table 1. Parameters for strike slip fault.

Fault length, L × width, W	15 km × 10 km
Asperity length, L_a × width, W_a	1.8 km × 1.8 km
Seismic moment, M_0	1.74×10^{18} N·m
Dip angle, δ	90°
Average stress drop, $\Delta\tau_s$	2.3 MPa
Stress drop in asperity, $\Delta\tau_a$	15.6 MPa
Slippage in asperity, D_a	70.3 cm
Slippage in back ground, D_b	29.0 cm

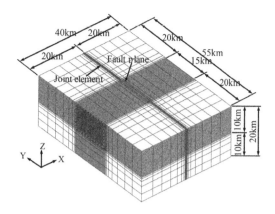

Figure 3. 3D-FEM mesh for strike slip fault.

centrated only at the hypocenter. Residual strength τ_r makes 10.0 MPa sufficiently larger than dynamic stress drop $\Delta\tau_d$ and peak strength is set by adding stress drop in asperity in Table 1 to residual strength. Excess strength $\Delta\tau_e$ in the mountain type was 1.6 times the static stress drop $\Delta\tau_s$ (Andrew 1976), and $\Delta\tau_d$ was 6.0 MPa, as shown in Figure 2. The shape of the peak stress distribution is a square, and the area of peak strength is calculated so that the volume of the truncated square pyramid shown as a blue solid line is equal to the volume of the cube shown as a dashed line. The shear stress at the hypocentre is assumed to be slightly larger than the peak strength τ_y.

Seismic response in the 3D-FEM is dependent on shear stiffness ks of the joint element (Iwata et al. 2018); therefore, we compared ΔCFF for the case of $ks = 1.0$ GN/m³ with that of $ks = 1.0 \times 10^4$ kN/m³.

3.3 FEM model

Figure 3 shows the 3D-FEM model for the strike slip fault. Joint elements are set up from the left-lateral boundary to the right lateral boundary and the strength of joint elements located inside or outside of the fault plane have sufficient strength not to fail. The distance from the fault edge to the boundaries is greater than the fault length and viscous dampers were introduced at the lateral and bottom boundaries to absorb scattering wave energy. When acceleration response is estimated using FEM, it is necessary to change the mesh height according to the natural frequency of object facilities, however, displacement and stress response do not depend on the mesh height (Iwata et al. 2018). In this study, the mesh height

around the fault plane was 250 m in order to express the changes in slippage and stress distribution.

4 ANALYSIS RESULTS BASED ON ELASTICITY THEORY OF DISLOCATION

Figure 4 shows the ΔCFF distribution at 5 km depth corresponding to the earthquake focal depth for the strike slip fault. The fault plane slipped in the direction indicated by white arrows in Figure 4. The positive domain spread radially towards the outside from the fault edge. The negative domain appeared on both sides of fault plane. The positive domain spreading from the asperity edge occurred because of extreme difference between the slip amount of the asperity area and that of background area.

Figure 5 shows the ΔCFF distribution at 5 km depth for the trust fault. The positive domain spreads from the fault edge and through the frontal domain, as illustrated by the white arrow. The domain of ΔCFF > 0.1 MPa where seismicity is enhanced becomes smaller than that of the strike slip fault.

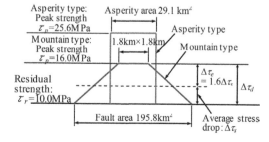

Figure 2. Schematic diagram of initial stress distribution.

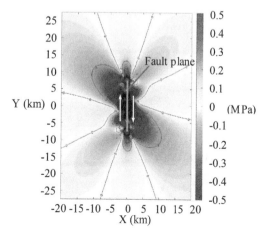

Figure 4. ΔCFF distribution at 5 km depth for strike slip fault.

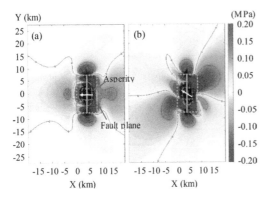

Figure 5. ΔCFF distribution at 5 km depth for the trust fault: (a) λ = 90°, (b) λ = 120° (trust fault with left-lateral slip).

5 ANALYSIS RESULTS BASED ON 3D-FEM

5.1 *Influence of initial stress distribution*

We carried out fault rupture simulations using different initial stress distributions: mountain type and asperity type, and compared slip distributions on the fault plane and ΔCFF distributions at 5 km depth. The fault type was strike slip and the spring stiffness of the joint element was 1 GN/m³. Figure 6 shows the slip distributions and Figure 7 shows the ΔCFF distributions. The maximum slip amount for the mountain type was 1.26 m and the rupture front reached the ground surface. The maximum slip

amount for the asperity type was 2.08 m, i.e. larger than that of the mountain type, although the rupture front did not reach the ground surface. The positive domain along the extending direction of the fault plane in the mountain type occurred from the fault edge, whereas that of the asperity type occurred from halfway across the fault plane because sliding stopped in the middle of the fault plane. The ΔCFF > 0.1 MPa domain in the mountain type spread farther than it did for the asperity type. The positive domain and quantity of ΔCFF on both sides of the fault plane in the asperity type were larger than that of the mountain type because of the larger amount of slip.

5.2 *Influence of spring stiffness of the joint element*

We conducted fault rupture simulations for the strike slip fault using different spring constants: $ks = 1.0$ and 0.01 GN/m³. The fault type was strike slip and the initial stress distribution was mountain type. Figures 8 and 9, respectively, show the slip and ΔCFF distributions for $ks = 0.01$ GN/m³. The slipping area and maximum slip amount for $ks = 0.01$ GN/m³ were smaller than those of $ks = 1.0$ GN/m³, as shown in Figures 6(b) and 7(b). If shear failure occurs in a joint element, stress drop occurs and the released stress spreads to nearby elements. When a shear spring constant is set to a smaller value, the shear stress of the fault transferred from a yield element becomes smaller, as does fault sliding; ΔCFF also becomes smaller for those reasons.

5.3 *Influence of fault type*

We performed fault rupture simulations for different fault types: a strike slip fault and a trust fault with λ=90° and 120°. The initial stress distribution was mountain type and the spring stiffness of the joint element was 1 GN/m³. Figure 10 shows the slip distribution for the trust fault with λ=90°. The maximum slip amount in the mountain type was 1.74 m and the rupture front reached the ground surface.

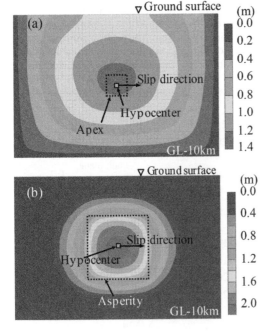

Figure 6. Comparison of slip distributions in different initial stress distributions for strike slip fault: (a) mountain type, (b) asperity type.

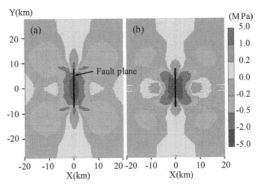

Figure 7. Comparison of ΔCFF distribution at 5 km depth for different initial stress distributions of the stroke slip fault: (a) mountain type, (b) asperity type.

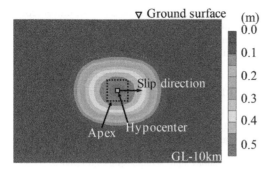

Figure 8. Slip distributions using joint spring constant of $ks = 0.01$ GN/m³ for strike slip fault.

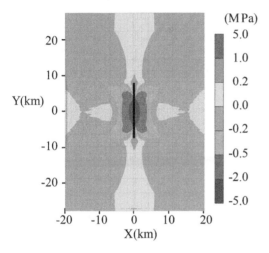

Figure 9. ΔCFF distribution at 5 km depth using joint spring constant of $ks = 0.01$ GN/m³ for strike slip fault.

When the rake angle was 120°, the rupture front spread upwards diagonally and the maximum slip amount became 1.65m; i.e. smaller than for λ=90°.

Figure 11 shows the comparison of ΔCFF distributions at 5 km depth. In case of λ=90°, the ΔCFF > 0.2 MPa domain shown in orange in Figure 11(a) distributes radially at both edges of the fault plane.

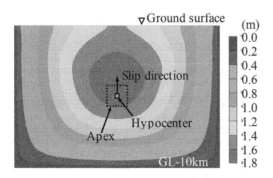

Figure 10. Slip distributions for trust slip fault (λ = 90°).

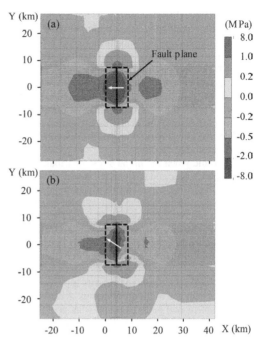

Figure 11. Comparison of ΔCFF distribution at 5km depth in different rake angles for trust fault: (a) λ = 90°, (b) λ = 120°.

A positive domain before and after the sliding direction seen in the strike slip fault did not occur because the rupture front reached the ground surface and the shear stress was released. The positive and negative domain for λ=120° extended to the sliding direction and the area became smaller than that of λ=90°.

6 COMPARISON OF ANALYTICAL METHODS

In this study, we estimated fault parameters based on Recipe. The slip amount calculated by 3D-FEM was larger than that of Recipe and ΔCFF based on 3D-FEM was larger than that of the elasticity theory of dislocation.

The shape of sliding distribution on the fault plane by 3D-FEM varied according to analysis conditions; nevertheless, in all cases, the amount of sliding changed smoothly. Unlike the assumption in Recipe, sliding did not concentrate only on asperity. For the elasticity theory of dislocation, shear stress was concentrated around the outer circumferential portion of asperity where the difference in slippage was large and ΔCFF became large. However such a concentration was not seen for 3D-FEM, and the shape of the ΔCFF distribution based on 3D-FEM differed from the discrepancy in the elasticity theory on both sides of the fault plane.

When rupture simulation was conducted using the initial stress distribution of the mountain type and a joint stiffness of $ks = 1.0$ GN/m³, the maximum slip amount was more than 1.2 m and seismic moment calculated from the slip distribution was estimated to be greater than 3.0×10^{18} N·m. These values are considerably larger than those assumed by Recipe. When we used the initial stress distribution of the asperity type, the maximum slip amount increased to 2.08 m but the estimated seismic moment was smaller than 3.0×10^{18} N·m because the sliding area was smaller.

On the other hand, with a joint stiffness of $ks = 0.01$ GN/m^3 and the initial stress distribution of the mountain type, the maximum slip amount became 0.52 m and seismic moment was estimated as 1.43×10^{18} N·m. These results are slightly smaller than those assumed by Recipe.

As a result, when the initial stress distribution was of the asperity type and the joint stiffness ks was set to around 0.01 GN/m^3, the calculated slippage and seismic moment values were closer to those of Recipe. In previous studies on the 2014 Northern Nagano Earthquake, for which the shear spring constant was $K_s = 0.01$ GN/m^3, the rupture did not propagate and only the immediate vicinity of the hypocentre was ruptured (Iwata et al. 2018). Moreover, the surface displacement when using initial stress distribution of the asperity type was markedly larger than that of the observations. A challenging future problem will be to examine what combination of parameters best reproduces the observations.

7 CONCLUSIONS

In this study, we calculated static ΔCFF for simple fault rupture models using the elasticity theory of dislocation and 3D-FEM. We compared the distribution shape and quantity of ΔCFF. The findings obtained from this study can be summarised as follows:

1. The slippage and seismic moment calculated by 3D-FEM using fault parameters based on Recipe become considerably larger than those assumed by Recipe (except for a joint stiffness of $ks = 0.01$ GN/m^3).
2. The distribution shape of ΔCFF in 3D-FEM is similar to that given by the elasticity theory of dislocation outside of the fault edge, but it is markedly different around the asperity owing to the difference in displace distribution.
3. The sliding domain in the initial stress distribution of the asperity type is smaller than that of the mountain type, but the amount of slide is larger. Therefore, the distribution range of the asperity type becomes smaller and the value of ΔCFF around the fault plane becomes larger.
4. The slip amount and ΔCFF become smaller as the joint stiffness is set smaller.
5. Regarding trust faults, the positive domain of ΔCFF distributes radially at both edges of the fault plane and extends towards the sliding direction.

Even if fault type and seismic magnitude such as fault geometry and moment magnitude are the same, the distribution domain and quantity of ΔCFF differ greatly according to the combination of analytical parameters. Therefore, it is necessary to examine the methods used for setting parameters

in the model to reproduce actual fault movements. We continue to examine the validity and applicability of fault rupture simulations using 3D-FEM to evaluate rupture propagation and fault activity.

REFERENCES

Andrew, D.J. 1976. Rupture velocity of plane strain shear racks, *J. of Geo. Res.*, Vol. 81, No. 32, 5679–5687.

Gomberg, J., Bodin, P. & Reasenberg, P. A. 2003. Observing earthquakes triggering near-field by dynamic deformations, *Bull. Seis. Soc. Am.*, 93: 118–138.

Hashimoto, M. 1996. Static stress changes associated with the Kobe earthquake: Calculation of changes in Coulomb failure function and comparison with seismicity change. *J. Seismol. Soc. Jpn.*, 48: 521–530.

Irikura, K. 2006. Predicting strong ground motions with "Recipe", *Bull. Earthq. Res. Inst. Univ. Tokyo*, 81: 341–352.

Iwata, N., Kiyota, R., Adachi, K., Takahashi, Y., Aydan, Ö., Miura, F. & Ito, T. 2018. Simulation of strong motions and surface rupture of the 2014 Northern Nagano Earthquake. *Proc. of the 3rd inter. conf. on Rock dynamics and applications (ROCDYN-3), Norway*: 541–546.

Iwata, N., Kiyota, R., Aydan, Ö., Ito, T. & Miura, F. 2019. Effects of Fault geometry and subsurface structure model in strong motion and surface rupture induced by the 2014 Kamishiro Fault Nagano Earthquake. *Proc. of 2019 Rock Dynamic Summit in Okinawa (2019RDS)* (under submission).

Kilb, D., Gomberg, J. & Bodin, P. 2002. Aftershock triggering by dynamic stresses, *J. Geophys. Res.* 107(B4): 2060, doi:10.1029/2001JB000202.

King, G. C. P., Stein, R. S. & Lin, J. 1994. Static stress changes and the triggering of earthquakes, Bull. Seis. Soc. Am., 84: 935-953.

Mizumoto, T., Tsuboi, T. & Miura, F. 2005. Foundamental study on failure rupture process and earthquake motions and near a fault by 3D-FEM. *Journal of Japan Society of Civil Engineers* 780/I-70: 27–40 (in Japanese with English abstract).

Okada, Y. 1992. Internal deformation due to shear and tensile faults in a half-space. Bull. Seismol. Soc. Am. 82: 1018–1040.

Stein, R. S., Barka, A. A., & Dieterich, J. H. 1997. Progressive failure on the North Anatolian fault since 1939 by earthquake stress triggering. *Geophys. J. Int.* 128: 594–604.

Toda, S., Stein, R.S., Reasenberg, P.A., Dieterich, J.H. & Yoshida, A. 1998. Stress transferred by the 1995 Mw = 6.9 Kobe, Japan, shock: Effect on aftershocks and future earthquake probabilities. *J. Geophys. Res.* 103: 24543–24565.

Toda, S., Stein, R.S., Sevilgen, V. & Lin, J. 2011. Coulomb 3.3 graphic-rich deformation and stress-change software for earthquake, tectonic, and Volcano Research and Teaching—user guide. *USGS Open-File Report* 2011–1060.

Toki, K. & Sawada, S. 1988. Simulation of the fault rupture process and near filed ground motion by the three-dimensional finite element method. *Proc. of 9th World Conf. on Earth. Eng., Japan*, Vol. II: 751–756.

Tsuboi, T. & Miura, F. 1996. Simulation of stick-slip shear failure of rock masses by a nonlinear finite element method. *Journal of Japan Society of Civil Engineers* 537/I-35: 61–76 (in Japanese with English abstract).

2019 Rock Dynamics Summit– Aydan et al. (eds)
© *2019 Taylor & Francis Group, London, ISBN 978-0-367-34783-3*

Analysis of surface fault displacements in 2014 Nagano-ken-hokubu earthquake by high performance computing

M.Sawada
Central Research Institute of Electric Power Industry, Abiko, Japan

K. Haba
Taisei Corporation, Tokyo, Japan

M. Hori
The University of Tokyo, Tokyo, Japan

ABSTRACT: Continuum mechanics-based numerical simulations are potential methods for evaluating surface fault displacements. We developed a parallel finite element method to evaluate these displacements. In this study, we applied a numerical method to the simulation of the 2014 Nagano-ken-hokubu Earthquake, in which surface faulting was observed. We modeled a 5 km × 5 km × 1 km domain around the northernmost region of the surface faults, which included secondary faults. We applied forced displacements on the bottom surface of the model based on the slip distribution on the primary fault and the elastic theory of dislocations. As the input slip increased, surface slips appeared on the primary fault and a secondary fault. The calculated surface slips were in good agreement with the measured values.

1 INTRODUCTION

Since the occurrence of massive earthquakes in Taiwan and Turkey in 1999, there have been growing concerns about the potential damage of various infrastructures and buildings caused by surface fault ruptures. For on-site fault assessment in nuclear power plants, it is important to estimate fault displacements and their impact on the safety functions of the facilities. It is necessary to reliably estimate the possible displacements.

Numerical simulations based on continuum mechanics are potential evaluation methods for surface fault displacements. However, there are major difficulties in simulating the fault rupture process. One difficulty is that it requires a large amount of numerical computations to simulate the fault rupture process of a target area of only a few hundreds of meters. Another difficulty is the loss of stability in the initial boundary value problem to which the numerical analysis is applied. Stability implies that a solution does not change when a small disturbance is added to the problem, and stability loss leads to drastic changes in the solution due to small disturbances. We overcame these difficulties by applying high performance computing methods. We developed a finite element method involving two functions: 1) a symplectic time integration explicit scheme to properly conserve the energy of the system and 2) rigorously formulated high-order joint elements. The finite element method was enhanced with parallel computing capabilities (Sawada et al., 2017).

In this paper, we applied this numerical method to simulate the 2014 Nagano-ken-hokubu Earthquake, in which surface faulting was observed, to validate the method.

2 METHOD

2.1 *Estimation of fault displacement*

For the fault displacement simulations, we constructed a continuum model of the ground and faults (see Fig.1). An input slip Δ was applied on the bottom of the primary fault (main fault), and the surface slip δ was obtained after calculating the spread and dispersion of the slip on the fault plane. For on-site fault assessment in nuclear power plants, surface slips on branch or secondary faults (sub-faults) must be estimated in addition to slips on the main fault.

branch fault or secondary fault (-> sub-fault)

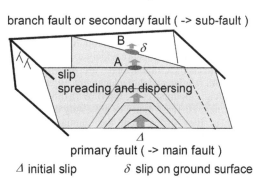

Δ initial slip δ slip on ground surface

Figure 1. Spreading and dispersing of a slip on faults.

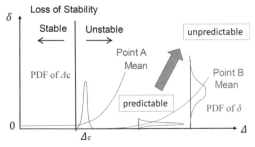

Figure 2. Prediction of the surface fault displacement.

A fundamental difficulty of simulating the fault rupture process is the loss of stability in the initial boundary value problem, which leads to significant changes in the solution due to small disturbances.

The treatment of uncertainties due to the limitations in the quality and quantity of the available relevant data for the underground structures, stress states, and source fault dynamics is important. Many simulations should be conducted with different conditions (capacity computing).

When considerable amounts of uncertainty exist, the calculated surface fault displacement δ by capacity computing are distributed over a wide range. In this case, δ is unpredictable in practice. It is important to evaluate the critical input slip Δ_c, which is the minimum input slip that causes a surface slip, before evaluating δ.

Quasi-static simulations allow a relationship between the input and surface slips to be obtained. This is extremely important for the estimation of Δ_c. Sawada et al. (2018) showed the calculated Δ_c and δ in the quasi-static simulations were close to those calculated in the dynamic simulations with damping and proposed the use of quasi-static simulations for capacity computing. Quasi-static simulations were also applied in this study.

2.2 FEM for the fault-displacement problem

We implemented a rigorous joint element and symplectic time integration method with a Hamiltonian formulation in the open-source FEM program FrontISTR (FrontISTR Commons, 2018) to develop a numerical tool for simulating fault displacements. In the joint elements that are widely used in the field of rock mechanics, nodal forces are directly calculated from relative nodal displacements on a discontinuity (e.g. Goodman et al., 1968). With this method, the element stiffness matrices are easily calculated. However, they differ from the standard method for constructing finite elements. Joint elements for fault-displacement simulations should be obtained to ensure proper convergence for the chosen element sizes and the applicability to curved faults. We proposed a rigorous joint element in which the element stiffness matrices are derived from the Lagrangian

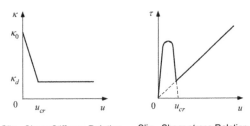

Figure 3. Frictional characteristics and slip-shear stress relation.

on the discontinuous surface with an isoparametric formulation (Sawada et al., 2017).

We implemented the joint elements with a non-linear spring-type constitutive equation to represent the fault movement. The following is the constitutive equation for shear movement of the joint element:

$$\tau = \kappa u \tag{1}$$

where τ denotes the shear stress and u denotes the slip on the fault plane. The spring coefficient per unit area (shear stiffness) κ is described by the following function of the slip u:

$$\kappa(u) = \begin{cases} \kappa_0 - \dfrac{\kappa_0 - \kappa_d}{u_{cr}}u & (u \le u_{cr}), \\ \kappa_d & (u > u_{cr}), \end{cases} \tag{2}$$

where κ_0 denotes the initial shear stiffness, κ_d denotes the final shear stiffness, and u_{cr} denotes the critical slip (see Fig.3). It should be noted that we did not use any sliders or dashpots in the model for κ.

We considered the effect of the normal stress on the shear stiffness of the faults, and we define the following equation:

$$\kappa_0 = a\sigma_n + b \tag{3}$$

where a and b are constants and σ_n denotes the normal stress on the fault plane.

3 MODEL

3.1 Target earthquake

We applied the numerical method to the 22 November 2014 Nagano-ken-hokubu earthquake of Mj = 6.7 (Mw = 6.2). The 9 km-long surface rupture was observed on the previously mapped Kamishiro fault. Figure 4 shows the mapped location of the rupture in the northernmost section. Sub-faults were also observed, including Sub-E, Sub-N, and Sub-S in Fig.4. Figure 5 shows the net slip along the surface rupture estimated by airborne LiDAR data obtained at two different times (Aoyagi, 2016). The maximum

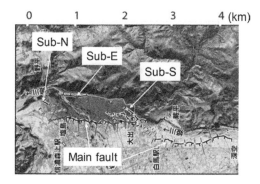

Figure 4. Distribution of surface ruptures in the northernmost section, modified after Aoyagi (2016).

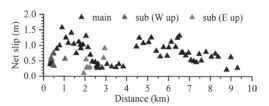

Figure 5. Surface fault dislocation from Aoyagi (2016). The horizontal axis corresponds to the distance from the northern end of the surface rupture, the scale of which is equivalent to Fig.4.

net slip values on the main fault and sub-faults were about 1.5 and 0.6 m, respectively.

3.2 Model

We determined the mechanical properties of the ground and strike angle of the main fault based on the J-SHIS database, which is provided by National Research Institute for Earth Science and Disaster Resilience (NIED) and covers all of Japan.

Figure 6 shows a finite element mesh used for the simulation. The target was a 5 km × 5 km region around the northern end of the surface rupture. The depth of the domain was about 1 km.

We included one main fault and three sub-faults. The dip angle of the main fault was 40°. Sub-E had the same strike angle as the main fault and a dip angle 40° in the opposite direction. Sub-N and Sub-S had strike angles of 45° from the main fault and 45° from the northside and southside of dip angles, respectively. The ground was discretized by second-order tetrahedral elements, and the fault planes were discretized by second-order triangle joint elements. The sizes of the elements on the fault planes were about 50 m, and the total number of degrees of freedom was about 2.17 million.

The Geospatial Information Authority of Japan (GSI) derived a slip distribution on the main fault from the crustal movements estimated by differential interferometric synthetic aperture radar

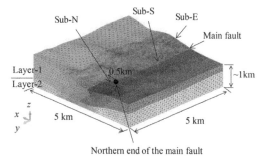

Figure 6. Finite element mesh.

(DInSAR). GSI's main fault had dip angles of 40° and 60° in the shallow and deep parts, respectively. We re-evaluated the slip distribution of the shallow part of the main fault based on the report of Aoyagi (2016). We used dislocation theory (Okada, 1985) with the slip distribution as the input and determined the forced displacements at the bottom of the model. The maximum slip on the bottom of the fault was 3.0 m, which was twice the estimated slip. The other outer boundaries of the model were traction free.

The J-SHIS database includes elevation data about the ground surface and boundaries of the elastic wave velocity and density. Based on this information, we considered the shape of the land and two layers of the ground. The Young's modulus and Poisson's ratio of each layer were determined from the elastic wave velocity and density. We assumed that the first peak of the slip-shear stress relation in Fig. 3 was identical to the shear strength. The initial shear stiffness κ_0 and coefficients were determined such that the shear strength satisfied Coulomb's friction law with a cohesion of 0.025 MPa and a 25° friction angle. The ratio between the initial and final shear stiffnesses, κ_d/κ_0, was assumed to be 0.01, and the critical slip u_{cr} was assumed to 0.1 m.

4 RESULT

4.1 Slip distribution

Figure 7 shows the displacement norm contour and deformation of the analytical domain. Figures 7 a) and b) show the results at $\Delta = 1.5$ and 2.7 m, respectively. The deformation is displayed with 300-times amplification. Large deformation occurred at the center of the model where the input slip was large. There was a displacement gap along the Sub-E fault, which indicates the appearance of surface slip.

Figures 8 a) and b) show contour plots of the slip at $\Delta = 3.0$ m on the main and Sub-E faults, respectively. These figures show views from the $+x$ (east) side. There was a surface slip of over 1.0 m on the main fault, and a surface slip in the north section of Sub-E fault.

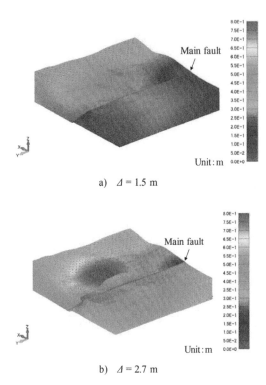

Main fault

Unit : m

a) $\Delta = 1.5$ m

Main fault

Unit : m

b) $\Delta = 2.7$ m

Figure 7. Contours of displacement norm and deformation.

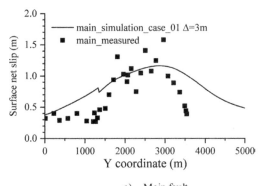

a) Main fault

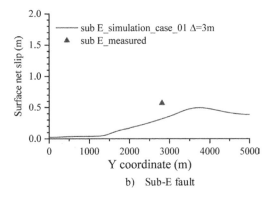

b) Sub-E fault

Figure 9. Comparison of surface slip distribution between in the simulation and multi LiDAR-DEM data.

Figures 9 a) and b) show the comparison between the calculated and measured net surface slips at $\Delta = 3.0$ m. The slips on the main fault were similar (see Fig.9 a). The simulated slip distribution was smoother than the measured distribution. In the simulation, the surface slip appeared in the northern section, $y > 3854.77$ m, where the surface rupture was not observed. There was only one slip measurement on the Sub-E fault, and the value was about 0.6 m. In the simulation, the maximum surface slip was about 0.5 m, similar to the measured value. However, the calculated surface slip was a small value of 0.3 m at the exact location of the measured data. Overall,

the simulated surface slip distribution at $\Delta = 3.0$ m agreed well with the measured distribution.

4.2 Surface slip-input slip relation

Figure 10 shows the evaluation points for the surface slip. In the legend, "main" and "sub" are the abbreviations of the main and sub-faults, and "Y2000" is the abbreviation for the y-coordinate of 2000 m.

Figure 11 shows the variation of the surface net slip δ as the input slip Δ increases. There was a steep increase in the surface slip at $\Delta = 0.705$ m at the evaluation points on the main fault. A steep increase

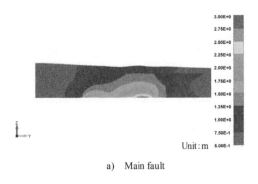

Unit : m

a) Main fault

Figure 8. Contours of the slip on the main and Sub-E faults.

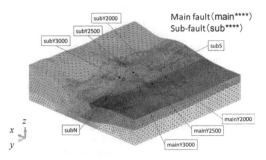

Figure 10. Evaluation points.

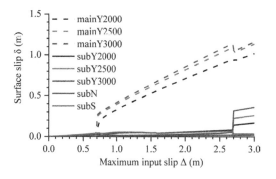

Figure 11. Surface slip–input slip relation.

in the surface slip on the Sub-E fault began at $\varDelta = 2.700$ m. Therefore, the critical slip $\varDelta_c$ was 0.705 m on the main fault and 2.700 m on the Sub-E fault.

After the appearance of the surface slip on the Sub-E fault, the net slips on the main and Sub-E faults were about 1.0 and 0.3 m, respectively. These agreed well with the measured values. However, surface slips did not appear on the Sub-N or Sub-S faults.

Figure 12 shows the vertical and strike-slip components of the surface slip at representative evaluation points of the main fault at Y2500 and sub-Y2500. These slip components were relative displacements of the $+x$ side of the faults against the $-x$ side. The slip on the main fault was positive in the vertical and strike-slip directions. This means that the $+x$ (east) side of the ground was lifted and moved north. However, the slip on the Sub-E fault was negative in the vertical direction and positive in the strike-slip direction. The Sub-E fault moved as though it was a back-thrust of the main fault. This movement agreed qualitatively with the observations. However, the vertical component of the slip on the Sub-E fault was smaller than the measured value.

4.3 *Discussion*

We constructed 5 km × 5 km × 1 km analytical model and derived the boundary conditions from the elastic

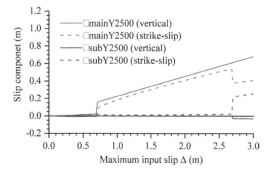

Figure 12. Vertical and strike-slip component of the surface slip.

theory of dislocations using the slip distribution on the main fault obtained from the inverse analysis of DInSAR and multi-LiDAR-DEM data.

When we introduced twice the inversely-analyzed slip on the main fault to the bottom of the analytical domain, we obtained a steep increase of the surface slip on the main fault and a sub-fault. The final surface net slips were in good agreement with the measured values. The vertical component of the slip on the Sub-E fault was qualitatively identical to the back-thrusted movement observed at the site. Since the simulation reproduced the main features of the observed surface rupture, the proposed numerical method is applicable for the predictive simulation of surface fault displacements.

However, some phenomena were not reproduced in the simulation. For instance, surface slips did not occur at the Sub-N or Sub-S faults. One possible reason is that slips in the region $y > 3855.77$ m on the main fault were not restricted. It may be necessary to consider spatial distributions of the fault friction properties to reproduce the surface slip on the Sub-N and Sub-S faults.

As described above, stability was lost during the simulations of the fault displacements, and thus, the simulations were affected by disturbances of the input conditions. When applying this numerical method, we must determine the input conditions by consulting with experts and conduct many simulations with different input conditions to evaluate the effect of uncertainty.

5 CONCLUSION

We developed a parallel finite element method program for the estimation of surface fault displacements. We applied the numerical method to simulate the 2014 Nagano-ken-hokubu Earthquake, in which surface faulting was observed. We modeled a 5 km × 5 km × 1 km domain around the northernmost region of the surface faults, which included secondary faults. We applied forced displacements on the bottom surface of the model based on a slip distribution on the primary fault and the elastic theory of dislocations. As the input slip increased, surface slips appeared on the primary fault and a secondary fault. The calculated surface slips were in good agreement with the measured values. Since the simulations reproduced the main features of the observed surface rupture, the proposed numerical method is applicable for predictive simulations of surface fault displacements.

ACKNOWLEDGEMENTS

The study was supported by commissioned projects from the Agency for Natural Resources and Energy, Ministry of Economy, Trade and Industry, Japan.

REFERENCES

Aoyagi, Y. 2016. Fault displacement distribution of the 2014 Nagano-ken Hokubu earthquake based on a differential analysis of multi LiDAR-DEM data. *Abstracts, Japan Geoscience Union Meeting*, SSS31–18. (in Japanese)

FrontISTR Commons. 2018. https://www.forntistr.org/, retrieved 14 December 2018. (in Japanese)

Geospatial Information Authority of Japan. 2015. https://www. jisihn.go.jp/main/chousa/14dec_nagano/p.29. htm, retrieved 5 June 2018. (in Japanese)

Goodman, R. E., Taylor, R. L. and Brekke, T. L. 1968. A model for the mechanics of jointed rocks. *J. ASCE* 94(SM3): 637–659.

National Research Institute for Earth Science and Disaster Resilience. Japan Seismic Hazard Information Station. http://www.j-shis.bosai.go.jp/en/, retrieved 5 June 2018.

Okada, Y. 1985. Surface deformation due to shear and tensile faults in a half-space. *Bull. Seism. Soc. Am.* 75(4): 1135–1154.

Sawada, M., Haba, K. and Hori, M. 2017. High performance computing for fault displacement simulation. *J. JSCE* 73 (2):I_699–I_710. (in Japanese)

Sawada, M., Haba, K. and Hori, M. 2018. Estimation of surface fault displacement by high performance computing. *Journal of Earthquake and Tsunami* 12(4): 1841003 (22 pages).

2019 Rock Dynamics Summit– Aydan et al. (eds)
© 2019 Taylor & Francis Group, London, ISBN 978-0-367-34783-3

Ground motion observation in Padang, West Sumatra, Indonesia

Y. Ono & T. Noguchi
Tottori University, Tottori, Japan

J. Kiyono
Kyoto University, Kyoto, Japan

T. Suzuki
Toyo University, Kawagoe, Saitama, Japan

Ö. Aydan
University of the Ryukyus, Nishihara-cho, Okinawa, Japan

R.P. Rusnardi
Padang State University, Padang, West Sumatra, Republic of Indonesia

A. Hakam
Andalas University, Padang, West Sumatra, Republic of Indonesia

ABSTRACT: Padang is the capital city of West Sumatra Province located on the west coast of Sumatra Island, Indonesia. The seismicity around Padang is quite active, and more than 1000 casualties occurred due to the M 7.6 earthquake of 30th October 2009. Besides, the existence of a seismic gap in the off-shore of Padang, where the plate boundary between the Eurasian and the Indo-Australian plates, has been pointed out and occurrence of an M 8.0 class earthquake is expected. In 2008, Engineers Without Borders Japan (EWB-JAPAN), a nonprofit organization (NPO) set up three earthquake ground motion observation sites in Padang and one in Bukittinggi approximately 80 km northward from Padang. Subsequently, one more site was set up in the campus of Padang State University. Currently, two sites in Padang are non-operational. So far, 86 earthquake ground motions have been recorded in total. In the present paper, 36 records whose horizontal PGA exceed 5 cm/s/s were selected, and Fourier analyses of them were conducted. The dominant frequencies of analyzed earthquake ground motion records coincided approximately the peak frequencies of the microtremor horizontal-to-vertical spectrum ratio (HVSR) obtained by the past study. Furthermore, the time-frequency analysis of two records obtained at two sites on the deep sedimentary ground was carried out. The results showed that surface waves of 2 Hz were dominant in the flat area along the coast of Padang. For more than ten years have elapsed since the installation of the observation sites, urgent maintenance works for the observation equipment are strongly required.

1 INTRODUCTION

Padang, the capital city of Western Sumatra Province, Republic of Indonesia, is located on the west coast of the Sumatra islands where the seismic activity is very high (Figure 1). The earthquake of M 7.5 occurred at the offshore Padang on 30th September 2009 and resulted in more than 1,000 casualties. It is expected that an M8 class earthquake will take place at the plate boundary of the offshore Padang shortly and will cause severe damage in Padang (Natawidjaja et al., 2006).

In 2008, Engineers without borders Japan (EWB-JAPAN), which is a nonprofit organization (NPO), installed accelerometers at three locations in Padang and one in Bukittinggi approximately 80 km far from the center of Padang. Subsequently, one accelerometer was added at Padang State University in Padang. At present, no seismic observations are made at two sites, and the observation continues at the remaining three sites. No borehole survey has been conducted at any observation site, and the detailed ground structures are unknown. However, in the four observation sites in Padang, the microtremor array observation was carried out, and the ground structures have been estimated (Ono et al. 2012).

In this paper, earthquake ground motions recorded at these observation sites are introduced, and their characteristics are analyzed.

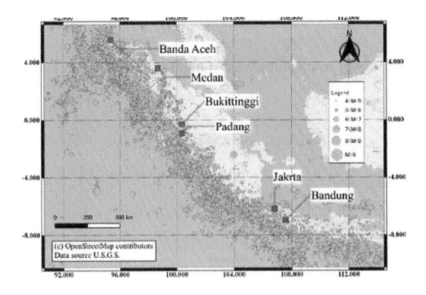

Figure 1. Location of Padang, West Sumatra, Indonesia and seismicity of the region.

2 OBSERVATION SITES

As described in the previous section, four sites in Padang (ADS, SMO, GVO, UNP) and one site (BKT) in Bukittinggi about 80 km away from the center of Padang were set up. The locations and elevations of each site are summarized in Table 1. At present, observations are being conducted at three sites excluding SMO and GVO.

As shown in Figure 2, ADS is installed in a quiet place inside the campus of Andalas University in eastern Padang city. The campus of Andalas University is located in a mountain area with an altitude of 228 meters. According to the ground structure estimation based on the microtremor array observation (Ono et al., 2012), the site of ADS is on hard soil or bedrock.

SMO was located in the middle of the line connecting ADS and GVO. According to the ground structure estimated from microtremor array observations, a slightly hard soil is on the soft sedimentary ground. SMO is located beside a roadwith heavy traffic. The building where the equipment was installed was a county office at the time of

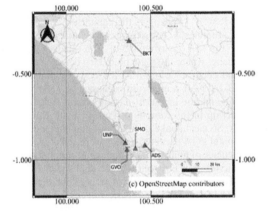

Figure 2. Location of observation sites.

installation, but now it is used as a fire department. At present, observation with SMO is not carried out due to mechanical problems.

GVO was installed on the flat land near the coast of the west side of Padang city. According to microtremor array observations, GVO is on a deep sedimentary ground. The GVO was set up in one of the provincial government buildings. The building was damaged by the 2009 Offshore Padang Earthquake and demolished. The GVO observation device was removed during the demolition of the building. The place of GVO is a region with high population density and high traffic volume. However, the building, where the equipment was placed, was in the area of the premises of the provincial government, and the surroundings was a quiet environment.

Table 1. Location and elevation of observation sites

site	latitude (°)	longitude (°)	elevation (m)
ADS	−0.913	100.465	288
SMO	−0.932	100.409	54
GVO	−0.938	100.360	7
UNP	−0.899	100.350	3
BKT	−0.307	100.373	917

UNP is placed near the coast of the western side of Padang, and the ground structure is same as GVO according to the microtremor survey. UNP is located inside of one school building of Padang State University. The area around Padang State University, like GVO, has high population density and high traffic volume. However, the observation equipment is installed in a quiet area in the center of campus.

BKT is installed in the tunnel in Bukittinggi, about 80 km from the center of Padang city. Although the estimation of the ground structure by the microtremor array observation is not carried out, according to the geological map, this zone is sedimentary rock. This tunnel was dug by the Japanese army during the Second World War and is now used as a tourist destination. Observation instruments are installed in areas where tourists are prohibited from entering. However, at present repair work of the tunnel is being carried out, and the influence of noise is concerned.

For all observation sites, SES60 manufactured by Azbil Co. (formerly Yamatake Co.) is used. The sampling frequency of observation is 100 Hz.

3 CHARACTERISTICS OF RECORDED EARTHQUAKE GROUND MOTIONS

3.1 Accelerograms and Fourier spectra

A total of 86 earthquake ground motion records were obtained so far. Among them, 36 records with a horizontal PGA exceeding 5 cm/s/s were selected and shown in Table 2. Four sites of ADS, SMO, GVO, BKT were in operation at the time of the offshore Padang earthquake on 30th September 2009. However, a power failure occurred, and the earthquake record was not preserved. The uninterruptible power supply was not installed at that time. The M6.7 earthquake on August 16, 2009, the foreshock of the 2009 Offshore Padang earthquake, was recorded at SMO and GVO (14: 36: 07 and 14: 36: 46 records, respectively). Among all records, the maximum value of PGA is approximately 80 cm/s/s of EW component recorded at 09/07/02 15:48:22 at ADS.

Among the records obtained at each site, the record with the largest horizontal PGA was chosen, and the acceleration waveform is plotted in Figures 3 to 7. For the sites of ADS and BKT on hard ground, high-frequency components around 10 Hz are dominant while low-frequency components of 0.2 to 1.0 Hz are small. On the other hand, for the sites of GVO and UNP on a deep sedimentary ground, the Fourier spectrum has a flat from 1 Hz to 3 Hz, and no clear dominant period is observed. The Fourier spectrum of SMO has a flat in a wide frequency range from 0.1 Hz to 1 Hz.

Table 2. List of recorded earthquake ground motions (PGA > 5 cm/s/s)

SITE	DATE	TIME	PGA(NS)	PGA(EW)	PGA(UD)
ADS	09/05/25	08:54:47	10.25	11.0	3.0
ADS	09/07/02	15:48:22	30.875	79.625	16.75
ADS	09/08/17	17:31:09	9.25	5.75	1.875
ADS	09/08/17	20:52:06	7.75	7.25	1.875
ADS	09/08/17	21:10:35	15.125	17.75	6.75
ADS	09/08/19	18:31:48	8.875	6.25	2.125
ADS	10/01/23	18:21:19	5.0	5.625	1.5
ADS	10/02/05	11:33:13	16.25	16.5	4.125
ADS	10/02/14	10:57:12	8.75	13.375	4.125
ADS	10/03/13	00:21:27	43.875	47.25	13.5
ADS	10/03/16	02:24:39	30.25	35.375	9.75
ADS	10/06/30	17:47:06	20.5	36.0	9.125
ADS	10/06/30	21:49:10	16.375	14.75	3.5
ADS	10/07/24	22:09:50	14.125	9.25	3.875
SMO	09/07/02	15:49:51	9.625	13.625	4.75
SMO	09/08/16	14:36:07	45.5	39.25	18.0
SMO	09/08/16	17:19:21	10.25	7.25	2.5
SMO	09/08/19	09:52:46	8.125	10.25	6.125
SMO	09/10/19	23:31:39	4.5	6.0	3.0
SMO	09/12/23	08:07:35	12.0	17.375	6.5
SMO	10/02/05	11:33:36	3.875	6.125	2.875
SMO	10/03/13	00:21:24	7.75	7.625	2.875
SMO	10/03/16	02:24:38	14.625	21.375	6.625
SMO	10/06/30	17:46:48	5.25	10.125	4.25
GVO	09/07/02	15:50:25	13.125	14.75	10.25
GVO	09/08/16	14:36:46	77.75	53.25	49.25
GVO	09/08/16	17:20:05	12.625	6.875	4.875
GVO	09/08/16	17:43:52	10.125	4.875	3.375
GVO	09/08/16	19:47:24	12.875	10.625	11.25
GVO	09/08/17	01:48:35	5.375	2.0	1.5
GVO	09/08/17	20:53:59	7.5	4.75	2.625
GVO	09/08/18	16:27:16	5.5	3.25	2.625
GVO	09/08/19	09:53:32	13.5	10.875	14.25
GVO	09/08/24	00:36:43	10.0	5.0	4.375
GVO	09/09/02	06:45:51	5.5	6.25	5.5
UNP	15/03/03	17:31:11	5.5	5.75	2.75
BKT	14/10/29	20:06:14	9.0	6.625	4.625

* The unit of PGA is cm/s/s.

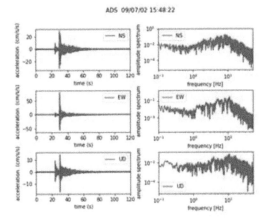

Figure 3. Observed accelergram at ADS.

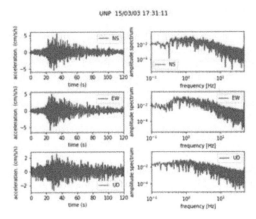

Figure 6. Accelereogram recorded at UNP.

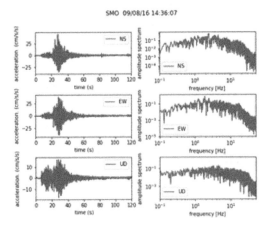

Figure 4. Accelereogram recorded at SMO.

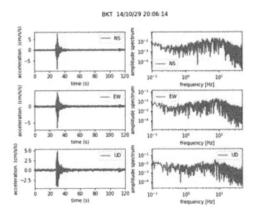

Figure 7. Accelereogram recorded at BKT.

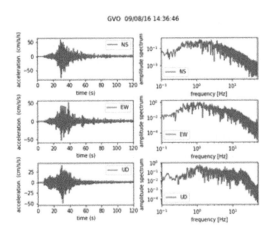

Figure 5. Accelereogram recorded at GVO.

Among the records shown in Figures 3 to 7, ADS, SMO and GVO records were able to identify the corresponding earthquake by the USGS earthquake catalog based on the recorded time. The ADS 09/07/02 record is the M5.1 earthquake occurring at 1.053 °S, 100.326 °E and the epicenter distance is approximately 22 km. SMO and GVO 09/08/16 record is M6.7 earthquake occurred at 1.479 °S, 99.490 °E, and the epicenter distance is about 115 km for both sites.

3.2 *Relationships between the dominant frequency of earthquake ground motion and peak frequency of HVSR of microtremor*

Microtremor array observations were conducted at four sites (ADS, SMO, GVO, and UNP) by Ono et al. (2012). Figure 8 shows a comparison of the dominant

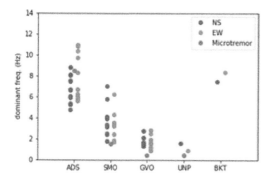

Figure 8. Dominan frequencies of recorded earthquake ground motions and comparison to peak frequencies of micrtremor horizontal-vertical-spectrum-ratio.

frequency of the seismic record shown in Table 2 and the peak frequency of the HVSR for each site. At any site where the peak frequency of the microtremor HVSR was obtained, the peak frequency of the HVSR showed good correspondence with the dominant frequency of seismic records. This result suggests that both microtremor observation and earthquake ground motion observation were appropriately performed.

More specifically, the peak frequency of HVSR of the microtremor of ADS corresponds to the mean value of the dominant frequency of earthquake ground motion records. On the other hand, in SMO, GVO, and UNP, the peak frequency of the microtremor HVSR corresponds to the lower limit of the dominant frequency of the earthquake ground motion records. This result suggests that the ground structure can influence the relationship between the peak period of microtremors and dominant period of seismic waves.

3.3 Time-frequency analysis of GVO and UNP records

Since GVO and UNP are located on a deep sedimentary ground, it is considered that surface wave reflecting the characteristics of the sedimentary ground is included in the recorded earthquake ground motion. Therefore, the time-frequency analysis was performed on the records of GVO and UNP. Figures 9 and 10 show the obtained evolutionary spectrum. In either the evolutionary spectrum of GVO or UNP, the characteristics of the frequency component contained in the ground motion record vary at near 50 s. Frequency components up to 10 Hz are contained before 50 s, whereas 2 Hz components are dominant after 50 s. This 2 Hz component is considered to be corresponding to a surface wave. This result indicates that surface waves of 2 Hz are dominant in the sedimentary ground where GVO and UNP are placed.

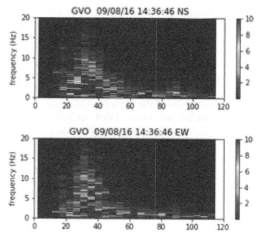

Figure 9. Evolutionaly spectra of recorded earthquake ground motion at GVO.

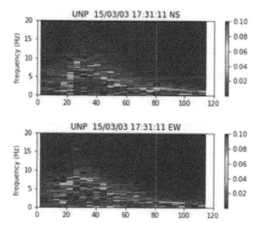

Figure 10. Evolutionaly spectra of recorded earthquake ground motion at UNP.

4 CONCLUSIONS

Earthquake observations are currently being conducted at two sites in Padang and one site at Bukittinggi. At the beginning of the observation, there were two more sites in Padang city. However, they have been abolished. At the observation sites in Padang, the ground structures had been estimated using the microtremor array observation. The sites of GVO and UNP, which are located close to the western coast of Padang city, are placed on the deep sedimentary ground. On the other hand, the site of ADS located at the high elevation place of the east side of Padang is on hard ground. The site of BKT in Bukittinggi is in the tunnel in the sedimentary rock.

So far 86 earthquake ground motion records have been obtained. Among them, there are 37 records in which PGA of the horizontal component exceeds 5 cm/s/s. As a result of spectral analysis of these 37 earthquake records, the dominant frequency of the seismic motion and the peak frequency of the microtremor HVSR coincided at each site. Time-frequency analysis of the records of GVO and UNP were carried out and revealed that surface waves of 2 Hz were dominant in the deep sedimentary ground where GVO and UNP were placed.

More than ten years have elapsed since the start of the observation in Padang and Bukittinggi, and it is necessary to maintain the equipment in order to continue observation in the future. Securing funds for the maintenance works is an urgent issue.

REFERENCES

Natawidjaja, D.H., Sieh, K., Chlieh, M., Galetzka, J., Suwargadi, B.W., Cheng, H., Edwards, R. L., Avouac, J.P. and Ward, S.N. 2006. Source parameters of the great Sumatran megathrust earthquake of 1797 and 1833 inferred from coral microatolls, *Journal of Geophysical Research*, 111, B06403.

Ono, Y., Noguchi, T., Rusnardi, R.P., Uemura, S., Ikeda, T. and Kiyono, J. 2012. Estimating subsurface shear wave velocity structure and site amplification characteristics of Padang, Indonesia, *Journal of Japan Society of Civil Engineers, Ser.A1 (Structural Engineering & Earthquake Engineering)*, 68(4):227–235 (in Japanese).

2019 Rock Dynamics Summit– Aydan et al. (eds)
© 2019 Taylor & Francis Group, London, ISBN 978-0-367-34783-3

Ground motion estimation at Kabul city for Mw 7.5 Hindu Kush earthquake

N.Z. Nasiry
Graduate School of Engineering and Science, Universityt of the Ryukyus, Okinawa, Japan

Ö. Aydan
University of the Ryukyus, Department of Civil Engineering, Nishihara, Okinawa, Japan

ABSTRACT: Within the last 16 years, Kabul, the capital and largest city of Afghanistan witnessed a development it had never seen. Many commercial and private high rise buildings in the mud-brown landscape of this city represent a new modernity. However, the seismological literature indicates that the safety assessment of the existing structures and those to build in future, with respect to ground motion variation is probably not studied. In this study, we used a stochastic point source modeling to estimate the ground motion variation at Kabul city by simulating the Mw 7.5 Hindu Kush earthquake that occurred on 2015 October 26 at a depth of 210 km. The earthquake was due to thrust faulting. The impact of this high magnitude earthquake was predominantly felt in Kabul and surrounding areas that killed around 267 people and damaged about 11,389 houses. We calculated the base-rock and surface acceleration at several sites in the epicentral region. The peak ground accelerations are compared with the acceleration records taken in Afghanistan. Although the strong motion instrumentation of Afghanistan is none, this study might be an important contribution of the seismic vulnerability of major settlements in Afghanistan.

1 INTRODUCTION

Kabul the capital of Afghanistan (34.55° N, 69.20° E), and the most populated (about 4.6 million, Kabul, World Population Review, 2017) and fastest growing city of the country is one of the largest cities (64th) in the world. Kabul city is located in the southern east earthquake region of Afghanistan, the region with most severe seismicity (Wheeler and others, 2005) in the country near where three main faults meet and form a junction (Figure 1). This region lies near the Eurasia and Indian plates boundary considered one of the most seismically active regions worldwide. This region has experienced great earthquakes and has caused immense damage to lives and properties. Pamir and Hindu Kush where continental deep earthquakes take place also lies near this region in about 300km on the North of Kabul. The M7.5 Hindu-Kush earthquake of October 26 2015 with the epicenter 45 km north of Alaqahdari-ye Kiran wa Munjan, Afghanistan at a depth of 212 km with a death tool of approximately 400 people, mostly in Pakistan. This earthquake was recorded at several regional accelerographic strong ground motion stations.

This earthquake caused tremendous shake to the nearest crowded city of Kabul. Damages to the old and new structures in Kabul city has been unofficially reported. Building damage due to foundation failures due to liquefaction were also reported in several parts of the city. The MMI intensity of his event at Kabul city has been estimated to be VI.

Although, Kabul city is located far away from the epicenter of this event and other large earthquakes (about 300 km), non-structural damages have been reported in many parts of this city.

With the fast increasing density of population and large industrial establishments in recent times, the vulnerability of Kabul city to damaging earthquakes is increasing day by day. Occurrence of another great earthquake in/near this region, might be highly devastating for Kabul city. Therefore it is of great need to have a reliable estimation of seismic hazard due to probable damaging earthquakes

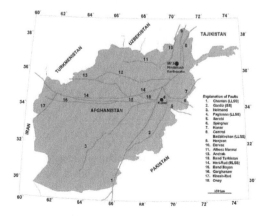

Figure 1. Active faults in Afghanistan, location of Kabul and M7.5 Hindu Kush earthquake. Faults map modified from (Wheeler and others, 2005).

occurring in this region. This requires good regional coverage of strong motion observations as well as detailed micro zonation studies for urban planning and design of infrastructures. There is lack of information on ground motion data of strong earthquakes as its application in engineering is concerned. The available quantified information on seismic hazard is sparse that earthquake engineers face problems in estimating the design ground motion at Kabul city. The lack of strong motion data can be remedied up to some extend by using analytical source mechanism models.

This study is performed to evaluate the ground motion at Kabul city for an earthquake of magnitude Mw 7.5 in Hindu Kush area of Afghanistan. The fault location and source parameters for this event is taken from the information released by the USGS. A modified version of stochastic point source model of Boore (1983) is used for simulating ground motion. First, acceleration time histories are simulated for bedrock conditions. Then the simulated response spectrum at bedrock and surface level are presented.

2 TECTONIC SETTING AND SOURCE PARAMETERS OF MW 7.5 EARTHQUAKE

The tectonic setting of Afghanistan involves fundamentally the northward motion of the Indian plate sub-ducting beneath Eurasia plate in the Tibet and the movement of the Indian plate is accommodated in Afghanistan and Pakistan by the Chaman Fault through mainly sinistral faulting (Fig. 2). However, the tectonic setting is more complex in the vicinity of Hindu Kush Mountains. Fig. 3 shows a cross-section depicting the subduction of the Indian plate beneath the Eurasia plate in the vicinity of the Hindu Kush Mountains. It is clearly noted that the Indian plate is steeply bended and it is probably in the process of the detachment from the Indian Plate and sinking into the upper mantle, which would definitely affect the seismicity of the region for decades from now on.

The M_w 7.5 Hindu Kush earthquake of Afghanistan took place on 26 October 2015 and the instrumental data of the earthquake are given in Table 1 together with some estimations. The focal mechanism of the earthquake was due to thrust faulting with a dip direction and dip angle of 16° and 70°, respectively according the USGS. The rake angle was 273°, which implies almost thrust faulting with a slight sinistral component. The hypocenter depth of the earthquake was 210-213 km.

The M_w 7.5 Hindu Kush earthquake of Afghanistan was widely felt in northeastern Afghanistan, northwestern Indian, and northern Pakistan. It killed at least 115 people in Jalalabad and destroyed more than 4,000 homes in Jarm of Afghanistan. The event was more strongly felt in Pakistan and killed at least 289 people and destroyed more than 29,230 house there. The Intensity of this event was (VI) at

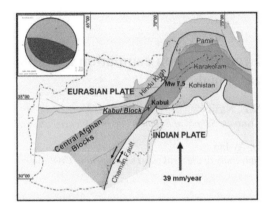

Figure 2. Tectonic setting of Afghanistan and focal mechanism of the Mw 7.5 Hindu Kush earthquake (modified from Kafarsky et al, 1975).

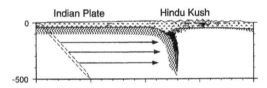

Figure 3. Illustration of subduction of Indian Plate beneath Hindu-Kush Mountains (modified from Pavlis and Das, 2000).

Table 1. Instrumental and estimated Data

Institute	Strike/dip/ rake	Slip(cm)	Rupture Duration (s)	Fault length(km)
USGS	106/70/87	615	20+5	40
HAR-VARD	104/69/91		27.2	
GFZ	106/65/95			
This study*	-	380	31.5	88

*Based on Aydan (2012) empirical formulas for Mw 7.5

Bagrami and Jalalabad, (V) at Kabul and Mahmud-e-Raqi, (VII) at Abbottabad and Wah of Pakistan; (VI) at Rawalpindi, (V) at Amritsar, Badambagh and Palwal of India and at Dushanbe of Tajikstan. It was felt (III) at Tashkent, Uzbekistan. It was also felt (II) at Ghorahi and Kathmandu, Nepal, Doha of Qatar. The intensity and inferred strong motions data of this event are presented in Table 2. Although there is no acceleration data for this earthquake, accelerations recorded in Peshawar, Nilore (Islamabad) and DI Khan in Pakistan (Ahmad, 2015; Ismail and Khattak, 2016). The accelerations at these stations are listed in Table 3 and the acceleration records and their response spectra are shown in Figs. 4 and 5.

The epicentral distance of Peshawar City in Pakistan is quite similar to the epicentral distance of Kabul City. The earthquake caused damage to

Table 2. Intensities and inferred strong motions for M7.5 Hindu Kush Earthquake, 26 Oct 2015.

Locality	Net-work	Intensity (MMI)	PGA (cm/s²)	PGV (cm/s)	Distance (km)
Charikar	DYFI	V	43	5.44	277
Kabul	DYFI	VI	70	9.74	306
Kabul	DYFI	V	51	6.77	309
Kabul	DYFI	V	42	5.44	316
Kabul	IU	VI	22	1.7	315
Dushanbe	DYFI	V	42	5.44	303
Garm	TJ	IV	53	1.9	315
Tashkent	DYFI	III	4	-	540
Peshawar	DYFI	V	45	5.85	339
Peshawar	DYFI	VI	70	9.74	341
Islama-bad	DYFI	VII	139	21.71	391
Srinagar	DYFI	V	54	7.28	489
Islama-bad	DYFI	VI	108	16.22	393
Islama-bad	DYFI	V	37	4.7	384

Note: Data in this table is retrieved from USGS website.

Table 3. Strong motion data (from N. Ahmad)

Institute	Epicentral distance (km)	EW	NS	UD
Peshawar	280	0.05 g	0.053 g	0.038 g
Nilore	430	0.021 g	0.020 g	0.017 g
D.I. Khan	514	0.026 g	0.036 g	0.015 g

mud-brick or stone masonry with earthen mortar in Peshawar and Kabul Cities. The analyses of collapsed structures imply that the acceleration might be up to 0.158 g in Kabul City, which may correspond to spectral acceleration rather than the base acceleration. The heavy damage in Peshawar also support the high spectral accelerations shown in Figure 5. It should be also noted that the hypocenter of the earthquake is very deep. It is expected to high frequency waves would attenuate and long-period components would be more dominant. As the earthquake was felt at far-distant locations, the observations support this conclusion.

3 SIMULATION METHOD AND MODEL PARAMETERS

The stochastic seismological model originally proposed by Hanks and McGuire (1981) and later generalized by Boore (1983) for simulating synthetic acceleration time histories is used in this study. This model is a good alternate for simulating synthetic acceleration time histories with few known source

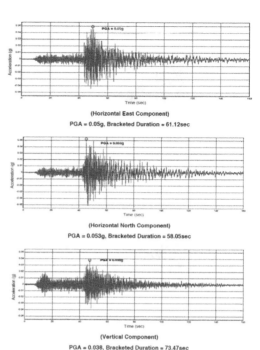

(Horizontal East Component)
PGA = 0.05g, Bracketed Duration = 61.12sec

(Horizontal North Component)
PGA = 0.053g, Bracketed Duration = 58.05sec

(Vertical Component)
PGA = 0.038, Bracketed Duration = 73.47sec

Figure 4. Acceleration records at Peshawar (from Ahmad, 2015).

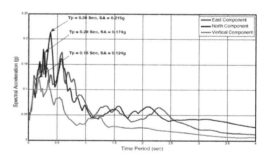

Figure 5. Response spectra of acceleration records at Peshawar (from Ahmad, 2015).

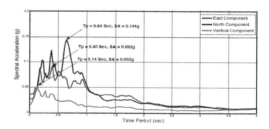

Figure 6. Response spectra of acceleration records at Peshawar (from Ahmad, 2015).

and medium parameters. Here, ground motion is modeled as a band limited finite duration Gaussian white noise in which the radiated energy is assumed to be distributed over a specified duration.

252

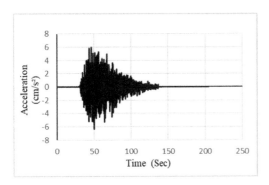

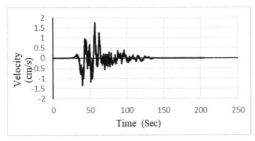

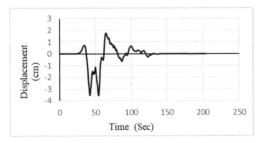

Figure 7. Acceleration, velocity and displacement time series, resulted from stochastic modeling for M = 7.5, R = 300 km.

Table 4. Model parameters used in simulation

Acceleration Fourier Amplitude Spectrum	$A(f,M_0,R)=S(f,M_0)D(f,R)P(f)$
Source Spectrum- *Brune Source Spectrum*	$S(f,M_0)=C(2\pi f)^2 M_0*1/(1+f/f_c)^2$ $C=R_\emptyset FV/(4\pi\rho\beta^3 R_1)$ $R_\emptyset=0.55, F=2, V=\sqrt{2}$ $\rho=2.8g/cm^3$
Corner Frequency	$Fc=4.9*10^6\beta\Delta_\sigma/M_0^-$ $\beta=3.5$ km/s $\Delta_\sigma=100$ bars
Attenuation	$D(f,R)=Dg(f,R)D(f)$
Geometric spreading	$Dg(R)=1/R$ (R<70 km) $Dg(R)=1/70; 70<R<130$ km $Dg(R)=1/70(130/R)^{0.5}$ R>130 km
Q model-Anelastic Attenuation	$Q(f)=220f^{0.52}$
Low-Cut Filter	$P(f) = \exp(-\pi\kappa f)$ Kaapa (κ)=0.006 hard rock

In this method, various factors affecting ground motion source, path, and site factors are put together into a physically determined algorithm and is used to predict ground motion. This method is widely used in different regions of the world and is proved to be reliable for the seismic regions of the world where strong ground motion occurs. The model parameters used for simulation are given in Table 4.

The simulated acceleration, velocity and displacement time series from the stochastic modeling are shown in Figure 7. The PGA at Kabul city is about 6 cm/s² and may be due to its big distance from the source. Maximum of this value reported in one of the four stations in Kabul is 70 cm/s² and by another of the four stations equal to 22. The simulated result shows a smaller number comparing it to the least recorded acceleration. The simulated response spectra for Kabul city at bedrock is presented in Figure 8.

The simulated response spectra for Kabul city at ground surface is presented in Figure 9. The acceleration response spectra particularly flat up to 2 seconds.

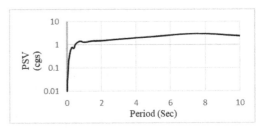

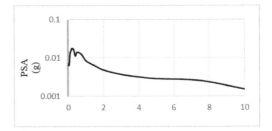

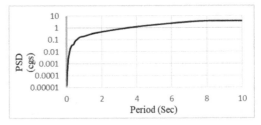

Figure 8. Response spectra at bedrock

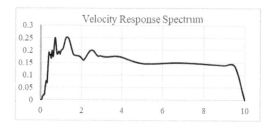

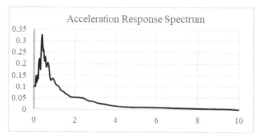

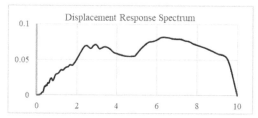

Figure 9. Surface level response at 5% damping

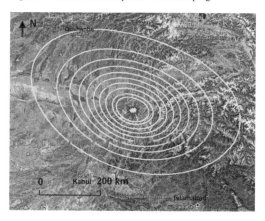

Figure 10. Estimated maximum ground accelerations for the Mw 7.5 earthquake.

4 STRONG MOTION ESTIMATION BY EMPIRICAL APPROACH

The method presented previously utilizes Green Function Method while its parameters are based on some empirical relations. Here, we utilize the method proposed by Aydan (2012) and Aydan et al. (2009a,b) to estimate the distribution of maximum strong motions. The method itself here is used for estimating the maximum ground acceleration using

the empirical relations for inter-plate earthquakes as the 2016 October 26 earthquake occurred at the plate boundary between the Indian Plate and Eurasia plate. The estimations of the maximum ground acceleration at the ground surface at the epicenter is only 95 gals and it attenuates to 4 gals beyond 250 km. Nevertheless, the directional effects in the attenuation of maximum ground acceleration are well evaluated.

Although, the damaging effects of the 2015 Hindu Kush earthquake was felt widely, the strong motions in Kabul City due to larger earthquakes are much more important. The major fault near Kabul City is the famous Chaman fault, which passes at about

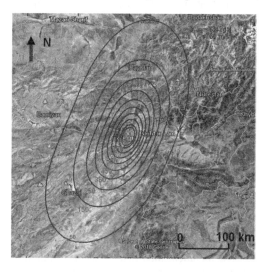

Figure 11. Estimated maximum ground acceleration contours for an earthquake of Mw 7.24 on the Chaman Fault near Kabul.

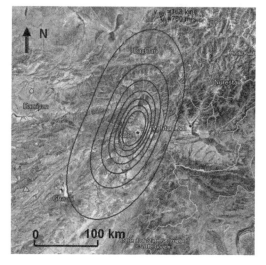

Figure 12. Estimated maximum ground velocity contours for an earthquake of Mw 7.24 on the Chaman Fault near Kabul.

Table 5. Characteristics of anticipated earthquake

Length (km)	Strike/ dip/rake	Mw	Slip (cm)	Rupture Duration (s)
90	206/90/0	7.24	274	23

16 km NW of the city. This fault joins to the Hari-Rud fault to the north of Kabul City. The Chaman fault is segmented and the nearest segment to Kabul City is 90 km long (see Figure 1). Figure 11 and 12 show the estimated contours of maximum ground acceleration and maximum ground velocity for bedrock by assuming that the fault has a strike, dip and rake angle with a hypocentral depth of 20 km as given in Table 5. The shear wave velocity of bedrock is taken as 760 m/s.

As noted from the figures, the ground accelerations could be greater than 0.9 g for Kabul City. Similarly, the maximum ground velocity would be about 153 kines at the epicenter and it is about 104 kines at Kabul City. In view of recent earthquakes with dense strong motion coverage in Japan, USA, Italy and Turkey, these values should be such that they must be the basis for the earthquake-resistant of structures in Afghanistan.

5 DISCUSSION AND CONCLUSION

In this study, we simulated strong ground motions induced in Kabul city by the Mw = 7.5 Hindu Kush earthquake and source to site distance of 300 km using SMSIM, a FORTRAN program that account for attenuation and local site conditions. The response spectra at base and surface level is calculated for Kabul city. The simulated spectral pseudo-acceleration with 5% damping is presented in Figure 9. The maximum acceleration at the surface is about 25% and is matching the recordings taken in Kabul city. Although the simulated results are smaller compared to the actual recordings, other construction factors might also contribute to the significant damage of buildings in Kabul city.

In addition, strong motion records taken in Peshawar and D.I. Khan induced by the 2015 Hindu Kush earthquake have been summarized. These results may be also of great significance to check the estimations. This signifies the importance of a thorough multi-hazard assessment on buildings in Afghanistan. Although the strong motion instrumentation of Afghanistan is none, the results presented here can be used in seismic analysis and design of infrastructure facilities in Kabul. Nevertheless, the most important aspect is to be well-prepared against the worst scenarios such as the possibility of a great earthquake on the Chaman fault. The strong ground motions can be quite high

and these would have an important implications in the earthquake preparedness of Afghanistan with a hope that the devastating 2005 Kashmir earthquake would not be repeated again. Furthermore, it is no need to say that the establishment of a strong motion network in Afghanistan is a must in order to check the assumptions as well as to have instrumental data of strong ground motions for the seismic design of structures.

REFERENCES

Ahmad, N. 2015. A Note on The Strong Ground Motions and Behavior of Buildings During 26th Oct. 2015 Afghanistan–Pakistan Earthquake. Earthquake Engineering Center, Department of Civil Engineering, UET Peshawar, 15p.

Aydan, Ö. (2012): Ground motions and deformations associated with earthquake faulting and their effects on the safety of engineering structures. Encyclopedia of Sustainability Science and Technology, Springer, R. Meyers (Ed.), 3233–3253.

Aydan, Ö., Hamada, M., Itoh, J, Ohkubo, K. 2009a. Damage to Civil Engineering Structures with an Emphasis on Rock Slope Failures and Tunnel Damage Induced by the 2008 Wenchuan Earthquake. Journal of Disaster Research, Vol.4, No.2, 153–164.

Aydan, Ö., Y. Ohta, M. Hamada, J. Ito, K. Ohkubo, 2009b.The response and damage of structures along the fault rupture traces of the 2008 Wenchuan Earthquake. Int. Conf. on Earthquake Engineering: the 1st Anniversary of Wenchuan Earthquake, Chengdu, pp. 625–633.

Boore, David M. 1996. SMSIM Fortran Programs for Simulating Ground Motions from Earthquakes: Version 2.0 A Revision of OFR 96-80-A.

Boore, D. M. 1983. Stochastic simulation of high-frequency ground motions based on seismological models of the radiated spectra". Bulletin of the Seismological Society of America 73.6A, 1865–1894.

Hanks, T. C., and R. K. McGuire 1981. The character of high-frequency strong ground motion". Bulletin of the Seismological Society of America 71.6, 2071–2095.

Ismail, N. Khattak, N. 2015. Building Typologies Prevalent in Northern Pakistan and Their Performance during the 2015 Hindu Kush Earthquake. Earthquake Spectra, 32(4),2473–2493,

Kabul - World Population Review (http://worldpopulation review.com/world-cities/kabul-population/)

Kafarsky AKh, Chmyriov VM, Stazhilo-Alekseev KF, Abdullah Sh, Saikovsky VS 1975. Geological map of Afghanistan, scale 1:2, 500,000.

Pavlis, G. L., Das, S. 2000. The Pamir-Hindu Kush seismic zone as a strain marker for flow in the upper mantle. Tectonics, **19, (1), 103–115**.

United States Geological Survey Earthquake Hazard Program (2015)., M 7.5-45 km E of Farkhar, Afghanistan.

Wheeler, R. L., Bufe, C. G., Johnson, M. L., Dart, R. L., & Norton, G. A. 2005. Seismotectonic map of Afghanistan, with annotated bibliography. US Department of the Interior, US Geological Survey.

2019 Rock Dynamics Summit– Aydan et al. (eds)
© *2019 Taylor & Francis Group, London, ISBN 978-0-367-34783-3*

Application of core-based stress measuring method in the vicinity of earthquake source fault: Diametrical core deformation analysis

T. Sugimoto, W. Lin & A. Lin
Kyoto University, Kyoto, Japan

ABSTRACT: Stress distribution around a source fault results from the rupture process of an earthquake. Therefore it can be key information for clarifying the rupture process of the fault. Moment tensor inversion is a powerful technique to determine principal stress orientations in the vicinity of a source fault. However, this technique determines only the orientations and regimes of principal stresses on the fault plane and has no ability to measure the stress far from the fault plane. In this study, we present one of the stress measuring methods which can determine stress far from source fault, diametrical core deformation analysis (DCDA). We have applied this method to the boring cores from a borehole penetrating the Futagawa fault, Kumamoto, Japan, which is one of the source faults of the 2016 Kumamoto earthquake sequence. Preliminary results of the DCDA curried out for limited core samples show core diameter deformation data is effective to figure out stress data.

1 INTRODUCTION

There is strong relationships between earthquake and in-situ stress of the rocks around the source fault (Hardebeck, 2004, 2012; Stein, 1999). In particular, stress distribution around a source fault after an earthquake results from the rupture process of the source fault. Therefore information on in-situ stress of rocks around the source fault becomes a clue to clarifying the rupture process of the fault.

Moment tensor inversion is a powerful technique to determine principal stress orientations in the vicinity of a source fault. However, this technique determines only the orientations and regimes of principal stresses on the fault plane and has no ability to measure the stress far from the fault plane. In this study, we present one of the stress measuring methods which have the ability to determine one dimensional stress distribution (i.e. depth profile of stress) and the magnitude of differential stress: diametrical core deformation analysis (DCDA).

The purpose of this study is evaluating the in-situ stress of rocks around the source fault of 2016 Kumamoto earthquake sequence after the main shock of the sequences. To this end, we conducted stress measurements by DCDA around Futagawa fault, Kumamoto, Japan. This fault is one of the source faults of 2016 Kumamoto earthquake sequence. In this area, stress measurements using moment tensor inversion were conducted by Yoshida et al. (2016).

We used several rock core samples retrieved from a drilling hole penetrating the Futagawa fault in order to evaluate the stress distribution around the fault.

2 2016 KUMAMOTO EARTHQUAKE SEQUENCE

The 2016 Kumamoto earthquake sequence occurred in Kumamoto, Japan. This earthquake sequence included two large earthquakes: a M6.5 foreshock which occurred on 14 April and a M7.3 main shock which occurred on 16 April (Asano & Iwata, 2016; Yoshida et al., 2016). Figure 1 shows the surface movement caused by the 2016 Kumamoto earthquakes.

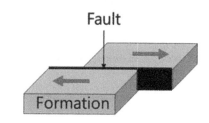

Figure 1. Surface movement around the source fault and a schematic diagram of right lateral strike slip fault.

Yoshida et al. (2016) conducted in-situ stress evaluations around the Futagawa fault using moment tensor inversion. This previous study reported the change of the stress state around the source fault before and after the earthquakes.

Yoshida et al. (2016) only determined the stress state in the vicinity of the fault plane. In this study, we aimed to determine the stress distribution along the drilling hole and to evaluate the more detailed stress condition around the source fault.

3 METHODOLOGY: DCDA

Diametrical core deformation analysis (DCDA) is one of the core-based methods for stress measurements (Funato & Ito, 2017).

This method is based on the elastic strain recovery which occurred on a drilled rock core. When a core is retrieved from a drilling hole, elastic strain recovers immediately after the drilling. We can measure the elastic strain recovery on the core sample by measuring the diameter of the core's section. Figure 2 shows a schematic diagram of the concept of DCDA. After drilling the core's section changes elliptically according to the two-dimensional in-situ stress state of rocks perpendicular to the drilling hole. DCDA theory says that the orientation of the major axis of the section corresponds to that of the maximum stress(Funato & Ito, 2017). If the drilling hole is vertical, we can determine the horizontal maximum stress orientation.

The diameter of the core is measured by the apparatus in Figure 3. The apparatus consists of an optical micrometer, a pair of motor-driven rollers and a data processing system. In the micrometer, the uniform parallel light emitted from a LED irradiates on a core sample. The image of a shadow created by the core sample is projected on a CCD, and the distance between the detected edges of the shadow is output as the core diameter. This measurement is carried out continuously while a core sample is rotated on two rollers at a constant speed (Funato & Ito, 2017).

According to the theory of DCDA, differential stress parallel to the core section is calculated from Equation 1

$$S_{H\,max} - S_{h\,min} \tag{1}$$
$$= E(d_{max} - d_{min}) / d_0 (1 + v).$$

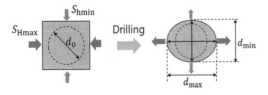

Figure 2. A schematic figure showing a core diameter change caused by drilling (modified from Funato & Ito, 2017).

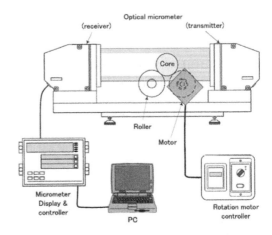

Figure 3. Apparatus used for core diameter measurement

4 BORING CORES USED FOR DCDA

In this study, we used several boring cores retrieved from a borehole penetrating to Futagawa fault. In this paper, we show information and results of the two of the cores. Figure 4 shows the picture of samples named FDB-13, FDB-19. Table 1 shows the depth information of the two samples.

5 RESULTS OF DCDA

Figure 5 shows the results of core diameter measurement of FDB-13 and FDB-19. These results show the core diameter change caused by in-situ stress

FDB-13

FDB-19

Figure 4. Two of the boring cores used for DCDA: FDB-13, FDB-19.

Table 1. Sample information

sample ID	Depth (m)
FDB-13	557.00
FDB-19	666.80

Table 2. Directions of S_{Hmax}

sample ID	direction of S_{Hmax}
FDB-13	N70°E
FDB-19	N70°E

FDB-13

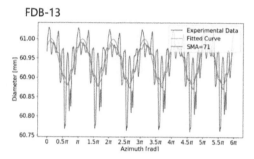

FDB-19

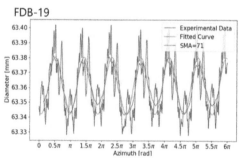

Figure 5. Results of core diameter measurement of FDB-13 and FDB-19

anisotropy occurred clearly. The result of FDB-13 has noise which caused by core surface roughness.

Table 2 shows the S_{Hmax} directions determined by the DCDA results. In the next section, the horizontal stress orientations evaluated using this results are shown.

6 STRESS STATE AROUND THE SOURCE FAULT

Figure 6 shows the Schmidt net diagram of horizontal maximum stress orientations evaluated with DCDA and moment tensor inversion by Yoshida et al. (2016). These three results of maximum horizontal stress orientation correspond well. Therefore the measurement with DCDA was evaluated with high accuracy.

These results say that the horizontal maximum stress is subparallel to the Futagawa fault under 550 m below the surface.

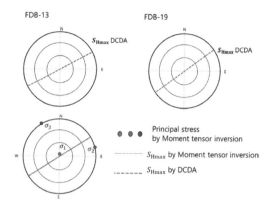

Figure 6. Schmidt net diagram of horizontal maximum stress orientations evaluated by DCDA and moment tensor inversion (Yoshida et al., 2016).

7 CONCLUSION

In this study, to determine the stress state around the source fault of 2016 Kumamoto earthquake, we conducted one of the core-based methods for stress measurement called DCDA. This method can determine the orientation of maximum stress parallel to a core section and the magnitude of differential stress. The results of DCDA correspond to that of moment tensor inversion. This fact says that the measurement with DCDA was successful.

ACKNOWLEDGMENTS

This work was supported by Grants-in-Aid for Scientific Research 16H04065 of the Japan Society for the Promotion of Science (JSPS), and a contract included in the Drilling into Fault Damage Zone (DFDZ) project of the Secretariat of Nuclear Regulation Authority of Japan.

REFERENCES

Asano, K., & Iwata, T. 2016. Source rupture processes of the foreshock and mainshock in the 2016 Kumamoto earthquake sequence estimated from the kinematic waveform inversion of strong motion data 2016 Kumamoto earthquake sequence and its impact on earthquake science and hazard assess. *Earth, Planets and Space*, *68*(1), 1–11.

Funato, A., & Ito, T. 2017. A new method of diametrical core deformation analysis for in-situ stress measurements. *International Journal of Rock Mechanics and Mining Sciences*, *91*, 112–118.

Hardebeck, J. L. 2004. Stress triggering and earthquake probability estimates. *Journal of Geophysical Research B: Solid Earth*, *109*(4), 1–16.

Hardebeck, J. L. 2012. Coseismic and postseismic stress rotations due to great subduction zone earthquakes. *Geophysical Research Letters*, *39*(21), 1–6.

Stein, R. S. 1999. The role of stress transfer in earthquake occurrence. *Nature*, *402*(6762), 605–609.

Yoshida, K., Hasegawa, A., Saito, T., Asano, Y., Tanaka, S., Sawazaki, K., … Fukuyama, E. 2016. Stress rotations due to the M6.5 foreshock and M7.3 main shock in the 2016 Kumamoto, SW Japan, earthquake sequence. *Geophysical Research Letters*, *43* (19),10,097-10,104.

2019 Rock Dynamics Summit– Aydan et al. (eds)
© *2019 Taylor & Francis Group, London, ISBN 978-0-367-34783-3*

Overview of seismic activity and historical earthquake disasters in Okinawa region

Y. Kamada

Miyakojima Meteorological Office, Japan Meteorological Agency, Hirara, Simozato, Miyakojima, Okinawa, Japan

A. Kamiya, S. Arakaki, T. Isikawa & Y. kawajo

Seismology and Volcanology Division, Okinawa Regional Headquarters, Japan Meteorological Agency, Higawa, Naha, Okinawa, Japan

ABSTRACT: Okinawa, which is located in the southwest part of Japan, is tectonically very active. Ryukyu subduction zone is characterized by the subducting Philippine Sea plate beneath the Eurasian plate and by crustal extension at the back of the arc in the Okinawa Trough. Historically, Okinawa region have been ex-perienced the severe damage of earthquake and tsunami several times. At the present day, a great num-bers of seismographs are deployed on Ryukyu arc islands by Japan Meteorological Agency (JMA) and other au-thorities to monitor the seismic activities for the purpose of disaster mitigation and various researches.

1 INTRODUCTION

To comprehend the topographic feature in Okinawa, it is essential to elucidate the tectonics of the Ry-ukyu subduction zone. In this paper, we describe the overview of the seismic activity associated with the tectonics and the stress field in the Ryukyu sub-duction zone briefly.

2 SEISMIC ACTIVITY

2.1 *Overview of the seismic activity in Okinawa region.*

Seismic activity in Okinawa region is very active over the whole area. More than 15,000 earth-quakes occurred in or around the Okinawa region in 2017 according to Japan Meteorological Agency (JMA) and it accounted for about 10% of total number of earthquake occurred in Japan (Fig.1). The distri-bution of hypocenters demonstrate that the earth-quakes occurred the shallow portion (depth<30km) are located toward the Ryukyu trench side. On the other hand, the earthquakes occurred the consider-ably deep portion is visible as going to continental side. It suggests that the Philippine Sea plate is sub-ducting beneath the Eurasian Plate and the earth-quakes occur along the subducting plate. The seismic stations in the Okinawa region are limited on the islands arc due to the topographical features. Considering that large numbers of earthquakes are determined in the Okinawa region is fully related to extreme seismic activities in spite of poor detection capability.

2.2 *The damaged earthquakes in Okinawa region.*

According to the historical documents and the field surveys, Okinawa region has been experienced many times earthquake and tsunami damages (Fig.2). Especially, Yaeyama Islands, the southwest part of Ryukyu arc, is subject to tsunami disaster due to the topographical factors. The prominent tsunami dis-aster is "Meiwa-Tsunami" (the Yaeyama earthquake Tsunami) in 1771. The devastating tsunami struck the coast of Yaeyama - Miyakojima islands, reach-ing a maximum runup height estimated over 30m, resulting in significant damage and approximately

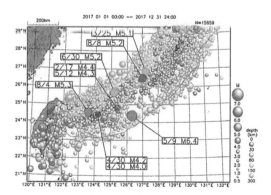

Figure 1. Distribution of hypocenters in or around Okinawa region in 2017 by JMA. Color circle indicates focal depth and its size is proportional to the M (magnitude). The flag are attached that seismic intensity with more than 3 are observed in the domestic.

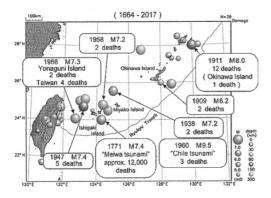

Figure 2. Map of the damaged earthquakes along the Ryukyu trench since 1664 to 2017. Data is based on JMA and number of the casualties is cited from Usami (2003).

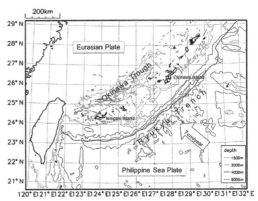

Figure 3. Outline of the topographic structure along the Ryukyu subduction zone.

12,000 casualties (Makino 1968). Several models of the "Meiwa-Tsunami" source have been proposed (Imamura et al. 2001, Nakamura 2009). However the source of "Meiwa-Tsunami" is still somewhat ambiguous. Recently some studies infer that the Ryukyu subduction zone has a high potential for large tsunami (Hsu et al.2012, Arai et al. 2016), and the great tsunami struck to the coast of Yaeyama Islands at least four times in the past repeatedly (Ando et al. 2018).

3 TECTONICS AND STRUCTURE IN THE RYUKYU SUBDUCTON ZONE

3.1 *Outline of the tectonics in the Ryukyu subduction zone.*

The tectonics of the Ryukyu subduction zone is characterized by mainly two factors, firstly sub-ducting the oceanic plate beneath the continental plate, secondly the existence of the active back arc basin called "Okinawa Trough". The Ryukyu Arc extends over 1,400 km from Kyusyu, Japan to Taiwan, which runs along the Ryukyu trench (Fig.3). The Philippine Sea plate subducts beneath the Eurasian plate northwestward. The Okinawa Trough located northwestward of the Ryukyu Arc is typical back-arc basin in continental margin which the seafloor depth exceeds 2000m, is now active opening (Sibuet et al. 1998, Arai et al. 2016). Associated with the continental rifting, it caused active seismic activity, especially southern part of the Okinawa Trough. Also southwest part of the Ryukyu subduction zone is characterized by extreme fast velocity of the plate convergence rate. Some studies that utilize a Global Positioning System (GPS) show the rate of subducting plate motion is about 7cm/yr, which the rate increase as going to southwestward (Nishimura et al. 2004). In addition, the southern Okinawa Trough is active opening to southward, eventually total relative plate convergence rate result in approximately 12cm/yr (Ando et al. 2009). There is no such subduction zone

in the world (Heki et al. 2008). On account of the reason that such the extreme plate convergence rate and no interplate great earthquakes (M>8) have been observed in the last few hundred years, the plate coupling rate in the Ryukyu subduction zone is assumed to be weak (Peterson et al. 1984). On the other hand, recent studies revealed the noteworthy events have discovered in the Ryukyu subduction zone, for example, "Slow-Slip Events" (Heki et al. 2008), "Low-Frequency Earthquakes" (Arai et al. 2016), "Very Low Frequency Earthquakes" (Nakamura et al. 2015). Such of the advanced research could elucidate the complex structure of this area.

4 EARTHQUAKE OBSERVATION NETWORK IN OKINAWA REGION

4.1 *JMA earthquake observation network.*

To monitor the earthquake, JMA operates earthquake observation network comprised of about 800 seismographs and seismic intensity meters. When an earthquake occurs, JMA issues information quickly on its hypocenter, magnitude and observed seismic intensity. In case of a damaging tsunami is expected in a coastal region, JMA issue the Tsunami Warning/Advisory for each region immediately. In Okinawa region, JMA deploy total 30 seismographs and seismic intensity meters (Fig.4). All the observed data are sent to central processing system in real time. It is inevitable that the distribution of the seismic stations is limited on islands due to the topographical condition.

4.2 *NIED and local government earthquake observation Network.*

At present day, National Research Institute for Earth Science and Disaster Resilience (NEID) install a nationwide broadband seismograph network, it's called "F-net" (Full Range Seismograph Network of Japan). The F-net data is quite useful and now 11

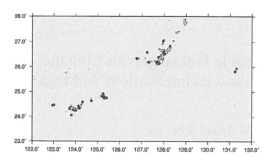

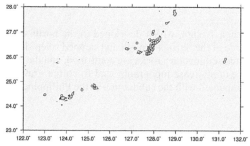

Figure 4. (top) The map of the installation of seismographs and seismic intensity meters by JMA. (bottom) The map of the F-net stations (NEID) and seismic intensity meters installed by Okinawa prefecture.

stations are installed in Okinawa region, which data is telemetered to JMA in real-time basis. Okinawa local government also installed total 44 seismic intensity meters in the prefecture. Seismic intensity information is quite important for relevant authorities because of it plays vital role as a trigger for the initial emergency operations.

5 CONCLUSION

In Okinawa region, the complicated topographical structure and stress field are associated with active seismic activities. Preparing for a disaster, It is important for us to realize the factor of earthquake occurrence. Further researches of the various points of view are required.

REFERENCES

Ando, M. Tu, Y. Kumagai, H. Yamanaka, Y. & Lin, C.H. (2012). Very low frequency earthquakes along the Ryukyu subduction zone. *Geophy. Res. Lett. 39, L04303.*

Ando, M. Kitamura, A. Tu, Y. Ohasi, Y. Imai, T. Nakamura, M. Ikuta, R. Miyairi, Y. Yokoyama, Y. Shishikura, M. (2018). Source of high tsunami along southernmost Ryukyu trench inferred from tsunami stratigraphy. *Tecnophysics. 722.* 265–276.

Arai, R. Takahashi, T. Kodaira, S. Kaiho, Y. Nakanisi, A. Fujie, G. Nakamura, Y. Yamamoto, Y. Isihara, Y. Miura,S. Kaneda, Y. (2016). Structure of the tsunami genic plate boundary and low-frequency earthquakes in the southern Ryukyu Trench.*Nat.Commun.712255, https://doi.org/10.1038/ncomms12255.*

Heki, K. Kataoka, T. (2008). On the biannually repeating slow-slip events at the Ryukyu Trench, southwestern Japan. *J. Geophys.Res113, B11402.http://dx.doi. org/10.1029/2008JB005739.*

Hsu, y-j. Ando, M. Yu, S-B. Simons, M. (2012). The potential for a great earthquake along the southernmost Ryukyu subduction zone, *Geophys.Res. Lett,*Vol.39,*L14302,doi:10.1029/2012GL052764.*

Imamura, F. Yoshida, I. Moore, A. (2001). Numerical study of the 1771 Meiwa tsunami at Ishigaki Island, Okinawa and the movement of the tsunami stones. *Proc. Coastal Eng. Jpn. Soc. Civ. Eng. 48*, 346–350 *(in Japanese).*

Makino, K. (1968). The Great Yaeyama Tsunami. *(printed privately, 462pp, in Japanese).*

Nakamura, M. (2009). Fault model of the 1771 Yaeyama earthquake along The Ryukyu Trench estimated from the devastating tsunami. *Geophys. Res. Lett. 36*, L19307.

Nakamura, M. Sunagawa, N. (2015). Activation of very low frequency earthquakes by slow slip events in the Ryukyu Trench. *Geophys.Res.Lett. 42*, 1076–10082.

Nishimura, S. Hashimoto, M. Ando, M. (2004). A rigid block rotation model for the GPS derived velocity field along the Ryukyu arc. *Phys. Earth Planet. Inter. 142*, 185–203.*http://dx.doi.org/10.1016/j.pepi.2003. 12.014.*

Peterson, E.T. Seno, T. (1984). Factors affecting seismic moment release rates in subduction zones. *J. Geophys. Res. 89*, 10233–10248.

Sibuet, J.C. Deffontaines, B. Hsu, S.K. Thareau, N. Formal, L. Liu, C.S. (1998). Okinawa trough backarc basin: Early tectonic and magmatic evolution. *J. Geophys. Res. 103*, 30245–30267.

Usami, T. (2003). New Comprehensive Summary of Historical Destructive Earthquakes in Japan. *University Tokyo, Press (in Japanese).*

2019 Rock Dynamics Summit– Aydan et al. (eds)
© 2019 Taylor & Francis Group, London, ISBN 978-0-367-34783-3

Geological and geomorphological features in broken gravels from the North Senoumi-bank, Suruga-bay, Japan and its implications on Mega-earthquakes

I. Sakamoto, S. Tomita, M. Fujimaki, Y. Yokoyama, N. Azumi & M. Yagi
Faculty of Marine Science and Technology, Tokai University, Shizuoka, Japan

ABSTRACT: In the central part of Suruga Bay, the Suruga Trough, which developed in the north-south direction, exists. Senoumi-bank is located on the western side of the Suruga Trough and stepped steep slopes are observed on the eastern slope. On this steep slope, crushed sedimentary rocks are confirmed, and the 90% of the collected gravels have a broken shape. The presence of staircase topography and fractured gravel is thought to be caused by the activity of the thrust fault accompanied with the subduction of the Philippine Sea Plate beneath Eurasian Plate.

1 INTRODUCTION

Suruga Bay is located in the center of the Honshu arc, its east-west width near the mouth of the bay is 60 km wide, its length is over 80 km in the north-south direction and the depth of the bay exceeds the 2,500 m (Fig. 1 a).

In the central part of Suruga Bay there is a Suruga Trough that developed in a narrow width to the north and south. The Suruga Trough is located at the boundary between the Philippine Sea Plate and the Eurasia Plate, and in the past several large earthquakes such as the Meiou Earthquake, the Hoei Earthquake, and the Ansei Tokai Earthquake occurred. Senoumi-bank (SB) is located on the side of Shizuoka (Eurasia Plate) in the central part of Suruga Bay (Fig. 1b), and on the east side of it, steep slopes ranging from water depth of 50 m to water depth of 1800 m are developed. On this steep slope, Multi-stepped topography

are developed, and the existence of a thrust faulting due to subduction is presumed.

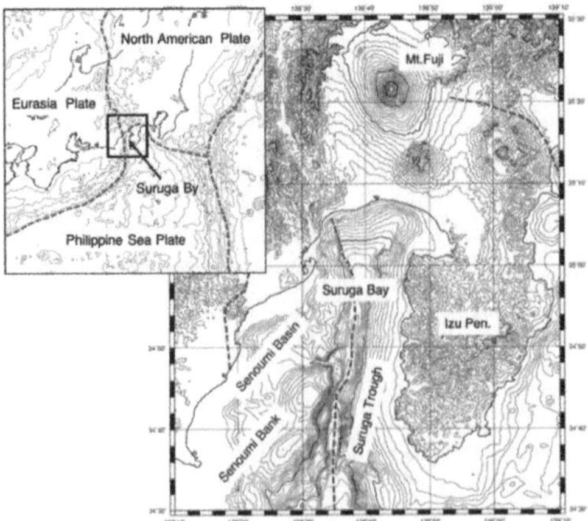

Sandstone and conglomerate are collected from the eastern side slope of Senoumi-bank. Many of these conglomerate rocks are fractured and it is presumed that shear or splitting failure by fault movement occurred due to seismo-geological conditions. In this paper, we have discussed the origin of fractured gravel from the geomorphological, geological features and sediment descriptions around the Senoumi-bank.

2 TOPOGRAPHIC FEATURES OF THE SENOUMI-BANK (SB)

The SB located near the center of Suruga Bay is the twin mountain with flat mountain peaks (Northern SB: -50 m · South SB: -170 m). The N-SB has a conical shape, but S-SB has a block shape with axes in the northeast southwest direction. On the western side of the SB, a Senoumi basin with a maximum depth of about 900 m spreads (Fig. 2).

2.1 *West side slope of SB*

In the northern SB, a gentle slope (800/6000 m) has developed across the Senoumi basin located west of the summit (Fig. 2). On the western side slope of the SB, a horseshoe shaped concave terrain (width 3000 m, length 6000 m) has developed from water depth 250 m to 800 m, and these are presumed to be landslide marks caused by the Ansei earthquake (Magnitude 8). Due to this landslide, a tsunami was caused at Yaizu city in the west side of Suruga Bay.

On the S-SB, steep slopes (200/1000 m) in the northwest direction are developed from -100 m to -300 m in depth, but they change to gentle (300/7000 m) slope at depths of -300 m or more (Fig. 2). On the

Figure 1. Location of Suruga Bay and topography detailed.

Figure 2. Location of sampling and seabed survey lines.

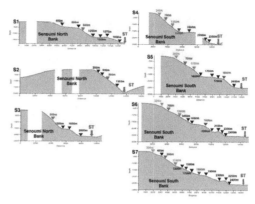

Figure 3. Seabed profiles along survey lines and sampling locations.

steep slope at the water depth of -100 m to -300 m on the west side of the S-SB, several gullies developed. However, they have not developed on gentle slopes deeper than that.

2.2 East side slope of the SB

On the northeastern slope of N-SB, three distinct flat planes are developed at the top of the mountain at a depth of less than 100 m, a water depth of 400 m to 600 m, and a water depth of 1100 m (Fig. 3: S1). In addition, steep slopes develop around the flat surfaces at depths of 100 m to 400 m, depths of 600 m to 1100 m, and depths of 1200 m to 1600 m, and stepwise terrains are developing. In addition, steep slopes develop around the flat surfaces at depths of 100 m to 400 m, depths of 600 m to 1100 m, and depths of 1200 m to 1600 m, and step-like terrains are observed. On the eastern side slope of the N-SB, a steep linear slope (1700 m/4000 m, Fig. 3: S2. 3) is observed from the top of the mountain (depth of about 100 m) to the bottom of the Suruga trough (depth of about 1800 m). Also, on the southeastern slope of N-SB, a steep slope develops around the flat surface around the water depth of 1550 m to 1600 m (1700 m/4000 m, Fig. 3: S 3). For the southeastern slope in Mt. Fuji, the topography change point is confirmed at water depths of 300 m, 800 m, 1150 m, 1500 m, 1800 m (Fig. 3: S4 - S7). Especially in the vicinity of the east side slope, a flat surface is observed at water depths of 2200-2300 m, 1400-1800 m, 700 - 1200 m.

On the east side slope of the SB, a flat surface is formed around the water depth of 1400-1800 m (called "upper flat surface": confirmed at the N-SB and S-SB) and near the water depth of 2200-2300 m (called "lower flat surface": confirmed at the S-SB), indicating the staircase topography.

3 GEOLOGICAL FEATURES OF SB

Dredge surveys at 8 sites (D1 - D8) was conducted around the SB (Fig. 2). From dredge surveys, fractured siltstone to black sandstone (volcanic) were collected around the flat upper surface of the northern eastern slope (Fig. 4, D6: water depth of 1800 m). Cliffs due to crushed sedimentary rocks were found also from the upper flat surface of the SB (Fig. 4, D8: depth of 1400 m), black sandstone (fine grained sandstone ~ coarse sandstone) and grayish siltstone (inclusive trace fossil) were collected.

On the other hand, on the upper slope of N-SB, a cliff with a huge amount of crushed gravels was observed (Fig. 4, D5: depth of 1000 m), and a large amount of gravel was collected. From the D7 site on the Northeast slope of the S-SB, gravels up to 30 cm has been collected.

4 MORPHOLOGICAL CHARACTERISTICS OF COLLECTED GRAVEL

A large amount of gravels (2 cm to 20 cm) was collected from the upper slope of the SB (Fig. 4, D5: depth of 1000 m). These gravels are classified into 1) circular gravel type, 2) broken rounded gravel type, 3) angular type, and 4) ruptured gravel type on the

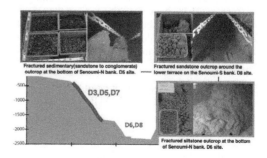

Figure 4. Views of sampling and seabed outcrops.

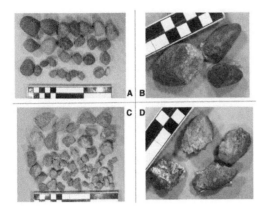

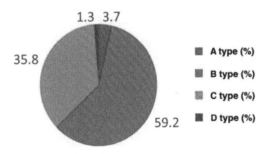

Figure 5. Views of broken boulders.

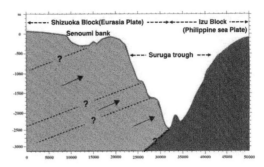

Figure 7. Tectonic model for the crashing of boulders in Suruga Bay.

1.3 3.7

35.8

59.2

- A type (%)
- B type (%)
- C type (%)
- D type (%)

Figure 6. Percentage of sampled boulder classes.

basis of their morphological features (Fig. 5). The rounded gravel type had no corners, had a spherical shape to an elliptical shape, no deformation such as fracture was observed, and it was a 3.7% sample in the gravel sample (5847 pieces) at the D5 site (Fig. 6). Although the broken rounded gravel type shows a circular congruent shape on a part of the surface, a straight fracture surface is also observed in the same sample (Fig. 5-B). At the D5 site, 59.2% of samples belongs to this type. For the angular type, all surfaces are square gravel and rounded surfaces are not observed (Fig. 5-C), and 35.8% belong to this type at the D5 site (Fig. 6). In the fractured gravel type, jigsaw cracks are formed on the surface, and solidified fine gravel matter is confirmed in the conjugate fracture, and 1.3% belongs to this type in the D5 site (Fig. 6). In addition, about 96% of the gravels collected from the upper slope of N-SB is a gravel that develops fresh rupture marks.

5 TOPOGRAPHY AND GEOLOGICAL FEATURES OF SB AND FORMATION OF RUPTURED GRAVEL

In the surrounding areas of the Suruga Trough and the Nankai Trough, huge earthquakes occur every several hundred years, and interest in earthquake

disaster prevention in Shizuoka prefecture is quite high. On the east slope of the SB, a stepped terrain consisting of a combination of a steep slope and a flat surface has been developed. Generally, thrust faults accompanying accretionary complexes have developed in continental slope areas of subduction zones such as the Nankai Trough (Shimamura, 1986), and the multi-stepped topography has developed. In the Suruga Trough, the Philippine Sea Plate is subducting obliquely (northwestward) relative to the Eurasian Plate, so that the thrust fault develops in the Eurasian side basement, and it is estimated that the multi-stepped topography developed on the SB slope (Fig. 7). 92% of the gravels collected from the eastern slope of the SB are sedimentary rocks and 8% volcanic rocks are included (Fig. 6), similar to the gravel description from the slope of the SB north side (Shiba et al., 1991) It is presumed that it was supplied from the Abe River. About 96% of the gravels collected from the upper slope of N-SB are the gravels with fresh rupture marks. About 70% of breaking gravels were collected from the slope of the northern slope of Mt. Fuji (Izu, 1991), and also on the slope of the southern upper slope (D7). Therefore, it was revealed that the breaking gravel was distributed over a wide area on the slope on the east side of SB. Furthermore, crushed basement rocks outcrops have been confirmed in the lower east slope area of the SB (Fig. 2). From this, the following process concerning the cause of fractured gravel is presumed. 1) In the vicinity of the SB, gravels fed from the Abe River accumulated. 2) The multi-stepped topography developed due to the thrust faulting activity accompanying the formation of the Suruga trough. 3) The sedimentary rocks, which constitutes the slope, were deformed due to the fault movement, and the gravels were fractured. (Fig. 7).

These broken gravels are considered to record past fault (earthquake) activity. From now on, we plan to conduct detailed analysis of topographical data and compression tests of gravels, and investigate past and present tectonics of this region and rock breaking mechanism.

6 CONCLUSION

Precise topography and bottom sediment sampling of the sea area around the SB were carried out. As a result, it was revealed that the multi-stepped terrain consisting of the steep slope and the flat surface developed, and crushed sedimentary rocks are distributed on the east slope of the SB. Since 90% of the gravels collected from the upper slope are fractured gravels, it was presumed that the multi-stepped terrain was formed by thrust faulting that characterize the continental slope area, and the fractured gravel was formed by faulting activity.

REFERENCES

Izu S. (1991): Geological significance of the sheared gravels dredged from the Senoumi-Bank, Suruga Bay. Proceedings of the Tokai University Junior Colleges, 25, 1–10. (in Japanese with English abstract)

Shiba M., Izu S., and Nemoto K. (1991): Origin of the Gravels on the North Senoumi Bank in Suruga Bay. Uplifting of the Fossa Magna, Monograph, 38, 11–18. (in Japanese with English abstract)

Shimamura K (1986): Topography and Geological structure in the Bottom of the Suruga Trough. Journal of Geography, 95, 1–22. (in Japanese with English abstract)

2019 Rock Dynamics Summit– Aydan et al. (eds)
© 2019 Taylor & Francis Group, London, ISBN 978-0-367-34783-3

Tectonics and crustal stresses in Yatsushiro Sea and its relation to the causative faults of the 2016 Kumamoto earthquakes.

M. Yagi
GEOSYS. Inc

I. Sakamoto
Tokai University, School of Marine Science and Technology, Tokyo, Japan

Ö. Aydan
University of the Ryukyus, Dept of Civil Engineering, Nishihara, Japan

ABSTRACT: Seismic reflection survey is used as an effective method to evaluate the activity of submarine active faults. By combining sediment core samples, activity intervals and average displacement rate are clarified with accuracy comparable to that of onshore trench survey. However, seismic reflection survey sees the underground information as a two-dimensional profile in the vertical direction. So, it is difficult to detect the displacement of strike-slip fault as compared with the normal fault or the reverse fault. The authors have been studying the detection of horizontal displacement at high density of high- resolution geologic exploration using sub bottom profiler at the Yatsushiro-sea which located southwest section of Hinagu fault zone in western part of Kyushu. In this paper, we consider the geological structure of this area by using the data obtained so far. Also, in the 2016 Kumamoto earthquake, a part of the fault zone was active, and an earthquake of Mw 7.0 scale occurred, and the epicenter distribution expanded in the NE-SW direction. In addition to these aftershocks, surface deformation is observed along with the main shock. As a result of comparing such observed data with the data obtained by exploration of the marine area and examining the relationship, although there is a difference in scale between the two structures, morphological similarity is recognized.

1 INTRODUCTION

On April 14, 2016, an earthquake of Mw 6.2 occurred at a depth of about 10 km in Kumamoto Prefecture, and this earthquake was named "The 2016 Kumamoto Earthquake" by the Japan Meteorological Agency (Fig. 1). Thereafter, Mw 6.0 and Mw 7.0 earthquakes occurred in the same area. The earthquakes of Mw 6.2 and Mw 6.0 are estimated to be due to the Hinagu Fault Zone while the Mw 7.0 earthquake is due to the activity of the Futagawa Fault Zone (HERP, 2016). The epicenter distribution after the Kumamoto earthquake in 2016 spreads along both fault zones, but noticeable activities are not observed in the middle part of the Hinagu Fault Zone to the southwestern part. From now on, it is necessary to understand the geological structure including the marine area in the southwest section in order to evaluate the activity and linkage of the Hinagu Fault Zone. Yagi et al. (2016) performed high-resolution seismic survey to reveal the activity history of the southwest section of Hinagu Fault Zone and also, challenged three-dimensionalization using total 277 reflection profile records in the Yatsushiro-sea which the junction of the central section to the south-west section of the Hinagu Fault Zone. As a

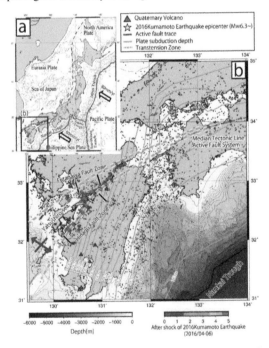

Figure 1. Plate boundary and subduction velocity surround Japan(a), Tectonic setting of Shikoku to Kyushu(b).

result, it is pointed out that there were, at least, four activities since the last glacial period in view of the presence of ridge-like rise, channel-shaped features, local dragging deformation of sedimentary layer, etc. developed along the seabed active fault.

In this paper, based on the results of Yagi et al. (2016), we will examine the horizontal deviation and examine the geological structures of this area and their relation to the seismicity of this area.

2 GEOLOGICAL SETTING OF KYUSHU AND 2016 KUMAMOTO EARTHQUAKES

The Kyushu area is characterized by east-west transpression accompanying the subduction of the Philippine Sea plate in the northwest direction and transtension of Beppu-Shimabara Graben in central Kyushu to the north and south. In Shikoku and Kyushu, Ikeda et al (2009) pointed out pull-apart basin formation by step-over along Median Tectonic Line Active Fault System (MTLAFS), and almost MTLAFS stress condition has a transpression sense.

The Hinagu Fault Zone is in west-central Kyushu, strikes NE–SW, and extends for a total distance of approximately 81 km beginning southeast 7.5 km from the Kumamoto to the Yatsushiro-sea (ERC, HERP, 2013 b; Fig. 2b).

The Futagawa Fault Zone lies north of the Hinagu Fault Zone and it strikes ENE–WSW from the western side of Mount Aso to the Uto Peninsula and has a total length of 64 km or more. The Hinagu Fault Zone is divided into three active segments: northern section (Takano–Shirohata segment), Central

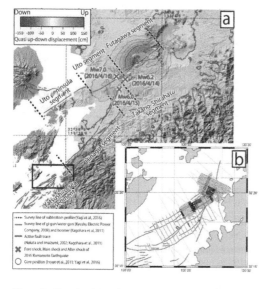

Figure 2. Foreshocks, mainshocks and ground deformations of the 2016 Kumamoto earthquake along the Hinagu Fault Zone and Futagawa Fault Zone (a). Exploratory survey conducted at the extension of the Hinagu Fault Zone (from Yagi et al., 2016) (b).

section (Hinagu segment), and southern section (Yatsushiro- sea segment).

The Futagawa Fault Zone is likewise divided into three active segments: Futagawa segment, Uto segment, and Uto Peninsula north coast segment. Among these active segments, many of ruptures generated during the 2016 earthquake were associated with the surface earthquake faults in the Futagawa Fault Zone (Futagawa segment) and the Hinagu Fault Zone (Takano–Shirohata segment and part of the Hinagu segment). In Mashiki town, Douzon (11km east from Kumamoto), a maximum right-lateral strike-slip displacement of 2.2 m was observed at the surface earthquake fault along the Futagawa fault, and a vertical displacement of about 2.0 m was observed at the Idenokuchi fault. Slip partitioning occurred with small vertical displacement and prevalent lateral displacement in the former case and predominately vertical displacement in the latter case (Toda et al., 2016). In addition, immediately following the Kumamoto earthquake in 2016, surface deformation measurements were carried out using Synthetic Aperture Radar (Daichi No. 2 ALOS/PALSAR) (Himematsu and Furuya, 2016; Fig. 2): A subsidence that fans out in an arc-shape on the NW of the Futagawa fault was identified.

3 METHOD AND DATA

In the seismic reflection survey method targeting the seabed active fault of the sea area, the activity history has been estimated by interpolating the dating values by the sediment cores sampled with two-dimensional reflection recording profiles and fault. In addition, Clarke et al (1985) and McNeilan et al (1986) attempted to detect horizontal displacement by multi-channel seismic reflection survey using boomer source and sub bottom profiler and numerous coring in the waters. In recent years, a 3D exploration system for coastal areas has been developed (e.g. Matsumura et al., 2013; Murakami et al., 2016), and it has been found that channels under the ocean floor are recognized by high resolution imaging and its flexion (For example, Soto and Escalone, 2007).

In the Yatsushiro-sea, Kyushu Electric Power Co., Inc. (2007) clarified the geological structure up to the Quaternary basement. Kagohara et al (2011) revealed geological deformation after 20 Ma. Yagi et al (2016) revealed detailed stratification of upper Pleistocene (after 20,000 yBP). In addition, sedimentary sample with a piston core (Inoue et al., 2011; Yagi et al., 2016) has been carried out, and the aged values of sediment by the 14C dating of dozens of points are calculated to be in Upper Pleistocene (Fig. 3). According to Yagi et al. (2016), from a total of 277 reflection records obtained by the sub bottom profiler, a reflecting surface, which can be tracked in a wide range with comparatively strong amplitude and good continuity, is selected and XYZ

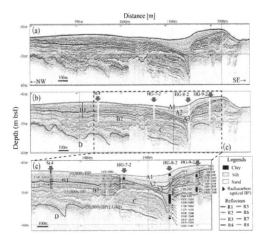

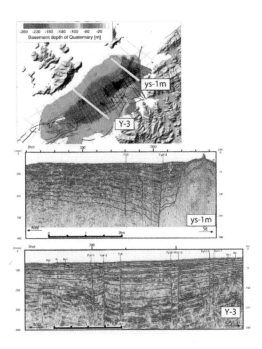

Figure 3. Seismic reflection profile (14AL-17) with NW-SE direction. (a) is core description and radio carbon ages of HG7-2, 8-2, 9-2 (Inoue et al., 2011) and St.4 (Yagi et al.,2016).

(Position, Depth) three-dimensional coordinate data are extracted.

Figure 4. Contour map of quaternary basement depth by reflective recorded section of Kyushu Electric Power Co. Ltd (2008).

4 EXTENSIONAL FEATURES OF YATSUSHIRO SEA

In Yatsushiro-sea, Kyushu Electric Power Co., Ltd (2007) conducted seismic reflection surveys using water-gun and air-gun. According the data, the Quaternary basement of the Yatsushiro-sea forms a sedimentary basin at the extension of the Hinagu Fault Zone, and towards the south the depth of the basement becomes shallow (Fig. 4). The presence of numerous submarine active faults of NNE-SSW or NE-SW direction has been recognized in this area, and the displacement of Holocene deposit is mainly concentrated in the north area. (e.g. Kagohara et al., 2011; ERC, HERP, 2013). Yagi et al. (2016) studied the distribution of these faults in more detail by high-density exploration mainly in the northern sea area. According to these, in the northern part of the Yatsushiro-sea, it mainly consists of faults extending in the NE - SW direction and, faults extending to the SE side and multiple sub-faults formed between them. In the south, the normal fault group with vertical fault displacement is dominant, but the amount of displacement is smaller than that in the northern part. Yagi et al. (2016) pointed out the drag deformation in the vicinity of the FA3 fault (a range of about 200 m). And, on the southwest side of the deformation zone, the existence of the channel-shaped features was revealed. These have a width of 50 m and arranged at an 200 m interval. (Fig. 5).

Fig. 6 shows the reflective surfaces R1 to 5 and the surface of the current seafloor created using XYZ data based on Horizon picking by Yagi et al. (2016). On the reflecting surface R2, an increase is

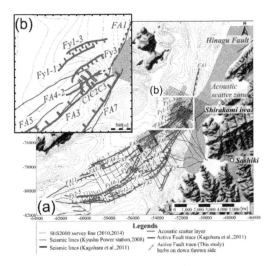

Figure 5. Positional relationship of fault distribution and channel-shaped features where displacement was recognized in the target area (Modified from Yagi et al., 2016).

observed near the fault, and this rise is cut by three channel-shaped buried landforms. These features extend generally in the NW-SE direction and oblique to the FA 3 fault. For convenience, these features are tentatively named C1 to C3 in order from the west. C1 to C3 bend in the east direction across the FA4 fault, and its extension direction changes toward the northwest margin. These channel-shaped features have the following features. 1) It is not clearly recognized inside the D layer, it is clearly observed in the B1 layer from the formation of the C layer, but it is not recognized on the current seafloor. 2) It is formed on the upper surface of the C layer, and thereafter also exists in the same place for a certain period, suffering a cumulative displacement. Therefore, assuming that C1 to C3 cumulatively suffered horizontal deviation, the horizontal features deviation amount was measured by extending features on both sides of the FA4 fault (Fig.7). As a result, C1 About 27 m for C2, about 56 m for C2, about 33 m for C3.

At the surface of about 20,000 yBP, subsidence area spread of about 2 km was observed. In this settling area, the depth on the lower side was averaged and the difference was calculated from the uniform surface (-48.5 m), and this was taken as the total subsidence amount (Fig. 7). However, it should be noted that this value does not indicate the actual displacement of the fault displacement. As a result of calculating the displacement through the surface, subsidence of maximum 9 m was estimated. Also,

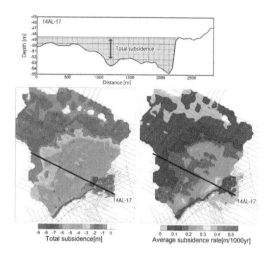

Figure 8. Amount of total subsidence and subsidence rate of the surface formed by 20,000 yBP.

the total subsidence amount from 20,000 yBP to the present time was calculated to be 0.45 m/kyr at the maximum when the average subsidence rate was considered.

The crustal deformation amount based on the interference SAR measured by ALOS 2 immediately after the Kumamoto earthquake was 1.5 m at the maximum, and it was harmonious with the displacement amount observed in the sea area from the ratio with the fault length. Also, in the case of the 2016 Kumamoto earthquake, the Hinagu fault occurs in the vicinity of diagonally intersecting with the Futagawa Fault Zone, whereas it branches towards the south in the marine area, and similarities were observed.

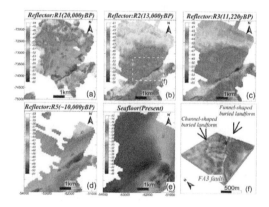

Figure 6. Results of 3D mapping from Last Glacial Maximum to current seabed (a-e). Each surface formation ages were estimated from 14C dating result near the reflectors. Reflector R1 to R4, the development of a subsidence area spreading in an arc circular shape on the northwest side of the FA3 fault is recognized. Three channel-like buried landform growing in the vicinity of the FA3 fault (f).

5 DISCUSSIONS AND CONCLUSIONS

On the bases of investigations presented in the previous sections, the following findings can be summarized including some discussions:

1. With respect to the estimation of the horizontal offset, measurement was carried out using the channel-shaped features recognized at the surface of about 13,000 yBP as a displacement reference. The result was about 27 m for C1, about 56 m for C2, about 33 m for C3. Considering the features formed on the surface about 13,000 years ago, the average displacement rate is calculated to be 2.1 to 4.3 m/kyr. However, the average displacement rate greatly differs from the Hinagu fault of on-land section (0.7 m/kyr; ERC, HERP, 2013). In the Kumamoto earthquake in 2016, a horizontal offset of about 2.2 m at the maximum was observed, and if it was 2.2 m in a single activity, in order to reach 27 to 56 m, at least 13 activities or more are required. In that case, the average activity interval is about 1000 years, and the interval estimated in the

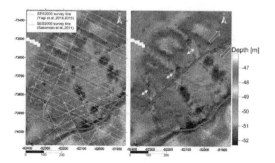

Figure 7. Three channel-shaped features of reflector R2 (formed about 20,000 yBP).

Yatsushiro-sea in the past is 1,100 to 6,400 years according to ERC, HERP (2013), 5,445 (Yagi et al., 2016) and 3,584 (Yagi, 2017), which implies that the fault can be categorized as a very active fault. However, the confirmed length of the FA4 fault that horizontally displaces the features terrain is short, and it is unlikely that it will result in activities of this scale.

2. The average subsidence rate calculated based on the subsidence region recognized about 20,000 years ago is 0.45 m/kyr at the maximum. In addition, this subsidence area occurs before the fault branches toward the south, and there is a morphological similarity although there is a difference in scale with that observed on ground surface at the time of the 2016 Kumamoto earthquake.

3. Ikeda et al (2009) pointed out the formation of sedimentary basins along the MTLFAS along the step-over of the fault. It is also considered that the stress condition changes to trans-tension from Shikoku to the Okinawa trough (Ikeda et al, 2009). The formation of numerous normal faults and sedimentary basins, which is recognized in the Yatsushiro-sea, is thought to be caused by such wide-area tectonic setting.

4. In this paper, we attempted to detect the horizontal offset displacements using a compact method compared with conventional 3D exploration but calculated a larger value than the average displacement rate estimated in the existing Hinagu fault. However, by creating the surface of the reflecting surface, we could estimate the average subsidence rate as 0.45 m/kyr maximum. The shape of the subsidence area recognized on the surface of about 13,000 yBP and 20,000 yBP was different from the ground deformation observed at the 2016 Kumamoto earthquake but morphological similarity was observed. In the coastal area, it is difficult to apply 3D exploration technique to explore faults because of difficulties such as fishing gear and shallow water depth. Therefore, this method is expected to be an effective tool as a method to evaluate the underground geologic structures in the shallow water region in the future.

REFERENCES

Clarke, S. H., Greene, H. G., Kennedy, M. P.1985. Identifying potentially active faults and unstable slopes offshore. In Evaluating Earthquake Hazards in the Los Angeles Region-An Earth-Science Perspective (J. I. Ziony, ed), U.S. Geol. Surv. Prof. Pap. 1360, 347–373.

Earthquake research Committee, Headquarters for Earthquake Research Promotion, 2013, Long-term evaluation of the Futagawa and Hinagu fault zones. http://jishin.go.jp/main/chousa/katsudansou_pdf/93_futagawa_hinagu_2.pdf, 66p (in Japanese).

Earthquake research Committee, Headquarters for Earthquake Research Promotion, 2016, Evaluation of the 2016 Kumamoto Earthquake. 1–23 www.static.jishin.go.jp/resource/monthly/2016/2016_04.pdf, 234823p (in Japanese).

Himematsu, Y., Furuya, M., 2016, Fault source model for the 2016 Kumamoto earthquake sequence based on ALOS-2/PALSAR-2 pixel-offset data: evidence for dynamic slip partitioning. Planets and Space, doi:10.1186/s40623-016-0545-7.

Ikeda, Michiharu., Toda, shinji., Ohno, Yuki., Nishizaka, Naoki., 2009. Tectonic model and fault segmentation of the Median Tectonic Line active faylt system on Shikoku, Japan. Tectonics, vol.28, doi:10.1029/2008TC002349.

Inoue, N., Kitada, N., Echigo, T., Kubo, T., Kazui, N., Hayashida, A., Sakamoto, I., Takino, Y., Kagohara, K., 2011, Piston coring survey of the Futagawa-Hinagu Fault Zone, Yatsushiro Sea, southwest Japan. Annual Report on Active Fault and Paleoearthquake Researches, 11, 295–308 (in Japanese).

Kagohara, K., Aiko, T., Adachi, I., Sakamoto, I., Takino, Y., Inoue, N., Kitada, N., 2011, High-resolution multi-channel seismic reflection imaging of the Futagawa-Hinagu Fault Zone, Yatsushiro Sea, southwest Japan. Annual Report on Active Fault and Paleoearthquake Researches, 11, 273–294 (in Japanese).

Kyusyu Electric Power Company, 2008, Handout of Nuclear Safety Commission Earthquake and Seismic motion evaluation committee and Institution soundness evaluation committee "WG3 No.40-5 Geology and Geological structure of Sendai Nuclear Power Station surrounding site/vicinity site: part 2". http://www.nsc.go.jp/senmon/shidai/taishin_godo_WG3/taishin_godo_WG3_40/taishin_godo_WG3_40.Htm (in Japanese).

Matsumura, K., Kamoshita, T., Miyamoto, K., 2013. P-cable: High-resolution 3D Marine Seismic Acquisition System - Feature and Applicability to New Fields-. OYO corporation technical report, No.32, 77–87(in japanese).

McNeilan, T. W., Rockwell, T. K., Resnick, G. S., 1996. Style and rate of Holocene slip, Palos Verdes fault, southern California. J. Geophys. Res. 101(B4), 8317–8334.

Murakami, F., Furuya, M., Kochi, Eijiro., Maruyama, K., Hatakeyama, K., Takeda, N., Sato, M., Baba, H., 2016. Development of the high-resolution three-dimensional seismic survey system for shallow water and the survey of active fault in the nearshore waters of the northern Suruga Bay using the system. Active Fault Research, No.44, 40–59(in japanese).

Shirahama, Y., Yoshimi, M., Awata, Y., Maruyama, T., Azuma, T., Miyashita, Y., Mori, H., Imanishi, K., Takeda, N., Ochi, T., Otsubo, M., Asahina, D., Miyakawa, A., 2016, Characteristics of the surface ruptures associated with the 2016 Kumamoto earthquake sequence, central Kyushu, Japan. Earth, Planets and Space, doi:10.1186/s40623-016-0559-1.

Soto David, M., Mann, Paul., Escalona, Alejandro., Wood, Leli, J., 2007. Late Holocene strike-slip offset of a subsurface channel interpreted from three-dimensional seismic data, eastern offshore Trinidad, Geology, vol. 35, Issue 9, 859–862.

Toda, S., Kaneda, H., Okada, S., Ishimura, D., Mildon, K., Z., 2016, Slip-partitioned surface ruptures for the Mw 7.0 16 April 2016 Kumamoto, Japan, earthquake. Earth, Planets and Space, doi:10.1186/s40623-016-0560-8.

Yagi, M., Sakamoto, I., Tanaka, H., Yokoyama, Y., Inoue, T., Mitsunari, K., Aydan, O., Fujimaki, M., Nemoto, K., 2016, Identification of faulting history of active faults in coastal area using high-resolution seismic survey and piston coring-A case study on the offshore extension of the Hinagu Fault Zone in the Yatsushiro Sea-. Active Fault Research, 45, 1–19 (in Japanese).

Yagi, Masatoshi, 2017, Application of the 3D geological structure analytical technique using seismic survey to sea section in the Hinagu Fault Zone. https://opac.time.u-tokai.ac.jp/webopac/TD00000231 (in japanese).

Wessel, P., W. H. F. Smith., 1998. New, improved version of the Generic Mapping Tools released, EOS Trans. AGU, 79, 579.

2019 Rock Dynamics Summit– Aydan et al. (eds)
© 2019 Taylor & Francis Group, London, ISBN 978-0-367-34783-3

Prediction of near fault ground motion by dynamic rupture simulation

M. Yamada
NEWJEC Inc., Osaka, Japan

R. Imai & K. Takamuku
Mizuho Information & Research Institute, Inc., Tokyo, Japan

H. Fujiwara
National Research Institute for Earth Science and Disaster Resilience, Tsukuba, Japan

ABSTRACT: The dynamic fault rupture simulation was conducted using the three dimensional finite dif-ference method without giving a priori rupture starting area and rupture stopping area by changing the coef-ficient of friction or changing the frictional constitutive law. It was conducted on condition that the shear rigidity was changed along the fault plane and that the shear stress on the fault plane was loaded by forced displacement. In this study, induced shear stress in the fault layer by the forced displacement was larger at the center part because of higher shear wave velocity and stress was smaller at the outskirts part because of lower shear wave velocity. The ununiform shear stress distribution caused spontaneous fault rupture starting from the central part in spite of constant frictional condition in this layer. The fault rupture was spontane-ously stopped after the rupture spread up to around 10km in the x-direction without giving a priori rupture stopping area by frictional condition. It was consistent with previous studies in the fault rupture process that the rupture velocity was 3~3.5km/s at most. It was not incompatible that the stress drop was about 2MPa. We could estimate the near fault ground motion for our model fault by using this dynamic rupture simulation. The slip velocity time function directly calculated by the dynamic fault rupture simulation was consistent with previous study.

1 INTRODUCTION

In recent years, the dynamic fault rupture simulation is used widely for the solution of the physical phe-nomenon of the fault dislocation and so on.

Tsuda (2016) tried to reproduce the rupture pro-cess of mega-thrust earthquakes such as the 2011 off the Pacific coast of Tohoku Earthquake. Kase (2016) tried to explain the source process of the interseg-mental rupture propagation for the 2014 northern Nagano earthquake. Kase et al. (2002) simulated the earthquake rupture process on the Uemachi fault system. Irie (2014) explored to be clear for the source properties of large inland strike-slip faults for strong motion prediction based on dynamic rup-ture simulation. However, in these simulations, the initial rupture area that was given smaller static fric-tion coefficient and the rupture stop area in which rupture was not permitted were indispensable.

Kame (1997) developed the calculation method of the analysis of spontaneous rupture growth with geometrical complexity. But it was different from our study that Kame (1997) assumed an initial fault plane having a length of critical crack.

In this study, the dynamic fault rupture simula-tion was conducted without giving a priori rupture starting area and rupture stopping area by changing the coefficient of friction or changing the frictional constitutive law. It was conducted on condition that the shear rigidity was changed along the fault plane and that the shear stress on the fault plane was loaded by forced displacement.

2 PROCEDURE AND CONDITION OF THE DYNAMIC FAULT RUPTURE SIMULATION

2.1 *Simulation model*

Three-dimensional finite difference method (Kase 2010) was used for the dynamic fault rupture simulation.

A rectangular parallelepiped model was pre-pared with 60km long for x-direction, which was strike direction of the fault plane, 20km long for y-direction, which was orthogonal to strike direc-tion of the fault plane and 8km high for vertical direction z. The model was assumed horizontal lay-ered structure with 4 layers and thickness of each layer was 2km. The second layer was the fault plane layer. The first layer was surface layer above the upper end of the fault plane layer. The third layer was the lower from the bottom of the seismogenic layer. The fourth layer was the transition layer to the viscous bound-ary of the bottom of the model which didn't dislocate. Simulation model section (x-z section) is shown in Figure 1.

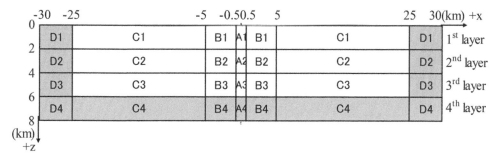

Figure 1. Simulation model section (x-z section) with 4 layers. Rupture is not permitted in gray shaded areas.

2.2 *Boundary condition and simulation model properties*

The top of the model was the stress-free boundary, and the bottom and the both edge of x-direction were the viscous boundary condition. The forced shear displacement was given at the edge of y-direction and induced the shear stress in the model body.

For the second layer of the fault plane layer, the 20% higher shear wave velocity was given at central part "A2" of x=−0.5km~x=+0.5km than the velocity for standard bedrock such as neighbor parts "B2". The induced shear stress by the displacement at the central part was larger than other parts. On the other hand, the 20% lower shear wave velocity was given at outskirts part "C2" of x>±5.0km for the fault plane layer. The static friction coefficient was set constant in each layer. Detailed model properties for each area were shown in Table 1.

3 RESULT OF THE DYNAMIC FAULT RUPTURE SIMULATION

3.1 *Share stress before rupture start*

Figure 2 shows spatial distribution of share stress before rupture start. It was found that the stable

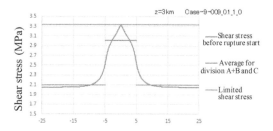

Figure 2. Spatial distribution of share stress before rupture start.

Table 1. Simulation model properties for each area (Rupture is not permitted in gray shaded areas)

Layer	Area	V_p (km/s)	V_s (km/s)	ρ (t/m³)	μ_s	μ_d	D_c (m)	Rupture
1st surface	A1	2.100	0.700	2.00	0.02	0.0	4.58	permitted
	B1	2.100	0.700	2.00	0.02	0.0	4.58	
	C1	2.100	0.700	2.00	0.02	0.0	4.58	
	D1	2.100	0.700	2.00	1	0.0	4.58	not permitted
2nd fault	A2	7.200	4.157	2.67	0.0364	0.0	0.10	permitted
	B2	6.000	3.464	2.67	0.0364	0.0	0.14	
	C2	4.800	2.771	2.67	0.0364	0.0	0.22	
	D2	4.800	2.771	2.67	1	0.0	0.22	not permitted
3nd under fault	A3	7.000	4.000	2.90	0.2	0.0	0.10	permitted
	B3	7.000	4.000	2.90	0.2	0.0	0.10	
	C3	7.000	4.000	2.90	0.2	0.0	0.10	
	D3	7.000	4.000	2.90	1	0.0	0.10	not permitted
4th transition for boundary	A4	7.000	4.000	2.90	1	0.0	0.10	not permitted
	B4	7.000	4.000	2.90	1	0.0	0.10	
	C4	7.000	4.000	2.90	1	0.0	0.10	
	D4	7.000	4.000	2.90	1	0.0	0.10	

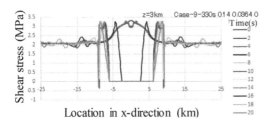

Figure 3. Spatial distribution of share stress after rupture start.

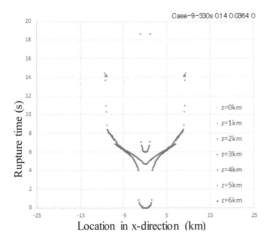

Figure 4. Distribution of rupture time in x-direction.

shear stress of the central part induced by the forced displacement was larger because higher shear wave velocity was given for this area.

3.2 *Share stress after rupture start*

Figure 3 shows spatial distribution of share stress after rupture start. Spontaneous fault rupture was occurred from the central part, although the static friction coefficient was set constant in this layer. The fault rupture starting at the central part was spontaneously stopped after the rupture spread at around 10km in the x-direction. It caused that the induced shear stress by the displacement at the part was smaller because the lower shear wave velocity was given at the corresponding outskirts part for the fault plane layer. In this figure, the stress drop was about 2MPa.

Figure 4 shows distribution of rupture time in x-direction. Figure 5 shows snap shots of x-component dislocation for x-y section at the depth of z=3km from 5s to 10s after rupture start. Figure 6 shows distribution of rupture time on the fault plane. It was found that the spontaneous fault rupture started at the central part was stopped after the rupture spread at around 10km in the x-direction as same as shown in Figure 3.

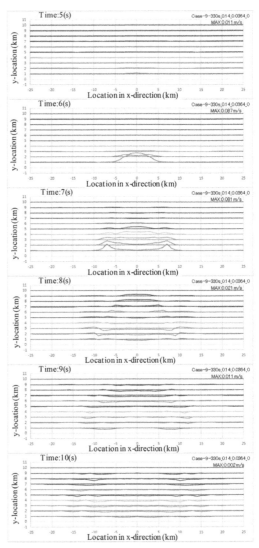

Figure 5. Snap shots of dislocation for x-direction at the depth of z=3km from 5s to 11s after rupture start.

3.3 *Rupture velocity*

Figure 7 shows distribution of rupture velocity on the fault plane. After rupture starts, the rupture velocity got faster up to 3-3.5km/s at around x=±4~7km while propagating and drastically slowed down to stop at x=±9km.

3.4 *Ground motion near the fault line*

Figure 8 shows velocity waveform on ground surface (z=0km) at 0.1km in y-direction apart from the fault line. Y-component of ground motion was larger than x-component of ground motion because rupture was not propagated to the 1st layer. Y-component of ground velocity was especially large at around x=±5~10km.

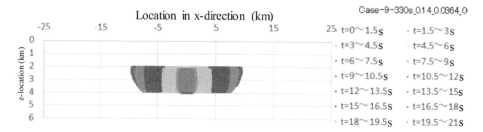

Figure 6. Distribution of rupture time on the fault plane.

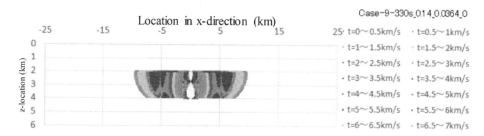

Figure 7. Distribution of rupture velocity on the fault plane.

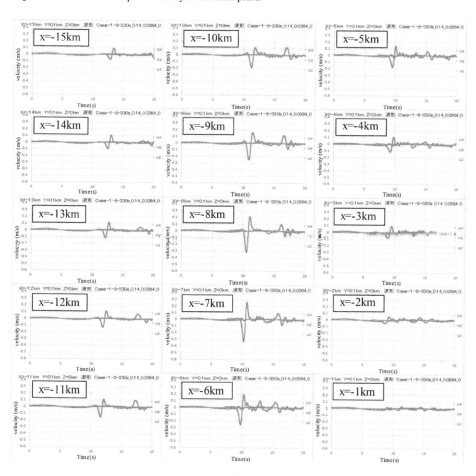

Figure 8. Velocity waveform on ground surface (z=0km) at 0.1km apart from the fault line for y-direction.

276

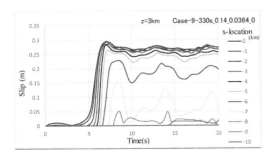

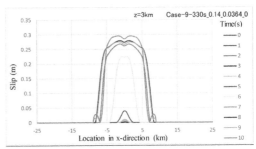

Figure 9. Time history of slip on the fault plane at each location in the x-direction.

Figure 10. Distribution of slip on the fault plane in the x-direction.

3.5 *Slip on the fault plane*

Figure 9 shows time history of slip on the fault plane at each location in the x-direction. Figure 10 shows distribution of slip on the fault plane in the x-direction. It was found that the slip on the fault plane was about 0.25m at the center part between x=-5km and x=+5km in the x-direction.

3.6 *Slip velocity*

Figure 11 shows comparison of slip velocity time function between by this study and by Nakamura & Miyatake (2000). In Figure 11, the slip velocity time function by this study was directly calculated at the points of (x,y,z)=(3,0,3) and (5.5,0,3) by the dynamic fault rupture simulation. The slip velocity time function by Nakamura & Miyatake (2000) was estimated by using rupture velocity, stress drop and slip that was obtained by the simulation of this study at the same points. The point of (x,y,z)=(3,0,3) was corresponding to strong motion generation area (SMGA), both slip velocity time functions were reasonably coincided each other. The point (x,y,z)=(5.5,0,3) was not corresponding to SMGA, slip velocity time function by this study was similar to that by Nakamura & Miyatake (2000) in its shape but was smaller in its peak slip velocity than that.

4 CONCLUSION

The dynamic fault rupture simulation was conducted using the three dimensional finite difference method without giving a priori rupture starting area and rupture stopping area by changing the coefficient of friction or changing the frictional constitutive law. It was conducted on condition that the shear rigidity was changed along the fault plane and that the shear stress on the fault plane was loaded by forced displacement.

In this study, induced shear stress in the fault layer by the forced displacement was larger at the center part because of higher shear wave velocity and stress was smaller at the outskirts part because of lower shear wave velocity. The ununiform shear stress distribution caused that fault rupture spontaneously starting from the central part in spite of constant frictional condition in this layer. The fault rupture was spontaneously stopped after the rupture spread up to around 10km in the x-direction without giving a priori rupture stopping area by frictional condition.

It was consistent with previous studies in the fault rupture process that the rupture velocity was 3~3.5km/s at most. It was not incompatible with previous studies that the stress drop was about 2MPa. We could estimate the near fault ground

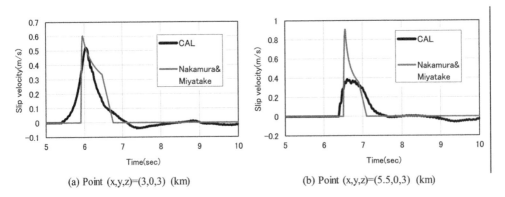

(a) Point (x,y,z)=(3,0,3) (km)

(b) Point (x,y,z)=(5.5,0,3) (km)

Figure 11. Comparison of slip velocity time function between by this study and by Nakamura & Miyatake (2000).

motion for our model fault by using this dynamic rupture simulation. The slip velocity time function directly calculated by the dynamic fault rupture simulation was consistent with previous study.

We intend to conduct this dynamic fault rupture simulation for the more realistic model with which the actual conditions and properties of underground can be illustrated such as model size, stress condition, friction coefficient and so on.

REFERENCES

Irie, K. 2014. Source properties of inland strike-slip faults for strong motion prediction based on dynamic rupture simulation, *Doctoral thesis of Hirosaki University.*

Kame, N. 1997. Theoretical study on arresting mechanism of dynamic earthquake faulting -A new method of the analysis of spontaneous rupture growth with geometrical complexity, *D. Sc. Thesis, the University of Tokyo.*

Kase, Y. 2010. Slip-length scaling low for strike-slip multiple segment earthquakes based on dynamic rupture simulations, *Bulletin of the Seismological Society of America* 100(2): 473–481.

Kase, Y. 2016. Dynamic rupture model of the 2014 northern Nagano, central Japan, earthquake (Part 3), *Japan Geoscience Union meeting.*

Kase, Y. & Horikawa, H. & Sekiguchi, H. & Satake, K. & Sugiyama, Y. 2002. Simulation of earthquake rupture process on the Uemachi fault system, *Annual report on active fault and paleoearthquake researches* 2: 325–340.

Nakamura, H. & Miyatake, T. 2000. An approximate expression of slip velocity time function for simulation of near-field strong ground motion, Zisin2 53: 1–9.

Tsuda, K. 2016. Dynamic Rupture Simulations Constrained by Experimental Data to Understand the Rupture Process of Mega-Thrust Earthquakes, *Technical research report of Shimizu Construction Co., Ltd.* 93: 82–88.

2019 Rock Dynamics Summit– Aydan et al. (eds)
© 2019 Taylor & Francis Group, London, ISBN 978-0-367-34783-3

Source modeling of the mid-scale crustal earthquake by forward modeling using the empirical Green's function method

T. Ikeda
Nagaoka University of Technology, Nigata, Japan

Y. Kojima
National Institute of Technology, Nagaoka College, Niigata, Japan

ABSTRACT: Recently, structural damage caused by the middle-scale earthquake were reported. These damage were not so serious, but middle-scale earthquake is often generated. It is important that study of the source modeling the middle-scale earthquake. In this study, we attempted to make a source model of middle-scale earthquake by the empirical Green's function method. The target earthquake is inland crustal earthquake with Mj5.6 which occurred in southern Nagano prefecture in Japan on 25th June 2017. Resultantly, we proposed source model with single strong motion generation areas located in near hypocenter.

1 INTRODUCTION

Characteristic strong ground motion which include semi-large pulse with a duration of 1 to 3 second was generated in the 1995 Kobe earthquake. It caused serious damage to many civil structures and many building structures. The characteristic ground motion was caused by forward rupture directivity effect from directional characteristics of source rupturing. After this earthquake, research on strong ground motion prediction including source modeling was conducted by many researchers energetically. As a result, "Recipe for strong ground motion prediction" was constructing based on several research findings. Effectiveness of the recipe was verifying by strong ground motion simulation against a large-size earthquake. Recently, structure or lifeline system damages caused by the mid-scale earthquake were reported. Generally, damage scale by mid-scale earthquake was not so serious, but that earthquake is often generated. So it is important that study of the source modeling the mid-scale earthquake. In this study, we attempted to make a source model of mid-scale earthquake by the forward modeling (Kamae and Irikura, 1998) using empirical Green's function method (Irikura, 1986).

2 THE 2017 SOUTHERN NAGANO PREFECTURE EARTHQUAKE

The earthquake of $M_w 5.2$ (Mj5.6) which occurred in southern Nagano prefecture in Japan on 25th June 2017 was selected for target earthquake. The fault plane of this earthquake was estimated a reverse fault type with the strike of NNE–SSW direction from CMT solution (NIED) and after-shocks distribution. Table 1. shows parameter of the southern Nagano prefecture earthquake. Figure 1 shows loca-

tion of the epicenter of the main-shock and after-shocks. These after-shocks were occurred within

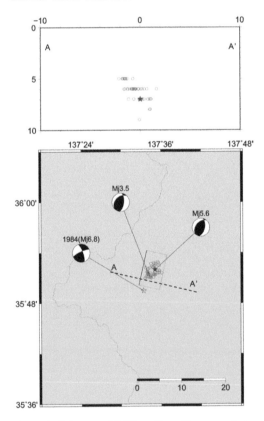

Figure 1. Epicenter of 2017 southern Nagano prefecture earthquake (Mw5.2) and after-shocks distribution which occurred within 24 hours and Mj ≥ 2. Mj6.8 is the earthquake occurred in 1984. Mj3.5 is the earthquake as used for empirical Green's function event.

Table 1. Source parameter of the 2017 southern Nagano prefecture earthquake

Origin time (JST)[a]	2017-6-25 7:02:15.3
Epicenter[a]	35°52.0' N 137°35.1' E
Depth (km)[a]	7
Magnitude	Mj5.6[a], Mw5.2[b]
Seismic moment[b]	$6.89×10^{16}$ Nm
STR/DIP/RAK[b]	219;13;40;53;111;73

[a]JMA, [b]F-net

24 hours from main-shock and its magnitude more than 2.

Strong ground motions were generated near source area in this earthquake and suffered to many houses.

Fire and Disaster Management Agency, Ministry of Internal Affairs and Communications was reported that 27 houses suffered partial damage and many roads was closed by landslide and rock fall.

A large scale earthquake of Mj6.8 is generated at the near epicenter in 1984 and it caused heavy damages in near source area. The epicenter of this earthquake was shown in Figure 1. It is located in very near with epicenter of target earthquake but fault mechanism is deferent.

3 STRONG GROUND MOTION WHICH OBSERVED AROUND SOURCE AREA

Strong ground motions in the wide area including near source area were observed at K-NET, KiK-net and local government seismic observation network. K-NET and KiK-net are seismic observation network of NIED, it was called "Kyoshin network". Figure 2 shows epicenter of main-shock and seismic stations.

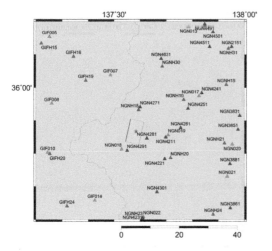

Figure 2. Location of the epicenter of the 2017 southern Nagano prefecture earthquake and seismic observation stations of NIED and local government.

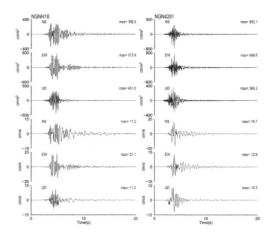

Figure 3. Acceleration and velocity waveform at NGNH18 and NGN4281.

Red triangle mark means KiK-net, green triangle mark means K-NET and blue triangle mark means local government station. Figure 3 shows acceleration waveform and velocity waveform which were observed at NGNH18 and NGN4281. Both records were observed on the ground surface. Velocity waveform was calculated from 0.1-10Hz band-pass acceleration waveform by Fourier integration procedure. NGNH18 was located in the strike direction of the fault plane. NGN4281 was located in the transverse direction of the fault plane.

Duration of principal motion at each waveform were not so long. NGN4281 of the nearest station is about 2 seconds. Pulse wave was included in waveform at NGN018 and NGN4281.

4 SOURCE MODELING

4.1 Method

We attempted making a source model of the 2017 Nagano prefecture earthquake by forward modeling (Kamae and Irikura, 1998) using empirical Green's function method (Irikura, 1986).

In this method, at first we make an initial source model which consist of single or multiple strong ground motion generation area (SMGA). Location and number of SMGA are referred from waveform inversion result. Secondary, parameters of SMGA such as location, size, stress drop and rise time were tuned based on agreement between observation and synthesis. The source model made by this procedure is called SMGA model.

Effectiveness of SMGA model was verified by previous researches (Ikeda et al., 2011) even though SMGA has a rectangular shape and homogenized physical property. In this study, we do not consider backward of source area because the strong ground motion was comprised of only elastic wave generated from SMGA. NGNH18, NGNH20, GIFH24,

GIFH19, GIFH20 and NGNH30 were used as target site that compared synthetics result with observed one. Because to remove the influence of the soil characteristics nonlinearity, we used underground records. Figure 2. shows location of each stations.

4.2 Assuming the fault plane

A fault plane that includes the JMA hypocenter with a length of 7.8 km and width of 7.8 km was assumed by after-shock distribution. The strike and dip angles were set to 13° and 54°, respectively, referring to the F-net moment tensor solution (NIED). Latitude and longitude of reference point of fault plain were 35.836N and 137.545E. Figure 2. shows fault plain.

4.3 Select of the element earthquake to use as empirical Green's function event

The empirical Green's function method (EGFM) synthesize strong ground motion of large earthquake using similarity law between large earthquake and small earthquake. So we have to very carefully select an element earthquake as use empirical Green's function event.

In this study, we selected element earthquake from 14 earthquakes which satisfied following condition.

1. Epicenter: 35.8N - 35.9N, 137.5E - 137.65E
2. Magnitude: Mj3.5 – Mj4.5
3. Focal mechanism of F-net: Evaluated

As a result, earthquake with Mj3.5 were selected as an empirical Green's function event. Figure 1 shows epicenter of element earthquake and focal mechanism of F-net. Table 2 shows source parameter of element earthquake.

Area and stress drop were evaluated by Brune method (Brune, 1970 and Brune, 1971) and circular crack method (Eshelby, 1957). Equation (1) and equation (2) shows Brune method and circular crack method. Corner frequency of element earthquake (f_{ca}) to use Brune method was evaluated by source spectral ratio fitting method (SSRM) by Miyake et al. (1999). SSRM method fit a source spectral function based on ω^{-2} source spectral theory into an observed source spectral ratio and calculate f_{ca}, N and C. Here, N is number of synthesis and C is ratio of stress drop between large earthquake and element earthquake. Equation (3) shows source spectral function (SSRF(f)).

$$r_e = \frac{2.34\beta}{2\pi f_{ca}} \quad (1)$$

$$\Delta\sigma_e = \frac{7}{16}\frac{m_0}{r_e^{3}} \quad (2)$$

Table 2. Source parameter of the element earthquake to use as empirical Green's function event

Origin time (JST)[a]	2017-6-25 9:48:42.7
Epicenter[a]	35°51.5' N 137°34.1' E
Depth (km)[a]	6
Magnitude	Mj3.5[a], Mw3.6[b]
Seismic moment[b]	3.04×10^{14} Nm
STR/DIP/RAK[b]	227;7/37;60/124;67
Corner frequency	2.37Hz
Stress drop	0.8MPa
Area	0.95km²

[a]JMA, [b] F-net

$$SSRF(f) = \frac{M_0}{m_0}\frac{1+\left(\dfrac{f}{f_{ca}}\right)^2}{1+\left(\dfrac{f}{f_{cm}}\right)^2} \quad (3)$$

Here, f is frequency, r_e is equivalent radius (km), β is shear wave velocity of rock basement (km/s), $\Delta\sigma_e$ is stress drop of element earthquake (MPa), M_0 and m_0 are seismic moment of large earthquake and element earthquake respectively (Nm). f_{cm} and f_{ca} are corner frequency of large earthquake and element earthquake respectively.

In this study, we use main-shock to use as large earthquake. Observed source function ratio was calculated by broadband velocity records which observed at KNM, NAA, FUJI, TTO and SRN. Figure 4 shows location of broadband seismic observation stations.

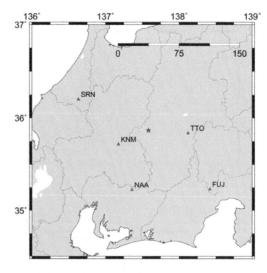

Figure 4. Location of broadband seismic observation station of F-net.

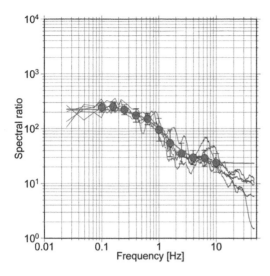

Figure 5. Curve fitting result between observed source spectral ratio and source spectral function.

Figure 5. shows result of curve fitting. Range of curve fitting is 0.1 – 10 Hz. Read curve is SSRF and blue circle are target. We evaluated corner frequency (f_{ca}) to be 2.37 Hz from curve fitting.

4.4 Source modeling

We set single SMGA because observation waveform does not see multi event. As a result of forward modeling, we make a source model with single SMGA which located in near hypocenter. Figure 6 shows source model of the 2007 Nagano prefecture earthquake. Table 3 shows parameters of the source model.

Figure 7 shows the synthesized acceleration waveforms, velocity waveforms and displacement waveform of EW component at each target stations compared with the observed ones. Figure 8 shows the pseudo velocity response spectrum with damping factor 0.05 (response spectrum) and acceleration Fourier spectrum (Fourier spectrum). Effective frequency range is 0.2–10.0 Hz.

Synthesized acceleration waveform and velocity waveform at NGNH18 which is the nearest station were good agreement with observed ones. Especially, synthetic waveform can reproduce pulse waveform. Also response spectrum and Fourier spectrum were good agreement with observed ones.

Table 3.Parameter of the source model

Area	8.56 km² (2.925 km × 2.925 km)
Synthesis number	3 × 3 × 3
depth (upper)	2.925 km
Stress drop	6.4 MPa
Rupture velocity	2.7 km/s
Rise time	0.2 s

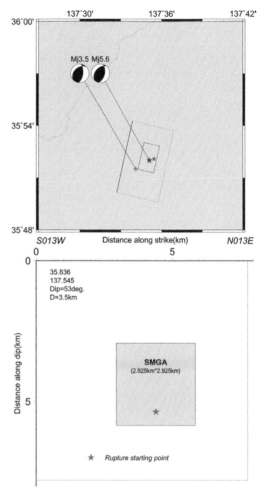

Figure 6. The source model of the 2007 Nagano prefecture earthquake.

Synthesized response spectrum and Fourier spectrum at GIFH24 is underestimated in more than 1Hz range.

Figure 9 shows relationship between seismic moment and short period source spectrum. Short period source spectrum was calculated by equation (4). Scaling law of Dan et al. (2001) was written in this figure. Equation (5) shows scaling law.

$$A = 2.46 \times 10^{10} \times \left(M_0 \times 10^7 \right)^{1/3} \tag{4}$$

$$A = 4\pi r \Delta\sigma_a \beta^2 \tag{5}$$

Here, A is short period source spectrum (Nm/s²), r is radius of SMGA (km) and $\Delta\sigma_a$ is stress drop of SMGA. Shear wave velocity assumed 3.5km/s.

Relation between seismic moment and short period source spectrum can be express scaling law.

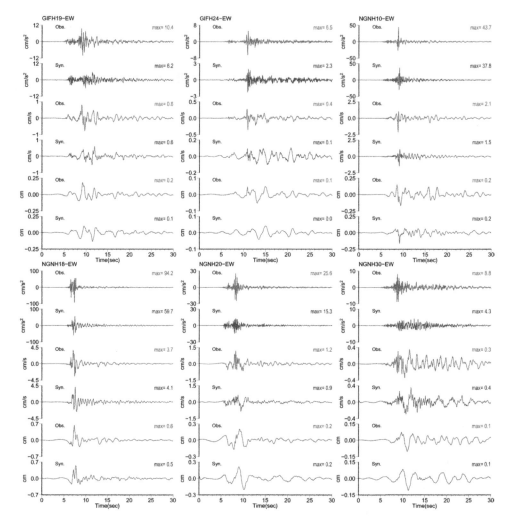

Figure 7.Comparison of the observed and synthetic waveforms of EW component at target station. The red color lines are the observed waveforms.

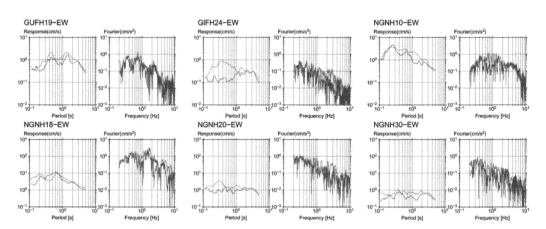

Figure 8.Comparison of the observed and synthetic pseudo velocity response spectrum and acceleration Fourier spectrum of EW component at target station. The red color lines are the observed waveforms.

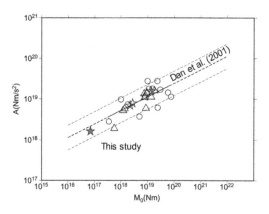

Figure 9. Relation between seismic moment and short period source spectrum.

5 CONCLUSION

We constructed the source model of the 2007 Nagano prefecture earthquake with Mj5.6 by the forward modeling using empirical Green's function method. The source model has a single SMGA. Relation between seismic moment and short period source spectrum can be express scaling law.

ACKNOWLEDGEMENT

The authors appreciate the National Institute of Earth Science and Disaster Prevention Research (NIED), Japan for providing the K-NET and the KiK-net strong motion data and F-net focal mechanism. This study was partially supported by JSPS KAKENHI (Grant Number: 18H01519)

REFERENCES

Brune, J. N. 1970. Tectonic stress and the spectra of seismic shear waves from earthquakes, *Journal of Geophysical Research*, Vol.75, 4997–5009.

Brune, J. N. 1971. Correction, *Journal of geophysical research*, Vol.76, 5002.

Dan, K., Motofumi, W., Sato, T. and Ishii, T. 2001. Short-period source spectra inferred from variable-slip rupture models and modeling of earthquake faults for strong motion prediction by semi-empirical method, *Journal of Structural and Construction Engineering*, AIJ, No.545, 51–62. (in Japanese with English abstract)

Eshelby, J.D. 1957. The determination of the elastic field of an ellipsoidal inclusion, and related problems, *Proceedings of the Royal Society of London. Series A, Mathematical and Physical Sciences*, Volume 241, Issue 1226, 376–396.

Ikeda, T., Kamae, K., Miwa, S. and Irikura, K. 2011. Source modeling using the empirical Green's function method and strong ground motion estimation considering nonlinear site effect, An application to the 2005 west off Fukuoka prefecture earthquake and the 2007 Noto hanto earthquak, *Journal of Structural and Construction Engineering*, AIJ, Vol. 76, No.665, 1253–1261. (in Japanese with English abstract)

Irikura, K. 1986. Prediction of strong acceleration motion using empirical Green's function, *Proceedings of the 7th Japan Earthquake Engineering Symposium*, 151–156.

JMA, Seismic intensity database (in Japanese), https://www.data.jma.go.jp/svd/eqdb/data/shindo/index.php

Kamae, K. and Irikura, K. 1998. Source model of the 1995 Hyogo-ken Nanbu earthquake and simulation of near-source ground motion, *Bulletin of the Seismological Society of America*, Vol.88, 400–412.

Miyake, H., Iwata, T. and Irikura, K. 1999. Strong ground motion simulation and source modeling of the Kagoshima-ken Hokuseibu earthquakes of March 26 (M_{JMA}6.5) and May 13 (M_{JMA}6.3), 1997, using empirical Green's function method, Journal of the seismological society of Japan. 2nd ser., Zisin, 51, 4, 431–442. (in Japanese with English abstract)

NIED, Strong motion seismograph networks (K-NET and KiK-net), http://www.kyoshin.bosai.go.jp/

NIED, Broadband seismograph network (F-net), http://www.fnet.bosai.go.jp/top.php?LANG=en

T3: Dynamic Response and Stability of Rock Foundations, Underground Excavations in Rock, Rock Slopes, and Stone Masonry Historical Structures

2019 Rock Dynamics Summit– Aydan et al. (eds)
© *2019 Taylor & Francis Group, London, ISBN 978-0-367-34783-3*

Centrifugal model tests on the seismic stability of rock foundations under critical facilities

A. Sekiguchi, M. Ishimaru, & T. Okada
Central Research Institute of Electric Power Industry, Chiba, Japan

K. Hiraga
CERES, Inc., Chiba, Japan

H. Morozumi
Kansai Electric Power Company, Inc., Osaka, Japan

ABSTRACT: In Japan, a slip safety factor based on an equivalent linear analysis is conventionally used to evaluate the stability of rock foundations under critical facilities in terms of the sliding motion during an earthquake. In this study, dynamic centrifugal model tests were performed to assess the seismic stability evaluation method for rock foundations. The results confirmed the feasibility of the method. In addition, the displacement of rock masses because of sliding was observed to be limited even when the slip safety factor had a value of less than 1. This confirms that, in the event of an earthquake, rock foundations do not become unstable spontaneously. Therefore, evaluating the seismic stability based on ground displacement is considered to be an effective approach.

1 INTRODUCTION

The occurrence of fatal and large-magnitude earthquakes in the recent past has increased attention on considering earthquake ground motions during the design phase of the construction of modern structures. Accordingly, the quantitative assessment of the seismic resistance of critical facilities to the earthquake-induced failure of rock foundations has become important.

In Japan, the seismic stability of rock foundations has conventionally been evaluated in terms of their bearing capacity, inclination, and sliding (JEAG 4601-1987 1987). With regard to the sliding motion during an earthquake, a slip safety factor based on an equivalent linear analysis is conventionally used to evaluate the stability of rock foundations. However, a slip safety factor value of less than 1 does not necessarily indicate immediate ground instability.

In this study, therefore, dynamic centrifugal model tests were performed to assess the applicability of conventional slip safety factor evaluation methods to the seismic stability of rock foundations.

2 CENTRIFUGAL MODEL TEST

A rock foundation model with a reduction ratio of 1:50 was constructed with artificial rock material and a weak layer. Vibration tests were performed in a centrifugal force field under a centrifugal acceleration of 50 g.

2.1 Rock foundation model

The rock foundation model and instrument arrangement are shown in Figure 1. The model was 200 mm (10 m upon real-scale conversion) in height and 300 mm in depth. The boundary surfaces had cutouts measuring 100 mm × 100 mm to avoid interference with the rigid box. The building model dimensions were 60 mm (width) × 40 mm (height) (3 m × 2 m upon real-scale conversion), and the density of the building material was 1200 kg/m³.

The measured variables included accelerations produced under and on the ground surface along with the corresponding displacements induced in the building model and on the ground surface. A relative displacement gauge was installed at a position straddling the weak layer. For comparison, a second relative displacement gauge was installed on the ground surface immediately adjacent to the weak layer. Three pressure receiving plates were installed on the bottom of the building model, and the horizontal and vertical stresses were measured. The pressure receiving plates and ground surface were fixed with an adhesive.

2.2 Properties of the rock foundation model

Table 1 lists the physical properties of the materials used to construct the artificial rock model and weak layer. The properties were obtained from various physical and mechanical tests.

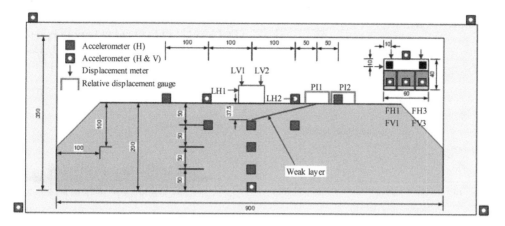

Figure 1. Rock foundation model and instrument arrangement.

Table 1. Physical properties of the artificial rock materials and weak layer (σ_m: mean stress).

	Rock	Weak layer
Unit weight	20.3 kN/m³	20.6 kN/m³
Peak shear strength	$c_p = 267.1$ kN/m² $\varphi_p = 34.7°$	$c_p = 0.0$ kN/m² $\varphi_p = 28.6°$
Residual shear strength	$a = 4.61, b = 0.70$ $(\tau_r = a \times \sigma_m{}^b)$	$c_r = 0.0$ kN/m² $\varphi_r = 19.3°$
Tensile strength	$\sigma_t = 41.4$ kN/m²	$\sigma_t = 0.0$ kN/m²
Initial elastic shear modulus	933000 kN/m²	2800 kN/m²
Poisson's ratio	0.42	0.49

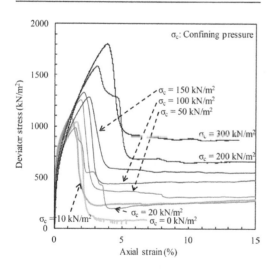

Figure 2. Stress–strain relationships obtained from plane-strain compression tests.

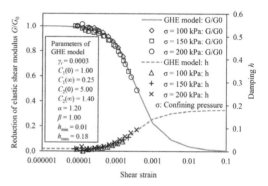

Figure 3. Dynamic deformation characteristics of the artificial rock material obtained from cyclic triaxial tests.

high early strength Portland cement, 370 kg of distilled water, 817 kg of crushed limestone sand, 817 kg of limestone fine powder, and 1 kg of admixture.

Figure 2 shows the stress–strain relationships obtained from plane-strain compression tests. Figure 3 shows the dynamic deformation characteristics obtained from cyclic triaxial tests.

2.2.2 Properties of the artificial weak layer

Based on the work by Ishimaru & Kawai (2011), the weak layer within the rock mass was reproduced by installing a 0.2 mm thick Teflon sheet within the rock foundation model before the artificial rock material started hardening. The resultant artificial weak layer had constant degrees of roughness, bite, etc. Prior examination confirmed that the cohesion between the post-hardening artificial rock material and Teflon sheet was very small. Under this condition, the shear resistance of the artificial weak layer can be considered to be equal to the frictional force generated between the artificial rock material and Teflon sheet.

The frictional force generated between the artificial rock material and Teflon sheet under normal-stress loading was examined through a

2.2.1 Properties of the artificial rock materials

Because the physical properties of different natural rocks vary considerably, the rock foundation model in this study was created from cement-modified soil with a curing period of 7 days. For a soil volume of approximately 1 m³, the formulation was 82 kg of

288

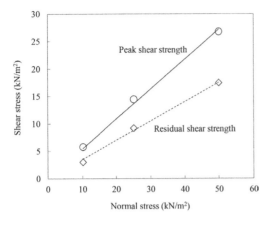

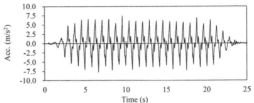

(a) Horizontal acceleration.

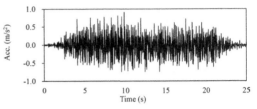

(b) Vertical acceleration.

Figure 4. Shear stress–normal stress relationships obtained from single-plane shearing tests.

Figure 5. Input acceleration (vibration step d04).

Table 2. Maximum values of the acceleration amplitude at different vibration steps.

Vibration step	Frequency	Horizontal acc. m/s²	Vertical acc. m/s²
d01	1.2	0.57	0.13
d02	1.2	3.47	0.42
d03	1.2	5.72	1.15
d04	1.2	7.77	0.91
d05	1.2	9.16	1.22
d06	1.2	10.40	1.50
d07	1.6	8.68	1.87
d08	1.6	10.04	2.88
d09	1.6	11.53	3.84
d10	1.6	11.25	3.39

single-plane shearing test. Figure 4 shows the test results; the maximum and residual shear resistances increased in proportion to the normal stress.

2.3 *Input acceleration*

The input acceleration was provided in the form of a sinusoidal wave with a wavenumber of 20 (frequencies of 1.2 and 1.6 Hz upon real-scale conversion) in the main part with four tapers before and after. During the test, the acceleration amplitude was increased for each vibration step. A horizontal movement was the only input. However, the vertical motion, which was considered to be caused by the shaking table rocking, was also measured during vibration. Figure 5 shows the input acceleration of vibration step d04, and Table 2 lists the maximum acceleration amplitudes at different vibration steps. The 1.6-Hz excitation produced a greater vertical motion than the 1.2-Hz excitation owing to the characteristics of the experimental apparatus.

2.4 *Test results*

Figure 6 shows the maximum values during vibration and the accumulated residual values for the inclination of the building model at different vibration steps. The maximum values during vibration were calculated by assuming a zero value at the start of each excitation step. Figure 7 shows the accumulated residual values for the differences between the stresses of the left and right pressure receiving plates at different vibration steps. Similarly, Figure 8 shows the accumulated residual values of the horizontal displacements of the building model and ground at different vibration steps, and Figure 9 shows the accumulated residual values of the displacements measured by the relative displacement gauge at different vibration steps. These figures confirm that the residual values rapidly increased after vibration step d09.

Figure 10 shows the strain distribution calculated from images captured by a high-speed camera at vibration step d10. Cracks connecting the lower end of the weak layer and the left side of the building model were generated, although they were not yet clear in images captured at vibration step d09. Owing to the occurrence of these cracks, the upper part of the weak layer was estimated to move.

3 EVALUATION OF THE SLIP SAFETY FACTOR

The results of the dynamic centrifugal model test were used to evaluate the applicability of the slip safety factor evaluation method based on the equivalent linear analysis. The properties of the rock foundation model used for the equivalent linear analysis are listed in Table 1. The dynamic deformation characteristics of the artificial rock material were set (Figure 3) by using the general hyperbolic equation (GHE) model (Tatsuoka & Shibuya 1992). In contrast, the artificial weak layer was modeled to represent linear elastic-joint elements. The unit weight of

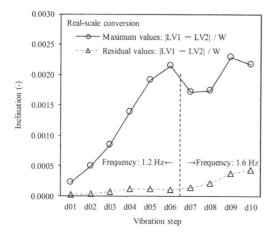

Figure 6. Maximum and accumulated residual values for the inclination of the building model at different vibration steps.

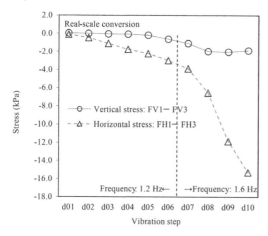

Figure 7. Accumulated residual values for the difference in stresses at different vibration steps.

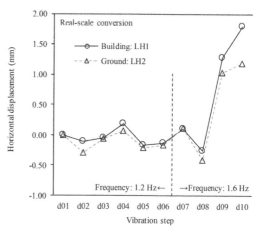

Figure 8. Accumulated residual values for the horizontal dis-placements of the building model and ground at different vi-bration steps.

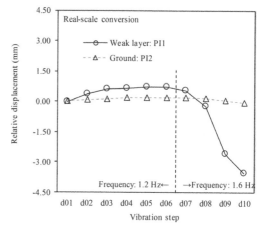

Figure 9. Accumulated residual values for the displacements obtained by the relative displacement gauge at different vibra-tion steps.

the artificial weak layer was 20.6 kN/m³, which was equal to that of the Teflon sheet, and the corresponding Poisson's ratio was 0.49 based on the assumption of no volume change. The pseudo shear modulus of elasticity, which was induced by modeling the artificial weak layer as linear elastic-joint elements, was set as 2800 kN/m² from the gradient up to the maximum shear resistance during the single-plane shearing tests.

Equivalent linear analyses were performed with the same input accelerogram as that measured in the centrifugal model test. The stresses used to cal-culate the slip safety factor were obtained by add-ing the stresses from the self-weight stresses and induced during an earthquake. Figure 11 shows the procedure for calculating the slip safety factor.

Table 3 lists the minimum slip safety factor values calculated during the different vibration steps, and Figure 12 shows the slip-line shapes

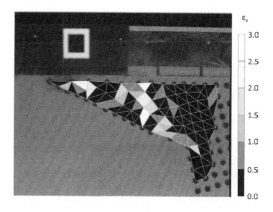

Figure 10. Horizontal strain distribution calculated from im-ages taken with a high-speed camera at vibration step d10.

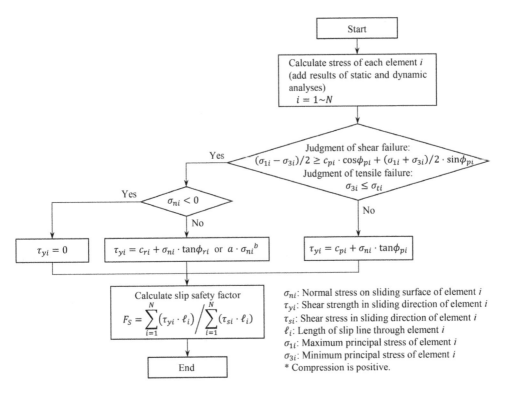

Figure 11. Flowchart for calculating the slip safety factor.

Flowchart content:

Start

Calculate stress of each element i
(add results of static and dynamic analyses)
$i = 1 \sim N$

Judgment of shear failure:
$(\sigma_{1i} - \sigma_{3i})/2 \geq c_{pi} \cdot \cos\phi_{pi} + (\sigma_{1i} + \sigma_{3i})/2 \cdot \sin\phi_{pi}$
Judgment of tensile failure:
$\sigma_{3i} \leq \sigma_{ti}$

Yes — $\sigma_{ni} < 0$

Yes — $\tau_{yi} = 0$

No — $\tau_{yi} = c_{ri} + \sigma_{ni} \cdot \tan\phi_{ri}$ or $a \cdot \sigma_{ni}^{\ b}$

No — $\tau_{yi} = c_{pi} + \sigma_{ni} \cdot \tan\phi_{pi}$

Calculate slip safety factor
$$F_S = \sum_{i=1}^{N}(\tau_{yi} \cdot \ell_i) \Big/ \sum_{i=1}^{N}(\tau_{si} \cdot \ell_i)$$

End

σ_{ni}: Normal stress on sliding surface of element i
τ_{yi}: Shear strength in sliding direction of element i
τ_{si}: Shear stress in sliding direction of element i
ℓ_i: Length of slip line through element i
σ_{1i}: Maximum principal stress of element i
σ_{3i}: Minimum principal stress of element i
* Compression is positive.

Table 3. Slip safety factors for different vibration steps.

Vibration step	Minimum slip safety factor		Slip safety factor obtained from model test	
	Slip line		Slip line	
d01	No. 6	24.78		24.78
d02	No. 6	8.38		8.38
d03	No. 5	5.10		5.61
d04	No. 4	3.02		3.33
d05	No. 1	2.12	No. 6	2.67
d06	No. 6	1.40		1.40
d07	No. 6	1.76		1.76
d08	No. 3	0.86		0.98
d09	No. 2	0.39		0.72
d10	No. 2	0.20		0.45

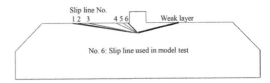

Slip line No.
1 2 3   4 5 6   Weak layer

No. 6: Slip line used in model test

Figure 12. Slip line shapes for the minimum slip safety factor.

4 CENTRIFUGAL MODEL TEST ASSUMING SLIDING FROM THE BEGINNING

In the centrifugal model test described above, even when the slip safety factor was less than 1, the amount of displacement due to sliding was limited, and the rock foundation did not become rapidly unstable. In order to confirm this more clearly, a centrifugal model test was performed assuming the occurrence of slip clump under the building model, as shown in Figure 13. Although only the surroundings of the building model are shown in this figure, the other model shapes and instrument arrangement were the same as that in Figure 1. The slip clump and weak layers were made of the same artificial rock material and Teflon sheets as above.

Table 4 lists the maximum accelerations measured at the bottom of the rock foundation model and the minimum slip safety factors at different vibration steps. Figure 14 shows the maximum

corresponding to these values. The minimum slip safety factor was less than 1 after vibration step d08, although the residual displacement rapidly increased at vibration step d09 during the test. Therefore, the slip safety factor evaluation method can be considered conservative. Although the slip safety factor of the slip line generated during the tests does not represent the minimum value, it is similar in that it was less than 1 before the residual displacement rapidly increased. In addition, even when the slip safety factor was less than 1, the amount of displacement that could be caused by sliding was limited. This indicates that, in the event of an earthquake, rock foundations do not spontaneously lose their seismic stability.

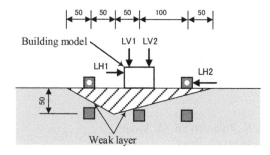

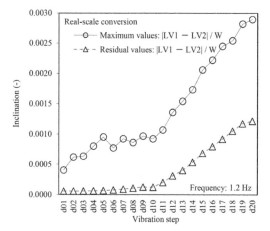

Figure 13. Rock foundation model for the centrifugal model test assuming the occurrence of slip clump under the building model.

Figure 14. Maximum and accumulated residual values for the inclination of the building model at different vibration steps.

Table 4. Maximum accelerations measured at the bottom of the rock foundation model and minimum slip safety factors at dif-ferent vibration steps.

Vibration step	Frequency	Horizontal acc. m/s²	Vertical acc. m/s²	Minimum slip safety factor
d01		0.39	0.09	1.85
d02		1.29	0.29	0.59
d03		1.81	0.38	0.41
d04		2.01	0.51	0.25
d05		2.46	0.72	0.19
d06		2.47	0.44	0.19
d07		2.86	0.63	0.19
d08		3.39	0.57	0.17
d09		4.07	0.87	0.14
d10	1.2	4.61	0.80	0.14
d11		5.09	0.89	0.14
d12		5.64	0.84	0.15
d13		5.92	1.02	0.12
d14		6.24	1.09	0.12
d15		6.53	1.17	0.11
d16		7.41	1.12	0.10
d17		7.79	1.26	0.09
d18		8.10	1.37	0.10
d19		8.79	1.65	0.09
d20		9.54	1.26	0.10

values during vibration and the accumulated residual values for the inclination of the building model at different vibration steps. The results once again confirmed that, even when the slip safety factor was less than 1, the rock foundation did not become unstable spontaneously.

5 CONCLUSION

The centrifugal model tests in this study confirmed the feasibility of the slip safety factor evaluation method. In addition, the displacement of rock masses because of sliding was observed to be limited even when the slip safety factor was less than 1. This confirms that, in the event of an earthquake, rock foundations do not become unstable spontaneously. Therefore, evaluating the seismic stability based on ground displacement is considered to be an effective approach.

6 ACKNOWLEDGMENTS

This work is part of a collaborative research project that was carried out jointly by Kansai Electric Power Co. Inc., Hokkaido Electric Power Co. Inc., Tohoku Electric Power Co. Inc., Tokyo Electric Power Company Holdings, Inc., Chubu Electric Power Co. Inc., Hokuriku Electric Power Co. Inc., Chugoku Electric Power Co. Inc., Shikoku Electric Power Co. Inc., Kyusyu Electric Power Co. Inc., The Japan Atomic Power Company, Electric Power Development Co., Ltd., and Japan Nuclear Fuel Limited.

REFERENCES

Ishimaru, M. and Kawai, T. 2011. Centrifuge model test on earthquake-induced failure behaviour of slope in discontinuous rock mass. *Proc. 12th International Congress on Rock Mechanics*, Beijing: 1919–1922.

JEAG4601-1987. 1987. Technical Guidelines for Aseismic Design of Nuclear Power Plants, Japan.

Tatsuoka, F. and Shibuya, S. 1992. Deformation characteristics of soils and rocks from field and laboratory tests. *Proc. 9th Asian Regional Conference on Soil Mechanics and Foundation Engineering*, Bangkok, Vol. 2: 101–170.

2019 Rock Dynamics Summit– Aydan et al. (eds)
© 2019 Taylor & Francis Group, London, ISBN 978-0-367-34783-3

A fundamental study on the foundations in Ryukyu Limestone Formation and the shear properties of interfaces and discontinuities under static and dynamic loading conditions

Y. Araki, T. Ito, J. Tomiyama, Y. Suda, K. Horiuchi, N. Tokashiki & Ö. Aydan
Department of Civil Engineering, University of the Ryukyus, Okinawa, Japan

Y. Shuri
Asahi Consultants Co., Urasoe, Okinawa, Japan

ABSTRACT: The Ryukyu limestone formation is not assumed to be a suitable load bearing foundation in Ryukyu Archipelago. The characteristics of Ryukyu limestone with various porosity under static and dynamic conditions have been investigated by the authors. Furthermore, the behavior of the interface between piles and Ryukyu limestone are tested using large-scale dynamic shear testing device. Some model piles founded on Ryukyu limestone are subjected to static and dynamic loads to check their deformation response and their load-bearing capacity. The authors will explain fundamental studies on foundations on Ryukyu Limestone Formation under static and dynamic loading conditions and present the outcomes of these studies and discuss their implications on bridge piles.

1 INTRODUCTION

Ryukyu limestone formation, which is broadly divided into coral limestone and sandy limestone, is widely distributed in Ryukyu Archipelago. As limestone is solvable due to percolating groundwater and accumulation of corals and as well broken pieces, the porosity of limestone is large. Furthermore, they may contain some large scale cavities due to combined action of groundwater and tectonic movements. Therefore, the Ryukyu limestone formation is considered to be not suitable as a foundation rock for large-scale engineering structures. When the thickness of the Ryukyu limestone formation is quite thick, it results in the non-economical foundation design. This is a serious engineering issue in Okinawa Prefecture, Japan and some fundamental studies are necessary to clarify whether it is really unsuitable or suitable as foundation rock. The authors investigate the characteristics of Ryukyu limestone with various porosity under static and dynamic conditions. The behavior of the interface between piles and Ryukyu limestone are tested using large-scale dynamic shear testing device. Furthermore, some model piles founded on Ryukyu limestone were subjected to static and dynamic loads to check their deformation response and their load-bearing capacity. In addition some analytical and numerical studies are proposed. The authors explain these fundamental studies on foundations on Ryukyu Limestone Formation under static and dynamic loading conditions and present the outcomes of these experimental studies and discuss their implications in foundation design.

2 CHARACTERISTICS OF RYUKYU LIMESTONE FORMATION

2.1 Geological Characteristics

The main islands of Ryukyu Archipelago are Okinawa, Amami-Oshima, Miyako, Ishigaki, Iriomote and Yonaguni. The islands are situated on Ryukyu arc bounded by Okinawa trough and Ryukyu trench. The environment is tropical. Ryukyu limestone is widely distributed in Ryukyu Archipelago. It is broadly defined as coral, gravely and sandy or sandy-gravel limestone. Geoengineering issues associated with Ryukyu limestone formation are cliff collapses, sinkholes due to karstic caves and their effect on super structures (Fig. 1).

Figure 1. Some examples of engineering problems associated with Ryukyu Limestone Formation.

2.2 Physico-mechanical Characteristics of Ryukyu Limestone

The mechanical properties of Ryukyu limestone have been investigated by Tokashiki (2010) and his colleagues (Tokashiki & Aydan, 2003, 2010) in details. Tokashiki (2010) utilized the stereology technique to evaluate the porosity of Ryukyu limestone and the shape of pores. This technique requires a number of slices of rock sample for the digitization and data processing to determine the porosity and the shape of pores. This is a quite cumbersome procedure.

A sample of Ryukyu limestone with a height of 100 mm and 50 mm in diameter was prepared and it was scanned using the inspeXio SMX-225CT FPD. Fig. 2 shows visual and scanned images of the Ryukyu limestone (Aydan et al. 2016). It is quite interesting to notice the porous structure of the sample can be easily visualized without any physical disturbance to the sample. Furthermore the shape, distribution and physical positions of pores in rock sample can be easily evaluated. For example, if such data is imported for some numerical simulations, the equivalent properties of porous rocks could be evaluated without any assumption of some models adopted in averaging techniques such as mixture theories, micromechanics, micro-structure models or homogenization technique. Fig. 3 shows a 3D visualization of porous structure of Ryukyu limestone and number of porosities with different volumes.

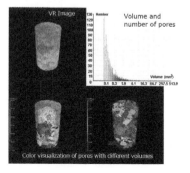

Figure 2. Comparison of actual and scanned images of a Ryukyu limestone sample.

Figure 3. A 3D visualization of porosities in Ryukyu Limestone.

2.3 Dynamic Mechanical Characteristics of Ryukyu Limestone

In the experiments, coral limestone is tested under uniaxial compression and Brazilian shock tests. Figs. 4 and 5 show the force and acceleration responses of Ryukyu limestone samples. The strength of Ryukyu limestone depends upon the porosity and the static UCS ranges between 20.0 and 33.3 MPa. Similarly the Brazilian tensile strength of Ryukyu limestone depends upon the porosity and it ranges between 2.4 and 5.3 MPa. Fig. 6 compares the failure state under static and dynamic conditions.

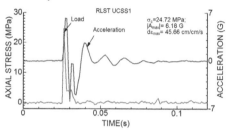

Figure 4. Axial stress and acceleration response of Ryukyu limestone sample under uniaxial compression shock test.

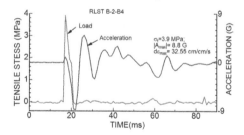

Figure 5. Axial stress and acceleration response of Ryukyu limestone sample under Brazilian shock test.

Table 1. Comparison of Static and Dynamics strength of Ryukyu limestone

Condition	Static (MPa)	Dynamic (MPa)
UCSS	20-33.3	24.72
BRS	2.4-5.3	3.90
Coral Finger-UCSS		27.94

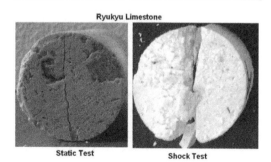

Figure 6. Comparison of fracturing of Ryukyu limestone samples under static and dynamic conditions.

2.4 Direct Shear Tests on Ryukyu Limestone

A direct shear is carried out to investigate the shear response of coral limestone stone with honeycomb-like structure (Fig. 7). Figs. 8 and 9 show the responses measured during the direct shear experiment. Once peak load exceeded, the deformation rate increases as noted from the figure.

2.5 Tilting, Stick-slip and Cyclic Tests on Limestone discontinuities

Tilting test technique is one of the cheapest techniques to determine the frictional properties of rock discontinuities and interfaces under different environmental conditions (Barton and Choubey, 1977; Aydan et al. 1995; Aydan 1998). This technique can be used to determine the apparent friction angle of discontinuities (rough or planar) under low stress levels. It definitely

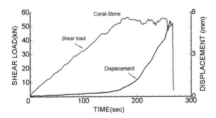

Figure 7. A view of direct shear testing by OA-DSTM testing machine

Figure 8. Shear displacement-shear load relation of coral-stone.

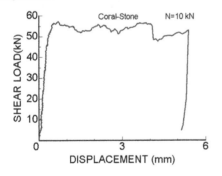

Figure 9. Shear displacement-shear load response of coral-stone

gives the maximum apparent friction angle, which would be one of the most important parameters to determine the shear strength criteria of rock discontinuities as well as various contacts. Therefore, the data for determining the parameters of the shear strength criteria for rock discontinuities should utilize both tilting test and direct shear experiment. Experimental results on saw-cut discontinuity planes of Ryukyu limestone samples are shown in Fig. 10, which shows responses measured during a tilting experiment on a saw-cut plane of Ryukyu-limestone. The static and dynamic friction coefficients of the interface were calculated from measured displacement response explained in previous section and they were estimated at 28.8-29.6° and 24.3-29.2° respectively.

A series of stick-slip experiments are carried out on the saw-cut discontinuities of Ryukyu Limestone. The peak (static) friction angle can be evaluated from the T/N response while the residual (kinetic) friction angle is obtained from the theoretical relation (Aydan et al. 2018). Fig. 11 show the stick-slip responses of discontinuity planes shown. The residual (kinetic) friction angle of saw-cut discontinuity plane of Ryukyu limestone obtained from stick-slip experiments are very close to those obtained from tilting experiments. Nevertheless, the kinetic or residual friction angle is generally lower than those obtained from the tilting experiments.

A multi-stage (multi-step) direct shear test on a saw-cut surface of sandy Ryukyu limestone sam-

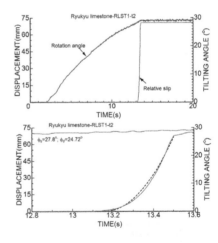

Figure 10. Responses of saw-cut discontinuity planes of Ryukyu limestone samples during a tilting test.

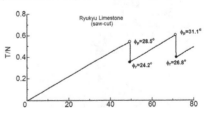

Figure 11. Stick-slip response of saw-cut plane of Ryukyu limestone

295

ple, which consist of two blocks with dimensions of 150x75x37.55mm, was varied out (Figs. 12 and 13). The initial normal load was about 17 kN and increases to 30, 40, 50, 60 and 70 kN during the experiment. Fig. 14 shows the shear displacement and shear load responses during the experiment. As noted from the figure, the relative slip occurs between blocks at a constant rate after each increase of normal and shear loads. This experiment is likely to yield shear strength of the interface two blocks under different normal stress levels. Fig. 14 shows the peak and residual levels of shear stress for each level of normal stress increment. Tilting tests were carried out on the same interface and the apparent friction angles ranged between 35.4 and 39.6 degrees. Tilting test results and direct shear tests are plotted in Fig. 15 together with shear strength envelopes using the shear strength failure criterion of Aydan (Aydan 2008; Aydan et al. 1966).

As noted from the figure, the friction angles obtained from tilting tests are very close to the initial part of the shear strength envelopes. However, the friction angle becomes smaller as the normal stress level increases. In other words, the friction angle obtained from tilting tests on saw-cut surfaces can not be equivalent to the basic friction angle of planar discontinuities and interfaces of rocks. The basic friction angle of the planar interface of sandy

limestone blocks is obtained as 27.5 degrees for the range of given normal stress levels.

A series of two-ways direct shear tests were carried out saw-cut discontinuity planes of Ryukyu limestone. Fig. 16 shows the normalized shear resistance (T/N) response of the saw-cut planes and acceleration during the two-ways shearing with ±15 mm forced displacement amplitude. As noted the acceleration response is quite different than that anticipated and it has a very irregular response with an amplitude less than 0.42g.

Fig. 17 shows the relative displacement-T/N relation. In the same figure, the trajectory of the first cycle is distinguished. It is quite interesting to note that the shear resistance of the discontinuity plane gradually changes after each cycle of shearing. At the end of the experiment, a thin powder is recognized on the discontinuity plane.

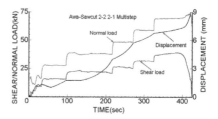

Figure 14. Shear stress-shear load response of the interface of sandy limestone blocks during the multi-stage(step) direct shear experiment.

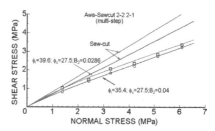

Figure 15. Comparison of shear strength envelope for the interface of sandy limestone blocks with experimental results from the multi-stage(step) direct shear experiment.

Figure 12. A view of direct shear test on a saw-cut discontinuity plane.

Figure 13. A view of a saw-cut discontinuity plane after a direct shear test.

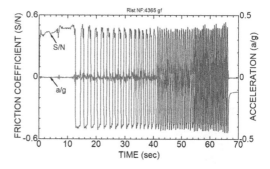

Figure 16. The response of the normalized S/N and acceleration response of a saw-cut plane during two shearing experiment.

3 EXAMPLES OF FOUNDATION DESIGN ON RYUKYU LIMESTONE FORMATION

The design of foundation on Ryukyu Limestone Formation (RLF) is always a major problem. Despite the general tendency to avoid foundation design on the RLF, there are some examples, in which the foundations are located in the RLF. These examples are Kouri Bridge (Fig. 18) and New Ishigaki Airport Protection structures (Fig. 19).

The foundations of the bridge along Gushikawa By-Pass roadway was designed to be having end-bearing on the phyllite formation below the Ryukyu limestone formation (Fig. 20). The distribution deformation and axial stress of the pile and the shear stress along the pile and surrounding ground was analyzed using the solution given by Aydan (2018) (Fig. 21). As noted from the figure, the applied load is not fully transferred to the tip of the pile at depth and it is only a small fraction of the total load. These result imply that the current concept of the design of pile foundation in Okinawa Prefecture requires substantial revision.

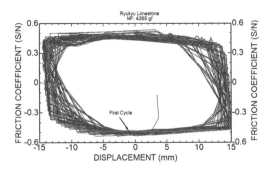

Figure 17. Shear response of a saw-cut discontinuity plane of Ryukyu limestone shown in Figure 18 in the space of relative displacement and T/N.

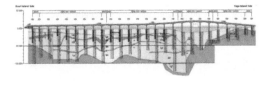

Figure 18. The design of bridge foundation piles of the Kouri Bridge.

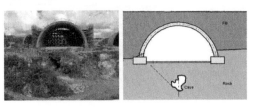

Figure 19. The design of arch protection structure on karstic caves at New Ishigaki Airport.

(a) Drilling and embedment of caisson

(b) Casting concrete piles   (c) Construction completed

Figure 20. Stages of construction of piles and the bridge.

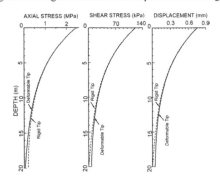

Figure 21. Distribution of axial and shear stresses and displacement with depth.

4 EXPERIMENTS ON BEARING CAPACITY FOUNDATIONS ON RYUKYU LIMESTONE FORMATION CONTAINING CAVITIES

4.1 Unfilled Cavities

The authors have performed laboratory experiments under static and dynamic conditions on the influence area above the caves (Fig. 22). Furthermore, bearing capacity experiments were performed on the limestone blocks obtained from the construction site. The influence line to estimate the effect of caves beneath foundations and structures is specified by taking a tangential line to the cavity with inclination of 45 degree in Japanese regulation. However, the experiments clearly showed that the regulation is not appropriate for evaluating the influence of cavities on the separation of discontinuities is not true and appropriate.

4.2 Filled Cavities

Some experiments were carried out to investigate the effect of backfilling on the bearing capacity of Ryukyu limestone formation. Fig. 23 compares the average strain stress relations for unfilled and pillars backfilled with granular and cohesive backfill materials. It is interesting to note that the bearing capacity of backfilled pillars is increased about 1.2-1.3 times compared with that of the unfilled sample at the same strain level. Furthermore, the behavior of backfilled pillars is elasto-plastic without any softening after the yielding of the pillar. When the limestone pillar is backfilled with NSK backfilling material, the overall strength of pillars greatly increased and the experiments were terminated in order to prevent the bursting of the acrylic cell resulting in undesirable accidents.

5 A PROPOSAL FOR FOUNDATION DESIGN ON RYUKYU LIMESTONE FORMATION

As pointed previously, there is high possibility of caves of different sizes in Ryukyu Limestone Formation (RLF). The existence of these caves may result in non-uniform settlement and/or collapse of foundations of the superstructures. For this reason, it is common to drill boreholes to a depth of rock layer such as Shimajiri formation or phyllite rock mass at each pier of the super structures. These boreholes can be used to check if there is any cave beneath the foundation. The spread foundation or very shallow foundations can be used if there is no cavity. If there is any cavity, the possibility of the backfilling should

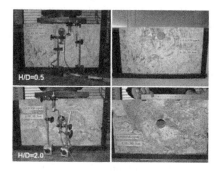

Figure 22. Experiment on the effect caves on footing separation.

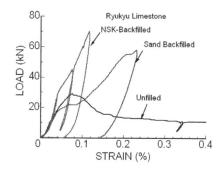

Figure 23. Comparison of strain-stress responses of unfilled pillar, backfilled pillars with granular and cohesive backfill materials during cyclic compression.

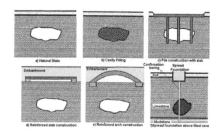

Figure 24. Various alternatives to deal cavities during the construction on Ryukyu Limestone Formation

be explored. The filling of such caves is possible and limited area type backfilling may also be utilized. Besides these suggestion, there may be other alternatives as illustrated in Fig. 24.

6 CONCLUSIONS

The utilization of Ryukyu Limestone Formation (RLF) as the foundation rock of superstructures is explored and the characteristics of the RLF are presented and the applications of various methods for the characterization of the RLF are explained. If sufficient explorations and counter-measures are implemented at a given foundation location, it is possible to utilize the RLF as foundation rock. Therefore, the present design philosophy is not correct and it is resulting in uneconomical foundation design and it is overconservative.

REFERENCES

Aydan, Ö. (2018): Rock Reinforcement and Rock Support, CRC Press, ISRM Book Series, No.7, 473 pages.
Aydan, Ö. and Tokashiki, N. 2007. Some damage observations in Ryukyu Limestone Caves of Ishigaki and Miyako islands and their possible relations to the 1771 Meiwa Earthquake. Journal of The School of Marine Science and Technology, 5(1), pp. 23–39.
Aydan, Ö., N Tokashiki, M. Edahiro (2016). Utilization of X-Ray CT Scanning technique in Rock Mechanics Applications. ARMS2016, Bali.
Tokashiki, N., 2010. Study on the Engineering Properties of Ryukyu Limestone and the Evaluation of the Stability of its Rock Mass and Masonry Structures. PhD Thesis, Waseda University, Tokyo.
Tokashiki, N. and Aydan, Ö., 2003, Characteristics of Ryukyu Limestone and its utilization as a building stone in historical and modern structures, International Symposium on Industrial Minerals and Building Stones, Istanbul, pp. 311–318.
Tokashiki, N. and Ö. Aydan, 2010. The stability assessment of overhanging Ryukyu limestone cliffs with an emphasis on the evaluation of tensile strength of Rock Mass. Journal of Geotechnical Engineering, JSCE, Vol. 66, No. 2, 397–406.

2019 Rock Dynamics Summit– Aydan et al. (eds)
© 2019 Taylor & Francis Group, London, ISBN 978-0-367-34783-3

Bingham flow model by fully implicit SPH and its application to reinforce underground caves

D.S. Morikawa & M. Asai
Kyushu University, Fukuoka, Japan

Y. Imoto
Tohoku University, Sendai, Japan

M. Isshiki
Ehime University, Matsuyama, Japan

ABSTRACT: The present work shows the application of the Smoothed Particle Hydrodynamics (SPH) on non-Newtonian fluids for simulating the injection of a cementitious material into underground caves to reinforce the soil. In special, it presents two main improvements over the already established SPH formulation: an implicit time integration scheme to overcome the problem of impracticable small time step restriction in highly viscous fluid simulation, and the introduction of air ghost particles (AGP) to fix problems on the free-surface treatment. This project utilizes the Incompressible SPH (ISPH) as a basis for the implementation of such improvements, which guarantees a stable and accurate pressure distribution. We validate the proposed implicit time integration scheme with pipe flow simulations and the free-surface treatment with a simple hydrostatic problem. Also, dam break numerical simulations using the proposed method resulted in very good agreement with experimental data. At last, we demonstrate the potential of this method with the highly viscous vertical jet flow over a horizontal plate validation test, which shows a complex viscous coiling behavior.

1 INTRODUCTION

In the city of Mitake, Gifu, Japan, there was an abundant supply of mineral coal underground in the 20th century. Since it was considered a great opportunity to generate income, the authorities decided to explore it as much as possible, leaving behind the empty underground caves. As a result, the current buildings, roads and other infrastructures built over these underground caves are in serious danger of problems such as soil settlement and ground collapse.

In this context, the authorities have proposed to fill some of these underground caves with a cementitious material called kira, which promotes great fluidity and fast hardening. In collaboration with them, the authors intend to utilize the Smoothed Particle Hydrodynamics (SPH) to simulate the behavior of this fluid during the injection process. The present study will show some improvements on the SPH to address this simulation accurately.

The SPH is one of the Lagrangian mesh-free particle methods which was first proposed simultaneously by Lucy (1977) and Gingold & Monaghan (1977), and it is widely used on fluid dynamics problem. Although it is already well recognized for dealing with problems of low viscosity fluids (eg., water), the SPH needs some modifications in order to accurately simulate highly viscous fluids, which includes non-Newtonian fluids as cementitious materials.

2 ISPH BASED ON THE PROJECTION METHOD

A stabilized ISPH method was proposed by Asai et al. (2012) with the objective of relaxing the density invariance condition on the pressure calculation procedure. This framework is based on the semi-implicit time integration scheme of the Moving Particle Semi-implicit (MPS) method and adapted to the SPH. In the following section, we aim to summarize the main aspects of this method.

2.1 Governing Equations

The ISPH is designed to numerically solve the main equations for hydrodynamic problems of incompressible flows: continuity equation and Navier-Stokes equation.

$$\nabla \cdot \mathbf{v} = 0 \tag{1}$$

$$\frac{D\mathbf{v}}{Dt} = -\frac{\nabla P}{\rho} + \nu \nabla^2 \mathbf{v} + \mathbf{g} \tag{2}$$

In the above equations, v is the velocity vector, D/Dt is the time derivative, P is the pressure field, ρ is the density of the fluid, ν is the kinematic viscosity, g is the external forces vector, and t represents time.

2.2 SPH approximations

The SPH method is a space integration method which smoothly approximates the value of functions and its derivatives integrating the contribution of the neighbor particles varying its influence according to a weight function W.

First, the space is discretized into a finite number of particles $\{\hat{\mathbf{x}}_i\}_{i=1}^N$, where N is the total number of particles, and $\mathbf{x}$ is the position vector, and the weight function is chosen between a wide range of possibilities. In this paper, we selected the quintic spline function as proposed by Schoenberg (1946).

Then, one can approximate the value of a generic function ϕ for a given target particle and its derivatives as

$$\langle \varphi_i \rangle \cong \sum_j \frac{m_j}{\rho_j} \varphi_j W\left(\mathbf{r}_{ij}, h\right) \tag{3}$$

where the subscripts i and j represents target particle and neighboring particle, respectively, m is the mass, $\mathbf{r}_{ij} = \mathbf{x}_i - \mathbf{x}_j$, and symbol $\langle \rangle$ represents the SPH approximation. Also, we define h, the smoothing length, as 1.2 times the particle distance d.

2.3 Original stabilized ISPH semi-implicit time integration scheme

The original time integration scheme (Asai et al., 2012) is based on a projection method which updates the velocity according to equation (2) in two steps: predictor and corrector steps.

First, the contribution of the viscous term and the external forces results in a predicted velocity field calculated explicitly.

$$\mathbf{v}^* = \mathbf{v}^n + \Delta t\left(\nu \nabla^2 \mathbf{v}^n + \mathbf{g}\right) \tag{4}$$

Then, the pressure is calculated through a Poisson Pressure Equation.

$$\nabla^2 P^{n+1} = \frac{\rho_0}{\Delta t} \nabla \cdot \mathbf{v}^* + \alpha \frac{\rho_0 - \rho^n}{\Delta t^2} \tag{5}$$

Finally, we add the contribution of the pressure field to calculate implicitly the updated velocity field.

$$\mathbf{v}^{n+1} = \mathbf{v}^* + \Delta t\left(-\frac{1}{\rho_0} \nabla P^{n+1}\right) \tag{6}$$

In the above equations, Δt is the time increment, ρ_0 is the reference density of the fluid, ρ^n is a SPH approximation of the density based on equation (3), α is a relaxation coefficient, the superscripts n and $n+1$ refers to the current and next iterations, and the superscript * represents the predictor step. Particularly, equations (4) and (5) are referred as predictor and corrector steps, respectively. For more details in the derivation of the pressure Poisson equation and the α coefficient, see Asai et al. (2012).

3 A FULLY IMPLICIT TIME INTEGRATION SCHEME FOR NEWTONIAN AND NON-NEWTONIAN FLUID INCLUDING FREE SURFACE

3.1 Newtonian fluid

In order to solve a highly viscous fluid with a moderate time increment, a fully implicit time integration has been introduced into the ISPH based on the projection method. Although the whole procedure is the same as the above mentioned ISPH, velocity is updated implicitly here to avoid the Courant-Friedrichs-Lewy condition (Violeau & Leroy, 2014) which becomes one of the most severe condition to select a larger time increment especially for the highly viscous fluid simulation.

The calculation procedure is also formulated in two steps, the main difference from the original stabilized ISPH is that the update of the intermediate velocity $\mathbf{v}^*$ is implicit in the predictor step as follows

$$\mathbf{v}^* = \mathbf{v}^n + \Delta t\left(\nu \nabla^2 \mathbf{v}^* + \mathbf{g}\right) \tag{7}$$

To illustrate this procedure, lets utilize the SPH approximation to discretize the equation (7) according to the target particle i.

$$\nu \nabla^2 \mathbf{v}_i^* \cong \sum_j B_{ij}\left(\mathbf{v}_i^* - \mathbf{v}_j^*\right) \tag{8}$$

where

$$B_{ij} = m_j \frac{2\nu}{\rho_i} \frac{\mathbf{r}_{ij} \cdot \nabla W\left(\mathbf{r}_{ij}, h\right)}{\mathbf{r}_{ij}^2} \tag{9}$$

Finally, equation (7) can be rewritten as

$$\left(1 - \Delta t \sum_j B_{ij}\right)\mathbf{v}_i^* + \left(\Delta t \sum_j B_{ij} \mathbf{v}_j^*\right) = \mathbf{v}^n + \Delta t \mathbf{g} \tag{10}$$

which is an $3N \times 3N$ coefficient matrix in the above linear equations for three dimensional problems.

The remaining time integration scheme follows exactly the same procedure as the original stabilized ISPH.

3.2 Non-Newtonian fluid

3.2.1 Rheological model

As opposed to Newtonian fluids, non-Newtonian fluids promote a not constant value for the viscosity parameter, and its value varies according to the rheological model chosen to evaluate it. In this research, we aim to simulate Bingham fluids, including the Bingham plastic and Bingham pseudoplastic.

To model the Bingham plastic fluid, we selected the cross model with the Papanastasiou regularization model as (Papanastasiou, 1987) which results in the following equation:

$$\tilde{v} = v_0 + \frac{\tau_y}{\dot{\gamma}p}\left(1 - e^{-m\dot{\gamma}}\right) \qquad (11)$$

Also, we selected the Herschel-Bulkley model to simulate Bingham pseudoplastic fluids as

$$\tilde{v} = \frac{K\dot{\gamma}^{n-1}}{\rho} + \frac{\tau_y}{\dot{\gamma}p} \qquad (12)$$

In equation (11) and (12), $\tilde{v}$ represents the approximated viscosity, v_0 the initial viscosity, τ_y the yield stress, $\dot{\gamma}$ the equivalent strain rate, m is a numerical parameter, and K is called consistency index. $\dot{\gamma}$ is defined as

$$\dot{\gamma} = \sqrt{\mathbf{S} : \mathbf{S}} \qquad (13)$$

$$\mathbf{S} = \frac{1}{2}\left(\left(\nabla \otimes \mathbf{v}\right) + \left(\nabla \otimes \mathbf{v}\right)^T\right) \qquad (14)$$

The Papanastasou regularization model is important to guarantee stability in the numerical modeling of the viscosity, since it avoids large gaps of viscosity in small space regions. In accordance with Cotela-Dalmau et al. (2017) we selected m to be 3000s.

In this study, the equivalent strain rate is calculated using a formula proposed by Violeau & Issa (2017), as follows

$$\dot{\gamma}_i^2 = \frac{1}{2}\sum_j m_j \frac{\rho_i + \rho_j}{\rho_i\rho_j} \frac{\mathbf{r}_{ij} \cdot \nabla W\left(\mathbf{r}_{ij}, h\right)}{\mathbf{r}_{ij}^2}\left|\mathbf{v}_i - \mathbf{v}_j\right|^2 \qquad (15)$$

Additionally, to avoid numerical problems of dividing by zero, we propose the following equation to finally evaluate the viscosity of each particle.

$$v = \begin{cases} v_{MAX}, & if \ \tilde{v} > v_{MAX}, \\ \\ \tilde{v}, & otherwise \end{cases} \qquad (16)$$

3.2.2 Time integration scheme for non-Newtonian fluids

The time integration of the predictor step for non-Newtonian fluid is very similar to the scheme proposed for Newtonian fluid. The only difference is the calculation of the viscosity term, which is dependent on the velocity. The following equation shows the proposed evaluation of the viscosity in an implicit manner:

$$\mathbf{v}^* - \Delta t\left(v^{\,n}\nabla^2\mathbf{v}^*\right) = \mathbf{v}^n + \Delta t\,\mathbf{g} \qquad (17)$$

Then, the viscous term can be computed from (8). However, in this case, the B term is defined as

$$B_{ij} = m_j \frac{v_j + v_i}{\rho_i} \frac{\mathbf{r}_{ij} \cdot \nabla W\left(\mathbf{r}_{ij}, h\right)}{\mathbf{r}_{ij}^2} \qquad (18)$$

to maintain the matrix symmetry.

3.3 Solid wall boundary treatment with Fixed Wall Ghost Particle (FWGP)

3.3.1 Pressure wall boundary treatment

Avoiding fluid penetration results in a Neumann boundary condition stating that the acceleration of the water particles close to the boundary surface in the direction of the wall should be zero. This boundary condition is used to calculate the value of the pressure of the wall particles. Then, we formulate this condition from the Navier-Stokes (2), and applying zero to the acceleration **a**.

$$\frac{\partial P}{\partial \mathbf{n}} = \rho(v\nabla^2\mathbf{v} + \mathbf{g} - \mathbf{a}) \qquad (19)$$

To apply the SPH approximation into equation (19), we will define the velocity on the virtual marker as the SPH average (3) of the velocities of wall particles using the virtual marker as the reference. Other variables of the virtual marker, such as viscosity and pressure are also calculated using equation (3). Then, manipulating equation (19) leads to the next equation.

$$P_i = P_I - \left|\mathbf{x}_i - \mathbf{x}_I\right|\rho(v\nabla^2\mathbf{v}_I + \mathbf{g}) \cdot \mathbf{n} \qquad (20)$$

which can be applied into equation (5).

3.3.2 Velocity wall boundary treatment

In this research, we propose a method to reinforce three slip conditions based on a single parameter γ_{slip}. This parameter physically corresponds to the proportion of allowed slip to occur on the wall boundary. Therefore, $\gamma_{slip} = 0$ represents the no-slip and $\gamma_{slip} = 1$, the free-slip condition.

First, lets decompose the velocities components into an orthogonal and a normal direction

$$\mathbf{v}_{f,n} = (\mathbf{n}_w \cdot \mathbf{v}_f)\mathbf{n}_w \qquad (21)$$

$$\mathbf{v}_{f,t} = \mathbf{v}_f - \mathbf{v}_{f,n} \qquad (22)$$

where the subscripts f, w, t and n are fluid, wall, orthogonal direction and normal direction, respectively. From the definition of the slip condition:

$$\mathbf{v}_{slip} = \gamma_{slip}\mathbf{v}_{f,t} \qquad (23)$$

As part of the no penetration wall boundary condition,

$$\mathbf{v}_{w,n} = -\mathbf{v}_{f,n} \qquad (24)$$

Then, we derived the following geometrical relationship

$$\mathbf{v}_{w,t} = C\mathbf{v}_{f,t} \tag{25}$$

$$C = \gamma_{slip} - \frac{(1-\gamma_{slip})d_w}{\left|\mathbf{x}_f - \mathbf{x}_w\right|\cos\theta - d_w} \tag{26}$$

Equations (21), (22), (24) and (25) lead to

$$\mathbf{v}_w = C\mathbf{v}_f - (1+C)(\mathbf{n}_w \cdot \mathbf{v}_f)\mathbf{n}_w \tag{27}$$

which can be applied into equation (10), the implicit predictor step.

3.4 Free-surface treatment with air ghost particles (AGP) implementation

In the original SPH method, there is no transition between inner fluid particles and the void space outside the fluid domain which causes some instability on the free-surface. Here, we propose the implementation of AGP in order to create fictitious mass around the free-surface and overcome this problem.

The algorithm described here is inspired by the SPP implementation (Tsuruta et al., 2015) as follows. First, during the neighboring search procedure, we attached the label of "free-surface particle"160 for 3D simulations. Then, with an algorithm proposed by Marrone et al. (2010), we reaffirm the label of "free-surface particle" for those which have no particles in the cone plus a semi-sphere region (3D) defined in the normal direction of each particle. Lastly, we create AGP in the normal direction of each free-surface particle.

Fig. 1 schematically illustrates the process of creating AGP adopted in this study.

If necessary, one could repeat processes (b) and (c) illustrated in Fig. 1 as many times as necessary to entirely fulfill the influence domain of all free-surface particles. In the case of this study, for a quantic spline weight function, three layers of AGP are necessary to achieve the highest accuracy.

During the solving procedure of (10), the Neumann boundary condition, null divergence of

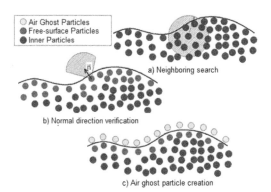

Figure 1. Process of creating AGP

the velocity field on the free-surface particles leads to a simplification that the velocities of neighboring AGP are equal to the velocity of the fluid target particle, and the Dirichlet boundary condition specifies that the pressure of all AGP are defined as zero.

4 NUMERICAL EXAMPLES

4.1 Pipe flow problem

4.1.1 Poiseuille flow
We conducted the Newtonian pipe flow problem, also known as Poiseuille flow problem, to verify our implicit time integration technique of the viscous term and the wall boundary treatment. The theoretical value of the 3D Newtonian pipe flow velocity is defined as showed in Munson et al. (2005), and Fig. 2 illustrates the geometrical parameters of this problem. We utilized the following parameters during the simulation: $R = 0.1325$m, $P_1 = 2000$Pa and $P_2 = 1000$Pa ($\Delta P = 1000$Pa), $L = 0.6$m, $v_0 = 0.1$m^2/s, and d = 0.005m. Fig. 3 is a graphs comparing the numerical results with the theoretical solution.

4.1.2 Non-Newtonian pipe flow
Additionally, we conducted the same problem for a non-Newtonian case of a Bingham plastic fluid using the theoretical solution showed in Mattiusi (2007). The parameters presented in the previous

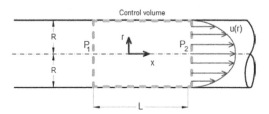

Figure 2. Geometrical parameters of the Poiseuille flow problem

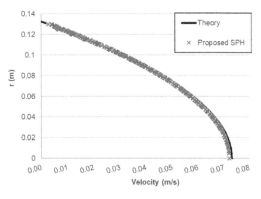

Figure 3. Comparison between numerical results and the theoretical solution of the Poiseuille flow simulation

302

Poiseuille flow problem are maintained. In addition, we utilized v_{max} = 1000 m²/s and τ_y = 50 Pa.

Fig. 4 shows the comparison with the expected theoretical values. Once more, the simulated results are very consistent with the theoretical values.

4.2 Hydrostatic problem

The objective of this numerical example is to test the ability of AGP to stabilize the free-surface and avoid the popping behavior on the top surface. Fig. 5 shows the result before and after the introduction of AGP. As illustrated, this implementation clearly improves the free-surface boundary treatment. Total fluid height of this numerical example is 0.5 meters, so the resulting 5000 Pa of pressure at the bottom is in accordance with the theoretical value (ρ = 1000kg/m³ and g = 9.8m/s²).

4.3 Dam break problems

In this section, we conducted a series of dam break validation tests. First, we selected the well-known experimental study from Martin & Moyce (1952) to verify the proposed SPH method applied on simple low viscosity case. To maintain the same notation as utilized on Martin and Moyce (1952), lets define the following non-dimensional quantities:

$$T = t \left(\frac{g}{a} \right)^{1/2} \tag{28}$$

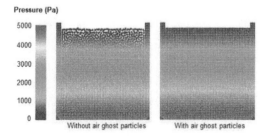

Figure 4. Comparison between numerical and theoretical results of the non-Newtonian pipe flow simulation

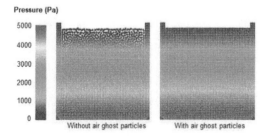

Figure 5. Hydrostatic problem with and without AGP

$$Z = \frac{z}{a} \tag{29}$$

where t is the time after the dam collapse, g is the gravity acceleration, a is the base length of the fluid, and z is the distance of the surge front from the initial wall.

The chosen parameters in this validation test are: H = a = 0.056 m, width = 0.056 m, d = 0.002 m, ρ_0 = 1000 kg/m³, v_0 = 1.4E-7 m²/s, and Δt = 1e-4 s.

Fig. 6 shows the simulation results graphically for the proposed SPH method. As expected, the results are very accurate.

Next, we analyzed the dam break of a Bingham pseudoplastic fluid comparing with experimental results from Minussi & Maciel (2012). The following parameters were used in this verification test: height = 0.13 m, a = 0.5 m, width = 0.32 m, ρ_0 = 1e3 kg/m³, τ_c = 49.179 Pa, k = 7.837 Pa sⁿ, n = 0.442, v_{max} = 100 m²/s, and Δt = 1e-4 s.

Fig. 7 shows the particle distribution of the non-Newtonian daybreak verification test after 1.5 s of simulation, and Fig. 8 shows the comparison between the results with different particle resolutions (d = 0.01 m and d = 0.005 m). The results become more accurate as the particle resolution is finer, which results in results more accurate than the reference numerical solution (Minussi & Maciel, 2012).

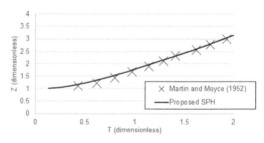

Figure 6. Newtonian dam break simulation compared with experiments on Martin & Moyce (1952)

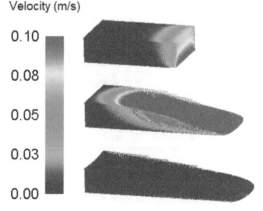

Figure 7. Non-Newtonian dam break particle positioning after 1.5 s of simulation

303

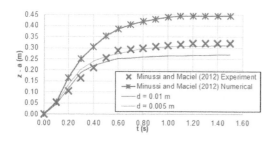

Figure 8. Comparison of the non-Newtonian dam break results with Minussi & Maciel (2012)

Figure 9. Vertical jet flow example without AGP

Figure 10. Vertical jet flow example with AGP

4.4 *Vertical jet flow over a horizontal plate*

The viscous coiling behavior is a widely used validation test for highly viscous fluid simulation, since it is not possible to generate it without special treatment (Violeau & Rogers, 2016). It is expected that viscous coil behavior occurs on a vertical jet flow over a horizontal plate experiment, with a high enough viscosity and H/D (height over diameter) proportion (Tomé et al., 2004).

The following numerical test demonstrates the robustness of the proposed method. The objective is to represent the behavior of honey falling over a bread. First, Fig. 9 reveals a rendered visualization of the vertical jet flow without the application of AGP. Since the free-surface condition is not well verified, coiling could not occur, and the result is a radial motion. As oppose to that, Fig. 10 shows the same simulation solved with the introduction of AGP. In this example, coiling occurs as a result of

the better free-surface treatment, and the movement appears to be very natural.

5 CONCLUSION

We proposed some improvements on the highly viscous fluid simulation for the SPH method. The main improvements are related to an implicit time integration scheme with a special boundary treatment of both free-surface and wall boundary treatment using ghost particles. Additionally, our proposed wall boundary approach (FWGP) can apply both slip and no-slip conditions.

Pipe flow with Newtonian and non-Newtonian model and hydrostatic verification tests, as well as the dam break validation tests show that these improvements can increase the accuracy of the simulation as well as alleviate the time step restriction. Furthermore, we demonstrated the robustness of our improvements with a coiling behavior test of high viscous fluid.

In conclusion, the authors consider that the proposed method is suitable to evaluate the behavior of the kira cementitious material injection in underground caves.

ACKNOWLEDGEMENTS

The first author is supported by Japan International Cooperation Agency (JICA) under the program Scholarship for Japanese Emigrants and their Descendants in Latin America and the Caribbean: Program for Developing Leaders in Nikkei Communities. This work was supported by Japan Society for the Promotion of Science (JSPS) KAKENHI Grant Number 17H02061 and "Joint Usage/Research Center for Interdisciplinary Large-scale Information Infrastructures" in Japan (Project ID: jh180060-NAH and jh180065-NAH).

REFERENCES

Asai M, Aly AM, Sonoda Y, Sakai Y (2012). A Stabilized Incompressible SPH Method by Relaxing the Density Invariance Condition. International Journal for Applied Mathematics No. 20130011.

Cotela-Dalmau, J, Rossi, R, Larese, A (2017). Simulation of two- and three-dimensional viscoplastic flows using adaptive mesh refinement. Int J Numer Meth Engng. Vol 112, pg. 1636–1658. DOI 40.1002/nme.5574.

Gingold RA, Monaghan JJ (1977). Smoothed Particle Hydrodynamics: Theory and Application to Non-Spherical Stars. Mon. Not. R. Astron Soc., 181, 375–389.

Lucy LB (1977). A Numerical Approach to the Testing of the Fusion Process. Astron J., Vol.88, pp.1013–1024.

Marrone S, Colagrossi A, Le Touzé D, Graziani G (2010). Fast Free-surface Detection and Level-set Function Definition in SPH Solvers. Journal of Computational Physics 229. 3652–3663.

Martin J, Moyce W (1052). An experimental study of the collapse of liquid columns on a rigid horizontal plane. Philos Trans R Soc Lond 24Ç312–324.

Mattiusi EM (2007). Escoament Laminar de Fluid Newtonianos Generalizados em Tubos de Seção Transversal Elíptica. Doctoral Dissertation of the Federal Technological University of Paraná.

Minussi RB, Maciel GF (2012). Numerical experimental comparison of dam break flows with non-Newtonian fluids. J. of the Braz. Soc. of Mech. Sci. & Eng. April-June 2012, Vol. XXXIV, No. 2/167.

Munson BR, Young D F, Theodore H O (2005). Fundamentals of Fluid Mechanics Fifth Edition. Willey & Sons, Incorporated, John.

Papanastasiou TC (1987). Flows of Materials with Yield. J Rhoel (1978-present). 14(6):385–302.

Schoenberg IJ (1946). Contributions to the problem of approximation of equidistant data by analytic functions. Quarterly of Applied Mathematics. Vol. 4, pp. 45–99.

Tomé MF et al (2004). A Numerical Method for Solving Three-dimensional Generalized Newtonian Free Surface Flows. Journal of Non-Newtonian Fluid Mechanics. Vol. 123, pp. 85–103.

Tsuruta N, Khayyer A, Gotoh H (2015). Space Potential Particles to Enhance the Stability of Projection-based Particle Methods. International Journal of Computation Fluid Dynamics. DOI: 10.1080/10618562.2015.1006130.

Violeau D, Issa R (2017). Numerical modelling of complex turbulent free-surface flows with the SPH method: an overview. International Journal for Numerical Methods in Fluids, Wiley, 2007, 53, pp.277–304. <10.1002/d.1292>. <hal-01097824>

Violeau D, Leroy A (2014). On the maximum time step in weakly compressible SPH. Journal of Computational Physics, Elsevier, 256, pp.388–415. <hal-00946833>

Violeau D, Rogers BD (2016). Smoothed Particle Hydrodynamics (SPH) for Free Surface Flows: Past, Present and Future. Journal of Hydraulic Research, 54:1, 1–26.

2019 Rock Dynamics Summit– Aydan et al. (eds)
© 2019 Taylor & Francis Group, London, ISBN 978-0-367-34783-3

Design of the tsunami protection wall against mega earthquakes and huge tsunamis

Masaaki Wani
Chubu Electric Power Co., Inc., Nagoya, Japan
Graduate School of Engineering and Science, University of the Ryukyus, Japan

Yoshihito Sato
Chubu Electric Power Co., Inc., Nagoya, Japan

Koto Ito
Chubu Electric Power Co., Inc., Nagoya, Japan

ABSTRACT: Chubu Electric Power Company has been implementing countermeasures in Hamaoka Nuclear Power Station (NPS) against tsunami and severe accidents following the disaster at Tokyo Electric Power Company's Fukushima Daiichi NPS that was caused by the 2011 Tohoku Earthquake. For this purpose, the L-shaped Tsunami protection wall 22 m high above sea level was constructed along coastline around the site. The total length of the protection wall was approximately 1.6 km and it was fixed to reinforced concrete underground walls, which were embedded in rock mass. Foundation rock consists of intercalated mudstone and sandstone. The tsunami protection wall has to withstand anticipated huge tsunamis such as the Nankai Trough Giant Tsunami, furthermore it has to withstand against the strong ground shaking of anticipated mega earthquakes to occur before the tsunami. The design seismic force is evaluated by the dynamic response analysis that can consider an interaction between structure, ground and bedrock, whereas the design wave force is set as hydrostatic pressure. Therefore, the tsunami protection wall must be designed reasonably based on the difference of seismic force and tsunami wave force. In this paper, the authors describe the concept of structural design on the basis of the above conditions and consider support performance of the protection wall and bedrock against the anticipated mega earthquake and huge tsunami.

1 INTRODUCTION

1.1 *Overview of Hamaoka NPS*

Chubu Electric Power supplies electric power to the central part of the main island of Japan facing the Pacific Ocean. The Hamaoka NPS is located in Shizuoka prefecture, and along the Pacific coast (Figure 1).

The Hamaoka NPS is Chubu Electric Power's only nuclear power station which has five nuclear power plants. Unit 1 and 2 are under decommissioning since 2009, and other 3 units are now waiting to

Figure 1. Location of Hamaoka NPS

restart. The total output of the remaining Units, 3, 4, and 5, is 3,617 MW.

1.2 *Tsunami countermeasures at Hamaoka NPS*

The 2011 off the Pacific coast of Tohoku Earthquake (hereinafter referred to as the 2011 Tohoku Earthquake), the most massive earthquake that Japan has received in the past, and a huge tsunami that followed caused extensive damage to the Pacific coastal area of eastern Japan. Furthermore the 2011 Tohoku Earthquake and subsequent tsunami had a major impact on nuclear power stations along the Pacific coast, and caused the accident at Tokyo Electric Power's Fukushima Daiichi NPS where radioactive materials were discharged.

When the 2011 Tohoku Earthquake occurred, the nuclear reactors of Fukushima Daiichi NPS sensed massive seismic ground motions and automatically shut down. However, after the earthquake, tsunami waves higher than the station site arrived, flooding the site and buildings. Key facilities were made unusable, including seawater intake pumps for cooling and emergency generators. When batteries ran out, the power station lost its "cooling function". Consequently this led to a severe incident escalating to a massive discharge of radioactive materials.

Fukushima Daiichi NPS was not fully prepared for the arrival of the tsunami nor the subsequent accident. To prevent a similar accident, we had promptly started safety improvement measures work, including tsunami countermeasures, after the accident.

We have applied a three phase strategy to tsunami countermeasures in the Hamaoka NPS (Table 1); "flooding prevention measures 1", "flooding prevention measures 2", and "enhanced emergency measures".

Firstly, "flooding prevention measures 1" are designed to prevent a tsunami flooding the station site. (Figure 2). We constructed "tsunami protection wall" measuring 22 m above sea level, stretching approximately 1.6 km along the front side of the station on the ocean side. In order to prevent a tsunami from entering the station site from the sides, "cement-mixed soil embankments" with a height of 22-24 m above sea level are also constructed on the eastern and western edges of the site (Figure 3-4).

Table 1. Three phase strategy to tsunami countermeasures

Flooding Prevention Measures 1	Prevention of tsunami inundation of the station site
Flooding Prevention Measures 2	Prevention of tsunami flooding of buildings on the site
Enhanced Emergency Measures	Adopting multiple alternative means of electric power supply, water injection, and heat sink

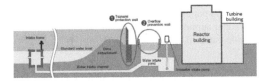

Figure 2. Flooding Prevention Measures 1

Figure 3. Perspective view of the tsunami protection wall and the cement-mixed soil embankments

Figure 4. Tsunami protection wall and cement-mixed soil embankment (eastern side)

In addition, we built "overflow prevention walls", approximately 4 m high, around water intake ponds that are linked to the sea via water intake tunnels.

Secondary, "flooding prevention measures 2" are designed to prevent buildings from flooding even if there is inundation in the station site (Figure 5). For preparedness against a tsunami higher than the tsunami protection wall, the pressure resilience and waterproof performance of exterior doors are reinforced by replacing reactor buildings' waterproof doors with watertight doors and combining them with new tsunami protection doors for dual protection (Figure 6). Watertight doors are also installed at rooms that contain important facilities.

Finally, "enhanced emergency measures" referring to ensure the cooling function will work even if there is a situation similar to that at the Fukushima Daiichi NPS, as there will be multiple alternative means of cooling the reactor such as electric power supplies (Figure 7), water injection, and heat sink.

In this paper, we explain the tsunami protection wall which is a major pillar of tsunami countermeasures of the Hamaoka NPS.

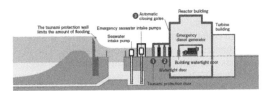

Figure 5. Flooding Prevention Measures 2

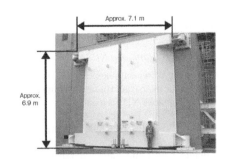

Figure 6. Reinforced protection door

Figure 7. Gas turbine generator Building

2 DESIGN OF TSUNAMI PROTECTION WALL

2.1 *Requirements for design*

There is a dune embankment with a height of 12-15 m above sea level on the ocean side of the Hamaoka NPS, and it had played a significant role as a natural protection against past-tsunamis until the Fukushima Daiichi accident occurred. However, we had promptly started safety improvement measures work, including the tsunami protection wall, after the accident.

When we started to design the tsunami protection wall, there were three requirements given below in consideration of the lessons in the disaster caused by the 2011 Tohoku Earthquake and the local conditions of the Hamaoka NPS.

- To withstand megaquakes and huge tsunamis, which may exceed paleo-quakes and paleo-tsunamis,
- To prevent large deformation against external forces far beyond the design force, and
- To be a slim structure that can be installed at the place with a limited width.

2.2 *Structural overview of the tsunami protection wall*

We had considered a structure satisfying above requirements and reached a conclusion that a combination of a wall, which had enough strength and resiliency, and a foundation, which supported the wall with high stability would be the most suitable.

As a result, we adopted a new structural system for seawalls. An L-shaped composite wall consisting of steel and steel-framed reinforced concrete was fixed to foundation of two underground walls of reinforced concrete that were embedded into solid bedrock (Figure 8). This structure provides an extra safety margin to seismic-resistant and tsunami-resistant design.

2.2.1 *L-shaped wall (upper structure)*

L-shaped wall consists of vertical wall and floor slab. To withstand huge tsunamis, it must have enough strength, but on the other hand, it is desirable to be lightweight to reduce the inertial force of megaquakes. As a result, the vertical wall was designed as steel structure which had high strength and resiliency and was lightweight. Moreover, to enhance seismic resistance, the structurally critical lower part of the vertical wall is filled with concrete.

The L-shaped wall stands 14-16 m high above the site, which is situated at an elevation of 6-8 m above sea level. A total of 109 blocks, each 12 m long, were constructed.

2.2.2 *Underground wall (foundation)*

The size of the underground wall is 7 m in width, 1.5 m thick, and approximately 10-30 m deep according to depth from ground level to bedrock surface.

To withstand megaquakes and huge tsunamis, Large-diameter reinforcing steel, such as D51 steel, is mainly used, and the underground wall is embedded into bedrock consisting of intercalated mudstone and sandstone (Figure 9).

218 underground walls in total were constructed at 6 m intervals and arranged so that they were perpendicular to the vertical wall. Special excavators were used to drill to the designated depth. After erecting reinforced frames assembled at the site, high-fluidity concrete was cast.

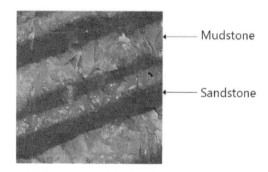

Figure 9. Bedrock (mudstone and sandstone)

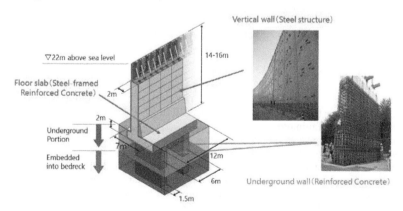

Figure 8. Structural overview of the tsunami protection wall

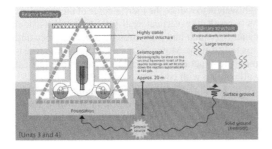

Figure 10. Basic seismic design of Hamaoka NPS

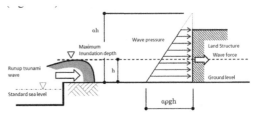

Figure 11. Assessment of wave force acting on a land structure

2.3 Concept of seismic-resistant design

As the Hamaoka NPS is within the hypocentral region of the anticipated Tokai Earthquake, the station has been built with a conservative seismic design from the very beginning of its construction (e.g. highly stable pyramid-like structure, built directly on bedrock) (Figure 10). Similarly, about the tsunami protection wall, seismic resistance was enhanced by embedding the foundation into bedrock.

The seismic structural design of the protection wall was based on a response analysis against the design earthquake ground motions for the Nankai Trough Megaquake that is expected to be even greater than the triple megaquake (the Tokai, Tonankai, and Nankai) that had been anticipated before the 2011 Tohoku Earthquake occurred.

2.4 Concept of tsunami-resistant design

When the 2011 Tohoku Earthquake occurred, many seawalls or other structures were destroyed by subsequent huge tsunamis. With respect to the damage, two failure mechanisms explained are regarded as the principal factors. One is sliding and falling the caisson by tsunami wave force, and the other is scouring the mound.

As mentioned above, the foundation of the tsunami protection wall is embedded into bedrock so that it has high resistance to sliding and scouring. Therefore, it is important to secure enough safety margin for wave force generated by anticipated huge tsunamis.

To calculate tsunami wave force, the appropriate assessment formula is used after taking into account the location where the structures to be assessed are located (undersea or on land).

The formula for assessing tsunami wave force acting on undersea structures is proposed by Tanimoto et al. (1984) and is mainly used in a working level.

On the other hand, regarding the wave force acting on land structures such as the tsunami protection wall, various assessment formulas are proposed (JSCE (2017)). Among these formulas, the major formulas such as Asakura et al. (2000) regard tsunami wave pressure as a hydrostatic pressure

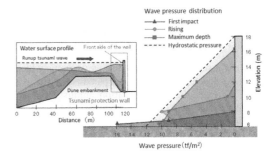

Figure 12. Wave pressures acting on the protection wall

equivalent to α times the maximum inundation depth of a runup tsunami wave in a state without a structure (Figure 11).

Based on the geographic characteristic of the site which has a dune embankment on the ocean side of the wall, hydraulic experiments were carried out to confirm applicability of the concept mentioned above. The nearshore topographical were set up in a 205m long 2-dimensional wave-generating channel to investigate tsunami wave force acting on the wall. The scale of models were 1/40. As a result, it was found that the maximum tsunami wave force acting on the wall was measured when tsunami overflew it, and the vertical distribution of pressures at the maximum tsunami wave force could be expressed by a linear relationship, which depend on the maximum water level at the wall as below Figure 12 (Matsuyama et al. (2012)).

Based on the result of experiment, the design wave force was set by integrating the vertical distribution of wave pressures acting on the wall equivalent to hydrostatic pressures of the wall height.

3 NUMERICAL ANALYSIS ABOUT THE STABILITY OF PROTECTION WALL AGAINST MEGAQUAKE AND HUGE TSUNAMI

The tsunami protection wall has to withstand huge tsunamis such as the Nankai Trough Giant Tsunami, furthermore it has to withstand against the strong ground shaking of anticipated mega earthquakes to occur before the tsunami. As mentioned above, seismic force acts on whole structure and ground dynamically. On the other hand, tsunami wave force acts on mainly the vertical wall statically. Therefore, the wall must be designed reasonably based on the difference of seismic force and tsunami wave force.

Based on the above, numerical analyses were carried out to confirm the differences in the behavior of the protection wall and support performance of the bedrock against the anticipated mega earthquake and huge tsunami.

3.1 Outline of analysis

3.1.1 Seismic response analysis

Firstly, a seismic response analysis of the structure-ground coupled system was conducted using two-dimensional FEM. The analysis model is shown in Figure 13. Specifically, ground including bedrock and floor slab were modelled by plane strain element, and vertical wall and underground wall by beam element. In addition, ground reaction springs to consider sliding and separation were set between the structure and the ground. The depth of the model was set 6m for one underground wall.

Main properties of the structure model were unit weight and Young's modulus. Regarding the bedrock, unit weight, initial shear modulus and deformation characteristics obtained from tests were used. The nonlinearity of deformation characteristics of the ground model during shaking was considered by the modified General Hyperbolic Equation (GHE) model. As an example, the dynamic deformation property of the bedrock is shown in Figure 14.

The input acceleration on bedrock for the analysis is shown in Figure 15.

3.1.2 Tsunami response analysis

Secondary, a tsunami response analysis was conducted. The analysis model was the same one of the seismic analysis basically, however the ground model was linear because it is static analysis. The initial state of the analysis was just after the earthquake, so shear modulus of the ground model was set in consideration of the decrease caused by the earthquake. Specifically, it was set in reference to the $G/G0$ - γ relations of the modified GHE model according to the maximum shear strain of each element.

The tsunami wave force to act on the vertical wall was set as hydrostatic pressures equivalent to 16m height, from ground surface to the top of the wall, based on the above-mentioned information.

3.2 Analysis result

Firstly, responses of the protection wall in case of earthquake and tsunami are shown in Table 2. The deformation of the vertical wall due to the tsunami is small with approximately 5mm at the same level as the maximum deformation due to the earthquake. However, in regard to the bending moment and shear force acting on the underground wall, the responses due to the tsunami are both smaller than those due to the earthquake, especially the difference in shear forces is remarkable.

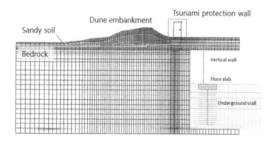

Figure 13. Analysis model (non-linear FEM)

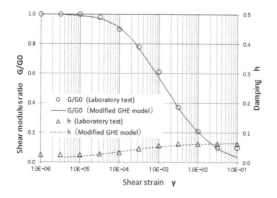

Figure 14. Dynamic deformation property of bedrock

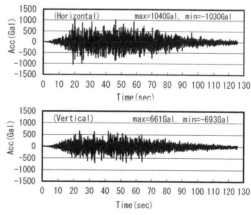

Figure 15. Input acceleration on bedrock

Table 2. Response of the structure (maximum value)

Event	Earthquake	Tsunami
Deformation of vertical wall	4.9mm	4.8mm
Bending moment of underground wall	93.2MNm	74.8MNm
Shear force of underground wall	13.4MN	7.6MN

The vertical distributions of shear force acting on the underground wall are shown in Figure 16. Though both direction and quantity of deformation of the vertical wall are mostly the same, positions of the peak of shear force distribution are different. Specifically, the maximum shear force due to tsunami occurs near the top of the underground wall, whereas the one due to earthquake occurs at lower elevation; near the border of sandy soil and bedrock. Behavior of the underground wall during earthquake is greatly affected by dynamic deformation of ground around the wall. Therefore, this is regarded as the main reason for the difference.

Secondly, with regard to responses of the bedrock, strain components ε_x, ε_y and γ_{xy} of the ground around the underground wall are shown in Figure 17. In the both case, locations of the peak of each strain distribution are mostly the same and they are either near the interface between of sandy soil and bedrock or near the bottom of the underground wall. However, a difference is noted in the distribution shape of the shear strain. Large shear strain is distributed mainly on the right side in case of tsunami, whereas it is distributed not only on the same side but on the other side in case of earthquake. In any way, the maximum

values of these strains on the bedrock due to both earthquake and tsunami are less than the deformability limit of the bedrock, so that it indicates that the underground wall is safely supported by the bedrock.

4 CONCLUSION

In this paper, the fundamental concept of seismic and tsunami-resistant design of the tsunami protection wall at the Hamaoka Nuclear Power Station against mega-earthquakes and huge tsunamis was described.

The evaluation method of the design seismic force and the design tsunami wave force acting on the protection wall in consideration of the local conditions of the site was discussed, moreover, numerical analyses were carried out to confirm the behavior of the protection wall and support performance of the bedrock during anticipated mega-earthquake and associated tsunami. As a result, the following conclusions are drawn.

1. The deformation of the vertical wall of the tsunami protection wall due to the tsunami is as small as the maximum deformation due to the earthquake. However, the loads acting on the underground wall due to the tsunami are smaller than those due to the earthquake.
2. Though both direction and quantity of deformation of the vertical wall are mostly the same, positions of the peak of shear force on the underground wall are different.
3. Behavior of the underground wall during earthquake is greatly affected by dynamic deformation of ground around the wall.
4. The underground wall is safely supported by the bedrock on the basis of the difference of seismic force and tsunami wave force.

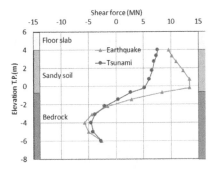

Figure 16. Distribution of shear force on the underground wall

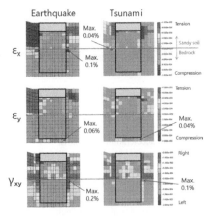

Figure 17. Normal and shear strain distribution in ground around the underground wall

REFERENCES

Tanimoto, K. & Tsuruya, H. & Nakano, S. 1984. Study of Wave Force in 1983 Japan Sea Central Region Earthquake Tsunami and Causes of Damage to Bulkheads. *31st Coastal Engineering Committee Collected Papers*, pp. 257–264 (in Japanese).

Japan Society of Civil Engineers. 2017. Tsunami Assessment Method for Nuclear Power Plants in Japan 2016 (in Japanese).

Asakura, R. & Iwase, K. & Ikeya, T. & Takao, M. & Kaneto, T. & Fujii, N. & Ohmori, M. 2000. Experimental Study on Wave Force Acting on Onshore Structures due to Overflowing Tsunamis. *Proc.of Coastal Engineering, Japan Society of Civil Engineers*, Vol.47: pp. 911–915 (in Japanese).

Matsuyama, M. & Uchino, D. & Hashi, K. & Tanaka, Y. & Sakakiyama, T. & Nakamura, J. & Inaba, D. 2012. Experimental Study on the Effect of Barrier against Tsunami Flowing over a Mound. *Journal of JSCE, B2 (Coastal Engineering), Japan Society of Civil Engineers*, Vol.68: pp. 236–240 (in Japanese).

2019 Rock Dynamics Summit– Aydan et al. (eds)
© 2019 Taylor & Francis Group, London, ISBN 978-0-367-34783-3

Analytical and numerical calculation of stresses and displacements around rectangular tunnel under SV waves in half space

P. Yiouta-Mitra
National Technical University of Athens, Greece

E. Pelli
Earthquake Planning & Protection Organization, Greece

A.I. Sofianos
National Technical University of Athens, Greece

ABSTRACT: The purpose of this research is the calculation of the stress field as well as the displacements occurring around tunnels of unlined rectangular shape, in a linear elastic ground due to SV seismic excitation. The numerical analyses are carried out with the finite-differences Code FLAC_3D. Then, the results are compared to those obtained from an analytical method. The analytical model is based on infinite series expansion of Bessel & Hankel type for the estimation of the wave potentials of incident, reflected and scattered waves, in combination with conformal mapping. This mapping transforms the area between the free surface and a tunnel of depth H in plane w onto two unit circles in the plane ζ, the centers of which are at vertical distance b the one from the other. The analytical calculation of the stress field as well as the deformations of the sections of underground structures, subjected to seismic excitation of shear SV waves, is achieved with the use of MATLAB. A series of dynamic analyses are performed the thickness of the soil layer above the structure in a half space. The model is subjected to seismic excitation simulated by a harmonic vertically propagating shear wave in terms of stress history. SV waves cause racking motion of the cross section of the tunnel. The results of the analyses showed that the sectional stresses and deformations depend on the structure depth, i.e. the deeper the structure the smaller the effect. To sum up, the comparison between the analytical and the numerical method in the elasticity range provided satisfactory results.

1 INTRODUCTION

1.1 *General*

The increasing use of underground facilities requires, in the case of countries within seismic zones, safety provisions against motions caused by earthquakes. Generally, underground structures are less vulnerable than surface structures in case of an earthquake. In recent years, however, the enhanced awareness of seismic hazards for underground structures has prompted an increased understanding of factors influencing their seismic behavior. Nevertheless, due to their importance, even slight damages can be harmful, e.g. leakage of hazardous materials incurring environmental disasters. Further, serious damage to transportation networks may endanger human life as well as hinder transport.

1.2 *Background Knowledge*

The dynamic response of unlined cylindrical cavities in a half-space to transverse SV waves has previously been investigated by many researchers. The stress-free boundary conditions at the free surface are among the most complicated boundary conditions

to be satisfied in physics (Lin et al., 2010). Up to now the problem has been solved approximately.

Lee and Karl, 1992 approached the problem of the application the zero-stress boundary conditions at the free surface by the assumption that, the half space boundary near the cylindrical cavity is approximated by an almost flat convex circular boundary of infinite radius b, while Davis et al. (2000), by the assumption that, the half space boundary near the cylindrical cavity is approximated by an almost flat concave circular boundary of infinite radius. Neither of these approximations did not yield satisfactory results.

Lin et al., 2010, suggested that, without a method to satisfy exactly the stress-free boundary conditions at the half-space surface, it might be better, to simply relax the stress-free conditions without taking into account the scattered SV & P wave from the surface. This proposal yielded better results although it did not solve the problem exactly, as the two ignored potentials, though small, they contribute significantly to the stress field.

In the current research, the analytical methodology proposed by Pelli and Sofianos, 2018 is examined. It solves exactly the problem of the calculation of the stress field around cavities in case of half space under seismic loading of SV waves. They proposed a mapping which transforms the region in the plane w consisting of the half space with a cavity of various shapes (circular, square, horse-shape) is mapped onto two vertical circles in the plane ζ. With the application of this mapping the problem of the free stress boundary conditions at the surface was solved. In order to simulate the depth where the underground structures are founded (deep foundation – full space, shallow foundation-half space), relevant transformation relationships in combination with the implementation of proper boundary conditions are proposed.

2 ANALYTICAL METHODOLOGY

2.1 *Full Space With Cavity*

The mapping Function which transforms the infinite region surrounding the square cavity in the complex plane w (i.e. $w = x + i\,y$) onto the exterior of the unit circle in the complex plane ζ ($\zeta = r\,e^{i\text{th}}$, r=1), applied herein, is given by Figure 1 (Nomikos, 2004)

$$w = R\left(\zeta + \left(\frac{c_3}{\zeta^3}\right) + \left(\frac{c_7}{\zeta^7}\right)\right) \qquad (1)$$

$c_3 = -0.16667$, $c_7 = 0.017857$

After separating real and imaginary parts of $w = x + i\,y$, the parametric equations become

$$x = R\left(cos\theta + c_3cos3\theta + c_7cos7\theta\right)$$
$$y = R\left(sin\theta - c_3sin3\theta - c_7sin7\theta\right) \qquad (2)$$

where R is a scaling factor depending on square geometry. If a_{sq} is the side length of the square, then R is given by:

$$R = \frac{0.5a_{sq}}{1 + c_3 + c_7} \qquad (3)$$

2.2 *Half Space with cavity*

The transformation is developed in three stages. In the first stage, the region in the plane w consisting of the half space with a square cavity of side length a_{sq}, buried at H m below the surface is mapped conformally onto a region consisting of the half space with a unit circle in the plane ξ, buried at «an equivalent depth h» (Figure 2). The transformation is provided

in equation (4) where the "equivalent depth h" of a unit circle onto which the square cavity is mapped, is calculated from equation (5):

$$w = \left(R((\zeta + (ih)) + \left(\frac{c_3}{(\xi + (ih))^3}\right)\right.$$
$$\left. + (\frac{c_3}{\left(\xi + (ih)\right)^{\wedge 7}})\right)) - (iH) \qquad (4)$$

$$R\left(ih + \left(\frac{c_3}{(ih)^3}\right) + \left(\frac{c_7}{(ih)^7}\right)\right) - (iH) = 0 \qquad (5)$$

where c_3, c_7 and R are calculated according to (1) and (3).

In the second stage, the plane ξ is transformed into a full space plane ζ containing two vertical circles, at a distance of their centers b. The second circle is the transformed boundary of the half-space in the previous mapping. The synthesis of all transformations is depicted in Figure 3 and performed according to equations (6-8).

$$w(\zeta_i) = R\left[\left(\xi(\zeta_i) + (ih)\right) + \left(\frac{c_3}{\left(\left(\xi(\zeta_i) + (ih)\right)^{\wedge} 3\right)}\right)\right.$$
$$\left. + \left(\frac{c_7}{\left(\left(\xi(\zeta_i) + (ih)\right)^{\wedge} 7\right)}\right)\right] - iH \qquad (6)$$

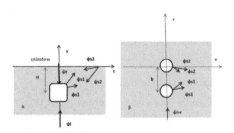

Figure 2. Mapping the half space plane w with square cavity onto the half space plane ξ with unit circle cavity and equivalent depth h.

$H = 10$

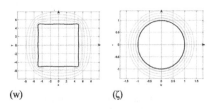

(w)   (ζ)

Figure 1. Exterior of square cavity in plane w onto the exterior of unit circle in plane ζ.

Figure 3. The entire transformation according to equations (6-8) and the respective waves

313

where

$$\xi(\zeta_i) = -ik \left(\frac{1 + \left(\frac{(e^{0.5lpi}\zeta_i) - a}{(e^{0.5lpi}\zeta_i\alpha) - 1} \right)}{1 - \left(\frac{(e^{0.5lpi}\zeta_i) - a}{(e^{0.5lpi}\zeta_i\alpha) - 1} \right)} \right) \quad (7)$$

$$\zeta_i : i = 1,2; \ \zeta_1 = r_1 e^{i\theta} - ib; \zeta_2 = r_2 e^{i\theta} \quad (8)$$

In order to implement the transformation, the curve must be expressed by parametric equations.

2.3 *Calculation of the stress field in half space*

Achenbach (1973) and Tselenti (1997) examined the problem of the calculation of the stress field and the deformations in the sections of underground structures subjected to seismic excitation of shear wave of SV type. Thus they solved the wave equation, by considering the "wave potentials" Φ & Ψ, for SV waves. The total displacement in terms of derivatives of potentials, is defined by the relation (9):

$$u = \nabla\Phi + \nabla \times \Psi \quad (9)$$

The calculation of the wave potentials is achieved by the expansion in finite number of terms, of Bessel-Hankel Fourrier Series, in combination with the implementation of the p roper boundary conditions on the tunnel section. According to Achenbach (1973), elastic stresses and displacements, in terms of potentials "Φ & Ψ", in polar coordinates can be written as:

$$t_{rr} = \lambda\nabla^2\varphi + 2\mu\left[\frac{\partial^2\varphi}{\partial r^2} + \frac{\partial}{\partial r}\left(\frac{1}{r}\frac{\partial\psi}{\partial\theta}\right)\right]$$

$$t_{\theta\theta} = \lambda\nabla^2\varphi + \frac{2\mu}{r}[\frac{\partial\varphi}{\partial r} + \frac{1}{r}\frac{\partial^2\varphi}{\partial\theta^2} + \frac{1}{r}\frac{\partial\psi}{\partial\theta} - \frac{\partial^2\psi}{\partial r\partial\theta}$$

$$t_{r\theta} = \mu\left\{2\left(\frac{1}{r}\frac{\partial^2\varphi}{\partial r\partial\theta} - \frac{1}{r^2}\frac{\partial\varphi}{\partial\theta}\right)\right. \quad (10)$$

$$\left. + \left[\frac{1}{r^2}\frac{\partial^2\psi}{\partial\theta^2} - r\frac{\partial}{\partial r}\left(\frac{1}{r}\frac{\partial\psi}{\partial r}\right)\right]\right\}$$

$$u_r = \frac{\partial\varphi}{\partial r} + \frac{1}{r}\frac{\partial\psi}{\partial r}$$

$$u_\theta = \frac{1}{r}\frac{\partial\varphi}{\partial\theta} - \frac{\partial\psi}{\partial r}$$

Mapping an arbitrary hole in the physical plane w onto a unit circle in the plane ζ, via the mapping function w (ζ), the equations (10) become equations (11):

$$t_{rr} = -k_a^2(\lambda + \mu)\varphi$$

$$+ \frac{1}{r^2}\frac{2\mu\zeta^2}{w'(\zeta)}\left[\frac{\partial}{\partial\zeta}\left(\frac{1}{w'(\zeta)}\frac{\partial(\varphi + i\psi)}{\partial\zeta}\right)\right]$$

$$+ \left[\frac{1}{r^2}\frac{2\mu\bar{\zeta}^2}{\overline{w'(\zeta)}}\left(\frac{\partial}{\partial\bar{\zeta}}\left(\frac{1}{\overline{w'(\zeta)}}\frac{\partial(\varphi - i\psi)}{\partial\bar{\zeta}}\right)\right)\right] \quad (11)$$

$$t_{r\theta}$$

$$= -\frac{1}{ir^2}\frac{2\mu\zeta^2}{w'(\zeta)}\left[\frac{\partial}{\partial\zeta}\left(\frac{1}{w'(\zeta)}\frac{\partial(\varphi + i\psi)}{\partial\zeta}\right)\right]$$

$$+ \left[\frac{1}{ir^2}\frac{2\mu\bar{\zeta}^2}{\overline{w'(\zeta)}}\left(\frac{\partial}{\partial\bar{\zeta}}\left(\frac{1}{\overline{w'(\zeta)}}\frac{\partial(\varphi - i\psi)}{\partial\bar{\zeta}}\right)\right)\right]$$

$$t_{\theta\theta}$$

$$= -k_a^2(\lambda + \mu)\varphi$$

$$- \frac{1}{r^2}\frac{2\mu\zeta^2}{w'(\zeta)}\left[\frac{\partial}{\partial\zeta}\left(\frac{1}{w'(\zeta)}\frac{\partial(\varphi + i\psi)}{\partial\zeta}\right)\right]$$

$$- \left[\frac{1}{r^2}\frac{2\mu\bar{\zeta}^2}{\overline{w'(\zeta)}}\left(\frac{\partial}{\partial\bar{\zeta}}\left(\frac{1}{\overline{w'(\zeta)}}\frac{\partial(\varphi - i\psi)}{\partial\bar{\zeta}}\right)\right)\right]$$

$$u_r = \frac{\zeta}{r}\frac{1}{|w'(\zeta)|}\frac{\partial(\Phi + i\Psi)}{\partial\zeta} + \frac{\bar{\zeta}}{r}\frac{1}{|w'(\zeta)|}\frac{\partial(\Phi - i\Psi)}{\partial\bar{\zeta}}$$

$$u_\theta = \frac{-\zeta}{ir}\frac{1}{|w'(\zeta)|}\frac{\partial(\Phi + i\Psi)}{\partial\zeta} + \frac{\bar{\zeta}}{ir}\frac{1}{|w'(\zeta)|}\frac{\partial(\Phi - i\Psi)}{\partial\bar{\zeta}}$$

The wave potentials, for a tunnel at depth H below the free surface under seismic loading of SV waves are: the incident, the reflected, two scattered wave potentials from the cavity and two scattered wave potentials from the free surface (see Figure 3). The boundary conditions in this case are:

(a) The stress-free boundary conditions at the cavity surface in plane ζ, $t_{r1r1} = t_{r1\theta1} = 0$
(b) The stress-free boundary conditions at the free surface in plane ζ, $t_{r2r2} = t_{r2\theta2} = 0$

By solving the system of equations from the implementation of (a) and (b) the calculation of the wave potentials and thus the stress field is achieved.

2.4 *Boundary Conditions*

The potentials must satisfy the stress-free boundary conditions both at the surfaces of the cavity and of half space in the plane ζ:

$$t_{r1r1} = t_{r1\theta1} = t_{r2r2} = t_{r2\theta2} = 0$$

314

where r_1, r_2 are the radii of the circles in the ζ plane corresponding to the cavity and the half space after mapping respectively. The stress-free boundary conditions at the cavity surface involve the following system of equations:

$$\psi_i = |\psi_0| e^{ik_b b} \frac{1}{2} \sum_{n=0}^{\infty} \varepsilon_n i^n J_n\left(k_b |w(\zeta_1)|\right)$$

$$\left(\left(\frac{w(\zeta_1)}{|w(\zeta_1)|}\right)^n i^{-n} + \left(\frac{w(\zeta_1)}{|w(\zeta_1)|}\right)^{-n} i^n\right)$$

$$\psi_r = |\psi_0| e^{-ik_b b} \frac{1}{2} \sum_{n=0}^{\infty} \varepsilon_n (-i)^n J_n\left(k_b |w(\zeta_1)|\right)$$

$$\left(\left(\frac{w(\zeta_1)}{|w(\zeta_1)|}\right)^n i^{-n} + \left(\frac{w(\zeta_1)}{|w(\zeta_1)|}\right)^{-n} i^n\right)$$

$$\psi_{s1} = \frac{1}{2} \sum_{n=0}^{\infty} B_n H_n\left(k_b |w(\zeta_1)|\right)$$

$$\left(\left(\frac{w(\zeta_1)}{|w(\zeta_1)|}\right)^n i^{-n} + \left(\frac{w(\zeta_1)}{|w(\zeta_1)|}\right)^{-n} i^n\right) \quad (12)$$

$$\varphi_{s1} = \frac{1}{2i} \sum_{n=0}^{\infty} A_n H_n\left(k_a |w(\zeta_1)|\right)$$

$$\left(\left(\frac{w(\zeta_1)}{|w(\zeta_1)|}\right)^n i^{-n} - \left(\frac{w(\zeta_1)}{|w(\zeta_1)|}\right)^{-n} i^n\right)$$

$$\psi_{s2} = \frac{1}{2} \sum_{n=0}^{\infty} D_n J_n\left(k_b |w(\zeta_2 f(\zeta_1))|\right)$$

$$\left(\left(\frac{w(\zeta_2 f(\zeta_1))}{|w(\zeta_2 f(\zeta_1))|}\right)^n i^{-n} + \left(\frac{w(\zeta_2 f(\zeta_1))}{|w(\zeta_2 f(\zeta_1))|}\right)^{-n} i^n\right)$$

$$\varphi_{s2} = \frac{1}{2i} \sum_{n=0}^{\infty} C_n J_n\left(k_a |w(\zeta_2 f(\zeta_1))|\right)$$

$$\left(\left(\frac{w(\zeta_2 f(\zeta_1))}{|w(\zeta_2 f(\zeta_1))|}\right)^n i^{-n} - \left(\frac{w(\zeta_2 f(\zeta_1))}{|w(\zeta_2 f(\zeta_1))|}\right)^{-n} i^n\right)$$

$$\zeta_1 = r_1 e^{i\theta}, \quad \zeta_2 = f(\zeta_1) = \zeta_1 + ib$$

$$\varphi_{tot} = \varphi_{s1} + \varphi_{s2} \quad (13)$$

$$\psi_{tot} = \psi_i + \psi_r + \psi_{s1} + \psi_{s2}$$

and the coefficient $\varepsilon_n = 1$ and $\varepsilon_n = 2$ for $n \geq 2$.

The stress-free boundary conditions at the half-space surface involve the following system of equations:

$$\psi_{s1}$$

$$= \frac{1}{2} \sum_{n=0}^{\infty} A_n J_n\left(k_b |w(\zeta_1 f(\zeta_2))|\right) \left(\left(\frac{w(\zeta_1 f(\zeta_2))}{|w(\zeta_1 f(\zeta_2))|}\right)^n i^{-n}\right.$$

$$\left. + \left(\frac{w(\zeta_1 f(\zeta_2))}{|w(\zeta_1 f(\zeta_2))|}\right)^{-n} i^n\right)$$

$$\varphi_{s1}$$

$$= \frac{1}{2i} \sum_{n=0}^{\infty} B_n J_n\left(k_a |w(\zeta_1 f(\zeta_2))|\right) \left(\left(\frac{w(\zeta_1 f(\zeta_2))}{|w(\zeta_1 f(\zeta_2))|}\right)^n i^{-n}\right.$$

$$\left. - \left(\frac{w(\zeta_1 f(\zeta_2))}{|w(\zeta_1 f(\zeta_2))|}\right)^{-n} i^n\right) \quad (14)$$

$$\psi_{s2} = \frac{1}{2} \sum_{n=0}^{\infty} D_n H_n\left(k_b |w(\zeta_2)|\right) \left(\left(\frac{w(\zeta_2)}{|w(\zeta_2)|}\right)^n i^{-n}\right.$$

$$\left. + \left(\frac{w(\zeta_2)}{|w(\zeta_2)|}\right)^{-n} i^n\right)$$

$$\varphi_{s2} = \frac{1}{2i} \sum_{n=0}^{\infty} C_n H_n\left(k_a |w(\zeta_2)|\right) \left(\left(\frac{w(\zeta_2)}{|w(\zeta_2)|}\right)^n i^{-n}\right.$$

$$\left. - \left(\frac{w(\zeta_2)}{|w(\zeta_2)|}\right)^{-n} i^n\right)$$

$$\zeta_1 = r_1 e^{i\theta}, \quad \zeta_2 = f(\zeta_1) = \zeta_1 + ib$$

$$\varphi_{tot} = \varphi_{s1} + \varphi_{s2} \quad (15)$$

$$\psi_{tot} = \psi_{s1} + \psi_{s2}$$

and the coefficient $\varepsilon_n = 1$ and $\varepsilon_n = 2$ for $n \geq 2$.

In both systems, $w(\zeta)$ is derived from equations (6-8) and ψ_{i+r} already satisfies the stress free conditions at the half space ground surface. The wave potentials φ_{tot} and ψ_{tot} in equations (13) and (15) are then differentiated twice, first for the cavity and second for the half space respectively and thus the boundary conditions can be applied by substituting the results into (12) & (14). Solving the resulting system the unknown coefficients A_n, B_n, C_n, D_n, are obtained.

2.5 Example Cases

An underground structure considered to be buried in linear elastic ground is selected as the example case. The cavity is subjected to incident plane SV waves having the potential ψ_i, angle of incidence 90°, angular frequency ω and transverse wave number k_b. Waves propagating upward through horizontal layers of successively lower transmission velocity, as it is common near the earth's surface, will be refracted closer and closer to a vertical path, (Kramer, 1996). Thus, the shear seismic waves practically propagate in the vertical direction.

The compilation of the transformations to square shape as well as the boundary conditions and the solutions of the system of the equations for the calculation of the wave potentials, have been achieved by composing the relevant computing codes in the language of technical computing MATLAB.

Two example cases have been analyzed, a deep buried tunnel and a shallow one, by use of the full and half space transformations respectively. The selected properties for the elastic soil are given in Table 1.

2.5.1 Deep Structure
The case of the square cavity (Figure 4), with side length $a_{sq} = 10$m in plane w under seismic loading of SV waves was examined and the relevant analytical solution was calculated. The square cavity in the plane w was transformed onto the unit circle in the plane ζ, with the implementation of (1) and R = 5.874149.

The absolute values of the wave potentials, the stress $t_{\theta\theta}$ and the displacements u_r & u_θ, are drawn in Figure 5, for theta $[\pi/2, 2\pi+\pi/2]$.

2.6 Shallow structure

In this case a square cavity of 10m side, at depth H = 10m below the free surface is considered. The area between the square cavity and the free surface in the plane w, is mapped onto the area outside two vertical unit circles in the plane ζ (Figure 6). The ground is mapped on to the circle at (0,0) and the cavity is mapped at distance -b according to the mapping function of equations (6-8). From the geometry of the square cavity:

R = 5.874149, h = 1.733964 m, b = 2.3383603 m

Table 1. Selected properties of elastic soil.

ρ: density	1800kg/m³
V_s: velocity of SV-wave propagation through medium	400 m/sec
V_p: velocity of P-wave propagation through medium	800 m/sec
v:Poisson ratio	0.33
G: shear modulus	2.88·10⁸Pa
$k_b = \omega/V_s$:Transverse wavenumber for S waves	0.6283
$k_a = \omega/V_p$:Transverse wavenumber for P waves	0.3141

With the use of MATLAB, a source code is compiled in order to calculate the coefficients of the system A_n, B_n, C_n, D_n and the wave potentials. Substituting the above into equations (11) the stress field is obtained.

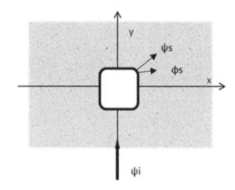

Figure 4. Example of incident SV wave on deep square cavity

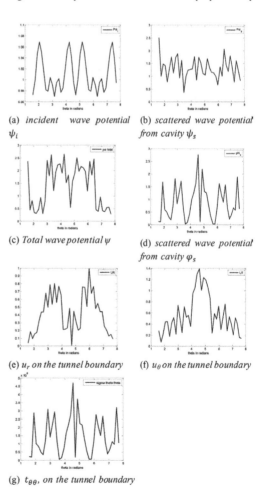

(a) incident wave potential ψ_i

(b) scattered wave potential from cavity ψ_s

(c) Total wave potential ψ

(d) scattered wave potential from cavity φ_s

(e) u_r on the tunnel boundary

(f) u_θ on the tunnel boundary

(g) $t_{\theta\theta}$, on the tunnel boundary

Figure 5. Absolute value of wave potentials, stress and displacement fields, for square deep cavity

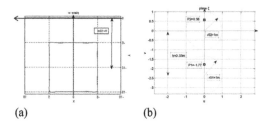

(a)   (b)

Figure 6. The composite transformation (6-8) for the example case of a specific shallow tunnel.

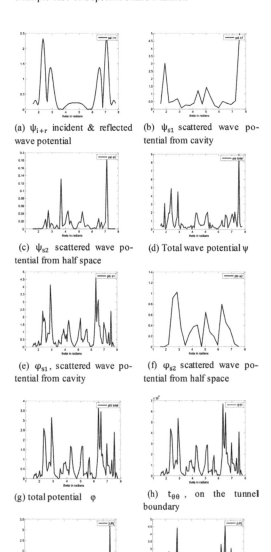

(a) ψ_{i+r} incident & reflected wave potential

(b) ψ_{s1} scattered wave potential from cavity

(c) ψ_{s2} scattered wave potential from half space

(d) Total wave potential ψ

(e) φ_{s1}, scattered wave potential from cavity

(f) φ_{s2} scattered wave potential from half space

(g) total potential φ

(h) $t_{\theta\theta}$, on the tunnel boundary

(i) u_r, on the tunnel boundary

(j) u_θ on the tunnel boundary

Figure 7. Absolute value of wave potentials, stress and displacement fields, for square shallow cavity

The absolute values of the wave potentials, the stress $t_{\theta\theta}$ and the displacements u_r & u_θ, are drawn in Figure 7, for theta $[\pi/2, 2\pi+\pi/2]$.

3 NUMERICAL ANALYSES

3.1 *Model*

In order to simulate the dynamic response of the underground structures, the numerical code of finite differences FLAC 3D, is used. This code is widely used in geotechnical engineering for the solution of the differential equation of motion. The soil-structure model is considered elastic and is subjected to seismic excitation, simulated by a harmonic vertically propagating shear wave SV. In addition the code incorporates the language FISH, for further programming and for the amelioration of the model. The boundary conditions applied at the artificial boundaries were:

a) Lateral dashpots to minimize wave reflections and achieve free-field conditions and

b) Absorbing boundaries at the bottom i.e. normal and shear dashpots of coefficient $c=\rho\cdot C_s\cdot V_s$ where, C_s is speed of s-wave propagation through medium, ρ is the mass density, and V_s the shear particle velocity (Kramer, 1996) to represent the effect of the bedrock presence in case of half space and absorbing boundaries both at the bottom and the surface in case of full space.

A stress boundary condition is used to input seismic motion at absorbing boundaries. The stress is calculates by converting a wave to an applied stress using the formula: $\sigma_{xz}=2\rho V_s C_s sin\omega t$ (Pelli et al, 2004). The soil-structure interaction effect is expressed by the flexibility ratio F, a measure of the soil stiffness relative to the flexural stiffness of the liner.

3.2 *Parameters*

Dynamic analyses were performed, in the elastic region taking into account:

− The thickness of the soil layer H above the structure. When the foundation is deep it is a full space case, in the opposite it is a half space case. In this study, the structure depth H is the distance between the structure and the free surface measured from canter to center, two «equivalent radii» in all other cases.

− The relative stiffness between the soil and the cavity, simulated by the flexibility ratio F (unlined or lined cavity rigid or flexible). In these analyses, unlined tunnel is considered.

− The hysteretic damping of Rayleigh type, independent of the frequency. In this study damping is considered equal to 2%.

The aforementioned factors play a crucial role in the calculation of the stress field. All the analyses verified the properties and the boundary conditions of SV waves in full and half space.

Finally, the properties of the elastic soil are according to Table 1. Both the square cavities were subjected to seismic wave of SV type with value of $\sigma_{xz} = 1.44$ MPa.

3.3 Results

3.3.1 Deep Cavity

The contours for displacements u_x & u_z and stresses $\sigma_{yy}, \sigma_{xx}, \sigma_{xz}, \sigma_{zz}$, for the deep square structure of 10m opening are drawn in Figure 8. A harmonic vertically propagating shear wave, causes racking deformation on the square cross section.

Results confirm that the cross section of the cavity has an important effect in the distribution of the internal forces.

Figure 9. depicts the displacements after they have been converted to polar coordinates and their maximum values plotted with respect to theta $[-\pi, +\pi]$.

In order to compare the results of the numerical analysis to the analytical, we must take into account the fact that in the analytical methodology, the displacements u_r & u_θ are multiplied by $|y_0 k_b|$, where y_0 is the wave amplitude (Lee & Karl, 1992). The value of y_0 corresponding to the incident wave is considered to be the unit, i.e. the displacements are normalized with respect to the wave amplitude. On the other hand, in the numerical solution, in order to calculate the stress corresponding to the incident wave, the value of V_0 was considered to be the unit, i.e. the displacements are normalized with respect $V_0 = \omega y_0$. As a result, the values of u_r and u_θ, from the analytical methodology must be multiplied by

$$\frac{V_0}{\omega} k_b = \frac{V_0}{\omega} \frac{\omega}{V_s} = \frac{1}{400}$$

where: $V_s = 400$m/sec, $k_b = 0.6283$

Then it becomes obvious that there is agreement between the numerical and the analytical solution.

Figure 10 depicts the superposition of the numerical results on the analytical solution for the deep cavity where it can be seen how they compare.

3.3.2 Shallow cavity

The contours for displacements u_x & u_z and stresses $\sigma_{yy}, \sigma_{xx}, \sigma_{xz}, \sigma_{zz}$, for the soil-structure system when the structure is a shallow square cavity of 10m opening are depicted in Figure 11. Displacements

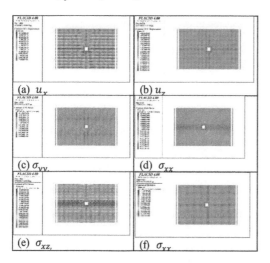

(a) u_x (b) u_z

(c) σ_{yy} (d) σ_{xx}

(e) σ_{xz},   (f) σ_{xx}

Figure 8. Displacements and stresses for square deep cavity from numerical analyses

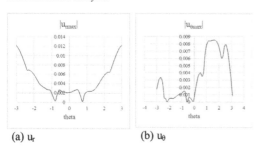

(a) u_r (b) u_θ

Figure 9. Maximum displacements in polar coordinates for square deep cavity from numerical analyses

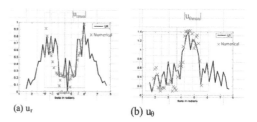

(a) u_r (b) u_θ

Figure 10. Superposition of maximum radial displacements numerical results on the analytical solution for the case of square deep cavity.

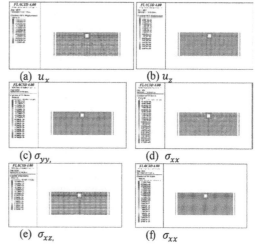

(a) u_x (b) u_z

(c) σ_{yy},   (d) σ_{xx}

(e) σ_{xz},   (f) σ_{xx}

Figure 11. Contours of displacements and stresses for square shallow cavity from numerical analyses

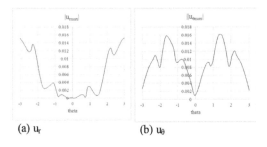

| (a) u_r | (b) u_θ |

Figure 12. Maximum displacements in polar coordinates for square shallow cavity from numerical analyses

exist in x and z direction on the cavity boundary while the displacements along the axis of the cavity are zero. The maximum displacements on the tunnel boundary have been converted to polar coordinates and are drawn in Figure 12 with respect to theta [-π, +π].

It can be seen that the shallow cavity can give up to 50% greater displacements as compared to the deep cavity. There is also general agreement between the respective analytical solution and the numerical solution, taking into account the difference in the two approaches as mentioned in the previous paragraph.

4 CONCLUSIONS

An analytical methodology for the effects of SV waves on square cavities has been examined that is based on the principles of conformal mapping. The general idea was based on the selection of a proper mapping function, which transforms the region surrounding the cavity in a complex z-plane onto the interior or exterior of the unit circle in another complex ζ-plane. Then, the boundary conditions of the problem given in the z-plane are simultaneously transformed into an appropriate form for the ζ-plane. The cases examined involved an investigation of the effects of depth on the displacements and stresses on the cavity.

Results showed that the structure depth plays a crucial role and is interrelated to the frequency of the dynamic excitation and the natural frequencies of the soil-structure system. It has been found that the deeper structure is not affected by the seismic excitation and thus previous related but less precise research is confirmed. The case presented for comparison to the full space, is a square cavity of 10 m side under seismic loading of waves of SV type situated at depth 10m and simulating shallow constructions (half space). Analyses indicate that

displacements and stresses on the cross section of the tunnel are about 50% greater than those of the cavity in full space.

A very important difference between the analytical and the numerical method is that the latter considers soil structure as an integrated system. It is a holistic approach, while on the other hand, the analytical method considers the tunnel individually

Finally, the cross section of the square cavity has an important effect in the distribution of the internal forces as compared to results from research on circular cavities.

ACKNOWLEDGEMENTS

The authors would like to acknowledge the support received from the European Commission Research Executive Agency via a Marie Skłodowska-Curie Research and Innovation Staff Exchange project (689857-PRIGeoC-RISE-2015).

REFERENCES

Achenbach J., (1973). "Wave Propagation in Elastic Solids», North Holland Publishing.

Davis C.A., Lee V.W., Bardet J.P., (2000). "Transverse response of underground cavities and pipes to incident SV waves", Earthquake Engineering and Structural Dynamics 30, p.383–410

Itasca Consulting Group, (2008). Manual FLAC Version 6.0, Fast Lagrangian Analysis of Continua, Inc. Minneapolis, Minnesota USA

Itasca Consulting Group, (2009). Manual FLAC3D Version 4.0, Fast Lagrangian Analysis of Continua in 3 Dimensions, Inc. Minneapolis, Minnesota USA

Kramer S., (1996). "Geotechnical Earthquake Engineering", Prentice Hall Upper Saddle River, New Jersey 07458.

Lee V., Karl J. (1992). "Diffraction of SV waves by underground, circular cylindrical cavities", Soil Dynamics and Earthquake Engineering,11 p.445–456.

Lin Chi Hsin, Lee VW, Todorovska M, Trifunac MD (2010). "Zero-stress, cylindrical wave functions around a circular underground tunnel in a flat, elastic half-space: Incident P-waves", Soil Dynamics and Earthquake Engineering 30, 879–894.

Pelli E, Yiouta-Mitra P, Sofianos AI (2006). "Seismic Behaviour of Square Lined Underground Structures", Tunnelling and Underground Space Technology, V.21, 3–4, May – June, 441.

Pelli E., Sofianos A., (2018). «Analytical calculation of the half space stress field around tunnels under seismic loading of SV waves », Tunnelling and Underground Space Technology 79, 150–174)

Pelli E. (2018): «Analytical relations and numerical code for the calculation of the stress field around tunnels of different shapes under seismic loading of S waves », PHD Thesis, National Technical University of Athens.

2019 Rock Dynamics Summit– Aydan et al. (eds)
© 2019 Taylor & Francis Group, London, ISBN 978-0-367-34783-3

Consideration of structural stability for Oya underground quarry with dynamic response

T. Seiki
Utsunomiya University, Tochigi, Japan

K. Takahashi
Utsunomiya City Hall, Tochigi, Japan

T.K.M. Dintwe
Kyusyu University, Fukuoka, Japan

S. Noguchi & T. Ohmura
Oya observation centre, Kawasaki Geological Engineering Co.,Ltd., Utsunomiya, Japan

Ö. Aydan
The University of Ryukyus, Okinawa, Japan

ABSTRACT: Oya tuff, which is categorized as soft rock, has been excavated in Oya area of Utsunomiya City, Tochigi Prefecture, Japan, for building stone and retaining walls for several decades. It is one of the most commonly used rock materials in Japan. As a result, there are many abandoned underground quarries existing in this area. After the 2011 East Japan Great Earthquake, the consideration of the seismic responses of underground quarries has become very important. Almost all of Oya underground quarries have been excavated using room and pillar method except some, which are long wall type. In the study, the quarry under investigation is an active room and pillar type and this paper presents the field measurement techniques used to study the seismic response and micro tremor of the quarry. Through the seismic response analysis, this study tried to clear underground quarries. A numerical model was then developed using a FDM to assess the dynamic stability of the quarry.

1 INTRODUCTION

There are over 200 abandoned underground quarries located in Oya town, Utsunomiya city and Tochigi prefecture, Japan. Those underground quarries have been excavated by mining Oya tuff as a building stone for retaining walls and decoration plate for interior walls of houses. In the past there has been sinkholes caused by the underground mines of Oya tuff and in 1989 there was a huge land subsidence which resulted in some property such as houses being damaged (Oyagi and Hungr, 1989). Recently, another collapse occurred in Oya town during the 2011 Great East Japan Earthquake; however, this time, it was collapse of a semi underground mine pillar in Oya town (Aydan, 2015). With occurrence of this events, underground quarries are considered as unstable, therefore, several study was conducted to evaluate the stability (Katayose et. al, 2008, Seiki et. al, 2007 and 2016).

Previous study tried to understand the pillar movement in seismic wave and micro tremor. And it shows the movement of a part of the underground quarry (Seiki et. al, 2018).

This study tried progress the characteristics of seismic response of Oya tuff underground quarry through field monitoring and numerical analysis. At first the author carried out selecting seismic record and calculating amplitude ratio via FFT technique to clarify site effect of seismic response in Oya tuff underground. Secondly, we also measured micro tremor in/ above the active underground quarries and calculated H/V spectrum (Nakamura, Y.,2008) to understand the relation of seismic response in/above the quarries deeply instead of the seismic record. Finally, we carried out the numerical analysis of Finite Differential Method to simulate the seismic response to understand the dominant frequency of the underground quarries and the ground in Oya.

2 SEISMIC RESPONSE OF UNDERGROUND QUAERRY

2.1 *Introduction to seismic analysis*

In this study, the authors carried out field seismic measurement at underground quarry and above ground. We also considered the dynamic stability of seismic response of the underground quarry with Fourier spectrum and natural frequency. We refereed the data measured by Observation center for

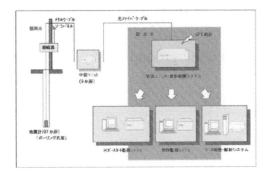

Figure 1. Concept for observation system in Oya.

(a)　　　　　　(b)　　　　　　(c)

Figure 2. Monitoring system for Oya underground quarries (a) solar panel and pole for observation point with amplitude,(b) seismology meter (SM-4-3D), (c) transmitting unit .

ruined underground quarry. Whereas, Observation system for Oya tuff ruined underground quarries which consisted in 97 velocity meters, has measured tremor by cracking and spalling the roof and walls of those underground quarries and monitored the structural stabilities integrally for the those safety. The system is also monitoring ground water depth. The acquired data is analytically used for evaluating the collapse of the ruined quarries. Figure 1 show the system diagram. Velocity meters of 3 components, i.e. one vertical and two horizontal direction or 1 component, vertical direction only, are equipped at individual measuring point. Velocity meter for 1 components are settled because it is enough to catch the tremor induce by cracking and spalling. However,only at concerning point, the 3 component one are equipped. The instruments are bullied at the bottom of the borehole of 66 mm in diameter installed under over 20m from the ground surface due to neglect artificial noise on the ground. And amplifier was equipped at individual point to get precise tremor wave-shape by solar panels (See Fig. 2). Measured data was transferred into optical digital signal and reached at the observation centre.

2.2 Analytical condition

In previous study mainly carried out analysis, it focuses on apparently huge seismic wave. However as the huge seismic wave may largely affect to the seismic response, the author adopt middle range seismic data to the numerical analysis. On the other way, the authors also apply seismic wave data recorded near epicentre on the bottom of the geological model. It gave the exaggerate results against the collops possibility. In this study, the author apply the seismic wave data recorded in Oya area. It is more realistic way to simulate the seismic response. To neglect the difference among the epicentre location of seismic data as small as possible the authors chose the seismic data from the proper location, of epicentre, south-west part in Ibaragi prefecture for the analysis. To neglect the difference of transmitting feature, we chose the seismic data of epicentre which are deeper

Table 1. Factor f or selecting seismic wave data.

Items	Option
Area of epicenter	Southwest of Ibaraki Prefecture Lat.: 36.019 N – 36.358 N Long: 139.685 E – 140.109 E
Depth of epicenter	Greater than 50 km
Magnitude Mj	Greater than 3.0

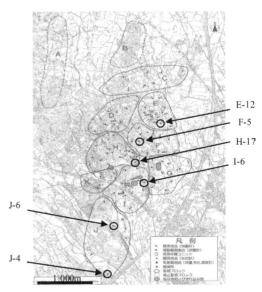

Figure 3. Observation points in Oya

than 50 km. And to get characteristics of the seismic wave in the analysis, we chose the magnitude of seismic waves, Mj, which are greater than 3.0 (See Table 1). Additionally the authors selected the seismic waves which are able to distinguish among P-wave and S-wave in those records. And we categorize the data group into the data before rising ground water or no-water in the quarries and the data of the quarry filled with ground water. Later, we will name those groups as no ground water and filled ground water, respectively. We chose every

16 monitoring point in the two groups. Figure 3 shows 6 observation points for analysis. In this analysis, we normalize the results of FFT analysis on every points by the results on the point J-4 and calculate the amplitude ratio of FFT, because the observation point J-4 locates most south part in Oya observation region and the seismic wave from southwest part in Ibaraki Prefecture may reach the top points. The amplitude ratio helps us to understand the characteristics of transmission for the seismic wave in this field. And we add the observation point J-6 which has no water in the ruined quarry beneath the observation point. We used the seismic wave from the beginning of P-wave to subsiding S-wave without the part of noise on the wave-shape. Figure 4 shows typical seismic wave for the FFT analysis.

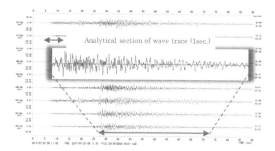

Figure 4. example of seismic record measured in Oya.

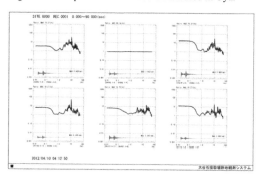

Figure 5. example of FFT spectrum on an observation point

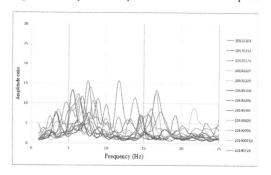

Figure 6. Example of Fourier spectrum of seismic wave on individual observation point.

2.3 *Results of FFT analysis*

We carried out Fourier spectrum analysis near the underground quarry with no ground water and filled water at all observation points. Figure 5 shows the example one. Figure 6 shows the result of FFT analysis of the all seismic wave data on the all observation points. As those spectrum includes dispersion widely in 16 seismic wave data before and after raising ground water level even though we neglect the characteristics of transmitting path of those waves, we have averaged those FFT spectrum to simplify those figures and show the results in Figure 7 and 8.

Transmitting feature shows that there is peak value between about 5.0Hz and 10 Hz shown in Figure 7 and 8. Additionally, there is peak value from about 15.0 Hz to 17.0 Hz after rising ground water in the ruined quarry shown in Figure 8. Comparing the amplitude ration before rising ground water with that after rising ground water, the amplitude ratio after rising ground water level is smaller than that before one. The value of observation point J-6 where the quarry has no ground water took apparently high amplitude ratio in all observation points shown in Figure 8. Based on those results it has been clear that underground quarries shows the peak of seismic response in about 5.0Hz -10Hz. Additionally, amplitude ratio after rising ground water tend to go down. Especially this tendency is

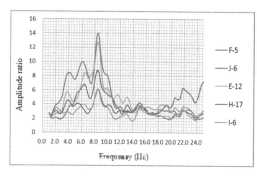

Figure 7. transmitting feature of underground quarries before rising ground water.

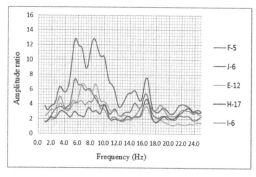

Figure 8. transmitting feature of underground quarries after rising ground water.

remarkably shown in the observation points H-17 and I-6. The amplitude reduction may indicate that the ground water inside of the ruined quarries affect to settle and develop the structural stability against seismic event. However, even though peak value of amplitude ratio around 15.0 Hz – 17.0 Hz is shown after rising ground water, this tendency is not seen before rising ground water and the observation point J-6 around the underground quarry with no ground water and some of observation points shows the peak value around 17.0Hz before and after rising ground water, the authors think that the peak amplitude ration around 17 Hz may be the characteristics of epicentres.

3 MICRO TREMOR ANALYSIS

Even though micro tremor and seismic wave are essentially different each other, relativity among them are recognized. So it is regarded that analysis of micro tremor gives same results by seismic wave. In this chapter, the authors carried out FFT analysis of observation data of micro tremor.

3.1 Introduction to the analysis

In this study, we used micro tremor measuring instrument, which has 2 components for horizontal velocity and 1 component of vertical one in a box. We settled the instrument and surveyed micro tremor at the base floor in the active underground quarries for Oya tuff. (J-5 and E-16 in Figure 9). Every measurement took 15 minutes for measuring micro tremor in three components. And we also compared the data and the data measured on the ground. We analyzed 3 components on FFT and calculated H/V spectrum.

3.2 Result of analysis for micro tremor

Figure 10.and 11 shows the results of active underground quarry A (it is a room and pillar type quarry, shown in J-5 in Figure 9) and B (it is a long wall type quarry shown in E-16 in Figure 9). We think those figures shows no special aspect of the underground quarries. The results above ground shows that the peak values are around 9.0 Hz - 11 Hz at the active underground quarry A and it is 6.0 Hz at the quarry B.

On the other hand, it is difficult to recognize the peak value in both underground quarries because the data include the effect of rock mass of Oya tuff instead of structure of the quarries as the measurement s carried out the on the base floor. However, the data on the ground includes the effect of the underground.

In the case of underground quarry A, the peak value of amplitude ratio shows 5.0 Hz -10.Hz as the seismic wave response shown in the Chapter 2. The peak value, around 11.0 Hz, by micro tremor on the ground may include the effect of soil layer,

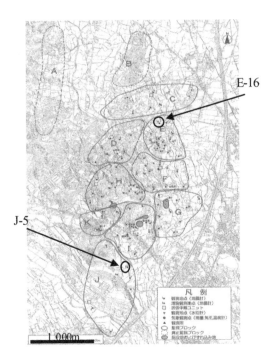

Figure 9. survey point of micro tremor in the active quarries.

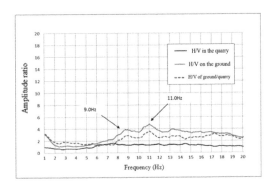

Figure 10. H/V spectrum in the underground A and its above ground.

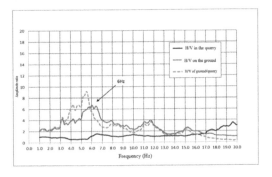

Figure 11. H/V spectrum in the underground B and its above ground.

323

i.e. Kanto roam layer and gravel layer laying among the ground surface and the Oya tuff layer. And the structure of underground quarry caused the peak difference. Additionally the difference of peak value between the underground quarry A and B should cause the structural difference each other.

4 THREE DIMENSIONAL DYNAMIC ANALYSIS FOR UNDER-GROUND QUARRY

4.1 Introduction

In this chapter, the authors carried out the numerical analysis to calculate deformation and seismic transmission for evaluating dominant frequency by on 3D-FDM software, FLAC3D. At first the authors generated three dimensional formation with proper boundary condition and initial condition of Oya area and calculated initial stress. Secondly, after excavating three dimensional structure of active underground quarry B, we applied seismic wave at the bottom of the model. Finally we have got the seismic response of the quarry.

4.2 Numerical modelling process

This study referred the geometry of an underground quarry for numerical analysis. It is an active quarry excavating Oya tuff and has long wall structure. Figure 12 and 13 show geological model for numerical analysis 500 m * 500 m * 100 m in this area. At first the model consisted in 1,875,000 rectangular solid shape zone. Around of the quarry, we employ fine zone which have 2 m in each side. The other zone have 10 m in each side. And we decrease height of the geological model to reduce the calculation time and apply distribution instead of overburden. To do that it is important to check the effect the inertia force. Finally we reduced the initial number of zone to 193,640. For dynamic analysis, we set free field boundary by the dash pot to regard the infinite boundary toward far from the sides

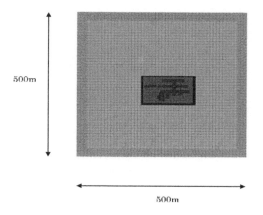

Figure 12. Overview of 3D Geological model for underground quarry (Dark blue area indicate fine zone and underground quarry.

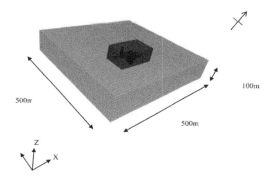

Figure 13. Plane view of 3D Geological model for underground quarry (Dark blue area indicate fine zone and underground quarry.

Table 2. Material parameter for numerical analysis.

Bulk modulus K (MPa)	1.38×10^3
Shear modulus G (MPa)	0.91×10^3
Cohesion C (MPa)	2.10
Internal friction angle ϕ (degree)	30
Tensile strength σ_t (MPa)	1.08
Density ρ_t (kg/m³)	1.77×10^3

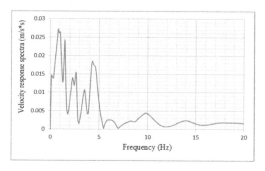

Figure 14. Seismic response at base floor in Oya underground quarry.

from geological model. We apply the apply the one of seismic waves, M4.8 in May 18, 2012 of which epicentre was located in South west of Ibaraki Prefecture and mentioned in chapter 2. Table 2 shows material parameters for the numerical analysis.

4.3 Results of numerical analysis

Figure 15.shows seismic response by FFT at base floor in the underground quarry and it explain the skew and gentle peak from about 4.0 Hz to10Hz except less than 4.0 Hz. It is same as micro tremor on FFT. In addition, results of numerical analysis show the other skew peak from about 1.0 Hz to 4.5Hz, it may be the structural effect of the underground quarry. It is considered that the dominant frequency of Oya area may be from about 5.0Hz to 10Hz and that of an Oya underground quarry may be from about 1.0 Hz to 4.5 Hz.

5 SUMMARY AND FUTURE WORK

5.1 *Summary*

This study has been clear the following knowledge.

Dominant frequency around Oya underground quarries is from about 5.0 Hz to 10 Hz by Fourier spectrum of seismic wave data which was recorded in Oya observation system. As amplitude ratio of those quarries after raising ground water level is smaller than those before rising ground water, it may tend that submerged underground quarries improve its structural stability.

This study measured micro tremor in 2 active underground quarries and those above ground and employ H/V process on the Fourier spectrum of 3 directional components. Roughly the results of micro tremor measured in active underground quarries also gave same tendency.

The authors also carried out three dimensional dynamic analysis by Flac3D to evaluate deformation and transmission feature in the state applying seismic wave. The peak frequencies are among about from 1.0 Hz to 4.5 Hz and from 4.0 Hz to 10 Hz at the seismic response on the base floor in the quarries. The dominant frequency of underground quarries may be regarded between about 1.0 Hz and 4.5 Hz. And the frequency between about 4.0 Hz and 10 Hz should be the dominant frequency in the Oya area. Because of attenuation in short frequency surrounding the quarries, the dominant frequency may not be observed.

5.2 *Future work*

As the dominant frequency of the velocity meter may cause the difficulty of measuring small frequency, other suitable velocity meters should be installed.

ACKNOWLWGEMENT

The authors thank to the Oya Area Development Co. for providing valuable data for this research. Additionally we also extend special thanks to the Coordinating Committee for Earthquake Prediction and the Chuden Engineering Consultants Co., Ltd. for financial support of some parts of this study.

REFERENCES

Aydan, Ö., Crustal stress changes and characteristics of damage to geo-engineering structures induced by the Great East Japan Earthquake of 2011, *Bull Eng Geol Environ* 74: 1057–1070, p. 1064, 2015.

Katayose, T. and Seiki, T. 2008, Trail for structural stability soundness using simple field test in Oya underground space, *Proc. of 37th Rock Mechanics symposium*, JSCE, CD-ROM (in Japanese).

Nakamura, Y., 2008, Basic concept of H/V spectrum ratio, Proc. of Seismic disaster symposium, *SEGJ*, (in Japanese).

Oyagi, N. and Hungr, O., Large rock collapse at Utsunomiya City near Nikko, Japan: *Landslide News*, 3, pp.11–12, 1989.

Seiki T., Nishi, J. and Nishida, Y., 2007, Renovation Challenge of Underground Quarries for Oya tuff, in *11th ACCUS conference: Underground Space: Expanding the Frontiers*.

Seiki, T.Ishii, K. Takashi, Noguchi, S. and Ohmura, T., Seismic response of numerical analysis and field measurement in Oya tuff quarry, Eurock 2016, 2016.

Seiki, T., Dintwe, T.K.M., Yamaguchi, R., Noguchi, S. and Ohmura, T.: Seismic characteristics of field measurements and numerical analyses of an underground quarry in Oya, *Proc. of 3rd Int. Conf. on Rock Dynamics and Applications*, Trondheim, Norway, 6ps., 2018.6.

2019 Rock Dynamics Summit– Aydan et al. (eds)
© 2019 Taylor & Francis Group, London, ISBN 978-0-367-34783-3

An experimental study on the effects of earthquake faulting on rock engineering structures

Y. Ohta
Graduate School of Science and Technology, Tokai University, Shizuoka, Japan (Presently with Nuclear Energy Regulation Authority), Shizuoka, Japan

Ö. Aydan
Department of Civil Enineering, University of the Ryukyus, Okinawa, Japan

ABSTRACT: The authors performed some laboratory model experiments on faulting and various rock engineering structures subjected to thrust, normal and strike-slip faulting movements under dynamic condition. Several experiments were carried out to investigate the effect of faulting on the slopes, underground openings, bridges and its foundations, and embedded linear structures. Model experiments were carried out using breakable blocks and layers. The experiments showed that the faulting mode and discontinuity orientation have great affect on the failure mode of various structures.

1 INTRODUCTION

Ground motion characteristics, deformation and surface breaks of earthquakes depend upon the causative faults. Their effects on the seismic design of engineering structures are almost not considered in the present codes of design although there are attempts to include in some countries (i.e. USA, Japan, Taiwan, Turkey (Aydan et al. 1999, 2009a,b,c).

The experimental device is developed amd used for investigating the effect of faulting under gravitational field. The orientation of faulting can be adjusted as desired (Ohta 2011; Ohta & Aydan 2010). The maximum displacement of faulting of the moving side of the faulting experiments was varied between 25 and 100mm. The base of the experimental set-up can model rigid body motions of base rock and it has a box of 780 mm long, 250 mm wide and 300 mm deep. This experimental device is used to investigate the effect of forced displacement due to faulting on rock slopes and underground openings. The displacement and accelerations were measured simultaneously.

Several experiments were carried out to investigate the effect of faulting on bridge and its foundations. Model experiments were carried out using either breakable blocks and layers or non-breakable blocks. The authors have been also performing some model experiments on the effect of faulting on the stability and failure modes of shallow undergound openings. In this study, the authors carried out some laboratory experiments to simulate the motions during normal and thrust faulting and their effects on model structures to investigate the effects of faulting due to earthquakes. They summarize the findings from this experimental study and their implications in practice.

2 MODEL EXPERIMENTS

2.1 Experimental Device

An experimental device shown in Figs. 1 and 2 was developed for model tests on faulting under dynamic condition (Ohta, 2011). One side of the device is moveable in a chosen direction to induce base movements similar to normal or thrust faulting with a different inclination. The device can simulate from 45 degrees normal faulting movement to 135 degrees thrust faulting.

The box is 780 mm long, 300 mm high and 300 mm wide. The length of the moveable side is 400 mm. The motion of the moveable side of the device

(a) A view of a faulting test with sand layer above the hard-base

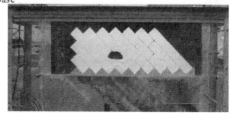
(b) A view of a faulting test on a model tunnel in soft rock

Figure 1. Views of some of faulting experiments

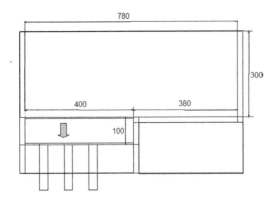

Figure 2. A drawing of the faulting test apparatus

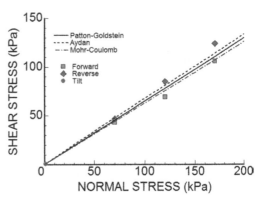

Figure 4. Shear strength envelopes of the quartz sand.

is achieved through its own weight by removing a stopper. The amount of the vertical movement of the moving base can be up to 200 mm and it can be set to a certain level as desired. The device equipped with non-contact laser displacement transducers and contact type accelerometers with three components. Besides the continuous monitoring of movements of the model ground through laser displacement transducers and accelerometers, the experiments were recorded using digital video cameras.

2.2 Materials

Two kinds of ground material were considered. The first type ground model material was granular material simulating soft ground on a rigid base. The second type ground material was a rock-like material to model layered and jointed rock mass. Granular ground material on the rigid movable base is dry quartz sand. Direct shear experiments on ground material were carried out. Fig. 3 shows some of displacement vs shear stress responses. The behaviour of sand is elasto-plastic without any softening. Fig. 4 shows some yield criteria for experimental results and the friction angle of sand ranges between 32-34 degrees with an average of 33 degrees.

Layers or blocks used for physical models of layered or jointed rock mass were created through the compression of a special mixture consisting of $BaSO_4$, ZnO and Vaseline oil under a chosen pressure. Various researchers determined the properties of this solid material. Frictional characteristics of discontinuities, unit weight, tensile and compressive strength of these samples have been measured in the laboratory as seen in Fig. 5.

(a) View of model material

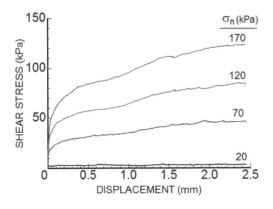

Figure 3. Shear response of sand under different normal stress.

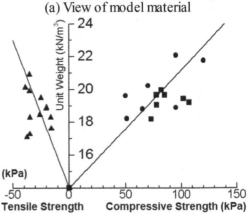

(b) Unit weight and strength relations

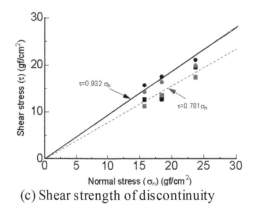

(c) Shear strength of discontinuity

Figure 5. View of rock-like model material and its mechanical properties.

3 EXPERIMENTS ON GRANULAR GROUND

The sand was poured into the soil box without any compaction. Therefore, the relative density of the sand ranges between 35-45 % with an average of 40%. Black dyed sand marker lines were set at an interval of about 50mm. However, the procedure for marker lines is extremely tedious and it is generally dificult to get perfect straight lines.

Once the soil model was prepared, the monitoring devices consisting of laser displacement transducers and accelerometers to observe ground motions on ground surface were set at three locations, specifically, movable and stationary blocks and just above the fault. Each experiment was recorded through digital video recorders and some pictures were taken during the experiment. Furthermore, variations of electric potential or electrical resistivity in relation to faulting were also measured using electrodes embedded into the soil and associated monitoring equipments for electric current and voltage in some of experiments.

The maximum displacement of faulting of the moving side of the faulting experiments was varied between 25 and 100mm. The vertical displacements of the fault was 25, 50, 75 and 100 mm. Due to the nature of the problem, the vertical component of accelerations becomes maximum among other components. Fig. 6 shows the vertical accelera-tion and displacement measured simultanenuously in the experiment with 200 mm thick soil deposit and 90 degrees normal faulting and Fig. 7 shows its motion at several time steps.

As seen from Fig. 6, the maximum acceleration of the movable side is greater than that of the stationary side. Furthermore, the maximum acceleration is observed when the movement of the movable side is restrained and the acceleration response is entirely unsymmetric while the acceleration response of the stationary side is almost symmetric. Although it is

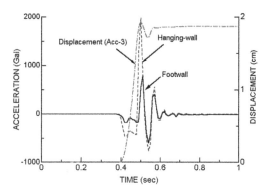

Figure 6. Acceleration responses of 200 mm thick soil deposit for 90° normal faulting.

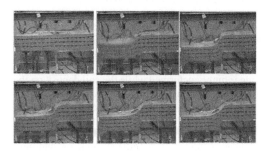

Figure 7. Motion of 200 mm thick soil deposit for 90° normal faulting.

not the purpose of this manuscript to discuss and compare with observations in actual earthquakes, the responses measured during these experiments are quite similar to the observations in actual earthquakes as well as in rock fracture experiments (i.e. Ohta 2011; Ohta and Aydan, 2010). The variation of soil deposit thickness has a certain effect on the resulting accelerations. However, if the displacement of the fault is same, its effect would be small as compared with that of the variation of the fault displacement.

The observations on deformation and slip-lines in experiments carried out for inclinations of 45, 90 and 135 degrees are shown in Fig. 8. Comparisons done for three different inclinations of faulting for the same amount of vertical displacements. The most extensive studies are carried out on experiments

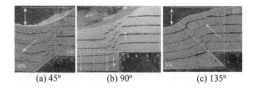

(a) 45°   (b) 90°   (c) 135°

Figure 8. Views of faulting experiments.

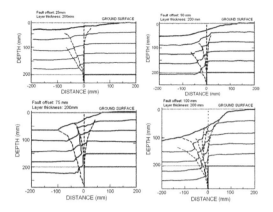

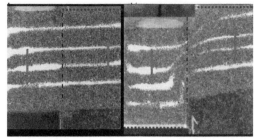

Figure 11. Negative images of several stages in a strike-slip faulting experiment in granular medium.

Figure 9. Sliplines in experiments with varying vertical displacement for faulting inclination of 90°.

with the faulting inclination of 90 degrees, in which the effects of allowable vertical displacements and soil thickness were investigated (Fig. 9).

The top soil deposit on the hanging wall (mobile) side is highly deformed while the soil deposit on the footwall (stationary) side is much less. Furthermore, the number slip-lines on the hanging-wall (mobile part) of the soil layer is greater than that in the foot-wall side. This probably associated with the amount of displacement of the mobilized soil in the hanging-wall side. It is also interesting to note that thrust type slip-lines occur at the hangingwall side while normal type slip-lines develop in the footwall side. Such slip-lines may be of great significance when ground deformation and slip-lines interpreted for faults in-situ.

The inclination of the thrust fault was set to 45 degrees and the amount of vertical displacement was varied between 40 mm and 100mm for a 200 mm soil layer. Views of soil layer after faulting are shown in Fig. 10. Similar to previous experiments, the number of slip-lines increases with the amount of fault offset. Several slip-lines develop sub-parallel to eact other. When the amount of fault offset is small, the slip-line does not reach ground surface and the ground surface configuration shows a flexural bend. However, if the fault offset is large enough, the top part of the soil on the hanging-wall side moves towards the footwall side and the ground deformation induces a slip-line similar to normal faults. In other words, the wedge-like body just

above the tip of the fault can not remain stable under gravitaional field and moves towards the foot-wall side and it becomes stabilized at a surface inclina-tion equivalent to its dynamic repose angle. This is an important observation as the interpretation of ground deformation and slip-lines may be wrongly interpreted as normal faulting despite that it is a thrust faulting. Furthermore, the ground surface deformation may resemble to the ground deforma-tion profiles resulting from classifical slope failures. This observation also implies that slope failures aligned should be interpretted as the surface expres-sion of earthquake faulting as happened in 2005 Kashmir earthquake (Aydan et al. 2009).

In addition, some strike-slip experiments are car-ried out under constant velocity condition using base friction apparatus. Fig. 11 shows negative images of several stages of a strike-slip faulting experiment on a granular media. Although slip-lines were not apparent from the figure, a wide deformation band with a thickness equivalent to the amount of relative displacement developed on both sides of the pro-jected fault-line. As noted from the figure, streching strain occur at certain direction while compressive straining occur perpendicular to the streching axis.

4 TESTS ON THE EFFECT OF FAULTING ON ENGINEERING STRUCTURES

Several model experiments were carried out to see the effect of faulting on rock engineering structures in/on rock-like ground as reported by Aydan et al. (2011). The experimental results are briefly explained in this section. Fig. 12 shows the

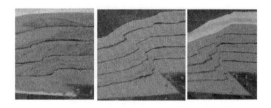

Figure 10. Views of 250 mm thick soil layer for various 45 degrees thrust fault offsets

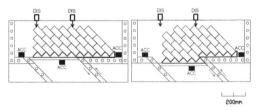

Figure 12. The instrumentation (DIS: laser displacement transducer; ACC:Accelerometer).

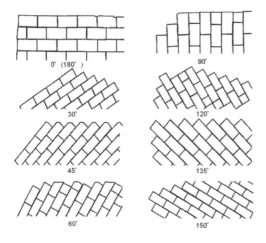

Figure 13. Illustration of experimental models of rock mass

instrumentation. Fig. 13 shows the sketches of rock mass models tested in experiments described herein.

4.1 Underground Openings

The authors performed some model experiments on the effect of faulting on the stability and failure modes of shallow undergound openings (Aydan et al. 2010). Fig. 14 shows views of some model experiments on shallow undergound openings subjected to the thrust faulting action with an inclination of 45°. Underground openings are assumed to be located on the projected line of the fault.

In some experiments three adjacent tunnels were excavated. While one of the tunnels was situated on the projected line of faulting, the other two tunnels were located in the footwall and hanging wall side of the fault. As seen in Fig. 14, the tunnel completely collapsed or was heavily damaged when it was located on the projected line of the faulting. When

the tunnel was located on the hanging wall side, the damage was almost none in spite of the close approximity of the model tunnel to the projected fault line. However, the tunnel in the footwall side of the fault was subjected to some damage due to relative slip of layers pushed towards the slope. This simple example clearly shows the damage state may differ depending upon the location of tunnels with respect to fault movement.

4.2 Slopes

The authors have initiated an experimental program on the effect of faulting on the stability and failure modes of rock slopes (Ohta 2011). The first series of experiments were carried out on rock slope models with breakable material under a thrust faulting action with an inclination of 45° (Fig. 15). When layers dip towards valleyside, the ground surface is tilted and the slope surface becomes particularly steeper.

As for layers dipping into mountain side, the slope may become unstable and flexural or columnar toppling failure occurs. Although the experiments are still insufficient to draw conclusions yet, they do show that discontinuity orientation has great effects on the overall stability of slopes in relation to faulting mode. These experiments clearly show that the forced displacement field induced by faulting has an additional destructive effect besides ground shaking on the stability of slopes.

4.3 Foundations of Bridges

Several experiments were carried out to investigate the effect of faulting on a bridge and its foundations. The bridge was a truss bridge just over the projected fault line. Fig. 16 shows truss bridge models above the jointed rock mass foundation. Fig. 16(a) shows views of the bridge model before and after the experiments, subjected to the forced displacement field of vertical normal faulting mode. Bridge foundations were pulled apart and tilted. The vertical offset was 0.37 times the bridge span. Similarly, Fig. 16(b) shows the bridge model before and after the experiments subjected to the forced displacement field of 45°

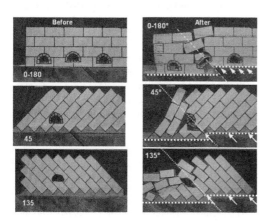

Figure 14. Effect of faulting on underground openings.

Figure 15. Effect of faulting on rock slopes.

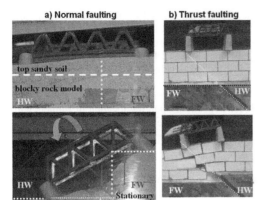

Figure 16. Effect of faulting on foundations of bridges.

thrust faulting mode. Bridge foundations were also pulled apart at the top and compressed at the bottom, and tilted as seen in Fig. 16(b).

4.4 *Embedded Linear Structures (such as Pipelines)*

A series of experiments were carried out on the effect of thrust faulting and normal faulting regarding the embedded linear structures in jointed breakable rock mass model such as pipelines etc. (Figs. 17 and 18). The model rock mass was layered with vertical cross joints. Furthermore, a house model was put on top of the projected tip of the fault. When

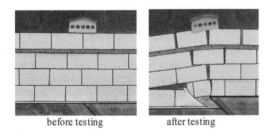

before testing   after testing

Figure 17. Deformation of embedded linear structure subjected to 45 degree thrust faulting.

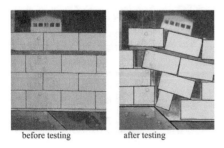

before testing   after testing

Figure 18. Deformation of embedded linear structure subjected to 45 degree normal faulting.

the thrust faulting is considered, the embedded linear structure deforms to accommodate faulting displacement. In addition, inter-slip among blocks occurs and some are separated and rotated. The block close to the fault tip is fractured. When the normal faulting is considered, the embedded linear structure deforms to accommodate faulting displacement. In addition, inter-slip among blocks and linear embedded structure occurs and some of blocks are separated and rotated. The block close to the fault tip is also fractured.

5 CONCLUSIONS

The following conclusions may be drawn from this experimental study as follow:

1. Dynamic component of faulting is an important factor governing the shape of deformed ground, ground surface structures and slip-lines compared with those observed in faulting experiments using constant velocity type experimental devices.
2. The volume of deformed ground is larger in experiments with dynamic component compared with those from the constant velocity experiments.
3. If the vertical component of faulting is large enough to induce ground failure on the stationary side, the ground surface would be stablised at an inclination equivalent to the dynamic repose angle of non-cohesive material.
4. The damage state may differ depending upon the location of tunnels with respect to fault movement. The damage of tunnels may be quite heavy if they are projection line of motion.
5. The experiments on slopes indicated that discontinuity orientation of rock mass has great effects on the overall stability of slopes in relation to faulting mode. The forced displacement field induced by faulting has an additional destructive effect besides ground shaking on the stability of slopes.
6. Bridge foundations may be pulled apart and tilted. Bridge foundations subjected to the forced displacement field of 45° thrust faulting mode. were also pulled apart at the top and compressed at the bottom, and tilted.
7. The embedded linear structures deform to accommodate faulting displacement and inter-slip among blocks occurs and some are separated and rotated. The block close to the fault tip is fractured.

REFERENCES

Aydan, Ö., 2003. Actual observations and numerical simulations of surface fault ruptures and their effects engineering structures. *The Eight U.S.-Japan Workshop on Earthquake Resistant Design of Lifeline Facilities and Countermeasures Against Liquefaction.* Technical Report, MCEER-03-0003, 227–237.

Aydan Ö, 2012. Ground motions and deformations associated with earthquake faulting and their effects on the safety of engineering structures. Encyclopedia of Sustainability Science and Technology, Springer, R. Meyers (Ed.), 3233–3253.

Aydan Ö., Ulusay, R., Hasgür, Z., and Hamada, M. (1999): The Behavior of structures built on active fault zones in view of actual examples from the 1999 Kocaeli and Chi-Chi earthquakes. ITU International Conference on Kocaeli Earthquake, Istanbul, pp.131–142, 1999.

Aydan, Ö., Y. Ohta, Hamada, M. 2009a. Geotechnical evaluation of slope and ground failures during the 8 October 2005 Muzaffarabad earthquake in Pakistan. Journal Seismology, Vol.13, No.3, pp.399–413.

Aydan, Ö., Y. Ohta, M. Hamada, J. Ito, K. Ohkubo, 2009b. The characteristics of the 2008 Wenchuan Earthquake disaster with a special emphasis on rock slope failures, quake lakes and damage to tunnels. Journal of the School of Marine Science and Technology, Tokai University, Vol.7, No.2,pp.1–23,2009.

Aydan, Ö., Hamada, M., Itoh, J, Ohkubo, K., 2009c. Damage to Civil Engineering Structures with an Emphasis on Rock Slope Failures and Tunnel Damage Induced by the 2008 Wenchuan Earthquake. Journal of Disaster Research, Vol.4, No.2, pp.153–164,2009.

Aydan, Ö., Ohta, Y., Genis, M., Tokashiki, N. and Ohkubo, K. 2010. Response and stability of underground structures in rock mass during earthquakes. Rock Mechanics and Rock Engineering, 43(6), 857–875.

Ohta, Y. (2011): A fundamental research on the effects of strong motions and ground deformations in the neighborhood of earthquakes faults on Civil Engineering structures. Doctorate Thesis, Tokai University, Graduate School of Science and Engineering, 285 pages.

Ohta, Y. and Aydan, Ö. 2010. The dynamic responses of geomaterials during fracturing and slippage. Rock Mechanics and Rock Engineering, Vol.43, No.6, 727–740.

2019 Rock Dynamics Summit– Aydan et al. (eds)
© *2019 Taylor & Francis Group, London, ISBN 978-0-367-34783-3*

Site characteristics of a rock tunnel based on field-monitored seismic response

Ya Chu Chiu
National Taiwan University, Taipei, Taiwan
National Chung Hsing University, Taichung, Taiwan

Chien Lun Kung
United Geotech, Inc., Taipei, Taiwan

Tai Tien Wang
National Taiwan University, Taipei, Taiwan

ABSTRACT: Most studies investigate the mechanism of tunnel seismic damages due to earthquakes by numerical simulation and physical model experiments. However, limited by inevitable simplifications, assumptions and boundary problems, the above methods deviate more or less from the real situation. Therefore, this research acquire the seismic response of a rock tunnel through in-*situ* monitoring. Five accelerometers were installed in a highway tunnel in southeast Taiwan, four on the same tunnel section, one on another. Since the monitoring started from 2014, there are four earthquake events exceed Richter magnitude 4.0. The acceleration records show that in general, the predominant frequencies of accelerometers installed in different locations inside the tunnel are similar with the predominant frequencies of the two neighboring seismic stations. Such similarity in predominant frequencies possesses the best consistency along the slope dip direction, and the worst along the vertical direction.

1 INTRODUCTION

Cases of tunnels damaged in earthquake had been reported worldwide, including scientific articles on a single earthquake event (Kawakami 1984, Asakura & Sato 1998, Wang et al. 2001, Yashiro et al. 2007, Wang et al 2009, Shen et al. 2014), or research papers that include multiple events (Dowding & Rozen 1978, Sharma & Judd 1991, Ulusay et al. 2002, Aydan et al. 2011, Asakura et al. 2008). It is known that under the attack of earthquakes with moment magnitude larger than 6.0, underground structures may, but not bound to, display local or complete collapse (Aydan 2017). Aydan (2017) classified tunnel damages into three types, 1.portal damage, 2.shaking induced damage and 3.permanent ground deformation induced damage. Part of portal damages and almost all of the permanent ground deformation induced damages are caused either by faulting or slope movements. This research concerns the shaking induced damages of tunnels.

Barred by the complexity of wave scattering, superposition of wave source with multiple frequency, or the composition of waves of different types, the universal analytical solution for seismic response of a tunnel in a general earthquake event is still absent. Therefore, the authors embarked on acquiring coseismic records of a mountain tunnel. Five accelerometers were installed in a highway tunnel in southeast Taiwan, four on the same tunnel

section, while one on another (Kung 2015, Chiu et al. 2015). Four seismic events produced more than 80 gal (cm/s^2) Peak Ground Accelerations (PGAs) in the two neighbouring seismic stations, which are less than 10 km to the case tunnel, since 2014. The acceleration records were then processed, and transformed into Fourier frequency spectra. Base on comparison between the seismic responses of the seismic stations and the accelerometers in the tunnel, a preliminary conclusion is given.

2 FIELD MONITORING AND SIGNAL PROCESSING

2.1 *Seismic monitoring of a mountain tunnel*

A highway tunnel in southeast Taiwan is chosen as the case in this research. The case tunnel is a two-lane highway tunnel that passes through a ridge (Fig.1). The tunnel is horseshoe shaped, 310 m long, 6.5 m high and 9 m in width (Fig.2b). The monitored location is about 5.5 m from ground surface. Quartz mica schist is the major rock type, investigation shows no adjacent large scale geologic structure. Micro tremor measured and analyzed by Lo (2013) indicate that the predominant frequency of this site is within 0.5-5.0 Hz. There are two nearby surface seismic stations established by the Central Weather Bureau, Taiwan. First, TTN051 (No.51, at 121.0251°E, 23.1870°N) which is 6.8 km northwest from the tunnel with height 1040 m. The other is TTN041 (No.41, at 121.1252°E, 23.1323°N), 10 km southeast from the tunnel, at 443 m elevation.

The main consideration in arranging monitoring instruments is the seismic response of different locations on the identical tunnel section. Four three-dimensional accelerometers were installed at section A-A'(Fig. 2a), one three-dimensional accelerometers were mounted at section B-B'. The defined coordinate system is shown on Figure 3a. Accelerometer A4 is located at the hillside spring line, where the arch transits to the vertical sidewall (Fig. 3b). A2 and A3 are installed on the arch vault, about 30 degree and 45 degree from the zenith to creek side and hill side, accordingly. A1 is placed on the creek side wall, about 1 m from the pavement. A5 is installed at about the same height as A4, but on section B-B'(Figs. 2-3).

2.2 Seismic response processing

The accelerometers were Type 2460-002 Micro Electro Mechanical Systems (MEMS) triaxial accelerometers made by the Silicon Designs, Inc. The sampling frequency of three-dimensional accelerometers were set to be 100 Hz. This research follows the procedure in Figure 4 to process the monitored acceleration. First, the mean value of acceleration is calculated and subtracted, the acceleration then shifted to a baseline of mean. Second, Discrete Fast Fourier Transform (DFFT) is performed to transfer acceleration from time domain to frequency domain, determine predominant frequencies in three directions (X, Y and Z). Third, since the noise of the instrument itself falls at around 20 Hz, a band pass filter 0-20 Hz is determined (Kung, 2015). Eventually, the frequency spectra were transformed back to the time domain by inverse Discrete Fast Fourier Transform (DFFIT).

The preliminary results of this section is based on an earthquake in May 25, 2014 (UTC+8), No.75 earthquake numbered after the Central Weather Bureau, Taiwan. The earthquake occurred at 121.16°E, 23.06°N, 12.8 km from the case tunnel with a Richter magnitude 5.0. Data of seismic stations came from the Geophysical Database Management System by Central Weather Bureau, Taiwan (Shin et al. 2013).

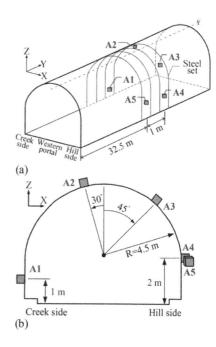

(a)

(b)

Figure 3. The arrangement of accelerometers installed in the tunnel in (a) perspective view and (b) cross section.

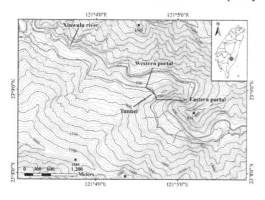

Figure 1. Topographic map of the case tunnel (Wang, 2010).

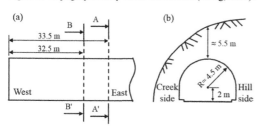

Figure 2. The (a) longitudinal map and (b) cross section A-A'of the monitoring location in the case tunnel.

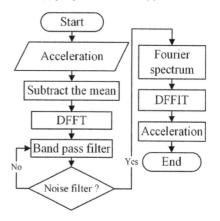

Figure 4. Flow chart of signal processing.

334

During earthquake No.75, seismic station TTN051 is 19.4 km from the epicenter with the fourth degree intensity scale (Tables 1-2). PGA is 18.00 gal in X direction, 18.16 gal in Y direction, and 18.70 in Z direction, as listed in Table 1. The Fourier spectrum in three directions display peaks at 4.8 Hz with amplitude 1.43 gal in X direction; 4.7 Hz and 7.4 Hz with amplitude 1.55 gal and 0.81 gal in Y direction; 5.0, 6.6 and 7.8 Hz with amplitude 0.94, 0.89 and 0.46 gal in Z direction.

From the recordings of accelerometer A1, the peak acceleration along X direction is 34.87 gal at the 4.98 second; the peak acceleration in Y direction is 12.16 gal at the 4.88 second, and the peak acceleration along Z direction is 19.05 gal at the 5.24 second. DFFT returns spectrum peaks on 5.0 Hz, 2.48 gal in

X direction; 1.0, 5.0 and 6.4 Hz in correspondence with 0.43, 0.85 and 0.54 gal in Y direction; 4.8, 6.9 and 8.6 Hz with 0.98, 0.60 and 0.63 gal in Z direction, accordingly.

Table 1 contains the analysis results of all monitored data during earthquake No.75. The peak X-acceleration of accelerometers A1 and A2, installed at creek side sidewall and vault, occurred at about the same time, but their peak Y- and Z-acceleration do not. Accelerometers A3 and A4, mounted on hill side sidewall and vault, reach peak acceleration in three dimension simultaneously. The peaks in Y- and Z- Fourier spectrum for seismic station 41, all three directional Fourier spectrum for seismic station 51 and accelerometers A1-A5 fall on about 5 Hz. The second peak in Z-direction of A1

Table 1. Analysis results of in-situ monitored acceleration.

Accelerometer No.	Direction	Peak ground acceleration (gal)	Time(s)	Peak amplitude (gal)	Predominant frequency f_1 (Hz)	Second peak f_2 (Hz) (gal)		Third Peak f_3 (Hz) (gal)	
	X	196.22	-	6.84	11.5	6.53	5.3	5.73	3.8
41	Y	203.62	-	9.37	4.4	9.07	5.5	8.85	2.8
	Z	134.22	-	9.24	5.0	3.73	10.4	-	-
	X	18.00	-	1.43	4.8	-	-	-	-
51	Y	18.16	-	1.55	4.7	0.81	7.4	-	-
	Z	18.70	-	0.94	5.0	0.89	6.6	0.46	7.8
	X	34.87	4.98	2.48	5.0	-	-	-	-
A1	Y	12.16	4.88	0.85	5.0	0.54	6.4	0.43	1.0
	Z	19.05	5.24	0.98	4.8	0.63	8.6	0.60	6.9
	X	37.38	4.99	2.82	5.0	-	-	-	-
A2	Y	22.22	5.25	0.85	6.0	0.63	8.8	0.50	1.4
	Z	14.23	5.07	1.05	4.8	0.56	6.4	0.53	8.5
	X	34.50	4.90	2.50	5.0	-	-	-	-
A3	Y	18.42	5.26	0.79	5.7	0.47	8.9	0.44	1.0
	Z	18.00	5.23	1.31	4.8	0.56	11.4	-	-
	X	31.08	4.89	2.07	5.0	-	-	-	-
A4	Y	14.50	5.25	0.61	5.9	0.46	13.18	0.45	1.0
	Z	18.00	5.24	1.23	4.8	0.44	11.7	-	-
A5	Y	10.80	4.82	0.54	5.9	0.53	3.9	0.42	1.0
	Z	18.28	5.18	1.05	4.8	0.48	9.18	-	-

Table 2. Top four earthquake events around case tunnel.

Event No.	Date	Time	Richter magnitude scale	Location	Seismic intensity scale*	
					TTN041	TTN051
01	2014/02/25	20:41	5.0	23.06N, 121.16E	5	4
02	2016/04/25	10:58	4.2	23.17N, 121.02E	4	5
03	2016/04/28	02:19	5.5	23.28N, 121.23E	5	4
04	2016/05/02	10:27	4.2	23.19N, 121.02E	4	5

*Seismic intensity scale after Central Weather Bureau, Taiwan. Peak Ground Acceleration (PGA) falls in 25-80 gal as seismic intensity scale = 4, PGA between 80-250 gal as seismic intensity scale = 5.

and A5, Y-direction of A2 and A3, Z-direction of A3 and A4 are also close with each other.

3 CHARACTERIZING SEISMIC RESPONSE OF A MOUNTAIN TUNNEL

3.1 *Earthquake events*

Since the beginning of monitoring of the case tunnel in 2014, there are 24 earthquake events that made the acceleration of the two neighbouring seismic stations exceed 8 gal. Among the 24 events, four induced more than five the seismic intensity scale at the seismic station, as listed in Table 2. The four events occurred in 2014-2016. Event No.1 is described in section 2.2, while Event No.2-4 took place within seven days in 2016, and all of them have Richter magnitude more than 4.0.

In Event No.2, the PGA of seismic station 41 in X-direction is 29.2 gal, and that of seismic station 51 in the same direction is 76.08 gal. For accelerometers in case tunnel, PGA values in X-direction are 14.82 gal, 14.82 gal, 16.66 gal and 16.5 gal for accelerometers A1-A4, accordingly (Figs. 5-6). PGA in Y-direction is close to which in X-direction.

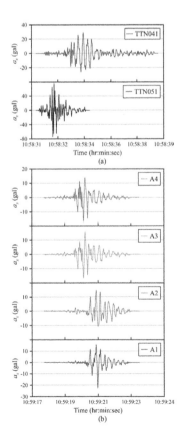

Figure 5. X-directional acceleration records of Event No.2, (a) seismic stations and (b) accelerometers in case tunnel. Note that the start and end time in (a) and (b) are different.

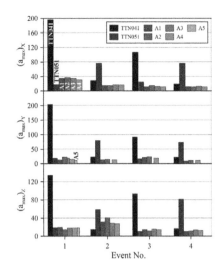

Figure 6. Peak ground accelerations in X-, Y- and Z-directions of all four events.

Seismic stations 41 and 51 have PGAs of 22.39 gal and 79.64 gal. PGAs of accelerometers A1-A2 and A4 are 12.91 gal, 14.25 gal, 12.71 gal, when the accelerometer in Y-direction of A3 is out of order. Acceleration of the seismic stations in Z-direction is much smaller than in X- and Y-directions. PGAs in Z-direction of seismic stations 41 and 51 are 14.61 gal and 58.55 gal. In the tunnel, PGAs of accelerometers A1-A5 in Z-direction went to 31.32 gal, 40.37 gal, 28.11 gal and 27.03 gal (Fig. 6).

3.2 *Characteristics of seismic response of a mountain tunnel*

Predominant frequencies of seismic stations and accelerometers in the tunnel are shown in Figure 7. X-direction of A5 malfunctioned before Event No. 1, and completely failure in all three directions before Event No. 2. The same tragedy happened on Y-direction of A3 before Event No. 2. In Figure 7, from above to below are X-, Y- and Z-directional predominant frequencies, from left to right predominant frequencies from Event No. 1 to Event No. 4. Each column group comprises the predominant frequencies f_1 of values for seismic station 41, 51, accelerometers A1-A5 in a sequence of left to right.

Figure 7 shows that regardless of the locations in the tunnel, the predominant frequencies are usually the same or very close to that of one of the seismic stations, e.g. X-directional predominant frequencies in Event No. 1, 2 and 4, and Z-directional predominant frequencies in Event No. 1 and 3. Such similarity in predominant frequencies possesses a diversity in different directions. Consider Event No.3 as a special case and exclude it, predominant frequencies of the tunnel and the seismic stations have the best coherence in X-direction, which is about parallel to the slope dip direction. Along the tunnel axis, which is the Y-direction, predominant frequencies of seismic stations and tunnel accelerometers are mainly close to each other but deviate slightly. Regarding Z-direction, seismic stations and tunnel accelerometers may have consistent predominant frequencies, or the coherent predominant frequencies of accelerometers in tunnel is close to the second frequency f_2 or the third frequency f_3 of a seismic station, or vice versa.

In Event No.1, the $(f_1)_X$ of accelerometers are all 5.0 Hz, similar with $(f_1)_X$ of station 51, which is 4.8 Hz. X-directional predominant frequencies of accelerometers all equals to 4.2 Hz, close to 4.0 Hz of station 51. In Event No.4, $(f_1)_X$ of accelerometers on creek side of the tunnel, A1 and A2, are 4.33 Hz; $(f_1)_X$ of accelerometers on hill side of the tunnel, A3 and A4, are 4.17 Hz. Accelerometers on both sides are close to the $(f_1)_X$ 4.4 Hz of station 51. In Event No.3, $(f_1)_X$ of accelerometers A1, A2 on creek side are 2.91 Hz and 5.76 Hz, and $(f_1)_X$ of accelerometers on hill side are both 2.91 Hz, presenting very little relevance with 4.0 Hz of station 41, and 7.8 Hz of station 51.

The coherence of f_1 between seismic stations and the tunnel is related to the similarity of their waveforms of frequency spectra. In most of the cases, the X-directional frequency spectra of seismic stations and the tunnel demonstrate similar waveforms. For Event No.1, 2 and 4, a single eminent peak can be found, as shown in Figure 8. Although multiple almost equal peaks exist in spectra of station 51, the highest peak locate at the same frequency as others.

Variation in Y- and Z- directional frequency spectra were found between the two seismic stations (Figs. 9-10). Frequency spectra of the tunnel possess characteristics of the two stations. For example, in Event No. 2 and 4 the Y-directional frequency spectra of accelerometers are similar to one seismic station, but their predominant frequencies are close to the other seismic station. Accelerometers A1, A2 and A4 exhibit similar waveform with station 51 in Y-direction, both reach peak values at about 4 Hz and 6 Hz (Fig. 9). However, the highest peak of the tunnel accelerometers fell in 3-4 Hz, which is closer to that of station 41. Furthermore, in Event No.4 the Y-directional frequency spectra of tunnel accelerometers have similar waveform with station 41, but their predominant frequencies 3.67 Hz are higher than 3.0 Hz of station 41. In the meanwhile, there is a f_2 of station 51 located at 4.6 Hz. While frequency spectra were formed with numerous adjacent peaks, such as Event No.2 in Z-direction (Fig. 10), the two seismic stations peak at 3 Hz, 5 Hz, 7 Hz, 9 Hz, 12 Hz and 16 Hz, but tunnel accelerometers assembled within 3-9 Hz only. Under this waveform, f_1 may alter with locations in the tunnel, but anyhow, f_1, f_2 or f_3 of the tunnel will conform with f_1 of seismic stations (Event No.2 in Fig. 7 and Fig. 10); or f_1 is consistent in the tunnel, but is closed to f_2 or f_3 of a seismic station. Moreover, the consistency of frequency spectra in the hill side, i.e. accelerometers A3 and A4, is always better than that in the creek side, accelerometers A1 and A2.

4 CONCLUSIONS

Based on records of five accelerometers installed in a mountain tunnel in Taiwan, this research analyzed the seismic responses of the tunnel and two neighboring

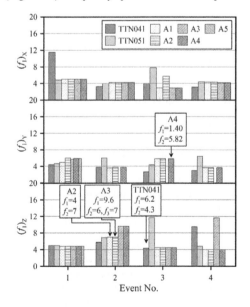

Figure 7. Predominant frequencies in X-, Y- and Z-directions of all four events.

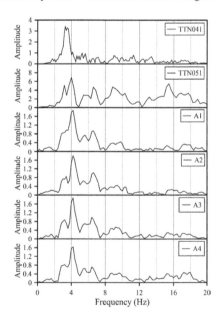

Figure 8. Frequency spectrum of X-direction acceleration in event No.2.

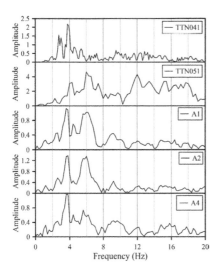

Figure 9. Frequency spectrum of Y-direction acceleration in event No.2.

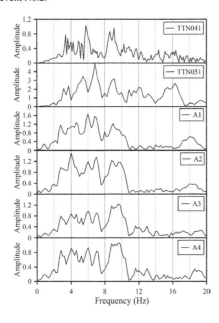

Figure 10. Frequency spectrum of Z-direction acceleration in event No.2.

seismic stations in four earthquake events with Richter magnitude larger than 4.0 occurred in 2014-2106.

Overall, the predominant frequencies of accelerometers installed in different locations inside the tunnel are similar with the predominant frequencies of the two neighboring seismic stations. In some events, few locations in the tunnel may have a double predominant frequency. Seismic responses of accelerometers in hill side are almost identical, while a larger differences can be found between two accelerometers in creek side. It is presumed that such phenomena have something to do with near surface effect, which awaits further investigation.

REFERENCES

Asakura, T. & Sato, Y. 1998. Mountain tunnels damage in the 1995. Hyogo-ken Nanbu Earthquake. *Railway Technical Research Institute*, 39(1): 9–16.

Asakura, T., Shiba Y., Matsuoka S., Oya T., Yashiro K. 2008. Damage to Mountain Tunnels by Earthquake and its Mechanism. *Taiwan Rock Engineering Symposium 2008*. October, 30–31. Taipei, Taiwan: 20–36.

Aydan, Ö., Ohta, Y., Daido, M., Kumsar, H., Genis, M., Tokashiki, N., Ito, T. & Amini, M. 2011. Chapter 15: Earthquakes as a rock dynamic problem and their effects on rock engineering structures. In Y. Zhou & J. Zhao (eds.), *Advances in Rock Dynamics and Applications*: 341–422. Boca Raton, FL, CRC Press, Taylor and Francis Group.

Aydan, Ö. 2017. *Rock Dynamics*. Leiden, the Netherlands, CRC Press/Balkema.

Chiu, Y.C., Kung, C.L., Liou, J.C., Cheng, D.S., Wang, T.T. 2015. Seismic response of a rock tunnel and associated crack variation. *East Asia Forum on Radwaste Management*. October 25–28. Taichung, Taiwan.

Dowding, C.H. & Rozen, A. 1978. Damage to rock tunnel from earthquake shaking. *Journal of the Geotechnical Engineering Division*. 2: 175–19.

Kawakami, H. 1984. Evaluation of deformation of tunnel structure due to Izu-Oshima Kinkai earthquake of 1978. *Earthquake Engineering & Structural Dynamics*, 12(3): 369–383.

Kung, C.L. 2015. *A rock tunnel under earthquake impact: in-situ monitoring and numerical simulation on harmonic incident wave excitation*. Master thesis. Department of Civil Engineering. Taipei: National Taiwan University.

Lo, B.C. 2013. *Evolution of a deep-seated creeping slope and associated characteristics: A case study*. Master thesis. Institute of Mineral Resources Engineering. Taipei: National Taipei University of Technology.

Sharma, S. & Judd W.R. 1991. Underground opening damage from earthquakes. *Engineering Geology*. 30: 263–276.

Shen, Y.S., Gao, B., Yang, X.M., Tao, S.J. 2014. Seismic damage mechanism and dynamic deformation characteristic analysis of mountain tunnel after Wenchuan earthquake. *Engineering Geology*. 180(2014): 85–98.

Shin, T.C., Chang, C.H., Pu, H.C., Lin, H.W., Leu, P.L. 2013. The Geophysical Database Management System in Taiwan. *Terrestrial, Atmospheric and Oceanic Sciences*. 24: 11–18.

Ulusay, R., Aydan, Ö. & Hamada, M. 2002. The behavior of structures built on active fault zones: Examples from the recent earthquakes of Turkey. *Structural Engineering/Earthquake Engineering*, 19(2): 149–167.

Wang, T.T. 2010. Characterizing crack patterns on tunnel linings associated with shear deformation induced by instability of neighboring slopes. *Engineering Geology*. 115(1-2): 80–95.

Wang, W.L., Wang, T.T., Su, J.J., Lin, C.H., Seng, C.R. & Huang, T.H. 2001. Assessment of damage in mountain tunnels due to the Taiwan Chi-Chi earthquake. *Tunneling and Underground Space Technology*, 16: 133–150.

Wang, Z.Z., Gao, B., Jiang, Y.J., Song, Y. 2009. Investigation and assessment on mountain tunnels and geotechnical damage after the Wenchuan earthquake. *Science in China Series E: Technological Sciences*. 52(2): 546–558.

Yashiro, K., Kojima, Y. & Shimizu, M. 2007. Historical earthquake damage to tunnels in Japan and case studies of railways tunnels in the 2004 Niigata-ken Chuetsu earthquake. *Quarterly Report of RTRI*, 48(3): 136–141.

2019 Rock Dynamics Summit– Aydan et al. (eds)
© *2019 Taylor & Francis Group, London, ISBN 978-0-367-34783-3*

AE and vibration monitoring on underground LPG storage caverns

Y. Tasaka, H. Kurose, T. Suido & C.S. Chang
Tokyo Electric Power Services Co.,Ltd, Japan

H. Fujii
Lazoc Inc., Japan

K. Toyoda
Japan Oil, Gas and Metals National Corporation, Japan

ABSTRACT: Namikata and Kurashiki underground LPG storage caverns apply natural groundwater with artificial hydraulic containment system to preserve the LPG at high pressure and normal temperature (e.g. Propane, 0.73MPaG at 20°C). The water curtain galleries and water curtain boreholes were designed to maintain the surrounding pore pressure higher than the storage pressure and to ensure the groundwater flow toward the storage caverns constantly. Therefore, the constitution of storage caverns under high water pressure has to be discussed throughout the construction phase, commissioning and operation phase. Pore pressure sensors, acoustic emission sensors, vibration sensors and accelerometers were installed in the caverns to ensure the rock mechanical stability. These monitoring and measurement are launched from the water curtain compression test, the access tunnel water filling and the cavern air tightness test, and are continued even in the operation phase. Based on the measurement results, the storage caverns have been evaluated the integrity.

1 INTRODUCTION

Namikata and Kurashiki are two of the national LPG stockpile bases in Japan. Both of the two sites utilize underground storage caverns to confine the LPG gas by relative higher water pressure at depth (Figure 1). Grouting works were applied to homogenize and to reduce the rock permeability and seepage at the cavern vicinity. Furthermore, water curtain system composed by an array of boreholes and water curtain tunnel, was constructed to enclose the storage caverns and to ensure the air-tightness by permanent groundwater flow toward the storage caverns

Figure 2 illustrates the layout of Kurashiki base. The Kurashiki base is composed by four 488~640m length underground storage caverns at EL.-160m within dimension of 24m(height) and 18m(width) to contribute 800,000m³ stockpile capacity. After the cavern excavation was completed, the water curtain system was submerged and confirmed the rock mechanical stability by maximum operation hydraulic pressure. Then, the access tunnel was also submerged and the storage caverns were compressed to maximum design pressure to examine the air-tightness.

In construction phase, groundwater level, pore pressure at cavern vicinity and seepage rate were continuously monitored to ensure air-tightness of storage cavern. Moreover, considering the rock mechanical behaviors under high water pressure belong to hydro-mechanical coupling mechanism, settlement and displacement at the cavern crown/walls were monitored to ensure the cavern stability under high water pressure. However, the settlement

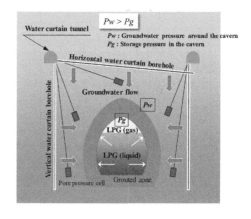

Figure 1. Concept of hydraulic containment type underground LPG storage cavern.

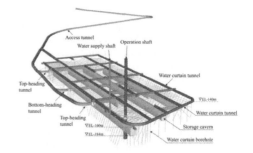

Figure 2. Layout of Kurashiki underground LPG storage cavern.

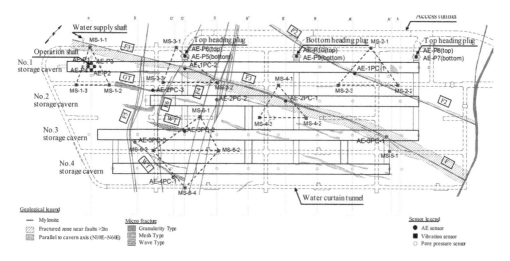

Figure 3. Horizontal geological map (EL.-167.5m, Level of LPG storage cavern top) and sensor arrangement

and displacement monitoring have difficulty in operation phase, therefore, aiming on the long term rock mechanical stability of storage caverns in operation phase, the authors attempted to install the acoustic emission (AE) sensors, vibration sensors and accelerometers at the geological structures and the vicinity of important facilities. The AE monitoring was utilized in laboratory tri-axial compression test to grasp the characteristics with dynamic mechanical behaviors, and then, was verified its applicability by comparison with the monitored displacement in compression of water curtain system.

This study aims to introduce the monitoring aspects on hydraulic behaviors and rock mechanical stability of the underground LPG storage caverns, also to report the monitoring results and observed characteristics.

2 THE INTEGRITY OF UNDERGROUND LPG STORAGE CAVERNS

2.1 *Geology of Kurashiki Site*

Gravel and sand-silt inter-bedding layers distribute below the reclaimed land to EL.-70m. From EL.-70 to EL.-120m, the main lithology represents weathered cretaceous granite, and then, fresh granite distributes below. Considering the maximum operation pressure of the cavern (0.95MPa) and the safety in construction, the authors have designed to construct the storage caverns at El.-160m. Regarding to the physical properties of host rock, the average permeability is 3×10^{-6}cm/s and the fracture density is 0.5-2/m (Kikui et al. 2014).

As the geology map illustrated in Figure 3, five fault zones were identified in the site area, F2 and F3 faults cross the storage cavern obliquely with 50-70 degree, meanwhile F4, F5 and F6 faults orthogonally cross the storage caverns with high dip angle. Metamorphic rock and clay are observed in all the

fault zones and micro-fracture structure composed by numerous tiny cracks are observed.

2.2 *Water curtain compression test*

After the cavern excavation was completed, the water curtain tunnel, operation shaft and water supply shaft were submerged, together with the water curtain boreholes were submerged, then, compressed from 1.1MPa(EL.-28m) to 1.35MPa(EL.-2.4m), meanwhile the hydraulic potential of the water curtain tunnel is 1225kPa (EL.-15m). Since the storage caverns are atmosphere pressure (0kPaG), the different pressure between water curtain system and storage caverns is maximum in experience, thus, the risk on rock mechanical stability together with the accompanied seepage increasing and pore pressure dropping are concerned.

2.3 *Access tunnel filling*

Another risk on rock mechanical stability was also indicated at access tunnel filling. A concrete plug was constructed at the conjunction (EL.-182m) between the storage caverns and access tunnel, after the access tunnel was submerged to EL+4m, the hydraulic head at concrete plug is equal to 186m. In consideration, new crack may be generated at the contact between concrete plug and host rock under high hydraulic head and harm the tightness of storage cavern.

2.4 *Air-tightness test*

Air-tightness test compressed the storage caverns from atmosphere pressure (0kPaG) to the designed maximum pressure (960kPaG, EL.-62m) meanwhile the water level of water curtain tunnel was EL.-15m. The storage caverns were examined the air-tightness by checking the variation of cavern pressure in 72 hours (Maejima et al. 2014). Since the hydraulic gradient is minimum in experience, the risk scenario is considered that the cavern pressure may exceed the hydraulic potential and compressed air may leak into cavern vicinity. Also, rock mechanical stability and hydraulic behaviors accompanied were concerned in cavern compression.

3 ROCK DYNAMIC BEHAVIOR MONITORING

3.1 *Sensors of dynamic behaviors monitoring*

The layout of sensors for dynamic behaviors monitoring is illustrated in Figure 3 and the specifications of the monitoring sensors are listed as shown in Table 1.

3.2 *Acoustic emission monitoring*

Aiming on evaluating the micro-scale failure and stress variation in rock mass, AE monitoring sensors were installed at the vicinity of concrete plugs at the conjunctions of operation shaft/access tunnel with storage caverns and in the geological structures which contact with storage caverns (Figure 4). Based on the preliminary experience of rock block test (see Section 4.2) and in-situ chamber test (Tasaka et al. 2010). AE sensors were also installed at the neigh-

Table 1. Summary of the monitoring sensors

Sensor	Type	Frequency	N.of sensors
AE	Optical	10kHz – 200kHz	19
Vibration	Optical	150Hz – 1kHz	18
Accelerometer	Electrical	1Hz – 10kHz	4

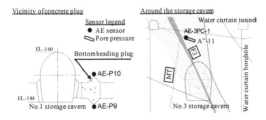

Figure 4. AE sensors installed in the vicinity of the concrete plug and around the storage caverns

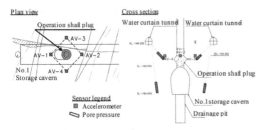

Figure 5. Arrangement of accelerometers

bor of pore pressure cells to monitor the hydro-mechanical coupling behaviors.

3.3 *Vibration monitoring*

Vibration monitoring aims on the key-block and rock falling in storage caverns. In consideration, massive and continuous rock falling events may indicate the propagation of rock failure at the storage cavern vicinity. For the purpose, preliminary falling tests was conducted to optimize the location of sensors for accurate identification ability on magnitude and location of rock falling. As the results, three vibration sensors were installed to enclose the geological sensitive area at cavern bottom (EL.-184m).

Moreover, regarding to the LPG storage status in operation phase, it is recognized that correction on vibration decay in LPG shall be concerned for estimation on the rock falling magnitude.

3.4 *Acceleration monitoring*

Operation shaft leads from the surface down to one of the storage caverns and contains pipelines for seepage withdrawing, LPG loading/unloading and cavern status measurements. Acceleration monitoring aims on evaluation of the mechanical stability of this important facilities. Four accelerometers were installed above the concrete plug of operation shaft at EL.-150m to identify the tiny vibration induced by crack generation (Figure 5). Also the accelerometers are capable to measure the acceleration under earthquake and evaluate with the designed value in seismic-resistant aspect.

4 MESSURED BEHAVIORS

4.1 *AE characteristics using granite specimens*

Figure 6 shows the AE characteristics (tani et al. 2007) in the tri-axial compression test with granite specimens sampled at the vicinities of Namikata storage caverns. As illustration, till the failure was observed in granite specimens, AE events rapidly increased and wave frequency, m-value dropped (whereas, amplitude increased) in accompany with deviator stress increasing.

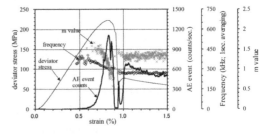

Figure 6. Fracture and AE Characteristics of Granite Specimen (tani et al. 2007)

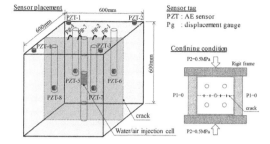

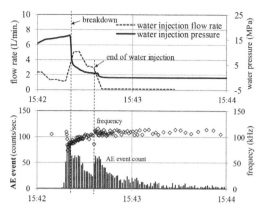

Figure 7. Water and air injection test using granite block Granite block was kept wetting during water/air injection tests.

4.2 *AE characteristics using granite block test*

In order to grasp the accompanied AE characteristics with mechanical and hydraulic behaviors in fractured rock mass, the author conducted a hydraulic fracturing to generate an artificial crack at center of the granite block to characterize the AE events in water/air injection tests (Figure 7).

Figure 8 illustrates the AE characteristics in hydraulic fracturing. The AE events increased with pressurization, then, dropped gradually after the failure was observed in granite block. The AE frequency concentrated in the range of 80 to 100 kHz and recovered to 100 kHz at the tail with crack closing.

In water injection test, several AE events within frequency ranged in 60-100 kHz were observed at the beginning of the crack opened at 2.0MPa. As the crack opened, numerous AE events within frequency ranged in 70-80 kHz were observed. As the crack closed, AE frequency became 100 kHz (Figure 9).

Regarding to the air injection test, since the injection pressure was only compressed to 0.7MPa and the crack opening is constant throughout the test, hence, it is considered the increasing air injection rate is due to air leakage and the 0.7 MPa maximum air injection pressure is correspond to the capillary entry pressure. After the gas leaked, the air injection pressure dropped (Figure 10).

Regarding to the AE events, 60 kHz AE frequency was characterized in air leakage phase in comparison with the observed AE characteristics in hydraulic fracturing and water injection test.

4.3 *Behaviors during water curtain compression test*

Figure 11 shows the AE behaviors and hydraulic pressure during water curtain compression test. The linearly increase of seepage and of pore pressure was observed in accordance with pressurizing water curtain tunnel and water curtain borehole (Fujii et al. 2014).

Most AE events were observed at AE-3PC-1 sensor during water curtain compression. The

Figure 8. AE behavior by hydraulic fracturing

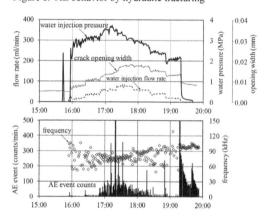

Figure 9. AE behavior during water injection test

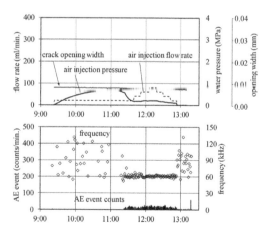

Figure 10. AE behavior during air injection test

AE-3PC-1 sensor is located in F3 fault zone, the AE frequency, amplitude and cumulated AE events are illustrated with pore pressure of neighbored A"-11 sensor, the injection pressure of water curtain system and seepage rate in storage caverns for integrated evaluation on cavern stability. As illustrated in Figure 11, continuous AE events were observed even after A"-11 reached to maximum hydraulic head (EL.-70m), and then, tended to convergence after EL.-77m.

Regarding to the AE frequency and amplitude, 100 kHz AE events were observed at the initial compression stage, then the AE frequency tended to decrease with water curtain compression. After A"-11 reached to maximum hydraulic head (EL.-70m), again, 100 kHz AE events are significant in water curtain decompression. Considering the observed AE frequency is similar as the monitored characteristics in water injection of granite block test and there was no abnormal AE characteristics (AE amplitude increasing and AE frequency decreasing) and hydraulic behaviors (rapid seepage rate increasing and pore pressure dropping) which may reflect to the macro failure on storage caverns. The rock mechanical stability was evaluated in water curtain compression.

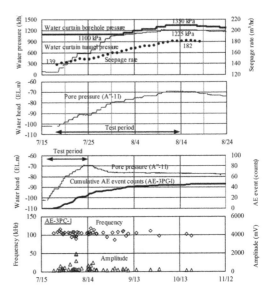

Figure 11. AE behavior during water curtain compression test

4.4 Behaviors during access tunnel filling

The water level of access tunnel was raised to EL.+4m in stepwise. Figure 12 illustrates the AE monitoring results of AE-P10 sensor at the top of concrete plug during access tunnel filling. AE events were observed after water level exceeded EL.-80m. From EL.-20m, rapid AE events increasing rate was observed and this trend continued till the maximum water level EL.+4m. After the water level reached, the triggered AE events decreased to quiet gradually.

The increasing AE events accompanied with water filling, and then decreased with the water level maintenance, is recognized as an elastic behaviors at the vicinity of concrete plug. Moreover, there was no AE frequency decreasing, amplitude increasing and m-value dropping was monitored during access tunnel filling (Mori et al. 2013). Based on the AE monitoring results, the rock mechanical stability of concrete plug in the access tunnel was evaluated in access tunnel filling.

4.5 Behaviors during air-tightness test

Figure 13 gives the evolutions of cavern pressure, seepage rate and pore pressure of A"-11 sensor. In air-tightness test, the storage caverns were compressed to the designed maximum pressure. After stabilization, the cavern pressure was corrected with consideration on the variation of temperature and gaseous volume in storage caverns, then, was examined the variation of cavern pressure in a range of criterion with non-decay trend to validate the air-tightness of storage caverns. (Maejima et al. 2014). During the storage cavern compression, the seepage decreased and pore pressure increased as linear correlation with pressure variation in storage cavern (Fujii et al. 2014). Moreover, regarding to the AE monitoring results, there was almost no AE event observed throughout the storage cavern compression till the air-tightness test. Consequently, the AE characteristics related to gas leakage and crack

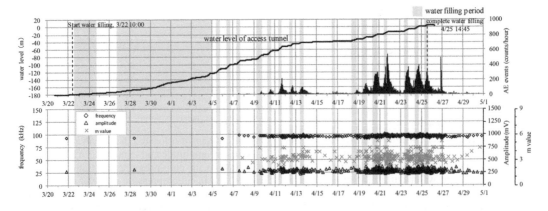

Figure 12. AE behavior of bottom heading plug part during access tunnel filling (Mori et al. 2013)

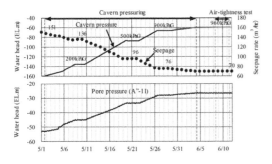

Figure 13. Hydraulic behavior during air tightness test. AE events were not observed in all sensors at top of the storage caverns.

opening/closing behaviors observed in water/gas injection of granite block test and in in-situ chamber test, were not observed and the air-tightness and rock mechanical stability were guaranteed.

5 BEHAVIORS DURING OPERATION PHASE

Figure 14 illustrates hydraulic behaviors and AE monitoring data during operation. The AE events of AE-P10 sensor express periodical fluctuation and are recognized as response with water level in access tunnel, it should be noted that the AE events tends to decrease as the water level of access tunnel tends to convergence. Regarding to the wave characteristics of the AE events, the frequency is ranged in 80 to 100 kHz without descending trends, meanwhile the amplitude decrease gradually.

Also, similar fluctuated hydraulic behaviors and AE characteristics are observed at pore pressure cell A"-11 and AE-3PC-1 sensor. The frequency of AE-3PC-1 sensor is in range of 80 to 100 kHz and no amplitude variation was identified. Considering the observed AE frequency, AE events are caused by crack opening/closing.

On 14th, March of 2014, an earthquake (M6.2) was occurred at 150km distance to Kurashiki base. The accelerometers installed in Kurahsiki storage cavern expressed the 11.3 gal maximum value, far less than the designed value 198 gal. Moreover, as illustrated in Figure 15, there is no abnormal seepage increasing and pore pressure dropping, together with AE event and vibration event related to earthquake was observed. Consequently, the integrity of the storage caverns was evaluated and validated under earthquake.

6 CONCLUSIONS

In this paper, the authors applied AE sensors, vibration sensors and accelerometers and installed in the caverns to ensure the rock mechanical stability. These monitoring and measurement are launched from the water curtain compression test, the access

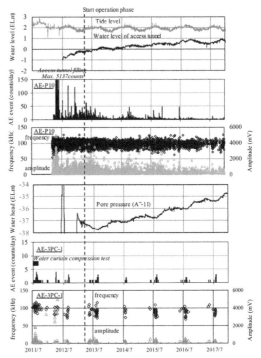

Figure 14. Hydraulic and AE behavior during operation phase

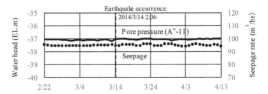

Figure 15. Hydraulic behavior at the earthquake there were no AE events and vibration events related to the earthquake.

tunnel water filling and the cavern air-tightness test, are continued even in the operation phase. Based on the measurement results, the storage caverns was evaluated its integrity consequently.

REFERENCES

Fujii, K., Kobuchi, T., Nishi, T., Kaneto, T., & Maejima, T., 2014. Establishment of water curtain system on the Kurashiki LPG storage cavern and evaluation on hydraulic behaviors during cavern air-tightness test, *The 42th Japan symposium on Rock Mechanics*: 131–136.

Kikui, T., Tasaka, Y., Uneda, A., Soya, M., Ogawa, T., Maejima, T., 2014, Rock mechanical evaluation and management on excavation of hydraulic containment LPG storage cavern: A case of Kurashiki LPG storage cavern, *The 42th Japan symposium on Rock Mechanics*: 137–142.

Maejima, T., Okazaki, Y., Kaneto, T., Mori, T., Soya, M., Kurose, H., 2014. Evaluation and methodology of air tightness test in the Kurashiki LPG underground storage cavern, *The 42th Japan symposium on Rock Mechanics*: 44–49.

Mori, T., Tezuka, Y., Miyajima, Y., Murakami, K., Takagishi, T., Fujii, H., Machijima, Y., Kaneto, T., 2013. Application of optical fiber type AE monitoring to rock underground cavern under high depth and high water pressure: Kurashiki national LPG stockpiling base: *Proceeding of the 68th Annual Conference of the Japan Society of Civil Engineering* VI-204: 407–408 (in Japanese).

Tani, T., Aoki, T., Hirai, H., 2007. Measurement of AE at tri-axial compression test of granite specimen, *Proceeding of the 62th Annual Conference of the Japan Society of Civil Engineering* III-195: 389–390 (in Japanese).

Tasaka, Y., Chang, C.S., Kurose, H., Shimo, M., Maejima, T., 2010. In-situ borehole air-tightness test and the evaluation by hydro-mechanical coupling analysis method, *The 39th Japan symposium on Rock Mechanics*: 255–260.

2019 Rock Dynamics Summit– Aydan et al. (eds)
© 2019 Taylor & Francis Group, London, ISBN 978-0-367-34783-3

Attempt of lignite pit exploration by seismic tomography using directional drilling borehole

K. Kiho
Present; Kawasaki Geological Engineering Co., Ltd., Tokyo, Japan
Former; Central Research Institute of Electric Power Industry, Abiko, Japan

K. Hase
Sumiko Resources Exploration & Development Co., Ltd., Tokyo, Japan

M. Nakadai
Geophysical Surveying Co., Ltd., Nagaoka, Japan

S. Abe
Japan Petroleum Exploration Co., Ltd., Tokyo, Japan

Y. Ohtsuka
Earth Scanning Association, Yokohama, Japan

A. Shidai
FUJITA Corporation, Tokyo, Japan

T. Kato
Kawasaki Geological Engineering Co., Ltd., Tokyo, Japan

ABSTRACT: The occurrence of depression on the ground surface due to lignite pit has been observed at central southern area of Gifu prefecture, Central Japan. In order to prevent the ground surface depression caused by collapse of the tunnel due to the shaking of the Tokai and Tonankai earthquakes, which is expected to occur near future, the work on filling fluent earth materials to the underground cavity is proceeding. To clarify the distribution of lignite pits and to monitor the filling condition of fluent earth materials in the cavity, the seismic tomography is planned between ground surface and the borehole several meters below and parallel to the lignite layer, which is drilled by the directional drilling technology. We report the current status of the site, drilling and survey plan, and outline of the 3D simulation results of seismic tomography.

1 INTRODUCTION

1.1 Background

Since 2000, CRIEPI (Central Research Institute of Electric Power Industry) has been conducting the project on the directional drilling and logging/measurement technologies in its boreholes (Kiho et al. 2009). This project was implemented as part of the development of an effective investigation method at the site selection stage of geological disposal in Japan. This project is almost completed in 2013 as planned and was scheduled to contribute to the site selection afterwards, but the geological disposal project has not started yet.

Until starting the geological disposal project, we have established a CD (Controlled Drilling) workshop consisting of members including authors in order to inherit this technology and started searching for the application of this technology, in order to apply this drilling method to civil engineering investigation and civil engineering construction.

On the other hands, currently, there are many artificial underground cavities such as abandoned mines and shelters and natural underground cavities such as limestone caves, nationwide. The ground surface depression and the ground facility disasters such as settlement and inclination are frequently occurring due to the modification of the ground surface such as the reduction of the soil covering thickness by excavation, the increase of the load by the embankment, the degradation and aging of the rock, the external forces such as the earthquake. In particular, cases where the abandonment of the lignite mine causing depression due to rock deterioration and earthquakes has become conspicuous. In Mitake Town, Gifu Prefecture, surveys and countermeasures are being carried out in order to suppress the risks of abandoned lignite mines ahead of the whole country, but a fixed method has not been established for the method of investigation and the evaluation of countermeasure effect.

1.2 Objectives

In order to confirm the distribution of abandoned lignite mines and to monitor the filling condition of the fluent earth-material to the cavity, we propose the seismic tomography between the borehole drilled by the directional drilling method and the ground surface. We also confirm the effectiveness of seismic tomography by three dimensional simulation..

2 LIGNITE MINE IN MITAKE TOWN

2.1 Geography and Geology

Mitake Town is located in the southeastern part of Gifu prefecture (Fig. 1), facing the Kiso River in the northeastern part and the western part of the town. The urban area spreads to the flat area between the hills of the north and the hills of the south side, roughly extending in the east-west direction and between the altitudes of 100 to 150 m.

The Mizunami group is one of the miocene strata deposited in the sea called "Koseto Inland Sea" in southwestern Japan, distributes from the central and southeastern part of Gifu prefecture. The Mizunami group is classified into the Hachiya Formation, the Nakamura Formation and the Hiramaki Formation from the lower level in the Mitake Town. The Nakamura Formation is about 120 m thick, divided into a lower layer composed of conglomerate or tuffaceous sandstone and siltstone and an upper layer consisting of tuffaceous sandstone and siltstone sandwiching the brown coal (lignite) layer. It is confirmed that at least three lignite layers belonging to the upper layer of Nakamura Formation are distributed in shallow underground around Mitake Town (Fig. 2).

These layers have east-west strike and gentle south dip, so the lower lignite layer distributes in the northern part and the upper lignite layer distributes in the southern part of the town.

2.2 Lignite mine

Since the discovery of the coal vein in 1869, the surroundings of Mitake Town became a major production area of the lignite, supporting important energy of Japan. Mining of the lignite was energetically carried out from the World War II to the postwar period and mining was done about 40 m in depth in the early stage and after the blasting was adopted in 1952, the mining was operated about 100 m deep. In the lignite mine in Mitake Town, the mining was carried out by a method of expanding the mining space from the tunnel while leaving lignite column at regular intervals. In 1956, the mining peaked, production volume rose nearly a quarter to 410,000 tons out of 1.55 million tons nationwide, Mitake Town prospered as a coal mine town. Since then, with the rise of better quality energy sources, in 1968 all the lignite mines closed.

3 SURFACE DEPRESSION AND FILLING

3.1 Depression on the ground surface

Lignite was mined by a method called "residual column method" (Fig. 3) that supports the stability of the gallery by leaving a part of the lignite layer in a columnar shape. The remaining column part gradually deteriorate and collapse, so the ground surface depression has come to occur (Fig.4).

Figure 1. Locality map

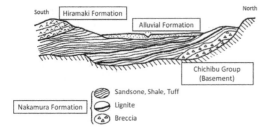

Figure 2. Geological section in Mitake Town (modified from Mitake Town history)

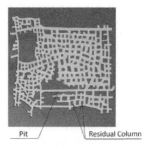

Figure 3. Lignite pit excavation (Hama, 2017)

347

Figure 4. Case example of Surface depression (Hama, 2017)

In Mitake Town, at least 246 surface depressions caused by abandoned mine were confirmed between 1959 and 2001, and 19 cases were confirmed in only 4 years from 2011 to 2014.

3.2 Filling to abandoned lignite mine

Even now, there are surface depressions caused by lignite galleries, but many and large scale surface depressions caused by shaking of the Tokai and Tonankai earthquakes that are expected to occur near future are concerned. In Mitake Town, from the viewpoint of "disaster prevention bases, places where evacuation facilities concentrate and depression frequently occurred in the past", two filling work areas of total 60,270 m2 were selected among the vast area where the lignite abandoned mine is supposed to exist underground. In these selected areas, filling work to the abandoned mine is in progress. In the filling work, a method of stabilizing the ground is adopted by injecting slurry kneaded fixation agent from the ground through the borehole into the underground cavity and closing it.

In the filling work, it is necessary to grasp the three dimensional distribution of the target lignite gallery. There is a location map of the lignite mine at that time in Mitake-Town, but it is difficult to grasp the detailed distribution. Geophysical exploration from the ground surface, for example, radar exploration, electric exploration, etc. are being carried out, but it has not yet been confirmed accurately the cavities existing at several tens of meters underground. For this reason, currently borehole is drilled according to a specific arrangement from the ground surface, and filling work is carried out by using the borehole which reached the cavity at the assumed depth. Also, after completion of filling work, check borehole is drilled, and filling into the cavity is judged according to the state of filling material amount in the borehole.

4 LIGNITE PIT EXPLORATION

The borehole is drilled parallel to the lignite layer at the lower level (about 10 m separation depth) of the target mining site (Fig. 5). By carrying out seismic (Vp and Vs) tomography between the borehole and several survey lines parallel to the borehole at the surface, three dimensional distribution of the lignite gallery can be detected and this can contribute to the decision for selecting the area and arranging borehole location. Also, tomography is performed before, during and after filling, and the filling situation (filling place and filling rate) is monitored by cross section analysis.

In addition, the seismic tomography will be performed between one borehole and multi-survey lines parallel to the borehole on the ground surface, so that the survey range becomes linear. For this reason, as a structure to be surveyed, it is assumed to be a road (national highway) which is a linear important structure.

In this chapter, we introduce the outline and actual results of directional drilling which is the key technology of the survey for lignite gallery which we plan to propose in the future. In addition, we introduce a part of simulation results using geological model of Mitake Town that was conducted to confirm the effectiveness of the seismic tomography.

4.1 Outline of directional drilling system

The directional drilling system is based on the wire line drilling principle, but conventional wire line tools can't insert the drilling and measurement assembly in a gently sloping or horizontal drill hole. The water pressure inserting apparatus (so called pump-in system) which can insert the downhole tools by using pressure of drilling fluid was selected.

This drilling system comprised downhole tools, casing pipe following the downhole tools in order to case the borehole wall protecting borehole wall collapse, and the armored cable with inserting apparatus which can push down and pull up the downhole tools easily and can transmit the data of the downhole tools by telemetric line installed inside the armored cable (Fig. 6).

The mud fluid flows down inside the casing pipe and also inside LWD and MWD, and provides the hydraulic pressure to rotate the DHM. Most of the fluid then flows up to the surface through the

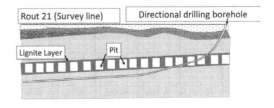

Figure 5. Layout of lignite pit exploration

annulus, which is the aperture between the borehole wall and outside diameter of the casing pipe.

The tools for drilling can be pulled up to the surface by using a wire-line cable in each 3m core runs; to collect core sample, and if a test is deemed necessary, the drilling tools will be exchanged for the testing tools.

4.2 *Drilling results at Kami-Horonobe site*

In 2005, the seismic reflection method was used at Kami-Horonobe site where the outcrop of the Omagari fault is located, and the fault lineament is well defined. Taking into consideration the fault profile deduced from the results of seismic reflection, a borehole trace was proposed to intersect the fault. In 2006, directional drilling was started to verify the applicability of the drilling and measuring system. The borehole was drilled on a bear-

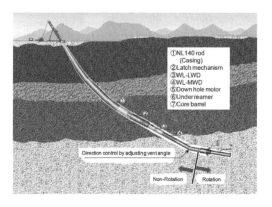

Figure 6. Outline of directional drilling system (Kiho, et al., 2016)

ing of S40W, with a stable inclination of 35° from 0 m to a length of 200 m. From 200 m onward, the directional drilling was relaxed to a shallower inclination of 3° per 30 m, and the borehole reached a horizontal attitude at a length of 720 m. Drilling was terminated at a length of 1000 m after horizontal drilling for 280 m, as of 2011 (Fig. 7). From 0 m to 200 m, drilling was conducted without coring, and from 200 m, cored drilling was performed, and the core recovery was almost 98% as of 2011, even in the fractured zone (Kiho et al. 2016).

4.3 *Waveform modeling and prestack imaging for the distribution of lignite pit*

We investigated capability of imaging the small-scale subsurface structure of abandoned lignite mines at shallow depth using synthetic seismogram calculated by 2D finite-difference acoustic modeling. The size of our model was 500 meters horizontally by 100 meters vertically, with the ground surface at the top (Fig. 8). The model consisted mainly of shale, with a thin layer of abandoned lignite mines at the depth of approximately 20 to 30 meters as the imaging target. Thickness of the target layer was set to 3 meters. It was configured with alternating pillars of lignite and water, whose horizontal widths are both 3 meters, to simulate the possible conditions of the abandoned mine. The synthetic seismic data was created by deploying the surface receivers at 1 meter intervals. While multiple sources were to be used for the survey in practice, calculated shot gathers from a single source at the center of the model was firstly investigated using Ricker wavelet as the source wavelet, to check the imaging capability of the target layer.

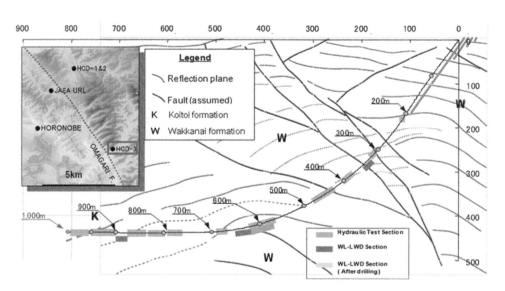

Figure 7. Drilling results at Kami-horonobe site in Hokkaido (Kiho, et al., 2016)

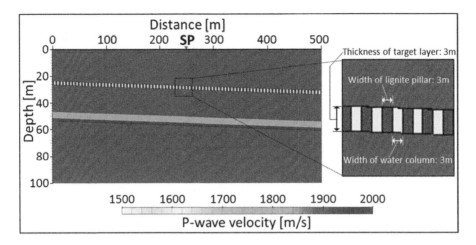

Figure 8. Two dimensional model for P-wave velocity, with an enlarged view of the target layer (Twice vertical exaggeration)

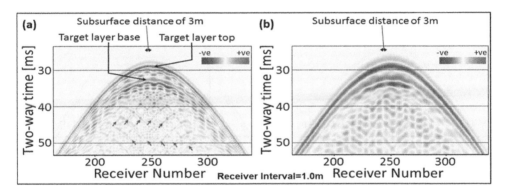

Figure 9. Shot gather created using Ricker wavelet with peak frequency of (a) 500Hz, (b) 250Hz for comparison. Diffracted waves from the target layer are indicated by gray arrows

We analyzed the shot gathers calculated with a range of peak frequency from 65Hz up to 1000Hz to determine how high the frequency should be to image the alternating pillars of lignite and water at the target layer. Diffracted waves from the target layer were observed on the shot gathers (Fig. 9). Individual pillars of lignite and water was delineated by applying Kirchhoff pre-stack time migration in common shot domain, when the dominant frequency of the source wavelet was increased to above 500Hz (Fig. 10).

5 CONCLUSION

- In Mitake Town, Gifu Prefecture, the lignite was mined until 1960's, the ground surface depressions caused by the abandoned lignite mines have been occurring, which is a social problem. Furthermore, there is concern that large-scale depressions may occur due to the shake of the Tokai and Tonankai earthquakes that are expected to occur near future.

- Currently, filling work of the fluent slag to the abandoned mines just beneath important structures is keenly underway, but since it is not possible to grasp the three dimensional distribution of underground cavity, multi boreholes are drilled and filling work is carried out from boreholes hit on underground cavity. Also, the method to confirm the filling situation in the cavity has not been established.

- Synthetic experiments indicate that effective high-frequency component above 500Hz is a prerequisite to detect the small-scale subsurface structure of abandoned lignite mines.

6 ACKNOWLEDGEMENT

The directional drilling project was performed under contracts awarded from the Agency of Natural Resources and Energy (ANRE) of Ministry of Economy, Trade and Industry (METI), and the in-situ drilling and survey was conducted as a collaboration with the Horonobe Underground

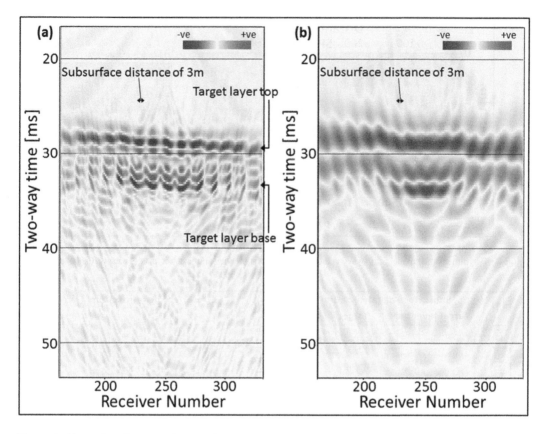

Figure 10. Shot gather after pre-stack time migration in common shot domain, created using Ricker wavelet with peak frequency of (a) 500Hz, (b) 250Hz for comparison.

Research Center of JAEA (Japan Atomic Energy Agency, formerly JNC).

REFERENCES

Hama, Y., 2017. Basic research on estimation of cavity filling condition at the lignite pit. Master's Thesis of Engineering, Ryukyu University. 70P. (Written in Japanese)

Kiho, K. et al. 2009. Development of Controlled Drilling Technology and Measurement Method in the Borehole (Phase 2). CRIEPI Report, N03, 97P. (written in Japanese with English abstract)

Kiho, K. et al. 2016. Directional Drilling Technology for HLW Disposal - Outline of System and its Application-, WM2016 Conference, 16078

Mitake Town. 1985. Mitake Town history, 781P. (written in Japanese)

2019 Rock Dynamics Summit– Aydan et al. (eds)
© 2019 Taylor & Francis Group, London, ISBN 978-0-367-34783-3

Aftershocks of Hyogo-ken Nanbu earthquake (M=7.3) whose epicenter was near the city of Kobe, Japan

S. Sakurai
Kobe University, Kobe, Japan

ABSTRACT: This paper deals with the lessons learned from the Hyogo-ken Nanbu earthquake (M = 7.3) which occurred in Kobe, Japan in 1995 and whose epicenter was shallow, namely, about 15 km in depth. Right after the earthquake, field observations were carried out to investigate the failure mechanism of structures. The results showed that certain structures, particularly a tunnel, were damaged by an unusual failure mechanism which might have been caused by high-frequency impulsive vertical seismic waves. However, in the results of measurements taken by accelerometers, there was no evidence of such high-frequency impulsive vertical seismic waves. Thus, in February of 1995, measurements of the ground vibrations caused by the aftershocks of the earthquake were initiated, and were continued for about two years. To measure the aftershocks, servo-type accelerometers (DC~100 Hz), were used; they differ from conventional earthquake accelerometers (SMAC-MDU) whose measurement range is limited to DC~30 Hz.

1 INTRODUCTION

The Hyogo-ken Nanbu earthquake (M=7.3) occurred in Kobe, Japan on January 17, 1995. Its epicenter was shallow, namely, about 15 km in depth. Numerous structures, such as bridges, buildings, tunnels etc., located along the earthquake fault, were destroyed. Many people believed that the failure mechanism of the structures might have been due to high-frequency vertical vibrations. However, in the published results of measurements taken by accelerometers, there was no evidence of such high-frequency impulsive vertical seismic waves. Therefore, it was decided that the ground vibrations caused by the aftershocks of the earthquake should be measured. To measure the aftershocks, servo-type accelerometers (DC~100 Hz) were used; they differ from conventional earthquake accelerometers (SMAC-MDU) whose measuring range is limited to DC~30 Hz. The measurements of the aftershocks were taken from February 1995 to May 1997. In this paper, the results of the two-year measurement period are shown and the ground vibrations of the aftershocks are discussed.

2 MEASUREMENT OF AFTERSHOCKS

The measuring point was on the floor of an abandoned belt conveyer tunnel located very close to the earthquake fault. The tunnel was situated at a depth of approximately 10 m. The geology of the ground consists of weathered grano-diorite. Eleven seismic events were recorded during the two years, as shown in Table 1. The time, the latitude and longitude of the epicenter, the region of the epicenter, the depth of the epicenter and the magnitude of each aftershock are indicated in Table 1.

Table 1. Collected data on aftershocks.

No	Occurrence time	Latitude	Longitude	Depth	M
①	'95.02.03 20:36'55.4"	N34° 43.7'	E135° 16.0'	12.3km	3.4
②	'95.02.18 21:37'33.9"	N34° 26.2'	E134° 49.0'	15.9km	4.8
③	'95.03.05 10:04'28.8"	N34° 44.3'	E135° 14.5'	13.2km	3.2
④	'95.03.30 14:24'48.3"	N34° 45.2'	E135° 17.8'	12.6km	3.6
⑤	'95.04.22 08:19'02.9"	N34° 43.7'	E135° 16.5'	11.9km	3.0
⑥	'95.05.04 05:53'16.7"	N34° 41.7'	E135° 11.2'	14.7km	3.6
⑦	'95.05.08 02:36'13.0"	N34° 42.7'	E135° 12.9'	13.8km	3.3
⑧	'95.06.16 07:55'50.6"	N34° 45.8'	E135° 17.7'	12.5km	3.8
⑨	'95.06.23 22:19'22.7"	N34° 45.5'	E135° 17.4'	13.2km	3.7
⑩	'95.10.14 02:04'05.7"	N34° 37.0'	E135° 06.0'	17.0km	4.8
⑪	'97.05.14 02:37'	N34° 42.0'	E135° 12.0'	13.0km	3.5

3 MEASUREMENT RESULTS

The epicenters of the aftershocks of the Hyogo-ken Nanbu earthquake were mainly located in the southeastern part of Hyogo Prefecture, as shown in Fig. 1.

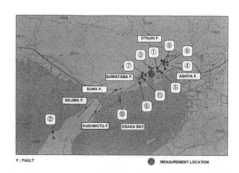

Figure 1. Locations of epicenters of aftershocks and major faults.

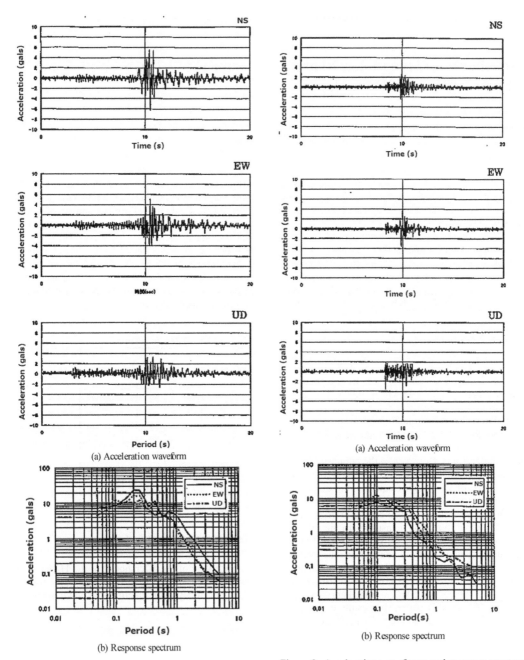

(a) Acceleration waveform

(b) Response spectrum

Figure 2. Acceleration waveforms and response spectrum of record ②.

(a) Acceleration waveform

(b) Response spectrum

Figure 3. Acceleration waveforms and response spectrum of record ③.

They are found approximately along the line of the earthquake fault. The magnitudes of the aftershocks fell in the range of approximately 3–5 gal. The epicenters of the aftershocks occurring closest to the measuring point were located about 2.5 km from the measuring point and are shown as ③ ⑤ and ⑦ in

the figure, while the epicenter of the aftershock occurring farthest from the measuring point was located on Awaji Island and is shown as ② in the figure. The distance between epicenter ② and the measuring point is approximately 47 km. The seismic waves recorded for epicenters ② and ③ are shown in Figs. 2 and 3, respectively.

4 DISCUSSIONS ON THE MEASUREMENT RESULTS

The acceleration waveforms for the nearest epicenters, i.e., ③, ⑤ and ⑦, approximately 2.5 km from the measuring point, are similar to each other in that high-frequency impulsive vertical vibrations exist for them. On the other hand, the acceleration waveform for the farthest epicenter, i.e., ②, 47 km from the measuring point, shows an acceleration waveform that is similar to that of usual earthquakes, for which the largest magnitude appears to be a horizontal vibration and the smallest a vertical vibration. This may be due to the fact that high-frequency seismic waves are easily attenuated during their propagation over a long distance, resulting in high-frequency vibrations for the closer epicenters that did not exist for the more distant ones, such as case ②.

On the basis of the above measurement results, it should be noted that high-frequency impulsive vertical vibrations exist along an earthquake fault within an area 2.5 km apart from the fault. In other words, high-frequency impulsive vertical vibrations exist in the region having a width of 5 km along the earthquake fault. The epicenter of aftershocks was located at a depth of approximately 10-20 km as shown in Table 1, which is almost the same as the depth of the earthquake fault.

The acceleration response spectrum indicates that the highest frequency was about 20 Hz, which is the highest measurable frequency because of the sampling time of the measuring system. In other words, more than 20 Hz could not have been detected even if it had existed. It should be noted that the sound of the aftershock could be heard when the epicenter was nearby; it was just like underground blasting. This proves that aftershocks have P-waves, and that the frequency of aftershocks must be more than 20 Hz. This is because the lowest frequency of the hearing range for human beings is 20 Hz.

5 CONCLUSIONS

The conclusions drawn from this study are as follows:

1. The aftershocks of the Hyogo-ken Nanbu earthquake were measured by servo-type accelerometers placed on the floor of an abandoned belt conveyer tunnel located very close to the earthquake fault. The geology of the ground consists of weathered granodiorite. Eleven seismic events were recorded during the two years. The nearest epicenter of aftershock was 2.5 km from the measuring point, while the farthest epicenter was 47 km.

2. The acceleration waveforms for the nearest epicenter i.e. ③ ⑤ and ⑦ which are approximately 2.5 km from the measuring point indicate almost similar to each other, namely, high frequency impulsive vertical vibration exists. On the other hand, for the farthest epicenter, i.e. ② 47 km from the measuring point, the acceleration waveform is similar to one of a usual earthquake, where the largest magnitude appears in horizontal vibration and the smallest being vertical vibration.

3. If the epicenter being very far, the horizontal vibration becomes largest and the vertical being smallest. This may be due to the fact that the high frequency seismic wave is easily attenuated during propagating a long distance, resulting in that the high frequency vibration disappears as shown in the case of ② 47 km.

4. According to the results of two-year measurements of the ground vibration caused by aftershocks, we can conclude that high-frequency impulsive vertical vibrations exist along an earthquake fault within an area 2.5 km apart from the fault. In other words, high-frequency impulsive vertical vibrations exist in the region having a width of 5 km along the earthquake fault. The epicenter of aftershocks was located at a depth of approximately 10-20 km as shown in Table 1, which is almost the same as the depth of the earthquake fault.

5. It should be noted that we could hear sound of aftershock when the epicenter is nearby just like underground blasting. This also proves that aftershock is high frequency P-wave, which must be more than 20 Hz. because the hearing range of lowest frequency of human being is higher than 20 Hz.

ACKNOWLEDGEMENTS

The author would like to sincerely appreciate Mr. E. Saitou for his great contribution to measurements of aftershocks of the Hyogo-ken Nanbu earthquake.

REFERENCE

Sakurai, S., 2014, Case Studies on the Dynamic Behavior of Tunnels Caused by Hyogoken- Nanbu Earthquake Whose Epicenter was Very Close to the Tunnels, 8th Asian Rock Mechanics Symposium (ARMS8), Sapporo, Japan.

2019 Rock Dynamics Summit– Aydan et al. (eds)
© *2019 Taylor & Francis Group, London, ISBN 978-0-367-34783-3*

Numerical analysis to evaluate repair work of swelling-rock damaged tunnels in the mountains

Keiichi Ota
Research and Development Center, Nippon Koei Co., LTD., Japan

Akio Arai
Highways & Bridges Department., Transportation & Urban Development Division, Nippon Koei Co., LTD., Japan

Yuji Ozaki
Engineering Department 1, Sendai Branch Office, Nippon Koei Co., LTD., Japan

Yoshio Nakamura
Transportation Infrastructure Management Department, Transportation & Urban Development Division, Nippon Koei Co., LTD., Japan

Masahiro Watanabe
Road Department, Fukushima Prefecture Kennan Construction Office, Japan

Takashi Kyoya
Infrastructural Materials Division, Department of Civil and Environmental Engineering, Tohoku University, Japan

ABSTRACT: When a tunnel is assumed to receive swelling pressure from rocks, it is necessary to estimate the swelling pressure properly and to set the external force condition for the tunnel's repair design. For the mountain tunnel under service in Fukushima prefecture, there was a deformation at the roadbed and the tunnel lining due to swelling rock. Therefore we developed a swelling-rock model based on onsite monitoring data and laboratory test data. The model was used to estimate the external forces for design. Several monitoring devices were installed in the site to obtain rock deformation data around the tunnel and evaluate the mechanism of swelling. In addition, swelling pressure tests using site materials were conducted to grasp the swelling characteristics of the rock. A numerical simulation model was made for the interaction between the swelling pressure of the rock and the tunnel structure using the rock deformation with the results of the tests The simulation model, which was made in consideration of past data, estimates the external forces, which are necessary for the reconstruction of tunnels in the future. This paper outlines the method of setting the swelling model by numerical analysis, the features of the model, and the results of applying it to a tunnel.

1 INTRODUCTION

1.1 *Overview of the tunnel*

This tunnel was constructed through a mountain with an altitude of about 2000 meters to improve the convenience of local transportation in Fukushima Prefecture. It is 4 km long and serves some of the local arterial roads.

The geology around the tunnel consists of lava, volcanic clastic rocks and intrusive rocks associated with volcanic activity of the Quaternary period.

According to the observation data of tunnel face during the construction, basaltic crushed lava, basalt lavas, tuff gravel rock was confirmed. Especially basaltic self-crushed lava is susceptible to alteration by crushing and is a geologically vulnerable.

1.2 *Tunnel deformation*

In this tunnel, uplift of the pavement surface was confirmed after service was started. After that, field surveys, onsite monitoring and laboratory tests were carried out, and one of the causes of the deformation was pointed out: in addition to progressive deterioration of the ground due to tunnel drilling, the influence of groundwater infiltration from the around tunnel and absorption swelling due to groundwater supply.

Therefore, we understood the observation record at the construction and the boring column. It confirmed that there was distribution of altered basaltic autogenous lavas and self-crushed conglomerate, a strong alteration zone, and the existence of smectite which was one of the swelling clay minerals.

1.3 Consideration on tunnel deformation

Based on such tunnel deformation and geological condition around the tunnel, we made a model to simulate the damage mechanism of the tunnel to help us consider the method of repair and the reconstruction design. We also developed a numerical simulation model that considered the absorption swelling of the ground, calculated the assumed external force by swelling, and examined repair and reconstruction design based on the calculation result.

2 SETTING SWELLING MODEL

2.1 Overview of swelling model

There are several existing swelling models for tunnels. One model that requires setting of multiple model parameters (Yashiro K., Shimamoto K., Kojima Y., Takahashi K., Matsunaga T., Asakura T. [2009]), and other models that require assumption of the swelling range before modeling (Okui Y., Tsuruhara T., Ohta H., Sakuma S., Nakata T. [2009]). For practical use, it is desirable that the model is easy to handle, and does not require assumptions in advance.

The model we developed uses the results of onsite monitoring data and laboratory test data. The mechanism of model is based on the elastic module of ground decreases with increasing in the swelling strain of the ground. This mechanism of this model is based on the process where the sampling material of survey indicates gradually swelling and leads to vulnerable nature after collecting the material. When using the model, the user can easily set parameters by onsite monitoring data and laboratory test data. Also, in this model, setting of the swelling range is unnecessary.

2.2 Feature of swelling model

The mechanism of this model is that the elastic module of the ground decreases with increasing in the swelling strain caused by the swell of the ground. Therefore, using the swelling stress and the absorption swelling ratio of the swelling test shown in Figure 1, the swelling strain is calculated from the volume change of the sample collected from the swelling. Using the swelling stress at that time, the elastic module, which is associated with the change in the swelling strain, is set.

In this model, the relationship between the stress after swelling, the elastic module, and the swelling strain is shown in Equation 1.

$$\{\sigma\} = [D_0] \exp(-a\varepsilon_v)(\{\varepsilon\} - \{\varepsilon_0\})$$ (1)

where,
$\{\sigma\}$:swelling stress after swelling, $[D_0]$:elastic module, a :any value, ε_v: volumetric swelling strain ($\varepsilon_v = \varepsilon_x + \varepsilon_y$), $\{\varepsilon\}$:actual strain, $\{\varepsilon_0\}$:swelling strain that given value

As shown in Equation 2, $[D_0]\exp(-a\varepsilon_v)$ which is the part of the elastic module corresponding to the volumetric swelling strain of Equation 1, can be obtained from the results by using the curve of Figure 2 arranged from the result of swelling test. Therefore, the elastic module decreases exponentially in accordance with the volumetric swelling strain, and the ratio of the reduction is in accordance with parameter a.

$$[E] = [D_0] \exp(-a\varepsilon_v)$$ (2)

Substituting Equation 2 into Equation 1 yields Equation 3.

$$\{\sigma\} = [E](\{\varepsilon\} - \{\varepsilon_0\})$$ (3)

According to Equation 3, the elastic module corresponding to the swelling strain commensurate with the actual strain is set by the curve shown in Figure 2, and as a result, the swelling stress considering swelling strain is calculated.

3 INVESTIGATION OF BASIC MECHANISM OF SWELLING MODEL

To confirm the basic behavior of the swelling model, we set the same conditions as the swelling test and examined it using numerical simulation.

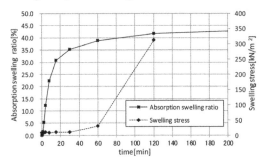

Figure 1. The results of swelling test

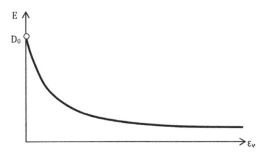

Figure 2. Relationship between the volumetric swelling strain and the elastic module

3.1 Setting swelling method

FLAC, which is one of the finite difference method analysis code was used for this numerical simulation. In the simulation, the relationship between the swelling strain and the elastic module shown in Figure 3 was used. Figure 2 shows the relationship between the swelling strain and the elastic module as a curve., However, to make the handling easier, Figure 3 set the swelling ratio and the reduction rate of elastic module as a straight line. Here, the elastic module gradually decreases as the swelling progresses, and the elastic module before swelling is set to about half of the initial value with swelling ratio of about 40%.

3.2 Simulation result under free swelling

Figure 4 shows actual strain when swelling strain is given to the specimen. Here, actual strain is the volumetric swelling strain obtained by simulation. The horizontal axis in Figure 4 is the cumulative value of the number of divisions of swelling strain given to the specimen during simulation.

As a result, when swelling strain of about 70% is set as the cumulative value, the swelling ratio of about 40% occurs in the specimen as the elastic module decreases. This swelling ratio is almost the same value as in Figure 1.

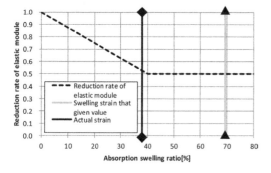

Figure 3. Relationship between the swelling ratio and the reduction rate of elastic module

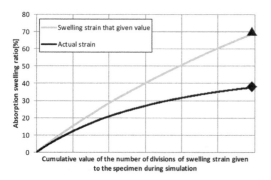

Figure 4. Relationship between the swelling strain that given value and the swelling ratio

As shown in Figure 4, as the swelling strain increases, the difference from the actual strain increases. This is because the elastic module gradually decreases as the swelling rate increases and becomes approximately half of the initial elastic module at about 40%. Even if the swelling strain is applied to the specimen, the generated swelling stress is multiplied by the elastic module after the reduction. Therefore, the difference between the swelling strain and the actual strain become large. Eventually, as shown in the result of the swelling test shown in Figure 1, the swelling ratio tends to converge gradually.

3.3 Simulation results under constraint conditions

Figure 5 shows the swelling stress in the vertical direction of the specimen when the upper surface of the specimen is constrained. The vertical stress gradually increases and converges to about 400 kN/m². The reason why vertical stress converges is the effect of restraint. As shown in Figures 2 and 3, after the beginning of swelling, the elastic module decreases as the expansion progresses but the specimen is restrained. Also, the swelling ratio and elastic module do not change although the swelling strain is set. As a result, the swelling stress was constant.

Thus, by using this swelling model, it is possible to obtain an external force due to swelling by setting the elastic module and swelling strain before swelling.

4 APPLICATION TO TUNNEL

This section covers the result of applying to the actual tunnel using the created swelling model.

4.1 Parametric analysis due to differences in swelling range and supportive work

Three cases were set for parametric analysis.

In Case 1, the swelling area was set around the tunnel.

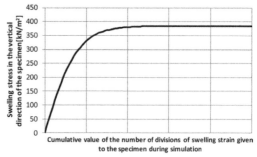

Figure 5. Relationship between the swelling strain that given value and the swelling stress

In Case 2, the swelling area was set a part of the tunnel periphery. Each case does not have tunnel structure.

In Case 3, a support structure and invert concrete were set in the tunnel and a part of invert concrete was set with a scratch, so the inverted concrete was deformable to the inner side.

The upper part of Figure 6 shows the amount of displacement after swelling and the lower part shows the ratio of volumetric swelling.

As a result of the analysis, Case 1 shows a substantially uniform displacement distribution around the entire circumference of the tunnel. Case 2 has a displacement distribution centered on the swelling range. There is no difference in the volume swelling ratio in each case, but the maximum displacement is larger in Case 2 than in Case 1. As a reason of the difference between the two cases, the uniform displacement distribution occurs in the whole circumference of the tunnel in Case 1; on the other hand, since the swelling range of Case 2 is smaller than Case 1 and localized, as a result, maximum displacement in Case 2 is larger than Case1. Also, in Case 3, maximum displacement at the scratch point is larger than Case 1 with the same swelling range and distribution of the volume swelling ratio is larger than Case 1 around the scratch point.

In this way, even if the swelling range is the same, when it has a tunnel support and local scratches, both the maximum displacement and the ratio of volumetric swelling are larger than when there is no tunnel support.

This is one of the important features of this model. The behavior of this model is consistent with the general mechanism where the stress concentrates on the weak part of the tunnel and is considered to be a feature indicating the validity of the developed model.

4.2 *Application to remodeling repair design*

Using the swelling model, we examined the swelling pressure and set the external force condition of the tunnel's design

In this tunnel, there was a deformation at the roadbed and the tunnel lining due to the swelling rock. Also, several measuring devices were installed to get rock deformation data around the tunnel. By using this monitoring data, the relationship between the swelling ratio and the elastic module was set. Figure 7 shows the result.

We modeled the tunnel shape and applied the swelling pressure to the tunnel using Figure 7. As a result, the distribution of the section force generated in the tunnel was obtained as shown in Figure 8, and this section force was used to decide the repair design. The dotted line shown in Figure 8 indicates the reconstruction tunnel specification.

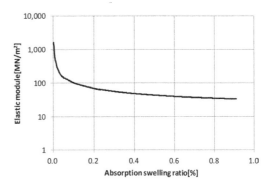

Figure 7. Relationship between the absorption swelling ratio and the elastic module

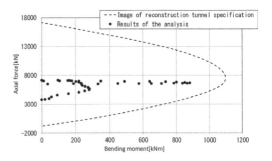

Figure 8. The distribution of the section force and the reconstruction tunnel specification

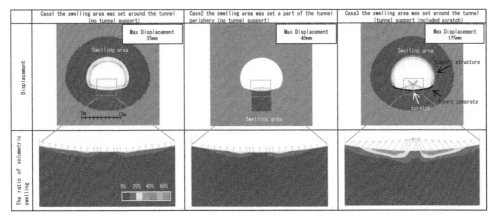

Figure 6. The results of parametric analysis

5 SUMMARY OF RESULTS

From the examination using the created swelling model, the following conclusions were drawn.

1. We developed a numerical simulation model, which considered the interaction between the swelling pressure of the rock and the tunnel structure using the rock deformation due to swelling characteristic of the rock. It was confirmed that this model showed the basic behavior as observed in the laboratory test.
2. By applying the model to the actual tunnel and examining the characteristics of the model, it was shown that this model was consistent with the general mechanism that the swelling pressure concentrated on the weak part of the tunnel due to the influence of the swelling pressure.
3. In addition, we examined the reconstruction design of the mountain tunnel under service. We investigated swelling pressure by using the model. As a result, we decided on the repair design that included the distribution of section force obtained by analysis

6 FUTURE DIRECTIONS

In this study, the adequacy of the swelling mechanism assumed from the field survey results was verified by using the developed swelling model and simulation model.

Since the developed model was able to reflect the site situation, we plan to improve the model using monitoring data and to proceed with the reconstruction tunnel design using the model in the future.

After the Great East Japan Earthquake, aftershocks have been followed around the tunnel occasionally. Generally speaking, tunnels behave together with mountain during earthquake, and therefore, the influence of the seismic motion to the tunnels would be small. However, ground expansion caused by hydraulic condition change due to the seismic motion may cause tunnel deformation. As the ground around the tunnel has a characteristic of the expansion, we plan to research about the relationship between the seismic motion and the tunnel deformation.

REFERENCES

Yashiro K., Shimamoto K., Kojima Y., Takahashi K., Matsunaga T., Asakura T. [2009]. "Study on simulation analyses of deformed tunnels due to earth pressure by ground strength deduction model and it's applicability to long term prediction", *Journal of Japan Society of Civil Engineering*, Vol.65, No.1. (In Japanese)

Okui Y., Tsuruhara T., Ohta H., Sakuma S., Nakata T. [2009]. "Analysis of heaving behavior in Sakazukiyama road tunnel under use", *Tunnel Engineering* Journal, *Vol.19*. (In Japanese)

2019 Rock Dynamics Summit– Aydan et al. (eds)
© 2019 Taylor & Francis Group, London, ISBN 978-0-367-34783-3

Effect of excavation of drives with blasting in paste-filled stopes on the mined-out zone

Sripad. R. Naik, Aditya Mishra & H.S. Venkatesh
National Institute of Rock Mechanics, Bengaluru, India

Aman Soni
Department of Mining and Minerals Engineering, Virginia Polytechnic and State University, USA

ABSTRACT: This paper evaluated stability of drives excavated in paste-filled stopes for drilling long holes for further blasting in underhand mining. The practical implementation of the process has been carried out at the Rampura Agucha underground zinc mine in India. Simulation for effect of excavation and blasting vibration on the surrounding area was carried out using dynamic analysis in 3 dimensions using FLAC3D. The actual geometrical profiles of the stopes along with the original rock mass properties were incorporated in the model. The stopes below the blasted stopes are prepared for ore extraction after excavation drives in paste fill. Rayleigh damping condition was applied to the model along with application of dynamic blast load with viscous boundary conditions. The vertical component of peak particle velocity (PPV) at various points in the model were measured and were compared to the values observed at the corresponding locations in the field.

1 INTRODUCTION

According to different mine conditions, mining methods are chosen to increase productivity. The activities are planned, however, by keeping the safety of personnel and equipment as a primary concern. The paper discusses the effect of vibration from blasting on mined-out stopes in the Rampura Agucha underground lead-zinc mine in India. The mining method applied is underhand mining method with paste backfill. To undergo underhand mining, the blast boreholes required to carry out blasting activities need to be drilled downwards from the crown level of the stope. Hence, paste filled drives are excavated by drilling and blasting operations. Vibrations generated from blasting affect the stability of the backfilled paste which has low strength compared to ore body or host rock. Therefore, it is necessary to ensure the stability of these drives until the drilling and blasting activities of the stope are completed. A similar method for underhand mining in India has not been implemented in the past.

Dynamic modeling in Fast Lagrangian Analysis of Continua (FLAC3D) can be used to analyze the stability of the drives after blasting activities. Rock and material properties were measured using in-situ geotechnical borehole and laboratory testing. Damping phenomenon for blast vibrations was defined using the Rayleigh damping equation. The damping parameters were calibrated using the field vibration read from blasting in a model stope. Peak particle velocity was classified as an effective parameter in calibrating and studying the damage from blast loading to the mined-out stopes. Effect of underground blasting on structures has been done by Drake & Little (1983), Kumar et al, (2016) and many authors in the past. Results of the numerical analysis have been compared to blasting in the other stopes in the mine. The peak particle velocities obtained from the field results and those obtained from the numerical model are quite comparable. Assessment of stability in the excavated drives has been provided to understand the effect of excavation using blasting in the underground mine.

2 DESCRIPTION OF SITE AND MINING METHOD

Rampura Agucha Underground (RAUG) mine, belonging to Hindustan Zinc Ltd. is one of the largest underground lead and zinc mines in India. The mining method adopted is Long Open Stopes with paste filling. In this method, the stopes are developed with 25m level intervals. At each level, the stopes are developed by drives and cross cuts. The stoping operation is carried out in a sequential manner so that the void created after extraction of the stope is immediately filled with paste while other stopes are kept ready for extraction. In a recent study conducted by Anon (2017), it was suggested that underhand mining is a feasible method which can increase the productivity of the mining activities, with due regard to safety. Subsequently, the underhand mining with paste fill was adopted at RAUG mine. In underhand mining, the blast boreholes required to carry out blasting activities need to be drilled downwards from the crown level of the stope. To access the crown level, the drives need to be excavated in paste filled stope by drilling and blasting operations. Therefore, it is necessary to ensure the stability of these drives until the drilling and blasting activities of the stope are completed. The stability of these drives was assessed by three-dimensional continuum models in FLAC3D.

3 DYNAMIC ANALYSIS IN FLAC-3D

3.1 *Methodology*

In this study, the behavior of the paste fill after blasting and excavation of drives is examined using 3D numerical models. The three-dimensional analysis of the excavated drives in paste fill and a typical model, pattern and sequence provided by the mine, was carried out using dynamic analysis. Mohr-Coulomb constitutive model was used to carry out simulation in FLAC3D. This constitutive model closely represents rock behaviour.

Blasting vibrations are characterized by their low peak amplitude and high dominant frequencies than earthquakes as the source of energy is nearer and propagation distance is also smaller. Many researchers have studied the effect of blasting on the cemented hydraulic fill and paste fill using numerical modelling tools. Lilley et al. (1998) studied damage due to production blasting adjacent to cemented hydraulic fill (CHF) stopes. McNearny & Li (2005) and Li et al. (2003) modelled backfilled stopes in the sublevel stoping method with delayed backfill. Gool et.al (2004) did experimental and numerical modeling and field work at Cannington mine in Australia. Variety of loading conditions for static and dynamic loading due to blasting conditions were studied using ABACUS to observe the changes in paste fill behavior due to different blasting conditions.

The dynamic analysis option in FLAC3D permits three-dimensional dynamic analysis. Explicit finite difference scheme is used to solve the full equations of motion, using grid point masses derived from the real density of surrounding zones. This formulation, coupled to the structural element model, permits analysis of ground and structure interaction brought about by ground vibrations. There are three aspects that one should consider when preparing a FLAC3D model for a dynamic analysis: (1) dynamic loading and boundary conditions; (2) wave transmission through the model; and (3) mechanical damping.

To study the response of drives to blast stress through numerical modeling, many factors such as fixity conditions, constitutive models, damping parameters need to be calibrated. Blast pressure is applied, and peak particle velocity time-history is obtained. Peak Particle Velocity (PPV) is monitored at various points away from the blast. The FLAC3D result has been compared with field data for calibration and results are obtained for evaluation of the response of paste drives subjected to blast pressure.

For dynamic analysis, the damping in the numerical simulation should reproduce in magnitude and form the energy losses in the natural system when subjected to dynamic loading. In this study, Rayleigh damping is used which, in time-domain programs, is commonly used to provide damping that is approximately frequency-independent over a restricted range of frequencies. A damping matrix, C, is used, with components proportional to the mass (M) and stiffness (K) matrices:

$$C = \alpha\, M + \beta\, K \qquad (1)$$

where α = the mass-proportional damping constant; and β = the stiffness-proportional damping constant.

Although both terms are frequency-dependent, an approximately frequency-independent response can be obtained over a limited frequency range, with the appropriate choice of parameters. These parameters were determined by the calibration of results using field data. In this study, field PPV's are compared with the values of modelling results to arrive at suitable mesh size and dynamic time step required to get a response of paste filled stopes subjected to blast excavations. The input to FLAC3D is given by free field pressure estimated by Kumar (2015) and Bulson (1997) which is established based on published data of various researchers. After the blast, PPV is calculated from the presently developed model.

3.2 *3D Model at RAUG Mine*

To conduct dynamic numerical modelling using FLAC3D, the following steps were followed:

Preparation of 3D model using plans and sections of the mine

– Application of boundary conditions
– Initialisation of the model with in-situ stresses and simulate upto equilibrium
– Calibration of the model with current mining activity
– Application of blast load around the tunnel in the paste drive
– Analysis of the results

The model is a close representation of the geometries and conditions at the Rampura Agucha underground mine and the stopes are located at approximately 580 m below the surface as shown in Figure 1. The level in which the stopes are simulated for blasting falls below the level of post-extraction paste filled stopes. The stopes below the blasted stopes are prepared for ore extraction after excavation of paste-filled drives. Peak Particle Velocity (PPV) is monitored at various points away from the blast as shown in Figure 2.

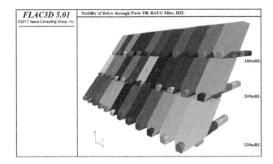

Figure 1. FLAC 3D Model showing the stopes and the crosscuts.

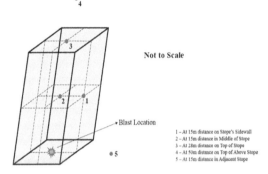

Figure 2. Location of points for Peak Particle Velocity observation from the blast location.

The rock mass properties for all geomaterials used in the model were measured using geotechnical borehole measurements and laboratory testing. The rock mass properties were calculated at every 100 m depth. From the Hoek and Brown theory (Hoek, 2006), the variation in depth causes the cohesion and friction angle to change. However, there was no change incorporated in the deformation modulus and tensile strength with depth.

3.3 In-situ Stress

The in-situ stresses measured in the RAUG mine are as follows:

$$SH = 0.034 \times z + 5.65 \quad (2)$$

$$Sh = 0.017 \times z + 2.82 \quad (3)$$

where, SH = Maximum Horizontal Stress (MPa);
Sh = Minimum Horizontal Stress (MPa); and
z = depth from the surface (m) (Surface of mine is at 392 mRL).

The major principal stress (Maximum horizontal Stress) lies 32° from Mine North towards Mine

East (Mine North is at N50°E from actual North) at the mine site. The mine North-South direction lies along the Y-axis of the model.

3.4 Applying Boundary Conditions

The FLAC3D model as shown in Figure 1 was enclosed by boundaries which allocated certain properties. The boundaries should be large enough so as not to affect the numerical simulations, yet not so big to occupy too much program memory. Viscous boundary is applied at all the sides except the top face, which is kept free. As described earlier, Rayleigh stiffness damping is applied. Damping ratio is taken as 0.04 at 55 Hz frequency.

3.5 Calibration using field data

To study the response of paste filled stopes and excavated drives to blast stress through numerical modeling, many factors such as boundary conditions, constitutive models, damping parameters need to be calibrated. Blast pressure is applied, and peak particle velocity time-history is obtained. The FLAC3D result was calibrated and compared with the field data as shown in Table 1 for calibration and results are obtained for evaluation of the response of paste drives subjected to blast pressure.

After calibration of the numerical model, the resultant peak particle velocity was noted at Location 4 and Location 5 which are corresponding to original reading locations at RAUG underground mine. Only the vertical component of peak particle velocity is calibrated against the model as it is the most important component amongst the PPV in all directions.

It can be seen from Figure 3 and Table 1, that the results achieved from the dynamic numerical modeling are in conformity with the values taken by field observations. The vertical PPV at Location 4 (corresponding to -130L S127) is coming out to be 62.5 mm/s in FLAC3D which is close to a value of 62.61 mm/s attained during the field blast monitoring. Similarly, the vertical PPV at Location 5 (corresponding to -105L S172) is coming out to be 30.0 mm/s in FLAC3D which is close to a value of 25.03 mm/s attained during the field blast monitoring. This shows that the model is calibrated accurately according to provided field measurements.

The displacements in the calibrated model show that after excavation of drives, high displacements are observed in the brow and shoulders of the blast plugs and tunnels and the maximum magnitude was around 112mm. The strength to stress ratios at the drive level at the paste fill before and after the blasting operation is shown in Figure 4. The results show that the drive in the paste fill showed failure with the strength to stress ratio falling below 1.

The strength to stress ratio contour plots was analysed to check for the possibility of shear failure in

Table 1. Field blast vibration data used for calibration of a numerical model.

Location of Blast	Monitoring Location	Total Explosive Kgs	Radial Distance. m	PPV mm/s	Frequency. Hertz	PPV mm/s	Time Sec	MCPD kg	No of Free face
S-112-130L	Open Pit 90mRL	4254	197	9.048	66.25	13.46	0.525	301	2
S-112-130L	-105L XC S 172	4254	50	25.03	158.9	25.54	0.443	301	2
S-112-130L	-130L XC S 127	4254	15	62.61	45	82.01	0.437	301	2

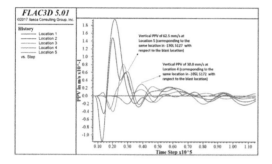

Figure 3. Peak particle velocity (Vertical) vs Time step for the blast cycle in Stope 1

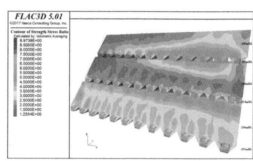

Figure 5. Strength to stress ratio contours before blasting and excavation of stopes (Front view)

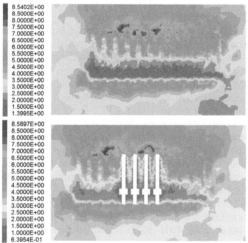

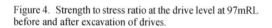

Plan View at 97mRL

Figure 4. Strength to stress ratio at the drive level at 97mRL before and after excavation of drives.

any zone. The contours are shown in figures 5, 6, 7, and 8. These figures show the factor of safety prior to blasting and excavation activities, as well as after the drive excavation. These contours indicate the deterioration of paste material especially in the brow and shoulder region of the blasted plugs and tunnels.

A drive through the paste fill is shown in Figure 9. The damage caused in the paste column was minimum due to blasting as shown in Figure 10.

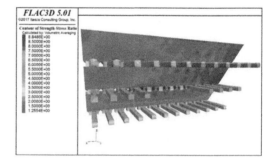

Figure 6. Strength to stress ratio contours before blasting and excavation of stopes (Back view)

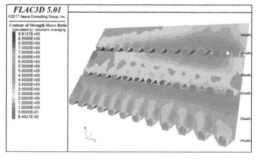

Figure 7. Strength to stress ratio post blasting and excavation of stopes (Front view)

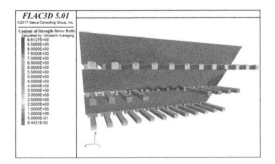

Figure 8. Strength to stress ratio post blasting and excavation of stopes (Back view)

Figure 9. A drive through the paste fill.

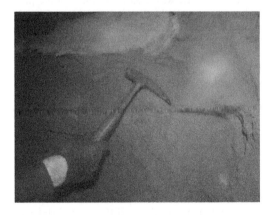

Figure 10. Minimal damage to paste fill column due to blasting.

4 CONCLUSIONS

The effect of blasting on mined out stopes was studied using dynamic analysis in FLAC3D. The parameters required for dynamic modelling was estimated using the measured peak particle velocity data. From the numerical modelling results, it can be predicted that the drives excavated through paste fill would be stable after drilling and blasting operations to further carry out downward drilling for underhand mining even when high displacements are observed in the brow and shoulders in the paste filled drive. The paste fill plays an important role in transferring stress from hangwall to footwall.

ACKNOWLEDGEMENT

Authors would like to thank Director, NIRM for permitting to publish the paper. Hindustan Zinc Limited. India is also thankfully acknowledged for their support and data during the study.

REFERENCES

Anon 2017. Scientific Studies to Assess Stability of Drives through Paste fill at Rampura Agucha Underground Mines, HZL. NIRM Report submitted to HZL.

Bulson, P. S. 1997. Explosive loading of Engineering Structures, 1st ed., E & FN Spon Publishers, New York, N, 236.

Drake, James L. & Charles D. Little Jr. 1983. Ground shock from penetrating conventional weapons. In Army engineer waterways experiment station Cicksburg MS.

Gool, V. Bronwyn, Karunasena, W. and Sivakugan, N. 2004. Modelling the effects of blasting on paste fill. In: *1st International Conference on Computational Methods (ICCM04)*, 15–17 December, Singapore.

Hoek E, Diederichs, M.S., 2006, Empirical Estimation of rock mass modulus, *International Journal of Rock Mechanics and Mining Sciences.*, 43, 203–215.

Kumar, R. Choudhury, D. & Bhargava, K. 2015. Simulation of rock subjected to underground blast using FLAC3D. In *The 15th Asian Regional Conference on Soil Mechanics and Geotechnical Engineering*, Nov 9–15, Fukuoka Prefecture Japan, 508–511

Li, Q. McNearny, R.L, & MacLaughlin, M.M. 2003. Stability of Sublevel Fill Stops Under Blasting Load In *Fourth International Conference in Computer Applications in Minerals Industries (CAMI 2003)* Calgary, Alberta, Canada.

Lilley, C.R. & Chitombo, G.P.F. 1998. Development of a near field damage model for cemented hydraulic fill. *Minefill'98*, Brisbane, p. 191–196.

McNearny, R.L. & Q. Li, 2005. Numerical Study of Stope Backfill Behavior in an Underground Mine, in AlaskaRocks. *Rock Mechanics for Energy, Mineral and Infrastructure Development in the Northern Regions.* ARMA. p. paper 824.

2019 Rock Dynamics Summit– Aydan et al. (eds)
© 2019 Taylor & Francis Group, London, ISBN 978-0-367-34783-3

Correlation between seismic damages to mountain tunnel and ground deformation: Case study on Tawarayama tunnel under the 2016 Kumamoto earthquake

Y.J. Jiang
School of Engineering, Nagasaki University, Nagasaki, Japan

X.P. Zhang
State Key Laboratory of Mining Disaster Prevention and Control Co-founded by Shandong Province and the Ministry of Science and Technology, Shandong University of Science and Technology, Qingdao, China

Y. Cai
School of Civil Engineering, Beijing Jiaotong University, Beijing, China

ABSTRACT: The spatial correlation of the ground deformation at Mt. Tawarayama in Kumamoto City in Japan with the seismic damages of Tawarayama tunnel was developed to explore whether the seismic damages of underground structures are related to the ground deformation. A pair of Digital Elevation Model (DEM) datasets were captured from the high-density airborne light detection and ranging (LiDAR) data before and after the 2016 Kumamoto earthquake. A new variant of Iteratively Closest Point (ICP) algorithm named Combination and Classification ICP (CCICP) was introduced to detect the three-dimensional (3-D) ground deformation field. The results indicated that the strong ground deformation can reflect the seismic performance of the tunnel to some extent. Furthermore, the results of the ground deformation direction validated the assumption of seismic wave propagation along the tunnel. It gives a clear explanation for the mechanism of the seismic damages under the earthquake force, especially lining cracks, pavement damage, and construction joint damage.

1 INTRODUCTION

Ground shaking and deformation induced by earthquakes may cause tremendous forces on long and/or large structures such as rock engineering structures (Aydan 2017). It can yield relevant information about the evolution of the temporal and spatial distribution of the ground deformation. Therefore, it is essential to understand the causes, triggering factors, and mechanisms, in order to delineate the most affected areas and achieve accurate assessment and mitigation of natural and anthropogenic hazards.

Inland earthquake with a magnitude greater than 7 can cause ground shaking and deformation (Park et al. 2018). Record of RTK-GNSS (Real-time Kinematic-Global Navigation Satellite System) time series showed that strong shaking and deformation caused by the 2016 Kumamoto earthquake could be observed at the stations within 100 km from the epicenter. Subjected to the 2016 Kumamoto earthquake, Tawarayama tunnel in Kumamoto City was severely damaged.

The present study aims at detecting the ground deformation characteristics and studying its correlation with the seismic damages of Tawarayama tunnel during the 2016 Kumamoto earthquake.

Airborne LiDAR was used to obtain both the pre- and post-event DEM (Digital Elevation Model) datasets. The 3-D deformation field of the ground above Tawarayama tunnel was calculated with the datasets using the Combination and Classification Iteratively Closest Point (CCICP) algorithm. The ground deformation was spatially compared with seismic damages of Tawarayama tunnel to explore whether the seismic damages of underground structures are related to the ground deformation.

2 STUDIED AREA AND DATA DESCRIPTION

A moderate-size earthquake with a magnitude 6.5 (Mj) struck Kumamoto City in Japan at a depth of about 11 km on April 14, 2016 (21:26 JST-Japan Standard Time) (Asian Disaster Reduction Center 2016) (Fig.1a). The fault rupture originated from the northern segment of the Hinagu fault. Subsequently, at 01:25 on April 16, 2016 (JST), a larger earthquake with magnitude 7.3 (Mj) occurred along the Futagawa fault at a depth of about 12 km.

Tawarayama tunnel with a total length of 2,057 m was excavated under Mt. Tawarayama on the Takamori Line of Kumamoto Prefectural Route 28.

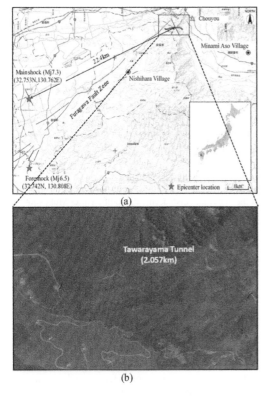

(a)

Tawarayama Tunnel
(2.057km)

(b)

Figure 1. Location of Tawarayama tunnel and the epicenter of the 2016 Kumamoto earthquake (a); Post-event earth image around Tawarayama tunnel captured on 31st May 2016 (Geospatial Information Authority of Japan 2016) (b).

The height of Mt. Tawarayama is 1,095 m. Most of the ground surface is covered with dense forests (Fig. 1b). The tunnel is located at about 22.4 km away from the epicenter of the mainshock (Mj7.3).

The studied area was set to the ground of Mt. Tawarayama above Tawarayama tunnel. An additional reference site station named Chouyou (location: 32.8707N, 130.9962E; code: EL04930274901) was investigated as well. It is applied to verify the accuracy of the detection method for the ground deformation.

Airborne LiDAR acquires a pair of DEM datasets before and after the 2016 Kumamoto earthquake with the ability to filter out reflections of forests. The DEM dataset before the earthquake was surveyed from January to February 2013 and has a resolution of 1.0 m. It is provided by the Geospatial Information Authority of Japan (GSI). The DEM dataset after the earthquake was surveyed from April to July 2016 and has a resolution of 0.5 m. It is provided by the Forestry Agency of Japan. Since the results of the detected deformation depend on the lower resolution, the resolution was unified into 1.0 m. The precision of the detected deformation can be up to 0.1 m with a resolution 1.0 m of the DEM datasets. Therefore, the horizontal deformation with magnitude larger than 0.1 m is deemed to be

effective. For the sake of brevity, we would call the DEM datasets acquired in 2013 and 2016 as pre-event DEM and post-event DEM, respectively.

3 METHODOLOGY

In the present study, an integrated system, which includes (i) data acquisition using airborne LiDAR system, (ii) 3-D ground deformation (horizontal and vertical displacement vectors) detection using the CCICP algorithm, and (iii) investigation on seismic damages of mountain tunnel, is developed for the correlation of the ground deformation with the seismic damages of the underground structures. Figure 2 presents the schematic diagram of the integrated system for the correlation analysis.

3.1 *Airborne LiDAR System*

Airborne LiDAR system is composed of an inertial navigation system (INS), and a laser scanner. The system sends pulses of laser light towards the ground and records the return time for calculating the distance between the sensor and the ground surface (Lillesand et al. 2007; Moya et al. 2017). Currently, it is the most detailed and accurate method for creating digital elevation models. One of the major advantages compared with photogrammetry is the ability to filter out reflections from vegetation due to its high penetrative abilities. The model can represent the actual situation of the ground surfaces such as rivers, paths, cultural heritage sites, etc., concealed by trees (Gigli et al. 2011; Riquelme et al. 2014; Riquelme et al. 2015). For rock mechanics, the system has been widely applied to rock characterization, detection of slope stability, and estimation of ground deformation (Muller & Harding 2007; Nissen et al. 2014; Moya et al. 2017). Usually, application of airborne LiDAR system is limited since only the post-event LiDAR dataset is available in most cases.

3.2 *CCICP Algorithm*

The CCICP algorithm was originally developed aiming at the accurate registration of MMS (Mobile Mapping System) point clouds since point clouds often have differences when the same areas are scanned several times by MMS (Takai et al. 2013). In the present study, the pre-event DEM is marked with S as a source point cloud. The post-event DEM is marked with T as a target point cloud. Firstly, each set of points sampled from both point clouds is categorized into linear points, planar points and scatter points based on the results of the Principle Component Analysis (PCA) method (Demantke et al. 2011). Local distributions of the points are evaluated by the eigenvalues λ_1, λ_2 and λ_3 of the variance-covariance matrix. The eigenvalues are calculated

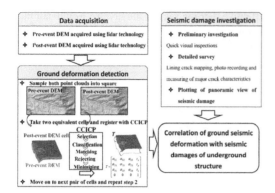

Figure 4. Vector of the horizontal displacement by Mukoyama et al. (2017) (blue arrow denotes the vector of the horizontal displacement, modified after Mukoyama et al. 2017).

Figure 2. Schematic diagram of the integrated system for correlation analysis of the ground seismic deformation with the seismic damages of the underground structure.

from the position of each point and its neighbors. Linear-planar and scatter-planar correspondences are rejected as incorrect correspondences for accurate registration. Execute the calculation for minimizing point-to-plane and point-to-point distance. We applied point-to-plane distance minimization to planar-planar correspondences and point-to-point distance minimization to the other correspondences (Takai et al. 2013; Oda et al. 2016).

The algorithm computes the 3-D displacements and rotations by iteratively minimizing the sum of square difference of corresponding coordinates of points between local subsets of the pre-event ('source') and post-event ('target') data as illustrated in Figure 2.

4 RESULTS AND DISCUSSION

4.1 Comparison between the observed and detected displacement

Figure 3 presents the horizontal displacement (combination of the West-East (EW), North-South (NS) components) detected by the CCICP algorithm. Although southwestward vectors accounted for a large portion of the whole area, westward vectors

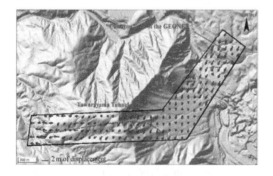

Figure 3. Vector of the horizontal displacement detected by the CCICP algorithm (blue arrow denotes the vector of the horizontal displacement).

were dominant in the west side of Tawarayama tunnel. Mukoyama et al. (2017) also detected the ground deformation using the 3D-GIV (Geomorphic Image Velocity) method. Figure 4 shows the deformation distribution detected using the 3D-GIV method. The western ground of the studied area moved westwards, and the eastern ground moved southwestwards. It is found out that the result of the CCICP method shows a good agreement with that of the 3D-GIV method.

The displacement at the reference Chouyou station is detected. The Real-time GEONET Analysis (REGARD) system successfully observed and recorded the coseismic displacement of each GEONET station (Kawamoto et al. 2016a, b). The station was observed to move in 0.97 m horizontally in the southwest direction. The horizontal displacement around Chouyou station was detected to be 0.97 m (an average value of the six points around Chouyou station) in the southwest direction using the present method. The horizontal displacement detected by the CCICP method is in good consistent with that observed by the REGARD system. The consistency between the observed and detected deformation validates the accuracy of the CCICP method.

4.2 Correlation between the ground deformation and the seismic damages of Tawarayama tunnel

To illustrate the correlation of the ground deformation with the seismic damages of the tunnel, a spatial combination between the ground deformation and the seismic damages of Tawarayama tunnel was performed, as shown in Figure 5. The ground deformation is composed of NS, EW, and vertical components. Detection lines in the south-north direction were set at an interval 100 m along the axis of the tunnel to illustrate the ground deformation and were numbered from L_1 to L_20, as shown in Figure 5.

At the ground near the western entrance of the tunnel ~L_6 in Figure 5a, Nishihara Village side), strong deformation occurred with significantly large horizontal displacement. The corresponding spans of Tawarayama tunnel are in the range S001~ S050. The maximum and average value of the horizontal displacements are 1.75 m (L_2) and 1.15 m, respectively. The strong deformation resulted from the activities of the Futagawa fault zone under the earthquake. It locates at the western side of the tunnel. The fault strikes northeast-southwest while the axis of the tunnel strikes west-east (Fig. 1a). The strike of the fault crosses obliquely with that of the tunnel axis. The large dislocation of the fault with the maximum displacement of 2.2 m (in Mashiki Machi) contributed to the strong deformation. Site investigation on the tunnel showed that the spans near the Nishihara Village side (S001~S050 in Fig. 5c) were seriously damaged. The seismic damages included lining cracks, concrete spalling, construction joint damage and pavement failure. As a result, there is a possibility that the ground horizontal deformation and shaking is related to the seismic response of the tunnel. In contrast with the horizontal deformation, vertical displacement revealed a small and smoothly varying pattern of uplift and subsidence (Fig. 5b). The ground at the lines L_1, L_2, and L_3 subsided with a maximum displacement 0.62 m and an average displacement 0.40 m, while the ground at the other lines (L_4~L_21) uplifted with a maximum displacement 0.46 m and an average displacement 0.16 m. No special significance for the seismic response of the tunnel was observed.

To highlight the correlation between the ground horizontal deformation and the seismic response of the tunnel, two detection lines in the east-west direction named north line (N_L) and south line (S_L) along the axis of Tawarayama tunnel were set. Both of the detection lines are illustrated in Figure 5a.

The N_L detection point is marked with a red solid square, and the S_L detection point marked with a black solid circle. In addition, we defined the acute angle (≤90°) between the horizontal displacement vector and the axis of the tunnel as angle θ. Figure 6 presents two components (NS and EW components) of the horizontal displacement on both of the detection lines and the angle θ along the tunnel axis.

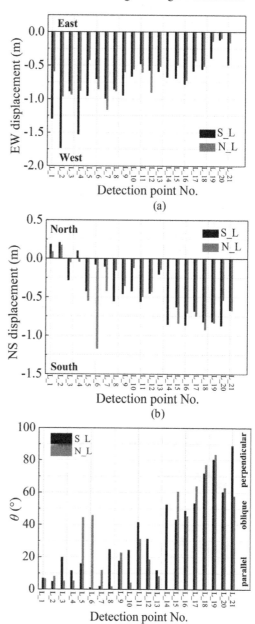

(a)

(b)

(c)

Figure 6. Three components of 3-D deformation on both detection lines and the variation of angle θ along the axis of the tunnel. (a) EW displacement; (b) NS displacement; (c) variation of angle θ along the axis of the tunnel.

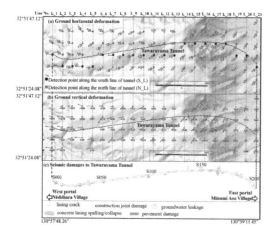

Figure 5. Horizontal (a) and vertical (b) displacement of the ground combined with seismic damages of Tawarayama tunnel (c) subjected to the 2016 Kumamoto earthquake (blue arrow denotes the vector of the horizontal displacement; for the vertical displacement, positive value denotes uplift of the ground, vice versa.).

At the lines in the range of L_1 ~ L_6 (Nishihara Village side), EW displacement (Fig. 6a) was larger than NS displacement (Fig. 6b). This contributes to the fact that the angle θ was less than 10° (Fig. 6c) in most cases. This kind of deformation pattern indicates that ground in the range of L_1~L_6 moved westwards, approximately parallel to the axis of the tunnel. Generally, seismic wave nearly parallel to the tunnel axis results in kinds of axial deformations, such as axial tension and compression. Once the compression or tensile strength of the concrete is reached, the damage or failure would take place. Tunnel section in the spans S001~S050 (corresponding to the ground of L_1~L_6) suffered from widespread damages. Most of the construction joint damages and pavement failure occurred in these spans were tensile/compressive failure with large displacement. The construction joint was compressed to fail circumferentially between S012 and S013 and opened due to tension between S001 and S002 (Fig. 7a). The pavement in the span S001 and at the portal deformed in tension and compression patterns, respectively (Fig. 7b). Furthermore, the tunnel portal (Nishihara Village side) moved westwards in 10 cm. The portal pavement was compressed horizontally to uplift in 15 cm (Fig. 8). These facts coincide with the observation of the ground westward deformation of Mt. Tawarayama.

At the lines L_1 and L_2, the ground moved northwards in 0.19 m (average value at the S_L) and 0.13 m (average value at the N_L) (Fig. 6b), while the ground at the other lines moved southwards. The change in the direction of the NS displacement contributed to ground dislocation. On the other hand, the lining between S001 and S002 of the tunnel dislocated in the north-south direction at the construction joint. Aforementioned results show that strong horizontal ground deformation can reflect the seismic performance of the tunnel to some extent.

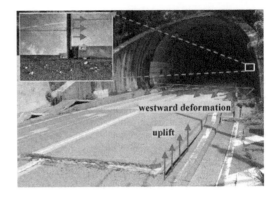

Figure 8. Portal pavement damage of Tawarayama tunnel (Nishihara Village side).

The horizontal deformation at the lines L_7~L_21 (Minami Aso Village side) decreased compared with that in the western part (L_1~L_6). The maximum and average value of the horizontal displacements are 1.24 m (L_7) and 0.87 m, respectively. However, seismic damages of the tunnel still occurred frequently in the spans S50~S200 under the ground in the range of L7~L_21. The previous study noted that geological condition, such as rock mass quality and fault zone etc. and tunnel lining conditions play a significant influence on the seismic response of the tunnel. From the viewpoint of the ground deformation, EW displacement decreased gradually with the increase of distance from the Futagawa fault segment (Fig. 6a), but NS displacement increased, even larger than EW displacement (Fig. 6b). The variation of the two components of the horizontal deformation contributes to the fact that the angle θ became larger (Fig. 6c). It indicates that the ground turned to move southwestwards. The direction of the ground deformation crosses the axis of the tunnel obliquely. It can be stated that the tunnel was subjected to seismic wave obliquely crossing the axis of the tunnel. The observation validates the assumption of seismic wave propagation (Zhang et al. 2018). As a result, the tunnel underwent axial tension or compression. When the actual strain in the lining exceeds the material ultimate strain, the crack will occur. Moreover, the dense Andesite and crushed Andesite appear in tilt alternately along the axis of the tunnel with space between 10 m and 20 m. Relative motion between the soft and hard ground contributes to the special phenomenon that ring cracks distributed along the axis of the tunnel regularly. The direction of the ground deformation can be taken as an intuitive index to reveal the seismic wave propagation direction for better understanding the seismic response of the underground structures.

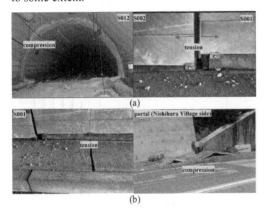

(a)

(b)

Figure 7. Site investigation on seismic damages of Tawarayama tunnel. (a) construction joint damage; (b) pavement failure.

5 CONCLUSIONS

An integrated system including data acquisition using airborne LiDAR system, ground deformation detection, and investigation of the seismic damages of mountain tunnel was proposed in this study. It aims at exploring the possible correlation between the ground deformation and the seismic damages of the underground structures. The CCICP (Combination and Classification Iterative Closest Point) algorithm was introduced to detect the 3-D deformation of the ground. The detected 3-D deformation field agreed well with the permanent displacements, which were detected using the 3D-GIV method and observed at the site station of Chouyou. The spatial comparison was carried out between the ground deformation and seismic damages of the tunnel.

Results show that strong horizontal deformation of the ground can reflect the seismic damage of the tunnel to some extent. Furthermore, the direction of the ground deformation can be taken as an intuitive index to reveal the seismic wave propagation direction for better understanding the seismic response of the underground structures. Ground deformation was observed to be parallel to or obliquely crossing the axis of the tunnel, which contributes to axial compression and tension deformation of the concrete lining and pavement. It provides a further explanation for the phenomenon that the ring cracks distributed regularly along the axis of the tunnel. On the other hand, there is no special significance for the seismic response of the underground structure for the vertical ground deformation.

REFERENCES

Asian Disaster Reduction Center. 2016. *Assessment of the Heisei 28-nen (2016) Kumamoto Earthquake Survey Report* (Preliminary).

Aydan, O. 2017. *Rock dynamics*. Balkema: CRC Press.

Demantke, J., Mallet, C., David, N., Vallet, B. 2011. Dimensionality based scale selection in 3D LiDAR Point Cloud. *Int Arch of the Photogramm Remote Sens Spatial Inform Sci* 38 (Part 5): W12.

Geospatial Information Authority of Japan. 2016. The 2016 Kumamoto Earthquake. http://www.gsi.go.jp/BOUSAI/H27-kumamoto-earthquake-index.html.

Gigli, G. & Casagli, N. 2011. Semi-automatic extraction of rock mass structural data from high resolution LIDAR point clouds. *International Journal of Rock Mechanics and Mining Sciences* 48(2): 187–198.

Kawamoto, S., Hiyama, Y., Ohta, Y. et al. 2016a. First result from the GEONET real-time analysis system (REGARD): the case of the 2016 Kumamoto earthquakes. *Earth, Planets and Space* 68(1): 190.

Kawamoto, S., Hiyama, Y., Kai, R. et al. 2016b. Crustal deformation caused by the 2016 Kumamoto earthquake revealed by GEONET. *Bulletin of the Geospatial Information Authority of Japan* 64: 27–33.

Lillesand, T.M., Kiefer, R.W., Chipman, J.W. 2007. *Remote sensing and image interpretation (5th ed)*. New York: John Wiley & Sons.

Moya, L., Yamazaki, F., Liu, W. 2017. Calculation of coseismic displacement from lidar data in the 2016 Kumamoto, Japan, earthquake. *Natural Hazards and Earth System Sciences* 17(1): 143.

Muller, J.R. & Harding, D.J. 2007. Using LIDAR surface deformation mapping to constrain earthquake magnitudes on the Seattle fault in Washington state, USA. *Urban Remote Sens Joint Event, IEEE*: 1–7.

Mukoyama, S., Sato, T., Takami, T. et al. 2017. Estimation of coseismic surface displacement in the Aso Caldera area before and after the 2016 Kumamoto earthquake by topographical data analysis from differential LiDAR DEM analysis. *Report of the 2016 Kumamoto, Oita Earthquake Disaster Research Mission, Japan society of engineering geology* 55–63.

Nissen, E., Maruyama, T., Arrowsmith, J.R. et al. 2014. Coseismic fault zone deformation revealed with differential lidar: Examples from Japanese Mw~7 intraplate earthquakes. *Earth and Planetary Science Letters* 405: 244–256.

Oda, K., Hattori, S., Takayama, T. 2016. Detection of slope movement by comparing point clouds created by SfM software. *Int Arch of the Photogramm Remote Sens Spatial Inform Sci* XLI-B5: 553–556.

Park, S.C., Yang, H., Lee, D.K. et al. 2018. Did the 12 September 2016 Gyeongju, South Korea earthquake cause surface deformation? *Geosciences Journal* 22(2): 337–346.

Riquelme, A.J., Abellán, A., Tomás, R. et al. 2014. A new approach for semi-automatic rock mass joints recognition from 3D point clouds. *Computers & Geosciences* 68:38–52.

Riquelme, A.J., Abellán, A., Tomás, R. 2015. Discontinuity spacing analysis in rock masses using 3D point clouds. *Engineering Geology* 195:185–195.

Takai, S., Date, H., Kanai, S. et al. 2013. Accurate registration of MMS point clouds of urban areas using trajectory. ISPRS Ann. Photogramm. *Remote. Sens. Spat. Inf. Sci.* 277–282 *November* 2013. Turkey: Antalya.

Zhang, X.P., Jiang, Y.J., Sugimoto, S. 2018. Seismic damage assessment of mountain tunnel: A case study on the Tawarayama tunnel due to the 2016 Kumamoto Earthquake. *Tunnelling and Underground Space Technology* 71: 138–148.

2019 Rock Dynamics Summit– Aydan et al. (eds)
© *2019 Taylor & Francis Group, London, ISBN 978-0-367-34783-3*

Seismic response analysis of the underground cavern type disposal facility

Y. Ito, Y. Tasaka & Y. Suzuki
Tokyo Electric Power Service Company, Japan

K. Niimi & N. Sugihashi
Shimizu Corporation, Japan

H. Fujihara & Y. Hironaka
Radioactive Waste Management Funding and Research Center, Japan

ABSTRACT: Seismic vibrations of the underground cavern type disposal facility during construction and operation phase were evaluated by considering the influence of ground motion on engineered barriers such as concrete pit, low diffusion layer and buffer. Two-dimensional FEM seismic response analyses with three different earthquake ground motions were carried out. The results indicated the cementitious materials (i.e., concrete pit and low diffusion layer) could withstand large earthquake motions, while the soil-based buffer would have some shear failure with small residual displacement after the earthquake. Even such failure, its influence on the permeability of the buffer was negligible because the increase of the permeability coefficient of the buffer was estimated to be small.

1 INTRODUCTION

Among the underground cavern type disposal facility (hereafter referred to as "underground cavern facility"), relatively large disposal cavern excavated in ground deeper than 70 m below the ground surface is planned for intermediate-level disposal. Fig.1 shows an example of the concept of the underground cavern facility in which engineered barriers composed of cement-based and soil-based materials are used to isolate the wastes. The concrete pit is made of steel-reinforced concrete for ensuring mechanical stability, radiation shielding, and migration resistance with nuclide sorption capability. Low diffusion layer is made of mortar for shielding, migration resistance with low diffusivity and high sorption, and bentonite buffer is also used for suppression of groundwater intrusion and migration resistance. These properties are required for the performance of the facility.

Underground structures are known to be relatively stable against earthquake, in general, and many studies have been conducted so far to evaluate the stability of rock mass and rock support for disposal tunnels in underground cavern facilities. Even though, the influence of earthquake loading on the stability of underground cavern facility that include engineered barriers is less examined. In this paper, we evaluate the influence of earthquake loading on the underground cavern facility during construction and operation phase with presumed ground motions, and the resulting deformation behaviors

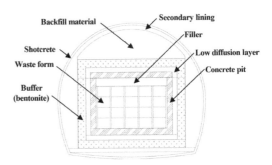

Figure 1. Conceptual cross-sectional view of underground cavern type disposal facility.

of the main components of the engineered barriers such as concrete pit, low diffusion layer, and buffer are reported.

This paper is based on the research report on the underground cavern facility (Radioactive Waste Management Funding and Research Center, 2012, 2013 and 2014).

2 OVERVIEW

Three earthquake motions with different maximum accelerations are assumed and two-dimensional FEM seismic analyses are performed. In the assumptions, the earthquake ground motions used for the conventional design are chosen basically and

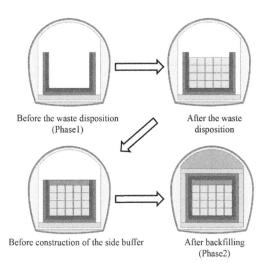

Before the waste disposition
(Phase1)

After the waste
disposition

Before construction of the side buffer

After backfilling
(Phase2)

Figure 2. Status of facility under construction and operation.

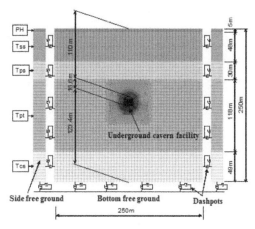

Figure 3. Simulation model.

some elevated values are set for comparison because the aim of the analyses is to reveal the characteristic behavior of the facility and to provide suggestions for design and construction.

At first, static analysis of the stress field before earthquake is carried out with considering the steps from cavern excavation to the construction of each engineered barrier member as shown in Fig.2. Then, seismic response analysis is performed. Analysis code is Soil Plus Dynamic (CTC). Stress from both analyses is superimposed to evaluate the response of the facility during earthquake.

The boundaries between the buffer and the surrounding cementitious members are considered as rigid boundaries (sharing nodes) basically (Basic analysis is described in Chapter 4). Additional analyses (in Chapter 5) are conducted as more realistic condition because local tensile stresses are found in the buffer during earthquake. In this case, joint elements are placed at the boundaries between the buffer and the surrounding materials, which allows analysis of boundary separation and sliding.

In addition, seismic response analysis considering elasto-plastic behavior of the buffer is conducted to obtain local plastic deformation of the buffer for more precise response of the buffer. Analysis code is FLAC (Itasca).

3 SIMULATION CONDITIONS

3.1 Model structure

The model structure for the simulation is shown in Fig.3. The geological structures around the cavern are assumed to be horizontal stratification based on the results of geological survey. To account for the dissipation of seismic waves in a semi-infinite ground, the side and the bottom boundaries of the model are set

as viscous boundaries using dashpots. Fig.4 presents the detail simulation model of the underground cavern facility immediately after backfilling.

3.2 Earthquake ground motion

The waveform of earthquake ground motion used in this study is shown in Fig.5. Three earthquake ground motions with the maximum accelerations in the horizontal direction adjusted to 250 Gal (Lv 1), 450 Gal (Lv 2), and 900 Gal (Lv 3) on the free rock surface are used. Case Lv 1 is set such that the maximum horizontal acceleration at the cavern position is about 200 Gal. The maximum accelerations of cases Lv 2

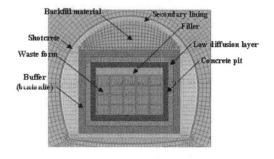

Figure 4. Detailed simulation model with the details of the underground cavern facility.

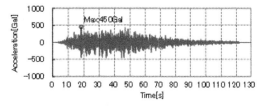

Figure 5. Earthquake wave form of acceleration at the free rock.

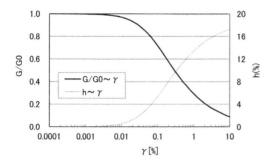

Figure 6. Strain-dependent curves of shear modulus and damping ratio of the buffer.

and Lv 3 are two and four times larger than that of case Lv 1, respectively, which are the extreme cases to examine seismic vibrations of the underground cavern facility.

The surface of the free rock is the upper part (depth 53 m) of Tps shown in Fig.3. The wave pulled back to the lower end of the model is taken from the theory of one-dimensional multiple reflection as the input of the ground motion. The amplitude of the vertical ground motion is assumed 2/3 that of the horizontal direction, and the horizontal and vertical vibrations are excited simultaneously.

3.3 Model properties

Material properties of the ground and the facility structures for the static and dynamic analyses are taken from the values in past report (JNFL. 2010). The modified Ramberg-Osgood model is applied to consider strain-dependent dynamic properties of the rock mass and the buffer, and elastic models are used for other members, instead. The strain-dependent material behavior of the buffer is shown in Fig.6.

4 BASIC ANALYSIS RESULT

4.1 Earthquake response of the entire facility

Fig.7 presents the seismic responses of the entire facility before the waste disposition (Phase 1 in Fig.2) and after backfilling (Phase 2). In Phase 1, the shotcrete and the secondary lining deform together with the surrounding rock mass, and the concrete pit and the low diffusion layer vibrate at a frequency different from that of the surrounding rock mass. In Phase 2, the entire facility including the concrete pit and the low diffusion layer deform together with the surrounding rock mass.

4.2 Stability of the cementitious members during earthquake

In Phase 1, where the concrete pit and the low diffusion layer are susceptible to the influence of earthquake loading, the relations between the

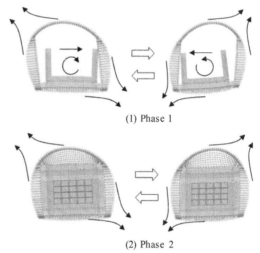

(1) Phase 1

(2) Phase 2

Figure 7. Seismic response of the entire facility (Lv2, displacement magnified 200 times).

minimum safety factor (SF) for tensile stress (tensile strength/historical maximum tensile stress) and the ground acceleration in each member are obtained and shown in Fig.8. The tensile strengths of the concrete pit and the low diffusion layer are 3.53 and 4.68 MPa, respectively. For both the concrete pit and the low diffusion layer, the smallest SF values appear at the lower end of the left side wall. The maximum acceleration on the free rock surface at which the tensile stress reaches the tensile strength are estimated to be about 500 and 1000 Gal for the concrete pit and the low diffusion layer, respectively. In the case of Lv 3, SF is less than 1 at the inner corner of the concrete pit (green circle in Fig.8). However,

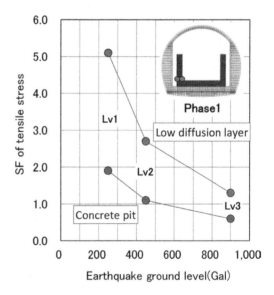

Figure 8. Relation between safety factor of tensile stress and earthquake ground level.

an inspection of the cross-section of the Reinforced Concrete (RC) indicates that even when cracks occur, the members (the reinforcing bar on the tensile side and the concrete on the edge of the compression side) have sufficient load carrying capacities.

4.3 *Dynamic response of the buffer*

The distributions of the minimum local safety factor SF of shear failure through seismic history in the buffer at each seismic motion level in Phase 1 and Phase 2 are shown in Fig.9 and the distributions of the maximum tensile principal stress through seismic history are shown in Fig.10 as well. The Mohr-Coulomb failure criterion is applied to calculate SF,

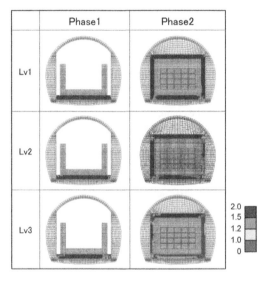

Figure 9. Local safety factor of shear failure in the buffer (minimum value through seismic history).

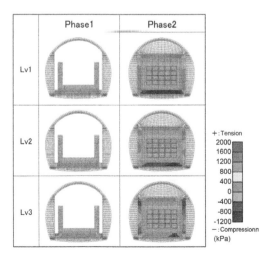

Figure 10. Maximum principal stress in the buffer (tensile side, history maximum).

and a cohesion of C = 370 kPa and an internal friction angle of φ – 2.64°, are determined from laboratory test prior to the analysis. Because the cavern deforms in a mode stretched to the left and right oblique directions, the buffer is in tension; hence in Phase 1 it is easy to have low SF and tensile stress zones at the left and right ends of the bottom part of the buffer. In Phase 2, the whole facility deforms together with the surrounding rock mass. The safey factor (SF) decreases and tensile stress is observed in a large area in the buffer depending on the seismic motion level.

As the seismic ground motion increases, more areas where SF decreases and tensile stress occurs are observed . This is obvious in Phase 2 for ground motions greater than Lv 2. Additional investigations of the seismic response of the buffer are thus carried out in Phase 2 using the earthquake motion of Lv 2.

5 ADDITIONAL ANALYSES OF THE BUFFER RESPONSE

5.1 *Simulation considering separation and sliding*

As pointed out in Chapter 4, when the buffer and the surrounding concrete member share nodes at their common boundary, the buffer is pulled by the peripheral member during earthquake and locally high tensile stresses are generated in the buffer. Complex phenomena such as separation and sliding may occur at the boundary between the ground and the underground structures. This may due to the type of earthquake motion used in the analysis and the difference of shear stiffness between the ground and the structures. Because the stiffness of the buffer and the surrounding members are very different, separation and sliding can be expected at the boundary. To consider this in the analysis, joints are placed at the boundary between the buffer and the surrounding members as shown in Fig.11.

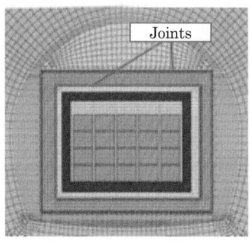

Figure 11. Locations where joints are placed.

Separation and sliding are considered by lowering the stiffness in the direction perpendicular and parallel to the boundary interface, respectively.

The distributions of the minimum SF and the maximum tensile principal stress in the buffer are depicted in Fig.12 and Fig.13, showing the results without and with joints. As shown in Fig.13, the area where the tensile stress occurs is smaller when separation and sliding are considered. As separation occurs, the compressive stress increases on the opposite side where separation does not occur. When joints are considered, the area with SF < 1 is larger than that in the case without considering the joints as can be seen in Fig.12.

5.2 Elasto-plastic analysis

It was found from the above analysis that the buffer can be sheared locally and showed plastic deformation depending on the magnitude of the earthquake. When plastic deformation occurs, there is stress redistribution and the plastic region increases with associated increase of strain and deformation as well as residual deformation after earthquake. This in turn can affect the layer thickness of the buffer and its water permeability property.

Therefore, seismic response analysis that takes elasto-plastic behavior of the buffer into account is adopted and additional simulations are conducted. The present modified Ramberg-Osgood model is improved in the way to redistributes the stresses exceeds the estimate with the Mohr-Coulomb failure criterion, and the elasto-plastic deformation behavior of the buffer is analyzed. Fig.14 presents the relation between shear stress and shear strain.

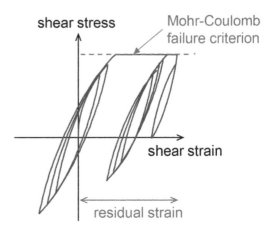

Figure 14. Relation between shear stress and shear strain.

The historical shear stress–shear strain relations at the lower right corner of the buffer are shown in Fig. 15. When the stress state reaches the Mohr-Coulomb failure criterion and enters the plastic deformation stage, the change of the shear strain is larger than the change of the shear stress. After that, the historical loop of the modified Ramberg-Osgood model seems to deviate significantly from its original position.

Fig.16 presents the plastic state at the end of the simulation. Compared with the SF distribution evaluated by the elastic analysis (Fig.12, with joints), the

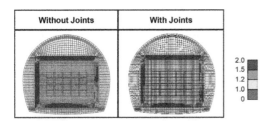

Figure 12. Comparison of local SF (minimum value through seismic history).

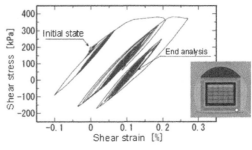

Figure 15. History curve of shear stress and shear strain of the buffer.

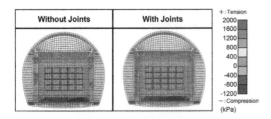

Figure 13. Comparison of maximum principal stress (tensile side, maximum through seismic history).

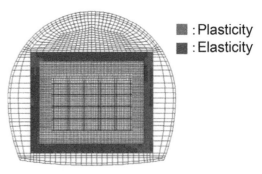

Figure 16. Plastic state of the buffer.

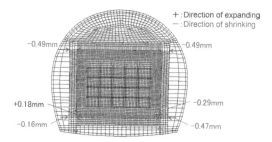

Figure 17. Residual displacements in the buffer.

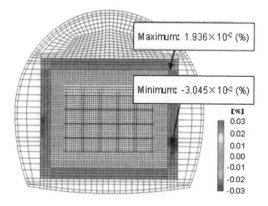

Figure 18. The increment of the volumetric strain of the buffer.

plastic region is larger than the region defined by SF < 1 in the elastic analysis. This is probably attributed to the fact that the plastic region increases due to stress redistribution after failure. On the other hand, in the elastic analysis result that shows regions at the upper right side inner corner and the lower left corner where SF < 1, there is no plastic deformation in the regions in the elasto-plastic analysis result.

The residual displacements of the buffer after the earthquake are shown in Fig.17. The residual displacements are in the order of -0.5 to 0.2 mm, which are very small compared with the original member thickness of 1000 mm. The maximum increment of the volumetric strain of the buffer at the end of the analysis is 1.936×10^{-2} (%) as shown in Fig.18, which corresponds to 0.131% increase of the permeability coefficient. Hence, the effect of seismic ground motion (Lv 2) on water permeability of the buffer is small even if plastic deformation of the material is considered.

6 CONCLUSION

Two-dimensional FEM seismic response analyses are conducted to evaluate the influence of seismic vibration on the overall behavior of the underground cavern facility under construction and operation phases. The findings by the analyses are summarized below.

1. Before the disposition of the waste, the shotcrete and the secondary lining deform together with surrounding rock mass. However, the concrete pit and the low diffusion layer vibrate at a frequency different from that of the surrounding rock mass. After the backfilling, the entire facility including the concrete pit and the low diffusion layer deform together with the surrounding rock mass.

2. Cementitious members such as the concrete pit and the low diffusion layer have sufficient strength to withstand large seismic vibrations.

3. The bottom of the buffer tends to have shear and tensile stresses at the left and right ends before the waste is dispositioned. Because the entire facility deforms together with the surrounding rock mass after the backfilling, shear and tensile stresses occur in an area larger than that before the waste is dispositioned.

4. When considered the elasto-plastic deformation and separation and sliding of the buffer relative to the surrounding members, the maximum residual displacement at the end of the analysis in the direction of the buffer depth is about 0.5 mm, which is very small compared with the original thickness (1000 mm) of the buffer.

5. The maximum increase in the permeability coefficient of the buffer material after the earthquake is 0.131%. This implies that the influence on the water permeability of the buffer is very small even if plastic deformation occurs due to earthquake loading.

ACKNOWLEDGEMENTS

This paper is based on a study, supported by the Ministry of Economy, Trade and Industry of Japan, "Verification test on performance of underground cavern type disposal facility (2012)" and "Verification test on closing technology of underground cavern type disposal facility (2013 and 2014)".

REFERENCES

Radioactive Waste Management Funding and Research Center, 2012. Verification test on performance of underground cavern type disposal facility.

Radioactive Waste Management Funding and Research Center, 2013. Verification test on closing technology of underground cavern type disposal facility.

Radioactive Waste Management Funding and Research Center, 2014. Verification test on closing technology of underground cavern type disposal facility.

CTC (ITOCHU Techno-Solutions Corporation), Soil Plus Dynamic.

Itasca Consulting Group, FLAC—Fast Lagrangian Analysis of Continua.

JNFL (JAPAN NUCLEAR FUEL LIMITED), 2010. Reprocessing facility and specified waste storage facility, Seismic safety evaluation report related to revision of "Seismic design review guideline of nuclear power facility", Replies to comments (The stability of the foundation ground) (in Japanese).

2019 Rock Dynamics Summit– Aydan et al. (eds)
© *2019 Taylor & Francis Group, London, ISBN 978-0-367-34783-3*

A study on the seismic behavior of underground tunnels, considering dynamic fault rupture, through large-scale finite element analysis

S. Rajasekharan, H. Sun, Y. Mitsuhashi & J. Shimabuku
Kozo Keikaku Engineering Inc., Tokyo, Japan

H. Noguchi & M. Maruta
Shizuoka Institute of Science and Technology, Shizuoka, Japan

G. Hashimoto & H. Okuda
The University of Tokyo, Tokyo, Japan

ABSTRACT: Recently, analyzing the effects of fault displacements on a fault-crossing tunnel has been in focus, in regard to vulnerability assessment, as well as the design of safe underground structures. The current study involves a large-scale seismic rupture simulation of an entire fault, tunnel and the surrounding crustal rock area. The parallel finite element program FrontISTR, was modified to include RC nonlinearity and fault modelling in the current study. The model was prepared using solid elements for the crust and joint elements for the fault surface. The tunnel is modelled using non-linear concrete solid elements and anisotropic plane-stress reinforcement steel elements. Through numerical simulation, the influence of considering dynamic fault rupture and tunnel non-linearity are investigated. This study highlights the importance of considering dynamic fault rupture with respect to the damages incurred by the tunnel, realized through its displacement mode, material degradation and crack patterns.

1 INTRODUCTION

In recent years, the evaluation of fault displacement is required for evaluating the soundness of underground structures during an earthquake. Fault displacements occurs as the result of the rupture of the earthquake source fault. Previous studies have been primarily conducted using the finite difference method (Dan et. al 2007), where the dynamic rupture simulation is performed through the spontaneous rupture process of a fault caused by a slip-weakening model. Few studies have also been done using the finite element method (Mizumoto et al. 2005), but the large computation requirement involved restricted the complexity of the model.

The present study used the 3D nonlinear finite element method to perform the simulation of the Kamishiro Fault earthquake in Nagano Prefecture on November 22, 2014(JAEE 2015). A 40 km × 40 km × 20 km model that included the earthquake source fault is modelled using solid elements for the crust and joint elements for the fault (Mitsuhashi 2017). A 100m tunnel crosses through the fault.

The rupture in the model is generated by applying an initial stress to the joint elements, which initiates the rupture process, as it uses a nonlinear constitutive law with a stress drop beyond failure. The simulation is performed through the FrontISTR, a massively parallel finite element program, which can perform large-scaled FEA with relative ease (Front ISTR 2018).

Through this program a wide area of the crust could be analyzed using a relatively fine mesh. The results obtained from the simulation are compared with results obtained from the static analysis, to investigate the influence of performing dynamic fault rupture simulation on the underground fault-crossing tunnel. The static method, which is based on the elastic theory of dislocations, is a very common method used to analyze soil deformation by nuclear agencies in Japan (Japan Nuclear Safety Institute 2014).

The previous study conducted by the authors (Mitsuhashi et al. 2017), assumed the tunnel to be linear as the research focus was on crustal/fault behavior. In the current study, the concrete tunnel is modelled using solid elements which incorporate a smeared-crack model with nonlinear constitutive relations (Noguchi et al. 2009). The effects of incorporating nonlinearity into the tunnel structure is observed. The steel reinforcement is modelled using anisotropic plane-stress elements. The influence of the introduction of steel reinforcement on the tunnel behavior is examined.

The first section of the paper describes the overall analysis model. The results obtained from the analysis are discussed in Section 3. The observations from the various simulations are summarized into the last section of this paper.

2 ANALYSIS MODEL

2.1 *Overall dynamic fault rupture model*

The model used in the analysis is shown in Figure 1, and the fault parameters are given in Table 1.

The fault shape was established by referring to AIST (2015) active fault data-base and aftershock distribution. A strike angle of 12 degree, dip angle of 50 degree, length of 18 km, and depth of 10 km (width of 12.2 km) were established based on the same literature. Typical physical values are assumed; Poisson's ratio, ν = 0.25, shear modulus of stiffness, μ = 30 GPa and unit weight γ = 2.5 t/m³.

Referring to the records on the hypocenter of this earthquake, the hypocenter was set to a position at a 5-km depth, and a large value was used for the shear yield stress τ_y of the joint elements. The Kamishiro fault earthquake was a reverse-fault earthquake. However, the earthquake was also indicated to have been accompanied by a left-lateral slip. So the direction of initial stress λ was assumed to be 60 degree. By the setting of an initial stress, rupture would occur as soon as the analysis begins. Since ground surface displacement was not observed on the north side of the fault, the shape of the fault surface might not have been a simple rectangle, but due to convenience, this study assumes a rectangular fault.

The stress drop was uniformly set to 1 MPa, the condition in which the moment magnitude Mw, calculated from the fault displacement Δu of each element in the final state was close to the value observed in the actual earthquake. τ_y was established according to Andrew (1976).

For damping, the stiffness-proportional damping was used. From the study by Mizumoto et al. (2005), it was established that there would be 2% damping at 1 Hz. In addition, the effect of the reflecting waves was eliminated by setting the viscous boundary at the model periphery that excluded the crust surface. Due to the stress drop incurred by the joint elements, this simulation problem is highly non-linear, hence a convergent calculation was performed by the Newton-Raphson method. The dynamic analysis used the Newmark-β method (parameter β = 0.25, γ = 0.5). The integration time step of the analysis Δt = 0.01 s, and the total duration of the analysis, T = 20 s.

2.2 *Non-linear Concrete Modelling*

A smeared-crack, hypo-elastic model is used to model the underground concrete tunnel (Noguchi et al. 2009). The concrete models in Table 1 has been introduced into FrontISTR as a part of this study. This enables the non-linear simulation of very large concrete structures in a parallel computing environment.

The method essentially computes an equivalent uniaxial strain function, and then using nonlinear constitutive relationship as shown in Figure 2, a stress update scheme based on the hypo-elastic incremental formulation, calculates the predicted stress at the integration points of the elements. The concrete model incorporates a detailed loading/unloading stress path and also consider the post-cracking reduction of compressive strength (Fig. 3) and shear modulus (Fig. 4). The models used have been experimentally verified in the past research, and can capture the complex state of stress in structural concrete with a good degree of accuracy (Noguchi et al. 2009). The program used in

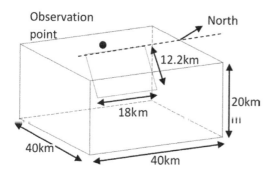

Figure 1. Overall analysis model

Table 1. Fault Parameters

Parameter	Symbol	Value	Unit
Fault width	W	12.2	km
Fault length	L	18.0	km
Strike angle	θ	12	degree
Dip angle	δ	50	degree
Stress drop	Δτ	1.00	MPa
Fault shearing stiffness	ks	1.20×10^4	kN/m/m²
Fault vertical stiffness	kv	1.20×10^4	kN/m/m²

Table 2. Models used for concrete modelling

Mechanism	Model
Concrete Cracking model	Multi-directional fixed cracking
Compressive stress-strain ascending curve	Saenz
Post-peak compressive stress-strain descending curve	Kent-Park
Pre-cracking stress-strain curve	Linear
Post-cracking tension stiffening	Shirai
Post-cracking compressive strength reduction	Noguchi-Hamada
Shear stress-strain relationship	AL-Mahaidi
Failure criteria	William-Warnke 5 parameter model

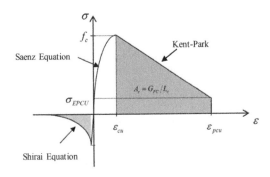

Figure 2. Concrete stress-strain relationship

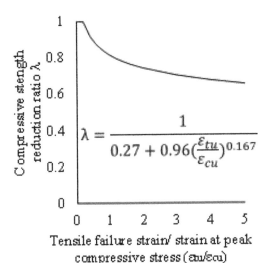

$$\lambda = \frac{1}{0.27 + 0.96(\frac{\varepsilon_{tu}}{\varepsilon_{cu}})^{0.167}}$$

Tensile failure strain/ strain at peak compressive stress (εtu/εcu)

Figure 3. Post-cracking compressive strength reduction model

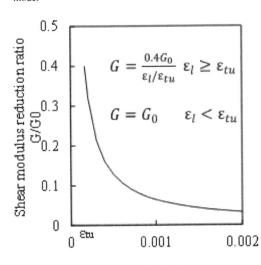

$$G = \frac{0.4G_0}{\varepsilon_l/\varepsilon_{tu}} \quad \varepsilon_l \geq \varepsilon_{tu}$$

$$G = G_0 \quad \varepsilon_l < \varepsilon_{tu}$$

Figure 4. The shear model used in simulation

the current study also has provisions to perform the analysis based on rotating crack model.

2.3 Reinforcement modelling

In most non-linear analysis methods of reinforced concrete, the steel reinforcement are modeled as truss or beam elements. Although a simple solution is necessary to reinforcement modeling, the modelling dependency on the reinforcement location presents a challenge when it comes to large-scaled simulation of concrete structures with a complex mesh. In addition, the modelling of closely spaced reinforcement, may lead to an unrealistically fine mesh. To overcome this problem, the current study employs the use of 2-D elements with plane stress formulation. The element can have 1-D anisotropic material properties in any given direction.

The steel elements share nodes and overlap with the concrete solid elements as shown in Figure 5, which displays the elements individually.

For the current study, 25mm diameter steel at a spacing of 200 mm were introduced both in the radial and axial direction of the tunnel; reinforcement ratio of about 0.2%. The thickness of the steel plate element was calculated such that the cross sectional area per unit area plan area was equal to the c/s area of steel provided (Fig. 6a). A bilinear stress-strain material model as shown in Figure 6b was used.

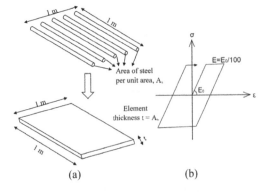

Figure 5. Modelling of steel reinforcement

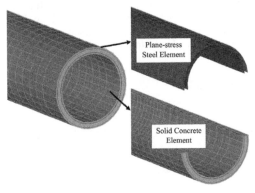

(a)   (b)

Figure 6. (a) Plane stress reinforcement model and (b) its bi-linear stress-strain relationship

Table 3. Reinforced Concrete Parameters

Parameter	Value	Unit
Concrete Young's Modulus, E_c	1.53×10^5	kgf/cm²
Poisson's Ratio, v	0.2	-
Compressive strength of concrete, f_{cu}	230.5	kgf/cm²
Tensile strength of concrete, f_{tu}	25.5	kgf/cm²
Strain at peak compressive stress, ε_{cu}	0.0025	-
Maximum compressive strain, ε_{cmax}	0.0125	-
Stress at max. compressive strain	23.05	kgf/cm²
Steel Young's Modulus, E_s	2.22×10^6	kgf/cm²
Steel Yield Stress, f_s	5660	kgf/cm²

Table 2 lists the values used for various parameters used in the simulation of the tunnel. Typical values of concrete and steel reinforcement were used for this study.

2.4 Fault representation by joint elements

In the present study, a model of the fault was created by the using joint elements. These joint elements, which were initially proposed by Goodman (1976), has been expanded to include triangle or quadrilateral joints elements (Mitsuhashi 2017). Usage of such elements into the FrontISTR allows high-precision analysis to be performed, even with a distorted mesh.

The relationship between the shearing stress and relative displacement of the joint element used in this study is shown in Figure 7. As shown in the figure, sliding rupture occurs when the shearing stress τ of joint elements reaches the yield stress τ_y, and a stress drop to τ_0 occurs. For convenience sake, the current study used a model in which the shearing stress has a sudden drop to τ_0 as soon as sliding rupture occurs. The vertical stiffness k_v was made to be linear with a sufficiently rigid value.

2.5 Static Analysis model

In order to investigate the influence of performing a dynamic fault rupture simulation, the analysis

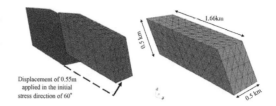

Figure 8. Static analysis model and its loading condition

results are compared to the results obtained from static analysis. Static analysis is performed by considering a smaller model as shown in Figure 8, where the input displacement is directly applied to the soil model near the tunnel. The input displacement is computed from the elastic theory of dislocation; a method commonly used while assessing the safety of important structure over a fault (Japan Nuclear Safety Institute 2013). Through these calculations an input displacement of 0.55m is applied along the fault surface with an initial stress angle of 60 degree.

3 ANALYTICAL RESULTS

3.1 Influence of dynamic fault rupture

Results from static and dynamic analysis are compared. For the comparison, the tunnel is assumed to be linear. Figure 9 shows the time history of maximum shear strain among all the elements in the tunnel. Although both the simulation results converge to similar strain levels, at the initial stage of loading maximum shear strain observed in the tunnel during dynamic analysis was almost 20% more than strain observed through static analysis, which is represented by the dotted line in Figure 9.

This increase in the maximum shear strain will have adverse effects in the damage incurred by the tunnel when considering concrete non-linearity, which is discussed in the next section.

3.2 Nonlinearity of concrete

Concrete is known for its softening behavior and post peak stress reduction due to energy dissipation through crack propagation. Figure 10 shows

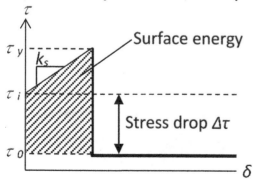

Figure 7. Stress-displacement relationship of joint element

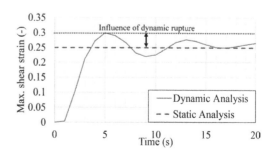

Figure 9. Comparison of maximum shear strain time history

380

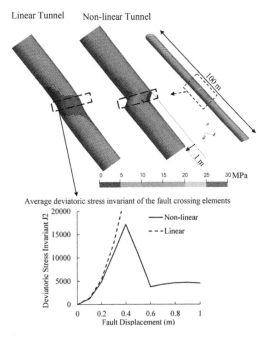

Linear Tunnel   Non-linear Tunnel

100 m

1 m

0   5   10   15   20   25   30 MPa

Average deviatoric stress invariant of the fault crossing elements

— Non-linear
--- Linear

Fault Displacement (m)

Figure 10. Linear vs. non-linear tunnel stress contour and deviatoric stress invariant time history

the stress reduction in the central part of the tunnel through which the fault crosses.

The reduction in stresses can also be observed in the graph in Figure 10, where it can be seen that the average deviatoric stress invariant of the fault-crossing elements of the tunnel reduces with an increase in fault displacement.

This reduction in stress is accompanied by an increase in strain in the elements where cracks have occurred. When the demand on the tunnel increases, the strain tends to concentrate at the location of cracked elements, which leads to the reduction of strain in the neighboring elements. This phenomenon of strain localization can be seen in Figure 11. In the linear case, the strain is low and well spread out, whereas the strain in the non-linear case is high and localized at the center of the tunnel. The tunnel deformation to be concentrated at the center of the tunnel.

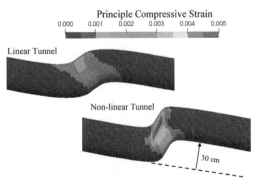

Principle Compressive Strain
0.000   0.001   0.002   0.003   0.004   0.005

Linear Tunnel

Non-linear Tunnel

30 cm

Figure 11. Strain localization; fault displacement

Hence, the displacement mode is a sudden shear-like guillotine shaped displacement as seen in Figure 11. In the current concrete model the strain softening model depends on the size of the mesh, such that the fracture energy released per unit area is constant.

The current simulation uses a multi-direction fixed-cracking model. This allows the simulation model to follow the crack propagation under complex loading conditions. Figure 12 shows the concrete elements where the cracks have occurred and the number of planes along which crack have been formed.

3.3 Effect of introduction of steel reinforcement

Figure 13 shows the comparison of results obtained from numerical simulation of plain concrete and reinforced concrete.

The introduction of steel increases the post-cracking capacity of the concrete. In Figure 13, even though the cracking pattern is almost similar in both cases, the strain incurred by the concrete elements reduces due to the introduction of steel. The overall failure of the tunnel arises from the cross-sectional shear failure of the tunnel passing through the fault and the introduction of steel delays this failure mode until the yielding of steel.

4 CONCLUSION

A dynamic rupture simulation analysis of the Kamishiro fault earthquake in Nagano Prefecture was performed using the finite element method. The effects of the fault rupture on a fault-crossing tunnel was investigated.

In conventional methods of assessment of a fault-crossing tunnel subjected to fault displacement, static analysis of the tunnel considering only the near surrounding soil, are usually performed. The current study highlighted the importance considering the dynamic fault rupture on the tunnel.

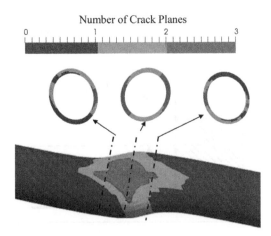

Number of Crack Planes
0   1   2   3

Figure 12. Cracking of Tunnel

381

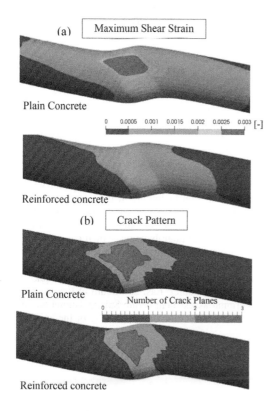

(a) Maximum Shear Strain

Plain Concrete

0   0.0005   0.001   0.0015   0.002   0.0025   0.003 [-]

Reinforced concrete

(b) Crack Pattern

Plain Concrete

Number of Crack Planes

0   1   2   3

Reinforced concrete

Figure 13. Change in (a) maximum shear strain and, (b) cracking due to introduction of steel reinforcement

Through the introduction of non-linear concrete material models, the response that are associated with the non-linearity concrete modelling was well observed through these simulations which includes (i) material degradation (ii) material softening (iii) tension stiffening (iv) strain localization (v) steel yielding (vi) crack propagation. The current model uses smeared-crack based models, which are based on the experimentation performed on fixed size concrete samples. The elements sizes used In the current study is within the limit prescribed by these models. Further study is required to investigate the changes required in the material models when larger element sizes are used and the related size effects.

A convenient method of modelling the reinforcement steel using 2-D plane stress elements with anisotropic material properties has been introduced. The usage of these elements simplifies the process of modelling the reinforcement, as they can be overlapped on the concrete solid elements. Unlike the usage of truss elements, there is no requirement to adjust the node arrangement to model steel reinforcement.

The crack response and strain distribution observed was similar to the patterns observed in past research, where static simulation of a fault-crossing underground tunnel was performed (Sakashita &Hata 2016).

The current study assumed a continuous tunnel with perfect bonding between steel and concrete. Further study is required to model the tunnel as segments, with joint elements representing the separation. The effects of inclusion of bond slip mechanisms between the concrete and steel elements also needs to be investigated.

Through the introduction of non-linear concrete material into the FrontISTR program, it has become convenient to perform large-scaled simulation in a high performance-computing environment, allowing a large analysis model to be simulated in parallel on multiple nodes, thereby drastically reducing the computation time.

ACKNOWLEDGMENTS

This study used the K-NET strong motion seismograms from the National Research Institute for Earth Science and Disaster Prevention.

REFERENCES

Andrews, D.J. 1976. Rupture velocity of plane strain shear cracks, *Journal of Geophysical Research*, Vol.81, pp. 5679–5687.

Dan, K., Muto, M., Torita, H., Ohashi, Y. & Kase, Y. 2007. Basic examination on consecutive fault rupturing by dynamic rupture simulation, *Annual Report on Active Fault and Paleo-earthquake Researches*, No.7, 259–271.

FrontISTR. 2018. Workshop home page: *https://www.frontistr.com/*Browsed on 22/11/2018(In Japanese).

Goodman, R.E. 1976. *Methods of geological engineering in discontinuous rocks*, West Publishing Company, Ch. 8, pp. 300-368.

Japan Association for Earthquake Engineering. 2015. *Report about the Earthquake at the Nagano Prefecture North in 2014*(In Japanese).

Japan Nuclear Safety Institute On-site Fault Assessment Method Review Committee. 2013. *Assessment Methods for Nuclear Power Plant against Fault Displacement*.

Mitsuhashi, Y. 2017. *Study on large-scale parallel analysis considering stochastic parameters for fault crossing structures*, Doctoral thesis, the University of Tokyo.

Mitsuhashi, Y., Hashimoto, G., Okuda H. & Uchiyama, F. 2017. Study on estimation of the fault model of Kamishiro Earthquake using dynamic fault rupture simulation, *SMiRT-24*.

Mizumoto, G., Tsuboi, T. & Miura, F. 2005. Fundamental study on fault rupture process and earthquake motions on and near a fault by 3D-FEM, *Journal of JSCE*, No.780, I–70, 27–40 (In Japanese).

National Institute of Advanced Industrial Science and Technology (AIST). 2015. *Active fault data base*, https://gbank.gsj.jp/activefault/index_gmap.html, Browsed on 01/08/2015(In Japanese).

Noguchi, H., Kashiwazaki, T. & Miura, K. 2009. : Finite Element Analysis of Reinforced Concrete Joints Subjected to Multi-Axial Loading, *Proc. of the Thomas T.C. Hsu Symposium on Shear and Torsion in Concrete Structures*, SP-265, American Concrete Institute, pp.223–244.

Sakashita, K. & Hata, A. 2016. Fundamental study on behavior of underground linear structure subjected to fault displacement, *Journal of Japan Society of Civil Engineers, Ser. AI, Structural engineering &Earthquake engineering*, JSCE.

2019 Rock Dynamics Summit– Aydan et al. (eds)
© 2019 Taylor & Francis Group, London, ISBN 978-0-367-34783-3

Key technical considerations on rehabilitation of existing Salang Tunnel – Afghanistan

N. Malistani & M. N. Nejabi
Faculty of Transportation Engineering, Kabul Polytechnic University, Kabul, Afghanistan

ABSTRACT: The existing Salang Tunnel is the first 2.6km long roadway tunnel in Afghanistan, which was constructed 55 years ago through conventional tunnel construction technique. It is located at an altitude of 3400 m so that the portals and mountains are subjected to severe environmental conditions such as freezing and thawing process. Unfortunately, the tunnel suffered from explosions during the civil war and it was blocked for the few times to traffic during the 1990s due to the huge blasting, which caused heavy damages inside the tunnel. The government of Afghanistan is planning to carry out a major rehabilitation and upgrading the existing Salang Tunnel and associated structures (snow galleries and road pavements). The Salang highway has a major socioeconomic importance for Afghanistan since it crosses over the Hindukush mountain range to connect Afghanistan's northern provinces to the Center, East and South of the country. This paper describes the technical/engineering and the functional assessment of the Salang Tunnel and as well the key technical considerations on works for the repairs, rehabilitation and upgrading of the existing Salang Tunnel.

1 INTRODUCTION

The existing two-lanes, two way Salang Tunnel is the first 2.6 km long roadway tunnel in Afghanistan, which was built almost 55 years ago by a conventional tunnel construction technique. It is located at an altitude of 3400 m at the Salang Pass in the Hindukush, starting the boundary between Parwan and Baghlan provinces of Afghanistan (Fig. 1). The tunnel portals and mountains are subjected to severe environmental conditions such as freezing and thawing process. Unfortunately, the tunnel suffered from explosions during the civil war and it was blocked for the few times to traffic during the 1990s due to huge blasting, which caused heavy damage in the tunnel. In 2010, it was observed that about 6,000 vehicles passed the Salang Tunnel daily. While the tunnel was originally designed for 1,000 vehicles a day, it is now handling seven to ten thousand vehicles a day.

Currently, the rehabilitating of the existing Salang Tunnel is a priority component of the Government of the Islamic Republic of Afghanistan. The Ministry of Public Works (MPW) of Afghanistan is planning to carry out a major rehabilitation and upgrading of the existing Salang Tunnel and of associated structures (snow galleries and road pavements). The tunnel and the snow galleries are part of the Salang Pass highway from Jabal Saraj to the town of Khenjan, with a total length of about 86 km. The planned rehabilitation of the Salang highway is part of the Trans – Hindukush Road Connectivity Project (THRCP) for which the Government has obtained financing from the World Bank.

Figure 1. South part of the Salang Tunnel

2 PROBLEM STATEMENT

In 2002, the Ministry of Public Works (MPW) of Afghanistan decided to reconstruct the Salang Tunnel which finally in 2012, the pavement of Salang road and tunnel, including lighting, installing fans for ventilation system, and tunnel lining was repaired. But unfortunately, it was not effective and could not solve the problems. It seems that the grouting on tunnel lining had not been conducted based on geological investigation along the alignment of tunnel. Despite all previous efforts and maintenance on the tunnel, it still faces to the following serious technical difficulties:

- Groundwater and water inflows
- Drainage system
- Lighting system
- Ventilation system
- Road pavement
- Support structural disrepair (concrete lining)
- Traffic control
- Safety
- Emergency exit

3 SITE GEOLOGY

The Salang Tunnel areas, both south and north portals, are topographically, considered to be gentle to steep sloping with an angle ranging 5° to 35° in the southern portal and 5° to 25° in the northern portal. Three stream channels are present at the southern portal and five stream channels are in the northern portal, small amount of water flowing permanently from the top (small lake located at the top of the north portal), and streams join each other at downstream.

The existing Salang Tunnel area consists of one rock type as observed at the outcrops in both southern and northern portals and it is granitic/granodiorite/plagiogranitic rock belonging to early to late Triassic age. Rock is characterized by coarse grains

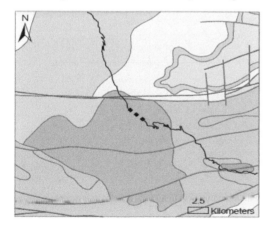

Figure 2. Local geology in the vicinity of the Salang tunnel.

of quartz, feldspar and mica (dominantly biotite) crystals with mafic enclaves of country rock. The outcrop seems to be massive plutonic body in a local scale, but it is highly disturbed tectonically in regional scale.

4 SEISMO-TECTONICS AND MAXIMUM GROUND MOTIONS

There are two major faults bounding the plutonic rock mass, in which the Salang Tunnel is located. The Andarab fault is in the north and it is reported to be a dextral fault. The total length of the Andarab fault is about 150-160 km. However, it is segmenetd and the maximum segment length is about 85 km, which can produce an earthquake with moment magnitude of 7.4 and 255 cm slip according to the empirical relations proposed by Aydan (2012).

The fault in the south of the Salang Tunnel is called the Hari-Rod fault (also known as Herat Fault) and this fault joins with the sinistral Chaman fault in the close vicinity of the Jabal Saraj and it has the dextral sense. The total length of this fault is more than 800 km long. However, it is highly segmented and the length of the segment in the vicinity of the Salang tunnel is about 64 km. The estimated moment magnitude of this segment is about 7.0 with a relative slip of 180 cm.

The contours of maximum ground surface accelerations for a rocky ground are shown in Fig. 3 provided that the earthquake occurs at the Andarab fault with a rupture length of 85 km and moment magnitude of 7.4. The maximum ground acceleration is expected to be ranging between 330-390 gals at the Salang Tunnel according to the relations proposed by Aydan (2012). However, the Hari-Rod fault with a rupture length of 64 km would induce maximum ground acceleration ranging between 220 and 260 gals near the tunnel. Therefore, the earthquake on Andarab Fault is more critical for the Salang Tunnel.

Aydan et al (2010) compiled case histories of tunnels affected by earthquakes and proposed some empirical relations for assessing the possibility of

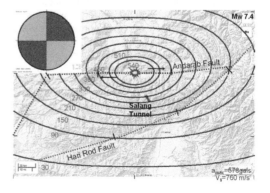

Figure 3. Estimated maximum ground accelerations.

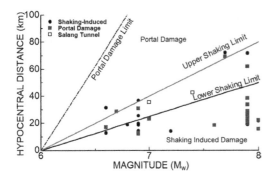

Figure 4. Assessment of damage possibility to Salang Tunnel by earthquakes (arranged from Aydan et al. 2010).

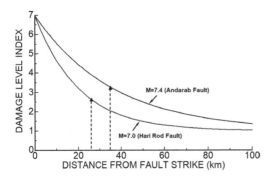

Figure 5. Estimated damage level index (DLI) at Salang Tunnel for hypothetical earthquakes with a moment magnitude of 7.0 and 7.4 (arranged from Aydan et al. 2010).

damage to the tunnel. The empirical relations imply that shaking-induced damage would be almost none or slight while the portals may suffer some damage due to the earthquake as seen in Fig. 4.

Aydan et al. (2010) also suggested more detailed damage level index (DLI). Fig. 5 shows the relation between the distance from the fault strike (Rf) and Damage Level Index (DLI). As seen from the figure, some damage is possible due to the earthquakes on Andarab fault and Hari-Rod fault. The damage level index may range between 2 and 3. When the damage level is about 3, visible cracking of concrete lining, shotcrete, noticeable plastic deformation of rockbolt platens and steel ribs, slight invert heaving may be observed. On the other hand, hair cracking of concrete lining and shotcrete, non-noticeable deformation of rockbolt platens and steel ribs, no invert heaving would be observed for the damage level index of 2.

5 IN-SITU STRESS

The stress state of Afghanistan is unknown and there are no in-situ measurements yet. However, the constructions of underground openings such as tunnels and underground caverns definitely require

the information on the stress state for stability assessments.

The Hindu Kush region of Afghanistan is seismically very active (according to earthquake data from USGS). There were two major earthquakes at a distance of about 100 km from the tunnel. Malistani et al. (2016) previously reported some in-situ stress evaluations for Salang Tunnel using the focal mechanism solutions obtained by Abers et al. (1988). The direction of p-axis solutions of earthquakes in the Hindu Kush mountain range is generally NW-SE, implying that maximum horizontal stress acts perpendicular the axis of the mountain range and parallel to the axis of the tunnel.

The estimated stress state for the earthquakes on Andarab Fault and Hari-Rod Faults using the method of Aydan (2000) are shown in Fig. 6. The maximum horizontal stress acts in NW-SE direction for the Andarab fault while it is almost E-W for Hari-Rod fault. Nevertheless, the lateral stress acting on the tunnel cross-section is almost close to the unity as estimated previously.

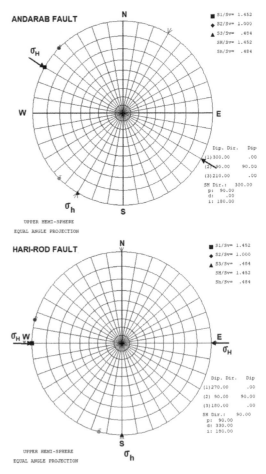

Figure 6. Estimated in-situ stresses from Andarab and Hari-Rod faults according to Aydan's method.

6 ENVIRONMENTAL MONITORING

The existing tunnel was reported to be suffering some ventilation problems. Environmental monitoring involving air pressure, temperature and humidity inside and outside the tunnel and CO_2 were carried out during August, 2015. The measurements were carried out at south, north portals and middle of the tunnel (Fig. 7). As seen in Fig. 7. the temperature fluctuates between 10 and 20 Celsius and humidity is more than 40% and sometimes reaches up to 100%. There is a difference between air-pressure at North portal and south portal so that natural ventilation takes place. Fig. 8 shows the CO_2 fluctuations in the middle of the tunnel. The CO_2

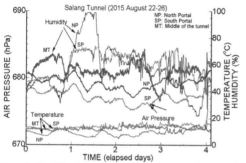

Figure 7. Humidity, temperature and air pressure change at the middle of the tunnel during August 22-26, 2015.

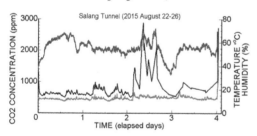

Figure 8. Carbon dioxide ($CO2$), Humidity and temperature variations at the Salang Tunnel during August 22-26, 2015 (from Malistani 2017, Malistani et al. 2016b)

concentration is about 600 - 700 ppm. However, the CO_2 concentration may be up to the level of 2800 ppm, and the air-quality is quite poor. Particularly high CO_2 concentration may be quite detrimental for the long-term performance of the concrete lining and steel members as observed in the existing tunnel.

7 TRAFFIC ANALYSES

The current average daily traffic obtained by analysis of traffic count surveys at the Salang Highway is shown in Table 1. While the tunnel was originally designed for 1000 vehicles per a day, it is now handling seven to ten thousand vehicles per a day.

8 STABILITY ANALYSES OF SURROUNDING ROCK MASS OF TUNNEL

In the Salang Tunnel, there are no sufficient investigation data from the site except some rock index investigations using the point load test at south and north portals of tunnel. As there is no appropriate investigation data, in-situ measurement and investigation on rock mass as well as in-situ stress measurements are required to analyze and to solve the current challenges of the tunnel.

It is indicated that some instability problems may appear at the surrounding rock mass of tunnel during the replacing of concrete lining in damaged sections. Therefore, it is recommended to conduct the stability analyze in damage sections and also it would be useful if such stability analyses are lead for the situation of support structures along the alignment of tunnel. Therefore, for this purpose, the properties of intact rock along the tunnel are required. Although, there are some Point Load Test data such as unconfined compressive strength, and uniaxial tensile strength are available from portals (south and north) of tunnel, but it is not a reliable data to design the project. Therefore, it requires for more investigation. The properties of intact rock listed below are required to analyze the stability of surrounding rock of tunnel to suggest the support structures.

1. Uniaxial compressive strength
2. Uniaxial tensile strength
3. Elastic modulus
4. Passion's ratio
5. Friction angle
6. Cohesion of intact rock

The geoengineering aspects of the tunnel such as some evaluations on local geology and geomechanical characterization of rocks, discontinuities and rock mass and their utilization for assessing

Table 1. Average daily traffic (ADT), 2018

	Direction	S to N	N to S	Total
	Car	1081	1087	2159
	Pick up	83	88	170
	Bus Light	52	61	113
	Bus Medium	86	91	178
Vehicles Type	Bus Big	150	156	305
	Truck 2 Axle	199	191	390
	Truck 3 Axle	391	380	771
	Truck 4 Axle	10	15	25
	Truck 5 Axle	82	85	167
	Truck 6 Axle	163	127	290
	Motor Cycle	3	4	7
	Total	2301	2275	4575

Source: Based on MPW's consultant volume count survey, Sep. 2018; N = North, S = South

the tunnel stability and support design are essential. Malistani (2017) and Malistani et al. (2016a,b) have already carried out some detailed studies on these aspects. The rock mass conditions were evaluated and some laboratory experiments on surrounding rock and discontinuities were carried out in the rock mechanics laboratory of the University of the Ryukyus. In addition, some dynamic shear experiments of Salang granite were carried out as shown in Fig. 9. The peak friction angle is about 42 degrees while the kinctic friction angle is about 26 degrees.

Large granite block samples were gathered from the site and some laboratory experiments were carried out on a large granite block sample having a circular tunnel with a diameter of 50 mm. Fig. 10 shows some of the responses measured during the experiment. During the experiment, the sample was instrumented to observe its multi-parameters responses such as strains, loads, acoustic emissions, accelerations, and infra-red thermographic imaging was also carried out. As noted from the figure, various interesting observations could be made. At the end of the experiment, permanent strains at the sidewalls, roof and total strains were observed. Kaiser effect can be also clearly distinguished during the first loading-unloading cycle. The sample was not subjected to a total failure.

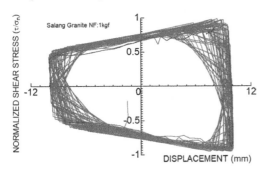

Figure 9. Dynamic shear behaviour of discontinuities of Salang granite.

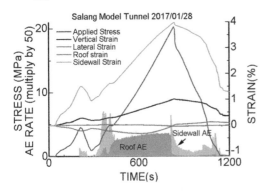

Figure 10. Multi-parameter responses of the model tunnel in large granite rock sample.

9 VENTILATION

In case of ventilation system, some environmental monitoring such as air pressure, CO_2, temperature and humidity, etc. inside and outside the tunnel are necessary to evaluate the ventilation problems quantitatively as explained in Section 6. This kind of environmental monitoring could provide quantitative data for the design of ventilation system for the existing tunnel. Nevertheless, such monitoring results should be performed for longer periods in order to see the effect of climatic variation throughout the year.

10 WATER INFLOW

Groundwater and seepage are a serious problem in Salang Tunnel. Some monitoring on groundwater conditions and measurement of permeability of the rock mass with an emphasis on the fracture zones are necessary. Such measurements should be quite useful for design and implementation of the drainage system for the tunnel. Techniques listed below are recommended for solution of the water inflow problems:

1. Investigation for finding the paths between tunnel and lake.
2. Investigation inside and above the tunnel to find the paths between lake and rock masses.
3. Grouting (shotcrete or another material) to surrounding rock mass of tunnel to reduce permeability.
4. Collecting groundwater by drilling relatively deep holes.
5. Drilling to rock masses in the seepage areas and spraying shotcrete.
6. Filling the voids by grouting at interface between ground and support structure (concrete lining).
7. It could also be effective to drain the ground water of the lake located above the tunnel temporarily, and then by spraying of specific materials at the surface of the lake the pores will be filled.

However, it is urgently required to carry out seepage analysis and measurements of permeability of rock masses. Furthermore, monitoring on ground water-level, snowfall, rainfall, etc. and water inflow into tunnel are necessary.

11 SUPPORT STRUCTURES

For the stability of surrounding rock mass of existing Salang Tunnel, it is anticipated that it would be more beneficial to analyze the support structures for the damaged sections, or even for the long alignment of tunnel. For the support system of the existing tunnel, it is recommended to evaluate it according to various updated rock classifications as given in Table 2. As noted from the Table 2, the Q-System provides very light support system as compared with those from the

RMQR and NEXCO (Malistani et al. 2016a,b). The RMQR rock classification provides a little conservative support system. However, the computations reported by Malistani (2017) and Malistani et al. (2016a,b) indicated that the yielding of surrounding rock mass is likely when overburden becomes greater than 150 m, which may require a support system as suggested by the RMQR. Rockbolts and shotcrete are not used in the excavation of the existing Salang tunnel and 650 mm thick concrete lining was constructed as a permanent support system. In addition, the suitability of the support systems should be evaluated by numerical analyses as done by Malistani (2017).

12 CONCLUSIONS AND RECOMMENDATION

The narrow width and poor situations of the tunnel adversely affect roadway safety and capacity. Multiple quick emergency type repairs have been attempted over the last 15 years, all of which have resulted in relative failure. Thus, after almost 55 years of service and little maintenance, the existing Salang Tunnel requires extensive structural, electrical and mechanical repairs along with renovations, in order to provide safe passage and adequate capacity for current traffic demand. Due to the lack of an appropriate and sufficient geological and geotechnical data along the alignment of tunnel, the following key considerations are required for the rehabilitation of the existing Salang Tunnel.

(i) Investigation on the sound principles of geomechanics and geoengineering

1. Investigations of geological structure of the site
2. *Rock classifications along the alignment of the tunnel*
3. *Properties of intact rock along the tunnel.*
4. *Geology of tunnel*
   - Faults (dip angle & dip direction, rake), shear zone
   - Folds, joints (spacing, filling, orientation data, etc.)
5. Rock stress measurement

6. Major structural and stability problems during concrete lining replacing, operation and vibration events such as earthquakes and bombing, etc.
7. Laboratory and in-situ tests
   - Mechanical properties of intact rock
   - Joints and seismic wave velocities

(ii) A general evaluation of properties and structure of rock, experiments and in-situ testing with the consideration of past case histories and experiences.
- Evaluation of properties and structure of rock mass
- *Major discontinuity sets*

(iii) Stability assessment (damaged sections)
- Support structures (concrete lining, rock-bolt and shotcrete) based on analysis of rock properties

REFERENCES

Abers, G, Bryan, C., Roecker, S. & McCaffrey, R. 1988. Thrusting of the Hindu Kush over the Southeastern Tadjik Basin, Afghanistan: Evidence from two large earthquakes. Tectonics, 7(1),41–56.

Aydan, Ö. 2000. An stress inference method based on structural geological features for the full-stress components in the earth' crust, *Yerbilimleri*, 22, 223-236.

Aydan, Ö. (2012): Ground motions and deformations associated with earthquake faulting and their effects on the safety of engineering structures. Encyclopedia of Sustainability Science and Technology, Springer, R. Meyers (Ed.), 3233-3253.

Aydan, Ö., Y. Ohta, M. Geniş, N. Tokashiki, K. Ohkubo 2010. Response and stability of underground structures in rock mass during earthquakes. Rock Mechanics and Rock Engineering, Vol.43, No.6, 857-875.

Aydan, Ö, Tokashiki, N. and Geniş, M, 2012. Some Considerations on Yield (Failure) Criteria in Rock Mechanics ARMA 12-640, 46th US Rock Mechanics/ Geo-mechanics Symposium, Chicago, Paper No. 640, 10 pages (on CD).

Aydan, Ö., Ulusay, R. and Tokashiki, N., 2014. A new Rock Mass Quality Rating System: Rock Mass Quality Rating (RMQR) and its application to the estimation of geomechanical characteristics of rock masses, Rock Mech. and Rock Eng., 47:1255-1276.

Malistani. N., 2017. A Fundamental Study on Integrated Tunnel Design System in Rock Masses with a Special Emphasis on Salang Tunnels, Afghanistan. Thesis, University of the Ryukyus.

Malistani, N., Aydan, Ö., Tomiyama, J. 2016a., Preliminary Rock Engineering Assessment of Salang Tunnel, Afghanistan. Paper361, EUROCK2016 Symposium Capadocia, Turkey, Rock Mechanics and Rock Engineering: From the past to the future. CRC Press, Pages 763-367, 2nd volume,

Malistani, N., Aydan, Ö. 2016b., Preliminary Rock Engineering Assessment of New Planned Salang Tunnels, Afghanistan. 9th Asian Rock Mechanics Symposium (ARMS9), 18th to 20th Oct, Bali, Indonesia. ARMS9, Paper126.

Table 2. Support system design according to the classifications.

Support Member	RMQR	Q-System	NEXCO
Shotcrete	150	40	100
Rockbolt	L=4-5m1.5x1.5 m	L=3m2x2	L=3m1.5x1.5
Steel ribs	Light(H125),1.5m	-	H125,1.2m
Lining	300 mm	-	300 mm
Invert	300 mm	-	400 mm

2019 Rock Dynamics Summit– Aydan et al. (eds)
© *2019 Taylor & Francis Group, London, ISBN 978-0-367-34783-3*

An integrated study on the risk assessment of Abuchiragama karstic underground shelter (Okinawa, Japan) under static and dynamic conditions

H. Inoue
Nanjyou consultants, Haebaru-city, Okinawa, Japan

N. Tokashiki and Ö.Aydan
Department of Civil Engineering., University of the Ryukyus, Nishihara, Okinawa, Japan

ABSTRACT: Abuchiragama is karstic cave in Ryukyu limestone. There is a increasing tendency that ground water is seeping in from top of cave after rainfalls. For the safety of entrants, it is necessary to evaluate the stability of the cave. This study is concerned with the risk managment of this cave utilizing RMQR as a new evaluation method of the quality of rock mass, and the stability evaluation of the cave was carried out through an empirical method and analytic methods. It is understood that it is necessary to examine in details the areas accessible to entrants, in particular, the entrance and exit areas. These areas are now being investigated in details and we have been considering to review the counter-measures master plan for rehabilitation.

1 INTRODUCTION

Itokazu Abuchiragama karstic underground cave is in Nanjyo-city, in south part of Okinawa Island (Figure 1). This cave was used as underground shelter during the battle of Okinawa. Now this cave is used for peace education for trips of the schools, and the number of entrants is about 110 thousand people.

From around 2012, there is an increasing tendency that groundwater infiltrates into the cave from its top of cave after rainfalls. For safety of entrants, it was necessary to evaluate safety and stability of this cave.

We have conducted various investigations. These involves discontinuity surveying, monitoring the crack displacements, acoustic emission measurements and accelerations at ground surface and inside the cave during seismic events. RMQR proposed by Aydan et al. (2014) as a new rock mass classification system is used to assess the quality of rock mass surrounding the cave and the stability assessment of the cave was carried out using empirical method and analytic methods.

2 GEOLOGICAL OVERVIEW

Ryukyu limestone is relatively young formation formed during Pleistocene. This formation is widely distributed in middle south area of Okinawa island.

Abuchiragama is located within the west plateau of Ryukyu limestone formation. There are faults around this underground shelter, There is a high possibility that they played a role on the formation of the cave (Figure 2).

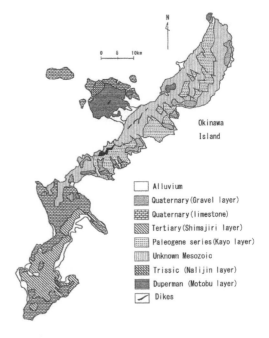

Figure 1. The geology of Okinawa Island (Okinawa Earth Science Society, 1997).

3 THE PRESENT SITUATION OF THE SHELTER

Abuchiragama cave is 200m long and its depth ranges between 6~17m from ground surface and its width is in between 12 and 30m.

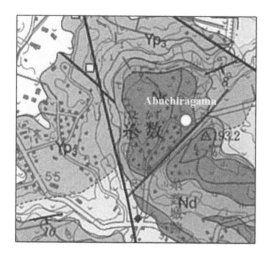

Figure 2. Close vicinity geology (National Institute of Advanced Industrial Science and Technology (2006)). Legend: Nr,Nd:Ryukyu limestone, Yp₃: Shimajiri Mudstone, Black Line: Fault

Figure 4. Status of the crown of the cave near exit (Area 7), note the movable block.

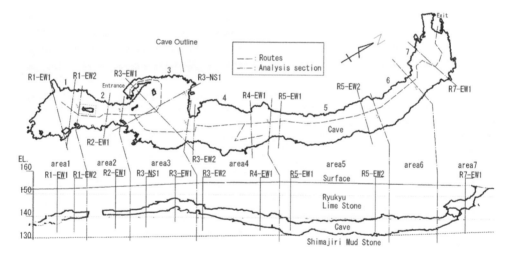

Figure 3. Areas in plan and cross section of the cave

Figure 5. Status of the top of the cave in Area 3.

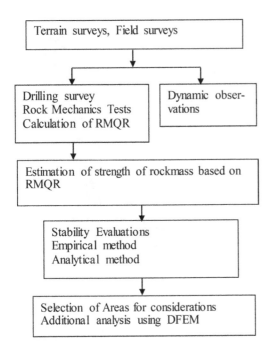

Figure 6. Flow chart of consideration

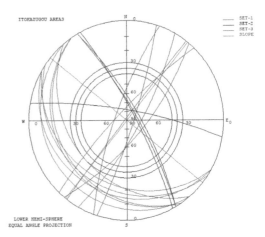

Figure 7. Stereo projection of discontinuities in Area 3.

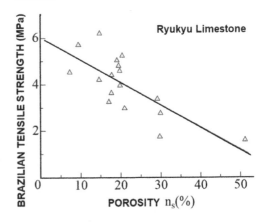

Figure 8. Boring cores and their evaluation and sample selection.

During wartime, almost all areas of this cave were used as shelter. For this reason, almost all areas of this cave had become routes for entrants.

To analyze, this cave was subdivided into 7 areas with the consideration of segmentation conditions such as shape, rock cover thickness etc. We have analyzed each area as illustrated in Figure 3. Some potentially unstable blocks were observed in the roof of the cave in Area 7.

4 FLOW CHART OF CONSIDERATIONS

We are currently conducting a series of studies. The method employed in this study has been already utilized in some projects such as "New Ishigaki Airport","Gushikawa Castle Remains", and "Himeyuri Peace Park". In each project, the existence of the caves is noted and they were considered in structural stability evaluations. The flow chart of this method is illustrated in Figure 6.

Discontinuity surveys were carried out through the cave. The investigations indicated that there are at least 3 discontinuity sets together with some random cross-joints. Figure 7 shows a stereo projection of discontinuities measured in Area 3.

Borings were drilled and the rock mass conditions were evaluated using RMQR rock classifications system as well as other rock mass classification systems (Figure 8). Rock samples have been prepared and the unit weight of rocks and elastic wave velocities were measured. The samples are used in uniaxial compression, Brazilian tensile tests. Figure 9 shows the variation of Brazilian tensile strength of samples with the ratio of porosity.

Figure 9. the variation of Brazilian tensile strength of Ryukyu limestone

5 STABILITY EVALUATION OF CAVE

5.1 *Empirical Evaluation Method*

Aydan (1989), Tokashiki (2011) and Aydan and Tokashiki (2011) developed some empirical and analytical methods to analyze the stability of natural caves. The rock mass conditions were evaluated using the RMR rock classification system. Aydan et al (2014) proposed a new rock mass classification which is named as Rock Mass Quality Rating (RMQR). This new rock classification system was utilized in this study to assess the state of the underground cave and the previously proposed method was extended to the use of RMQR together with further subdivision of stability conditions around the caves in view of the conditions observed in Abuchiragama cave. (Table 1, Figure 11). Meaning of H_f, H_r, B_s and B are shown to Figure 10. We plotted the width of

Table 1. State of stability classification (from Aydan 2018)

Category	State	$\frac{H_f}{H_r}$	$\frac{B_s}{B}$	Comments
I		0.0	0.0	Opening locally and globally stable
II		0.0-0.1	0.0	Some rock block falls from roof. Opening globally stable
III		0.1-0.3	0.0	Block falls from roof and sidewalls into the opening occur and the failure zone increases in size. Roof height is higher than opening width. The failure zone may increase in size with time.
IV		0.3-0.7	0.0	Considerable scale of falls and sliding of rock blocks from the roof and sidewall of openings occur and the failure zone larger in size. Roof height is much higher than opening width. The failure zone may increase reach ground surface in long-term
V		0.7-1.0	0.8-1.0	Failure zone reaches to ground surface and a small size crater develops at ground surface. The possibility of collapse zone may increase in size and shoulder may fall into opening in long-term
VI		>1.0	>1.0	Opening globally unstable. In other words, it is in a total collapse state. Deep Sinkhole appear on the ground surface.

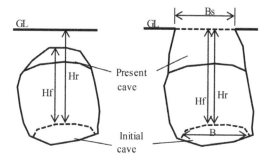

Figure 10. State of cave

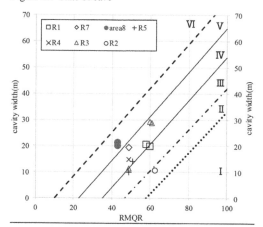

Figure 11. Comparison of stability categories of areas with estimations from the empirical method.

cave and RMQR shown in Figure 11. R1 to R7 are section lines in Areas 1 to 7. As a result, we concluded that Area 3 and Area 7 need further detailed considerations

5.2 Analytical Evaluation Method

The rock mass strength for analytical stability evaluation method is necessary. In this study, RMQR system is used to estimate rock mass properties through the utilization of RMQR value and intact rock properties. The properties of rock mass is obtained from the Eq. (1) given below (Aydan et al. 2014).

$$\alpha = \alpha_0 - (\alpha_0 - \alpha_{100})\frac{RMQR}{RMQR + \beta(100 - RMQR)} \quad (1)$$

The strength reduction was calculated by applying the RMQR values in Table 2 to Eq.(1). Calculation examples are given in Table 4.

Aydan (1989), Kawamoto et al. 1991, Tokashiki (2011) and Aydan and Tokashiki (2011) showed the following three patterns for stability of the roof of the cave (Figure 12). The following relationships are used to estimate the maximum width of the cave for the stability under no-support condition:
(Formula for simple beam and built-in beam)

Table 2. Results of empirical stability evaluations

Area	Section	Cavity width (m)	Value of RMQR	Stability Category
1	R1-EW1	20.8	58	IV
	R1-EW2	20.2	60	III
2	R2-EW1	10.5	63	II
3	R3-EW1	28.8	61	IV
	R3-EW2	29.3	60	IV
	R3-EW3	11.5	49	III
4	R4-EW1	15.0	49	IV
5	R5-EW1	10.0	49	III
	R5-EW2	14.3	51	III
7	R7-EW1	19.5	49	IV

Table 3. Values of parameters used in the Eq. (1)

Property (α)	α_0	α_{100}	β
Deformation modulus	0.0	1.0	6
Poisson's ratio	2.5	1.0	0.3
Uniaxial compressive strength	0.0	1.0	6
Tensile strength	0.0	1.0	6
Cohesion	0.0	1.0	6
Friction angle	0.3	1.0	1.0

Table 4. Example of calculation used for estimating rock mass strength using RMQR

Strength of Intact rock			
Uniaxial compressive strength	σ_{ci}	12.34	MPa
Tensile strength	σ_{ti}	4.0	MPa

Constants used in Eq. (1)

	α_0	α_{100}	β
Uniaxial compressive strength	0.0	1.0	6
Tensile strength	0.0	1.0	6

Strength of rock mass (RMQR Value is 61)			
Uniaxial compressive strength	σ_{cm}	2.55	MPa
Tensile strength	σ_{tm}	0.83	MPa

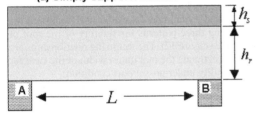

(a) Simply supported beam

h_s

h_r

A ◄──── L ────► B

Figure 12. Models for assessing roof stability

$$\frac{L}{h_r} = \sqrt{\beta \frac{\sigma_t}{(\gamma_r h_r + \gamma_s h_s)}} \qquad (2)$$

Simple beam:$\beta=2/3$, Built-in beam: $\beta=2$
(Formula of Arching model)

$$\frac{L}{h_r} = \sqrt{\beta \frac{\sigma_c}{(\gamma_r h_r + \gamma_s h_s)}} \qquad (3)$$

In case of no crack: $\beta=4/3$

Eqs. (2) & (3) were applied to each section and the stability state were evaluated. An example of analyzed cross section used for evaluation is shown in Figure 13. The evaluation results are given in Table 5. As the roof situation is close to the built-in beam condition, the cave is stable in many section except Area 3 and Area 7. The estimations imply that cracking may occur. However, the arching model implies that the roof should be stable even in Area 3 and Area 7 under static conditions.

Table 5. Results of stability evaluation based on analytical method

Area	Section	Simple Beam	Built-in Beam	Arching Model
1	R1-EW1	Potential	OK	OK
	R1-EW2	Potential	OK	OK
2	R2-EW1	OK	OK	OK
3	R3-EW1	Potential	Potential	OK
	R3-EW2	Potential	Potential	OK
	R3-EW3	OK	OK	OK
4	R4-EW1	Potential	OK	OK
5	R5-EW1	OK	OK	OK
	R5-EW2	OK	OK	OK
7	R7-EW1	Potential	OK	OK

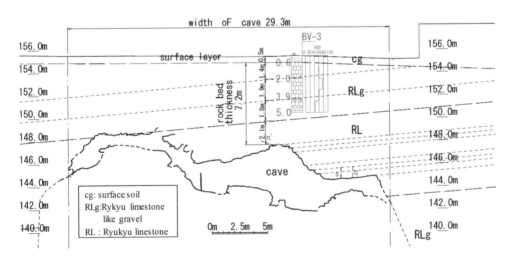

Figure 13. Analysis section for Stability evaluation, (R3-EW2)

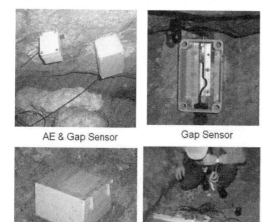

AE & Gap Sensor   Gap Sensor

Accelerometer   Micro-tremor Measurement

Figure 14. Instrumentation for multi-parameter monitoring

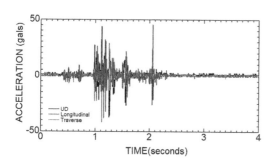

Figure 15. An acceleration record on ground surface on February 7, 2018.

6 MULTI-PARAMETER MONITORING

The long-term monitoring is essential for structures of great importance. Particularly for structures of this kind, it is very rare to see any monitoring worldwide. However, the monitoring studies in the Gushikawa Castle remains is probably the first attempt for such a purpose.

The multi-parameter monitoring of groundwater seepage into the cave, the crack displacements, acoustic emission measurements and accelerations at ground surface and inside the cave during seismic events has been initiated at the Abuchiragama cave since May 2017 (Figure 14). In addition some microtremor measurements were implemented inside and the surface of the underground shelter. Figure 15 shows an example of acceleration record on the ground surface taken on Feb. 7, 2018.

7 CONCLUSIONS

The following conclusions may be drawn from this study:

1. The stability category of Area 3 is estimated form the analytical method lower than that estimated from the empirical method.
2. It is estimated that the Area 3 should be stable if the arching model is valid. In other words, such a condition may be violated in case of large scale seismic events.
3. Further studies are necessary for Areas 3 and 7. Such studies are currently underway such as using the Discrete Finite Element Method (DFEM) (Aydan et al. 1996).

ACKNOWLEDGEMENTS

This work is carried out as a part of project with the consignment of the Nanjo City. We gratefully acknowledge the authorities of the Nanjo City for the permission to publish the content of the project in this conference.

REFERENCES

Aydan, Ö., 1989. The stabilization of rock engineering structures by rockbolts. Doctorate Thesis, Nagoya University, 204 pages.

Aydan, Ö., 2018. Some Thoughts on the Risk of Natural Disasters in Ryukyu Archipelago. International Journal of Environmental Science and Development, 9(10),282-289.

Aydan, Ö., Tokashiki, N. (2007). Some damage observations in Ryukyu limestone caves of Ishigaki and Miyako Islands and their possible relations to the 1771 Meiwa Earthquake. J. of The School of Marine Sci. and Tech.,Tokai University, 5(1),23-39.

Aydan, Ö. and Tokashiki, N. (2011): A comparative study on the applicability of analytical stability assessment methods with numerical methods for shallow natural underground openings. The 13th International Conference of the International Association for Computer Methods and Advances in Geomechanics, Melbourne, Australia, pp.964-969.

Aydan, Ö., I.H.P Mamaghani, T. Kawamoto (1996). Application of discrete finite element method (DFEM) to rock engineering structures. NARMS'96, 2039-2046.

Aydan, Ö., Ulusay, R. & Tokashiki, N. 2014. A new rock mass quality rating system: Rock Mass Quality Rating (RMQR) and its application to the estimation of geomechanical characteristics of rock masses. Rock Mech Rock Eng 47: 1255–1276.

Kawamoto, T., Ö. Aydan, and S. Tsuchiyama, 1991. A consideration on the local instability of large underground openings. Int. Conf., GEOMECHANICS'91, Hradec, 33-41.

Tokashiki, N. (2011). Study on the Engineering Properties of Ryukyu Limestone and the Evaluation of the Stability of its Rock Mass and Masonry Structures. PhD Thesis, 221p, Waseda University, Engineering and Science Graduate School.

Tokashiki, N. Aydan, Ö. (2010): The stability assessment of overhanging Ryukyu limestone cliffs with an emphasis on the evaluation of tensile strength of Rock Mass. Journal of Geotechnical Engineering, JSCE, 66(2),397-406.

2019 Rock Dynamics Summit– Aydan et al. (eds)
© *2019 Taylor & Francis Group, London, ISBN 978-0-367-34783-3*

An integrated study on the large-scale arch structure for protection of karstic caves at New Ishigaki Airport

H. Minei & Y. Nagado
Iwashita Kengi Consultants Co., Urasoe, Okinawa, Japan

Y. Ooshiro
Okinawa Prefectural Government, Civil and Architectural Division, Naha, Japan

Ö. Aydan & N. Tokashiki
Department of Civil Engineering, University of the Ryukyus, Nishihara, Okinawa, Japan

M. Geniş
Department of Mining Engineering, Bülent Ecevit University, Zonguldak, Turkey

ABSTRACT: The site of the Ishigaki New Airport had several karstic caves, which crosses the runway at several locations. The bats in these caves were claimed to be precious species, the Okinawa Prefectural Government decided to construct the reinforced protection arch structures above the caves against their possible collapses. The authors carried out a series of computations using 2D and 3D numerical methods to check the static and dynamic stability assessment of natural underground openings with its protection arch structure. In this article, the authors would present the outcomes of the numerical studies and discuss their implications.

1 INTRODUCTION

The natural underground openings are generally caused by the dissolution and/or erosion of rocks by sea waves, winds, river flow or percolating rain water and they may present some engineering problems especially in urbanized areas above karstic regions. The stability problems may arise in the form of sinkholes. However, the studies are very few for evaluating the stability of such natural underground openings in literature.

Ryukyu limestone is widely distributed all over Ryukyu Islands. Large engineering projects have been recently increasing in the Ryukyu Islands and natural underground openings present some stability problems to the superstructures on the ground surface. Although the numerical methods are superior over empirical methods and simplified analytical methods, the authors also explored the applicability of quick stability assessment techniques based on either empirical methods or simplified analytical methods for natural caves.

The site of the Ishigaki New Airport had several karstic caves, which crosses the runway at several locations. The bats in these caves were claimed to be precious species, the Okinawa Prefectural Government decided to construct the reinforced protection arch structures above the caves against their possible collapses. The construction of the protection structures was quite unusual in civil engineering and their design was a quite challenging problem. The width of the protection structures was decided on the basis of past case histories, model test under static and dynamic conditions. Analytical and numerical studies were performed to check the stability of the caves beneath and protection arch structures using 2D and 3D static and dynamic analyses. During construction and afterwards, continuous real-time monitoring was undertaken to check the response of the structures. Furthermore, strong motion measurements for the response of the structure during earthquakes were also implemented and it is still continued. The authors explain this integrated study on this very unique structure and present the outcomes. Furthermore, they discuss their implications in rock dynamics field.

2 OUTLINES OF PROTECTION STRUCTURES

New Ishigaki Airport is built on a Ryukyu Limestone formation which includes some solution cavities of various sizes in the form of karstic caves. Some precious bat species live in these caves and there is environmental concern to protect these bats living in the caves. The existing airport had a short runway of about 1500 m and there was a huge demand to build a new airport with a longer runway. The construction of embankment up to 14 m high was required over the limestone formation. As the load due to the embankment as well as airplanes etc., it was required to build some arch-like reinforced concrete structures with a 22 m span at some locations. Furthermore, these arch structures had to follow the karstic caves below.

Fig. 1 shows a view of arch structures during the construction stage. The cave numbered A1 was the largest and a huge semi-circular arch structure with a span of 22.4 m and length of 100 m was built over the karstic cave. The overall structure divided into 5 blocks and the geometry of the protection arch structure is quite complex in three dimensions. The overall embankment height was 14 m from the base and the depth of the embankment was about 4 m above the crown of the arch structure (Fig. 2). The thickness of reinforced concrete arch is about 1.2 m. The new airport was open in March 2013 with more than 50 flights per-day. Boeing 777 type airplanes were considered for the design of the arch protection structure and the runway.

3 TWO-DIMENSIONAL DFEM ANALYSES

First a series of discrete finite element analyses (Aydan et al. 1996; Mamaghani et al. 1999) were carried out for four different steps by considering actual construction stages as illustrated in Fig. 3. Constraint conditions were introduced at the bottom and sides of the finite element meshes. Material properties used in finite element analyses are given in Table 1. At Step 0, the initial stress stage was established before the construction of arch structure by applying

Figure 1. An aerial view of arch protection structures during construction.

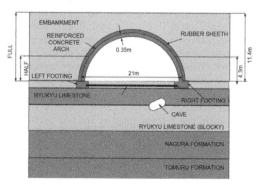

Figure 2. A cross section view of the arch protection structure for A1 Cave.

body force under the assumed constraint conditions. At Step 1, the arch structure was built under its own weight. At Step 2, the embankment to the half-height of the arch structure was constructed with the consideration of its self-weight. At Step 3, which is the final stage, the embankment was constructed to its full height with the consideration of the self-weight of the embankment material of the model.

Figs. 4, 5 and 6 show the maximum principal stress contours, principal stresses and deformed configuration at Step 1, Step 2 and Step 3, respectively. In Figs. 4, 5 and 6, tensile stress regions

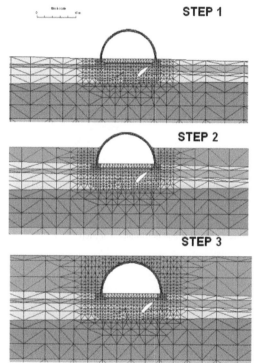

Figure 3. Finite element meshed for each embankment step.

Table 1. Material properties used in the analyses.

Formation (layers)	Υ kN/m³	c MPa	ϕ (°)	E MPa	v
Tomuru Form. (Phyllite-schist)	30	10.0	0	5900	0.20
Nagura Form. (Fractured limestone)	19	0.0043	35	177	0.29
Ryukyu limestone (Clayey limestone)	19	0.006	35	113	0.33
Ryukyu limestone (Solid limestone)	23	0.409	30	500	0.18
Enbankment	22	0.059	28	16	0.33
Concrete	24	8.0	35	10000	0.15

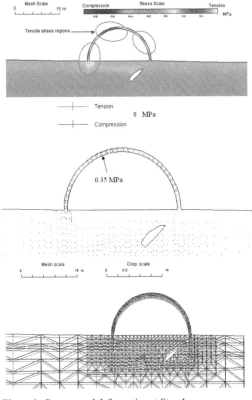

Figure 4. Stresses and deformation at Step 1.

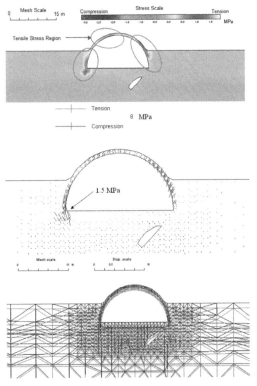

Figure 5. Stresses and deformation at Step 2.

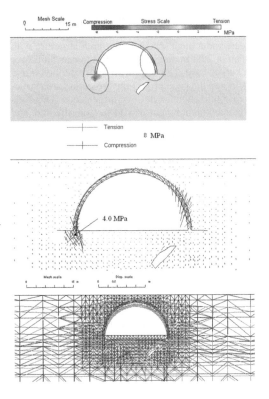

Figure 6. Stresses and deformation at Step 3.

are also pointed out. As noted from the computed results, tensile stresses develop in the arch structure. Although tensile stress decreases in magnitude when the full embankment load is imposed (Step 3), there are some still tensile stress regions in the arch structure. These locations almost coincide with observed cracking in the actual arch structure.

4 THREE DIMENSIONAL ANALYSES

Some preliminary static elasto-plastic three-dimensional numerical analyses together with the same material properties were carried out. Three dimensional analyses were carried out using finite difference code (FLAC3D) (Itasca, 2005). The constructions stages were fully considered in computation stages similar to those if the 2D analyses. Figure 7 shows various views of computational models of three-dimensional analyses without the karstic cave beneath the arch structure.

Figs. 8 and 9 show the displacement contours under fully constructed stage of the arch structure and embankment. It is interesting to note that the displacements are large in the area above the footings.

Figures 10 and 11 show the maximum and minimum principal stress contours in the arch concrete structure, respectively. As noted from the Figs. 10 and 11, some tensile stresses develop in the crown of the arch concrete structure. In addition some

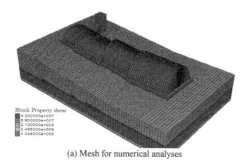

Block Property shear
4.200000e+007
8.900000e+007
2.120000e+008
2.488000e+009
4.348000e+009

(a) Mesh for numerical analyses

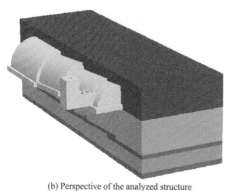

(b) Perspective of the analyzed structure

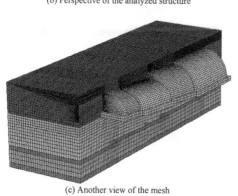

(c) Another view of the mesh

Figure 7, Three-dimensional view and finite difference mesh of the numerical models.

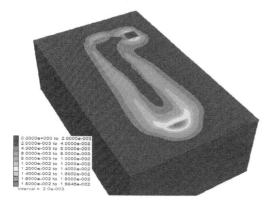

0.0000e+000 to 2.0000e-003
2.0000e-003 to 4.0000e-003
6.0000e-003 to 8.0000e-003
8.0000e-003 to 1.0000e-002
1.2000e-002 to 1.2000e-002
1.2000e-002 to 1.4000e-002
1.4000e-002 to 1.6000e-002
1.8000e-002 to 1.8000e-002
1.8000e-002 to 1.9846e-002
Interval = 2.0e-003

Figure 8. Settlement of the embankment at the ground surface (unit:m).

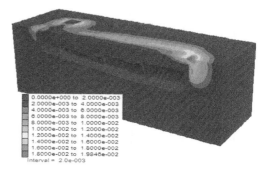

0.0000e+000 to 2.0000e-003
2.0000e-003 to 4.0000e-003
4.0000e-003 to 6.0000e-003
6.0000e-003 to 8.0000e-003
8.0000e-003 to 1.0000e-002
1.0000e-002 to 1.2000e-002
1.2000e-002 to 1.4000e-002
1.4000e-002 to 1.6000e-002
1.6000e-002 to 1.8000e-002
1.8000e-002 to 1.9846e-002
Interval = 2.0e-003

Figure 9. Displacement contours of the model.

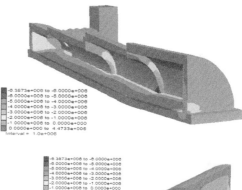

-6.3873e+006 to -6.0000e+006
-6.0000e+006 to -5.0000e+006
-5.0000e+006 to -4.0000e+006
-4.0000e+006 to -3.0000e+006
-3.0000e+006 to -2.0000e+006
-2.0000e+006 to -1.0000e+006
-1.0000e+006 to 0.0000e+000
0.0000e+000 to 4.4733e+006
Interval = 1.0e+006

-6.3873e+006 to -6.0000e+006
-6.0000e+006 to -5.0000e+006
-5.0000e+006 to -4.0000e+006
-4.0000e+006 to -3.0000e+006
-3.0000e+006 to -2.0000e+006
-2.0000e+006 to -1.0000e+006
-1.0000e+006 to 0.0000e+000
0.0000e+000 to 4.4733e+006
Interval = 1.0e+006

Figure 10. Maximum principal stress contours in the concrete arch structure (Negative values are compressive stress).

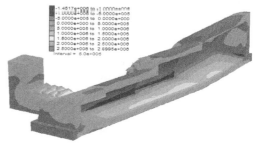

-1.4517e+006 to -1.0000e+006
-1.0000e+006 to -5.0000e+005
-5.0000e+005 to 0.0000e+000
0.0000e+000 to 5.0000e+005
5.0000e+005 to 1.0000e+006
1.0000e+006 to 1.5000e+006
1.5000e+006 to 2.0000e+006
2.0000e+006 to 2.5000e+006
2.5000e+006 to 2.8995e+006
Interval = 5.0e+005

Figure 11. Minimum principal stress contours in the concrete arch structure (Positive values are tensile stress).

tensile regions appear at the floor of the opening. It should be noted that these tensile stresses are taken by tie beams used in the actual structure. Compared with two-dimensional analyses results, there are some differences regarding the maximum tensile stress regions in the arch structure. The difference

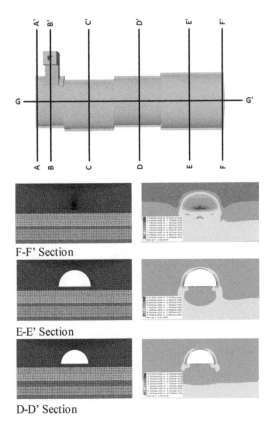

F-F' Section

E-E' Section

D-D' Section

Figure 12. Maximum principal stress contours in the arch and surrounding ground for different sections.

among the maximum tensile stress regions might be due to non-consideration of the cave beneath the arch structure. Fig. 12 shows the maximum principal stress contours at three different sections of the arch concrete structure. As noted from the figure the crown of the arch structure is subjected to tensile stresses. Furthermore, the stress state is quite different at the end wall of the structure, which is just beneath the center-line of the runway.

5 EFFECT OF IMPACT VIBRATIONS AT THE GROUND SURFACE ON ARCH STRUCTURE

One of the major concern is the impacts induced by landing or take-off of the airplanes on the underground arch structure. Although it was difficult to do such a monitoring using actual planes, some tests on the impact vibrations at the ground surface on arch structure were carried out using a 10 tonf truck and 1 tonf sand-bag dropped from a height of 100-180 cm. the accelerations induced by the passing of truck over 10 cm high barrier was quite small and attenuated very rapidly. Therefore, it was decided to use 1 tonf sand bag falling from a height of 100 cm or 180 cm as shown in Fig. 13. Besides accelerometers on the ground surface, 5 accelerometers were

installed in the underground arch structure (Fig. 14). As shown in Fig. 15, 1 tonf sand bag was dropped at various locations projected on the ground surface and induced vibrations were measured.

Fig. 16 shows the attenuation of maximum accelerations measured at various points in relation to the distance from the source point. The attenuation of maximum accelerations are fitted to the following equation:

$$a_{max} = 20000e^{0.8*W*h}\frac{1}{1+r^b} \qquad (1)$$

where W (tonf) is weight of dropped bag, h (m) is the drop height and r (m) is the distance from the drop location. The coefficient b is an empirical value and

Figure 13. A view of the in-situ experiment and set-up of accelerometers.

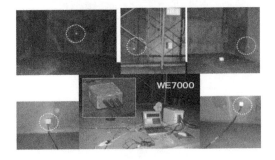

Figure 14. Locations of accelerometers in the underground arch structure.

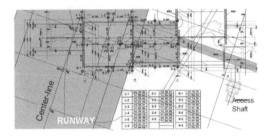

Figure 15. Locations of measurement points and sand-bag drop.

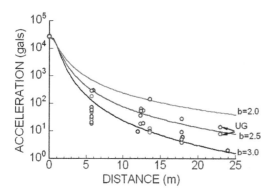

Figure 16. The attenuation of maximum accelerations with distance and its comparison with the attenuation relation.

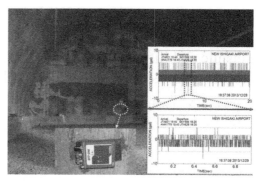

Figure 17. Induced accelerations in the arch structure by landing and take-off of the airplanes.

it was found that it ranges between 2 (cylindrical) and 3 (spherical). Including the measurements in the underground arch structure, if its value is 2.5, it fits the in-situ measurements. As noted from Fig. 16, the vibrations drastically reduced as a function of the distance from the source area. Therefore, the dynamic impact effects of the airplanes during landing or take-off would be quite small in the underground actual structure.

The authors installed some accelerometers in the underground arch structure for long-term monitoring of ground motions due to earthquakes or other sources. One of the device installed at the end-wall of the arch structure was triggered by the airplanes as shown in Fig. 17. The vibrations are due to Boeing 737-800 or 737-500 airplanes and the maximum amplitude in the order of 8 gals. This value is quite small compared to the maximum accelerations during the landing and take-off of the airplanes, which are generally in the order of 1300-1800 gals.

6 CONCLUSIONS

Some two-dimensional and three dimensional numerical analyses of the arch structure beneath the new Ishigaki airport runway are presented in this article. 55 different cases were considered in actual two dimensional numerical analyses were carried out. In this article, some of these numerical analyses were presented in relation to the actual monitoring of the arch structure. The two-dimensional numerical analyses carried out in details were capable of simulating the actual observations with due considerations of possible geological and geotechnical situations in-situ. The results clearly showed that some tensile fracturing was likely in the arch concrete structure due to existence of the karstic cave beneath the structure. The three-dimensional analyses are still preliminary. Nevertheless, the consideration of karstic cave beneath the arch structure is necessary to simulate the actual conditions. Furthermore, this area subjected to earthquakes from time to time. Therefore, some dynamic analyses are necessary and such analyses are planned to be carried out. Furthermore, the dynamic impacts loads due to landing airplanes is also of great concern for the structure and such studies are also planned to be carried out.

REFERENCES

Aydan, Ö., Mamaghani, I.H.P, Kawamoto, T. 1996. Application of discrete finite element method (DFEM) to rock engineering structures. *NARMS'96*, 2039–2046.

Itasca, 2005. FLAC3D-Fast Lagrangian Analysis of Continua-User Manuel (Version 2.1). Itasca Consulting Group Inc., Minneapolis.

Mamagahani, I.H.P, Aydan, Ö., Kajikawa, Y. 1999. Analysis of masonry structures under static and dynamic loading by discrete finite element method. JSCE, Journal of Structural and Earthquake Engineering Division, Vol.16, No. 626 75–86.

2019 Rock Dynamics Summit– Aydan et al. (eds)
© 2019 Taylor & Francis Group, London, ISBN 978-0-367-34783-3

Seismic stability evaluation of the existing rock slope subject to ground motion records of 2011 Tohoku earthquake

Y. Kikuchi, G. Itoh & Y. Benno
Tohoku Electric Power Company, Inc., Miyagi, Japan

M. Ishimaru
Central Research Institute of Electric Power Industry, Chiba, Japan

ABSTRACT: This study verified a slip safety factor evaluation method, as developed subsequent to equivalent linear analysis, for the existing rock slope subject to earthquake ground motion records of the 2011 Tohoku earthquake off the Pacific coast of Japan. Results demonstrated that, although a large deformation was not confirmed via slope measurements taken after the earthquake, the minimum slip safety factor was determined to be lower than 1.0. Therefore, it was found that a certain margin is included in the slip safety factor evaluation method. Next, for the purpose of developing a more practical seismic stability evaluation method, the slip safety factor was calculated based on seismic response analysis in the time domain while taking into account the stress redistribution. The analysis revealed the results to be consistent with the existing rock slope.

1 INTRODUCTION

Since the great earthquake of 2011 struck the northeastern part of Japan, evaluation of the stability and safety of slopes situated in the vicinity of important structures when a large-scale earthquake strikes have been called for all the more fervently.

The stability evaluation of a rock slope against potential sliding failure during earthquakes, typically conducted as an evaluation of its safety factor against slipping (hereinafter, "slip safety factor"), has been carried out through examination of stresses in the slope based on seismic response analysis in the frequency domain using the equivalent linearization method (hereinafter, "equivalent linear analysis") (Nuclear Standards Committee of Japan Electric Association, 2015) (Nuclear Power & Civil Engineering Committee of Japan Society of Civil Engineers, 1985). However, since only the equilibrium between forces is considered in the evaluation of a slip safety factor, it has usually been the case that a slope is judged to be subject to failure if its safety factor is evaluated to be below 1.0 due to the fact that the equilibrium has broken down at that time momentarily. However, it cannot be said that a slope is likely to fail only because the equilibrium condition is not satisfied momentarily. Conventional slip safety factor evaluations, it has to be said, tend to be very conservative in nature.

In this study, the establishment of a new method for evaluating the slip safety factor, which is a calculation method of stresses for determining it, is attempted, focusing on seismic response analysis in the time domain, where redistribution of the stresses in a slope occurs every time a failure incident takes place in it, unlike in the case of equivalent linear analyses which cannot consider the influence of such failures, in order to eliminate their conservative nature.

2 ROCK SLOPE AND INPUT SEISMIC GROUND MOTION

In this study, the seismic response of a rock slope situated near an important structure located in Onagawa-cho, Miyagi Prefecture was examined in a simulation study, using the seismic ground motion observed during the 2011 Tohoku Earthquake off the Pacific Coast near the slope.

2.1 *Rock slope*

A layout drawing of the slope and its rock classification cross-section drawing are shown in Figure 1 and Figure 2, respectively. At the three points shown in the cross-section, boring surveys and P-and-S wave velocity loggings were conducted. The important structure, an underground facility, is situated near the foot of the rock slope.

The slope was an excavated one, with a height of approx. 43 m and a gradient of 1:1.0. The geological structure is a Jurassic system of the Mesozoic Group consisting of alternating layers of sedimentary sandstone and shale, with porphyritic rock partially distributed therein. In the deep part of the slope is a fault, and seams parallel to the bedding planes have been detected in the part between sandstone and shale. Although the slope was struck by the 2011 Tohoku Earthquake off the Pacific Coast, no particular change was observed thereto in an inspection carried out after the earthquake.

Shown in Figure 3 are the analytical model of the slope. The ground water levels are shown in Figure 4.

The material properties used in the analysis are shown in Table 1 and Figure 5. The properties of the rocks were assumed to be linear. As for the properties of the D-Class rocks, backfill, fault and seams, their dependencies on strain were considered. The fault and seams were modeled as joint elements.

2.2 *Input seismic motion*

The seismic ground motions used in this study are the actual motions observed during the 2011 Tohoku Earthquake off the Pacific Coast at a seismic activity observation point near the study site.

The input seismic motions are shown in Figure 6. The maximum acceleration of horizontal motion

was 4.65 m/s² (at 90.12 s), and that of vertical motion, 3.11 m/s² (at 40.72 s).

3 SLIP SAFETY FACTOR EVALUATION BASED ON EQUIVALENT LINEAR ANALYSIS

3.1 *Slip safety factor evaluation*

The slip lines to be considered in this study is were those in D-Class rock strata (approx. 10 m thick) whose strength was relatively low.

The shear strength properties of D-class rocks (with zero tensile strengths) are shown in Table 2.

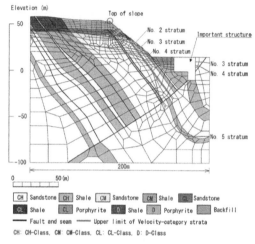

Figure 3. Analytical model of the slope.

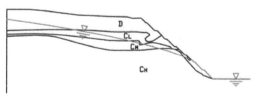

Figure 4. Underground water level.

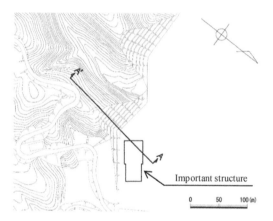

Figure 1. Layout plan.

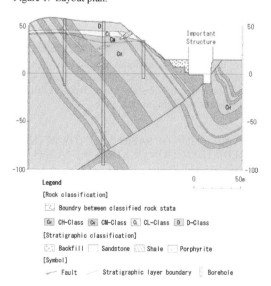

Figure 2. Rock classification cross-section.

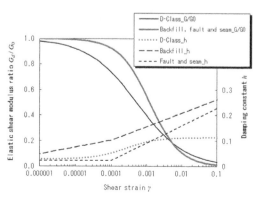

Figure 5. Dynamic deformation property.

402

Table 1. Material properties used in analysis.

	Unit weight γ (kN/m³)	Modulus of static elasticity E_s (MN/m²)	Static Poisson's ratio v_s	Dynamic shear modulus G_d (MN/m²) Velocity-category		Dynamic Poisson's ratio v_d	Damping constant h (%)
CH-Class sandstone	26.2	1770	0.24	No. 1 stratum	200	0.48	3
				No. 2 stratum	1500	0.44	
				No. 3 stratum	5900	0.40	
				No. 4 stratum	13200	0.36	
				No. 5 stratum	16500	0.35	
CH-Class shale	26.6	2160	0.25	No. 2 stratum	1600	0.44	3
				No. 3 stratum	6000	0.40	
				No. 4 stratum	13500	0.36	
				No. 5 stratum	16700	0.35	
CM-Class sandstone	25.2	980	0.26	No. 1 stratum	200	0.48	3
				No. 2 stratum	1500	0.44	
				No. 3 stratum	5700	0.40	
				No. 4 stratum	12700	0.36	
				No. 5 stratum	15800	0.35	
CM-Class shale	25.8	980	0.20	No. 1 stratum	200	0.48	3
				No. 2 stratum	1500	0.44	
				No. 3 stratum	5900	0.40	
				No. 4 stratum	13000	0.36	
				No. 5 stratum	16200	0.35	
CL-Class sandstone CL-Class shale CL-Class porphyrite	24.1	400	0.31	No. 1 stratum	200	0.48	3
				No. 2 stratum	1400	0.44	
				No. 3 stratum	5500	0.40	
D-Class sandstone D-Class shale D-Class porphyrite	20.2	80 40 80	0.38	No. 1 stratum $G_0=255\sigma^{0.26}$		0.48	Refer to Fig. 5
				No. 2 stratum Refer to Fig. 5		0.44	
Backfill	21.8	$198\sigma^{0.60}$	0.40	$G_0=404\sigma^{0.70}$ Refer to Fig. 5		0.48	Refer to Fig. 5
Fault & seam	19.9	$74.6\sigma^{0.81}$ (Compression) $26.6\sigma^{0.81}$ (Shear)	0.40	$G_0=162\sigma^{0.55}$ Refer to Fig. 5		0.46	Refer to Fig. 5
Structure	8.2	2254	0.1667	966		0.1667	3

σ representing consolidation stress, and, for the structure, material properties equivalent to it are considered in the analysis model.

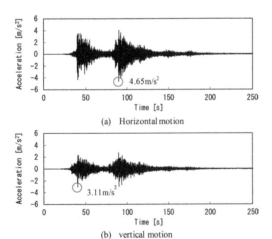

(a) Horizontal motion

4.65m/s²

(b) vertical motion

3.11m/s²

Figure 6. Input seismic motion.

The calculation results of the slip safety factor are shown in Figure 7. From this figure it is evident that the value of the safety factor was below 1.0 for slip line No.6. However, no significant change was observed in the rock slope after the earthquake. Therefore, it can be concluded that these results

Table 2. Shear strength properties of D-class rock.

	Peak strength		Residual strength
	Cohesion c (kN/m²)	Internal friction angle φ (°)	τ_r (kN/m²)
D-Class sandstone			$13.89\sigma^{0.49}$
D-Class shale	100	24.0	$28.13\sigma^{0.21}$
D-Class porphyrite			$13.89\sigma^{0.49}$

σ representing mean stress (kN/m²).

show that the conventional evaluation method allows some margin for the slip safety factor as an evaluation index of the stability of structures under the influence of earthquakes.

4 EVALUATION OF SAFETY FACTOR WITH REDISTRIBUTION OF STRESSES TAKEN INTO ACCOUNT

The reason why no significant change was observed in the slope during the earthquake despite the fact that the minimum slip safety factor was calculated to be below 1.0 in the equivalent linear analysis (in other words, an imbalance of forces in the slope created at the time) might have been that the imbalance condition occurred just momentarily while alternating stresses were generated in the rock slope. Or, it might have been the case that the imbalance condition transitioned to the equilibrium condition when redistribution of stresses took place therein. Due to these possibilities, a seismic response analysis in the time domain with stress redistribution taken into account (to be referred to as stress redistribution analysis) as a method for calculating stresses to determine the slip safety factor was attempted in order to make our evaluation more reasonable.

4.1 Overview of stress redistribution of analysis

In our study, linearization in the time domain convergent material properties obtained in the equivalent linearization was used because emphasis was placed on keeping our study consistent with the result of the equivalent linear analysis. A point of difference of our study from conventional linear analyses was that the judgments on whether a failure of the

Slip line	Safety factors
No. 1	1.96 (40.94s)
No. 2	1.79 (40.95s)
No. 3	1.67 (40.94s)
No. 4	1.68 (40.94s)
No. 5	1.64 (40.95s)
No. 6	0.77 (40.95s)
No. 7	1.11 (40.95s)

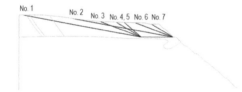

Figure 7. Slip safety factors of equivalent linear analysis.

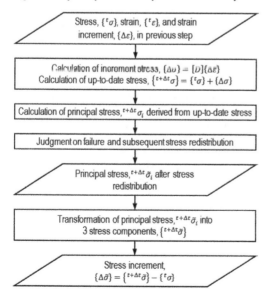

Figure 8. Processing flow of constitutive model.

element had taken place were made from moment to moment. When a failure occurred at a point, stresses were redistributed to nearby elements. Figure 8 shows the processing flow of the constitutive model used in this analysis method. Figure 9 shows the processing flow of stress redistribution (in this paper, the compression side is defined as positive).

Failure judgment is made with regard to principal stresses. When the minimum principal stress σ3 exceeds the tensile strength of the rock σt tensile failure is assumed to have occurred. When the maximum sheer stress τmax exceeds the shear strength of the rock τf shear failure is assumed to have occurred.

After the shear failure the shear strength is assumed to be equal to the residual strength, and the tensile strength, zero. However, after the tensile failure, the shear strength is assumed to retain its peak strength. This assumption stems from our understanding of the origin of the shear strength of a rock stratum, i.e., the conditions of rocks being engaged with one another in the stratum. Although some rocks may become disengaged momentarily, the shear strength can be expected to be preserved as long as no sliding takes place in the stratum.

4.2 Conditions of stress redistribution analysis

We conducted the stress redistribution analysis incorporating the aforementioned constitutive model in TDAP III as a user nonlinear model. As for the material property values and input motions, the descriptions are provided in the foregoing.

With respect to the Rayleigh damping used in our analysis, the reference frequency of stiffness-proportional damping is set to 8 Hz (common to all material properties) so that the slip safety factor obtained in equivalent linear analyses is roughly equal to that obtained in the linear analysis (without stress redistribution) in the time domain where convergent material properties in equivalent linearization (shear modulus and damping constant) are used.

4.3 Evaluation of slip safety factor

Using the stresses obtained from the stress redistribution analysis, slip safety factors are calculated. The calculation results for all the slip lines shown in Figure 7 are listed in Table 3. Figure 10 shows an example of the failure situation at the time when the slip safety factor is the smallest. In Table 3, the results obtained in the equivalent linear analysis are listed together. When the results are compared, it can be said that the deeper the slip line, the lower the value of its safety factor, and conversely, the shallower the slip line, the higher its safety factor.

These trends may be attributed to the fact that many tensile failures occurred around the top of the slope where the slip safety factors were the smallest.

404

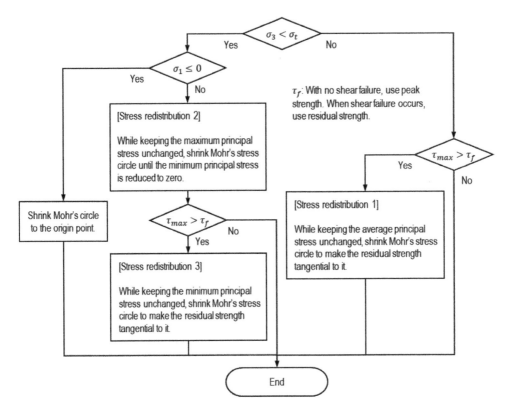

Figure 9. Processing flow of stress redistribution.

Table 3. List of minimum slip safety factors(time of occurrence in parentheses).

Slip line	Equivalent linear analysis	Stress redistribution analysis
No. 1	1.96 (40.94s)	1.72 (40.52s)
No. 2	1.79 (40.95s)	1.63 (40.52s)
No. 3	1.67 (40.94s)	1.44 (40.53s)
No. 4	1.68 (40.94s)	1.47 (40.53s)
No. 5	1.64 (40.95s)	1.56 (98.28s)
No. 6	0.77 (40.95s)	1.29 (40.95s)
No. 7	1.11 (40.95s)	1.19 (40.94s)

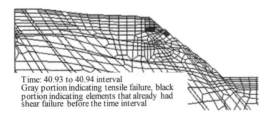

Time: 40.93 to 40.94 interval
Gray portion indicating tensile failure, black portion indicating elements that already had shear failure before the time interval

Figure 10. Failure condition at the time when the minimum slip safety factor resulted.

Excessive stresses at the points where tensile failures occurred may well have been distributed downwards, creating the observed results.

Focusing on the values of the slip safety factors, the smallest safety factor shown in Table 3 is 1.19 for slip line No.7 (elapsed time at 40.94 s). Figure 11 shows the results of an exhaustive search among circular slip lines (cross-sections of spherical slip surfaces) for the one with the smallest slip safety factor. Here, the smallest value is 1.06 (elapsed time at 40.54 s), which is smaller still than 1.19, but greater than 1.0 nonetheless.

In the case where the lowering of the shear strength to the residual strength in the stress redistribution is considered, the safety factors cannot be lower than 1.0. However, for the method used in this paper, it can be lower than 1.0 because the shear strength was not lowered if there was a concomitant tensile failure, and because the residual strength was used when there was a tensile failure in the slip safety factor evaluation. Therefore, in such a case as shown in Figure 10 where tensile failures are dominant, it can be concluded that the stability of a slope can be secured with a certain margin of safety even if the slip safety factor is greater than 1.0. Considering the results described in the foregoing, the conclusion of this study that no slip incident occurred in its slip safety evaluation can be said to be reasonable, and this result is consistent with the fact that no particular change had been observed in the actual slope.

5 SUMMARY

In this study, using seismic motion observed during the 2011 Tohoku Earthquake off the Pacific Coast, a slip safety factor evaluation method based on equivalent linear analysis was verified with an actual rock slope. The minimum safety factor for the slope turned out to be less than 1.0 despite the fact that no significant change had been observed in the slope after the earthquake. This indicated to us that the conventional evaluation method allows some margin for the slip safety factor.

Following this result, we proposed a method for calculating slip safety factors based on a seismic response analysis in the time domain with stress redistribution taken into account, and its applicability was examined in order to establish a more reasonable method for evaluating the slope's stability during earthquakes. As a result, a minimum slip safety factor of 1.06 was obtained, indicating consistency of this result with the fact that there was no change observed in the slope.

In the future, we intend to continue to verify this slip safety evaluation method with the stress redistribution taken into account in other observations, model experiments and so on. It is our opinion that it is important to prove the validity of the proposed method through more cases.

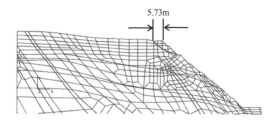

5.73m

Figure 11. Circular slip line with minimum slip safety factor.

REFERENCES

Nuclear Power & Civil Engineering Committee of Japan Society of Civil Engineers. 1985. Anti-Seismic Stability Evaluation Method for Slopes in Vicinity. Ed. 5 of Reports on Surveying and Testing Methods for Geological Features and Grounds for Nuclear Power Plants and Evaluation Method for Stability of Ground during Earthquakes. Japan Society of Civil Engineers

Nuclear Standards Committee of Japan Electric Association. 2015. Technical Guidelines for seismic design nuclear power plants (JEAG4601-2015). The Japan Electric Association

2019 Rock Dynamics Summit– Aydan et al. (eds)
© 2019 Taylor & Francis Group, London, ISBN 978-0-367-34783-3

The dynamic and multi-parameter responses of the Taru-toge tunnel during excavation

M. Imazu
Geo-space Engineering Center, Engineering Advancement Association of Japan, Tokyo, Japan

Ö. Aydan
Department of Civil Engineering and Architecture, University of the Ryukyus, Okinawa, Japan

H. Tano
Department of Civil Engineering, Nihon University, Koriyama, Japan

ABSTRACT: The authors monitored the dynamic response of the surrounding rock mass at the excavation face of Taru-toge tunnel at the prefectural border of Shizuoka and Yamanashi prefectures, Japan built in connection route between New Tomei and Chuo Expressways. This monitoring study in the immediate vicinity of the tunnel face is probably the first one in the world so far within the knowledge of the authors. In this paper, the authors describe the several monitoring studies and the results obtained so far. They also discuss the implications of this study within the integrated framework of tunnel construction.

1 INTRODUCTION

The monitoring of the real-time response of underground strutcures during excavation is of paramount importance for the safety evaluation and stability assessment. Drilling and blasting (DB) are commonly used for the excavation of rock engineering structures. The variation of stress state around the underground opening during excavation is a dynamic process and the stress state would be generally different from that under the static condition for the geometry particularly soon after blasting operation. One can not find any monitoring study on the dynamic response of underground excavations in literature except few studies by the authors. The authors have developed a multi-parameter monitoring system for monitoring the response and stability of the rock engineering structures.

Drilling-Blasting is the most commonly used excavation technique in mining and civil engineering applications. Blasting induces strong ground motions and fracturing of rock mass in rock excavations. The second author has developed a monitoring system for measuring blasting induced vibrations (Aydan et al. 2016, Imazu et al. 2014, 2015).

In this study, the authors attempted to monitor the blasting-induced vibrations and multi-parameter responses at the very close vicinity of the tunnel face in order to investigate if it is possible to evaluate its real time stability as well as the characteristics of surrounding rock masses in a Taru-Toge tunnel in Central Japan (Fig. 1). The multi-parameters measured involve electric potential variations, acoustic emissions, rock temperature,

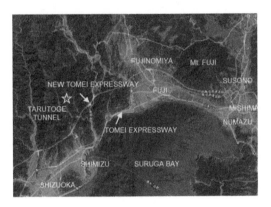

Figure 1. Location of Taru-toge tunnel.

temperature and humidity of the tunnel during the face advance. The authors describe the outcomes of these unique monitoring studies in this paper and discuss the implications in tunneling.

2 GEOLOGY OF MONITORING LOCATION

Taru-toge tunnel is being constructed as a part of expressway project connecting New-Tomei Expressway and Chuo Expressway at the boundary of Shizuoka and Yamanashi Prefectures in the Central Japan. The tunnel pass through a series of mudstone, sandstone and conglomerate layers with folding axes aligned north-south (Fig. 2). Furthermore it is bounded by Fujikawa Fault in the east and Itoikawa-Shizuoka Tectonic Line. Mt. Fuji

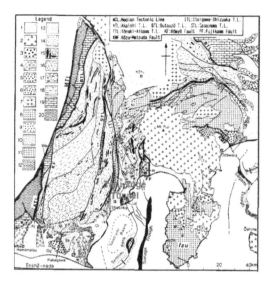

Figure 2. Regional geology in the vicinity of the Taru-toge tunnel.

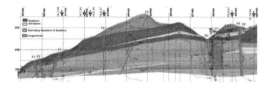

Figure 3. Local geology of at the monitoring locations in Taru-Toge tunnel (south portal side)

is to the east of the tunnel. Fig. 3 shows the local geology of southern part of the tunnel where blasting operations are carried. In the section, nine major faults (F1-F9) exist.

3 MONITORING SYSTEMS

The monitoring systems for vibration, elastic wave velocity, AE emissions and environmental measurements are described in this section. So far the monitoring was performed 6 times for certain durations.

3.1 Vibration and Wave-velocity monitoring

Two attempts to measure the vibrations and acoustic emissions at the tunnel face were done in August 31-September 1 and October 5-8, 2015 (Fig. 4). The elastic wave velocity measurements were done on March 1, 2014 and Sep. 1, 2015.

The total number of accelerometers was 10 and one of them was installed in the sidewall at a distance of 1.5-2.5 m from the tunnel face during August 31-September 1, 2015 monitoring (Fig. 4(a)). The remaining 9 accelerometers were fixed to the plates of the rockbolts or steel ribs.

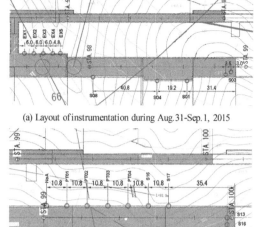

(a) Layout of instrumentation during Aug. 31-Sep. 1, 2015

(b) Layout of instrumentation during Oct. 5-8, 2015

Figure 4. Layouts of instrumentation and position of the tunnel faces of evacuation and main tunnels (EX1-EX5; NoA, PT01-PT04, S16-S17: Accelerometers; THP1-THP3: Temperature, Humidity and Air Pressure Sensors; CO2: CO2 sensor).

16 accelerometers were used and two of them were installed in the roof and sidewall at a distance of 2 m from the tunnel face (Fig. 4(b), Fig. 5). The remaining 14 accelerometers were fixed to the plates of the rockbolts or steel ribs (Fig. 6). One of the accelerometers could measure accelerations up to 16G.

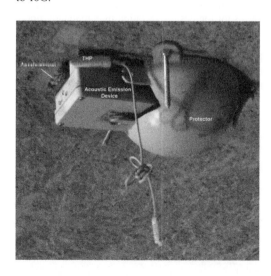

Figure 5. A view of accelerometer, AE device and THP device during October 5-7, 2015 monitoring in the roof at the face.

Figure 6. Fixation of accelerometers on a rockbolt platen.

The accelerometer is named QV3-OAM-SYC and it has a storage capacity of 2 GB and it is a stand-alone type. The trigger threshold and the period of each record can be set to any level and chosen time as desired. The accelerometers can be synchronized and they can be set to the triggering mode with a capability of recording pre-trigger waves for a period of 1.2s. It can operate for two-days using its own battery and the power source can be solar-light if appropriate equipments are used. Fig. 4 shows the location of accelerometers in the August 31-September 1 and October 5-8, 2015 monitoring.

Wave velocity measurements was first attempted in March 1, 2014 using three accelerometers attached the plates of rockbolts at the passage tunnel between the main tunnel and the evacuation tunnel for wave velocity measurements during the blasting operation at the main tunnel. As the data was so huge to handle, the main concept of the data recording was changed and the operation and data recording scheme was modified to record the acceleration waves for the desired duration by a manually operated a 'start-stop' switch. This modification was successful and the data handling became much easier. This new system was first applied during measurements September 1-3, 2015. Fig. 7 shows one of the accelerometer attached to the rockbolt platen.

3.2 Acoustic Emission System

Acoustic emissions (AE) were measured using a compact acoustic emission system, which consists of AE sensor, amplifier and pulse-counting logger (Tano et al. 2005). The system operates using batteries. The protection of instrumentation devices during blasting operation is extremely difficult issue. In previous studies, the devices were installed in larger holes at the sidewall and covered by some protection sheaths (Aydan et al. 2005). In this study, the instruments were put in an aluminum box attached to rockbolt head and protected by a semi-circular steel cover fixed to the tunnel surface by bolts. Figure 4 illustrates the installation of the instruments near the close vicinity of the tunnel face. The instruments installed at the crown were about 1.5 m from tunnel face while the instruments installed at the mid-height of the sidewall were about 2.5 m from the tunnel face during September 1-3, 2015 and 2m during the October 5-7, 2015 monitoring. AE counts were recorded for 1 second interval and the recording was started and stopped by a remote-operated switch device.

3.3 Environmental Monitoring System

Three locations, which were about 1.5-2.5, 31 and 76 m from the tunnel face, were chosen for temperature, humidity and air-pressure variations (Figs. 4,5 and 9). The air-pressure measurements can be also

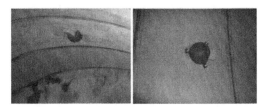

Figure 8. Views of AE device during Aug. 31 -September 1, 2015 monitoring in the roof and sidewall at the tunnel face.

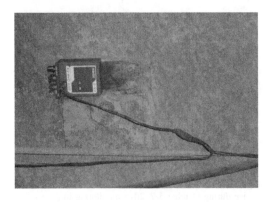

Figure 7. Fixation of an accelerometer for wave-velocity measurement.

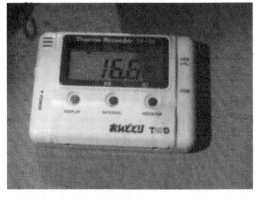

Figure 9. Temperature, humidity and air pressure device nearby the blasting switch hut.

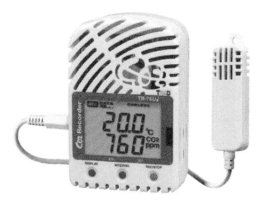

Figure 10. Views of CO2 measurement device.

used for investigating the blasting pressure wave propagation in the tunnel besides identifying the exact blasting times.

Blasting operations also cause Carbon dioxide (CO2) emissions. A CO_2 monitoring device produced by T&D was used for this purpose and set just in front of the mobile ventilation equipment. The CO_2 measurement was done during October 5-7, 2015 monitoring (Fig. 10).

4 MONITORING RESULTS

Because of some malfunctioning of some devices at one of the monitoring stations, we present the results of the station where all devices worked properly. It should be noted the devices were set at locations as close as possible to the tunnel face.

4.1 Monitoring during August 31-September 1, 2015

Monitoring during the blasting events at 20:36, 22:20 August 31 and at 9:20, 9:27 on September 1, 2015 was carried out. This was the first attempt to monitor acoustic emissions and acceleration in the immediate vicinity of the tunnel face (Fig. 11). Total amounts of explosives were 174.5 and 155.8 kg with 10 rounds having a 0.25 s delay, respectively at 22:20 (Aug. 21) and 9:20 (Sep.1) in the main tunnel. The total duration of each blasting was 2.5s.

Fig. 12 shows the acoustic emissions counts obtained from the acoustic emission devices with a sampling rate of 1s in this particular monitoring. The blasting with about 155 kgf explosives was carried out at 9:20 in the main tunnel. As noted from Fig. 12(a) AE counts terminated within 5 seconds. The ventilation system was started about 60-70 seconds after blasting in order to eliminate noises from the operation of construction and ventilation equipments. The blasting at evacuation tunnel (pillar width is about 20m) with a similar amount of explosives was also carried about 7 minutes after

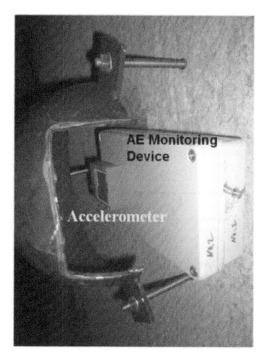

Figure 11. A close-up view of the instrumentation at the immediate vicinity of the tunnel face.

that in the main tunnel. The response is quite similar to that as seen in Fig. 12(b). Nevertheless, some further acoustic emission events occurred in the main tunnel. When the tunnel is stable, the acoustic emission counts decrease soon after blasting. This is in accordance with previous observations by Aydan et al. (2005), Imazu et al. (2014) as well as those presented in the previous section.

The acceleration records of 4 accelerometers (S03: 1.5m; S01: 30m; S04: 40m; S08: 50m from the face of the main tunnel) installed on the sidewall adjacent to the evacuation tunnel in the main tunnel during the blasting at 20:36 on Aug. 31, 2015 in the evacuation tunnel are shown in Fig. 13. As noted from the figure, the accelerations are higher as they become closer to the blasting location. The closest accelerometer S03 could not unfortunately record the accelerations due to malfunctioning following

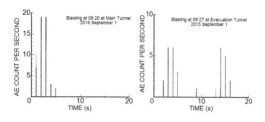

Figure 12. Acoustic emissions responses (a) blasting at 9:20 in the main tunnel; (b) blasting at 9:27 in the evacuation tunnel.

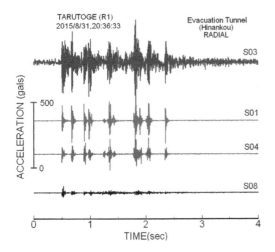

Figure 13. Acceleration records during the blasting at the evacuation tunnel at 20:36 on Aug. 31, 2015.

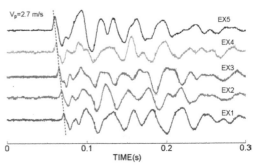

Figure 15. Acceleration records during the arrival of P-waves induced by blasting.

sive strength (UCS) of intact rock ranges between 40-70 MPa and its p-wave velocity is in the range of 3 and 4 km/s. This result is also consistent with previous measurements by Imazu et al. (2014).

4.2 Monitoring during October 5-7, 2015

The second attempt was done to measure both acoustic emissions (AE) and acceleration responses in the immediate vicinity of the tunnel face. Although two accelerometers were installed nearby the tunnel face, the acceleration measurements were not successful due to mal-functioning of the accelerometers due to extreme adverse conditions although accelerometers at distances greater than 32 m could successfully record the blasting induced accelerations. The results were quite similar to those given in the previous sub-section and earlier reports by Aydan et al. (2016) and Imazu et al. (2014, 2015). Therefore, the discussion the acceleration responses would be skipped in this section.

Although four acoustic emission sensors were installed in the roof and the sidewall in the immediate vicinity of the tunnel (i.e. Fig. 5), the acoustic emission sensor installed in the roof could only response properly due to adverse conditions at the tunnel face. The AE counts are shown in Fig. 16 together with air-pressure records also shown in Fig. 17. The spikes on the air-pressure records correspond directly to the blasting events in the main

the recording of the blasting at the evacuation tunnel. Fig. 14 shows the three acceleration records for radial component. The maximum acceleration was 800 gals at the accelerometer S01. As also noted previously (Aydan et al. 2016), acceleration records are not symmetric with respect to time axis.

The wave velocity of rock mass was measured during the blasting operations August 31 and September 1, 2015 by 5 accelerometers separated from each other by 6 m with a sampling rate of 0.05 ms. Fig. 15 shows the acceleration records induced by the blasting at 9:20 on Sep.1, 2015 in the main tunnel. It is possible to distinguish the p-wave on the acceleration records. P-wave arrival indicates that the P-wave of rock mass is about 2.7 km/s in the vicinity of the blasting point. The uniaxial compres-

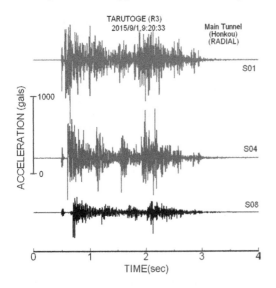

Figure 14. Acceleration records during the blasting at the main tunnel at 9:20 on Sep. 1, 2015.

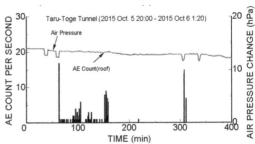

Figure 16. Acoustic emissions responses of the active AE sensor installed in the roof of the main tunnel.

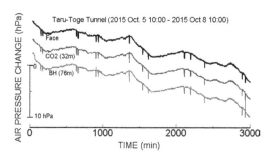

Figure 17. Air-pressure variations at three locations in the tunnel during October 5-7, 2015.

and evacuation tunnels. Blasting operations were carried out at 20:45 and 21:05 on Oct. 5 and 1:20 and 1:40 on Oct. 6 in the evacuation and main tunnels during the AE measurements shown in Fig. 16. The acoustic emissions are directly related to blasting events and disappear after a certain time interval. Nevertheless, the AE counts are somewhat prolonged following the blasting at 21:05 on Oct. 5.

The variations of air-pressure, temperature, humidity and CO_2 concentrations during October 5 and 8 are shown in Figs. 18, 19 and 20, respectively. The air pressures measured at three locations were shown by an offset of 3 hPa for better illustrations. The air pressure fluctuations are within 3 hPa for a 170 kg explosives. Fig. 21 shows the air pressure variation in the tunnel during a measurement on March 1, 2014.

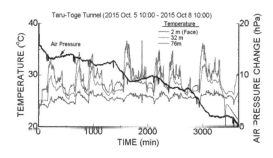

Figure 18. Temperature variations at three locations in the tunnel during October 5-7, 2015.

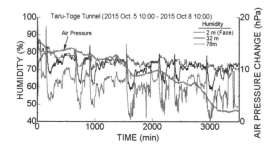

Figure 19. Humidity variations at three locations in the tunnel during October 5-7, 2015.

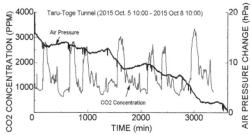

Figure 20. CO_2 concentration in the tunnel during October 5-7, 2015.

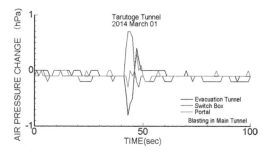

Figure 21. Air-pressure variation at several locations in the tunnel during a blasting on March 1, 2014.

The temperature and CO_2 concentration increase and humidity decreases after each blasting event as the ventilation system is activated. The increase in temperature and decrease in humidity may be partially related to the stoppage of the ventilation system before blasting. The temperature increase may be up to 10 degrees while the change of the humidity is within the range of 30%. The increase of CO_2 concentration may be up to 3300 and it decreases to about 700 PPM due to ventilation.

4.3 Electric potential and Acoustic Emissions Monitoring Results during Sep. 20-26, 2014

Two locations were selected for multi-parameter responses. Station 1 and Station 2 are approximately 45 and 25 m away from the face, respectively. The system consists of AE and EP monitoring devices. The blasting times at the main tunnel (honkou) and evacuation tunnel (hinankou) are shown in the figure and they are closely associated with rapid air pressure changes. Fig. 22 shows the variations of acoustic emissions and electric potential variations at Station 1 (St1). The acoustic emission counts were higher after each blasting operation although the counts decreased after a certain period of time. The electric potential variations showed very high-spike-like responses during blasting and disappeared after a certain period time, implying the tunnel response is stable. Nevertheless, the high variations

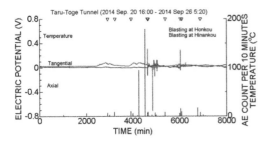

Figure 22. Electrical potential and acoustic emission variations at Station 1 (St1) in the main tunnel

may be associated with sudden increase of pore-pressure and seepage conditions in the vicinity of the tunnel face.

5 CONCLUSIONS

In this study, the authors applied this portable acceleration and multi-parameter systems are used to monitor acoustic emissions, electrical potential variations, the blasting-induced vibrations and wave-velocity characteristics of surrounding rock masses in a Taru-Toge tunnel in Central Japan. The portable acceleration monitoring system can be effectively used to obtain the vibrations caused by blasting.

The acoustic emission counts were higher after each blasting operation although the counts decreased after a certain short period of time. The acoustic emission measurements with a sampling interval of 1s at the immediate vicinity of the tunnel face during and after each blasting during Aug. 31 – Sep. 1, 2015 and Oct. 5-7, 2015 were quite similar to previous measurements in the Taru-toge tunnel as well as in the Shizuoka 3rd tunnel. Therefore, the monitoring as close as possible to tunnel face should be quite useful for engineers for the real-time stability assessment of underground excavations.

The multi-parameter monitoring indicated that the temperature, humidity, CO_2 concentration and electric potentials fluctuate after each blasting event. The activation of the ventilation system results in the recovery of the suitable working conditions.

REFERENCES

Aydan Ö., Daido M., Tano H., Tokashiki N. & Ohkubo K., 2005. A real-time multi-parameter monitoring system for assessing the stability of tunnels during excavation. *ITA Conference*, Istanbul, 1253–1259.

Aydan Ö., Tano H., Imazu M., Ideura H. & Soya M., 2016. The dynamic response of the Taru-Toge tunnel during blasting. *ITA WTC 2016 Congress and 42st General Assembly*, San Francisco, USA.

Imazu M., Ideura H. & Aydan Ö., 2014. A Monitoring System for Blasting-induced Vibrations in Tunneling and Its Possible Uses for The Assessment of Rock Mass Properties and In-situ Stress Inferences. *Proc. of the 8th Asian Rock Mechanics Symposium*, Sapporo, 881–890.

Imazu M., Soya M., Ideura H. & Aydan Ö., 2015. A monitoring system for blasting-induced vibrations in tunneling and its possible uses for the assessment of rock mass properties and in-situ stress inferences. *ITA WTC 2015 Congress and 41st General Assembly*, Dubrovnik, Croatia.

Tano, H., Abe, T., and Aydan, Ö., 2005. The development of an in-situ AE monitoring system and its application to rock engineering with particular emphasis on tunnelling. *ITA Conference*, Istanbul, 1245–1252.

2019 Rock Dynamics Summit– Aydan et al. (eds)
© *2019 Taylor & Francis Group, London, ISBN 978-0-367-34783-3*

An integrated study on the response of unsupported underground cavity to the nearby construction of piles of Gushikawa By-Pass Bridge

T. Tomori
Shibaiwa Consulting Co., Urasoe, Okinawa, Japan

K. Yogi
Okinawa Prefectural Government, Naha, Okinawa, Japan

Ö. Aydan & N.Tokashiki
Department of Civil Engineering, University of the Ryukyus, Nishihara, Okinawa, Japan

ABSTRACT: There was an unsupported underground tomb very close to the foundations of the bridge along Gushikawa By-Pass roadway. The effect of constructing bridge piles on the nearby underground tomb was a concern to the owner of the tomb. As a result, the authors were asked to check the effect of the construction on the tomb utilizing various non-destructive testing techniques and multi-parameter monitoring scheme. The authors describe an integrated study on the response of underground tomb undertaken before and during and after the construction of bridge piles. The authors describe the outcomes of laboratory test, in-situ explorations, rock mass evaluation, numerical analyses and multi-parameter monitoring undertaken in this study and discuss their implications. By virtue of this integrated study, it was possible to construct the bridge foundations without any harming effects on the underground tomb.

1 INTRODUCTION

The use of unsupported cavities as tombs was quite common in Okinawa Prefecture, Japan. Some of them are still used tombs and there was a case that an unsupported underground cave was very close to the foundations of the bridge along Gushikawa By-Pass roadway. The owner of the cavity was extremely worried for the stability of the cavity during the construction of the bridge piles having end-bearing on the phyllite formation below the Ryukyu limestone formation. The authors were asked to check the effect of the piles on the response and stability of the underground cavity, which was excavated about 400 years ago, during the construction process. Various laboratory on rock samples, in-situ explorations, seismic wave propagation studies and multi-parameter monitoring system to measure AE and displacement together with environmental monitoring and strong motion measurements were utilized to investigate the effect of pile construction on the nearby unsupported underground cavity. The new Rock Mass Quality Rating (RMQR) was utilized to characterize the surrounding rock mass. Furthermore, 2D elasto-plastic finite element simulation were also performed to check the stress changes and deformation response in relation to the construction stages. The authors will explain the outcomes of this unique integrated study on the response of unsupported undergound cavity to the nearby construction of piles of Gushikawa By-Pass Bridge. It will be also shown that various laboratory

Figure 1. An aerial view of the site after the construction.

on rock samples, in-situ explorations, seismic wave propagation studies and multi-parameter monitoring system and strong motion measurements were quite effective to evaluate the effect of pile construction on the nearby unsupported underground cavity.

2 GEOLOGY AROUND TOMB

2.1 *Preliminary Geological Invesigations*

The geologic explorations involve surface geology and boring. Except the highly weathered soil-like layer near the ground surface, the rock mass consists of Ryukyu formation, which consists of sandy, sandy-gravely and gravelly limestone and it is underlain by Kunigami formation, which consists of phyllite, mainly. The Ryukyu formation is about 42 m thick. The Kunigami phyllite was chosen as the load-bearing rock unit for pile design of the bridge.

2.2 New Borehole Drillings

As the preliminary geotechnical investigations were mostly based on soil-mechanics types of investigations and evaluations, they were not appropriate. For this reason, several new boreholes were drilled. The rock mass was classified as Ryukyu limestone (Rlst) Some of them were drilled for the purpose of installation of steel-bars as waveguides for Acoustic Emission (AE) sensors. The diameter of the new boreholes was 76 mm and the core-recovery was quite high (Fig. 2). These new borehole drilling yielded cores with a diameter of 70 mm. Furthermore, they resulted in a different perspective of rock mass conditions. While the earlier investigations implied that ground conditions as gravely soil, the new boreholes implied that the ground conditions correspond to rock mass. This new finding implied that the diameter of boreholes should be larger, say, 76 mm or more. In addition, it is quite common to utilize N-values (number of blow values to per 30 cm penetration) from Standard Penetration Tests (SPT) in Okinawa Prefecture to evaluate ground conditions even though the ground corresponds to rock mass conditions.

Fig. 3 and Fig. 4 show the core recovery of for No1 horizontal borehole and its internal view, respectively. Despite the very poor core recovery for the first 1m from the borehole head, the horizontal borehole is in stable condition. This fact clearly indicates the disturbance caused by borehole drilling on surrounding rock mass is tremendous In other words, borehole cores may give very wrong interpretation of ground conditions.

(a) View of cores of No. 2 new Borehole

(b) View of cores of No. 3 new Borehole

Figure 2. Views of cores of new borehole drillings.

Figure 3. Views of cores of No.1 horizontal borehole.

Figure 4. Internal view of the new Borehole No.1.

3 ROCK MECHANICS TESTS AND EXPLORATIONS

3.1 Rock Mechanics Tests

The cores obtained from the new boreholes were used to determine the physical and mechanical properties of rock units. For this purpose, unit weight, elastic wave velocity, tensile strength and uniaxial compression tests were carried out. The physico-mechanical properties of rock units are given in Table 1 and Mohr-coulomb yield criterion for each rock unit are shown in Figs. 5-7. These

Table 1. Physico-mechanical properties of rock units

Rock unit	UW kN/m³	Vp km/s	EM GPa	BT MPa	UCS MPa
Sandy	19.4	2.6	0.43	0.4	2.4
Sandy-Gravely	19.5	3.0	0.96	0.9	4.3
Gravely	23.3	3.4	1.9	1.9	7.6
Coral	20-24	4.5-6.3	8-27	2.5-4.1	20-34

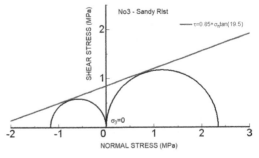

Figure 5. Sand Ryukyu Limestone (Rlst).

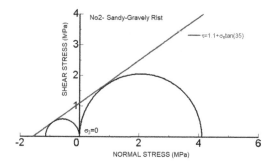

Figure 6. Sandy-gravely Ryukyu Limestone (Rlst).

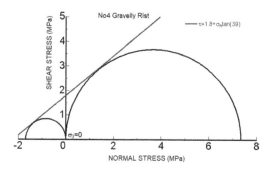

Figure 7. Gravely Ryukyu Limestone (Rlst).

evaluations clearly indicated first time that the rock units should be regarded as rock mass units rather than soil-like ground in Okinawa Prefecture.

3.2 *In-situ Explorations*

The assessment of ground conditions are of great importance to assess the response of ground during construction and in-service of piles including the 400 old years underground tomb. The authors utilized some dynamic vibration measurements to assess the ground conditions (Aydan et al. 2005). The initial vibration measurements were carried out using five QV3-OAM-EX/W portable accelerometers developed by Aydan et al. (2016), which was utilized for different purposes, vibration measurements system consisting of five TOKYO SOKKI AR-10TF accelerometer, Yokogawa WE7000 measurements station and lap-top computer, and TOKYO SOKUSHIN SPC-51 microtremor device.

As the source of vibration, a 1 tonf sand bag was dropped from a height of 1m above the ground (Fig. 8(a)). As the bag was torn when it hit the ground, it was decided to use the backhoe bucket as the vibration source (Fig. 8(b)). Although it is difficult to adjust the vibration level as desired, it proved to be quite useful as the vibration source. Fig. 9 shows the layout of accelerometer sensors A1-A5, EX/W1-EX/W5 and SPC51.

The main purposes of vibrations were to measure the wave-velocity of the ground (Fig. 10) and the transmission and attenuation of the Acoustic

(a) 1 tonf sand bag   (b) Backhoe bucket strike

Figure 8. Views of vibration sources.

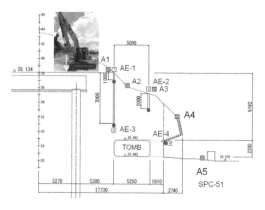

Figure 9. Layout of sensors for measuring vibrations.

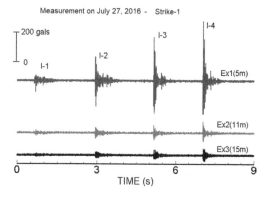

Figure 10. Acceleration records induced by the backhoe bucket striking the ground.

Emission (AE) signals with distance (Fig. 11). Particularly the attenuation of AE signals between Sensor AE-1 (top), Sensor AE-3 (above the tomb) and Sensor AE4 (outer side wall of the tomb) was of great importance. The results indicated that the amplitude of the vibrations drastically reduced as a function of distance as shown in Fig. 12. Fig. 13 shows the attenuation of AE count numbers at AE-1, AE-3 and AE4. The result indicated that AE signals were also drastically reduced with distance.

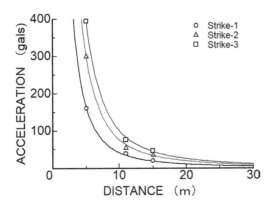

Figure 11. Attenuation of maximum acceleration with distance.

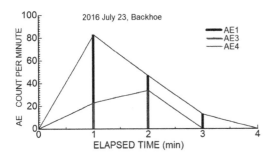

Figure 12. Attenuation of AE counts.

3.3 Characterization and Properties of Rock Mass

The new rock mass classification system called Rock Mass Quality Rating (RMQR) was applied to assess the rock mass conditions at the construction site. Four different rock mass conditions were recognized and the RMQR value of the each rock mass condition are given in Table 2. In the same table, p-wave velocities of in-situ rock mass are also given.

4 NUMERICAL ANALYSES

4.1 Analyses by Analytical Method

The original solution developed by Aydan (1989) for bolts and anchors under pull-out condition was recently modified by Aydan (2018) to analyze the distribution deformation and axial stress of the pile and the shear stress along the pile and surrounding ground. Fig. 13 illustrates the modeled geology, pile

Table 2. Characteristics of rock mass units

Rock mass unit	RMQR	Vp (km/s)
Heavily weathered Rlst	10-25	0.35-0.47
Sandy Rlst	34-43	0.68-0.93
Sandy-Gravely-Rlst	52-60	1.28-1.74
Gravely Rlst	70-74	1.87-2.52

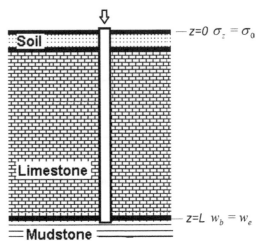

Figure 13. Modeled ground, pile and boundary conditions

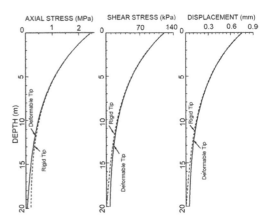

Figure 14. Distribution of axial and shear stresses and displacement with depth.

and boundary conditions. As noted from Fig. 14, the applied load is not fully transferred to the tip of the pile at depth and it is only a small fraction of the total load. These result imply that the current concept of the design of pile foundation in Okinawa Prefecture requires substantial revision.

4.2 Analyses by Finite Element Method

Numerical methods such as the finite element method would be generally desirable to simulate various load, ground and boundary conditions. In this particular case, the existence of the tomb and the effect of construction steps on the responses of the tomb and the surrounding ground to the varying loading conditions require a sophisticated numerical simulations. The authors utilized GTS-NX finite element software of MIDAS to analyse this structure. The computation involves the following steps

STEP 1: Applying gravity loading and excavation of the tomb (Fig. 15),

STEP 2: Excavation of the top layer,
STEP 3: Construction of the pile and
STEP 4: Applying the design load of 280 tonf to the top of the pile.

Fig. 16 shows the finite element models at each step and Table 3 gives material properties used. Figs. 17-19 show the safety factor distributions at each step while Fig. 20 shows the maximum shear distribution at the final stage. The results indicated that the surrounding rock mass should behave elastically without any yielding.

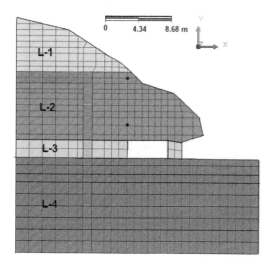

Figure 15. Ground and boundary conditions

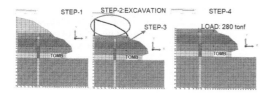

Figure 16. Finite element models for each step

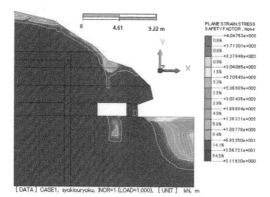

Figure 17. Safety factor distribution for STEP-1

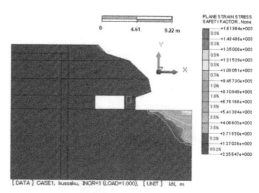

Figure 18. Safety factor distribution for STEP-2 & STEP-3

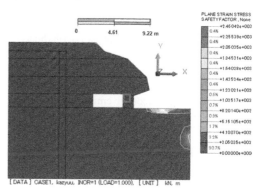

Figure 19. Safety factor distribution for STEP-4

Table 3. Mechanical properties of rock mass units

Rock unit	UW kN/m³	EM MPa	υ	Cm MPa	φ (°)
Layer-1(L-1)	19.4	137	0.3	0.03	19.5
Layer-2 (L-2)	19.5	345	0.3	0.39	29.4
Layer-3 (L-3)	23.0	468	0.3	0.64	32.7
Layer-3 (L-4)	23.6	724	0.3	1.8	39.0
Pile	20.0	20000	0.2	-	-

5 CONSTRUCTION AND MONITORING

5.1 Construction

The construction was carried out at several stages. First ground was excavated down to the design level and then piles were drilled, caissons were sank and filled with concrete together with reinforced cage (Fig. 20). Finally the super-structure was built and the construction was complete as seen in Figure 20 and Figure 1.

5.2 Monitoring

A monitoring program was undertaken to check that there was no negative effect of the construc-

(a) Drilling and embedment of caisson

(b) Casting concrete piles (c) Construction completed

Figure 20. Stages of construction of piles and the bridge.

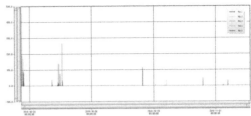

Figure 22. AE counts observed during Oct. 21 - Dec.02, 2016.

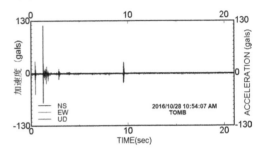

Figure 23. An acceleration record at the tomb

tion of piles on the nearby underground tomb. The monitoring involved a multi-parameter monitoring system consisting of acoustic emission(AE) sensors installed at 4 locations, displacement measurements between layers L2-L3, inclination at the top of excavation and at the tomb entrance and vibration due to machinery or earthquakes were monitored as illustrated in Fig. 21. Some of results are presented in this section.

5.2.1 Acoustic Emission (AE) Monitoring
Acoustic emission measurements were used for real-time monitoring of the safety of the tomb during construction (Tano et al. 2005). Four acoustic emission sensors used for the monitoring the rock mass responses and 1 AE sensor was used dummy sensor for emissions resulting from rain and other major sources. Fig. 22 shows an example of acoustic emissions recorded at 5 AE sensors. The accelerometer at the ground surface was also used to check the vibrations caused by the machinery operations. In addition, major construction operations and their timing were also recorded. Most of the acoustic emissions were due to the construction

activities during the day time. However, some special attentions were made to the acoustic emissions after 18 PM and before 8 AM during a typical working day. The results indicated that there was no acoustic emissions and they were due to the construction induced vibrations during the working hours.

5.2.2 Displacement and Inclination Measurements
The measurements of displacement and inclinations indicated that there was almost no effect of construction. The inclination variations were between 0.1 and 0.2 degrees. Similarly displacement variations was less than 0.1 mm.

5.2.3 Vibration and Strong Motion Monitoring
During the monitoring period, there was no major earthquake to trigger the accelerometers. However, some accelerations were recorded after commencement (after 9 AM) and before the termination (before 6 PM) of the construction work. One of the accelerometer was at the top of the tomb while the second accelerometer was installed at the entrance of the tomb. Fig. 23 shows an example of an acceleration record taken at the tomb.

6 CONCLUSIONS

An integrated study was undertaken during the construction of the foundations of the bridge along Gushikawa By-Pass roadway, nearby which there was an unsupported underground tomb. The authors checked the effect of the construction on the tomb

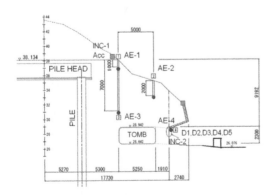

Figure 21. An illustration of instruments

by utilizing various non-destructive testing techniques and multi-parameter monitoring scheme. Furthermore, a series of analyses on the stress distribution were carried out using analytical and numerical methods. By virtue of this integrated study, it was possible to construct the bridge foundations without any harming effect on the underground tomb, which was initially the major concern.

REFERENCES

Aydan, Ö. (1989). The stabilisation of rock engineering structures by rockbolts. *Doctorate Thesis, Nagoya University*, 204 pages.

Aydan, Ö. (2018): Rock Reinforcement and Rock Support, CRC Press, ISRM Book Series, No.7, 473 pages.

Aydan Ö., Daido M., Tano H., Tokashiki N. & Ohkubo, K. 2005. A real-time multi-parameter monitoring system for assessing the stability of tunnels during excavation. *ITA Conference*, Istanbul, 1253-1259.

Aydan Ö., Tano H., Imazu M., Ideura H. & Soya M. 2016. The dynamic response of the Taru-Toge tunnel during blasting. *ITA WTC 2016 Congress and 42st General Assembly*, San Francisco, USA.

GTS-NX (2016). Geotechnical Sofware, MIDAS, Tokyo.

Tano, H., Abe, T., and Aydan, Ö. (2005). The development of an in-situ AE monitoring system and its application to rock engineering with particular emphasis on tunnelling. *ITA Conference*, Istanbul, 1245-1252.

2019 Rock Dynamics Summit– Aydan et al. (eds)
© 2019 Taylor & Francis Group, London, ISBN 978-0-367-34783-3

A study on the dynamic and multi-parameter responses of Yanbaru Underground Powerhouse

Ö. Aydan, N. Tokashiki, J. Tomiyama & T. Morita
Department of Civil Engineering, University of the Ryukyus, Okinawa, Japan

M. Kashiwayanagi, T. Tobase & Y. Nishimoto
Electric Power Company Co., Ltd, Chigasaki, Kanagawa, Japan

ABSTRACT: The Yanbaru Pumped Storage Power Station is the unique hydro-electric power station utilizing seawater, located in Okinawa Island, Japan. The construction of underground house was completed in 1999 and it is located at a depth of about 132 m below the ground surface. This underground powerhouse provides a unique chance to monitor its dynamic and multi-parameter responses during earthquakes and long-term period. The authors installed three accelerometers in the underground powerhouse and one accelerometer at the ground surface to observe its acceleration response during earthquakes. In addition, some displacement gap sensors, AE sensors at the adjacent connection tunnel, rock temperature and environmental parameters such as temperature, humidity and CO2 sensors in the powerhouse and ground surface were installed to monitor its multi-parameter responses. In this study, the authors describe this instrumentation, some numerical analyses and the results obtained so far.

1 INTRODUCTION

The Yanbaru Seawater Pumped Storage Power Station is the first experimental hydro-electric power station utilizing seawater, located in Okinawa Island, Japan (Fig. 1). The construction was completed in 1999. The underground powerhouse is located at a depth of about 132 m below the ground surface (Fig. 2).

This underground powerhouse provides a unique chance to monitor its dynamic and multi-parameter responses during earthquakes and long-term period. The authors initiated a collaborative research since July 2017. Four accelerometers were installed in the underground powerhouse to observed its acceleration response during earthquakes. Specifically, one accelerometer at the bottom level, two accelerometers at the mid-level and one accelerometer at the ground surface at an elevation of 132 m were installed. Furthermore, micro-tremor measurements were carried out in the powerhouse, access tunnel and the ground surface.

Some displacement gap sensors, AE sensors at the adjacent connection tunnel and environmental parameters such as temperature, humidity and CO2 sensors in the powerhouse and ground surface were installed to monitor its multi-parameter responses. Recently, some temperature sensors installed to observe the temperature of rock mass in relation to the temperature variation in the powerhouse.

Besides this monitoring program, 2D and 3D finite element dynamic and Eigen-value analyses have been performed to evaluate the dynamic response of the powerhouse as well as its close vicinity and vibration amplification in the powerhouse as well as at the ground surface. Furthermore, some rockanchors and rockbolts installed in surrounding ground in the powerhouse and the adjacent tunnels were selected and their responses to shock waves were measured with the purpose of evaluating their

Figure 1. An aerial view of the Yanbaru pumped-storage power station.

Figure 2. A cross section of the Yanbaru pumped-storage power station.

soundness since the beginning of the construction. The authors would explain the details of this unique instrumentation and the outcomes obtained from the monitoring program and numerical analyses and discuss their implementations.

2 DESCRIPTION OF POWERHOUSE

2.1 Geological characteristics

Powerhouse is located within the Kunikami zone of Nago Metamorphic rocks and it is regarded as the southern extension of Shimanto zone (Sato et al. 1994). Rocks are subjected to pressure metamorphism rocks and the overall schistosity plane dips at angle of 40-50 degrees to SW. However, the rocks contain numerous micro-folds (Fig. 3) and thrust-type faults of different scale. Fig. 4 shows cores of a borehole in the powerhouse. The boring was done about 30 years ago and the disintegration of cores along the schistosity (foliation) planes are observed. In the cores, some fracture zones are also noticed perpendicular to the schistosity plane.

2.2 Rockmass characteristics

The observation of the outcrops, cores of a borehole indicated that the rock mass has more than two discontinuity sets. The rock mass according DENKEN classification system is generally classified as CH with CM type zones, occasionally. The rock mass was classified using the new rock classification system called "Rock Mass Quality Rating - RMQR). Table 1 gives some rating of outcrops.

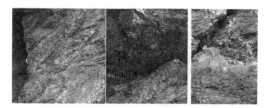

Figure 3. Views of rock mass conditions outside of the powerhouse.

Figure 4. View of cores of a borehole drilled about 30 years ago.

2.3 Geometry and characteristics of powerhouse

The crown of the powerhouse at 0 m elevation and it is 31.8m high, 16.4 m wide and 40.4 m long (Fig. 5). There is an access shaft next to the powerhouse for materials transport and accessible through an elevator and a stair from the ground surface. The cavern was supported through the use of rockbolts, rock anchors and 300 mm thick shotcrete with a wire mesh.

3 INSTRUMENTATION AND INSTALLATION

3.1 Strong motion instrumentation

The first author has developed a portable accelerometer, which can be used under 4 different modes. For strong motion observations during earthquakes, every accelerometer should have a triggering level to start and stop recording for a given time interval, sampling rate and store in a digital format. The minimum sampling rate is 1Hz. The device is named as QV3-OAM-XXX and it has the ability with a storage capacity of 2 GB. The power of the accelerometer can be an internal battery, external

Table 1. RMQR rating of outcrops.

Parameter	Description	Natural	Description	Disturbed
DD	Fresh	15	slight-weathering	11
DSN	1-3	12-16	DSN (3sets+)	4-8
DS	0.3-1.2m	8-12	DS	1-4
DC	rough	15-22	Rough	15-22
GWSC	Wet	5	drip	3
GWAC	Highly-absorptive	5-6	H. Absorptive	4-5
RMQR		60-76	RMQR	38-53

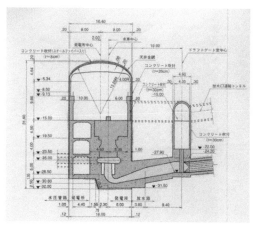

Figure 5. A cross-section of the powerhouse.

battery, solar energy or ordinary 100-240 V electricity. In case solar energy or ordinary electricity, the power is stored into an external battery through an adapter from a power supply such as solar panels or electricity. The system adopted at the powerhouse is designed to utilize the ordinary electricity (100V) or two external batteries. Currently, the system utilize electricity available in the powerhouse.

We installed four accelerometers. Three accelerometers installed in the powerhouse as illustrated in Fig. 6. One accelerometer is at the Underground 4F, which is at the bottom of the powerhouse, two accelerometers at Underground 1F, which is the mid-level of the powerhouse. One of the accelerometer is fixed to the sidewall at the penstock side and the other accelerometer is fixed to the middle of the end-wall of the cavern. The fourth accelerometer is installed at the surface. Fig. 7 shows some views of the installed accelerometers. The main purposes of the installation is to observe the seismic response of the cavern during earthquakes and to evaluate the ground motion amplifications as pointed out by previous pioneering researchers (Nasu 1931; Komada and Hayashi, 1980).

3.2 *Multi-parameter Instrumentation*

The monitoring of underground caverns is essential for the long-term performance and stability of the underground powerhouses. Aydan et al. (2005) have initiated multi-parameter monitoring system to observe the behaviour of structures during excavations as well as during their service life. The multi-parameter system fundamentally covers all measurable quantities such as displacement, acoustic emissions, electric potential or electrical resistivity, water level changes, climatic parameters such as temperature, humidity and CO2, temperature changes of rock. In this study, a thoroughgoing open crack at the access tunnel next the powerhouse cavern was selected to monitor its movement (displacement), acoustic emissions (AE) together with climatic changes (temperature and humidity) (Fig. 8). The power supply is fundamentally battery operated.

The locations of climatic parameters such as temperature, humidity and air pressure (THP) are measured at the ground surface (132 m elevation). Two CO2 sensors are installed at the ground surface and Underground 1F. The monitoring of CO2 is to check the air quality as well as the condition for the carbonation environment for concrete. Recently, two temperature sensors are installed in a short borehole to monitor rock temperature and air temperature around its vicinity. This measurement is expected to yield some information of cyclic thermally induced deformations of the powerhouse.

3.3 *Micro-tremor measurement*

The vibration characteristics in the powerhouse and ground surface are essential to assess the seismic response and stability of underground as well as surface structures. Micro-tremor measurements using

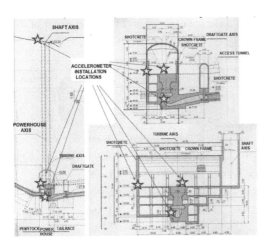

Figure 6. Installation locations of accelerometers.

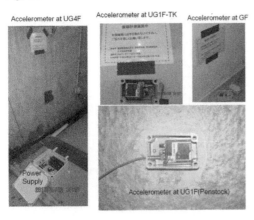

Figure 7. Views of installed accelerometers.

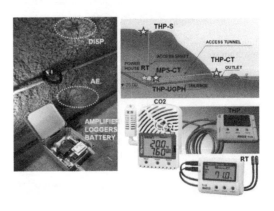

Figure 8. Views and locations of installed multi-parameter monitoring system.

SPC51A micro-tremor device produced by TOKYO-SOKUSHIN was utilized and vibration characteristics in the powerhouse (UG1F, UG4F, Access Tunnel and ground surface were evaluated (Fig. 9). At each location, at least, 3 records for 100 second periods were taken.

3.4 *Impact tests on rockanchors and rockbolts*

The non-destructive testing of support members such as rockanchors, rockbolts (Aydan 2017, 2018) are essential for the maintenance of underground and Surface rock engineering structures from time to time. With this in mind, rockanchors and rockbolts installed in the penstock side of the powerhouse and the access tunnel were selected for non-destructive testing. Three rockbolts in the access tunnel and two rockanchors were tested using a non-destructive testing device produced by IMV. The device is quite compact and a single person can apply the impact force and record the acceleration/velocity wave and its Fourier spectra. The sensor

can be easily attached to the rockbolt/rockanchor head using a magnet. We essentially utilize one of three sensors with the consideration of the testing object. Particularly, the STD-S sensor produced by IMV is quite suitable for non-destructive testing of bar-type rockbolts and rock anchors.

4 NUMERICAL ANALYSES

The use of numerical analyses is essential for evaluating and understanding behaviour of underground structures for their dynamic and long-term performance. In this study, the fundamental modes of underground powerhouse with the consideration of the surface topography were analyzed using 3D MIDAS-FEA software. In addition, 2D response analyses of the powerhouse were carried out using FEM (Kashiwayanagi 2018). Material properties used in this study are given in Table 2. One of the main purposes of the 3D analyses was to see the effect of mountain configuration (Fig. 11). For that purpose, Eigen value analyses were carried out for the domain with and without underground opening and shaft (Figs. 12 & 13). Table 3 gives the natural frequencies obtained from the mode analysis of

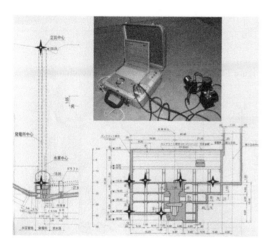

Figure 9 A view of micro-tremor measurement in the powerhouse.

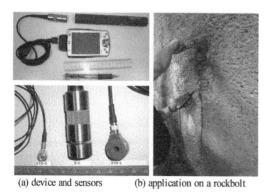

(a) device and sensors   (b) application on a rockbolt

Figure 10. Non-destructive test device and its utilization on a rockbolt in the access tunnel.

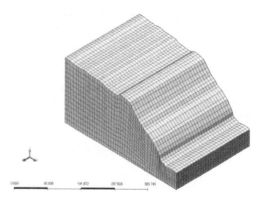

Figure 11. 3D Finite element Mesh

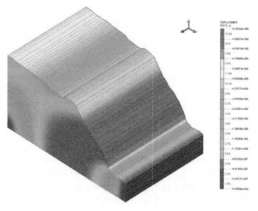

Figure 12. Mode 1 displacement response (no cavern).

Figure 13. Mode 1 displacement response (with cavern).

Table 2. Material Properties

Material	UW(kN/m³)	E(GPa)	Poisson Ratio
Rock mass	27.0	4.00	0.25
Concrete	23.5	11.042	0.20

Table 3. Eigen values for each mode

	Mode 1(s)	Mode 2(s)	Mode 3(s)
No Cavern	0.898	0.651	0.646
With Cavern	0.898	0.653	0.646

the numerical model. There is almost no difference for the dominant period due to the existence of the cavern.

5 MEASUREMENTS AND DISCUSSIONS

5.1 *Strong motion measurement*

The measurements have been continuing since 2017 July 26. All devices were set to the trigger value of 19 gals. Except some man-made vibrations, no accelerations recorded due to earthquakes. We recently reduced the trigger value of accelerometers to 4 gals, which is the lowest level of triggering for the accelerometers. The three devices recorded one event at 13:39 on 2018 March 23. Fig. 14 shows the acceleration records. Although the event was due to man-made activity, it was interesting to note that three devices were triggered.

5.2 *Micro-tremor Measurements*

Micro-tremor measurements were carried out at several locations as shown in Fig. 8. Fig. 15 compares the H/V spectra of micro-tremor measurements at different locations. The dominant frequency for the surface is about 0.976Hz, which implies that the natural period of the ground surface is about 1.025s.

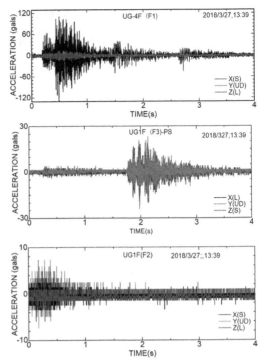

Figure 14. Acceleration records of three accelerometers.

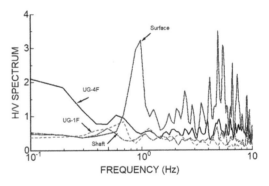

Figure 15. Comparison of Micro-tremor measurements

Although this value is slightly different from the eigen value analyses, it may be said that the computational results may estimate the vibration characteristics of the Yanbaru Sewater Pumped Storage Power Station.

5.3 *Multi-parameter Measurements*

In this subsection, the multi-parameter measurements, specifically, the responses of crack displacement, acoustic emissions and climatic response of the cavern, access tunnel and ground surface are presented. responses of crack displacement, AE, Temperature, Air Pressure, CO2 concentration, and rock temperature are shown in Figs 16 to 21.

425

As noted from Fig. 16. residual crack displacement occurs after one year, implying that some time-dependent deformation taking place. The AE response also confirms the time-dependency of the deformation behaviour of the crack. Figs. 18 & 19 show that temperature and air pressure fluctuates and their amplitude depends upon the location. Similarly, CO_2 concentration and temperature response of rock mass (20cm from sidewall) and

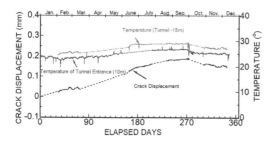

Figure 16. Displacement and temperature responses of the crack

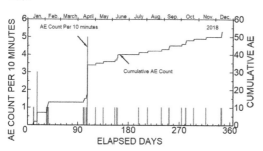

Figure 17. Acoustic emission response of the crack

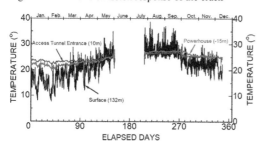

Figure 18. Temperature variations at various locations

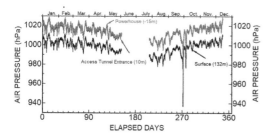

Figure 19. Air pressure variations at various locations

inner temperature of the powerhouse varies with time. Particularly the rock temperature implies some thermal stress variations as reported by Aydan et al. (2013).

5.4 Non-destructive tests on rockanchors and rockbolts

The non-destructive tests on rockanchors and rockbolts are essential to evaluate the performance of support members and their maintenance. Three rockbolts in the access tunnel and two anchors at the penstock side of the powerhouse were selected for non-destructive testing and some non-destructive tests were carried out. Fig. 22 shows one example of acceleration response together with computed response using the procedure proposed by Aydan (2017, 2018). These non-destructive tests and computations have been continued.

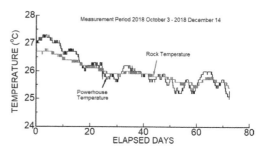

Figure 20. Temperature response of rock mass to powerhouse temperature.

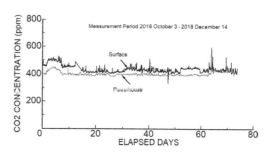

Figure 21. CO_2 variations in the powerhouse and ground surface.

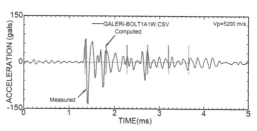

Figure 22. Temperature variations at various locations

426

6 CONCLUSIONS

The authors explained the strong motion instrumentation and multi-parameter monitoring system at the Yanbaru underground powerhouse. In this study, the authors described this instrumentation, some numerical analyses and the results of monitoring obtained so far. Altough the recorded and monitored results are not that so long, it still provides some insight views on the vibration and multi-parameter responses of the underground powerhouse. It is expected that the continuation of this observation and monitoring system would be quite valuable for the dynamic response of large underground structures and their long-term performances.

REFERENCES

Aydan, Ö. (2017): *Rock Dynamics*, CRC Press, ISRM Book Series, No.3, 464 pages.

Aydan, Ö. (2018): *Rock Reinforcement and Rock Support*, CRC Press, ISRM Book Series, No.7, 473 pages.

Aydan, Ö., Daido, M., Tano, H., Tokashiki, N., Ohkubo, K. (2005). A real-time multi-parameter monitoring system for assessing the stability of tunnels during excavation. *ITA Conference*, Istanbul, 1253–1259.

Aydan, Ö., Uehara, F. and Kawamoto, T. (2012): Numerical Study of the Long-Term Performance of an Underground Powerhouse Subjected to Varying Initial Stress States, Cyclic Water Heads, and Temperature Variations. *Int. Journal of Geomechanics*, ASCE, Vol. 12, No.1, 14–26.

Kashiwayanagi, M. 2018. A report on micro-tremor monitoring of underground powerhouse of Okinawa Seawater Pumped Storage Station. *JPower Chigasaki Research Institute*, 37 pages (in Japanese, unpublished).

Komada H., Hayashi M. 1980. Earthquake observation around the site of underground power station. CRIEPI Report, E379003, *Central Research Institute of Electric Power Industry*, Japan, 1–34.

Nasu, N. (1931) Comparative studies of earthquake motions above ground and in a tunnel. *Bull Earthquake Res. Inst*, Tokyo University, 9, 454–472.

Sato, S., Fukuhara, A., Koyama, K. 1994. The construction of underground powerhouse as a part of Okinawa Seawater Pumped Storage Pilot Plant (in Japanese). *J. of Electric Power Civil Engineering*, 253, 69–76.

2019 Rock Dynamics Summit– Aydan et al. (eds)
© 2019 Taylor & Francis Group, London, ISBN 978-0-367-34783-3

An integrated system for the cavity-filling of an abandoned lignite mine beneath Kyowa Secondary School in Mitake, Japan against an anticipated mega-earthquake

K.Sugiura & A.Sakamoto
Tobishima Co., Nagoya Branch, Nagoya, Japan

H. Tano
Nihon University, Koriyama, Japan

Ö. Aydan
Department of Civil Engineering, University of the Ryukyus, Okinawa, Japan

ABSTRACT: There is a great concern in Japan how to deal with abandoned room and pillar mines, quarries as well as karstic caves following the 2011 Great East Japan Earthquake with a moment magnitude 9.0, which caused the collapses abandoned lignite and coalmines and underground stone quarries and associated damage to super structures at 316 localities in Tohoku Region of Japan. The authors have been involved the backfilling of an abandoned lignite mine beneath Kyowa Secondary School in Mitake Town, Gifu Prefecture, Japan against an anticipated mega earthquake. In this study, the authors present an integrated study on the backfilling of abandoned mine beneath the secondary school These studies involve the geological and geotechnical conditions, static and dynamic numerical analyses, monitoring before and after backfilling operations and some attempts of the verification of the effect of backfilling of the abandoned mine. The authors explain this unique integrated study and outcomes and findings from this study.

1 INTRODUCTION

The extraction of lignite using the room and pillar mining technique was extensive in various parts of Japan until 1960s. These mines are abandoned and some of these areas have become urbanized since then. Tokai region in Central Japan is a well-known example for such a case. The Great East Japan Earthquake with a moment magnitude 9.0 caused the collapses abandoned lignite and coalmines and underground stone quarries and associated damage to super structures at 316 localities in Tohoku Region of Japan. There is now a great concern in Japan how to deal with potential damage resulting from the collapse of abandoned room and pillar mines, quarries as well as karstic caves in relation to the anticipated Nankai-Tonankai-Tokai mega earthquake. The anticipated magnitude would be similar to that of the Great East Japan earthquake, which occurred along the Tohoku region of Japan in 2011.

The authors have been involved the backfilling of an abandoned lignite mine beneath Kyowa Secondary School in Mitake Town, Gifu Prefecture, Japan against an anticipated mega earthquake. The integrated study involves the surveying of mine layout beneath the secondary school, the geological and geotechnical conditions, preliminary 1D, 2D and 3D static and dynamic numerical analyses, monitoring before and after backfilling operations

and some attempts of the verification of the effect of backfilling of the abandoned mine. The authors presents the features of this unique integrated study and outcomes and findings from this study. Furthermore, the implications of this study on similar projects are discussed.

2 INTEGRATED BACK-FILLING TECHNIQUE

The cavity filling or backfilling technique involves several fundamental steps such as the exploration of dimensions of cavities to be filled, establishing the filling plant facilities, filling and constructional and environmental monitoring and surveying. All these steps must be performed with due considerations of environmental restrictions and regulations. Fig. 1 shows the schematic flow-chart of the integrated cavity filling technique. The details of the filling technique are given in the following sub-sections.

2.1 *Exploration and Surveys*

Explorations and surveys involve the determination of layouts and spatial distributions of cavities to be filled and underground water conditions. Although some mining plans submitted to the authorities at the time of mining operations are available, these documents do not generally coincide with actual mining layouts. Furthermore, the geometry may change with partial degradation and collapses of pillars and roof layers in abandoned mines as time goes by. In order to perform underground filling works effectively, the following exploration and surveys are conducted.

1. Site investigations for cavities to be filled, plant-yards, offices, etc.,
2. Investigations for pre-existing cracks in exterior walls, ground floors and elevations of nearby buildings and of the presence of wells,
3. Examination of groundwater and waters of rivers and wells,
4. Investigation of buried structures,
5. Collection and analysis of existent
   - data on topography, strata, and geology,
   - Treatises and documents issued by public offices,
   - Legwork.

The authors utilize several exploration techniques such as borings, gravity exploration, high-density electrical resistivity technique, sonar exploration technique depending upon the geological and environmental conditions. The most efficient method is the boring technique together with sonar exploration technique through drillholes. Figs. 2 and 3 illustrate the sonar exploration technique utilizing a drillhole and its typical output. This system can be used under submerged underground environments for determining the geometry of cavities along the project routes.

2.2 Design of Grouting and Monitoring Drillholes

The next step is to design the layout of drillholes for grouting and piping. Grouting drillholes (Fig. 4) should be drilled at points about 0.5m away from exploration drillholes, where the existence of caverns is confirmed. Although the spacing between

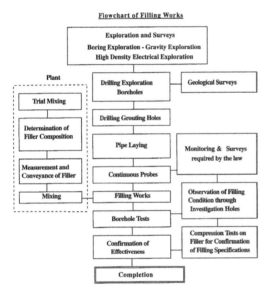

Figure 1. A schematic illustration of the integrated cavity filling technique.

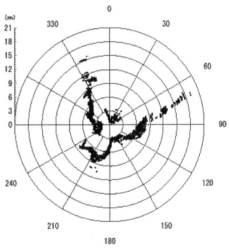

Figure 3. An application of the Sonar exploration technique to an abandoned lignite mine

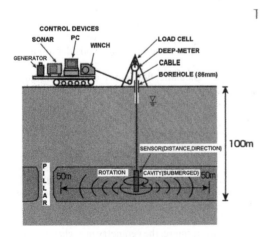

Figure 2. Sonar exploration technique

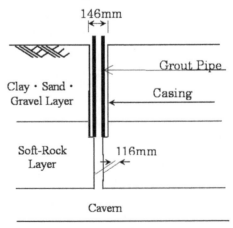

Figure 4. An illustration of dimensions of drillholes.

429

Figure 5. A monitoring probe for grouting.

drillholes depends upon the characteristics of filling material, an area with a radius of about 20m can be filled through a single grouting drillhole with use of filling materials used by the authors. Therefore the spacing distances between the grouting boreholes can be chosen as approximately 40m. However, if there are much deposits in caverns or the area is densely urbanized, the spacing of grouting boreholes should be less than 40m in order to control grouting pressure.

The water level in other grouting drillholes should be continuously monitored through probes after infusing water into some grouting boreholes. Through the monitoring of water level, it is possible to confirm if the drillholes are linked with each other. These probes also serve to know whether cavities are connected to each other or independent, and it is possible to determine the filling order. The ground settlement or uplift is also measured continuously through probes as shown in Fig. 5.

2.3 Filling Plant

A typical filling plant basically consists of suppliers of scums, agitators, mixers and slurry pumps as illustrated in Fig. 6 and Fig. 7. The capacity of the plant depends upon the size of the site, the period of filling and the usable space. For example, a plant with a maximum mixing capacity of 300m³ per day should have an area of about 2000m², through the consideration of the filling materials, which are transported into the plant by 10 to 20 dump

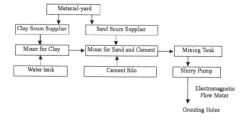

Figure 6. Flowchart of the filling plant

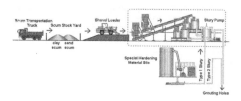

Figure 7. An illustration of the filling plant and filling operation

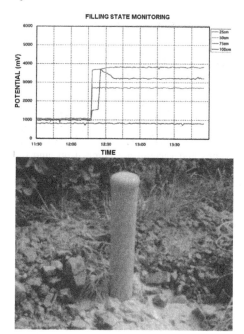

Figure 8. Monitoring result for cavity filling state and the inference from the overflow of wells.

trucks with 11 tonnages per day, and the materials, which should be stocked for a few days.

Mixers are used to mix the various scums before pumping filling material in cavities. Mixers can dissolve even viscous materials like clay scum in a short time and mix them easily. The maximum mixing capacity ranges between 20 and 550m³ per day depending upon the plant size. However, the usual mixing capacity ranges between 10 and 300m³ per day. The amount of filling per day also depends on cavern conditions.

Slurry pumps are used to pump the mixed slurry of filling material and they have the capability of conveying slurry with a density ranging between 1.3 and 1.5g/cm³ even though grouting drillholes may be about 1000m far away from the location of the pumps.

2.4 Safety measures and Monitoring

The environmental monitoring involves the ground deformations due to filling operations, the solvents

Table 1. Criteria of Quality Control

A List of Tests	Criteria of Quality Control
Flow Value	9 - 14 seconds
Bleeding Rate	3% or less than
Uniaxial Compressive Strength after 28 Days	Uniaxial strength cured samples in water should be 50 kPa or more
	Uniaxial strength from the boring cores should be 20 kPa or more
Analysis of Toxic Materials	Less than the statutory criteria

in underground water migration as well as groundwater levels. Protuberances caused by overfilling are prevented by monitoring the pore water pressure, the tape measurement of the slurry level in grouting boreholes, and monitored changes through tilt-meters and crack-meters on the ground.

The spurts of water from grouting boreholes after neutralizing are drained. The spurts of slurry with sand pumps and vacuum are pumped up, and the wastes are discarded after required processing. pH value of ground water, solvents are monitored to assess the effects of grouting on the environmental conditions of ground water. To ensure the quality of grouting and filling state, several tests and measurements are carried as listed in Table 1.

3 APPLICATION KYOWA SECONDARY SCHOOL

3.1 *Geology and Material Properties*

2D geological cross section of the Kyowa Secondary School is shown in Fig. 9. The base layer is chert of Paleozoic period and it is overlain by intercalated mudstone, sandstone and lignite belonging to Mizunami formation of Tertiary period. Top soil is underlain by intercalated mudstone and sandstone of Quaternary period. The average width and height

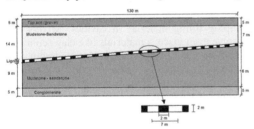

Figure 9. A geological cross section of the site.

Table 2. Material properties of layers

Layer	γ kN/m³	E MPa	v	c MPa	$\varphi(°)$
Topsoil	19	270	0.35	0.0	38
Upper Mst-Sst	19	750	0.3	0.7	25
Lignite	14	400	0.3	0.66	45
Lower Mst-Sst	19	1073	0.3	1.00	45
Chert	19	3647	0.3	3.00	45

of pillars are 2m with an excavation ratio of 70%, respectively. Material properties of layers are given in Table 2. The material properties determined using laboratory experiments, borehole pressure-meter tests and in-situ geophysical exploration methods.

3.2 *2D Finite Element Analyses*

First a series of 2D finite element analyses were carried out. Fig. 10 shows the safety factor contours for short-term properties of rock mass for static case. As expected, the safety factors are lower in pillars and the minimum safety factor was observed in deepest

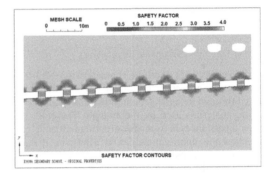

Figure 10. Safety factor contours for short-term properties and static condition.

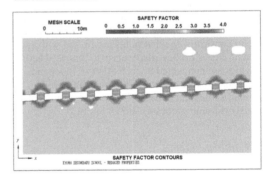

Figure 11. Safety factor contours for long-term properties and static condition.

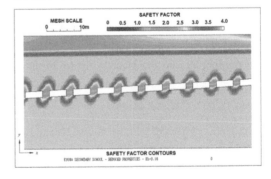

Figure 12. Safety factor contours for long-term properties for pseudo-dynamic condition (Lateral seismic coefficient is 0.16).

pillar. Similarly, if the long-term properties are used, the safety factor values become much smaller (Fig. 11). Nevertheless, minimum safety factors are observed at the pillars and the safety factor of the deepest pillar is the lowest. If pseudo-dynamic condition with a lateral seismic coefficient of 0.16 is imposed, the safety factors becomes much smaller (Fig. 12). Similar to the previous cases, the deepest pillar has the lowest safety factor implying that the yielding would occur at the deepest part and propagate towards to pillars in shallower part.

3.3 1D Amplification Analyses

Series of numerical analyses using the transfer matrix method were carried for the fundamental vibrations of the ground for a 40m deep ground profiles as shown in Fig. 13. The computational results implies that the natural period of the ground is more than 0.2 seconds and less than 0.4 s. However, if the depth of mined level becomes shallower, shorter natural periods also become apparent.

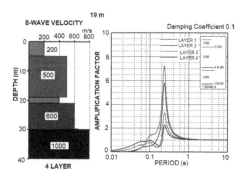

Figure 13. Amplification analyses of ground beneath the Kyowa Secondary School using the matrix transfer method.

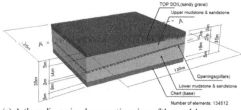

(a) A three-dimensional perspective view of the model

(b) A three-dimensional perspective view of the room and pillar abandoned mine

Figure 14. A three-dimensional views of the numerical model and abandoned room and pillar mine.

3.4 3D Elasto-plastic Response Analyses

A three-dimensional elasto-plastic numerical analyses of abandoned lignite mine beneath the Kyowa Secondary school (Fig. 14) using the estimated ground motion record (Fig. 15), based on the methods developed by Sugito et al. (2000) and Aydan (2012) for the anticipated Nankai-Tonankai-Tokai mega earthquake. The Nankai earthquake terminates at 43 seconds and Tonankai earthquake starts and terminates at 75 seconds. The last earthquake is Tonkai earthquakes and it terminates at about 125 seconds.

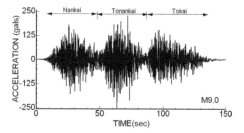

Figure 15. Input base acceleration record

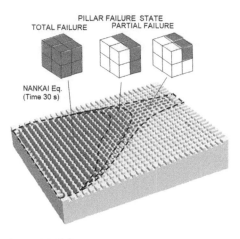

Figure 16. Failure state at time 30 seconds (Nankai Eq.)

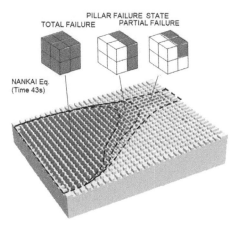

Figure 17. Failure state at time 43 seconds (Nankai Eq.)

The failure of the pillars starts at the deepest site and propagates towards to shallower parts as estimated from 2D numerical analyses (Figs. 16 to 19). The failure state of pillars can be broadly classified as total failure and partial failure as illustrated. When the Nankai earthquake terminates, about 60 percent of the pillars were already totally or partially yielded. The Tonankai earthquake is the nearest one to the Mitake town and the all pillars are in total failure state. The Tokai earthquake has no further effect on the failure state.

3.5 Backfilling and Monitoring

The computational results clearly indicated that the abandoned mine beneath the Kyowa Secondary School would totally collapse and have tremendous effect on the ground surface and superstructures. Therefore, the Ministry of Education, Science, Sports have decided to backfill the abandoned room and pillar lignite mine beneath the Kyowa Secondary School. The abandoned mine was backfilled using the integrated back-filling technique described in Section 2. In this particular project, the area limited to the school ground was backfilled and backfilling process was monitored through multi-parameter monitoring system involving the Acoustic emission technique developed by Tano et al. (2005). This project was

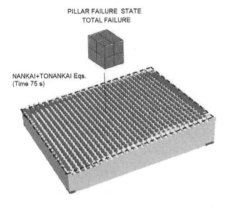

Figure 18. Failure state at time 75 seconds (Nankai & Tonankai Eqs.)

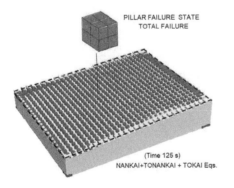

Figure 19. Failure state at time 125 seconds (Nankai & Tonankai & Tokai Eqs.)

also very unique for the first time to implement counter-measures before the disaster occurs.

4 CONCLUSIONS

The authors presented an integrated study on the backfilling of abandoned mine beneath the secondary school involving the geological and geotechnical conditions, static and dynamic numerical analyses, monitoring before and after backfilling operations and some attempts for the verification of the effect of backfilling of the abandoned mine. This project was successfully carried out and it was very unique for the first time to implement counter-measures before the disaster occurs.

ACKNOWLEDGEMENTS

We thank Emeritus Prof. Dr. Toshikazu KAWAMOTO of Nagoya University and Emeritus Prof. Dr. Masanori HAMADA of Waseda University for their invaluable suggestions and encouragement for the development and application of this integrated cavity filling technique utilizing also recycling materials. Three dimensional computations were carried out by Prof. Dr. M. Geniş of Bülent Ecevit University, Zonguldak, Turkey using FLAC3D. We sincerely acknowledge his help with the computations.

REFERENCES

Aydan, Ö. (2012): Ground motions and deformations associated with earthquake faulting and their effects on the safety of engineering structures. Encyclopedia of Sustainability Science and Technology, Springer, R. Meyers (Ed.), 3233–3253.

Sakamoto, A., Yamada, N., Sugiura, K., Kawamoto, T. (2005a). Some examples of the cavity filling along transportation routes above abandoned room and pillar lignite mines in Tokai Region. Post-Mining 2005, Nancy.

Sakamoto, A., Yamada, N., Iwaki, K., Kawamoto, T. (2005b). Applicability of re-cycling materials to cavity filling materials. Zairyo, Japan Society of Material Science (in print).

Itasca, 2005. FLAC3D-Fast Lagrangian Analysis of Continua-User Manuel (dynamic option) (Version 2.21). Minneapolis: Itasca Consulting Group Inc.

Sugiura, K. and Yamada, N. (2003). On the completion of the cavity filling works of Tokai By-Pass Expressway (in Japanese). Juten, Journal of Japan Grouting (Juten) Association. No.44, 8–17.

Sugiura, K., Ishiai, N., Wada, S. (2005). Cavity filling in urbanized area: Changes and state of construction environment in cavity filling works (in Japanese). Juten, Journal of Japan Grouting (Juten) Association. No.47, 8–13.

Sugito, M., Furumoto, Y. and Sugiyama, T.: Strong Motion Prediction on Rock Surface by Superposed Evolutionary Spectra, 12th World Conference on Earthquake Engineering, 2111/4/A, CD-ROM, 2001.

Tano, H., Abe, T., and Aydan, Ö. (2005). The development of an in-situ AE monitoring system and its application to rock engineering with particular emphasis on tunneling. ITA Conference, Istanbul, 1245-1252

2019 Rock Dynamics Summit– Aydan et al. (eds)
© 2019 Taylor & Francis Group, London, ISBN 978-0-367-34783-3

The effect of cave-filling of abandoned lignite mines in Tokai Region, Japan against an anticipated mega-earthquake

T.Ito & Ö. Aydan
Department of Civil Engineering, University of the Ryukyus, Okinawa, Japan

K.Sugiura & A. Sakamoto
Tobishima Corporation, Nagoya Branch, Nagoya, Japan

ABSTRACT: There is now a great concern how to deal with potential damage resulting from the collapse of abandoned room and pillar mines in relation to the anticipated Nankai-Tonankai-Tokai mega earthquake. The authors have initiated some analytical, numerical and monitoring studies on stability and performance abandoned cavities as well as their performance before and after back-filling. In this study, the authors present the outcomes of a series of experimental studies on the supporting effect of backfilling on the response and stability of abandoned mines. Furthermore, the authors discuss their implications on backfilling against anticipated mega earthquakes through a series of numerical analyses with the considerations of experiments.

1 INTRODUCTION

The Great East Japan Earthquake with a moment magnitude 9.0 caused gigantic tsunami waves, which destroyed many cities and towns along the shores of Tohoku and Kanto Regions of Japan. Besides the structural damage on ground surface, this earthquake caused the collapses abandoned lignite and coalmines and underground stone quarries and associated damage to super structures at 329 localities (Fig. 1). Similar events occurred in the previous 1978 Off-Miyagi earthquake, 2003 Miyagi-hokubu earthquake and 2008 Iwate-Miyagi intraplate earthquake. There is now a great concern in Japan how to deal with potential damage resulting from the collapse of abandoned room and pillar mines and quarries in relation to the anticipated Nankai-Tonankai-Tokai mega earthquake. The anticipated magnitude would be similar to that of the 2011 Great East Japan earthquake occurred along the Tohoku region of Japan (Fig. 2).

The authors have involved with the performance and responses of abandoned mines in Tokai Region of Japan during earthquakes and have initiated a continuous measurement and monitoring program for investigating the stability and performance of several abandoned room and pillar lignite mine in Mitake town since April 2004. The authors have also been carrying out some monitoring studies on the response of abandoned cavities before and after back-filling. In this study, the authors present the outcomes of a series of experimental studies on the supporting effect of backfilling on the response and stability of abandoned mines and quarries and report the results of a series of numerical analyses.

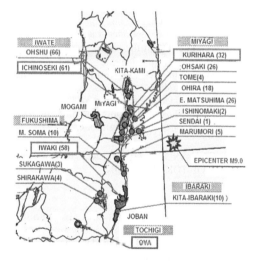

Figure 1. Locations of damage to abandoned mines and quarries after the 2011 Great East Japan Earthquake (from Aydan and Tano, 2012).

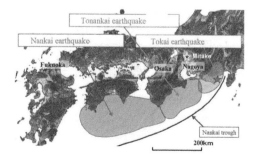

Figure 2. Anticipated M9 Nankani-Tonankai-Tokai earthquake and location of Mitake town.

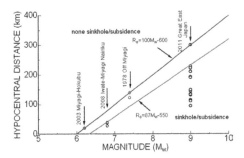

Figure 3. Comparison of case history data with an empirical relations for assessing the possibility of sinkhole/subsidence occurrence as a function of earthquake magnitude (from Aydan and Tano, 2012)

Aydan and Tano (2012) evaluated the seismic vulnerability of abandoned lignite mines with the consideration of the recent earthquakes and they plotted the case history data described in the previous section together with those from other earthquakes occurred in Tohoku region in the space of earthquake moment magnitude versus hypocentral distance of the locality where sinkhole or large subsidence occurred as shown in Fig. 3. The data was fitted to a linear function whose coefficients are shown in the same figure. The line in the figure should be interpreted as a limiting line between surface damage and non-damage on the ground surface. This figure may serve as a guideline to assess the risk of sinkhole or large subsidence due to abandoned mines and quarries exploited using the room and pillar method as a function of earthquake magnitude.

2 GEOLOGY AND CHARACTERISTICS OF ROCKS OF TOKAI REGION

2.1 Geology

The geological age of soft rocks associated with lignite deposits differ from location to location. Fig. 4 shows the distribution of lignite deposits in Tokai region. While the age of lignite field within Nakamura unit of Mizunami formation at Mitake belongs to the

Miocene era, the lignite field belonging to Yadagawa unit of Seto formation at Nagakute does to Pliocene. Therefore, rocks of Mitake are older than those at Nagakute. There are four lignite seams in Mitake and the second and the third seams were extracted. The thickness of the third seam is 1-3m thick. Although the number of lignite seams at Nagakute is four, the seams are much thinner (less than 1m). Sedimentary rocks at Mitake are broadly classified as lignite, sandstone, siltstone and mudstone with conglomerate layers at the base of the formation above the pre-Tertiary rocks. Although sedimentary rocks of Nagakute can be geologically classified in the same manner, the cementation is poor and sometimes non-cemented sand layers are found below lignite seams. Rock samples were obtained through boring at Mitake town.

2.2 Characteristics of Rocks

Rock samples were obtained from Tokai Region, mainly in Mitake town, Nagakute and Akaike City. Various short term experiments were carried out. Table 1 gives major properties of rocks in Mitake town.

The mechanical properties of soft sedimentary rocks are influenced by the water content. The compressive strength and elastic modulus of soft rocks generally decrease with the increase of water content. During some experiments, electric potential, electrical resistivity, magnetic force, acoustic emission (AE) were measured besides conventional load and displacement. The main purpose of such measurements was to establish some experimental bases for the real-time monitoring of multi-parameters for the stability assessment of an abandoned mine in Mitake (Aydan et al. 2005a,b).

The experiments were mainly carried out by using specially designed creep-loading devices. Fig. 5 shows the responses of sandstone sample measured in one of the creep tests. Fig. 6 shows the long-term strength of rocks of Mitake town together with other rocks.

Table 1. Physical and mechanical properties of soft rocks from Mitake site

Rock	UW (kN/m³)	UCS (MPa)	EM (GPa)	Vp (km/s)
Lignite	10.6-14.1	4.0-6.0	1.8-1.9	1.6-2.3
Sandstone	16.8-19.2	2.3-7.0	0.3-0.5	1.6-2.7
Mudstone	15.6-16.5	1.0-1.5	0.1-0.3	1.3-1.5

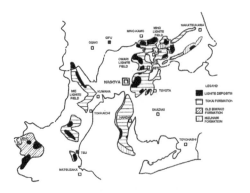

Figure 4. Lignite deposits in Tokai region and locations of sites.

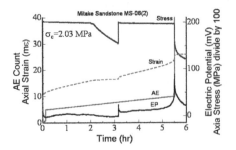

Figure 5. Creep test result on sample MS-06.

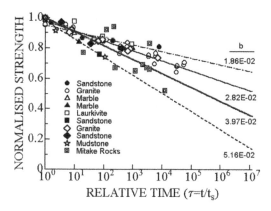

Figure 6. Comparison of normalized long term strength of Mitake soft rocks with other rocks.

Figure 7. Examples of disintegration of soft rocks of Mitake.

2.3 Degradation of Lignite and Surrounding Rocks

The surrounding rocks and lignite itself in abandoned lignite mines are sedimentary rocks and very prone to degradation due to cyclic variations of water content during absorption and desorption. Such variations cause flaking of surrounding rocks from roof and pillars of lignite seam and results in the reduction of the support area of pillars and thinning of roof layers. Soft rocks in Mitake exhibit such behavior as seen in Fig. 7. The same phenomenon observed in abandoned mines.

3 EXPERIMENTS ON BACKFILLING OF ABANDONED LIGNITE MINES

Aydan et al. (2013a) have carried out large-scale short-term experiments on lignite samples of abandoned mines of Nagakute and Mitake towns and of Oya tuff under both unfilled and back-filled state using granular backfilling and cohesive backfilling materials (Fig. 8). Cohesive backfilling material is a mixture of clayey and sandy residue from ceramic factories and cement. Experiments were carried out at the age of 28 days with the consideration of hydration process of cement. This backfill material is commonly used in the backfill of abandoned lignite mines in Tokai Region in Japan and it is named as NSK backfilling material. These experiments repeated recently and performed both under static and dynamic conditions. Figs. 9-11 show results of some experiments carried out under static condition. Figs. 9, 10 and 11 compares the average strain

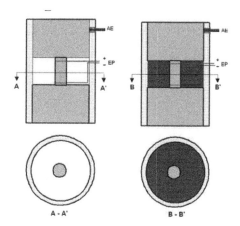

Figure 8. Set-up unfilled and backfilled samples.

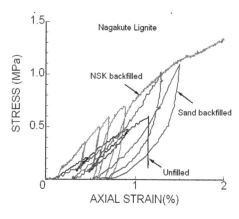

Figure 9. Comparison of strain-stress responses of unfilled pillar, backfilled pillars with granular and cohesive backfill materials during cyclic compression.

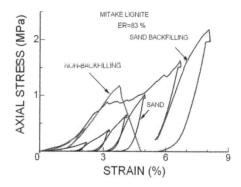

Figure 10. Comparison of strain-stress responses of unfilled pillar, backfilled pillars with granular backfill materials during cyclic compression.

stress relations for unfilled and pillars backfilled with granular and cohesive backfill materials for lignite samples. It is interesting to note that the bearing capacity of backfilled pillars is increased about 1.3-1.5 times compared with that of the unfilled sample at the same strain level. Furthermore, the behavior of backfilled pillars

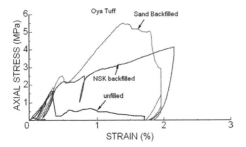

Figure 11. Comparison of strain-stress responses of unfilled pillar, backfilled pillars with granular and cohesive backfill materials during cyclic compression.

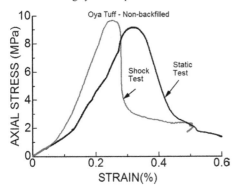

Figure 12. Effect of loading condition on strain-stress responses of the pillars for non-backfilled situation.

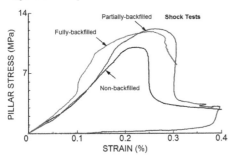

Figure 13. Strain-stress responses of the pillars for different backfilling condition subjected to shock loads

is elasto-plastic without any softening even after the failure of the pillars. Furthermore, the bearing capacity of the backfilled pillars was of great significance. The bearing capacity of the pillar backfilled with NSK backfill material is greater than that of the pillar backfilled with granular backfill material.

Fig. 11 compares the average strain stress relations for unfilled and Oya tuff pillars backfilled with granular and cohesive backfill materials for lignite samples. It is interesting to note that the bearing capacity of backfilled pillars is increased about 1.3-2.5 times compared with that of the unfilled sample at the same strain level. Furthermore, the behavior of backfilled pillars is elasto-plastic without any softening even after the failure of the pillars. The bearing capacity of the pillar backfilled with NSK backfill material is greater

than that of the pillar backfilled with granular backfill material.

4 SHOCK TESTS

Aydan et al. (2018) reported a new series of experiments to investigate the response of Oya tuff pillars under three different backfilling state under shock tests. The back-filling situation were the same as those of the second series experiments, which were:

i. Non-backfilled,
ii. Partially-backfilled up to 70% of the pillar height,
iii. Fully backfilled.

The load was imposed as shocks on samples. Fig. 12 compares the response of Oya tuff without backfilling under static and shock loads. As noted from the figure, the overall stiffness and the strength of the pillar sample subjected to shock loading are slightly higher than those of the sample tested under static loading. Furthermore, the residual strength of the sample subjected to shock loading is also slightly higher than that of the sample tested under static condition.

Fig. 13 shows the strain-stress responses of the Oya tuff pillar samples with different degree of backfilling subjected to shock loads. It is also noted that the overall stiffness increases as a function of degree of backfilling similar to those of samples tested under static case.

The acceleration response of samples at the time of failure for non-backfilled and fully-backfilled situations are shown in Fig. 14. As noted from the figures, the accelerations are higher for the non-backfilled sample as compared with that of the fully-backfilled sample. The maximum amplitude of the acceleration is almost four times that of the fully-backfilled sample. Furthermore, the accelerations are not symmetric with respect to time axis as noted previously by Aydan et al. (2011).

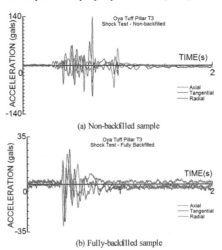

Figure 14. Acceleration responses of the non-backfilled and fully-backfilled samples.

5 ASSESSMENT OF EFFECT OF BACK-FILLING DURING MEGA EARTHQUAKES

Aydan (Aydan 2017; Aydan et al. (2006, 2011, 2013b) proposed a practical method for evaluating the stability of pillars and roof of abandoned room and pillar lignite mines using the seismic coefficient approach. Fig. 15(a,b) shows the assumed models for the dynamic stability of pillars and roofs. In this approach, the pillar was assumed to fail under the maximum compressive stress. The load condition under horizontal shaking is assumed to consist of gravitational load inducing the bending stresses and linearly varying axial stress along the roof axis from tension to compression due to horizontal shaking. On the basis of this assumption, the seismic coefficient at the time of roof layer failure by bending is obtained. Fig. 16 shows computed diagrams for the relation between overburden ratio and seismic coefficient for various failure modes for the chosen parameters shown in the same figure. These results indicate that the shallow mines are prone to roof failure while the deeper mines are prone to pillar failure for actual strong ground motions during earthquakes.

As shown in Sections 3 and 4, the backfilling drastically increases the resistance of the pillars against collapses. In other words, the resistance of the backfilled abandoned mines is the same as that under unexploited condition, provided that the abandoned mines are fully backfilled. If the abandoned mines are partially backfilled, the resistance of the backfilled abandoned mines against earthquakes would be also partial.

Another approach was also proposed by Aydan et al. (2012) to evaluate the response and stability of abandoned room and pillar mines with the use of dynamic tributary area concept. Fig. 17 shows the basic concept of this method. One can evaluate the safety of pillars subjected to ground shaking due to earthquakes.

Aydan et al. (2012) used this model to evaluate the response of the abandoned mines in Mitake town with the use of ground acceleration due to M9 class anticipated mega earthquake. Fig. 18 shows the acceleration responses of the abandoned mines at ground surface and lignite seam together with the input base acceleration. As noted from the figure,

the amplification of the ground acceleration at the ground surface is very high above the abandoned mines and it may be up to 5 times while the amplification at the lignite seam level is 3 times.

Fig. 19 shows the safety factor and shear stress variation at the lignite seam level (19m below the ground surface) for abandoned lignite mines in Mitake town subjected to anticipated base acceleration of the mega earthquake. As noted from the figure, when short term strength is considered the safety factor is about 1.5. On the other hand, the safety factor may be drastically reduced for long-

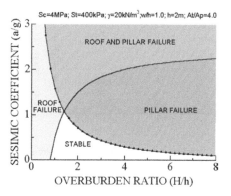

Figure 16. An example of computation for the stability of pillars and roof layers under dynamic conditions.

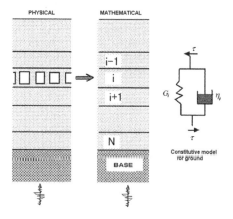

Figure 17. Mathematical model for dynamic response of pillar during earthquakes (from Aydan et al. 2013b).

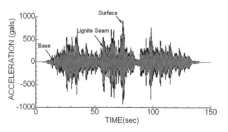

Figure 18. Computed acceleration responses at lignite seam level and ground surface for the anticipated M9 class mega earthquake.

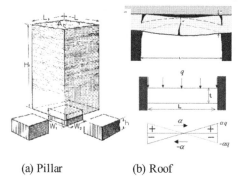

(a) Pillar   (b) Roof

Figure 15. Mathematical models for stability assessment of pillars and roof layers under dynamic conditions

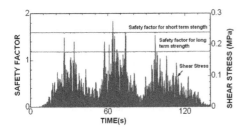

Figure 19. Variation of the safety factor and shear stress at the lignite seam level of the abandoned room and pillar during the mega earthquake shaking.

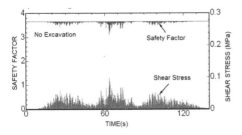

Figure 20. Variation of the safety factor and shear stress at the virgin lignite seam level during the mega earthquake shaking.

term strength of the lignite pillar. As discussed in Sections 3 and 4, the backfilling increases the pillar strength and the ground condition would be quite close to the unexploited state. For such case, the variation of the safety factor would be quite close that under virgin state as shown in Fig. 20. The safety factor of the lignite seam level under backfilled state is expected to be more 3.6.

6 CONCLUSIONS

There are many abandoned mines in various parts of Japan and the recent earthquakes showed these mines are quite vulnerable to the formation of sink-holes and subsidence. Therefore, there is a growing concern on the safety of areas above such abandoned mines in relation to the anticipated Nankai-Tonankai-Tokai earthquake. The authors have been involved with the research on the short and long-term stability of ground above these abandoned mines. In this study, the authors present the results of a series of experimental studies on the supporting effect of backfilling on the response and stability of abandoned mines under static and dynamic situations. Furthermore, the authors discussed their implications on backfilling against anticipated mega earthquakes in view of the experimental and numerical studies. It is shown that the backfilling has a great stabilizing effect on the abandoned room and pillar mines provided that the fully backfilled.

REFERENCES

Aydan, Ö., 2017. Rock Dynamics. CRC Press, Taylor and Francis Group, 462p, ISRM Book Series No. 3, ISBN 9781138032286.

Aydan, Ö., 2004. Damage to abandoned lignite mines induced by 2003 Miyagi-Hokubu earthquakes and some considerations on its possible causes. J. of School of Marine Science and Technology, Vol. 2, No.1, 1-17.

Aydan, Ö., 2007. Monitoring. In Mine Closure and Post-Mining Management – International State of Art, C. Didier (Ed.). International Commission on Mine-Closure, ISRM, 111–123.

Aydan Ö, Kawamoto T., 2004. The damage to abandoned lignite mines caused by the 2003 Miyagi-Hokubu earthquake and some considerations on its causes. *3rd Asian Rock Mechanics Symposium*, Kyoto, 525–530, 2004.

Aydan, Ö., Geniş, M., 2007. Assessment of dynamic stability of an abandoned room and pillar underground lignite mine, 23–44. Turkish Journal of Rock Mechanics Bulletin, Turkish National Rock Mechanics Group, ISRM, No. 16,

Aydan, Ö., Tano, H. 2012. The observations on abandoned mines and quarries by the Great East Japan Earthquake on March 11, 2011 and their implications. Journal of Japan Association on Earthquake Engineering, Vol.12, No.4, 229–248.

Aydan, Ö., Sakamoto, A., Yamada, N., Sugiura, K. and Kawamoto, T., 2005a. The characteristics of soft rocks and their effects on the long term stability of abandoned room and pillar lignite mines. Post Mining 2005, Nancy.

Aydan, Ö., Sakamoto, A., Yamada, N., Sugiura, K. and Kawamoto, T., 2005b. A real time monitoring system for the assessment of stability and performance of abandoned room and pillar lignite mines. Post Mining 2005, Nancy.

Aydan, Ö., Daido, M., Ito, T., Tano, H. and Kawamoto, T., 2006. Instability of abandoned lignite mines and the assessment of their stability in long term and during earthquakes. 4th Asian Rock Mechanics Symposium, Singapore, Paper No. A0355 (on CD)..

Aydan Ö, Ohta, Y., Genis, M., Tokashiki, N., Ohkubo, K., 2011. Response and Stability of Underground Structures in Rock Mass during Earthquakes. Rock Mech Rock Eng (2010) 43:857–875

Aydan Ö, Genis, M., Sakamoto, S. and Sugiura, K., 2012. Characteristics and amplification of ground motions above abandoned mines. International Symposium on Earthquake Engineering, JAEE, 75–84.

Aydan Ö, Tokashiki N, Tano H, 2013a. An experimental study on the supporting effect of back-filling on abandoned room and pillar mines, quarries and karstic caves, and its in-situ verification. ARMA 13-379. 47th US Rock Mechanics/Geomechanics Sym., San Francisco, 10pages.

Aydan Ö, K. Sugiura, A. Sakamoto 2013b. Dynamic characteristics of ground above abandoned mines. 13th *Japan Rock Mechanics Symposium*, 519–524 (in Japanese).

Ito, T., Ö. Aydan: Abandoned Lignite Mines in Mitake and Issues 2019. Post-Mining and Environmental Issues Themed Issue, Environmental Geotechnics, Institute of Civil Engineers (in press).

Aydan, Ö., Sugiura, K., Sakamoto, A. 2018. The quality of backfilling of abandoned lignite mines on ground settlement under static and dynamic conditions. AusRock2018, 134–141.

2019 Rock Dynamics Summit– Aydan et al. (eds)
© *2019 Taylor & Francis Group, London, ISBN 978-0-367-34783-3*

The numerical studies for fault displacement damage of Shih-Gang Dam in Chi-Chi Earthquake

Y. Nikaido & Y. Mihara
Kajima Corporation, Tokyo, Japan

H. Tsutsumi & K. Ebisawa
Central Research Institute of Electric Power Industry, Tokyo, Japan

R. Haraguchi
Mitsubishi Heavy Industries, Ltd, Kobe, Japan

ABSTRACT: In Japan, interests on the impact of fault displacement on nuclear facilities have increased and been recognized as one of the urgent issues for nuclear safety. However, the damage process of the structures under the event of fault displacement remains to be elucidated due to the lack of observed and test data. In order to confirm the feasibility of the numerical procedures to simulate the damage process of the structure above the fault displacement, the authors conducted the 3D-FE analysis of Shih-Gang Dam in Chi-Chi Earthquake (Taiwan, 1999). Throughout this study, we acquired a prospect of the feasibility of the procedures.

1 INTRODUCTION

Fault displacement have caused serious damage to civil infrastructures and buildings above it in several earthquakes such as Chi-Chi Earthquake (1999, Taiwan) and Izmit Earthquake (1999, Turkey). Even though the risk from such event is anticipated to be low enough compared to other disasters, it is still important to understand the mechanism of such events in consideration of the potential risk to an infrastructure.

Recently in Japan, interest on the impact of principal and secondary fault displacement on nuclear facilities has increased, and it is currently recognized as one of the urgent issues for nuclear safety. Under this context, the authors have conducted examination of fault displacement risk assessment methodology framework and identification of technical issues (Tsutsumi et al, 2018). As a part of the study, the authors conducted the verification analysis of Shih-Gang Dam that is the fault displacement damaged structure in Chi-Chi Earthquake (Taiwan, 1999) using the finite element method to confirm the feasibility of the numerical procedures to simulate the damage of civil infrastructures against fault displacement.

2 OVERVIEW OF THE DAMAGE SITUATION

2.1 Chi-Chi Earthquake

At 1:47 am (local time) on September 21, 1999, an earthquake with magnitude 7.7 occurred in 150 km southwest Of Taipei. As of Oct 6, 1999, the casualties, number of injured and missing persons were 2413 and 8700.(JMA,2018)

The number of collapsed houses was 9 thousand and several hundreds. Because the epicenter was located near Chi-Chi in Nan-Tou prefecture in Taiwan this earthquake was named the 1999 Chi-Chi earthquake. This earthquake reportedly caused by the movement of the Chelongpu fault (Figure 1). This is a thrust type fault with dip angle about 30 degrees. The west side of the Chelongpu fault was the foot wall and rose up 1-4 m to the west side. The damage in the region from Dong-Shyh to Chao-Tuen located on the hanging wall was severe. The length and width of this earthquake fault are 80 km and 40 km. The maximum dislocation measured at ground surface was 6.5 m. More than 700 strong motion observation stations are distributed and maintained by the Central Weather Bureau, Ministry of Transportation and Communications, Taiwan. About 70 percent of observation station was

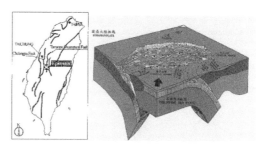

Figure 1. Plate Geometry of Taiwan and epicenter and Chelongpu faults in Chi-Chi earthquake

triggered. Based on recorded earthquake motions the detail study on the rupture process of the earthquake fault will be done because spatial distribution of predominant periods and durations of strong motions along the Chelongpu fault shows a possibility to exist rather big asperity compared to the 1995 Hyogoken Nambu Earthquake. The average distance of observation stations near the Chelongpu fault is several kilometers. Observed records at these stations will be used extensively to investigate mechanisms of damage to structures.(JSCE,1999)

2.2 Damage of Shih-Gang Dam

Shih-Gang Dam is one of the valuable samples for actual damaged structure due to the fault displacement. Fig.2 shows the topological relationship between the Chelongpu Fault and Shih-Gang Dam. The end of Chelongpu Fault nearby Shih-Gang Dam is branched into three lines stated A, B, and C fault.

The length and height of Shih-Gang Dam are about 290 meters and about 27 meters respectively (Fig.3). As shown in Fig.4, the vertical slip fault struck at the point about 20 meters from the right side of the dam and 8 meters differential occurred at the point. As a result, the embankment body was destroyed and reservoir function was lost

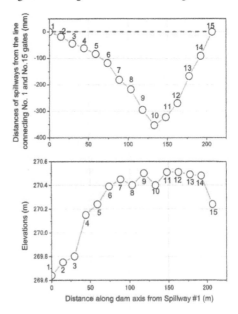

Figure 4. Damage overview of Shih-Gang Dam

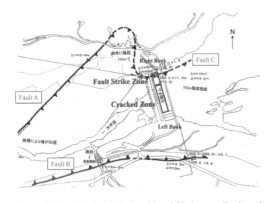

Figure 2. Topological Relationship of Chelongpu Fault and Shih-Gang Dam

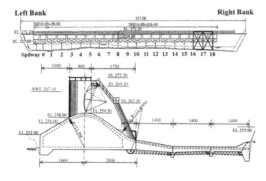

Figure 3. Structural Overview of Shih-Gang Dam

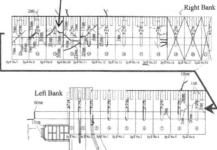

Figure 5. Observed movement of Shih-Gang Dam (Ohmachi, 2000)

(Konagai et al, 2000) and also, the crack occurred over the 270 meters of left embankment body due to the shortening of dam length of 6 meters in the axial direction and 2 meters in the downstream direction. Additionally, Over 3 meters of deformation occurred at the center of water sluices and 75 meters of peeling due to uplift was estimated from the restoration work after the disaster.

3 NUMERICAL SIMULATION FOR GLOBAL BEHAVIOR OF SHIH-GANG DAM

3.1 Numerical model

Firstly, we studied with the simple numerical simulation to capture the whole behavior of the dam body such as uplift and bending.

3.1.1 Finite element model

To simplify the problem, two types of finite element models are developed for both vertical and horizontal movements. Both models are composed of beam elements which represent twenty dam bodies, gap element that judges contact between dam body and soil, and soil springs. Figure 6 shows the diagram of the two models.

Model 1 is developed to study the relationship between soil-structure interaction and uplift. Forced vertical displacement is given by the blue side of fixed points.

On the other hand, Model 2 is developed to study the relationship between the peeled area and bending behavior in horizontal direction. Considering the eccentricity in the cross sectional area of dam body, axial forced displacement is applied to one point.

3.1.2 Material properties

Major material properties for the study are shown in Table 1.

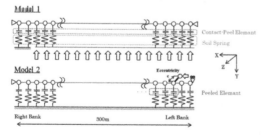

Figure 6. Model diagrams

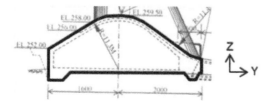

3.2 Study Cases

The combinations for the study cases are listed in Table 2 and Table 3.

Index A in Table 2 is defined as the equation below that is the well-known value as the indicator of deformation mode for horizontal load in pile design.

$$A = \sqrt[4]{EI_y / G_{Soil}} \qquad (1)$$

$$G_{soil} = \frac{\gamma}{g} V_s^2 \qquad (2)$$

Where, E is Young's Modulus and G_{Soil} is Shear Modulus of Soil.

Table 1. Material Properties

	Young's Modulus	1.25, 6.33, 20.0, 101, 320 (GPa)
	Poisson Ratio	0.2
Concrete	Section Area	305 m²
	Iy	15,130 m⁴
	Iz	110,600 m⁴
Soil	Shear Wave Vel.	50, 100, 200, 316, 1000 (m/s)
	Unit Weight	

Table 2. Study Cases for Model 1

Case No.	Index A	Young's Modulus of Concrete(GPa) 5 levels (1.25~320) 1	2	3	4	5	Shear Wave Vel. of Soil (m/s) 5 levels (50~1000) 1	2	3	4	5	Weight (MN/m)
1	62.1			✓			✓					10
2	34.9			✓						✓		10
3	19.6			✓							✓	10
4	31.0	✓					✓					10
5	46.5		✓				✓					10
6	93.0				✓		✓					10
7	124.1					✓	✓					10
8	43.9			✓						✓		10
9	87.8			✓				✓				10
10	62.1			✓			✓					5
11	34.9			✓						✓		5
12	19.6			✓							✓	5
13	31.0	✓					✓					5
14	46.5		✓				✓					5
15	93.0				✓		✓					5
16	124.1					✓	✓					5
17	43.9			✓						✓		5
18	87.8			✓				✓				5

Table 3. Study Cases for Model 2

Case No.	Eccentricity (e in Fig.6)	Peel Area(from left bank)			
		-	30~105m	75~150m	105~180m
1	0m	✓			
2	2m	✓			
3	2m		✓		
4	2m			✓	
5	2m				✓

For Model 2, the length of peeled area is fixed as 75m and the location was studied.

3.3 Study Results

Figure 7 shows the examples of study results of Model 1. The red colored elements indicate the peeled area. In addition to the closest point to the fault shown at the left side of Figure, the various contact peel behavior appears in the rest of area.

Figure 8 shows the relationship of Index A and ratio of peeled area in the study results of Model 1. As can be seen on the figure, the linear relationship and the reasonable range of the stiffness of dam body and supporting soil are assumed.

Figure 9 shows the examples of study results of Model 2 and Table 4 shows the summary of study. As can be seen on Figure and Table, Consideration of both peeling and eccentricity in axial compression enable to reproduce the observed behavior of dam body in horizontal direction.

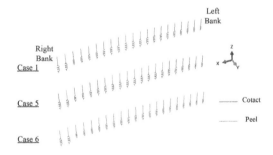

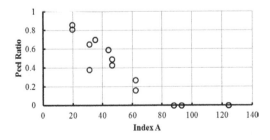

Figure 7. Results of numerical simulation (Model 1)

Figure 9. Results of numerical simulation (Model 2)

Table 4. Study Results for Model 2

Case No.	Eccentricity	Peel Area(from left bank)				Maximum Deformation
		-	30~105m	75~150m	105~180m	
1	0m	✓				0.00m
2	2m	✓				0.42m
3	2m		✓			2.58m
4	2m			✓		2.73m
5	2m				✓	2.63m

4 NUMERICAL SIMULATION FOR DETAILED BEHAVIOR OF SHIH-GANG DAM

4.1 Numerical model

We also performed the nonlinear 3D-FE simulation to understand the detailed behavior of the structure such as crushing or cracking. Based on the damage situation observed, severe damages of the dam body are mainly absorbed by the deformation of the ground and the dam body directly above the C fault, and the dam body somewhat away from the C fault remains slightly cracked and deformed. Therefore, as the damaged part/mode, the following two models were set separately for damage of the right bank cutoff dam body and crack of the left bank dam body.
- Model A: Cracking mode of the dam wall body with minor damage (deformation about 1 m)
- Model B: Functional loss mode of the dam body and directly above the fault(deformation about 10 m)

4.1.1 Finite element model

Figure 10 and Table 5 show the summary of finite element models. Since it involves large deformation, destruction and contact-peel, an explicit method that can proceed stably and get solution even with a problem of strong nonlinearity is adopted.

4.1.2 Material properties

Assume that the concrete constitutive law follows equation (3) to (8) from Japanese standard for concrete design(JSCE, 2015).

$$\sigma_c = E_0 K \left(\varepsilon_c - \varepsilon_p \right) \qquad (3)$$

$$E_0 = \frac{2 \cdot f_c}{\varepsilon_{peak}} \qquad (4)$$

Figure 8. Relationship of Index A and Peel Ratio

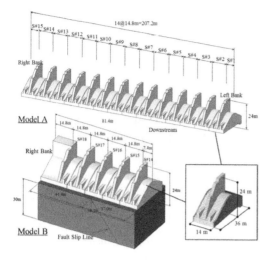

Model A

Model B

Figure 10. Finite element models for detailed evaluation

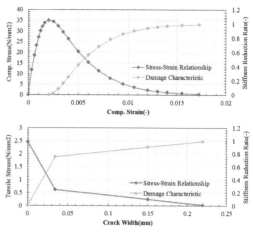

Figure 11. Material Characteristic of concrete

maximum aggregate size (mm), f_t is tensile strength, and w_t is crack width (mm). Figure 11 shows the compressive stress - compressive strain and the tensile - crack displacement relationship.

To take the joint of each dam body into consideration, tensile strength is reduced to half of the other concrete.

As material properties of the supporting soil, Young's modulus is 320 MPa set from the studies in Chapter 3. Poisson's ratio and mass density are assumed to be 0.25 and 2.0 ton/m³ as soft rock, friction coefficient between the dam body and ground is 0.4.

4.1.3 Procedure

For the calculation of crack damage (Model A), forced displacement obtained by global analysis and observed data was given to the bottom and the side of the body part shown in Figure 12. On the other hand, for the calculation of fault struck damage (Model B), the fault slip rate is 7.6 meters for vertical upward,

Table 5. Specifications of FE model and method

Model	Model A	Model B
Element	10737290965	8727773640
Node		
Element Type	C3D8R (Reduced Integration)	
Constitutive Low	Concrete Isotropic Plasticity Damage Model	
Method	Stress Deformation Analysis (Explicit/ Large Deformation)	
Code	Abaqus/Explicit 3DEXPERIENCE R2017x	

$$K = \exp\left\{-0.73\frac{\varepsilon_c}{\varepsilon_{peak}}\left(1-\exp\left(-1.25\frac{\varepsilon_c}{\varepsilon_{peak}}\right)\right)\right\} \quad (5)$$

$$\varepsilon_p = \varepsilon_c - 2.86\cdot\varepsilon_{peak}\left\{1-\exp\left(-0.35\frac{\varepsilon_c}{\varepsilon_{peak}}\right)\right\} \quad (6)$$

$$G_{ft} = 10\left(d_{max}\right)^{1/3}\cdot f_c^{1/3} \quad (7)$$

$$w_t = 5G_f / f_t \quad (8)$$

Where, σ_c is average compressive stress of long term property of dam concrete (35 N/mm²) (Kondo et al, 2016), ε_c is compressive strain, ε_{peak} is strain at compressive strength (assumed to be 0.002), ε_p is compressive plastic strain, E_0 is the initial elastic modulus, G_f is tensile fracture energy (N/m), d_{max} is

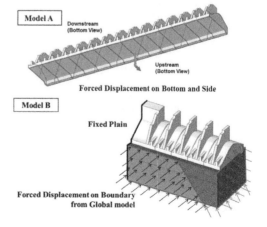

Figure 12. Boundary conditions of computation

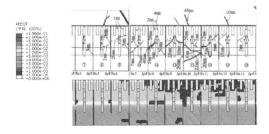

Figure 13. Comparison between study result of plastic tensile strain (Model A) and observed crack distribution

Figure 14. Comparison between study result of plastic tensile strain (Model B) and damage state after the earthquake (SGRDF, 1999)

6 meters toward the dam axial direction (left bank to right bank), 2 meters toward downstream direction, and the slip speed of 1 m/s forced displacement obtained by global analysis was given to the side of the dam body and supporting soil.

4.2 *Study Results*

Figure 13 shows the comparison between the study result of plastic tensile strain of Model A and observed crack distribution.

Figure 14 shows the comparison between the study result of plastic tensile strain of Model B and flyover view picture. Regarding the numerical analysis of Model B the deformation and damage situation of the dam body are roughly grasped. Also, regarding the numerical analysis of Model A, occurrence positions of cracks are caught.

5 CONCLUSION

In order to confirm the feasibility of the numerical procedures to simulate the damage process of the structure above the fault displacement, the authors

conducted the 3D-FE analysis of Shih-Gang Dam in Chi-Chi Earthquake (Taiwan, 1999). Throughout this study, we got a prospect of the feasibility of the procedures.

At first, we performed the simple springs and beam FE analysis to grab the global behavior of dam body.

From the numerical analysis by Model 1, the relationship and the reasonable range of the stiffness of dam body and supporting soil are assumed. From the numerical analysis by Model 2, we pointed out that consideration of both peeling and eccentricity in axial compression enable to reproduce the observed behavior of dam body in horizontal direction.

Secondly, we performed the nonlinear 3D-FE simulation to simulate the detailed behavior of the structure such as crushing or cracking. As the damaged part/mode, the two models were developed separately for cracked area and crashed area. Regarding the numerical analysis of Model B the deformation and damage situation of the dam body are roughly grasped. Also, regarding the numerical analysis of Model A, occurrence positions of cracks are caught.

REFERENCES

Japan Meteorological Agency(JMA). 2018. Major Earthquake in the world. *https://www.jma.go.jp/jma/press/1304/08a/1303eq-world.pdf* (In Japanese).

Japan Society of Civil Engineering(JSCE). 1999. THE 1999 JI-JI EARTHQUAKE,TAIWAN -Investigation into Damage to Civil Engineering Structures.

Japan Society of Civil Engineering(JSCE). 2015. Standard Specifications for Concrete Structure Design. Vol.10, Ch.2

Konagai, K et al. 2000. About the damage of the Shin-Kang Dam in Chi-Chi Earthquake (preliminary report). *Daidam*; No.171. (In Japanese)

Kondo, M. 2016. A Basic Survey and Analysis of Long-Term Properties of Dam Concrete (In Japanese). *Dam Engineering*; Vol.26. 56–68 (In Japanese)

Ohmachi, T. 2000. On damage to dams in Taiwan due to the ChiChi Earthquake. *Dam Engineering*; Vol.10. 138–150 (In Japanese)

Sino-Geotechnics Research and Development Foundation (SGRDF). 1999. Bird's Eye View of Cher-Lung Pu Fault, Sino-Geotechnics Research and Development Foundation

Tsutsumi, H. 2018. The Concept of Validation Strategy about Fault Displacement Fragility Evaluation Methodology and its Application to Actual Damaged Structure. *Proceeding of Probabilistic Safety Assessment and Management (PSAM 14)*: #354.

2019 Rock Dynamics Summit– Aydan et al. (eds)
© *2019 Taylor & Francis Group, London, ISBN 978-0-367-34783-3*

Numerical simulation of rock slope failure with dynamic frictional contact based on co-rotational technique

S. Suzuki & K.Takashi
School of Engineering, Tohoku University, Sendai, Japan

K.Terada & S.Moriguchi
International Research Institute of Disaster Science, Tohoku University, Sendai, Japan

N. Takeuchi
Department of Art and Technology, Hosei University, Tokyo, Japan

ABSTRACT: Based on the co-rotational formulation to deal with large displacements and rotations of a solid, we propose a method of numerical simulations of rock slope failure involving dynamic frictional-contact behavior. In order to simulate failure process of a rock slope subjected to external loading such as an earthquake, we divide the phenomena into multiple stages. The proposed method enables us to represent all of these deformation and failure stages in a continuous fashion. Some simple numerical examples are presented to demonstrate the capability of the proposed method to simulate multi-stage failure processes.

1 INTRODUCTION

In order to simulate failure process of a discontinuous rock slope subjected to external loading such as an earthquake, we divide the phenomena into multiple stages. The stages we are concerned with in this study are the following four:

−   At the first stage: a slope deforms in response to dynamic or static excitations.
−   At the second stage: the material is degraded due to micro-cracks distributed in the rock. These micro-cracks gradually evolve and increase in size, so that the load-carrying capacity is reduced.
−   At the third stage: the slope collapses and breaks into many segments. Several sets of segments start to move dynamically.
−   At the fourth stage: moving segments contact each other and some of them break up into several blocks. These blocks act as new external force against other structures.

The proposed method is designed to simulate all of these deformation and failure stages in a continuous fashion. In this study, we try to simulate the collapse of a slope involving collision followed by segmentation by cracking. In order to represent dynamic frictional contact with large rotations after the crack propagation, we incorporate the effects of dynamics, gluing, and contact within the framework of the co-rotational theory.

2 OUTLINE OF THE PROPOSED METHOD

2.1 *Three-dimensional co-rotational formulation*

The idea of the co-rotational formulation (Moita & Crisfield 1996) is to decompose the motion into a rigid body motion and a strain-producing motion. Thanks to this decomposition, we can follow the standard small deformation theory for material characterization on the local coordinate frame that rotates with a finite element. The geometrically nonlinearity is induced by the rotation of the local frame. In order to obtain the rotation matrix with respect to the global frame, we compute a deformation gradient at the centroid of the element F_c as

$$F_c = \sum_{a=1}^{n} x_s^a \otimes \frac{\partial N_a}{\partial X_s}\bigg|_{\xi=0} \tag{1}$$

and then the rotation matrix R_c is determined in terms of polar decomposition

$$R_c = U F_c^{-1} \tag{2}$$

where x_s^a is the global current configuration at a node a on an element s and $\partial N_a / \partial X_s|_{\xi=0}$ is the partial derivative of a shape function at centroid of the element. Equation (2) above leads to the transformation matrix T_s with respect to the element s as

$$T_s = z_s w_s^T + \left[\text{diag} R_s^T \right] \tag{3}$$

where z_s and w_s are both of the 24×3 matrix involving R_c. In order to calculate the variables such as the internal force vector F_s and tangent stiffness matrix K_s^t for the global coordinate system, we transform local variables to corresponding global variables by using T_s as below:

$$F_s = T_s^T F_{\ell s} \tag{4}$$

$$K_s^t = T_s^T K_{\ell s} T_s + K_\sigma^t \tag{5}$$

Note that the vector $F_{\ell s}$ and the matrix $K_{\ell s}$ are local internal force vector and stiffness matrix respectively, and K_σ^t is the geometric stiffness matrix. The equations of motion are solved by use of these global variables so that the deformation with large displacement and rotations can be updated.

2.2 Formulation for mid-point dynamics

The discrete energy-momentum method (Simo & Tarnow 1992) and the mid-point algorithm (HYPERLINK "0113.docx" \l "LinkManagerBM_REF_TKZ6TfdxCrisfield et al. 1997) are adopted to deal with the dynamic behavior. These two approaches impose the equivalence between the change of total momentum of the system and the impulse of the external forces acting on the system during each time step. In addition, both of these methods exhibit the proper energy balance within a time interval so that unconditional stability in time can be realized. In the present study, we formulate the three-dimensional co-rotational dynamic problem based on both the energy-momentum method and the mid-point algorithm.

Based on these theory, velocities and accelerations at the middle configuration are defined by convex combination

$$x_m = \frac{1}{2}\left(x_{n+1} + x_n\right) \tag{6}$$

$$V_m = \frac{1}{2}\left(V_{n+1} + V_n\right) \tag{7}$$

where the variables with subscript $n+1$, n, and m indicate the quantity corresponding to the value at $n+1$, n, and m configuration respectively. Then, we define the mid-point inertial force vector F_{ms}^{mass} and the internal force vector F_{ms} that satisfy equivalence between the change of total momentum of the system as

$$F_{m_s}^{mass} = \frac{1}{\Delta t} M_s \left(V_{n+1_s}^a - V_{n_s}^a\right) \tag{8}$$

$$F_{m_s} = \left(\frac{T_{n+1_s} + T_{n_s}}{2}\right)^T \left(\frac{\bar{F}_{n+1_s} + \bar{F}_{n_s}}{2}\right) \tag{9}$$

Note here that the matrix M_s is a consistent mass matrix and Δt is the time interval. Differentiation of

Equation (9) with respect to the displacement leads to mid-point tangent stiffness matrix as

$$K_{m_s}^t = \left(\frac{T_{n+1_s} + T_{n_s}}{2}\right)^T \frac{1}{2} K_{\ell s} T_{n+1_s} + \frac{1}{2} K_{\sigma_s}^t \tag{10}$$

2.3 Formulation for gluing/breakoff

For the sake of simplicity, we employ the classical discrete crack model and the traction-separation law for a brittle fracture to represent the crack initiation and propagation. The penalty method is applied to connect the interface between neighboring segments, which are assumed to be potential discontinuities equipped with the employed traction-separation law. Each of these set of segments is supposed to move and deform independently after complete separation without the cohesive force

To obtain the relative displacement for the traction-separation law, we define the gap functions in the normal and tangential directions below as

$$\dot{g}_{T_m}^\alpha \equiv -\left[\left(x_{n+1}^{(2)} - x_{n+1}^{(1)}\right) - \left(x_n^{(2)} - x_n^{(1)}\right)\right] \cdot v_m$$

$$\dot{g}_{T_m}^\alpha \equiv \left[\left(x_{n+1}^{(2)} - x_{n+1}^{(1)}\right) - \left(x_n^{(2)} - x_n^{(1)}\right)\right] \cdot x_m^\alpha \tag{11}$$

Here, the subscripts N and T represent the normal direction and the tangential direction, respectively, and the superscript is used to denote a quantity which is associated with the regions between crack path. Also, v_m and χ_m^α are the unit normal vector and the contravariant basis, which are defined as

$$v_m = \frac{\partial \langle x_m \rangle}{\partial \eta^1} \times \frac{\partial \langle x_m \rangle}{\partial \eta^2}; \quad x_m^\alpha = g_m^{\alpha\beta} \frac{\partial \langle x_m \rangle}{\partial \eta^\beta} \tag{12}$$

with $g_m^{\alpha\beta}$ being the metric tensor. Here, $\langle \cdot \rangle$ indicates an average quantity over the whole domain.

In this study, to represent brittle fracture often observed in discontinuous rock mass, we employ the following traction-separation law below

$$\hat{t}_N \equiv \in_N \dot{g}_{N_m}; \quad \hat{t}_{T_\alpha} \equiv \in_T g_{\alpha\beta_m} \dot{g}_{T_m}^\beta \tag{13}$$

where $\in_N$ and $\in_T$ are penalty parameters. When the normal component of traction reaches tensile strength, both of the traction values are set to be zeros as.

$$\hat{t}_N \equiv 0; \quad g_{T_\alpha} \equiv 0 \tag{14}$$

Then, we obtain the internal force vector and the tangent stiffness matrix associated with the penalty method as follows:

$$F_{m_s}^{dis} = \int_{\Gamma_d} \left(\hat{t}_N N_m - \hat{t}_{T_1} \hat{D}_m^1 - \hat{t}_{T_2} \hat{D}_m^2\right) dA \tag{15}$$

447

$$K_{m_s}^{fD} = \int\limits_{\Gamma_d}^{m_s} \left(2\hat{t}_N N_m N_m^T \right.$$

$$\left. + 2 \in_T g_{\alpha\beta_m} \hat{D}_m^\alpha \hat{D}_m^{\beta T} \right) dA \qquad (16)$$

Here, N_m and $\hat{D}_m^\alpha$ are the vectors involving shape function. By introducing these variables into co-rotational dynamics based on the mid-point rule, we can properly update the deformation with large rotations involving cracks.

2.4 Frictional Contact formulation

The so-called node-to-surface approach is adopted for a discretization technique for non-matching meshes of contact blocks (Armero & Petocz 1998, Laursen 2002). In this method, a contact pair is composed of a vertex node of a slave element and an element surface of a master element, on which the contacting point must be determined. To find this point, the projection is made to provide the minimum distance with the slave node.

$$\left(x^2 - x^1\left(\xi^1, \xi^2\right) \right) \cdot \frac{\partial x^1\left(\xi^1, \xi^2\right)}{\partial \xi^\alpha} = 0 \qquad (17$$

Here, we can obtain the projection point $\tilde{x}^1 = x^1\left(\tilde{\xi}^1, \tilde{\xi}^2\right)$ by solving the above equation. The projection point leads to dynamic normal gap function associated to the mid-point algorithm as follows:

$$\dot{g}_m \equiv -\left[\left(x_{n+1}^2 - \tilde{x}_{n+1}^1 \right) - \left(x_n^2 - \tilde{x}_n^1 \right) \right] \cdot \tilde{n}_m^1 \qquad (17)$$

with normal traction vector $\tilde{n}_m^1$ on master element. In addition, the augmented Lagrangian

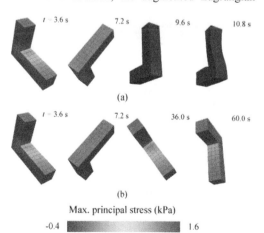

(a)

(b)

Max. principal stress (kPa)

-0.4　　　　　　　　　1.6

Figure 1. Example 1, deformed configuration with maximum principal stress distribution via (a) standard Newmark method and (b) present method.

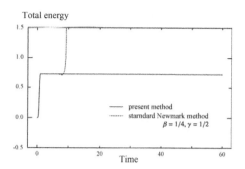

Figure 2. Example 1, total energy vs time for the two methods.

method is incorporated with the energy-momentum scheme to impose contact constraints. To determine the contact force, we define the contact force-gap function law as

$$\hat{t}_N \equiv \left\langle \bar{\lambda}_N^{(k)} + \in_N \left\langle \dot{g}_m \right\rangle_+ \right\rangle_+ \qquad (18)$$

$$\hat{t}_{T_\alpha}^{\text{trial}(k)} \equiv \bar{\lambda}_{T_\alpha}^{(k)} + \in_T m_{\alpha\beta_m} \left(\xi_m^{\beta(k)} - \bar{\xi}^\beta \right) \qquad (19)$$

where $m_{\alpha\beta m}$ is the metric tensor and $\bar{\xi}^\beta$ is the stick point in the β direction. Here, superscript (k) indicates the number of iterations with Uzawa algorithm. Also $\bar{\lambda}_N$ and $\bar{\lambda}_{T\alpha}$ are the Lagrange multipliers in the normal direction and tangential direction, respectively. Note that the superscript denotes trial state in return mapping method associated with friction. For the tangential direction, we define the slip function of Coulomb's friction law as

$$\Phi^{\text{trial}} = \left\| t_T^{\text{trial}(k)} \right\| - \mu t_N^{(k)} \qquad (20)$$

Then, we can compute the contact force as

$$\hat{t}_{T_\alpha}^{(k)} = \begin{cases} \hat{t}_{T_\alpha}^{\text{trial}(k)} & \text{if} \quad \Phi^{\text{trial}} \leq 0 \\[2mm] \mu \dfrac{\hat{t}_{T_\alpha}^{\text{trial}(k)}}{\left\| t_T^{\text{trial}(k)} \right\|} & \text{otherwise} \end{cases} \qquad (21)$$

The contact forces $\hat{t}_N$ and $\hat{t}_{T_\alpha}^{(k)}$ become involved in the internal force vector $F_{m_s}^{\text{con}}$ and the tangent stiffness matrix $K_{m_s}^{tC}$. Then, the linearized equation in this study is derived as follows:

$$\left[\frac{2}{\Delta t^2} M_s + \left(\frac{T_{n+1_s} + T_{n_s}}{2} \right)^T \frac{1}{2} K_{\ell s} T_{n+1_s} \right.$$

$$\left. + \frac{1}{2} K_{\sigma_s}^t + \frac{1}{2} K_{m_s}^{fD} + \frac{1}{2} K_{m_s}^{tC} \right] \Delta d_{n+1} \qquad (22)$$

$$= -\left(F_{m_s}^{\text{mass}} + F_{m_s} + F_{m_s}^{\text{dis}} + F_{m_s}^{\text{con}} - F_{m_s}^{\text{ext}} \right)$$

3 NUMERICAL ANALYSIS EXAMPLE

In order to confirm the performance of the proposed method, three simple applications are conducted.

3.1 Example 1: dynamic analysis for an L-shaped block

In order to demonstrate the numerical stability in dynamic analysis of an L-shaped block, we compare the present method to the standard Newmark method whose parameters are set at $\beta = 1/4$ and $\gamma = 1/2$. The time interval in an increment is chosen as $\Delta t = 0.01$s for both of the methods and the total time is set to be 60 seconds. The L-shaped block is subjected to no displacement boundary restrictions and the external force applied on the top of surface of the block within one second.

Figure 1 shows the deformed configuration with the maximum principal stress distribution obtained by these two approaches. The result with the present method properly represents large rotation, while the result with the standard Newmark method seems to suffer from locking as shown in this figure. Figure 2 shows that the Newmark procedure causes a sudden energy growth. On the other hands, the present approach conserves the total energy all the time during the analysis.

3.2 Example 2: Freefall problem with contact

To see the performance of the augmented Lagrange method for dynamic contact, we carry out a freefall problem in comparison with penalty method. The example considers the situation, in which a cubic block falls down on the base block.

Figure 3 shows the results obtained with the augmented Lagrangian method and the penalty method. The upper block's motion with the penalty method does not restore the potential energy after contact with the base block, while the augmented Lagrangian method exhibits the conservation of total energy.

3.3 Example 3: Demonstration of rock slope failure

The last example is provided to simulate a slope sliding failure of a rock mass as shown in Figure 4. In order to confirm that the proposed method is capable of representing the large rotation with frictional contact behavior, we carry out numerical simulation of the rock slope in consideration with potential discontinuities. Here, the slope is subjected to the seismic motion caused by the enforced displacement to the bottom of the model and is considered to fail at a certain time after the excitation. This displacement wave is obtained by transforming from the acceleration waveform observed in the 2016 Kumamoto earthquake.

Figure 5 shows the sequence of the deformed configurations of the rock slope subjected to the excitation. The blocks that are initially glued together separate and the resulting crack surfaces contact with each other. It is safe to conclude that the rock slope failure could be successfully simulated.

4 CONCLUSION

We have proposed a new analysis method for numerical simulations for slope failure of discontinuous rock masses. In order to simulate the large rotation of the structure after the collapses due to excitation, we have formulated the method based on the three dimensional co-rotaional technique. Also, to carry out numerical simulation without numerical instability, the mid-point co-rotational dynamics was

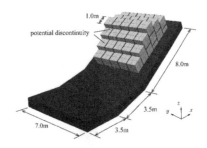

Figure 4. Example 3, geometry and finite element mesh for analysis of a rock slope in consideration of potential discontinuity.

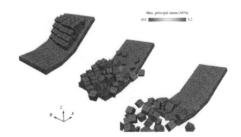

Figure 5. Example 3, sequence of deformed configuration with maximum principal stress distribution

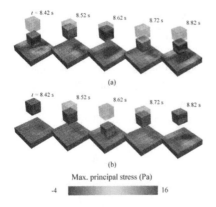

Figure 3. Example 2, deformed configuration with maximum principal stress distribution via (a) penalty method and (b) augmented Lagrangian method.

developed for three dimensional continuum elements. The penalty method has employed for gluing of separate blocks composed of finite elements and each joining surface is regarded as a potential discontinuity. In addition, the augmented Lagrangian method was adopted for constrained optimization problem associated with frictional contact.

According to the numerical verification, the proposed method is capable of representing large rotations with frictional contact behavior. The linearized equation obtained the present method seems to be complicated beyond necessity. However, the multi-stage failure of a rock slope failure is essentially a long time phenomenon, especially when material degradation of the rock mass is considered. Inevitably therefore, the implicit method is intrinsically needed for the numerical simulations. Incorporation with inelastic constitute models for rock masses must be done to realize actual material responses and would be our future work.

REFERENCES

Armero, F. & Petocz, E. 1998. Formulation and analysis of conserving algorithms for frictionless dynamic contact/ impact problems, *Comput. Methods Appl. Mech. Engrg.*, Vol.158, pp.269–300.
Crisfield, M.A., Galvanetto, U., & Jelenić, G. 1997. Dynamics of 3-D co-rotational beams, *Comput. Mech.*, *Vol.20, pp.*507–519.
Laursen, T.A. 2002. *Computational Contact and Impact Mechanics: Fundamentals of Modeling Interfacial Phenomena in Nonlinear Finite Element Analysis*, Springer, 2002.
Moita, G.F. & Crisfield, M.A. 1996. A finite element formulation for 3-D continua using the co-rotational technique, *Int. J. Numer. Methods Engrg.*, Vol.39, pp.3775-3759.
Simo, J.C. & Tarnow, N. 1992. The discrete energy-momentum method. Conserving algorithms for nonlinear elastodynamics, *Z. Agnew. Math. Phys.*, Vol.43, pp.757–792.

2019 Rock Dynamics Summit– Aydan et al. (eds)
© *2019 Taylor & Francis Group, London, ISBN 978-0-367-34783-3*

A study of formations of a sliding surface during an earthquake

T. Kawai
Tohoku University, Japan

S. Nakamura
Nihon University, Japan

M. Shinoda
National Defense Academy

K. Watanabe
Tokyo University, Japan

S. Nakajima & K. Abe
Railway Technical Research Institute

H. Nakamura
Japan Nuclear Energy Safety Organization

ABSTRACT: Evaluating the seismic stability of a rock slope typically involves searching for the minimum value of calculated safety factors (SF) for each supposed sliding block. Because only the transient equilibrium is evaluated, the likelihood of any slope failure can be deemed negligible if all the calculated SFs are greater than unity. However, even if some of the calculated SF are less than unity, it cannot be assumed that all such slopes will collapse. Recently, in the wake of extremely large earthquakes in Japan, the design earthquake standards for nuclear power plants (NPP) have been extended. After the experience of the 2011 off the Pacific coast of Tohoku Earthquake, the designer is expected to consider beyond design basis earthquakes to determine whether more can reasonably be done to reduce the potential for damage, especially where major consequences may ensue (IAEA, 2011). With this in mind, the method employed to evaluate the seismic performance of the slopes surrounding an NPP needs to be capable of doing more than determing the likelihood of failure: it must also to consider the process toward failure in the event of an earthquake beyond the design basis. In this paper, a detailed study of formations of a sliding surface was conducted by using a series of model shaking table tests, and then some issues to be considered toward inplementation of a new evaluation method on seismic performance of a slope are indicated.

1 INTRODUCTION

In response to the extremely large earthquakes experienced in recent history in Japan, Indonesia and Chile, the design earthquake level for nuclear power plants (NPP) has been extended. More specifically, after the experience of the 2011 off the Pacific coast of Tohoku Earthquake, the designer is expected to consider the consequences that may arise from such an earthquake. That is, the designers must consider if there is anything more than can be done to reduce the damage potential above and beyond the design basis for earthquakes (IAEA, 2011). In other words, there is no earthquake magnitude limitation when considering what kind of phenomena occur in the event of an earthquake.

The evaluation of the seismic stability of a rock slope involves calculating the safety factor (SF) of a supposed sliding block. In the method, the SF is defined as the ratio of the resistant force derived from the rock strength divided by the transient driving force due to the earthquake, and the minimum value for this calculation over the entire slope for the duration of the earthquake is determined. Because the principle of the method is both simple and valid, the method is adequate to distinguish safe slopes from 'potentially' dangerous slopes when subject to a design earthquake. However, since the method is focussed only on the transient equilibrium, it has one clear limitation: it is not capable of predicting what would happen in the event that the resultant force exceeds the strength for a short time. That is, because it only evaluates the transient equilibrium, it can only conclude that the likelihood of any slope failure is negligible when all the calculated SFs are greater than unity. It is theoretically pointed out and then confirmed by the experimental results and simulations that some sliding block can remain on the slope even after the least SF becomes less than unity (Kawai & Ishimaru, 2010).

Shinoda et al. (2015) proposed a evaluating flow chart to consider this, in which the evaluation of SF during an earthquake is regarded as initiation of 'unstable' and then followed by the evaluation of SF with the residual strength of the material as an indicator to distinguish failure modes of the slope as shown in Figure 1. In their flow, there is a possibility of 'OK' only for the slope which is guaranteed 'gradual' movements after being judged as 'unstable' and the calculated displacements remain tiny. On the other hand, if the likelihood of a sudden failure, that is a rapid progress of failure and immediate falling down of the sliding block, is highly likely, rather larger displacements will be checked only for evaluating the influence of falling rock mass on the important facilities at the bottom of the slope. Therefore, there are two kinds of displacement evaluation in their flow chart; one is for the safety evaluation, the other is for the evaluation of influences of falling rock mass. For the former displacement evaluation, the method proposed by Newmark (1965) is a practical option.

The method is based on the assumption that the behaviour of a potential sliding mass in an embankment is similar to the behaviour of a sliding block on an inclined plane. The block slides only if the earthquake acceleration exceeds the yield acceleration of the block. The yield acceleration is assumed to be constant, e.g. Tsai et al. (2015), and constant values of the strength parameters, such as the friction angle, are employed to calculate the yielding acceleration and to consider the stress change in the normal direction to the plane caused by both the horizontal acceleration and the vertical acceleration (Malla 2017). Kan et al. (2017) studied the relationships between the tuning ratio and the deformation of dams under earthquake loading and proposed a practical framework to avoid the risk involved in Newmark-type methods, that is, their failure to predict conservative displacement for some dams. All these research works focus on the improvement in the precision of small estimated deformations rather than predicting collapse. However, in order to determine the limitation of such displacements in the safety design, more detailed evaluations of failure process seem to be important.

In this paper, in order to support the concept included in Figure 1, the obserbations of the failure process are summarized through analyzing the formerly reported shaking table tests (Shinoda et al. 2015, Nakajima et al. 2015, Nakamura et al. 2015), and then some issues to be considered toward inplementation of the new evaluation method on seismic performance of a slope are indicated.

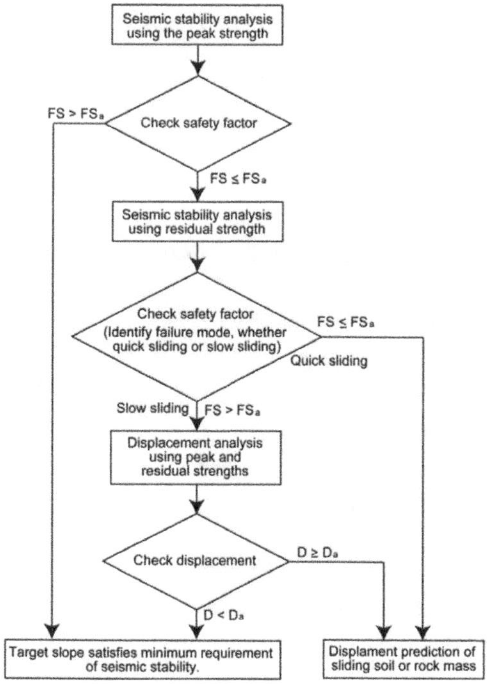

2 OVERVIEW OF PREVIOUSLY CONDUCTED SHAKING TABLE TESTS

A large number of series of shaking table tests (JNES, 2013), including various size models and both at 1 G and a centrifugl acceleration were

Figure 1. Flow chart for evaluating the seismic stability of a slope (Shinoda et al. 2015).

Table 1. Experimental cases.

Case	Size	Num. Layer	Weak Layer Angle (degree)	Input	Actual Failure Amp. m/s²	Seismic Coef. k_h	Mat.
#1	S						A
#2	S						A
#3	S			NG			A
#4	S						A
#5	S	3	45	S	1	0.400	A
#6	S	3	40	S	5	0.5	A
#7	S	3	40	S	6	0.434	A
#8	S	3	35	S	6	0.61	A
#9	S	3	35	E	9		A
#10	M	3	35	S	3.5		A
#11	M	2	35	S	4.5		A
#12	M	3	45	E	2	0.156	A
#13	M	3	45	E	3	–	B
#14	S	3	40	S	5	–	B
#15	S	1	–	S	–	–	A
#16	S	1	–	S	6	0.6	A
#17*	S	1	–	S	7	1.022	A
#18*	S	1	–	S	7	1.2	A
#19	M	1	–	E	9	1.05	C
#20	M	3	45	E	9	1.33	A
#21	L	1	–	S, E	7	0.862	C
#22	L	3	45	S, E	8	0.941	C
#23*	M	3	45	S, E	–	0.634	A
#24	C	3	45	–	–	–	D
#25	C	3	45	S	–	–	E
#26*	C	3	45	S	–	–	D

*Reinforced by ground anchors

(a)Three layers model.　(b)Two layers model.　(c)One layer model.
(main target)

Figure 2. Typical model shapes used in the series of shaking table tests.

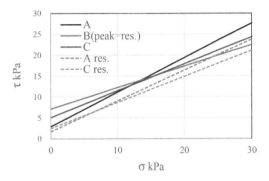

Figure 3. The material strength for the selected cases in this report.

conducted intensively by NRA and RTRI. The cases conducted in that project is summarized in Table 1 and the detailed experimental conditions were shown in the literature mentioned above.

In those tests, three types of model slopes were adopted as shown in Figure 2 and the materials to compose those models are summarized in Figure 3: the material A and C have a different residual strength from the peak one and, on the other hand, the material B has a constant strength after reaching its peak value, it can retain the same strength even after largely sheared. All tests were intended to cause sliding failure in the slope by increasing the input accelerations step by step for each shaking stage, and the moments at the failure were recorded by a high-speed video camera. A sinusoildal wave form was mainly used to analyze the mechanisms at the moment of failure, and in some cases, a synthetic earthquake motion was used to evaluate the perfomance of the slope against an actual input of an irregular form wave.

In this report, some seleted cases are picked up to evaluate the validity of the flow chart shown in Figure 1, especially to check the use of the two different safety factors.

3 OVSERVATIONS OF THE FAILURE PROCESS OF SOME SELECTED CASES

3.1 Difference between 'sliding' and 'falling'

In Figure 4, typical forming processes of a sliding surface from local small slips up to a unified large sliding surface passing through a whole slope are

shown. As shown in Figure 4(a), before forming a final whole sliding surface, several local slips occur simultaneously or one after another in the same shaking event. In this Case 9, two different shaking

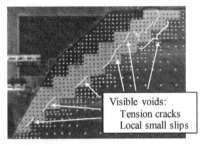

Visible voids:
Tension cracks
Local small slips

(a)After the former shaking events.

(b)During the latter shaking events.

Figure 4. A typical process of forming a whole sliding surface after occuring several local slips (Case 9 in Table 1).

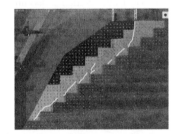

(a)After the former shaking event of the input acceleration amplitud of 3.5 m/s².

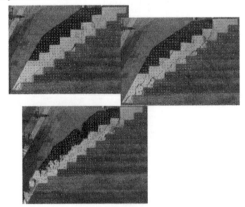

(b)During the latter shaking event of the input acceleration amlitude of 3.0 m/s².

Figure 5. Another demonstration of showing the difference between the 'sliding' and 'falling' (Case 10 in Table 1).

453

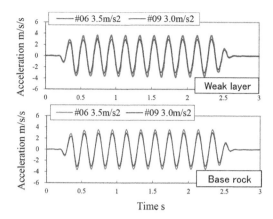

Figure 6. Input and response accelerations of the two shaking events.

events having the same maximum acceleations of a synthetic earthquake motion were conducted, and in the former shaking event many local slips were recoginzed as narrow voids in the weak 2nd layer, but the upper block of the slope was rested on the original position without falling down even after forming those local slips. Since the sum of the length of those voids was rather long, the minimum safety factor during the shaking must be less than unity. It proves that the transient safety factor being less than unity during an earthquake does not always mean the slope falling down perfectly. That is, the 'sliding' is a different phenomenon from the 'falling', and it should be evaluated separately.

Finally, the slope fell down by the latter shaking event of the same maximum acceleration through the process that the local slips were connected together to form a large sliding surface as shown in Figure 4(b). Further, in Case 10 shown in Figure 5, the less input accelerations of the next shaking event were sufficient (Figure 6) to form a whole sliding surface and then make the slope fall down totally after the larger former shaking event, in which only unconnected local slips were observed.

3.2 Influence of dynamic input motions on the 'falling' of the slope

Consider the 'sliding' and 'falling' separately, the influence of dynamic input motions on the 'falling', which is needed while evaluating the safety of the structures located below the slope, should be reconsidered to grasp the actual behavior of the slope during an earthquake.

In Figure 7, the results of Case 11 are shown to reconsider the effect of additional dynamic forces caused by a shaking event. In this case, a large sliding surface was formed at the bottom of the surface layer parallel to the every conrners of the base rock steps, but as same as the other cases mentioned above, the upper block did not move largely just

after forming the sliding surface. Instead of that, in the lower part, a new sliding block was formed and fell down to the bottom. From the geometry of the lower falling part, authers must acknowledge that it might be difficult to distinct forming of a new sliding surface from breaking apart in one sliding block on the way to fall down. But in this case, since the upper block remained on the sliding surface with only a slight movement even after the whole shaking, it is possible to regard this as a new sliding surface formation.

Considering above, the main influences of the additional dynamic stress seem to be the degradations of material strength and the following redistributions of stress. It also suggests that the main cause of the 'falling' is not the dynamic stress but the self-weight of the block. Therefore, there is possibility of misleading while detecting the actual falling part only by drawing attention to the least safety factor sliding surface.

It is also supported by the results from Case 21 comprised of one material for the whole slope as shown in Figure 8. As shown in Figure 9 (JNES, 2013), the supposed sliding surface of the least safety factor calculated by using the conventional modified Fellenius (1927) method with seismic coefficients was located much deeper than the actual sliding surface observed during the shaking table test. The reason of the difference of the two sliding surfaces was summarized that it seemed to attribute the difficulties of preliminary strength evaluations and dynamic response evaluations. In

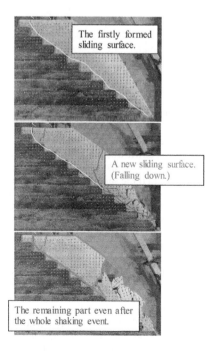

Figure 7. A new sliding surface was formed after the deeper and former sliding surface (Case11 in Table 1).

Figure 8. The results from the one-layer slopes (Case 21).

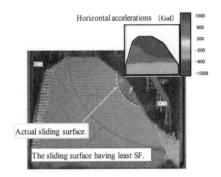

Figure 9. Two different sliding surfaces.

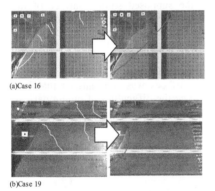

(a)Case 16

(b)Case 19

Figure 10. Failure patterns of the other one-layer slopes.

that report, it was also explained that by using the material strength obtained after the shaking table test and the actual shape of the sliding surface, the event of falling down was reasonably evaluated.

However, there is a possibility of degradations of material strength if the samples were taken from the model after the shaking events. Besides this, Kawai and Ishimaru (2018) explained that an actual falling part of a slope may be possible to differ from the supposed sliding surface having the least safety factor with considering the additional transient stress caused by an earthquake.

Case 16 and Case 19 were also the cases comprised of one material for the whole slope as shown in Figure 10. Because of the reinforcement of the right-hand side of the models, the responces of those models were complicated and many tension cracks were observed. Despite the deep tension cracks, especially in Case 19, only shallow sliding blocks fell down. Since the model shape was the same as Case 21, the sliding surface of the least safety

factor might be as deep as the one in Case 21 and the locations of the actual fallen part were affected by the degradations of the strength and related redistributions of stress.

4 EXPECTED PROBLEMS TO BE SOLVED FOR IMPLEMENTATION OF EVALUATIONS OF SLOPE DEFORMATION IN THE NEW FLOW CHART

In the above sections, the validities of using two different safety factors; the one is for the evaluation of initiation of 'unstable' and the other is for the evaluation of possibility of 'falling down', are partly proved. This is more inportant to evaluate seismic performance of a slope in the beyond design basis, in which the actual phenomenon of slope failure should be evaluated properly to think the scenario for an improvement of safety. Therefore, the flow chart, shown in Figure 1, is expected to be used in practical design process for the evaluation of seismic safety of important facilities.

On the other hand, if the event 'falling down' is considered separately from the event 'partial slide' or the initiation of 'unstable' without any prominent movements, the following two issues should be taken into consideration to determine the limitation of displacement being compared to the estimated value in the last process of the flow chart in Figure 1.

The first thing is that the stress degradations may be realized without any displacements toward the direction of 'falling down'. If the supposed sliding block moves even toward upward, from damage point of view, the effect of moving upward is equivalent to that of moving downward. The sketch shown in Figure 11 also shows the same apprehension. It should be noted that the both have the same effect, but the latter cannot be evaluated from the movements parallel to the direction of sliding. Therefore, while considering the limit state of the slope or perhaps conducting the calculation of

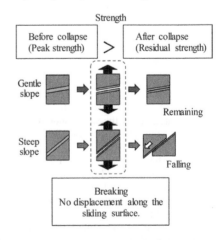

Figure 11. The possibility of no slip before falling down.

455

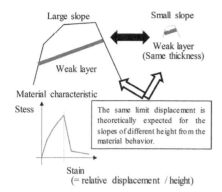

Figure 12. The necessity of evaluation of failure process.

displacements of a sliding block by using a method like Newmark's type method, the degradation of slope materials should be carefully evaluated. Those are brought not only by the movements parralel to the direction of sliding but also by the total responsec of the slope.

The second thing is shown in Figure 12. That is, there is a possibility if an idealized behavior of the material composing the slope is only considered uniformly withing a weak layer, the limitation of displacement of the upper block is dominated only by the thickness of the weak layer. It means that the allowable displacement is indipendent of the size of the whole slope. It may lead to a too strict evaluation for the larger slope. Therefore, in order to evaluate the seismic safety of a slope practically, it is important to know the failure process withing the weak layer, in general, behaviors of the neighbor parts of the final sliding surface.

In the series of shaking table tests above mentioned, it is confirmed from the detailed observations of the images obtained by a high-speed camera that all the emerging local slips were attributed to the movement of the upper supposed sliding block. Therefore, nonlinear FEM analysis, or perhaps more advanced simulation method (e.g. Abe et al. 2012), are effective to proceed the seismic evaluation of a slope with considering failure process upto the actual limit of falling down.

5 SUMMARY

In order to proceed the implementation of a new evaluation flow of seismic safety of a slope shown in Figure 1, a detailed obserbations of failure process of various slopes reported previously in the leterature were conducted. The summaries of the analysis and the related problems to be solved in the future are follows.

1. The least safety factor during shaking, which means 'tentative' or 'in short time', becoming less than unity does not always mean that the supposed sliding block fall down perfectly. It is partly confirmed by the fact that a sliding block was moved

during a smaller shaking event after the shaking event conducted by the maximum input motion.
2. From the above-mentioned test case and the other, the main influence of an earthquake to the seismic performance of a slope seems to be degradations of the material.
3. From damage point of view, just considering the slope behavior along a sliding surface seems to be insufficient, and the whole responses of a slope including failure process of the elements should be evaluated at the same time.

REFRENCES

Abe, K., Shinoda, M., Watanabe, K., Sanagawa, S., Nakamura, S., Kawai, T., Murata, M. and Nakamura, H. 2012. Numerical simulation of landslides after slope failure using MPM with SYS Cam-clay model in shaking table tests, Proc. of 15th World Conference on Earthquake Engineering (CD-ROM): 1999

Fellenius, W., 1927. Erdstatische Berechnungen mit Reibung ind Kohaesion (Adhasesion) und unter Annahme kreizilindrischer Gleiflaechen. W. Ernst und Sohn Verlag, Berlin.

IAEA. 2011. IAEA international fact finding expert mission of the Fukushima dai-ichi NPP accident following the Great East Japan Earthquake and Tsunami, Mission report, IAEA.

JNES. 2013. Guideline for design and risk evaluation against the seismic stability of the ground foundation and the slope, ii-2-1 A series of shaking table tests of small and medium size slope model, *JNES-RE-Report Series*, JNES-RE-2013-2037.

Kan, E.M., Taiebat, H.A. and Taiebat, M. 2017. Framework to assess Newmark-type simplified methods for evaluation of earthquake-induced deformation of embankments, *Can. Geotech. J.* 54:392–404.

Kawai, T. and Ishimaru, M. 2010. A development of an evaluation flow chart for seismic stability of rock slopes based on relations between safety factor and sliding failure, CRIEPI report, N09030(in Japanese).

Kawai, T. and Ishimaru, M. 2018. The implementation of the gradual movement of sliding blocks to a new method for evaluating the seismic stability of slopes reinforced by geotextile, *J. Earthquake and Tsunami*, 12 (4): 1841010–1–19.

Nakajima, S., Watanabe, K., Shinoda, M., Abe, K., Nakamura, S., Kawai, T. and Nakamura, H. 2015. Consideration on evaluation of seismic slope stability based on shaking table model test, *The 15th Asian Regional Conference on Soil Mechanics and Geotechnical Engineering*: JPN–100.

Nakamura, S., Abe, K., Shinoda, S. and Kawai. T. 2013. Study on evaluation of behavior for slope stability and failure during earthquake for using seismic risk of NPP, *SMiRT-23, Manchester*, UK: 10–14.

Newmark, N.M. 1965. Effects of earthquakes on dams and embankments, Geotechnique. 15:139–160.

Shinoda, M., Watanabe, K., Sanagawa, T., Abe, K., Nakamura, H., Kawai, T. and Nakamura, S. 2015. Dynamic behavior of slope models with various slope inclinations, *Soils and Foundations*, 55(1): 127–142.

Tsai, C.C. and Chien, Y.C. 2015. A simple procedure to directly estimate yield acceleration for seismic slope stability assesment, *Porc. The 15th Asian Regional Conference on Soil Mechanics and Geotechnical Engineering*: TWN–08.

2019 Rock Dynamics Summit– Aydan et al. (eds)
© 2019 Taylor & Francis Group, London, ISBN 978-0-367-34783-3

Some considerations on the failure of Güney Waterfall, Denizli, Turkey

H. Kumsar
Department of Geological Engineering, Pamukkale University, Denizli, Turkey

Ö. Aydan
Department of Civil Engineering and Architecture, University of the Ryukyus, Okinawa, Japan

ABSTRACT: The Güney Waterfall in Denizli was ruptured by the The 21st July 2003 Buldan earthquake with magnitude 5.6. The ruptured block toppled after 10 years and traveled towards the reservoir of Cindere Dam more than 132 m. The authors attempt to clarify the mechanism and process of the toppling failure of the ruptured block. The authors attempt to analyse the failure process in fundamentally in three stages. The first stage is the initial failure caused by the earthquake. The second stage is the growth of the ruptured block and progress of degradation at the toe of the failed block. Third stage is the final collapse and movement in the valley towards the Cindere Dam reservoir. The pre-rupture stage before the Buldan earthquake and post-rupture and final collapse conditions are analysed through some analytical and numerical techniques to understand the fundamental causes of the failure, which took more than 10 years. This is study is unique in a sense it involves slow and rapid dynamic processes. The authors present the results of the various studies and investigations undertaken so far and and they discuss their implications.

1 INTRODUCTION

The 21st July 2003 Buldan earthquake with magnitude 5.6 ruptured the Güney Waterfall in Denizli, which is the 23rd natural beauty of Turkey. However, the ruptured block of the Güney Waterfall did not toppled at that time. The ruptured block toppled after 10 years and traveled towards the reservoir of Cindere Dam more than 132 m. The initially ruptured block has a prism-like shape with 34m width, 20 m length and 12 m height. Since the 2003 Buldan earthquake, several earthquakes with a moment magnitude more than 5 occurred in the vicinity of the Güney Waterfall, which may further cause at the toe of degradation and additional damage.

The authors attempt to clarify the mechanism and process of the toppling failure of the ruptured block. The failure process can be divided into three stages. The first stage is rupturing of the waterfall due to the 2003 Buldan earthquake. The second stage is the growth of the ruptured block in size and the degreadation and erosion of the toe of the block. This geometrical changes of the ruptured block resulting from the growth and degradation prepared the conditions of the toppling failure. In the third stage, the block toppled and traveled to the Cindere Dam reservoir and disintegrated into several large blocks. The authors investigated the area during the reconnaissance of the Buldan earthquake and after the failure. Besides site investigations and geometrical measurements, samples were gathered and cored to carry out some physico-mechanical properties of travertine. The pre-failure and post-failure conditions are analysed through

some analytical and numerical techniques to understand the fundamental causes of the failure, which took more than 10 years. Besides the clarification of the mechanism and the causes of the failure, this is study is unique in a sense it has both slow and rapid dynamic processes. The authors explain the results of these studies and investigations undertaken so far and discuss the implications of this failure event.

2 GEOLOGICAL SETTING

Tufa and travertine are common carbonate deposits in Quaternary and present-day depositional systems (Chafetz and Folk 1984). Tufa deposition in the left abutment of Büyük Menderes River in Güney District in NW part of Denizli Province (western Turkey) has been accumulating for more than thousands of years. The tufa deposition in the area overlays marble and schist of Menderes Masif metamorphic bed rock Rocks belongs to Paleozoic age (Özkul et al, 2010). The tufa deposits cover an area of about 20 hectares (Fig. 1). The Güney waterfall site is an area of unique geological/natural heritage that is visited by many people, especially during spring and summer seasons. There are mainly four springs, having total flow rate about 80 l/s and discharge at 18.7–18.8 8°C. Spring waters in the tufa site are type of Ca–HCO$_3$, and have formed coalescent tufa bodies that occur at elevations between 220–400 m above the Büyük Menderes river bed. The springline tufa body contains numerous primary cavities up to cave size in some cases. There are two main caves in

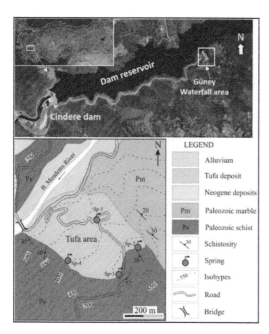

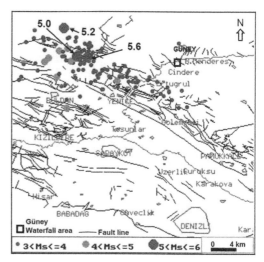

Figure 2. Fault lines and epicenter distribution of 23-26 July 2003 Buldan earthquakes (Kumsar et al, 2008)

Figure 1. Location map and geological map of Güney waterfall tufa depositional area (After Özkul et al, 2010)

the tufa site. ^{14}C age in the range from 2000 yr BP to 5800 yr BP (Özkul et al, 2010).

Fault geometry of Denizli basin consists of segments of different length along the major fault zones of Pamukkale, Babadağ, Honaz, Laodikeia and also secondary-fault zones in Denizli basin. Pamukkale fault is the major active fault bounding the basin in the North (Fig. 2) and it is about 32-35 km long (Aydan et al. 2002; Kumsar et al, 2015). The total length of Babadağ fault zone is about 40 km extending from Tekkeköy in the east to northwest of Babadağ district in the west (Fig. 2), and it consists of several segments, whose lengths change between 5 km and 10 km and trend E-W or NW-SE (Bozkuş et al, 2001; Kumsar et al, 2015)

3 RUPTURE PROCESS OF TUFA BLOCK UNDER DYNAMIC LOADING

The seismic activity in Buldan and surrounding in Denizli basin started with a 5.2 magnitude at 07:56 AM on 23rd of July 2003. Another earthquake with a magnitude of 5.6 occurred at 11:26 AM on 26th July 2003 (Fig. 2). The last shock caused damages on masonry structures, slopes failures and tension cracks on roads and slopes (Kumsar et al, 2003). Güney waterfall, which is located at about 15 km east the earthquake epicenter location, was also ruptured due to 5.6M earthquake. 13 cm wide tension crack developed at 8 m back side from the outer most of the of the waterfall tufa head (Fig. 3), and this crack continued to the base of the tufa (Kumsar et al., 2008). An illustration, showing rupture of tufa block, is given in Fig. 4a. After the earthquake, some part of spring water flowed through the cracks

Figure 3. a) Overhanging tufa block of Güney waterfall, b, c) tension cracks developed during 2003 Buldan earthquake with a magnitude of 5.6 (Kumsar et al, 2008).

and, the caves within the tufa block were filled up with water (Fig. 4b). The maximum ground acceleration caused by the earthquake at the waterfall site was estimated to be more than 120 gals.

The tufa block stayed stable for about 10 years. A big block from the most outer part of the tufa block that had been overhanging for more than 10 years, toppled on 13.05.2013 (Kumsar et al, 2013). The unstable tufa block was broken into many parts and partly moved into Cindere Dam reservoir (Fig. 5).

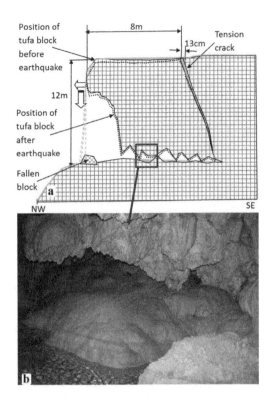

(a)   (b)

Figure 6. Views of the model before and after the experiment.

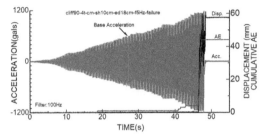

Figure 7. Input acceleration waves and measured displacement and acoustic emission responses.

particular experiment. Fig. 7 shows the input acceleration waves to the shaking table together with displacement and acoustic emission responses of the failed part of the cliff. As noted from the figure, the overhanging block fails in tension after certain acceleration level and moves towards valley side. Acoustic emission events starts long before the breakage of the overhanging part.

5 ANALYSES OF RUPTURE AND FAILURE PROCESS OF THE TUFA BLOCK

5.1 Analyses of rupture process

5.1.1 Analyses by theoretical methods

Aydan and Kawamoto (1992) proposed an analytical model to analyses the stability of layered rock slopes against flexural toppling failure under static and dynamic conditions. We utilize the method of Aydan and Kawamoto (1992) for a single layer with the consideration of shape of the overhanging body and horizontal and vertical seismic coefficients. Based on this concept, the force condition acting on the overhanging cantilever beam can be modeled as shown in Fig. 8. Thus, the formula for the outermost fiber tensile stress, which is maximum, can be written as follows:

$$\sigma = k_h \gamma h_b L \left(\frac{1+\alpha}{2} \right) + 6 \frac{M_o}{h^2} \qquad (1)$$

where

$$\alpha = \frac{h_s}{h_b}; \quad M_o = (1+k_v)\gamma h_b L^2 \left(\frac{1}{2} - \frac{(1-\alpha)}{3} \right)$$

x, h_b, h_s, γ and L are distance from the base, beam height at the base and at the far end, unit weight of

Figure 4. a) Schematic drawing of tension crack development within the tufa block of Güney waterfall, b) view from a cave within the tufa block.

Figure 5. A view of the collapsed waterfall tufa and disintegration from the Cindere Dam reservoir.

4 A MODEL STUDY FOR THE RUPTURE OF TUFA BLOCK UNDER DYNAMIC LOADING

The second author have performed many studies on the static and dynamic stability of the overhanging cliffs. The situation of the waterfall cliffs resembles to the overhanging cliffs (Aydan 2015; Aydan et al. 1989b, 2007, 2011; Tokashiki and Aydan 2010). Fig. 6 shows some views of a model experiment on a cliff model with toe erosion using a shaking table (Aydan et al. 2011). The height of the overhanging part was 100 mm high and the erosion depth was 180 mm in this

459

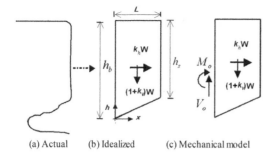

(a) Actual   (b) Idealized   (c) Mechanical model

Figure 8. Modeling of overhanging cliffs.

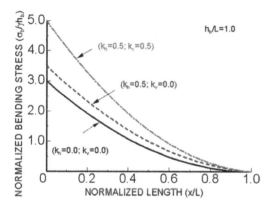

Figure 9. Distribution of bending stress at the outermost fiber of beam subjected to different seismic coefficients.

rock mass and erosion depth, respectively. k_h and k_v are horizontal and vertical seismic coefficients.

As discussed by Aydan and Kawamoto (1992), the cantilevers fail immediately once the tensile stress exceeds the tensile strength of rock mass. In addition to gravitational load, the seismic loads would make the cliffs more vulnerable to failure during earthquakes as seen in Fig. 9. As noted from Fig. 9, the vertical seismic loading has greater effects on the stability, compared to that of the horizontal seismic loading.

The height and length of the overhanging part were measured to be 10 m and 8 m, respectively in view of site observations and investigations as shown in Fig. 4(a). Under gravitational conditions with a unit weight of 20 kN/m³ for rock mass, the maximum tensile stress at the top of the overhanging waterfall tufa would be 384 kPa.

Although there was no strong motion stations near the waterfall, the ground motions were recorded at the Sarayköy strong motion station of Turkish National Strong Motion Network (DAD-ERD, 2003). The maximum ground acceleration was 155 gals at Sarayköy with a vertical acceleration of 154 gals (Fig. 10). As the distances of Sarayköy and the waterwall from the epicenter of the 2003 Buldan earthquake were quite similar, the earthquake is likely to increase the maximum tensile stress to 467 kPa with the consideration of

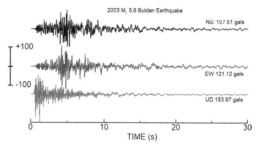

Figure 10. Acceleration records at Sarayköy.

horizontal and vertical components of the acceleration records. In view of the mass tensile strength of tufa deposits, the rupture of the overhanging part of the waterfall tufa would be possible as observed in the case of this waterfall.

5.2 Analyses of toppling stage

Aydan et al. (1989) analyzed various modes of failure of blocky rock mass slopes with the consideration dynamic limiting equilibrium method, which also includes the those of a single block. Although the initiation can be easily estimated from that approach, the utilization of numerical integration techniques is necessary to evaluate the displacement and rotation of the falling tufa block during the motion stage. This approach is used herein. When the overhanging tufa was ruptured by the earthquake, it was separated and fallen over the base and it has became an individual single block as illustrated in Fig. 4a. The mass center of the fallen block was such that it was stable against toppling failure. The bottom side of the fallen block was jagged. The protruding parts of the jagged bottom side of the fallen block could had been damaged during the falling stage. Furthermore, the high stress concentration should also occur at protruding parts. The seeping water from cracks, high stress concentration may lead further deterioration,

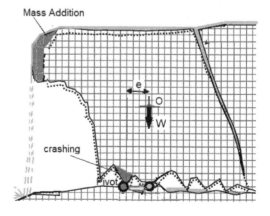

Figure 11. An illustration of the mechanism leading to the toppling of the fallen tufa block.

460

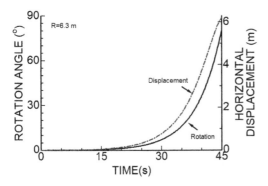

Figure 12. Rotation angle and horizontal responses of the fallen rock block during toppling stage.

degradation and crashing of the tufa at the protruding parts. As a result of this process, the pivot of the fallen tufa block against rotation should have started in space. Furthermore, the mass addition to the already fallen tufa block due to the sedimentation from water flow having dissolved Ca–HCO$_3$ content would occur. These two processes should had prepared the conditions suitable for toppling failure of the fallen block after a period of 10 years. This situation is conceptually illustrated in Fig.11. The horizontal rate of the movement of the pivot in the space has been estimated to be about 20 mm/year. Furthermore, the mass addition of 10 cm thick and 3-4 m high would had been taken place. This is a slow dynamic process taking place in years.

The block would stay almost in the same position until the horizontal location of the pivot is becomes beyond the horizontal position of the fallen block provided that the vertical position of the mass center does not change with time. Once the location of the pivot becomes beyond the horizontal location of the mass center, the fallen block would start to topple and it would be a very rapid dynamic process. Until the blocks hits the ground, the process is quite simple and it can be easily evaluated as a rigid body motion. However, the process would become quite complicated if the block starts to disintegrate after hitting the ground and/or fracturing during motion in space. Fig. 12 shows the computed rotation angle and horizontal displacement responses of the block shown in Fig. 11. As it is noted from the figure, the toppling rapidly occurs within 45 seconds while the preparation of the fallen block to topple took more than 10 years.

5.3 *Analyses of post-toppling stage*

Models, based on the motion of the failed body, which may be a monolithic mass or several slices, are most commonly used models to obtain estimations of travel distances and the reach angle of slope failures (e.g. Aydan et al. 1992; Aydan 2016). The motion of the failed body on a prescribed basal plane is estimated from the equation of motion with

the consideration of effective sliding resistance or rolling frictional resistance. Despite very complicated nature of the problem, the reach angle (α_c) for describing the motion of the failed body can be obtained from the following relation:

$$\tan \alpha_c = \frac{H_g}{L_g} = (1 - r_u) \tan \varphi_s \qquad (2)$$

where H_g: vertical drop of the failed mass center; L_g: horizontal travel distance of the failed mass center; r_u: pore-water pressure coefficient; φ_s: sliding friction angle. There are different concepts to estimate the sliding friction angle. One of them is to use the rolling friction coefficient. Eq. (2) can be used as it is by just replacing the sliding friction coefficient by the rolling friction coefficient. If the motion path has a V-shape, the equivalent friction angle may be greater than for the planar surface and the wedge factor of the V-shape path can be introduced as (e.g. Kumsar et al. 2000; Aydan and Kumsar, 2010)

$$\tan \varphi_{eq} = \frac{\cos \omega_1 + \cos \omega_2}{\sin(\omega_1 + \omega_2)} \cdot \tan \varphi_s \qquad (3)$$

where ω_1 and ω_2 are the inclinations of the side walls of the path from vertical and they may be distance dependent. It should be noted that the following relation holds:

$$2\omega = \omega_1 + \omega_2 \qquad (4)$$

The tangent of the reach angle of the fallen tufa block is about 0.61. The friction angle between the fallen tufa block and side-walls of the valley is about 32 degrees and the wedge angle is about 150 degrees. As the water flows through the creek, which constitutes the path of block motion, the water may affect the motion of the block. As it is a small creek with small amount of water, the water pressure coefficient (r_u) was estimated to be about 0.05 in view of water height and partial submergence of the block after it stopped the motion. The reach angle from Eq.(2) can be obtained as 31.4 degrees, which is almost the same as that observed.

6 CONCLUSIONS

The authors attempted to clarify the mechanism and rupture and collapse processes of the Güney waterfall in Denizli Province of Turkey. The process involves both slow and rapid dynamics. It was shown that the ruptude was caused by 2003 Buldan earthquake in view of presented model experiments and theoretical considerations. The second stage was a slope process involving the growth of the ruptured block in size and progress of degradation at the toe of the failed block, which took more than 10 years. This process was simulated by taking into account the facts in computational models. The third stage was the final collapse and movement in the valley towards the Cindere Dam reservoir. This is study is a unique contribution to the dynamics of rockfall involving slow and rapid processes.

REFERENCES

Aydan, Ö., 1989. The stabilization of rock engineering structures by rockbolts. Doctorate Thesis, Nagoya University, 204 pages.

Aydan, Ö. 2015. Large Rock Slope Failures induced by Recent Earthquakes. Rock Mechanics and Rock Engineering, Special Issue on Deep–seated Landslides. 49(6), pp. 2503–2524.

Aydan Ö., Kawamoto T (1992) The stability of slopes and underground openings against flexural toppling and their stabilisation. Rock Mech Rock Eng 25(3):143–165.

Aydan Ö., Shimizu Y, Ichikawa Y. 1989a. The effective failure modes and stability of slopes in rock mass with two discontinuity sets. Rock Mech Rock Eng 22(3):163–188.

Aydan, Ö., T. Kyoya, Y. Ichikawa, T. Kawamoto and Y. Shimizu, 1989b. A model study on failure modes and mechanisms of slopes in discontinuous rock mass. *The 24ᵗʰ Annual Meetings of Soil Mechanics and Foundation Eng. of Japan*, Miyazaki, 415, 1089–1093.

Aydan, Ö., Shimizu Y., Kawamoto T. 1992. The reach of slope failures. *The 6th Int. Symp. Slope failures, ISL 92*, 1, 301–306, Christchurch.

Aydan, Ö., Kumsar, H., Ulusay, R., 2002. How to infer the possible mechanism and characteristics of earthquakes from the striations and ground surface traces of existing faults. JSCE, Earthquake and Structural Engineering Div. 19 (2): 199–208.

Aydan Ö, Tano, H., Watanabe, H., Ulusay R. & Tuncay, E. 2007. A rock mechanics evaluation of antique and modern rock structures in Cappadocia Region of Turkey. *Proceedings of Symposium on the Geology of the Cappadocia Region*, Niğde, Turkey, October 2007, 13–23.

Aydan, Ö., Ohta, Y., Daido, M., Kumsar, H. Genis, M., Tokashiki, N., Ito, T., Amini, M. 2011. Chapter 15: Earthquakes as a rock dynamic problem and their effects on rock engineering structures. *Advances in Rock Dynamics and Applications*, Editors Y. Zhou and J. Zhao, CRC Press, Taylor and Francis Group, 341–422.

Bozkuş, C., Kumsar, H., Özkul, M., Hançer, M., 2001. Seismicity of Active Honaz fault under an extensional tectonic regime. Int. Earth Sciences Collq. On the Aegean Region, Izmir, Turkey, IESCA–2000, Dokuz Eylul Univ., pp 7–16

Chafetz, H.S. and Folk, R.L. 1984. Travertines: depositional morphology and the bacterially constructed constituents. Journal of edimentary Petrology, 54, 289–316.

Kumsar, H., Aydan, Ö. and Ulusay, R. 2000. Dynamic and static stability of rock slopes against wedge failures. *Rock Mechanics and Rock Engineering*, Vol.33, No.1, 31–51.

Kumsar, H., Çelik, S. B., and Aydan, Ö., 2003, Some Characteristics of recent earthquakes in western Turkey: Seferihisar (Izmir) and Buldan (Denizli) Earthquakes in 2003. An International Colloquium on The Instrumentation and Monitoring of Landslides and Earthquakes in Japan and Turkey, November, 8, 2003, Japan, pp 109–120.

Kumsar, H., Aydan. Ö., Tano, H., Çelik, S.B., 2008. An investigation of Buldan (Denizli) Earthquakes of July 23–26, 2003 (in Turkish), Ekin Basın Yayın Dağıtım, Bursa, 118 s.

Kumsar, H., Aydan, Ö., Şimşek, C., D'Andria, F., 2016. Historical earthquakes that damaged Hierapolis and Laodikeia antique cities and their implications for earthquake potential of Denizli basin in western Turkey, Bulletin of Engineering Geology and the Environment, 74, 1037–1055.

Kumsar, H., Özkul, M., Çobanoğlu, İ., Çelik, S.B. 2013. Güney şelalesi kaya düşmesi sahasina ait gözlemsel içerikli jeolojik ön değerlendirme raporu, Unpublished report (in Turkish), Pamukkale University, 14 p.

Özkul, M., Gökgöz, A., Horvatinčić, N., 2010. Depositional properties and geochemistry of Holocene perched springline tufa deposits and associated spring waters: A case study from the Denizli province, Western Turkey. In: Pedley, H.M. (Ed.), Tufas and Speleothems: Unravelling the Microbial and Physical Controls. The Geological Society, London. Special Publications 336, pp. 245–262.

Tokashiki, N. Aydan, Ö. 2010. The stability assessment of overhanging Ryukyu limestone cliffs with an emphasis on the evaluation of tensile strength of rock mass. JSCE, Vol. 66, No.2, pp. 397–406.

2019 Rock Dynamics Summit– Aydan et al. (eds)
© *2019 Taylor & Francis Group, London, ISBN 978-0-367-34783-3*

Numerical simulation on progressive failure in rock slope using a 3D lattice spring model

Tsuyoshi Nishimura & Masanori Kohno
Tottori University, Tottori, Japan

ABSTRACT: In slope stability analyses, the failure surface is often assumed to be predefined as a persistent flat or circular concave plane and the slide resistance along the plane is evaluated. Though this procedure gives the safety factor based on the limit equilibrium theory, the existence of such plane is highly unlikely, and a complex interaction between pre-existing flaws, stress concentration and resulting crack generation, these are not modeled. The development of advanced numerical methods is the key issues of importance. This paper attempts to develop a numerical procedure providing means to analyze the kinetic failure process and the state of stability using a discrete approach. The ground is modeled with an assembly of mass points connected by a pair of springs, normal and tangential direction and the translational motion of each mass point is calculated by solving the equation of motion. The stress state is evaluated on each mass point.

1 INTRODUCTION

Slope instability occurs in many parts of urban and rural areas and causes damages to housing, roads, railways and other facilities. Slope engineering has always involved some form of risk management and this has led to the process of the identification and the characterization of the potential slope failure together with evaluation of their frequency of occurrence. An essential part of the hazard (slope failure) identification is the prediction in terms of the character of failure (type, volume), the post-failure motion (travel distance, velocity) and the state of activity (Fell, et al. 2008). The literatures of slope stability analysis using the limit equilibrium method (LEM) and the finite element method (FEM) were reviewed by Duncan (1996), and a number of valuable lessons concerning the advantages and limitations of the methods for use in engineering problems were presented. Jing and Hudson (2002) presented a review of the techniques, advances problems and future development directions in numerical modeling for rock mechanics and rock engineering. The expanded version of the brief review was presented by Jing (2003) and he has suggested that computer methods available can be still inadequate when facing the challenge of practical problems, especially when representation of rock fracture systems and fracture behavior are a pre-condition for successful modeling. Despite of all the advances in both continuum and discrete approaches, the development of advanced numerical methods is the key issues of importance. This paper attempts to develop a numerical procedure providing means to analyze the kinetic failure process and the state of stability as a function of a trial gravitational acceleration to a lattice spring model. This procedure is also able to explain a possible depth and volume of failure at the site.

Most rock slopes are inhomogeneous structures comprising anisotropic layers of rock characterized by different material properties, and they are often discontinuous because of jointing, bedding and fault. In rock slope stability analyses, the failure surface is often assumed to be predefined as a persistent plane or series of interconnected planes, where the planes are fitted to the surfaces based on the structural observation. Such assumptions are partly due to the constraints of the analytical technique employed (e.g. limit equilibrium method, the distinct element method, etc.) and can be valid in cases in which the response of single discontinuity or a small number of discontinuities is of critical importance on the stability. However, especially on a large scale slope, it is highly unlikely that such a system of fully persistent discontinuous planes exists a priori to form the failure surface. Instead, the persistence of the key discontinuities may be limited and a complex interaction between pre-existing flaws, stress concentration and resulting crack generation, is required to bring the slope to failure. In small engineered slopes, excavation gives significant changes in stress distribution in the slopes and may generate fully persistent planes keep propagating with stress redistribution. Larger natural rock slopes seldom experience such a disturbance and have stood in relatively stable features over the period of thousands of years. This does not imply that in natural rock slopes a system of discontinuities may not be interconnected developing the portion of where the failure surface will be formed. Strength degradation may occur in rock mass with time-dependent manner and drive the slope unstable state. Thus, rock slope instability problem requires the progressive failure modeling to drive the slope to catastrophic events.

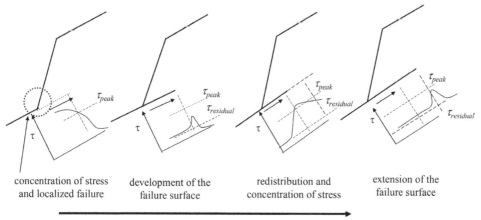

concentration of stress and localized failure

development of the failure surface

redistribution and concentration of stress

extension of the failure surface

progressive failure sequence

Figure 1. Failure surface initiation and its propagation (after Bjerrum (1967))

2 MODELING OF THE FAILURE SURFACE

Based on the coulomb shearing strength criterion, both the shearing resistance depending on the normal stress value (i.e. the frictional strength) and the cohesion of intact rock between discontinuous joints resist to shear failure. At the tip of the joints, stresses would increase and subsequent failure in rock would occur. Progressive failure in rock mass would involve the failure of intact rock as their strength is exceeded. There have been a number of investigations focused on the failure process of rock slope using the finite element method and the boundary element method.

Kaneko et al. (1997) used the displacement-discontinuity method (DDM) and fractures' principles to model the progressive development of shear crack in rock slope. In their analysis, rock material was assumed to be homogeneous and any pre-existing cracks were not considered. They compared the DDM results with the conventional limit equilibrium method (LEM) and discussed the allowable slope height under the given strength parameters and the slope angle. Eberhardt et al. (2004) discussed some aspects of the modeling of progressive

failure surface development linking initiation and degradation to eventual catastrophic slope failure, using a hybrid method that combines both continuum and discontinuum numerical techniques to model fracture propagation. They concluded that the use of the hybrid modeling technique helped to provide important insight as to the underlying mechanism, focusing on the example of a rockslide in the Swiss Alps.

Developments in the field of rock slope analysis were reviewed by Stead et al. (2006). They have attempted to illustrate the wide range of tools available with particular emphasis on emerging powerful modeling of hybrid techniques that allow realistic simulation of rock slope failure. Generally, there are two choices of hybrid technique; technique of hybrid continuum-/discrete-element method and technique based on a discrete modeling. In the first choice, fracture plane is aligned in two ways shown in Figure 2.

1. a process known as intra-element fracturing where a series of new nodal points and elements are systematically created as shown Figure 2(b).
2. a process known as inter-element fracturing where a series of new nodal points are systematically created but no new element is generated as shown Figure 2(c). This process is usually preferred from the computational standpoint and the discrete fracture orientation is aligned with the best oriented element boundary attached to the node considered.

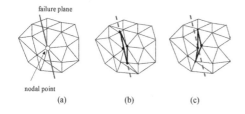

failure plane

nodal point

(a)   (b)   (c)

Figure 2. Modeling of crack generation: (a):element discritization and failure plane, (b)continuous discritization of elements along the direction of new cracks, (c) a priori placement of crack element along the possible direction of the crack propagation.

The presence of contact phenomena during post fracturing requires an effective contact simulation procedure. One fundamental advantage of the discrete modeling is that the phenomena can be incorporated into the method by the rigid body or the mass point contact with the spring-dashpot system. This method has been used to investigate a wide variety of rock failure mechanisms. In these analyses, cracks propagate with the path calculated by the failure criterion under the stress state. This paper presents a numerical trial to simulate the progressive development of failure surface in slope.

3 DISCRETE ELEMENT MODELING FOR ELASTICITY

The body to be analyzed is divided into small portions and is represented by a system of points linked together with the neighboring points. The particles and bonds form a network system representing the material. For this system, the equation of motion can be expressed as

$$\mathbf{m\ddot{u} + c\dot{u} + ku = f} \tag{1}$$

where $\mathbf{u}$ represents the vector of particle displacement, $\mathbf{k}$ is the stiffness, $\mathbf{m}$ is the mass matrix, $\mathbf{c}$ is the damping matrix and $\mathbf{f}$ is the vector of external force. Equation (1) is solved by using the explicit finite difference scheme.

$$\mathbf{\ddot{u}}_t = \frac{1}{\mathbf{m}}\left[\mathbf{f} - \mathbf{c\dot{u}}_t - \mathbf{ku}_t\right] \tag{2}$$

The displacement of particle at time $t+\Delta t$ can be expressed as

$$\Delta\mathbf{u}_t = \Delta\mathbf{u}_{t-\Delta t} + \mathbf{\ddot{u}}_t\Delta t^2 \tag{3}$$

where $\Delta\mathbf{u}_t = \mathbf{u}_{t+\Delta t} - \mathbf{u}_t$. The new position of particle is also expressed as

$$\mathbf{u}_{t+\Delta t} = \mathbf{u}_t + \mathbf{\dot{u}}_t\Delta t + \mathbf{\ddot{u}}_t\Delta t^2/2 \tag{4}$$

The particle velocity at t is given by

$$\mathbf{\dot{u}}_t = \frac{\mathbf{u}_t - \mathbf{u}_{t-\Delta t}}{\Delta t} + \mathbf{\ddot{u}}_t\Delta t/2 \tag{5}$$

To keep the computation stable, the time step could be chosen as

$$\Delta t < \min\left(\frac{d_0}{c_\mathrm{p}}\right) \tag{6}$$

where d_0 is the contact length, that is the distance between particles, and c_p is the P-wave velocity. For static simulation, the equations of motion are damped to reach an equilibrium state under given boundary conditions as quickly as possible. In this modeling, the damping effect is incorporated as written in the following.

$$\mathbf{\ddot{u}}_t = \frac{1}{\mathbf{m}}\left(\sum_t \mathbf{f} - \mathrm{sgn}(\mathbf{\dot{u}}_{t-\Delta t})\cdot\alpha\left|\sum_t \mathbf{f}\right|\right) \tag{7}$$

where α is the damping constant which is independent of mechanical properties of the material.

We assume that the medium is loaded from zero condition to an initial condition defined by the strain e_{ij} and the stress σ_{ij}. For such a system, one can write the displacement for a particle (or element) p with position x_i as follows

$$u_i^\mathrm{p} = e_{ij}x_j^\mathrm{p} \tag{8}$$

where e_{ij} should be a symmetric tensor which is calculated by removing an asymmetric tensor from the displacement-gradient tensor. The rotation-related term is removed from the relative translational displacement between these two points, then this method ensures that the calculated strain is independent of rotational displacement (Nishimura, T. et. al. 2014).

The following contact law, which relates normal and shear forces $F_{(\mathrm{n})}$, $F_{(\mathrm{s})}$ to normal and shear relative displacements $U_{(\mathrm{n})}$, $U_{(\mathrm{s})}$ holds at the contact

$$F_{(\mathrm{n})} = k_\mathrm{n}U_{(\mathrm{n})}, \quad F_{(\mathrm{s})} = k_\mathrm{s}U_{(\mathrm{s})} \tag{9}$$

Let us assume that a contact m connects two particles p1 and p2, then the normal and shear relative displacements can be written as

$$U_{(\mathrm{n})}^m = \Delta u_i^m I_i^m \tag{10}$$

$$U_{(\mathrm{s})i}^m = \Delta u_i^m - U_{(\mathrm{n})}^m I_i^m \tag{11}$$

where the relative displacement at the contact Δu_i^m is given as

$$\Delta u_i^m = e_{ij}\left(x_j^{\mathrm{p}1} - x_j^{\mathrm{p}2}\right) = e_{ij}d_b^m I_j^m \tag{12}$$

where I_i^m is the normal unit vector. The total force f_i at contact m can be written as

$$f_i^m = k_\mathrm{n}^m\Delta u_j^m I_j^m I_i^m + k_\mathrm{s}^m\left(\Delta u_i^m - \Delta u_j^m I_j^m I_i^m\right) \tag{13}$$

$$= \left(k_\mathrm{n}^m - k_\mathrm{s}^m\right)\left(e_{kl}I_k^m I_l^m\right)I_i^m d_b^m + k_\mathrm{s}^m e_{ij}I_j^m d_b^m$$

Notice that Einstein summation convention with dummy subscript i, j, k, l is used in the preceding equations. The total strain energy stored per unit volume is

$$\Pi = \frac{\Pi_b}{V} = \frac{1}{V}\sum_{m=1}^{N_c}\frac{1}{2}\left(e_{ij}d_b^m I_j^m f_j^m + e_{ji}d_b^m I_i^m f_i^m\right) \tag{14}$$

where N_c is the number of contacts inside the medium, V is the medium volume. The stress tensor of the continuum can be obtained through the classical elastic theory, and it can be written as (assuming symmetrical stress) (Walton, K. 1987, Richard, J. B. & Leo, R. 1988).

$$\sigma_{ij} = \frac{1}{2V}\sum_{m=1}^{N_c}\left(d_b^m I_i^m f_j^m + d_b^m I_j^m f_i^m\right) \tag{15}$$

From Equations (13) and (15), we end up with

$$\sigma_{ij} - \frac{1}{V}\sum_{m=1}^{N_c}\frac{1}{2}\left(k_s^m e_{jl}I_l^m I_i^m d_b^{m2} + k_s^m e_{il}I_l^m I_j^m d_b^{m2}\right.$$

$$\left. + \left(k_n^m - k_s^m\right)e_{kl}I_i^m I_j^m I_k^m I_l^m d_b^{m2}\right) \quad (16)$$

The constitutive matrix C_{ijkl} in the classical elasticity theory is expressed as

$$\sigma_{ij} = C_{ijkl}e_{kl} \quad (17)$$

and finally, by substituting Equation (17) into Equation (16), C_{ijkl} can be given as

$$C_{ijkl} = \frac{1}{V}\sum_{m=1}^{N_c}\left(\frac{k_s^m d_b^{m2}}{4}\left(I_j^m I_k^m \delta_{il} + I_i^m I_k^m \delta_{jl} + I_j^m I_l^m \delta_{ik}\right.\right.$$

$$\left.\left. + I_i^m I_l^m \delta_{jk}\right) + \left(k_n^m - k_s^m\right)d_b^{m2}I_i^m I_j^m I_k^m I_l^m\right) \quad (18)$$

where δ_{ij} is the Kronecker's delta. Then, the relationship the micro-mechanical parameters k_n, k_s and the macro material constants the Young' modulus E_0 and the Poisson's ratio v_0 can be obtained by comparing Equation (18) to the classical elastic matrix. As seen in this equation, the values of the stiffness are influenced by the normal unit vector, therefore, the lattice structure should be carefully modeled. A 18-bond cubic lattice geometry as shown in Figure 3 is prepared. The micro-mechanical parameters k_n, k_s are obtained by

$$k_n = \frac{E_0 d_0}{5(1-2v_0)} \quad (19)$$

$$k_s = \frac{(1-4v_0)E_0 d_0}{5(1+v_0)(1-2v_0)} \quad (20)$$

Equation (20) shows that the spring stiffness of negative occurs when v_0 is greater than 0.25. For this condition, we have failed to obtain the quasi-static state under a given static boundary condition and the model shows unstable behavior. Therefore, numerical results shown in this paper are carried out for $v_0 < 0.25$. This negative effect of the Poisson's ratio in such lattice modeling has also been reported (e.g. Zhao, G. F. Fang, J. & Zhao, J. 2011).

4 NUMERICAL SIMULATION ON PROGRESSIVE FAILURE IN SLOPE

Figure 3 shows the slope model which is formed by the discrete lattice spring model. The height of slope h is 100m and the bottom width w is 200m. Slope inclination is expressed with $\tan\beta=2$ ($\beta=63.7°$). The Young's modulus and the Poisson's ratio are set 1000MPa and 0.24 respectively. Figure

4 shows the results of τ_{xy} for g=9.8m/s² (1G) and the stress concentration is recognized around the slope toe. Then, the gravitational acceleration is assumed to increase as nG (n>1). This condition is to load the slope model and to simulate the progressive development of failure surface and slope instability. We have introduced the Mohr-coulomb failure criterion written as;

$$F = (\sigma_1 - \sigma_3) + (\sigma_1 + \sigma_3)\sin\phi - 2c\cos\phi \quad (21)$$

σ_1: the major principal stress, σ_3: the minor principal stress, c: cohesion, ϕ: frictional angle. When $F=0$ the stress state fulfills the failure criterion and $F>0$ represents by points outside the failure surface. Plastic strains may develop during increment from an initial state belonging to the yield surface ($F=0$). The plastic strain is governed by the plastic flow rule and then the corresponding stress increment to the elastic strain is evaluated so as to hold the stress state on/inside the yield surface. This procedure is the same to the ordinal numerical modeling of plasticity.

Figure 5 demonstrates the transition from stable slope conditions to those of shear failure by showing the evolution of failure points. The parameter related to the strength are set to c=35kPa, $\phi=50°$ and these values keep constant and no reduction is assumed during the loading. Though the assumptions are introduced, this numerical modeling can be used to examine the evolution of stresses strain and failure surface development within the rock slope. From

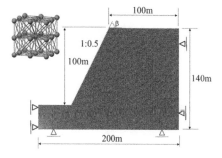

Figure 3. Rock slope model with constant inclination using the assembly of circular element.

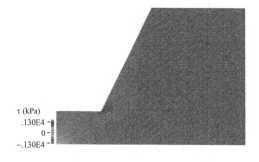

Figure 4. Distribution of τ_{xy}

466

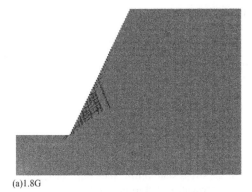

(a)1.8G

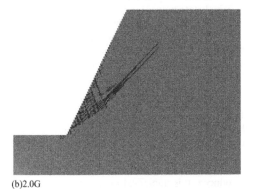

(b)2.0G

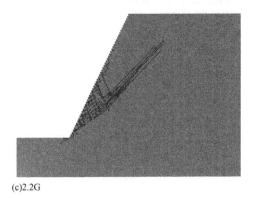

(c)2.2G

Figure 5. Progressive development of failure surface.

these figures, the shear-based Mohr-coulomb fracture surface has propagated from the toe of the slope model to the inner part. These figures also involve the progressive fracturing cross to the shear-based surface mentioned the above. To more closely examine these mechanisms for the development, it can be recognized that firstly the shear-based fracture surface has propagated from the toe of the slope and secondly the tensile crack propagate in the slide mass as illustrated in Figure 6. The failure criterion used in this simulation is set up to explain tensile failure without the inclusion of a predefined failure element and the stress state on the mass point calculated by Equation

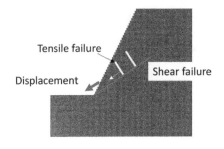

Figure 6. Development of shear surface and propagation of tensile damage upwards through slide mass, dividing the slide mass into blocks.

(15) and examined by Equation (21), the stress state is inside the failure criterion or outside including tensile fracturing. Results show that a zone of yield due to shear damage develops near the toe of slope and this development transforms into a tensile failure in the slide body. This transformation of failure continues through the body and divides the body into blocks. These progressive fracturing must lead the mass to collapse of the frontal region of the slope. This model results agree with the analysis by Eberhaldt (2004).

5 CONCLUSION

Various numerical methods (continuum and discontinuum methods, and hybrid methods which combine both continuum and discontinuum techniques to simulate fracturing process) have been applied to demonstrate the evolution of failure in rock slope. In this paper, a numerical modeling of progressive failure in rock mass using the discrete lattice spring model is introduced to simulate the rock slope failure. The modeling could give a possible failure volume of rock material based on the geometrical data and strength properties, such as cohesion and internal friction angle. No decisive statement about the effect of the mechanical parameters can be made because the analysis was performed under the limited input values. On going work should be done incorporating the effects of the macro parameters and the initial stress condition.

ACKNOWLEDGEMENT

This work is partly supported by Japanese Society for the Promotion of Sciences (JSPS), Grant-in-Aid for Science Research, No.17K06554.

REFERENCES

Bjerrum, L. 1967. Progressive failures in slopes of overconsolidated plastic clay and clay shales, *Journal of Soil Mechanics and Foundation Division, ASCE,* 93(SM), 1–49.

Duncun, J. M. 1996. State of the art: Limit equilibrium and finite element analysis of slopes. *Journal of Geotechnical Engineering, ASCE*, 122(7),577–596

Eberhardt, E., Stead, D. & Coggan, J. S. 2004. Numerical analysis of initiation and progressive failure in natural rock slopes - the 1991 Randa rockslide. *International Journal of Rock Mechanics and Mining sciences*, 41(1),69–87.

Fell, R., Corominas, J., Bonnard, C., Cascini, L., Leroi, E. & Savage, W. Z. 2008. Guideline for landslide susceptibility, hazard and risk zoning for land use planning, *Engineering Geology*, 102(3/4), 85–111.

Kaneko, K., Otani, J., Noguchi, Y. & Togashiki, N. 1997. Rock fracture mechanics analysis of slope failure. *Defomation and Progressive failure in Geomechanics*, Nagoya, Japan, 671–676.

Ling, L. & Hudson, J. A. 2002. Numerical methods in rock mechanics. *International Journal of Rock Mechanics and Mining sciences*, 39(4),409–427.

Ling, L. 2003. A review of techniques, advances and outstanding issues in numerical modelling for rock mechanics and rock engineering. *International Journal of Rock Mechanics and Mining sciences*, 40(3),283–353.

Nishimura, T., Fumimura, K., Kohno, M. & Mitsuhashi, D. 2016. 3D Distinct Element-Based Model for Deformation and Failure of Rock, *Proceedings of 2016 ISRM Interna-tional Symposium - 9th Asian Rock Mechanics Symposi-um*, PO4–P74.

Richard, J. B. & Leo, R. 1988. Note on a random isotropic granular material with negative Poisson's ratio, *International Journal of Engineering Science*, 26(4),373–383.

Stead, D., Eberhardt E. & Coggan, J. S. 2006. Developments in the characterization of complex rock slope deformation and failure using numerical modeling techniques, *Engineering Geology*, 83(1/3), 217–235.

Walton, K. 1987. The effective elastic moduli of a random packing of spheres, *Journal of the Mechanics and Physics of Solids*, 35(2),213–226.

Zhao, G. F., Fang, J. & Zhao, J. 2011, A 3D distinct lattice spring model for elasticity and dynamic failure, *International Journal for Numerical and Analytical Methods in Geomechanics*, 35(8),859–885.

2019 Rock Dynamics Summit– Aydan et al. (eds)
© 2019 Taylor & Francis Group, London, ISBN 978-0-367-34783-3

Failure patterns of granular slopes subjected to dynamic impact: Experimental observations

K. Uenishi
Department of Advanced Energy, The University of Tokyo, Kashiwa, Japan

T. Goji
Department of Aeronautics and Astronautics, The University of Tokyo, Tokyo, Japan

ABSTRACT: Contrary to the firmly established mechanics of continuum media, the mechanical behavior of particles in granular media, especially that under dynamic load, has not been fully understood. Here, as an initial investigation into wave and fracture propagation inside granular media, experimental technique of dynamic photoelasticity is utilized. Penny-shaped birefringent particles are prepared and placed on a rigid horizontal plane and two-dimensional dry model slopes with some inclination are formed. Using a high-speed digital video camera, the transient stress and fracture development due to dynamic impact on the top free surface of the model slope is recorded. It is found that there exist at least two failure patterns depending on the energy profile associated with the impact: (i) total collapse of the slope or mass flow resulting from one-dimensional force-chain-like stress transfer; and (ii) dynamic, toppling failure-like separation of the slope face induced by widely spread multi-dimensional wave propagation.

1 INTRODUCTION

Besides the disastrous power of Tsunamis, one of the puzzling phenomena noticed on the occasion of the devastating 2011 off the Pacific coast of Tohoku (Great East Japan) earthquake (moment magnitude M_w 9.0) is the unique dynamic failure in slopes in the city of Sendai where open cracks, located some meters away from the upper edge (crest) in the top surface, extended parallel to that edge. It should be noted that similar tensile cracks were generated precisely in the identical slopes by the 1978 Miyagi-ken-oki earthquake (M_w 7.5; Fig. 1a), and using the conventional countermeasures that consider the influence of body waves, the slopes were reinforced with steel pipe piles after the 1978 event. This earthquake-induced failure pattern, not widely recognized, has been found also at other rockier places including the South Island, New Zealand, in 2011 (M_w 6.2) (Hancox et al. 2011) and California in 1906 (M_w 7.8), 1957 (M_w 5.7) (Sitar et al. 1980) and 1989 (M_w 6.9) (Ashford & Sitar 1997). However, the mechanical analyses of body wave interaction with continuum model slopes (e.g. Sitar et al. 1980, Ashford & Sitar 1997, Ashford et al. 1997) as well as dynamic study of granular slopes, often assumed also in the field of rock mechanics, does not seem to be able to straightforwardly reproduce the failure pattern. In typical granular models (see e.g. Cleary & Prakash (2004)), usually, the effect of wave propagation on dynamic fracture is neglected, and only granular mass flow, namely, nearly simultaneous, total collapse of the slope face, edge and top surface, is depicted.

Recently, based on a mechanical analysis of continua (Uenishi 2010: before the 2011 seismic events; Fig. 1b) as well as by a numerical particle method (Uenishi & Sakurai 2015; Fig. 2), it has been theoretically suggested that the dynamic roles played by Rayleigh surface waves are more influential than those by body waves in inducing the tensile cracks at positions away from the slope edges. According to the numerical simulations employing the original version of the moving particle semi-implicit (MPS) method (Fig. 2), body waves, not only in a conventional low frequency range but also in a higher frequency range, cannot produce such tensile cracks, and the investigation only treating body wave interaction with slopes is not sufficient to completely describe the dynamics and stability of slopes. The fact that the same slopes in Sendai, reinforced using the traditional countermeasures against body waves after the 1978 quake, were affected by the 2011 event may support this viewpoint.

Experimentally, the two-dimensional propagation of Rayleigh waves in vertical and inclined continuum model slopes made of birefringent polycarbonate has been observed by utilizing the technique of dynamic photoelasticity in conjunction with high-speed cinematography (Uenishi & Takahashi 2014), but the effect of such dynamic disturbances on fracture generation near the top surface has not been visibly confirmed so far (Fig. 3). In this contribution, therefore, instead of continua, wave and fracture propagation in a two-dimensional dry granular slope is experimentally investigated.

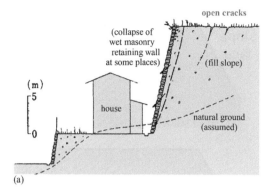

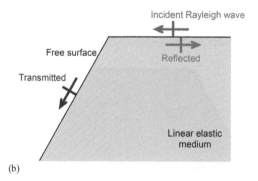

Figure 1. (a) Typical dynamic slope failure (open or tensile cracks) found in Sendai, Japan, on the occasion of the 1978 Miyagi-ken-oki earthquake (modified after Uenishi (2010)). Although the slopes were reinforced after the 1978 event utilizing the conventional countermeasures, similar open cracks were caused in the same slopes by the 2011 off the Pacific coast of Tohoku (Great East Japan) earthquake. (b) A simplified two-dimensional linear elastic model slope employed in the continua-based study. The superposition of the incident and reflected Rayleigh waves induce strong dynamic stress amplification in tension in the top surface at positions away from the edge.

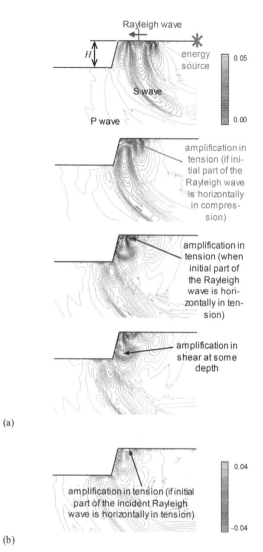

Figure 2. (a) Numerically generated, archetypal dynamic linear elastic wave field related to the Rayleigh wave interaction with a slope with $V_R T_R/H = 1$ (V_R: Rayleigh wave speed; At the energy source on the free surface, a vertical normal stress $A_R \sin(\pi t/T_R)$ (for $0 \le t \le T_R$; zero otherwise) is applied, with t, A_R and T_R being the time, the amplitude and duration of the stress pulse). The time-dependent evolution of the maximum in-plane shear stress (τ_{max}/A_R) is drawn. The slope inclination angle is 75 degrees and Poisson's ratio of the medium is 0.25. The time elapsed from the energy source excitation is $V_R t/H = 2.762$, 3.399, 3.930 and 4.037 (parts of the figures modified after Uenishi & Sakurai (2015)). (b) Distribution of the principal stress σ_1 (tension positive) at $V_R t/H = 3.930$, showing large amplification in tension in the horizontal direction owing to the interaction between the incident, reflected and transmitted Rayleigh waves. The initial (latter) part of the incident Rayleigh wave here is horizontally in tension (compression) on the free surface, respectively. If the initial part of the incident Rayleigh wave is originally horizontally in compression, superposition of the initial part of the reflected (now in tension) and the latter (originally tensile) part of the incident Rayleigh waves takes place earlier at $V_R t/H = 3.399$. See Figure 2a.

2 DYNAMIC FRACTURE IN A DRY GRANULAR MODEL SLOPE

Here, mechanical properties of particles in dry granular media under dynamic impact are studied by looking into possible relation between waves and fracture in such granular media. For this purpose, besides the basic examination into transient gravity-induced granular mass flow from a (semi-) cylindrical column filled with dry glass beads (Uenishi & Tsuji 2008, Uenishi et al. 2009), dynamic photoelastic study has been performed regarding the stress transfers inside granular media due to impact loading (Uenishi et al. 2017). In this preliminary work, the granular media are composed of penny-shaped birefringent particles (diameter 20 or 40 mm, thickness 10 mm) that are made of epoxy resin, and it has been shown that in a single particle system, i.e. if only one particle is placed on a horizontal rigid plane, dynamic wave propagation can be recognized

470

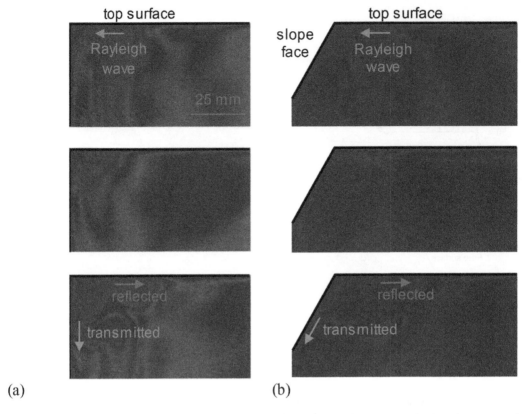

(a)　　　　　　　　　　　　　　(b)

Figure 3. Experimentally obtained typical isochromatic fringe patterns (contours of the maximum in-plane shear stress τ_{max}) associated with dynamic Rayleigh wave-slope interaction for (a) vertical and (b) inclined (inclination angle 60 degrees) model slopes made of polycarbonate plates. For both models, the temporal interval between each photograph is some 27 microseconds (modified after Uenishi & Takahashi (2014)). Reflected and transmitted Rayleigh waves can be clearly identified in both cases, but the initial dynamic impact given to the top surface by the airsoft gun-launched projectile is not sufficient to visibly induce fracture in the polycarbonate slopes.

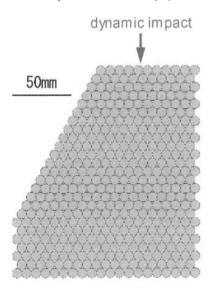

Figure 4. Archetypal two-dimensional model slope made of penny-shaped polycarbonate particles and used for the more recent series of dynamic fracture experiments.

inside that single particle. On the other hand, in a layered multi-particle system, if loosely packed, stress seems to be rather quasi-statically transferred from one particle to another without clear existence of multi-dimensional wave propagation.

Therefore, in the more recent experiments, using a laser cutter, smaller, easier-to-pack penny-shaped particles (diameter 8 mm, thickness 3 mm) are cut out from birefringent polycarbonate plates, and for instance, 407 out of them are stacked on a rigid horizontal plane and a two-dimensional model slope with certain inclination angle (60 degrees in Fig. 4) is formed. Dynamic impact is given to the top free surface of the model slope by free-falling button-shaped aluminum (Fig. 5) or by an airsoft gun-launched projectile (Fig. 6). The time-dependent variations of stress and development of fracture network are recorded with a high-speed digital video camera at a frame rate of 50,000 frames per second. The snapshots taken by the video camera indicate the presence of at least two failure patterns in the granular slope, which depends on the profile (time history) of energy imparted to the slope by the impact. One is mass flow related to one-dimensional stress

Figure 5. Experimentally recorded movement of each photoelastic particle in a two-dimensional model slope. The particle motion and the isochromatic fringe patterns inside the particles are induced by free-fall of button-shaped aluminum (diameter 20 mm, thickness 10 mm, mass 8.4 grams, impact velocity 3.5 m/s). The time elapsed after the topmost snapshot, which shows the slope just before dynamic impact, is 600, 1200, 1800 and 4000 microseconds (modified after Uenishi & Goji (2018)).

Figure 6. Now the particle movement and the isochromatic fringe patterns in the model slope are depicted for the case where the initial impact energy is given by a projectile launched using an airsoft gun (sphere of diameter 6 mm, mass 0.2 grams, impact velocity 76 m/s). The time elapsed after the topmost snapshot is 300, 600, 900 and 2000 microseconds (modified after Uenishi & Goji (2018)).

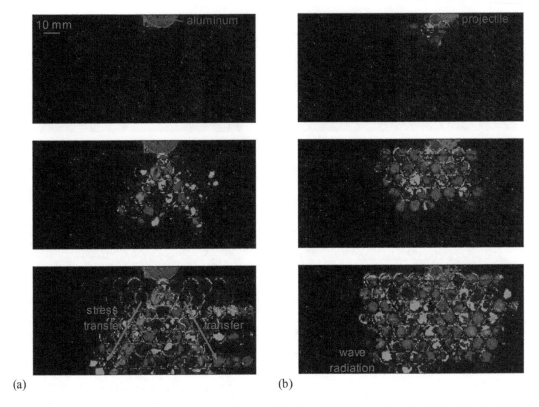

(a)   (b)

Figure 7.(a) One-dimensional (straight) stress transfer in two directions inside the granular slope induced by free-falling aluminum (Fig. 5) and (b) broadly spread two-dimensional waves generated by the impact of the projectile (Fig. 6). The stress change in each particle is evaluated by the difference of colors between the experimentally taken photographs (like those in Figures 5 and 6) before and after the dynamic impact. The time after the topmost image is (a) 160 and 320 microseconds and (b) 40 and 80 microseconds, respectively (modified after Uenishi & Goji (2018)).

transmission that is akin to force chains observed in quasi-static models (Figs 5, 7a; hereafter pattern (i)). The other is separation of the slope face only, which is similar to toppling failure and seems to be owing to widely spread two-dimensional waves (Figs 6, 7b; pattern (ii)). In the pattern (i), tensile fracture initially develops just beneath the point of impact while in (ii) severe opening occurs at the slope face, that is, at a distance from the point of impact, and seems to be caused more by wave interaction than by translational movement of particles itself. The actual earthquake-induced failure pattern in slopes, Figure 1a, has only open cracks and basically no mass flow, and in this sense it is rather similar to the above pattern (ii), which seems to be, as mentioned, caused by multi-dimensional wave propagation and also akin to the analytical models in Figures 1b and 2. However, the existence of surface waves and possible roles of body and surface waves in generating fracture in granular media have not been fully identified, and more systematic observations and quantitative measurements are required to give any conclusions regarding the dynamic stability of dry granular slopes. Note that numerical simulations based on particle models tend to produce fracture just beneath the point of impact similar to the

experimental pattern (i) in Figure 5, and usually will not cause toppling-like failure, i.e. calculation conditions should be re-examined in order to obtain more reliable and closer-to-reality numerical outcome.

3 CONCLUSIONS

Utilizing the experimental technique of dynamic photoelasticity in conjunction with high-speed cinematography, stress transfer as well as wave and fracture propagation in two-dimensional dry granular slopes that are under impact loading is investigated. The experimentally noticed mechanical characteristics of slopes are compared with those inferred from theoretical models and actual slope failures under dynamic loading conditions like earthquakes. The experiments suggest the existence of two totally dissimilar patterns of energy transmission inside the granular slopes: (i) one-dimensional, unidirectional straight stress transfers similar to force chains in the quasi-static analyses, resulting in mass flow or total collapse of the slopes; and (ii) wide spread of energy in the form of wave propagation, possibly with wave-induced separation of slope faces akin to toppling failure;

473

Whichever occurs seems to depend on the energy profile associated with the impact. Currently, more systematic experimental observations by controlling the impact velocity, etc., are being conducted in order to clarify the governing factors in the development of wave and dynamic fracture in rock-like and soil-like granular media.

ACKNOWLEDGEMENTS

The financial support provided by the Japan Society for the Promotion of Science (JSPS) through the "KAKENHI: Grant-in-Aid for Scientific Research (C)" Program (No. 16K06487) is kindly acknowledged.

REFERENCES

Ashford, S. A. & Sitar, N. 1997. Analysis of topographic amplification of inclined shear waves in a steep coastal bluff. *Bulletin of the Seismological Society of America* 87: 692–700.

Ashford, S. A., Sitar, N., Lysmer, J. & Deng, N. 1997. Topographic effects on the seismic response of steep slopes. *Bulletin of the Seismological Society of America* 87: 701–709.

Cleary, P. W. & Prakash, M. 2004. Discrete-element modelling and smoothed particle hydrodynamics: Potential in the environmental sciences. *Philosophical Transactions of the Royal Society A* 362: 2003–2030.

Hancox, G., Perrin, N. & Van Dissen, R. 2011. Report on landslide reconnaissance flight on 24 February 2011 following the Mw 6.3 Christchurch earthquake of 22 February 2011. *GNS Science Immediate Report* LD8 (NZTopo50–BX24)/941–947.

Sitar, N., Clough, G. W. & Bachus, R. C. 1980. *Behavior of Weakly Cemented Soil Slopes under Static and Seismic Loading.* Stanford: Stanford University.

Uenishi, K. & Tsuji, K. 2008. The dynamics of collapsing granular columns and its implications in earthquake mechanics. In *Proceedings of the Japan Geoscience Union Meeting 2008, Chiba, 25–30 May 2008*: S142–P013. Tokyo: Japan Geoscience Union.

Uenishi, K., Tsuji, K. & Doi, S. 2009. Dynamic deformation and collapse of granular columns. *Eos Transactions of AGU* 90 (52): Fall Meeting Supplement, T41A–2004.

Uenishi, K. 2010. On a possible role of Rayleigh surface waves in dynamic slope failures. *International Journal of Geomechanics* 10: 153–160.

Uenishi, K. & Takahashi, T. 2014. Rayleigh waves and dynamic tensile cracking in slopes. In *Proceedings of the 17th U.S. National Congress of Theoretical and Applied Mechanics, East Lansing, 15–20 June 2014*: S–04–341. East Lansing: Michigan State University.

Uenishi, K. & Sakurai, S. 2015. Dynamic tensile cracking in slopes possibly induced by Rayleigh surface waves. *Geomechanics and Geoengineering* 10: 212–222.

Uenishi, K., Goji, T. & Debski, W. 2017. Photoelastic study of dynamic stress transfers in granular media. In *Proceedings of the Joint Scientific Assembly of the International Association of Geodesy and the International Association of Seismology and Physics of the Earth's Interior, Kobe, 30 July – 4 August 2017*: S13–6–01. Tokyo: The Geodetic Society of Japan and Seismological Society of Japan.

Uenishi, K. & Goji, T. 2018. Dynamic fracture and wave propagation in a granular medium: A photoelastic study. Accepted for publication in *Procedia Structural Integrity*.

2019 Rock Dynamics Summit– Aydan et al. (eds)
© 2019 Taylor & Francis Group, London, ISBN 978-0-367-34783-3

Physical model study on microtremor characteristics of rock block on slope

T.T. Wang
National Taiwan University, Taipei, Taiwan

T.T. Lee & K.L. Wang
National Taipei University of Technology, Taipei, Taiwan

ABSTRACT: The stability of rock block on slope is an important evaluation index for the prevention of rockfall disaster. This study aims to provide an evaluation method other than stereographic projection and conventional static analyses. A physical model is manufactured to measure and analyze the microtremor characteristics of blocks on slopes designed with different conditions of joint roughness. The results show that when the rock block is stable, the frequency spectrum of microtremor signal is dispersed, and the predominant frequency is not prominent. When the rock block tends to be unstable due to the increase of slope angle, the frequency spectrum is concentrated, and the amplitude of predominant frequency increases in vertical direction and in the direction of potential movement. With smoother surface between slope and block, the frequency spectra of microtremor display a more concentrated figure, and the amplitude of predominant frequency also increase.

1 INTRODUCTION

Rockfall is one of the most difficult to predict slope instability, and it is also a slope disaster which is difficult to prevent and control effectively at present. The downhill movement may be carried out in the form of falling, slipping, rolling, or even bouncing once the rock mass is unstable and detached from the slope. It is difficult to predict accurately since motion trajectory is affected by many factors, such as slope, geotechnical material characteristics, vegetation, and rock block collision rupture in motion process. The stability of rock block on slope is related to internal factors such as rock strength and mechanical properties of discontinuous surface, and is influenced by external factors such as rainfall, earthquake or other vibration, so the occurrence of rock-fall disaster not only has the dispersion of space, but also has the sudden nature of time.

The stability analysis of rock block was paid attention to in the late 1960, and a series of analytical methods were proposed, including kinematic analysis method (Hoek 1973, Kogure & Matsukura 2012) and deformation analysis method (Zhu et al. 2016, Wu et al. 2016), and the simulation analysis methods of motion trajectory also moved forward in the 1980 (Azzoni et al. 1995, Chen et al. 2013). The occurrence position of rock-fall is the necessary information of motion trajectory analysis. Therefore, the identification of potential unstable rock block on slope is the first work of disaster prevention and control of rock-fall. Rock wedge

or rock block is bounded by discontinuous surface and slope. The most direct and reliable method to assess their stability at present is field investigation through personal contact by geologists or engineers. In the near future, geometrical morphology can be obtained by non-contact mapping method, and the results may help derive the probability of instability through stereographic projection, hydrostatic analysis and empirical methods (Royán et al. 2014, Yang et al. 2014). It is also possible that vibration characteristics can be utilized as the signs of instability (Du et al. 2015, 2016, Chao et al. 2017).

In this study, a physical model is made considering slope angle, joint roughness and mechanical characteristics of rock material. Microtremor signals of slope and rock block under different internal and external conditions are processed to reveal the effects of these factors. The results can provide a reference for the stability assessment method of rock block in addition to the traditional stereographic projection and statics analysis.

2 PHYSICAL MODEL AND EXPERIMENTAL PLANNING

This study designs a physical model to measure the microtremor signal of slope and rock block. Considering that the stability of rock block is mainly affected by factors such as slope, mechanical characteristics of rock and discontinuous surface roughness, the physical model and following experiments are designed for these three factors.

2.1 Model Design

The physical model is made of concrete, and the compressive strength of concrete design (after curing for 28 days) is 21 MPa. After concrete pouring, the Uniaxial Compressive Strength (UCS) and P wave velocity (V_p) change with the curing time. After 7 days of curing, the average of UCS and V_p are 12.9 MPa and 4243 m/s. After 14 days of curing, the value increase to 14.3 MPa and 4305 m/s respectively. After 28 days of curing, the indexes reach 20.4 MPa and 4827 m/s. The average friction angle of the concrete is 43° in tilt test after curing for 28 days.

Figure 1 shows a design diagram and a photograph of the physical model. The length, width and height of the whole model are 160 cm, 120 cm and 102 cm respectively. The slope of the model is decided according to the average friction angle (43°) of concrete, which is poured into 3 different slopes including 0°, 28° and 42°. Each slope area is divided into 3 sub-zones. After the concrete is poured for about the initial setting time, the left and right areas are brushed by a sparse ruler parallel and vertical to slope direction, addressed below as "tendency roughness" and "strike roughness" respectively.

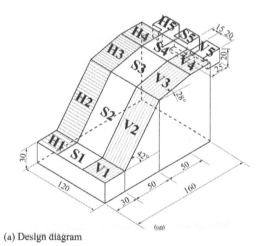

(a) Design diagram

(b) photograph

Figure 1. Physical Models

This provides the parallel and vertical concave and convex stripes, producing the effect of rough surface. The central part retains the original pouring surface, addressed below as "smooth roughness". The codes for different areas on the physical model are shown in Figure 1. The influence of rock materials is simulated with the measured results in 7, 14 and 28 curing days. In addition, 20 cm × 20 cm × 20 cm concrete cubes are used as the rock blocks on slope. A reverse groove is made on the cube to keep accelerometer level.

2.2 Experimental planning

Table 1 is the experimental planning of this study and the corresponding discussion topic. In M01, the microtremor signals of ground, low and high platforms (S1 and S4) are measured, and the background noise and vibration characteristics of the platform are obtained. M02 is used to measure the microtremor signals of H1, H2 and H3 to compare the effects of different slopes. M03 measures H3, S3 and V3 microtremor signals to investigate the effect of surface roughness. M04 was used to compare the measurement results of H5, exploring the influence of concrete curing for 7, 14 and 28 days as rock blocks with different strengths.

Microtremor signal measurement uses portable triaxial accelerometer, size specification of 25×25×21 mm, weight for about 0.02 kg. The accelerometer can detect the vibration of ±20 m/s² (±2000 gal), and is valid within frequency response 0-300 Hz. The sensitivity is 2000 mV/g, and is paired with a data logger which has eight channels and the sampling frequency can reach up to 10^6 Hz.

In the experimental process, the V direction of the accelerometer is set to the vertical direction, the south and North (SN) are dip direction, and the East and West (EW) are parallel with the contour of the slope. During the measurement, the rock block is placed on the slope to be measured, the accelerometer is fixed and backward glued to the small groove, the sampling frequency is 500 Hz, the measurement is 120 seconds, and a total of 60000 data is obtained. Each measurement is carried out 2 times to obtain

Table 1. Experimental planning and content.

No.	Station	Purpose
M01	Ground、S1、S4	Measure background noise and select base stations
M02	H1、H2、H3	Compare microtremor characteristics of different slopes (0°, 28°, 42°)
M03	S3、V3、H3	Microtremor characteristics of the smooth surface and the rough surface
M04	H5	Comparison of microtremor characteristics of different material strengths

duplicate data to provide filtering and Fourier transform and other signal processing.

3 RESULTS AND DISCUSSIONS

Figure 2 shows that the measurement results of the three-direction acceleration records of the ground station, the EW (Fig. 2a), the SN (Fig. 2b) and the V axis (Fig. 2c) are within +0.5 gal. Firstly, the signal processing is carried out by moving average method and band-pass filter. Considering that the accelerometer characteristics and the main vibration frequency of the experimental location site is 1 Hz, the band-pass filter range is set to 0.2-10 Hz, and then the spectrum diagram is obtained by Fast Fourier transform. Figure 3 is the three-directional Fourier spectrum of the ground station, in which the EW amplitude is between 2.0×10^{-7} and 5.0×10^{-7} (Fig. 3a), the SN (Fig. 3b) and the V axis (Fig. 3c) amplitude are between 3.0×10^{-7} and 7.0×10^{-7}. In this study, the predominant frequencies of the first three were selected as the predominant frequencies of the model. The top three frequencies of EW direction are 4.0 Hz, 4.2 Hz and 4.6 Hz; the top three frequencies along SN direction are 4.2 Hz, 3.6 Hz and 4.6 Hz; and the top three frequencies along V direction are 3.8 Hz, 3.6 Hz and 2.8 Hz sequentially. Overall, the predominant frequency of the ground is between 2.8 Hz and 4.6 Hz, and the amplitude is less than 2.0×10^{-7}, therefore it is not easy to distinguish the predominant frequency from the microtremor signal.

3.1 Slope

Table 2 shows the measured results of microtremor signals. A and F represent the amplitude and the frequency band respectively.

Figure 4 shows the three-directional Fourier spectrum on H1(0°), H3(28°) and H2(42°), it can be seen that the amplitude of EW and SN direction with a slope of 0° is between 1.0×10^{-7} and 2.0×10^{-7},

the amplitude along the V axis is between 1.0×10^{-7} and 2.5×10^{-7}, and the frequency bands are between 3.0 Hz and 9.2 Hz (Fig. 4a).

As the slope is 28°, the amplitude of EW direction is between 3.0×10^{-7} and 7.0×10^{-7}, the amplitude of SN direction is between 1.0×10^{-7} and 2.0×10^{-7}, the amplitude along the V axis is between 1.0×10^{-7} and 2.5×10^{-7}, and the frequency bands are between 1.8 Hz and 7.0 Hz (Fig. 4b).

As the slope is 42°, the amplitude of EW direction is between 3.0×10^{-7} and 8.0×10^{-7}, the amplitude of the SN direction is between 3.0×10^{-7} and 8.0×10^{-7}, the amplitude along the V axis is between 4.0×10^{-7} and 1.0×10^{-6}, and the frequency bands are between 0.8 Hz and 5.5 Hz (Fig. 4c).

In Figure 4, the vibration characteristics of rock blocks on low platform (0°), gentle slope (28°) and medium slope (42°) slope are obviously different, and the maximum amplitude of the frequency bands of the three-directional amplitude on the low platform is mostly 2.0 to 2.4×10^{-7}. The spectrum measured on the gentle slope has the tendency of moving to the left, the predominant frequency band distribution range is lower than the platform result concentration, the corresponding maximum amplitude amplification is about 3-4 times to 7.0×10^{-7}. The medium slope is close to the friction angle of the material used this study. On the smooth surface (S3), it is difficult to ensure the stability of the rock block. The results show that when the rock block is in an unstable state, the microtremor signal has the phenomenon of increasing the frequency concentration and the maximum amplitude.

3.2 Roughness

Figure 5 is the three-directional Fourier spectrum of S3, V3 and H3 stations, the EW amplitude of the smooth surface (S3) is between $3.0 \times e^{-7}$ and $6.0 \times e^{-7}$, the frequency band is between 1.2 Hz and 4.0 Hz; SN amplitude falls between 1.5×10^{-7} and 3.5×10^{-7}, and the frequency band is between 1.2 Hz and 4.6

Table 2. Measurement results of the influence factors of microtremor signals.

Influence factor		EW		SN		V axis	
		A ($\times 10^{-7}$)	F (Hz)	A ($\times 10^{-7}$)	F (Hz)	A ($\times 10^{-7}$)	F (Hz)
Slope	0°	1.0-2.0	3.0 -9.2	1.0-2.0	3.0-9.2	1.0-2.5	3.0- 9.2
	28°	3.0-7.0	1.8-7.0	1.0-2.0	1.8-7.0	1.0-2.5	1.8-7.0
	42°	3.0-8.0	0.8-5.5	3.0-8.0	0.8-5.5	4.0-10.0	0.8-5.5
Roughness	Smooth	3.0-6.0	1.2-4.0	1.5-3.5	1.2-4.6	2.5-5.0	2.2-6.2
	Tendency roughness	2.5-5.0	2.2-6.2	3.0-6.0	2.2-5.8	3.0-6.0	2.8-7.0
	Strike roughness	3.5-7.0	1.8-6.0	1.0-2.0	2.2-6.8	1.0-2.5	2.4-6.8
Strength	7 day	3.0-6.0	2.0-5.5	2.5-5.0	2.0-5.5	2.5-4.8	2.0-5.5
	28 day	4.5-3.5	3.0-7.5	2.0-4.0	3.0-7.5	2.0-3.8	3.0-7.5

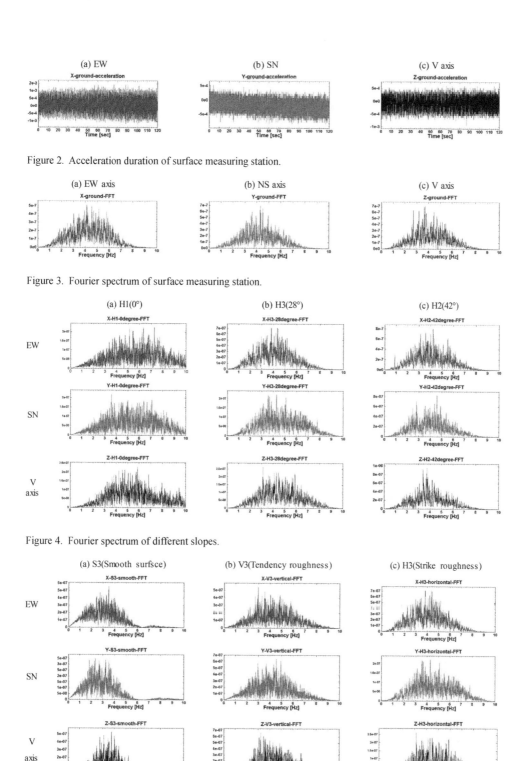

Figure 2. Acceleration duration of surface measuring station.

Figure 3. Fourier spectrum of surface measuring station.

Figure 4. Fourier spectrum of different slopes.

Figure 5. Fourier spectrum of different roughness.

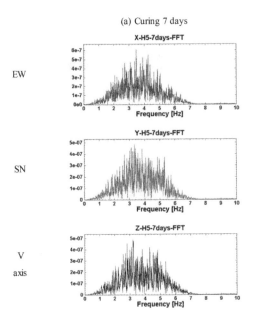
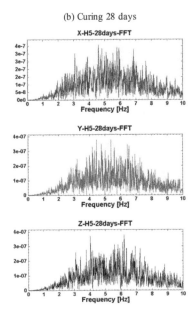

(a) Curing 7 days   (b) Curing 28 days

Figure 6. Fourier spectrum of different rock block strength

Hz; V axis amplitude ranges from 2.5×10^{-7} to 5.0×10^{-7}, and the frequency band is between 2.2 Hz and 5.2 Hz (Fig. 5a).

The EW amplitude of the tendency roughness (V3) is between 2.5×10^{-7} and 5.0×10^{-7}, the frequency band is between 2.2 Hz and 6.2 Hz; the SN amplitude is between 3.0×10^{-7} and 6.0×10^{-7}, the frequency band is between 2.2 Hz and 5.8 Hz; the V axis amplitude is between 3.0×10^{-7} and 6.0×10^{-7}, the frequency band between 2.8 Hz and 7.0 Hz (Fig. 5b).

The EW axis amplitude of the discontinuous surface for the strike roughness (H3) is between 3.5×10^{-7} and 7.0×10^{-7}, the frequency band is between 1.8 Hz and 6.0 Hz; the NS axis amplitude is between 1.0×10^{-7} and 2.0×10^{-7}, and the frequency band is between 2.2 Hz and 6.8 Hz; and the V axis amplitude between 1.0×10^{-7} and 2.5×10^{-7}, the band is between 2.4 Hz and 6.8 Hz (Fig. 5c).

In Figure 5, when the discontinuous surface between the rock block and the slope is a smooth surface, the frequency is significantly concentrated compared to the experimental results of the strike roughness (H3) and the tendency roughness (V3), which is higher than the frequency band range (>4.6 Hz) of the experimental position, and the amplitude is almost gone. However, the maximum amplitude corresponding to the predominant frequency increases, but the increase amount is less than twice times. By further comparing the spectrum obtained

from the strike roughness (H3) and the tendency roughness (V3), it is observed that the trend of the former spectrum moving to the left, but the variation in predominant frequency and the corresponding maximum amplitude are not significant.

3.3 *Rock block Strength*

Figure 6 is the three-directional Fourier spectrum of different rock block strength, it can be seen that the EW amplitude after curing seven days is between 3.0×10^{-7} and 6.0×10^{-7}, SN amplitude is between 2.5×10^{-7} and 5.0×10^{-7}, the V axis amplitude is between 2.5×10^{-7} and 4.8×10^{-7}, and the frequency bands are between 2.0 Hz and 5.5 Hz (Fig. 6a). The EW amplitude of curing 28 days is between 1.5×10^{-7} and 3.5×10^{-7}, the NS amplitude is between 2.0×10^{-7} and 4.0×10^{-7}, the V axis amplitude is between 2.0×10^{-7} and 3.8×10^{-7}, and the frequency bands are between 3.0 Hz and 7.5 Hz (Fig. 6b).

In Figure 6, the lower the concrete strength is, the lower the wave velocity is. As the trend of the microtremor signal spectrum towards the low-frequency concentration is more obvious, the maximum amplitude corresponding to the predominant frequency increases by about twice times.

Table 3 listed the predominant frequency of the integrated factors and their corresponding amplitudes.

Table 3. Integration of the predominant frequencies and their corresponding amplitudes.

Influence factor		No.	Amplitude (×e^-7)	Predominant frequency (Hz)
Slope	0°	1	2.2	7.2
		2	2.0	6.2
		3	2.4	5.8
	28°	1		4.4
		2	7	4.2
		3		3.8
	42°	1	10	3.6
		2	9	3.2
		3	8.2	2.8
Roughness	Smooth	1	5.2	3.4
		2	5.0	2.2
		3	4.0	1.8
	Tendency roughness	1	6.2	4.6
		2	6.0	3.2
		3	5.0	2.8
	Strike roughness	1		4.4
		2	7	4.2
		3		3.8
Strength	7-day	1	5.5	4.2
		2	5.0	3.8
		3	6.0	3.2
	28-day	1	3.2	6.8
		2	3.4	6.0
		3	3.8	4.2

4 CONCLUSION

In this study, a physical model was made to test if microtremor signal can be used as an index for rock block stability on slope. The following conclusions were reached:

1. When the rock block is stable, the spectrum of the microtremor signal is dispersed, and the predominant frequency is not prominent. As the rock block tends to be unstable due to the increase of slope, the spectrum has a more concentrated trend, and the maximum amplitude of the predominant frequency increases by several times.

2. The phenomena that the frequency spectrum concentrated and the maximum amplitude increased is obvious in the vertical direction and the potential moving direction of the rock block.

3. When the discontinuous surface between rock block and slope is smooth surface, the stability of rock block is worse than that the strike roughness and the tendency roughness, the predominant frequency of microtremor is more concentrated, and the amplitude of the predominant frequency band range (>4.6 Hz) almost disappeared. In addition, concentration of frequency spectrum of the strike roughness moves to the left.

4. Comparing the spectrum of horizontal stripes (H3) and parallel stripes (V3), it is observed that joint orientation can influence the trend of spectrum concentration moving to the left, and the maximum amplitude corresponding to the predominant frequency increases by about twice.

REFERENCES

Hoek, E., Bray J.W. & Boyd J.M. 1973. The stability of a rock slope containing a wedge resting on two intersecting discontinuities. *Quarterly Journal of Engineering Geology and Hydrogeology* 1: 1–55.

Azzoni A., La Barbera G. & Zaninetti A. 1995. Analysis and prediction of rockfalls using a mathematical model. *International Journal of Rock Mechanics and Mining* 32: 709–724.

Kogure, T. and Matsukura Y. 2012. Threshold height of coastal cliffs for collapse due to tsunami: theoretical analysis of the coral limestone cliffs of the Ryukyu islands, Japan. *Marine Geology* 323: 14–23.

Chen G.Q., Zheng L., Zhang Y.B., Wu J. 2013. Numerical Simulation in Rockfall Analysis: A Close Comparison of 2–D and 3–D DDA. *Rock Mechanics and Rock Engineering* 46: 527–541.

Zhu H., Wu W., Chen J., Ma G., Liu X. & Zhuang X. 2016. Integration of three dimensional discontinuous deformation analysis (DDA) with binocular photogrammetry for stability analysis of tunnels in blocky rockmass. *Tunnelling and Underground Space Technology* 51: 30–40.

Wu L.Z., Li B., Huang R.Q. & Wang Q.Z. 2016. Study on Mode I–II hybrid fracture criteria for the stability analysis of sliding overhanging rock. *Engineering Geology* 209: 187–195.

Barton, N. 1971. A relationship between joint roughness and joint shear strength. *Rock Faracture; Proc. int. symp. rock mech., Nancy, 4–6 October 1971.*

Du Y., Xie M.W., Jiang Y.J., Liand B. & Gao Y. 2015. Methods for determining early warning indices based on natural frequency monitoring. *Yantu Lixue/Rock and Soil Mechanics* 36(8): 2284–2290.

Du Y., Xie M.W., Jiang Y.J., Liand B., Gao, Y. & Liu Q.Q. 2016. Safety monitoring experiment of unstable rock based on natural vibration frequency. *Yantu Lixue/Rock and Soil Mechanics* 37(10): 3035–3040.

2019 Rock Dynamics Summit– Aydan et al. (eds)
© 2019 Taylor & Francis Group, London, ISBN 978-0-367-34783-3

Simulation-based optimal design approach for rockfall protection walls

S. Moriguchi, H. Kanno, K. Terada & T. Kyoya
Tohoku University, Sendai, Japan

ABSTRACT: This study presents a new design approach which can optimize position and size of a rockfall protection wall with the help of a numerical simulation. In the proposed framework, sufficient numbers of rockfall paths with information of kinematic energy are accumulated by rockfall simulation. A protection wall is then virtually overlapped on the simulated results, and values of a safety function are checked under the different combinations of design parameters. A response surface of the safety function is then approximately constructed, and a cost function is defined to express the penalty in unreasonable situation. Once the two functions are obtained, an optimization problem can be formulated. In order to demonstrate the capability of the proposed optimal design approach, a numerical example is presented in this study. The obtained result indicates that the proposed framework has the possibility of developing optimal design of rockfall protection walls.

1 INTRODUCTION

Once rockfalls reach to roads or housing area, houses and infrastructures are easily damaged or destroyed. The construction of a protection wall is one of the typical measures, which can withstand rockfall impacts with few meters in diameter. In order to construct safety and reasonable protection walls, we have to estimate the path and the kinetic energy of rockfall. It is, however, still hard to predict the movement of rockfall with high accuracy, because rockfall movement strongly depends on a local situation. In addition, strong variability is generally involved in rockfall movement and its path. Under such uncertainties, there are a lot of difficulties in the design of rockfall protection walls. Thus, the information of past rockfall records and engineering knowledge has been widely utilized in the design procedure. The problems are that the design of protection walls strongly depends on the designers' ability and experience, and that quantitative design procedures have not been established completely.

Rockfall simulations are powerful tools that enable us to obtain large amounts of quantitative information for prediction and risk assessment of rockfalls. Simulation methods are roughly classified into two types; the first one is the lumped mass approach, and the other one is the discrete modeling approach. The lumped mass approach (e.g. Stevens 1998, Crosta & Agliardi 2004) can analyze the motion of a falling block that has simple shape or neglects the shape effects. The approach has the advantage of extremely quick calculation. However, it is necessary to use assumptions for transition between movement modes in addition to solve different governing equations for each movement

mode. On the other hand, the discrete modeling approach, such as Discrete Element Method (Cundall 1971, Cundall & Strack 1979) or Discontinuous Deformation Analysis (Shi & Goodman 1985), can express the complicated geometric shape effects of rocks and slopes directly. Furthermore, the transition of movement forms can be naturally described without any assumption because the equations of rigid body motion are directly solved.

This study aims to propose a new optimal design approach for rockfall protection walls based on DEM simulation. The width of protection walls and the horizontal distance between the toe of a slope and protection walls are selected as key design parameters, and these values are optimized in the proposed method. In this paper, the method is briefly explained, and the performance of the method is discussed based on result of an simple numerical example.

2 DISCRETE ELEMENT METHOD

2.1 *Contact force model*

As mentioned before, DEM is employed to simulate rockfall movement in this study. Because DEM can calculate the contact force between rigid bodies, the method has been widely applied to rock fall simulations. In this study, a spherical element is employed as a basic element. The shape of a rock block model is represented by clumps of spherical elements. Figure 1 describes a contact force model used in this study. The contact force model consists of spring, dashpot, and slider, and is popularly used in DEM simulations. The contact forces affecting on elements can be obtained by solving the following equations;

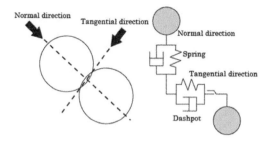

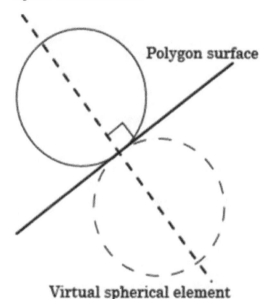

Figure 1. Contact force model

Figure 2. Image of virtual spherical element

$$\mathbf{Fn} + \mathrm{Cn\dot{u}} + \mathrm{Knu} = \mathbf{0} \tag{1}$$

$$\mathbf{Fs} + \mathrm{Cs\dot{v}} + \mathrm{Ksv} = \mathbf{0} \tag{2}$$

Here, **Fn** and **Fs** are the normal and tangential contact force vector. **u** and **v** are the relative displacement vector in the normal and tangential directions, respectively. Kn and Cn are the spring constant and the damping coefficient in the normal direction, and Ks and Cs are those in the tangential direction. The upper bound of **Fs** is given by the Coulomb friction criterion with the friction angle.

2.2 Modeling of the contact between a rockfall and slope surface

Configuration of slope surface is represented by assembly of polygons. As shown in Figure 2, a virtual spherical element is set on the opposite side when a spherical element collides onto polygon surface. This indicates that the contact force can be calculated by considering only the interactions between the spherical elements even in the contact between a spherical element and a polygon element.

3 OPTIMIZATION METHOD

Although there are a lot of parameters we have to consider in the design of rockfall protection wall, we especially focus on two key parameters, such as the width of a protection wall and the horizontal distance between the toe of a slope and a protection wall. By using those parameters, the safety and the construction cost of designed walls can be expressed quantitatively. We also specify the acceptable energy of protection wall for the safety constraint. Consequently, there are three parameters: the horizontal distance **x**, the width of a protection wall **l**, and the acceptable energy e_d. These are the parameters for an optimal design problem considered in this study.

3.1 Safety function

In the rockfall simulations, a number of calculation cases are carried out to obtain enough number of rockfall paths with the information of kinetic energy. Based on these simulated results, the safety of a rockfall protection wall is quantitatively defined. Figure 3 shows an image of a flow for evaluating the safety of a protection wall. The protection wall is virtually placed on the simulated results, and it is checked whether the protection wall has enough capability against each rockfall path. When the rockfall path crosses over the protection wall and the associated rockfall energy at collision point is

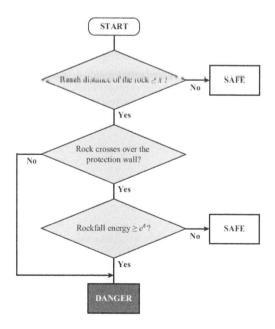

Figure 3. Flow for evaluating the safety of a protection wall

larger than the acceptable energy, the protection wall is assumed to be destroyed. In addition, it is also assumed to be dangerous situation when the rock passes through the sides of wall. On the other hand, when the rock stops in front of the protection wall or the rockfall energy at collision point is smaller than the acceptable energy, it is regarded that the protection wall has enough capability. For simplification, the wall is assumed to be a rectangular shape and it has infinite height. This post-productive operation enables us to perform contact simulation virtually with low computational cost.

Based on the assumption mentioned above, the safety of the protection wall is quantified as,

$$g(x,l) = n^s / (n^s + n^d) \tag{3}$$

Here again, x is the horizontal distance between the toe of the slope and the wall, and l is the width of the wall. n^s and n^d are the number of the "SAFE" and "DANGER" paths, respectively. Values of the safety index g can be defined for each combination of x and l. Because the values are discrete data, a continuous function $\hat{g}$ is considered to solve an optimal design problem easily. To be more specific, a relational expression of $\hat{g}$ is approximately described by employing the polynomial function. The polynomial function model has advantages such that it offers a flexible variable selection and can be simply estimated by the linear least square method. In this study, the maximum likelihood values of coefficients of the polynomial function are estimated based on the backward elimination approach in which Akaike Information Criterion (Akaike 1974) is chosen as a model fit criterion. That procedure is performed by R, the statistical software (R Core Team 2017).

3.2 Cost function

The total cost is assumed to be the sum of two factors: the construction cost that is proportional to the wall width, and the land cost of the rectangle area in front of protection wall. Therefore, the total cost f(x; l) is calculated as follows:

$$f(x,l) = c_1 l + c_2 lx \tag{4}$$

Here, c_1 is the construction cost of rockfall protection wall per unit length, and c_2 is the land cost per unit area.

3.3 Design optimization problem

A design optimization problem that enables us to find the best combination of x and l is expressed by making use of the safety function and the cost function as below.

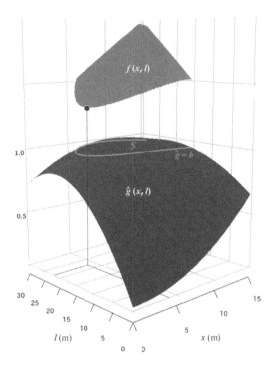

Figure 4. Image of the design optimization problem

$$\min \ f(x,l) \tag{5}$$

$$\text{s.t.} \ (x,l) \in S \tag{6}$$

$$b \leq \hat{g}(x,l), \ \ 0 \leq b \leq 1 \tag{7}$$

Here, S is the feasible set of parameter combinations, and b is the lower safety limit that the rockfall protection wall must satisfy. Figure 4 is an image of the optimization problem. The lower surface expresses the safety function $\hat{g}$, the upper surface indicates the cost function f, the feasible set S is described as area surrounded by the curved line defined by (7).

4 NUMERICAL EXAMPLE

4.1 Calculation condition

The proposed method was applied to an example of the optimal design problem on a simple slope model. Figures 5 and 6 show images of the numerical example. Rock brock is initially placed on the top of the slope, then the rock brock falls down due to the gravity. Input parameters used in the DEM simulation are summarized in Table 1. As shown in Figure 5, rockfall paths including and kinematic energy were accumulated, and we tried to design a rockfall protection wall using the proposed method.

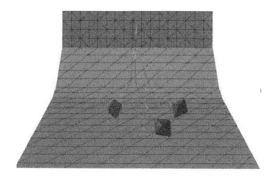

Figure 5. Image of the numerical example

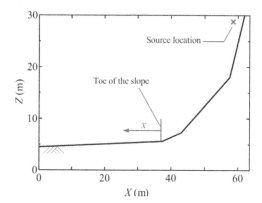

Figure 6. Longitudinal profile of slope model used in the numerical example

Table 1. Input parameters of DEM simulation

Rock volume	38	m^3
Rock mass	1.1×10^5	kg
Particle radius	0.60	m
Restitution coefficient	0.30	-
Friction angle	30	deg
Spring coefficient	1.0×10^8	N/m
Time step	1.0×10^{-6}	sec

In order to check performance of the proposed framework, three kinds of protection walls are considered. Information of the protection walls are summarized in Table 2. Type 1 has high acceptable energy, but is most expensive protection wall. Type 2 is middle, and Type 3 has lowest acceptable energy, and is cheapest one. The lower safety limit b is set to 0.95 (95%). Figure 7 shows cost functions that are cut inside of the feasible

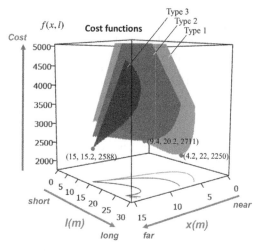

Figure 7. Results of the design optimization

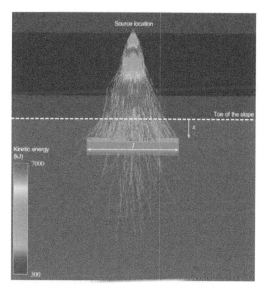

Figure 8. Optimized result of the numerical example

domain. As we can understand from the results, Type 1 is the most suitable one, and optimized values of The design parameters, **x** and **l**, can be automatically obtained. Figure 8 shows an image of the obtained result.

Table 2. Parameters of protection walls

		Type1	Type2	Type3
Strength	e^d (kJ)	7,000	5,500	4,000
Constr. cost	c_1 (Kyen/m)	600	400	200
Land price	c_2 (Kyen/m^2)	100	100	100

5 CONCLUSION

This study proposed a new design approach which enables us to optimize design parameters of rockfall protection walls based on the result of three-dimensional DEM rockfall simulations. In this approach, two mathematical functions are defined based on results of DEM rockfall simulation to quantify the safety and risk of rockfalls, and the method can optimize design parameters of rockfall protection walls with low calculation cost. The numerical example was solved to discuss performance of the proposed method. Based on the result, high flexibility and capability of the method was confirmed.

REFERENCES

Akaike, H. 1974, A new look at the statistical model identification, IEEE transactions on automatic control 19(6),pp.716–723.

Crosta, G. B. & Agliardi F. 2004. Parametric evaluation of 3D dispersion of rockfall trajectories. Natural Hazards and Earth System Sciences, 4, pp.583–598.

Cundall, P. A. 1971, A computer model for simulating progressive, large scale movement in blocky rock systems. Proceedings of ISRM Symposium, pp.11–18.

Cundall, P. A. & Strack, O. D. L. 1979. A discrete numerical model for granular assemblies, G´eotechnique, 29(1),pp.47–65.

R Core Team 2017, R: A Language and Environment for Statistical Computing. Vienna, Austria: R Foundation for Statistical Computing.

Shi, G. H. & Goodman, R. E. 1985, Two dimensional discontinuous deformation analysis. International Journal for Numerical & Analytical Methods in Geomechanics, 9(6),pp.541–556.

Stevens, W. D. 1998, A tool for probabilistic analysis, design of remedial measures and prediction of rockfalls, Master thesis, University of Toronto, 1988.

2019 Rock Dynamics Summit– Aydan et al. (eds)
© 2019 Taylor & Francis Group, London, ISBN 978-0-367-34783-3

Solid and water interaction analysis by NMM-DDA and MPS methods applied to large-scale landslide triggered by earthquake

S. Miki
Kiso-Jiban Consultants Co. Ltd., Tokyo, Japan

Y. Ohnishi
Kyoto University, Kyoto, Japan

T. Sasaki
Suncoh Consultants Co. Ltd., Tokyo, Japan

ABSTRACT: Discontinuous Deformation Analysis (DDA) and Distinct Element Method (DEM) has been used as analyses methods of the slope disasters. However, these methods cannot handle the mixed body of fluid and solid material appropriately. In the past decades, the meshless MPS method has been developed within the Lagrangian framework. In MPS, the fully Navier–Stoke equation can be solved without using computational grids or meshes, and MPS is suitable for simulating motion and fragmentation of fluid. In this study, NMM-DDA and MPS have been newly combined to simulate landslide behavior in the earthquake. In the coupling method of NMM-DDA and MPS, rock block movement is solved by NMM-DDA and water movement inside the rock mass is solved by MPS method separately. The formulation of the coupling method and the simulation for the large-scale landslide triggered by the earthquake will be presented.

1 INTRODUCTION

Topographically, the mountainous area in Japan occupies about 70% of the whole country and the earthquake and volcanic activities is very dynamic. The disaster due to earthquake and heavy rain has occurred frequently. The analytic method also has been applied to the estimation of the hazards and damage. However, it is not always successful. One of the reason is considered that continuum method such as FEM is applied to analyze the phenomenon with large movement and a breakdown occurring in slope disasters. On the other hand, discontinuous method such as Discontinuous Deformation Analysis (DDA) (Shi & Goodman 1989) and Distinct Element Method (DEM) (Cundall 1971) has also been used as analyses methods of the slope disasters. However, these methods cannot handle the mixed body of fluid and solid material appropriately. It is known that the water explicitly affect stability of rock slope in earthquake.

Recently, a particle method such as MPS (Moving Particle Semi-implicit) (Koshizuka & Oka 1996) and SPH (Smoothed Particle Hydrodynamics) (Lucy 1977), which is one of mesh free methods, has been used to analyze water movement. The particle method is simple but powerful method to solve the problem of fluid mechanics. The authors introduce the coupling analysis by a discontinuous DDA and a particle MPS method to solve complex solid and fluid interaction problems such as collapse of natural landslide dam and debris flow (Miki et al. 2017). The proposed method shows the potential ability for

the application in landslide, debris flow and other related fields.

For the dynamic response analysis of discontinuous rock slopes by DDA, seismic forces are commonly applied to a basement using a single block. However, it is often necessary to consider the local variation of seismic forces and stress conditions, especially when the size of slopes is large and/or the slope geometry becomes complicated. It is difficult for DDA to consider the local displacements of the single block for the basement due to the fact that the strain in the single block is assumed to be uniform. On the other hand, the Numerical Manifold Method (NMM) (Shi 1991) can simulate both continuous and discontinuous deformation of blocks with contact and separation. However, the rigid body rotation of blocks, which is one of the typical behaviors for rock slope failure, cannot be treated properly because NMM does not deal with the rigid body rotation in explicit form. For the numerical simulations of the dynamic behavior of slopes during earthquakes, it is necessary and preferable to consider both continuous and discontinuous deformations of fractured rock masses appropriately. According to the above mentioned features and drawbacks, the authors have introduced the coupling analysis of DDA and NMM (Miki et al. 2010), and this method (NMM-DDA) was applied to the simulation of the large-scale landslide trigged by Iwate-Miyagi inland earthquake (M7.2) occurred in 2008 (Miki et al. 2013). In the simulation, the basement of the landslide was divided by NMM elements and the landslide body consisted of DDA blocks. The seismic forces were given to the basement NMM elements to consider the local variation of seismic forces. However, in order to simulate the large displacement of the landslide body in the earthquake, the friction angles between the DDA blocks should choose low values in the simulation. It has considered that the lack of ground water affection caused the low friction angles.

In this study, NMM-DDA and MPS have been newly combined, and the newly propose method is applied to the simulation of the large-scale landslide triggered by Iwate-Miyagi inland earthquake. The coupling method of NMM-DDA and MPS will be presented, and the simulation results for the large-scale landslide will be considered.

2 OUTLINE OF NMM-DDA AND MPS

2.1 NMM-DDA

In this study, solid part is analyzed by NMM-DDA. NMM-DDA, which is coupling analysis of NMM and DDA, can deal with large movement of blocks with deformation and rigid body rotation.

NMM-DDA is formulated with the kinematic equations based on Hamilton's principle expressed as:

$$M\ddot{D} + C\dot{D} + KD = F \tag{1}$$

where M = mass matrix; C = viscosity matrix; K = stiffness matrix; and F = external force vector. D, $\dot{D}$ and $\ddot{D}$ are displacement, velocity and acceleration vector, respectively. The total displacement (u, v) at the point (x, y) inside i-th DDA block can be calculated by:

$$\begin{pmatrix} u \\ v \end{pmatrix} = \left[T_i^d(x,y) \right]\left[D_i^d \right] \tag{2}$$

$$\left[T_i^d \right] =$$

$$\begin{pmatrix} 1 & 0 & -(y-y_0) & (x-x_0) & 0 & \dfrac{(y-y_0)}{2} \\ 0 & 1 & (x-x_0) & 0 & (y-y_0) & \dfrac{(x-x_0)}{2} \end{pmatrix} \tag{3}$$

$$\left[D_i^d \right] = \begin{pmatrix} u_0 & v_0 & r_0 & \varepsilon_x & \varepsilon_y & \gamma_{xy} \end{pmatrix}^T \tag{4}$$

where $[T_i^d]$ = the DDA block deformation matrix (displacement function) for i-th block; and (x_0, y_0) = the location of gravity center of block i. In Equation (4), (u_0, v_0) is the rigid body transformation, r_0 is the rigid body rotation of the block at the gravity center (x_0, y_0), and ε_x, ε_y, γ_{xy} are the normal (in the x- and y-directions) and shear strains of the block, respectively.

For the NMM, assuming that the shape of the cover mesh is triangle and the coordinates of three nodes of the triangle, C_1, C_2 and C_3 are (x_1, y_1), (x_2, y_2) and (x_3, y_3), respectively, the displacement (u, v) at the point (x, y) in the element defined by C_1, C_2 and C_3 can be given by:

$$\begin{pmatrix} u \\ v \end{pmatrix} = \left[T_j^m(x,y) \right]\left[D_j^m \right] \tag{5}$$

$$\left[D_j^m \right] = \begin{pmatrix} u_1 & v_1 & u_2 & v_2 & u_3 & v_3 \end{pmatrix}^T \tag{6}$$

where $[T_j^m]$ = the element deformation matrix (displacement function) for j-th element and (u_k, v_k) $(k=1, 2, 3)$ in Equation (6) means the displacements at the triangle nodes C_k. The deformation matrix $[T_j^m]$ can be extended as following equation.

$$\left[T_j^m(x,y) \right] = \begin{pmatrix} f_1 & 0 & f_2 & 0 & f_3 & 0 \\ 0 & f_1 & 0 & f_2 & 0 & f_3 \end{pmatrix} \tag{7}$$

Where,

$$\begin{pmatrix} f_1 & f_2 & f_3 \end{pmatrix} = \begin{pmatrix} 1 & x & y \end{pmatrix}$$

$$\dfrac{\begin{pmatrix} x_2 y_3 - x_3 y_2 & x_3 y_1 - x_1 y_3 & x_1 y_2 - x_2 y_1 \\ y_2 - y_3 & y_3 - y_1 & y_1 - y_2 \\ x_3 - x_2 & x_1 - x_3 & x_2 - x_1 \end{pmatrix}}{\begin{vmatrix} 1 & x_1 & y_1 \\ 1 & x_2 & y_2 \\ 1 & x_3 & y_3 \end{vmatrix}} \tag{8}$$

The total potential energy Π_{sys} of the block system, which includes the DDA blocks and NMM elements, can be expressed as the following equation.

$$\Pi_{sys} = \Pi_{sys}^m + \Pi_{sys}^d + \sum_{B,i}\sum_{E,j}\Pi_{i,j}^{PL} \tag{9}$$

In Equation (9), the first and second term on the right side are the potential energy for the DDA blocks and NMM elements, respectively. The last term on the right side of Equation (9) represents the potential energy for the contacts between DDA block i and NMM element j. The matrices and vector in kinematic equations based on Hamilton's principle (Equation (1)), can be also obtained by minimizing the potential energy expressed as Equation (9). However, the potential energy for DDA part, Π_{sys}^d and NMM part, Π_{sys}^m are minimized with respect to the block displacement $[D^d]$ and displacement $[D^m]$, respectively.

2.2 MPS

Discretization of continuum medium (usually fluid) by particle method is categorized into roughly two groups. Both SPH and MPS are used

487

in the field of fluid dynamics and geomechanics. In this study, the MPS method is adopted as combined analysis with NMM-DDA because stable solution of water pressure is easily obtained for larger time span in MPS calculation. MPS, which has been proposed by Koshizuka & Oka 1996, is suitable for simulating motion and fragmentation of incompressible fluid. MPS solves the Navier-Stoke equation of Equation (10) explicitly and implicitly:

$$\frac{Du}{Dt} = -\frac{1}{\rho}\nabla P + \frac{\mu}{\rho}\nabla^2 u + F \qquad (10)$$

where u = velocity vector; ρ = density; P = pressure; μ = viscosity; and F = external force. D/Dt means Lagrangian derivative. First term on the right side of Equation (10) means pressure gradient, and the second term means viscosity. The external force term and viscosity term are solved explicitly, and the pressure gradient term is solved implicitly. The gradient and Laplacian model for i-th particle are expressed as:

$$\left\langle \nabla \phi_i \right\rangle = \frac{d}{n^0} \sum_{j \neq i} \left[\frac{\phi_j - \phi_i}{\left| \mathbf{r}_j - \mathbf{r}_i \right|^2} \left(\mathbf{r}_j - \mathbf{r}_i \right) w \left(\left| \mathbf{r}_j - \mathbf{r}_i \right| \right) \right] \qquad (11)$$

$$\left\langle \nabla^2 \phi_i \right\rangle = \frac{2d}{n^0 \lambda^0} \sum_{j \neq i} \left(\phi_j - \phi_i \right) w \left(\left| \mathbf{r}_j - \mathbf{r}_i \right| \right)$$

$$\lambda^0 = \frac{\sum_{j \neq i} \left| \mathbf{r}_j^0 - \mathbf{r}_i^0 \right|^2 w \left(\left| \mathbf{r}_j - \mathbf{r}_i \right| \right)}{\sum_{j \neq i} w \left(\left| \mathbf{r}_j - \mathbf{r}_i \right| \right)} \qquad (12)$$

where ϕ is the variable, r_i and r_j are the position vector of i-th and j-th particle, respectively. d is the number of the spatial dimension, and n^0 is the initial particle number density. The function w is the weighted function and expresses the degree of influence to the particle j that exists inside the influential radius r_e of the particle i as given by Equation (13).

$$w \left(\left| \mathbf{r}_j - \mathbf{r}_i \right| \right) = \frac{r_e}{\left| \mathbf{r}_j - \mathbf{r}_i \right|} \qquad \left(0 \leq r < r_e \right)$$

$$w \left(\left| \mathbf{r}_j - \mathbf{r}_i \right| \right) = 0 \qquad \left(r_e \leq r \right) \qquad (13)$$

3 COUPLING METHOD NMM-DDA AND MPS

3.1 Modeling DDA block and NMM element

In an actual joints between sound rock mass, flow water does not go through a solid rock because permeability of rock is usually very low comparing to soil or void of rock mass. Therefore, water particles

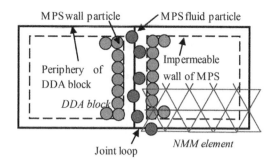

Figure 1. Modeling of joint between DDA blocks.

in MPS are analyzed in conditions that they do not enter the interior of solid parts, such as DDA block and NMM element. To realize this condition in the analysis, wall particles of MPS, which are used for setting impermeable condition in MPS, are arranged inside the DDA block as shown Figure 1. The wall particles arranged inside the block act as an impermeable wall because particles of MPS avoid closing to each other in the principle of the analysis. At the time of analysis, the wall particles move with the movement of the block in the NMM-DDA analysis.

On the other hand, in the above mention method, water particles cannot go through joint and void between blocks or elements under contact. It means that water cannot flow along the joint and no water pressure act along the joint. To avoid this condition, the wall particles of MPS are arranged inside the DDA block to enable the flow of water particles as shown Figure 1. For NMM elements, the wall particles of MPS are arranged inside the joint loop, which is the periphery of NMM elements. Water pressure and water flow along the joints between DDA blocks, NMM elements, and DDA block and NMM element are performed by above mentioned method.

3.2 Force acting form the water particles to DDA block and NMM element

Acting force from the MPS water particles to the DDA block and/or NMM elements is calculated by water pressure. Water pressure is caused by the water particles located in the range of influence to the DDA block (Fig. 2). Water pressure acts on periphery of the block and a direction of the force is normal to the periphery of the block. For the DDA block, water pressure act along the edge of the block as line distributed force. Assuming that the water pressure acting on 2 points (x_1, y_1) and (x_2, y_2) along the edge of DDA block is $F_1(f_{x1}, f_{y1})$ and $F_2(f_{x2}, f_{y2})$ as shown in Figure 3, the sub-matrices for the line distributed force is easily derived.

For NMM elements, water pressure act along the wall particle periphery inside the joint loop of NMM elements as line distributed force. The sub-matrix for the line distributed force in NMM element is easily derived by the same procedure as that of DDA block.

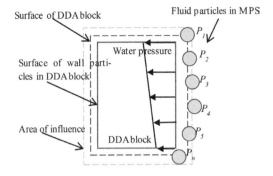

Figure 2. Conception of interaction from fluid particles to blocks.

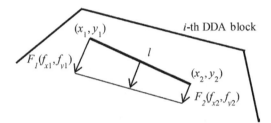

Figure 3. Line load along the point (x_1, y_1) to the point (x_2, y_2).

3.3 Procedure of coupling analysis

At the beginning of the coupling analysis, the initial state must be set both in MPS analysis also in NMM-DDA analysis. In NMM-DDA, an initial block alignment should be obtained under the condition of own weight and contact forces between blocks and/or elements. In MPS, an initial velocity and water pressure of the particles should be constructed. An initial setting analysis of MPS and NMM-DDA is individually performed at first. Next, the coupling model is constructed by superposing the results of the initial setting analysis of MPS and NMM-DDA.

Coupling analysis is performed by alternatively performing MPS analysis and NMM-DDA analysis. The flow chart of coupling analysis is shown in Figure 4 schematically.

4 APPLICATION TO THE LANDSLIDE

The Large-scale landslide (Aratozawa landslide) trigged by Iwate-Miyagi inland earthquake (M7.2) occurred in 2008. The length of the moved slope is about 1.3km, the volume of moved landslide body amount to 6.7×107 m³, and the maximum displacement of moved masses in the middle and lower part of the landslide is about 300m. The dip angle of main sliding surface is 0-2°. The formation around the landslide, which is Neogene sedimentary rocks, consists of welded tuff, pumice tuff and alternation of sandstone and siltstone (Ohno et al. 2010, Irie et al. 2009).

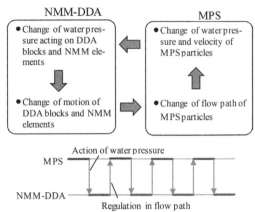

Figure 4. The flow of coupling analysis.

The simulation was attempted to ascertain applicability of the newly coupling method to the large scale and large displacement problem during earthquake.

4.1 Analytical model and condition

Figure 5 shows the analytical models. The length and height of the model are 1900 m and 325 m, respectively. The basement of the landslide was divided by NMM elements, and the landslide body consists of DDA blocks.

The material properties and analytical conditions are summarized in Table 1. The seismic forces, which are acceleration records during Iwate-Miyagi inland earthquake observed at AKTH04 (NIED in Japan), were given to the basement NMM elements as a dynamic body force. The maximum acceleration of horizontal motion was beyond 1000gal. In the simulation, the viscous boundary was applied along both right and left sides, and the displacements along bottom of the analytical model were fixed.

In the simulation, three cases were performed. The geometries, physical properties, seismic force and boundary conditions without groundwater conditions were equal in three cases. As shown in Figure 5, Case 1 was NMM-DDA analysis alone. In Case 2, the void was arranged along the basement of landslide body

Number of NMM elements: 1551
Number of DDA blocks: 180

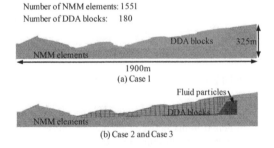

(a) Case 1

(b) Case 2 and Case 3

Figure 5. The analytical model of the large-scale landslide.

Table 1. Physical properties and parameters in NMM-DDA.

Unit mass (kg/m³)		1600
Unit weight (kN/m³)		16
Young's modulus (MN/m²)		800
Poisson's ratio		0.3
Penalty spring constant(GN/m)	Normal	2.0
	Shear	0.1
Friction angle of joint (deg)		3.0
Cohesion of joint (kN/m²)		0.0
Tensile strength of joint (kN/m²)		0.0
Max. time step (s)		0.01
Number of time step	Initial	1000
	Combined	80,000

of DDA blocks. However, the artificial void between blocks was not arranged. The initial groundwater condition in Case 2 is shown in Figure 5(b). Fluid particles in MPS were distributed in right side of the NMM basement to apply water pressure to DDA blocks. In the analysis, interaction between fluid particles in MPS and NMM elements were ignored, and interaction between fluid particles and DDA blocks were considered. In Case 3, the void was arranged along the basement of landslide body of DDA blocks and the artificial void between block was arranged. The width of artificial void between DDA blocks, where a water pressure is given to the DDA blocks, was set about twice size of representative diameter of water particles in MPS. The initial groundwater condition in Case 3 was equal to that in Case 2. Table 2 shows physical properties and parameters used in MPS analysis.

Before the coupling analysis, the initial setting analysis was carried out to build the contact forces between the DDA blocks, where 1000 iterations in NMM-DDA analysis was performed before applying seismic force.

4.2 Results

Figure 6 shows the simulation results after 40,000 steps. The displacement of DDA blocks in Case 1 and Case 3 was nearly equal. The displacement in Case 2 was different from that in Case 3.

Table 2. Physical properties and analytic parameters in MPS

Properties and parameters	MPS
Initial distance among particles (m)	5.0
Density of fluid (kg/m³)	1000
Coefficient of kinematic viscosity (m²/s)	1.0×10^{-6}
Compressibility (1/Pa)	0.45×10^{-7}
Gravitational acceleration (m/s²)	-10.0
Time step (s)	0.01

Comparing Case 2 with Case 3, the block displacement in upper part of the landslide body in Case 2 was larger than that in Case 3. However, the block displacement along the basement in Case 2 was slightly larger than that in Case 3. In Case 3, fluid particles moved through the artificial void between blocks, and the groundwater surface after 40,000 steps was almost flat as shown in Figure 5(c). On the other hand, the groundwater surface of Case 2 was not flat because fluid particles could not move between blocks. The water pressure along the basement in Case 2 was larger than that in Case 3 consequently. From this reason, it is consider that the block displacement along the basement in Case 2 showed large displacement.

Figure 7 shows the simulation results, which are the block displacement in Case 2 after the earthquake. The large displacement of DDA blocks appeared in the center and upper part of the landslide body, and maximum displacement was beyond 300m. The water pressure acting on the bottom of

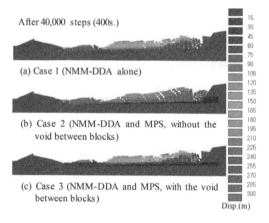

(a) Case 1 (NMM-DDA alone)

(b) Case 2 (NMM-DDA and MPS, without the void between blocks)

(c) Case 3 (NMM-DDA and MPS, with the void between blocks)

Disp.(m)

Figure 6. Displacement of DDA blocks after 40,000 steps.

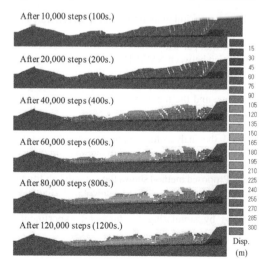

Figure 7. Displacement distribution of DDA blocks in Case 2.

Figure 8. Comparison between simulation results and ground surface line after the earthquake.

the landslide body was about 800,000Pa. In early stage of the simulation after the earthquake, the subsidence appeared in the head part of the landslide, which is the right side in the model. The toe part of the landslide in the left side of the model was thrust up to the left side slope in late stage of the simulation. A remarkable thrust up in Case 2 did not appear in Case 1 and Case 3. It is clear that a water pressure affected the landslide movement.

Figure 8 shows the simulation results after 120,000 steps in Case 2. The ground surface line by geological survey after the earthquake is also shown in Figure 8. The surface line of the simulation in Case 2 is similar to that of the geological survey. It is consider that the simulation results are consistent with the geological survey.

These results show the potential ability to simulate the large-scale landslide movement during earthquake. However, the simulation results also show complex relations among strength of sliding surface, water pressure and permeability of joint.

5 CONCLUDING REMARKS

The authors introduced coupling analysis of the discontinuous models using NMM-DDA and particle MPS to solve a complex solid and fluid interaction problems. In the coupling analysis of NMM-DDA and MPS, the wall particles of MPS are arranged inside DDA blocks or joint loops along NMM elements, and these wall particles prevent the water particles from invading into the DDA block and NMM elements. The effect of water inflow into the joint is incorporated by making artificial void between the block and alignment of wall particles inside the block. Moreover, the interaction between NMM-DDA and MPS analysis is performed as distributed forces along the surface of the DDA block and the joint loops surrounding NMM elements, and the forces are evaluated by the water pressure in MPS analysis.

In order to examine the ability of the proposed method, the analysis for the movement of large-scale landslide triggered by the earthquake were attempted. The results agreed with the common landslide movement. Proposed method shows the potential ability for the application to landslide movement in earthquake.

The validity of the method should be additionally verified by simulating fundamental experimental results such as laboratory liquefaction test. In order to simulate landslide movement in earthquake practically, groundwater conditions before and after earthquake should be made clear also.

ACKNOWLEDGEMENT

In this study, the seismograms recorded by National Research Institute for Earth Science Disaster Prevention (NIED) were used. The authors would like to thank to NIED.

REFERENCES

Cundall, P.A. 1971. A computer model for simulation progress, large scale movement in block system. *Proceedings of ISRM Symp., Nacncy, France*, 11–18.

Irie, K., Koyama, T., Hamasaki, E., Nishiyama, S., Shimaoka, K., & Ohnishi, Y. 2009. DDA simulations for huge landslides in Aratozawa area, Miyagi, Japan caused by Iwate-Miyagi Nairiku earthquake. *Proceedings of ICADD-9, Singapore*, 153–160.

Koshizuka, S. & Oka, Y. 1996. Moving particle semi-implicit method for fragmentation of incompressible fluid. *Nuclear Science and Engineering*, 123, 421–434.

Kuno, M., Miki, S., Ohnishi, Y., & Sasaki, T. 2016. Coupling analysis of rock mass and water for debris flow on a rock slope by DDA (Discontinuous De-formation Analysis) and MPS (Moving Particle Simulation) Method. *Proceedings of 9th Asian Rock Mechanics Symp., Bali, Indonesia*.

Miki, S., Sasaki, T., Koyama, T., Nishiyama, S., & Ohnishi, Y. 2010. Development of coupled discontinuous deformation analysis and numerical manifold method (DDA-NMM). *International Journal of Computational Methods*, Vol.7, No.1: 131–150.

Miki, S., Sasaki, T., Ohnishi, Y., Nishiyama, S., & Koyama, T. 2013. Application of NMM-DDA to earthquake induced slope failure and landslide, *Proceedings of 47th US Rock Mechanics Geomechanics Symposium*, ARMA13–492.

Miki, S., Ohnishi, Y., & Sasaki, T. 2017. Water flow and rock mass coupling analysis of debris flow on a rock slope by DDA and MPS (Moving Particle Simulation) method, *Proceedings of 51st US Rock Mechanics Geomechanics Symposium*, AR-MA17–630.

Ohno, R., Yamashina, S., Yamasaki, T., Koyama, T., Esaki, F., & Kasai, S. 2010. Mechanism of a large-scale landslide triggered by the earthquake in 2008 – A study of Artosawa landslide. *Journal of the Japan Landslide Society*, Vol.47, No.2: 8–14. (in Japanese)

Shi, G.H., & Goodman, R.E. 1989. Generalization of two-dimensional discontinuous deformation analysis for forward modeling. *Int J. Numer. Anal. Meth. Geomech*, 13, 359–380.

Shi, G.H. 1991. Manifold method of material analysis, *Transactions of the 9th Army Conference on Applied Mathematics and Computing*, Report No.92-1, U.S. Army Research Office.

2019 Rock Dynamics Summit– Aydan et al. (eds)
© 2019 Taylor & Francis Group, London, ISBN 978-0-367-34783-3

Model tests on rock slopes prone to wedge sliding and some case histories from recent earthquakes

H. Kumsar
Department of Geological Engineering, Pamukkale University, Dept. of Geological Engineering, Denizli, Turkey

Ö. Aydan
Department of Civil Engineering and Architecture, University of the Ryukyus, Okinawa, Japan

ABSTRACT: The wedge failure is one of the common forms of slope failures. In this study, the authors investigate the sliding responses of rock wedges under dynamic loads rather than the initiation of wedge sliding. Firstly some laboratory model tests are described. On the basis of these model tests on rock wedges, the theoretical model proposed previously is extended to compute the sliding responses of rock wedges in time domain. The proposed theoretical model is applied to simulate the sliding responses of rock wedge model tests and its validity is discussed. In the final part, the method proposed is applied to actual wedge failures observed in 1995 Dinar earthquake, 2007 Çameli earthquake and 2005 Pakistan-Kashmir earthquake, and the results are discussed.

1 INTRODUCTION

The stability of rock slopes under dynamic loading in mining and civil engineering depends upon the slope geometry, mechanical properties of rock mass and discontinuities, and the characteristics of dynamic loads with time (Aydan 2015). The wedge failure is one of the common forms of rock slope failures (Fig. 1).

In this study, the sliding responses of rock wedges under dynamic loads rather than the initiation of wedge sliding are investigated. Firstly some laboratory tests on model wedges are described. On the basis of these model tests on rock wedges, the theoretical models developed by Aydan and Kumsar (2010) are used to compute the sliding responses of rock wedges in time domain. In the final part, the method is applied to actual wedge failures observed in 1995 Dinar earthquake, 2007 Çameli earthquake and 2005 Pakistan-Kashmir earthquake, and the results are discussed.

2 DYNAMIC MODEL TESTS

2.1 *Preparation of Models*

Six special moulds were prepared to cast model wedges (Kumsar et al. 2000). For each wedge configuration, three wedge blocks were prepared. Each base block had dimensions of 140x100x260mm. Base and wedge models were made of mortar and their geomechanical parameters were similar to those of rocks.

The composition of the mortar used for the preparation of the models is 1781 kgf/m³ of fine sand,

Figure 1. Some examples of wedge failure of rock slopes

360 kgf/m³ of cement with a water-cement ratio of 0.5. The cement used in mortar was rapid hardening type and samples were cured for about 7 days in a room with a constant temperature. The wedge angles and the initial intersection angles of wedge blocks are listed in Table 1.

In addition, several mortar slabs were cast to measure the friction angle of sliding planes. A number of tilting tests were performed. The inferred friction angle measured in tilting tests ranged between 30° and 34° with an average of 32°.

Table 1. Geometric parameters of wedges (also see Figure 9).

Wedge Number	Intersection inclination - i_a (°)	Half-wedge Angle, $\omega_1 = \omega_2$ (°)
TB1(Swedge120)	29	61.5
TB2(Swedge100)	29	51.5
TB3(Swedge90)	31	47.8
TB4(Swedge70)	27	40.0
TB5(Swedge60)	30	33.8
TB6(Swedge45)	30	26.0

Figure 2. A view of a wedge model

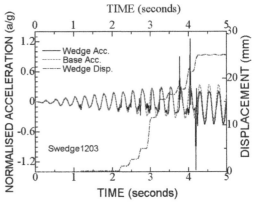

Figure 3. Dynamic response of Wedge Model TB1

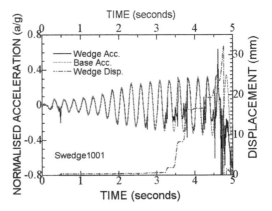

Figure 4. Dynamic response of Wedge Model TB2

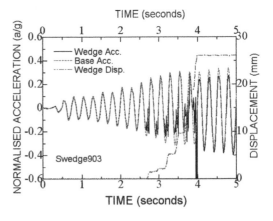

Figure 5. Dynamic response of Wedge Model TB3

Each wedge base block was fixed on the shaking table to receive same shaking with the shaking table during the dynamic test. The accelerations acting on the shaking table, at the base and wedge blocks were recorded during the experiment, and saved on a data file as digital data (Fig. 2). Furthermore, in the second series of dynamic tests, a laser displacement transducer was used to record the movement of the wedge block during the experiments. The reason for recording accelerations at three different locations is to determine the acceleration at the moment of failure as well as any amplification from the base to the top of the block. In fact, when the amplitude of input acceleration wave is increased, there is a sudden decrease on the wedge block acceleration records during the wedge failure, while the others are increasing. A barrier was installed at a distance of 20-30 mm away from the front of the wedge block to prevent their damage by falling off from the base block.

Dynamic testing of the wedge models were performed in the laboratory by means of a one-dimensional shaking table, which moves along horizontal plane. The applicable waveforms of the shaking table are sinusoidal, saw tooth, rectangular, trapezoidal and triangle. The shaking table has a square shape with 1m-side length. The frequency of waves to be applicable to the shaking table can range between 1 Hz and 50 Hz. The table has a maximum stroke of 100 mm and a maximum acceleration of 6 m/s² for a maximum load of 980.7 N.

2.2 Shaking Table Tests

Three experiments were carried out on each wedge block configuration and dynamic displacement responses of wedge blocks in addition to the acceleration responses were measured. Figs. 3-8 shows typical acceleration-time and displacement-time responses for each wedge block configuration. As it is noted from the responses shown in Figs. 3-8, the acceleration responses of the wedge block indicate some high frequency waveforms on the overall trend of the acceleration imposed by the shaking table. When this type waveform appears, the permanent displacement of the wedge block with respect to the base block takes place. Depending upon the amplitude of the acceleration waves as well as its direction, the motion of the block may cease. In other words, a step-like behavior occurs.

The motion of the block starts when the amplitude of the input wave acts in the direction of the

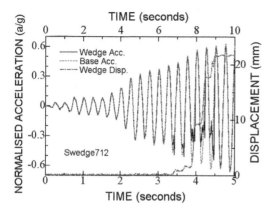

Figure 6. Dynamic response of Wedge Model TB4

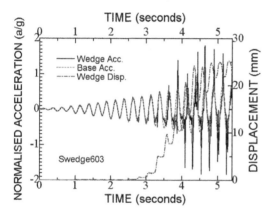

Figure 7. Dynamic response of Wedge Model TB5

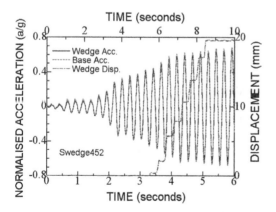

Figure 8. Dynamic response of Wedge Model TB6

downside and exceeds the frictional resistance of the wedge block. When the direction of the input acceleration is reversed, the motion of the block terminates after a certain amount of relative sliding. As a result, the overall displacement response is step-like.

Another important observation is that the frictional resistance between the wedge block and base-block

limits the inertial forces acting on the wedge block and the base block even though the base-block may undergone higher inertial forces. The sudden jumps in the acceleration response of the wedge block as seen in Figs. 3-7 are due to the collision of the wedge block with the barrier. The initiation of the sliding of the wedge blocks was almost the same as those measured in the first series of the experiments.

3 THEORETICAL MODELLING

The authors have advanced the method of stability assessment proposed by Kovari and Fritz (1975) for wedge failure of rock slopes under different loading conditions and confirmed its validity through experiments (Kumsar et al. 2000). Aydan and Kumsar (2010) extended to evaluate sliding responses of rock wedges under dynamic loading conditions under submerged conditions with viscos resistance. Let us consider a wedge subjected to dynamic and water loading as shown in Figure 9. One can easily write the following dynamic equilibrum conditions for the wedge during sliding motion on two basal planes in a coordinate system $Osnp$ shown in Fig. 9.

$$\sum F_s = (W - E_v)\sin i_a - E_i \cos i_a - S = m\frac{d^2 s}{dt^2} \quad (1)$$

$$\sum F_n = (W - E_v)\cos i_a - E_i \sin i_a - N = m\frac{d^2 n}{dt^2} \quad (2)$$

$$\sum F_p = -N_1 \cos\omega_1 + N_2 \cos\omega_2 + E_p = m\frac{d^2 p}{dt^2} \quad (3)$$

Where $N = N_1 \sin\omega_1 + N_2 \sin\omega_2$; W: weight of wedge; E_v: dynamic vertical load; E_i: dynamic force in the direction of intersection line; E_p: dynamic load

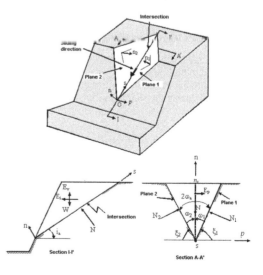

Figure 9. Illustration of mathematical model for wedge failure.

494

perpendicular to intersection line. Other parameters are shown in Fig. 9. Although the dynamic vectorial equilibirum equation are written in terms of its component, they correspond to a very general form for wedges sliding along the intersection line while being in contact with two basal planes. Furthermore, the earthquake force is decomposed to its corresponding components in the chosen coordinate system.

One can easily obtain the following identity from Eq.(3) by assuming that there are no motions upward and perpendicular to the intersection line:

$$N_1 + N_2 = \left[(W - E_v)\cos i_a - E_i \sin i_a \right] \lambda_i - E_p \lambda_p \quad (4)$$

Where

$$\lambda_i = \frac{\cos \omega_1 + \cos \omega_2}{\sin(\omega_1 + \omega_2)}; \quad \lambda_p = \frac{\sin \omega_1 - \sin \omega_2}{\sin(\omega_1 + \omega_2)} \quad (5)$$

If the resistance is assumed to obey Mohr-Coulomb criterion (Aydan and Ulusay, 2002; Aydan et al. 2008) one may write the following:

$$T = (N_1 + N_1)\mu ; \quad \mu = \tan \varphi \quad (6a)$$

Following the initiation of sliding, the friction angle can be reduced to the kinetic friction angle as given below

$$\mu = \tan \varphi_r \quad (6b)$$

Where φ_r are residual cohesion and friction angle. Under frictional condition, it should be noted that normal force $(N_1 + N_2)$ can not be negative (tensile). If such a situation arise, normal force $(N_1 + N_2)$ should be set to 0 during computations. Let us introduce the following parameters:

$$\eta_v = \frac{E_v}{W} = \frac{a_v}{g}; \quad \eta_i = \frac{E_i}{W} = \frac{a_i}{g}; \quad \eta_p = \frac{E_p}{W} = \frac{a_p}{g} \quad (7)$$

where $a_v; a_i; a_p$ are acceleration components resulting from dynamic loading.

The following dynamic equilibrium equation must be satisfied during the sliding motion of the wedge.

$$S = T \quad (8)$$

If the relations given by equations (1),(4),(6),(7) are inserted in Eq. (8), one can easily obtain the following differential equation

$$\ddot{s} = \frac{d^2 s}{dt^2} = g\left[(1 - \eta_v)A + \eta_i B + \eta_p C \right] \quad (9)$$

Where

$$A = (\sin i_a - \cos i_a \mu \lambda_i); \quad B = (\cos i_a + \sin i_a \mu \lambda_i);$$
$$C = \mu \lambda_p$$

Since dynamic loads are very complex in time domain, the solution of Eq. (9) is only possible through numerical integration methods. The time-domain problems in mechanics are generally solved by finite difference techniques. For this purpose, there are different finite difference schemes. In this article the solution of Eq. (9) based on linear acceleration finite difference technique (i.e Aydan and Ulusay, 2002; Aydan et al. 2008). One can write the velocity ($\dot{s}$) and displacement of wedge (s) for a time step $n+1$ as follows:

$$\dot{s}_{n+1} = \dot{s}_n + \frac{\ddot{s}_n}{2}\Delta t + \frac{\ddot{s}_{n+1}}{2}\Delta t \quad (10)$$

$$s_{n+1} = s_n + \frac{\dot{s}_n}{1}\Delta t + \frac{\ddot{s}_n}{3}\Delta t^2 + \frac{\ddot{s}_{n+1}}{6}\Delta t^2 \quad (11)$$

Provided that resulting dynamic shear force exceeds the shear resistance of the wedge at time $(t = t_i = i\Delta t)$, one can easily incorporate the variation of shear strength of discontinuities from peak state $(\mu = \tan \varphi_p)$ to residual state $(\mu = \tan \varphi_r)$.

4 COMPARISONS WITH EXPERIMENTS

The theoretical model presented in the previous section has been applied to experiments to compare the computed responses with those from experiments. Figs. 10 and 11 compare computational results for some of wedge blocks with measured sliding responses. The detailed comparisons for all wedge blocks are reported elsewhere (Aydan and Kumsar 2010). In computations, the peak friction angle of discontinuity planes reported in the previous publication (Kumsar et al. 2000) was reduced by 3° in order to take into account the slight damage to surfaces due to multiple utilization of wedge blocks. As noted from the measured and computed displacement responses, the results are remarkably similar to each other. Although some slight differences exist, these may be associated with the variation of the non-linear surface friction between the base-block and wedge block and the negligence of viscous effects in computations.

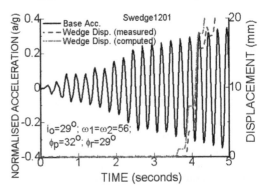

Figure 10. Comparison of computed responses with experimental response for TB1 wedge model.

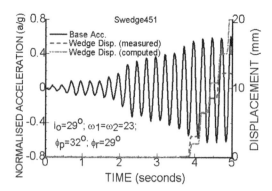

Figure 11. Comparison of computed responses with experimental response for TB6 wedge model.

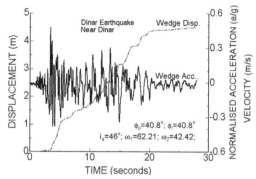

Figure 13. Computed displacement response of chosen wedge failure nearby Dinar.

5 APPLICATIONS TO WEDGE FAILURES CAUSED BY RECENT EARTHQUAKES

5.1 *Dinar Earthquake*

Dinar earthquake with a magnitude of 6.0 occurred on October 1, 1995 (Aydan and Kumsar, 1997). Many rock slope failures observed along the surface trace of the earthquake fault. At several locations on the eastern slope of the graben adjacent to the fault scarps, where the rock mass shows up, there were some rock slope failures. Rock mass is karstic conglomerate and bedding planes dip towards south with an inclination of 20°-25°. Most of rock slope failures were of small scale and associated with existing joints in rock mass. Slope failures were observed on eastern side of the graben next to Dinar-Çivril fault. The slope failures shown in Figure 12 were wedge failures nearby Dinar.

Tilting tests indicated that the friction angle of joint surfaces was about 40°. Fig. 13 shows the dynamic response of the wedge blocks nearby Dinar computed using the acceleration record obtained at Dinar strong motion station (DAD-ERD, 1995). As noted from the figure, wedge-sliding failure is initiated. Nevertheless, the motion terminates after a certain amount of displacement (4.5m). Such an amount of relative displacement was sufficient to displace the blocks and the blocks were fallen on the flatter ground as seen in Fig. 13.

5.2 *The Kashmir Earthquake*

On October 8, 2005 at 8:50 (3:50UTC), a large devastating earthquake occurred in Kashmir region of Pakistan. The depth of the earthquake was estimated to be about 10 km and it had the magnitude of 7.6 (Aydan, 2006). The large slope failure occurred at Hattian (Dana Hill) and it was an asymmetric wedge sliding (Fig. 14). The wedge sliding failure at Hattian was quite large in scale. The sliding area was 1.5 km long and 1.0 km wide. Rock mass consisted of shale and sandstone and it constituted a syncline (Aydan et al. 2009). The estimated wedge angle was about 100° and it was asymmetric. The friction angle of shale from tilting test was more than 35° with an average of 40°.

The limiting equilibrium analysis for wedge sliding failure (Kumsar et al., 2000) indicated that the safety factor of the slope would be 1.55 under dry static conditions (Aydan and Kumsar, 2010). However, the mountain wedge becomes unstable when ground acceleration is equivalent to the horizontal seismic coefficient of 0.3 and the safety factor reduces to 0.9 under such a condition.

Using the acceleration record of Abbotabad (Ohkawa 2005) and multiplying the record by an

Figure 12. Wedge failures nearby Dinar.

Figure 14. A view of Hattian slope failure.

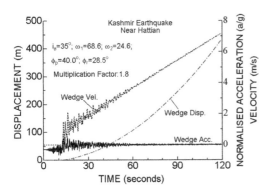

Figure 15. Computed displacement response of failed body of Hattian slope.

amplification factor of 1.27 so that the seismic coefficient value of 0.3 was achieved, a dynamic simulation of the wedge failure was carried out. The residual friction angle was reduced to 28.5° from the peak friction after yielding. The results are shown in Figure 15. The slope becomes unstable by the earthquake-induced ground shaking, the motion of the failed body increases with time.

Figure 16. A view of wedge failures at Taşçılar village.

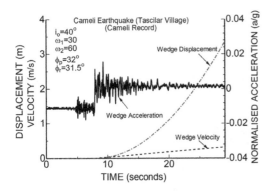

Figure 17. Computed displacement response of wedges failed at Taşçılar village due to the 2007 Çameli earthquake.

Unless the geometrical profile of the sliding surface changes, the sliding motion would continue with a constant velocity.

5.3 *The Çameli Earthquake*

The Çameli earthquake occurred at at 11:23 on TST (9:23 UTC) on October 29, 2007. The estimated magnitude of the earthquake varies between 4.9 and 5.4 depending upon the seismological institute. The earthquake caused some planar and wedge sliding failures in rock slopes consisting of marn. Figure 16 shows the computed dynamic response of the rock wedge for the assumed parameters shown in Figure 17. The wedge was assumed to be under dry conditions. As noted from the figure, the strong motion records taken at Çameli are sufficient to induce rock wedge sliding failure.

6 CONCLUSIONS

A series laboratory shaking table tests were carried on wedge models under dynamic excitations for the assessment of the validity of the limiting equilibrium method as well as to evaluate their sliding responses during shaking. The shaking table experiments on the wedge models were performed under dry conditions. Then, a method was presented to evaluate the dynamic sliding response of wedge blocks and the estimated sliding responses from the method presented were compared with experimental results. The results show that the estimated results are in a good agreement with the experimental results. In addition, the wedge failures induced by the 1995 Dinar earthquake, Çameli earthquake and the 2005 Kashmir earthquake were back analyzed and discussed. Although the wedge failures can be generally in small scale, it may sometimes be quite large in scale as observed at Hattian in Kashmir region of Pakistan.

REFERENCES

Aydan, Ö. (2006): Geological and Seismological Aspects of Kashmir Earthquake of October 8, 2005 and A Geotechnical Evaluation of Induced Failures of Natural and Cut Slopes. Journal of Marine Science and Technology, Tokai University, Vol.4, No.1, 25–44.

Aydan, Ö., (2015). Large Rock Slope Failures induced by Recent Earthquakes. Rock Mechanics and Rock Engineering, Special Issue on Deep-seated Landslides. 49(6),2503–2524.

Aydan, Ö. and H. Kumsar (1997). A site investigation of Oct. 1, 1995 Dinar Earthquake. Turkish Earthquake Foundation, TDV/DR 97–003.

Aydan, Ö., Ulusay, R. (2002): Back analysis of a seismically induced highway embankment during the 1999 Düzce earthquake. Environmental Geology, 42, 621–631.

Aydan, Ö., Kumsar, H. (2010). An Experimental and Theoretical Approach on the Modeling of Sliding Response of Rock Wedges under Dynamic Loading. Rock mechanics and Rock Engineering, Vol.43, No.6, 821–830.

Aydan, Ö., Ulusay R., Atak, V.O. (2008): Evaluation of ground deformations induced by the 1999 Kocaeli earthquake (Turkey) at selected sites on shorelines. Environmental Geology, Springer Verlag, 54, 165–182.

Aydan, Ö., Ohta, S., Hamada, M. (2009): Geotechnical evaluation of slope and ground failures during the 8 October 2005 Muzaffarabad Earthquake, Pakistan. Journal of Seismology 13(3),399-413. (DOI 10.1007/s10950-008-9146-7).

DAD-ERD, (1995). Earthquake Research Department of Turkey, http:www.deprem.gov.tr

Kovari, K., Fritz, P. (1975). Stability analysis of rock slopes for plane and wedge failure with the aid of a programmable pocket calculator. 16th US Rock Mech. Symp., Minneapolis, USA, 25–33.

Kumsar, H., Aydan, Ö. and Ulusay, R. (2000). Dynamic and static stability of rock slopes against wedge failures. Rock Mechanics and Rock Engineering, Vol.33, No.1, 31-51.

Okawa, I. (2005): Strong earthquake motion recordings during the Pakistan, 2005/10/8, http://www.bri.go.jp/

2019 Rock Dynamics Summit– Aydan et al. (eds)
© 2019 Taylor & Francis Group, London, ISBN 978-0-367-34783-3

Failure mechanism and causes of Ergenekon landslide (Turkish Republic of Northern Cyprus - TRNC)

C. Atalar
Faculty of Engineering, Near East University, Lefkoşa, TRNC

H. Kumsar
Department of Geological Engineering, Pamukkale University, Denizli, Turkey

Ö. Aydan
Department of Civil Engineering and Architecture, University of the Ryukyus, Nishihara, Okinawa, Japan

ABSTRACT: The mechanism and causes of a landslide, which blocked a road to transportation and took place in 500 m NE of Ergenekon village of Serdarlı District located on the southern slopes of Beşparmak Mountain Range in Turkish Republic of Northern Cyprus, are investigated in this study. The failure surface was within the melange and resulted in buckling of the road up to 2 m. Different limiting equilibrium methods and discrete finite element method (DFEM) were used for the assessment of the slope instability. The factor of safety of the slope ranges between 1.22 and 1.26 when the slope is dry. The results showed that the slope failure may occur when the pore water pressure has a value ranging between 0.25 and 0.35. A failure surface, which is a combination of discontinuities with tectonic origin, was assumed for the stability assessment of the slope by using a pseudo-dynamic version of DFEM and the results indicated that the slope becomes unstable when pore water pressure in the slope forming material becomes 0.4.

1 INTRODUCTION

Beşparmak Mountain Range of Northern Cyprus is made up Miocene aged flysch at the bottom that was overlain by the Cretaceous aged melange. The top of the mountains are mainly composed of Jurrasic – Triassic aged limestones. A landslide occurred in the melange in Ergenekon village of Serdarlı District at 28 km northeast of Lefkoşa, Turkish Republic of Northern Cyprus (Fig. 1) and it is named as Ergenekon landslide.

The melange is composed of radiolorite, wheathered tuff, serpentinite, limestone and pillow lava. The asphalt pavement was buckled at the toe of the slope and blocked the road due to this landslide (Fig. 2). In this study, a series of stability analyses were carried out using different limiting equilibrium methods of analysis and a pseudo-dynamic version of DFEM. The results are presented and discussed.

2 GEOLOGY

In the study area, Miocene aged flysch formation is located at the bottom (Fig. 3). The strata dips into the mountain and the slope of the road cut is about 42°. This formation is overlain by the Cretaceous aged mélange that consisting of weathered serpentinite, radiolorite, tuff and limestone blocks. The Cretaceous aged pillow lavas overly this formation with an unconformity.

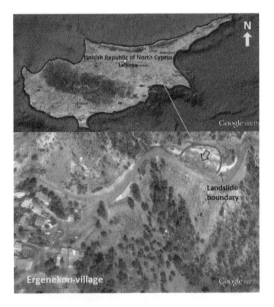

Figure 1. Google view showing the location of the Ergenekon landslide.

In order to investigate the failure mechanism and the factors triggering the slope failure, 8 geotechnical boreholes were drilled in the landslide area (Fig. 3). Their depths ranged between 4.33 m and 7.50 m. Since the mélange was highly weathered,

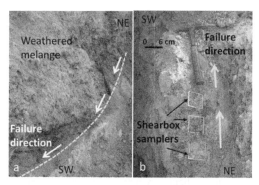

Figure 4. (a) Failure surface within the weathered melange observed in Trench-5, (b) undisturbed sampling on the failure surface in Trench-6.

Figure 2. a) General view of the slope failure, b) buckled road pavement at the toe of the slope.

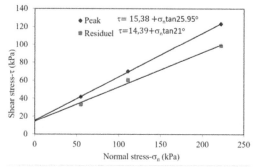

Figure 5. Failure envelopes for the weathered melange samples collected from Trench-6.

weathered serpentinite and tuff in the mélange. It was observed that the failure surface occurred along the boundary between mélange and flysch formations at the toe of the slope and it caused the buckling of the asphalt pavement of the road.

Direct shear test of the weathered melange samples obtained from Trench-6 were tested in the laboratory under consolidated-drained conditions. The test results indicated that the peak cohesion (c_p) is 15.4 kPa, peak internal friction angle (ϕ_p) is 26°, residual cohesion (c_r) is 14.4 kPa and residual internal friction angle (ϕ_r) is 21° (Fig. 5). Average unit weight of the slope, 23 kN/m³, was used for calculations.

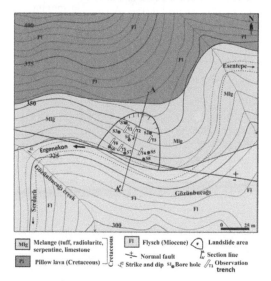

Figure 3. Engineering geological map of the Ergenekon landslide area (Atalar et al., 2016).

tectonized, and water was used during borings, the failure surface could not be clearly observed in the borehole sections. Six observation trenches were excavated with depths ranging between 1.5 m and 3.7 m, and lengths between 3 m and 12 m (Fig. 3).

A circular failure surface was observed at 0.9 m in Trench-4, 0.5 m in Trench-6, 1.5 m in Trench-5 (Fig. 4). The failure surface occurred through the

3 STABILITY ASSESMENTS

In order to assess the instability, a slope profile was drawn along Section line A-A' of the pre-failure topographic map as shown in Fig. 3. SLOPAC2D computer program (Kumsar, 1993) was used for the definition of the failure surface using the failure surface data collected from the observation pits. The slope failure was assumed to occur along a circular and planar surface passing through the toe of the slope. The circular part occurred within the weathered mélange in the upper part of the slope, and the planar failure surface was occurred between the boundary of the weathered mélange and flysch at the toe of the slope where the buckling of the asphalt pavement of the road occurred (Fig. 6).

The stability of the slope was carried out by using conventional limiting equilibrium methods (LEM) and also a pseudo-dynamic version of discrete finite element method (DFEM) (Aydan et al. 1996, 1997, 2001).

3.1 *Stability assessment based on limiting equilibrium methods*

The vertical slice interval of the sliding mass from its top to the boundary between weathered melange and flysch at the toe was selected as 5m. Two smaller slices were defined at the toe of the slope. The first once is in weathered melange and the second one is along its boundary with flysch (Fig. 7). Pore water pressure coefficient (r_u) was assumed to be zero as the slope was stable in dry season and the factors of safety were calculated using Bishop (1954), Aydan et al (1992) and Kumsar (1993) methods. The calculated factors of safety (F) ranged between 1.23 and 1.26 under dry conditions (Table 1) indicating the slope was stable. As the slope failure occurred in a rainy season in March, it was likely that the r_u value would be increased. Taking this fact into consideration, the pore pressure coefficient was ranged between 0 and 0.4, and the factor of safety of the slope was calculated for each r_u value as shown in Fig. 7. Slope failure occurs when r_u value is greater than 0.25 depending on the limiting equilibrium method.

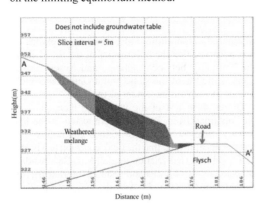

Figure 6. Failure surface defined on the cross-section A-A' given in Figure 3.

Table 1. Stability assessments for the sliding mass

Method	∑Disturbing Forces (kN)	∑Resisting forces (kN)	Factor of safety (FS)
Bishop (1954)	1803.61	1449.59	1.24
Aydan et al. (1992)	1790.21	1415.76	1.26
Kumsar (1993)	1765.41	1440.49	1.23

3.2 *Stability assessment based on discrete Finite element method (DFEM)*

The tectonic zones, which were observed at Ergenekon landslide site, consist of crushed rock between the Cretaceous and Miocene units. Therefore, this zone has its intrinsic structural weakness surfaces as well-known in structural geology. The situation of ground conditions corresponds to crushed/shattered rock masses and it deserves to be evaluated using rock mechanics approaches rather than soil-mechanics type of evaluations (Fig. 8). In this section, three rock mass classifications systems were used for evaluating the state of rock mass (Tables 2-4). The RMQR value ranges between 13-19 while the Q-values ranges between 0.0127-0.0625. Although the use of RMR is not apropriate to apply to fracture zones, the value of RMR is estimating to be ranging between 15-16.

Needle penetration tests (Ulusay et al., 2014; ISRM, 2015) on rocks in the tectonic zone were carried out. The needle penetration index (NPI) values ranged between 1 and 5 for the decomposed/shattered tuff and 20 and 60 for the intact greenish or whitish tuffs. The shear strength characteristics are assumed to be the same as those used in LEM analyses and Lame coefficients (λ, μ) given in Table 5 were computed from the relations between the NPI and mechanical properties proposed by Aydan (2012) and Aydan et al. (2014b) and used in DFEM analyses described in Section.

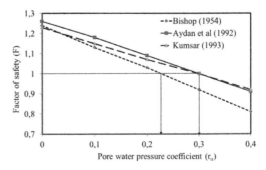

Figure 7. Variation of the safety factor with pore water pressure (r_u) coefficient.

Figure 8. Excavated Pit 2 in the failure mass, b) a close up view of crushed/shattered rock mass in Pit 2.

Table 2. Rock mass quality rating classification (RMQR; Aydan et al., 2014a) of the melange.

Parameter		Description	Rate
Degradation Degree (DD)		Heavy degradation/decomposed	1-3
Discontinuity set number (DSN)		Crushed/shattered	1
Discontinuity spacing (DS)		0.3m>DS>0.07 / 0.07>DS	1-4
Discontinuity condition (DC)		separation>10mm; thick infill t>60 mm; shear band	3
Groundwater Condition (GWC)	Seepage	wet	5
	Absorption	Highly/extremely absorptive	1-2
Total rating		Very poor / Weak	13-19

Table 3. Rock mass classification (Q-System; Barton et al., 1974) of the melange.

Parameter	Description	Value
RQD	0-25	10-25
Joint set number (Jn)	Crushed rock	20
Joint Roughness parameter (Jr)	G: slickensided, planar	0.5
Joint alteration parameter (Ja)	M: thick continuous bands	10-13
Joint Water Reduction factor (Jw)	A: Minor inflow or B. Medium inflow	1-0.66
Stress Reduction Factor (SRF)	No description for slopes	1.0
Quality	Q	0.0127-0.0625

Table 4. Rock mass rating classification (RMR; Bieniawski 1989) of melange formation.

Parameter	Description	Rate
UCS	0-5 MPa	0-1
RQD	<25%	3
Discontinuity spacing (DS)	<60 mm	5
Discontinuity condition (DC)	Soft, thick, gouge, decomposed	0
Groundwater condition (GWC)	wet	7
Total Basic RMR	Very poor rock	15-16

Table 5. Shear strength parameters of the weathered tuffs in the melange used in DFEM analyses.

Unit	λ (MPa)	μ (MPa)	γ (kN/m³)	c (kPa)	ϕ (°)
Body	1000	1000	23	-	-
FP-S1	150	15	-	15.38	26.0
FP-S2	150	15	-	28	17.15

FP: Failure Plane; S1 are S2: segments; γ: unit weight; c: cohesion; ϕ: friction angle; Body: unstable mass

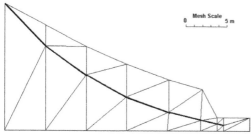

Figure 9. DFEM mesh used for circular sliding.

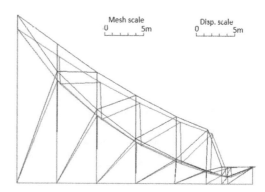

Figure 10. Deformed configuration of the slope for r_u=0 at step 12.

The stability assessment of the sliding body was carried out using the pseudo dynamic version of DFEM for circular sliding and structurally controlled failure surfaces. Thus the results were presented for two different failure surfaces, namely, circular sliding failure surface and structurally controlled failure surface in this section.

3.2.1 Stability assessment for circular failure surface by Discrete Finite Element Method (DFEM)

The finite element mesh shown in Figure 9 was used for assessing the stability of the slope for circular surface, which is the same as that used in limiting equilibrium analyses (LEM). The pore pressure coefficient r_u was varied between 0 and 0.4. Figures 10 and 11 show the deformed configuration of slope for pore pressure coefficient r_u of 0 and 0.4 at the pseudo dynamic time step 12, respectively. Figure 12 shows the relative displacement between two nodes at the toe of the slope.

As noted from the figure, when the slope is dry, it implies that the slope should have been stable as estimated from the limiting equilibrium methods. However, if the pore-pressure coefficient increases

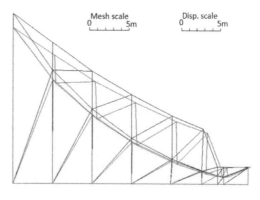

Figure 11. Deformed configuration of the slope for r_u=0.4 at step 12.

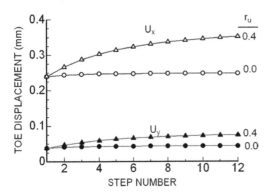

Figure 12. The variation of the relative displacement between nodes at the toe of the slope for $r_u=0.0$ and 0.4 in relation to pseudo dynamic time steps.

the slope becomes unstable as it was also estimated from the limiting equilibrium methods of analysis.

3.2.2 Stability assessment for structurally controlled failure surface by Discrete Finite Element Method (DFEM)

As the failure occurred within the tectonically disturbed zone, a failure surface was determined using the distribution of various fracture patterns anticipated within the tuffeceaous contact zone between the upper limestone and lower phyllite zone (Aydan and Kawamoto, 1990; Aydan et al. 1993). The failure surface follow of R'and T fracture paths and contact between phyllite and tuffeceaous melange. Figure 13 shows the mesh incorporating the structurally controlled failure surface, which is modeled using the contact elements.

Mechanical properties used in the pseudo dynamic version of DFEM analyses are those given in Table 5. The values of pore pressure coefficient (r_u) were assumed to be 0 and 0.4. Figures 14 and 15 show the deformed configuration of the slope for pore pressure coefficient r_u of 0 and 0.4 at the pseudo dynamic time step 12, respectively.

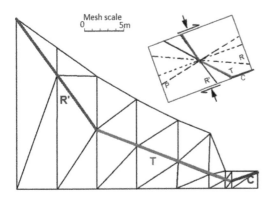

Figure 13. Finite element mesh used for structurally controlled failure surface.

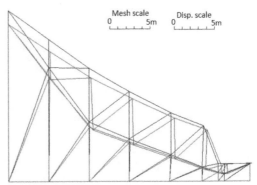

Figure 14. Deformed configuration of meshed slope for $r_u=0.0$ at pseudo dynamic time step 12.

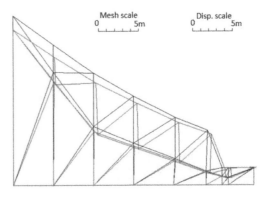

Figure 15. Deformed configuration of the slope for $r_u=0.4$ at pseudo dynamic time step 12.

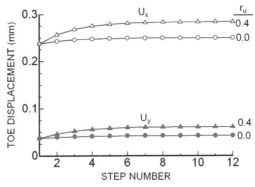

Figure 16. The variation of the relative displacement between nodes at the toe of the slope for r_u of 0.0 and 0.4 in relation to pseudo dynamic time steps.

Figure 16 shows the relative displacement between two nodes at the toe of the slope. As noted from the figure, when the slope is dry, it implies that the slope should have been stable as estimated from the limiting equilibrium methods of analysis.

However, if the pore-pressure coefficient increases, the slope becomes unstable as it was also estimated from the limiting equilibrium methods of analysis as well as from the pseudo dynamic version of DFEM for the circular failure surface.

4 CONCLUSIONS

In this paper, the results of the investigation of mechanism and reasons of a landslide, which blocked a road and took place in 500 m NE of Ergenekon village of Serdarlı Disctrict located on the southern slopes of Beşparmak Mountain in Turkish Republic of Northern Cyprus, are presented. The outcomes of this study may be summarized as

- Failure occurs when the contact zone was fully saturated. This result confirms why the failure occurred following a heavy rainy period.
- The degradation of cohesion and friction angle of the failure surface with the amount of water content should be expected and it may be easily incorporated in the models described herein.
- Failure surfaces based on circular sliding model and tectonic model yielded similar results.
- The failure surface was determined from the consideration of a tectonic model utilizing intrinsic surfaces of weaknesses in the tectonic melange zone of shattered tuffecaous rocks and it was consisted of R' and T surfaces of weakness of shear zones in structural geology. This landslide points out that there is a high potential of similar events in the future within the weak tectonic zones along Beşparmak Mountain Range particularly following rainy periods as well as during earthquakes in/around the Turkish Republic of the Northern Cyprus.

ACKNOWLEDGEMENT

This project NEU/1-05-8 is supported with a grant from the Turkish Republic/Turkish Republic of Northern Cyprus Scientific Research Projects (BAP). The authors are grateful for the financial support. The authors also thank to Prof. Dr. R. Kılıç, Prof. Dr. Baki Varol of Ankara University, Prof. Dr. R. Ulusay of Hacettepe University and Asst. Prof. Dr. Ali Denker of Near East University for their help and suggestions during the investigation and preparation of this paper.

REFERENCES

Atalar C., Kumsar H., Aydan Ö., Ulusay R., Kılıç R. 2016. Investigation of the mechanism and causes of Serdarlı District landslide (Turkish Republic of Northern Cyprus – TRNC) *Proceedings of The 2016 ISRM International Symposium, EUROCK 2016. Rock Mechanics and Rock Engineering: From the Past to the Future* Volume 1, Taylor & Francis Group, pp 635–640.

Aydan, Ö. 2012. The inference of physico–mechanical properties of soft rocks and the evaluation of the effect of water content and weathering on their mechanical properties from needle penetration tests ARMA 12–639. *American Rock Mechanics Association, 46th US Rock Mechanics/Geomechanics Symposium held in Chicago, IL, USA, 24–27 June 2012*, pp 1–6, on CD.

Aydan, Ö. & Kawamoto, T. 1990. Discontinuities and Their Effect on Rock Masses. *Int. Conf. Rock Joints*, 149–155, ISRM.

Aydan, Ö., Shimizu, Y., and Kawamoto, T. 1992. The stability of slopes against combined shearing and sliding failures and their stabilization. *Asian Regional Symposium on Rock Slopes*, pp 1105–1117, Oxford & IBH Publ., New Delhi.

Aydan, Ö., Ito, T., and Ichikawa Y. (1993). Failure phenomena and strain localisation in rock mechanics and rock engineering: A phenomenological description. *Int. Symp. Assessment and Prevention of Failure Pheomena in Rock Engineering*, 119–128, Istanbul, Balkema.

Aydan, Ö., Mamaghani, I.H.P., & Kawamoto, T. 1996. Application of discrete finite element method (DFEM) to rock engineering structures. *NARMS'96*, 2039–2046.

Aydan, Ö., Kumsar, H., Ulusay, R., & Shimizu, Y. 1997. Assessing limiting equilibrium methods (LEM) for slope stability by discrete finite element method (DFEM). *IACMAG*, Wuhan, 1681–1686.

Aydan, Ö., Tokashiki, N., Shimizu, Y., & Mamaghani, I.H.P. 2001. A stability analysis of masonry walls by Discrete Finite Element Method (DFEM). *10th IACMAG Conference, Austin*, 1625–1628.

Aydan, Ö., Ulusay, R., and Tokashiki, N. 2014a. A new Rock Mass Quality Rating System: Rock Mass Quality Rating (RMQR) and its application to the estimation of geomechanical characteristics of rock masses, *Rock Mech. and Rock Eng.*, 47:1255–1276.

Aydan, Ö., Sato, A., & Yagi, M. 2014b. The inference of geomechanical properties of soft rocks and their degradation from needle penetration tests. *Rock Mechanics and Rock Engineering*, 47(5),1867–1890.

Barton, N., Lien, R., and Lunde, I. 1974, Engineering classification of rock masses for the design of tunnel supports, *Rock Mech.*, 6 (4), 189–239.

Bieniawski, Z.T. 1989. *Engineering Rock Mass Classifications*. New York, John Wiley & Sons.

Bishop, A.W. 1955. The use of slip circle in the stability analysis of slopes. *Geotechnique*, 5, 7–17.

ISRM 2015. *The ISRM Suggested Methods for Rock Characterization, Testing and Monitoring: 2007–2014.* Suggested Methods Prepared by the Commission on Testing Methods, International Society for Rock Mechanics, Springer.

Kumsar, H. 1993. Mine Slope Stability Assessment by Using Inter–slice Force Transmission, *PhD Thesis*, 251 p., Nottingham University, UK.,

Ulusay, R., Aydan, Ö., Ergüler, Z.A., Ngan–Tillard, D.J.M., Seiki, T., Verwaal, W., Sasaki, Y., & Sato, A. 2014. ISRM Suggested Method for the needle penetration test. *Rock Mechanics and Rock Engineering* 47: 1073–1085.

2019 Rock Dynamics Summit– Aydan et al. (eds)
© 2019 Taylor & Francis Group, London, ISBN 978-0-367-34783-3

An experimental study on the dynamic stability of overhanging cliffs

K. Horiuchi & Ö. Aydan
Department of Civil Engineering, University of the Ryukyus, Okinawa, Japan

ABSTRACT: Overhanging rock cliffs may generally become unstable due to toe erosion resulting from the wave action. The critical erosion depth depends upon the height of cliff and the strength (tensile or shear) of the rock, and if the resulting stress state exceeds the strength of rock, it will collapse. Recently, it is reported that overhanging rock slopes may also fail during recent earthquakes. This experimental study was undertaken to investigate the stability of model rock cliffs under dynamic loading conditions in order to clarify the governing factors associated with their failure. The rock mass was modeled as continuum, layered, and blocky model. Shaking table (OA-ST1000X) has a size of 1 m × 1 m and the allowable maximum load is 100 kg under the conditions of maximum horizontal displacement of ±50 mm, maximum velocity of 0.56 m/s and maximum acceleration of 600 gals. The outcomes of the experimental studies are presented and discussed in this study. It is found that the failure modes of overhanging cliffs very much depend upon the number of discontinuity sets, tensile and shear strength of rock mass and their geometrical configuration.

1 INTRODUCTION

There are many cliffs with toe-erosion along shorelines of around the world and it may cause stability problems especially to historic structures and lighthouses. Such an example is the Gushikawa castle constructed on Ryukyu limestone cliffs. Sea waves, winds and percolating rain water are the main causes of toe-erosion leading to the failure. To prevent collapses, reinforcement for improving the stability of eroded cliffs using filling materials with the consideration of landscape has been introduced. numerous collapses of the overhanging cliffs have been reported due to earthquakes (Aydan 2013; Aydan and Amini, 2008; Aydan et al. 2012). In the vicinity of Japan, there are the four plates, namely "Eurasia Plate", "North American Plate", "Pacific Plate", "Philippine Sea Plate", interacting with each other in a complex manner (Figure 1). While Japan constitutes less than 1% of the world's area, earthquakes of about 10% of the world occurs in Japan. Considering the Philippine sea plate as an example, which is subducting in the northwest direction at a rate of 3-5 cm per year beneath the south east side of Ryukyu Archipelago. Also, on the continental side, there is a seabed topography called Okinawa trough. Following the formation of fractures the stability of the cliffs declines, and the occurrence of an earthquake may cause their collapse. But, there are few studies on the estimation of the collapse mechanism of the eroded cliffs during the earthquakes and the effect of discontinuity orientation and erosion on the collapse mode. It is necessary to examine the different failure form and the stability of rock cliffs depending on the discontinuous nature of rock mass at the time of the earthquakes. In this study, the failure modes of eroded cliffs are examined using rock

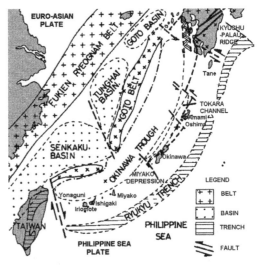

Figure 1. Ryukyu Islands and their tectonics (modified from Kizaki)

mass models having different number of discontinuity sets, namely, continuum, layered and blocky. Then, the authors classified each failure mode into three categories based on strength and discontinuous nature of rock mass.

2 MATERIALS PROPERTY TEST

The model materials used in this study is in powder form obtained by mixing barium (BaSO$_4$), zinc oxide (ZnO) and Vaseline in a weight ratio of 70: 21: 9 and it can be formed into various

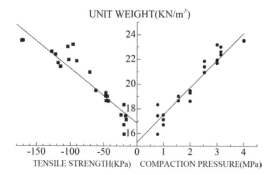

Figure 2. Relationship between tensile strength and compressive strength by unit volume weight.

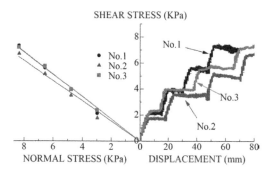

Figure 3. Shear characteristic of model materials.

shapes by compacting in a mold. The strength of the model block mostly depends on its unit weight and strength of the model block can be easily changed by varying the compaction force and it can return to its original powder form after model tests. In this study, two types of experiments were conducted to investigate the tensile properties of the model material and shear strength characteristics of interfaces between model materials. The cantilever test was carried to determine the tensile properties. Figure 2 shows the relationship between unit weight and tensile strength and compaction pressure. Experimental results confirmed that the unit volume the tensile strength increase are proportional to compressive strength. As the compaction pressure increases, compressive strength of the model material also increases. The shear test was used to obtain shearing characteristics of interfaces between blocks by gradually increasing normal load as 0 gf, 500 gf, 1000 gf, 1500 gf. Figure 3 shows the results of shear tests on interfaces between the blocks. The shear strength increases as the normal force increases.

Table 1. experimental condition.

	Acceleration(gal)	Frequency(Hz)
Sweep	50	3-50-3
Failure	0 - 400	5

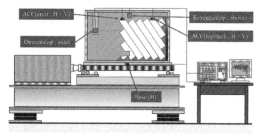

Figure 4. Illustration of the set-up of shaking table model tests.

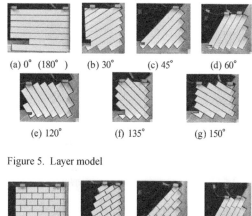

Figure 5. Layer model

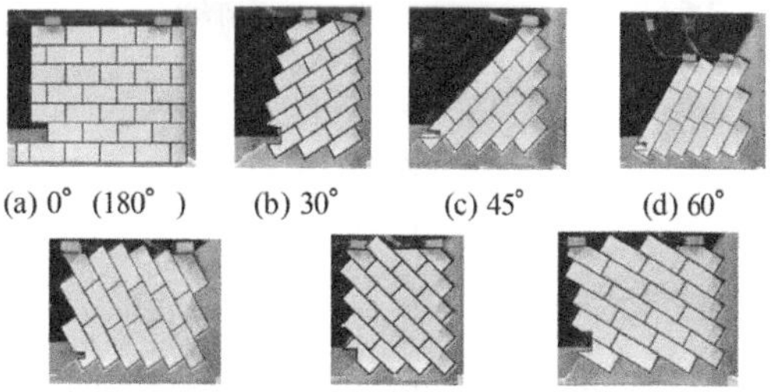

(a) 0° (180°)   (b) 30°   (c) 45°   (d) 60°

(e) 120°   (f) 135°   (g) 150°

Figure 6. Blocky model

3 SHAKING TABLE TESTS

3.1 Dynamic experiment method

Two types of shaking table experiments were conducted, specifically, natural period characteristic test (Sweep test) and failure experiment. In some model experiments for layered and blocky rock mass models, FFT analyses were also performed. Experimental conditions are given in Table 1. In experiments, model blocks was laid into a model frame of 25 cm or 50 cm after the blocks were prepared under a compaction force of 2.5 tonf or 5 tonf respectively, and compaction was performed. The compaction pressure was selected on the assumption that the model itself didn't fail in a static state and could preserve its shape after molding. The model was subjected to vibration by the shaking table, and the acceleration and displacement responses were measured. Figure 4 shows the set-up of models and

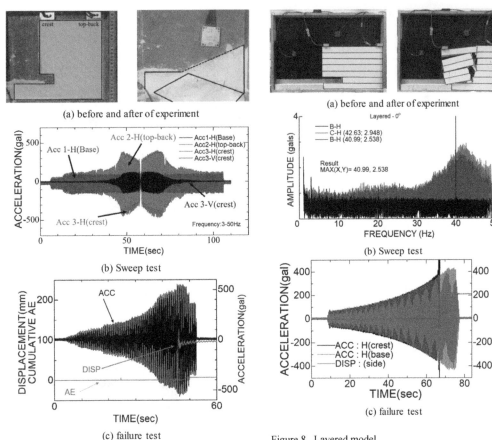

(a) before and after of experiment

(b) Sweep test

(c) failure test

Figure 7. Continuous model

(a) before and after of experiment

(b) Sweep test

(c) failure test

Figure 8. Layered model

instrumentation. We installed three accelerometers and two non-contact laser displacement transducers in the model of cliff. The model experiment of the eroded cliff was carried out three times. A model experiment is shown in Figure 5&6. As seen from the figures, the slope angle was 90 degrees (except for 45 degree and 60 degree). The erosion was introduced into the model and sweep test was first carried and results are shown in Figure 7 (b). Then, the model was subjected to shaking until failure under a chosen frequency 3-5Hz. The slope angle and erosion depth is chosen such a way that the model slope is stable under static condition and it may fail under dynamic conditions.

3.2 Result of continuous model

The collapse of continuum model entirely depends upon the strength of material constituting slope. The collapse of the continuum model may be due to tensile failure or combination of tensile and shear failure.

First a sweep test was carried out and it is found that the acceleration at the crest of the slopes was more than three times that on the shaking table.

Regarding the acceleration response in the vertical direction, an amplification was observed in a certain frequency band.

Regarding the test resulting in the failure of the model slope, the results of the acceleration response, displacement and AE indicated that the model was greatly displaced when the acceleration response was around 500 gals. In addition, when the acceleration response at the crest of the slope was compared with that of the shaking table, it was about 1.5 times (Figure 7 (c)). From the above results, it was found that the crest of the slope shows larger acceleration response. The failure mode of the continuum cliff with toe erosion, a crack was initiated near the toe of the model slope, it was found that the collapse depended on the strength of the rock.

3.3 Result of layered model

The model experiment was conducted with a erosion depth being 50 mm. But, the model didn't collapse. As our purpose was to investigate the collapse mode, we increased the erosion depth of the model to 100 mm. Figure 8 show the experimental results before and after the layered model with toe erosion of 100 mm. The layer inclination of cliff model with toe

(a) before and after of experiment

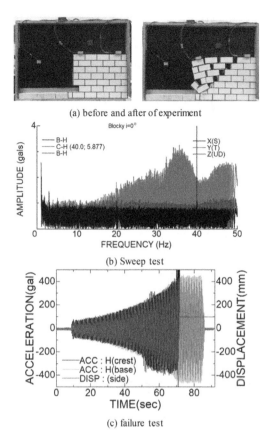

(b) Sweep test

(c) failure test

Figure 9. Blocky model

| (a) layered
30 degree | (b) blocky
30 degree | (c) blocky
135 degree |

Figure 10.Example of flow model and received model

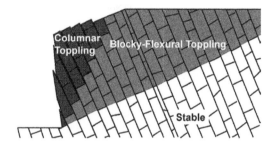

Figure 11.Blocky flexural toppling failure

| (a) bending failure | (b) shearing failure |

Figure 12.Special failure modes

erosion was 0 degree and the failure occurred in the form of bending failure, which is just above the erosion tip. From the result of the acceleration response, it turned out that the model had collapsed at about 380 gals. There was almost no big difference regarding the amplitude of the acceleration at the failure in three experimental results. The movements was slightly different during the test, although the final collapse mode was same.

3.4 Result of blocky model

Although the layered model, did not collapse when the erosion depth was 50 mm, the blocky model collapsed when the erosion depth was 50 mm.

Figure 9 shows the views of the models before and after the experiment. For cliffs with toe erosion and 0 degree thoroughgoing discontinuity set, failure occurred near the vicinity of erosion tip and resulted in the toppling of blocks above a stepped failure surface. From the result of the acceleration response, it turned out that the model had collapsed at about 400 gals. There was almost no big difference regarding the amplitude of the acceleration at the failure in the three experimental results. The failure mode was slightly different during the test, but the final collapse mode was same.

3.5 Evaluation of experimental results

In this study, an investigation on the acceleration response and failure modes of the cliff models with toe erosion having different number of discontinuity sets was undertaken. From the comparison of experimental results, it may be stated that.

1. The blocky collapses when the erosion depth is less than of the layered model.
2. The results of acceleration response for three experiments showed almost the same result.

4 FAILURE MODES OF ERODED CLIFF

4.1 Example of cliff failure after the experiments

1. In the layered model, the model material itself was broken. However failure mode was affected by the discontinuities in blocky model.
2. The collapsed region in the blocky model collapses was larger than the layered model as the discontinuities of the rock mass has a large influence (Figure 10).

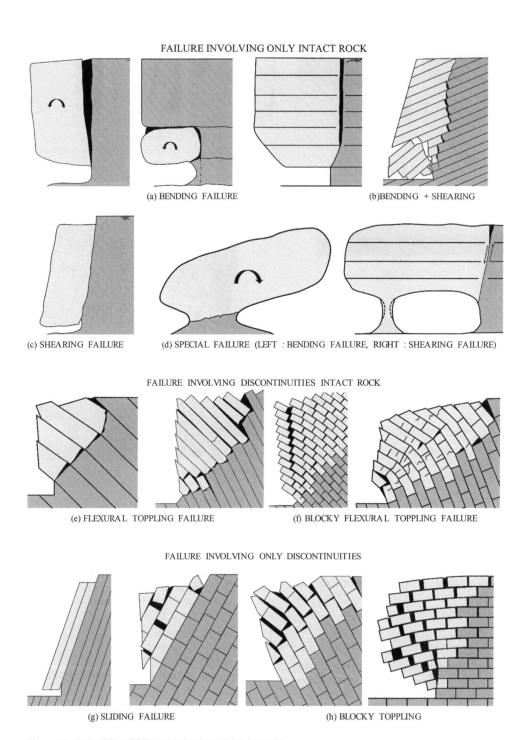

Figure 13. FAILURE MODE OF EROSION ROCK SLOPES

4.2 Blocky flexural toppling failure

Blocky flexural toppling failure is the failure of flexural failure of some of blocks while the others topple. (Figure 11). Blocky toppling failure related to purely to discontinuities. Flexural toppling failure caused by tensile stresses in layers of the rock mass (e.g. Aydan and Kawamoto, 1992).

4.3 Proposal of failure modes of cliffs with erosion

The authors have been surveying case histories of many cliffs with toe erosion in major islands of the Ryukyu Archipelago. From dynamic experiments and survey of cliffs (e.g. Aydan and Amini, 2009; Horiuchi et al. 2018; Tokashiki and Aydan 2010) the failure modes of the cliffs with toe erosion are classified as shown in Figure 13. The area shown in yellow in the figure corresponds to the area of actual collapse. The area shown in blue is in a stable state. This failure modes can be roughly categorized into three classes. The mechanism of detailed failure form is described in a study conducted by Aydan et al. (1989 a), b)) and Shimizu et al.

1. Failure involving only intact rock
   a) Bending failure
   b) Bending failure + Shear failure
   c) Shear failure
   d) Special failures (Figure 12)
2. Failure involving discontinuity and intact rock
   e) Flexural toppling failure
   f) Blocky flexural toppling failure
3. Failure involving only discontinuities
   g) Planar sliding
   h) Blocky toppling

5 CONCLUSIONS

In this study, the authors investigated dynamic stability of cliffs with toe erosion by varying discontinuity inclination and the number of discontinuity sets. The model material used in the experiment can be repeatedly used and it is most suitable for model tests of discontinuous rock slopes. In this experimental study, we also investigated the failure modes of cliffs with toe erosion as well as the acceleration response. The findings obtained in this research may be listed as.

1. The failure modes of the continuum, layered, blocky model cliffs are greatly affected by the number of discontinuity sets and their distributions.
2. It was difficult to compare the acceleration response of the model cliffs as the geometry of each model was different. In the future studies, we need to reconsider the model of the sea cliffs and the stability evaluation in relation to acceleration response.

REFERENCES

Aydan, Ö. (2013).: The effects of earthquakes on rock slopes. The 47th US Rock Mechanics/Geomechanics Symposium, San Francisco, ARMA pp. 13–378.

Aydan, Ö. (2015).: Large Rock Slope Failures induced by Recent Earthquakes. Rock Mechanics and Rock Engineering, Special Issue on Deep-seated Landslides. 49(6), pp. 2503–2524.

Aydan, Ö. & Amini, M.G. (2009): An experimental study on rock slopes against flexural toppling failure under dynamic loading and some … Journal of School of Marine Science and Technology, Tokai University, No.55, 53–66.

Kizaki, K. (1986). : Geology of tectonics of the Ryukyu Islands Tectonophysics, Vol. 125, pp. 193–207.

Aydan, Ö., T. Kyoya, Y. Ichikawa, T. Kawamoto and Y. Shimizu (1989a). :A model study on failure modes and mechanisms of slopes in discontinuous rock mass. The 24th Annual Meetings of Soil Mechanics and Foundation Eng. of Japan, Miyazaki, 415, 1089–1093.

Aydan, Ö., Y. Shimizu, Y. Ichikawa (1989b). :The Effective Failure Modes and Stability of Slopes in Rock Mass with Two Discontinuity Sets. Rock Mechanics and Rock Engineering, 22(3),163–188.

Aydan, Ö., Ulusay, R., Hamada, M. and Beetham, D. (2012): Geotechnical aspects of the 2010 Darfield and 2011 Christchurch earthquakes of New Zealand and geotechnical damage to structures and lifelines. Bulletin of Engineering Geology and Environment, 71, pp. 637–662.

Horiuchi, K., Aydan, Ö, Tokashiki, N. (2018). : Recent failures of limestone cliffs in Ryukyu Archipelago and their analyses. Proc. of 45th Rock Mechanics Symposium of Japan, JSCE, 131–136.

Shimizu, Y., Aydan, Ö., Tsuchiyama, S., Ichikawa, Y (1990).: An Integrated Method for Slope Stability in Discontinuous Rock Mass. JSCE, VI–12, No. 415, pp. 109–118.

Tokashiki, N. and Aydan, Ö. (2010).: The stability assessment of overhanging Ryukyu limestone cliffs with an emphasis on the evaluation of tensile strength of rock mass. JSCE, Vol. 66, No.2, pp. 397–406.

Aydan, Ö. and Kawamoto, T (1992). :The stability of slopes and underground against flexural toppling and their stabilization. Rock Mechanics and Rock Engineering, pp-143–165.

2019 Rock Dynamics Summit– Aydan et al. (eds)
© *2019 Taylor & Francis Group, London, ISBN 978-0-367-34783-3*

Dynamic model tests on the Babadağ-Gündoğdu landslide (Denizli-Turkey)

S.B. Çelik & H. Kumsar
Department of Geological Engineering, Pamukkale University, Denizli, Turkey

Ö. Aydan
Department of Civil Engineering, University of the Ryukyus, Nishihara, Japan

ABSTRACT: Long term creep like Gündoğdu landslide which is moving in various annual rates along the dip direction of bedding planes of sandstone-marl layers is seen in Babadağ town of Denizli city. In this study, the sliding behavior of Gündoğdu landslide was investigated by means of laboratory tests on a slope model which was prepared in accordance with in-situ geological conditions. For this purpose, series of laboratory tests including tilting tests and dynamic shaking table tests were carried out. According to tilting test results average values of measured static and calculated kinetic friction angles were determined as 33.8° and 31.13° respectively. Recorded total AE activity for block slide in tilting and dynamic tests were determined as 3.57 for 68.4 mm and 54.2 for 101.2 mm as average values respectively. Critical acceleration values for planar sliding of upper block were determined as 0.45g and 0.22g for the dip angles 16° and 24° of sandstone and marl succession respectively. Block displacements which were developed under dynamic cyclic loads were calculated by a computer program and compared with measured displacements. If obtained critical accelerations develop in the study area in case of a possible earthquake, ongoing mechanism of the Gündoğdu landslide could be affected or triggered.

1 INTRODUCTION

Landslides are widely known major geologic hazards and threatening people all over the world. Babadağ town is located in 40 km northwest of Denizli city and it has been exposing to creep like slope movement since 1940 (Figure 1). Effects of the movement in the Gündoğdu district can be classified as; tilting and settling of the buildings, damages on buried lifeline systems, deformation of roads and pavements. Moreover, crackings of walls are very typical and clearly observed.

The mechanism and factors, which contribute to the Gündoğdu slope movement, have been investigated by Tano et al. 2003; Aydan et al. 2003; Kumsar et al. 2004; Çevik 2003; Çevik and Ulusay 2005; Tano et al. 2006a; Tano et al, 2008; Çelik et al. 2011; Kumsar et al. 2012, Çelik 2012; Kumsar et al. 2016a.

Figure 1. View of Gündoğdu landslide in Babadağ town.

Aydan et al. (2003) stated that marl under dry conditions has higher strength than under wet and saturated conditions. Mechanical properties of marl decrease under wet and saturated condition. Aydan (2003) concluded that sudden and rapid movement through Gökdere valley may occur after a rainy period in case of an earthquake with a magnitude 6 or bigger within an area having a 20 km radius around Babadağ.

Causes and mechanism of the mass movement were investigated by Çevik (2003) and Çevik and Ulusay (2005), they stated that annual displacement values were ranging between 3.8 and 15 cm according to field measurements. They also reported the factor of safety values decreases with changing the groundwater level for the limit equilibrium condition is provided. Tano et al. (2006a) stated that Gündoğdu slope movement continues through N27E direction towards the Gökdere valley according to pipe strain data. Amount of displacements tend to increase after precipitations. Rainfall and infiltration affected the movement, especially in excessive rainy periods sudden changes in pore water pressure reduces the effective stress which is reduces the strength and deformability properties of materials. In this way resisting forces of the movement reduces and movement rate increases. Çelik (2012), presented all monitored field data representing the unstable nature of Gündoğdu landslide. Experiments were carried out on rock samples in order to investigate deformability properties of marls and a model was proposed which is based on deformability of marls by saturation from groundwater level fluctuations. Kumsar et al. (2016a) summarized the studies about the Gündoğdu landslide since 2000. The ongoing mechanism and the characteristics of the Gündoğdu landslide were also given. Under the risk of potential landslide, the landslide area was designated as "Natural Disaster Area for Life" by governmental institutions and the people living in this part of the town were re-settled to a new area.

Babadağ town is situated on Babadağ sandstone-marl member of Kolonkaya formation in the south margin of Denizli basin which is bounded by normal faults. Layer thicknesses of the member vary from 1 to 20 cm. The slope movement has been developing in this member along the dip direction of marl sandstone intercalation since 1940s (Kumsar et al, 2016). Main strike of bedding planes is N30E and dipping to SE with 16°-20° (Çelik 2012).

In this study a slope model was prepared by considering layered geological structure of the Gündoğdu landslide in Babadağ. Series of laboratory tests including tilting and dynamic shaking table tests were carried out. In these tests, static friction angle at which upper block started to slide was determined. From recorded time dependent response of the mobile block, kinetic friction angles were determined by using a method proposed by Aydan et al. (2014, 2016). Dynamic model tests were carried out in order to assess block sliding with increasing base acceleration by using a unidirectional shaking table. The dip angles of marl layers in-situ vary between 16-24°. By considering these values, models using actual materials were prepared and tested in different inclination angles. For each inclination value, critical acceleration was measured and relationship between base angles and critical accelerations were determined. Block movements were calculated by a dynamic limiting equilibrium method proposed by Aydan (2003) and Aydan and Ulusay (2002) then compared with measured displacements, similarity between measured and calculated displacement values were compared. During the tests, AE activities were monitored and relationship between AE activities and slippage was also presented.

2 PREPARATION OF A SLOPE MODEL

Physical models were prepared by using natural marl and sand material taken from the field. Marl layers were transported to laboratory as two blocks, however sand material of weak sandstone layers could not be collect as undisturbed sample. Marl was easily dimensioned however modeling of weak sandstone layers under laboratory conditions was quite difficult. In order to overcome of this difficulty, a binder material was used. Acrylic polymer was mixed with sand in order to prepare artificially bounded sandstone layer shear strength parameters were obtained

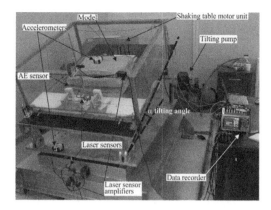

Figure 2. Shaking table, slope model and testing equipment (Çelik, 2012).

for the samples with a 1 % mixing ratio. When residual shear strength parameters are taken into consideration, cohesion and internal friction angles were obtained as 6.9 kPa and 26.5°, respectively and these values were found to be closer to the parameters obtained from natural material. Physical model and testing system was given in Figure 2. Test system is consisted of unidirectional electro-motor driven shaking table with adjustable tilting upper surface, acceleration recording unit, laser displacement sensor, acoustic emission system and their equipment.

All tests were carried out in Geological Engineering laboratories at Pamukkale University. In geotechnical engineering shaking tables are widely used to analyze seismic behavior of natural or manmade earth structures such as slopes, embankments etc. (i.e. Kumsar et al. 2000; Park et al. 2006; Aydan and Kumsar 2010). In this study, small scaled, low cost, unidirectional, electric motor driven shaking table was used. Table dimension is 1.5 m and 1.0 m and load capacity of the table is up to 500 kgf, and produced accelerations are up to 0.7g (Çelik 2012; Çelik et al. 2013).

3 SLOPE MODEL TESTS

3.1 Static tilting tests

In various geoengineering investigations both in design and practice, determination of frictional resistance between discontinuities has great importance. The friction has been utilized since the human started to build structures. Historical masonry structures of past such as pyramids, bridges, stadiums walls etc. utilize the frictional resistance of rock and brick interfaces and rock discontinuities in their structural design. The friction is an important element of well-known yield criteria of Mohr-Coulomb. As a result, many articles and textbook are inclined to see the friction is an integral part of the yield criterion of Mohr-Coulomb. There are many methods to measure the friction coefficient of interfaces and

Table 1. Measured sliding angles and related parameters in tilting tests.

Test no	Measured friction angle (°)	Travel time(s)	Displacement (mm)	Total AE counts	Average velocity (cm/s)
1.1	35.69	0.43	81.80	3	19.02
1.2	32.94	0.41	81.70	6	19.93
1.3	31.87	0.48	38.50	4	8.02
1.4	32.78	0.84	82.35	3	9.80
1.5	34.04	0.51	78.50	3	15.39
1.6	34.81	2.10	36.80	4	1.75
1.7	32.89	0.80	73.40	2	9.17
1.8	35.41	0.78	74.25	NA	9.52

discontinuities. These methods are described in various sources (Bowden and Tabor 1964; Jaeger and Cook 1979; Goodman 1989). One of the well-known methods is tilting tests (Barton 1973, Barton and Choubey, 1977) and it has become one of the convenient tools to measure the frictional characteristics of rock discontinuities in laboratory and in-situ under particularly low normal stress or load conditions. Portable tilting devices were developed to measure the friction angles of discontinuities both in laboratory and in situ under low normal stresses (Xian-Qin and Cruden 1992; Aydan et al. 1995; Kumsar et al. 2000; Alejano et al. 2012; Ulusay and Karakul 2016; Aydan et al. 2016).

Tilting tests were repeated eight times. At the end of each test the upper block was set to initial position and test was repeated. In all the tests, tilting rate was tried to keep constant as possible as 0.3°/s. Average duration of tests was about 107 s. In all tests, tilting values, AE activity counts, and sliding block movement were recorded. All collected data analyzed and results were presented. Determined sliding angles, block sliding durations, displacements, AE counts and block velocities from repeated tilting tests were also given in Table 1.

Block movements were recorded in 10 ms time interval and this provided precise monitoring of the block movements. From eight repeated tilt tests, average values of static friction angle, sliding duration, displacement and velocity values were determined as 33.8°, 0.8 s, 68.4 mm and 11.58 cm/s, respectively. It is observed that, AE activities started to increase simultaneously as block start to slide and it can be said that AE activities were directly related to sliding. In average 68.4 mm block displacement, recorded average total AE activity determined as 3.57.

In tilting tests, block movements are controlled by the static friction angle until upper block starts to slide, and thereafter block movement is controlled by kinetic friction angle. In this study, kinetic friction angle values of the sliding blocks for the eight tests were determined from time dependent block

movement data by a method which was proposed by Aydan et al. (2014, 2016). At the initiation of sliding, the inertia terms are zero so that the following relation is obtained (Aydan et al. 1989).

$$\tan \alpha = \tan \varphi_{static} \tag{1}$$

$$\frac{W \sin \alpha}{W \cos \alpha} = \tan \varphi_{static} \tag{2}$$

Where, α is the base inclination and W is the dead weight of the upper block. For dynamic equilibrium one can write (Aydan et al. 2014, 2016);

$$\frac{d^2 s}{dt^2} = g(\sin \alpha - \cos \alpha \tan \varphi_{kinetic}) \tag{3}$$

Where, $g(\sin \alpha - \cos \alpha \tan \varphi_{kinetic}) = A$ then the integration of Eq.3 yields the following form;

$$s = A\frac{t^2}{2} + C_1 t + C_2 \tag{4}$$

Since the followings hold at the initiation of sliding as; displacement and velocity are zero and $t = T_{static}$ integration coefficients in Eq. (4) are obtained as,

$$C_2 = -A\frac{t_s^2}{2} - C_1 t_s = A\frac{t_s^2}{2} \quad \text{and} \quad \frac{ds}{dt} = At_s + C_1,$$

$C_1 = -At_s$ Thus, the relative movement of the upper block is given by,

$$s = \frac{A}{2}(t - T_{static})^2 \tag{5}$$

As the measured values of relative displacement are many, the least square technique is applied to Eq. (5) and coefficient A is obtained (Aydan et al. 2016). Then kinetic friction angle from the following equation (Eq. 6) can be calculated by using the coefficient A as.

$$\varphi_{kinetic} = \tan^{-1}\left(\tan \alpha - \frac{1}{\cos \alpha}\frac{A}{g}\right) \tag{6}$$

This method was used to determine to kinetic friction angles for repeated tilting tests, time dependent block movements analyzed and "A" coefficient for each test was determined. And then, kinetic friction angles were back-calculated. Block movements and fitted curves for a test was given in Figure 3. Calculated kinetic friction angles were ranging between 26.61° and 34.38°, and average value as 31.13°.

3.2 Dynamic shaking table tests

There are many studies on investigation of slopes under dynamic loads by using shaking tables (e.g. Aydan 2003, 2015; Aydan and Ulusay 2002; Aydan and Kumsar 2010; Kumsar et al, 2000; Park et al. 2006).

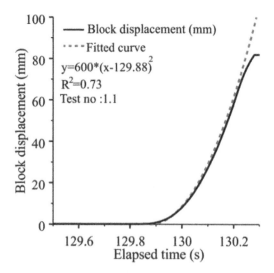

Figure 3. Block movements and fitted curve for the tilting test no 1.1.

Sliding response of block under cyclic loads was investigated by using unidirectional, electric motor driven shaking table. Tests were carried out at the inclination angles of 16, 18, 20, 22 and 24°. In each test, acceleration values were increased gradually until the block started to slide. Moving block was monitored by using a laser sensor as in tilting tests and table acceleration was recorded by the portable accelerometer. It is observed that, AE activities started to increase simultaneously during sliding of block and, it can be said that AE activities were directly related to displacement of the block. In dynamic tests, average value of total AE counts was 54.2 for 10.12 cm of block displacement. In shaking table tests, recorded AE counts for 1 mm displacement were found to be ten times greater than the tilting tests. As an example shaking table test results were given in Figure 4. In shaking table tests, critical acceleration values, at which the sliding was initiated, were determined. As a result, critical acceleration was decreased from 0.45 g to 0.22 g as the table inclination was increased from 16° to 24°.

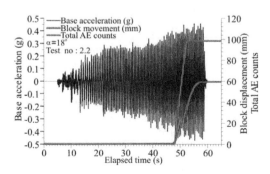

Figure 4. Shaking table test results at the inclination of 18°.

4 STABILITY ANALYSES OF GÜNDOĞDU SLOPE MOVEMENT

Gündoğdu landslide was investigated on a cross section (Figure 5) which was taken parallel to moving direction of the landslide based on previous studies (Tano et al. 2003, Tano et al. 2006 a, b; Aydan et al. 2003). For the determination of the major sliding surface, pipestrain data of SK-3 borehole was evaluated. In the borehole, strain concentration at 28.5 m depth was observed (Çelik 2012; Kumsar et al. 2016). This depth was assumed to be the major sliding surface. In SK-2 borehole groundwater level was monitored, although groundwater level is changing seasonally it was determined that it fluctuates on the sliding surface. Gündoğdu landslide has been moving at various rates and it was considered that instability was controlled by residual shear strength parameters. Cohesion and internal friction values along the failure surface were selected as 5 kPa and 20° respectively. In static limit equilibrium analyses, factor of safety values were calculated between 1.17 and 1.28 (Çelik 2012). Decreasing of shear strength parameters of the failure surface during a rainy season can cause the decrease of FOS to 1. A possible earthquake near the study area can affect or trigger the Gündoğdu landslide. For this reason stability assessment of the Gündoğdu landslide under dynamic conditions is very important. The Denizli basin is seismically active and it is known that strong earthquakes occurred in ancient times. These earthquakes are commonly related to basin bounding Pamukkale, Honaz and Babadağ fault systems, this is supported by areal distribution of the earthquake epicenters in the basin.

The greatest magnitude of the earthquake near study area is 5.7 (Ms) since 1900. Strong ground motion recording station in the Babadağ town was established in 2007. Therefore, peak ground acceleration values in the study area caused by previous earthquakes cannot be known. There is also no information about the contribution of the past earthquakes to Gündoğdu landslide in terms of slope displacements. Earthquakes occurred in Denizli basin are mostly related to active Pamukkale fault with a 35 km length. Aydan (2007) proposed a relationship between fault rupture length and earthquake magnitude (Figure 6), according to this relationship Pamukkale fault could produce an earthquake

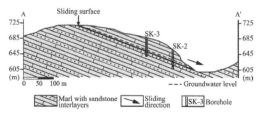

Figure 5. Location and cross section of the Gündoğdu landslide.

514

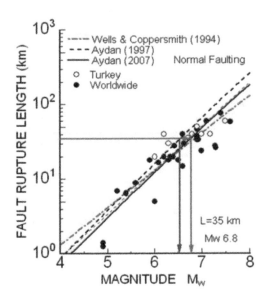

Figure 6. Magnitude-fault rupture relationship (Aydan, 2007, 2012).

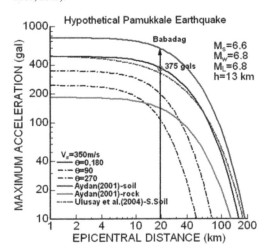

Figure 7. Estimated acceleration values for hypothetical 6.8 earthquake in Babadağ.

with magnitude of 6.8 magnitude (Kumsar et al. 2015). In case of a 6.8 magnitude earthquake in the Denizli basin, acceleration values in Babadağ were estimated by using attenuation relationships (Aydan 2001, 2007, 2012; Ulusay et al. 2004), (Figure 7).

Estimated acceleration values were between 350-375 cm/s². Average of these values is 86.8 times greater than the peak acceleration value which were recorded in Babadağ on 5th of February 2011 Akköy earthquake with local magnitude (M_L) of 4. In dynamic stability analyses acceleration time history of Akköy earthquake was used by multiplying by a ratio of 86.8. Slope displacements were calculated by the same dynamic limiting equilibrium method of Aydan (2003, 2015), which was used to estimate

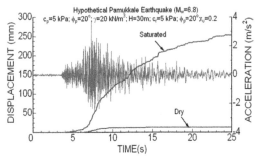

Figure 8. Calculated slope displacement values in hypothetical 6.8 (M_w) earthquake.

the block displacements in shaking table tests. Slope displacements caused by ground acceleration were calculated by using the method proposed by Aydan and Ulusay (2002), Aydan et al. (2014, 2015).

Unit weight, cohesion and internal friction angle values were taken as 18.6 kN/m³, 5 kPa and 20° respectively in dry and partly saturated (r_u = 0.2) conditions. In Figure 8, calculated displacement values were given both for dry and partly saturated conditions. According to results, in case of a hypothetical earthquake, relative displacement for dry conditions is computed to be very small. On the other hand, a total of 25 cm relative displacement occurs under partially saturated condition.

5 CONCLUSIONS

In this study, creep type Gündoğdu landslide in Babadağ town (SW, Turkey) was investigated by a prepared slope model tests. Model composed of marl and weak sandstone layer.

Under static condition, tilting tests were carried out in order to determine the static friction angle. From time dependent block movement data, kinetic friction angles were calculated. According to tilting test results average values of static and kinetic friction angles were determined as 33.8° and 31.13° respectively. In both tilting and shaking table tests, AE counts which correspond to sliding response were also monitored. As a result, average 3.57 AE counts were recorded at 68.4 mm average block displacement in tilting tests. In dynamic tests, average value of 54.2 AE counts were recorded for 101.2 mm block displacement. In shaking table tests, the recorded AE counts for 1 mm displacement were found to be ten times greater than the AE counts in tilting tests.

The dip angles of sandstone marl unit ranges from 16° in town center to 24° in Gökdere valley. According to shaking table test results, critical acceleration values for sliding were obtained as 0.22g and 0.45g for the inclinations of 24° and 16° respectively. In case of a possible earthquake, which may produce the critical accelerations near the study area, Gündoğdu landslide could be affected or triggered.

In the actual scale, Gündoğdu slope was investigated for a representative cross section. Under static condition, factor of safety values was calculated to be ranging between 1.17 and 1.28. Ongoing instability can be explained by the factor of safety values are close to 1. In dynamic condition, in case of a hypothetical 6.8 magnitude earthquake occurs, a total of 25 cm relative displacement takes place under partially saturated condition. This implies that the degree of saturation is crucial when the seismic forces occur for the motion of the landslide body.

ACKNOWLEDGEMENT:

The shaking table used in this study was financially supported by an institutional infrastructure project of Pamukkale University (2009KRM001) this support is gratefully acknowledged. The authors are also grateful to Prof. Dr. Hisataka Tano (Emeritus) from Nihon University for his support and providing AE testing system.

REFERENCES

Alejano, L.R., González, J., Muralha, J. 2012. Comparison of Different Techniques of Tilt Testing and Basic Friction Angle Variability Assessment. *Rock Mech Rock Eng* 45:1023–1035.

Aydan, Ö. 2003. The Mechanism of the Creep-Type Landslide at Babadağ. In H. Tano, H. Watanabe and Ö. Aydan (eds.), *International Colloquium on the Instrumentation of Landslides and Earthquakes in Japan and Turkey*, Koriyama, Japan, 39–50.

Aydan, Ö. 2007. Inference of seismic characteristics of possible earthquakes and liquefaction and landslide risks from active faults. *Conference of the 6th National Earthquake Engineering*, İstanbul, 563–574 (In Turkish).

Aydan, Ö. 2012. Ground motions and deformations associated with earthquake faulting and their effects on the safety of engineering structures. In R. Meyers (Ed.) *Encyclopedia of Sustainability Science and Technology*. Springer, 3233–3253.

Aydan, Ö. 2015. Some considerations on a large landslide at the left bank of the Aratozawa Dam caused by the 2008 Iwate-Miyagi intraplate earthquake. *Rock Mech Rock Eng Special Issue on Deep-seated Landslides* 49(6):2525–2539.

Aydan, Ö. & Ulusay, R. 2002. Back-analysis of a seismically induced highway embankment failure during the 1999 Düzce earthquake. *Environ Geol* 42:621–631.

Aydan, Ö. & Kumsar, H. 2010. An experimental and theoretical approach on the modeling of sliding response of rock wedges under dynamic loading. *Rock Mech Rock Eng* 43:821–830.

Aydan, Ö., Shimizu, Y., Ichikawa, Y. 1989. The Effective Failure Modes and Stability of Slopes in Rock Mass with Two Discontinuity Sets. *Rock Mech Rock Eng* 22(3):163–188.

Aydan, Ö., Shimizu, Y., Kawamoto, T. 1995. A portable system for in-situ characterization of surface morphology and frictional properties of rock discontinuities. *The 4th International Symposium on Field Measurements in Geomechanics*, Bergamo, 463–470.

Aydan, Ö., Ulusay, R., Kumsar, H., Çevik, S.Y. 2003. Laboratory and in–situ tests on rock and bedding planes and machinery induced vibrations of Babadağ landslide area. In H. Tano, H. Watanabe and Ö. Aydan (eds.), *International Colloquium on the Instrumentation of Landslides and Earthquakes in Japan and Turkey*. Koriyama, Japan, 91–100.

Aydan, Ö., Tokashiki, N., Iwata, Takahashi, K., Adachi, K. 2016. Determination of static and dynamic friction angle of rock discontinuities from tilting tests. In *14th National Rock Mechanics Symposium of Japan*, Paper No: 041, 6p.

Barton, N. 1973. Review of a new shear strength criterion for rock joints. *Eng Geol* 7:287–332.

Barton, N. & Choubey, V. 1977. The shear strength of rock joints in theory and practice. *Rock Mech* 10:1–54.

Bowden, F.P. & Tabor, D. 1964. The friction and lubrication of solids. Clarendon Press, Oxford.

Çelik, S.B. 2012. Investigation of the slope instability in Babadağ town (Denizli) using multi parameter monitoring techniques, physical model tests and mathematical methods. Unpublished PhD dissertation, Pamukkale University (In Turkish).

Çelik, S.B., Kumsar, H., Aydan, Ö. 2011. The investigation of the mechanism of Babadağ Gündoğdu landslide through static and dynamic physical model tests. *Xth Regional Rock Mechanics Symposium*, Ankara, 161–170 (in Turkish).

Çelik, S.B., Kumsar, H., Aydan, Ö. 2013. Motion parameters of a mechanical shaking table and an application. *Pamukkale University Journal of Engineering Sciences*, 19 (5):224–230 (in Turkish).

Çevik, S.Y. 2003. An investigation on the causes, mechanism and modelling of slope instability at Babadağ town (Denizli). Unpublished MSc dissertation, Hacettepe University (in Turkish).

Çevik, S.Y. & Ulusay, R. 2005. Engineering geological assessments of the repeated plane shear slope instability threatening Babadağ (Turkey) and its environmental impacts. *Environ Geol* 47(5):685–701.

Jaeger, J.C. & Cook, N.G.W. 1979. *Fundamentals of rock mechanics*. Chapman and Hall, London.

Kumsar, H., Aydan, Ö., Ulusay, R. 2000. Dynamic and static stability assessment of rock slopes against wedge failures. *Rock Mech Rock Eng* 33(1):31–51.

Kumsar, H., Aydan, Ö., Tano, H., Atak, O. 2004. Investigation of long-term Babadağ (Denizli) landslide in terms of rock mechanics. In *7th Rock Mechanics Symposium*, Sivas.

Kumsar, H., Çelik, S.B., Aydan, Ö., Tano, H., Ulusay, R. 2012. Investigation of Babadağ Gündoğdu landslide using multi parameter monitoring techniques and its evaluation within natural hazard scope. In *65th Years Mahir Vardar - Geomechanics, Tunneling and Design of Rock Structures - Special Sessions*, Istanbul, 311–337 (in Turkish).

Kumsar, H, Aydan, Ö., Şimşek, C., D'Andria F. 2015. Historical earthquakes that damaged Hierapolis and Laodikeia antique cities and their implications for earthquake potential of Denizli basin in western Turkey, *B Eng Geol Environ*, 74, 1037–1055.

Kumsar, H., Aydan, Ö., Tano, H., Çelik, S.B., Ulusay, R. 2016a. An integrated geomechanical investigation, multi-parameter monitoring and analyses of

Babadağ-Gündoğdu creep-like landslide. *Rock Mech Rock Eng* 49:2277–2299.

Kumsar, H., Aydan, Ö., Şimşek, C., D'Andria, F. 2016b. Historical earthquakes that damaged Hierapolis and Laodikeia antique cities and their implications for earthquake potential of Denizli basin in western Turkey. *B Eng Geol Environ* 74:1037–1055.

Park, B.K., Jeon, S., Lee, C.S. 2006. Evaluation of dynamic frictional behavior of rock joints through shaking table test. *Tunn Undergr Sp Tech* 21:427.

Tano, H., Kumsar, H., Aydan, Ö., Ulusay, R. 2003. The assessment of the Babadağ landslide behavior by a simple field measurement system. In H. Tano, H. Watanabe and Ö. Aydan (eds.), *International Colloquium on the Instrumentation of Landslides and Earthquakes in Japan and Turkey*, Koriyama, Japan, 1–9.

Tano, H., Aydan, Ö., Ulusay, R., Kumsar, H. 2006a. Joint research related to the effects of geo-environmental factors on slope movements: Investigation of the creep type Babadağ (Turkey) landslide. *Interim report*, 74p (in Turkish).

Tano, H., Abe, T., Aydan, Ö., Kumsar, H, Kaya, M., Çelik, S.B., Ulusay, R. 2006b. Long-term monitoring of Babadağ landslide through an integrated monitoring system with an emphasis on AE responses and ground straining. In *Symposium on Recent Applications in Engineering Geology*, Pamukkale University, Denizli, 131–141.

Tano, H., Aydan, Ö., Kumsar, H., Kaya, M., Çelik, S.B., Ulusay, R., Abe, T. 2008. Investigation of Babadağ landslide and its implications for hazard mitigation before natural disaster. In *First Collaborative Symposium of Turk-Japan Civil Engineers*, 5 June 2008, Istanbul, 65–73.

Ulusay, R., Tuncay, E., Sönmez, H., Gökçeoğlu, C. 2004. An attenuation relationship based on Turkish strong motion data and iso-acceleration map of Turkey. *Eng Geol*, 74: 265–291.

Ulusay, R. & Karakul, H. 2016. Assessment of basic friction angles of various rock types from Turkey under dry, wet and submerged conditions and some considerations on tilt testing. *B Eng Geol Environ* 78(4):1683–1699.

Xian–Qin, H. & Cruden, D.M. 1992. A portable tilting table for on–site tests of the friction angles of discontinuities in rock masses. *Bulletin of the International Association of Eng Geol* 46:59–62.

2019 Rock Dynamics Summit– Aydan et al. (eds)
© 2019 Taylor & Francis Group, London, ISBN 978-0-367-34783-3

Seismic stability evaluation verification of slopes reinforced with prevention piles

H. Kobayakawa, M. Ishimaru, K. Hidaka A. Sekiguchi & T. Okada
Central Research Institute of Electric Power Industry, Japan

S. Mori
NEWJEC, Incorporated, Japan

K. Hiraga
Civil Engineering Research & Environmental Studies, Incorporated, Japan

H. Morozumi
Kansai Electric Power Company, Incorporated, Japan

ABSTRACT: In order to verify the stability evaluation flow of slopes with prevention piles installed, a model test of a slope was conducted. Prevention piles were designed to be installed on a slope with a distribution of successive weak layers, ensuring the safety factor of 1.2 for the design maximum seismic coefficient (K_H) equal to 0.6, based on past stability evaluation flows. A slope model installed with prevention piles based on the design was created and a dynamic centrifugal force model experiment was conducted. As a result, it was confirmed that the prevention piles functioned even when the maximum horizontal acceleration (6.31 m/s²) equivalent to the designed seismic coefficient worked, and that no noticeable change occurred on the slope. Based on these facts, the validity of the existing stability evaluation flow was demonstrated.

1 INTRODUCTION

In order to improve the seismic resistance of slopes around important structures, countermeasure construction is sometimes carried out (Nuclear Standards Committee of JEA 2016). The use of prevention piles is one effective measure. An earthquake stability evaluation method has been proposed for slopes with prevention piles (Toda et al. 2013). However, the effectiveness of the flow of stability evaluation has not been sufficiently verified, since it is difficult to ascertain the behavior of the actual slopes with the prevention piles installed at the time of the collapse. In order to verify the flow, a dynamic centrifugal model experiment was carried out on a model slope reinforced with prevention piles (Kobayakawa et al. 2017). In this paper, the effectiveness of the proposed stability evaluation flow is demonstrated from the results of the dynamic centrifugal model experiment.

2 PREVENTION PILE DESIGN BASED ON THE STABILITY EVALUATION FLOW

2.1 The outline of the flow

The proposed flow is shown in Figure 1. Prevention piles for landslides are usually designed according to the guidelines for the steel pipe pile for landslide (Board of directors of the guidelines for the steel pipe pile for landslide 2013). In this guideline, the concept of design differs depending on the pile deterrence mechanism. However, since the stress state of the piles and the surrounding ground during the earthquake is complicated, it is difficult to presume the deterrence mechanism in advance. Therefore, in this flow, the specifications of the prevention piles were decided based on the section force of the piles and the destruction state of the surrounding ground by a

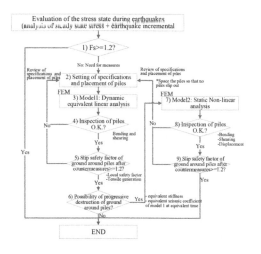

Figure 1. Flow of the seismic stability evaluation for slopes reinforced with prevention piles (Toda et al. 2013).

parameter study of dynamic equivalent linear analysis. Specifically, the section force of the piles was set so as to be within the elastic limit. With respect to the destruction of the ground around the piles, we checked the slip safety ratio of the ground surrounding the piles and determined the setting of the piles so as to secure an allowable safety factor of 1.2.

2.2 The model slope and pile design

2.2.1 Outline of the model slope

The cross-section diagram of the target slope is shown in Figure 2. The slope is a model used for centrifuge shaking tests. Since the size of the model slope used for the experiment was on a scale of 1 to 50, and the apparent gravity of the centrifugal acceleration set at 50G, the real scale of the height of the slope was 10m. The materials used for the slope consists of artificial soft rock and discontinuous plane, simulating a weak layer. The soft rock was compounded as shown in Table 1 in consideration of the homogeneity and strength of the material. The rock was left to sit for one week. The properties of the rock are as follows: The wet density (ρ_t) = 2.07 Mg/m³, the initial shear modulus of rigidity (G_0) = 933MPa, tensile strength (σ_t) = 41.4kPa and the peak parameter for cohesion (c) = 267kPa and the angle of sheer resistance (ϕ) = 34.7° and. A Teflon sheet (thickness: 0.2mm) was used for the weak layer, in consideration of the ease of making a model and the reproducibility of physical properties. The mechanical properties of the weak layer are ρ_t = 2.10 Mg/m3, c = 0, ϕ = 28.6° (at peak) and 19.3° (at residual).

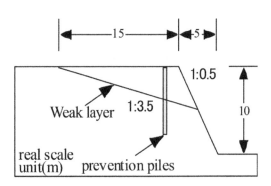

Figure 2. Shape of the slope

Table 1. Compound of the artificial soft rock.

Material	composition (kg/m³)
Cement	82
Crushed sand	817
micronized powder	817
admixture	1
Water	370

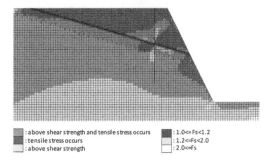

: above shear strength and tensile stress occurs   : 1.0<= Fs<1.2
: tensile stress occurs   : 1.2<=Fs<2.0
: above shear strength   : 2.0<=Fs

Figure 3. Distribution chart of local safety factor

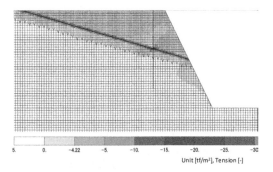

5.   0.   -4.22   -5.   -10.   -15.   -20.   -25.   -30
Unit [tf/m²], Tension [-]

Figure 4. Distribution chart of tensile stress.

Based on these strengths, when the relationship between the slip safety factor and the horizontal seismic coefficient (K_H) of the slope is calculated by the limit equilibrium method, the slip safety factor falls below 1.0 when K_H = 0.3.

2.2.2 The design of the piles

Based on the design philosophy of wedge piles and shear piles according to the guidelines for the steel pipe pile for landslide, the specification of prevention piles was determined so that the safety factor would be 1.2 or more for K_H = 0.6 (Flow procedure 1)). As a result, on the real scale, the outer diameter of the piles was 0.4m, the thickness 15mm, the spacing between the piles 2m, the length 8m, and it was determined that the position of the piles would be set at 13.75m from the intersection of the slip surface and the ground surface (Flow procedure 2)).

Next, based on these specifications, an equivalent linear analysis was carried out on the slope model with the prevention piles. The cross-sectional force of the piles and the stress on the ground were calculated from the equivalent linear analysis, and the slip safety factor in front of the prevention piles and the possibility of progressive destruction were evaluated. The evaluation results are shown in Table 2. Procedures 3) and 4) of the flow meet the inspection reference value. An analysis of the local safety distribution diagram (Fig. 3) and the tensile stress distribution diagram (Fig. 4), however, shows that a stress exceeding

Table 2. Evaluation results of analyses.

Inspection items	Inspection result
Equivalent linear analysis	
Sectional force of prevention piles (allowable stress ratio < 1.0) (procedure 4))	Shear 0.13 < 1.0: O.K.Bending 0.038 < 1.0: O.K.
Slip safety factor in front of prevention piles (Fs > 1.2) (procedure 5))	Fs = 5.26 > 1.2: O.K.
Possibility of progressive destruction (procedure 6))	Possible occurrence of destruction in the weak layers (Fig.5). Tensile stress exceeding the tensile strength occurs around the prevention piles (Fig.6).
Static non-linear analysis	
Sectional force of prevention pile (allowable stress ratio < 1.0) (procedure 8))	Shear 0.27 < 1.0: O.K.Bending 0.116 < 1.0: O.K.
Slip safety factor in front of piles (Fs > 1.2) (procedure 9))	Fs = 2.59 > 1.2: O.K.

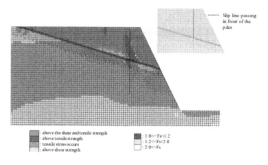

Figure 5. Distribution chart of local safety factor and failure element.

the tensile strength of the rock was generated at the intersection of the piles and the weak layers, which made it clear that there was a possibility of a progressive destruction.

Therefore, a static non-linear analysis shown in the flow procedure 7) was conducted. The evaluation results are shown in Table 2, and the distribution of destruction elements and the local safety factor in the static non-linear analysis are shown in Figure 5. In the static non-linear analysis, there was a tendency for the tensile failure to progress towards the front of the prevention piles. Although the slip safety factor in front of the prevention piles became lower than that of the equivalent linear analysis, the section force and the slip safety factor of the piles satisfy the predetermined criteria. These results show that based on the flow shown in Figure 1, the specification of the piles installed on the slope was appropriately determined.

3 CENTRIFUGAL MODEL TEST OF THE SLOPE

3.1 Outline of the test

Stainless-steel pipes with an outer diameter of 8.0 mm and an inner diameter of 7.4 mm were used at the 1/50 scale of the piles studied in the design were used in the experiment. Seven piles were installed at intervals of 40 mm in the depth direction of the slope. The lower end of the model of the pile was fixed to the bottom of the soil tank, and strain gauges were attached to the inside of the prevention piles at the center of the model in the axial direction and the strain was measured.

The slope model and sensor placement are shown in Figure 6. Two types of experimental case were conducted: case1 without prevention piles and case 2 with prevention piles installed. The slope height was 200 mm (10 m in real scale). The slope model was installed in a soil tank of 400 mm in length × 700 mm in width × 300 mm in depth, and two transparent silicone rubber sheets coated with petrolatum in between was placed between the side of the slope model and the wall of the soil tank to reduce friction. For the input acceleration waveform, a sinusoidal wave of 1.2 Hz at the actual conversion frequency was set to 20 waves in the main part, and four tapers were provided before and after the main part. The input acceleration was started at a small level, and a step shaking method was carried out in which the input level was incrementally increased after the shaking was completed at each step.

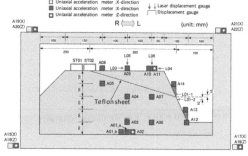

(1) Case1 (Unreinforced slope)

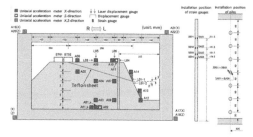

(2) Case2 (Slope reinforced with prevention piles)

Figure 6. Outline of the models

3.2 Result (case1)

The maximum horizontal acceleration (a_h) and the visual observation results at each step are shown in Table 3. In shaking step d06 where the horizontal acceleration exceeds 3 m/s², it was confirmed for the first time that the rock mass at the upper part of the weak layers moved downward during the shaking. The residual displacement accumulated as the steps proceeded since the downward displacement remained after the end of the shaking at each step. The residual displacement after the shaking at each step are shown in Figure 7. In all the measurement results of the displacement, we can see how the rock mass from the upper part of the weal layers moves down the slope as the shaking step advances. The results of calculating the slip safety factor by the limit equilibrium method are also shown in Figure 7. The horizontal acceleration (about 3 m/s²) where the slip safety factor by the limit equilibrium method was less than 1.0 and the horizontal acceleration of step d06 where the accumulative displacement was remarkably observed were almost in agreement.

3.3 Result (case2)

The maximum a_h and the visual observation results at each step are shown in Table 4. Although a shaking exceeding the input acceleration of the design was made for the first time in step d06, no deformation could be confirmed upon visual observation. The accumulative residual displacement captured by each displacement gauge are shown in Figure 8. In the displacement gauge ST02, the residual displacement transitioned at 1 mm or less until step d08, while about 50 mm of displacement occurred at step d09. In L01 at the front of the prevention piles, the residual displacement was 1 mm or less until step d08, as was the case with ST02, whereas displacement exceeding the measurement range of the sensor occurred at step d09.

4 EXAMINATION OF EXPERIMENTAL RESULTS AND VERIFICATION OF STABILITY EVALUATION FLOW

4.1 Verification of the design of piles based on the results of the equivalent linear analysis

As shown in Figure 1, in the stability evaluation flow of the pile design, the specification of the piles is first set based on the equivalent linear analysis. This is because it is considered that the behavior of the piles can be grasped by equivalent linear analysis in the range of design. Therefore, the validity of determining the specification of the piles based on the equivalent linear analysis is shown by comparing the experiment results of response of the piles and the surrounding ground with the analysis results.

Table 3. Maximum horizontal acceleration (a_h) and observation at each step.

Step	a_h m/s²	Observation results
d01	0.30	d01~d05: No deformation confirmed visually.
d02	0.63	d06~d07: rock mass on the weak layer moves slightly downward along the weak layer.
d03	1.48	
d04	2.01	
d05	2.35	d08: rock mass on the weak layer moves 1 mm (at model scale) downward along the weak layer.
d06	3.02	
d07	3.09	d09: rock mass on the weak layer moves 1 mm (at model scale) downward along the weak layer. d10: rock mass on the weak layer moves 3 mm (at model scale) downward along the weak layer.
d08	3.50	
d09	3.92	
d10	4.27	

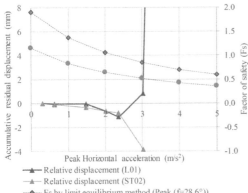

Figure 7. Residual displacement at each shaking step and slip Fs by limit equilibrium method.

Legend:
—▲— Relative displacement (L01)
—▲— Relative displacement (ST02)
··◆·· Fs by limit equilibrium method (Peak (f=28.6°))
··●·· Fs by limit equilibrium method (Residual (f=18.3°))

Table 4. Maximum horizontal acceleration (a_h) and observation at each step.

Step	a_h m/s²	Observation results
d01	0.72	d01~d08: No deformation confirmed visually.
d02	1.64	
d03	2.65	d09: rock mass above the weak layers vibrates back and forth after the shaking, then a crack will generate on the surface of the installation of the prevention piles. As the vibration continues, the crack propagates to the weak layer, and the rock mass at the front of the prevention piles is greatly displaced.
d04	3.42	
d05	4.76	
d06	6.31	
d07	8.64	
d08	9.32	
d09	11.47	

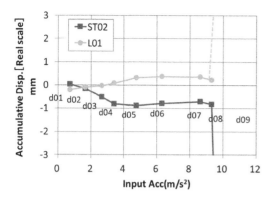

Figure 8. Relationship between input acceleration and accumulative displacement.

Table 5. Mechanical and physical properties of the prevention piles.

Properties	Per pile	Per unit depth
Shear modulus (GPa)	74.49	74.49
Poisson ratio	0.3	0.3
Sectional area (m²)	0.01814	9.466×10^{-3}
Second moment of area (m⁴)	3.367×10^{-4}	1.571×10^{-4}
Unit weight (kN/m³)	77.8	36.3

4.1.1 Analysis condition

An analysis model diagram is shown in Figure 9. The slope ground was modeled by plane strain elements considering the non-linear characteristics of the material, and the weak layer part adopted a linear joint element. The prevention pile was modeled as a beam element and rigidly connected to the ground. The number of the piles was converted into the number per unit depth and then modeled. The boundary conditions were as follows: The bottom of the soil tank was completely constrained. A joint element was provided between the ground and the soil tank, and the value of the tangential spring of the joint element was set to 0, and the value of the vertical spring was set to be rigid so as to make it equivalent to the vertical roller. The properties of the soft rock and weak layer mentioned in 2.2 were used. The mechanical and physical properties of the prevention piles are shown in Table 5. For the input acceleration, the horizontal acceleration (A01-a) and the vertical acceleration (A02) measured at the bottom of the model were uniformly input to the bottom of the model. Input waves were given to every 20 seconds of each shaking step in the experiment.

4.1.2 Analysis result and examination

The horizontal maximum acceleration for each step by experiment and analysis is shown in Figure 10.

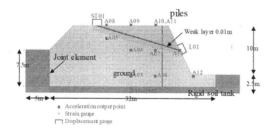

Figure 9. Analytical model. The model slope with a weak layer.

The analysis results of the maximum horizontal acceleration distribution are almost consistent with the experimental value, except for measurement point A 08. The reason why the behavior of measurement point A 08 was different was that this measurement point was located in the immediate vicinity of the weak layer, and the upper rock mass of the weak layer was thinned, and it can be viewed that it may have been affected by the local behavior. In the acceleration time history, which will be described later, the peak acceleration of the measurement point is also larger than the other measurement points, but considering the behavior of the entire slope, it is thought to be a local behavior.

We focus on step d06 where the acceleration response value became the maximum within the range of design. As a representative example of the acceleration response at that step, the responses at A06, 07, 10, 11 around the piles are shown in Figure 11. The acceleration response values of the analysis results are either equal to or slightly larger than the experimental results, but have been calculated almost appropriately. In the range of design where displacement of the rock mass does not occur, the acceleration response has been appropriately evaluated. The time history of the axial force and the bending moment of the piles obtained from the measurement results of the strain gauge attached to the prevention piles is shown in Figure 12. The experimental and analytical values are generally consistent with respect to the section force of a piles in the rock mass above G.L. - 3.425 m near the weak layer. At the position of G.L. - 3.625 m, although the experimental and analytical values are consistent for the axial force, the bending moment is slightly larger in the analysis value. It is considered that the experimental value of the sectional force generated in the piles were smaller than the analysis value due to the local non-linearity of the surrounding ground where the piles and weak layer intersect. From the aspect of pile design, however, since the section force generated by the shake is estimated to be larger than the actual force, it is a setting on the side of prudence. For this reason, the evaluation is

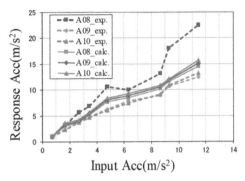

(1) Top part of the model.

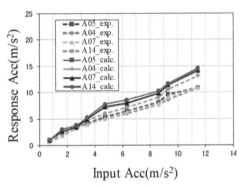

(2) Middle part of the model

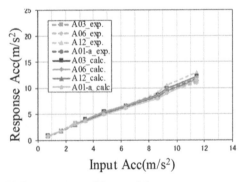

(3) Bottom part of the model

Figure 10. Change in maximum acceleration at each shaking step.

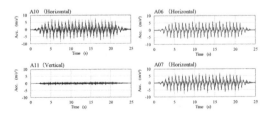

Figure 11. Time histories of the response acceleration (step d06)

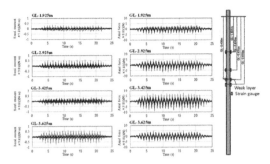

Figure 12. Time histories of the sectional force of pile (step d06).

appropriate as it becomes a setting on the safe side for piling design.

From the above results, it was confirmed that the response of the ground and piles obtained by the equivalent linear analysis is consistent with or slightly larger than the results of the experiment in the range of the horizontal seismic intensity set at the pile design. Therefore, it is considered that the design of the prevention piles based on the equivalent linear analysis is appropriate.

4.2 The validity of the earthquake stability evaluation flow for slopes with prevention piles

A shaking experiment was carried out on the slope model created based on the specification of the prevention piles set by the flow shown on Figure 1. In the shaking step d06 (maximum $a_h = 6.31$ m/s²) equal to or greater than the design seismic coefficient $K_H = 0.6$, it was confirmed that both the slope and the prevention piles were healthy without significant change occurrences. From these facts, the validity of the flow was confirmed. Furthermore, no noticeable variations such as slip or local failure of the slope were observed up to shaking step d08 (maximum $a_h = 9.32$ m/s²) about 1.5 times the design seismic coefficient, and at shaking step d09 (maximum $a_h = 11.47$ m/s²), about 1.8 times the design seismic coefficient, the rock mass at the front of the prevention piles collapsed (Table 4). This result shows that there is some allowance in the evaluation by this flow.

5 CONCLUSION

In order to investigate the validity of the stability evaluation flow of a slope on which prevention piles were installed, shaking tests were conducted in the centrifugal force field of the slope model with a weak layer. Based on the flow, the specification of the piles was determined so as to secure a slip safety factor of 1.2 or more against $K_H = 0.6$, and the piles were installed on a slope with a weak layer.

The shaking tests of the model slope confirmed that no noticeable change could be observed in either the prevention piles or the slope at the assumed input ground motion level. Based on these facts, the validity of the flow was demonstrated. For future earthquake risk assessment, it is necessary to establish a non-linear time-history analysis that can evaluate the behavior of the ground and piles in the range beyond the design input earthquake ground motions.

ACKNOWLEDGMENTS

This paper is the result obtained by the common research on nuclear risk center (FY2016) by 9 electric power companies, the Japan Atomic Power Company, Electric Power Development Co., Ltd. and Japan Nuclear Fuel Ltd.

REFERENCES

Board of directors of the guidelines for the steel pipe pile for landslide. 2013. The guidelines for the steel pipe pile for landslide, Association for slope disaster management, 215pp (in Japanese).
H. Kobayakawa, M. Ishimaru, A. Sekiguchi, T. Okada, T. Taniguchi, T Hiraga, S. Mori and H.Nakamura. 2017. Earthquake stability evaluation of slope with prevention pile (part 1) - dynamic centrifugal force model experiment on prevention pile slope model -, *Japan Society of Civil Engineers 2017 Annual Meeting, Japan Society of Civil Engineering*, VII-20, pp.30–31 (in Japanese).
Nuclear standards committee of JEA. 2016. Technical Guidelines for seismic design nuclear power plants. (JEAG4601-2015), The Japan electric association (in Japanese).
T. Toda, H. Kobayakawa and K. Haraguchi. 2013. Dynamic stability of the rock slopes with prevention piles, *Abstract of the 52th annual meeting of the Japan landslide society, The Japan Landslide Society*, pp.126–127 (in Japanese).

2019 Rock Dynamics Summit– Aydan et al. (eds)
© *2019 Taylor & Francis Group, London, ISBN 978-0-367-34783-3*

Characteristics and mechanisms of earthquake-induced landslides according to recent events and studies

S. Komata

Technology Headquarters, Nippon Koei Co., Ltd, Tokyo, Japan

ABSTRACT: According to events and studies on landslides of rock slopes due to recent earthquakes mainly in Japan, the author remarks some differences of the characteristics and mechanisms among landslide phenomena. Rockfalls occur by the seismic inertia force on steeper reverse-dip slopes as topples and falls, which consist of loosen rock or jointed rock material in shallow portions. On the other hand, rockslides occur on gentler dip slopes as translational or rotational slide, which develop surface gravitational deformation, consist of mainly bed rock and intercalate sandy, silty or clayey strata of slip plane. Furthermore, rockslides are caused not by the seismic inertia force but by the cyclic shear load which results in both decrease of shear strength and liquefaction with excess pore water pressure of slip plane in earthquake duration.

1 INTRODUCTION

Many earthquake-induced landslides have occurred on rock slopes in the seismically active areas of the world. In Japan, earthquake-induced landslides have caused hundreds of deaths and billions of dollars in economic losses in recent years. Since Japan situates on the subduction of both the oceanic Pacific plate and the oceanic Philippines plate beneath both North American plate and Eurasian plate.

Characteristics and mechanisms of earthquake-induced landslide have been discussed based on their numerous historical records induced by seismic ground motions. For example, Voight & Pariseau (1978) introduced the distance at which earthquakes can trigger landslides is subject to the following: the stability of the potential slide mass, the orientation of the earthquake in relation to the slide mass, earthquake magnitude, focal depth, seismic attenuation, and aftershock distribution.

Wyllie & Mah (2004) introduced the performance of rock slopes during earthquakes by Keefer (1984, 1992) that the following five slope parameters have the greatest influence on stability during earthquakes:

- *Slope angle* Rock falls and slides rarely occur on slopes with angles less than about 25˚;
- *Weathering* Highly weathered rock comprising core stones in a fine soil matrix, and residual soil are more likely to fail than fresh rock;
- *Induration* Poorly indurated rock in which the particles are weakly bounded is more likely to fail than stronger and well-indurated rock;
- *Discontinuity characteristics* Rock containing closely spaced or open discontinuities are more susceptible to failure than massive rock in which the discontinuities are closed and healed; and

- *Water* Slopes in which the water table is high, or where there has been recent rainfall, are susceptible to failure.

However, these discussions are mainly about effects of shaking acceleration on the site. There were few discussions about the slope instability in earthquake duration except, for example, Newmark (1965) and Gucwa & Kehle (1978).

Newmark (1965) proposed that in the case the displacement on slip plane increases, the slope should be unstable with decrease of slip plane, because the repeated inertia of the sliding mass by downward acceleration pulse results in residual strength of slip planes during earthquake motions.

Gucwa & Kehle (1978) mentioned that earthquake loading is suggested as a cause of elevated pore fluid pressure or an equivalent loss of strength during earthquake duration, and catastrophic slope failures accompanying earthquakes are the result of soil liquefaction caused by the cyclic loading of the soil; this load occurs with the upward passage of seismic shear waves.

Previous discussions above are almost proved by recent events. However, according to events and studies on recent earthquake-induced landslides of rock slopes mainly in Japan, the author remarks that there are several differences in performance between types of earthquake-induced landslides: rockfall and rockslide, and some rockslides occur on the slopes with angles less than about 25˚. We have investigated their mechanisms concerning topography, geology, groundwater condition and soil strength under dynamic stress, and discussed particularly rockslide occurrence during cyclic shaking duration of earthquake. As a result, we suggest that rockslides occur owing to both excess pore water pressure by liquefaction and decrease of shear strength by particle breakage of slip plane.

In this paper, earthquake-induced landslides of rock slopes are classified into rockfalls and rockslides based on the principles and terminology of Varnes (1978). Rockfalls are defined as rock descending of individual boulders or disrupted masses on slopes by bounding, rolling, free fall or toppling. Whereas rockslides are defined as rock mass movements by rotational slump, translational slide including deep-seated block slide, and lateral spread.

2 EARTHQUAKE GROUND MOTION RELATING TO ROCKFALLS OR ROCKSLIDES

Recently, many landslides have occurred in Japan and its adjacent area as follows: East Nagano Earthquake (Eq. ; 1984, M6.8), Hokkaido south-western offshore Eq. (1993, M7.8), South Hyogo Eq. (1995, M7.2), Taiwan-Chi chi Eq. (1999, M7.6), Chuetsu Eq. (2004, M6.8), Noto peninsula Eq. (2007, M6.9), Chuetsu offshore Eq. (2007, M6.8), Iwate/Miyagi inland Eq. (2008, M7.2), China-Bunsen Eq. (2008, M7.9), East Japan Eq. (2011, M9.0), Kumamoto Eq. (2016, M7.3), and East Iburi of Hokkaido Eq. (2018, M6.7).

According to studies of both these events and earthquake-induced landslides and to previous discussions above, the author remarks the performance of earthquake-induced landslides as follows:

2.1 *Moment magnitude and intensity of earthquakes relating to landslide occurrence*

The intensity of earthquake at specific location depends on several factors in recent events of Japan: (1) the total amount of energy released, (2) the distance

M: Moment Magnitude, L: length of earthquake source fault, Li: extent of surface deformation in the direction of fault strike, B & D: Mode extent of JMA's intensity 6 from earthquake source fault, and D max: Maximum extent of JMA's intensity 6 from earthquake source fault from the epicenter, (3) the site topography, and (4) the type of rock and degree of consolidation (seismic velocity of materials).

1. *The total amounts of energy released* (moment magnitudes), which concerned almost all earthquake-induced landslides, are more than M6.1 of offshore (inland) earthquakes, and more than M7.9 of onshore (submarine trench) earthquakes.

Similarly, ground motions relating to earthquake-induced landslides are that minimum earthquake intensity is 5 minus of Japan Meteorological Agency scale (JMAs) almost same as IV of Modified Mercalli Intensity scale (MMIs). Furthermore, the number of landslide occurrence increases more than 6 plus of JMAs almost same as IX of MMIs.

Maximum accelerations of earthquake ground motion concerning landslide occurrence range from 200 gal to 1000 gal. Especially, rockfalls are likely to occur with more than 200 gal, and larger rockslides with more than 500 gal.

2. *The distance from the epicenter* is that the longer the distance is, the weaker the intensity is due to seismic attenuation. For example, Figure 1 shows the correlation between moment magnitude of earthquake and extent of intensity more than 6 of JMAs from earthquake source fault in Japan (Kuwahara, 2008); the intensity is likely to trigger landslides.

3. *The site topography* of convex landforms causes amplification of seismic waves as shown in Figure 2a.

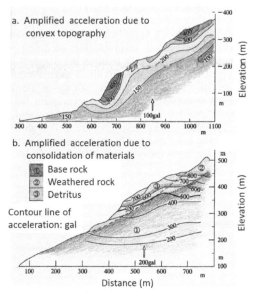

Figure 2. Amplified acceleration of seismic wave by response analysis in mountain area (Yamaguchi, 1998)

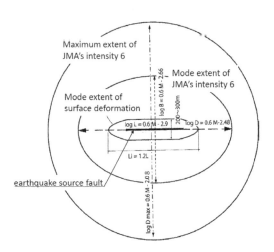

Figure 1. Assumed extents of JMA's intensity 6 and surface deformation from earthquake source fault (Kuwahara, 2008)

a: amplified acceleration of seismic wave on convex topography, and b: amplified acceleration of seismic wave in weak consolidated materials

4. *The type of rock and degree of consolidation* (seismic velocity of materials) causes amplification of seismic waves as shown in Figure 2b. The lower the seismic velocity of materials (less consolidated materials) is, the lager the amplification of seismic waves becomes.

2.2 Landslide distribution from earthquake source faults

The correlation between maximum distance of disrupted slides or falls from epicenter and earthquake magnitude has been discussed by Keefer (1984). In addition to this, we remark on the distribution of landslides from recent events in Japan as follows:

1. Landslides by offshore(inland) earthquakes on reverse fault (thrust) are concentrated on the slopes of hanging wall by so-called "hanging wall effect" within about 15 km far from earthquake source faults (Fig. 3);
2. Landslides by offshore(inland) earthquakes on normal fault occur on the slopes of hanging wall, however they are distributed less than about 2 km near earthquake source faults (Inagaki, 2015); and
3. Landslides in extension of the earthquake source faults occur owing to so-called "Doppler effect" by accumulation of seismic waves to the direction of fault rapture.

3 TYPES OF EARTHQUAKE-INDUSED LANDSLIDES CORRELATING WITH TOPOGRAPHY AND GOELOGY

Earthquake-induced landslides of rock slopes are classified into rockfalls and rockslides according to types of movement. Table 1 shows the diagram of these types combining slope gradients and materials.

Rockfalls are relatively free falling of newly detached segments of bed rock especially on steep slopes (Table 1-1, 2a, 2b, 2c, 3) or on reverse-dip slopes (Table 1-5, 6, 8); these types are shallow slides, topples and falls, and are likely to result in debris flows or sediments on the foot of cliff after collapse.

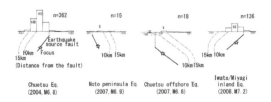

Figure 3. Distribution of earthquake-induced landslides on the section parallel to the dip of the reverse earthquake source faults

Table 1. Types of earthquake-induced landslides (Komata, 2015)

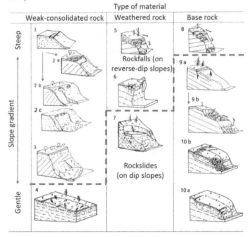

: Maximum extents of landslides are almost all within 15 km from the faults.

: Rockfalls are defined as rock descending of individual boulders or disrupted masses on slopes by bounding, rolling, free fall or toppling, and rockslides also as rotational slump, translational slide including deep-seated block slides, and lateral spreads.

Rockslides of larger rock mass occur on slopes with convex ridge in topography or on gentle slopes behind the shoulder of cliff eroded by river, sea or man-made so-called "knick point" (Table 1-9a, 9b).

Furthermore, rockslide slopes have developed gravitational deformations in head portion of the sliding mass under tensional stress, such as steps, trenches or double ridges before earthquake (Table 1-9a). They also consist of downward-dipping bedding strata: dip slope (Table 1-9a, 9b, 10a, 10b).

In fractured rock slope, rotational slides occur owing to earthquake (Table 1-7). In the case of slopes of soft rock, for example, newly volcanic sediment, translational slides occur owing to earthquake (Table 1-4).

4 MECHANISM OF EARTHQUAKE-INDUSED ROCKFALLS

Ground motion (an earthquake tremor) increases extremely on cliff shoulders, and the amplified tremor exists decreasing backwards of the cliff shoulder within the distance of two times of the cliff height shown in Figure 4. The ground motion is greater in perpendicular to the cliff face (towards the free cliff face) than in parallel to it. Therefore, rockfalls occur easily from cliff shoulders so-called "knick point" and its adjacent potions.

Rockfalls of small mass are caused by the inertia force and the amplified tremor with repeating (cyclic) ground motion of earthquake depending on site

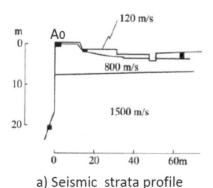

a) Seismic strata profile

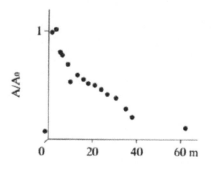

b) Ratio of shaking amplitudes(A) to cliff shoulder amplitude(A0)

Figure 4. Measurement of ground motion amplification on cliff shoulder (Komaki & Toida, 1980)

: Ground motion: an earthquake tremor increases extremely on cliff shoulders, and the amplified tremor exists decreasing backwards of the cliff shoulder within the distance of two times of the cliff height.

conditions. They are convex portion or "knick point" of topographical condition, which consists of jointed or loosen rock material with lower seismic velocity than fresh bed rocks. Yagi *et al.* (2007) mentioned that 80 % of rockfalls triggered by Chuetsu Eq. were distributed on steep reverse-dip slopes of more than 35°gradient and its mode were between 45°and 50°gradient.

The amplified tremor causes shear or tensile stress along joints to increase and rock materials to loosen. In the case that the shear or tensile stress becomes larger than joint strengths, rockfalls such as strip type rock-fall or wedge failure occur. The ground motion which triggers rockfalls almost has the intensity of more than 5 minus of JMAs and the duration of about 10 seconds.

In general, the natural shaking period of the slopes around cliff shoulders is from 0.1 to 0.3 seconds about 10m below the surface. If the period of the slope resonates with the period of input earthquake ground motion, which is in the case of short period and large amplification of the input

earthquake, then rock-falls and rock failures may occur easily by the earthquake ground motion.

5 MECHANISM OF EARTHQUAKE-INDUCED ROCKSLIDES

5.1 *Effect of the inertia force to rockslide mass*

Rockslides of large rock mass could not be unstable only by the earthquake ground motion. Since the motion of earthquake is the cyclic shaking of frequency with about 10 to 0.1 Hz, the downward acceleration pulse does not work to the whole sliding mass at the same time during earthquake motions. Consequently, it is unlikely to move the large rock mass by only the inertia force. Furthermore, the dynamic shear strength is larger than the static one by the effect of shear velocity especially in the case of cohesion soil; the effect is that the soil strength become stronger as the shear velocity rise during earthquake with high frequency.

In addition, the deeper the depth of slip plane is, the larger the confined pressure on slip plane becomes. For example, the confined pressure in the case of a rockslide of 50m deep is assumed about 1MPa. Landslide mass could not move easily under high confined pressure, because the slip plane strength is unlikely to decrease owing to the small displacement by earthquake inertia force. Relating to this, Kokusho (1980) experimented on the confined pressure effect concerning the relation between shear strain rise and shear strength fall; the experiment suggests that the decrease of slip plane during earthquake under high confined pressure of deep rockslides needs more strain than under low confined pressure of shallow rockslides.

5.2 *Liquefaction of slip plane materials*

The slip planes of earthquake-induced landslides of Chuetsu Eq. consist of saturated sandy or silty soil. Sasa et al. (2007) have experimented with these materials of slip plane by dynamic and cyclic ring shear test shown as Figure 5, which continued the vibration of shear stress raise by a sine wave of 1 Hz of frequency until 15 second (15 cycles), then stopped the vibration. It was found that the shear resistance increases with the dynamic effect until 10 second and decreases after 10 second as the pore-water pressure rising. Pore-water pressure rises to almost same as normal stress at 20 second and after, consequently the shear displacement increases after stopping the vibration; the liquefaction of slip plane occurs at 20 second and after.

In Japan, many large rockslides have recently occurred after rainfall or in snow melting season, when the water table of landslide slopes becomes higher and the materials of slip plane are saturated. Therefore, earthquake-induced rockslides are suggested to occur by liquefaction of slip plane.

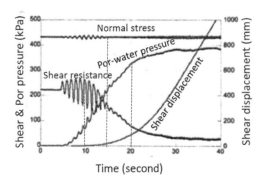

Figure 5. Undrained cycle shear test of sandy material (Sasa et al., 2007)

: vibration of shear stress is continued until 15 second (15 cycles) by a sine wave of 1 Hz of frequency. Then stopped the vibration. Shear resistance increases until 10 second, then it was decreasing as the Pore-water pressure rising after 10 second. Pore-water pressure rose to almost same as normal stress at 20 second, consequently the shear displacement increases after stopping the vibration.

5.3 Particle breakage of slip surface materials

In addition to the liquefaction, the particle breakage of slip plane materials occurs owing to earthquake ground motion. Figure 6 shows the decrease of slip plane material strength by cyclic shear loading test.

The strength of test piece decreases slightly in shear velocity (rate of displacement) of 50mm/min (0.8 mm/s) and is reduced rapidly to about 1/3 of the peak strength in 500mm/min (8mm/s). The reduction may occur not only on the slip plane but also in the side boundary materials of the slide mass, therefore the friction on these portions decreases and results in occurrence of landslide.

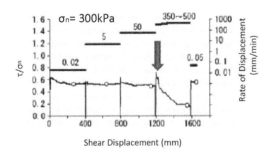

Figure 6. Shear strength decrease of fractured sample relating to shear velocity increase (Kinoshita et al. 2009)

: Slip plane material decreased by cyclic shear loading test with normal stress σ_n of 300 kPa. The strength of the test piece decreases slightly in 50mm/min of shear velocity (rate of displacement) and are reduced rapidly to about 1/3 of the peak strength in 500mm/min (8mm/s).

Furthermore, in the case that slip plane materials are saturated, the particle breakage occurs and results in the reduction of void volume so that excess pore water pressure rise and landslides are more likely to occur.

6 DISCUSSION

It is said that rockfalls occur owing to earthquake of lightening acceleration type with high frequency and in short duration, while rockslides occur on dip down-ward slopes (dip slope) owing to earthquake of energy type (of greater wave amplitudes than rockfall slopes) with low frequency and in long duration (Keefer, 1984).

The author discussed earthquake-induced landslides of recent events and previous discussions above and remarked the performance of them as follows (Fig. 7):

Rockfalls occur on steep reverse-dip slopes as toppling and fall, which consist of loosen rock material of sheeting, columnar joint or gravitational deformation in shallow portions. Since earthquake acceleration of ground motion (the intensity of an earthquake) can be magnified on the slopes both with topography such as convex slopes, shoulder of cliff ("knick line": erosion front) in upper slopes and along ridges, and with unconsolidated materials such as talus deposits, detritus and loosen rock structure. It is said that rockfalls occur on these portions by the seismic inertia force.

Whereas rockslides occur on gentle dip slopes as translational or rotational slide, which develop surface deformation structure such as cracks,

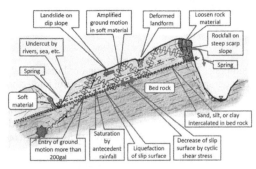

Figure 7. Mechanism of earthquake-induced landslides

: Rockfalls occur on steep reverse-dip slopes as toppling and fall, which consist of loosen rock material in shallow portions result from sheeting, columnar joint or gravitational deformation. Rockslides occur on gentle dip slopes as translational or rotational slide, where develop surface deformation structure such as cracks, depression, uphill-facing scarp and double ridges, and consist of mainly bedrock and intercalate a thin saturated sandy, silty or clayey bed.

depression, uphill-facing scarp and double ridges. The slopes consist of mainly bedrock and intercalate a thin saturated sandy, silty or clayey bed. If rivers, sea or man-made undercut into those unstable slopes, large rockslides might occur owing to earthquake.

Furthermore, in the case there are springs or antecedent rainfall add to saturating slope body, rockslides are more likely to occur than unsaturated condition.

When earthquake wave locally reaches these slopes in more than about 200 gal, the intensity of earthquake is magnified with greater wave amplitudes and longer durations than that in solid bedrock. Shear resistance of slip plane (layer of rupture) decreases by shear repeating and by liquefaction with excess porewater pressure, resulting in the rockslide occurrence. Under these conditions, rockslides are likely occur even on gentle slopes of lower than 25°.

Deep rockslides on gentle gradient portions are unlikely to occur only by the seismic repeating inertia force. They result from slope instability caused by factors of increasing shear stress or lowering shear resistance on rupture of slip plane in earthquake duration.

Furthermore, these rockslides may occur in distance less than about 15km from earthquake fault of shallow direct hit; especially they occur on slopes to overlie the fault (hanging wall of the fault) or to extend in the fault direction.

7 CONCLUSIONS

The author reviewed previous discussions about performances of earthquake-induced landslides, and then studied events and investigations of recent earthquake-induced landslides on rock slopes mainly in Japan. We have recognized that there are several differences in performance between the types of earthquake-induced landslides: rockfall and rockslide. Some rockslides occur on the gentler slopes with angles less than about 25°. The suggestions from studies are as follows:

Rockfalls occur by the seismic inertia force, to say the primary effect of earthquake, on adjacent portion of so-called "knick line (cliff shoulders)" in steeper reverse-dip slopes, which consist of loosen rock or jointed rock materials and are likely to amplify ground motion. They rupture in shallow portions as topples and falls.

Rockslides occur on gentler dip slopes, which develop the surface gravitational deformation, consist of mainly bed rock and intercalate a saturated sandy, silty or clayey strata of slip plane. On these slopes, rockslides occur as translational or rotational slide on slip planes owing to both excess pore water pressure with liquefaction and shear strength decrease by particle breakage of slip planes during earthquake cyclic shaking; that is the secondary effects of earthquake.

REFERENCES

Gucwa, P. R. & Kuhle, R. O. 1978. Beapaw Mountains rockslide, Montana, USA, *Rockslides and Avalanches* (ed. Voight, B.), Elsevier: 393–422.

Inagaki, H. 2015. Relation between the distance from active fault and the scale of mass movement, *Journal of the Japan Society of Engineering*, 56–1: 10–15. (in Japanese)

Keefer, D. L. 1984. Landslides caused by earthquakes, *Geol. Soc. of America Bull.*, 95–4: 406–421.

Keefer, D. L. 1992. The susceptibility of rock slopes to earthquake-induced failure, *Proc. 35th Annual Meeting of the Assoc. of Eng. Geologists* (ed. Stout, M. L.), Long Beach, CA: 529–538.

Kinoshita, A. et al. 2009. Study on geo-technical characteristics of the slip surface of a large-moved landslide triggered by the 2004 Mid-Niigata Prefecture earthquake, *Journal of the Japan Landslide Society*, 45–6: 6–15. (in Japanese)

Kokusho, T. 1980. Cyclic Triaxial test of Dynamic Soil Properties for Wide Strain Range, *Soils and Foundations*, 20(2): 45–60.

Komata, S. 2015. Failures and landslides of slopes, *Geomorphology and geology concerning disaster management, environment and maintenance*, Japan geotechnical society: 34–79. (in Japanese)

Komaki, S. & Toida, M. 1980. Measurement of ground motion amplification on cliff shoulder and its adjacent area in Izu Peninsula, no. 1, *Proc. 17th Symposium of Natural disaster scence, 198.* (in Japanese)

Kuwahara, K. 2008. Guard yourself from geo-disasters—Knowledges of safety for life—, Kokon-shoin, 131p. (in Japanese)

Newmark, N. M. 1965. Effects of earthquakes on dams and embankments. *Geotechnique*, 15(2),139–160.

Sasa, K. et al. 2007. Sliding mechanism of the 2004 Mid-Niigata Prefecture Earthquake-triggered-rapid landslides occurred within the past landslide masses, *Journal of the Japan Landslide Society*, 44–2: 1–8. (in Japanese)

Varnes, D. J. 1978. Slope movement types and processes, *Landslides—Analysis and control* (eds. Schuster, R. L. & Krizek, R. J.), Transportation Research Board Special Report 176, National Academy of Sciences: 12–33.

Voight, B. & Pariseau, W.G. 1978. Rockslides and Avalanches: an introduction, *Rockslides and Avalanches* (ed. Voight, B.), Elsevier: 1–67.

Wyllie, D. C. & Mah, C. W. 2004. Seismic analysis of rock slopes, *Rock slope engineering 4th ed.*, Spon Press: 141–148.

Yagi, K. et al. 2007. GIS analysis on geomorphological features and soil mechanical implication of landslides caused by 2004 Niigata Chuetsu earthquake, *Journal of the Japan Landslide Society*, 43–5: 44–56. (in Japanese)

Yamaguchi, I. 1998. Earthquake–induced disasters in mountainous area and their countermeasures—Lessons from South Hyogo Eq.—(ed. Preservation technology research center of mountainous area), *Report of Japan Association of Mountain and River Preservation*, 7–24: (in Japanese)

2019 Rock Dynamics Summit– Aydan et al. (eds)
© 2019 Taylor & Francis Group, London, ISBN 978-0-367-34783-3

The dynamic response and stability of discontinuous rock slopes

Ö. Aydan
Department of Civil Engineering, University of the Ryukyus, Okinawa, Japan

Y. Ohta
Graduate School of Science & Technology, Tokai University, Shizuoka, Japan (Presently with Nuclear Energy Regulation Authority, Tokyo, Japan)

M. Amini
Department of Mining Engineering, Tehran University, Tehran, Iran

Y. Shimizu
Department of Civil Engineering, Meijo University, Nagoya, Japan

ABSTRACT: The dynamic response and stability of rock slopes during earthquakes are of great concern in rock engineering works such as highway, dam and nuclear power plant constructions. The main objectives of the study are to investigate the dynamic response of slopes and likely forms of instability of the slopes in relation to the number and orientation of discontinuity sets with respect to the slope geometry under various kinds of acceleration waves. The dynamic response of the model slopes were measured using accelerometers installed at various points in the slope. In the tests, various parameters such as the effect of the frequency and the amplitude of input acceleration waves are investigated in relation to discontinuity patterns and their inclinations and the slope geometry. Finally, the model slopes were forced to fail by increasing the amplitude of input acceleration waves and the forms of instability were investigated. In this study, the authors describe the results of the model tests on the dynamic response and stability of rock slopes and discuss their implications.

1 INTRODUCTION

The dynamic response and the stability of rock slopes associated with rock engineering works such as highway, dam and nuclear power plant constructions during earthquakes are of great concern. Fig. 1 shows some slope failure examples observed during recent earthquakes (e.g. Aydan 2016). As rock masses in nature generally have discontinuities due to various physical, chemical phenomena they underwent in the geologic past, there is a need to take into account the discontinuous nature of rock masses for evaluating the dynamic response and the stability of rock slopes.

An experimental study was undertaken to investigate the stability of rock-cut excavations during earthquakes. The main objectives of the study were to investigate the dynamic response and instability modes of the rock slopes in relation to the number and orientation of discontinuity sets with respect to the slope geometry under various kinds of acceleration waves. Blocks with two kinds of materials are used. The blocks with unbreakable material was intended to study the effects of disontinuities while the breakable blocks were intended to study the effect of block failure on the response and stability of discontinuous rock slopes. The results obtained from these experimental study on model rock slopes with various discontinuity sets are presented and discussed.

Figure 1. Examples of slope failures induced by earthquakes.

2 DEVICES AND MATERIALS

2.1 Shaking Table Devices

Two shaking test (ST) devices were used. The shaking table at Nagoya University(NU) was used for studying the response of models slopes for unbreakable material and the shaking table at University of the Ryukyus(UR) was used to study the response of models slopes with breakable material. Main features of the shaking table apparatusses are given

Table 1. Specifications of shaking tables

Parameters	NU Shaking Table	UR Shaking Table
Vibration Direction	Uni-axial	Uni-axial
Operation Method	Electro-oil servo	Magnetic
Table size	1300x1300	1000x1000
Load	30 kN	6 kN
Stroke	150 mm	100 mm
Amplitude	5G	0.6G
Wave Form	Harmonic, triangular, rectangular, random	Harmonic, triangular, rectangular, random

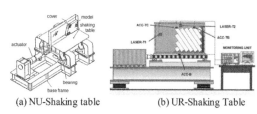

(a) NU-Shaking table   (b) UR-Shaking Table

Figure 2. Illustration of shaking tables and intsrumentation.

in Table 1. Fig. 2 shows a sketches of the devices together with the model mounted upon and instrumentations. Slope models were two-dimensional and were mounted upon the table through metal frames. The metal frame at NU-ST was 1200 mm long and 800 mm high while it was 1000 mm long and 750 mm high at UR-ST. The frame width was 100 mm wide at two shaking table experiments. Acceleration responses of the slope at several locations and input waves were measured using the accelerometers.

2.2 Model materials

2.2.1 Non-breakable materials
Blocks with dimensions of $10 \times 10 \times 100$ mm and $10 \times 20 \times 100$ mm were made of wood and used to simulate the discontinuity sets in rock masses.

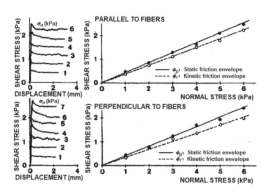

Figure 3. Shear tests on interfaces between wood blocks.

Direct shear tests were carried out on discontinuities between wood blocks and results together with shear strength envelopes are shown in Fig. 3.

2.2.2 Breakable materials
Breakable blocks are made of $BaSO_4$, ZnO and Vaseline oil, which is commonly used in base friction experiments (Aydan and Kawamoto, 1992). Properties of materials of blocks and layers are described in detail by Aydan and Amini (2009) and Egger (1983). Fig. 4 shows the variation of the strength of the model material with respect to compaction pressure. The material can be powderized and re-used after each experiment. The friction angle of interfaces between blocks are tested and shown in Fig. 5.

2.3 Testing procedure

The metal frames have some special attachments to generate different discontinuity patterns. The models were subjected to some selected forms of acceleration waves through a shaking table. The acceleration responses of model slopes were

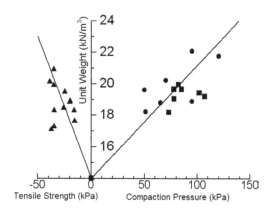

Figure 4. Variation of tensile strength of model material.

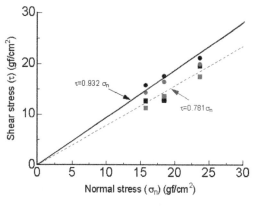

Figure 5. Shear tests on interfaces between breakable blocks.

measured using accelerometers installed at various points in the slope.

2.3.1 Non-breakable blocks

Model slopes were prepared by arranging wood blocks in various patterns to generate discontinuity sets with different orientations in space. Slope angles were 45, 63 and 90 degrees and the height and base width of model slopes were 800 mm and 1200 mm respectively. The intermittency angle ξ of cross joints were 0 and 45 degrees (Aydan et al. 1989) and one discontinuity set was always continuous as such sets in actual rock masses always do exist.

The inclination of the thoroughgoing (continuous) set was varied from 0 to 180 degrees by 15 degrees. At some inclinations, model slopes were statically unstable and at such inclinations no tests were done. Besides varying the inclination of the continuous set, the following cases were investigated:

CASE 1: Frequency was varied from 2.5 Hz to 50 Hz while the amplitude of the acceleration is kept at 50 or 100 gal.

CASE 2: The amplitude of the acceleration waves was varied until the failure of the slope occured, while keeping the frequency of the wave at 2.5 Hz.

2.3.2 Breakable blocks

The inclination of thoroughgoing discontinuity set was selected as 0, 45, 60, 90, 120, 135 and 180 degrees. Before forcing the models to failure in each test, vibration responses of some observation points in the slope were measured with the purpose of investigating the natural frequency of slopes and amplification through sweep tests with a frequency range between 3-40 Hz. Also, deflection of the slope surface was monitored by laser displacement transducers and acoustic emission sensors.

3 DYNAMIC RESPONSE AND STABILITY OF MODEL SLOPES

Various parameters such as the effect of the frequency and the amplitude of input acceleration waves are investigated in relation to discontinuity patterns and their inclinations and the slope geometry for the model slopes with non-breakable and breakable models. The model slopes were finaly forced to fail by increasing the amplitude of input acceleration waves and the forms of instability were investigated.

3.1 Natural frequency of model slopes

3.1.1 Model slopes with non-breakable blocks

Fig. 6 shows the amplification of waves measured at selected points in relation to the variation of input wave frequency The inclination of the throughgoing set for both discontinuity patterns

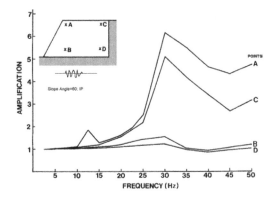

Figure 6. Variation of amplification with respect to frequency of model slopes and measurement locations.

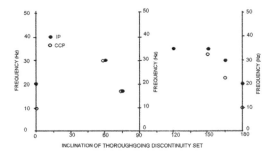

Figure 7. Variation of natural frequency of model slopes with respect to the inclination of thoroughgoing discontinuity set.

was 75 degrees.. The letter on each curve indicates the selected points within the model slopes. It is noted that if the natural frequency of the slopes exist, it varies with the spatial distributions of the sets and the structure of the mass.

In the followings, the frequency responses are discussed and compared for each respective inclination of the throughgoing discontinuity set for the point A (see Fig.6 for location) as shown in Fig. 7. The slope angle was 63 degrees in the sweep tests shown in Fig. 7. The results for each discontinuity set pattern are indicated in the figure for intermittent pattern as IP and for cross-continuous pattern as CCP.

Inclination 0: The natural frequency of the slope is 10 Hz for cross-continuous pattern and 20 Hz for intermittent pattern, respectively. Therefore, the natural frequencies of the slopes for intermittent pattern and cross-continuous pattern are different even the slope geometry and intact material are same. This may be related to the resulting slender columnar structure of the mass in the case of cross-continuous pattern.

Inclinations 15, 30, 45: The slopes for these inclination of the througoing set could not be tested as they were statically unstable for the slope angle of 60 degrees.

Inclination 60: From the figure, the natural frequencies for both patterns coincide and they have a value of 30 Hz. This may be attributed to the similarity of the structure of the mass for this inclination of the throughgoing discontinuity set.

Inclination 75: Natural frequences of the slopes for both patterns are almost same and it appears to have a value of 17.5 Hz. Similar reasoning as in the case of inclination of 60 can be stated for this case.

Inclination 90: No tests for this inclination could be made.

Inclination 120: Slopes having intermittent pattern were only tested as slopes having cross-continuous pattern could not be tested as they were statically unstable. For this inclination of the throughgoing set, the natural frequency of the slope has a value of 35 Hz.

Inclination 150: Natural frequencies of the slopes for both patterns are almost the same and it has a value of 35 Hz.

Inclination 165: The natural frequency of the slope is 22.5 Hz Hz for cross-continuous pattern and 30 Hz for intermittent pattern, respectively. In addition, the natural frequencies of the slopes for intermittent and cross-continuous pattern are different.

3.1.2 *Model slopes with breakable blocks*

Fundamentally, the vibration response of model slopes are quite similar to those of model slopes made with unbreakable blocks. Fig. 8(a) shows the input and measured wave forms at selected two points on the slope. The amplification of the vibration response is highest at the slope crest and the

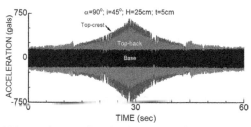

(a) input and measured acceleration responses in sweep tests

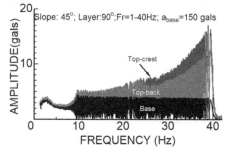

(b) Fourier spectra of acceleration responses in sweep tests

Figure 8. Acceleration responses of selected points on model slopes made with breakable blocks.

ampification at the top-back (ACC-TB) are a bit smaller than that at the slope crest as seen in Figure 8(b). From this figure, we can clearly state that amplification of the acceleration waves increases towards the slope (free) surfaces. In addition to this, the amplifications are larger at the top and have the maximum value at the crest of the slope (ACC-TC) as it was also noted in Figs. 6 and 7 for model slopes made with non-breakable blocks.

3.2 *Stability of model slopes: Failure tests*

When rock slopes are subjected to shaking, passive failure modes occur in addition to active modes (Aydan et al. 2009a,b, 2011; Aydan and Amini, 2009). Figs. 9 and 10 show examples of failures of some model slopes consisting of non-breakable and breakable blocks and/or layers. The experiments also show that flexural toppling failure of passive type occur when layers (60° or more) dip into valleyside.

The records of base acceleration and deflection of slope surface of the model are shown in Fig. 11 for a layer inclination of 90 degrees as an example. The acoustic emissions are also shown in the figure. Acoustic emissions starts to increase long before the displacement start to increase. This observation may also have an important implications for the monitoring of rock slopes. These responses were observed

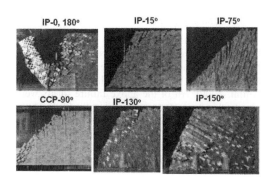

Figure 9. Failure modes of rock slope models with non-breakable material.

Figure 10. Failure modes of rock slope models with breakable material.

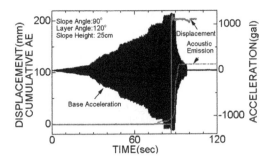

Figure 11. Acceleration, displacement and acoustic emission responses of a model slope.

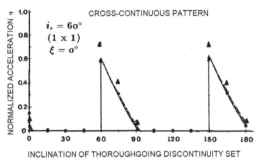

Figure 12. Comparison of theoretical estimations with experimental results (cross-continuous pattern).

in experiments on layered and blocky model slopes made with breakable blocks.

4 FAILURE MODES OF MODEL SLOPES

4.1 *Failure mode observations*

When the blocks of the mass are strong, failure modes involve only discontinuity sets (Aydan, 1989, 2015; Aydan et al. 1989, 1990). When failures involving only discontinuity sets, they are:

1. Sliding failure,
2. block toppling failure,
3. Combined toppling and sliding failure, and
4. Block buckling failure.

However, if the model slopes were made with breakable layers or blocks, flexural toppling or flexural block toppling failure are observed (Aydan and Kawamoto, 1992; Aydan et al. 2011; Aydan and Amini, 2009). All of the above failure forms were observed in our tests as seen in Figs. 9 and 10. These tests also showed that the above failure forms can also be subdivided into active and passive modes according to the vertical component of the displacement vectors with respect to that of the gravity.

4.2 *Analysis*

The authors developed a method of analysis for the stability of slopes under dynamic loadings using the limiting equilibrium concept. As the details of this method were described elsewhere (Aydan et al. 1989, 1991; Shimizu et al. 1986, 1988; Aydan and Kawamoto, 1992). Altough the method is not described herein, we compare the estimations by the method with the results of the tests. Figs. 12 and 13 show a comparison of the estimations by the method with the tests results for the slopes with an inclination of 60 degrees for both patterns. We note that predictions by the method are quite similar to the experimental results. However, the experimental results are above the predicted stable-unstable transition curves. This discrepancy is due to non-consideration of the amplification due

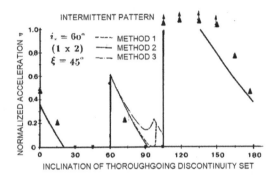

Figure 13. Comparison of theoretical estimations with experimental results (intermittent pattern).

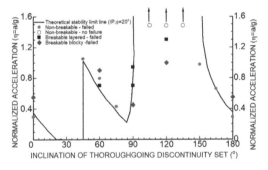

Figure 14. Comparison of theoretical estimations with experimental results (breakable layers and blocks).

to frequency content of the acceleration waves in the method as well as the inertial effects.

As explained in previous sections, model slopes are tested using breakable layers or blocks. The experimental results are plotted together with theoretical line computed for the stability of slopes having intermittent pattern and non-breakable material, which was published in a previous work by Shimizu et al. (1986). The theoretical line was computed for a slope with an inclination of 45 degrees and discontinuities to have a friction angle of 20 degrees. The stability line should act as a guideline for both layered and blocky rock mass with intermittent pattern for the upper limit of seismic resistance of the rock slopes. In other words, if the material constituting the slope fails, the slope will fail at lower acceleration levels. In accordance with this statement, the seismic resistance of the model slopes with breakable layers or blocks was either equal to theoretical lines or less than that. It should be noted that the friction angle for breakable material is much higher than the friction angle assumed in computations. Therefore, the resistance for sliding behaviour is expected to be higher than the theoretical line shown in Figure 14.

The model slopes indicated that flexural toppling or blocky toppling failure did occur. To analyse such failures, the methods proposed by Aydan and Kawamoto (1992) and Aydan et al. (1989) could be usefull. Aydan and Kawamoto (1992) have already proposed a method for analyzing the flexural toppling failure under dynamic loading condition and it can be easily used for active as well as passive mode of flexural toppling failure. Therefore, the available methods based on the bending theory of cantilever methods (i.e. Aydan and Kawamoto 1992; Amini et al. 2008) may be used for analyzing the experimental results in this study. Nevertheless, the most critical issue with these methods is how to designate the inclination of failure plane, above which layers are subjected to bending. Aydan and Kawamoto (1992) suggested the failure plane should be equal to the normal to layers for active flexural toppling failure and it was confirmed by experiments. The experiments presented in this study showed that inclination of the failure plane ranges between 0 and 15 degree above normal to discontinuities for active flexural toppling failure under dynamic condition. However, the inclination of the failure plane should be chosen through a minimization technique for the passive flexural toppling failure of rock slopes under dynamic condition.

5 CONCLUSIONS

The main objectives of the study were to investigate the dynamic response of the slopes and likely forms of instability of the slopes in relation to the number and orientation of discontinuity sets with respect to the slope geometry under various kinds of acceleration waves. In this particular study, some emphasis was put upon the effect of breakage of the layers/blocks used in creating model slopes.

The dynamic response of the model slopes were measured using accelerometers installed at various points in the slope. In the tests, various parameters such as the effect of the frequency and the amplitude of input acceleration waves are investigated in relation to discontinuity sets patterns and their inclinations and the slope geometry. Finally, the model slopes were forced to fail by increasing the magnitude of the input acceleration waves and their form of instability were investigated.

Test results show that slopes in discontinuous rock mass have a natural frequency, that depends largely upon the inclination of discontinuity sets rather than the properties of rock material. The failure modes slopes involving only discontinuities are:

1. sliding,
2. toppling and
3. combined sliding and toppling

However, each failure mode could be further subdivided to active mode and passive modes.

When layers or blocks constituting the model slopes are breakable under the induced stress state, some additional conclusions are as follow:

- If dip direction of layers discontinuities is approximately the same as that of the slope, the rock mass had a potential to fail in passive flexural toppling failure under dynamic condition.
- Flexural toppling failure under dynamic condition can also be classified into two modes: active and passive.
- As shown by Aydan and Amini (2009), experiments indicated that the required seismic coefficient is much higher for single column subjected to passive toppling as compared with that of columns subjected to active toppling.
- The experiments on model rock slopes also indicated that the required seismic coefficient is much higher for slopes subjected to passive toppling as compared with that of slopes subjected to active toppling.

REFERENCES

Aydan, Ö. (1989): The stabilisation of rock engineering structures by rockbolts. Doctorate Thesis, Nagoya University, 204 pages.

Aydan, Ö. (2008). New directions of rock mechanics and rock engineering: Geomechanics and Geoengineering. 5th Asian Rock Mechanics Symposium (ARMS5), Tehran, 3–21.

Aydan Ö (2016). Large Rock Slope Failures Induced by Recent Earthquakes. Rock Mech. Rock Eng., 49: 2503–2524.

Aydan 2017. Rock Dynamics. CRC Press, ISRM Book Series, No.6, 468 pages.

Aydan, Ö. and Kawamoto, T., (1992): Stability of slopes and underground openings against flexural toppling and their stabilization, Rock Mech. Rock Engng. 25 (3), 143–165.

Aydan, Ö., Amini, M. G. (2009). An experimental study on rock slopes against flexural toppling failure under dynamic loading and some theoretical considerations for its stability assessment. Journal of Marine Science and Technology, Tokai University, Vol. 7, No. 2, 25–40.

Aydan, Ö., Y. Ichikawa, Y. Shimizu and K., Murata (1991): An integrated system for the stability analysis of rock slopes, Proc. 7th IACMAG Conference, Cairns, 1, 469–474.

Aydan Ö., Ohta Y., Daido M., Kumsar H. Genis M., Tokashiki N., Ito T. & Amini M. (2011): Chapter 15: Earthquakes as a rock dynamic problem and their effects on rock engineering structures. *Advances in Rock Dynamics and Applications*, Editors Y. Zhou and J. Zhao, CRC Press, Taylor and Francis Group, 341–422.

Egger, P. (1983):A new development in the base friction technique. Coll. Phys. Geomech. Models, ISMES, 67–87, Bergamo.

Shimizu, Y., Aydan, Ö., Y. Ichikawa and T. Kawamoto (1986): A model study on dynamic failure modes of discontinuous rock slopes. Proc. Int. Symp. on Complex Rock Formations, Beijing, 183–189.

Shimizu, Y., Aydan, Ö., Y. Ichikawa and T. Kawamoto (1988): Dynamic stability and failure modes of slopes in discontinuous rock mass (in Japanese). Geotechnical Journal, JSCE, No.400–10, 189–198.

2019 Rock Dynamics Summit– Aydan et al. (eds)
© 2019 Taylor & Francis Group, London, ISBN 978-0-367-34783-3

Shake table experiments on masonry retaining wall reinforced with nails

Tsuyoshi Takayanagi
Disaster prevention Division, Railway Technical Research Institute, Tokyo, Japan

Naoki Sawada
Taisei Corporation, Tokyo, Japan (Formerly, Tokyo Institute of Technology, Tokyo, Japan)

Osamu Nunokawa
Disaster prevention Division, Railway Technical Research Institute, Tokyo, Japan

Akihiro Takahashi
Department of Civil and Environmental Engineering, Tokyo Institute of Technology, Tokyo, Japan

ABSTRACT: Many masonry walls have been constructed on slope along the Japanese railway. These walls are used as a surface protection on cut slopes and as retaining walls on embankments. Since these masonry walls are more brittle than the concrete retaining walls, the risk of collapse of masonry walls due to earthquake is high. For this reason, in order to improve railway safety against earthquakes, the reinforcement of masonry walls is required. Therefore, we develop a seismic reinforcement method for masonry walls and confirmed its effectiveness by centrifuge model experiments. The reinforcement method mainly consists of the penetration of the nailing into the slope and the solidification of the gravel backfill by the grout injection. As the result of the shake table experiments in the centrifuge, we confirm that the reinforced masonry wall does not collapse even under the shaking with a maximum acceleration greater than 1000 gal.

1 INTRODUCTION

1.1 Background

There are a lot of masonry walls (Fig. 1) that have constructed on slopes along the railway in Japan. A masonry wall is one of the structures commonly used as a slope protection work in early 20th century before reinforced concrete retaining walls becoming popular. Many of the old masonry walls remain not only in the local railways but also in urban areas including the metropolitan area of Japan. In the current railway design standards, use of reinforced concrete is required for the slope protection work. The sectional view of the standard masonry wall presented by the Ministry of Railways of Japan in the past is shown in Fig. 2 (Ota, 2008). Masonry walls have been applied to both cut slopes and embankment slopes and the main purpose of their construction is to prevent erosion, weathering and collapse of the slope surface. In the case of embankment, the thicker gravel backfill has been recommended as shown in the figure. These specifications in the standard masonry wall design have been empirically determined and no design calculation has been made.

The masonry wall consists mainly of the masonry blocks (stone blocks) and gravel backfill. The stone block which is shaped to a certain size is called "*Kenchi-ishi*", and the gravel backfill is called "*Uraguri*". The stone block generally has a shape close to a quadrangular pyramid, and the quadrangular face is placed in front. The gravel backfill is

an important structure that functions as a drainage layer for discharging groundwater in the slope.

There are two types in the gravel backfill. One is "*Neri-zumi*" in which gravels behind the stone blocks are mixed with mortar. The other one is "*Kara-zumi*" in which no mortar is used in the backfill. Since *Kara-zumi* is more brittle compared o *Neri-zumi*, the masonry walls with *Kara-zumi* are more vulnerable to earthquake. In the 2001 Geiyo Earthquake (magnitude of 6.7 in Richter scale), many *Kara-zumi* retaining walls in residential area were collapsed (Fig. 3, JSCE, 2001). In the large-scale shake table experiments conducted by Ota (2008), large bulging of the wall and subsidence in the gravel backfill were observed in the case of the *Kara-zumi* masonry wall.

Figure 1. Example of masonry wall remaining along the railway in urban area

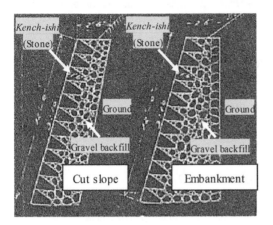

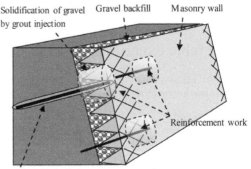

Figure 2. Sectional view of standard masonry walls (Ota, 2008)

Figure 4. Conceptual diagram of proposed improvement of pin-up method

Figure 3. Example of collapsed masonry wall (Geiyo Earthquake), (JSCE, 2001)

To reinforce existing masonry walls, it is possible to build a reinforced concrete wall in front of the existing wall. However, in the urban railways, the space available for seismic reinforcement is very limited and construction of additional structure in front of the existing wall is not feasible. Therefore, we have developed the pin-up method (RTRI, 2008) that does not require additional space in front of the existing masonry wall to reinforce the structure. In this method, a short reinforcing bar is installed to pin the wall to the ground (Fig.4). Furthermore, to increase the stiffness of the wall, a grout is injected into the gravel backfill around the reinforcing bar. Since the primal aim of this reinforcement method is to stabilize the masonry wall only, the standard length of the reinforcing bar is as short as about 1 m. However, when the soil behind the masonry wall is not stable against earthquake, the required length of the reinforcing bar can be longer.

1.2 *Objective*

In this research, we develop a new seismic reinforcement method that can be applied to the masonry

wall retaining unstable soil against earthquake. Effectiveness of the reinforcement method is examined by centrifuge model experiments. The type of masonry wall to which this reinforcement method shall be applied is *Kara-zumi* composed of *Kenchi-ishi*. Through this study, we intend to make scope of application of the pin-up method wider. The difference between the proposed method and the original pin-up method is the length of the reinforcing bar. In the proposed one, to stabilize the soil behind the masonry wall, the longer reinforcing bars are used, which is the same as the soil nailing (NEXCO, 2007). To examine the effectiveness of the reinforcement, response of the reinforced walls are compared to that without reinforcement.

2 MODEL EXPERIMENT

2.1 *Centrifuge model experiment*

In this study, we conducted shake table experiments in a centrifugal acceleration field. The centrifuge model experiment is a physical model test on a 1/n-scaled model in a centrifugal acceleration field whose gravitational acceleration is n times. In this experiment method, the stress state corresponding to the prototype is simulated in the small-scaled model. Table 1 summarizes major scaling factors in the centrifuge model experiment. The Mark III Centrifuge at Tokyo Institute of Technology was used in the experiments.

2.2 *Small-scaled model of masonry wall*

Cross-sectional view of the model is shown in Fig. 5. Since Ota (2008) reported that the height of the existing masonry walls for railways are typically less than 5 m, the target height of the masonry wall is set 4.5m in prototype scale. The centrifuge model experiments were carried out at 25 G centrifugal acceleration field and dimensions of the model were scaled down to 1/25 of the prototype. The model

Table 1. Correspondence between the centrifugal acceleration ratio and each physical property

Physical property	Model	Prototype
Length L	1	n
Area A	1	n^2
Volume V	1	n^3
Weight of unit volume ρ	1	$1/n$
Stress σ	1	1
Strain ε	1	1
Displacement d	1	n
Acceleration α	1	$1/n$
Flexural rigidity of reinforcing EI	1	n^4

Table 2. Properties of soil used

	Dry Density γ_d (kN/m³)	Compaction degree Dc (%)	Water content W (%)	Friction angle Φ (°)	Cohesion c (kN/m²)
Soil behind wall	14.1	85	18	36.8	2.0
Foundation ground	15.8	95	18	—	

Table 3. Masonry wall conditions in prototype

Slope Height	4.5 m
Slope gradient	1:0.3
Size of *Kenchi-ishi*	0.3 m × 0.3 m × 0.5 m
Thickness of gravel backfill	0.3 m
Particle size of gravel	About 0.08 m
Structure of wall	*Kara-zumi* (not mixed mortal)
Soil behind wall	Low stability

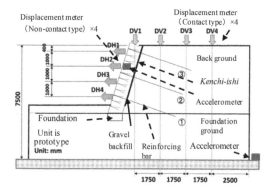

Figure 5. Sectional view of masonry wall model

ground was constructed in a rigid steel box having a length L = 600 mm, a width W = 122 mm and a height H = 400 mm.

The soil retained by the masonry wall was constructed by using partially saturated *Edosaki* sand (particle density ρ_s = 2.72 Mg/m³, mean particle diameter D_{50} = 0.29 mm, effective particle diameter D_{10} = 0.14 mm, maximum dry density ρ_{dmax} = 1.66 Mg/m³, optimum water content w_{opt} =17.8%) with compaction degree Dc of 85%. The foundation ground was constructed using the same material with compaction degree Dc of 95%. Table 2 summarizes properties of the soils used.

The target *Kenchi-ishi* (stone block) for the masonry wall was the quadrangular pyramid stone with 0.3 m × 0.3 m quadrangular in front and 0.5 m in depth. To adjust the density of the stone, the cement mixed with iron powder was used as a material. The dry density of the *Kenchi-shi* model was 2.4 Mg/m³. For the gravel backfill, cobble stones with large diameter with no adhesive strength are often used. Therefore, in this experiment, Silica sand No. 3 (soil particle density ρ_s = 2.65 Mg/m³, mean particle diameter D_{50} = 1.72 mm, effective particle diameter D_{10} = 1.37 mm) was used to model the material for the gravel backfill. Table 3 summarizes the masonry wall conditions in prototype.

2.3 Reinforcing bar

Table 4 summarizes specifications of the reinforcing bar in prototype scale. According the scaling laws for the centrifuge model experiments, the flexural rigidity EI of the model reinforcing bar shall be $1/n^4$ of that in prototype scale. Since it was not feasible to model it as a round bar, the steel strip of 0.5 mm in thickness, 3 mm in width and 200 mm in length was used to model the reinforcing bar (see Fig. 6).

To measure the axial force acting on the reinforcing bar, three pair of strain gauges were attached to the steel strip at three points of 47 mm, 93 mm and 149 mm from the back of the stone block in model scale. In the improved pin-up method, as the reinforcing bar and stone blocks shall be fixed by the grout, the steel strip and model stone blocks were fixed by using plaster. Before inserting the model reinforcing

Table 4. Specifications of the reinforcing bar in prototype scale

Deformed reinforcing bar	Nominal diameter: 21 mm
Material of nailing core	SD345
Length of Reinforcing bar	4 m

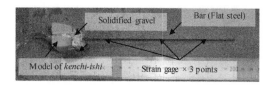

Figure 6. Model reinforcing bar

bar into the soil, 4 mm diameter pre-boring and grouting using plaster were conducted.

Here, prior to the model experiment, a pull-out tests were carried out in order to examine the pull-out resistance (maximum tensile force T_{max}) of the model reinforcing bar. The results of the pull-out tests are shown in Fig. 7. In the pull-out test, three patterns of surcharge pressure (p=5 kPa, 10 kPa, and 15 kPa) were applied to the ground and the reinforcing bar was pulled-out at 2.0 mm/min, and the relation between the maximum tensile force T_{max} and the surcharge pressure p was obtained. Although the resistance did not linearly increase with the surcharge pressure, it could be confirmed that a certain resistance is mobilized along the model reinforcing bar.

2.4 Experiment conditions

In this study, we conducted three experiments in total, with different reinforcement on the masonry wall. The outline of the reinforcing patterns of the masonry wall is shown in Fig. 8. No reinforcement was applied in Case 1 to observe collapse behavior on of the non-reinforced masonry wall. In Cases 2 and 3, the masonry wall was reinforced by the improved pin-up method. In Case 2, the spacing between the reinforcement was 1.7 m, while that was 1.2 m in Case 3 in prototype scale.

2.5 Experiment procedures

In the experiment, we first build the foundation ground and the soil retained by the wall with tar-

geted compaction degree Dc, using the *Edosaki* sand with the target water content. Then excavated the ground to the shape shown in Fig. 5. Next, the masonry wall and the gravel backfill were constructed from the lower position of the cut slope. In the cases with reinforcement, the reinforcing bar was inserted into the ground at the predetermined height during construction of the masonry wall. The installation procedure of the reinforcing bar is described in Subsection 2.3.

After constructing the model, we placed displacement transducers in the model box (Fig. 5). After setting all the measurement sensors, we spun up the centrifuge to 25 G and applied earthquake motion shown in Subsection 3.1.

3 EXPERIMENT RESULTS

3.1 Applied earthquake motion

We applied earthquake motion used in the seismic design of railway structures. The earthquake motion used was G2 ground motion for spectrum II specified in "Design Standard for Railway Structures "in Japan (RTRI, 2006). In the experiment, three motions having the same waveform but different amplitude were applied to the model. In STEP 1, the target maximum acceleration was about 100 gal in prototype scale. In STEP 2, that was about 500 gal. And in STEP 3, the maximum acceleration was around 1400 gal. The acceleration waveform in STEP 3 of Case 1 is shown in Fig. 9.

Table 5 summarizes the actual maximum acceleration α_{max} and velocity V_{max} of the motion applied to the model in prototype scale. In Table 6, examples of α_{max} and V_{max} in the major earthquakes in Japan are summarized (GIROJ, 2009).

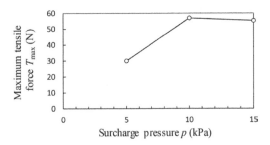

Figure 7. Pull-out resistance of model reinforcing bar

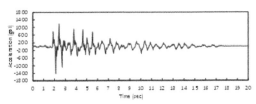

Figure 9. Acceleration waveform in STEP 3 of Case 1 in prototype scale

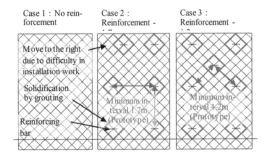

Figure 8. Reinforcing patterns of masonry wall (front view)

Table 5. Maximum acceleration and velocity of applied motion in prototype scale

STEP of vibration		STEP 1	STEP 2	STEP 3
Case 1	α_{max}(gal)	99	488	1410
	V_{max}(cm/sec)	7.5	26	51
Case 2	α_{max}(gal)	84	361	1394
	V_{max}(cm/sec)	6.7	20	50
Case 3	α_{max}(gal)	98	435	1419
	V_{max}(cm/sec)	7.4	25	50

Table 6. Examples of maximum acceleration and velocity in major earthquakes in Japan (GIROJ, 2009)

Earthquake	α_{max} (gal)	V_{max} (cm/sec)
Miyagioffshore (2003) Ounato	1106	32
Geyo (2001) Yuki	830	31
Hyogo (1995) Ashiya	891	112
Grate East Japan (2011) Tsukidate	2933	106
Kumamoto (2016) Mashiki	1362	133

Comparison between Tables 5 and 6 reveals that the maximum acceleration applied in the experiment was comparable to those in the past earthquakes, while the maximum velocity was rather smaller because the applied motion was rich in high frequency components.

3.2 Deformation of non-reinforced masonry wall (Case 1)

In Case 1 in which no reinforcement was applied, deformation of the masonry wall cumulatively progressed with application of the earthquake motion. Particularly in STEP 3, marked settlement occurred in the gravel backfill and the masonry wall showed bulging (see Fig. 11). Furthermore, after STEP 3, a slip surface was confirmed in the soil behind the wall. Figure 10 shows the masonry wall before and after STEP 3 in Case 1. It can be confirmed that large permanent deformation of the masonry wall and settlement of the ground surface, which is comparable to those observed in actual earthquake-induced damage of masonry walls.

3.3 Mitigation of permanent deformation of masonry wall with reinforcement (Cases 2 and 3)

Figure 12 shows side view of the masonry wall before and after STEP 3 in Case 2. In addition, Fig. 13 shows time histories of the horizontal wall displacement and ground surface settlement in STEP 3 of Case 2 in prototype scale. It is confirmed that the horizontal wall displacement dH at DH2 in Case 2 is

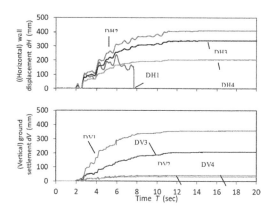

Figure 11. Time histories of horizontal wall displacement and ground surface settlement in STEP 3 of Case 1 in prototype scale

reduced to about 1/40 of that in Case 1. Furthermore, the ground surface settlement dV at DV1 in Case 2 is also reduced to about 1/6 of that in Case 1. It can be said that the improved pin-up method can restraint the permanent deformation of the masonry wall and can prevent formation of the slip deformation in the soil behind the wall due to earthquake.

The relationships between the cumulative horizontal displacement ΣdH at DH3 and the maximum acceleration α_{max} of the motion applied are shown in Fig. 14. Comparisons between Case 2 and Case 3 reveal no big difference between the two cases. However, the accumulated horizontal displacement ΣdH of Case 3 is slightly larger than Case 2 whose reinforcement density is slightly low. This reason is thought that the seismic force is slightly larger in Case 3 and/or the possibility that the variation of grouting for reinforcing bar has an influence.

3.4 Mobilization of tensile force in reinforcing bar

Figure 15 shows the axial force distributions of the reinforcing bar (Reinforcing bar ② shown in Fig. 5) installed in the middle of the masonry wall in STEP 3 of Case 3. Mobilized axial force during earthquake is larger than that before shaking. This suggests that the

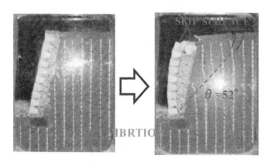

Figure 10. Side view of masonry wall before (left) and after (right) STEP 3 in Case 1

Figure 12. Side view of masonry wall before (left) and after (right) STEP 3 in Case 2

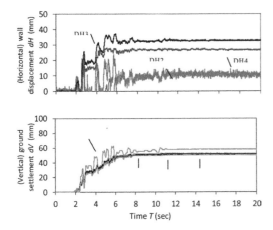

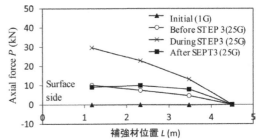

Figure 15. Axial force distributions of reinforcing bar ② in STEP 3 of Case 3 in prototype scale

Figure 13. Time histories of horizontal wall displacement and ground surface settlement in STEP 3 of Case 2 in prototype scale

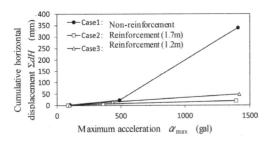

Figure 14. Relationships between cumulative horizontal displacement at DH3 and maximum acceleration of motion applied

reinforcing bars tied the masonry wall the soil behind and the deformation of the soil was restrained by the reinforcing bars during earthquake. From previous research, it is known that the axial force distribution could change according to the stiffness of the slope protection work connected to the reinforcing bars and, in the case where the stiffness of the slope protection work is particularly large, the axial force near the surface tends to increase (NEXCO, 2007). The axial force of the reinforcing bar near the surface is also large in this experiment, which suggests that the stiffness of the wall becomes larger due to the solidification of the gravel backfill by the improved pin-up method.

4 CONCLUSIONS

In this study, a centrifuge model experiments were conducted to confirm effectiveness of the reinforcement of the masonry wall by the improved pin-up method. The conclusions drawn are summarized as follows:

1. The centrifuge model experiment on the non-reinforced masonry wall reveals that (a) the non-reinforced masonry wall exhibits noticeable deformation during earthquake whose maximum acceleration is about 500 gal and (b) deformation becomes large once the maximum acceleration exceeds about 1000 gal.

2. When stability of the soil retained by the masonry wall is not high as in the case without reinforcement, marked earthquake-induced permanent deformation of the soil behind the wall occurs with formation of slip surface.

3. No marked deformation is observed in the masonry wall reinforced with the improved pin-up method even under the earthquake whose maximum acceleration exceeds 1000 gal.

4. No marked difference can be seen between the case with 1.7 m reinforcing bar spacing and that with 1.2 m spacing. But, at least, it can be said that the reinforcement with 1.7 m spacing performs better even under the earthquake whose maximum acceleration exceeds 1000 gal.

5. Observed distribution of the axial force of the reinforcing bar is comparable to the case with the large stiffness slope protection work in the previous research. This indicates that the reasonable increase in the stiffness of the wall can be expected when the wall is reinforced by the improved pin-up method.

REFERENCE

Naoyuki Ota. 2008. Study on stability evaluation of masonry wall using pyramidal stone and reinforcement method against earthquake. *Kokushikan University, Doctoral thesis*
Geiyo Earthquake Damage Investigation Team of Japan Society of Civil Engineers. 2001. March 4th 2001 Geiyo Earthquake Damage Survey Report. *Japan Society of Civil Engineers*
Railway Technical Research Institute. 2008. Design and construction manual of Pin-Up method. *Railway Technical Research Institute*
NEXCO. 2007. Design and construction manual of cut soil reinforcing method, *Nippon Expressway Company Limited.*
Railway Technical Research Institute. 2006. Design Standards for Railway Structures and Commentary (Seismic Design), *Railway Technical Research Institute*
General Insurance Rating Organization of Japan. 2009. A Study on Earthquake Damage Prediction Method of Household Properties, Part 2, Chapter 2. *General Insurance Rating Organization of Japan*

2019 Rock Dynamics Summit– Aydan et al. (eds)
© 2019 Taylor & Francis Group, London, ISBN 978-0-367-34783-3

Study on the stability of stone wall in earthquake by discontinuous deformation analysis

G.C. Ma
Sabo Disaster Prevention Division, OYO Corporation, Saitama, Japan

M. Nakashima
Zivil Survey design corporation, Fukui, Japan

K. Arai
NPO regional geotechnical disaster prevention institute in Fukui, Fukui, Japan

ABSTRACT: Old-time stone walls are important cultural assets in Japan. There are many cases of large deformation or collapse of stone walls caused by earthquakes in the past. In order to maintain the seismic stability of stone wall, good repair of large deformed wall is an essential project. In the repair, it is a big issue whether the reinforcement of wall is employed or not. If reinforcement is not applied, a risk remains that the stone wall deforms largely again when encountering a big earthquake. If reinforcement is applied, it is required how to enhance the seismic resistance by the reinforcement. To examine these issues, we carry out a numerical analysis of stone wall by the discontinuous deformation analysis (DDA). In the analysis, a large deformed stone wall is reproduced, and the material parameters for analysis are determined. Based on the results, the seismic stability of the stone wall is examined for the case of employing reinforcement by soil improvement.

1 INTRODUCTION

In Japan, there are many classical stone walls that experienced hundreds of years after construction. In order to maintain and manage these traditional structures in the future, it is important to properly evaluate the stability of structure. Furthermore, in Japan where earthquakes occur frequently, there are many stone walls that cause large deformation and lead to collapse due to earthquake. In order to maintain the long-term stability of traditional structures, a method to accurately evaluate the stability on earthquake is required.

In recent years, many studies on the method for evaluating the stability of stone wall have been carried out. As the example of past research, an exit index which is an empirical index (Nishida, 1998), a method based on earth pressure theory (Ichioka, 1996) and a numerical analysis (Noma, 2013), have been reported. However, at the present stage it is difficult to establish a method for quantitatively evaluating the stability of stone wall based on the deformation of wall during earthquake which considers the waveform of seismic motion.

In this paper, we investigate the necessity of reinforcement when repairing or rebuilding a stone wall where collapse occurs due to past earthquakes and large deformation exudes due to soil pressure, and we examine the method for reinforcement for stone wall.

2 STRUCTURE OF STONE WALL

2.1 Current status

Figure 1 shows the current deformation of the stone wall. Point A in Fig. 1 (b) is sinking, and point B is staring. From this fact, it can be easily inferred that the ground between point A and B is loose, and a slip surface may exist between point A and B, the shape of which is unknown. For the stone wall shown in Figure 1, the rebuilt repair work was planned. It became an issue whether reinforcement should be carried out or not, when rebuilding. We perform a numerical analysis with consideration for this stone wall.

2.2 Collapse due to earthquake

On June 28, 1948, an earthquake of magnitude 7.1 with epicenter of Fukui Plain occurred. The earthquake ground motion was intense, and in the vicinity of the epicenter almost 90% of houses were destroyed. Due to this earthquake, collapse occurred in several stone walls adjacent to the stone wall shown in Fig. 1. Figure 2 shows an example of the collapse. By reproducing this collapse, we confirm the applicability of our numerical analysis later.

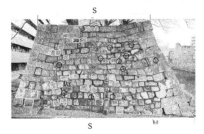

(a) Front view

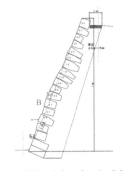

(b) Cross section of S-S

(c) Enlarged view of section S-S

Figure 1. Deformation of stone wall.

Figure 2. Stone wall collapsed in Fukui earthquake, 1948.

3 METHOD OF ANALYSIS

3.1 Method

The analysis method is required to express the state of stone wall at all times, i.e., stability during earthquake or collapse at earthquake. The analysis is also required to represent the process from the stable state to the collapse of stone wall during earthquake. Since the discontinuous deformation analysis (DDA) is considered to satisfy these conditions (Monma, 1994), we reproduce the collapse of stone wall during earthquake, and examine the effect of reinforcement by soil improvement.

3.2 Self-weight analysis and earthquake analysis

We perform the analysis at three stages of self-weight static analysis, self-weight dynamic analysis and earthquake analysis.

Self-weight analysis is separated to static analysis and dynamic analysis. The self-weight static analysis is performed by stepwise analysis where the initial block velocity at each step reaches zero.

In the self-weight dynamic analysis, the initial block velocity at each step is inherited from the final speed at the preceding step.

Earthquake analysis inherits the stress and deformation of the self-weight dynamic analysis, and performs the analysis by applying the seismic force for each step.

3.3 Dynamic analysis

The external force in the seismic vibration analysis may be "own weight and seismic force". There are two kinds of method considering seismic force. One is the static seismic intensity method and the other is the dynamic seismic force method. In the static seismic intensity method, a constant inertia force is uniformly applied to all blocks in a horizontal or a vertical direction at a fixed time. In the dynamic seismic force method, the observed acceleration waveform of actual earthquake is applied to the box-shaped block outside the model. In this paper, the latter method is adopted in order to analyze the deformation of the stone wall during earthquake.

4 REPRODUCTION ANALYSIS

4.1 Analysis model

Figure 3 shows the analysis model used for the reproduction analysis. In Fig. 3, area A is a stone wall made of the shape shown in Fig. 1 (b), but the surface shape is restored to the shape before ingestion. Area B is a chestnut stone, and is assumed a discontinuous surface which coincides with the backfill ground and constitutes blocks. Since the chestnut stone part is finer than the stone wall part, the constituent block is divided into smaller parts than C. Area C is the backfill ground behind stone walls, and the block is divided with an arcuate discontinuity. Area D is the same ground as C, but it is set as a different category from C because the ground surface is a road. Area E is a box type that surrounds the ground of the model and that is a block simulating the ground around the model.

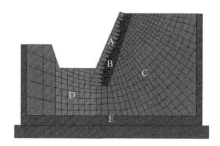

Figure 3. Division of stone wall area. A: stone wall, B: chestnut stone, C: backfill behind stone wall, D: same ground as C (ground surface is a road), E: box type that surrounds the ground of the model.

Table 1. Material property values used for reproduction analysis.

Category	Unit volume weight (kN/m³)	Elastic coefficient (MPa)	Poisson's ratio (-)	Friction angle (°)	Adhesive force (kN/m²)
A	26	1,000	0.25	35	0
B	26	100	0.25	35	25
C	19	100	0.3	30	25
D	19	100	0.3	30	50
E	19	10,000	0.3	30	0

During earthquake analysis, we apply seismic force to the center of these blocks as inertia force. The block indicated by E is a slide, and constitutes a horizontal roller.

4.2 Material properties

Table 1 shows the material property values used for analysis. Since the reproduction analysis is a basic study, the material property values are set by referring to the past literature.

4.3 Seismic waveform

In the analysis, inertia force is applied in the X direction of the box-shaped block from outside of the model. As the seismic waveform, we used the observed acceleration waveform (maximum value 500 gal) in the Kobe earthquake.

4.4 Result of reproduction analysis

Figure 4 shows the collapse process of the stone wall during the earthquake, which is obtained by the DDA. Figure 5 shows the time history of the maximum displacement rate obtained by the analysis.

As shown in Fig. 4, (a) starts to collapse, (b) shows large deformation, (c) - (e) shows collapse. In Fig. 4 (b), point A is settled and point B is impregnated, which is consistent with Fig. 1 (b). Fig. 4 (e) is consistent with the collapse mode shown in Fig. 2.

Figure 5 (a) shows the time history of the maximum displacement rate obtained by analysis, and

(b) shows the input seismic waveform. In Fig. 5 (a), time 0 - 2.5 s is the self-weight static analysis, and time 2.5 - 2.8 s is the self-weight dynamic analysis, both of which are self-weight analysis. In self-weight analysis, the maximum displacement rate converged to 10⁻³ or less. After 2.8 s, the earthquake

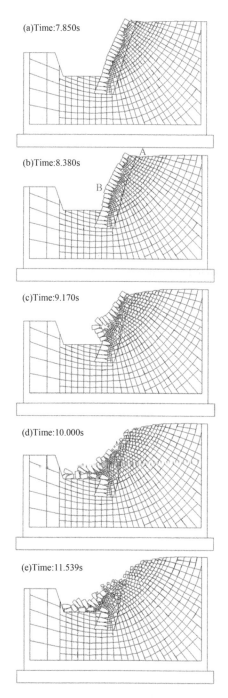

(a)Time:7.850s

(b)Time:8.380s

(c)Time:9.170s

(d)Time:10.000s

(e)Time:11.539s

Figure 4. Collapse process of stone wall during earthquake obtained by DDA

546

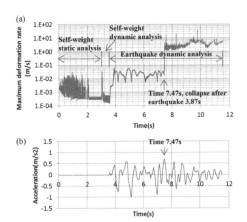

Figure 5. Time history of maximum displacement rate obtained by analysis. (a) time history, (b) input seismic waveform.

dynamic analysis is shown. At time 7.47 s (3.87 s after the start of earthquake) the maximum displacement rate increased sharply. This time may be considered the start time of collapse. This time matches the maximum value of the acceleration in the plus direction in the input seismic waveform shown in Fig. 5 (b).

Based on the above results, it seems that the analysis reproduces the constant stability of stone wall and the collapse due to earthquake.

5 REINFORCEMENT ANALYSIS

From the results of reproduction analysis, there is a possibility that the stone wall shown in Fig. 1 will collapse in the future due to earthquake, if the wall is rebuilt only by restoring the surface shape.

Referring to the collapse process shown in the reproduction analysis, the collapse of stone wall during earthquake is considered due to the failure of the backfill ground behind the stone wall. Then improvement of the ground behind the stone wall is required.

Figure 6 shows the analytical model of the ground improvement scheme. In Fig. 6, area A: stone wall, B: chestnut stone part, C-G: ground, H: a box type block that surrounds the ground of the model, and I: ground improvement part. Distribution and material property values of C - G are set from the results of the laboratory soil test of the boring core and the surface wave survey of the site. Table 2 shows the material property values of the block and discontinuous surface used for analysis.

According to the residential land disaster prevention manual (Residential disaster prevention study group, 2007.), if the ground surface acceleration of a large earthquake is assumed the order of 400 to 500 gal, the design horizontal seismic intensity is considered to be about 0.25. In the reinforcement analysis, considering

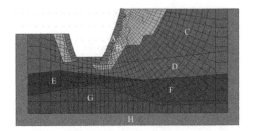

Figure 6. Analytical model of soil improvement plan. A: stone wall, B: chestnut stone, C-G: ground, backfill, H: box type block that surrounds the ground of the model, and I: soil improvement part.

Table 2. Material property values of block and discontinuous surface used for reinforcement analysis.

category	Unit volume weight (kN/m3)	Elastic coefficient (MPa)	Poisson's ratio (-)	Friction angle (°)	Adhesive force (kN/m2)
A	22	100	0.25	45	0
B	20	30.0	0.25	40	10
C	18	16.8	0.3	15	30
D	18	14.0	0.3	20	35
E	18	14.0	0.3	20	75
F	18	16.8	0.3	20	80
G	18	47.6	0.3	30	30
H	18	1,000	0.3	0	0
I	18	28.0	0.3	15	50

the safety side, the maximum acceleration of the input ground motion is set to 500 gal.

For the analysis, we use the observed acceleration waveform in the Kobe earthquake, which is shown in Fig. 7. In the analysis, horizontal waveform is NS and vertical is UD. As shown in the recorded waveform, the maximum acceleration in the vertical direction is 0.41 to 0.54 times smaller in the

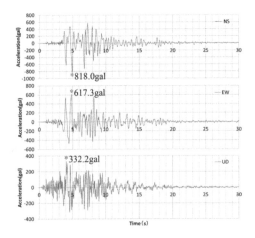

Figure 7. Observed acceleration waveform of Kobe earthquake.

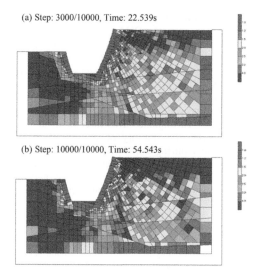

Figure 8. Distribution of real-time local safety factor obtained by reinforcement analysis. (a): After self-weight analysis, (b): After earthquake dynamic analysis.

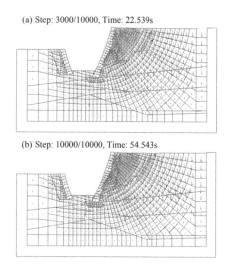

(a) Step: 3000/10000, Time: 22.539s

(b) Step: 10000/10000, Time: 54.543s

Figure 9. Distribution of cumulative displacement vector obtained by reinforcement analysis. (a): After self-weight analysis, (b): After earthquake dynamic analysis.

horizontal direction. In the analysis, maximum value 300 gal in the vertical direction is used compared to 500 gal in the horizontal direction.

Reinforcement analysis was carried out in the same order as the reconstruction analysis, self-weight static analysis, self-weight dynamic analysis and earthquake dynamic analysis.

Figure 8 shows the distribution diagram of Real-time local safety factor obtained by reinforcement analysis. In Fig.8, (a): after self-weight analysis, and (b): after earthquake dynamic analysis.

After the earthquake, the safety factor of the lower part of soil improvement section decreases constantly (after self-weight analysis), but there is no continuous decrease. The stone wall and the backfill ground are stable during earthquake.

Figure 9 shows the distribution of cumulative displacement vector obtained by reinforcement analysis. In Fig.9, (a): after self-weight analysis, (b): after earthquake dynamic analysis. The cumulative displacement vector distributes in a direction similar to the bulge form observed in the actual stone wall. This result is consistent with the actual deformation of stone wall. The cumulative displacement vector distributes regularly, and there is no point where the vector becomes extremely large. As the result, it can be inferred that both the stone wall and ground are stable after the earthquake.

After the earthquake, the cumulative displacement and displacement rate in the lower part of stone wall are sorted out to confirm the cumulative amount of displacement of the stone wall. Figure 10 shows the cumulative displacement and displacement rate of block 21. In Fig.10, (a): the position of block 21, (b): cumulative displacement, (c): displacement rate. In Fig.10 (a), U_x is the cumulative displacement in

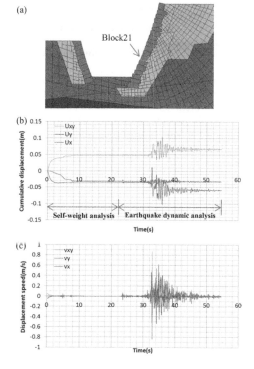

Figure 10. Cumulative displacement and displacement rate of block 21 obtained by reinforcement analysis. (a): The position of block 21, (b): Cumulative displacement, (c): Displacement rate.

horizontal direction, U_y is the cumulative displacement in vertical direction, and U_{xy} is the cumulative displacement vector. The cumulative displacement due to the earthquake increases by about 26.2 mm in horizontal direction before the earthquake (after self-weight analysis). There is a decrease of about 2.3 mm in vertical direction. It is confirmed that the cumulative displacement after the earthquake clearly converges.

In Fig.10 (b), vx is the displacement rate in horizontal direction, vy is the displacement rate in vertical direction, and vxy is the displacement rate vector. As similar with the cumulative displacement, it is confirmed that the displacement rate after the earthquake converges clearly.

It is possible to confirm that the stone wall and ground are stable during the earthquake, from the distribution of local safety factor, the distribution of cumulative displacement vector, the cumulative displacement, and displacement rate of the lower part of stone wall. As the result, we confirm the effectiveness of soil improvement.

6 CONCLUSION

It is an important task of disaster prevention to evaluate the stability of ground structure during earthquake. Numerical analysis is a powerful method for the evaluation. It is necessary to grasp accurately ground condition, to construct an appropriate model and to select an appropriate numerical method.

In ground structures like stone wall, stone wall is discontinuous body, and ground is continuum. Numerical methods targeting continuum are difficult to apply to stone wall. It is difficult to apply numerical methods targeting discontinuous object to ground. In this paper, we employ DDA. It is important to divide the ground into discontinuous blocks when applying DDA. Therefore, the stress state obtained by FEM is grasped beforehand, and the ground block is divided by arc discontinuity line with the slip line considered possible as the potential crack.

By using DDA, we can reproduce the deformed state of stone wall and the collapse of stone wall during the earthquake. Reinforcement analysis also confirms the stability of stone wall by soil improvement during earthquake. As the result, it is confirmed that DDA is effective for the stability analysis of stone wall during earthquake, and it is shown how to apply DDA to stone wall structure.

REFERENCES

Nishida, K. 1998. Engineering proposal on the stability evaluation method of the castle stone wall as historical heritage, symposium on maintenance, repair and management related to the ground, pp27-32.
Ichioka, T., 1996. Mechanical devises of traditional sedimentary technique, Kozo Kogaku Ronbunshu. A (Journal of Structural Engineering. A) of the Japan Society of Civil Engineers, Vol.42A, 519-526.
Noma, Y., 2013. Study on seismic deformation prediction and dynamic stability evaluation of castle masonry wall, Journal of Japan Society of Civil Engineers, Ser. C (Geosphere Engineering), Vol.69, No.4, 444-456.
Monma, K., 1994. Study on the application of discontinuous deformation analysis to evaluate the mechanism of rock slope instability, Doboku Gakkai Ronbunshu, No.757/III-66, 45-55.
Residential disaster prevention study group, 2007. explanation of residential disaster prevention manual, 96-97.

2019 Rock Dynamics Summit– Aydan et al. (eds)
© *2019 Taylor & Francis Group, London, ISBN 978-0-367-34783-3*

Seismic response analysis using by 4-node iso-parametric NMM and 3D-DDA

Y. Ohnishi
Kyoto University, Kyoto, Japan

R. Hashimoto
Hiroshima University, Hiroshima, Japan

T. Sasaki
Suncoh Consultants Co. Ltd., Tokyo, Japan

S. Miki and N. Iwata
Kiso-Jiban Consultants Co. Ltd., Tokyo, Japan
Chuden Engineering Consultants, Hiroshima, Japan

ABSTRACT: In the discontinuous models as DDA and NMM, the earthquake wave is propagating from the base block to the upper blocks by the friction between blocks in theoretical. The experiments of response on the artificial structure models has been applied by prescribed displacement on the actuator of excitation apparatus. The authors report some factors in earthquake motions affects in modeling described above for the discontinuous models by using 4-node iso-parametric NMM (Sasaki & Ohnishi, 2001) and 3D-DDA (Shi, 2001).

1 INTRODUCTION

The characteristics of structures on resonance in the earthquake response analysis is depends on the frequency of the input wave motions and the own natural period of the structures. The motion of the earthquake is input by the accelerations or the prescribed displacements on time histories in general. The earthquake motion is recorded by seismographs on accelerations on site in usual. The response of the foundation model is dependent on the stratum of elastic modulus and the boundary condition of analytical regions. The viscous dampers models are introduced around the boundaries in the analytical regions in some times. In this case, the response of earthquake motion of foundation model is dependent on its boundary condition and the boundaries are not clear and ambiguous in physically. The other hand, the experiments of response on the artificial structure models has been applied by prescribed displacement on the actuator of excitation apparatus. In this case, the boundary condition is clear and in depends. In the discontinuous models as DDA and NMM, the earthquake wave is propagating from the base block to the upper blocks by the friction between blocks in theoretical. And the superfluous force is disappearing by the slip motions between blocks. In this time, the precision of fiction forces depending penalty should be calculating on the strength of yield surface (Hashimoto et al., 2016). The authors report some factors in earthquake motions affects in modeling described above for the discontinuous models by using 4-node iso-parametric NMM (Sasaki & Ohnishi, 2001) and 3D-DDA (Shi, 2001).

2 FOUR NODE ISO-PARAMETRIC ELEMENT NUMERICAL MANIHOLD METHOD

2.1 *Outline of the theory*

In the Manifold method, the finite covers which is define the displacement are overlapped on the material blocks which contain discontinuous planes in an analytical area, as shown in Figure 1. The finite covers are shape functions, which are disconnected on the two sides of the discontinuous planes. The equilibrium equation is solved dynamically with contact problem by Penalty method.

The total potential energy of the global elastic block system in large deformation on dynamic analysis with contacts Π^{sys} is explained by Eq. (1).

$$\Pi^{sys} = \sum_{i=1}^{n} \Pi^{(block)i} = \sum_{i=1}^{n} \left(\Pi^i + \sum_{j=1}^{m} \Pi_{PL}^{i,j} \right) \quad (1)$$

The first term of the right hand of Eq. (1) shows the potential energy of continuum of each block and the second term shows the potential energy of contacts between block *I* due to block *j* respectively. And the first term is explained Eq. (2).

The first term of Eq. (2) shows the strain energy, the second term shows surface traction energy, the third

$$\Pi^i = F(x,y) \int_{VI} \frac{\rho^c}{0 \, \rho} [\tau_{ij}^* \delta \varepsilon_l] dV_l \quad (2)$$

$$- \int_{\Gamma} \bar{t} \cdot u d\Gamma - F(x,y) \int_{VI} [\rho(\dot{b} - \ddot{u}) - c\dot{u}] dV_l$$

term shows inertia and viscosity energy of a block.

where, σ: the stress tensor, ε: the strain tensor, u: the displacement vector, $\ddot{u}$: the acceleration vector, $\dot{u}$: the velocity vector, ρ: the unit mass, b: the body force vector, c: the dumping coefficient, $\bar{t}$: the surface traction force, V: the volume of block, Γ: the area of block surface.

The strain increment of each step is expressed multiplying the local strain matrix $[B_l]$ and local displacement increment as shown Eq. (3).

$$\{\Delta \varepsilon_l\} = [B_l]\{\Delta u_l\} \tag{3}$$

where, $\{\Delta \varepsilon_l\}$ is Green strain (Malvern, 1969) as defined before deformation co-ordinate (c°). 2nd Piola Kirchhoff stress increment is defined Eq. (4).

$$\{\tau^*\} = [C]\{\Delta \varepsilon_l\} \tag{4}$$

where, $[C]$ is 2nd Piola Kirchhoff stress - Green strain constitutive law defined in reference co-ordinate. Generally, Cauchy stress - Almansi strain constitutive law in small deformation analysis is defined in after deformation co-ordinate and as coincide of experiment. $[C]$ could not defined correctly, therefore, Cauchy stress-Almansi strain constitutive law is use approximately (McMeeking, Rice,1975). Snice, Cauchy stress is expressed by the ratio of before and after deformation as expressed Eq. (5).

$$\{\Delta \sigma_l\} \cong \frac{\rho^c}{\rho^0}\{\tau^*\} \tag{5}$$

The function of the finite covers is expressed by Eq. (6).

$$F(x,y) = \sum_{i=1}^{n} w_i(x,y) f_i(x,y) \tag{6}$$

where, $w_i(x,y)$: Weight function, $f_i(x,y)$: displacement shape function, x, y: generalized co-ordinates.

The weight function is given the following characteristics when if the mathematical mesh and the physical mesh are overlapped, the weight function has a certain value.

$$w_i(x,y) \geq 0 \quad (x,y) \in U_i \tag{7}$$

And the non-overlapping area is zero in Eq. (6).

$$w_i(x,y) = 0 \quad (x,y) \notin U_i \tag{8}$$

Total value of the weight function in the over-lapped area is 1.

In the case of FEM, the physical region and the mathematical region coincide and DDA is a special case in

$$\sum_{(x,y) \in U_j} w_j(x,y) = 1 \tag{9}$$

which one block has one cover. Hence, Manifold method is a generalized method covering both FEM and DDA. Significance of Eq. (6) is to deal with mathematical meshes independently on the shape of physical blocks.

The second term represents the surface traction energy. The third term represents the inertia and viscosity energy of blocks. On the other hand, the potential energy for the contact between discontinuous blocks can be expressed using penalty as follow.

$$\Pi_{PL}^{i,j} = \frac{1}{2}k_N[(u^j - u^i)\cdot n]^2 - \frac{1}{2}k_T[u_T^j - u_T^i]^2 \tag{10}$$

where, k_N and k_T are the penalty coefficients in the normal and shear directions, respectively, $(u^j - u^i)\cdot n$ is the amount of penetration between two blocks in the normal direction, u_T is the amount of slip in the shear direction, and n is the direction cosine of the contact plane. In the case of the contacts of three-dimensional analyses, the basic contact is assumed between the vertex and the polygon of the polyhedron (Shi, 2001).

2.2 Shape function of element

The displacement in arbitrary point of element using four node iso-parametric shape function is defined by Eq. (11).

$$\begin{Bmatrix} u_l \\ v_l \end{Bmatrix} = [N]\{\Delta u_l\} \tag{11}$$

where, $[N]$ is the shape function of normal local co-ordinate.

$$[N] = \begin{bmatrix} N_1 & 0 & N_2 & 0 & N_3 & 0 & N_4 & 0 \\ 0 & N_1 & 0 & N_2 & 0 & N_3 & 0 & N_4 \end{bmatrix} \tag{12}$$

The displacement increment in nodal points is,

$$\{\Delta u_l\}^T = \{u_1, v_1, u_2, v_2, u_3, v_3, u_4, v_4\} \tag{13}$$

The shape functions in normal co-ordinate is expressed Eq. (14).

$$N_1 = \frac{1}{4}(1-\xi)(1-\eta), N_2 = \frac{1}{4}(1+\xi)(1-\eta)$$

$$N_3 = \frac{1}{4}(1+\xi)(1+\eta), N_4 = \frac{1}{4}(1-\xi)(1+\eta) \tag{14}$$

Figure 1 shows the finite cover on joints. The stiffness matrix of the element is integrated only on the physical element area by Gauss integration by using weight function as Eq. (6). Figure 2 shows four node iso-parametric element in local normal co-ordinate as same as standard FEM. The contact

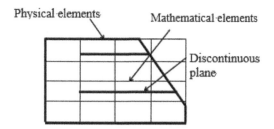

Figure 1. Finite cover on joint

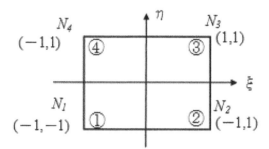

Figure 2. Local co-ordinate of iso-parametric element

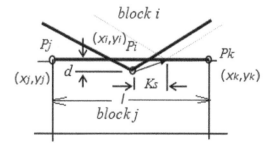

Figure 3. The contact mechanism between blocks

conditions at the boundaries of discontinuous planes are judged by Eq. (10), as shown in Figure 3.

The contact conditions at the boundaries of discontinuous planes are judged by Eq. (10), as shown in Figure 3. In this case, the square penetrating energy on a contact point between two elements is converged toward zero by Penalty method.

The strain energy of the contact spring by penetration between elements i and j in Fig.3 is evaluated by Eq. (13), where d is expressed by Eq. (14) with the coordinate functions $\{H\}$ and $\{G\}$, and the unknowns $\{D_i\}$ and $\{D_j\}$ of elemental nodes.

$$d = (\frac{S_0}{l} + \{H\}^T\{D_i\} + \{G\}^T\{D_j\})^2 \qquad (15)$$

where, S_0 is the penetrating area of physical elements, $\{H\}$, $\{G\}$ are the function of nodal coordinates x_i, y_i. The contact matrix $\{H\}$ and $\{G\}$ are

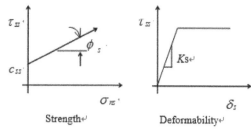

Strength   Deformability

Figure 4. The friction conditions of the element surfaces

expressed by function of $N(\xi_i, \eta_i)$ and block coordinates of x_i and y_i.

$$\{H\} = \frac{1}{l}[N(\xi_i, \eta_i)]^T \begin{Bmatrix} y_j - y_k \\ x_k - x_j \end{Bmatrix} \qquad (16)$$

$$\{G\} = \frac{1}{l}[N(\xi_k, \eta_k)]^T \begin{Bmatrix} y_k - y_i \\ x_i - x_k \end{Bmatrix} \qquad (17)$$

$$+ \frac{1}{l}[N(\xi_j, \eta_j)]^T \begin{Bmatrix} y_i - y_j \\ x_j - x_i \end{Bmatrix}$$

The normalized coordinates ξ_i, η_i are determined using relation Eq. (16) and Eq. (17) by iteration.

$$\begin{Bmatrix} x_s \\ y_s \end{Bmatrix} = \begin{bmatrix} \sum_{i=1}^{4} N_i(\xi_s, \eta_s) \\ \sum_{i=1}^{4} N_i(\xi_s, \eta_s) \end{bmatrix} \begin{Bmatrix} x_i^e \\ y_i^e \end{Bmatrix} \qquad (18)$$

Figure 4 shows the friction conditions of the contact element surfaces.

2.3 The governing equation for NMM and DDA

The governing equation for MM and DDA in the matrix form can be derived from the kinematic equations based on Hamilton's principle by minimizing the total potential energy for the block system as follows (Shi, 2001):

$$[M]\{\ddot{u}\} + [C]\{\dot{u}\} + [K]\{u\} = \{F\} \qquad (19)$$

where, M is the mass matrix; C is the viscosity matrix; K is the stiffness matrix; F is the external force vector; $\ddot{u}$ is the acceleration; $\dot{u}$ is the velocity and u are the displacement at the center of the block.

The kinematic equation of motion expressed as Eq. (19) is solved by Newmark's β and γ methods (Doolin & Sitar, 2002) with β is 0.5 and γ is 1.0, and the simultaneous algebraic equations for the increment in displacement is solved for each time increment,

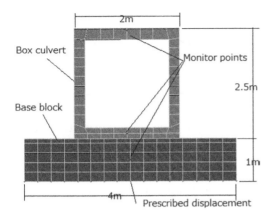

Figure 5. Model of 4-node Manifold Method

$$\tilde{K} \cdot \Delta u = \tilde{F} \qquad (20)$$

$$\tilde{K} = \frac{2}{\Delta t^2} M + \frac{2}{\Delta t} C + \frac{\rho^c}{\rho^0}[K_e + K_s] \qquad (21)$$

$$\tilde{F} = \frac{2}{\Delta t} M \cdot \dot{u} + (\Delta F - \sum \int \sigma \, dv) - M\alpha(t) \qquad (22)$$

where, Δu is the incremental displacement; M is the mass matrix; C is the viscosity matrix; K_e is the stiffness matrix of the linear term; K_s is the initial stress matrix caused by rigid body rotation; $\dot{u}$ is the velocity tensor of the center of gravity of a block; ρ^0 is the volume before deformation; ρ^c is the volume after deformation and $\alpha(t)$ is the acceleration history of external forces such as an earthquake.

The incremental prescrived displacement inputed each time step in 4-node manifold and 3D-DDA.

2.4 Numerical model and results

Figure 5 shows the model of 4-node manifold method. The reindocement box culvert is constructed on the base rock. Figure 6 shows the initial horizontal stress distribution caused by gavity force. The distributions on the border between the base block and the box culvert are not show

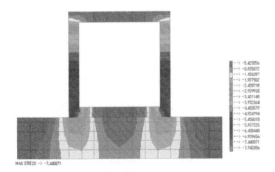

Figure 6. Initial horizontal stress disribution

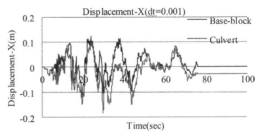

Figure 7. Time histries of monitor points

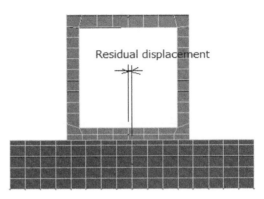

Figure 8. Final position of the box culvert after earthquake

uniform in NMM. The elastic modulus of the box culvert is 22Gpa and the base block is 4Gpa. The time increment is adopted 0.001 second. Figure 7 shows the time histories of the base block and box culvert at the monitor points. The input prescribed displacement of the base block is assumed by integration of the accseleration of actural earthquake record by an accelometer. The displacement time history of the box culvert in Figure 7 as shown as the red line, reached different point from the base block the blue line.from the base block it self as the blue line. The result shows the residual displacement of the box culvert is occured after earthquake as shown in Figure 8.

3 THREE DIMENSIONAL DDA

3.1 Outline of the theory

Figure 9 shows the coordinate system and the unknowns for 3D - DDA as also shown in Eq. (23). The unknowns are defined at the center of gravity of the blocks using first order shape functions (Shi, 2001).

$$[D_i] = \begin{pmatrix} u_c & v_c & w_c & r_x & r_y & r_z \\ & \varepsilon_x & \varepsilon_y & \varepsilon_z & \gamma_{yz} & \gamma_{zx} & \gamma_{xy} \end{pmatrix}^T \qquad (23)$$

553

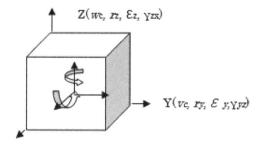

Z(w_c, r_z, ε_z, γ_{zx})

Y(v_c, r_y, ε y,γ_{yz})

X (u_c, r_x, ε_x, γ_{xy})

Figure 9. The coordinate system and the unknowns in 3D-DDA

where, $\{D_i\}$ is the unknown vector at the block center of gravity, u_c, v_c, w_c are the rigid body displacements, r_x, r_y, r_z are the rigid body rotations, ε_x, ε_y, ε_z, γ_{yz}, γ_{zx}, γ_{xy} are the strains of block i.

The displacement of arbitrary points in block i is expressed by Eq. (24).

$$\begin{Bmatrix} u \\ v \\ w \end{Bmatrix} = [T_i(x,y,z)][D_i] \tag{24}$$

where, the deformation function matrix is defined by Eq. (25).

$$\left[T(x,y,z)\right] = \begin{pmatrix} 1 & 0 & 0 & 0 & Z & -Y & X & 0 & 0 & 0 & \dfrac{Z}{2} & \dfrac{Y}{2} \\ 0 & 1 & 0 & -Z & 0 & X & 0 & Y & 0 & \dfrac{Z}{2} & 0 & \dfrac{X}{2} \\ 0 & 0 & 1 & Y & -X & 0 & 0 & 0 & Z & \dfrac{Y}{2} & \dfrac{X}{2} & 0 \end{pmatrix} \tag{25}$$

The structure viscosity matrix of the second term on the left side of Eq. (19) is expressed by Eq. (26) by the viscosity η and the mass matrix M.

$$C = \eta M \tag{26}$$

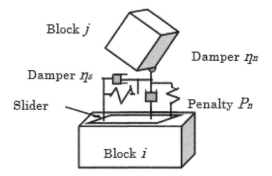

Figure 10. The viscosity at the contact of two blocks i and j

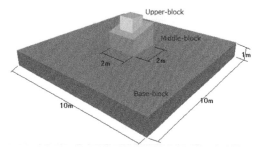

Figure 11. 3D-DDA Model

The physical meaning of the structure damping η is the dissipation of energy through air resistance and the vegetation on the surface of the slope.

The authors also introduced the contact viscosity used by the Voigt type model as shown in Figure 8 similar to that in a two-dimensional analysis (Sasaki et al., 2005).

$$\Pi_{p\eta} = \frac{1}{2}(p + \frac{\eta_p}{\Delta t})d^2 \tag{27}$$

where, p is the penalty coefficient Shi (2001) and is the contact viscosity (Sasaki, et al, 2011).

3.2 Numerical model and results

Figure 11 shows 3D-DDA model.The elastic modulus of the base, middle and upper block is 2Gpa and the friction angle of the block surface is assumed 35 degrees. The angle between the middle block and the upper block of the model is 20 degrees as shown in Figure 11. The penalty coficient of the contact calculation is adopted 2Gpa/m. The time increment is adopted 0.001 second for the steps.

Figure 12 shows the input prescrived displacement of Nort-West, East-West and Up-and down directions given at the four points of the coners of the base block. Figure 13 and Figure 14 show the displacement responces of the middle and the upper block. The displacement responce of the upper block show sliped on the between middle and upper blocks at the slope plane. Figure 15 shows the final position of the upper block after accelerations. The residual displacement is ocurred between the midle and the upper block.

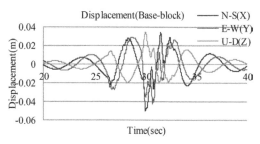

Figure 12. Input displacement of 3D-DDA Model

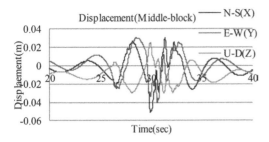

Figure 13. Displacement of middle block

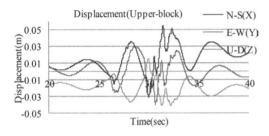

Figure 14. Displacement of middle block

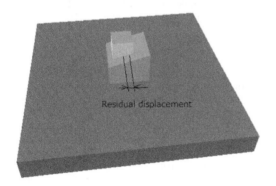

Figure 15. Final position of the upper block

4 CONCLUSION

The authors applied the prescribed displacement of earthquake motions at the base block of 4-node manifold and 3D-DDA models. The recommended method is very simple and avoided to determine the boundary conditions of base block in discontinuous model. The input motion of 4-node manifold method is indicated acceleration on the site and trance formed to the displacement. And the displacement history for 3D-DDA is made using statistical Green function of the active fault model. The results of NMM and 3D-DDA is good much in physical

meanings, especially, the contact fiction distribution of four node iso-parametric NMM. The actual earthquake motion is three dimensional in general, therefore, the three-dimensional DDA is effective for the practical problems. The prescribed displacement input motions for the NMM and DDA is very simple and could be avoid complex boundary conditions as relation with circumference analytical regions as a damper of boundaries, for example. The reliability between the field measures and the calculation models would be the future problems.

REFERENCES

Doolin, D. M. & Sitar, N., 2002. Displacement accuracy of discontinuous deformation analysis method to sliding block, *Journal of Engineering Mechanics*, ASCE, 1158–1168.

Malvern, L. E., 1969. Introduction to the mechanics of a continuous medium, Prentice-Hall.

McMeeking, R.,M. & Rice, J.,R. 1975. Finite element formulation for problems of large elastic-plastic deformations, *International Journal of Solids Structure*, Vol. 11, pp.601–616.

Miki, S. Sasaki, T., Koyama, T. Nishiyama, S.& Ohnishi, Y. 2010. Development of Coupled Discontinuous Deformation analysis and Numerical Manifold Method (NMM-DDA), International Journal of Computational Methods, Volume 7, Issue 1, pp.131–150.

Miki, S. Ohnishi, Y. Sasaki, T. & Okuda, S. 2016. Stability assessment for the stone-block retaining wall by the NMM-DDA, WCCM12 & APCOM 2016, Seoul.

Sasaki, T.& Ohnishi, Y. 2001. Analysis of the discontinuous rock mass by four node iso-parametric Manifold method, *Fourth International Conference for Analysis of Discontinuous Deformation*, Glasgow, Scotland, pp.369–378, UK.

Sasaki, T. et al. 2004.: Earthquake response analysis of a rock falling model by Discontinuous Deformation Analysis, *ISRM Symposium & 3rd Asia Rock Mechanics Symposium*, Millpress, pp1267–1272.

Sasaki, T. Hagiwara, I. Sasaki, K. Horikawa, S. Ohnishi, Y. Nishiyama, S. & Yoshinaka, R. 2005. Earthquake response analysis of a rock falling by Discontinuous Deformation Analysis, *Seventh International Conference on the Analysis of Discontinuous Deformation*, pp.137–146.

Sasaki, T. Hagiwara, I. Sasaki, K. Ohnishi, Y. & Ito, H. 2007. Fundamental studies for dynamic response of simple block structures by DDA, *Eighth International Conference on the Analysis of Discontinuous Deformation*, Beijing, pp. 141–146.

Sasaki, T., Hagiwara, I. Sasaki, K. Yoshinaka, R., Ohnishi, Y. Nishiyama, S.& Koyama, T., 2011. Stability analysis of ancient masonry structures by using DDA and Manifold method, *International Journal of Computational Methods*, Volume 8, Issue, 2, pp.247–275.

Shi, G. H. 2001. Three-dimensional discontinuous deformation analyses. *Proceedings of Fourth International Conference on Analysis of Discontinuous Deformation*, pp. 1–21.

2019 Rock Dynamics Summit– Aydan et al. (eds)
© 2019 Taylor & Francis Group, London, ISBN 978-0-367-34783-3

Fundamental study on the dynamic behavior of Japanese castle masonry wall using NMM-DDA

T. Sueoka & R. Hashimoto
Hiroshima University, Hiroshima, Japan

M. Kikumoto
Yokohama National University, Kanagawa, Japan

T. Koyama
Kansai University, Osaka, Japan

ABSTRACT: There are a lot of historic castle masonry walls all over Japan. Those are regarded as the important cultural properties to be conserved. However, in recent years, the masonry walls frequently suffered the collapse due to the strong earthquake such as the 2016 Kumamoto Earthquake, and the investigations on the mechanical characteristics and the effective reinforcement method of the masonry walls have become the urgent task. In this study, to study the influence of the vertical excitation on the stability of the masonry wall, a series of shaking analysis was carried out using the coupled Numerical Manifold Method and Discontinuous Deformation Analysis (NMM-DDA), a discontinuum-based numerical method. The simulated results showed that the vertical excitation changes the apparent gravity of the soil and the stones, and consequently affect the amount of the deformation during the excitation.

1 INTRODUCTION

There exists a lot of castle masonry walls all over Japan, and the conservation of them become the important task because they have aesthetic landscape and sophisticated construction techniques. However, in recent years, the masonry walls frequently suffered the collapse due to the strong earthquakes such as the 2011 Great East Japan Earthquake and the 2016 Kumamoto Earthquake (see, Fig. 1). After these earthquakes, the investigations on the reinforcement method have become the urgent task for the conservation of their historical values and the prevention of the human damages. To propose a proper measure, the mechanical characteristics of the masonry walls during the earthquakes should be revealed.

In order to understand the behaviors of the masonry walls, the mechanical interaction among the stones and the backfill soil should be considered. Since the masonry walls sustain the backfill soil by the friction between the stones, the characteristics of the friction and the earth pressure during the earthquake is a key factor to evaluate the stability of the masonry walls. Moreover, the previous studies pointed out that the vertical excitation may induce the changes in contact conditions and friction strength among the stones, and the phase difference between horizontal and vertical waves also influences the stability of the masonry wall (Ikemoto et al. 1995, Koyama et al. 2016). In this study, therefore, a series of shaking analyses of a masonry wall with vertical excitation is conducted.

Figure 1. A collapsed masonry wall in Kumamoto castle (after the 2016 Kumamoto Earthquake).

To analyze the behavior of the masonry wall, which is a composite structure of stones and soil, a numerical method that can deal with the mechanical interaction of discontinua and continuum is required. In this study, as a method applicable to the dynamic analysis of the masonry walls, a coupled method of Numerical Manifold Method (NMM) and Discontinuous Deformation Analysis (DDA) proposed by Miki et al. (2010) is focused on. Both of the NMM (Shi 1991) and the DDA (Shi & Goodman 1989) are the numerical methods for the dynamic contact problems among the deformable blocky system, and the contact treatment algorithms are

556

identical for two methods. NMM can analyze the detailed deformation of continua considering the mutual contact. On the other hand, the DDA can properly simulate the dynamic behaviors of blocks with large rotation. Therefore, the NMM-DDA is a suitable method for the simulation of mechanical interaction among the stones and the ground, and has been applied to the quasi-static stability problems of the masonry structures (Hashimoto et al. 2017). In addition, the NMM-DDA has been improved for dynamic analysis introducing the implicit updating algorithm of the friction law (Sueoka et al., 2018). This study adopts this fully implicit NMM-DDA.

Using the above numerical method, shaking analyses of a simple masonry wall model with and without the vertical excitation are carried out. In the analyses with the vertical wave, the phase difference with the horizontal wave is considered as well. From the simulated results, the influence of the vertical excitation on the stability of the masonry wall is discussed.

2 THEORETICAL OUTLINES OF THE FULLY IMPLICIT NMM-DDA

2.1 Governing equations and weak form

The NMM-DDA solves the equation of motion of the system consists of the multiple continua considering their mutual contacts (Fig. 2). Hereafter, a multibody system consisting of n continua (Ω_1, Ω_2, …, Ω_n) is assumed and the kinematic problem of each object is described with the equation of motion:

$$\rho_i \ddot{u}_i - \nabla \cdot \sigma_i - \rho_i \bar{b}_i = 0 \quad (1)$$

the strain compatibility condition:

$$\varepsilon_i = \frac{1}{2}\left\{ \nabla u_i + \left(\nabla u_i\right)^{\mathrm{T}} \right\} \quad (2)$$

the incremental form of the constitutive equation:

$$\Delta \sigma_i = D_i : \Delta \varepsilon_i \quad (3)$$

where the subscript i indicates the physical quantity of object Ω_i, the dot over the symbols means the material derivative, ρ is the density, u is the displacement vector, σ is the Cauchy stress tensor, $\bar{b}$ is the known body force vector, ε is the infinitesimal strain tensor, D is the constitutive relations tensor. Moreover, the displacement field must satisfy the displacement boundary conditions (Eq. (4)) and the stress boundary conditions (Eq. (5)).

$$u_i = \bar{u}_i \text{ on } \Gamma_{iu} \quad (4)$$

$$t_i = \sigma_i . n_i = \bar{t}_i \text{ on } \Gamma_{iu} \quad (5)$$

here, t is the traction vector, and n is the unit normal vector to Γ. Assuming the virtual displacement

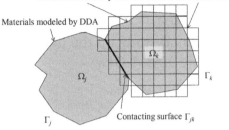

Figure 2. Modeling in NMM-DDA.

δu, which is a vector function satisfying $\delta u = 0$ on Γ_u, the weak form of the above boundary value problem expressed with Eqs. (1)-(5) will be derived as follows,

$$G_i^{\text{ext, int}}\left(u_i, \delta u_i\right) = \int_{\Omega_i} \delta u_i \cdot \rho_i \ddot{u}_i \mathrm{d}\Omega + \int_{\Omega_i} \delta \varepsilon_i : \sigma_i \mathrm{d}\Omega$$

$$- \int_{\Omega_i} \delta u_i \cdot \bar{b}_i \mathrm{d}\Omega - \int_{\Gamma_{i\sigma}} \delta u_i \cdot \bar{t}_i \mathrm{d}\Gamma \quad (6)$$

where $G_i^{\text{ext, int}}$ means the virtual work by external and internal forces for Ω_i.

2.2 Contact condition

In case of a contact between Ω_j and Ω_k along the interface Γ_{jk} (Fig. 2) is assumed hereafter. The normal gap g_N between two objects has to satisfy the non-penetration condition:

$$g_N \geq 0 \text{ on } \Gamma_{jk} \quad (7)$$

and the contact force, which is the tractions acting against each other on Γ_{jk}, must balance:

$$t_j + t_k = 0 \text{ on } \Gamma_{jk} \quad (8)$$

The contact force on Γ_{jk} was decomposed into the normal component t_N and the tangential component t_S, the virtual work by the contact force $G_{jk}^c(t_N, t_S, g_N, g_S)$ is obtained as follows:

$$G_{jk}^c\left(t_N, t_S, g_N, g_S\right) = \int_{\Gamma_{jk}} t_N \delta g_N \mathrm{d}\Gamma + \int_{\Gamma_{jk}} t_S \delta g_S \mathrm{d}\Gamma \quad (9)$$

where δg_N and δg_S are the virtual displacement corresponding to g_N and g_S, which is the shear displacement along Γ_{jk}, respectively. Since the total virtual work by the external force, the internal force, and the contact force must balance during the contact, the weak form is consequently expressed as the following equation.

$$\sum_{i=1}^{n} G_i^{\text{ext, int}}\left(u_i, \delta u_i\right) + G_{jk}^c\left(t_N, t_S, g_N, g_S\right) = 0 \quad (10)$$

557

2.3 *Penalty regularization and return mapping*

Eq. (10) includes the unknown contact force t_N and t_S. Therefore, it is necessary that Lagrange multiplier method or the penalty method is applied to solve the Eq. (10). In the NMM-DDA, the penalty method is adopted. In the penalty method, the non-penetration condition is weakened, and permitting small inter-penetration between the objects, the normal contact force is expressed with a function of as follows:

$$t_N = p_N g_N \tag{11}$$

where p_N is the penalty coefficient for the normal direction. On the other hand, the tangential component t_S increases with the progress of the tangential displacement g_S, and t_S shows nonlinearity when t_S reaches the friction strength. If the following Coulomb's friction law is adopted as the friction yield criterion:

$$f = t_S \mathrm{sgn}(g_S) - |t_N| \tan\phi \tag{12}$$

t_S will be updated by follows:

$$t_S = \begin{cases} p_S g_S & \text{if } f < 0 \text{ (elastic state)} \\ |t_N| \tan\phi \cdot \mathrm{sgn}(g_S) & \text{if } f = 0 \text{ (slipping state)} \end{cases} \tag{13}$$

where p_S is the penalty coefficient for the tangential direction. For the accurate expression of this nonlinear behavior, an implicit integration scheme of the friction law, called return mapping method (Wriggers, 2006) should be employed.

In the conventional NMM-DDA, t_S is updated incrementally as the tangential displacement progresses based on the first equation of Eq. (13). Then, if t_S exceeds the friction strength determined by the yield condition (Eq. (12)), t_S will be drawn back to the friction strength at the beginning of the 'next' time step (Fig. 3(a)). Since this scheme temporarily overestimates the friction, the error of the shear displacement on the material interface is unavoidable. Moreover, the error accumulates in case of the seismic analyses that induce cyclic shear along the discontinuity.

For this drawback, Sueoka et al. (2018) improved the NMM-DDA by introducing an implicit updating algorithm of friction force so-called return mapping method. In this method, after solving the global equilibrium equation, t_s is once estimated assuming that the tangential displacement increment is totally elastic component. This procedure is called 'trial elastic step' (see Fig. 3(b)). Then, if the updated t_S exceeds the yield criterion, t_S will be pulled back to the correct friction strength (see Fig. 3(b)). This process is called 'return mapping'. Since this scheme modifies the tangential force immediately after violating the yield condition, the friction strength can be precisely reproduced. However, it is necessary that iterative solution scheme is incorporated into the

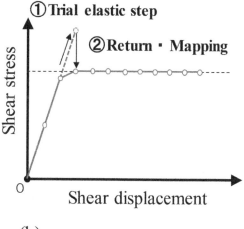

(a)

(b)

Figure 3. Updating procedure od friction force: (a) conventional NMM-DDA and (b) return mapping method.

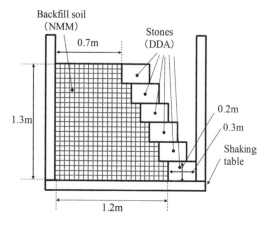

Figure 4. Simple masonry wall model.

558

NMM-DDA since the residual force is caused due to pulling back the shear force in the return mapping of the tangential force. Therefore, in this study, the Newton-Raphson iteration is also implemented the NMM-DDA to make converged the residual force.

By discretizing the weak form consisting of Eqs. (10), (11) and (13), the contact problem of continua can be solved. The displacement variables in the weak form are discretized by NMM or DDA, and the acceleration term is discretized by Newmark's β method. Detailed description about spatial and temporal discretization can be seen in Hashimoto et al. (2017).

3 SHAKING ANALYSIS OF A MASONRY WALL MODEL

3.1 Analytical conditions

A series of the shaking analyses of masonry wall model considering vertical excitation is conducted here using the fully implicit NMM-DDA. The simple masonry wall model which consists of the shaking table and the stones modeled by DDA and the backfill soil modeled by NMM is shown in Figure 4. In order to investigate the stress transmission properties between the stones under different shaking conditions, the rectangular shaped stones were used although the actual masonry walls constructed with the tapered stones. This model enables to compare the effect of vertical excitation easily because its direction coincides with that of the normal contact force between the stones. Material properties and surface parameters for a shaking analysis of a masonry wall model are summarized in Table 1 and 2, respectively. The stones are modeled as linear elastic material and the soil is assumed Drucker-Prager elasto-perfectly plastic material with the internal friction angle $\phi' = 30°$. The initial stress conditions were set by a self-weight analysis, and subsequently the acceleration was input to the shaking table. Figures 5 and 6 show the used input waves in horizontal and vertical directions, respectively. There are three cases with different vertical excitation conditions under same horizontal wave condition shown in Figure 5. Figure 6(a) is the case without vertical excitation, and (b) and (c) are the cases with vertical excitation under different phase lags (case (b): 0° and

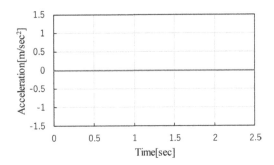

(a)

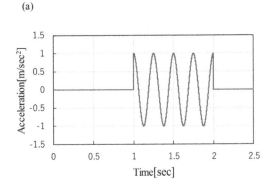

(b)

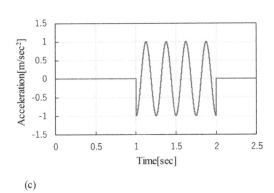

(c)

Figure 6. Input vertical waves: (a) without vertical wave, (b) with vertical wave (phase lag is 0°), (c) with vertical wave (phase lag 180°).

case (c): 180°) to the horizontal wave, respectively. Both of horizontal and vertical waves are the sinusoidal wave which frequency is 4Hz and amplitude is 1.0m/sec², and duration is 1.0sec.

3.2 Simulation results

Firstly, the results of the case (a) (without vertical excitation) will be shown. Figure 7 shows the distribution of the horizontal displacement at the final state. The upper four stones moved horizontally and the discontinuity of the displacement that means slip was observed among the stones. The third stone from the bottom showed the largest displacement (0.87mm). It should be noted that

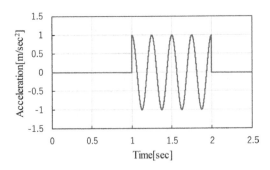

Figure 5. Input horizontal wave

Table 1. Material properties for a shaking analysis of a masonry wall model.

	Stones (Elastic)	Ground (Drucker-Prager)
Unit weight (kN/m³)	29.4	14.7
Young's modulus (MPa)	10000	100
Poisson's ratio	0.2	0.3
Internal friction angle(°)	N/A	30.0
Dilatancy angle (°)	N/A	0.0
Cohesion (kPa)	N/A	2.0

Table 2. Material surface parameters for a shaking analysis of a masonry wall model.

Friction angle (°)	30.0
The normal penalty coefficient (GN/m)	10.0
The tangential penalty coefficient (MN/m)	1.0

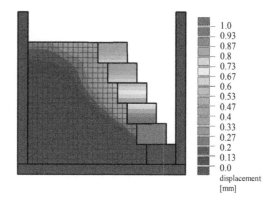

Figure 7. Final distribution of horizontal displacement (without vertical wave).

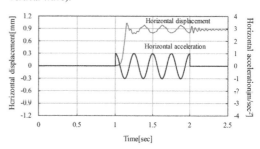

Figure 8. Final distribution of deviator strain (without vertical wave).

the displacement of the masonry wall model was very small because the maximum amplitude of the input waves was 1.0m/sec². Figure 8 shows the distribution of the deviator strain at the final state. The figure shows that shear deformation was caused in the backfill soil due to the displacement of the stones.

Figure 9 shows the time history of the horizontal displacement of the third stone from the bottom and the horizontal input wave. The residual displacement occurred in the first cycle of the horizontal wave, but did not progress during the following cycles. For discussion of this phenomenon, the time history of the vertical normal stress of the third stone from the bottom, which affects the friction strength along the stone interface, is shown in Figure 10. In this figure, the vertical stress increased during the first cycle of the excitation because the stones move rightward and the weight transferred from the upper stones. Therefore, the residual displacement did not progress from the second cycle of the excitation due to the increase of the friction strength along the stone interface.

Figure 11 and 12 show the time history of the horizontal displacement and vertical acceleration of the third stone obtained from the cases (b) and (c). Comparing with Figure 8 (without vertical wave), the amount of the displacement is different among each case. The smallest one is the case (b), the largest one is the case (c), and the case (a) without vertical motion showed intermediate displacement.

The above difference among the simulated results can be discussed considering the relation between the vertical excitation and the horizontal displacement. As shown in Figure 11, the phase of the vertical acceleration is reverse to that of the horizontal displacement in case (b). This means that the stone moved when the wall model was shaken downward. Thus, the apparent gravity of the wall model decreased when the stone moved. In Figure 12 (case (c)), on the other hand, the phase of the vertical acceleration is

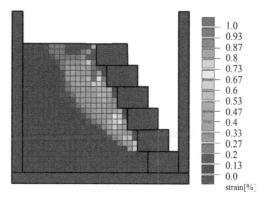

Figure 9. Time history of the horizontal displacement and the input acceleration

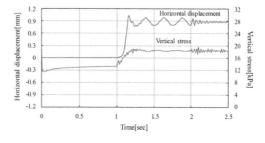

Figure 10. Time history of the horizontal displacement and the vertical stress (without vertical excitation).

560

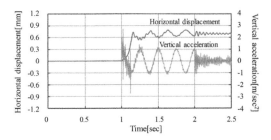

Figure 11. Time history of the horizontal displacement and the vertical acceleration (phase lag 0°).

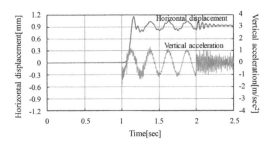

Figure 12. Time history of the horizontal displacement and the vertical acceleration (phase lag 180°).

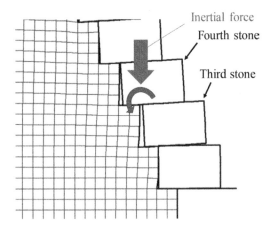

Figure 13. Deformation of the masonry wall model at 1.15sec (phase lag 180°)

consistent with that of the horizontal displacement. This means that the stone moved when the wall model was shaken upward, and the apparent gravity of the model increased during the stone movement.

Now, to investigate the influence of the apparent gravity, the configuration around the third stone from the bottom of case (c) at the moment of deformation progress (1.15sec in Figure 12) is shown in Figure 13. The figure shows that the upper four stones dropped into backfill soil, and the lower stone was pushed out due to the rotated upper stone. Therefore, it is concluded that the case (c) showed the largest horizontal displacement of the stone because of the increase of the apparent gravity due to upward excitation.

4 CONCLUSIONS

In this study, to investigate the influence of the vertical excitation on the stability of the masonry wall, a series of shaking analysis was carried out using the coupled Numerical Manifold Method and Discontinuous Deformation Analysis (NMM-DDA), a discontinuum-based numerical method. The simulated results showed that the vertical excitation changes the apparent gravity of the soil and the stones, and consequently affect the amount of the deformation during the excitation.

In the future study, the validation study should be carried out by simulating a shaking table experiments of a masonry wall model with the NMM-DDA. Then, from the parametric studies on various construction conditions will be performed under various external force conditions to find the vulnerable structural condition of the masonry walls. Furthermore, the influence of the inner structure of the masonry walls should be investigated. Although the actual masonry walls in Kumamoto castle are constructed using river gravel for the backfill, this study assumed the entire backfill is soil to investigate the influence of the vertical excitation. Therefore, seismic analysis of the masonry wall model considering the river gravel should be carried out.

REFERENCES

Hashimoto, R., Kikumoto, M., Koyama, T. & Mimura, M. 2017. Method of deformation analysis for compare structures of soils and masonry stones. *Computers and Geotechnics.* 82: 67–84.

Ikemoto, T., Kitaura, M., Nishida, Y. & Yoshida, R. 1995. Dynamic behavior of masonry considering vertical motion. *Geotechnical Research Presentation Lecture Outline.* Vol. 23: 549–552.

Koyama, T., Kikumoto, M., Hashimoto, R. & Kuwajima, R. 2016. Damage situation of stone walls in Kumamoto castle caused by the 2016 Kumamoto earthquake. *Safety Science Review.* 7: 87–94 (in Japanese).

Miki, S., Sasaki, T., Koyama, T., Nishiyama, S. & Ohnishi, Y. 2010. Development of coupled discontinuous deformation analysis and numerical manifold method. *International Journal of Computational Methods.* Vol. 7. No. 1: 1–20.

Shi, G.H. & Goodman, R.E. 1989. Generalization of two-dimensional discontinuous deformation analysis. *International Journal of Numerical and Analytical Methods in Geomechanics.* 13: 359–380.

Shi, G.H. 1991. Manifold method of material analysis. *Trans. 9th Army Conf. on Applied Mathematics and Computing.* Report No. 92–1.

Sueoka, T., Hashimoto, R., Kikumoto, M. & Koyama, T. 2018. Introducing implicit integration of friction law into NMM-DDA and dynamic analysis of castle masonry wall model. In: *Proc. of The 17th International Conference on Civil and Environmental Engineering,* Paper No. HU2.

Wriggers, P. 2006. *Computational Contact Mechanics.* Springer.

2019 Rock Dynamics Summit– Aydan et al. (eds)
© 2019 Taylor & Francis Group, London, ISBN 978-0-367-34783-3

Seismic diagnosis of monastery of St. Stepanos in Iran

M. Miyajima & T. Ikemoto
Kanazawa University, Kanazawa, Japan

A. Fallahi
Azerbaijan Shahid Madani University, Tabriz, Iran

K. Yamaguchi
Kyushu University, Fukuoka, Japan

M. Yoshida
National Institute of Technology, Fukui College, Sabae, Japan

ABSTRACT: This paper deals with a seismic diagnosis of monastery of St. Stepanos that is UNESCO world heritage in east Azerbaijan, Iran. An oldest buildings of monastery of St. Stepanos was built in tenth century and most of buildings are masonry structures. Several kinds of in situ tests such as microtremor measurements, shear wave velocity measurements, etc. were conducted around the monastery to estimate dynamic characteristics of the site. Microtremor measurements of a church and tower of the monastery ware also performed to estimate dynamic characteristics of the buildings. Then, a strong ground motion of a scenario earthquake was predicted by using the stochastic green function method. Scenario earthquake fault is assumed as North Tabriz fault in this study. Finally, we evaluate safety of the monastery for the scenario earthquake.

1 INTRODUCTION

The monastery of St. Stepanos is located 15km west of Jolfa, near Darresham village in Aras Kenar rural district and 6km from Aras River in Iran (See Figure 1). The monastery was registered as the UNESCO world's heritage in 2008 as Armenian monastery structure group. An oldest buildings of Monastery of St. Stepanos was built in tenth century. The monastery of St. Stepanos is situated an earthquake prone area and has experienced several strong earthquakes in the past. Historical masonry structures such as the monastery of St. Stepanos have suffered damage from an earthquake occurred at some countries in the earthquake prone areas in the world. Since the historical masonry buildings must be survived from the future large earthquakes, vulnerability assessment to earthquake and its strengthening are indispensable. It is in importance to estimate not only the strength of buildings but also magnitude of possible ground motion due to a scenario earthquake in the vulnerability assessment.

The purpose of this study is to examine the dynamic characteristics of the surface ground at the monastery using shear wave velocity measurements method, and to discuss about the seismic characteristic of the monastery of St. Stepanos using microtremor measurements of the ground and structure. Furthermore, strong ground motion prediction at the monastery is tried.

Figure 1. Location of the monastery of St. Stepanos.

2 SHEAR WAVE VELOCITY MEASUREMENTS AROUND MONESTRY OF ST. STEPANOS

2.1 Refraction wave technique

It is difficult to obtain information of the in-situ soil properties below the uppermost layer of the earth by conventional sampling methods. Refraction wave technique is one of seismic methods which has been used in this study. Figure 2 shows a schematic diagram of refraction wave technique. S is an impulsive energy source and R1, R2, and R3 are three receivers in this figure. The interval of each receiver is 5m in this measurements. The measurement installment is McSEIC-3, Model 1817 (OYO Co. Ltd.). This figure also shows three travel-time curves recorded at each receiver station. Each line shows the first arrival for each type of wave. The slope of each travel-time curve indicates the velocity of the wave.

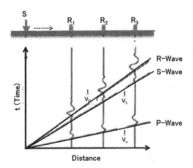

Figure 2. Schematic diagram of refraction wave technique and traveling-time curves.

2.2 *Estimated shear wave velocity*

Figure 3 show a plan of the monastery of St. Stepanos. Shear wave velocity measurements using refraction wave technique was conducted along an east-west line of northern and southern sites of the monastery. The shear wave velocity and depth of the upper layer of the ground and the natural frequency of surface ground in each station were estimated from the traveling-time curves. The natural frequency and the shear velocity of surface ground near the monastery of St. Stepanos is listed in Table 1.

3 MICROTREMORE MEASUREMENTS

3.1 *Outline of measurement*

Microtremor measurement is a very useful method for obtaining dynamic characteristics of structure (Enomoto et al. (2000)), including the natural frequency and the amplification factor of the church using H/H spectral ratios (H: horizontal direction). A portable highly sensitive vibration measurement machine of SPC-35N (Tokyo Sokushin Co. Ltd.) and seismometer with 8 pick-up sensors was used in microtremor measurements. The velocity amplitudes of microtremor were measured using a sampling frequency of 100 Hz and 0.1-Hz high pass-filter. Each recording time was 10 minutes. The record was divided into ten segments with 2048 data each after eliminating visible noise parts. The mean spectrum was smoothed by a Parzen window of 0.4 Hz band. After transform in the frequency domain, the ratio between the horizontal component at 2nd or 3rd floors and the horizontal one at 1st floor (basement) was calculated in order to estimate the natural frequencies of the structure. Table 2 shows the specification of church area at the monastery of St. Stepanos.

3.2 *Microtremor measurements of building*

Microtremor measurements were conducted at a church area in the monastery and a tower area (shown in Figure 3). The microtremor installations

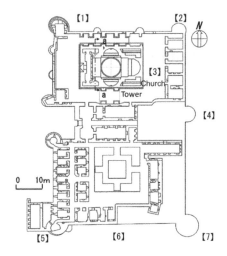

Figure 3. A plan of the monastery of St. Stepanos.

Table 1. Results of shear wave velocity measurement.

	Shear wave velocity (m / s)	Natural frequency (Hz)	The thickness of surface layer (m)
Measuring line on the north side	333	8.3	10.0
Measuring line on the south side	341	8.3	10.2

were two directions, EW and NS in the 1st, 2nd and 3rd floor level. Figure 4 shows each measurement point. The natural frequencies obtained from the H/H spectral ratio at each point are listed in Table 2.

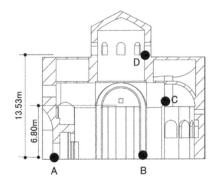

Figure 4. Section of church and measurement points.

Table 2. The Natural frequency.

Place		Direction	Natural frequency (Hz)		
			1st	2nd	3rd
Church		EW	4.2	-	-
		NS	5.4	-	-
Tower	North	EW	4.1	7.2	-
		NS	4.1	5.1	9.4
	South	EW	4.1	7.2	-
		NS	4.2	5.1	9.9
	East	EW	4.1	7.2	-
		NS	4.2	5.1	9.6
	West	EW	4.1	7.2	-
		NS	4.2	5.1	9.9

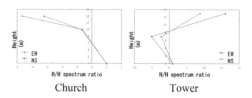

Church   Tower

Figure 5. Vibration modes of each measurement point.

Table 3. Natural frequency of surface ground.

Measurement point	Natural frequency(Hz)
1	8.0
2	8.2
3	8.2
4	8.0
5	8.3
6	8.3
7	8.3

In the church area, the natural frequency was obtained only first mode, and the modes beyond first mode were obtained at the tower area. Vibration mode on measured microtremor in each area is shown in Figure 5. The first mode was dominated in the church area, and secondary mode was dominated in the tower area. It was understood that predominant frequency of the church was about 4 Hz and that of the tower was 4-9Hz, respectively.

3.3 Microtremor measurements of ground

The microtremors of a surface ground around the monastery were also measured (Karashi (2015)). The measurement points of the surface ground were indicated as numbers in Figure 3. Natural Frequency and maximum spectral ratio of the surface ground obtained from the H/V spectral ratio of microtremor measurement (Andrej (2012)) in each number point are summarized in Table 3. The natural frequency of surface ground at the monastery is roughly 8.0 to 8.3 Hz.

4 STRONG GROUNG MOTION PREDICTION

Strong ground motions from future large earthquake at historical masonry building, that is, the monastery of St. Stepanos are predicted. The source, path, and site characteristics used in the prediction and the obtained results are introduced in this paper. Because there is no observed record from small earthquake occurred in North Tabriz fault, the stochastic Green's function method is adopted in the prediction.

The target fault is North Tabriz fault, its whole length is over 200km. The location of North Tabriz fault is decided by an active fault map (Hessami et al. (2003)). The fault closest distance to the monastery is about 54km. The fault is divided to six segments (from segment A to segment F) based on the strike angle. Four rupture scenarios according to rupture area are set here. One segment ruptures individually in two scenarios, and some segments rupture simultaneously in two scenarios. The rupture area and the other parameters of each scenario are shown in Table 4 and Figure 6. The moment magnitude of each scenario is in the range from 6.4 to 7.4. The outer and inner fault parameters are given according

to the recipe for predicting strong ground motions from future large earthquakes (Irikura et al. (2004), Irikura & Kurahashi (2008)). For example, the fault parameters of scenario No.2 is shown in Table 5. Figure 7 shows the location of asperities and rupture starting points of the scenario No.2. The asperity area is deployed the nearest location of fault plain to the target site, because the location is considered as most critical case for the target site. Several points are given as rupture starting points and the parametric study is done. For example, four points are given as rupture starting points in scenario No.2 (from R1 to R4).

Path characteristic is evaluated by spectral inversion method (Iwata & Irikura (1988)) using observed records of four small earthquakes (Table 6) at three sites, Basmanj, Khajeh, and Tabriz6. Q-factor is obtained as a function of frequency shown in Equation (1).

$$Q(f) = 17.4 \times f^{1.00} \tag{1}$$

where, f is frequency.

Site amplification factor is evaluated by an empirical equation proposed by Midorikawa et al. (1992). According to the empirical equation, a site amplification factor is simply estimated by using an average shear wave velocity above 10m from surface ground.

Predicted acceleration waveforms and response acceleration spectra (damping 5%) for the fault rupture scenario No.2 are shown in Figures 8, 9 and 10.

The fault rupture scenario No.2 is in case that only one segment which is the nearest segment to the monastery, segment A, ruptures individually. The predicted peak ground accelerations of EW component for scenario No. 2 are in the range from 92.0cm/s/s to 147cm/s/s. The peak response acceleration spectra are short period range (T<0.1 sec) is in the range from 400cm/s/s to 700cm/s/s.

According to the mictrotremor measurements at the monastery shows that the predominant frequency is about 8 Hz, that is, 0.12 sec. The resonance of strong ground motion of a scenario earthquake and surface ground, therefore, is possible. The strengthening of the monastery is necessary for this kind of the worst scenario.

564

Table 4. Each fault rupture scenario.

Scenario No.	No.1	No.2	No.3	No.4
Ruprure Area	Segment C	Segment A	Segment A Segment B	All Segment
Length (km)	22	30	74	218
Width (km)	15	15	15	15
Area (km^2)	330	450	1,110	3,270
Seismic moment (N·m)	5.69×10^{18}	1.13×10^{19}	3.55×10^{19}	1.36×10^{20}
Moment magunitude	6.4	6.6	7.0	7.4

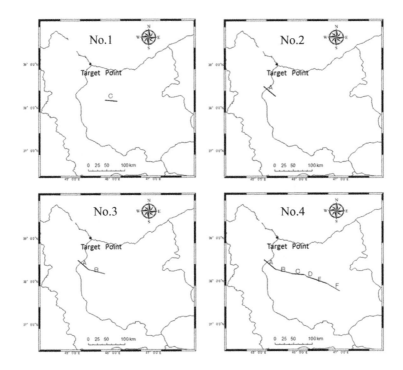

Figure 6. Location of each scenario fault.

Table 5. Fault parameters of scenario No.2.

Outer fault parameters			Inner fault parameters	
Upper depth (km)	2		Asperity area	
Lower depth (km)	17		Area (km2)	90
Fault length (km)	30		Moment (N·m)	4.53×10^{18}
Fault width (km)	15		Slip (cm)	152.1
Fault area (km^2)	450		Stress drop (MPa)	14.37
Strike angle (°)	134.9		Rise time (sec)	1.79
Dip angle (°)	90		Off-asperity area	
Rake angle (°)	172		Area (km2)	360
Seismic moment (N·m)	5.13×10^{19}		Moment (N·m)	6.74×10^{19}
Moment magnitude	6.6		Slip (cm)	56.6
Average slip (cm)	75.7		Stress drop (MPa)	2.87
Average stress drop (MPa)	2.87		Rise time (sec)	2.98

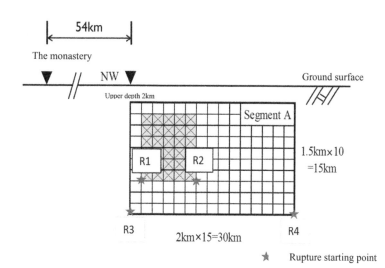

Figure 7. Location of asperity and rupture starting point (★) of the scenario

Table 6. Observed records used in spectral inversion analysis.

No.	Origin Time	Epicenter Lat.	Lon.	D	M
1	2007.09.16 05:20:01	38.100	46.360	24	4.1
2	2007.12.01 18:22:17	38.150	46.410	22	4.2
3	2007.12.01 18:45:11	38.130	46.430	22	4.8
4	2007.12.02 10:00:02	38.090	46.430	16	4.1

Lat.: Latitude (Degree), Lon.: Longitude (Degree)

D: Depth(km), M: Magnitude

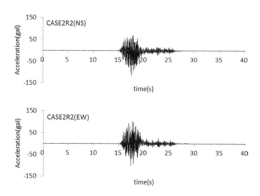

Figure 8. Predicted acceleration waveforms for scenario No.2. (Rupture starting point: R2).

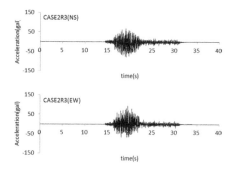

Figure 9. Predicted acceleration waveforms for scenario No.2. (Rupture starting point: R3).

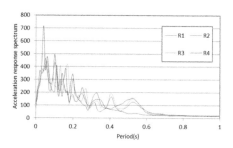

Figure 10. Response acceleration spectra (damping ration 5%) foe scenario No. 2.

5 CONCLUSION

Shear wave velocity and microtremor measurements were carried out at the monastery of St. Stepanos. Strong ground motion prediction was also conducted. Obtained results can be summarized as follows:

1. By refraction wave method on shear wave velocity measurements, the natural frequency and shear wave velocities of surface ground around the monastery were determined. It is 8.3 Hz.

2. By H/H Fourier spectral ratio analysis on microtremor measurements of the church area, the natural frequencies of the sites were determined. It is 8.0 to 8.3 Hz.

3. The natural frequency of the surface ground at the church and tower obtained by microtremor and shear wave velocity measurements showed good agreement.

4. According to the mictrotremor measurements at the monastery shows that the predominant frequency is about 8 Hz, that is, 0.12 sec. The resonance of strong ground motion of a scenario earthquake and surface ground, therefore, is possible. The strengthening of the monastery is necessary for this kind of the worst scenario.

ACKNOWLEDGEMENTS

This study was supported by JSPS KAKENHI Grant Number JP18KK0129 and Azerbaijan Shahid Madani University, Tabriz, Iran.

REFERENCES

Andrej., G. 2012. Determination of masonry building fundamental frequencies in five Slovenian towns by microtremor excitation and implications for seismic risk assessment, journal of Nat Hazards, Springer, No.62, pp.1059–1079.

Enomoto, T., Kuriyama T., Abeki N., Iwatate T., Navarro M. and Nagumo, M. 2000. Study on microtremor characteristics based on simultaneous measurements between basement and surface using borehole, Proc. of the 12tn world conference on Earthquake Engineering, Paper No. 0918.

Hessami,K., Jamali.,F, and Tabassi,H. 2003. Major Active Faults of Iran, International Institute of Earthquake Engineering and Seismology.

Irikura,K., Miyake,H., Iwata,T., Kamae,K., Kawabe,H., Dalguer,L.A. 2004. Recipe for predicting strong ground motions from future large earthquakes, Proc. of the 13th world conference on Earthquake Engineering, Vancouver, Canada:1371.

Irikura,K. and Kurahashi,S. 2008. Validity of strong motion prediction recipe for inland-crust earthquakes, Proc. of the 14th world conference on Earthquake Engineering, Beijing, China.

Iwata,T., and Irikura,K. 1988. Source parameters of the 1983 Japan Sea earthquake sequence, Journal of Physics of Earth, 36:155–184.

Karashi., J. 2015. Results of microtremor measurements to determine soil dynamic characteristics at Saint Stephano's Church-Jolfa, Master thesis of Azerbaijan Shahid Madani University, Tabriz, Iran.

Midorikawa, S., Matsuoka, M. and Sakugawa, K. 1992. Estimation of ground characteristics in PGA and PGV in the 1987 Chiba Toho-oki Earthquake, Journal of Architectural Institute of Japan, 442, 71–78.

2019 Rock Dynamics Summit– Aydan et al. (eds)
© *2019 Taylor & Francis Group, London, ISBN 978-0-367-34783-3*

The numerical analysis of response and stability of stone masonry bridges in Aizanoi Antique City in Kütahya Province of Turkey

J. Tomiyama, Y. Suda and Ö. Aydan
Department of Civil Engineering, University of the Ryukyus, Nishihara, Okinawa, Japan

H. Kumsar and E. Özer
Department of Geological Engineering, Pamukkale University, Denizli, Turkey

ABSTRACT: Aizanoi or Azan antique city was first established by Phrygians and later become a part of Roman Empire. This city was the first city in the world to construct the flood protection measures such as dams and embankment walls. Out of the four stone masonry bridges, two masonry bridges are remaining and they are still used today. The authors carried out some site investigations and surveying of these remaining stone bridges. In this study, the response and stability of stone bridges are investigated using finite element method under various loading conditions. Furthermore, some Eigen value analysis utilizing the FEM were also carried out, and the results of the investigations and analyses are presented and discussed.

1 INTRODUCTION

The antique city called Azan in Phrygian or Aizanoi in ancient Greek, located in Kütahya Province of Turkey, was established by Phrygians at 3000 B.C. and become a part of Roman Empire in 133 B.C. and then become of a part of Seljuk in 13th Century, later Ottoman Empire and finally Modern Turkey (Figure 1). One can find traces of various civilizations in this city. This city was the first to introduce the fundamentals of the modern stock exchange and modern civil engineering principles.

This city was also the first city in the world to construct the flood protection measures such as dams and embankment walls. In other words, it is the city of the birth of the hydraulic engineering of modern civil engineering. The city had four stone masonry bridges made of marble. The bridges have performed very well even under very heavy trucks of modern times. However, the traffic has been recently restricted to light vehicles such as cars and they are still in use.

The authors have performed some site investigations and surveying on two remaining stone bridges constructed during the Roman period. In this study, the response and stability of stone bridge denoted as Bridge-L are investigated using finite element method under various loading conditions in order to find out why these stone bridges performed very well so far since their construction about 1850 years ago. Furthermore, some modal analysis utilizing the FEM were also carried out. The results of the investigations and analyses are presented and discussed in this paper.

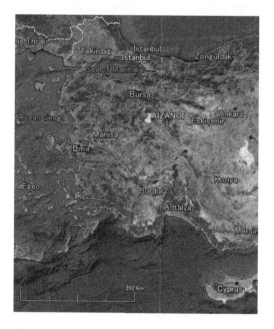

Figure 1. Location of Aizanoi

2 DESCRIPTION OF BRIDGES

There were four stone masonry bridges over Koca (Bedir) Çay in Turkish, which was called Penkalas in Roman period. However, it is very likely that the river could have been called with a different name by Phrygians as western historical sources generally

Figure 2. Views and locations of Bridges

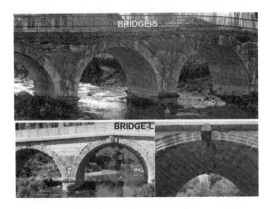

Figure 3. Relative separation and sliding observed in both bridges.

quote ancient Greek or Roman names rather than the original names. Two of these four bridges are remaining and they are still being used today (Figure 2). Two of the bridges are missing while the building stones of one of the missing bridges were recently retrieved during the rehabilitation works. The bridges were made of white marbles presumed to be extracted from nearby ancient quarries at Göynükören, which is about 12 km to the east.

There are no official or historical names for these remaining bridges. The authors would denote bridges as Bridge-L and Bridge-S with the consideration of their length. Some spalling and slight relative separation and sliding of blocks were observed at both bridges as seen in Figure 3. The bridges were subjected to freezing and thawing cycles for the last 1850 years at least. In this study, the bridge denoted Bridge-L, which is about 33m long, would be investigated in details.

The Bridge-L consists of semi-circular five-arches with different radii (Figure 4) built in 157 AD. The maximum arch width is 6.6m and foundation width is about 1m. It is known that wooden piles are employed in the foundations of Cybele (wrongly known as Zeus) temple. Therefore, it is very likely that the wooden piles might have been used in the

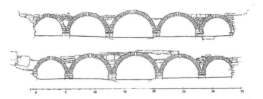

Figure 4. Drawing of Bridge-L (Hoffman and Rheidt, 1995).

Figure 5. A view of Bridge-L after dismantling the asphalt pavement.

foundations of the bridges against erosion as well as their uneven settlement in soft ground in river beds, which may eventually lead the collapse of the bridges. It must be also noted the region is prone to earthquakes, the use of wooden piles could be also as a counter measure against earthquakes (Aydan et al. 2003a,b), which is not well-understood by western archeologists from non-seismic countries (i.e. Rheidt 1990). The asphalt pavement of this bridge has been recently dismantled to observe the original structure as shown in Figure 5.

3 REGIONAL SEISMICITY

The region is no exception in Turkey and it suffers earthquakes from time to time. Figure 6 shows the earthquakes with a magnitude of 4 or more in the last 50 years. The most recent largest event was the 1970 Gediz earthquake with a magnitude of 7.2. The epicentral distance of this earthquake to

Figure 6. Seismicity in the vicinity of Aizanoi.

Figure 7. Damaged and partially restored section of the Cybele temple due to the 1970 Gediz earthquake.

Figure 8. Damage to columns of the Cybele temple and walls and stairs of the theatre.

the Aizanoi was about 10 km and four columns and one free standing column at the NE corner of the Cybele temple toppled in the direction towards the epicenter. Three columns were re-erected and two columns remain on the ground (Figure 7). One can also see various damage to columns and walls of this temple and other monumental structures in Aizanoi antique city (Figure 8).

Another large earthquake with a surface magnitude of 6.0 occurred in Simav on May19, 2011. However, this earthquake did not cause any damage to Aizanoi (Kumsar et al. 2015) as the epicentral distance was more than 50 km and the estimated acceleration was less than 30 gals (Aydan 2012).

4 MECHANICAL PROPERTIES OF MARBLES AND BLOCK INTERFACES

The mechanical properties of marbles and block interfaces have not been tested as sampling are not allowed from this historical bridges. The authors

(a) Göynükören antique quarries

(b) Örencik antique quarry

Figure 9. Views of antique quarries

located the ancient quarries for sampling on rock blocks and their interfaces. The ancient quarries are found to be at Göynükören and Örencik villages (Figure 9). The investigation at the ancient quarries indicated that the quarrymen at ancient times utilized rock discontinuities to their advantage to extract marble blocks. The authors have recently obtained rock samples for uniaxial and Brazilian

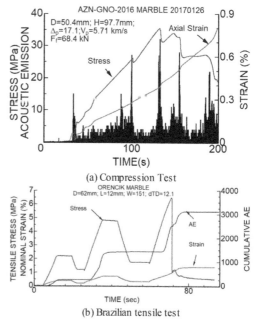

(a) Compression Test

(b) Brazilian tensile test

Figure 10. Strain-stress relations of marble samples of Aizanoi under uniaxial compression and Brazilian tests.

tensile tests as well as the rock plates for friction properties of block interfaces. Although the experiments have not been completed yet, some preliminary tests are shown in Figure 10.

The mechanical behaviour of contacts between blocks is the most important parameter governing the overall stability of the masonry structures. An experimental program has been initiated to test the mechanical behaviour of the contact conditions of marble blocks under static and dynamic direct shear loading conditions using a high capacity dynamic shear testing machine at the Civil Eng. Dept., the University of the Ryukyus. In addition of some tilting tests are performed on the marble and limestone blocks with surface conditions. Table 1 summarizes the friction angle obtained from tilting tests. The friction angle of dimension stones of marble and limestone is generally high unless polished and they approach to their intrinsic friction angles under high normal stresses. Figure 11 shows a shear strength criterion for conventional marble contacts using the criterion of Aydan et al (1996).

5 NUMERICAL ANALYSES

A series of numerical analyses were carried out using the non-linear finite element code MIDAS-Civil (2015). Four different conditions are analysed as listed in Table 2. The material properties used in the analyses are given in Table 3. Figure 12 shows the details of the finite element meshes for each part of the bridge. Figure 13 shows the boundary conditions. Subgrade stiffnesses were introduced

Table 1. Friction properties of marble contacts.

Surface condition	Friction angle (degrees)
Polished	22-27
Saw-cut	32-33
Rough	38-41

Table 2. Details of numerical analyses cases

Case No	Details
1	Modal analyses (Eigen values)
2	Self-weight only
3	Self-weight + 20 ton vehicle on the center arch
4	Self-weight + 20 ton vehicles on three arches

Table 3. Friction properties of marble and limestone blocks.

Parameter	Marble	Contact
Density (kN/m³)	26.5	
Elastic Modulus (GPa)	17.72	
Poisson's ratio	0.23	
Cohesion (MPa)		0
Friction angle (degree)		30
Normal stiffness (N/mm³)		10^{10}
Shear stiffness (N/mm³)		1.12

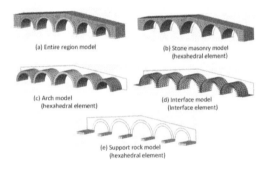

(a) Entire region model

(b) Stone masonry model (hexahedral element)

(c) Arch model (hexahedral element)

(d) Interface model (Interface element)

(e) Support rock model (hexahedral element)

Figure 12. Exploded view of finite element meshes for each part of the arch bridge

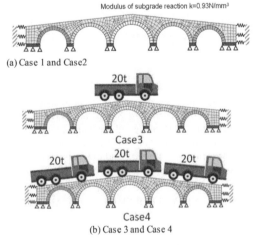

Modulus of subgrade reaction k=0.93N/mm³

(a) Case 1 and Case2

20t

Case3

20t   20t   20t

Case4

(b) Case 3 and Case 4

Figure 13. Illustration of boundary and loading conditions for different cases.

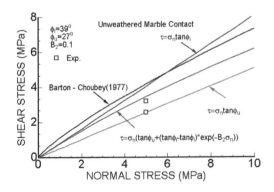

Figure 11. Shear strength envelope of marble block contacts

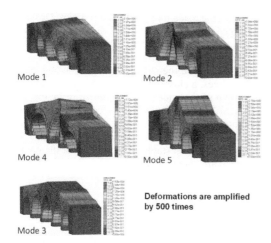

Mode 1   Mode 2

Mode 4   Mode 5

Deformations are amplified
by 500 times

Mode 3

Figure 14. Deformed configuration of the bridge for five
different modes

Table 4. Results of the modal analyses.

Mode No	Natural Frequency (Hz)	Natural Period (s)
1	11.82	0.084
2	18.40	0.054
3	24.36	0.041
4	27.64	0.036
5	28.90	0.035

in order to account the lateral ground reaction and
possible relative settlement of foundations.

5.1 Modal Analyses (Case 1)

Figure 14 shows the deformation of the bridge for
five different modes. Table 4 summarizes the eigen
values of the bridge for each mode. One of important
feature of this results is that the natural frequency is
quite different from the frequency content of earthquakes which ranges between 1-5 Hz.

5.2 Self-weight Loading (Case 2)

The stone masonry bridges are designed on the
concept of arch action, which induced compressive
stresses within the blocks without any inter-block
sliding. Different arch configurations are utilized
for this purpose. The original arching concept was
developed by Sumerians more than 5000 years ago
and it has been used widely since then (Aydan 2008,
2014). Figure 15 shows the deformation of the arch
bridge under its own self-weight. The maximum displacement occurs at the center of the widest arch.

Figure 16 shows the principal stress distributions
in the arch (tension is assumed to be positive). It
is interesting to note that the arch is subjected to

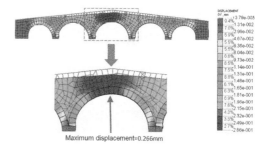

Maximum displacement=0.266mm

Figure 15. Deformed configuration of the arches under
their own self-weight.

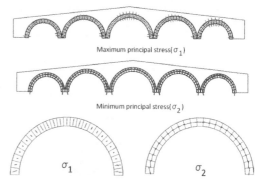

Maximum principal stress(σ_1)

Minimum principal stress(σ_2)

σ_1 σ_2

Figure 16. Principal stress distribution in the arches of the
bridge.

all around tangential compressive stresses, which
is of paramount importance for the validity of the
arching concept. Nevertheless, small radial tensile
stresses develop in the arch blocks. This may imply
that the blocks may split at the mid point if such
stresses exceed their tensile strength.

5.3 Self-weight and Vehicle Loading (Case 3 & 4)

The passage of the trucks and buses (generally more
than 10 tons and may be up-to 20 tons) is now forbid-
den. However, it was known that some heavy trucks
up to 20 tons passed over the bridge. Sometimes,
three trucks might have passed over the bridges.
These two possible conditions were carried out.
Following the application of the self-weight loading, loads resulting from the vehicles are applied to
the model and a series of non-linear finite element
analyses which allow the relative slip among blocks
along contacts, were carried out. Figure 17 shows
the deformed configuration of the arch bridge for the
final step of computation. The number of vehicles
did not have much influence on the overall deformation and stress state within the bridge. The numerical analyses indicated that the overall deformation
and stress state are influenced by the self-weight of
the arch bridge. Nevertheless, the consideration of
dynamic loading conditions may lead to the different conclusions.

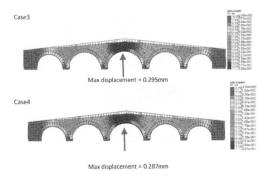

Case3

Max displacement = 0.295mm

Case4

Max displacement = 0.287mm

Figure 17. Deformed configuration of the bridge for Case 3 and Case 4 loading condition at the final step of the computation.

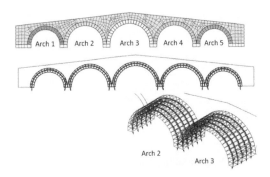

Arch 1   Arch 2   Arch 3   Arch 4   Arch 5

Arch 2

Arch 3

Figure 18. Compressive principle stresses in the arches of the bridge for Case 4 at the last step of computation.

Figure 18 shows the compressive principle stresses in the arches of the bridge for Case 4 at the final step of computation. As pointed above, the self-weight of the structure has much more influ-

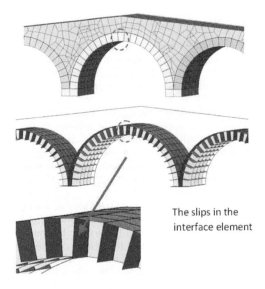

The slips in the interface element

Figure 19. Relative slips at the crown of the widest arch.

ence than the loads induced by the vehicles. Despite additional 60 tons load due to vehicles, the stresses in the arches are compressive, which implying the arch action is not disturbed by the vehicles.

In order to see the slip among blocks along contact surfaces, a close-up view of the deformation in the crown of the widest arch is shown in Figure 19 for CASE 4 at the final step of computation. The computations indicated that the relative slips along the contacts of the marble blocks are quite small and the computation is stable. In other words, the computations are convergent.

6 CONCLUSIONS

The authors investigated the response and stability stone masonry bridges constructed during the Roman period in Azan (Aizanoi) antique city under different loading conditions. This study on the stone masonry bridges existing in this ancient city is of the first kind, and the in-situ investigations, experiments and numerical analyses clearly showed why these bridges have performed very well so far since their construction about 1850 years ago. Nevertheless, further experiments on the mechanical characteristics of marble blocks and contacts are necessary.

The region is subjected to earthquakes from time to time. Further numerical analyses are necessary on the dynamics response and stability of the bridges in this ancient city.

Two bridges are missing. The causes of the collapse of these bridges need to be clarified by further investigations.

REFERENCES

Aydan, Ö. 2008. New directions of rock mechanics and rock engineering: Geomechanics and Geoengineering. Keynote Lecture, 5th Asian Rock Mechanics Symposium (ARMS5), Tehran, 3-21.

Aydan, Ö., 2012. Ground motions and deformations associated with earthquake faulting and their effects on the safety of engineering structures. *Encyclopaedia of Sustainability Science and Technology*, Springer, R.Meyers (Ed.), 3233-3253.

Aydan, Ö. 2014. Future advancement of rock mechanics and rock engineering (RMRE). Keynote Lecture, *ROCKMEC'2014-XIth Regional Rock Mechanics Symposium, Afyonkarahisar, Turkey*, 27-50.

Aydan, Ö., Y. Shimizu, T. Kawamoto 1996. The anisotropy of surface morphology and shear strength characteristics of rock discontinuities and its evaluation. *NARMS'96*, 1391-1398.

Aydan, Ö., Y. Ogura, M. Daido N. Tokashiki S. Irabu, K. Takara 2003a. Re-assessment of the seismic response and stability of stone masonry structures of Shuri castle. *Proceedings of Industrial Minerals and Building Stones, IMBS '2003*, pp.109-117.

Aydan, Ö., Y. Ogura, M. Daido N. Tokashiki 2003b. A model study on the seismic response and stability of masonry structures through shaking table tests. *Fifth National Conference on Earthquake Engineering, Istanbul, Turkey, Paper No:AE-041(CD)*

Barton, N. and Choubey, V. 1977. The shear strength of rock joints in theory and practice. Rock Mechanics, 10, 1-54.

Hoffman, A., and Rheidt, K. 1995. Die ausgrabungen in Aizanoi 1990. T.C. Kültür Bakanlığı Anıtlar ve Müze Genel Müdürlüğü, XIII Kazı Sonuçları Toplantısı, II, 323-340.

Kumsar, H., Ar, E., Aydan, Ö. 2015. The Characteristics of 2011 Simav Earthquake (Turkey) with an Emphasis on Geotechnical Damage. G. Lollino et al. (eds.), Engineering Geology for Society and Territory – Volume 5, 1059-1063, DOI: 10.1007/978-3-319-09048-1_201, Springer Int. Pub Switzerland..,

MIDAS-Civil 2015. Finite Element Code. MIDAS Information Technology, Tokyo Branch.

Rheidt, K. 1990. Pile Foundation in the Anatolian Mountains – Wrong Technique at the Wrong Place? Proceedings of the Third International Congress on Construction History, Cottbus.

2019 Rock Dynamics Summit– Aydan et al. (eds)
© 2019 Taylor & Francis Group, London, ISBN 978-0-367-34783-3

Analysis of the stone wall damage of Kumamoto Castle by 2016 Kumamoto earthquake using 3D laser scanner and ground survey

A.T. Hashimoto
Department of Civil Engineering, Kokushikan University, Japan

B. K. Ishidukuri & T. Matsuo
Nissoku Corporation, Japan

ABSTRACT: In the Kumamoto earthquake in 2016, the stone wall of Kumamoto Castle has reached 30% of the total, but the cause of the damage is not clear. Therefore, in this paper, the structure of each stone wall and the back ground were classified into three types of stone base, semi-stone base and non-stone base type, and the relationship between the shape and damage ratio was analyzed. As a result, it was revealed that the damage mechanism of Kumamoto Castle stone wall was the main cause of the collapse of the stone base type, and the shocking induced the collapse. Analysis of liquefaction damage using GIS in the liquefaction district of Kumamoto city by on-site survey and survey in the 2016 Kumamoto earthquake.

1 INTRODUCTION

The 2016 Kumamoto earthquake occurred in the whole area of Kumamoto prefecture with the M6.5 front shock and the M7.3 main shock. In Kumamoto Castle, 30% of the total stone walls caused extensive damage such as collapse and infestation, as shown in Photo 1 and Photo 2. However, the cause of the damage is not clear. Therefore, this paper analyzed the damage ratio by dividing the type of The castle stone wall into three types, stone base, semi-stone base and non-stone base type. As a result, we aim to clarify the mechanism of the cause of The castle stone wall's collapse.

2 THE CASTLE STONE WALL'S REPAIR AND DAMAGED

In the Kumamoto earthquake in 2016, the damage of Kumamoto Castle caused by the foreshock occurred in 10 places that have been restored so far as shown in Figure 2 (a). In the main shock, as shown in Figure 2 (b), most of the parts that have been repaired have been damaged.

(a)Hayaken stone wall (b) Kita Juhachiken Yagura

Figure 1. Collapse of Important Cultural Property

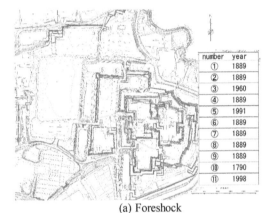

(a) Foreshock

number	year
①	1889
②	1889
③	1960
④	1889
⑤	1991
⑥	1889
⑦	1889
⑧	1889
⑨	1889
⑩	1790
⑪	1998

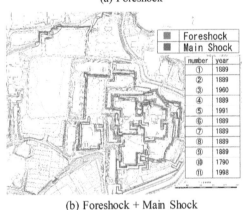

■ Foreshock
■ Main Shock

number	year
①	1889
②	1889
③	1960
④	1889
⑤	1991
⑥	1889
⑦	1889
⑧	1889
⑨	1889
⑩	1790
⑪	1998

(b) Foreshock + Main Shock

Figure 2. Damage to restoration points of The castle stone wall by the Kumamoto earthquake

3 PLACESAGE OF THE CASTLE STONE WALL CONSTRUCTION AND DAMAGED PLACES

Figure 3 is a plan view of the age of building The castle stone wall and its damage location. The pink color (A to Z) is the position 3) where Kuwabara measured the section of 23 stone walls in 1984 before the earthquake. Green (FK 1 - 28) is the position 4) Fukuda measured the section of the stone wall in 28 places of Kumamoto castle before the earthquake. Furthermore, we created 9 blue cross sections (KC 1 ~ 9) according to the report of Kumamoto Castle Research Center. From the above, the total is 49 cross sections

Figure 4 is a cross-sectional view showing the shape and damage of each The castle stone wall construction age.Hashimoto and Ishidukuri surveyed the same place before the earthquake measured by Kuwabara.I made a cross section of the stone wall before and after the earthquake occurred and investigated the state of the scattering deformation.

(a) Phase I (1599) is located in the center of the vicinity of the castle tower, many of which are not collapsed because they are made with a gentle slope. However, when it comes to a steep slope, it is causing a scatter.

(b) Phase II (1600), the steep gradient near the Iida Maru 5th Floor Oborg is causing scratches and collapse.

(c) Phase III (1601) has a gentle slope and no damage.

(d) Phase IV (1601-1607), many damaged stone walls such as the Udo Oborg, where the collapse occurred, the Kitaeighth tower, the Touzen-eighta oar, etc. were built There.

(e) Phase V (1607) has many corner parts such as many Bulging.

(f) Phase VI (1633 - 1820) is all Bulging.

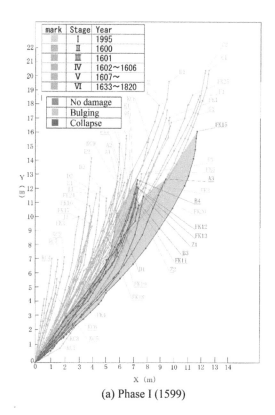

(a) Phase I (1599)

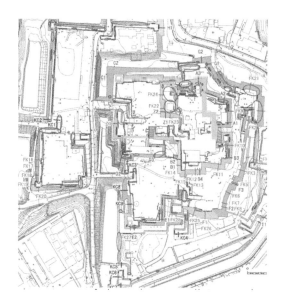

Figure 3. Age of The castle stone wall Construction and Damaged Places

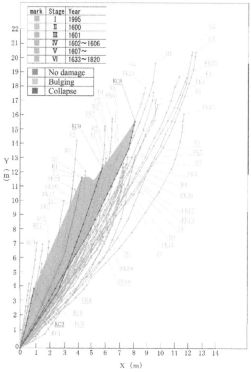

(b) Phase II (1600)

576

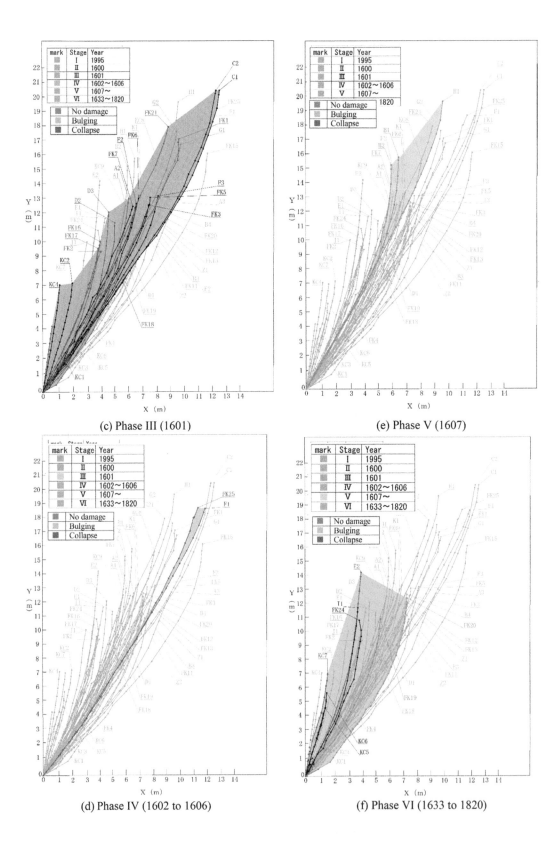

(c) Phase III (1601)

(e) Phase V (1607)

(d) Phase IV (1602 to 1606)

(f) Phase VI (1633 to 1820)

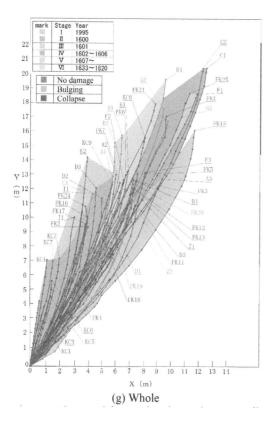

mark	Stage	Year
	I	1995
	II	1600
	III	1601
	IV	1602~1606
	V	1607~
	VI	1633~1820

	No damage
	Bulging
	Collapse

(g) Whole

Figure 4. Shape and damage by The castle stone wall construction age

4 TYPES OF THE CASTLE STONE WALL AND ANALYSIS OF DAMAGE CONDITIONS

4.1 *Type classification of The castle stone wall*

Figure 5(a) shows the position of the stone base type. The stone base type also has a core with a stone base. Figure 5(b) shows the position of the semi-stone base type. There are three types of semi-stone base type: loose ground from the bottom of the stone wall, loose ground behind the stone wall, solid ground behind the ground. Figure 5(c) shows the position of the non-stone base type. There are the same three as the semi-stone base type for the non-stone base type.

4.2 *Relationship between stone base type and shape and degree of damage*

Figure 6 shows the detailed damage of each type of stone wall. Figure 7 shows the shape (height and slope) of the stone wall and the damage situation,

summarized in a graph from the bottom of the stone wall.

The shape of The castle stone wall uses the result of the cross section actually measured by The castle stone wall before the earthquake by Kuwabara, Fukuda and Kumamoto Castle Research Center.

In this figure, we made three colors of collapsed part (red wire) and part where it was engulfed (orange wire), no damage (black wire).

The castle stone wall's giraffe part was extracted from the 3D survey 1), 2) after the earthquake by comparison before and after the earthquake.

Furthermore, the type of the stone wall was classified into a stone base type (yellow), a semi-stone base type (green), and a non-stone base type (light blue). The stone base type (yellow) in Figure 7 (a) has a low height but a steep gradient, so many are collapsing. The half-stone base type (green) in Figure 7 (b) has a high height and a gradual slack a little, many of which are congested. The non-stone base type (light blue) in Figure 7 (c) has a high height but a gradual slope, so it is less collapsed.

Figure 8 shows the classification of the stone base type obtained from the extension of Kumamoto Castle. The stone base type is 26%, including 21% only for stone base and 5% with core. Semi-stone base type is 11% of loose ground, 31% of slightly loose ground and 48% of 6% of hard ground. Non-stone base type is 13% of loose ground, 6% of slightly loose ground and 26% of 7% of hard ground.

Figure 9 shows the ratio of the degree of damage for each stone wall type. From this graph, it can be seen that many of the stone base types are collapsing even if the height is low, although the gradient is steep. In addition, it can be seen that there are many damage of collapse or pregnancy in semi-stone base type and non-stone base type embankment type. On the other hand, it can be seen that both the semi-stone base type and the non-stone base type ground type have minimal damage.

5 CONCLUSION

Most stone base types were collapsed due to the steep slope despite the low height. It was found that the embankment type of semi - stone base and non - stone base was damaged by collapse or infestation. On the other hand, it was found that the ground type of both the semi-stone base and the non-stone base are minor damage. In other words, the cause of the damage of Kumamoto Castle was revealed that the stone base type was the cause of disintegration, and the shocking event occurred as a disintegration trigger.

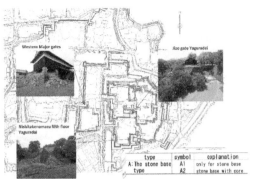

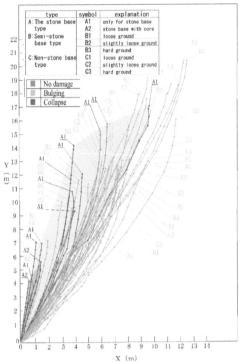

type		symbol	explanation
A:The stone base type		A1	only for stone base
		A2	stone base with core
B:Semi-stone base type		B1	loose ground
		B2	slightly loose ground
		B3	hard ground
C:Non-stone base type		C1	loose ground
		C2	slightly loose ground
		C3	hard ground

■	No damage
■	Bulging
■	Collapse

(a)The stone base type

type	symbol	explanation
A:The stone base type	A1	only for stone base
	A2	stone base with core

(a)The stone base type

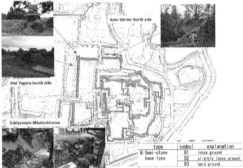

type	symbol	explanation
B:Semi-stone base type	B1	loose ground
	B2	slightly loose ground
	B3	hard ground

(b)The semi-stone base type

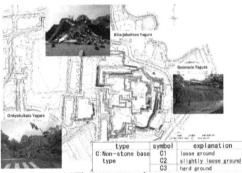

type	symbol	explanation
C:Non-stone base type	C1	loose ground
	C2	slightly loose ground
	C3	hard ground

(c)The non-stone base type

Figure 5. Position diagram of Stone type

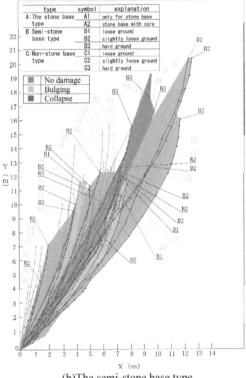

type		symbol	explanation
A:The stone base type		A1	only for stone base
		A2	stone base with core
B:Semi-stone base type		B1	loose ground
		B2	slightly loose ground
		B3	hard ground
C:Non-stone base type		C1	loose ground
		C2	slightly loose ground
		C3	hard ground

■	No damage
■	Bulging
■	Collapse

(b)The semi-stone base type

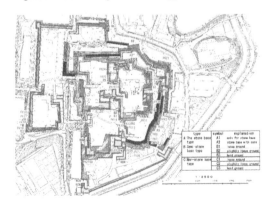

Figure 6. Degree of damage per detail stone wall type

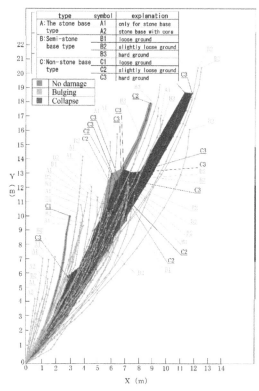

type	symbol	explanation
A:The stone base type	A1	only for stone base
	A2	stone base with core
B:Semi-stone base type	B1	loose ground
	B2	slightly loose ground
	B3	hard ground
C:Non-stone base type	C1	loose ground
	C2	slightly loose ground
	C3	hard ground

▨	No damage
▨	Bulging
■	Collapse

(c)The non-stone base type

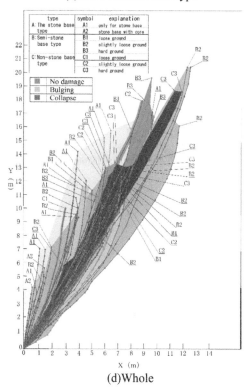

type	symbol	explanation
A:The stone base type	A1	only for stone base
	A2	stone base with core
B:Semi-stone base type	B1	loose ground
	B2	slightly loose ground
	B3	hard ground
C:Non-stone base type	C1	loose ground
	C2	slightly loose ground
	C3	hard ground

▨	No damage
▨	Bulging
■	Collapse

(d)Whole

Figure 7. Shape and damage situation of The castle stone wall

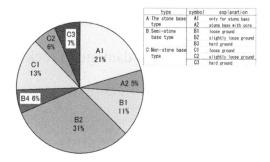

type	symbol	explanation
A:The stone base type	A1	only for stone base
	A2	stone base with core
B:Semi-stone base type	B1	loose ground
	B2	slightly loose ground
	B3	hard ground
C:Non-stone base type	C1	loose ground
	C2	slightly loose ground
	C3	hard ground

Figure 8. Classification of The castle stone wall type

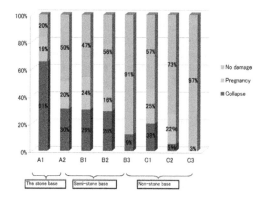

Figure 9. Percentage of Damage Degree by The castle stone wall Type

ACKNOWLEDGMENT

We cooperated with the staff of Kumamoto Castle Research and Research Center for preparing this paper. Thank you for writing.

REFERENCES

1) Takao Hashimoto, Katsuya Ishidukuri: Deformation analysis of Kumamoto Castle stone wall using 3 dimensional laser scanner, The 72nd Annual Scientific Lecture, Civil Engineering Society, 2017.
2) Takao Hashimoto, Sayaka Suzuki, Katsuya Ishidukuri: A study on damage survey of Kumamoto castle Ishigaki by the 2016 Kumamoto earthquake using infrared thermotracer and 3D laser scanner, bulletin of the Faculty of Science and Technology, Kokushikan University, peer review, No. 11, 2017. 11.
3) Fumio Kuwabara: The castle stone wall gradient of Kumamoto Castle, Nippon Institute of Technology research report Vol. 14, No. 2, 1984.
4) Public benefit organization corporate civil engineering association Kansai branch: research committee report on preservation and utilization of historical ground site, research committee on conservation and utilization of historic ground ruins, pp. 79-105, 2018.9.

2019 Rock Dynamics Summit– Aydan et al. (eds)
© *2019 Taylor & Francis Group, London, ISBN 978-0-367-34783-3*

Some examples of damage to rock masonry structures caused by recent earthquakes

N. Tokashiki & Ö. Aydan

Department of Civil Engineering, University of the Ryukyus, Nishihara, Okinawa, Japan

ABSTRACT: Historical structures are mainly masonry structures, and they are composed of blocks made of natural stones, bricks or both, and they are built in different patterns with or without mortar. As Japan is a seismically active country, an emphasis was given on the seismic response and stability of restored masonry structures such as arches and walls during earthquakes and natural rock structures. Furthermore, The deterioration by atmospheric agents, bombing and seismic shaking damaged some of these structures. The authors describes the outcomes of experimental, limiting equilibrium and numerical studies on the stability of historical masonry structures under dynamic loads such as those induced by earthquakes and their implications are discussed.

1 INTRODUCTION

Historical structures are mainly masonry type, and they are composed of blocks made of natural stones, bricks or both, and they are built in different patterns with or without mortar. The authors have been recently involved with the restoration of the famous historical structure in Okinawa Island such as Shuri Castle, Nakagusuku Castle, Katsuren Castle, Gushikawa Castle, an arch bridge in Iedonchi royal garden and Yodore royal mausoleum of the Ryukyu Imperial period, as well as the assessment of static and dynamic stability of some natural rock structures such as Wakariji.

As Japan is a seismically active country, an emphasis was given on the seismic response and stability of restored masonry structures such as arches and walls during earthquakes and natural rock structures. Furthermore, The deterioration by atmospheric agents, bombing and seismic shaking damaged some of these structures.

Different arch configurations used in Shuri Castle in Okinawa Island were tested. The stability of the dynamic arch bridge of Iedonchi Royal Garden and Wakariji natural rock structure were investigated using the physical models (Tokashiki et al. 2007). Furthermore, dynamic limiting equilibrium methods (D-LEM) as well as numerical methods (i.e. conventional finite element method (FEM), discrete finite element method (Aydan 1998; Aydan et al. 1996; Mamaghani et al. 1989).

The authors describes the outcomes of experimental, limiting equilibrium and numerical studies on the stability of historical masonry structures under dynamic loads such as those induced by earthquakes and their implications are discussed.

2 RETAINING WALL COLLAPSE OF KATSUREN CASTLE

An earthquake with a magnitude of 7.2 on Japan Meteorological Agency Scale occurred near Okinawa Island of Japan (Fig. 1). The focal depth of the quake, which occurred at 5:31 a.m. local time (2031 GMT Friday), was about 40 km under the sea 107 km east off Naha, capital of Okinawa.

The 2010 earthquake caused the collapse of some parts of castle walls at Katsuren Castle, which is designated as a world heritage site. The castle is located over a hill in Uruma City and the nearest

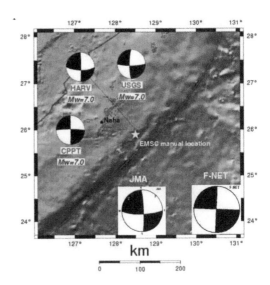

Figure 1. Location and focal mechanism of the 2010 earthquake

Figure 2. A views of the failed retaining wall

strong motion station of the K-NET strong motion network is Gushikawa. The NW corner of the castle wall with a height of 4m collapsed and there were numerous dislocations and rotation of blocks in the castle as seen in Fig. 2.

The authors carried out a series of analyses using the dynamic limit equilibrium method (D-LEM) and the acceleration records at Chinen and Gushikawa strong motion stations of the K-NET strong motion network. The typical size of the blocks ranges between 50 to 60 cm. Fig. 3 illustrates the geometry of retaining wall.

The authors used their method (Tokashiki et al. 2007) to back-analyze the collapse of the wall using the strong motion records taken at Gushikawa and Chinen. The wall is stable against toppling mode for strong motions recorded at Gushikawa and Chinen strong motion stations. If the records taken at Gushikawa are used, the relative slid-

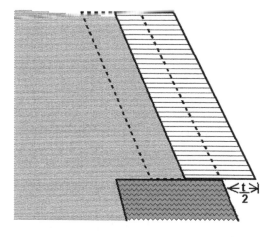

Figure 3. The idealized geometry of the failed retaining wall.

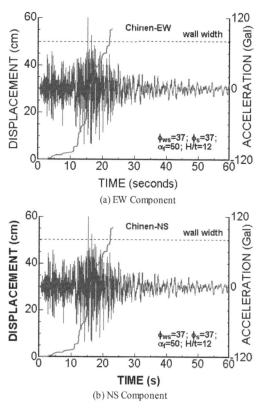

(a) EW Component

(b) NS Component

Figure 4. Sliding response of collapsed retaining wall for EW and NS components taken at the Chinen strong motion station

ing can not be greater than 10cm, which implies that the wall should be stable although some slip might take place. However, if the records taken for EW and NS components at Chinen are used the relative sliding can be greater than 60 cm for $\theta = 5°$, which exceeds the half size of the block and this implies that the wall should collapse (Fig. 4). The bulging of the wall and inclination of the foundation rock strongly supports that this condition would be prevailing at the location of the collapse. As the castle is situated on the top of the hill, it is likely that ground motions might have been amplified also.

The authors have initiated a multi-parameter monitoring program at the Katsuren Castle and micro-tremor measurements. The multi-parameter monitoring system was installed at the north east corner of the castle, where some spalling and separation of blocks were observed. The system involves the real-time monitoring of relative displacement, acoustic emissions, temperature, humidity, air pressure, geo-electrical potential and accelerations (Fig. 5). A OA-SYC stand-alone type three component accelerometer is installed. This accelerometer can be set into different modes. For long-time monitoring, the accelerometer is set to trigger mode

(a) Location of transducers and sensor

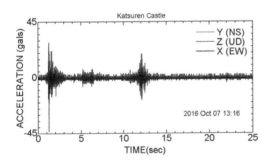

(b) monitoring box.

Figure 5. View of installed devices and monitoring box

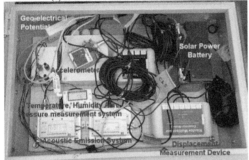

Figure 6. An acceleration record taken at the Katsuren Castle.

with a chosen threshold. Fig. 6 shows an example of acceleration record measured by the system in 2016, The system utilizes solar energy as a power source and it is an environment-friendly.

Some micro-tremor measurements are undertaken at the Katsuren Castle at different level. One of the purposes was to evaluate the frequency characteristics of the castle at various levels (Fig. 7). In addition, some measurements were also taken at the top of the retaining walls including the one collapsed and restored after the 2010 earthquake. The Fourier spectra of each channel at three levels were evaluated from the measurements and they

Figure 7. Locations of micro-tremor measurements

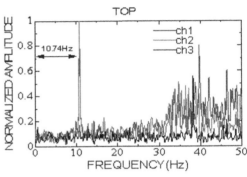

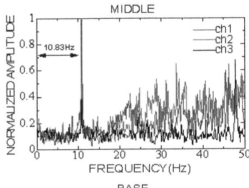

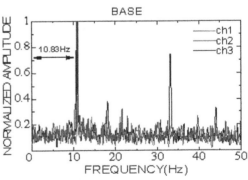

Figure 8. Fourier spectra of records at three levels.

are shown in Figure 8. The Fourier spectra was almost same for three levels and there was no big difference.

3 MEASUREMENTS AND OBSERVATIONS NAKAGUSUKU CASTLE

A multi-parameter monitoring system was also initiated by the authors at Nakagusuku Castle. The system at the castle was actually installed about 3 years before the one installed at the Katsuren Castle, which is probably the first attempt regarding the masonry structures in the world. The monitoring was initiated in December 2013 and it has been still continued. During the period of measurements, some earthquakes occurred and long-term creep-like separation of a huge crack in Ryukyu Limestone layer extending to the Shimajiri formation layer has been taking place. Fig. 9 shows the installation location.

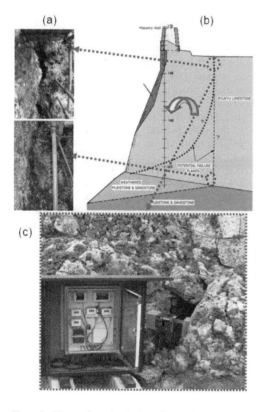

Figure 9. Views of monitoring location and instrumentation

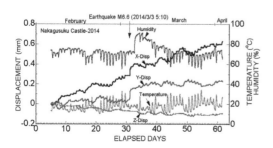

Figure 10. Monitoring results during Feb. to March, 2014.

An earthquake with a moment magnitude of 6.5 occurred at 5:10 AM on 2014 March 13 (JST) in East China Sea at a depth of 120 km on the western side of the Okinawa Island. Another earthquake occurred at 11:27 AM on the same day near Kumejima Island. Although the magnitude of the earthquake was intermediate and far from the location, some permanent displacement occurred as seen in Fig. 10.

A series of analysis using DFEM (Aydan et al. 1996) together with implementation of softening and hardening process of weathered rock mass in relation to rainfall (Aydan 2016) was carried out. Figure 11 shows the mesh used in analyses and comparison of computed results with measurements. The computational model can simulate both permanent displacement induced by the earthquake as well as rainfall induced softening process of weathered rock mass.

At this site, there is a monumental natural rock block used for religious purpose in historical times. According to archeologists, this natural block was toppled about 400 years ago during an earthquake. It was restored to its original position about 3 years ago (Fig. 12). It was also an important evidence of past earthquakes occurring in Okinawa Island or its close vicinity. A series of dynamic analyses on

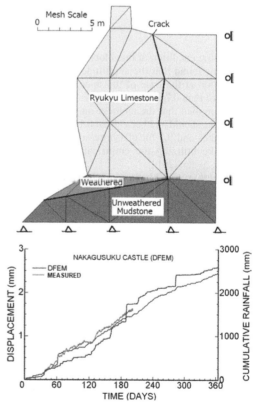

Figure 11. Mesh used DFEM analysis and comparison with measured responses.

Figure 12. Views of toppled block before and after restoration

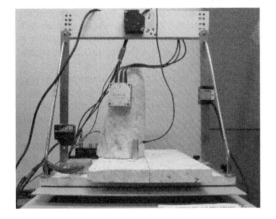

Figure 13. A view of shaking experimental set-up.

the toppling condition of this block was carried out and it was found that the amplitude of the ground acceleration at the base of this block should have been more than 0.7g. As the block seats on the rock base, it implied that the moment magnitude of the earthquake should had been about 8.1 for intra-plate earthquakes and 9.1 for off-shore subduction plate earthquake utilizing the empirical relations proposed by Aydan (2012).

A coral limestone block, which had a very similar geometrical configuration, was subjected to shaking and its stability was investigated. Fig. 13 shows a view of the coral limestone block together with the experimental set-up, which involves the measurement of accelerations, relative displacement and acoustic emissions. Fig. 14 shows the acceleration, acoustic emissions (AE) and relative displacement responses during the experiment. As the friction angle between the limestone and coral limestone was saw-cut, the block instability occurred due to sliding rather than toppling. Nevertheless, some rocking type movements was also observed.

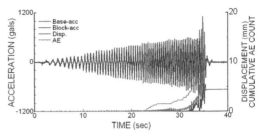

Figure 14. Measured responses of acceleration, relative slip and acoustic emissions.

4 THE DYNAMIC STABILITY OF WAKARIJI NATURAL ROCK BLOCK

There is a natural rock block, known as "Wakariji Stone" in Urasoe City of Okinawa Island, Japan, which is made of Ryukyu Limestone. It is also known as "Needle Rock" internationally. Wakariji stone is about 13m high with an eliptical base. The short axis is about 7.8m wide while the long axis is about 10m.

The authors were asked to evaluate the stability of this natural rock block under both static and dynamic conditions. As stability chart as shown in Fig. 15 was prepared for different modes of failure of rock block. The friction angle of the block with its base was estimated to be ranging between 40 and 46 degrees. As seen from the figure, the block is likely to topple rather than slide over of its base.

Besides site investigations and kinematic analyses of Wakariji Stone and kinematic analyses, a physical model of Wakariji Stone was prepared at a scale of 1/50 using a material having density similar to original rock material.

Two type of experiments were carried out. First several tilt tests were carried out in order to estimate the initiation of instability of the physical model and to check the validity of the stability chart. The second type experiment was shaking experiments. However, the base of the physical model was not fully corresponding to the actual situation and it was

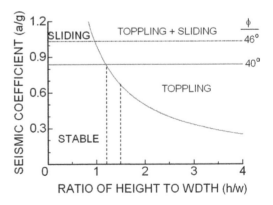

Figure 15. Stability chart for the Wakariji stone

585

Figure 16. A view of tilting experimental set-up.

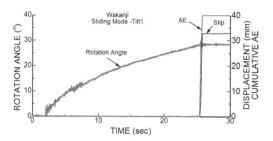

Figure 17. Measured responses of acceleration, relative slip and acoustic emissions.

Figure 18. A view of tilting experimental set-up.

build on a base with a width 0.923 time its height. Futhermore, the base was smooth with a friction angle of 30 degrees on the Ryukyu Limestone base plate. Fig. 16 shows the physical model in a tilting test while Fig. 17 shows the relation between the rotation angle and relative slip. The physical model slip over the base platen when the rotation angle reached the value of 28.5 degrees.

Next shaking table experiments were carried out. Fig. 18 shows the physical model during

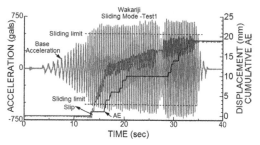

Figure 19. Measured responses of acceleration, relative slip and acoustic emissions.

shaking-table test while Fig 19 shows the acceleration, AE and relative slip responses as a function of time. The slip of the model occurred at the base acceleration estimated from friction angle.

5 CONCLUSIONS

Historical structures are mainly masonry structures and there are also monumental natural rocks. The historical structures, which are generally composed of blocks made of natural stones, bricks or both, are the ones remain over centuries or thousands years. In this study, the authors considered the stability of some masonry structures and monumental natural rock structures. The stability and response of these structures were investigated through an integrated approach involving multi-parameter in-situ monitoring system, dynamic physical tests, dynamic limiting equilibrium and numerical methods. It is shown that the stability and responses of various masonry structures and monumental natural rock structures can be estimated using the dynamic limiting equilibrium and numerical methods if appropriate considerations are given to the structure in hand.

ACKNOWLEDGEMENTS

The authors greatefull acknowledge Mr. M. Toguchi from Nakagusuku Village Education, Committee, Mr. K. Takara of Mahae Consultants Company for introducing the structures studied in this study and financial funding. Okinawa Shimatate-Kyokai is also greatefully acknowledged for providing financial funding for multi-parameter monitoring at Katsuren Castle.

REFERENCES

Aydan, Ö., 1998. Analysis of masonary structures by finite element method, *Proceedings of Prof. Dr. Rifat Yarar Symposium*, Istanbul, TDV, 141–150.
Aydan, Ö., Shimizu, Y., & Kawamoto, T. 1992. The stability of rock slopes against combined shearing and sliding

failures and their stabilisation. *Reg. Symp. on Rock Slopes*, India, 203–210.

Aydan, Ö., Mamaghani, I.H.P., Kawamoto, T. 1996. Application of dicrete finite element method (DFEM) to rock engineering structures. NARMS'96, Montreal.

Mamaghani, I.H.P, Baba, S., Aydan, Ö., Shimizu, Y. 1994. Dicrete finite element method for blocky systems. IACMAG, 843–850.

Aydan, Ö., Tokashiki, N., Shimizu, Y., Mamaghani, I.H.P., 2001. A stability analysis of masonry walls by Discrete Finite Element Method (DFEM). *10th IACMAG Conference*, Austin.

Aydan, Ö., Ogura, Y., Daido, M. and Tokashiki, N., 2003. A model study on the seismic response and stability of masonary structures through shaking table tests, *Fifth National Conference on Earthquake Engineering*, Istanbul, Turkey, Paper No: AE-041(CD).

Tokashiki, N., Aydan, Ö., Daido, M. and Akagi, T., 2007. Experiments and analyses of seismic stability of masonry retaining walls, *Rock Mechs. Symp., JSCE*, 115–120.

2019 Rock Dynamics Summit– Aydan et al. (eds)
© *2019 Taylor & Francis Group, London, ISBN 978-0-367-34783-3*

A dynamic model study on the dynamic response and stability of Perry Banner Rock in Nakagusuku, Okinawa, Japan

K. Takara
Mahae Consultants Co., Naha, Okinawa, Japan

M. Toguchi
Nakagusuku Village, Education Committee, Nakagusuku, Okinawa, Japan

Ö. Aydan & N. Tokashiki
Department of Civil Engineering, University of the Ryukyus, Nishihara, Okinawa, Japan

ABSTRACT: A natural rock block, which was visited by he famous Black Ship visited Japan in 1853, exists in Nakagusuku Village. This natural rock structure and its vicinity has been recently designated as a touristic spot and the authorities are concerned with its static and dynamic safety. The authors have been asked to investigate the stability of this natural monumental rock block. For this purpose, a physical model of this block with a similar density was prepared and its stability was investigated using a shaking table. The block become unstable when the base acceleration exceeds 350 gals. The authors also studied how to increase the dynamic resistance of the block against large earthquakes using an agent to bond the block to its base. The experimental results clearly indicated that the increase of the resistance was possible and the overall seismic resistance of the block increases. The outcomes of this experimental study has been described and their implications are presented and discussed.

1 INTRODUCTION

The recent earthquakes have recently shown that the monumental rock structures may be quite vulnerable to collapse besides other engineered structures. A natural rock structure exists in Nakagusuku Village of Okinawa Island. There is a drawing depicting the banner on the rock structure at the time that the famous Black Ship visited Japan in 1853 (Fig. 1a). The vicinity of the rock block has been now designated as a touristic spot. The elevation of the base of the rock structure is about 155 m and the rock itself is about 10 m high. The rock structure is locally known as Taachi (twin rock) and it was officially named as "Perry Banner Rock" by the Nakagusuku Village in 1997.

The authors visited this natural rock structure in April 2, 2011 and its vicinity was like a jungle. It was designated as a touristic spot in Nakagusuku Village and a project was initiated to arrange the site as a park. Geologic and rock mechanics investigations were carried out to check the static and dynamic stability of the natural rock structure and its vicinity.

There is a stepped discontinuity surface, on which the block sits on. The authors developed a physical model of this block with a similar density and studied its stability using a shaking table. In this study, the authors report the geological and rock mechanics investigations and the experiments and evaluate its static and dynamic stability.

(a) A drawing in 1853

(b) A view on 2012 December 22

Figure 1. Drawing and view of Perry Banner Rock

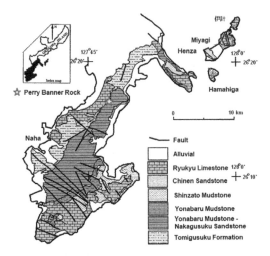

Figure 2. General geology and location of the Perry Banner Rock (modified from Kizaki 1986).

2 GEOGRAPHY AND GEOLOGY

The Perry Banner Rock is located in the Nakagusuku village in the central part of Okinawa Island (Fig. 2). The base of the natural rock structure is at the elevation of about 156 m and the rock structure is about 10m. There is an open crack with a width of 1 m at the base between two rock blocks.

The natural rock structure is made of Ryukyu limestone and the base rock is Shimajiri formation, which generally consists of mudstone, sandstone, siltstone and tuff from time to time. Shimajiri formation is subdivided into several sub-groups.

The rock block on the west-side is very close to the edge of slope and contains a thoroughgoing stepped discontinuity plane, which is probably a fault plane. The top rock block sits on two contacts and there is an open gap between the contacts. The rock at the lower contact (Contact-L) area is crushed and contains numerous cracks (Fig. 3).

To investigate foundation rock, three borings were performed. The boring B2 was 8 m deep while Boring B3 was 7 m deep. The rock was highly weathered up to a depth of 1.4-1.6 m. The rock below was fresh without any weathering. Fig. 4 shows the borehole logs for a depth of 5m for Boring Numbered B3. Fig. 5 shows the variation of the Needle Penetration Index (NPI) with depth.

3 ROCK MECHANICS INVESTIGATIONS

The natural rock structure mainly consists of Ryukyu limestone and its physical and mechanical properties are well known. Table 1 gives the physico-mechanical properties of Ryukyu limestone (Tokashiki and Aydan 2010). The frictional properties of Ryukyu limestone has been also well

Figure 3. A view of potentially unstable rock block

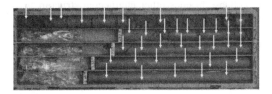

Figure 4. Views of borehole logs and needle penetration locations

investigated by Aydan et al. (2016). The physico-mechanical properties of rocks belonging to Shimajiri Formation depend upon the water content. Table 2 gives the physico-mechanical properties some rocks of Shimajiri formation at natural state. Figure 6 and 7 show the mechanical responses of rocks.

Aydan (2013) and Aydan et al. (2014) have studied the relations between NPI and various mechanical properties of soft rocks. Fig. 8 compares the estimations from relations proposed by Aydan et al. (2014) with experimental results on rocks of Shimajiri Formation. The results indicate that it is possible to estimate the mechanical properties from NPI tests, which is very practical for site investigations.

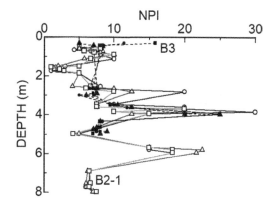

Figure 5.. Variation of NPI with depth

Table 1. Physico-mechanical properties of Ryukyu Limestone

ρ (kN/m³)	V_p(km/s)	E(GPa)	υ	σ_c (MPa)
19.6-23.4	4.5-6.3	8.1-27.8	0.15-0.3	20.0-33.9

Table 2. Physico-mechanical properties of Shimajiri Formation

Rock	γ(kN/m³)	σ_c (MPa)	φ_i
Sandstone	17.6-22.6	0.34	34-38
Mudstone	14.7-18.5	1.2-3.6	24-28

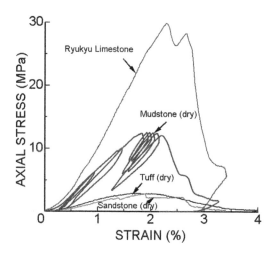

Figure 6. Uniaxial strain-stress relations of various rocks.

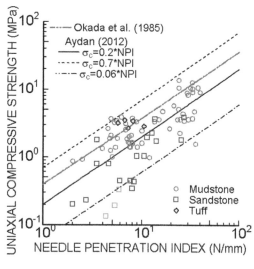

Figure 8. Comparison of the estimations from relations proposed by Aydan et al. (2014) with experimental results on rocks of Shimajiri Formation.

150 x 75 x 37.55 mm, was varied out using the newly upgraded direct shear testing machine (Aydan et al. 2016). The initial normal load was about 17 kN and increases to 30, 40, 50, 60 and 70 kN during the experiment. Fig. 9 shows the shear displacement and shear load responses during the experiment.

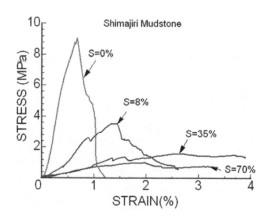

Figure 7. Uniaxial strain-stress relations at various water content.

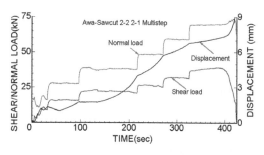

Figure 9. Shear stress-shear load response of the interface of sandy limestone blocks during the multi-stage(step) direct shear experiment.

A multi-stage (multi-step) direct shear test on a saw-cut surface of Ryukyu limestone sample, which consist of two blocks with dimensions of

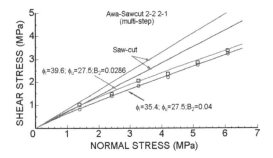

Figure 10. Comparison of shear strength envelope for the interface of sandy limestone blocks with experimental results from the multi-stage(step) direct shear experiment.

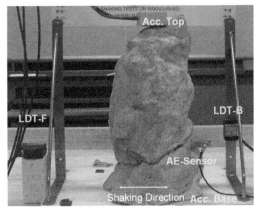

Figure 11. A view of the instrumentation of the physical model.

As noted from the figure, the relative slip occurs between blocks at a constant rate after each increase of normal and shear loads. This experiment is likely to yield shear strength of the interface two blocks under different normal stress levels. Fig. 10 shows the peak and residual levels of shear stress for each level of normal stress increment. Tilting tests were carried out on the same interface and the apparent friction angles ranged between 35.4 and 39.6 degrees (Aydan et al. 1995).

Tilting test results and direct shear tests are plotted in Fig. 10 together with shear strength envelopes using the shear strength failure criterion of Aydan (Aydan 2008, Aydan et al. 1966). As noted from the figure, the friction angles obtained from tilting tests are very close to the initial part of the shear strength envelopes. However, the friction angle becomes smaller as the normal stress level increases. In other words, the friction angle obtained from tilting tests on saw-cut surfaces can not be equivalent to the basic friction angle of planar discontinuities and interfaces of rocks. The basic friction angle of the planar interface of sandy limestone blocks is obtained as 27.5 degrees for the range of given normal stress levels.

4 PHYSICAL MODEL AND SHAKING TABLE

4.1 Physical Model

A physical model of the Perry Banner Rock was prepared using a model material, which has almost the same density of the original rock at a scale of 1/40 and the total model height was 300 mm. The model was equipped with two laser displacement transducers (LDT-F, LDT-B), one Acoustic Emission sensor (AE-Sensor) and two accelerometers (Acc. Top, Acc. B) as shown in Fig. 11.

4.2 Shaking Table And Testing Procedure

The applicable waveforms of the shaking table are sinusoidal, saw tooth, rectangular, trapezoidal, triangle and arbitrary earthquake wave form. The shaking table has a square shape with 1m-side length. The frequency of waves to be applicable to the shaking table can range between 1 Hz and 50 Hz for sweep tests. The table has a maximum stroke of 100 mm and a maximum acceleration of 6 m/s² for a maximum load of 980.7 N.

Model was fixed on the shaking table to receive same shaking with the shaking table during the dynamic test. The displacements, acoustic emissions (AE) and accelerations acting on the shaking table, at the base and model were recorded during the experiment, and saved on a data file as digital data (Fig. 12).

As the shaking table can apply uniaxial accelerations, the model was shaken in three directions namely, 0, 45 and 90 degrees in order to investigate the effect of the inclination of the thoroughgoing discontinuity plane (Fig. 13). The shaking direction angle is the acute angle between the strike of the thoroughgoing discontinuity plane and direction of shaking. Before the experiment leading to failure, the model was tested using the sweep testing

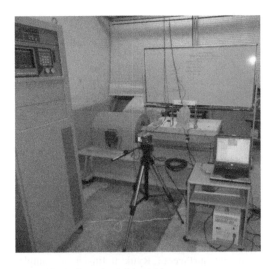

Figure 12. A view of a shaking table test.

Figure 13. Views of orientations of shaking

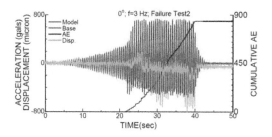

Figure 14. Responses for the shaking orientation of 0 degrees.

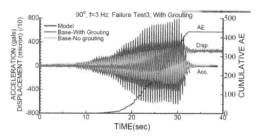

Figure 17. Responses of the model with counter-measures for the shaking orientation of 90 degrees.

the shaking the blocks starts to exhibit non-linear behaviour at about 100 gals for the direction of 90 degrees while it is about 230 gals for the direction of 0 degree. The blocks become unstable when the base acceleration exceeds 350 gals.

5.2 Models Tests with Counter-measures

Rockbolts and rockanchors are one of the effective way of dealing reinforcement issues in rock engineering (Aydan 1989, 2018). However, the utilization of such countermeasures may not be attractive due to the disturbance of the appearance especially in archeological structures. As the resistance of the model was minimum for the orientation of 90 degrees, an experiment was carried out by introducing a bonding resistance to the thoroughgoing discontinuity plane. The bonding agent was double sided bonding tape. Figure 17 shows the responses measured during the experiment. As noted from the figure and comparison with the response shown in Figure 17, the experimental results clearly indicated that the increase of the resistance was possible and the overall seismic resistance of the block increases. This experimental finding was taken into account and it was implemented during the remedial measures of the potentially unstable block.

procedure. The acceleration was fixed to 100 gals and the frequency of wave was varied from 1 to 50 Hz. The final experiment was concerned with the effect of grouting the gap in the model.

5 SHAKING TABLE TESTS

5.1 Models Tests without Counter-measures

Figs. 14, 15 and 16 shows the acceleration, AE and displacement responses measured for three directions respectively. Depending upon the direction of

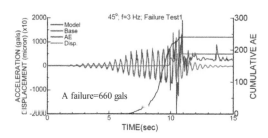

Figure 15. Responses for the shaking orientation of 45 degrees.

6 CONCLUSIONS

The authors performed a series of physical model experiments on a potentially unstable natural rock block, named as "Perry Banner Rock" using a material with a similar density of the natural rock and studied its stability using a shaking table. Depending upon the direction of the shaking the blocks starts to exhibit non-linear behaviour at about 100 gals for the direction of 90 degrees while it is about 230 gals for the direction of 0 degree. The blocks become unstable when the base acceleration exceeds 350 gals. The dynamic resistance of the natural rock block against large earthquakes can be increased using an agent to bond the block to its base. The experimental results clearly indicated that the increase of the resistance was possible and the overall seismic resistance of the block increases.

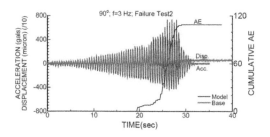

Figure 16. Responses for the shaking orientation of 90 degrees.

REFERENCES

Aydan Ö., 1989. The stabilisation of rock engineering structures by rockbolts. *Doctorate Thesis*, Nagoya University.

Aydan, Ö., 2008. New directions of rock mechanics and rock engineering: Geomechanics and Geoengineering. 5th Asian Rock Mechanics Symposium (ARMS5), Tehran, 3–21.

Aydan, Ö. 2018. Rock Reinforcement and Rock Support. CRC Press, Taylor and Francis Group, 486p, ISRM Book Series, No.6, ISBN 9781138095830 - CAT# K34886.

Aydan, Ö. 2012.The inference of physico-mechanical properties of soft rocks and the evaluation of the effect of water content and weathering on their mechanical properties from needle penetration tests. ARMA 12-639, 46th US Rock Mechanics/Geomechanics Symposium, Chicago, Paper No. 639, 10 pages (on CD), 2012.

Aydan Ö., Akagi T., Okuda H. & Kawamoto T. 1994.The cyclic shear behaviour of interfaces of rock anchors and its effect on the long term behaviour of rock anchors. *Int. Symp. on New Developments in Rock Mechanics and Rock Engineering*, Shenyang, 15–22.

Aydan Ö., Shimizu Y. & Kawamoto T. 1995. A portable system for in-situ characterization of surface morphology and frictional properties of rock discontinuities. *Field Measurements in Geomechanics. 4th International Symposium* 463–470.

Aydan Ö., Shimizu, Y. & Kawamoto T. 1996. The anisotropy of surface morphology and shear strength characteristics of rock discontinuities and its evaluation. *North American Roc Mechanics Symposium, NARMS'96*, 1391–1398.

Aydan, Ö., Sato A., Yagi. M., 2014. The Inference of Geo-Mechanical Properties of Soft Rocks and their Degradation from Needle Penetration Tests. Rock Mechanics and Rock Engineering, 47, pp.1867–1890.

Aydan, Ö., N. Tokashiki, J. Tomiyama, N. Iwata, K. Adachi, Y. Takahashi, 2016. The Development of A Servo-control Testing Machine for Dynamic Shear Testing of Rock Discontinuities and Soft Rocks, EUROCK2016, Ürgüp, 791–796.

Kizaki, K. 1986. Geology and tectonics of the Ryukyu Islands, Tectonophysics, 125, pp.193–207

Tokashiki, N. Aydan, Ö. 2010. The stability assessment of overhanging Ryukyu limestone cliffs with an emphasis on the evaluation of tensile strength of Rock Mass. Journal of Geotechnical Engineering, JSCE, Vol. 66, No. 2, pp.397–406.

T4: Induced Seismicity

T5: Dynamic Simulation of Loading and Excavation

2019 Rock Dynamics Summit– Aydan et al. (eds)
© 2019 Taylor & Francis Group, London, ISBN 978-0-367-34783-3

Sinkholes induced by dewatering in an open pit mine: Case study from a coal basin in Eastern Turkey

Y. Mahmutoğlu, R. Karagüzel, G. Şans & M. Erdoğan Topçuoğlu
Istanbul Technical University, Faculty of Mines, Ayazaga Campus, Turkiye.

ABSTRACT: Karstified Bedrock, and the weak cover on it, is most important boundary condition in sinkhole formation at site conditions susceptible to dewatering. Such contacts provide a balance between the unconfined and granular aquifers, fed by surface water and the confined aquifer located at the base. Excessive groundwater withdrawal from one of these aquifers causes interaction throughout their contacts. Six drainage borehole further operated about 20 days of after the dewatering process, which increased 2-3 times compared with the amount discharge in the same period of the previous year, in the Çöllolar Open-pit Mine. The first one, week after the new boreholes put into operation, four sinkholes formed at the edge of the coal basin, at a distance of 200-350 m to the borehole occurred until June 2015. Later, four more sinkholes were formed, one in the same month, the other two in December 2015 and one in March 2017 despite the reduction of groundwater withdrawal. In this study, the hydrological, the result of numerical analyses of a conceptual modal built by considering geological, hydrogeological, geo-mechanical boundary conditions and dewatering process in the region are presented. As a result of the numerical analysis, it was determined that there is a significant decrease in pore water pressure due to extreme water withdrawal from the karstic aquifer and a significant increase in the vertical effective stresses on the sinkhole forming area. Moreover, internal erosion develops between the limestone and coal-bearing gyttja contacts, and the weak upper cover composed of young sedimentary layers collapses into karstic cavities in the base rock due to the gravity.

1 INTRODUCTION

It is recorded that numerous karst-related sinkholes formed on the edge of the lowlands due to dewatering process. There are more than 4 000 sinkholes occurred due to dewatering on the coal mining sites between 1900 and 1980 years in Alabama, United States (Newton, 1984). Unconsolidated and weak sedimentary material on karstic cavities, migrate into the voids of underlying base rock by losing of buoyancy support. The increase in velocity of groundwater (internal erosion), lowering of groundwater level below unconsolidated cover, and surcharge are the most important reasons of sinkhole occurrence in similar areas. Susceptibility to sinkhole occurrence of an area depends, firstly, on the existence of carbonate base rocks containing karst structure and loose, weak and saturated cover on it. The elevation of groundwater level is generally related to the topography, geological characteristics, artificial recharge and water withdrawal. The geological, hydrological, geomorphological, hydrogeological and geo-mechanical conditions at the edge of the Afşin Elbistan Coal Basin, where the western slope of the Çöllolar Open Pit Box located resemble the features described above. In this study, the valid boundary conditions and the effects of dewatering for slope stability are explained and the

results of the numerical model are evaluated considering the karst dewatering during February 2015 in studied area.

2 BOUNDARY CONDITIONS AT THE EDGE OF THE BASIN, SINKHOLES OCCURED

2.1 Geological sequence

The study area is located between the western slope of the Çöllolar Open Pit Mine and surrounding region. It is next to the edge of the basin. The Çayderesi Formation having heavy karstic structure forms the base of Plio-Quaternary basin (Özgül, 1981, Yılmaz et al., 1997, Gökmenoğlu and Aslan, 2013). Young sedimentary sequence consisting of thick coal layers up to 60 m and rich organic matter content gyttja are unconformably overlain the basement rock, which also contains clay, silt clay, and river deposits on top (Fig.1). Neogene sequence generally starts with green clay and ends with blue clay on the top. The thickness of the blue-clay layer at the top with the green-gray clay at the bottom of the sequence decrease towards the basin side in the south west of the Çöllolar Coal Sector. The bottom clay having very low impermeability on base rock is either very thin (1-2 m) or not continuous in the area where the sinkholes are formed. The old alluvium (terrace), slope debris,

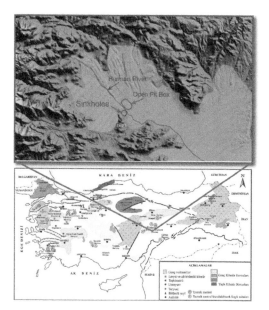

Figure 1. Locations and Afşin-Elbistan Coal Basin and site characteristics of sinkhole area.

and alluvium of Hurman River locate on top of the basin fill. The limestone ridge extending inside of the basin, near the sinkhole occurred area (Fig.1). In this area, the terrain elevations are 1165-1170 m and the base rock depth is 20-40 m (Fig. 2).

The green-gray clay (bottom clay) seen in deeper parts of the basin, but it does not continue to the edge of the basin (Karaguzel et al,, 2016).

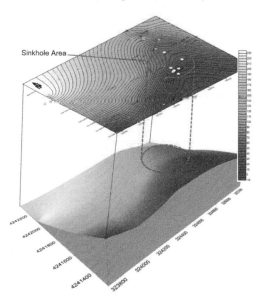

Figure 2. Topography of the bedrock (limestone), where sinkholes occurred during dewatering of karstic aquifer (Karaguzel et al., 2016).

2.2 Surface water

The K-dewatering wells operated in February 2015 were drilled at the top of southwestern slope and close to each other and the Hurman River runs throughout between sinkhole area and dewatering wells (Fig.1 and Fig.3). Therefore, the Hurman River is an important boundary condition. In order to determine the effect of surface water on formation of the sinkholes, the variation in river discharge around of study area is taken into account. For this reason, records during mine operation and dewatering in 2011, current observation stations located on the Hurman River and the side-branch Kurudere were correlated (Fig.3). It was concluded that surface water discharges totally at the stations OP3 and Q2 sometimes is higher than that of station OP1.

2.3 Hydrogeological systems

Geological units are classified by taking into consideration the stratigraphy, physical and hydraulic characteristics representing the Çöllolar Sector (Table 1). In this study, hydrogeological systems which have no significant difference between their hydraulic characteristics have been identified and the units defined as "coal-interbedded gyttja" in the previous studies, are incorporate into gyttja layer. In the same way, the zones in the previous studies, which are defined as "poor coal" with thin layer of gyttja stratums are included in the "coal" layer.

The hydraulic parameters valid in the area of the open-pit were determined by taking into account the previous studies. The average arithmetic and geometric hydraulic conductivity values (k) calculated for the lithological units from the pumping tests in investigation wells opened in the study area are given in Table 2 (MBEG, 2012). Hydraulic conductivities determined for geological units encountered

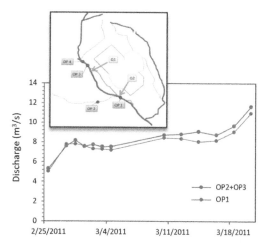

Figure 3. Locations of current observation stations and correlation of measured discharges in 2011 near to the studied area.

Table 1. Hydrogeological systems and their properties.

Formation	Lithology	Hydrogeological System	Types of Aquifers
Alluvium	Young and old Alluviums	Coarse (Poorly cemented)	Unconfined
			Unconfined
Sediments with Coal	Top Clay	Cohesive	Aquitard
	Marl	Consolidated	Aquitard
	Gyttja	Coarse/Cohesive	Poor Aquifer/Aqitard Unconfined/ Semi Confined
	Coal	Fissured/Jointed	
	Lower Gyttja	Coarse/Cohesive	
	Bottom Clay	Cohesive	Aquitard
Coarse Sediments (Eocene)	Inter-bedded Sandstone-Claystone and Sand-Coarse	Coarse Poorly Cemented Layered Jointed	Semi High-Confined
Base Rocks	Limestone Marble Schist	Karstified and Fractured	Extensive-High Confined

in ITU coded borehole in sinkhole area are also included in this table.

2.4 Dewatering of karstic aquifer

At the beginning of February 2015, K-9 and K-84 karst wells were operated in addition to existing TK5, TK6, TK-7 and TK-8 dewatering wells. Totally

Table 2. Hydraulic parameters calculated from the pumping wells near the sinkhole area (MBEG, 2012 and Karaguzel et al., 2016).

Model Layer	k (m/s)			
	MBEG, 2012		ITU, 2016	
	Arithmetic Mean	Geometric Mean	Arithmetic Mean	Geometric Mean
Alluvium	3.33E-05		3.09E-4	2.92E-4
Top Clay	4.31E-06	4.27E-6	–	–
Gyttja	8,80E-07	8.52E-7	3E-7	3E-7
Coal	1.15E-07	1.73E-7	1.58E-4	1.58E-4
Bottom Clay	1.15E-06	1.05E-6	3.55E-4	3.59E-4
Base Rock (Limestone)	1.00E-3		2.98E-4	2.97E-4

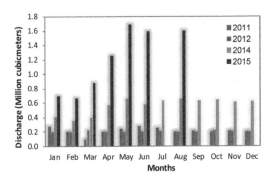

Figure 4. Monthly discharges from karstic aquifer during the period of January 2011 and August 2015.

654 458 m³ of groundwater was withdrawn by these wells from the karstic aquifer during the month of February 2015 (Fig. 4). Additionally, K-13, K-63 and K-73 dewatering wells (Fig. 5) were commissioned in April-2015 and groundwater withdrawal from the karstic aquifer gradually increased. During February 2015-June 2015, a total of about 6.54 million cubic meters of water was withdrawn from the karstic aquifer. The total amount of pumping water per month in May-2015 is 1.68 million cubic meters. This is 2.64 times more than the amount pumped in the same period of previous year. During this activity, on February 19, 2015, collapse in the first sinkhole began to develop at a distance of about 200–350 m from the southwestern slope of cut-box and dewatering wells newly operated. Simultaneously, some set of cracks growing in time were observed at sinkhole area. After noticing these failures, karst dewatering activity have been cancelled immediately. Later, in August 20, 2015, except K-84 dewatering well, others were deactivated and the underground water withdrawal from the karstic system was reduced. Also, in July, groundwater drainage was interrupted due to pump damage of dewatering well TK-7. After reducing total capacity of wells, considerably groundwater discharge from the springs in the southwestern slope of cut-box was observed. For the safety of western

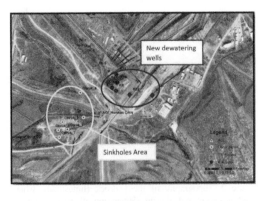

Figure 5. The locations of sinkhole area and new dewatering wells commissioned before sinkholes occurrence.

slope K-63 well put under operation instead of TK-7 pumping well. Later, monthly amount of discharge in karst wells were reduced to a range of 765–850 thousand cubic meters (Fig. 4).

From the dewatering process explained above, it is indisputable that the failures such as sinkholes and surface cracks were occurred depending on the groundwater withdrawal from the karstic system.

2.5 Geo-mechanical parameters

To find out the mechanism of landslides in western and eastern slopes of cut-box, additional geotechnical boreholes were drilled in February 2011 (MBEG, 2012). Taking into account the geological units encountered in guide boreholes drilled before, undisturbed samples were taken from the planned depths of these boreholes. Due to the apparent differences, zones of gyttja is distinguished into gray and beige (calcareous) gyttja layers. The void ratio obtained for all levels of these unit is high (1.5 <e<3.19). Numerous drained (CD) and undrained (CU) direct shear tests were performed and the results of these test statistically evaluated. The shear strength parameters determined for lithological units as a result of these evaluations are summarized in Table 3 and Table 4.

In order to determine the safe geo-mechanical parameters in the area where the sinkholes were formed, another three investigation boreholes were drilled for site characterization during this study. Disturbed (SPT) and undisturbed (UD) samples were also taken from different depths of these boreholes and they were subjected to sieve analysis, consistency limit tests, and undrained shear tests.

Table 3. Peak (ϕ_p, c_p) and residual (ϕ_r, c_r) shear strength parameters obtained from undrained (CU) tests (MBEG,2012).

Unit	Peak St.		Unit W.	Residuel St.	
	$\phi_p(°)$	c_p(kpa)	γn(kN/m³)	$\phi_r(°)$	c_r(kpa)
Alluvium	–	–	–	–	–
Loam	27	31	17.3	19	22
Marl	–	–	–	–	–
Top Clay	16	42	18.5	9	31
Black Clay	15	46	17.2	9	15
Gyttja	32	25	13.9	30	15
Biege Gyttja	31	32	11.5	30	27
Gyttja/ Coal	30	38	13.6	29	14
Coal	33	30	11.8	27	21
Clay interbed	17	32	16.6	7	15

Table 4. Peak (ϕ_p, c_p) and residual (ϕ_r, c_r) shear strength parameters obtained from drained (CD) tests (MBEG,2012).

Unit	Peak St.		Unit W.	Residuel St.	
	$\phi_p(°)$	c_p(kpa)	γn(kN/m³)	$\phi_r(°)$	c_r(kpa)
Alluvium	–	–	–	–	–
Loam	21	31	17.6	14	18
Marl	–	–	–	–	–
Top Clay	23	33	18.1	15	12
Black Clay	28	11	19.2	26	6
Gyttja	32	23	15.4	29	15
Biege Gyttja	34	27	13.1	30	16
Gyttja/ Coal	32	24	15	29	17
Coal	28	36	12.4	21	26
Clay interbed	20	19	15.9	12	9

Some of the test results are given in Table 5. In addition, alluvial samples taken from the walls of Sinkhole3, 5 and 7 were subjected to sieve analysis. As a result of these analysis, it was determined that the alluvium composed of coarse gravel with small amount of sand. Sand content did not exceed 25 percent. From the consistency tests, it was found that the thin layer of cohesive soil covering the alluvium in sinkhole area has medium plasticity (MI) and the samples taken from different depths of İTÜ coded boreholes (Fig.7) have high plasticity (CH, MH). The natural water content of the UD-3 sample taken at 37.50-38.00 m depth of the İTÜ-3 borehole exceeds the liquid limit.

As can be seen from Table 5, the porosity of the gyttja is usually over 60 %. In addition, experiments were carried out on six undisturbed UD specimens under undrained condition. The results of these tests

Table 5. Solid and natural unit weights (γ_s, γ_n), porosity (n) and undrained shear strength parameters of UD samples taken from İTÜ boreholes drilled sinkhole area.

Units	γ_s (kN/m³)	n	γ_n (kN/m³)	c_p (kPa)	ϕ_p (°)
Peat	27.0	0.48	18.1	20	31
Beige Gyttja	23.6	0.61	14.3	20	34
Gyttja	24.5	0.59	15.8	60	33
Beige Gyttja	24.3	0.63	14.5	5	35
Beige Gyttja	-	-	15.0	5	35
Coal/ Gyttja	25.1	0.71	13.6	35	32

show that the maximum shear strength parameters obtained for gyttja varies in a wide range.

If shear strength parameters compared to the MBEG (2012) data presented in Table 3, it is evident that the same layers of Plio-Quaternary sediments are weaker at the edge of the basin.

3 CONCEPTUAL MODEL AND NUMERICAL ANALYSES

The conceptual hydrogeological model based on the field survey and boring is depicted in Fig.6. The main geological, hydrological, hydrogeological boundary conditions are represented in this figure. The interaction between aquifers after groundwater drawdown induced by pumping the water from karst wells. Karstified base rock and shallow karstic cavities in it are also taken into account. When new dewatering K wells put into operation, the drawdown within the edge of basin is lowered and pore water pressure decreased sharply. As a result the turbulent flow has occurred and the groundwater flows turned to vertical direction around the sinkhole area. The water pumped from confined aquifer flowed throughout cavities being inside the base rock and its velocity has been considerably increased.

From the site survey on the southwestern slope and sinkhole walls behind the Çöllolar Open Pit, it is understood that the river alluvium and slope wash deposits have high permeability. The main branches of the drainage network on the surface, the exploration boreholes and the morphology indicate the young and unconsolidated alluvial deposits have considerable thickness in area where sinkholes occurred (Fig. 6).

In order to control the conceptual model shown in Fig. 6, a research pit (AÇ-1) was excavated to control the flow direction and level of the groundwater in area where the sinkholes were developed (Fig.7). The groundwater level measured in this pit was compared to Kurudere's water level between the pit and the K-63 dewatering well. It was noticed that the water level elevation in Kurudere (1157.23 m) is higher than that of AÇ-1 (1156 m). The same

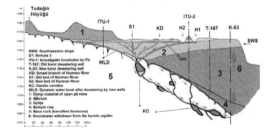

Figure 6. Schematic description of conceptual model, interaction between unconfined and confined aquifers at the edge of basin after dewatering operation (Mahmutoglu et al., 2017).

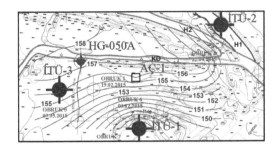

Figure 7. The dynamics water level around sinkholes occurred area after pumping the water from the karst aquifer (H1 and H2 old and new beds of Hurman River, locations and recognizing date of sinkholes (obruk) are marked by red writing).

comparison was also repeated for the level at İTÜ-3 and the nearest point of Kurudere.

Consequently, all measurements and comparisons have shown that maximum drawdown took placed around the sinkholes during the water withdrawal from the karst wells (Fig.7).

Karstic limestone indicated by number 5 in the conceptual model continues toward to the center of basin with a gentle inclination. But its shallow paleo-topographic surface under sinkholes indicates a ridge morphology lain beneath the young sediments. On the other side, during field survey around the vicinity of the basin, old sinkholes having equivalent dimeters exceeding 30 m were recognized near to the study area at south.

Depending on the observed site characteristic, it was ensured that there may be also karstic cavities under sinkhole area. Therefore similar cavities (KC in Fig.7) were taken into consideration for the construction of framework of conceptual model. the KC symbol in the conceptual model. The gyttja layers having high void ratio and low hydraulic conductivity are identified by the number 3 in this figure. Experimental studies indicate that this geological formation is mostly composed of high plasticity clay and silt, and occasionally sandy lens.

From the investigation boreholes around the study area, it was found out that the green clay (bottom clay) forming an impermeable boundary condition above karstic limestone does not continue toward to the edge of basin. Therefore, at the sinkhole area, there is interaction between the upper hydrogeological system and the karstic aquifer. The young alluvium of Hurman river having very high permeability is marked by the number 2 in figure 6. At the top of alluvium, there is a brownish clay silt with a thickness ranging from 0.8 to 1.5 m and having properties of swamp deposits. The number 1 in Figure 6 corresponds to ash and damping material of open-pit mine.

The conceptual model introduced above has been digitized and finite analyses for two different scenarios have been carried out by using the software "Rocscience RS2 v.9".

For the Plio-Quaternary units model parameters were selected from Table 2, Table 3 and Table 5, and for the base rock (limestone) were obtained according to the Hoek & Brown (1988) by taking into account the lowest unconfined compressive strength in laboratory (60 MPa) for intact rock. Hydraulic head belonging to unconfined and confined aquifers and the Hurman River are introduced separately. In the numerical model, the water level of the Hurman River feeding the upper alluvium aquifer is kept constant. At the last stage of modeling, karst dewatering process was introduced.

According to the Equation 1, the average filtration rate of karst wells which were under operation calculated.

$$V=Q/(F_k.H_f.R.\pi) \qquad (1)$$

Where;
Q: Total discharge from karst wells (March 2015: 865670 m³/month) = 0.32 m³/s
F_k: Filter coefficient = 0.1
H_f: Total length of filtered zones in dewatering wells under operation at the same period =104 m
R: Radius of wells = 0.5 m.

The average filtration rate in karst wells was obtained as $1.9.10^{-2}$ m/s.

The results of finite element analyses carried out for two different scenarios are explained separately under following topics. The difference between these scenarios is the location of karstic cavity. In scenario I, it is located shallow depth of karstic limestone under sinkhole area. Without any change in boundary conditions, it is replaced to the bottom of southwestern slope of open-pit mine in scenario II.

3.1 Scenario I

The karst cavity is modeled on the upper part of the basement rock, between the Hurman and under the area where the sinkholes are formed. The geometry of karstic cavity was randomly generated, with a maximum height of 15 m and a longest axis of 50 m approximately. After the model, has been run, no significant displacement can be obtained in the area where the failures are formed. However, considering the cavity in the bedrock, only the initial hydraulic conductivity of the karstic aquifer has been increased until the collapse, without changing the other model parameters until the vertical movement measured in sinkholes. As a result of this calibration; the hydraulic conductivity of karst aquifer was increased up to 2.32E-2 m/s in order to be able to migrate on the surface. Due to the karst dewatering, it is observed that there is a considerable decrease in pore water pressure under sinkhole area. The result of finite analysis are shown for different stages in Figure 8.

Therefore, it was determined that both hydraulic gradient and velocity of groundwater were increased. In addition, most of groundwater flow

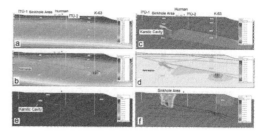

Figure 8. Variation of pore water pressure before and after the groundwater withdrawn from K wells (a, b), vertical flow velocities (c, d) and vertical displacements (e, f) for Scenario I.

lines were developed in vertical direction along the contact of alluvium and base rock and some of them were directed towards the edge of basin. Due to the migration of soil mass towards the cavity, the vertical displacement exceeding 3.5 m at surface was occurred (Figure 8).

3.2 Scenario II

In Scenario II, the same karstic cavity envisaged for Scenario I was shifted to a new location under southwestern slope of open-pit. The hydraulic conductivity of karstic aquifer was taken as the calibration value obtained in Scenario I (2.32E-2 m/s) and then the model was analyzed using the same boundary conditions. As a result of this solution (Fig. 9), similar variation in pore water pressure, flow line and groundwater velocity were observed around the cavity. In this case it has been shown that vertical displacement in southwestern slope reached up 1.5 m.

4 CONCLUSION

As a result, it was understood that dewatering of karstic aquifer at the edge of Afşin Elbistan Coal result in the sinkhole occurrence. It has been determined that the karst aquifer fed from the upper hydrogeological system during operation of dewatering wells, where the Hurman River continuously feeds the upper system and the interaction has devel-

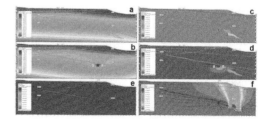

Figure 9. Variation of pore water pressure before and after the groundwater withdrawn from K wells (a, b), vertical flow velocities (c, d) and vertical displacements (e, f) for Scenario II.

oped along the boundary between the upper hydro-geological system and the karst aquifer in the region around the sinkholes. During karst dewatering work, it was determined that significant decrease in pore water pressure was observed at the edge of the basin. Thus, the vertical effective stress increases and the sedimentary mass weight on the base rock becomes heavier than before dewatering work.

REFERENCES

ÇÜ MMF, 2015. Integrated dewatering project for Çöllolar Coal Basin, Çukurova University, Faculty of Engineering and Architecture, Adana, p. 63 (in Turkish).

Gökmenoğlu, O., Aslan, M. 2013. Hydrogeological investigation project of Kahramanmaraş Afşin-Elbistan coal basin, Dewatering report for east side of Hurman River. MTA (General Directorate of Mineral Research and Exploration) Energy Depertment, Archieve No: 1346, 236 pp, Ankara (Unpublished); (in Turkish).

Hoek, E., Brown, E. T. 1988. The Hoek-Brown failure criterion –a 1988 update. Proceeding 15th Canadian Rock Mechanics Symposium (ed.) H. Curran, Toronto: Civil Engineering Department, University of Toronto, pp. 31–38.

.Karaguzel. R., Akyuz. S.H., Mahmutoglu, Y., Isseven. T., Bozkurtoglu. E. 2017.Geological research for sinkhole occurance at Afşin-Elbistan Coal Basin, ITU Mining Faculty Foundation, 217 pp. and appendixes, Istanbul (Unpublished Report); (in Turkish).

Mahmutoglu, Y., Karaguzel. R., Sans. G., Erdogan. M., Bozkurtoglu. E., Akyuz. S.H., Isseven. T., Tastekin. C., Ata. E. 2017. The effect of karst dewatering on sinkhole occurrence, A case study from Afşin-Elbistan Coal Basin, National Symposium of Engineering Geology and Geotechnics, proceeding book (Ed: Çan. T & Taga. H), pp 164–171, Adana (in Turkish)

MBEG, 2012. Consultancy and engineering services for mine operation in Çöllolar open cast mine (Afşin Elbistan coal basin), Phase V: Mine plan for the restart of coal mining & status report 06/2012, Subprojects: Geological modelling/Hydrogeology/Dewatering/Geotechnics and mine planning, p. 160 and Appendix.

Newton, J. G., 1984. Sinkhole resulting from groundwater withdrawals in carbonate terranes -an overview. Geological Society of America, Reviews in Engineering Geology, volume VI. pp. 195–202.

Özgül, N., 1981. Geology of Munzur Mountains. MTA (General Directorate of Mineral Research and Exploration) Report no: 6995, 136 p. Ankara (Unpublished); (in Turkish).

Park Teknik, 2011–2016 Montly Activity Reports (in Turkish).

Yılmaz, A., Bedi, Y., Uysal, Ş. Aydın, N. 1997. 1:100000 scaled geological maps of Turkey, book of Elbistan-İ23. MTA (General Directorate of Mineral Research and Exploration), Ankara (in Turkish).

2019 Rock Dynamics Summit– Aydan et al. (eds)
© 2019 Taylor & Francis Group, London, ISBN 978-0-367-34783-3

An X-FEM investigation of deflection/penetration of hydro-fractures at material discontinuities

M.Vahab
School of Civil and Environmental Engineering, UNSW Sydney, Sydney, Australia

Sh. Akhondzadeh & A.R. Khoei
Center of Excellence in Structural and Earthquake Engineering, Department of Civil Engineering, Sharif University of Technology, Tehran, Iran

N. Khalili
School of Civil and Environmental Engineering, UNSW Sydney, Sydney, Australia

ABSTRACT: In this study, a computational framework is extended for the simulation of the interaction of hydro-fractures with material interfaces. Using the extended finite element method (X-FEM), the equilibrium equation of the bulk is solved in conjunction with the hydro-fracture inflow and continuity equations using the staggered Newton method. The lubrication theory is applied to model the hydro-fracture inflow, where the hydro-fracture permeability is incorporated by means of the cubic law. Specific enrichment strategy is elaborated on the basis of the Eigen-function expansion method so as to achieve at high resolution for the stress field at singular zones. The competition between different interaction scenarios is explored through the application of the Ming-Yuan & Hutchinson's penetration/deflection criterion. Finally, the robustness of the proposed framework is investigated by means of numerical simulation.

1 INTRODUCTION

Hydraulic fracturing technique is a well-known engineering process in which permeable fractures are driven within low permeability formations by means of highly pressurized fluid injection. Since first introduced in 1950s, hydraulic fracturing has been the common practice in enhancing oil and gas recovery from tight reservoirs (Detournay & Cheng 1993). Geertsma & de'Klerk (1969), Spence & Sharp (1985), and Khristianovic & Zheltov (1955) were amongst the first who presented simplified analytical solutions to the hydraulic fracturing problem. Numerical approaches on the other hand have proven to be a versatile tool in tackling with the wide range of complicated mechanisms involved in hydraulic fracturing treatments, namely the hydro-mechanical coupling, matrix permeability, fluid-lag, and domain inhomogeneities (e.g. Detournay & Cheng 1993, Boone & Ingraffea 1990, Secchi & Schrefler 2012). A step forward in the numerical simulations of hydraulic fracturing has been due to the advanced mesh independent implementations in the context of X-FEM framework (e.g. Réthoré et al. 2007, Khoei et al. 2014, Vahab & Khalili 2018a,b).

Nearly all geological formations contain inhomogeneities, which affect their geomechanical responses. The study of the interaction of hydro-fractures with material interfaces is proven to be a formidable task, whereas researchers have

reached at contradictory conclusions. In this regard, Thiercelin et al. (1987) investigated hydro-fractures approaching material interfaces within rocks subjected to traction-free boundaries, and showed hydro-fractures can penetrate across such interfaces regardless of contrast in material properties. In a similar study by Eekelen (1982), it is shown neither changes in material stiffness nor in-situ stresses may prevent the hydro-fracture penetration across the material interfaces. In contradictory, Biot et al. (1983) implied unless a critical fluid pressure is exceeded any mechanical property differences may cease fracture penetration. Using a stress analysis, Gudmundsson & Brenner (2001) predicted hydro-fractures may penetrate across the material interfaces in soft to stiff configuration as a consequence of the stress built up throughout secondary domain. On the contrary, Wang (2015) showed that hydro-fractures tend to propagate into softer rocks which possess lower fracture energy and require smaller driving force. In consistency to Wang (2015), through an energy analysis, Behnia et al. (2014) showed that even minor inclination angle may lead to rapid diversion of the hydro-fracture from the interface in soft to stiff configuration.

The primary objective here is to present a robust energy based computational framework for the simulation of hydro-fracture evolution in naturally layered domains. Particular attention is given to the admissible variations in crack length through the second law of thermodynamics and minimization of the total potential energy. The penetration/deflection criterion introduced by Ming-Yuan & Hutchinson (1989) and Hutchinson & Sue (1991) is applied in order to determine the resultant hydro-fracture/material interface interaction scenarios. The interfacial fracture toughness is computed using the mode mixity parameter, defined as a function of the Stress Intensity Factors (SIFs) at the instant the fracture tip merges with the material interface. The stress singularity in the vicinity of crack-tips is obtained based on the X-FEM enrichment strategy, which ensures the high resolution of the stress field, as well as the precision and reliability of the model simulations. Finally, the robustness of the proposed computational framework is examined by means of numerical simulation.

2 MATHEMATICAL MODEL

2.1 Fractured bulk

Consider a two-dimensional impermeable domain Ω, bounded by the external boundary Γ, with the external normal unit vector $\mathbf{n}_\Gamma$, as depicted in Figure 1. The external boundary conditions are defined as the prescribed traction $\mathbf{t}$, and the prescribed displacement $\bar{\mathbf{u}}$, which are respectively imposed on Γ_t and Γ_u such that $\Gamma_t \cup \Gamma_u = \Gamma$ and $\Gamma_t \cap \Gamma_u = \varnothing$ hold. The domain is composed of two subdomains Ω_i ($i = 1, 2$) with different mechanical characteristics, which are fully bonded at the intersection of two subdomains indicated by Γ_M and the normal unit vector $\mathbf{n}_{\Gamma_M}$. An internal hydro-fracture interface Γ_{HF} is assumed within the domain with its unit normal vector denoted by $\mathbf{n}_{\Gamma_{HF}}$. The hydro-fracture is driven by the injection of the fracturing fluid at the constant rate of Q_{INJ}, which induces the fluid pressure p in direction normal to the hydro-fracture faces (i.e. $p^{HF} = p\mathbf{n}_{\Gamma_{HF}}$). Furthermore, the plausible overlap of the fracture faces is prevented by contact tractions $\mathbf{t}^{Cont}$. Bringing it all together, the boundary conditions associated with hydraulic fracturing problem in layered domains can be summarized as

$$\sigma\,\mathbf{n}_\Gamma = \mathbf{t} \quad \text{on } \Gamma_t,$$
$$\mathbf{u} = \bar{\mathbf{u}} \quad \text{on } \Gamma_u, \tag{1}$$

for the external boundaries, and

$$\sigma\,\mathbf{n}_{\Gamma_{HF}} = p\,\mathbf{n}_{\Gamma_{HF}} + \mathbf{t}^{Cont} \quad \text{on } \Gamma_{HF},$$
$$\mathbf{u}_{1/2} = 0 \quad \text{on } \Gamma_M, \tag{2}$$

over the internal boundaries, respectively. The linear momentum balance equation of the fractured domain in quasi static condition under the assumption of infinitesimal deformations in Lagrangian description is expressed as

$$\nabla \cdot \sigma + \rho\mathbf{b} = \mathbf{0}, \tag{3}$$

where ρ is the bulk density, $\mathbf{b}$ is the body force per unit mass vector, and σ is the Cauchy stress tensor. The constitutive law relating the Cauchy stress tensor to the resultant strain tensor is described as $\sigma = \mathbf{D} : \varepsilon$, where $\mathbf{D}$ is the Hook's elasticity tensor and ε is the strain tensor defined as $\varepsilon(\mathbf{u}) = \nabla^s\mathbf{u}$. Note that ∇^s is the symmetric part of the spatial gradient operator ∇. Using the test and trial functions of the displacement field, the weak form of the boundary value problem introduced through equations (1–3) is obtained as

$$\delta\Pi^{domain} = \int_\Omega \delta\varepsilon : \sigma \; d\Omega +$$
$$\int_{\Gamma_{HF}} [\![\delta\mathbf{u}]\!] \cdot p\,\mathbf{n}_{\Gamma_{HF}} \; d\Gamma \tag{4}$$
$$-\int_\Omega \delta\mathbf{u} \cdot \rho\mathbf{b} \; d\Omega - \int_\Gamma \delta\mathbf{u} \cdot \mathbf{t} \; d\Gamma = 0,$$

where the notation $[\![*]\!]$ indicates the difference between $*$ at the fracture faces (i.e. $[\![*]\!] = *^+ - *^-$).

2.2 Hydro-fracture inflow

The hydro-fracture inflow is governed by means of flow continuity equations. Consider the local Cartesian coordinate system (x', y'), in which x' and y' are aligned with the tangent and normal to the flow line, respectively. Assuming a one-dimensional incompressible Newtonian flow for the fracturing fluid within the cavity, and neglecting the variations of fluid pressure across the hydro-fracture (i.e. $\partial p/\partial y' = 0$), the continuity equation is expressed by

$$\frac{\partial Q}{\partial x'} + [\![\dot{u}_N]\!] + Q' = 0, \tag{5}$$

where Q is the flow rate, Q' is the leak-off flow given by Carter's relation (Carter 1957), and $[\![u_N]\!]$ is the normal component of the relative jump in displacement field (i.e. the crack opening) with the over-dot indicating the time derivative. Moreover, the hydro-fracture inflow is described by using the cubic law as

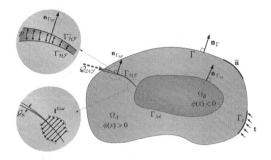

Figure 1. Hydro-fracture subjected to layered domain; problem definition and boundary conditions.

$$Q = \frac{1}{f} \frac{h^3}{12\mu} \frac{\partial p}{\partial x'} \equiv \kappa_{\Gamma\mathcal{HF}} \cdot \frac{\partial p}{\partial x'}. \qquad (6)$$

In the above relation, $h = [\![u_N]\!]$ is the hydro-fracture opening, μ is the fluid viscosity, and f is the modification factor suggested by Witherspoon et al. (1980) which accounts for the deviations from the ideal condition of the smooth crack faces. The boundary conditions for the flow continuity equations are defined by assuming the hydro-fracture inflow is vanished at the hydro-fracture tip (i.e. zero-lag assumption, see Vahab & Khalili 2018b), while the injection rate at the borehole is remained constant. It can be summarized as

$$\left. Q \right|_{x'=0} = Q_{\mathcal{INJ}},$$

$$\left. Q \right|_{x'=\ell_{\mathcal{HF}}} = 0, \qquad (7)$$

in which $\ell_{\mathcal{HF}}$ is the hydro-fracture length. Notably, the system of differential equations (5) and (6) governing the hydro-fracture inflow, together with the boundary conditions (7) is not well-posed (Boone & Ingraffea 1990, Khoei et al. 2014). To overcome this issue, the flow rate boundary condition at the borehole is replaced by the injection pressure $p_{\mathcal{INJ}}$, which is modified such that the prescribed injection rate $Q_{\mathcal{INJ}}$ is achieved at the fracture mouth.

3 DISCRETIZATION

X-FEM is elaborated in order to discretize the weak form of the equilibrium equation of the fractured domain given by equation (4). The X-FEM approximation of the displacement field $\mathbf{u}^h(x,t)$ is represented by

$$\mathbf{u}^h(x,t) = \mathbf{u}^{\text{std}}(x,t) + \mathbf{u}^{\mathcal{HF}}(x,t) + \mathbf{u}^{\mathcal{M}}(x,t), \qquad (8)$$

in which $\mathbf{u}^{\text{std}}$, $\mathbf{u}^{\mathcal{HF}}$ and $\mathbf{u}^{\mathcal{M}}$ are associated with the standard, hydro-fracture and bimaterial displacement field, respectively, defined as

$$\mathbf{u}^{\text{std}} = \sum_{I \in N^{\text{std}}} N_I(x) \mathbf{u}_I(t),$$

$$\mathbf{u}^{\mathcal{HF}}(x,t) = \sum_{I \in \mathcal{M}^{\text{Hev}}} N_I(x)(\mathcal{H}(\phi_{\Gamma_{\mathcal{HF}}}(x))$$
$$- \mathcal{H}(\phi_{\Gamma_{\mathcal{HF}}}(x_I))) \, \overline{\mathbf{a}}_I$$
$$+ \sum_{I \in \mathcal{M}^{\text{Tip}}} N_I(x) \sum_{\alpha=1}^{\mathcal{M}} (\mathcal{B}^\alpha(x) \qquad (9)$$
$$- \mathcal{B}^\alpha(x_I)) \mathcal{R}_{\text{Tip}}(x) \, \overline{\mathbf{b}}_{\alpha I},$$

$$\mathbf{u}^{\mathcal{M}}(x,t) = \sum_{I \in \mathcal{M}^{\text{Ridge}}} N_I(x) \mathcal{D}_m(\phi_{\Gamma_{\mathcal{M}}}(x)) \overline{\mathbf{c}}_I.$$

In the above relation, N_I is the standard shape function of node I corresponding to the standard DOFs $\mathbf{u}_I(t)$, and $\mathcal{N}^{\text{std}}$ is the set of nodal points consisting the whole domain. Moreover, $\mathcal{H}$ and $\mathcal{B}^\alpha$ are the Heaviside and asymptotic enrichment functions, respectively, with $\overline{\mathbf{a}}_I$ and $\overline{\mathbf{b}}_{\alpha I}$ being the DOFs associated with the Heaviside and asymptotic parts of the displacement field. Finally, $\mathcal{D}_m(x)$ is the ridge enrichment function, with $\overline{\mathbf{c}}_I$ denoting the DOFs corresponding to the ridge parts of the displacement field. Notably, the asymptotic enrichment functions for fracture tips terminated at material interface must be replaced by the terms manifest in the Airy stress function solution of the displacement field as (see Bouhala et al 2013):

$$\mathcal{B}_{\text{Tip}}^{\text{SSE}}(r,\theta) = \{\mathcal{B}_1, \mathcal{B}_2, \mathcal{B}_3, \mathcal{B}_4\} \quad \text{for } \theta_c = \pi/2$$
$$= \{r^\lambda \sin((\lambda \pm 1)\theta), \, r^\lambda \cos((\lambda \pm 1)\theta)\},$$

$$\mathcal{B}_{\text{Tip}}^{\text{SSE}}(r,\theta) = \{\mathcal{B}_1, \mathcal{B}_2, ..., \mathcal{B}_8\} \quad \text{for } \theta_c > \theta_p$$
$$= \{r^{\lambda_j} \sin((\lambda_j \pm 1)\theta), \, r^{\lambda_j} \cos((\lambda_j \pm 1)\theta)\},$$
$$\text{where } j = 1,2$$

$$\mathcal{B}_{\text{Tip}}^{\text{SSE}}(r,\theta) = \{\mathcal{B}_1, \mathcal{B}_2, ..., \mathcal{B}_{16}\} \quad \text{for } \theta_c < \theta_p$$
$$= \{r^{\lambda_{\text{Re}}} e^{\pm \lambda_{\text{Im}}\theta} \cos(\lambda_{\text{Im}} \ln(r)) \cos((\lambda_{\text{Re}} \pm 1)\theta),$$
$$r^{\lambda_{\text{Re}}} e^{\pm \lambda_{\text{Im}}\theta} \cos(\lambda_{\text{Im}} \ln(r)) \sin((\lambda_{\text{Re}} \pm 1)\theta),$$
$$r^{\lambda_{\text{Re}}} e^{\pm \lambda_{\text{Im}}\theta} \sin(\lambda_{\text{Im}} \ln(r)) \cos((\lambda_{\text{Re}} \pm 1)\theta),$$
$$r^{\lambda_{\text{Re}}} e^{\pm \lambda_{\text{Im}}\theta} \sin(\lambda_{\text{Im}} \ln(r)) \sin((\lambda_{\text{Re}} \pm 1)\theta)\},$$
$$(10)$$

In the above relations, θ_p is the power logarithmic angle defined as the angle where the transition between real and imaginary solutions for λ occurs. The singular enrichment function used in this work for the instance of crack terminating at the material interface is the modified version of the above enrichment functions, in which a weighted summation of the above functions, on the basis of the the Airy stress function coefficients is employed as the enrichment function. This approach, known as the modified singular enrichments (MSE), reduce the number of enrichment functions with increased accuracy (see Akhondzadeh et al. 2017).

Finally, it is noteworthy to add that the strong form of the hydro-fracture inflow given by Eqs. (5-7) is simply discretised by means of finite difference scheme. To this end, the first-order upwind scheme for the spatial and time domain discretization is employed. Further details in relation to the finite difference solution to the flow continuity equations within hydro-fractures can be found in Vahab et al. (2018).

4 PENETRATION/DEFLECTION CRITERION

Based on the literature (e.g. see Martinez & Gupta 1994), the interaction of hydro-fractures with material interfaces leads to three distinct scenarios; hydro-fracture may penetrate across the interface, as the successful scenario, or it can be deflected either along one or both wings of the interface, which is referred to as failure (Fig. 2). The resultant interaction scenario is determined by taking advantages from the energy release rate associated with each scenario. Based on Hutchinson & Suo (1991), the singly deflected case is always energetically favorable in comparison to the doubly deflected case. Accordingly, the doubly deflected case is disregarded in the present study. The penetration/deflection criterion is indicated based on Ming-Yuan & Hutchinson (1989) as

$$
\begin{cases}
\dfrac{\mathcal{G}_d}{\mathcal{G}_p^{max}} < \dfrac{\mathcal{G}_{intf}}{\mathcal{G}_2} & \text{penetration,} \\[2ex]
\dfrac{\mathcal{G}_d}{\mathcal{G}_p^{max}} > \dfrac{\mathcal{G}_{intf}}{\mathcal{G}_2} & \text{deflection.}
\end{cases}
\tag{11}
$$

In the above relations, $\mathcal{G}_2$ and $\mathcal{G}_{intf}$ are the fracture toughness of the secondary material and the interface, respectively, and $\mathcal{G}_p^{max}$ is the penetration energy release rate which is determined as the maximum value of the energy release rate related to the penetrated crack at all plausible penetration angles. Based on Hutchinson & Suo (1991), and Banks-Sills & Ashkenazi (2000) the interfacial fracture toughness

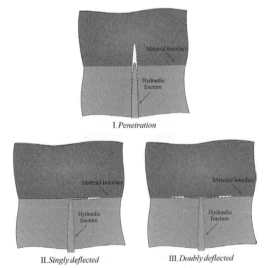

I. Penetration

II. Singly deflected   III. Doubly deflected

Figure 2. Schematic representations of possible interaction scenarios when the hydraulic fracture merges with the material interface.

associated with a bimaterial interface intersected by a fracture is estimated by

$$
\mathcal{G}_{intf} = \hat{\mathcal{G}}(1 + \tan^2 \Psi),
\tag{12}
$$

in which $\hat{\mathcal{G}}$ is the reference energy release rate, and Ψ is the mode mixity defined as

$$
\Psi = \tan^{-1}\left(\frac{\mathcal{K}_{II}^{intf}}{\mathcal{K}_I^{intf}}\right).
\tag{13}
$$

Evidently, due to difference in λ_1 and λ_2 for the generalized SIFs appearing in the above relation (see equation (10)), they do not contain the identical units. Thus, in order to use these SIFs in evaluation of the mode mixity a characteristic length factor of $l_0^{\lambda_1-\lambda_2}$ is applied to the nominator of the above relation so that the units be compatible.

The deflection energy release rate, $\mathcal{G}_d$, is calculated by adopting the values of the SIFs obtained via the solution of an imaginary debounding of the material interface with an arbitrary length occurring in a single side as

$$
\mathcal{G}_d = \frac{\eta[(\mathcal{K}_I^d)^2 + (\mathcal{K}_{II}^d)^2]}{E_1},
\tag{14}
$$

where η is a coefficient which depends on the material properties on either sides of the interface. In addition, $\mathcal{K}_I^d$ and $\mathcal{K}_{II}^d$ are the mode I and II generalized SIFs, which are calculated using the interaction integral method by incorporating the discontinuity within the solution field along the material interface (see Akhondzadeh et al. 2017).

Finally, $\mathcal{G}_p^{max}$ is computed through the auxiliary solution of the imaginary hydro-fracture penetration at the bimaterial interface using the same arbitrary growth length used in the singly deflected case as $\mathcal{G}_p = [(\mathcal{K}_I^p)^2 + (\mathcal{K}_{II}^p)^2]/E_2$. Note that it is required to solve the penetration problem at different penetration angles in order to determine the maximum value of the penetration energy release rate. At the end, all the ingredients appearing in relation (11) are elaborated to describe the resultant scenario developed within the interaction of the hydro-fracture with the bimaterial interface.

5 NUMERICAL SIMULATIONS

Numerical simulations here are performed by using an impervious $8\,m \times 8\,m$ rock block under plane strain condition. The block is composed of four different material configurations categorized into 1) soft to stiff ($E_1 = E_0$, $E_2 = 2E_0$ or $4E_0$, $v_1 = v_2 = v_0$) and 2) stiff to soft ($E_1 = 2E_0$ or $4E_0$, $E_2 = E_0$, $v_1 = v_2 = v_0$), in which $E_0 = 10$ GPa and $v_0 = 0.3$ are respectively the reference Young modulus of elasticity and Poisson's ratio. A hydro-fracture of negligible length is positioned at the left surface of the block with initial inclination angles of $\theta_i = 0°, 15°$ and $30°$. The hydro-fracture is permitted to evolve from the right tip until it merges with the material interface located at $x = 4$ m. The fracturing fluid is injected at the constant rate of $Q_{INJ} = 10^{-3}$ m^2/s with the dynamic fluid viscosity of $\mu = 0.001$ Pa.s. The solution is performed by adopting a FE mesh of 95×95 quadrilateral elements. The time increments of $\Delta t = 0.02$ s are utilized together with the characteristic length of $\Delta a = 0.1$ m for crack increment. The geometry, configuration together with the utilized boundary conditions are represented in Figure 3.

In all solutions the probability of hydro-fracture deflection or penetration is investigated by taking advantages from the penetration/deflection criterion (11). For this sake, in all cases the fracture toughness of the secondary domain (i.e. G_2) is set to 100 J/m^2. The interfacial toughness on the other hand is computed according to Eq. (12) through the assumption of $\hat{G} = 50$ J/m^2, where a characteristic length of $\ell_0 = 0.4$ m is adopted for the calculation of the mode mixity based on Eq. (13). This characteristic length is also utilized as the imaginary penetration/deflection crack increment for the evaluation of the penetration/deflection energy release rates G_d and $G_{\bar{d}}$. Finally, the possibility of the occurrence of deflection or penetration scenario in each case is determined by introduction of the deflection and penetration factors defined as $R_1 = G_d / G_{intf}$ and $R_2 = G_p^{max} / G_2$, respectively. Notably, the greater factor governs the response of the hydro-fracture subjected to the material interface.

5.1 The soft to stiff configuration

The injection of the fracturing fluid leads to the hydro-fracture evolution within the block. Using the

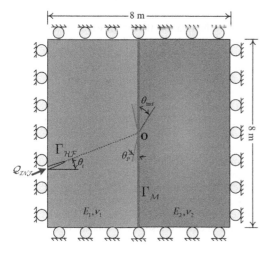

Figure 3. Hydro-fracture evolution throughout a bimaterial block; the problem definition and boundary conditions.

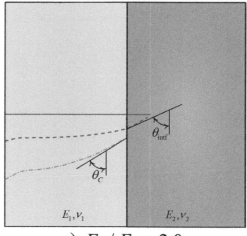

a) $E_2 / E_1 = 2.0$

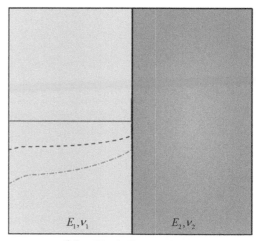

b) $E_2 / E_1 = 4.0$

Figure 4. The final crack path trajectory for hydro-fracture subjected to material interface in soft to stiff configuration;: $\theta_i = 0°$_____,: _____ $\theta_i = 15°$,: _____ $\theta_i = 30°$.

stress field at the instant the hydro-fracture is merged with the material interface, the SIFs are determined and applied in order to determine the fracture toughness of the material interface G_{intf}. The deflection energy release rate and deflection factor G_d and R_1 are calculated based on the values of K_I^d and K_{II}^d. Finally, the penetration energy release rate is calculated using the same crack increment (i.e. $\ell_0 = 0.4$ m). The maximum value of the penetration energy release rate versus the penetration angles is utilized in order to determine the governing penetration factor R_2. Our simulations show the penetration factors related to $\theta_i = 0°, 15°$ is slightly higher than the corresponding deflection factors; hence, for these cases the penetration scenario is more probable. However, by increasing the initial inclination angle to $\theta_i = 30°$

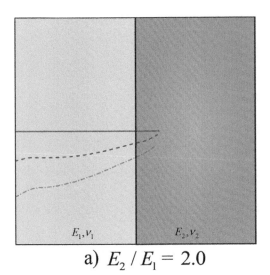

a) $E_2 / E_1 = 2.0$

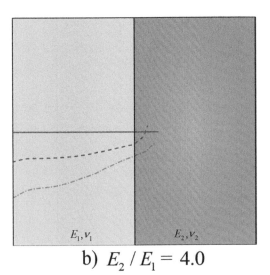

b) $E_2 / E_1 = 4.0$

Figure 5. The final crack path trajectory for hydro-fracture subjected to material interface in stiff to soft configuration;: $\theta_i = 0°$ _____,: $\theta_i = 15°$ _____,: $\theta_i = 30°$ _____.

the deflection scenario predominates. Interestingly, by increase in stiffness contrast from 2 to 4 the penetration factors are significantly decreased. This is justified by taking into account that the energy release rate is reversely proportional to the material stiffness. This yields to the unconditional occurrence of the deflection scenario for all the associated inclination angles. Based on the aforementioned discussion, the final crack path trajectory is represented for all cases in Figure 4. Results here clearly imply that the hydro-fracture is prone to be deflected at the material interfaces in soft to stiff configuration.

5.2 *The stiff to soft configuration*

In this example, the role of the material interfaces in stiff to soft configuration on hydro-fracture evolution is examined. Using the obtained stress field at the instant the hydro-fracture intersects with the material interface, the SIFs $\mathcal{K}_I^{\text{intf}}$ and $\mathcal{K}_{II}^{\text{intf}}$, mode mixity Ψ and interfacial toughness $\mathcal{G}_{\text{intf}}$ are calculated. In resemblance to the previous example, the penetration/deflection calculation is carried out. Based on the values obtained for the SIFs $\mathcal{K}_I^d$ and $\mathcal{K}_{II}^d$, and deflection energy release rate $\mathcal{G}_d$, the deflection factor $\mathcal{R}_1$ associated with the plausible deflection scenario is evaluated. The maximum values obtained for the penetration factor is again utilized as $\mathcal{R}_2$. Notably, the values of the deflection factors are observed to slightly increase in comparison to the first example; yet, the penetration factors are significantly increased. This can be attributed to the intensified stress field within the primary domain on one hand, and the appearance of the lower Young modulus in calculation of the energy release rate within the secondary domain ($\mathcal{G}_p \propto 1 / E_2$) on the other. Therefore, in contrast to the first example for all cases the penetration scenario is observed. In this fashion, by increasing the stiffness ratio the penetration factors grow much faster than the deflection factors; so, it can be concluded that the probability of the penetration scenario is intensified by increase in stiffness contrast. In Figure 5, the resultant hydro-fracture path is shown for all cases. Based on the discussions in this example, the material configuration of stiff to soft the probability of hydro-fracture penetration is higher in general.

6 CONCLUDING REMARKS

In the present paper, an extended finite element framework is developed to simulate hydraulic fracturing within layered formations. The viscous fluid flow within the hydro-fracture is modelled by using the lubrication theory in which the fracture permeability is determined based on the cubic law. Exploiting the enrichment strategy, the discretization of the displacement field is performed. The well-known penetration/deflection criterion introduced by Ming-Yuan and Hutchinson (1989) is utilized in order to predict the plausible penetration/deflection scenario throughout the interaction of hydro-fractures with material interfaces. Finally, the performance of the developed framework is explored by means of numerical simulation. In the first example, the effects of soft to stiff configuration of the material interfaces as well as the role of initial inclination of perforations are investigated. Based on the energy analysis performed, it is shown that by increase in either the stiffness contrast or the initial inclination angle the probability of hydro-fracture deflection at the material interface dramatically increases. The second example studies the interaction of hydro-fractures with stiff to soft configured material interfaces. Taking into account that the penetration factors are substantially greater than their corresponding deflection factors, it is concluded that in general for stiff to soft configurations of material interfaces the probability of hydro-fracture penetration is quite high. The results proposed in the current study are believed to give a better understanding from the interaction problem for being based on the high resolution for the stress field on one hand, and the use of the energy based penetration/deflection criterion on the other hand.

REFERENCES

Akhondzadeh, S., Khoei, A. R. & Broumand, P. 2017. An efficient enrichment strategy for modeling stress singularities in isotropic composite materials with X-FEM technique. *Engineering Fracture Mechanics*, 169, 201–225.

Banks-Sills, L. & Ashkenazi, D. 2000. A note on fracture criteria for interface fracture. *International Journal of Fracture*, 103, 177–188.

Behnia, M., Goshtasbi, K., Marji, M. F. & Golshani, A. 2014. Numerical simulation of crack propagation in layered formations. *Arabian Journal of Geosciences*, 7, 2729–2737.

Biot, M. A., Medlin, W. & Masse, L. 1983. Fracture penetration through an interface. *Society of Petroleum Engineers Journal*, 23, 857–869.

Boone, T. J. & Ingraffea, A. R. 1990. A numerical procedure for simulation of hydraulically-driven fracture propagation in poroelastic media. *International Journal for Numerical and Analytical Methods in Geomechanics*, 14, 27–47.

Bouhala, L., Shao, Q., Koutsawa, Y., Younes, A., Núñez, P., Makradi, A. & Belouettar, S. 2013. An XFEM crack-tip enrichment for a crack terminating at a bi-material interface. *Engineering Fracture Mechanics*, 102, 51–64.

Carter, R. 1957. Derivation of the general equation for estimating the extent of the fractured area. *Appendix I of "Optimum Fluid Characteristics for Fracture Extension," Drilling and Production Practice, GC Howard and CR Fast, New York, New York, USA, American Petroleum Institute*, 261–269.

Detournay, E. & Cheng, A. H.-D. 1993. Fundamentals of poroelasticity1. *Chapter 5 in Comprehensive Rock Engineering: Principles, Practice and Projects, II*, 113–171.

Geertsma, J. & De Klerk, F. 1969. A rapid method of predicting width and extent of hydraulically induced fractures. *Journal of Petroleum Technology*, 21, 1571–1581.

Gudmundsson, A. & Brenner, S. L. 2001. How hydrofractures become arrested. *Terra Nova*, 13, 456–462.

Hutchinson, J. W. & Suo, Z. 1991. Mixed mode cracking in layered materials. *Advances in applied mechanics*, 29, 63–191.

Khoei, A. R., Vahab, M., Haghighat, E. & Moallemi, S. 2014. A mesh-independent finite element formulation for modeling crack growth in saturated porous media based on an enriched-FEM technique. *International Journal of Fracture*, 188, 79–108.

Khristianovic, S. & Zheltov, Y. Formation of vertical fractures by means of highly viscous fluids. Proc. 4th world petroleum congress, Rome, 1955. 579–586.

Martinez, D. and Gupta, V., 1994. Energy criterion for crack deflection at an interface between two orthotropic media. *Journal of the Mechanics and Physics of Solids*, 42, 1247–1271.

Ming-Yuan, H. & Hutchinson, J. W. 1989. Crack deflection at an interface between dissimilar elastic materials. *International Journal of Solids and Structures*, 25, 1053–1067.

Réthoré, J., Borst, R. D. & Abellan, M. A. 2007. A two-scale approach for fluid flow in fractured porous media. *International Journal for Numerical Methods in Engineering*, 71, 780–800.

Secchi, S. & Schrefler, B. 2012. A method for 3-D hydraulic fracturing simulation. *International journal of fracture*, 178, 245–258.

Spence, D. & Sharp, P. Self-similar solutions for elastohydrodynamic cavity flow. Proceedings of the Royal Society of London A: Mathematical, Physical and Engineering Sciences, 1985. The Royal Society, 289–313.

Thiercelin, M., Roegiers, J., Boone, T. & Ingraffea, A. An investigation of the material parameters that govern the behavior of fractures approaching rock interfaces. 6th ISRM Congress, 1987. International Society for Rock Mechanics.

Vahab, M. & Khalili, N. 2018a. An X-FEM Formulation for the Optimized Graded Proppant Injection into Hydrofractures Within Saturated Porous Media. *Transport in Porous Media*, 121, 289–314.

Vahab, M. & Khalili, N. 2018b. Computational Algorithm for the Anticipation of the Fluid-Lag Zone in Hydraulic Fracturing Treatments. *International Journal of Geomechanics*, 18, 04018139.

Vahab, M., Akhondzadeh, S.H., Khoei, A. R. & Khalili, N. 2018. An X-FEM investigation of hydro fracture evolution in naturally-layered domains. *Engineering Fracture Mechanics*, 191, 187–204.

Van Eekelen, H. 1982. Hydraulic fracture geometry: fracture containment in layered formations. *Society of Petroleum Engineers Journal*, 22, 341–349.

Wang, H. 2015. Numerical modeling of non-planar hydraulic fracture propagation in brittle and ductile rocks using XFEM with cohesive zone method. *Journal of Petroleum Science and Engineering*, 135, 127–140.

Witherspoon, P. A., Wang, J., Iwai, K. & Gale, J. 1980. Validity of cubic law for fluid flow in a deformable rock fracture. *Water resources research*, 16, 1016–1024.

2019 Rock Dynamics Summit– Aydan et al. (eds)
© 2019 Taylor & Francis Group, London, ISBN 978-0-367-34783-3

A Verification of the multi-body dynamics based on impulse-based method

Yi Li & Mitsuteru Asai
Department of civil engineering, Kyushu University

ABSTRACT: In this paper, an impulse-based method is applied instead of penalty method to deal with the collision between rigid bodies. In penalty method, the interpenetration of the two related rigid bodies is used to calculate the contact force and several parameters like the coefficient of restitution, Young's module and Poission's ratio are needed to adjust for different materials. In addition, for colliding contact problem, whose interval is very short, a tiny time increment is needed, or the simulation result will be very unstable. However, as penalty method is easy to implement and understand, it is widely adopted in the engineering simulations. On the contrary, using impulse-based method can also allow us to use a larger time increment to improve the efficiency of the simulation with only one parameter (coefficient of restitution) to adjust. By only considering the relative velocity at the contact point, the interpenetration between the related rigid bodies is not needed and the process of the entire collision is simulated in one time increment to prevent a deeper interpenetration, which also mitigates the error from the unrealistic interpenetration.

1 INTRODUCTION

Multi-Body Dynamics (MBD) is widely utilized in the computer graphics, 3-D games and so on. The main purpose of the MBD is to provide a plausible simulation result of rigid body motion including contacts and articulated connections within a limited computational time as for interactive simulation. Then, fast calculation has the most priority on the MBD simulations and each rigid body is generally modelled as an equivalent polygon in order to reduce discretized points and computational costs. On the other hand, engineering purpose of simulation requires accuracy, numerical stability and reproducibility by others. In addition, engineering and industry requires applications of the rigid body simulation into complicated shapes of body and huge number of rigid bodies. Commonly, in the area of engineering penalty method has the dominant position as it calculates the contact force between the rigid bodies and update the motion of the relating rigid body by the sequence of force→acceleration→velocity→position. And the contact force is directly based on the interpenetration between the rigid bodies. It may work well for the situation where the normal contact force is constant. However, for the collision problem, penalty method needs a smaller time increment to track the interpenetration (position) of the relating rigid bodies and keep the velocity change continuous. In addition, the parameter-tuning work is also time-consuming in penalty method (Kurose, S., et al. 2009).

Instead of penalty method, impulse-based method (Mirtich, B et al. 1995) is popular in the area of CG or robotic engineering because it is fast and robust to get a physically plausible simulation result. In impulse-based method, the impulse is calculated based on the relative velocity and there is no need to calculate the force and acceleration. And in many physical engines, a branch of impulse-based method, sequential impulse is widely used, which will be discussed later. Otherwise, some similar methods are also applied as the alternatives like velocity constraint (Erleben, K et al. 2007) or acceleration constraint (Baraff, D. et al 1993) method. However, in these methods using polygon to represent the rigid body makes the contact detection complicated and computationally expensive. In this work, a particle-based rigid body which can be discretized into many particles is chosen to represent the rigid bodies in the simulation. The particle-based rigid body simulation commonly used in the Discrete Element Method (DEM) is referred as a clustered DEM which is always based on penalty method. Recently, the position-based dynamics (PBD), which is particle-based deformable bodies and fluids simulator proposed by (Müller, M et al. 2007), becomes popular because it is fast, robust and controllable, while no implicit time integration is required. The PBD can be applied easily for the rigid body simulation (Duel,C et al. 2016). (Tang, X. et al. 2014) simplifies the work of (Mirtich, B et al. 1995) and apply it to the area of rock mechanics. In this paper, we follow their ideas to rearrange an impulse-based method that can be applied to a particle discretized rigid body simulation. Then, we implement some verification tests to maintain reliability and accuracy of simulations.

2 RIGID BODY DYNAMICS

In this section, the basic knowledge of rigid body dynamics and the variables to represent the rigid body are introduced.

2.1 *Mass property for a rigid body*

In a particle-based model, the total mass M is simply calculated as:

$$M = \sum_{j=1}^{N} m_i \tag{1}$$

where N is the number of particles forming the rigid body and m_i is the mass of each particle.

The center of mass X_{rb} is decided by the mass distribution, so it is expressed as:

$$X_{rb} = \frac{\sum m_i x_i}{M} \tag{2}$$

where x_i means the position of each particle in the global coordinate.

For rotational motion of the rigid body, the inertia tensor I is the equivalent of the total mass M. To calculate the inertia tensor, the relative position r_i of each rigid body particle to the center of mass must be calculated as:

$$r_i = x_i - X_{rb} \tag{3}$$

Using the relative position, the initial inertial tensor can be expressed as:

$$I = \sum_{i=1}^{N} m_i \begin{bmatrix} r_{iy}^2 + r_{iz}^2 & -r_{ix} r_{iy} & -r_{ix} r_{iz} \\ -r_{iy} r_{ix} & r_{ix}^2 + r_{iz}^2 & -r_{iy} r_{iz} \\ -r_{iz} r_{ix} & -r_{iz} r_{iy} & r_{ix}^2 + r_{iy}^2 \end{bmatrix} \tag{4}$$

where the last subscript (x, y, z) means the x, y, z component of the relative position, and m is the mass value of one rigid body particle.

The orientation caused by rotation in each time step is expressed as:

$$I(t) = R(t) I R(t)^T \tag{5}$$

where $I(t)$ is the inertia tensor at t-th time step, the subscript T means transport matrix. And the rotation matrix of time step t $R(t)$ is calculated according to unit quaternion.

2.2 *Update the linear and angular velocity*

As the main content in this paper is for impulse-based method where the force and acceleration from external force are not needed, the part for acceleration is skipped. When a contact force or an impulse P is calculated, the linear and angular velocity of a rigid body are updated as:

$$v^+ = v^- + \frac{P}{M} \tag{6}$$

$$\omega^+ = \omega^- + I(r_i \times P) \tag{7}$$

where vector P is used to express the direction the impulse.

2.3 *Update the position of rigid body particles*

The position of the center of mass is updated as:

$$X_{rb}^+ = X_{rb}^- + v \Delta t \tag{8}$$

To update the rigid body particles, the rotation and the orientation of the rigid body are considered and expressed by rotation matrix and inertial tensor. The position of each particle is updated as:

$$x_i^+ = X_{rb}^+ + R r_i^0 \tag{9}$$

where i means the index of each rigid body particle, the subscript 0 indicates the initial step.

3 IMPULSE-BASED METHOD

Generally, impulse-based method is used in polygon-based model and a very complicated algorithm is needed to do the contact detection. However, in the paper, the model is built by many particles and the thanks to the development of computer and parallel programing, the contact detection based on SPH (smoothed particle hydrodynamics) method's neighboring particle searching algorithm makes it easy to implement and very efficient. Each particle has its own position and normal vector which are used to generate the surface of the rigid body. Whether collision happens is easily judged by the next equations:

$$\|x_A - x_B\| - d < 0 \tag{10}$$

$$(v_B - v_A) \cdot n < 0 \tag{11}$$

where x means position of each particle, d is the particle diameter, v is the velocity of each particle and n is normal at the contact point. When both equations (10) and (11) are satisfied, two objects are

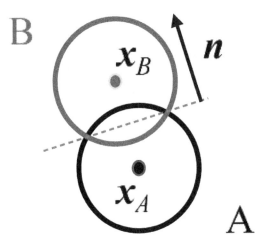

B

x_B

n

x_A

A

Figure 1. Contact detection model

regarded as being in collision. In this paper, the normal direction points from collider A to collider B for convenience as shown in Figure.1.

Traditionally, impulse-based method is applied to deal with the contact like collision whose interval is very short. And it is only suitable for the collision happening at one contact point. Therefore, usually the contact point with the deepest interpenetration is chosen to be the effective point. However, in the real world it is very common that a rigid body has many collisions at the same time, which could be treated as a multiple contact point problem. Generally, for multiple contact point problem, sequential impulse and velocity constraint method are used. Both of them follows Newton's impact law as:

$$u_i^+ = -eu_i^-$$
(12)

where the superscript + and − mean the post-collision and pre-collision, and here u is used to represent the relative velocity at the contact point i instead of the linear velocity of the rigid body v.

To calculate the relative velocity at the contact point i, which is expressed as $u_{A-i} - u_{B-i}$ needs the velocity of both sides of the rigid bodies' velocity at the contact point as:

$$u_{A-i} = v_A + \omega_A \times r_{A-i}$$
(13)

where the capital letter A is used to represent the id of rigid body A and A-i means the contact point i at rigid body A.

The impulse applied to the contact point is calculated as:

$$P = \frac{-(1+e)u_i^- \cdot \mathbf{n}}{\sum_{i=\{i,j\}} \left(M_i^{-1} + \left(I_i^{-1}(r_i \times \mathbf{n}) \times r_i \right) \cdot \mathbf{n} \right)}$$
(14)

where $\mathbf{n}$ is the unit normal vector at the contact point and M^{-1} is the inverse of mass of the rigid body. Note that if the rigid body is immovable like the boundary, the mass is infinite large.

After the calculation of impulse, the linear velocity and angular velocity of the rigid body is updated by Eq.(6) and Eq.(7). And the difference between velocity constraint method and sequential impulse is for the order of dealing with collisions when there are several collisions happening at the same time. In sequential impulse, the collision is solved to make the relative velocity at the contact point follow Newton's impact law Eq.(12) one by one. And the sequence of solving the collision may affect the simulation results. In velocity constraint method, the post-collision velocity at all the contact points are calculated before the calculation of impulses. The impulses at all the contact points are calculated iteratively to make all the contact points satisfy the post-collision velocity calculated. However, this method suffers from an erroneous impulse distribution when a face to face collision happens.

In the work of (Mirtich, B et al. 1995), the author follows Stronge's hypothesis (Stronge, W. J. 2018) instead of Newton's impact law because of its energy conservation. In Stronge's hypothesis, the process of a collision is divided into compression and restitution. During compression, the contact point absorbs energy from the impact. During restitution, the energy stored at each contact point is released to both sides of the contact points. The relationship between the energy absorbed and released is:

$$\left| W_{release} \right| = e^2 \left| W_{mc} \right|$$
(15)

where $W_{release}$ and W_{mc} are the energy released and absorbed respectively.

The energy in Eq.(15) is calculated as:

$$W = \frac{1}{2}\left(u_n^+ + u_n^- \right) P$$
(16)

where the subscript n means it is the normal relative velocity.

The energy stored at the contact point is only affected by the impulse in the normal direction, as the force in the tangential direction (friction) is dissipative to the kinematic energy of the rigid body. The compression part stops when the relative velocity becomes 0 and the restitution stops when energy storage is 0. The whole process of Stronge's hypothesis is shown in Figure 2.

When considering the effect of friction, usually a local coordinate is established at the contact point including the normal direction and two tangential direction to represent the velocity change as three components (Hahn, J. K. 1988). As the tangential impulse (friction) only works to stop the relative

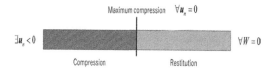

Maximum compression $\forall u_n = 0$

$\exists u_n < 0$ $\forall W = 0$

Compression   Restitution

Figure 2. The demonstration for Stronge's hypothesis

motion in the tangential direction, the velocity after collision is a vector as $[0, 0, u_n^+]$. The velocity before collision is a vector like $[u_t^-, u_q^-, u_n^-]$ and the difference between the pre- and post- collision velocity is Δu. A collision matrix K is built to calculate the impulse as:

$$P = K^{-1}\Delta u \qquad (17)$$

where P is also a vector to represent the impulses in three directions and K is expressed as:

$$K = \left(\frac{1}{M_A} + \frac{1}{M_B}\right)\mathbf{1} - \left(r_{A-i}I_A^{-1}r_{A-i} + r_{B-i}I_B^{-1}r_{B-i}\right) \qquad (18)$$

The magnitude of x and y component in P must be less than the z component according to Coulomb's law of friction. If not, the impulse in normal direction P_n must be recalculated as:

$$P_n = \frac{u_n^+ - u_n^-}{n^T K \left(n + \mu t\right)} \qquad (19)$$

where the vector t is the unit vector of the tangential relative velocity and the numerator is the relative velocity change in the normal direction. Then the tangential impulse is calculated as $P_t = t\mu P_n$.

However, as pointed out in (Kurose, S., et al. 2009), impulse-based method can only solve the collisions one by one. In the work of (Tang, X. et al. 2014), the priority of the collision is decided by the magnitude of the relative velocity at the contact points during compression and during restitution the points with maximum energy storage is solved first. Therefore, in this paper as the rigid body is formed by a lot of rigid body particle, which means the collision or contact may happen at the same time, many contact points having the largest relative velocity need to be dealt with at the same time, which is intuitively physically correct. The largest relative velocity is found numerically as u_{max} and other contact points with very similar relative velocity are chosen and put into a list. If apply an impulse to make the relative velocity at the contact points all 0 according to the local information, the influence of the impulses will be too large. Therefore, the velocity change is

sliced into pieces and the impulses are applied iteratively to make sure they don't push the rigid body too hard. In this work, a tolerance is set to find the contact points with the similar velocity as:

$$Tol = \delta |u_{max}| \qquad (20)$$

where δ is chosen as 1%-5%.

When the compression is over, the Stronge's hypothesis Eq.(15) is applied to all the contact points. In restitution, the simulation starts from the contact points with the largest energy storage. The maximum energy storage W_{max} is founded first and a tolerance like Eq.(20) is used to find the similar points. And limitation for velocity change in restitution is decided by the energy storage at each contact point and calculated by:

$$\Delta u_{lim} = -u_n + \sqrt{u_n^2 + 2Wn^T Kn} \qquad (21)$$

To keep the continuity of the velocity change in the simulation, the limitation velocity change is not directly used to calculate the impulse applied to the rigid body. Instead, it is used as the maximum change and sliced into small pieces to apply the impulse iteratively.

In this work, the friction calculation follows the equation used in the work of (Bender, J. et al. 2006) as:

$$P_t = \frac{u_t}{t^T Kt} \qquad (22)$$

This friction is calculated directly from the local tangential relative velocity to stop the relative motion in the tangential direction. And according to Coulomb's law of friction, it needs to be modified according to the normal impulse as:

$$P_t = \begin{cases} \mu P_n & \left(P_t > \mu P_n\right) \\ P_t & otherwise \end{cases} \qquad (23)$$

4 NUMERICAL TEST

4.1 *Free fall box-shaped rigid dody*

To test whether using Stronge's hypothesis can reproduce the simplest free fall simulation. A 2m x 2m x 1m box with 0.1m particle diameter is dropped freely to the ground. The coefficient of restitution is set as 0.5 and the time increment is 0.001s. From the simulation result (Figure.3), it can keep the energy conservation in the cases of free fall simulation.

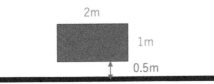

Figure 3. The demonstration for free fall simulation

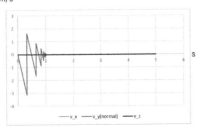

Figure 4. The demonstration for free fall simulation

4.2 *Friction test for a box-shaped rigid body*

The same rigid body in the free fall test is used and located on a slope with 30 degree as Figure 5. To make sure the box slides down the slope, the coefficient of restitution is set as 0 to keep the bottom of the rigid body as a resting contact. 4 cases are tested with different coefficient of friction as 0.2, 0.3, 0.4 and 0.5. These simulation results show that the friction of sliding condition can be reproduced correctly like Figure 6.

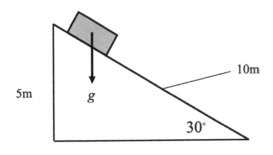

Figure 5. The demonstration for free fall simulation

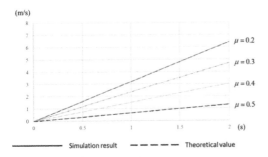

Figure 6. Simulation result for friction tests

Figure 7. Simulation result for Stanford Bunnies

4.3 *Demonstration of multiple rigid body*

This simulation is a test for multiple rigid body contact problem. 3cm x 3cm x 2cm Stanford Bunny with 0.1cm particle diameter is used as the rigid body with complicated shape. In the simulation, the time increment is 0.001s. The coefficient of restitution and coefficient of friction are chosen as 0.5 and 0.2 respectively. 45 bunnies drop freely into a hemisphere with the dimension of 15cm diameter. Each layer of bunnies is dropped with a small interval. From the simulation results, the motion of the bunnies looks natural and at last all of the bunnies stack static without interpenetration.

5 CONCLUSION AND FUTURE WORK

By applying Stronge's hypothesis and impulse-based method, the simple rigid body contact problem like the free fall collision and the silding friction contact can be easily solved and the energy conservation is also kept well. The test of multiple rigid body has no interpenetration.

As the future work, the resting contact (normal relative velocity is near 0) can be solve by using coefficient of restitution 0, but there is some problem to judge the if it is a resting contact automatically. And for multiple rigid body simulation, some experiment data is needed to do the validation to check if the algorithm is correct and robust for the multiple rigid body simulation. The fluid-rigid body interaction simulation is a challenge after the simulation for multiple rigid body can give a physically correct result.

REFERENCES

Mirtich, B., & Canny, J. (1995, April). Impulse-based simula tion of rigid bodies. In *Proceedings of the 1995 symposium on Interactive 3D graphics* (pp. 181-1ff). ACM.

Erleben, K. (2007). Velocity-based shock propagation for multi body dynamics animation. *ACM Transactions on Graphics (TOG)*, *26*(2), 12.

Baraff, D. (1993). Issues in computing contact forces for non-penetrating rigid bodies. *Algorithmica*, *10*(2-4), 292.

Müller, M., Heidelberger, B., Hennix, M., & Ratcliff, J. (2007). Position based dynamics. *Journal of Visual Communication and Image Representation*, *18*(2), 109–118.

Deul, C., Charrier, P., & Bender, J. (2016). Position-based rigid-body dynamics. *Computer Animation and Virtual Worlds*, *27*(2), 103–112.

Tang, X., Paluszny, A., & Zimmerman, R. W. (2014). An im pulse-based energy tracking method for collision resolution. *Computer Methods in Applied Mechanics and Engineering*, *278*, 160–185.

Stronge, W. J. (2018). *Impact mechanics*. Cambridge university press.

Hahn, J. K. (1988, August). Realistic animation of rigid bodies. In *Acm Siggraph Computer Graphics* (Vol. 22, No. 4, pp. 299–308). ACM.

Kurose, S., & Takahashi, S. (2009, April). Constraint-based sim ulation of interactions between fluids and unconstrained rigid bodies. In *Proceedings of the 25th Spring Conference on Computer Graphics* (pp. 181–188). ACM.

Bender, J., & Schmitt, A. (2006, July). Constraint-based colli sion and contact handling using impulses. In *Proceedings of the 19th international conference on computer animation and social agents* (pp. 3–11).

2019 Rock Dynamics Summit– Aydan et al. (eds)
© 2019 Taylor & Francis Group, London, ISBN 978-0-367-34783-3

Dynamic damage constitutive relationship of mesoscopic heterogeneous brittle rock under rotation of the principal stress axes

D.P. Guo
Deputy General Manager of Xuzhen Railway Co.,Ltd., Zigong, China

J.Z. Zhang
School of Civil Engineering, Chongqing University, Chongqing, China

ABSTRACT: A micro-mechanics-based model is developed to investigate microcrack damage mechanism at the stages of rapid stress drop and strain softening of brittle rock under rotation of the principal stress axes. The frictional sliding crack model is applied to analyze microcracks nucleation and propagation. Effects of rotation of the principal stress axes on microcracks nucleation and propagation and coalescence, as well as the inelastic strain increments and shear deformation localization of brittle rock, are investigated. It is shown that the dynamic damage constitutive relationship and the failure strength of brittle rock are sensitive to rotation of the principal stress axes.

1 INTRODUCTION

The sequencing and advancement of a tunnel face leads to the disturbance and redistribution of the primary in-situ stress field under excavation condition. This disturbance and redistribution involve both changes in magnitude and orientation of the stress-field tensor in the proximity of the tunnel boundary. Moving away from the tunnel boundary, the stress tensor eventually returns to its initial in-situ state. Given the controlling influence that stress magnitude and orientation results in the development of brittle fractures, strength degradation and instabilities of rock mass, the analysis of such changes has become standard practice in most rock excavation designs (Eberhardt 2001).

The analysis of excavation-induced stresses has, in the past, been primarily restricted to change in magnitude of the stress-field tensor (Abel & Lee 1973; Chen et al. 1999; Kyung & Yong 2006; Jaeger et al. 2007; Li & Michel 2009). However, it is inadequate that change in magnitude of the stress-field tensor is only taken into account if effects of the principal stress axes rotation are considered to be significant. In the case of an advancing tunnel face, effects of the principal stress axes rotation have been shown to be an important factor, especially with respect to induced stress concentrations and rock strength degradation. Better understanding of effects of the principal stress axes rotation promises benefit in many areas from rock mechanics to slope and underground engineering and earthquake prediction. It is essential and important to understand how microcracks nucleate and propagate and coalesce under rotation of the principal stress axes in order to provide better understanding of brittle rock fracture process that occur in the slope and underground engineering fields. In the paper, the analyses concentrate on effects of the principal stress axes rotation on brittle fracture propagation, induced damage and rock strength degradation.

2 DAMAGE MECHANISM AND MICRO-MECHANICS-BASED MODEL UNDER ROTATION OF THE PRINCIPAL STRESS AXES

In order to study shear failure of rock materials, periodically distributed array of cracks is adopted, as shown in Figure 1. In Figure 1, ξ_1 is the angle measured from the direction of mode II crack unstable growth to x_2-axis. The corresponding local coordinate system (o-$x_1'' x_2''$) is established. The magnitude of shear fault failure stress is sensitive to the distribution of the orientations and sizes of microcracks. Shear fault failure is due to some microcracks whose orientations are θ_0 and maximum initial size c_1 to become unstable. It is proved that the shear fault failure stress is the minimum applied axial stress when

$$\xi_1 = \tan^{-1}\left(\mu + \sqrt{\mu^2 + 1}\right).$$

The critical length of tensile cracks can be defined as

$$l_2 =$$

$$\frac{w\cos\delta - c_1\sin(\theta_0 - \beta) - \left[w\sin\delta - c_1\cos(\theta_0 - \beta)\right]\tan\xi_1}{\sin(\theta_0 - \beta + \varphi) - \cos(\theta_0 - \beta + \varphi)\tan\xi_1}$$

$$\tag{1}$$

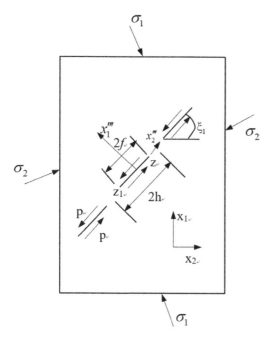

Figure 1. The shear fault model of rock under dynamic compression

where δ denotes the orientation of the array relative to the x_1-axis.

Following works by Tada (1973), the dynamic stress intensity factor of mode II crack array in Figure 1 can be written as:

$$K_{IID} = \frac{v_r - v}{v_r - 0.65v}\left[(\sigma_2 - \sigma_1)\cos\xi_1 \sin\xi_1 + \mu(\sigma_1 \cos^2\xi_1 + \sigma_2 \sin^2\xi_1)\right]\sqrt{2h\tan\frac{\pi f}{2h}}$$

(2)

where the mode II crack spacing $2h = 2f + \frac{2[w\cos\delta - r_i \sin(\theta_0 - \beta)]}{\sin\xi_1}$, $2f$ is the length of mode II crack.

The crack instability condition is then (Sih 1973,1991)

$$a_{22}K_{IID}^2 = S_c$$

(3)

where $S_c = r_c\left(\dfrac{dW}{dV}\right)_c = \dfrac{(1+v_0)(1-2v_0)}{2\pi E_0}\left(K_{ICC}^d\right)^2$ is

the critical strain energy density factor, K_{ICC}^d is mode I dynamic critical stress intensity factor.

The average mode II crack opening displacement can be determined by:

$$b_3 = \frac{1}{2f}\int_{-f}^{f} \frac{-4(1-v_0^2)p}{E_0}\sqrt{f^2 - x_2'''^2}\, dx_2''' = -\frac{\pi p f(1-v_0^2)}{E_0}$$

(4)

where f can be determined by (2) and (3), and $p = (\sigma_1 - \sigma_2)\cos\xi_1 \sin\xi_1 - \mu(\sigma_1 \cos^2\xi_1 + \sigma_2 \sin^2\xi_1)$

$\sqrt{2h\tan\dfrac{\pi f}{2h}}$.

Once σ_1 reaches σ_{1m}, mode II cracks satisfying the growth criterion (3) will experience unstable growth, which may cause shear localization, and a rapid stress drop. During this stage, only mode II cracks oriented at ξ_1 propagate further and other microcracks undergo elastic unloading or back-sliding. Under the condition of strain-controlled loading, the deformation which has received contribution from all microcracks concentrates gradually to the minority of microcracks experiencing unstable growth, which results in a localization of shear deformation.

During the stage of stress drop, the criterion should be satisfied by microcracks experiencing unstable growth. The relation between f and σ can be obtained approximately from the criterion (3) of unstable growth of a microcrack with orientation ξ_1. The inelastic portion of the specific complementary energy including the energy release rate in this phase can be written as

$$\Delta\psi(\sigma, H) = \frac{1}{A_0}\int_{-f}^{f}\int_0^{b_3(x_2')} \tau_{12}'''^r(\sigma, b)\, db\, dx_2'' + \frac{2}{A_0}\int_0^f G(\sigma, f)\, df$$

(5)

where the actual shear traction along zz_1 is

$$\tau_{12}''^r = \left[(\sigma_2 - \sigma_1)\cos\xi_1 \sin\xi_1\right]\sqrt{2h\tan\frac{\pi f}{2h}},$$

$$G(\sigma, l) = \frac{1 - v_0^2}{E_0}K_{IID}^2.$$

If $b_3(x_2') = b_3$, expression (4) can be rewritten as:

$$\Delta\psi(\sigma, b_3) = \frac{2f}{A_0}\int_0^{b_3}\tau_{12}''^r(\sigma, b)\, db + \frac{2}{A_0}\int_0^f G(\sigma, f)\, df$$

(6)

The inelastic change of ψ can be computed by:

$$d^i\psi = \frac{\partial(\Delta\psi(\sigma, b_3))}{\partial b_3}db_3 + \frac{\partial(\Delta\psi(\sigma, f))}{\partial f}df$$

$$= \frac{2}{A_0}\left[f\tau_{12}''^r db_3 + G(\sigma, f)\, df\right]$$

(7)

The increment of inelastic strain in the local coordinate system (0-$x_1''x_2''$) takes the following explicit form:

618

$$\begin{bmatrix} d\varepsilon_{11}^{\prime m5} \\ d\varepsilon_{22}^{\prime m5} \end{bmatrix} = \omega_1 \sqrt{2h\tan\frac{\pi f}{2h}} \begin{bmatrix} -\sin 2\xi_1 \\ \sin 2\xi_1 \end{bmatrix} d\tilde{b}_3$$

$$+ 2\omega p \left(\frac{v_r - v}{v_r - 0.65v}\right)^2 \sqrt{\frac{2\tilde{h}}{c}\tan\left(\frac{\pi f}{2h}\right)}$$ (8)

$$\begin{bmatrix} \mu\cos^2\xi_1 - \dfrac{1}{2}\sin 2\xi_1 \\ \mu\sin^2\xi_1 + \dfrac{1}{2}\sin 2\xi_1 \end{bmatrix} d\tilde{f}$$

where the normalized slip $\tilde{b}_3 = b_3 / c$, $\omega_1 = Nf^2 / A_0$, $\tilde{f} = f / c$, $\omega = Nc^2 / A_0$ is the initial crack density parameter in 2D case material, N is the number of microcracks.

From the relationship between the global coordinate system and the local coordinate system $(0\text{-}x_1''x_2'')$, the increment of inelastic strain in the global coordinate system takes the following explicit form

$$\begin{bmatrix} d\varepsilon_{11}^{m5} \\ d\varepsilon_{22}^{m5} \end{bmatrix} =$$

$$\omega_1 \sqrt{2h\tan\frac{\pi f}{2h}} \begin{bmatrix} \cos\beta & \sin\beta \\ -\sin\beta & \cos\beta \end{bmatrix} \begin{bmatrix} -\sin 2\xi_1 \\ \sin 2\xi_1 \end{bmatrix} d\tilde{b}_3 +$$

$$2\omega p \left(\frac{v_r - v}{v_r - 0.65v}\right)^2 \sqrt{\frac{2\tilde{h}}{c}\tan\left(\frac{\pi f}{2h}\right)} \begin{bmatrix} \cos\beta & \sin\beta \\ -\sin\beta & \cos\beta \end{bmatrix}$$

$$\begin{bmatrix} \mu\cos^2\xi_1 - \dfrac{1}{2}\sin 2\xi_1 \\ \mu\sin^2\xi_1 + \dfrac{1}{2}\sin 2\xi_1 \end{bmatrix} d\tilde{f}$$ (9)

On the other hand, during the stage of stress drop, as the applied axial stress is reduced, the opposing forces of friction acting on some preexisting crack faces change sign, so that backsliding may commence. The backsliding activation equation has same form as the forward sliding activation equation, except for the signs of frictional stresses. In consequence, the onset of backsliding take place when

$$\sigma_{1b} =$$

$$\frac{A_6 + \sigma_2 \sin(\theta - \beta)[\cos(\theta - \beta) - \mu\sin(\theta - \beta)] + 2\tau_c}{\sin(\theta - \beta)\cos(\theta - \beta) + \mu\cos^2(\theta - \beta)}$$ (10)

where

$$A_6 = (\sigma_{1m} - \sigma_{2m})\sin(\theta - \beta)[\mu\sin(\theta - \beta) + \cos(\theta - \beta)] -$$

$$\mu\sigma_{1m}$$

, and σ_{1m} and σ_{2m} are peak loads, respectively.

The orientation bounds $(\theta_{1s}, \theta_{2s})$ defining microcrack backsliding domain can be determined by Eq. (10). The inelastic part of strain increment in the

local coordinate system $(0\text{-}x_1''x_2'')$ due to backsliding of microcracks can be determined by

$$\begin{bmatrix} d\varepsilon_{11}^{\prime mb} \\ d\varepsilon_{22}^{\prime mb} \end{bmatrix} =$$

$$\rho \int_{\theta_{1s}}^{\theta_{2s}} \int_{c_{01}}^{c_1} p(\theta)p(c)c^2 \begin{bmatrix} -\sin 2(\theta - \beta) \\ \sin 2(\theta - \beta) \end{bmatrix} d\tilde{b}_m d\theta +$$

$$2\rho \int_{\theta_{1s}}^{\theta_{2s}} \int_{c_{01}}^{c_1} p(c)p(\theta)c^2 \begin{bmatrix} -\cos(\theta - \beta + \varphi)\sin(\theta - \beta) \\ \cos(\theta - \beta)\sin(\theta - \beta + \varphi) \end{bmatrix}$$

$$\tilde{l}_m d\tilde{b}_m d\theta$$ (11)

where the subscript m assigned to a variable denotes its maximum value recorded in the loading process.

From the relationship between the global coordinate system and the local coordinate system $(0\text{-}x_1''x_2'')$, the inelastic part of strain increment in the global coordinate system due to backsliding of microcracks can be deduced by

$$\begin{bmatrix} d\varepsilon_{11}^{mb} \\ d\varepsilon_{22}^{mb} \end{bmatrix} =$$

$$\rho \int_{\theta_{1s}}^{\theta_{2s}} \int_{c_{01}}^{c_1} p(\theta)p(c)c^2 \begin{bmatrix} \cos\beta & \sin\beta \\ -\sin\beta & \cos\beta \end{bmatrix} \cdot$$

$$\begin{bmatrix} -\sin 2(\theta - \beta) \\ \sin 2(\theta - \beta) \end{bmatrix} d\tilde{b}_m d\theta$$ (12)

$$+ 2\rho \int_{\theta_{1s}}^{\theta_{2s}} \int_{c_{01}}^{c_1} p(c)p(\theta)c^2 \begin{bmatrix} \cos\beta & \sin\beta \\ -\sin\beta & \cos\beta \end{bmatrix} \cdot$$

$$\begin{bmatrix} -\cos(\theta - \beta + \varphi)\sin(\theta - \beta) \\ \cos(\theta - \beta)\sin(\theta - \beta + \varphi) \end{bmatrix} \tilde{l}_m d\tilde{b}_m d\theta$$

During the stage of stress drop, strain remain constant, which can determine the magnitude of stress drop. Then, we have

$$d\varepsilon_{ij(m)} =$$

$$d\varepsilon_{ij(m)}^0 + d\varepsilon_{ij(m)}^{m1} + d\varepsilon_{ij(m)}^{m2} + d\varepsilon_{ij(m)}^{m3} + d\varepsilon_{ij(m)}^{m4} + d\varepsilon_{ij(m)}^{m5}$$

$$= d\varepsilon_{ij}^0 + d\varepsilon_{ij}^{m1} + d\varepsilon_{ij}^{m2} + d\varepsilon_{ij}^{m3} + d\varepsilon_{ij}^{m4} + d\varepsilon_{ij}^{m5} + d\varepsilon_{ij}^b$$ (13)

where $\varepsilon_{ij(m)}$ represents the total strain of rock material at peak loads, the subscript m when assigned to a variable denotes its maximum value recorded in the loading process. In Eq. (13), $d\varepsilon_{ij}^{m1}, d\varepsilon_{ij}^{m2}, d\varepsilon_{ij}^{m3}, d\varepsilon_{ij}^{m4}$ can be determined as follows:

$$\begin{bmatrix} \varepsilon_{11}^{m1} \\ \varepsilon_{22}^{m1} \end{bmatrix} = \rho \int_{\theta_{1f}}^{\theta_{2f}} \int_{c_0}^{c_1} p(\theta)p(c) \begin{bmatrix} \cos\beta & \sin\beta \\ -\sin\beta & \cos\beta \end{bmatrix} \cdot$$

$$\begin{bmatrix} -\sin 2(\theta - \beta) \\ \sin 2(\theta - \beta) \end{bmatrix} c^2 d\tilde{b}_1 d\theta$$ (14)

$$\begin{bmatrix} \varepsilon_{11}^{m2} \\ \varepsilon_{22}^{m2} \end{bmatrix} = \rho \int_{\theta_{1pp}}^{\theta_{2pp}} \int_{c_{01}}^{c_1} p(\theta)p(c) \begin{bmatrix} \cos\beta & \sin\beta \\ -\sin\beta & \cos\beta \end{bmatrix} \cdot$$

$$\begin{bmatrix} -\sin 2(\theta - \beta) \\ \sin 2(\theta - \beta) \end{bmatrix} d\tilde{b}_2 d\theta$$

$$+ 2\rho \int_{\theta_{1pp}}^{\theta_{2pp}} \int_{c_{01}}^{c_1} p(\theta)p(c) \begin{bmatrix} \cos\beta & \sin\beta \\ -\sin\beta & \cos\beta \end{bmatrix} \cdot$$

$$\begin{bmatrix} -\cos(\theta - \beta + \varphi)\sin(\theta - \beta) \\ \cos(\theta - \beta)\sin(\theta - \beta + \varphi) \end{bmatrix} \tilde{l}_1 d\tilde{b}_2 d\theta \tag{15}$$

$$\begin{bmatrix} \varepsilon_{11}^{m3} \\ \varepsilon_{22}^{m3} \end{bmatrix} = \frac{-8\rho(1 - v_0^2)(v_r - v)^2 \tau_{eff}}{E_0} \int_{\theta_{1pp}}^{\theta_{2pp}} \int_{c_{01}}^{c_1} p(\theta)p(c) \cdot$$

$$f_1(\tilde{l}_1) \begin{bmatrix} \cos\beta & \sin\beta \\ -\sin\beta & \cos\beta \end{bmatrix} [D_1] d\tilde{l}_1 d\theta$$

$$+ \frac{4\pi\rho(1 - v_0^2)(v_r - v)^2}{E_0(v_r - 0.5v)^2} \int_{\theta_{1pp}}^{\theta_{2pp}} \int_{c_{01}}^{c_1} p(\theta)p(c)D_0 \cdot$$

$$\begin{bmatrix} \cos\beta & \sin\beta \\ -\sin\beta & \cos\beta \end{bmatrix} [D_2] \tilde{l}_1 d\tilde{l}_1 d\theta$$

$$- \frac{\pi\rho(1 - v_0^2)(v_r - v)^2}{E_0(v_r - 0.65v)^2} \int_{\theta_{1pp}}^{\theta_{2pp}} \int_{c_{01}}^{c_1} p(\theta)p(c)D_4 \cdot$$

$$\begin{bmatrix} \cos\beta & \sin\beta \\ -\sin\beta & \cos\beta \end{bmatrix} [D_3] \tilde{l}_1 d\tilde{l}_1 d\theta \tag{16}$$

$$\begin{bmatrix} \varepsilon_{11}^{m4} \\ \varepsilon_{22}^{m4} \end{bmatrix} = -8\tau_{eff} A_5 \int_{\theta_{1pp}}^{\theta_{2pp}} \int_{c_{01}}^{c_1} \int_0^{\pi/2} p(\theta)p(c)f_2(\tilde{l}_1) \cdot$$

$$\begin{bmatrix} \cos\beta & \sin\beta \\ -\sin\beta & \cos\beta \end{bmatrix} [B_1] d\varphi dc d\theta$$

$$+ 2\pi A_5 \int_{\theta_{1pp}}^{\theta_{2pp}} \int_{c_{01}}^{c_1} \int_0^{\pi/2} p(\theta)p(c)\tilde{l}_1^2 \sin 2(\theta + \varphi) \cdot$$

$$\begin{bmatrix} \cos\beta & \sin\beta \\ -\sin\beta & \cos\beta \end{bmatrix} [B_2] d\varphi dc d\theta \tag{17}$$

After a certain magnitude of stress drop, the rapid stress drop will stop and the rock material will exhibit strain softening behavior in strain-controlled tests. During the stage of strain softening, the localization of deformation and damage further develop, the criterion (3) must be satisfied by microcrack oriented at ξ_1, which gives the relation between the crack length f and stress σ_{ij}. The constitution relation for the stage of strain softening is then

$$d\varepsilon_{ij} = d\varepsilon_{ij}^0 + d\varepsilon_{ij}^{m1} + d\varepsilon_{ij}^{m2} + d\varepsilon_{ij}^{m3} + d\varepsilon_{ij}^{m4} + d\varepsilon_{ij}^{m5} + d\varepsilon_{ij}^b \tag{18}$$

3 CONCLUSIONS

A micromechanical model is developed to investigate the microcrack damage mechanisms of brittle rock under rotation of the principal stress axes. Weibull distribution of orientation and size of microcracks is introduced into the dynamic damage constitutive relationship of brittle rock under rotation of the principal stress axes. Effects of rotation of the principal stress axes on microcracks nucleation and propagation and coalescence, as well as the inelastic strain increments and shear deformation localization of brittle rock, are investigated by using the frictional sliding crack model and the strain energy density factor approach. It is shown that the dynamic damage constitutive relationship and the failure strength depend on rotation of the principal stress axes, the distribution of orientation and size, strain rate and geometric parameters.

REFERENCES

Eberhardt, E., 2001. Numerical modelling of three-dimension stress rotation ahead of an advancing tunnel face. International Journal of Rock Mechanics and Mining Sciences 38(4),499–518.

Abel, J.F., Lee, F.T., 1973. Stress changes ahead of an advancing tunnel. International Journal of Rock Mechanics and Mining Sciences & Geomechanics Abstracts 10(6),673–697.

Chen, X., Tan, C.P., Haberfield, C.M., 1999. Solutions for the deformations and stability of elastoplastic hollow cylinders subjected to boundary pressures. International Journal for Numerical and Analytical Methods in Geomechanics 23(8), 779–800.

Li, L., Michel, A., 2009. An elastoplastic evaluation of the stress state around cylindrical openings based on a closed multiaxial yield surface, International Journal for Numerical and Analytical Methods in Geomechanics 33, 193–213.

Kyung, H.P., Yong, J.K., 2006. Analytical solution for a circular opening in an elastic–brittle–plastic rock. International Journal of Rock Mechanics and Mining Sciences 43(4),616–622.

Jaeger, J.C., Cook, N.G.W., Zimmerman, R.W., 2007. Fundamentals of rock mechanics, 4th ed. Oxford, Wiley-Blackwell.

Tada, H., Paris, P.C., Irwin, G.R., 1973. Stress analysis of cracks handbook, Del Research. Corporation, St. Louis.

Sih, G.C., 1973. A special theory of crack propagation: methods of analysis and solutions of crack problems In: G.C. Sih, Editors, Mechanics of Fracture I, Noordhoof, Leyden.

Sih, G.C., 1991. Mechanics of fracture initiation and propagation, Kluwer Academic Publishers, Netherlands.

2019 Rock Dynamics Summit– Aydan et al. (eds)
© 2019 Taylor & Francis Group, London, ISBN 978-0-367-34783-3

Energy transformations related to rockburst in underground excavation

C.C. Li

Department of Geoscience and Petroleum, Norwegian University of Science and Technology (NTNU), Norway

ABSTRACT: The intensity of a rock burst event is associated with the excess strain energy released from the ejected rock, the energy released from the surrounding rock and the energy of the seismic waves in the case that the burst event is triggered by a fault slippage nearby. The transformations of the energy in a rockburst event are analyzed in this paper. Three energy resources exist when a burst event occurs: the strain energy stored in the rock to be ejected, the energy released from the surrounding rock and seismic energy. All the three types of energy are transformed to fracture energy to damage the rock and kinetic energy to eject the rock during bursting. The energy transforms are first illustrated in a conceptual model. A circular tunnel is taken as an example to demonstrate the relationships of the released energies with the burst depth. Finally, quantitative relationships are established between the ejection velocity of the rock and the released energies. The study shows that the intensity of rockburst is related to the excess energy released for the ejected rock, the burst depth and the magnitude of the seismic event. The intensity of a strain is mainly dependent on the excess energy in the case of a shallow burst, but the role of the energy released from the surrounding rock becomes important when the burst depth is large.

1 INTRODUCTION

Transformation of energies in a rockburst event has been studied by many in the past decades (Hauquina et al. 2018, Li et al. 2017, Tarasov & Stacey 2017, Ai et al. 2016, Huang & Li 2014, Tarasov & Potvin 2013, Duvall & Stephenson 1965). The common sense is that the intensity of the event is related to the elastic energy released from the rock mass during the occurrence of the burst. The results of an analytical study in the transformation of energies are presented in this paper. In this study, the energy contributors are classified into three types: the energy stored in the rock to ejected, the energy released from the surrounding rock after bursting and the seismic energy in that case that the burst event is triggered by fault slippage nearby. It is thought that all the energies are dissipated for doing two things: to fracture the rock and to eject the rock. Therefore, they are transformed to fracture energy for rock damage and kinetic energy for the burst. This concept is illustrated in a concept model below. A circular tunnel is taken as an example to demonstrate the relationships of the released energies with the burst depth and the ejection velocity. The study shows that the intensity of rockburst is related to the excess energy released for the ejected rock, the burst depth and the magnitude of the seismic event. The intensity of a strain is mainly dependent on the excess energy in the case of a shallow burst, but the role of the energy released from the surrounding rock becomes important when the burst depth is large.

2 TYPES OF ROCKBURST

Rockburst is classified to two types according to the trigger mechanisms. The first type is called *strain burst* that is caused directly by the stress concentration in the immediate surrounding rock after excavation (Figure 1a). The highly stressed rock fails when the major principal stress is beyond the strength of the rock. The strain energy released after rock failure is composed of two parts. One part is the elastic strain energy stored in the rock just before the rock failure. The other part is the elastic energy released from the surrounding rock after the rock failure. A portion of the released energy is dissipated for rock damage and the other portion is transformed to kinetic energy to eject the failed rock. Strain burst occurs always in Class II rock, but conditionally in Class I rock. Strain burst is characterized by fine, thin and knife-sharp rock debris (Figure 1b). Strain burst does not involve any seismic activity prior to the rockburst event.

The second type of burst is called fault-slip or seismic burst that is triggered by a seismic event (Figure 2a). The change in the rock stresses caused by mining activities may activate slippage of pre-existing faults/discontinuities. A rockburst event may occur when fault-slip induced seismic waves reach the free rock surfaces of an underground opening. Fault slippage could release a significant amount of energy. Therefore, a fault-slip burst event could be stronger than a strain burst event. Rock debris from a fault-slip burst is composed of rock pieces of different sizes, ranging from finely fragmented debris to large blocks (Figure 2b).

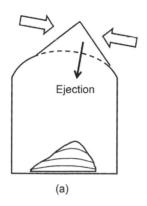

(a)

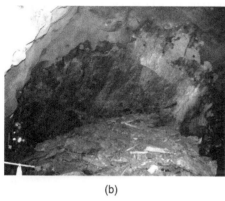

(b)

Figure 1. Strain burst. (a) sketch and (b) an example of strain burst in a metal mine.

3 A CONCEPT MODEL FOR ENERGY TRANSFORMATIONS

The transformation of energies during the occurrence of a rockburst event is schematically illustrated in Figure 3. The curve to the left hand side represents the stress-strain behaviour of the stressed rock that potentially bursts after point P. At point P, the elastic strain energy stored in the burst rock body is represented by the sum of W_r and W_a. The thick solid line to the right hand side is the response line of the surrounding rock after the occurrence of the burst. The elastic energy released from the surrounding rock in this period is represented by the sum of W_{m1} and W_{m2}. When the rockburst occurs, a portion of the strain energy transformed to fracture energy W_r to damage the rock and the rest to kinetic energy W_a to eject the failed rock. The surrounding rock converges in this period and contributes a portion of energy, represented by the sum of W_{m1} and W_{m2} in Figure 3. Part of this portion energy is transformed to kinetic energy W_{m1} and the rest to fracture energy W_{m2}. It will be demonstrated that the amount of the elastic energy released from the surrounding rock with respect to the burst depth in the example below.

Assume that a circular tunnel is excavated in a homogeneous and isotropic massive rock mass which is subject to a hydrostatic stress p_o. On the

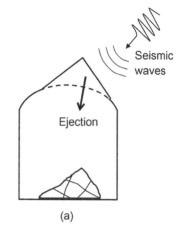

(a)

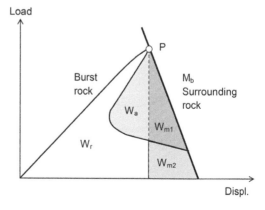

(b)

Figure 2. Fault-slip burst. (a) sketch and (b) an example of fault-slip burst event of 3.8 Mn in a deep metal mine (Counter 2014).

basis of the Kirsch equations, the elastic strain energy density is calculated as:

$$e = \frac{1+v}{E} p_o^2 \left(1 - 2v + \frac{a^4}{r^4} \right) \qquad (1)$$

Figure 3. A sketch illustrating the different energy components in a burst event.

622

where a stands for the radius of the tunnel, r the distance from the position to the tunnel center, E and v the Young's modulus and Poisson's ratio of the rock, respectively. It is assumed that the rock is ejected from radius a to b during the rockburst, Figure 4. The elastic strain energy stored in the ejected rock is

$$W_b = \int_a^b 2\pi redr$$

Substituting (1) into the above integral yields:

$$W_b = \frac{1+v}{E} \pi p_o^2 \left(b^2 - a^2\right)\left(1 - 2v + \frac{a^2}{b^2}\right). \qquad (2)$$

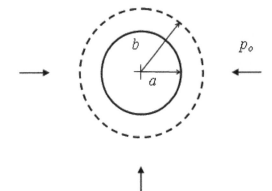

p_o

(a)

p_o

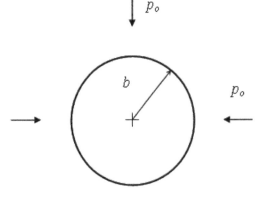

p_o

(b)

Figure 4. Sketches for rock ejection from radius a to b. (a) before the ejection and (b) after the ejection.

W_b equals the sum of W_r and W_a in Figure 3.

The radial displacement u in r in the surrounding rock is expressed as:

$$u = -\frac{1+v}{E} r p_o \left[1 - 2v + \left(\frac{a}{r}\right)^2\right] \qquad (3)$$

The radius displacement at radius b before rock ejection is obtained by letting $r = b$ in (3):

$$u_{b1} = -\frac{1+v}{E} b p_o \left[1 - 2v + \left(\frac{a}{b}\right)^2\right].$$

The radius displacement at radius b after rock ejection is obtained by letting $a = b$ and $r = b$ in (3):

$$u_{b2} = -\frac{1+v}{E} b p_o (2 - 2v).$$

The displacement increment after the rock ejection is:

$$\Delta u_b = u_{b2} - u_{b1} = \frac{1}{M_b} p_o \qquad (4)$$

where M_b is the radial stiffness of the tunnel at radius b, that is expressed by:

$$M_b = -\frac{E}{(1+v)} \frac{b}{\left(b^2 - a^2\right)}. \qquad (5)$$

The radial stress at radius b before burst, according to the Kirsch equations, is:

$$\sigma_{rb} = p_o \left(1 - \frac{a^2}{b^2}\right). \qquad (6)$$

The radial stress linearly drops to zero after a radial displacement of Δu_b. The energy released from the surrounding rock during this process is calculated as:

$$W_m = -\frac{1}{2}\left(2\pi b \sigma_{rb}\right)\Delta u_b.$$

Substituting (4) and (6) into the above expression yields

$$W_m = \frac{1+v}{E} \pi p_o^2 \frac{\left(b^2 - a^2\right)^2}{b^2} \qquad (7)$$

W_m is the sum of W_{m1} and W_{m2} in Figure 3.

Let tunnel radius $a = 4$ m, $E = 70000$ MPa, $v = 0.2$, $p_o = 30$ MPa. The stiffness of the surrounding rock is calculated according to (5). The variation of the stiffness with the burst depth is presented in Figure 5. It indicates that the stiffness is very high when the burst depth is small, for instance, less than 0.5 m. A high stiffness means that the M_b line in Figure 3 is steep. The stiffness becomes relatively low when the burst depth is large, such as more than 1.5 m.

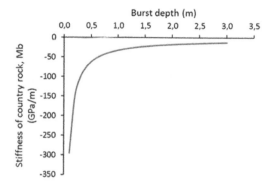

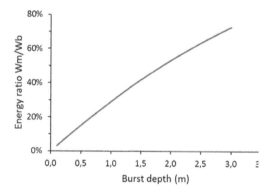

Figure 6. Ratio of W_m to W_b versus the average burst depth in an 8m circular tunnel.

To compare the energy released from the surrounding rock, W_m, with the strain energy in the ejected rock party, W_b, the ratio W_m to W_b is plotted versus the burst depth in Figure 6. The ratio is less than 7% when the burst depth is less than 0. 5 m. It implies that the elastic energy released from the surrounding rock is very small compared to the strain energy in the ejected rock. The released energy W_m becomes significant when the burst depth is large.

4 KINETIC ENERGIES IN A ROCKBURST EVENT

4.1 Strain burst

The total potential strain energy stored in the ejected rock prior to ejection is quantitatively expressed as

$$W_b = \int_0^v \frac{1}{2E} \sigma_i^2 \, dV \tag{8}$$

where V is the volume of the rock to be ejected, and σ_i are the principal stresses in the rock ($i = 1, 2$ and 3). As mentioned above, part of the total potential strain energy, denoted as W_a, is released in the form of elasticity. Let α represent the percentage of the elastically released energy in the total strain energy. The excess energy W_a, is then expressed as:

$$W_a = \alpha \int_0^v \frac{1}{2E} \sigma_i^2 \, dV \cdot \tag{9}$$

Note that α is not zero only in Class II rock and it is zero in Class I rock. Another part of the kinetic energy for rock ejection is W_{m1} which is quantitatively expressed as:

$$W_{m1} = \beta \int_0^A \frac{1}{2} \Delta u_{ij} \sigma_{ij} \, dA \cdot \tag{10a}$$

where β is the transformation coefficient, varying from 0 to 1, A is the area of the burst contour, σu_{ij} are the displacement increment at the contour after ejection and σ_{ij} are the stresses on the contour area. In the case of a circular ejection as illustrated in Figure 4, the W_{m1} is expressed as:

$$W_{m1} = \beta \frac{1+v}{E} \pi p_o^2 \frac{\left(b^2 - a^2\right)^2}{b^2} \tag{10b}$$

The ejection velocity of the strain burst event, denoted as v_{SB}, is then expressed as

$$v_{SB} = \sqrt{\frac{2}{m}\left(W_a + W_{m1}\right)} \tag{11a}$$

or

$$v_{SB} = \sqrt{\frac{2}{m}\left(\alpha W_b + \beta W_m\right)} \tag{11b}$$

where m is the mass of the ejected rock.

4.2 Fault-slip burst

Assume that a fault-slip event generates a sinusoidal seismic wave (Figure 7) which is expressed by (Li 2011).

$$u_x = A \sin\left(\omega t - kx\right) \tag{12}$$

where u_x: particle displacement at position x; A: the displacement amplitude; ω: angular frequency, $\omega = 2\pi f$; f: frequency; t: time; k: wave number, k = ω/C; C: wave velocity and x: position.

The seismic wave induces particle vibrations in the rock it passes through. The vibrations bring about a strain and stress in the rock so that a static strain energy density, w_s, is thus induced in the rock by the seismic wave. On the other hand, the wave propagation means that a kinetic energy component, w_k, is also induced by the seismic wave. Therefore, the total energy induced by the seismic wave, w, is the sum of the two components w_s and w_k, that is, $w = w_s + w_k$.

In the case of a longitudinal wave (i.e. the P wave), the normal strain induced by the wave is $\varepsilon_x = du_x/dx$ and the normal stress is $\sigma_x = E\varepsilon_x$. The static wave

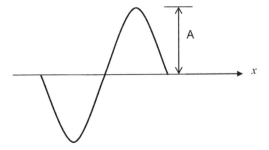

Figure 7. A sinusoidal seismic wave pulse.

strain energy density is expressed as $w_s = \sigma_x \varepsilon_x/2$. The average static strain energy density of the seismic wave is then obtained as

$$W_s = \frac{1}{4} \rho \left(PPV \right)^2 \tag{13}$$

where ρ is the density of the rock and PPV represents the Peak Particle Velocity, $PPV = A\omega$. The particle velocity is expressed by $\dot{u}_x = du_x/dt$. The kinetic energy density is calculated as $w_k = \rho_x/2$. The average kinetic energy density is then obtained as

$$W_k = \frac{1}{4} \rho \left(PPV \right)^2 \tag{14}$$

The total energy density in the rock, which is caused by the seismic wave, is thus

$$W = W_s + W_k = \frac{1}{2} \rho \left(PPV \right)^2. \tag{15}$$

Let v_{FS} represent the wave-induced velocity of the ejected rock. The following equilibrium must exist:
$$\frac{1}{2} m v_{FS}^2 = \gamma WV$$
where γ represents the percentage of the seismic energy that is transformed to kinetic energy for rock ejection. The ejection velocity v_{FS} is then obtained as:

$$v_{FS} = PPV \sqrt{\gamma}. \tag{16}$$

A fault-slip rockburst involves release of both seismic energy and the potential strain energy. The total ejection velocity of the burst rock body is then expressed as

$$v = \sqrt{v_{SB}^2 + v_{FS}^2} = \sqrt{\frac{2}{m} \left(W_a + W_{m1} \right) + \gamma \left(PPV \right)^2} \tag{17a}$$

or

$$v = \sqrt{\frac{2}{m} \left(\alpha W_b + \beta W_m \right) + \gamma \left(PPV \right)^2}. \tag{17b}$$

5 CONCLUSION

A theoretical solution to the intensity of a rockburst event is obtained by considering the released energy available for rock ejection. The contributions to the kinetic energy of the ejected rock are the excess elastic energy released from the burst rock, part of the elastic energy released from the surrounding rock and part of the seismic energy. Referring to Figures 5 and 6 and Equation (11), the kinetic energy of a strain burst is mainly contributed by the excess energy released from the burst rock as long as the burst depth is shallow, such as less than 0.5 m. The role of the surrounding rock becomes more important when the burst depth is large, such as deeper than 1.5 m. Shallow strain burst occurs only in Class II rock, but deep strain burst could occur in both Class II and Class I rocks.

Referring to Equation (17), the role of the seismic energy could be simply a trigger of the burst event or a major contributor of the burst energy in a fault-slip burst event. In the former, the burst volume and ejection velocity would be limited, while in the latter the burst volume could be vast and the burst intensity could be high.

REFERENCES

Ai, C., Zhang, J., Li, Y.-W., Zeng, J., Yang, X.-L. & Wang, J.-G. 2016. Estimation criteria for rock brittleness based on energy analysis during the rupturing process. *Rock Mech. Rock Eng.* 49: 4681–4698.

Counter, D. 2014. Kidd mine – dealing with the issue of deep and high stress mining – past, present and future. *Deep Mining* 2014 (eds by M Hudyma and Y Potvin). Australian Centre for Geomechanics, Perth, 3–22.

Duvall, W.I. & Stephenson, D.E. 1965. Seismic energy available from rockbursts and underground explosions. *Society of Mining Engineers*, September 1965: 235–240.

Hauquina, T., Gunzburgerb, Y. & Deckb, O. 2018. Predicting pillar burst by an explicit modelling of kinetic energy. *Int. J. Rock Mech. Min. Sci.* 107: 159–171.

Huang, D. & Li, Y. 2014. Conversion of strain energy in triaxial unloading tests on marble. *Int. J. Rock Mech. Min. Sci.* 66: 160–168.

Li, C.C. 2011. Chapter 18 Rock support for underground excavations subjected to dynamic loads and failure. In: *Advances in Rock Dynamics and Applications* (eds by Y. Zhou and J. Zhao). CRC Press, Taylor & Francis Group, Boca Raton, 483–506.

Li, D., Sun, Z., Xie, T., Li, X. & Ranjithb, P.G. 2017. Energy evolution characteristics of hard rock during triaxial failure with different loading and unloading paths. *Engineering Geology* 228: 270–281.

Tarasov, B. & Potvin, Y. 2013. Universal criteria for rock brittleness estimation under triaxial compression. *Int. J. Rock Mech. Min. Sci.* 59: 57–69.

Tarasov, B.G. & Stacey, T.R. 2017. Features of the energy balance and fragmentation mechanisms at spontaneous failure of Class I and Class II rocks. *Rock Mech. Rock Eng.* 50: 2563–2584.

2019 Rock Dynamics Summit– Aydan et al. (eds)
© 2019 Taylor & Francis Group, London, ISBN 978-0-367-34783-3

Study on amplification characteristics of seismicity at dam sites using monitored acceleration data, affected by disturbance of the nearby existing cavern

M. Kashiwayanagi
Electric Power Development Co., Ltd., Chigasaki, Japan

M. Yoda
JP Business Service, Tokyo, Japan

ABSTRACT: The earthquake monitoring for Kuzuryu dam has been conducted since its completion. Due to the aging deterioration and following update of the monitoring system, the current system is arranged to the underground powerhouse in addition to the dam crest and both dam abutments. Around 200 earthquake data accumulated so far clarify the dynamic behavior of the dam and its foundation. The dynamic amplification characteristics of the dam and its abutments reflect the characteristics of these materials and configurations. The possible influence on the monitored acceleration due to the existence of the underground powerhouse nearby the dam is examined referring to the numerical simulation of the seismic propagation as well as the frequency analysis. The disturbance due to nearby existing underground structure on the seismic wave amplification is confined in its surrounding area and negligible in other part of the dam foundation.

1 INTRODUCTION

Propagation characteristics of the dam foundation are critical parameters to evaluate the seismicity for the assessment of the anti-seismic performance of dams and these appurtenant structures. 3-D numerical model consisting of a dam, foundation and a reservoir is frequently utilized to examine its characteristics. However simplification of the model and/ or uncertainties of material properties always claim the verification on the applicability of the model. The reproduction analysis of the dam behavior during earthquakes is quite adequate for this purpose.

The earthquake monitoring is recently a standard menu for the dam. It has been scarcely conducted in the existing dams aged more than several decades. Devices for the earthquake monitoring have been newly arranged in some of these. The monitoring of the dam behavior is usually focused, resulting that the arrangement of the devices are limited only on the crest and lower elevation of the dam. A few attentions are paid to the amplification characteristics of the dam foundation. Therefore, very few data are available to the verification above-mentioned.

Kuzuryuu dam (Electric Power Development Co., Ltd.) located in the central Japan is an embankment dam of 129 m high with inclined impervious core for the hydropower. The earthquake monitoring has been conducted since its completion in 1968.

Many earthquake data and response of the dam and the dam foundation are provided in 50-year monitoring period. The amplification characteristics of the dam foundation and the dam are examined with an indicator of amplification ratio and transfer function of these monitored data. Recent monitoring in the dam foundation is conducted in the underground powerhouse close to the dam. The disturbance on the monitoring due to the nearby existing large underground structure is concerned and estimated by a 2-D dynamic simulation to clarify the influence on the propagation characteristics in the dam foundation.

2 SEISMIC MONITORING

The dam section of the maximum height has been focused for the seismic monitoring of the dam. The 3-D accelerometers had been aligned at the center section of the dam. Three sets had been embedded in the dam. Other sets had been arranged on the dam crest and both abutments. Due to the deterioration of these devices since 1968, the system update, providing digital record system as well, has been made for the externally arranged ones, not for the internal ones. The embedded devices have been malfunctioned after approximately-20-year working. As the remarkable update, the similar updated device has been additionally arranged in the gallery (hereinafter, referred to as P-location) of the underground powerhouse close to the dam at almost same elevation of the dam bottom instead of ones embedded at the dam bottom since 2011. Its location is, however, slightly out of the dam. The present monitoring system located at the dam crest (D1, D2 and D3), dam abutments (D4-D9) and the gallery of the

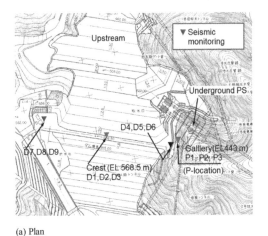

(a) Plan

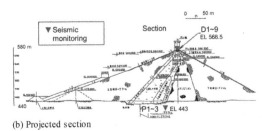

(b) Projected section

Figure 1. Kuzuryu dam and current seismic monitoring

underground powerhouse (P-location, P1, P2 and P3) is shown in Figure 1. All monitored seismic waves counted approximately 200 cases are summarized in Table 1, categorized corresponding to the period of the system update. 17 cases have been monitored in the underground powerhouse since 2011.

Table 1 Outline of earthquakes monitored as of 2018

Period: Year	Monitored locations	Number*	Recording media
A: 1968–1985	Dam foundation, Inside (2 elevations), Crest, Abutments	38	Photosensitive paper
B: 1986–1997	Dam foundation, Inside (2 elevations), Crest, Abutments	34	Magnetic tape
C: 1997–2011	Dam crest, Abutments	92	Digital recording
D: 2011–2018	Dam crest, Abutments, Underground powerhouse (P-location)	42(17)	Digital recording

*Numbers of earthquakes monitored (in underground powerhouse)

3 AMPLIFICATION CHARACTERISTICS

Propagation characteristics in the dam foundation are examined for the monitored data at the dam crest and both abutments respect to ones at the dam bottom and P-location.

An amplification ratio is calculated using the maximum acceleration at each monitoring locations as a comprehensive characteristics of the wave propagation in the dam foundation. 90 data in Period A, B and D (Table 1), providing both data at the dam bottom area and the dam crest, are used. To understand the propagation characteristics in detail, the transfer function corresponding to frequency characteristics is also examined using typical recent 3 sets of data in Period D, which were monitored by the current monitoring system.

The amplification ratio is shown in Figure 2 respect to the maximum acceleration at the dam bottom area. These are illustrated in each period of A, B and D. No obvious tendency is found associated to the system update or the periods in Table 1. It roughly indicates that the propagation characteristics have been stable and the updated monitoring system is consistent to the previous one. The latter is interpreted that both monitored data at the dam bottom and P-location are consistent.

The amplifications are high up to 10 at the dam crest, while these are at most 4 at both abutments, where similar tendency is seen. Such contrast reflects the response characteristics of the dam and the dam foundation that attribute to each mecanical property. Technical committee on the earthquake database of Japan Large Dam Commission (2015) summarized similar characteristics of major types of dams. One for embankment dams verifies the amplification characteristics of the dam above examined.

No nonlinear amplification is found in our data at the dam crest. High response could bring nonlinear response to embankment dams in general. Within our data, such high responses of the dam or intensive seismic waves have not been monitored so far.

As shown in Figure 3, the transfer functions are broken into 9 (nine) components that are three orthogonal ones cited by three orthogonal excitation. These are estimated by the matrix transfer function method (Cao & Kashiwayanagi 2018). The transfer function S_{ij} stands for the response characteristics in i direction against to the input in j direction. Ones of S_{ii} roughly correspond to usual transfer functions but eliminating the response by the input in j direction. These provide clearer propagation characteristics by extracting the directional interference. However, these estimation require three wave sets monitored.

The dominant frequency is identified at 2.4 Hz and 3 Hz. These are predominant frequency of the dam response. The similar dominant frequency of

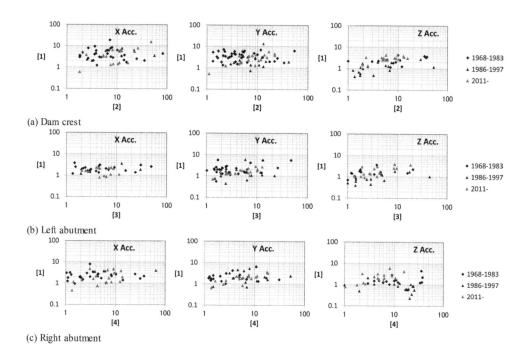

(a) Dam crest

(b) Left abutment

(c) Right abutment

Figure 2. Amplification characteristics of the dam and its abutments respect to the dam foundation

From left, ones for horizontally perpendicular to dam axis (X) and longitudinal direction (Y), and vertical direction (Z)
[1] Amplification of the maximum acceleration to dam bottom(EL 449 m) or cavern (EL 443 m), [2] Maximum acceleration at the center of the dam crest (EL 568.5 m) (cm/²),[3] Maximum acceleration at left abutment (EL 568.5 m) (cm/²), [4] Maximum acceleration at right abutment (EL 568.5 m) (cm/²)

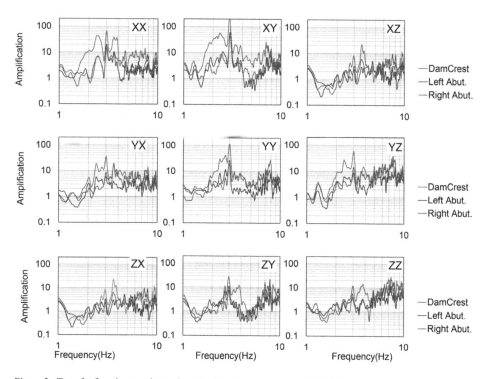

Figure 3. Transfer function matrix, evaluated using three wave sets in Period D

S_{xy} exhibits that the response in X direction of the dam crest is easily caused by the input in Y direction at the dam bottom. Checking ones of abutments, the similar dominant frequencies in the dam crest are found, but differences in 3 Hz to 6 Hz of S_{xx} and S_{yy} are found. The former is resulted from the response interaction between the dam and the abutments, while the latter indicates the response characteristics depending on local mechanical properties of the abutments. In addition reflecting the mechanical properties of the abutments and the dam, the amplification ratios of the abutments are at most 2 and less than ones of the dam crest.

4 EFFECT OF NEARBY CAVERN ON THE PROPARGATION OF EARTHQUAKES

4.1 Outline of the study

In this chapter, the effect of the nearby existing cavern on the propagation characteristics of the earthquake in the dam foundation is examined. The earthquake monitoring has been conducted in the cavern of the underground powerhouse since 2011 (See Table 1, Period D), which is located at 125 m below the left abutment of the dam as shown in Figure 1. The cavern is 20.3 m wide, 42 m high and 70 m long with 130 m overburden. It is concerned such large cavern would incur alteration of the wave propagation in the dam foundation. However, no apparent differences are found in Figure 2 among ones in periods A, B and D, of which the deepest data are monitored at the bottom of the dam, below the points D1, 2, and 3 and in the cavern, respectively, as shown in Figure 1. Even though the monitored locations are different from each other, these, however, suggest no influence of the cavern on the wave propagation in the dam foundation. In other words, the data monitored in the cavern may be equivalent to ones at the dam bottom. It is examined by the numerical simulation as follows.

4.2 Analysis

The simulation is conducted using the 2-D numerical model of the virtual cavern as shown in Figure 4. The foundation boundary is fixed with damping condition to simulate semi-infinite foundation. The material properties are defined referring to ones of existing caverns as shown in Table 2.

The acceleration responses under the seismic load on the model bottom are examined for the models with and without the cavern. Synthetic seismic wave is prepared to match the design acceleration response spectrum for the dam safety assessment (National Institute for Land and Infrastructure Management 2005), so called Level 2 earthquake. These are shown in Figure 5.

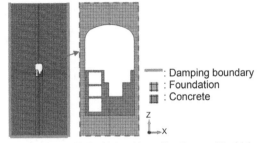

: Damping boundary
: Foundation
: Concrete

Model extent: 293.7m high, 130m wide, Cavern: 32m high, 16.4m wide, Numbers of nodes & element are 41,021and 40.520, respectively.

Figure 4. Numerical model

Table 2. Material properties for the numerical model

Material	*1)	*2)	*3)	*4)	*5)
Foundation	4000	1600	2.7	0.25	5
Concrete	26,501	11,042	2.35	0.2	5

*1) Elastic modulus (N/mm²), 2) Shear modulus (N/mm²), 3) Unit weight (g/cm³), 4) Poisson's ratio, 5) Damping (%)

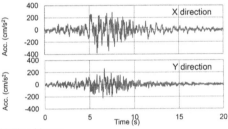

(b) Time history

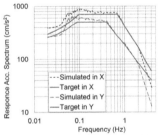

(b) Acceleration response spectrum in horizontal (X) and vertical (V) directions

Figure 5 Synthetic seismic wave for the study

4.3 Results of the analysis

The flash image of the horizontal acceleration distribution is shown in Figure 6 for the maximum values at the bottom and the ground surface. These

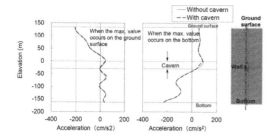

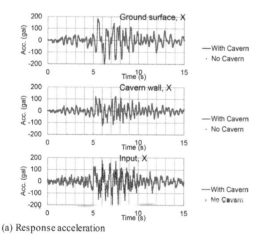

Figure 6 Instantaneous distribution of the horizontal acceleration at the center of the model

* The values on the cavern wall are substituted for the cavern center.

characteristics are shown in Figure 7. The transfer functions indicate some peaks as 0.6 Hz, 2 Hz that agree with the dominant frequencies of multiple reflection theory, taking the foundation depth to the bottom of the model. The example at the ground surface is shown in Table 3.

These characteristics indicate that the influence of the existence of the cavern on the wave propagation is limited in surrounding area of the cavern. It is neither on the ground surface nor in other parts of the foundation.

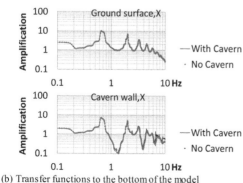

(a) Response acceleration

(b) Transfer functions to the bottom of the model

Figure 7. Characteristics of horizontal response

Table 3 Dominant frequency by multiple reflection theory on the ground surface

Item	Symbol	Value
Shear modulus (N/mm²)	G	1600
Unit weight (g/cm³)	ρ	2.7
Shear velocity (m/s)	V_s	769.8
Foundation depth (m)	H	293.7
Dominant frequency* (Hz)	f_p	0.66 (Primary)
		1.97 (Secondary)
		3.28 (Tertiary)

*Estimated by following equation, $f_p = V_s/4H$

5 DISCUSSION

Based on the above examination, the following discussion is made for the amplification characteristics in the dam foundation of seismic waves.

5.1 Disturbance due to nearby cavern on the seismic wave in the dam foundation

As illustrated as in Figure 2, the monitored data in the underground powerhouse appears roughly consistent to ones at the dam bottom in terms of the amplification ratio in the dam foundation. For the verification, the amplification characteristics are compared as the transfer functions corresponding to the foundation data monitored at the underground powerhouse and the real dam bottom as shown in Figures 3 and 8, respectively. The comprehensive figures of both are similar such as the predominant frequency and the contrast between both abutments. Any clear disturbances due to the existing cavern are not pointed out in the amplification characteristics in the dam foundation in Figures 3 and 8.

The uniform seismic wave must be found at a certain elevation in the rock foundation below the dam bottom, since the seismic wave approaches vertically into the dam foundation from the uniform stratum below. The disturbance on the seismic wave due to the nearby existing cavern is very limited only around the cavern as shown in Figure 6. These are reasons that very slight disturbances due to the existing cavern near the dam are found in the amplification characteristics at the ground surface.

From the standpoint of the earthquake monitoring at the dam site, the monitoring in the underground structures near the dam can be one of the options. The outcropped rock foundation downstream of the dam may be available for the location of the earthquake monitoring. In case of the embankment dams without the foundation galleries, such options are beneficial not to pose the extensive cost to renew the deteriorated earthquake monitoring devises in the foundation. It can be available also for dams with no earthquake monitoring system at present. In addition, it is considered that such options can enhance the existing earthquake monitoring system.

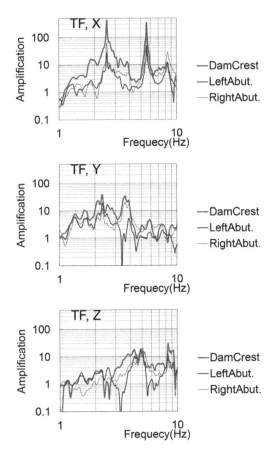

Figure 8. Transfer function of the seismic wave monitored in Period B (1987 June 6th)

The clear dominant frequency is easily found on the dam crest by resulting from its unique configuration. The surrounding rock including the abutment of the dam tends to behave like uniform stratum with less amplification than one of the dam. On the other hand, the interaction caused by the dam response is dominant in the abutments. It is typically demonstrated in the studied case, where the monitoring is conducted at the locations very close to the dam.

6 CONCLUSION

The dynamic behavior of the dam and its foundation are examined using many monitored earthquake records at the dam site. The conclusions drawn from the study are summarized below.

1) The dynamic amplification characteristics of the dam and its abutments reflect the characteristics of these materials and configurations. The dam behavior is recognized as a clear dominant frequency and high amplification ratio due to its unique configuration. For the dam abutments, the interaction by the dam response clearly imposes on.

2) Both monitoring data and the numerical simulation verify that the disturbance due to nearby existing underground structure as large as the underground powerhouse on the seismic wave amplification is confined in its surrounding area and negligible in other parts of the dam foundation. This fact may enhance the earthquake monitoring system in dams by arranging extra monitoring system at existing underground structures near the dam.

5.2 *Amplification characteristics of the dam foundation*

The seismic waves propagate in the uniform stratum as the multiple reflection waves in general. It is observed at the dam bottom or below and the neighboring underground structures. These appear on the ground surface with the alteration in frequency characteristics and intensity due to the disturbance of the configuration of the ground surface or the dam and the dependency on local mechanical properties. The disturbance of the underground structure, which is as large as the underground powerhouse, is negligible.

REFERENCES

Cao, Z. & Kashiwayanagi, M. 2018. Application of the transfer function matric method in dam engineering, *ICOLD Congress, Communication; Proc., Vienna, 4-6 July 2018*. CD

National Institute for Land and Infrastructure Management (ed.) 2005. Technical note on seismic performance evaluation of dams against large earthquake *Technical note*. 244 (in Japanese)

Technical committee on the earthquake database of Japan Large Dam Commission (ed.) 2015. Report on upgrading of earthquake database in Japan, *Large Dams*. 231: 4–54. (in Japanese)

2019 Rock Dynamics Summit– Aydan et al. (eds)
© *2019 Taylor & Francis Group, London, ISBN 978-0-367-34783-3*

Numerical modelling of dynamic indentation of rock with polygonal finite elements

T. Saksala
Laboratory of Civil Engineering, Tampere University of Technology, Tampere, Finland

ABSTRACT: A numerical method based on a damage-viscoplasticity model for rock implemented with the polygonal finite element method is presented in this paper. As many rocks display polygonal mineral texture, polygonal elements are naturally suited for modelling these rocks. A heterogeneity description based on random clusters of elements representing the constituent minerals of the rock is included. The numerical simulations of dynamic indentation of rock with a spherical indenter are carried out to demonstrate the method. The effect of surface roughness is accounted for in the simulation showing that it adds to the reliability of the results.

1 INTRODUCTION

The basic problem to be tackled in numerical modelling of purely percussive drilling is the dynamic indentation of rock by a single indenter (Saadati et al., 2014; Saksala, 2011). A numerical method with predictive capabilities in modelling this problem provides a valuable tool for the drill bit design. Therefore, this is the problem where a numerical model for percussive drilling is to be first tested.

In the two studies cited above, the continuum approach based on finite elements and damage-plasticity type of rock modelling were chosen. However, they used conventional hexahedral (Saadati et al. 2014) and axisymmetric triangular (Saksala, 2011) finite elements. In the present study, the polygonal finite elements are employed in numerical modelling of rock failure under dynamic indentation with a spherical tool.

Polygonal finite elements are rarely used in numerical analyses of rock failure. However, as many rocks exhibit polygonal mineral texture, it is natural to used polygonal elements to describe rock mineral mesostructure. Saksala (2018) used polygonal elements in numerical simulation of rock fracture under dynamic loading. The rock heterogeneity was described at the mesolevel by random clusters of elements representing the rock constituent minerals with their respective material properties.

Compared to the usual triangular and quadrilateral elements, polygonal elements offer, in many cases, greater flexibility in meshing arbitrary geometries, better accuracy in the numerical solution, better description of certain materials, and less locking-prone behavior under volume-preserving deformation (Sukumar & Tabarraei, 2004). However, due to the rational basis functions, the numerical integration is more involved.

In this paper, the polygonal finite element method is applied in numerical simulation of dynamic indentation of rock. The rock failure is accounted for by a damage-viscoplasticity model based on the Drucker-Prager criterion with the modified Rankine criterion as a tensile cut-off, and as a parabolic cap function as a compressive cut-off. In the numerical examples, the dynamic indentation of a granitic rock with a single spherical indenter under axisymmetric conditions is simulated for a demonstration of the method.

2 NUMERICAL MODEL

2.1 *Constitutive model for rock*

A modified version of the damage-viscoplasticity model originally presented by Saksala (2010) is briefly reviewed here for the convenience of the reader. In this model, the stress states leading to viscoplasticity and damage are indicated by the Drucker-Prager (DP) yield function with the Modified Rankine (MR) criterion and a parabolic cap surface as tension and compression cut-offs, respectively:

$$
\begin{aligned}
f_{\mathrm{DP}}(\sigma,\dot{\kappa}_{\mathrm{DP}}) &= \sqrt{J_2} + \alpha_{\mathrm{DP}}I_1 - k_{\mathrm{DP}}c(\dot{\kappa}_{\mathrm{DP}}) \\
f_{\mathrm{MR}}(\sigma,\dot{\kappa}_{\mathrm{MR}}) &= \sqrt{\sum_{i=1}^{3}\langle\sigma_i\rangle^2} - \sigma_{\mathrm{t}}(\dot{\kappa}_{\mathrm{DP}}) \\
f_{\mathrm{Cap}}(\sigma,c,p_{\mathrm{P}}) &= \sqrt{J_2} - C_1(c,p_{\mathrm{P}})I_1^2 - C_2(c,p_{\mathrm{P}})I \\
&\quad - C_3(c,p_{\mathrm{P}})
\end{aligned}
\tag{1}
$$

where I_1 and J_2 are the first and the second invariants of the stress tensor σ, σ_i is the *i*th principal stress, $\langle\rangle$ are the McAuley brackets, α_{DP} and k_{DP} are the DP parameters, *c* and σ_{t} are the dynamic cohesion and tensile strength depending on the rates of the internal variables κ_{DP} and κ_{MR}, respectively, p_p is the hydrostatic pressure that defines the location of the cap, and C_i are the cap fitting parameters. The DP parameters are expressed in terms of the friction angle φ so as to match the uniaxial compressive strength: $\alpha_{DP} = 2\sin\varphi/(3-\sin\varphi)$ and $k_{DP} = 6\cos\varphi/(3-\sin\varphi)$. A plastic potential, denoted g_{DP} hereafter, similar to the DP criterion (but depending on dilatation angle ψ instead of the friction angle φ) is chosen in order to account for the correct dilatational behaviour of the rock in compression.

The dynamic cohesion, tensile strength and the hydrostatic strength (pressure in compression) are defined as

$$c(\dot{\kappa}_{DP}) = c_0 + s_{DP}\dot{\kappa}_{DP}, \quad \sigma_t(\dot{\kappa}_{MR}) = \sigma_{t0} + s_{MR}\dot{\kappa}_{MR}$$
$$p_P(\varepsilon_V^p) = p_{P0} + \frac{1}{D}\ln(1+\frac{\varepsilon_V^p}{W}) \tag{2}$$

where s_{DP} and s_{MR} are the constant viscosity moduli in compression and in tension, respectively. Moreover, ε_V^p is the hydrostatic plastic strain, p_{P0} is the initial (intact) value of pressure, and D and W are parameters controlling the the maximum plastic volumetric strain (W) and the initial slope of the volumetric response curve (D). Thus, perfect viscoplastic behavior is assumed.

As rocks exhibit highly asymmetric behavior in tension and compression, the damage part of the model is formulated with separate scalar damage variables. The damage, both in tension and compression, is chosen to be driven by the viscoplastic strain. By this choice and the perfect viscoplasticity assumption made above, both the strength and the stiffness degradation both in tension and in compression are governed by the damage part of the model.

The damage part of the model is defined by equations

$$\omega_t = A_t\left(1-\exp(-\beta_t\varepsilon_{eqvt}^{vp})\right)$$
$$\omega_c = A_c\left(1-\exp(-\beta_c\varepsilon_{eqvc}^{vp})\right)$$
$$\dot{\varepsilon}_{eqvt}^{vp} = \sqrt{\sum_{i=1}^3\left\langle\dot{\varepsilon}_i^{vp}\right\rangle^2}, \dot{\varepsilon}_{eqvc}^{vp} = \sqrt{\frac{2}{3}\dot{\varepsilon}^{vp}:\dot{\varepsilon}^{vp}} \tag{3}$$
$$\dot{\varepsilon}^{vp} = \dot{\lambda}_{MR}\frac{\partial f_{MR}}{\partial\sigma} + \dot{\lambda}_{DP}\frac{\partial g_{DP}}{\partial\sigma}$$

where parameters A_t and A_c control the final value of the damage variables ω_t and ω_c in tension and in compression, respectively. Parameters β_t and β_c are determined based on the fracture energies G_{Ic} and G_{IIc} by $\beta_t = \sigma_{t0}h_e/G_{Ic}$ and $\beta_c = \sigma_{c0}h_e/G_{IIc}$ with h_e being a characteristic length of a finite element. The equivalent viscoplastic strain in tension and compression, $\dot{\varepsilon}_{eqvt}^{vp}, \dot{\varepsilon}_{eqvc}^{vp}$, respectively, are defined by the

rate of viscoplastic strain tensor $\dot{\varepsilon}^{vp}$ and its principal values $\dot{\varepsilon}_i^{vp}$. Finally, the nominal-effective stress relationship is written as

$$\sigma = (1-\omega_t)\bar{\sigma}_+ + (1-\omega_c)\bar{\sigma}_- \quad (\bar{\sigma} = \bar{\sigma}_+ + \bar{\sigma}_-)$$
$$\bar{\sigma}_+ = \max(\bar{\sigma},0), \quad \bar{\sigma}_- = \min(\bar{\sigma},0) \tag{4}$$

The stress integration process of this model is illustrated in Figure 1.

Few words on the model components are in order here. First, many rocks, such as granite, usually has a negligible porosity. For this reason, it is emphasized that the cap model is not primarily intended for modelling the behaviour of rock due to porosity. Instead, it is employed in order to model the phenomenological behaviour, i.e. the force-penetration response, of the crushing of rock just beneath the indenter where most of the available energy is consumed. The stress state under the indenter is highly confined compression so that the linear (in the stress invariant space) DP criterion is seldom violated there (depending on the calibration). Hence, the compression cut-off is needed in order to have response significantly differing from linear elastic, i.e. to have significant dissipation, see Saksala et al. (2014). Moreover, under the indenter, there are two competing processes, shear-induced flow or pressure induced densification (cataclastic flow) in action during indentation. Inclusion of the cap model allows thus for both of these processes to be accounted for: the cap model for the pressure induced densification (compaction) and the DP viscoplasticity model for the shear induced flow. Finally, the MR criterion is needed to account for tensile failures, and the viscosity is needed to accommodate the strain rate effects. Heterogeneity description is also included as explained in Section 2.4.

Given $\varepsilon_{t+\Delta t} = \mathbf{B}u_{t+\Delta t}^e$ Do:

1. Predict trial elastic stress state:
   $\bar{\sigma}_{trial} = \mathbf{E}:(\varepsilon_{t+\Delta t}-\varepsilon_t^{vp}) \rightarrow \bar{\sigma}_{trial}^{pr}$ (principal stresses)
   $f_{DP}^{trial} = f_{DP}(\bar{\sigma}_{trial}^{pr},\kappa_{DP}^t,\dot{\kappa}_{DP}^t), f_{MR}^{trial} = f_{MR}(\bar{\sigma}_{trial}^{pr},\kappa_{MR}^t,\dot{\kappa}_{MR}^t)$
   $f_{Cap}^{trial} = f_{Cap}(\bar{\sigma}_{trial}^{pr},c^t,p_P^t)$

2. If $\max(f_{MR}^{trial},f_{DP}^{trial},f_{Cap}^{trial}) > 0$
   Perform Viscoplastic correction $\rightarrow \bar{\sigma}^{t+\Delta t}, \varepsilon_{t+\Delta t}^{vp}$
   Otherwise elastic state correct $\rightarrow$ EXIT

3. Update damage variables:
   $\omega_t^{t+\Delta t} = g_t(\varepsilon_{eqvt}^{vp,t+\Delta t}), \quad \omega_c^{t+\Delta t} = g_c(\varepsilon_{eqvc}^{vp,t+\Delta t})$

4. Calculate nominal stress:
   $\sigma^{t+\Delta t} = (1-\omega_t^{t+\Delta t})\bar{\sigma}_+ + (1-\omega_c^{t+\Delta t})\bar{\sigma}_- \quad (\bar{\sigma}^{t+\Delta t} = \bar{\sigma}_+ + \bar{\sigma}_-)$

Figure 1. Solution process (stress integration algorithm) at the material (Gauss) point level.

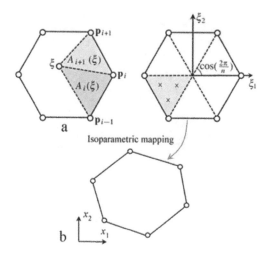

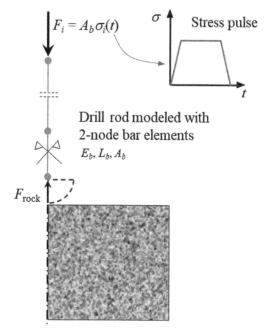

Figure 2. Illustration of the triangular areas used in the definition of Wachspress shape function (a), and the triangulation of the reference regular polygon with 3 integration points in each triangle, and the isoparametric mapping to a physical element (b).

2.2 Polygonal finite elements

Here, the implementation by Talischi et al. (2012b) of the polygonal finite element method based on Wachspress interpolation functions is chosen. It exploits the standard isoparametric mapping from a reference element to the physical element, as illustrated in Figure 2.

Mathematically, a Wachspress type of barycentric interpolant at node i of a reference n-gon reads

$$N_i(\xi) = \frac{\alpha_i(\xi)}{\sum_{j=1}^{n} \alpha_j(\xi)}, \alpha_i(\xi) = \frac{A(\mathbf{p}_{i-1}, \mathbf{p}_i, \mathbf{p}_{i+1})}{A(\mathbf{p}_{i-1}, \mathbf{p}_i, \xi) A(\mathbf{p}_i, \mathbf{p}_{i+1}, \xi)}$$

(5)

where $A(a, b, c)$ denotes the signed area of triangle a, b, c (Figure 1a). The numerical integration is based on a sub-division of the reference polygon into triangles and applying a three-point quadrature for each triangle (resulting $3n$ integration points for each n-gon), as illustrated in Figure 1b.

2.3 Method for indentation simulation

The modelling principle for the indentation simulation is illustrated in Figure 3. Axisymmetry is assumed here. The indentation setup imitates the principle of percussive drilling where a piston impacts the drill rod and the induced compressive stress wave enforces the bit, the indenter, to penetrate into the rock. Here, the drill rod is modeled using 2-node linear bar elements while the external stress pulse $\sigma_i(t)$, applied to the other end of the drill rod, simulates the compressive stress wave due to impact of the piston.

Figure 3. Axisymmetric model for dynamic indentation of rock with a spherical tool.

The dynamic indentation of the virtual and rigid bit into the rock is modeled by imposing kinematic (impenetrability) contact constraints between the indenter and the rock. These contact constraints are of form $u_{bit,z} - u_{i,z} = b_i$, where $u_{bit,z}$ and $u_{i,z}$ are the degrees of freedom in the z-direction of the bit node and ith rock contact node of the bit, respectively, and b_i is the corresponding distance.

The details of the solution procedure based on the Lagrange multipliers and the explicit Modified Euler time integrator can be found in Saksala (2010).

2.4 Rock heterogeneity description

Polygonal elements are suitable for modelling polycrystalline materials such as rocks. In the present approach, the rock constituent minerals are represented by random clusters of Voronoi cells which themselves are the physical polygonal finite elements. The polygonal element mesh is generated by the PolyMesher Matlab code by Talischi et al. (2012a). This code generates 2D Voronoi diagrams (tessellations) consisting of centroidal (or alternatively non-centroidal) Voronoi cells, which are used as polygonal finite element meshes in the present work.

The rock is Kuru granite, which consists of four main minerals with their corresponding percentages as: Quartz 35 %, Albite 31 %, Microcline 28 %, Biotite 6 %, which have their respective material properties. Thereby, the rock heterogeneity can be described. A numerical rock specimen generated by this method is illustrated in Section 3.

3 NUMERICAL SIMULATIONS

3.1 Material properties, model parameters, and the FE mesh

The mechanical properties of the minerals and model parameters used in the simulations are given in Table 1. These values are collected from various sources and may not match the actual ones when determined from a single specimen of actual Kuru granite.

The reader should also note that for most of the model parameters a single value valid for a homogenized material is set. Moreover, due to the lack of experimental data, the exact numerical replication of Kuru granite under dynamic indentation with a specific set up is not the target in this paper.

The axisymmetric finite element mesh with 10000 polygonal elements is shown in Figure 4.

A relatively large domain of rock is used for minimizing the rigid body movements due to the indentation loading. However, as damage due to the impact of the striker at velocity levels used in percussive drilling is very local, the mesh is refinement in the indentation area. The average element size in the refined area in Figure 4 is about 0.6 mm. This requires extremely short time step to meet the Courant stability limit. Indeed, the time used in the simulations was ~ 3E-8 s.

Table 1. Material properties and model parameters used in simulations.

Parameter/ mineral	Quartz	Albite	Microcline	Biotite
ρ [kg/m^3]	2650	2620	2570	3050
E [GPa]	94.5	69	89	33.8
v	0.08	0.28	0.25	0.36
σ_{t0} [MPa]	10	30	7	7
c_0 [MPa]	25	42	25	25
φ[°]	63	63	63	63
ψ[°]	5	5	5	5
G_{Ic} [J/m^2]	100	100	100	100
G_{IIc} [J/m^2]	1000	1000	1000	1000
A_t	0.98	0.98	0.98	0.98
A_c	0.9	0.9	0.9	0.9
S_{MR} [MPa·s]	0.01	0.01	0.01	0.01
s_{DP} [MPa·s]	0.01	0.01	0.01	0.01
D	0.003	0.003	0.003	0.003
W	$20\sigma_{c0}/E$	$20\sigma_{c0}/E$	$20\sigma_{c0}/E$	$20\sigma_{c0}/E$
P_{p0} [MPa]	$9\sigma_{c0}$*	$9\sigma_{c0}$	$9\sigma_{c0}$	$9\sigma_{c0}$

*Compressive strength = $2c_0\cos\varphi/(1-\sin\varphi)$

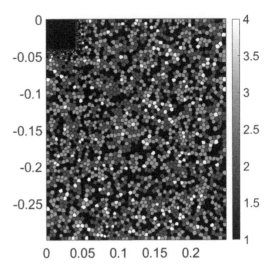

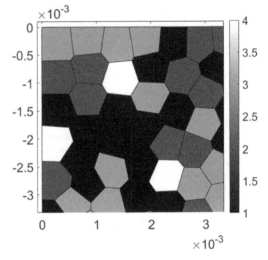

Figure 4. Polygon mesh with 10k elements for simulations (upper), and a detail (lower) (dimensions in [m], and the colobar codes are: Quartz = 1, Albite = 2, Microcline = 3, Biotite = 4).

3.2 Indentation simulations

Two representative indentation simulations are carried out here. In the first, the rock surface in the contact area is assumed perfectly smooth and in the second, the surface nodes of the mesh at the contact area are randomly perturbed in the z-direction so that the resulting surface roughness value is 0.5 mm. The indenter radius is 8 mm and the drill rod (Figure 3) diameter is 22 mm. A stress pulse with a magnitude of 280 MPa, and duration of 0.1 ms with the rise and descent times of 20 μs is applied. The results are shown in Figures 5 and 6.

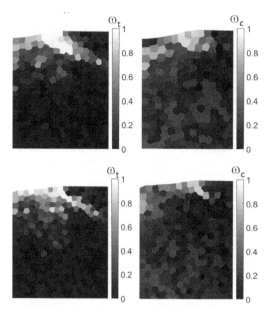

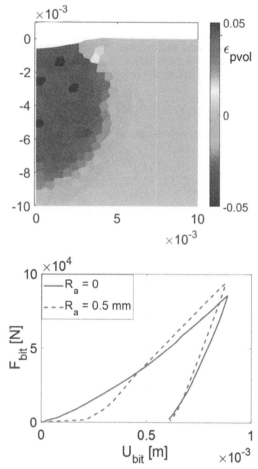

Figure 5. Simulations results: tensile and compressive damage patterns (an average value over the Gauss points in an element is shown) at the end of simulation for the smooth surface (upper), and for the $R_a = 0.5$ mm (lower).

According to the simulation results in Figure 5, the damage induced by the indenter is somewhat mild at this level of compressive stress pulse. The tensile damage patterns display the typical Hertzian cone shape with disrupted details due to the heterogeneity of the rock. The amount of compressive (shear) damage is surprisingly small, which is due the linearity of the Drucker-Prager criterion and its calibration to match the experimental confined compression test results at low confined region, see Saksala et al. (2014). As the stress state under the indenter is extremely highly confined, the rock does not properly fail according to linear DP criterion (the strength envelope of Kuru granite under confined compression is nonlinear in confining pressure). The rock, however, fails in the compression cap mode, as attested in plastic volumetric strain plot shown in Figure 6. Nevertheless, there are some compressive damage in the droplet shaped volume failed in the compression cap mode. Therefore, the two processes mentioned above, i.e. shear-induced flow and pressure induced densification, really are in action under the indenter.

The bit force-penetration curves in Figure 6 display typical non-linear response during the loading and off-loading stages (usually approximated by bi-linear curve). With the non-smooth surface, an initial stage a very low bit force is attested. This is clearly the experimentally observed phenomenon of crushing of the surface asperities.

Figure 6. Simulations results: volumetric plastic strain distribution (the average value over the Gauss points in an element is shown) at the end of simulation with the smooth surface (upper), and the force-penetration curves for both simulations (lower).

4 CONCLUSIONS

This paper presented a percussive drilling simulation method based on a damage-viscoplasticity model for rock implemented with polygonal finite elements. These elements are naturally suited for rock heterogeneity description as the mineral texture of many rocks exhibit polygonal structure.

The numerical simulations of the indentation of rock with a single spherical button under axisymmetric condition demonstrated that the present method has some predictive capabilities in the simulation of this percussive drilling. As the present constitutive model for rock has the Drucker-Prager shear failure criterion with tension and compression cut-offs, it can account for all the major failure types occurring in rock during the indentation process. These include the shear-induced flow and the pressure induced densification in the highly confined zone just under the indenter as well as the tensile side-crack and the Hertzian cone crack, which is of mixed shear-tensile nature.

ACKNOWLEDGEMENTS

This research was funded by Academy of Finland (Grant number 298345).

REFERENCES

Saadati, M., Forquin, P., Weddfelt, K., Larsson, P.L., Hild, H. 2014. Granite rock fragmentation at percussive drilling – experimental and numerical investigation. *International Journal for Numerical and Analytical Methods in Geomechanics* 38:828–843.

Saksala, T. 2010. Damage-viscoplastic consistency model with a parabolic cap for rocks with brittle and ductile behaviour under low-velocity impact loading. *International Journal for Numerical and Analytical Methods in Geomechanics*. 34:1362–1386.

Saksala, T. 2011. Numerical modelling of bit–rock fracture mechanisms in percussive drilling with a continuum approach. *International Journal for Numerical and Analytical Methods in Geomechanics* 35:1483–1505.

Saksala, T., Gomon, D., Hokka, M., Kuokkala, V-T. 2014 .Numerical and experimental studies of percussive drilling with a triple-button bit on Kuru granite. *International Journal of Impact Engineering*. 72: 56–66.

Saksala, T. 2018. Numerical modelling of rock fracture with a Hoek-Brown viscoplastic-damage model implemented with polygonal finite elements. In: Litvinenko, V. (ed.) *Proceedings of the 2018 European Rock Mechanics Symposium (EUROCK2018: Geomechanics and Geodynamics of Rock Masses)*, Saint Petersburg, Russia, 22–26 May, 2018. Vol1, pp. 903–908.

Sukumar, N., Tabarraei, A. 2004. Conforming polygonal finite elements. *International Journal for Numerical Methods in Engineering* 61:2045–2066.

Talischi, C., Paulino, G.H., Pereira, A., Menezes, I.F.M. 2012. PolyMesher: a general-purpose mesh generator for polygonal elements written in Matlab. *Structural and Multidisciplinary Optimization* 45:309–328.

Talischi, C., Paulino, G.H., Pereira, A., Menezes, I.F.M. 2012. PolyTop: a Matlab implementation of a general topology optimization framework using unstructured polygonal finite element meshes. *Structural and Multidisciplinary Optimization* 45:329–357.

2019 Rock Dynamics Summit– Aydan et al. (eds)
© 2019 Taylor & Francis Group, London, ISBN 978-0-367-34783-3

Rock stresses around active faults measured by using the high stiffness hydraulic fracturing technique

T. Yokoyama, M. Murakami T. Danjo K. Ogawa
OYO Corporation, Saitama, Japan

A.Lin W. Lin
Kyoto University, Kyoto, Japan

T.Ito
Tohoku University, Sendai, Japan

ABSTRACT: The hydraulic fracturing technique is a method to measure the stress states in the rocks based on the change of the water pressure observed when the induced fracture in the borehole wall generated by water pressure reopens or closes. A new standard of hydraulic fracturing method in Japan will have two important observation parameters as shown below. One parameter is the water pressure P_s (Shut-in pressure) at the time when the tip of the fracture begins to close after shut-in operation at stopping pressurization in a test interval. The other parameter is the water pressure P_r (Reopening pressure) at the time when the mouth of the fracture begins to open when the test interval is re-pressurized. The new hydraulic fracturing technique highlights that a compliance of the measuring system was adequate for correctly measuring Pr. The small compliance means that the capacity of the water supply system is extremely small and the water supply system has the high stiffness. Small compliance is synonymous with high stiffness, and an ideal measuring system is required to have smaller compliance. We got an opportunity to measure the crustal stress by the hydraulic fracturing technique around Nojima fault which appeared on the surface at the 1995 Southern Hyogo Prefecture Earthquake. The hydraulic fracturing test was conducted at the foot side of Asano fault which is derived from Nojima fault. The measurement depth is about 800 m. The magnitude of the measured maximum principal stress is smaller than the overburden stress, and its principal stress direction does not match the fault movement and the direction of the compression axis is greatly deviated. Therefore, the current stress state around the fault is considered to represent the stress relaxation state after the fault activity.

1 INTRODUCTION

Nojima fault in Awaji Island was active at the 1995 Southern Hyogo Prefecture Earthquake (hereinafter referred to as "the 1995 earthquake"). In the crustal stress measurement by the hydraulic fracturing test around this measurement point carried out immediately after this earthquake, it is reported that the small stress values and the NW-SE direction in maximum principal stress are measured (Tsukahara et al., 1998). Here, it was measured at a depth of about 1500 m at the upper side of Nojima fault, but no induced longitudinal fracture due to hydraulic fracturing was confirmed, and the principal stress direction was estimated from the borehole breakout phenomenon observed in the same borehole.

As a result of the hydraulic fracturing test in the northern part of Awaji Island measured at the same time after the 1995 earthquake, the lateral stress ratio K (maximum principal stress/vertical stress) is more than 2 in less than 900 m, but in near 1200 m, K is around 1, and the magnitude of the stress

is relatively small (Ikeda et al., 1998). Also, the direction of the maximum principal stress is in the NW-SE direction as well as above report.

Twenty three years after the 1995 earthquake, we got the opportunity to measure the crustal stress using a borehole penetrating Asano fault which is a derived fault from Nojima fault. The hydraulic fracturing technique similar to Tsukahara et al. was used for the measurement, but in this time a more accurate method called "high stiffness hydraulic fracturing technique" was used.

In this paper, we first introduce the geological setting around the Nojima and Asano faults. Next, outline the high stiffness hydraulic fracturing technique used for stress measurement is explained. In this section, observation equations and two important parameters obtained from actual pressure change data in hydraulic fracturing test are described, and a replica of the induced fracture are shown. Then, the crustal stress state around the fault analyzed from those data is clarified. Finally, the relation between the obtained crustal stress state and the fault activity is discussed.

2 GEOLOGICAL SETTING

The study area is located in the northern part of Awaji Island, which consists of a mountains area ranged in the NE-SW direction with an altitude of about 200 to 500 m and a hills-plateau-lowland area at the foot of the mountains area narrowly on both sides. The mountains area is formed as a horst between eastern and western escarps of about 200-300 m in height, and a small gently undulating terrain spreads with constant height at the summit. Active faults are developed in the steep escarps on both sides of the mountains area, and Nojima and Asano faults on the west side are formed in the area.

Nojima fault extends over about 10 km in the NE-SW direction on the west side of the mountain range. A systematic right flexion from 20 to 190 m is shown in river valleys and ridges near the fault. Asano fault derived from Nojima fault extends over about 6 km in the NE-SW direction, and is located in the slope conversion line of the mountain foot that forms the boundary between granite and sedimentary rocks.

These two faults were penetrated and sampled at depth in "Nojima Fault Drilling Project" from 2016 to 2018. The drilling was successful, and we collected the core samples at 350-1,000 m including Nojima fault (~530 m) and at 100-700 m including Asano fault (~540 m). Figure 1 shows the location of these faults and borehole for hydraulic fracturing test.

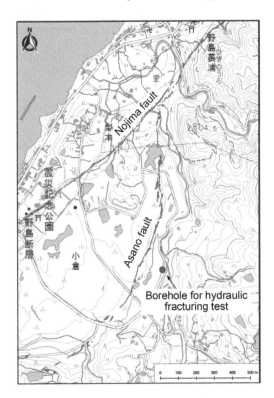

Figure 1. Location of Nojima fault, Asano fault and borehole for hydraulic fracturing test.

3 HIGH STIFFNESS HYDRAULIC FRACTURING

3.1 Observation equations

In the hydraulic fracturing technique, pressure is applied to the wall of an isolated section of a borehole. This technique is used to measure the applied pressures required to open and close new fractures in the borehole wall. When a borehole is drilled in a homogeneous and isotropic elastic rock mass, the distribution of stress in the two-dimensional plane perpendicular to the borehole axis will be as shown in Figure 2. The stress $\sigma_{\theta 1}$ at point A that crosses the S_H axis and stress $\sigma_{\theta 2}$ at point B that crosses the S_h axis can be described as follows:

$$\sigma_{\theta 1} = 3S_h - S_H \tag{1}$$

$$\sigma_{\theta 2} = 3S_H - S_h \tag{2}$$

where S_H is the maximum initial principal stress and S_h is the minimum initial principal stress on the plane. $\sigma_{\theta 2}$ will be smaller than $\sigma_{\theta 2}$ if the compressive stress is positive. Increasing the fluid pressure in the borehole, when the tensile stress at the point A will reach a tensile strength of the rock, a fracture will be generated at the point A. The breakdown pressure P_b can be described as

$$P_b = 3S_h - S_H + T - P_p \tag{3}$$

where T is tensile strength of the rock and P_p is the pore pressure of the rocks. The fracture is closed once with a decrease of pressure under the shut-in of a fluid valve. The fracture will reopen with a re-increase in pressure as shown in Figure 3. The reopening pressure P_r can be described as

$$P_r = (3S_h - S_H)/2 \tag{4}$$

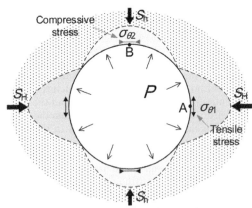

Figure 2. Stress state around a borehole induced by fluid pressure P in a hydraulic fracturing test.

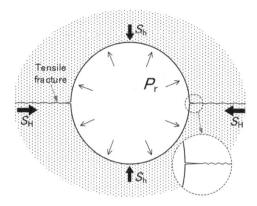

Figure 3. Tensile fractures in a borehole wall induced by fluid pressure P in a hydraulic fracturing test.

At reopening, it is assumed that the tensile strength of the rock is zero, and the pore pressure P_p at the fracture mouth is equal to P_r. If we assume that the minimum principal stress S_h is balanced by the Shut-in pressure P_s, the following equation is obtained.

$$P_s = S_h \qquad (5)$$

These equations are modified for the hydraulic fracturing technique. The maximum principal stress S_H and minimum principal stress S_h are obtained from two observed parameters P_r and P_s.

3.2 High stiffness system

The most reliable way to determine the fracture closing pressure P_s is to calculate it from a depressurization curve after shut-in (Hayashi and Haimson, 1991). However, the maximum principle stress S_H depends not only on P_s but also on the re-opening pressure P_r, and thus the reliability of P_r is also quite important. P_r is greatly influenced by the compliance of the pressurization system (Ito et al., 1999). Here, compliance is defined as the relationship between the pumping volume and the pressure in the closed pressurization system. Thus, a system with less compliance has greater stiffness, and is more sensitive to the change in pressure during the injection of fluid. Therefore, for the hydraulic fracturing test, it would be ideal to use a pressurization system that requires as low a volume of fluid as possible with a tubing system with a high stiffness. Such a system would enable us to obtain a more accurate re-opening pressure P_r from the relationship between the injection volume and pressure in the re-opening test.

Figure 4 shows a modified hydraulic fracturing test system (Yokoyama and Ogawa, 2016). The significant difference between this system and a conventional system is the stiffer stainless steel tubing for the fluid circuit and the use of a syringe pump for the injection of fluid. This modification results in a

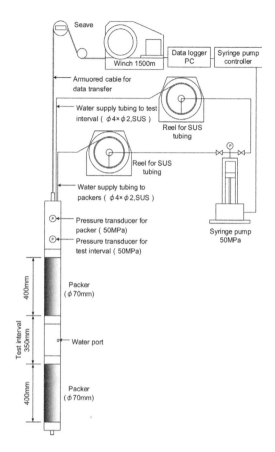

Figure 4. New hydraulic fracturing system with high-stiffness mechanism for 1200m depth.

system with significantly low compliance, and enables the extremely accurate control of the pressure and volume of the fluid. This system can be used in a borehole at a depth of up to 1200m with a diameter of 76mm. Figure 5 shows the hydraulic fracturing test apparatus used in this time. Figure 6 shows an example of a hydraulic fracturing test record for fracture initiation, reopening, and shut-in. The data indicate that the system can be used to inject a very small amount of fluid at a rate of as low as 5 ml/min,

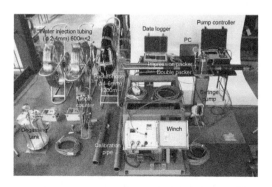

Figure 5. Hydraulic fracturing apparatus used in this study.

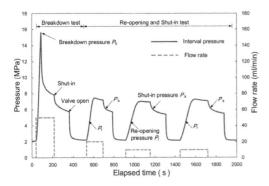

Figure 6. An example of a hydraulic fracturing test record for breakdown, reopening and shut-in.

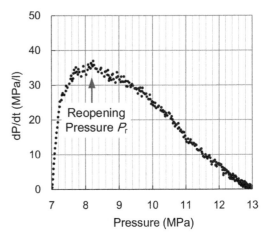

Figure 8. An example of interpretation result for reopening pressure.

and has extremely low compliance for detecting a change in pressure immediately after the injection starts. The compliance of this system in a borehole drilled in the granite is around 2.5ml/MPa. This value was determined from the first injection test for breakdown based on the relationship between the pressure and the flow rate.

4 TEST RESULTS

4.1 Reopening pressure

The reopening pressure P_r is defined as the pressure at the time when a mouth of induced fracture starts to be opened again by the water injection into the test interval. This restart pressure can be read as the pressure at the time when the pressure-time relationship deviates from the straight line. However, it is difficult to accurately read this pressure in the reopening test shown in Figure 7. Therefore, the differential coefficient dp/dQ (MPa/l) of pressure-flow data recorded at 0.1 second intervals was calculated and the reopening pressure was read from the peak value of the pressure increment for each unit flow rate. An example of the interpretation result is shown in Figure 8. It should be noted that since the differential coefficient of the pressure-flow data

actually uses numerical data obtained at intervals of 0.1 second, the variations occur in this graph. The differential coefficient of the pressure-flow data shown in Figure 8 was represented by 11 moving averages, and the peak value of the all data was taken as the reopening pressure.

4.2 Shut-in pressure

The shut-in pressure P_s is defined as the pressure at the moment when an induced fracture generated by hydraulic fracturing starts to close at the fracture tip after water supply stops. Since it is difficult to directly read this closing pressure from the pressure-time relationship, the method of Hayashi and Haimson

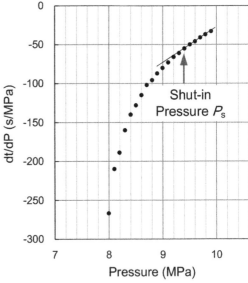

Figure 9. An example of interpretation result for shut-in pressure.

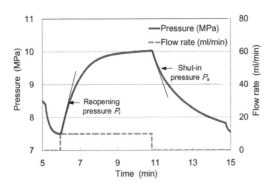

Figure 7. An example of read for reopening and shut-in pressure.

(1991) was used. The shut-in pressure according to this method is a pressure indicating the first inflection point from the high pressure side of the differential coefficient - dT/dP (s/MPa) of the pressure-time relation. An example of the interpretation result is shown in Figure 9. Data on the pressure-time relationship are discrete graphs, because they use the average value of the numerical data obtained at 0.1 second intervals for several seconds.

4.3 Replica of induced factures

Taking a shape of the fracture on the borehole wall was performed by using the oriented impression packer. First, the shape of the natural borehole wall is taken before the hydraulic fracturing test, and the shape at the same depth once again after hydraulic fracturing test. The sheet with traces of fractures obtained by these operations is called a replica. The induced fractures in the hydraulic fracturing test are read by comparing replicas before and after of the test. Since the induced fractures are an extremely

fine, a special plastic rubber attached to the surface of the impression packer is pressed in the similar pressure to the test interval, and the traces of the induced fractures are detected. Figure 10 shows the examples of the replicas of the induced fractures generated by the hydraulic fracturing tests. Lateral or inclined traces are natural fractures, vertical traces are induced fractures.

4.4 Stress states

The hydraulic fracturing tests were conducted at three depths of 725.40 m, 726.64 m, 769.44 m. From the pressure change curves in the test interval obtained in these tests, only one depth at 769.44 m shown in Figure 11 was judged to be an apparent induced longitudinal fracture by hydraulic fracturing. In the pressure change curve obtained at the other two depths, there is almost no difference

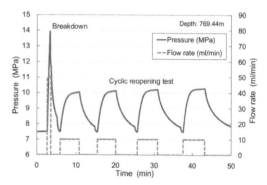

Figure 11. Pressure-time data obtained in a hydraulic fracturing test considered to be successful at 769.44m depth.

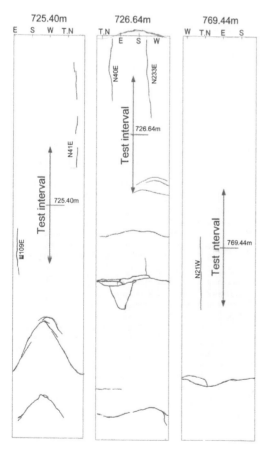

Figure 10. Replicas of natural and induced fractures after the hydraulic fracturing tests at 725.40m, 726.64m, 769.44m depth. Lateral or inclined traces are natural fractures, vertical traces are induced fractures.

Table 1.Measured parameters and analyzed results by the hydraulic fracturing tests at three depths.

Depth (m)	P_b (MPa)	P_s (MPa)	P_r (MPa)	P_p (MPa)	S_h (MPa)	S_H (MPa)	σ_z (MPa)	θ	Q (ml/min)
	-	7.5			-	-			10
	12.0	8.2			12.0	19.6			50
725.40	13.5	11.8	8.4	6.9	11.8	18.6	18.7	N41E	80
	11.8	8.2			11.8	19.0		N109E	50
	12.0	8.2			12.0	19.6			50
	Average				11.9	19.2			-
	14.7	8.1			14.7	27.9			50
	14.6	8.2			14.6	27.4			80
	15.3	9.1			15.3	27.7			100
726.64	17.2	15.0	9.0	6.9	15.0	27.0	18.7	N40E	100
	14.8	9.5			14.8	25.4		N233E	100
	14.8	10.2			14.8	24.0			150
	14.8	10.2			14.8	24.0			150
	Average				14.9	26.2			-
	9.1	7.7			9.1	11.9			10
	9.0	7.7			9.0	11.6			10
769.44	13.9	9.1	7.7	7.5	9.1	11.9	19.8	N21W	10
	9.1	7.7			9.1	11.9			10
	Average				9.1	11.8			-

Depth	: Depth of test interval
P_b	: Break down pressur
P_s	: Shut-in pressure
P_r	: Reopening pressure
P_p	: Pore pressure
S_h	: Minimum principal stress
S_H	: Maximum principal stress
σ_z	: Over burden pressure
θ	: Azimuth of principal maximum stress
Q	: Flow rate

between the peak pressure at breakdown and the peak pressure at the reopening test. Therefore, it is considered that the breakdown in the hydraulic fracturing test at these two depths is likely to have been caused by the opening of the pre-existing fracture.

Table 1 shows the reopening pressure, the shut-in pressure and the azimuth of the induced fracture read from the data of reopening tests obtained at three depth. In the most reliable stress state obtained at 769.44 m, the maximum principal stress is 11.8 MPa, the minimum principal stress is 9.1 MPa, and the azimuth of the maximum principal stress is N21W. The magnitude of the stress obtained at 769.44 m is considerably less than the estimated overburden pressure of 19.8 MPa. The lateral stress ratio K (S_H/σ_z) is 0.6 at 769.44 m.

5 DISCUSSION

The crustal stress state around Nojima fault before the 1995 Southern Hyogo Prefecture Earthquake is unknown. Tanaka et al. (1994) reported the results of continuous stress measurement at the Hiraki mine located about 25 km north of Kobe city. According to this paper, the shear stress ratio R ($(S_H - S_h)/(S_H + S_h)$) linearly increases in the crustal stress measurement results of 5 times from 1978 to 1993, and this value exceeded 0.5 in 1993.

On the other hand, the lateral stress ratios K and the shear stress ratios R obtained by the hydraulic fracturing tests in this time are as shown in Table 2. These shear stress ratios are in the range of 0.13 to 0.27, which are less than half of the values before the 1995 Southern Hyogo Prefecture Earthquake. The stress states at three depths around the Asano fault measured this time largely varies with respect to the direction of the principal stress and the magnitude of the crustal stress is small.

The cause of this stress state can be considered as follows. Assuming that the repetition period of the 1995 earthquake is around 1000 years, the current state of stress is still close to the state immediately after the earthquake and the magnitude of the crustal stress around the fault will be a state of stress drop. For this reason, it is considered that the

each azimuth of the principal stress varies largely depending on the heterogeneity of the rocks around the fault.

6 CONCLUSIONS

We conducted stress measurement around the Nojima fault in Awaji Island and obtained the following results.

- We developed a high stiffness hydraulic fracturing test apparatus capable of measuring stress at depths of over 1000 m.
- This test apparatus is expected to have high stiffness of the water supply system for hydraulic fracturing and to obtain more reliable test data.
- In the foot wall of Asano fault, which is a derived fault of Nojima fault, stress measurements were conducted at a depth of about 800 m.
- It was found that the magnitude of the stress was equal to or smaller than the overburden pressure.
- The direction of the maximum principal stress intersects the fault plane at a high angle and faces the direction opposite to the movement of the fault slip.
- It is inferred that the stress state around the present Nojima fault is the state of stress drop immediately after the seismic activity.

REFERENCES

Ikeda, R. Ilo, Y. and Omura, K. 1998. Stress measurement for active faults, *Earth Monthly 21 (Special issue)*, pp. 91–98. (in Japanese)

Ito T., Evans, K. Kawai, K. and Hayashi, K. 1999. Hydraulic fracture reopening pressure and the estimation of maximum horizontal stress, *Int. J. Rock Mech. Min. Sci. & Geomech. Abstr.*, 36, pp.811–826.

Hayashi, K. and Haimson, B. C. 1991. Characteristics of shut-in curves in hydraulic fracturing stress measurements and determination of in situ minimum compressive stress, *J. Geophys. Res.*, 96: 18, 311–321.

Tanaka, Y. 1993, Crustal stress states and stress changes in the northern part of Kinki district, *Annuals, Disas. Prev. Res. Inst., Kyoto Univ.*, No. 36 B-1, pp. 279–290. (in Japanese)

Tsukahara, H. Ikeda, R. and Yamamoto, K. 1998. Crustal stress measurement in the vicinity of the Nojima fault at the depth of 1500m: A small maximum horizontal compressional stress perpendicular to the fault, *Earth Monthly 21 (Special issue)*, pp. 66–69. (in Japanese)

Yokoyama, T. and Ogawa, K. 2016. New hydraulic fracturing system for in-situ stress measurement by using high stiffness mechanism, *7th International Symposium on In-Situ Rock Stress*, pp. 569–577.

Table 2. Principal stresses, K and R.

Depth (m)	S_H (MPa)	S_h (MPa)	σ_z (MPa)	K[※1]	R[※2]
725.40	19.2	11.9	18.7	1.0	0.23
726.64	26.2	14.9	18.7	1.4	0.27
769.44	11.8	9.1	19.8	0.6	0.13

K[※1] : Lateral stress ratio (SH/σz)
R[※2] : Shear stress ratio(SH−Sh)/(SH +Sh)

2019 Rock Dynamics Summit– Aydan et al. (eds)
© 2019 Taylor & Francis Group, London, ISBN 978-0-367-34783-3

Model tests and numerical simulations on evaluation method of earthquake induced failure of rock slopes

Hideki Nakamura
Chubu Electric Power Company, Inc., Japan

Makoto Ishimaru & Kosuke Hidaka
Central Research Institute of Electric Power Industry, Japan

ABSTRACT: It is satisfied with the Nuclear Regulation in Japan to estimate slope failures in case of a huge earthquake using an equivalent linear analysis. However, in term of the risk assessment and the accident management as an electric company, it is necessary to estimate which area of slopes collapses, and how much ground slides. Therefore, we compared the dynamic model tests using cut rock slope models with equivalent linear analysis and time-history nonlinear analysis from the viewpoint of reproducibility of the collapse behavior and the collapse area. As a result, the equivalent linear analysis proved to have a certain margin in the evaluation of the slip safety factor. However, the slip surface with the minimum slip safety factor did not coincide with the actual collapse range. On the other hand, time-history nonlinear analyses were able to reproduce the residual displacement of the slope model. Furthermore, the accumulation area of strain and the collapse range of the slope model generally agreed

1 INTRODUCTION

For important facilities of nuclear power plants, it is necessary to evaluate the stability of the surrounding slope at the time of earthquake. Because it is necessary to confirm that the safety function of the facility is not seriously affected when the surrounding slope collapses due to the earthquake.

In the guide(Nuclear Regulation Authority.2013) established with the new regulation for the nuclear power reactor, the stability of the slope is considered to confirm that the safety factor against sliding defined as the ratio of resistance of the slope material to sliding force of ground on the slip surface is 1.2 or more by dynamic analysis. And equivalent linear analysis by finite element method is generally used as dynamic analysis (Japan Electric Association Guide. 2015). However, a slip safety factor of less than 1 and a slope collapse are not necessarily equivalent (Ishimaru et al 2011).

From the viewpoint of risk assessment and accident management, it is necessary to estimate which area of slopes collapses and how much ground slides. However, equivalent linear analysis is not a method for quantitatively evaluating deformation. And the safety factor against sliding is evaluated from the instantaneous force balance in this method. Therefore, it is difficult to accurately evaluate the collapse behavior and the collapse area. As a method of obtaining the amount of deformation of the slope, the Newmark method (Newmark,

1965) etc. has already standardized in the field of railways (Railway Technical Research Institute. 2013) and roadways (Japan Road Association. 2010) in Japan. However, these methods are not suitable for evaluation of slope stability and collapsed area, but they are more suitable for evaluation of subsidence and irregularity of the ground against permissible displacement for considering traffic performance and recovery time after the disaster.

In order to calculate a highly reliable collapse area by analysis using large earthquake motions such as ground failure, it is desirable to simulate the process of sliding of the ground inside the slope. Therefore, dynamic analysis by time history nonlinear response analysis considering nonlinearity of ground material is considered useful, but it is necessary to verify applicability to various conditions such as slope geometry and input wave.

Wakai et al. (2007), Shinoda et al. (2015) and others provided examples of collapse cases due to earthquakes on real slopes and shaking table tests using model slopes by time history nonlinear analysis and verification of its applicability. They show the applicability of the time history nonlinear response analysis by simulating the slopes with the consideration of various factors such as sliding planes (e.g. bedding plane and weak layers). Ishimaru et al. (2018) showed the applicability of time history nonlinear analysis to the evaluation of slope failures due to earthquakes for centrifugal model experiments on rock slopes subjected to countermeasures. On the other hand, the sliding plane and the failure area due to the earthquake are thought to be dependent on the characteristics of the seismic motions such as frequency content, however, there are few cases showing the applicability of the time history nonlinear response analysis to the slope failure that is difficult to identify the slip plane before the earthquake.

In this study, therefore, we compared the dynamic model tests using cut rock slope models with equivalent linear analyses and time-history nonlinear response analyses from the viewpoint of simulating the collapse behavior and the collapse area.

2 OUTLINE OF THE SLOPE MODEL TESTS

2.1 Slope model

The slope model and instrumentation are shown in Figures 1 and 2. The slope model in this test was constructed to simulate a real soft rock slope composed of homogeneous sandstone, and its scale was 1/70 of the real slope. The height was 485mm (real-scale is 34m), the gradient was 1:1.5, and the depth was 800mm. The silicone rubber sheet and the lubricant (Vaseline) were used to reduce the friction at the boundary between the model side surface and the box.

2.2 Model ground materials

In model tests, it is difficult to satisfy the perfect similarity rule for the real slope under the test conditions, because these tests are at the 1G field. Therefore, materials were selected so as to simulate the relation between the generated stress and the material strength of the real slope as much as possible.

Specifically, first, for the purpose of simplifying the adjustment of the material strength, the materials having a density as high as possible are selected so that failure occurs within the range of the vibration capability of the shaking box. Next, in accordance with the model scale of 1/70, the material composition was determined so that the internal friction angle is equivalent to that of actual slope material and the cohesion is about 1/70 × (density ratio) of the real slope.

From the results of the material test, it was decided to use a mixed material containing stainless steel particles (diameter: 2 mm, height: 2 mm, cylindrical): sand iron (manufactured by New Zealand Taharoa): water at a weight ratio of 40: 30: 1.

Figure 3 shows the axial differential stress - axial strain relationship obtained from the plane strain compression test. As a result, the peak strength and the stress drop after reaching the peak stress level are clearly observed.

2.3 Input acceleration

Figure 4 shows the input seismic acceleration. A sinusoidal wave was input in the horizontal direction, and the maximum acceleration of the input wave was gradually increased by about 100 gal

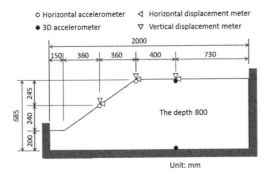

Figure 1.Slope model and instrumentation.

Figure 2. Slope model before the shaking test.

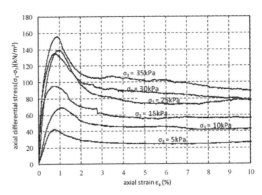

Figure 3. Plane strain compression test results

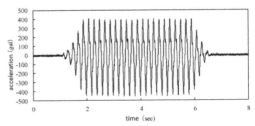

Figure 4. Example of input acceleration (400gal).

every vibration step. The test was carried out using sinusoidal wave with a frequency of 5Hz. According to the white noise excitation before the main excitation, the natural period of the slope model was about 38 to 39 Hz.

3 TEST RESULTS

Figure 5 shows the relationship between the maximum acceleration at each step and the residual displacement (horizontal direction) at each measurement position at the end of the step. The residual displacement rapidly increased when the input acceleration was about 710 gal, and the sliding surface was clearly observed when the input acceleration was about 910 gals. In this test, the sliding failure occurred due to the inertial force resulting from shaking, and the failure did not progressed after the shaking table stopped.

The slope model after all steps is shown in Figure 6, and after removing the failed model material, which corresponds to the approximate slip surface, is shown measurement lines in the vertical direction following the removal of the failed parts. As a result, it is understood that the failure occurred in an arc shape from the top edge of the slope to the bottom edge of the slope as shown in Figure 7.

4 NUMERICAL SIMULATION OF THE MODEL TESTS

In this section, equivalent linear analysis was carried out using material properties of the slope model test, and the minimum safety factor was evaluated and compared with the test results.

4.1 Analysis conditions

The analysis model is shown in Figure 8. The box was modeled using element width and physical properties considered to be rigid body, and the side and bottom surfaces of the slope model were rigidly connected with the box. Table 1 presents the physical properties of the slope, and Figure 9 shows the dynamic deformation characteristics. For the input acceleration in this analysis, the horizontal acceleration measured at the bottom of the slope test was used.

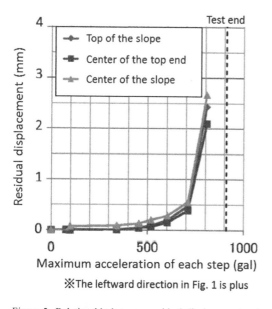

Figure 5. Relationship between residual displacement and maximum acceleration of each step.

Figure 6. The slope model after all steps.

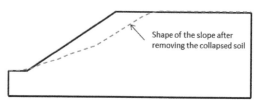

Figure 7. Shape of the slope after the removal of the collapsed soil.

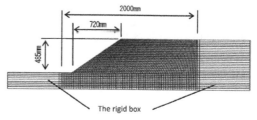

Figure 8. The analysis model.

Table 1. Physical properties and setting parameters

Unit weight γ	t/m³	4.2
Poisson's ratio ν		0.09
Static elastic modulus E	MPa	$1.360\ \sigma^{1.03}$
Initial shear modulus G_0	MPa	$34.44\ \sigma^{0.32}$
Tensile strength G_0	kPa	0.5
Shear strength (peak)	kPa	$7.0+\sigma\tan40.9°$
Shear strength (residual)	kPa	$2.05\ \sigma^{0.69}$

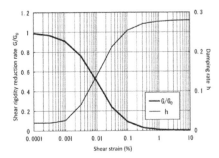

Figure 9. Dynamic deformation characteristics.

Table 2. Results of equivalent linear analysis.

Input acceleration	Minimum slip safety factor	Slip line of minimum slip safety factor
600gal	1.17	
710gal	0.99	810gal / 710gal / 600gal
810gal	0.63	

4.2 Safety factor evaluation and consideration

The stress used to calculate the safety factor were obtained by adding gravity-induced stress and stress caused by vibration. In consideration of the test result, the slip line in this analysis was assumed to be an arc shape passing through the bottom of the slope from the top end of the slope. Table 2 gives the minimum safety factor and its slip line in typical acceleration levels. The slope model in tests did not fail at the moment when the minimum safety factor falls below 1.0 in the analysis and it failed in a higher vibration level. Also, the slip surface with the minimum safety factor passes through a deeper position than the vibration test result shown in Figure 7, and it can not match with the test. This may imply that the design may result in an excessive countermeasures such as piles and anchors, if the slip surface is determined using the result of this evaluation.

5 NONLINEAR ANALYSIS

In this chapter, the applicability of a nonlinear analysis method to the slope model test was evaluated. Ishimaru et al. (2019) proposed a constitutive model that considers the influences of rock shear failure and tensile failure on the shear stress-shear strain relation based on the multiple shear spring model (Towhata et al., 1985) under two-dimensional plane strain state. First, we outline this nonlinear analysis method, and then describe numerical simulations of the model test.

5.1 Nonlinear analysis method

The multiple shear spring model can consider anisotropy if different stiffness and strengths are assigned to each spring. Since, the process becomes very complicated in dynamic analysis, isotropy was assumed herein.

5.1.1 Constitutive model of materials pre-failure
The shear stress-shear strain curve pre rock failure was modeled with the general hyperbolic equation (GHE) model (Tatsuoka and Shibuya, 1992), which is given below by Eq. 1.

$$\tau = \frac{G_0 \cdot \gamma}{\dfrac{1}{C_1(\gamma)} + \dfrac{1}{C_2(\gamma)} \cdot \dfrac{|\gamma|}{\gamma_r}} \tag{1}$$

where τ is the shear stress, γ is the shear strain, G_0 is the initial shear modulus, γ_r is the reference shear strain ($\gamma_r = \tau_{a0} / G_0$), and τ_{a0} is the initial reference shear strength. In addition, $C_1(\gamma)$ and $C_2(\gamma)$ are correction coefficients. The damping characteristics are assumed to follow the model given by Eq. 2, which uses a virtual shear stress-shear strain curve (Ishihara et al., 1985) and the maximum damping constant h_{max}.

$$h = h_{max} \cdot \left(1 - G_R / G_0\right)^{\beta_1} \tag{2}$$

Here, G_R is the shear modulus at the strain level of the unloading point, and β_1 is the adjustment parameter of the damping characteristics.

5.1.2 Definitions of failure
The shear strength is defined by Eq. 3 (in this study, the compression is considered as positive):

$$\tau_f = c_p \cdot \cos\phi_p + \frac{\sigma_1 + \sigma_3}{2} \cdot \sin\phi_p \tag{3}$$

where τ_f is the peak shear strength and C_p is the peak cohesion. In addition, ϕ_p is the peak internal friction angle, σ_1 is the maximum principal stress, and σ_3 is the minimum principal stress. Shear failure is assessed according to Eq. 4, and tensile failure is assessed according to Eq. 5, where σ_t is the tensile strength.

$$\left(\sigma_1 - \sigma_3\right)/2 \geq \tau_f \tag{4}$$

$$\sigma_3 \leq \sigma_t \tag{5}$$

5.1.3 Modeling post failure
The shear stress-shear strain curve after rock failure also employs the GHE model given in Eq. 1. However, the reference shear strain , and the reference shear strength $\gamma_r = \tau_a / G_0$ decreases from the

initial value to the residual shear strength (Eq. 6) after rock failure.

$$\tau_r = a \cdot \left(\frac{\sigma_1 + \sigma_3}{2} \right)^b \qquad (6)$$

where a and b are the adjustment parameters of the residual strength. In addition, the damping characteristics in post-failure are assumed to be the same as those in Eq. 2.

Next, modeling of the reference shear strength τ_a and tensile strength σ_t after rock failure are explained. First of all, as for tensile failure, we consider reducing the strength uniformly in all directions according to the number of tensile failure surfaces. More specifically, an analysis element is divided at a certain angle (N: division number), and the number of divided regions including the tensile failure surface defined by the direction of the principal stress surface is counted. Figure10 shows the concept of division number pre- and post-tensile failure. Then, τ_a and σ_t are gradually reduced by equations (7) and (8).

$$\tau_a = \tau_{as} \cdot \left(1 - n_f / N\right)^{\alpha_1} + \tau_r \cdot \left\{1 - \left(1 - n_f / N\right)^{\alpha_1}\right\} \qquad (7)$$

$$\sigma_t = \sigma_{ts} \cdot \left(1 - n_f / N\right)^{\alpha_1} \qquad (8)$$

where n_f is the number of divided regions including the tensile failure surface, and α_1 is the adjustment parameter of the strength reduction. In addition, τ_{as} and σ_{at} are reference shear strength and tensile strength considering the influence of shear failure. τ_{as} and σ_{at} are gradually reduced by the following equations according to the amount of strain generated after shear failure.

$$\tau_{as} = \tau_r + \frac{\left(\tau_{a0} - \tau_r\right)}{\left(A \cdot \gamma^p + 1\right)} \qquad (9)$$

$$\sigma_{ts} = \frac{\sigma_{t0}}{\left(A \cdot \gamma^p + 1\right)} \qquad (10)$$

where σ_{t0} is the initial tensile strength, γ^p is the maximum value of the maximum shear strain (incremental amount from the value at shear failure), and A is the strain softening coefficient, which is a parameter that determines the rate of decrease in τ_{as} and σ_{ts}.

5.1.4 Redistribution of stresses

The method of redistribution of stresses exceeding strengths is as follows:

i. For shear failure, shrink the Mohr's stress circle with average principal stress fixed.

ii. For tensile failure, shrink the Mohr's stress circle with maximum principal stress fixed. If it exceeds the shear strength even after this treatment, shrink the Mohr's stress circle with minimum principal stress fixed.

iii. Apply residual force calculated from external force and stress.

5.2 Numerical simulations of the model test

5.2.1 Analysis conditions

Table 3 presents the selected values for the parameters of the constitutive model (the other parameters are shown in Table 1). The model parameters were assigned according to the results of the physical and mechanical tests. The strain softening coefficient A was obtained from the relation of the deviatoric stress σ_d - the axial strain ε_a in the strain-softening process after the peak of the plane strain compression tests. In addition, case studies were conducted on the parameters N and α_1 related to strength reduction after tensile failure.

Figure11 shows the analysis model. For the boundary conditions in the numerical simulations of the model test, the bottom side of the model was fixed, and the joint elements (tension/shear spring: 0 kPa; compression spring: 1.0×10^8 kPa) were used on the vertical side of the model. First, an incremental gravity analysis was performed by dividing

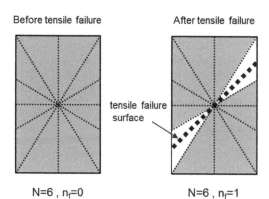

Before tensile failure   After tensile failure

tensile failure surface

N=6 , n_f=0   N=6 , n_f=1

Figure 10. Concept of division number.

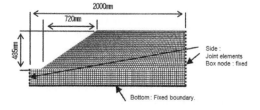

Figure 11. The analysis model.

Table 3. Setting parameters of the constitutive model.

	Case 1	Case 2	Case 3	Case 4
Number of the multiple shear spring in semicircle		12		
Parameters of strength reduction after tensile failre: N, α_1	1, 1.0	3, 1.0	12, 1.0	12, 0.5
Strain softening coefficient A		120		
Parameters of GHE model: γ_r, $C_1(0), C_1(\infty)$, $C_2(0), C_2(\infty)$, α, β		0.0001, 1.0, 0.25, 1.5, 2.5, 1.0, 1.0		
Parameters of damping characteristics: h_{max}, β_1		0.27, 1.0		

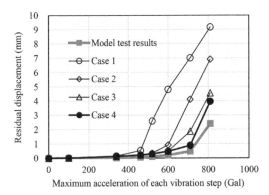

Figure 12. The residual values (accumulated values) of horizon-tal displacement of the top of the slope model for each vibration step.

gravity by 100. Then, in the earthquake response analyses, the accelerations of the shaking table were input. In the seismic response analyses, the stress and deformation state of the preceding step were carried to the next step considering 2 % stiffness proportional damping at 10 Hz.

5.2.2 Analysis results

Figure 12 shows the residual values of horizontal displacement of the top of the slope model for each vibration step. From this figure, it can be seen that the faster the strength reduction after tensile failure, the faster the vibration step in which the residual displacement rapidly increases, and the larger the amount of residual displacement. In particular, Case 1, in which the strengths after tensile failure is immediately reduced, is over conservative as

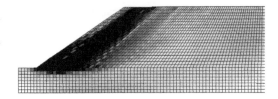

Figure 13. The maximum shear strain distribution of Case 4 in the final state.

compared with the model test results. Case 2 and Case 3 with a larger number of divisions than Case 1 tends to approach the test results. And, in Case 4, the vibration step in which the residual displacement rapidly increases is the same as that of the model test, and the amount of residual displacement is also comparatively close to the model test results.

Figure 13 shows the maximum shear strain distribution of Case 4 in the final state. From this figure, it can be concluded that the region with the largest maximum shear strain is almost the same as the collapsed region in the model test. As described above, it is considered that the residual displacement and the failure region of the slope can be evaluated with certain accuracy by the nonlinear analysis. From now on, the question is how to select the parameters related to strength reduction after tensile failure. In this test, it was found that reproducibility can be improved by increasing the division number N, which is equal to the number of the multiple shear springs. On the other hand, for another parameter α_1, conservative evaluation is possible if a large value is set. However, for quantitative evaluation of the amount of residual displacement, etc., detailed examination of the setting of α_1 is necessary.

6 CONCLUSION

In this study, model tests were conducted simulating cut slopes of homogeneous soft rocks. And the equivalent linear analysis and the time history nonlinear analysis were carried out, from the viewpoint of reproducibility of post-failure behavior. As a result, the following conclusions were obtained.

1. In the equivalent linear analysis, it was suggested that using the slip safety factor as an indicator of the stability of the slope is an evaluation on the safety side. However, the slip surface with the minimum slip safety factor did not coincide with the actual failure surface.

2. In the time history nonlinear response analysis, the reproducibility was better in the case of gradually lowering the shear strength and the tensile strength after the tensile failure than in the case of abrupt decreasing the shear strength and the tensile strength. From now on, it is necessary to study the reasonable selection method of related

parameters in detail in order to more accurately evaluate the collapse area and residual displacement amount.

REFERENCES

Ishimaru, M. and Kawai, T. 2011. Centrifuge model test on earthquake-induced failure behavior of slope in discontinuous rock mass. *ISRM 12th International Congress on Rock Mechanics, Beijing*, 1919–1922

Ishimaru, M., Hashi K., Oikawa K. and Kawai T. 2018. Nonlinear Analysis Method for Evaluating the Seismic Stability of Rock Slopes Considering Progressive Failure. *Journal of Earthquake and Tsunami* Vol. 12, No. 4 1841008.

Ishimaru, M. and Okada, T. 2019. Improvement of a nonlinear constitutive model of rock for dynamic analysis. *Proc. 46th Symposium on Rock Mechanics, Morioka*, 123–126 (in Japanese).

Ishihara, K., Yoshida, N. and Tsujino, S. 1985. Modeling of stress-strain relations of soils in cyclic loading. *Proc. 5th Int. Conf. Numerical Methods in Geomechanics, Nagoya, April 1-5*, Vol. 1, pp. 373–380.

Japan Electric Association Guide.4601-2015, 2015, Technical Guidelines for Aseismic Design of Nuclear Power Plants, pp. 226(in Japanese)

Japan Road Association. 2010. ISBN978-4-88950-417-0

Nuclear Regulation Authority. 2013. Review Guide on the Evaluation of the Stability of the Foundation and the Surrounding Slope, pp. 4–5

Newmark, N. M., 1965. Effects of earthquakes on dams and embankments, Geotechnique, Vol. 15, No.2, pp.139-160.

Railway Technical Research Institute. 2013. Design Standards for Railway Structures and Commentary (Earth Structures).

Shinoda, M., Watanabe, K., Sanagawa, T., Abe, K., Nakamura, H., Kawai, T., and Nakamura, S., 2015. Dynamic behavior of slope models with various slope inclinations. Soils and Foundations; 55(1): pp.127–142

Tatsuoka, F. and Shibuya, S. 1992. Deformation characteristics of soils and rocks from field and laboratory tests. *Proc. 9th Asian Regional Conf. Soil Mechanics and Foundation Engineering, Bangkok*, Vol. 2, pp. 101–170.

Towhata, I. and Ishihara, K. 1985. Modelling soil behavior under principal stress axes rotation. *Proc. 5th Int. Conf. Numerical Methods in Geomechanics, Nagoya, April 1-5*, Vol. 1, pp. 523–530.

Wakai, A., Ugai, K., Onoue, A., Higuchi, K., and Seiichiro, K., 2007. Finite element simulation for collapse of dip slope during earthquake induced by strain-softening behavior of bedding plane. *Landslides journal of the Japan Landslide Society*, Vol. 44, No.3, pp.145–155.

2019 Rock Dynamics Summit– Aydan et al. (eds)
© 2019 Taylor & Francis Group, London, ISBN 978-0-367-34783-3

Seismic response analysis of a tailing dam foundation composed of discontinuous rock masses and countermeasure design

Y. Ohara and S. Fukushima
Japan Atomic Energy Agency, Okayama, Japan

S. Horikawa, M. Koshigai and T. Sasaki
Suncoh Consultants Co., Ltd., Tokyo, Japan

ABSTRACT: The authors report an earthquake response analysis of a tailing dam with a foundation composed of jointed rock masses. The non-linear deformation characteristics of the rock joints are assumed to be combinations of different angles of joint sets using the Multiple Yield Model introduced by Sasaki et al. (1994). The non-linear deformations of a joint in the normal direction and in the shear direction are dependent on the confining pressure presented by Bandis et al. (1981) and Kulhawy (1975), respectively. The results indicate that the tailing dam is sound and that the water-gathering basin with a gravity-type wall should be collapsed. They also show that the additional small gravity dam is sound when the countermeasures considered are taken, and that acceleration at the gravity-type wall is reduced by one-third compared with the structure with no countermeasures taken.

1 INTRODUCTION

In 1955, the Japan Atomic Agency constructed a tailing dam to contain the debris and debris water from a mine in order to provide advanced water treatment of the mineral poison. Later, in 1980, the Agency reconstructed an extension of the dam to a height of 1 m. As of this writing, the volume of debris in the dam has reached 3.6×10^4 m³. According to the Headquarters for the Promotion of Earthquake Research under the Japanese Ministry of Education Culture, Sports, Science and Technology and the Division of Disaster Prevention in Okayama Prefecture, the existence of active faults near the dam site indicates that several earthquakes have occurred. The authors have therefore evaluated the stability of the dam using the Multiple Yield Model (MYM), a non-linear finite element analysis of joint sets in rock masses introduced by Sasaki et al. (1994). This paper reports the results of the analysis on the present state of the dam structure and state of the dam countermeasure design determined through earthquake response analysis.

2 GEOLOGY OF THE SITE

Figure 1 shows a plane view of the geology of the site. Figure 2 shows a section of the tailing dam along A-H-E-B on the plane view in Figure 1. The dam foundation consists of granite classified as C_H with two joint sets: one oriented at a low angle ranging from 4 to 14 degrees and one oriented at a high angle of 90 degrees from the horizontal plane.

And portions of weathered granite classified as C_M, C_L and D are distributed on the basic rock mass. Figure 3 shows a geologic map of the 2-D section for analysis. Table 1 shows the material properties of the tailing dam and foundations. Table 2 shows the material properties of joint sets in the granite.

3 OUTLINE OF THE MULTIPLE YIELD MODEL

Sasaki et al. (1994) proposed the Multiple yield model (MYM) as a tool for analyzing discontinuous rock under static loading. MYM can be considered the equivalent of FEM for modeling discontinuous rock, and adopts a compliance matrix method based on the theories of Hill (1963) and Singh (1973). The basic constitutive model is shown in Figure 4. The applicability of the method was verified under a static loading condition by comparing the results of deformation behavior directly measured during a large-scale excavation with the analytical results

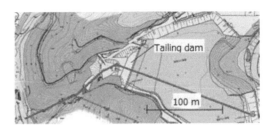

Figure 1. Geologic map along the tailing dam showing the position of the 2-D section.

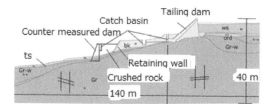

Figure 2. Section of the tailing dam along A-H-E-B on the plane view in Figure 1.

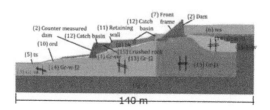

Figure 3. Geologic map of the 2-D section for analysis

Table 1. Material properties of the tailing dam foundation

Mat-No.	Material	E_s (MPa)	v_r	γ_t (kN/m³)	C (MPa)	ϕ (Deg.)	St (MPa)	Ed (MPa)	V_d
1	Gr-CH	2000	0.3	25.4	2	45	100	8000	0.354
2	Dam	22000	0.2	22	3	45	200	22000	0.2
3	Gr-w-CM	310	0.3	22	0.4	37.5	10	1240	0.473
4	Gr-sw-CL	200	0.3	14.2	0.014	26	0	800	0.462
5	ts	2.6	0.3	10.8	0	18	0	27	0.497
6	ws	5	0.3	10.5	0	25	0	26	0.497
7	Front-frame	1000	0.2	22	1	45	100	1000	0.2
8	bk	2.6	0.3	18.6	0.098	19	0	46	0.497
9	Catch basin	1000	0.2	5	10	45	100	1000	0.2
10	ord-gravel	5.3	0.3	13.5	0.01	28	0	53	0.496
11	Retaining wall	22000	0.2	22	3	45	200	22000	0.2
12	Catch basin	2000	0.2	5	1	45	100	2000	0.2
13	Gr-J	2000	0.3	25.4	2	45	100	8000	0.354
14	Gr-w-J	310	0.3	22	0.4	37.5	10	1240	0.473
15	Crushed rock	53	0.3	15	0	40	0	530	0.4

Table 2. Material properties of the joint sets

Joint No.	Angle (Deg.)	Space (m)	K_n (MP/m)	K_{au} (MP/m)	K_s (MP/m)	K_{s-u} (MP/m)	Thickness (m)	ϕ (Deg.)
J1-1	5.28	0.3	2*10⁵	2*10⁶	2*10⁵	2*10⁵	0.002	35
J1-2	90.0	3.0	2*10⁵	2*10⁶	2*10⁵	2*10⁵	0.002	35
J2-1	-13.24	0.3	2*10⁵	2*10⁶	2*10⁵	2*10⁵	0.002	35
J2-2	90.0	3.0	2*10⁵	2*10⁶	2*10⁵	2*10⁵	0.002	35

obtained by MYM (Yoshinaka et al., 2009; Iwata et al., 2011).

MYM supposes that the total strain of the rock mass $\{\varepsilon_T\}$ is given by the summation of the strain of the intact rocks $\{\varepsilon_R\}$ and of the joint systems consisting of joint sets $\{\varepsilon_I\}$, as expressed in Fig. 1 and in Eq. (1) from Sasaki et al (1994).

$$\{\varepsilon_T\} = \sum \{\varepsilon_I\} + \{\varepsilon_R\} \qquad (1)$$

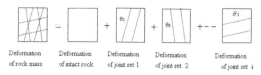

Figure 4. Concepts adopted in the Multiple Yield Model for discontinuous rock

If we assume that each joint set has a periodical distribution, that the volume of the joint set can take no account of the volume of the rock mass, and that the stresses of both the intact rocks and joint sets are equal, the total strain will be as follows in Eq. (2):

$$\{\varepsilon_T\} = [\Sigma[F_I] + [E]^{-1}] \cdot \{\sigma\} = [C] \cdot \{\sigma\} \qquad (2)$$

where $\{\sigma\}$ is the total stress vector, $[E]$ is the stress-strain matrix of the isotropic-elastic body, and $[C]$ is the compliance matrix arranged as a summation with the intact rock and joint sets expressed in Eq. (3).

$$[C] = [D]^{-1} \qquad (3)$$

where $[D]$ is the stress-strain matrix of the rock masses contained within the joint sets.

The compliance matrix of joint sets $[F_I]$ is expressed by Eq. (4) and Eq. (5).

$$[F_I] = [T_I]^T [C_I][T_I] \qquad (4)$$

$$[C_I] = [K_I]^{-1} / S_I \qquad (5)$$

where S_I is the joint spacing, $[T_I]$ is the coordinate transpose matrix, $[K_I]$ is the stiffness matrix of joint set I, and $[B_I]$ is the stress concentration matrix expressed in Eq. (6) and Eq. (7).

$$[T_I] = \begin{bmatrix} \sin^2\theta_I & \cos^2\theta_I & -2\sin\theta_I\cos\theta_I \\ -\sin\theta_I\cos\theta_I & \sin\theta_I\cos\theta_I & \cos^2\theta_I - \sin^2\theta_I \end{bmatrix} \qquad (6)$$

$$[K_I] = \begin{bmatrix} K_{nI} & 0 \\ 0 & K_{sI} \end{bmatrix} \qquad (7)$$

where θ_I is the angle of the joint set from the X axis of the anticlockwise, K_{nI} are the stiffness of joint set I in the normal direction and in the shear direction, respectively. The relations between the stress $\{\sigma_I\} = \{\sigma_{nI}, \tau_{sI}\}^T$ and the strain $\{\varepsilon_I\} = \{\varepsilon_{nI}, \gamma_{sI}\}^T$ of a joint set are expressed by Eq. (8) and Eq. (9).

$$\{\sigma_I\} = [T_I]\{\sigma\} \tag{8}$$

$$\{\varepsilon_I\} = [C_I][T_I]\{\sigma\} = [C_I]\{\sigma_I\} \tag{9}$$

The compliance matrix of the joint sets therefore introduces the anisotropy of the rock mass as variables independent of the intact rock. Hence, the strength of the intact rock and the joint sets can also be defined.

The Mohr-Coulomb failure criterion expressed by Eq. (10) is used for joint strength:

$$|\tau_s| = (c + \sigma_n \tan \phi_J) \tag{10}$$

where τ_s is the shear stress of the joint, σ_n is the normal stress of the joint, c is the cohesive strength of the joint, and ϕ_J is the friction angle of the joint.

Figure 5. shows the Mohr-Coulomb failure envelop of the intact rock and the joint sets in the MYM. The local safety factor for shear stress is expressed by Eq. (11).

$$f_s = \frac{\tau_{max}}{\tau_{max}} \tag{11}$$

The local safety factor for tensile stress is expressed by Eq. (12).

$$f_t = \frac{\sigma_t + (\sigma_x + \sigma_y)/2}{\tau_{max}} \tag{12}$$

The hyperbolic function of Eq. (13) proposed by Bandis et al. (1983) is used to express the normal stiffness K_n of the joint:

$$K_n = K_{ni}\left[1 - \frac{\sigma_n}{V_m K_{ni} + \sigma_n}\right]^{-2} \tag{13}$$

where K_{ni} is the initial normal stiffness, V_m is the maximum closure, and σ_n is the normal stress.

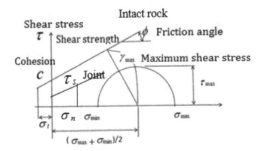

Figure 5. Mohr-Coulomb's failure envelope of the intact rock and the joint sets in MYM

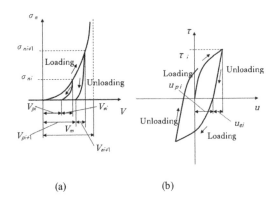

(a)   (b)

Figure 6. Model of joint deformation; (a) Normal deformation of joint under loading and unloading conditions, (b) Shear deformation of joint under a cyclic loading condition

The results of the cyclic loading test in the normal direction to the joint plane show the nonlinearity and residual deformation in the unloading process, as plotted in Figure 6(a). From this fact, Eq. (14) expresses the unloading deformation function, where the value of closure V_i under an arbitrary confining pressure σ_{ni} is given by the sum of the elastic deformation V_{ei}. The plastic deformation V_{pi} and V_{ei} in Eq. (15) are defined by the quadratic function of V_m. The deformation characteristics at unloading can be expressed by adding Bandis's equation to the elastic deformation V_{ei} described by Iwata et al. (2011).

$$v_i = v_{ei} + v_{pi} \tag{14}$$

$$v_{ei} = \left[1 - \frac{V_i}{V_m}\right]^{m_j} \cdot v_i \tag{15}$$

The hyperbolic function of Eq. (16) proposed by Kulhawy (1975) is used to express the shear stiffness K_{st} of the joint:

$$K_{st} = K_{si}\left(\frac{\sigma_n}{P_a}\right)^{n_j}\left(1 - \frac{\tau_s \cdot R_f}{\tau_p}\right)^2 \tag{16}$$

where k_{si} is the initial shear stiffness, σ_n is the normal stress, P_a is the atmospheric pressure, τ_p is the shear strength of the joint, τ_s is the shear stress, n_j is the coefficient of stiffness, and R_f is the material constant (about 0.6~0.9).

As shown in Figure 6(b), the residual deformation under the unloading process is due to the cyclic shear loading test. Considering the experimental result of Yoshinaka et al. (2009), we can formulate the shear deformation U_i can as arranged in Eq. (17), that is, as a sum of elastic and plastic deformations, in a manner much like normal deformation. The elastic deformation u_{ei} is defined by the quadratic function of failure severity τ_i / τ_p shown in Eq. (18). For the sake of simplicity, however, we can assume the stress-strain relation under the unloading process to be linear.

$$u_i = u_{ei} + u_{pi} \tag{17}$$

$$u_e = u_i \cdot \left[1 - \frac{\tau_i \cdot R_f}{\tau_p} \right]^2 \tag{18}$$

Figures 7 and 8 shows the scale effect of the joint area on normal stiffness and shear stiffness, respectively. The joint stiffness is inversely proportional to the joint area, a parameter that depends on the joint spacing on the two perpendicular cross joints shown in Table 2. In this analysis, the representative area of joints is assumed 9 m squares of both the low and high angle joints set determined by each joint spacing.

4 INITIAL STRESS ANALYSIS

Figure 9 shows the boundary condition of the initial stress analysis. The vertical external body force is applied for the 50th increment. Figures 10 and 11 show the horizontal stress distribution and vertical stress distribution of the initial condition, respectively. Figure 12 shows the local safety factor distribution of the initial condition. The safety factor is less than 1.0 around the point of the retaining wall foundation in the soil bank. The stability of

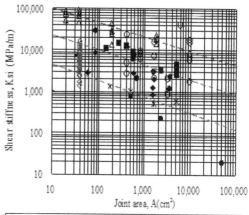

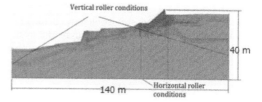

Figure 8. Scale effects of the joint area on shear stiffness

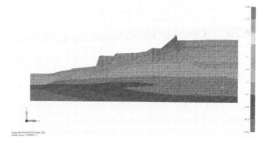

Figure 9. Boundary conditions of the initial stress analysis

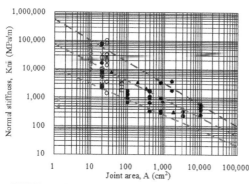

Figure 7. Scale effects of the joint area on normal stiffness

Figure 10. Horizontal stress distribution of the initial condition

the retaining wall is therefore unstable in the initial condition.

5 EARTHQUAKE RESPONSE ANALYSIS

The use of direct integration makes it impracticable to attempt a seismic response analysis of the equivalent continuous rock with discontinuities that show

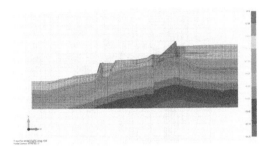

Figure 11. Vertical stress distribution of the initial condition

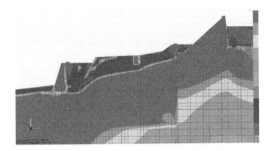

Figure 12. Local safety factor distribution of initial condition

strong nonlinearity and are dependent on the confinement of stress. The results of dynamic-cyclic direct shear tests, however, improve the constitutive equations of the joints, which has enabled the rapid development of the calculation techniques. The fundamental studies on seismic response analyses using MYM were reported by Yoshinaka et al. (2009) and Iwata et al. (2011).

The equation of motion according to Hamilton's principle is as follows:

$$[M]\{\ddot{u}\}+[C]\{\dot{u}\}+[K]\{u\}=$$

$$\int_{S2}[N]^T\{f\}ds-[M]\{\ddot{U}\}+\int_V[N]^T\begin{Bmatrix}0\\-\gamma\end{Bmatrix}dV \quad (19)$$

where $[M]$ is the mass matrix, $[C]$ is the damping matrix, $[K]$ is the stiffness matrix, $\{\ddot{u}\}$ is the acceleration matrix, $\{\dot{u}\}$ is the velocity, and $\{u\}$ is the displacement. The first term on the right-hand side of the equation is the surface external force, the second term is the external force of seismic acceleration, and the third term is the body force from its own weight. The body force can be obtained by integrating the volume V, as the surface external force is the surface area S. These relations are expressed in Eq. (19) in incremental form:

$$[M]\{\Delta\ddot{u}\}+[C]\{\Delta\dot{u}\}+[K]\{\Delta u\}=\{\Delta f\} \quad (20)$$

Eq. (20) can be obtained by solving the simultaneous equation Eq. (21) at each time interval and the Newmark's β, γ method, where $\beta = 0.25$, $\gamma = 0.5$:

$$\left[[K]+\frac{1}{\beta\Delta t^2}[M]+\frac{1}{\beta\Delta t}[C]\right]\{\Delta u\}$$

$$=-\alpha(\Delta t)[M]+[M]\left[\frac{1}{\beta\Delta t}\{\dot{u}(t)\}+\frac{1}{2\beta}\{\ddot{u}(t)\}\right] \quad (21)$$

$$+[C]\left[\frac{\delta}{\beta}\{\dot{u}(t)\}+\left\{\frac{\delta}{2\beta}-1\right\}\Delta t\{\ddot{u}(t)\}\right]$$

$$\{\Delta\ddot{u}\}=\frac{1}{\beta\Delta t^2}\left[\{\Delta u\}-\{\dot{u}(t)\}\Delta t-\frac{\Delta t^2}{2}\{\ddot{u}(t)\}\right] \quad (22)$$

$$\{\Delta\dot{u}\}=[\{\ddot{u}(t)\}+\delta\{\Delta\ddot{u}\}]. \quad (23)$$

Eq. (22) and Eq. (23) express the acceleration and velocity.

C, the damping matrix in Eq. (21), is expressed by Eq. (24) using the viscosity coefficient and mass matrix:

$$[C]=C_M[M]+C_K[K] \quad (24)$$

$$C_M=\frac{2\omega_i\omega_j(h_i\omega_i-h_j\omega_i)}{\omega_j^2-\omega_i^2} \quad (25)$$

$$C_K=\frac{2(h_i\omega_i-h_j\omega_i)}{\omega_j^2-\omega_i^2} \quad (26)$$

where C_M and C_K are damping constants concerning mass and stiffness, the affixing characters i and j express the mode order on the eigenvalue analysis, and $0\,\omega_i$ and ω_j show the circle frequencies. Because the damping factors h_i and h_j differ from the considered order of C_M and C_K, the lower order mode or/and larger simulation constants are practically adopted for general use in spite of the difference of actual physical phenomenon.

6 INPUT EARTHQUAKE WAVE

Figures 13 and 14 show the input acceleration of the Oodate N-S wave and up and down wave sing the statistic Green function and Fourier spectrum with the hypo-center model and active fault parameters presented by Okayama prefecture. The maximum acceleration of N-S wave is 572.77 gal and the up and down wave is 195.29 gal. Figure 15 shows the Fourier spectrum of the Oodate N-S wave. The natural frequency is distributed between 0.1 to 10Hz. The horizontal and up and down wave motions are input at the same time. The authors calculated the case of Oodate S-W wave also, in which the maximum acceleration is 527.56 gal. Therefore, we choose N-S wave for the representative case.

7 RESULT OF THE EARTHQUAKE RESPONSE ANALYSIS

Figure 16 shows the boundary condition of the earthquake response analysis. The fixed condition assumed on the bottom boundary and Lysmer damper (1969) are assumed both left and right

boundaries. Figures 17, 18, and 19 respectively show the acceleration time histories, horizontal stress time histories, and local safety factor time histories of the monitoring points. The maximum response of the acceleration of the top of counter measured dam is 2000 gal and as the bottom is 500 gal. The local safety factor at the dam crest of monitor No. 4750 is shows over 10 during earthquake as safe. The local safety factor at the dam foundation of monitor No. 4089 is shows under 1.0 for 3 second during earthquake. Figure 20 shows a comparison of acceleration at the retaining wall with and without the dam countermeasures. The countermeasures are effective in reducing the acceleration at the retaining wall. Figure 20 shows a one/third reduction in comparison with the original design.

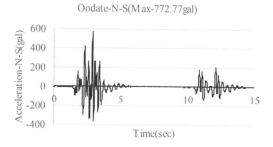

Figure 13. Input acceleration of the Oodate N-S direction wave

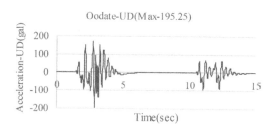

Figure 14. Input acceleration of the Oodate U-D direction wave

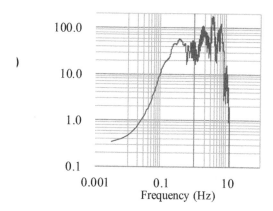

Figure 15. Fourier Spectrum of the Oodate N-S direction wave

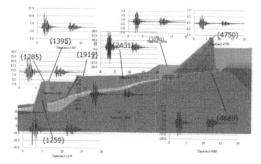

Figure 16. Boundary conditions of the earthquake response analysis

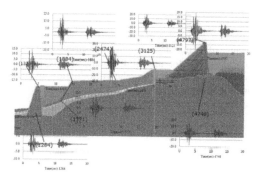

Figure 17. Acceleration time histories of the monitoring points

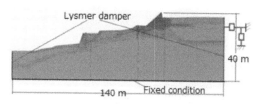

Figure 18. Horizontal stress time histories of the monitoring points

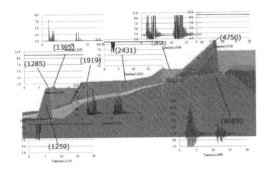

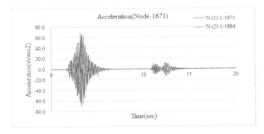

Figure 19. Local safety factor time histories of the monitoring points

Figure 20. Comparison of acceleration at the retaining wall with (red line) and without (blue line) dam countermeasures

8 CONCLUSION

MYM can be affectively applied to a jointed rock model in an earthquake response analysis for the design of tiling dam countermeasures. The geological survey for the joint system on site is very important in this study. The introduction of non-linearity of joint deformation should be necessary in large earthquake input motions. Acceleration during actual earthquakes will have to directly measured on-site to confirm the results of this study.

REFERENCES

Bandis S. C. Lumsden A. C. & Barton, N. R. 1983. Fundamentals of rock joint deformation, Int. J. Mech. Min. Sci. Geomech. Abstr., 6, 249–268.

Hill R. 1963. Elastic properties of reinforced solid, Some theoretical principles, J. Mech. Phys. Solids;11: 357–372.

Iwata, N., Sasaki, T., Sasaki, K., & Yoshinaka, R. 2011. Static and dynamic response analysis of rock mass considering joint distribution and its applicability, 11th ISRM Congress Beijing, pp. 233–236.

Iwata, N., Yoshinaka, R. & Sasaki, T. 2013. Applicability of the earthquake response analysis using Multiple Yield Model for discontinuous rock, International Journal for Rock Mechanics and Mining Science, 60, pp.196–207.

Kulhawy F. H. 1975. Stress deformation properties of rock and rock discontinuities, Eng. Geol., 9:327–350.

Lysmer & Kuhlemeyer, 1969. Finite dynamic model for infinite media, Journal of the Engineering Mechanics Division, 95(4),pp.859–877.

Sasaki T. Yoshinaka R. & Nagai F. 1994. A study of the multiple yield models on jointed rock mass by finite element method, J Geotech Eng., JSCE, 505, 59–68. (in Japanese)

Singh B. 1973. Continuum characterization of jointed rock mass, Part I – The constitutive equations, Int. J. Rock Mech. Min. Sci., 10, 311–335.

Souley M. Homand F. & Amadai B. 1999. An extension of the Saeb and Amadei constitutive model for rock joints to include cyclic loading paths, Int. J. Mech. Min. Sci. Geomech. Abst., 32, 2, 101–109.

Yoshinaka, R., Iwata, N., Sasaki, T., Sasaki, K. & Yoshida, J., 2009. Deformation behavior of discontinuous rock due to large scaled vertical excavation - Comparison between the prediction by numerical analysis and the measurement -, ISRM Symposium, SINOROCK2009, Hong-Kong, pp.547–551.

2019 Rock Dynamics Summit– Aydan et al. (eds)
© 2019 Taylor & Francis Group, London, ISBN 978-0-367-34783-3

Numerical simulations on the seismic stability of rock foundations under critical facilities via dynamic nonlinear analysis

M. Ishimaru, A. Sekiguchi, & T. Okada
Central Research Institute of Electric Power Industry, Chiba, Japan

H. Morozumi
Kansai Electric Power Company, Inc., Osaka, Japan

ABSTRACT: Evaluation of seismic stability of critical facilities to earthquake-induced failure of rock foundations based on ground displacement is considered to be crucial. In this study, a constitutive model was developed to consider the effects of both shear and tensile failure of rock on the stress–strain relation derived from the multiple shear spring model. This model was then used for dynamic nonlinear analysis that considers progressive failure. The applicability of this analysis method to a dynamic centrifugal model test of a rock foundation was evaluated. The amounts of residual displacements of the analysis results were comparatively close to the model test results although the vibration step, which begins after the occurrence of residual displacements, was slightly faster than that of the model test.

1 INTRODUCTION

The occurrence of fatal, large-magnitude earthquakes in the recent past has led to increased attention on earthquake ground motion during the design phase of modern structures. Accordingly, quantitative assessment of seismic resistance of critical facilities to earthquake-induced failure of rock foundations has become important.

In Japan, the seismic stability of rock foundations has conventionally been evaluated in terms of their bearing capacity, inclination, and sliding (JEAG 4601-1987 1987). In terms of the sliding motion during an earthquake, a slip safety factor based on an equivalent linear analysis is conventionally used to evaluate the stability of rock foundations. However, a slip safety factor value of less than 1 does not necessarily indicate immediate ground instability (Ishimaru et al. 2018a). Therefore, the evaluation of seismic stability based on ground displacement is considered to be a more effective approach.

In this study, therefore, the applicability of a nonlinear analysis method that considers progressive failure to evaluate the seismic stability of rock foundations (including post-earthquake residual displacement) was investigated. This paper explains the proposed nonlinear analysis method. The applicability of this nonlinear analysis method to a dynamic centrifugal model test of rock foundation is presented.

2 NONLINEAR ANALYSIS METHOD

The influences of rock shear failure and tensile failure need to be properly considered in a nonlinear analysis. Ishimaru et al. (2018b) proposed a constitutive model of materials that considers strain-softening characteristics after shear failure in the shear stress–shear strain relation derived from the multiple shear spring model (Towhata & Ishihara 1985) in the two-dimensional plane strain state. In this study, modeling of the strength characteristics after tensile failure was added to the above constitutive model, and the method of redistribution of stress after tensile failure was modified.

In addition, the multiple shear spring model can consider anisotropy if different values of hardness and strength are assigned for each spring. However, the process becomes very complicated in dynamic analysis, and thus, isotropy was assumed in this study.

2.1 *Constitutive model of materials before failure*

The shear stress–shear strain curve before rock failure was modeled with the general hyperbolic equation (GHE) model (Tatsuoka & Shibuya 1992), which is given in Equation 1. The GHE model can be fitted to experimental results over a wide strain range.

$$\tau = \frac{G_0 \cdot \gamma}{\dfrac{1}{C_1(\gamma)} + \dfrac{1}{C_2(\gamma)} \cdot \dfrac{|\gamma|}{\gamma_r}} \tag{1}$$

Here, τ is the shear stress, γ is the shear strain, G_0 is the initial shear modulus, γ_r is the reference shear strain ($\gamma_r = \tau_{a0}/G_0$), and τ_{a0} is the initial reference shear strength. $C_1(\gamma)$ and $C_2(\gamma)$ are correction coefficients expressed as follows:

$$C_1(\gamma) = \frac{C_1(0) + C_1(\infty)}{2} + \qquad (2a)$$

$$\frac{C_1(0) + C_1(\infty)}{2} \cdot \cos\left\{\frac{\pi}{\alpha / (|\gamma| / \gamma_r) + 1}\right\}$$

$$C_2(\gamma) = \frac{C_2(0) + C_2(\infty)}{2} + \qquad (2b)$$

$$\frac{C_2(0) + C_2(\infty)}{2} \cdot \cos\left\{\frac{\pi}{\beta / (|\gamma| / \gamma_r) + 1}\right\}$$

where $C_1(0)$, $C_2(0)$, $C_1(\infty)$, $C_2(\infty)$, α, and β are parameters. The damping characteristics are assumed to follow the model given in Eq. 3, which uses a virtual shear stress–shear strain curve (Ishihara et al. 1985) and the maximum damping constant h_{max}.

$$h = h_{max} \cdot \left(1 - G_R / G_0\right)^{\beta_1} \qquad (3)$$

Here, G_R is the shear modulus at the strain level of the unloading point, and β_1 is the adjustment parameter of the damping characteristics. The method proposed by Ozutsumi & Iai (2001) is used to set the damping constant in the multiple shear spring model.

2.2 Definitions of failure

The shear strength is defined in Eq. 4 (in this paper, the compression side is defined as being positive):

$$\tau_f = c_p \cdot \cos\phi_p + \frac{\sigma_1 + \sigma_3}{2} \cdot \sin\phi_p \qquad (4)$$

where τ_f is the peak shear strength, and c_p is the peak cohesion. In addition, ϕ_p is the peak internal friction angle, σ_1 is the maximum principal stress, and σ_3 is the minimum principal stress. Shear failure is estimated according to Eq. 5, and tensile failure is estimated according to Eq. 6, where σ_t is the tensile strength.

$$(\sigma_1 - \sigma_3) / 2 \geq \tau_f \qquad (5)$$

$$\sigma_3 \leq \sigma_t \qquad (6)$$

2.3 Modeling after failure

The shear stress–shear strain curve after rock failure also employs the GHE model shown in Eq. 1. However, the reference shear strain $\gamma_r = \tau_a / G_0$, and the reference shear strength τ_a decreases from the initial value τ_{a0} to the residual shear strength τ_r (Eq. 7) after rock failure.

$$\tau_r = a \cdot \left(\frac{\sigma_1 + \sigma_3}{2}\right)^b \qquad (7)$$

Here, a and b are the adjustment parameters of the residual shear strength. In addition, the damping characteristics after failure are assumed to be the same as those in Eq. 3.

Modeling of the reference shear strength τ_a and tensile strength σ_t after rock failure are explained below. Anisotropy should be considered for modeling the strength after rock failure, but the process becomes very complicated in dynamic analysis. Thus, we assume isotropy here. First, for tensile failure, we consider reducing the strength uniformly in all directions according to the number of tensile failure surfaces. More specifically, the analysis element is divided at a certain angle (N: division number), and the number of divided regions, including the tensile failure surface defined by the direction of the principal stress surface, is counted. Then, τ_a and σ_t are gradually reduced by Eqs. (8) and (9).

$$\tau_a = \tau_{as} \cdot \left(1 - n_f / N\right)^{\alpha_1} + \tau_r \cdot \left\{1 - \left(1 - n_f / N\right)^{\alpha_1}\right\} \qquad (8)$$

$$\sigma_t = \sigma_{ts} \cdot \left(1 - n_f / N\right)^{\alpha_1} \qquad (9)$$

where n_f is the number of divided regions including the tensile failure surface, α_1 is the adjustment parameter of the strength reduction, and τ_{as} and σ_{as} are the reference shear strength and tensile strength considering the influence of shear failure. τ_{as} and σ_{as} are gradually reduced by the following equations according to the amount of strain generated after shear failure, considering the gradient of strain softening.

$$\tau_{as} = \tau_r + \frac{(\tau_{a0} - \tau_r)}{(A \cdot \gamma^p + 1)} \qquad (10)$$

$$\sigma_{ts} = \frac{\sigma_{t0}}{(A \cdot \gamma^p + 1)} \qquad (11)$$

Here, σ_{t0} is the initial tensile strength, γ^p is the maximum value of the maximum shear strain (incremental amount from the value at failure), and A is the strain softening coefficient, which is a parameter that determines the rate of decrease in τ_{as} and σ_a.

2.4 Redistribution of stress

The method of redistribution of stresses exceeding strengths is as follows:

i. For shear failure, shrink the Mole's stress circle under the condition of fixed average principal stress.

ii. For tensile failure, shrink the Mole's stress circle under the condition of fixed maximum principal stress. If it exceeds the shear strength even after this treatment, shrink the Mole's stress circle

659

under the condition of fixed minimum principal stress.

iii. Apply residual force calculated from external force and stress.

3 DYNAMIC CENTRIFUGAL MODEL TEST

A dynamic centrifugal model test was performed to assess the seismic stability of rock foundations (Ishimaru et al. 2018a). The rock foundation model with a reduction ratio of 1:50 was constructed with artificial rock material and a weak layer. Vibration tests were performed in a centrifugal force field under a centrifugal acceleration of 50 g.

3.1 Rock foundation model

The rock foundation model and instrument arrangement are shown in Figure 1. The model was 200 mm (10 m upon real-scale conversion) in height and 300 mm in depth. The boundary surfaces had cutouts measuring 100 mm × 100 mm to avoid interference with the rigid box. The building model dimensions were 60 mm (width) × 40 mm (height) (3 m × 2 m upon real-scale conversion), and the density of the building material was 1200 kg/m³. In addition, the bottom of the building model and ground surface were fixed with an adhesive.

The measured variables included accelerations produced under and on the ground surface along with the corresponding displacements induced in the building model and on the ground surface.

3.2 Properties of the rock foundation model

The rock foundation model was created from cement-modified soil with a curing period of 7 d. For a soil volume of approximately 1 m³, the formulation comprised 82 kg of high early-strength Portland cement, 370 kg of distilled water, 817 kg of crushed limestone sand, 817 kg of fine limestone powder, and 1 kg of admixture. Table 1 lists the physical properties of the artificial rock materials. The properties were obtained from various physical and mechanical tests. Figure 2 shows the dynamic deformation characteristics obtained from cyclic triaxial tests.

Based on the work by Ishimaru & Kawai (2011), the weak layer within the rock mass was reproduced by installing a 0.2 mm thick Teflon sheet within the rock foundation model before the artificial rock material started hardening. The resultant artificial weak layer had constant degrees of roughness, bite, etc. Prior examination confirmed that the cohesion between the post-hardening artificial rock material and Teflon sheet was very small. Under this condition, the shear resistance of the artificial weak layer can be considered to be equal to the frictional force generated between the artificial rock material and Teflon sheet. Therefore, the frictional force generated between the artificial rock material and Teflon sheet under normal-stress loading was examined through a single-plane shearing test. Table 1 shows the maximum and residual shear resistances increased in proportion to the normal stress.

Table 1. Physical properties of the artificial rock materials and weak layer (σ_m: mean stress).

	Rock	Weak layer
Unit weight	20.3 kN/m³	20.6 kN/m³
Peak shear strength	$c_p = 267.1$ kN/m² $\varphi_p = 34.7°$	$c_p = 0.0$ kN/m² $\varphi_p = 28.6°$
Residual shear strength	$a = 4.61, b = 0.70$ ($\tau_r = a \times \sigma_m{}^b$)	$c_r = 0.0$ kN/m² $\varphi_r = 19.3°$
Tensile strength	$\sigma_t = 41.4$ kN/m²	$\sigma_t = 0.0$ kN/m²
Initial elastic shear modulus	933000 kN/m²	2800 kN/m²
Poisson's ratio	0.42	0.49

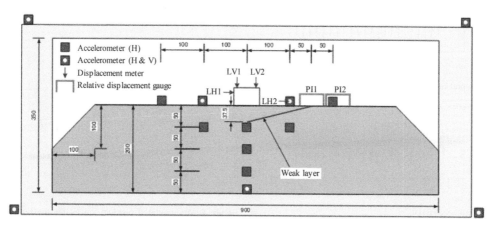

Figure 1. Rock foundation model and instrument arrangement (units: mm).

660

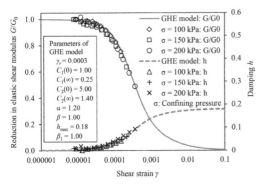

Figure 2. Dynamic deformation characteristics of the artificial rock material obtained from cyclic triaxial tests.

Table 2. Maximum values of the acceleration amplitude at different vibration steps (real-scale conversion).

Vibration step	Frequency	Horizontal acc. m/s²	Vertical acc. m/s²
d01	1.2	0.57	0.13
d02	1.2	3.47	0.42
d03	1.2	5.72	1.15
d04	1.2	7.77	0.91
d05	1.2	9.16	1.22
d06	1.2	10.40	1.50
d07	1.6	8.68	1.87
d08	1.6	10.04	2.88
d09	1.6	11.53	3.84
d10	1.6	11.25	3.39

3.3 Input acceleration

The input acceleration was provided in the form of a sinusoidal wave with a wavenumber of 20 (frequencies of 1.2 and 1.6 Hz upon real-scale conversion) in the main part with four tapers before and after that. During the test, the acceleration amplitude was increased for each vibration step. A horizontal movement was the only input. However, the vertical motion, which was considered to be caused by the rocking of the shaking table, was also measured during vibration. Figure 3 shows the input acceleration of vibration step d04, and Table 2 lists the maximum acceleration amplitudes at different vibration steps. The 1.6 Hz excitation produced a greater vertical motion than the 1.2 Hz excitation owing to the characteristics of the experimental apparatus.

3.4 Test results

Figure 4 shows the accumulated residual values of the horizontal displacements of the building model and ground at different vibration steps. This figure confirms that the residual displacements rapidly increased after vibration step d09.

Figure 5 shows the strain distribution calculated from images captured by a high-speed camera at vibration step d10. Cracks connecting the lower end of the weak layer and the left side of the building model were generated, although they were not yet clear in images captured at vibration step d09. Owing to the occurrence of these cracks, the upper part of the weak layer was considered to have moved.

4 NUMERICAL SIMULATIONS

4.1 Analysis conditions

The set values for the parameters of the artificial rock materials used for the nonlinear analyses are listed in Table 3.

The other parameters were set according to the results of the physical and mechanical tests

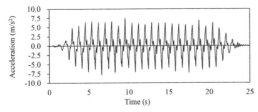

(a) Horizontal acceleration.

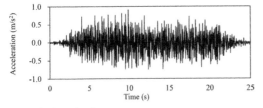

(b) Vertical acceleration.

Figure 3. Input acceleration (real-scale conversion).

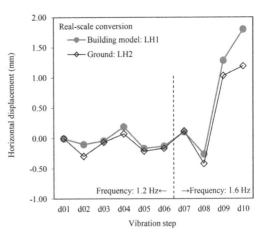

Figure 4. Accumulated residual values for the horizontal displacements of the building model and ground at different vibration steps.

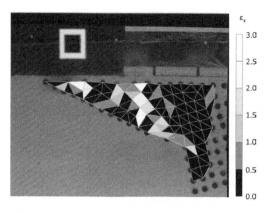

Figure 5. Horizontal strain distribution calculated from images taken with a high-speed camera at vibration step d10.

Table 3. Setting parameters of the constitutive model.

	Case 1	Case 2	Case 3
Number of multiple shear springs in the semicircle		12	
Parameters of strength reduction after tensile failure: N, α_1	$N = 18$, $\alpha_1 = 0.5$	$N = 18$, $\alpha_1 = 0.25$	$N = 18$, $\alpha_1 = 0.125$
Strain softening coefficient A		$A = 300$	

(Table 1 and Figure 2). The strain softening coefficient A was obtained from the relation of the deviator stress σ_d minus the axial strain ε_a in the strain-softening process after the peak of the plane strain compression tests. In addition, case studies were conducted on the parameters N and α_1 related to the strength reduction after tensile failure occurred.

In contrast, the artificial weak layer was modeled to joint elements. The unit weight of the artificial weak layer was 20.6 kN/m³, which was equal to that of the Teflon sheet, and the corresponding Poisson's ratio was 0.49 based on the assumption of no volume change. The pseudo shear modulus of elasticity, which was induced by modeling the artificial weak layer as joint elements, was set as 2800 kN/m² from the gradient up to the maximum shear resistance during the single-plane shearing tests.

For the boundary conditions in the numerical simulations of the model test, the bottom was fixed, and the joint elements (tension/shear spring: 0 kPa; compression spring: 1.0×10^8 kPa) were installed on the side. A self-weight analysis was performed by dividing the value of gravity by 100. The accelerations of the shaking table were then input for the earthquake response analyses. In the seismic response analyses, the stress and deformation state of the preceding step were

carried to the next step considering 1 % stiffness-proportional damping at 10 Hz. In addition, the time increment of the calculation was 2.0×10^{-4} s, and the convergence criterion was set to a residual force norm/external force norm of less than 1.0×10^{-4}. The iterative calculation reached 10x, and residual forces were carried over to the next calculation step.

4.2 Analysis results

Figure 6 shows the accumulated residual values for the inclination of the building model at different vibration steps. This figure confirms that the inclination amounts of the analysis results are smaller than those of the model test although the inclination amounts of the analyses rapidly increase at vibration step d08. This may be due to properties of the joint element for the artificial weak layer. In particular, the vertical rigidity of the joint element will be examined in detail.

Figure 7 and Figure 8 show the accumulated residual values of the horizontal displacements of the building model and ground at different vibration steps, respectively. These figures confirm that the faster the strength reduction after tensile failure, the larger the amount of residual displacements. However, the difference between Case 2 and Case 3 is small, and the amounts of residual displacements were comparatively close to the model test results although the vibration step, which begins after the occurrence of residual displacements, was slightly faster than that of the model test.

Figure 9 shows the tensile failure surface ratio n/N at the end of vibration step d08 for Case 2. This figure confirms that the tensile failures connecting the lower end of the weak layer and the left side of the building model were generated as the same of the model test result shown in Figure 5.

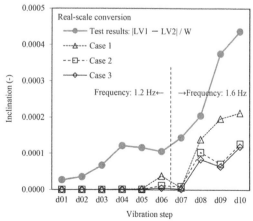

Figure 6. Accumulated residual values for the inclination of the building model at different vibration steps.

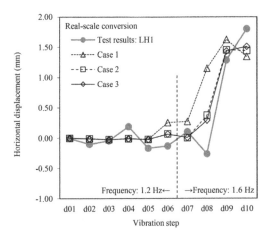

Figure 7. Accumulated residual values for the horizontal displacements of the building model at different vibration steps.

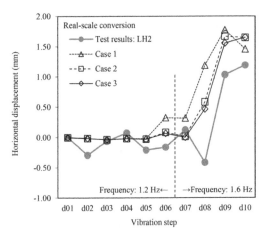

Figure 8. Accumulated residual values for the horizontal displacements of the ground at different vibration steps.

5 CONCLUSION

This study developed a constitutive model that considers the effects of both shear and tensile failure of rock on the stress–strain relation derived from the multiple shear spring model. This model was then used for dynamic nonlinear analysis that considers progressive failure.

The applicability of this analysis method to a dynamic centrifugal model test was evaluated. The vibration step, which begins after the occurrence of residual displacements, was slightly faster than that of the model test although the amounts of residual displacements were comparatively close to the model test results. This may be because

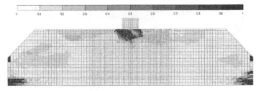

Figure 9. Tensile failure surface ratio at the end of vibration step d08 for Case 2.

anisotropy was not considered in this analysis. To evaluate the displacement more quantitatively, the proposed method needs to consider anisotropy, such as changes in strength and rigidity according to the failure surface direction.

Another problem concerns setting the parameters related to the strength reduction after tensile failure. Conservative evaluation is possible if large values are set. However, for a more accurate quantitative evaluation of the amount of displacement, a detailed examination for setting the parameters is necessary.

REFERENCES

Ishihara, K., Yoshida, N. and Tsujino, S. 1985. Modeling of stress-strain relations of soils in cyclic loading. *Proc. 5th Int. Conf. Numerical Methods in Geomechanics, Nagoya*, Vol. 1: 373–380.

Ishimaru, M., Hashi, K., Oikawa, K. and Kawai, T. 2018b. Nonlinear Analysis Method for Evaluating the Seismic Stability of Rock Slopes Considering Progressive Failure, *Journal of Earthquake and Tsunami*, Vol. 12, No. 4: 1–16.

Ishimaru, M. and Kawai, T. 2011. Centrifuge model test on earthquake-induced failure behaviour of slope in discontinuous rock mass. *Proc. 12th Int. Congress on Rock Mechanics, Beijing*, 1919–1922.

Ishimaru, M., Sekiguchi A., Okada, T., Hiraga, K. and Ozawa, K. 2018a. Experimental study on seismic stability of foundation rocks under critical facilities, *Rock Dynamics - Experiments, Theories and Applications; Proc. 3rd Int. Conf. on Rock Dynamics and Applications, Trondheim*, 569–574.

JEAG 4601-1987. 1987. Technical Guidelines for Aseismic Design of Nuclear Power Plants, Japan.

Ozutsumi, O. and Iai, S. 2001. Adjustment method of the hysteresis damping for multiple shear spring model. *Proc. 4th Int. Conf. Recent Advances in Geotechnical Earthquake Engineering and Soil Dynamics, San Diego, California*, Paper No. 1.68: 1–6.

Tatsuoka, F. and Shibuya, S. 1992. Deformation characteristics of soils and rocks from field and laboratory tests. *Proc. 9th Asian Regional Conf. Soil Mechanics and Foundation Engineering, Bangkok*, Vol. 2: 101–170.

Towhata, I. and Ishihara, K. 1985. Modelling soil behavior under principal stress axes rotation. *Proc. 5th Int. Conf. Numerical Methods in Geomechanics, Nagoya*, Vol. 1: 523–530.

2019 Rock Dynamics Summit– Aydan et al. (eds)
© *2019 Taylor & Francis Group, London, ISBN 978-0-367-34783-3*

The dynamic response of Horonobe Underground Research Center during the 2018 June 20 earthquake

T. Sato, K. Aoyagi & N. Miyara
Japan Atomic Energy Agency, Horonobe Underground Research Center, Hokkaido, Japan

Ö. Aydan, J. Tomiyama & T. Morita
Department of Civil Engineering, University of the Ryukyus, Okinawa, Japan

ABSTRACT: The Horonobe Underground Research Laboratory (URL) is assumed to be located in an aseismic region and earthquakes are rarely occur. An earthquake with a moment magnitude of 4 occurred in at 5:28 (JST) in early morning of June 20, 2018. The strong motions induced by this earthquake were recorded by the accelerometers installed in the URL as well as the Kik-Net and K-Net strong motions networks operated by the National Research Institute for Earth Science and Disaster Prevention of Japan (NIED). The authors explain the results of analyses carried out on the ground amplification and frequency characteristics of the acceleration records at the URL and those of the Kik-net strong motion station and the structural effect of the URL on the ground amplification and frequency characteristics. Furthermore, the implications of the results obtained from this study in practice and the safety of the nuclear waste disposal at depth are discussed.

1 INTRODUCTION

In order to cover the general geological environment in Japan, two URLs, one for sedimentary rock and another for crystalline rock have been planned, one is the Horonobe Underground Research Laboratory the other is the Mizunami Underground Research Laboratory (JNC, 2001; 2002). One purpose of this plan is to confirm the technical reliability of the geo-logical disposal methods for high-level radioactive waste, as indicated by the Second Progress Report (JNC, 2000).

The site of the URL project for sedimentary rock is located at Horonobe, in the northern part of Hok-kaido, north of Japan. The geology consists of Ter-tiary and Quaternary sedimentary rocks. Conceptual design for the Horonobe URL at present is as follow:

– Two 500 m access shafts and one Ventilation Shaft; and
– Experiment levels, at 140 m, 250 m, 350 m and 500 m depths.

In 2012, excavation of the Ventilation Shaft and East Access Shaft reached 350 m depth, and the ex-periment gallery at 350 m depth was excavated be-tween 2012 and 2014. In 2012, excavation of the West Access Shaft started, and reached 350 m depth in 2013. Countermeasures against rock bursts and large volume/high-pressure inflows of water or in-flammable gas are important issues to be addressed during excavation of shafts and research galleries. Figure 1 shows the layout of the Horonobe URL.

The shafts and galleries have been excavated through overlying Neogene sedimentary rocks named Koetoi formation (diatomaceous mudstones with opal-A) and into the Wakkanai formation (sili-ceous mudstones

with opal-CT). The Koetoi formation and upper part of Wakkanai formation have been faulted and have undergone sever-al episodes of uplift and subsidence from the Miocene to the Pliocene, indicated by the presence of lacustrine and marine sedimentary forma-tions unconformably overlying the basement. Table 1 shows properties of the formations and groundwater.

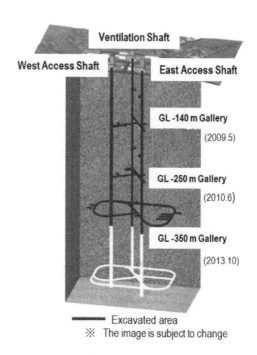

Figure 1. Layout of Horonobe URL.

It will be necessary to evaluate direct and/or indirect influence of large earthquakes to design and construction of repository, and safety of geological disposal system during long period in safety assessment. The Horonobe URL is assumed to be located in an aseismic region and earthquakes are rarely occur. But, an earthquake with a moment magnitude of 4.1 occurred in at 5:28 (JST) in early morning of June 20, 2018, which is also named as the 2018 June 20 Soya Region earthquake. The seismic intensity at Horonobe town was 3 on the Japan Meteorological Scale.

This paper shows the results of the analyses carried out on the ground amplification and frequency characteristics of the acceleration records at the Horonobe URL and those of the Kik-net strong motion station for the 2018 June 20 Soya Region earthquake and the structural effect of the URL on the ground amplification and frequency characteristics. Furthermore, the authors discuss the implications of the results obtained from this study in practice and the safety of the nuclear waste disposal at depth.

2 THE 2018 JUNE 20 SOYA REGION EARTHQUAKE

2.1 Characteristics of Soya Region Earthquake

The Soya Region Earthquake occurred on 2018 June 20 at 5:28 AM. The moment magnitude of the earthquake was 4.0 according to F-NET of NIED. The focal mechanism of the earthquake was estimated to be thrust fault (Figure 2), which is a consistent mechanism in view of tectonics of the Soya region. Figure 3 shows the inferred stress state for the earthquake. The maximum horizontal stress acts in almost in East-West (EW) direction.

2.2 Characteristics of Iburi Earthquake

Another major earthquake with a moment magnitude of 6.6 (Mj 6.7) occurred on September 6, 2018

Table 1. Properties of rock and groundwater.

Property	Koetoi F.	Wakkanai F.
UCS* (MPa)	1.44 - 9.80	8.90 - 34.9
Young's modulus (GPa)	0.38 - 1.03	1.41 - 6.35
Poison's ratio	0.22 - 0.47	0.17 - 0.34
Effective porosity (%)	60 - 65	40 - 50
Unit weight (kN/mm³)	14 - 16	15 - 19
Hydraulic conductivity	$10^{-8} - 10^{-9}$	$10^{-6} - 10^{-11}$
Swelling factor	<0.04	<0.03
Durability factor (Id²) (%)	>90	>95
Dissolved gas		Methane dominant
Groundwater		Saline water

*Unconfined Compressive Strength

at 3:08 in Iburi Region of Hokkaido Island, which is about 260 km away from Horonobe. The focal mechanism of this earthquake was due to the blind steeply dipping thrust fault. The earthquake was felt in Horonobe as recorded by Kik-Net network. However, the maximum ground acceleration was 3.3 gals.

2.3 Acceleration records at Horonobe URL

Accelerometers are set at the ground surface, GL.-250 m and GL.-350 m galleries in Horonobe URL (Figure 4). Figure 5 shows the Seismic records of the 2018 June 20 Soya Region earthquake. Table 2 compares the maximum acceleration at each strong motion station installed at various depths. The ground motions are amplified towards the ground surface. The data even in the same level is scattered, which may imply some local effects such as the geological

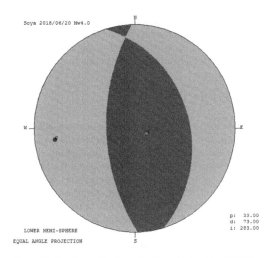

Figure 2. Re-drawn focal mechanism obtained by F-NET

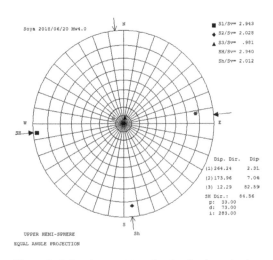

Figure 3. Inferred stress state for the focal mechanism obtained by F-NET by Aydan's method (2000).

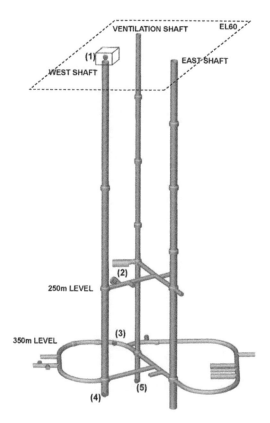

Figure 4. Locations of strong motion observation stations.

conditions, the geometry of the opening where the devices are installed.

2.4 *Kik-Net and K-Net data*

National Research Institute for Earth Science and Disaster Prevention of Japan (NIED) has been operating the Kik-Net and K-Net strong motions networks. There is a strong motion station of the Kik-net in Horonobe town and the accelerations recorded at the ground surface and at a depth of 100 m from the ground surface (-70 m). Figure 6 shows the acceleration records taken at the ground surface and at the base (100 m below the ground). Table 1 gives the maximum ground accelerations and their amplifications. Theoretically, the amplification is expected to be greater than 2 for an elastic ground (Nasu 1931). The comparison indicates that the amplification is more than 3 times. Compared to data from the Kik-Net, the measurements at the Horonobe URL are somewhat scattered.

3 FOURIER AND ACCELERATION RESPONSE SPECTRA ANALYSES

The Fourier and accelerations response spectra analyses have been carried out for each strong motion stations. We report some of them herein.

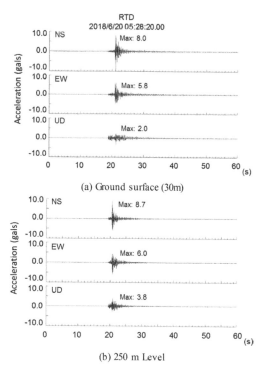

(a) Ground surface (30m)

(b) 250 m Level

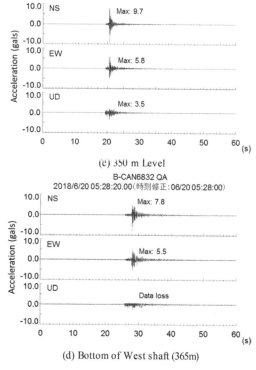

(c) 350 m Level

(d) Bottom of West shaft (365m)

Figure 5. Seismic records of the 2018 June 20 Soya Region earthquake (20 Jun. 2018, M=4.8) observed in Horonobe URL. There is a possibility that the recorder set at the bottom of West shaft has a problem.

Table 2. Maximum acceleration

Locations	NS (gals)	EW (gals)	UD (gals)
Surface (+60 m)	8.0	5.8	2.0*
250 m Level	8.7	6.0	3.8
350 m Level	9.7	5.8	3.5
West shaft (365 m)**	7.8	5.5	Data loss

*Data having low quality are possible.

**There is a possibility that the recorder set at the bottom of West shaft has a problem.

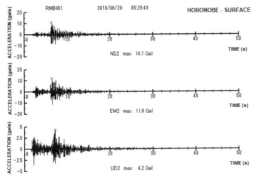

(a) Ground surface (30m)

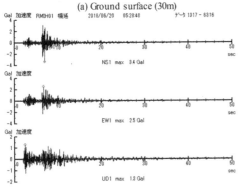

(b) Base (100 m below the ground surface)

Figure 6. The acceleration records taken at the ground surface and at the base of Kik-net.

3.1 Fourier spectra analyses

The Fourier spectra analyses of acceleration records measured by the Kik-Net (RMIH01) at the ground surface and base are shown in Figure 7. As noted from the figure, the dominant frequency ranges between 4-8 Hz and the Fourier Spectra characteristics do not change with depth, although the amplitude of the ground surface is at least 3 times that at the base.

Figure 8 shows the FFT of records taken at ground surface and at a depth of 365 m at the West Shaft bottom (No.4) in Horonobe URL. The FFT amplitude of the shaft bottom records are almost the same as that of the ground surface. The frequency characteristics are also quite similar and they resemble to those of the Kik-Net records.

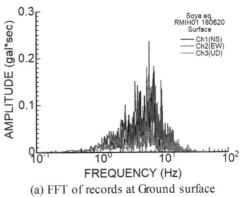

(a) FFT of records at Ground surface

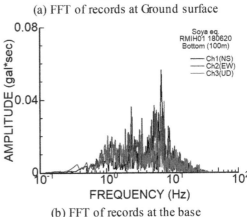

(b) FFT of records at the base

Figure 7. FFT of records at the ground surface and base.

Figure 9. shows the FFT spectra of the record taken at the bottom of Vent. Shaft (No.5) observation station during the 2018 Iburi Earthquake. Except UD component, the other components are quite similar to those of the Soya Region Earthquake shown in Figure 7, except amplitude. The normalized amplitude may be useful for comparison purposes.

3.2 Acceleration response spectra analyses

A series of acceleration response analyses are carried out. Figure 10 shows the acceleration response spectra for RHIM01 and Horonobe URL Surface (+60 m) strong ground motion stations (No.1). The amplitude and frequency characteristics are somewhat different. The ground conditions at the RHIM01 may be softer than those at the Horonobe URL site.

4 MODAL ANALYSES

A series of 3D finite element modal analyses were carried out for four conditions, which are namely, no shafts, single shaft, double shafts and triple shafts. The software used was 3D MIDAS-FEA. Table 4 gives the material properties used in numerical analyses while Table 5 compares the Eigen values for four different conditions and Figure 11 shows

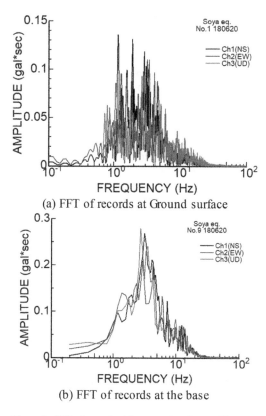

(a) FFT of records at Ground surface

(b) FFT of records at the base

Figure 8. FFT of records at the ground surface and base.

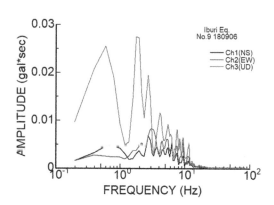

Figure 9. FFT of records at the ground surface for the records due to 2018 Iburi earthquake of 2018 Sep. 06.

displacement response for Mode 1. Detailed comparison of actual response and analyzed results is necessary for future work.

5 CONCLUSIONS

An earthquake with a moment magnitude of 4 occurred on June 20, 2018 in Soya Region and the strong motions induced by this earthquake were recorded by the accelerometers installed in the Horonobe URL

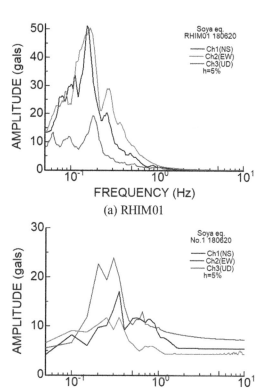

(a) RHIM01

(b) Horonobe URL

Figure 10. Comparison of acceleration response spectra for RHIM01 and Horonobe URL site.

Table 3. Maximum acceleration and amplification

	NS (gals)	EW (gals)	UD (gals)
Surface (+30 m)	10.7	11.8	4.2
Base (-70 m)	3.4	2.5	1.3
Amplification	3.15	4.72	3.23

Table 4. Material Properties

Material	UW (kN/m³)	E (GPa)	Poisson's ratio
Rock mass	26.5	0.600	0.37
Concrete	23.5	11.042	0.20

Table 5. Eigen values for Mode 1

	No shaft	Single	Double	Triple
Mode 1(s)	1.763	1.752	1.203	1.199
Mode 2(s)	1.645	1.635	1.889	1.172
Mode 3(s)	1.564	1.554	1.117	1.111

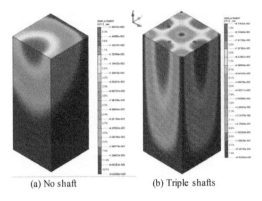

| (a) No shaft | (b) Triple shafts |

Figure 11. Displacement response for Mode 1.

as well as the Kik-Net and K-Net strong motions networks. The authors analyzed strong ground motions with an emphasis on the ground amplification and frequency characteristics of the acceleration records at the Horonobe URL and those of the Kik-net strong motion station and the structural effect of the URL on the ground amplification and frequency characteristics. Although the results show similar tendencies regarding the ground amplification and frequency characteristics, the structural effect of the Horonobe URL does exist on the recorded data. The results are probably affected by the geological and rock mass characteristics of the site. As the amplitude of the motions were less than 10 gals, the non-linear features could not observed. It is needless to say that the large amplitude records may reveal better insights on the dynamic responses of the URL. Nevertheless, these result clearly show that there is a ground amplification as the depth become shallower, which may have some important implications on the dynamic safety of the nuclear waste disposal at depth.

REFERENCES

Aydan, Ö. 2000. A stress inference method based on structural geological features for the full-stress components in the earth' crust. *Yerbilimleri*, 22, 223–236.

Aydan, Ö., Kumsar, H. & Ulusay, R. 2002. How to infer the possible mechanism and characteristics of earthquakes from the striations and ground surface traces of existing faults. *JSCE, Earthquake and Structural Engineering Division*, Vol. 19, No.2, 199–208.

F-Net. 2018. Focal mechanism solutions. http://www.fnet. bosai.go.jp/

Kik-Net and K-Net. 2018. Strong motion records of Soya region and Iburi Earthquakes. http://www.bosai.go.jp/

Japan Nuclear Cycle Development Institute. 2001. Horonobe Underground Research Laboratory Project, Plan of surface-based Investigation. *JNC Technical Report, JNC TN1410 2001–001.*

Japan Nuclear Cycle Development Institute. 2002. Master Plan of the Mizunami Underground Research Laboratory Project. *JNC Technical Report, JNC TN7410 2003–001.*

Japan Nuclear Cycle Development Institute. 2000. Second Progress Report on Research and Development for the Geological Disposal of HLW in Japan. *JNC Technical Report, JNC TN1410 2000–001–004.*

2019 Rock Dynamics Summit– Aydan et al. (eds)
© 2019 Taylor & Francis Group, London, ISBN 978-0-367-34783-3

Mechanical behaviour and characteristics of rocks subjected to shock loads

S. Kodate, J. Tomiyama, Y. Suda, K. Horiuchi & Ö. Aydan
University of the Ryukyus, Department of Civil Engineering, Okinawa, Japan

ABSTRACT: The dynamic mechanical behaviour and characteristics of rocks have been investigated using different techniques. One of the commonly used technique is the Split Hopkinson Pressure Bar technique. However, the measured responses are inferred from the strain gauges attached to intermittent and transmitting bars rather than directly from samples. The authors devised a new experimental apparatus to investigate the behaviour of rocks under shock waves. The device may be fundamentally categorized as the drop-weight apparatus and it is possible to evaluate the mechanical behaviour and characteristics of rocks subjected to shock waves during pre-failure as well as post-failure stages. The rocks tested are tuff, limetsone, granite, marble, gneiss, porphyrite, ranging from soft rocks to hard rocks. The testing conditions correspond to uniaxial compression test and Brazilian tensile test. The nominal impact velocity can be easily adjusted and it can be easily correlated with the measured responses and dynamic mechanical properties. The authors present the outcomes of this experimental study and discuss their implications in the field of rock dynamics.

1 INTRODUCTION

The dynamic mechanical behaviour and characteristics of rocks have been investigated using different techniques in relation to evaluate the engineering structures by dynamic loads such as those resulting from the impact of missiles or meteorites, One of the most commonly used techniques is the Split Hopkinson Pressure Bar technique, which was originally developed Hopkinson in 1914 to measure stress pulse propagation in a metal bar. Kolsky (1949) refined Hopkinson's technique by using two Hopkinson bars in series, now known as the Split-Hopkinson Pressure bar (SHPB), to measure stress and strain. However, the measured responses are inferred from the strain gauges attached to intermittent and transmitting bars. Furthemore, it is difficult to evaluate the post-failure characteristics of rocks.

The authors devised new experimental apparatuses to investigate the behavior of rocks under shock waves. The devices are fundamentally categorized as the drop-weight apparatus, it can be equipped with load-cell, non-contact type laser displacement transducers, accelerometers and infra-red thermo-graphic imaging. Therefore, it is possible to evaluate the mechanical behavior and characteristics of rocks subjected to shock waves during pre-failure as well as post-failure stages. The tested rocks are tuff, limestone, granite, marble, gneiss, porphyrite. In other words, rocks tested range from soft rocks to hard rocks. The testing conditions correspond to uniaxial compression test and Brazilian tensile test. Nevertheless, it is possible to do experiments such as punching tests, bending tests. The nominal impact velocity can also be

easily adjusted and it can be easily correlated with the measured responses and dynamic mechanical properties. The authors explain the newly developed experimental apparatus and experiments on various rock under shock loads and present the outcomes of this experimental study and discuss their implications in rock dynamics field.

2 SHOCK TESTING DEVICES AND ROCKS

2.1 Shock testing devices

The authors first developed a special shock testing device. The device consist of a steel cylinder with a weight of 8300gf having a diameter of 97 mm, a load cell, an accelerometer (Fig. 1). The plastic pipe container, in which the cylinder was dropped, had an internal diameter of 100 mm and height of 500mm. The steel cylinder was dropped from certain heights and acceleration and force were measured simultaneously using YOKOGAWA WE7000 data-acquisition system at a sampling rate of 1ms. No digital filtering was imposed on measured force and accelerometer records. The initially designed device shown in Fig. 1 was not equipped with non-contact type laser displacement transducers.

The authors recently improved their initial device and equipped with non-contact type laser displacement transducers to measure the displacement of the loading platen. The load cell was also improved, which is now capable of measuring much higher dynamic loads. The device is shown in Fig. 2 and it also enables to take infra-red thermo-graphic images during the shock tests. The displacement of the loading platen is allowed to move downward

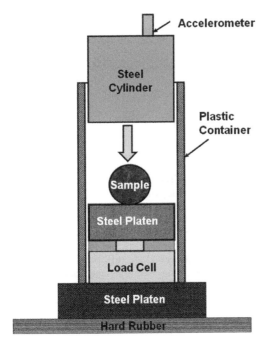

Figure 1. Schematic drawing of the initial shock testing device.

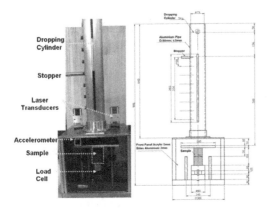

Figure 2. Schematic drawing of the new shock testing device.

up-to 20 mm in order to prevent the total destruction of samples upon failure. The cylindrical weight can be dropped from different heights up to 500 mm with an interval of 50 mm.

2.2 Samples

Samples of rocks are gathered from different locations in Japan and Turkey and the attention was given to those rocks with well-known mechanical properties. One of the main goal of this study is also to see the effect of velocity of shock load on the deformability and strength characteristics of rocks.

The rocks were tuffs from Cappadocia, Oya and Fukui, limestone from Ryukyu limestone formation and Bazda in south-western Turkey, mudstone belonging Shimajiri formation and Seyitömer in Turkey, andesite from Sinop in Turkey, porphyrite from Atera fault zone in Japan. The experiments were carried out under uniaxial compression and Brazilian test conditions.

2.3 Effect of velocity and location of impact

Before tests on actual rock samples, some preliminary impact experiments were carried out on paraffin samples. Fig. 3 shows the state of paraffin sample tested under uniaxial compression shock test. It was quite interesting that the ends of the sample deformed while the center part of the sample was less deformed. This is quite different situation when the deformation state of samples is compared under static testing. Furthermore, the conical failure surface was observed at the impact side while there was no such a failure surface at the restrained side. This fact implies that the propagation of the stress wave through the sample is not uniform.

Figs. 4 and 5 show the responses during the shock tests on the paraffin sample numbered UCS2. If the shock load is sufficient enough to yield the sample, the shock load looks has a triangular shape. The larger plastic zone developed in the sample near the

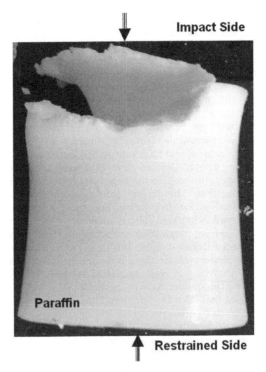

Figure 3. The deformation and failure state of sample under uniaxial compression shock test.

671

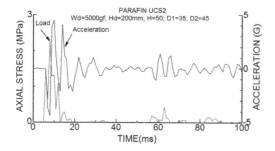

Figure 4. Time response of load and acceleration during shock test on Paraffin sample UCS2.

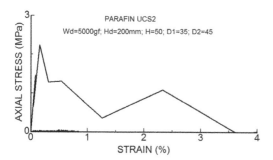

Figure 5. Strain-stress relation during shock test on Paraffin sample UCS2.

impact side. Furthermore, the post failure indicates strain softening behavior (Fig. 5).

Fig. 6 shows the deformed and failure state of two paraffin samples subjected to different impact velocity. The plastic deformation (whitened) zone is wider at the impact side than that at the restrained side, which also implies non-uniform stress wave propagation through the sample.

Another important feature is that the damage zone increase in size as the impact velocity increases. This observation has an important implication that the plastic deformation and fracturing would be much larger as the impact velocity increases. In other words, the failure zone or fractured volume is much larger in dynamic shock tests as compared with that under static condition. This may also imply that the plastic work done would be larger under dynamic conditions compared with that under static condition. This fact may also be interpreted that the apparent strength increase mentioned in many previous studies may be related to this situation in samples.

The maximum nominal velocity at the time of impact on samples can be computed from the following formula:

$$V_{max} = \sqrt{2gH_d} \qquad (1)$$

where g is gravitational acceleration and H_d is drop height. In this study, we also define maximum nominal strain rate by dividing the maximum nominal

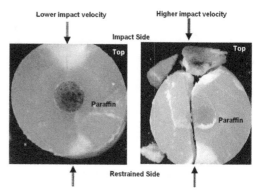

Figure 6. The deformation and failure state of sample under Brazilian shock test.

impact velocity by the sample height or sample diameter as given below:

$$d\dot{\varepsilon}_{max} = \frac{V_{max}}{L} \quad \text{or} \quad d\dot{\varepsilon}_{max} = \frac{V_{max}}{D} \qquad (2)$$

Maximum acceleration is obtained from the acceleration response during the experiment. It is expected to increase with the increase of strength of rock samples.

3 SHOCK TESTS

3.1 *Ryukyu limestone*

Ryukyu limestone is widely distributed in Ryukyu Archipelago. It is broadly defined as coral and sandy limestone (Tokashiki and Aydan 2010). In the experiments, coral limestone is tested under uniaxial compression and Brazilian shock tests. Figs. 7 and 8 show the force and acceleration responses of Ryukyu limestone samples. The strength of Ryukyu limestone depends upon the porosity and the static UCS ranges between 20.0 and 33.3 MPa. Similarly the Brazilian tensile strength of Ryukyu limestone depends upon the porosity and it ranges between 2.4 and 5.3 MPa. Fig. 9 compares the failure state under static and dynamic conditions.

From the comparison of experimental results given in Table 1, there is no remarkable strength increase

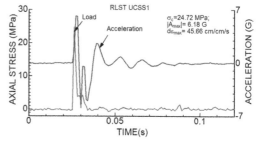

Figure 7. Axial stress and acceleration response of Ryukyu limestone sample under uniaxial compression shock test.

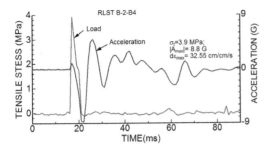

Figure 8. Axial stress and acceleration response of Ryukyu limestone sample under Brazilian shock test.

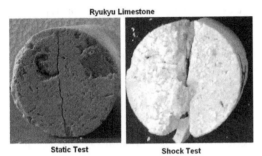

Figure 9. Comparison of fracturing of Ryukyu limestone samples under static and dynamic conditions.

under dynamic conditions for the given testing conditions. The finger-like corals are also widely distributed along the shores of Ryukyu Islands. An experiment was carried out on coral fingers samples with a diameter of 20 mm under compression shock test. Fig. 10 shows the axial stress and acceleration response during the experiment. It was quite interesting that the dynamic UCS of the coral finger is almost the same as that of Ryukyu coral limestone.

3.2 Tuff of Fukui (Shakudani-ishi)

Tuff of Fukui or locally known as Shakudani-ishi is a welded tuff and there are many abandoned underground quarries, which collapse from time to time. Table 2 compares the static and dynamic strength of Shakudani-ishi. Fig. 11 shows the tensile stress and acceleration response of the sample while Fig. 12 shows the fracturing state of Shakudani-ishi Brazilian samples tested under static and dynamic conditions. As noted from Fig. 12, the fracturing state is entirely different under dynamic condition than that under static condition. The damage is much more intense in dynamic shock test and it involves more energy dissipation in samples if the load exceeds the energy required to fracture under static condition.

3.3 Tuff of Derinkuyu

Antique Derinkuyu underground city in Cappadocia is excavated in tuff formation. The characteristics of

Table 1. Comparison of Static and Dynamics strength of Ryukyu limestone

Condition	Static (MPa)	Dynamic (MPa)
UCS	20-33.3	24.72
BRS	2.4-5.3	3.90
Coral Finger-UCS		27.94

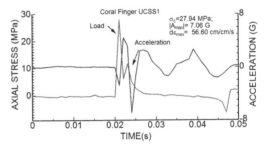

Figure 10. Axial stress and acceleration response of coral finger limestone sample under uniaxial compression shock test.

Table 2. Comparison of Static and Dynamics strength of Shakudani-ishi

Condition	Static (MPa)	Dynamic (MPa)
UCS	33.8-37.9	
BRS	3.8-	5.43

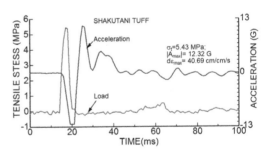

Figure 11. Tensile stress and acceleration response of Shakudani-ishi sample under Brazilian shock test.

Figure 12. Comparison of fracturing of Ryukyu limestone samples under static and dynamic conditions.

surround rock are well-investigated by Aydan and Ulusay (2014). Tuff of Derinkuyu constitutes main rock mass around the underground city. A series of experiments on Derinkuyu tuff under dynamic Brazilian and compression shock loading conditions were investigated. Fig. 13 shows the tensile stress and acceleration responses of the sample while Fig. 14 shows the views of the sample before and after testing. Table 3 compares the experimental results.

Fig. 15 shows the shows the axial stress and acceleration response during the experiment on a compression shock test on a Derinkuyu tuff sample. From the comparison of experimental results given in Table 3, there is no remarkable strength increase under dynamic conditions for the given testing conditions.

3.4 Oya tuff

Oya tuff is one of well-known rock in Japan and it has been used as construction material. A series of experiments on samples under Brazilian and compression shock loading conditions were carried out. Figures 16 and 17 show the dynamic response of

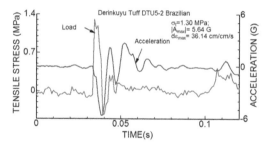

Figure 13. Tensile stress and acceleration response of Derinkuyu tuff sample under Brazilian shock test.

Figure 14. Comparison of fracturing of Derinkuyu tuff sample before and after Brazilian shock testing.

Table 3. Comparison of Static and Dynamics strength of tuff of Derinkuyu

Condition	Static (MPa)	Dynamic (MPa)
UCS	4.1-8.3	4.6
BRS	0.5-1.1	1.1

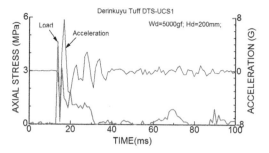

Figure 15. Axial stress and acceleration response of Derinkuyu tuff under uniaxial compression shock test.

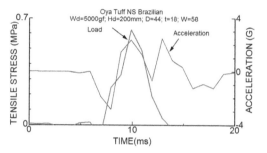

Figure 16. Tensile stress and acceleration response of Oya tuff sample under Brazilian shock test.

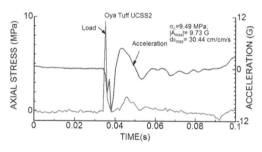

Figure 17. Axial stress and acceleration response of Oya tuff sample under uniaxial compression shock test.

Oya tuff samples under Brazilian and Compression loading conditions. Figure 18 shows the views of the sample before and after the compression shock experiment. Table 4 compares the strength of Oya tuff under static and dynamic conditions. Once again it is noted that there is no remarkable strength increase under dynamic conditions for the given testing conditions when the experimental results given in Table 3 are compared with each other. Furthermore, the fracturing state is more intense under dynamic conditions.

3.5 Sinop red andesite

Andesite is one of the widely distributed rock in the Sinop Nuclear Power Plant site. It has a grayish or reddish color. The grayish colored andesite has

Figure 18. Comparison of fracturing of Oya tuff sample before and after compression shock testing.

Table 4. Comparison of Static and Dynamics strength of Oya tuff

Condition	Static (MPa)	Dynamic (MPa)
UCSS	4.7-11.2	9.73
BRS	0.5-1.0	0.62

higher strength as compared with that of reddish colored andesite (Aydan et al. 2015). A series of experiments on samples under Brazilian and compression shock loading conditions were carried out. Figs. 19 and 20 show the dynamic response of Sinop reddish andesite samples under Brazilian and Compression loading conditions. Fig. 21 shows the views of the sample before and after the compression shock experiment.

4 CONCLUSIONS

The authors devised a new experimental apparatus, which may be categorized as drop-weight testing technique, to investigate the behaviour of rocks subjected to shock waves. Various rock samples

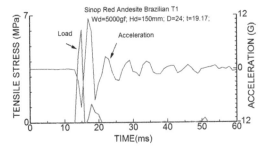

Figure 19. Tensile stress and acceleration response of Sinop reddish andesite under Brazilian shock test.

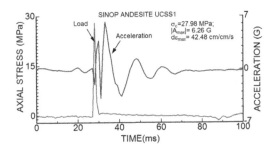

Figure 20. Axial stress and acceleration response of Sinop reddish andesite sample under uniaxial compression shock test.

Figure 21. Comparison of fracturing of Sinop Reddish andesite sample before and after compression shock testing.

having a sedimentary origin to igneous rock have been tested under Brazilian and compression testing conditions. The device is equipped with non-contact laser displacement transducers to observe the behaviour of rocks during pre-failure as well as post-failure stages. The nominal impact velocity can be easily adjusted and it can be easily correlated with the measured responses and dynamic mechanical properties. Some of important conclusions from this experimental study may be states as:

1. Under the given testing conditions, there is no remarkable strength increase under dynamic conditions as compared with that under static condition. If the load level under dynamic condition is higher than that to fracture the rock under static condition, the excess energy will be dissipated by intense fracturing as well as inertia forces. In other words, the so-called strength increase in SPHB experiments reported in literature may be due to this phenomenon (e.g. Kobayashi 1970; Aydan 2017), which cast a doubt on the truthness of the results of experiments using the SHPB technique.

2. The monitoring and high-speed video records indicated that the samples under Brazilian and compression shock test are first compressed at

675

the initial contact stage and they fully re-bound if they are not fractured. They will also partially rebound even they became fractured.

3. Acceleration responses are not symmetric with respect to time axis as noted previously by Aydan et al. (2011).

REFERENCES

Aydan, Ö., 2017. Rock Dynamics. CRC Press, Taylor and Francis Group, 462p, ISRM Book Series No. 3, ISBN 9781138032286

Aydan, Ö., Ulusay, R., 2013. Geomechanical evaluation of Derinkuyu Antique Underground City and its implications in geo-engineering. Rock Mechanics and Rock Engineering, Springer, Volume 46, Issue 4, pp.731–754.

Aydan Ö., Ohta Y., Daido M., Kumsar H. Genis M., Tokashiki N., Ito T. & Amini M., 2011. Earthquakes as a rock dynamic problem and their effects on rock engineering structures. In Advances in Rock Dynamics and Applications, Editors Y. Zhou and J. Zhao, Taylor and Francis Group, 341–422.

Aydan, Ö. Fuse, T. Ito, T., 2015. An experimental study on thermal response of rock discontinuities during cyclic shearing by Infrared (IR) thermography. Proc. 43rd Symposium on Rock Mechanics, JSCE, 123–128.

Hopkinson, B., 1914. A Method of Measuring the Pressure Produced in the Detonation of High Explosives or by the Impact of Bullets," Philosophy Trans. R. Soc. (London) A, 213, 437–456.

Kobayashi, R., 1970. On mechanical behaviours of rocks under various loading-rates. Rock Mechanics in Japan 1:56–58.

Kolsky, H., 1949. An Investigation of the Mechanical Properties of Materials at Very High Rates of Loading. Proc. Phys. Soc. London, B62, p.676–700.

Kolsky, H., 1953 Stress Waves in Solids. Clarendon Press, Oxford.

Tokashiki, N. Aydan, Ö., 2010. The stability assessment of overhanging Ryukyu limestone cliffs with an emphasis on the evaluation of tensile strength of Rock Mass. J. of Geotechnical Eng., JSCE, 66(2),pp.397–406.

2019 Rock Dynamics Summit– Aydan et al. (eds)
© 2019 Taylor & Francis Group, London, ISBN 978-0-367-34783-3

Numerical simulation of the rock cutting

M. Mohammadnejad & S. Dehkhoda
CSIRO, Minerals Resources Business Unit, Australia

H. Liu, & A. Chan
School of Engineering, University of Tasmania

D. Fukuda
Hokkaido University, Japan, Sapporo, Japan

ABSTRACT: Mechanical rock breakage has been of interest in modern mining and civil construction industries for decades now. However, the rock fracture mechanism associated with rock cutting processes is not yet well understood due to complex interaction between the cutter and the rock. With recent advances in numerical simulation methods, these techniques have become powerful tools for investigating the early rock fracturing process. However, not all numerical modelling methods are able to simulate this process correctly. This paper uses a new extension of combined finite-discreet element method to simulate the mechanical rock breakage with a simple drag cutting tool. The modelling technique takes into account the mix-mode I-II fracture criteria in addition to mode I and mode II for predicting the initiation and propagation of the cracks. After calibration of input parameters against uniaxial compressive and Brazilian tensile strength tests, tool-rock interaction is simulated in three different cutting velocities.

1 INTRODUCTION

During recent decades, the application of mechanical breakage tools in rock fragmentation has been widely extended to mining and civil engineering industries. Mechanical excavation machines provide a more flexible and environmentally friendly alternative to conventional blasting, especially in urban or non-ventilated environments.

Typically, there are two types of rock cutting methods classified based on the moving direction of the mechanical tool with respect to the rock surface: drag cutters (or fixed cutter bits) and indenters. While a drag cutter hits the rock in a direction parallel to the rock surface, the indenter penetrates normal into the rock surface (Hood and Alehossein, 2000). Rock cutting experiments are largely used to investigate the cutting process and the associated cutting forces; however, due to the extensive number of factors that influence the cutting process, the application of experiments to study the effective parameters can become expensive. Meanwhile, analytical and empirical methods developed so far suffer from too many simplistic assumptions. Over the years, different numerical techniques have been considered for investigation of rock fracture process with a cutting tool; however, not all the numerical techniques are able to model the entire rock cutting process (Menezes, 2016). Hence, despite many researches, the rock fragmentation mechanism with a mechanical cutter has not been well understood due to the complexity of the interaction between

mechanical tool and rock, and complex rock fracture process (Che et al., 2016). In recent years, however, new advancements in numerical modelling methods have significantly improved their potential for simulating the complex rock failure mechanisms. This paper aims to investigate the rock cutting process with a simple drag tool using a new implementation of combined finite-discrete element method (FDEM). The presented FDEM code uses parallel computation enabled by general purpose graphic processing units (GPGPU). This capability significantly speeds up the computations, which makes the code suitable for simulating time-consuming processes involved in rock cutting.

2 NUMERICAL COMBINED FINITE-DISCRETE ELEMENT METHOD

The principles of the FDEM are based on continuum mechanics, cohesive zone modelling and contact mechanics, all of which are formulated in the framework of explicit FEM (Munjiza, 2004). The continuum isotropic elastic behaviour of materials is modelled by an assembly of continuum 3−node triangular finite elements (TRI3s) (Figure 1(a)), while transition from continuum to discontinuum is modelled through cohesive zone model (CZM) with the concept of smeared crack (Liu et al., 2018).

This paper uses an implementation of FDEM, named as Y-HFDEM code, which is speeded up by the GPGPU parallelization scheme (Fukuda et al., 2018).

In the Y-HFDEM code, normal and shear cohesive tractions, (σ^{coh} and τ^{coh}, respectively), acting on each face of CE4 are computed using Eqs. (1) and (2) assuming tensile and shear softening behaviours:

$$\sigma^{coh}\begin{cases} \dfrac{2o}{o_{overlap}}T_s \text{ if } o<0 \\[2ex] \left[\dfrac{2o}{o_p}-\left(\dfrac{o}{o_p}\right)^2\right]f(D)T_s \text{ if } 0\le o\le o_p \\[2ex] f(D)T_s \text{ if } o_p<o \end{cases} \tag{1}$$

$$\tau^{coh}\begin{cases} -\sigma\tan(\phi)+\left[\dfrac{2|s|}{s_p}-\left(\dfrac{|s|}{s_p}\right)^2\right]f(D)c \text{ if } 0\le|s|\le s_p \\[2ex] -\sigma^{coh}\tan(\phi)+f(D)c \text{ if } s_p<|s| \end{cases} \tag{2}$$

where o_p and s_p are the elastic limits of the crack opening and sliding displacements (*o* and *s*), respectively, $o_{overlap}$ is the representative overlapping when *o* is negative, T_s is the tensile strength of CE4, *c* is the cohesion of CE4 and ϕ is the internal friction angle of CE4. Positive *o* and σ^{coh} mean crack opening and tensile cohesive traction. Eq. (2) corresponds to the Mohr-Coulomb (MC) shear strength model with tension cut-off.

The function, $f(D)$, in Eqs. (1) and (2) is the characteristic function for tensile and shear softening curves (Figure 1(b)) depending on a damage value *D* of the CE4. The following definitions of *D* and $f(D)$ are used to consider mode I, II and mixed mode I-II failure modes (Liu et al., 2018; Mahabadi et al., 2012):

$$D=\text{Minimum}\left(1,\sqrt{\left(\dfrac{o-o_p}{o_t}\right)^2+\left(\dfrac{|s|-s_p}{s_t}\right)^2}\right) \tag{3}$$

if $o\ge o_p$ or $|s|>s_p$, otherwise 0

$$f(D)=\left[1-\dfrac{A+B-1}{A+B}\exp\left(D\dfrac{A+CB}{(A+B)(1-A-B)}\right)\right]$$
$$\left[A(1-D)+B(1-D)^C\right](0\le D\le 1) \tag{4}$$

where *A*, *B* and *C* are the intrinsic rock properties that determine the shape of softening curves, o_t and s_t are the critical values of *o* and *s*, respectively, in

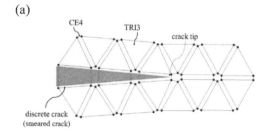

(a)

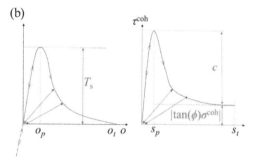

(b)

Figure 1. Transition from continuum to discontinuum in 2D Y-HFDEM IDE. (a) Assembly of TRI3s and CE4s, (b) Tensile/Shear softening curves

which a CE4 breaks and turns into a macro/explicit fracture. The o_t and s_t in Eq. (3) satisfy the Mode I and Mode II fracture energies, G_{fI} and G_{fII}, specified in Eqs. (5) and (6):

$$G_{fI}=\int_{o_p}^{o_t}\sigma^{coh}(0)\,do \tag{5}$$

$$G_{fII}+W_{res}=\int_{s_p}^{s_t}\left\{\tau^{coh}(|s|)\right\}d|s| \tag{6}$$

where W_{res} is the amount of work per area of CE4 done by the residual stress term in the MC shear strength model. This paper uses the same $f(D)$ functions for both mode I and II failure processes with *A = 0.63*, *B = 1.8* and *C = 6.0* (Liu et al., 2018).

3 NUMERICAL MODEL CALIBRATION

An appropriate determination of micromechanical parameters of FDEM model is essential for accurate simulation of any phenomenon of interest. In this study, the model parameters are calibrated against two standard laboratory testing methods used in rock mechanics: Brazilian tensile strength (BTS) and uniaxial compressive strength (UCS) tests. The physical and mechanical properties of the selected sandstone specimen for calibration and the input parameters used in the numerical simulations are listed in Table 1.

Table 1. Rock mechanical properties and numerical parameters

Parameter	Unit	Value
Density (ρ)	Kg/m3	1800
Young's modulus (E)	GPa	12.2
Poisson's ratio (v)	-	0.25
Tensile strength (T_s)	MPa	1.77
Cohesion (c)	MPa	5
Internal Friction angle of intact rock (ϕ)	°	25
Mode I fracture energy (G_{fI})	J/m²	16
Mode II fracture energy (G_{fII})	J/m²	160
Normal contact penalty number (P_{n_con})	GPa	1220
Tangent contact penalty number (P_{tan_con})	GPa/m	1220
Fracture penalty numbers (P_f, P_{tan}, $P_{overlap}$)	GPa/m	12200
Average element size (h_{ave})	mm	0.7

Figure 2 illustrates the geometry of the BTS and UCS tests simulations. The average edge length, h_{ave}, of TRI3s in both models is 0.7 mm. The rock specimens are placed between two rigid platens that load the numerical sample at a constant velocity of 0.05 m/s to satisfy quasi-static loading conditions. The friction coefficient μ_{fric} between the platens and the rock is assumed equal to 0.1, and within the rock is assumed equal to 0.5 based on (Fukuda et al., 2018).

Figure 3 depicts the tensile stress versus the vertical strain curve resulted from simulations along with the horizontal stress distribution, σ_{xx}, within the sample at three selected time snaps; in these figures the compressive stress is shown as negative (cold colour) and tensile stress is regarded as positive (warm colour).

Figure 3 (b) is the snapshot of the stress distribution taken at point A in Figure 3(a) and corresponds to the moment where the macroscopic splitting/tensile failure is initiated. When the induced tensile

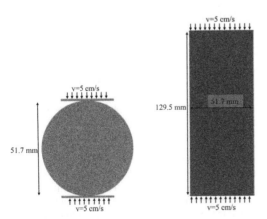

Figure 2. Model geometry and loading condition: a) BTS test, b) UCS test

stress along the central line of the disc reaches the peak value, point B in Figure 3 (a), tensile micro fractures are initiated approximately along the central line of the model, as shown in Figure 3(c). Figure 3(d), which corresponds to point (C) in Figure 3(a), illustrates the final splitting cracks along the central line of the Brazilian disc sample.

Obtained axial stress-strain curve for UCS test simulation is shown in Figure 4(a). Figures 4 (b)-(d) illustrate the minimum principal stress built-up and corresponding rock failure process. In these figures also, compressive stress is shown as negative (cold colour) and tensile stress is regarded as positive (warm colour). Figure 4(b) corresponds to the snapshot of stress distribution within the sample taken

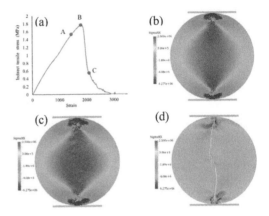

Figure 3. Numerical simulation of Brazilian test. a) indirect tensile stress versus axial strain, horizontal stress distribution b) before the peak stress, (c) at the peak stress, and (d) during the post failure

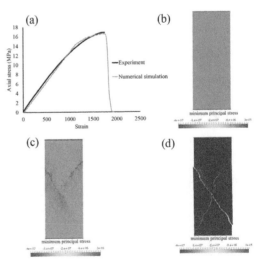

Figure 4. Numerical simulation of UCS test. a) Axial stress versus axial strain, minimum principal stress distribution b) before the peak stress, (c) at the peak stress, (d) during the post failure

at point A of Figure 4(a), when the microcracks appear in the model. As loading platens narrow further, microcrack coalescence and localization occur, which result in nonlinear stress-strain behavior from the point A to point B of Figure 4 (a). The minimum principal stress distributions corresponding to this is illustrated in Fig 4 (c). After point B, the coalescence and localization of the microcracks lead to initiation of macroscopic cracks, resulting in a diagonal fracture pattern as shown in Figure 4(d).

Figure 5 shows the final crack patterns in BTS and UCS tests. As shown, while the Mode I and Mode II damages are the dominant failure mechanisms to initiate fracture in BTS and UCS simulations respectively, the mixed-mode failure is the final failure mechanism in both simulations, leading to disintegration of the sample.

4 NUMERICAL SIMULATION OF ROCK CUTTING AND DISCUSSION

Once the model parameters were calibrated for the UCS and BTS tests as shown in previous section, the numerical simulations were extended to investigates the linear cutting process of a simple drag cutting tool e.g. a PDC cutter. Linear cutting test can be simulated under plane strain mode, if the cutting width is much bigger than the cutting depth (Jaime, 2011). The geometry of the model is, therefore, simplified as shown in Figure 6 (Menezes, 2016; Huang and Detournay, 2008). In these simulations the cutter is assumed rigid with a rake angle of 15°, which moves across the rock at a constant depth of cut of 0.3mm. The elements size near the cutting edge is 0.05 mm while larger elements are used further away from the cutter. The bottom and right boundaries of the rock slab are fixed, while the cutter is forced to move only in horizontal direction at a set speed. Three numerical

cutting speeds of 0.5m/s, 1m/s, and 2m/s have been investigated to best simulate the numerical ductile failure, which is expected at a 0.3mm depth of cut.

4.1 Failure Mechanism

As shown in Figure 6, the cutter is in full contact with the rock when simulation starts. Figure 7 shows the rock breakage process in the model with 0.5m/s cutting velocity. The corresponding cutting force with respect to the cutter location is depicted in Figure 7(a) for first chipping process. Under the loading action of the cutter, stress is concentrated at the contact region between the rock and the cutter. Compression is shown negative and tension is positive in Figure 7 (b).

As the cutter moves further into the rock, the reaction force reaches its first peak at point A in Figure 7(a). in this zone, which is known as crushed zone, different types of microcracks, mode I (Figure 7 (c)), mode II (Figure 7 (d)) and mix mode I-II (Figure 7 (e)) are initiated, with some propagated within the

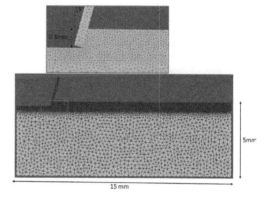

Figure 6. The rock cutting model geometry

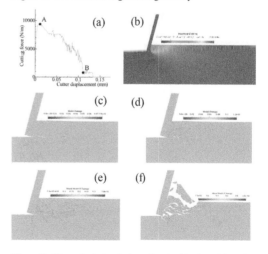

Figure 7. Numerical simulation of rock chipping. (a) cutting force vs. cutter displacement history, (b) induced stress field, (c) Mode I microcrack development in the model, (d) Mode II microcrack development in the model, (e) mixed Mode I-II fracturing, (f) chip formation due to mixed mode I-II fracturing

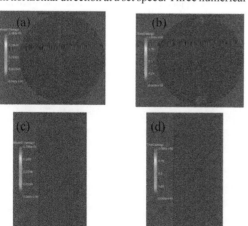

Figure 5. (a) Mode I micro-cracks in BTS model, (b) final mix-mode macro-cracks during the post failure in BTS model, (c) shear micro-cracks in UCS model, (d) mix-mode macro-cracks during the post failure in UCS model

compressed zone ahead of the cutter. As the cutter displacement further increases, due to coalescence of multiple cracks, explicit cracks unstably extend to free surface (point B in Figure 7(a)). Figure 7(f) shows that the final chipping form due to propagation of unstable mixed mode I-II fractures. Under perfect tool-rock contact condition, shearing failure seems to be the main failure mechanism, since the stress state is compression (Figure 7(b)), although the chipping is due to the mixed-mode failure mechanism.

The chipping process changes slightly when cutter-rock interface is relatively smaller, and the cutter is not fully in contact with the rock. Under such condition, both compressive (cold colour) and tensile stresses (warm colour) are induced into the rock ahead of the cutter, as shown in Figure 8(a). In such conditions, unstable mode I microcracks are more likely to contribute to the failure process (Figure 8(b)), while the final chipping is still due to the to the mix mode I-II fracture (Figure 8(c)).

4.2 Cutting velocity

In this section the effect of numerical cutting velocity on observed failure mode is investigated in 3 levels: 0.5 m/s, 1 m/s and 2 m/s. Figure 9 shows the first chipping process for all 3 levels of numerical cutting speed when there is perfect contact between the rock and the cutter. As can be observed, the chip morphology is almost identical at 3 levels of numerical cutting speeds. However, in second chipping the difference becomes obvious. As it can be seen in Figure 10, there are more cutting fragments in front of the cutter with the velocity of 2 m/s than the one with velocity of 0.5 m/s. This could be due the fact that the resulted fragments have less time to move away with higher velocity. Moreover, resulting fragments are finer in cutting with higher velocity, which

could be due to maximized cutter-rock contact facilitated by fragments and secondary breakage.

Figure 11 records the cutting force versus the cutter displacement history for three different cutting velocities. It is evident that the cutting force itself is not affected by the cutting velocity in first chipping process when there is full contact between the cutter

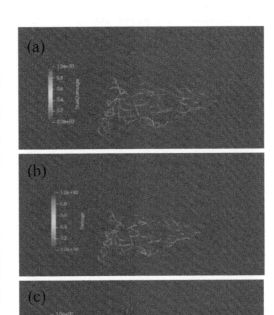

Figure 9. Chipping process under full contact between the tool and the rock in different cutting velocities; (a) 0.5 m/s, (b) 1 m/s (c), 2 m/s

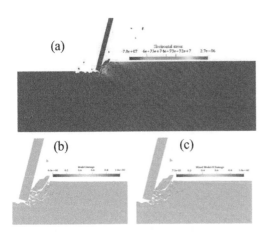

Figure 8. Chipping process at partial tool-rock engagement condition; (a) creation of stress field, (b) Mode I micro crack development in the model, (c) chip formation due to mixed mode I-II fracturing

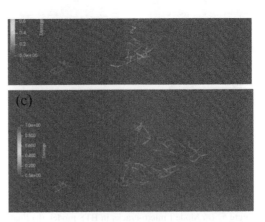

Figure 10. Chipping process of partial contact between the tool and the rock at different cutting velocities; (a) 0.5 m/s, (b) 1 m/s (c), 2 m/s

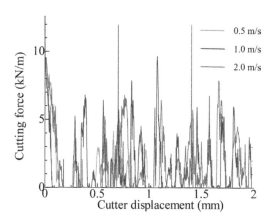

Figure 11. Cutting force vs. cutter displacement history in different cutting speeds

and the rock. However, the force associated with second major failure cycle is relatively lower for 0.5 m/s cutting speed than other two velocities, i.e. 1 m/s and 2 m/s. At higher speeds, although more cracks initiate and propagate into the rock, which itself require more energy, the resulted fragments also do not have time to move away from the front of the cutter. As a result, the accumulation of fragments ahead of the cutter provide better contact area between the tool and the rock, leading to better distribution of compressive stress and more crushing failure.

5 CONCLUSION

This study investigated the rock cutting process of a simple drag cutting tool using combined finite-discrete element method (FDEM). The input parameters of the numerical model were calibrated against standard Brazilian tensile strength (BTS) and uniaxial compressive strength (UCS) tests, prior to simulating the cutting process at three levels of cutting

speeds. Obtained results showed that the chipping at all cutting velocities is due to mix-mode I-II fracture. Moreover, at higher speeds accumulation of the resultant fragments ahead of the cutter provided a better contact between the tool and rock, leading to better distribution of stresses as well as secondary crushing of the fragments into finer products. Finally, the results showed ductile failure in shallow cutting depths could be modelled more realistically in 2 m/s cutting speed.

REFERENCES

Che D, Zhu W-L and Ehmann KF. 2016. Chipping and crushing mechanisms in orthogonal rock cutting. *International Journal of Mechanical Sciences* 119: 224–236.
Fukuda D, LIU H, Mohammadnejad M, et al. 2018. Development of 2-D Hybrid FEM/DEM Method Code Using GPGPU. In: ZHOU Y (ed) *10th Asian Rock Mechanics Symposium cum the ISRM International Symposium for 2018*. Singapore.
Hood M and Alehossein H. 2000. A development in rock cutting technology. *International Journal of Rock Mechanics and Mining Sciences* 37: 297–305.
Huang H and Detournay E. 2008. Intrinsic Length Scales in Tool-Rock Interaction1. *International Journal of Geomechanics* 8: 39–44.
Jaime MC. 2011. Numerical modeling of rock cutting and its associated fragmentation process using the finite element method. University of Pittsburgh.
Liu W, Zhu X and Jing J. 2018. The analysis of ductile-brittle failure mode transition in rock cutting. *Journal of Petroleum Science and Engineering* 163: 311–319.
Mahabadi OK, Lisjak A, Munjiza A, et al. 2012. Y-Geo: New Combined Finite-Discrete Element Numerical Code for Geomechanical Applications. *International Journal of Geomechanics* 12: 676–688.
Menezes PL. 2016. Influence of rock mechanical properties and rake angle on the formation of rock fragments during cutting operation. *The International Journal of Advanced Manufacturing Technology*: 1–13.
Munjiza A. 2004. *The Combined Finite-Discrete Element Method*: Wiley

T6: Blasting and machinery induced vibrations

T7: Rockburst, Outburst, Impacts

2019 Rock Dynamics Summit– Aydan et al. (eds)
© 2019 Taylor & Francis Group, London, ISBN 978-0-367-34783-3

Application of peridynamics to dynamic fracture process analysis of rock-like materials

D. Fukuda, J. Kodama & Y. Fujii
Hokkaido University, Sapporo, Japan

S.H. Cho
Chonbuk National University, Jeonju, South-Korea

H. Liu & A. Chan
University of Tasmania, Hobart, Australia

ABSTRACT: Peridynamics (PD) has recently attracted significant attentions from researchers in the field of computational mechanics because it can model complex fracture process with relative ease even for 3D dynamic fracture problems. However, the number of its applications to dynamic fracture problems of rock-like materials has been very limited. This paper presents the application of a self-devolved 2D/3D PD simulator based on mesh-free particle discretization scheme. To overcome the significant computational burden of the PD, we implemented a PD simulator with a parallelization scheme utilizing general purpose graphic processing unit (GPGPU). The developed code is verified first using some benchmark simulations. Then, by applying the code to 2D/3D dynamic fracture problems of rock-like materials assuming high loading rate such as detonation and deflagration phenomena, the applicability and future task of the developed PD simulator are demonstrated.

1 INTRODUCTION

Successful modeling of complex dynamic fracture process in rocks due to external impact loads, such as percussive hammer drilling and explosive blasting, is very important yet challenging task. In many existing computational mechanics approaches based on such as finite element method (FEM) including eXtended FEM, the target governing equation includes the spacial derivative of stress tensor and this causes the crack tip singularity of strain/stress. Silling (Silling, 2000,) developed the peridynamics (PD) as a new paradigm of continuum mechanics based on non-local theory. The spacial derivative of stress tensor is replaced by integration of force (state) in the PD which is characterized by length-scale parameter "horizon" and the crack is a part of the solution and not a part of the problem. Furthermore, no representation of the crack topology is needed. This feature makes the PD theory ideal for handling the problems with complex fracture process with relative ease. In fact, crack initiation/propagation/branching/coalescence even for 3D dynamics fracture problems can be easily modeled in the PD. At present, there have been three types of peridynamics formulations developed, namely, the bond-based PD (BB-PD), ordinary state-based PD (OSB-PD) and non-ordinary state-based PD (NOSB-PD). However, the PD is still a relatively new theory and the number of its applications to rock dynamics problems has been very limited.

This paper presents a self-devolved 2D/3D PD simulator based on mesh-free particle discretization scheme including a brief introduction of the PD theory. In addition, because of its non-local nature, its computational burden is significant. To overcome this, we implemented the BB-PD, OSB-PD and NOSB-PD based on a parallelization scheme using general purpose graphic processing unit (GPGPU) controlled by CUDA (Compute Unified Device Architecture) C/C++. The developed code is first verified using some benchmark problems. Then, by applying the code to 2D/3D dynamic fracture problems of rock-like materials assuming very high loading rate such as detonation and deflagration phenomena, the applicability and future task of the developed PD simulator are presented.

2 PERIDYNAMICS (PD)

In the classical continuum mechanics (CCM), the governing equation of a solid is based on a local theory as in the following equation:

$$\rho_0\left(\mathbf{X}\right)\ddot{\mathbf{u}}\left(\mathbf{X},t\right)=\nabla\mathbf{P}\left(\mathbf{X},t\right)+\mathbf{b}\left(\mathbf{X},t\right) \tag{1}$$

where $\mathbf{X}$ and t indicate the material points of a target solid in the initial configuration and time, respectively. In addition, ρ_0, $\ddot{\mathbf{u}}$, $\nabla\mathbf{P}$ and $\mathbf{b}$ are the initial density of the solid, the second-order time derivative of displacement **u**, the special derivative of the 1st Piola-Kirchhoff stress **P** and the body force, respectably, at **X**. Because of the inclusion of special derivative of the stress, crack-tip singularity is inevitable and thus appropriate treatment of the singularity is inevitable.

On the other hand, the PD is a new paradigm of continuum mechanics based on a non-local theory. As shown in Fig. 1, the PD considers bonds $\xi(=\mathbf{X'}-\mathbf{X})$ between a target material point **X** and its neighbor points **X'** within a finite distance δ from **X**, and the force interactions between **X** and **X'** are assumed to occur through each bond ξ. This δ is called "horizon". Each bond ξ existing within δ from **X** in the initial configuration is deformed to be $(\xi+\eta)$ by the force interaction where η is the relative displacement between **X'** and **X** due to bond deformation. This paper uses the dual-horizon PD concept (Ren et al., 2017) and the governing equation is given by the following equation:

$$\rho_0(\mathbf{X})\ddot{\mathbf{u}}(\mathbf{X},t) = \iiint_{\mathbf{X'}\in H'_{\mathbf{X}}} \mathbf{f}_{\mathbf{X}\mathbf{X'}}(\xi,\eta,t)dV_{\mathbf{X'}}$$
$$- \iiint_{\mathbf{X'}\in H_{\mathbf{X}}} \mathbf{f}_{\mathbf{X'}\mathbf{X}}(-\xi,-\eta,t)dV_{\mathbf{X'}} \quad (2)$$
$$+\mathbf{b}(\mathbf{X},t)$$

where $dV_{\mathbf{X'}}$ is the volume associate with **X'** and $\mathbf{f}_{\mathbf{X}\mathbf{X'}}$ is the bond force vector density (with the unit of [N/m^6]) exerting on **X** by **X'** and vice versa for $\mathbf{f}_{\mathbf{X'}\mathbf{X}}$. $\mathbf{X'}\in H'_{\mathbf{X}}$ indicates the union of points **X'** whose horizons include **X** (i.e. family of **X**), while $\mathbf{X'}\in H_{\mathbf{X}}$ indicates the union of the points **X'** within the horizon of **X**. i.e. family of **X**. The dual horizon PD makes it possible to use the variable horizon while Eq. (2) results in the original PD when constant horizon is assumed for every points. By comparing between

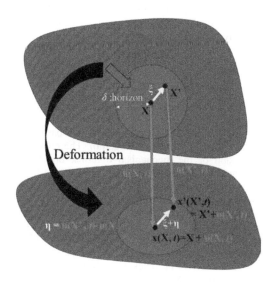

Figure 1. The concept of bond deformation in PD.

Eqs. (1) and (2), only the difference between the PD and CCM is clearly the replacement of special derivative of stress by the integration of force. Further detail can be found in the literature (Ren et al., 2017).

In the PD, the bond forces $\mathbf{f}_{\mathbf{X}\mathbf{X'}}$ and $\mathbf{f}_{\mathbf{X'}\mathbf{X}}$ are determined through the bond deformation depending on the material type, i.e. constitutive equation. In this aspect, three concepts called bond-based PD (BB-PD), ordinary state-based PD (OSB-PD) and non-ordinary state-based PD (NOSB-PD) have been actively used/developed. The BB-PD was historically the first generation of PD and each bond ξ in $H_{\mathbf{X}}$ is considered as independent spring. Thus, the bond force at **X** of a particular bond ξ in Eq. (2) is assumed to be the function of the bond itself and not affected by other bonds' deformation in the same family. This results in the requirement that directions of $\mathbf{f}_{\mathbf{X}\mathbf{X'}}$ and $\mathbf{f}_{\mathbf{X'}\mathbf{X}}$ must be parallel to the deformed bond. Thus, in the case of modelling isotropic linear elastic solid (ILES), Poisson's ratio can only take the value of 1/3 (2D) and 1/4 (3D) while the arbitrary Young's modulus can be used. To overcome this restriction of Poisson's ratio, Silling (Silling et al., 2007) also developed the OSB-PD by introducing the concept of "state" which is the generalization of tensor. In the OSB-PD, the force density at **X** is determined through the collective deformation of all the bonds in the family **X**, which allows to use arbitrary Poisson's ratio in the case of ILES. However, the OSB-PD is still subjected to the constraint in which directions of $\mathbf{f}_{\mathbf{X}\mathbf{X'}}$ and $\mathbf{f}_{\mathbf{X'}\mathbf{X}}$ must be parallel to the deformed bond. Furthermore, the number of available constitutive equations in the OSB-PD has been still limited. To overcome the constraints in the OSB-PD, the NOSB-PD has been developed by Silling (Silling et al., 2007) and this approach is not subjected to the constraint in the directions of $\mathbf{f}_{\mathbf{X}\mathbf{X'}}$ and $\mathbf{f}_{\mathbf{X'}\mathbf{X}}$. The NOSB-PD introduces a concept of non-local deformation gradient, which can be related to the constitutive equations developed for the CCM while overcoming the zero-energy mode due to the introduction of non-local deformation gradient must be solved and thus the NOSB-PD can be considered as "*not mature enough*". This paper only considers the BB-PD and OSB-PB.

In the BB-PD, $\mathbf{f}_{\mathbf{X}\mathbf{X'}}$ in the case of ILES is given as follows:

$$\mathbf{f}_{\mathbf{X}\mathbf{X'}} = C(\delta_{\mathbf{X'}})\cdot s_{\mathbf{X}\mathbf{X'}}\cdot\frac{(\xi+\eta)}{\|\xi+\eta\|}(^{\forall}\mathbf{X}\in H_{\mathbf{X'}}) \quad (3)$$

where C is the micro-modulus, which is a function of the horizon at **X'**, $\delta(\mathbf{X'})$, Young's modulus, E, and Poisson's ratio, v ($v=1/3$ in 2D, $v=1/4$ in 3D):

$$C(\delta_{\mathbf{X'}}) = \begin{cases} 3E/\left\{\pi\delta_{\mathbf{X'}}^3\cdot(1-v)\right\} & \text{(Plane Stress)} \\ 3E/\left\{\pi\delta_{\mathbf{X'}}^3\cdot(1+v)(1-2v)\right\} & \text{(Plane Strain)} \\ 3E/\left\{\pi\delta_{\mathbf{X'}}^4\cdot(1-2v)\right\} & \text{(3D)} \end{cases}$$

$$(4)$$

while $s_{\mathbf{XX'}}(=(\|\boldsymbol{\xi}+\boldsymbol{\eta}\|-\|\boldsymbol{\xi}\|)/\|\boldsymbol{\xi}\|)$ is the bond stretch.

On the other hand, in the OSB-PD, $\mathbf{f}_{\mathbf{XX'}}$ in the case of ILES is given as follows:

$$\mathbf{f}_{\mathbf{XX'}} = \underline{t}\langle\boldsymbol{\xi}\rangle \cdot \frac{(\boldsymbol{\xi}+\boldsymbol{\eta})}{\|\boldsymbol{\xi}+\boldsymbol{\eta}\|}(\forall \mathbf{X} \in H_{\mathbf{X'}}) \tag{5}$$

where $\underline{t}\langle\boldsymbol{\xi}\rangle$ is the force scalar "state" with the unit of [N/m^6] indicating the interaction force between $\mathbf{X}$ and $\mathbf{X'}$. Similar to tensor, the scalar "state" $\underline{t}$ operates on the bond $\langle\boldsymbol{\xi}\rangle$ and maps $\langle\boldsymbol{\xi}\rangle$ to a scalar. $\underline{t}\langle\boldsymbol{\xi}\rangle$ is given for 3D problem as follows:

$$\underline{t}\langle\boldsymbol{\xi}\rangle = \frac{3K\theta}{m}\underline{\omega}\langle\boldsymbol{\xi}\rangle\|\boldsymbol{\xi}\| + \frac{15G}{m}\underline{\omega}\langle\boldsymbol{\xi}\rangle\left(\underline{e}\langle\boldsymbol{\xi}\rangle - \frac{\theta}{3}\|\boldsymbol{\xi}\|\right) \tag{6}$$

where K and G are bulk modulus and shear modulus, respectively, $\underline{e}\langle\boldsymbol{\xi}\rangle (= \|\boldsymbol{\xi}+\boldsymbol{\eta}\| - \|\boldsymbol{\xi}\|)$ is the extension state of the bond, and m and θ are weighted volume and volumetric strain, respectively, given as follows:

$$m = \int_{H_{\mathbf{X}}} \underline{\omega}\langle\boldsymbol{\xi}\rangle\boldsymbol{\xi} \cdot \boldsymbol{\xi}\, dV_{\boldsymbol{\xi}}, \theta = \int_{H_{\mathbf{X}}} \underline{\omega}\langle\boldsymbol{\xi}\rangle\|\boldsymbol{\xi}\|\underline{e}\langle\boldsymbol{\xi}\rangle\, dV_{\boldsymbol{\xi}} \tag{7}$$

The scalar state $\underline{\omega}\langle\boldsymbol{\xi}\rangle$ has a role of weighting function and can control the influence of each bond $\boldsymbol{\xi}$ depending on $\|\boldsymbol{\xi}\|$, which is given as follows:

$$\underline{\omega}\langle\boldsymbol{\xi}\rangle = \omega\langle\|\boldsymbol{\xi}\|\rangle = \exp\left(-\|\boldsymbol{\xi}\|^2 / \delta_{\mathbf{X}}^2\right) \tag{8}$$

Excellent explanation about the concept of "state" can be found in the literature (Silling et al., 2007). Note that the computation of $\mathbf{f}_{\mathbf{X'X}}$ can be easily done by exchanging $\mathbf{X}$ and $\mathbf{X'}$ in Eqs. (3) or (5).

Fracture process in the framework of the PD is often expressed by the breakage of each bond $\boldsymbol{\xi}$, which is also used in this paper. When a particular bond $\boldsymbol{\xi}$ satisfies a given failure criterion, the bond is deleted. Hence, by introducing a scalar variable $\mu(\mathbf{X}, \boldsymbol{\xi}, t)$ to each bond $\boldsymbol{\xi}$ belonging to the family of $\mathbf{X}$, either $\mu=0$ or $\mu=1$ is given to the bond depending on whether it is broken or intact, respectively. Then the damage $D(\mathbf{X}, t)$ at $\mathbf{X}$ is defined as follows:

$$D(\mathbf{X},t) = 1 - \left(\int_{H_{\mathbf{X}}} \mu(\mathbf{X},\boldsymbol{\xi},t)\, dV_{\boldsymbol{\xi}}\right) \Big/ \left(\int_{H_{\mathbf{X}}} dV_{\boldsymbol{\xi}}\right) \tag{9}$$

where $D=0$, $0<D<1$ and $D=1$ indicate that the point $\mathbf{X}$ is intact, softening and completely broken state, respectively. With this approach, even complex 3D fracture processes can be handled with ease. Each bond is considered as broken, i.e. $\mu=0$, when the bond stretch $s_{\mathbf{XX'}}$ reaches its critical stretch s_0 given as follows:

$$s_0(\delta_{\mathbf{X'}}) = \begin{cases} \sqrt{(4\pi G_0)/(9E\pi\delta_{\mathbf{X'}})} & \text{(Plane Stress)} \\ \sqrt{(5\pi G_0)/(12E\pi\delta_{\mathbf{X'}})} & \text{(Plane Strain)} \\ \sqrt{(5G_0)/(6E\pi\delta_{\mathbf{X'}})} & \text{(3D)} \end{cases} \tag{10}$$

where G_0 is fracture energy.

Eq. (2) is just a governing equation in the PD and it should be solved by such as FEM or mesh free particle method. The mesh free particle approach is used and discretized form of Eq. (2) is given as follows:

$$\rho_0(\mathbf{X}_i)\ddot{\mathbf{u}}(\mathbf{X}_i,t) = \begin{bmatrix} \sum_{\mathbf{X}_j \in H'_{\mathbf{X}_i}} \mathbf{f}_{\mathbf{X}_i\mathbf{X}_j}\Delta V_{\mathbf{X}_j} \\ -\sum_{\mathbf{X}_j \in H_{\mathbf{X}_i}} \mathbf{f}_{\mathbf{X}_j\mathbf{X}_i}\Delta V_{\mathbf{X}_j} \end{bmatrix} + \mathbf{b}(\mathbf{X}_i,t) \tag{11}$$

where i stands for the ith PD particle among N PD particles in total in the system. The PD particles are distributed either regularly or irregularly depending on each problem, and each PD particle has its initial position $\mathbf{X}_i$ and the volume associate with the particle i. In the case of regular particle discretization, the value of horizon $\delta(\mathbf{X}_i)$ at $\mathbf{X}_i$ must be larger than $3\Delta x\sim4\Delta x$ to avoid the dependency of crack paths on the particle discretization. Explicit time integration scheme, velocity Verlet, is used for all the following simulation examples. Although the implementation of the PD simulator is relatively easy, its computational burden is tremendous especially for 3D problems. Thus, we utilized the GPGPU parallelization controlled with CUDA C/C++ to efficiently solve the PD simulation. It is worth mentioning that the OSB-PD (Eqs. (5)-(9)) requires more demanding computation than the BB-PD ((Eqs. (3)-(4) and (9))), and the NOSB-PD needs more computation than the OSB-PD.

3 RESULTS AND DISCUSSION

Since the PD code has been self-developed by the authors utilizing parallel computation by the GPGPU, extensive verifications must be necessary. Here some of the varication simulations are demonstrated. Then the applications of BB- and OSB- PDs to the dynamic fracture process analysis of rock-like materials are demonstrated.

3.1 Examples of verifications

A crack branching problem in a pre-notched Homalite slab (20 cm × 40 cm) with 20 cm notch investigated in Bobaru and Zhang (Bobaru and Zhang, 2016) is simulated (Fig.2(a)). The complete discussion of the mechanism of crack branching simulated by the BB-PD can be found in the literature. At $t=0$, tensile traction of 1MPa is suddenly applied to the top and bottom of the slab. Following in Bobaru and Zhang (Bobaru and Zhang, 2016), the 2D BB-PD under plane stress condition is used with the physical properties of $E = 4.55$GPa, $\rho_0 = 1230$ kg/m^3, and $G_0=38.46$ J/m^2. A regular particle discretization (83809 particles in total) with a constant particle spacing $\Delta x = 0.001$ m and a constant horizon $\delta = 4.015\Delta x$ is used. An example of the obtained result is shown in Fig. 2 where damage D in Eq. (9) is displayed. The warm color indicates the crack. The result shows a good agreement with the one presented in Bobaru and Zhang (Bobaru and Zhang, 2016). It is notable that even the BB-PD can easily capture the crack branching phenomena, contrary to FEM-based fracture simulations including XFEM.

Consideration of microscopic structure, i.e. material heterogeneity in rocks, is of fundamental importance. Here the fracture process in a solid rectangular slab with multiple circular inclusions subjected to prescribed velocity is simulated as another example (Fig. 3). The initial model consists of a 0.3 mm-notch (indicated in dotted boxes in Fig. 3(a)(b)), matrix of the solid material (ρ_1 = 1230 kg/m³, E_1 = 72 GPa, v_1=1/3, G_1 = 40 J/m²) and randomly distributed circular inclusions (ρ_2 = 1230 kg/m³, E_2 = 144 GPa, v_2 = 1/3, G_2 = 80 J/m²) following Ren et al. (Ren et al., 2017). The top and bottom of the pre-notched slab are subjected to prescribed constant velocities.

The constant particle spacing Δx=1.5×10⁻⁵ m and a constant horizon δ = 3.015Δx are used, resulting in

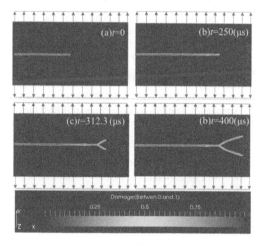

Figure 2. Verification simulation (1).

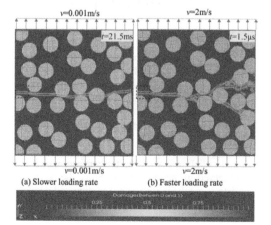

(a) Slower loading rate   (b) Faster loading rate

Figure 3. Verification simulation (2).

41607 particles in total. Two cases are considered for the prescribed velocities, i.e. 0.001m/s (Fig. 3(a)) and 2 m/s (Fig. 3(b)). Fig.3(a) with the slower loading rate shows that the single crack propagates in the way that it avoids the stiffer/stronger inclusions in the early stage. However, when the single crack meets one of the inclusions just in front of the crack, the crack propagation within the inclusion occurred under this loading rate. Fig. 3(b) with the faster loading rate shows that the crack became relatively thicker and multi-stage crack branching is also found through the interaction of crack-tip and inclusions. This condition was also investigated by Ren et al. (Ren et al., 2017) and the obtained result in Fig.3(b) shows the excellent similarity with the literature. Although the physical properties used here is different from those for rocks for the purpose of verification, it is expected that the PD can be used for the investigation of crack propagation in heterogeneous rock considering micro-structures of rock.

3.2 Dynamic fracture due to blast-induced load

In rock blasting, controlling the resultant fracture patterns and minimizing the blast induced-damage zone is of fundamental importance, which is yet challenging task. If the blast-induced dynamic rock fracture process can be reasonably simulated by numerical simulators, the benefits are considerable for such as optimization of blasts to specific geological settings without conducting costly field trials. Development of such a useful numerical simulator for blasting is a grand challenge in engineering since at least the numerical simulator needs to capture detonation process including gas flow into newly generated cracks and complex dynamic fracture, each of which is also very challenging. Since the PD can handle the complex dynamic fracture problems with relative ease, this subsection applies the developed BB-PD simulator to a simple single hole blasting problem to see how the obtained results look like. The 2D dynamic fracture process due to stress propagation by blast-loading is only considered and thus modelling gas flow into newly generated fractures is future task.

A 2D rock model with a single blast hole is prepared following the experimental setup used in (Dehghan Banadaki and Mohanty, 2012) in which a copper tube was placed between blasthole surface and explosive and thus no gas flow into the generated cracks occurred. The radii of the rock cylinder and cylindrical blast hole are 2.575 mm and 72 mm, respectively. The model is first discretized by 213,933 PD particles with a constant particle spacing Δx = 0.25 mm and a constant horizon δ = 3.015Δx. The blast-induced pressure $P(t)$ applied to the blasthole surface is assumed to be given as follows:

$$P(t) = P_0 \left(e^{-\alpha t} - e^{-\beta t} \right) / \left(e^{-\alpha t_0} - e^{-\beta t_0} \right) \qquad (12)$$

where maximum pressure P_0, rise time t_0 and pressure decay rate β/α are just assumed to be 150 MPa,

5 µs and 5, respectively. For the physical properties of rock, ρ_0 = 2660 kg/m³, E = 50.0 GPa and v = 1/3 are assumed to be constant for the entire rock. Since the BB-PD is used here, v = 1/3 is only possible while the value of E is assumed considering the measured wave speeds of Barre granite used in (Dehghan Banadaki and Mohanty, 2012). The material heterogeneity of rock is simply expressed by distributing fracture energy G_0 for each bond using Weibull's distribution with heterogeneity index m = 1.5 and the average value of G_0 is set to be 200 J/m² assuming intact rock. The outer surface of the model is free face.

Fig.4 shows the particle velocity magnitude and damage evolution, i.e. fracture process, for the BB-PD simulation at selected time intervals. The results at 1.25 µs show that the stress wave due to the application of the blast-induced pressure commenced from the surface of the blast hole. At t = 11.25 µs, the stress wave front further propagated toward the outer free surface. The significant fracture (crushing zone) is formed in the vicinity of the blast hole and extension of several tension induced fractures were found

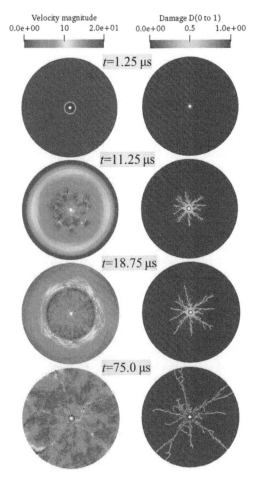

Figure 4. Numerical simulation of the dynamic fracture process of rock modelling blast-induced loading.

surrounding the crushed zone. Due to wave reflection upon impinging on the outer free face, the results at t = 18.75 µs shows the interaction between the reflected wave and the tips of radial fractures propagating from the blasthole, which resulted in further extension of these radial cracks (see the results at t =75 µs). By comparing the obtained result with the experimental result (Dehghan Banadaki and Mohanty, 2012), the fracture patterns show good similarity although further detailed modeling/calibration of the PD simulation is still necessary.

3.3 Dynamic fracture due to deflagration

This section considers more 3D PD dynamic fracture simulation of concrete due to deflagration, i.e. very fast combustion. An experiment conducted by (Yamachi, 2015) is modeled. A cylindrical concrete specimen with a single borehole is fragmented using the deflagration of nitromethane (NM) (Fig. 5(a)(b)). The NM filled in plastic container is set into a drilled borehole and it is initiated by electric discharge. According to (Fukuda et al., 2013), the peak pressure is achieved 140 µs after the initiation of the NM and thus the corresponding loading rate is much slower than that due to detonation. Further detail of this method can be found in (Fukuda et al., 2013). The 3D OSB-PD model is shown in Fig.5(c) and the number of total PD particles is 289504, which requires the significant computation. The model is first discretized by 289504 unstructured 3D tetrahedra similar to conventional FEM mesh. Then the PD particles are placed in all the vertexes of the tetrahedra and the volume of each tetrahedron is equally lumped to vertexes of each tetrahedron. For the applied pressure to borehole due to the deflagration of the NM, Eq.(12) with P_0 = 1 GPa, t_0 = 140 µs and β/α = 1.5. As physical/mechanical properties of concrete, v = 1/4, ρ_0 = 2320 kg/m³, E = 34.2GPa and average fracture energy G_0 = 25 J/m² with heterogeneity index m = 5. All the outer boundaries are free faces and stemming of the borehole is neglected for simplicity.

Fig. 6(a)(b) show the results just after the commencement of loading and resultant fracture pattern, respectably. For each figure, top left and bottom left indicate the damage, i.e. fracture, patterns on the surface and inside of the model, respectively. On the other hand, right hand side of the figure shows the velocity distribution for selected two cross-sections in the model. Fig. 6(a) clearly shows that the deflagration-induced deformation propagates from the borehole toward the outer boundary of the model. By comparing the fracture pattern of the surface in Fig. 5(b) and Fig. 6(b), it is found that the PD simulation can reasonably capture the similar "cross" cracks on the lateral surface and radial cracks on the top surface, which are also observed in the experiment. Since the no measured data is available for surface motion and detailed 3D fracture pattern in the experiment, further detailed discussion including the correct setting of material/physical properties is our future task. However, with the results of Fig. 6, the high potential of the PD simulation for the dynamic fracture of rock-like materials has been shown. To obtain the similar results by such as XFEM and FEM based on such as cohesive zone model, implementing such simulation code is not so easy task while the implementation of the PD itself is quite straightforward and in fact the difference in terms of implementing 2D/3D PD codes are very small.

4 CONCLUSIONS

This paper newly developed the GPGPU-based 2D/3D PD simulator. Some verification simulations such as crack branching and crack propagation in heterogeneous medium were shown, and the applicability of the PD to the 2D/3D dynamic fracture problems of rock-like materials due to detonation/deflagration were demonstrated.

As a future task, the authors will apply the developed PD simulator to modelling more and more fundamental rock fracture problems including modelling the conventional laboratory rock testing such as Brazilian test and compression test. In addition, by coupling with such as smoothed particle hydrodynamics which has also been implemented by the authors, modelling the full blasting process is another important task. Furthermore, only BB-PD and OSB-PD were tested in this paper and the development/application of NOSB-PD, which can utilize the conventional constitutive equations in the CCM, should be also considered. Last but not least, since the numerical implementation of the PD is computationally very demanding, multiple GPGPU-accelerators should be utilized to make the simulations of practical rock engineering problems with much larger scale possible. Since the application of PD to rock engineering problem has not been mature enough, more and more extensive investigations regarding the applicability/improvement of the PD are expected to be pursued in the rock engineering community.

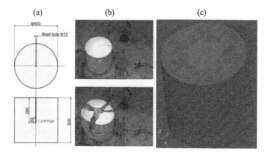

Figure 5. The overview of dynamic fracture test utilizing the deflagration of the NM and its numerical OSB-PD model. (a) The cylindrical concrete specimen used for the experiment, (b) Before and after the fracture due to deflagration ((a) and (b) are after (Yamachi, 2015)), (c)The OSB-PD model.

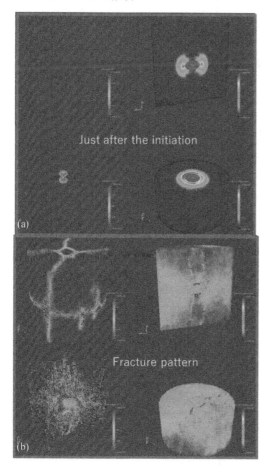

Figure 6. The result of numerical simulation modelling the dynamic fracture process of concrete due to deflagration of the NM.

REFERENCES

Bobaru, F. & Zhang, G. 2016. Why do cracks branch? A peridynamic investigation of dynamic brittle fracture. *International Journal of Fracture*, 196, 59–98

Dehghan Banadaki, M. M. & Mohanty, B. 2012. Numerical simulation of stress wave induced fractures in rock. *International Journal of Impact Engineering*, 40–41, 16–25.

Fukuda, D., Moriya, K., Kaneko, K., Sasaki, K., Sakamoto, R. & Hidani, K. 2013. Numerical simulation of the fracture process in concrete resulting from deflagration phenomena. *International Journal of Fracture*, 180, 163–175.

Ren, H., Zhuang, X. & Rabczuk, T. 2017. Dual-horizon peridynamics: A stable solution to varying horizons. *Computer Methods in Applied Mechanics and Engineering*, 318, 762–782.

Silling, S. A. 2000. Reformulation of elasticity theory for discontinuities and long-range forces. *Journal of the Mechanics and Physics of Solids*, 48, 175–209.

Silling, S. A., Epton, M., Weckner, O., Xu, J. & Askari, E. 2007. Peridynamic States and Constitutive Modeling. *Journal of Elasticity*, 88, 151–184.

Yamachi, H. N., Jun-Ichiro 2015. The Numerical Evaluation of Shock Wave Propagation Caused by Electric Discharge. *Reports of Technical Research Institute of Sumitomo Mitsui Construction Co., Ltd.* (In Japanese)

2019 Rock Dynamics Summit– Aydan et al. (eds)
© *2019 Taylor & Francis Group, London, ISBN 978-0-367-34783-3*

Predicting seismic velocity distribution in the ground ahead of tunnel face using drilling vibration of hydraulic rock drill

K. Tsukamoto
Okumura Corporation, Tsukuba, Japan

M. Shinji
Yamaguchi University, Ube, Japan

ABSTRACT: Accurate prediction of the geological conditions ahead of the tunnel face is essential for safe and economical tunnel construction. As an application of the drill logging, a new method for predicting the distribution of seismic velocity ahead of the tunnel face has been developed. The method makes it possible to determine the travel-time from transmission to reception by determining the transmission time at which the rock is hit by the drill bit and processing the wave form of the vibration received by the receivers on the tunnel face. Field test results have shown that the newly developed method is useful in predicting the seismic wave velocity distribution ahead of the tunnel face.

1 INTRODUCTION

Non-core drilling exploration (Kuwahara, 2012) is one of the exploration methods used for the ground ahead of the tunnel face. In this method, advancing drilling is carried out for 30 to 50 m ahead of the tunnel face, and we predict the ground condition based on the energy required for excavation per unit volume (specific energy) as an index. The specific energy used as an index varies depending on the change in the working pressure of the hydraulic rock drill for drilling, characteristics or discharge conditions of drilled rock fragments, etc.; thus, the accuracy of the specific energy may also vary. In addition, we can only determine the distribution of the specific energy in the direction of the advancing drilling, and it is hard to determine the ground conditions in a two-dimensional plane.

Considering the above circumstances, we have developed an exploration method for a hydraulic rock drill mounted on a drill jumbo used for mountain tunnel construction, as shown in Figure 1. In this method, we accurately obtain the propagation time (travel-time) for the vibration generated when the drill bit strikes the ground during drilling to reach a geophone placed at the tunnel face. We developed the non-core drilling exploration method to determine the distribution of elastic wave velocities in the ground ahead of the tunnel face.

Figure 2 shows the basic conceptual diagram of the exploration method, and Figure 3 shows the flow of the exploration. In the method developed in this study, advancing drilling is carried out from two or more positions in the tunnel face. When a bit strikes the ground, we obtain the generation time of the vibration from a pilot sensor set on the hydraulic drifter of a hydraulic rock drill, and then measure the

arrival time of the vibration to reach the tunnel face after propagating through the bedrock using multiple geophones set at the face. Using the travel-time, which is the difference between the generation time and arrival time, we perform tomography analysis (an analysis to obtain the distribution of elastic wave velocities in the exploration region from data on propagation distances and travel-times by setting multiple generation and receive points around the region) to obtain the distribution of elastic wave velocities of the ground ahead of the tunnel face.

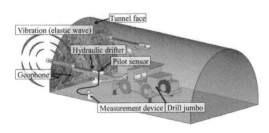

Figure 1. Exploration by a drill jumbo.

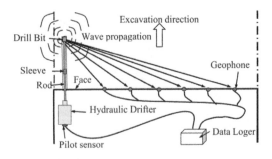

Figure 2. Conceptual diagram of the exploration method.

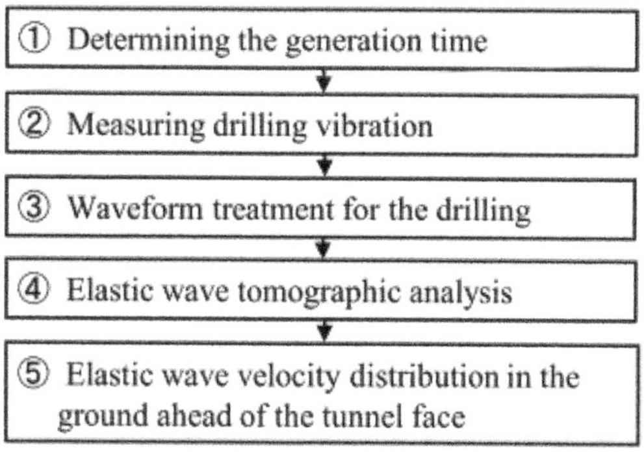

① Determining the generation time

↓

② Measuring drilling vibration

↓

③ Waveform treatment for the drilling

↓

④ Elastic wave tomographic analysis

↓

⑤ Elastic wave velocity distribution in the ground ahead of the tunnel face

Figure 3. Exploration flow chart.

2 OVERVIEW OF THE EXPLORATION METHOD

As shown in Figure 4, the vibration (elastic wave) is generated when the piston in the hydraulic drifter strikes the shank rod propagates from blow position A to the pilot sensor in position B through the hydraulic drifter. On the other hand, the wave generated in the shank rod due to the blow also propagates to position C on the rod at time T_{AB} measured by the pilot sensor. When the wave reaches position D at the tip of the bit, it breaks rock, thus generating wave in the ground. The time difference of the first break obtained from the waveforms measured by the geophone and pilot sensor corresponds to $T_{CD} + T_{DE}$. Based on the above, we obtained the travel-time T_{DE} from Eqs.(1).

$$T_{DE} = T_E - T_B - T_{CD} \qquad (1)$$

where T_{DE} is the travel-time of the bit and geophone; T_E is the arrival time at E; T_B is the arrival time at B; T_{CD} is travel time from C to D.

We mounted accelerometers on the hydraulic drifter and the bit as shown in Figure 5, and then generated wave by a blow to determine beforehand the time difference T_{CD} for the elastic wave to reach the hydraulic drifter and the bit. Figure 6 shows the measurement results for the hydraulic drifter vibration and the bit vibration. We notice that the time difference T_{CD} for the elastic wave to travel from the hydraulic drifter to the bit is 0.709 ms.

3 IN-SITU EXPERIMENT

3.1 Measurement site and instrumentation setup

We performed the experiment in a tunnel with a width of 8.5 m, a length of 980 m, having a cross-section of 65 m². The geological features are mainly shale, but also include sandstones, green rocks, and tuff.

In the measurement of drilling vibrations, as shown in Figure 7, geophones were placed at seven positions (the same height as the drilling position

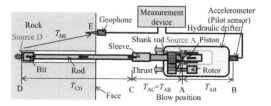

Figure 4. Conceptual diagram of the exploration method.

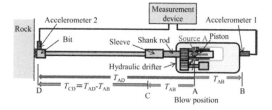

Figure 5. Experiment to determine T_{CD} time.

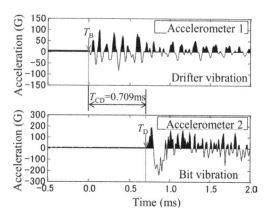

Figure 6. Relation between vibrations of a hydraulic drifter and a bit.

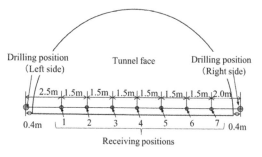

Figure 7. Configuration of measurement instruments.

with 1.5 m intervals) on the shotcrete surface at the tunnel face. At two positions on the left and right sides of the tunnel face, drilling was done from the side walls that were 1.5 m from the tunnel-face surface in the direction ahead of the tunnel face at an angle 15° from the tunnel axis. Figure 8 shows

692

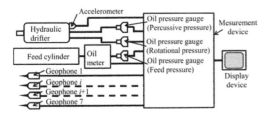

Figure 8. Composition of measurement instruments.

Table 1 Specifications of measurement instruments

Instrument	Specifications	
Accelerometer	Output	0.103 mV/G
	Resonance frequency	100kHz <
Geophone	Output	104 mV/cm/s
	Resonance frequency	28Hz
Oil meter	Range	0~80ℓ/min
	Output	12V(pulse)
Oil gauge	Range	0~20MPa
	Output	1~5V
Measurement device	AD resolution	24 bit
	Maximum sampling frequency	100kHz

the composition of the measurement instruments while Table 1 gives their specifications. We used a measurement device with 24-bit A/D resolution; the maximum sampling frequency was 100 kHz.

3.2 Measurement of drilling vibrations

We obtained a series of waveforms for each blow from the pilot sensor placed on the rock drill and geophones placed on the face surface. In the waveforms for each blow, when the bit hit the ground, the blow is sometimes not uniform because drilled rock fragments can be caught between the bit and the ground. After removing such waveforms, we performed stacking procedure (procedure of adding to the time series of waveform data) on 10 waveforms to enhance the S/N ratio.

3.2.1 Separating the waveform based on each blow
In the hydraulic rock drill, a large vibration is generated when the piston strikes the shank rod, and when the piston moves backward. As the number of blows of the piston per second during drilling ranges 50 to 60, a waveform is read at every 20 ms from the time series of measured data.

The maximum amplitude of a vibration due to a blow is 40–80% of the vibration induced when the piston moves backward. We set the trigger level of the pilot sensor at 20% of the maximum acceleration level and it is less than 40%. We further extracted the waveforms that were measured at 20 ms after the first breaks in cases where the bit struck the ground and where the bit did not strike the ground. Then, we distinguished these two cases based on whether the wave due to a blow was observed after the expected travel-time in the waveforms measured by the geophones placed on the tunnel face. The measured waveforms resulted from waves that propagated in the ground, and we distinguish them

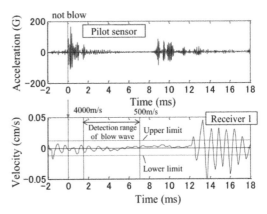

Figure 9. Vibration waveforms when the bit strike the ground.

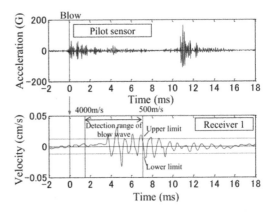

Figure 10. Vibration waveforms when the bit did not strike the ground.

by the S-waves, whose amplitudes are greater than those of the P-waves.

The S-wave velocity (V_s) of rock ranges from 0.5 km/s to 4 km/s in most cases; thus, to distinguish the waveform due to a blow, we set the range of the arrival times corresponding to the upper and lower velocity limits of 4 km/s and 0.5 km/s of the S-wave, respectively, as the confirmation range of waves, as shown in Figure 9. We determine that the bit strikes the ground when the maximum amplitude of the waveform is measured within the range. In contrast, as Figure 10 shows, the bit did not strike the ground when the maximum amplitude of the waveform was not measured within the confirmation range.

3.2.2 Waveform stacking procedure
A stacking procedure is a calculation method that adds up the time series data of waveforms aligned on the time of the first break. Using this procedure increases the value and random noises are cancelled out or decrease, thus enhancing the S/N ratio.

When the bit strikes the ground, drilled rock fragments can sometimes be stacked between the bit and the ground, and continuous blows cannot be performed. If stacking procedure is applied to original waveforms due to multiple blows, then the S/N ratio becomes low. Therefore, we obtain the cross-correlation functions between waveforms, and discard waveforms that have low correlation coefficients, instead of treating them as vibration sources. The cross-correlation function for a pair of waveforms is calculated from Eqs.(2)-(4).

$$\mu(i) = \frac{1}{N} \sum_{n=1}^{N} y_n(i) \qquad (2)$$

$$C_k(i,j) = \frac{1}{N} \sum_{n=k+1}^{N} \{y_n(i) - \mu(i)\}\{y_{n-k}(j) - \mu(j)\} \quad (3)$$

$$R_k(i,j) = \frac{C_k(i,j)}{\sqrt{C_0(i,i)C_0(j,j)}} \qquad (4)$$

where $y(i)$ is the amplitude of the vibration due to blow i; N is the number of data samples; $\mu(i)$ is the mean of the vibration due to blow i; k is the lag (the shift time between the vibration due to blow i and blow j), $C(i,j)$ is the cross-covariance function for the vibration due to blow i and blow j, and $R(i,j)$ is the cross-correlation function for the vibration due to blow i and blow j.

To reduce noise and enhance the S/N ratio, we obtain the lag of the cross-correlation function for each pair of waveforms, and perform stacking procedure by shifting the time on the lags added to the waveforms. Here, considering the effectiveness of the calculation process for cross-correlation functions, we perform stacking treatment on waveforms due to 10 blows by the hydraulic rock drill to improve the S/N ratio. The drilling length by 10 blows is approximately 2–5 cm, and its effect on the travel-time is small because the change in the drilling length is also small. We set the threshold as 20% of the maximum amplitude of the pilot sensor. Figure 11 shows the waveforms between the time 2 ms (pre-trigger) and 5ms. The waveforms from the pilot sensor are not in a pulse shape as the

vibration of the piston, generated when the piston in the rock drill strikes the shank rod, and the vibration due to the surge pressure, which is associated with switching valves, are transmitted to the sensor through the frame of the rock drill; the waveforms become a sweep shape that varies for each blow. There are differences in the first breaks that are read in the waveforms from the pilot sensor within a range of 0.32 ms, which will cause an error in the travel-time if we set the first break using the threshold of the maximum amplitude.

Figure 12 shows the waveforms for 10 blows measured at receiving point 1, whose triggers were the vibrations from the pilot sensor. We can see that noise is mixed in with these waveforms, such as in the first blow, where another subsequent waveform is seen at approximately 10 ms; this is attributed to drilled rock fragments (clast) caught between the bit and rock. In the third blow, the waveform is deformed. The P-waves of vibrations due to a blow start with anaseism, and then propagate in the ground. Therefore, when we read the first break in a waveform at a receiving point, we read the rising time of the anaseism (right side).

Figure 13 shows the zoomed-in waveforms around the first breaks at receiving point 1. As the first breaks from the pilot sensor do not occur at

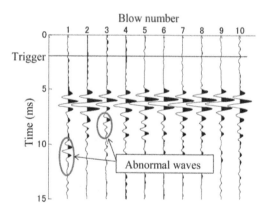

Figure 12. Waveforms that include abnormal parts at receiving point 1.

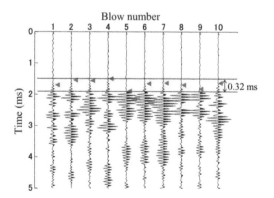

Figure 11. Around the first breaks of the waveforms from the pilot sensor (is the position of the first break).

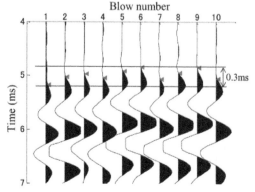

Figure 13. First breaks in waveforms at a receiving point 1.

the same time instant, there is an error of 0.3 ms in reading the travel-time. If we apply stacking procedure including vibrations with such noise, then the S/N ratio becomes low. Therefore, we used the cross-correlation function method, which is the easiest way to detect the time delay between waveforms.

Figure 14 shows an example of the stacking procedure of waveforms at a receiving point. We calculated the cross-correlation functions for waveforms due to 10 blows and removed waveforms with low correlation coefficients. Then, we performed stacking procedure by shifting the time for waveforms whose correlation coefficients were 0.9 and over, from which we consider that there is a strong correlation, and enhanced the S/N ratio. Usually, this treatment is performed using the waveforms from the pilot sensor, which were on the generation side.

However, we used the waveforms at the receiving points because it is difficult to perform the procedure using waveforms that vary for each blow.

For the waveforms where we applied stacking procedure, we obtained the time of the first break for each waveform. We read the first breaks from the stacked waveforms at positions with intervals 0.2 m; thus, the

number of readings is large, and the work becomes cumbersome. Therefore, we performed readings of the first arrivals of P-waves using an automated system to ensure mathematical objectivity.

To read the first arrivals, we used Akaike's information criterion (Akaike, 1973). We set the range for the first arrival reading as the time between vibration generation due to a blow and time when the vibration amplitude became the maximum.

Figure 15 shows an example of the first arrival obtained from the waveform recorded at the pilot sensor, and Figure 16 shows an example of the first arrivals obtained from the waveform at Receiving Point 1. By aligning wave-trains using cross-correlation functions and stacking, the start of the wave rise and accurate evaluation of the time of the first break by the AIC are clearly observed.

Figure 17 shows the result at Receiving Point 1, where we applied the stacking treatment to the waveforms received by the pilot sensor and the receiving point, and we aligned the waveforms using an interval of 0.2 m based on the generating time when the bit struck the ground. The figure shows

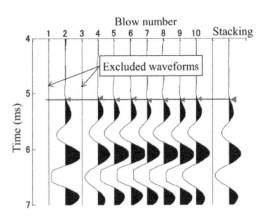

Figure 14. Stacking of waveforms at a receiving point 1.

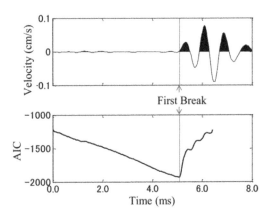

Figure 16. Detecting the first break in the waveform at receiving point 1.

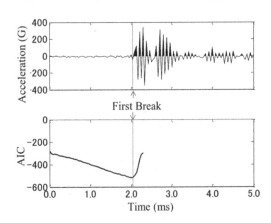

Figure 15. Detecting the first break in the waveform at the pilot sensor.

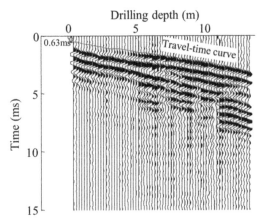

Figure 17. Received waveforms and the travel-time curve at receiving point 1.

the travel-time curve (representing the relationship between the time intervals when the elastic wave due to a blow reaches the receiving point and the drilling length), which is obtained by choosing the position (dot) of the first arrival for each waveform, and drawing a line passing through these dots. The travel-time for the drilling length of 0.2 m is 0.630 ms at Receiving Point 1, and 2.364 ms at Receiving Point 7. The travel-time increases as the receiving point moves further away from the drilling point.

4 TOMOGRAPHY ANALYSIS

In general, tomography is a method to obtain information on physical property values from multiple vibration generating and receiving points surrounding the target exploration region. By performing travel-time tomography analysis using travel-time data at each receiving point, we obtain the distribution of elastic wave velocities in the ground ahead of the tunnel face. For the analysis, we used a method where the velocity distribution is iteratively corrected so that the difference between the observed travel-time and theoretical travel-time obtained from the ray method for the velocity distribution at that time, becomes the minimum value. We set the cell size for the analysis as 1 m × 1 m; and we divide the analysis region, where all the generating and receiving points are included in 21 cells in the direction along the cross-section of the tunnel, and 16 cells in the direction ahead of the tunnel face. For the analysis, we use the least-squares method and perform the calculation iteratively until the relative RMS residue between the observed travel-time and calculated travel-time is within the error range defined in the convergence test.

Figure 18 shows the distribution of elastic wave velocities ahead of the tunnel face obtained by tomography analysis. Table 2 shows the interval mean values of specific energy. Around the tunnel face, elastic wave velocities in the range of 3.5 km/s to 4.5 km/s are distributed in the ground. The result of the distribution of elastic wave velocities predicts that from the position approximately 3 m ahead of the tunnel face, areas with elastic wave velocity of about 5 km/s appear, gradually expanding from the left to right. Grounds with such characteristics exist ahead of the tunnel face as a layer with thickness of approximately 5 m, inclined at an angle 20° from the direction along the cross-section. In the figure, the arrows show positions where the change of the ground is predicted from the change in specific energy. In the left side of the non-core drilling exploration, the elastic wave velocities range between 4.4 km/s and 5 km/s in the section with drilling distance between 2.8 m and 10.3 m. The ground with elastic wave velocity of 5 km/s appears in the right side of the non-core drilling exploration, with drilling distance of approximately 8.8 m. The results of specific energy and that of the distribution of elastic wave velocities are well correlated with each other.

In line graph, the elastic wave velocity around the tunnel side wall obtained by underground

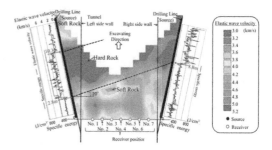

Figure 18. Analysis result of elastic wave tomography.

Table 2 Interval mean values of specific energy

Left side		Right side	
Drilling length (m)	Interval mean (J/cm³)	Drilling length (m)	Interval mean (J/cm³)
0–2.8	190	0–8.8	164
2.8–10.3	304	8.8–14.8	276
10.3–14.8	196	-	-

elastic wave exploration and the elastic wave velocity obtained by drilling vibrations are almost consistent.

5 CONCLUSION

The following conclusions may be drawn from this study:

1. It is possible to estimate the velocity distribution in the ground ahead of the tunnel face from machine induced vibration data during drilling with hydraulic rock drills used in actual construction, and from the travel-time data at the tunnel face surface.

2. The elastic wave velocity around the tunnel side wall obtained by underground elastic wave exploration and the elastic wave velocity obtained by drilling vibrations are almost consistent, which confirms the validity of the proposed prediction method

3. The developed exploration method utilizes rock drills in a drill jumbo for tunnel construction, when we perform non-core drilling exploration at a site. In addition, compared with an exploration method where drilling is temporarily halted at a specified depth to apply only blows to the ground to align the vibration source positions at the depth at regular intervals, the developed method is more practical because we can make measurements without interrupting the drilling process.

REFERENCES

Akaike, H, 1973. Information theory and an extension of the maximum likelihood principle, Second International Symposium on Information Theory, Akademiai Kiado, Budapest, 267–281.
Kuwahara, T, et al. 2012. International Journal of the JCRM Vol.8 pp.25–30.

2019 Rock Dynamics Summit– Aydan et al. (eds)
© 2019 Taylor & Francis Group, London, ISBN 978-0-367-34783-3

Effect of burden to hole diameter ratio on rock fragmentation by blasting using LS-DYNA

M.I. Shahrin, R.A. Abdullah, R. Sa'ari & M.N.A. Alel
School of Civil Engineering, Faculty of Engineering, Universiti Teknologi Malaysia, Johor, Malaysia

S. Jeon
Department of Energy Systems Engineering, Seoul National University, Seoul, Korea

B. Jeon
Research Institute of Energy and Resources, Seoul National University, Seoul, Korea

ABSTRACT: Rock fragmentation in which the fragment size distribution of blasted rock material used in the quarry and mining industry as an index to estimate the productivity of blasting activity. The aim of this study is to investigate the effect of burden to hole diameter ratio on rock fragmentation induced by blasting. Numerical investigation was carried out by utilizing the Discrete Element Method (DEM) and Particle Blast Method (PBM) approaches in LS-DYNA software, on the response of rock fragmentation. In bench blasting simulation, the rock media was modelled using DEM and the blast loading was modelled using PBM. Parametric study was carried out on the hole diameter and burden was kept constant. The comparison of rock fragmentation distribution has been made between numerical analysis and fieldwork data. It was observed that uniformity index increases with the decreases in burden to hole diameter ratio. It can be concluded that by increasing hole diameter without changing burden will increase the powder factor which in turn yields finer fragmentation. The numerical model has successfully showed the crack propagation of rock which can form the rock fragments and helps to predict fragment size distribution of rock.

1 INTRODUCTION

Fragmentation is one of the most important aspects on explosives engineering. The mean fragment size of rock should not too high or too low. This is because oversize rock and very fines rock will occur. Rock fragmentation plays a critical role in large-scale quarrying operations because of its direct effects on the costs of drilling, blasting, secondary blasting and crushing (Dershowitz, 1993; Goodman & Shi, 1985; Faramarzi et al. 2013). The optimum-blasting pattern to excavate a quarry efficiently and economically can be determined based on the minimum production cost, which generally estimated according to rock fragmentation.

Factors influence the blast result divided into controllable and uncontrollable factors. Blast design parameters and explosive charge characteristics grouped under controllable parameters. Whereas, properties of rock mass, discontinuities and geomechanical characteristics of the intact rock are under uncontrollable factors. Controllable parameters for blast results was listed by Bakhtavar (2014) which are the burden, spacing, bench height, stemming length hole diameter, delay sequence, firing pattern, explosive per hole and powder factor.

Burden is the distance between rows of holes and the nearest free face. The influence of the effective burden on rock fragmentation is related to the mechanism of rock fracture. Furthermore, Persson (1975) shows that increasing size of hole diameter will decreasing the cost of the cost of drilling and blasting which because of hole volume increase so volume of explosive became larger (Wyllie & Mah, 2017).

Cunningham (1983) states that high values of n are preferred and mentioned that the parameters n increase when the burden to hole diameter ratio decreases. Ouchterlony (2003) also agreed with the statement and was summarized in his review paper.

In this study, the effect of burden to hole diameter on rock fragment size distribution is investigated using numerical simulation.

2 NUMERICAL METHOD

2.1 Bonded particle model

Potyondy & Cundall (2004) mentioned that rock behaves like a cemented granular material of complex-shaped grains which the grains and the cement can deform and break. The mechanical behavior of rock is driven by the formation, growth and existing of microcracks. The mechanical properties of rock are determined by its constituent particles and its structure (Hallbauer et al. 1973; Jaeger et al. 2007).

Bonded particle model (BPM) was defined by Potyondy and Cundall as a model consist of a dense packing of particles either in non-uniform-sized circular or in spherical which the particles are bonded together at their contact points with parallel bond (Potyondy & Cundall, 2004). The mechanical behavior can simulate by DEM using commercialized programs PFC2D, PFC3D (Itasca, 2003) and LS-DYNA (Livermore Software Technology Corporation (LSTC), 2012). Bonded particle model can provide both a scientific tool to investigate micro-mechanisms and an engineering tool to predict macroscopic behaviour. The DEM was introduced by Cundall for analysis of rock mechanics problems and then applied to soils by (Cundall & Strack, 1979).

The rigid particles interact only at the soft contact which finite normal and shear stiffness. Force and moment acting at each contact bond between particles. Bonded particles in BPM where all of the particles are linked to their neighboring particles through bonds. Bonds represent the complete mechanical behavior of solid mechanics and bonds are independent of the DEM. Every bond between particles is subjected to tension, bending, shearing and twisting (Karajan et al. 2013).

2.2 Particle blast method

Continuum-based Eulerian approach is one of the most accurate technologies to stimulate blast loading. However, this approach has several difficulties in modelling of blast loading. One of the disadvantages is advection error relative to Lagrangian simulations. Both momentum and kinetic energy is not conserved at the same time when advection is used. Next, greater computational effort is needed over Lagrangian simulations due to the advection and geometrical complexities are hard to handle with continuum-based Eulerian approaches (Teng, 2016).

A corpuscular particle method (CPM) has been proposed for airbag deployment application to overcome those difficulties. This method is a coarse grained method based on kinetic molecular theory (KMT) which the study of gas molecules and their interaction that lead to ideal gas law. The CPM considers the effect of transient gas dynamics and thermodynamics by using a particle to represent a group of gas molecules. Each particle carries translational energy as well as spin energy (Teng, 2016).

Corpuscular particle method assumes that the system is always thermal equilibrium. This is a reasonable assumption for airbag simulation with moderate temperature and low pressure. However, for blast simulation where gas flow is extremely high, the assumption of thermal equilibrium is invalid (Teng & Wang, 2014).

A particle blast method (PBM) has been proposed to model the interaction between detonation products, air, and structure. This method improves corpuscular method to account for the thermally non-equilibrium behaviour. Furthermore, co-volume effects have been considered to represent gas behaviour at high temperature (Teng & Wang, 2014).

3 NUMERICAL MODELLING OF ROCK BLASTING

Calibration procedures have been done to represent the fieldwork granite rock. The aim of calibration is to meet elastic modulus and UCS tolerance of intact rock specimen form the laboratory test. The uniaxial compression strength, shear strength and uniaxial tensile strength were set to 90 MPa, 30 MPa and 20 MPa respectively. In discrete element modelling of geomaterials, there are microproperties that need to be specified. These properties cannot be measured in the laboratory test. Therefore, in order to calibrate the numerical model, some initial values for the microproperties must be examined to create model similar to the actual rock. Trial and error method are needed to calibrate the numerical model. Input parameters of calibrated models are shown in Table 1 below.

Bench blasting numerical simulation performed in this study using the commercial software LS-DYNA. A one-borehole model was constructed to model the field test. Figure 1(a) shows the model geometry and the sizes of rock mass. The bench height is 15 m and the borehole depth is 14.8 m. 2.75 m from the borehole depth is the length of stemming. The burden and spacing length are 3.05 m and 3.65 m. Parametric study was carried out on the hole diameter and burden was kept constant. The size of hole diameter used in this study are 76 mm, 89 mm, 102 mm, 115 mm and 127 mm. Bonded particle model combined with particle blast method were used to model the rock mass as shown in Figure 1(b).

Table 1. Input parameters of the calibrated model

Properties	Value
Particle radius/mm	100-125
Parallel-bond normal stiffness (PBN)/GPa	20.76
Parallel-bond shear stiffness (PBS)	0.25
Parallel-bond maximum normal stress (PBN_S)	0.03
Parallel-bond maximum shear stress (PBS_S)	0.02
Bond radius multiplier (SFA)	1.31
Numerical damping (ALPHA)	0.5
Maximum gap between two bonded spheres (MAXGAP)	-1.31
Normal damping coefficient (NDAMP)	0.7
Tangential damping coefficient (TDAMP)	0.01
Friction coefficient (Fric)	0.99
Rolling friction coefficient (FricR)	0.98
Normal spring constant (NormK)	0.1
Shear spring constant (ShearK)	0.4

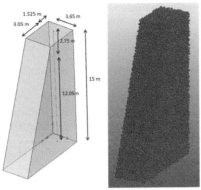

a) Geometry of bench blast model

b) Bench blast model using DEM/BPM

Figure 1. One borehole bench blast model

Table 2. Particle parameters of high explosive

Parameters	Value
Density, ρ (kg/m)	1200
Energy, E (GJ/m)	3.2
Detonation velocity, D (m/s)	4500
High Explosive fraction, γ	1.4
Co-volume, b	0.3

The numerical model to stimulate bench blasting is shown in Figure 1(b). The explosive was modelled by PBM using 50000 HE particles and the particle parameters of high explosive are listed in Table 2. The rock mass was modelled between 35019 to 37629 DEM vary with different hole size. The density of DEM is adjusted such that total mass of DEM equal to the total mass of rock.

4 RESULTS AND ANALYSIS

Numerical simulation was performed using LS-DYNA software and the running time duration taken to analyze for each model is between 24 to 42 hours. Figure 2 shows the rock fragmentation induced by blasting from section view of numerical simulation with varies size of hole diameter. It was observed that the larger the size of hole diameter tends to create more crack propagation thus more rock fragment will be produced. Small diameter holes with smaller burden and spacing produce smaller fragmentation. Reducing the hole diameter without changing burden and spacing will lowers the powder factor which in turn yields coarser fragmentation (Singh & H, 2012).

Digital image processing in fragmentation assessment techniques allows rapid and low cost of blast fragmentation size distribution (Siddiqui et al. 2009). There are several software namely Split

a) 76 mm hole diameter

b) 89 mm hole diameter

c) 102 mm hole diameter

d) 115 mm hole diameter

e) 127 mm hole diameter

Figure 2. Rock fragmentation from section view of numerical simulation with different size of hole diameter

Desktop, Wipfrag, GoldSize to obtain fragment size distribution. GoldSize software was used in this study to obtain fragment size distribution in Gemencheh Granite quarry.

Images of rock fragment in point cloud form were taken after blasting process. A total of 4 different images were used to obtain rock fragment distribution for fieldwork. Rock fragment from numerical simulation is shown in Figure 2 from one section view. A total of 4 different section views for each model were taken for analysis. To determine the size of rock particles, scaled object in the image were required as a reference. Then, the rock fragment from rock pile and particles that form a

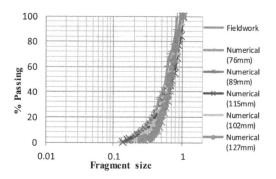

Figure 3. Comparison of rock fragment distribution of different size of hole diameter

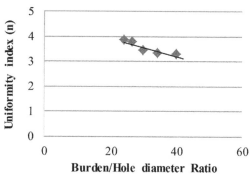

Figure 4. Burden to hole diameter ratio vs. Uniformity index

Table 3. Comparison of particle size distribution results

Burden (m)	Hole diameter (mm)	B/D Ratio	d_{50}	d_{80}	n
3.05	76	40.13	0.58	0.75	3.28
3.05	86	34.27	0.69	0.89	3.31
3.05	102	29.90	0.68	0.87	3.42
3.05	115	26.52	0.72	0.9	3.77
3.05	127	24.02	0.69	0.86	3.82

fragment need to be identify manually. The particle edge was outlined manually in a continuous line. Once the binary image was completely edited, computation of size distribution can be carried out. In this study, simple estimation was used which applied a fines correction formula to increase the amount of fine material at sizes below a specified threshold.

Figure 3 shows the comparison of rock fragment distributions of different hole size diameter. It can be seen that the curve trend for each hole size diameter was similar with fieldwork. Sudhakar (2005) suggested experimental equation (1) and also was used by Sereshki (2016), which the uniformity parameter (n) can be calculated and compared with the real value based on d_{50} (size at which the passing fraction is 50%) and d_{80} (size at which the passing fraction is 80%). Table 3 shows the values of d_{50} and d_{80} that were recorded from curve Figure 4 for fieldwork and numerical results. The uniformity parameter (n) was obtained from equation (1) below:

$$n = 0.842 \big/ \left(ln\ d_{80} - ln\ d_{50} \right) \qquad (1)$$

Figure 4. shows the graph of burden to hole diameter ratio versus uniformity index. It was observed that the lower the burden to hole diameter ratio will increase the uniformity index. Cunningham (1983) states that high values of n are preferred and mentioned that the parameters n increase when the burden to hole diameter ratio decreases.

5 CONCLUSION

Numerical simulation using combination of DEM-PBM was used in this study to simulate rock fragmentation induced by blasting. Rock fragment size distribution was obtained using image digital processing. Parametric study was carried out on the hole diameter and burden was kept constant. The comparison of rock fragmentation distribution has been made between numerical analysis and fieldwork data. It was observed that uniformity index increases with the decreases in burden to hole diameter ratio. It can be concluded that by increasing hole diameter without changing burden will increase powder factor which in turn yields finer fragmentation. The numerical model has successfully showed the crack propagation of rock which can form the rock fragments and helps to predict fragment size distribution of rock.

ACKNOWLEDGEMENT

This study is supported by Universiti Teknologi Malaysia and Ministry of Higher Education Malaysia under Research University Grant – Tier 1 (PY/2017/01396).

REFERENCES

Bakhtavar, E., Khoshrou, H. and Badroddin, M. 2014. 'Using dimensional-regression analysis to predict the mean particle size of fragmentation by blasting at the Sungun copper mine', *Arabian Journal of Geosciences*, 8(4), pp. 2111–2120.

Cundall, P. A. and Strack, O. D. L. 1979. 'A discrete numerical model for granular assemblies', *Géotechnique*, 29(1), pp. 47–65.

Cunningham 1983. 'The Kuz–Ram model for prediction of fragmentation from blasting.', in *In: Proceedings of the 1st international symposium on rock fragmentation by blasting. Lulea, Sweden*, p. 439–453.

Dershowitz, W. S. 1993. 'Geometric conceptual models for fractured rock masses: implications for groundwater flow and rock deformation', in *ISRM International*

Symposium - EUROCK 93, 21–24 June, Lisboa, Portugal. International Society for Rock Mechanics and Rock Engineering, pp. 71–81.

Faramarzi, F., Mansouri, H. and Ebrahimi Farsangi, M. a. 2013. 'A rock engineering systems based model to predict rock fragmentation by blasting', *International Journal of Rock Mechanics and Mining Sciences*. Elsevier, 60, pp. 82–94.

Goodman, R. E. and Shi, G. 1985. *Block theory and its application to rock engineering.*, Englewood Cliffs, N.J. : Prentice-Hall.

Hallbauer, D. K., Wagner, H. and Cook, N. G. W. 1973. 'Some observations concerning the microscopic and mechanical behaviour of quartzite specimens in stiff, triaxial compression tests', *International Journal of Rock Mechanics and Mining Sciences and*, 10(6), pp. 713–726.

Itasca 2003. 'PFC3D PFC2D User's Manual. Itasca Consulting Group Inc., Minneapolis.'

Jaeger, J. C., Cook, N. G. W. and Zimmerman, R. W. 2007. *Fundamentals of Rock Mechanics, 4th Edition, Blackwell Publishing.*

Karajan, N. *et al.* 2013. 'Interaction Possibilities of Bonded and Loose Particles in LS-DYNA', in *9th European LS-DYNA Conference*, pp. 1–27.

Livermore Software Technology Corporation (LSTC) 2012. 'LS-DYNA keyword user manual version 971', I(August).

Ouchterlony, F. 2003. 'Influence of blasting on the size distribution and properties of muckpile fragments, a state-of-the-art review', *MinFo proj. P2000–10*, pp.29–32, 97–102.

Potyondy, D. O. and Cundall, P. A. 2004. 'A bonded-particle model for rock', *International Journal of Rock Mechanics and Mining Sciences*, 41(8 SPEC.ISS.), pp. 1329–1364.

Sereshki, F., Hoseini, S. M. and Ataei, M. 2016. 'Blast fragmentation analysis using image processing', *International Journal of Mining and Geo-Engineering Blast fragmentation analysis using image processing*, pp. 211–218.

Siddiqui, F. I., Shah, S. M. A. and Behan, M. Y. 2009. 'Measurement of Size Distribution of Blasted Rock Using Digital Image Processing', *Engineering Science*, 20(2), pp. 81–93.

Singh, S. P. and H, A. 2012. 'Investigation of blast design parameters to optimize fragmentation', in *Fragblast 10, International conference on fragmentation by blasting*.

Sudhakar, J., Adhikari, G. R. and Gupta, R. N. 2005. 'Comparison of Fragmentation Measurements by Photographic and Image Analysis Techniques', *Rock Mechanics and Rock Engineering*, 39(2), pp. 159–168.

Teng, H. 2016. 'Coupling of Particle Blast Method (PBM) with Discrete Element Method for buried mine blast simulation', *14th International LS-DYNA Users Conference*, pp. 1–12.

Teng, H. and Wang, J. 2014. 'Particle Blast Method (PBM) for the Simulation of Blast Loading', *13th International LS-DYNA Users Conference*, pp. 1–7.

Wyllie, D. C. and Mah, C. 2017. *Rock Slope Engineering, Civil and Mining*. 4th edn. London and New York: Taylor & Francis Group.

2019 Rock Dynamics Summit– Aydan et al. (eds)
© 2019 Taylor & Francis Group, London, ISBN 978-0-367-34783-3

Application of advanced tunnel blasting technique for reducing vibration and optimizing the excavation advanced rate

T. Inuzuka, K. Iwano & Y. Tezuka
Kajima Corporation, Tokyo, Japan

ABSTRACT: Blasting is a relatively economical and efficient excavation method compared to other methods such as mechanical excavation. However, when tunnel sites are located near residential areas, the charge weight is often reduced to limit the impacts of vibration and noise. This makes it difficult to maintain the scheduled excavation advance rate. In recent years, however, advanced blasting technologies such as computerized jumbos and high-accuracy electronic detonators have allowed better control over tunnel blasting vibrations. Additionally, new vibration measurement systems have been introduced that enable the simultaneous monitoring and analysis of blast vibration results. Here, we applied these advanced technologies in a restricted blasting zone to reduce the environmental impact and enable the excavation advance rate to be maintained.

1 INTRODUCTION

Blasting is rapid and economical, making it advantageous compared to other tunnel excavation methods. However, it can sometimes cause environmental problems such as vibrations, especially when excavation sites are close to residential areas. To mitigate vibrations in such areas, the amount of charge is commonly reduced by shortening the drill length (advance distance) or by suspending blasting at night. However, these precautions can reduce the advance rate, thus delaying the construction schedule.

2 STUDY SITE

This paper discusses a case study in which attempts were made to control vibrations in a residential area. Newly developed advanced techniques, such as computer jumbo, advanced electronic detonators, and real-time vibration measurement systems (Takaaki Inuzuka, 2018), could be used to modify the blasting

conditions such that an appropriate advance rate could be maintained despite vibration mitigation in blasting-restricted areas. In this case study, the blasting-restricted area covers ~300 m of tunnel, as shown in Figure 1. In this area, several residential houses are located along the tunnel alignment. Figure 2 shows the relation between the tunnel distance (TD) and the overburden. At TD = ~1450 m, several residential houses are located above the tunnel route with a small overburden of ~45 m.

Further, the overburden rock is quite hard, with elastic wave velocity of ~5.0 km/s. Together, these geological conditions might cause more environmental impact on the residential area than in a conventional situation. Therefore, to confirm the vibration propagation properties, a blasting test was carried out just before blasting operations in the blasting-restricted area.

3 BLASTING TEST

Figure 3 shows a bird's-eye view of the test area, and Tables 1–2 list the blasting test conditions. In Japan, blasting vibration propagation is conventionally estimated by the following equation proposed by the explosive distributer (Japan Explosive Society, 2002).

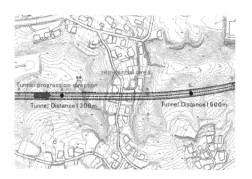

Figure 1. Tunnel site.

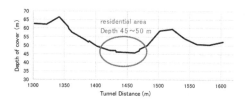

Figure 2. Relationship between TD and depth of cover.

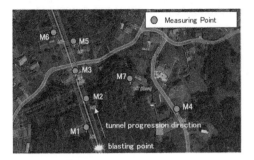

Figure 3. Measuring points.

Table 1. Blasting conditions (distance).

Measuring point	Distance (m)
M1	87
M2	90
M3	115
M4	155
M5	160
M6	174
M7	125

Table 2. Blasting conditions (holes, charge).

Test case	Number of holes	Charge amount	
		kg	kg/hole*
case 1	108	125.4	1.2
case 2	114	166.8	1.8
case 3	114	154.4	1.6
case 4	114	135.0	1.4

*MIC; maximum instantaneous charge

$$V = K \cdot W^{2/3} \cdot D^{-2} \qquad (1)$$

Here, K changes depending on the blasting conditions and rock properties, V is the vibration velocity (cm/s), W is the maximum instantaneous charge (MIC; kg/hole), and D is the distance of propagation (m).

Based on the blasting test results, K in Equation 1 was set to 2574 (Fig. 4, Table 3); this is much larger than the generally proposed value of 500–1000 (Table 4). Therefore, this tunnel site showed larger blasting vibrations than those at other sites.

The vibration effect on the blasting-restricted area was then considered. Figure 5 shows the relationship between the MIC and the vibration propagation distance. V values of 0.2 and 1.0 cm/s were used, and K was set to 2574. V = 0.2 cm/s is considered the threshold at which people feel vibrations, and

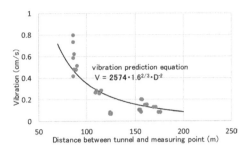

Figure 4. Test results.

Table 3. K-value (Test results).

Test case	Vibration (cm/s)							K
	M1	M2	M3	M4	M5	M6	M7	
case 1	0.6	0.5	0.3	0.1	0.2	0.1	0.1	2913
case 2	0.8	0.5	0.3	0.1	0.2	0.1	0.1	2200
case 3	0.7	0.5	0.3	0.1	0.2	0.1	0.1	2624
case 4	0.6	0.4	0.3	0.1	0.2	0.1	0.1	2559
Average								2574

Table 4. K values for different patterns (Japan Explosive Society, 2002).

Pattern	K
cut of tunnel blast	500–1000
perimeter of tunnel blast	200–500
open cast blasting	200–500

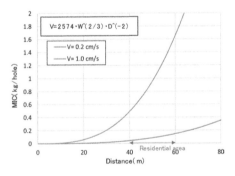

Figure 5. Relationship between distance and MIC.

V = 1.0 cm/s is considered the threshold at which building structures experience vibrations (Japan Explosive Society, 2002). Based on the blasting test results shown in Figures 4–5, only small MIC values should be used for blasting in the restricted area. However, this tunnel site needs to be in service as soon as possible, making it necessary to shorten the construction schedule.

4 CONSTRUCTION

Figure 6 shows the concept of the integrated construction cycle that was implemented at this site for controlling both the advance distance and the vibrations. Conventional vibration mitigation methods are based solely on the repetition of blasting and vibration measurements; an effective countermeasure is to simply reduce the MIC by shortening the advance distance. However, in this case study, as shown in the integrated construction cycle in Figure 6, drilling and blasting were modified using both a computer jumbo and an advanced electronic detonator. Furthermore, blasting vibration data were obtained instantly, and accordingly, blasting patterns were modified to mitigate vibrations under the appropriate advance rate. These processes involve the following steps:

(1) Drilling and blasting

A computer jumbo can not only provide accurate drilling alignment as designed but also record digital drilling data such as hole position and geological condition; these data are useful for optimizing the blasting patterns. Moreover, because the computer jumbo used at this site had four booms with high-performance drifters, the drilling time could be shortened even though the number of holes was increased due to the small MIC.

For blasting, the advanced electronic detonator has highly accurate delay time with error within 0.01% of the designed delay time. This enables the determination of the complete ignition order for each hole. The electronic detonator can also set different delay times for each hole, thus reducing the MIC.

(2) Vibration measurement

An advanced vibration measurement system was applied for vibration monitoring. This system has a remote monitoring function that interacts with a geographic information system (GIS) via a mobile phone,

and it can analyze the vibration data (Fig. 7). It also has a cloud database system; therefore, job site staff do not have to collect vibration data from the device. It also features both a solar panel and a battery; therefore, staff do not need to connect it to a power supply.

(3) Blasting pattern modification

The blasting conditions can be modified based on the data acquired in steps (1) and (2). As shown in Figure 8, vibration data obtained using the remote vibration measurement system can be synchronized with hole position data from the computer jumbo. A comparison between these data can then clarify the specific hole that is producing the strongest vibrations. In this case study, the blasting pattern was modified based on this comparison analysis. Then, the number of holes, burden, and MIC for each part of the drilling face, such as the cut hole and perimeter hole, were modified.

Figure 9 shows the relationship between vibrations and the number of holes at three monitoring points (M3, M4, and M5) and the TD in the blasting-restricted area. At the beginning of the

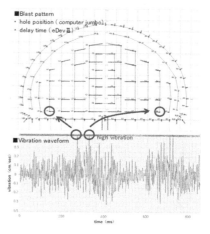

Figure 8. Modification of blast pattern.

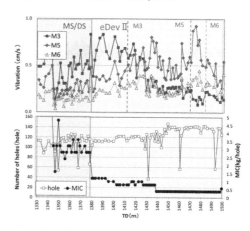

Figure 9. Vibration monitoring at M3, M5, and M6; number of holes, and MIC.

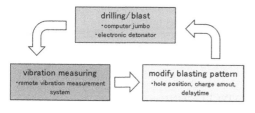

Figure 6. Control blast correction cycle.

Figure 7. Remote vibration measurement system.

restriction area, the MIC is reduced by a combination of two types of electric detonators, such as decisecond (DC) and millisecond (MS) detonators. As the vibrations increased up to monitoring point M3, the detonator was switched to the advanced electronic detonator that can ignite each hole separately. This electronic detonator performs blasting with quite short vibration durations, and its waveform has a high frequency (Keita Iwano, 2014). These vibrations tend to seem relatively small to the affected residents compared to those produced when using a conventional detonator for blasting. Therefore, even though a vibration increase was seen after switching to the advanced electronic detonator, there were no complaints from residents and excavation could be completed in the blasting-restricted area while maintaining an appropriate advance rate.

5 OPTIMUM DELAY TIME SETTING

As the MIC was reduced to its minimum value of 0.4 kg/hole at TD = ~1440 m, vibration increase was still seen at monitoring point M5 (Fig. 9). Therefore, an additional study was conducted to determine the optimum delay time to reduce vibrations. As shown in Figure 10, the produced waves are supposed to consist of a single overlapped waveform in each

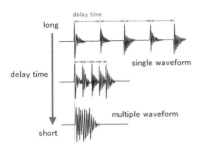

Figure 10. Single and multiple waveforms.

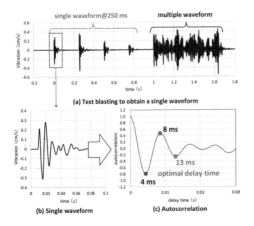

Figure 11. Determination of optimal delay time through autocorrelation of a single waveform.

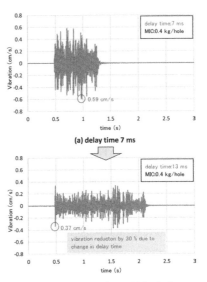

(a) delay time 7 ms

(b) delay time 13 ms (optimal delay time)

Figure 12. Example of vibration reduction through changing the delay time.

hole. Therefore, the optimum delay time to best cancel out the adjacent waveforms was determined. First, very large delay times were set to obtain the complete shape of a single waveform, and the autocorrelation of these single waveforms was estimated (Fig. 11) (Keita Iwano, 2018). The autocorrelation shows a similarity between the original waveform, $x(t)$, and the modified waveform, $x(t+\Delta t)$, that has a different time phase, Δt. The delay time that had the minimum autocorrelation was that for which adjacent waveforms were best canceled out, implying that the vibration of the produced wave was minimized. As shown in Figure 11, the current delay time of 7 ms at the site was found to be unfavorable for mitigating vibrations; the optimum delay time was seen to be 4 ms. As 4 ms is a quite small delay time, the delay time was instead set to 13 ms, for which the second-smallest autocorrelation was obtained. Changing the delay time to 13 ms reduced vibrations by 30% (Fig. 12).

6 CONCLUSION

The tunnel site featured in this case study needs to be in service as soon as possible. Therefore, adhering to the construction schedule was mandatory even in a blasting-restricted area. A blasting test preformed to characterize the restricted area revealed a low possibility of mitigating vibrations owing to the geological conditions, and more severe blasting conditions, such as low MIC and small drill length, were anticipated. Therefore, cutting-edge techniques such as a computer jumbo, an advanced electronic detonator, and a remote vibration measurement system were used in this area. Then, based on acquired digital data, the blasting conditions were modified. In addition, the optimum delay time to minimize vibrations was estimated, and blasting was continued with sufficient consideration of its environmental impact. Through the above analyses, an advance rate of 22 m per month was achieved even in the blasting-restricted area. This advance rate is comparable with those achieved in normal blasting areas.

The process and methods used in the present study to estimate the optimum blasting conditions would be quite effective for controlled blasting at any other tunnel site even though each tunnel has different excavation methods and geological conditions.

REFERENCES

Inuzuka, T., et al. 2018. Application of control blasting technique with electronic detonator and ICT. *Proc. of the 27th Tunnel Engineering Symposium*, I-46.
Japan Explosive Society. 2002. Blasting examples. *Eiko publishing*.
Iwano, K., et al. 2014. Environmentally low impact blasting with highly accurate electronic detonator. *Proc. of the 23rd Tunnel Engineering Symposium*, I-35.
Iwano, K., et al. 2018. Control blasting by information construction using computer jumbo and electronic detonator. *Proc. of the 73rd Annual Meeting of Japan Society of Civil Engineering*, 157–158.

2019 Rock Dynamics Summit– Aydan et al. (eds)
© 2019 Taylor & Francis Group, London, ISBN 978-0-367-34783-3

Fracture characterization and rock mass behavior induced by blasting and mechanical excavation of shafts in Horonobe Underground Research Laboratory

K. Aoyagi
Horonobe Underground Research Center, Japan Atomic Energy Agency, Japan

T. Tokiwa
Shinshu University, Japan

T. Sato, A. Hayano
Horonobe Underground Research Center, Japan Atomic Energy Agency, Japan

ABSTRACT: In high-level radioactive disposal projects, it is important to investigate the extent of the excavation-damaged zone (EDZ) for safety assessment because EDZ can provide a migration pathway for radionuclides from the facility. To investigate the quantitative differences between EDZs formed because of blasting and mechanical excavation, we studied the characteristics of fractures induced by excavation based on fracture mapping performed during shaft sinking (V- and E-Shafts). As a result, it was found that blasting excavation can lead to the formation of a large number of newly created fractures (EDZ fractures) compared with mechanical excavation. In addition, the seismic velocity (P-wave velocity) measured during blasting excavation (E-Shaft) was lower than that measured during mechanical excavation (V-Shaft). Furthermore, we found that the support pattern that reinforces forward rocks to be appropriate for limiting damage to the shaft wall.

1 INTRODUCTION

The construction of underground facilities induces fractures in the surrounding rock mass because of the resultant stress redistribution. This has particular implications for projects related to the disposal of high-level radioactive wastes (HLW), where fracture development creates an excavation-damaged zone (EDZ) that increases the permeability of the surrounding rock mass and potentially creates a pathway for the migration of radionuclides from storage facilities (Tsang et al. 2005). Therefore, in the construction of a repository for HLW disposal, it is important to understand EDZ development and the behavior of rock mass around the facility.

A HLW disposal repository is spread over a few square kilometers, and the associated tunneling length is around 300 km (JNC, 2000). Therefore, it is necessary to apply an efficient excavation method to reduce the time required for and the monetary cost of excavation.

To this end, blasting excavation is applied in various tunneling projects. However, it has been reported that the extent of EDZ in blasting excavation is considerably larger than that in mechanical excavation by using a roadheader or a tunnel-boring machine (e.g., Sato et al. 2000; Siren et al. 2015; Tokiwa et al. 2018). This is one of the disadvantages of blasting excavation in the construction of a repository for HLW disposal. Therefore, it is important to investigate an efficient excavation method that limits its EDZ development.

With this objective, we investigated the quantitative difference in the EDZs formed because of blasting and mechanical excavation in Horonobe Underground Research Laboratory located in Hokkaido, Japan. Furthermore, we investigated the efficiency of excavation and appropriate support patterns to identify an efficient excavation method that does not enlarge the EDZ.

2 HORONOBE URL

Figure 1 shows the layout of the Horonobe URL. The ventilation shaft (V-Shaft) and the east access shaft (E-Shaft) were excavated to 380 m below ground level, and the west access shaft (W-Shaft) was excavated to 365 m by the end of 2018. Excavations of horizontal galleries were also completed at depths of 140 m, 250 m, and 350 m.

The targets of this study are the V-Shaft and the E-Shaft. The target depth is 250–350 m. The rocks in this interval consist mainly of the siliceous mudstone. The representative properties of siliceous mudstone are indicated in Table 1 (Aoyagi and Ishii, 2018).

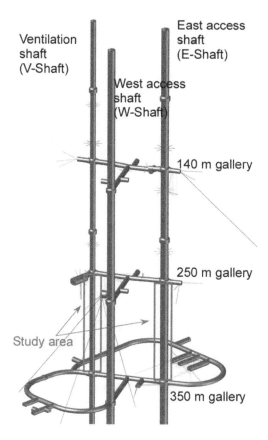

Figure 1. Layout of Horonobe URL.

Figure 2. Photograph of roadheader in V-Shaft.

Table 2. Major specifications of excavation equipment (roadheader).

Engine	49 kW (400 V)
Revolution speed	36rpm
Size of the cutter head	φ830 × 600 mm
Excavatavility	3.5 - 5.6 m^3/h
Weight	about 5 t
capacity of muck bucket	1.5 m^3

Table 1. Representative properties of siliceous mudstone (Aoyagi and Ishii, 2018).

Properties	Average value
Uniaxial compressive strength (MPa)	15.4
Elastic modulus (GPa)	1.82
Poisson's ratio	0.17
P-wave velocity (km/s)	2.08
Effective porosity (%)	41.6
Saturated density (g/cm^3)	1.84

The V-Shaft was excavated mechanically by using an excavator called roadheader (Fig. 2), and its diameter is 4.5 m. The specifications of the excavation equipment are listed in Table 2. By contrast, the E-Shaft was excavated by blasting, and its diameter is 6.5 m. The amount of charge was determined based on guidelines for blasting engineering (Japan Explosives Society, 2001). The average amount of charge was 0.24 g/m^3.

The shafts were excavated by following a four-step cyclical procedure. Figure 3 summarizes the construction procedure. Note that the excavation face was approximately 1 m below the lower edge of a concrete lining (Fig. 3a). First, the excavation proceeded in 1-m intervals. Then, steel arch ribs and rock bolts were installed after every 1 m of advance in excavation (Figs. 3b and 3c). After every 2 m of excavation, fracture mapping and seismic velocity measurement were conducted on the exposed shaft wall of the V- and E-Shafts (Inagaki et al. 2013; Inagaki, 2014).

After the excavation proceeded 2 m, the concrete lining was installed (Fig. 3d). The number of rock bolts and the span of the installed concrete lining were changed depending on geological conditions (Tsusaka et al. 2012; 2013).

3 IN SITU SURVEYS

3.1 Fracture observation and analysis

Detailed fracture observations during excavation of the V- and E-Shafts were carried out by Tokiwa et al. (2018). Therefore, we used the fracture data obtained by Tokiwa et al. (2018) in the present study. Fractures are observed frequently on the shaft wall, and these fractures can be divided into two main types, as shown in Figure 4 (Tokiwa et al. 2018). The first type is characterized with shear evidence. A slickenside can be recognized on the fracture planes, and the planes are commonly associated with slickenlines and/or slickensteps. Thus, first type of fractures is shear fracture. The second type is characterized by plumose structures on the fracture plane, and therefore, the second type of fracture is extension fracture. Based on studies in the literature (Tokiwa et al

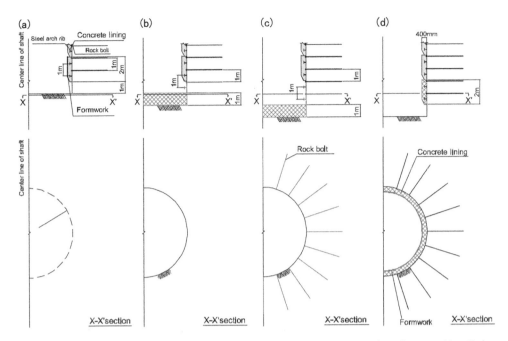

Figure 3. Procedure of shaft construction. (a) Initial state, (b) first one-meter excavation advance and installation of steel arch rib and rock bolts, (c) second one-meter advance and installation of steel arch rib and rock bolts, (d) installation of concrete lining.

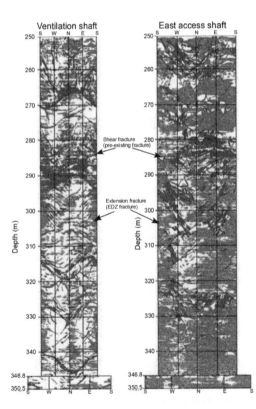

Figure 4. Fracture mapping of V- and E-Shafts.

2018), we assumed that shear fractures are natural pre-existing fractures, and extension fractures are induced as a result of stress redistribution during excavation, that is, EDZ fractures.

As a statistical characteristic of fractures, we analyzed the number of fractures (fracture frequency) of each type based on the result of fracture mapping (Fig. 4). The differences in damage induced on the wall in both shafts were discussed based on the frequency of EDZ fractures.

3.2 *Seismic velocity measurement*

In a previous study, it was reported that seismic velocity (P-wave velocity) decreases as fracture density increases (Aoyagi et al. 2014). Thus, we analyzed the results of seismic velocity measurements on the surface of theshaft wall to discuss the damage induced on the shaft wall.

Figure 5 shows a photograph of the measurement system (McSEIS-3, OYO Corporation) and an illustration of the measurement. Seismic velocity was measured along the horizontal and vertical directions. The measurement length was 1.5 m. As a seismic source, direct shot to the rock was made using a hammer. The measurement was performed five times, and average value was used for comparison with the velocity measured in each shaft.

As a result of fracture observation, the orientation of EDZ fractures was found to be almost parallel

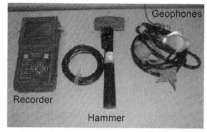

(a) Photograph of seismic velocity measurement system (McSEIS-3).

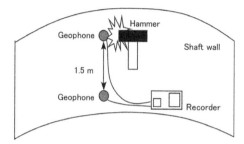

(b) Illustration of seismic velocity measurement.

Figure 5. Seismic velocity measurement on shaft wall.

to the respective shaft walls (Tokiwa et al. 2018). Thus, the seismic velocity measured along the vertical direction is effective for detecting the effects of the damage, that is, it is a suitable parameter for discussing the degree of damage to the wall.

4 RESULTS OF IN SITU SURVEYS

4.1 *Fracture analysis*

Figure 6 shows the total number of pre-existing and EDZ fractures contacting the north, east, south, and west scan lines. From this figure, it can be found easily that the EDZ fractures developed in the E-shaft are larger than those developed in the V-Shaft. In fact, the total numbers of pre-existing fractures were 561 in the V-Shaft and 693 in the E-Shaft, and the difference was only 1.24 times. By contrast, the total number of EDZ fractures was 660 in the V-Shaft and 1,204 in the E-Shaft, and the number of EDZ fractures in the E-Shaft was 2.38 times higher than that in the V-Shaft.

From the analysis result of fractures observed on the wall, it was quantitatively ascertained that blasting excavation (E-Shaft) induces a large number of EDZ fractures on the shaft wall compared with mechanical excavation (V-Shaft).

4.2 *Seismic velocity measurement*

Figure 7 shows the measured seismic velocity in the V- and E-Shafts along the depth direction. In addition, the results of velocity logging performed in the

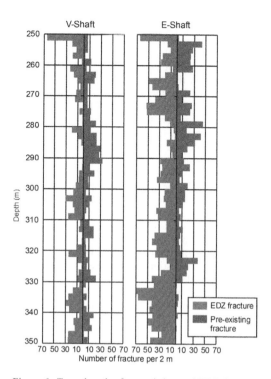

Figure 6. Trace length of pre-existing and EDZ fractures measured in V- and E-shafts.

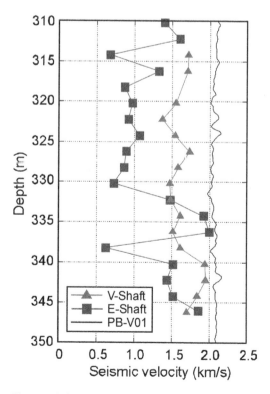

Figure 7. Seismic velocity measured on walls of shafts and borehole drilled before excavation of URL (PB-V01).

borehole drilled near the V-Shaft (borehole PB-V01) are shown in the graph to indicate the seismic velocity of the intact rock (Funaki et al. 2010). From the graph, the seismic velocities measured in both shafts are lower than that measured in borehole PB-V01. It is estimated that the stress redistribution during excavation can damage the wall. Then, the seismic velocity on the wall of the shaft decreases relative to the value measured for intact rock.

Furthermore, we compared the trends of seismic velocity measured in each shaft. The seismic velocity measured in the E-Shaft was lower than that measured in the V-Shaft, except for a few high value measured at the depths of 334 m and 336 m in the E-shaft. The average seismic velocity was 1.64 km/s in the V-shaft and 1.25 km/s in the E-shaft. Considering the relationship between seismic velocity and fracture density described by Aoyagi et al. (2014), it is estimated that blasting excavation induced considerably higher levels of damage than did mechanical excavation. This trend is consistent with the result of fracture observation.

5 DISCUSSION

As a parameter for discussion of excavation efficiency, we calculated the excavated volume per hour (excavation rate) based on construction work records.

Figure 8 shows the excavation rate as a function of excavation depth. The average excavation rates were 2.75 m³/h and 5.16 m³/h in the V-Shaft and the E-Shaft, respectively. That is, the excavation rate of blasting excavation (E-Shaft) was about 1.87 times higher than that of mechanical excavation (V-Shaft). Thus, blasting excavation is more efficient than mechanical excavation.

As described in chapter 4, blasting excavation can induce a greater number of EDZ fractures than mechanical excavation. This, as opposed to excavation efficiency, is one of the disadvantages from the viewpoint of its use in HLW disposal projects. However, it is estimated that the support pattern can affect the degree of damage on the shaft wall. We investigated the damage development depending on the support pattern by using the result of seismic velocity measurements.

Figures 9a to 9c shows the three support patterns applied at depths of 310–346 m. In the construction of these shafts, a few support options were designed to control excessive spalling of the shaft wall to prevent failure of the concrete lining (Tsusaka et al. 2012; 2013). At depths of 332–346 m, rockbolts were installed downward after every meter of advance in excavation to reinforce the forward rock, as shown in Figure 9c. In that section, the average seismic velocity was found to be 1.55 km/s. By contrast, the average seismic velocity at depths of 310–322 m was 1.07 km/s. The seismic velocity in the section in which the support pattern shown in Figure 9c was

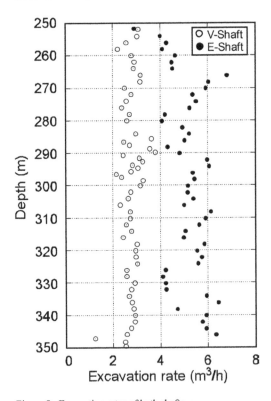

Figure 8. Excavation rates of both shafts.

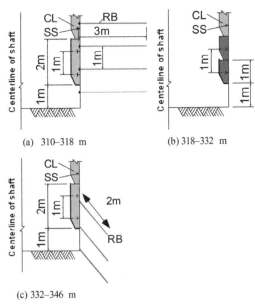

(a) 310–318 m

(b) 318–332 m

(c) 332–346 m

Figure 9. Support pattern applied at depths of 310 – 350 m of the East access shaft (cross-sectional view). RB denotes rock bolt, CL denotes concrete lining, and SS denotes steel ribs (after Tsusaka et al. 2013).

applied was higher than that in the section in which the support pattern shown in Figures 9a and 9b was applied. In addition, the difference in seismic velocity between both shafts shows a decreasing trend. Therefore, the effect of reinforcement of forward rock was validated from the measured seismic velocity. In addition, the support pattern shown in Figure 9c is estimated to be appropriate for limiting damage to the shaft wall.

In the present study, we investigated the development of damage only on the shaft wall. However, as described in chapter 1, EDZ can enhance the permeability of rock and thus create critical migration pathways for radionuclides. Therefore, it is necessary to investigate the extent of EDZ fracture development into the wall and hydraulic conductivity in the EDZ by applying techniques such as tomography and hydraulic testing to discuss the hydro-mechanical behavior of the EDZ in blasting and mechanical excavations.

In addition, the sizes of the two shafts considered herein are different, that is, the diameters of the V- and the E-Shafts are 4.5 m and 6.5 m, respectively. Thus, it is necessary to investigate failure behavior considering the scale effect of the shaft. We will investigate these topics in the future.

6 CONCLUSIONS

In this study, we investigated the quantitative differences between the EDZs formed because of blasting and mechanical excavations in Horonobe URL based on the result of fracture analysis and measured seismic velocity on the wall. Furthermore, we investigated the efficiency of excavation and appropriate support patterns to identify an efficient excavation method from the viewpoint of limiting the extent of EDZ.

The excavation rate of blasting excavation (E-Shaft) was found to be approximately 1.87 times higher than that of mechanical excavation (V-Shaft). However, the number of the EDZ fractures formed because of blasting excavation was approximately 2.38 times higher than those formed because of mechanical excavation.

The relationship between seismic velocity and support pattern in blasting excavation (E-Shaft) revealed that the support pattern that reinforces the forward rock, for example, installation of rockbolts downward, is appropriate for limiting damage to the shaft wall.

In this study, we could not determine the extent of EDZ because we did not employ a survey technique, such as seismic tomography, hydraulic testing, or rock core observation. In the future, we will investigate the extent and hydraulic conductivity of the EDZ to discuss the relationship between the EDZ and excavation efficiency.

REFERENCES

Aoyagi, K., Tsusaka, K., Kondo, K. & Inagaki, D. 2014. Quantitative assessment of an Excavation Damaged Zone from variations in seismic velocity and fracture distribution around a gallery in the Horonobe Underground Research Laboratory. *Rock Engineering and Rock Mechanics: Structures in and on Rock Masses.* 487–492.

Aoyagi, K. & Ishii, E. 2018. A method for estimating highest potential hydraulic conductivity in the excavation damaged zone in mudstone. *Rock Mech Rock Eng.* https://doi.org/10.1007/s00603-018-1577-z.

Funaki, Y. et al. 2010. Horonobe Underground Research Laboratory Project: The results of the pilot borehole investigation of the ventilation shaft (PB-V01) –. Geophysical loggings -. *Tech Rep JAEA-Research 2010-002* (in Japanese with English abstract).

Inagaki, D., Tokiwa, T., Murakami, H. 2013. Collection of URL Measurement Data in 2011 Fiscal Year at the Horonobe Underground Research Project. *Tech. Rep JAEA-Data/Code 2012-029* (in Japanese with English abstract).

Inagaki, D. 2014. Collection of URL Measurement Data in 2012 Fiscal Year at the Horonobe Underground Research Project. *Tech. Rep JAEA-Data/Code 2013-022* (in Japanese with English abstract).

Japan Explosives Society (ed). 2001. *Handbook of Blasting Engineering for Field Engineer.* Tokyo: Kyoritsu Shuppan. (in Japanese)

Japan Nuclear Cycle Development Institute. 2000. H12: Project to Establish the Scientific and Technical Basis for HLW Disposal in Japan – Project Overview Report -. *Tech. Rep. JNC TN1410* 2000–2001.

Ota, K., Abe, H. & Kunimaru, T. 2010. Horonobe Underground Research Laboratory Project Synthesis of Phase I Investigations 2001–2005 Volume "Geoscientific Research". *Tech rep. JAEA-Research* 2010–2068.

Sato, T., Kikuchi, T., & Sugihara, K. 2000. In-situ experiments on an excavation disturbed zone induced by mechanical excavation in Neogene sedimentary rock at Tono mine, central Japan. *Eng Geol.* 56: 97–108.

Siren, T., Kantia, P. & Rinne, M. 2015. Considerations and observations of stress-induced and construction-induced excavation damage zone in crystalline rock. *Int J Rock Mech Min Sci,* 73: 165–174.

Tokiwa, T., Tsusaka, K. & Aoyagi, K. 2018. Fracture Characterization and Rock Mass Damage Induced by Different Excavation Methods in the Horonobe URL of Japan. *Int J Civil Eng.* 16: 371–381.

Tsang, C., Bernier, F. & Davies, C. 2005. Geohydromechanical processes in the Excavation Damaged Zone in crystalline rock, rock salt, and indurated and plastic clays in the context of radioactive waste disposal. *Int. J. Rock Mech. Min. Sci.* 42: 109–125.

Tsusaka, K., Inagaki, D., Tokiwa, T., Yokota, H., Nago, M., Matsubara, M., Shigehiro, M. 2012. An Observational Construction Management in the Horonobe Underground Research Laboratory Project. *Proc. World Tunnel Congress* 2012.

Tsusaka, K., Inagaki, D., Nago, M., Kamemura, K., Matsubara, M. & Shigehiro, M. 2013. Relationship between rock mass properties and damage of a concrete lining during shaft sinking in the Horonobe Underground Research Laboratory Project. *Proc. World Tunnel Congress 2013.* 2014–2021.

2019 Rock Dynamics Summit– Aydan et al. (eds)
© 2019 Taylor & Francis Group, London, ISBN 978-0-367-34783-3

Consideration on setting of detonation time interval of control blast

Hiroaki Takamura, Yoshiharu Tanaka
Nishimatsu Construction Co. Ltd

Satoki Nakamura, Moriaki Arimitsu
Kayaku Japan Co., Ltd

ABSTRACT: In NATM tunnel construction, Noise, low frequency sound, vibration propagates to the surroundings by blasting excavation. Although these are instantaneous phenomena of several seconds a day, they have a large influence on neighboring houses, therefore, countermeasures are often used. In the problem of blasting vibrations, countermeasures are taken by control blasting to change that increase the number of detonations, but it is known that a high reduction effect can be exerted particularly by using an electronic delay detonator. Normally excavation is carried out by detonation of about 100 holes per 10 steps using an electric detonator. In the case of using an electronic delay detonator excellent in accuracy of control of time intervals, since it detonates in one hole per step, the influence of ground vibration can be suppressed by reducing the amount of explosive per step. In recent years, electronic delay detonators that can set and change the explosion time interval have been developed according to site geology such as ground geology and rock quality of vibration propagation routes. In this report, test blasting using an electronic delay detonator capable of changing the detonation time interval at the site was carried out, and the effect and control blasting utilization method were examined. From the results of test blasting, it has been shown that more effective vibration reduction is possible by setting the detonation time interval of the electronic detonator according to the ground surface hardness in the vicinity of the maintenance target and the geological condition of the propagation route. Furthermore, in order to improve the vibration reduction effect, we grasp that it is important to grasp the dominant frequency in the maintenance target and the propagation route. Therefore, we examined the method of grasping the dominant frequency and discussed the method of determining the detonation time interval and the method of determining the optimum range of application of control blast.

1 INTRODUCTION

Tunnel construction by drilling and blasting based on the New Australian Tunneling Method (NATM) generates instantaneous noise, low frequency sound, and repeated vibrations that can last for several seconds. Although this is instantaneous and occurs only several times per day, a significant sonic influence will be exerted on the neighboring inhabitants. Therefore, countermeasures are considered in several scenarios (Japan Explosives Industry Association 2002).

The blasting vibration can be reduced using controlled blasting based on an appropriate detonation method. For instance, if hole-by-hole detonation is possible, the amount of explosives that are required per blast can be reduced through the application of an electronic detonator (Tanaka 1992) over approximately ten precise detonation intervals. Recently, electronic detonators that can set and change the on-site detonation interval (Iwano 2014) have been developed, enabling the duration of blasting to be easily altered in response to the site conditions.

In this study, we assess the usage of controlled blasting based on the analysis of the results of test blasting in which an electronic detonator was used to alter the on-site detonation interval.

2 TEST BLASTING USING ELELCRONIC DELAY DETONATORS

2.1 *Site situation*

Test blasting was conducted during tunnel construction in the Fukuoka Prefecture. Although the sediment-like decomposed granite soil and severely weathered granite are distributed near the mouth of the tunnel, fresh granite (CM to CH class) is observed to fill the inner part of the tunnel. This tunnel exhibits several cracks with weathering in some parts; however, hard rocks that do not require steel support account for approximately 85% of the total volume. The tunnel face that was used for performing test blasting comprised fresh and hard granite and appeared to be tightly closed with only small cracks.

2.2 *Measurement conditions*

The vibration velocity was measured at the two locations (A and B) that are depicted in Fig. 1. Test blasting was conducted in the standard CI-pattern section depicted in Fig. 2 and involved the detonation of an electric detonator that was commonly used in the

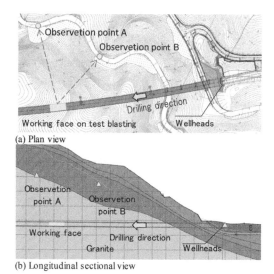

(a) Plan view

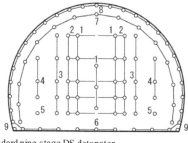

(b) Longitudinal sectional view

Figure 1. Vibration measurement positions

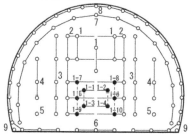

a) Standard nine-stage DS detonator

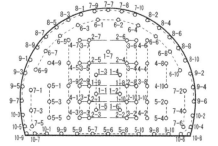

b) Center-cut one-shot waveform measurement: ten-stage with electronic detonator + nine-stage with DS detonator

c) Hole-by-hole detonation using an electronic detonator

Figure 2. Blast pattern diagram

field (hereafter referred to as the nine-stage DS detonator), single-shot detonations for evaluating the resulting waveforms, and hole-by-hole detonation using an electronic detonator. The detonations were applied in the pattern and positional relations that are depicted in Tables 1 and 2, respectively. As will be detailed in Section 2.4, the detonation interval of the electronic detonator was determined based on the vibration prediction results that were obtained from the single-shot detonation waveform.

2.3 Single-shot detonation test results

The single-shot detonations were conducted in four stages (0.6kg/hole) in an auxiliary center cut that was followed by detonations that were conducted in six stages (1.0kg/hole) in a deep center cut at 300ms intervals. Further, the vibration velocity waveforms that are obtained at observation points A and B are depicted in Fig. 3. The fourth stage vibrations are not recorded in Fig. 3 owing to the cutoff at which rocks were blown off around the blast holes. Detailed waveforms with large velocity amplitudes and frequency characteristics that are produced by the fifth and seventh stages are depicted in Figs. 4 and 5, respectively.

Although the velocity amplitudes that are received from the center-cut single shots vary significantly, the waveform properties at each measurement point are observed to be considerably consistent; further, the frequency characteristics also follow a similar trend. This demonstrated that the coincident measurement points and propagation paths resulted in high reproducibility of the waveform even though the velocity amplitudes that were produced at each

Table 1. Relation between the detonator type and charge dosage

Standard DS electric detonator			Electronic detonator								
			Center cut		32-ms intervals		8-ms intervals		20-ms intervals		13-ms intervals
Number of detonators	Number of holes	Amount of charge (kg)	Number of holes	Amount of charge (kg)	Number of holes	Amount of charge (kg)	Number of holes	Amount of charge (kg)	Number of holes	Amount of charge (kg)	Number of holes	Amount of charge (kg)
DS1	13	13	4	2.4	10	8.8	9	8.4	10	8.8	9	7.8
DS2	10	10	6	6	10	12	10	10.4	10	10	10	10
DS3	10	10			10	10	10	10	10	10	10	10
DS4	7	5.6			10	10	10	10	9	9	10	10
DS5	1	0.8			10	10	10	10	10	10	10	10
DS6	4	3.2			10	10	10	10	10	10	10	10
DS7	17	13.6			10	10	10	10	10	10	10	10
DS8	31	29.8			10	10	10	10	10	8.6	10	9.8
DS9	2	2			10	10	10	10	10	10	10	10
					10	10	10	10	10	10	10	10
Total	95	88	10	8.4	100	100.8	99	98.8	99	96.4	99	97.6

Table 2. Distance between the observation point and the working face

Blasting method	Tunnel distance (m)	Distance of observation point A	Distance of observation point B
Standard DS electric detonator	211.4	126.2	94.0
Center cut by the electronic detonator	218.9	124.9	98.8
Electronic detonator; 32ms, 8ms interval	224.9	124.1	102.9
Electronic detonator; 20ms, 13ms interval	232.4	123.5	108.3

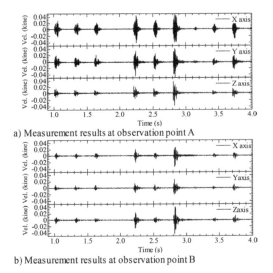

a) Measurement results at observation point A

b) Measurement results at observation point B

Figure 3. A time-series single-shot blasting waveform with ten-stage center cut

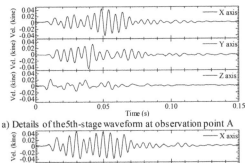

a) Details of the 5th-stage waveform at observation point A

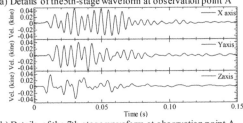

b) Details of the 7th-stage waveform at observation point A

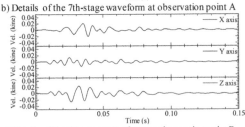

c) Details of the 5th-stage waveform at observation point B

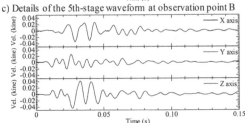

d) Details of the 7th-stage waveform at observation point B

Figure 4. A time-series single-shot blasting waveform

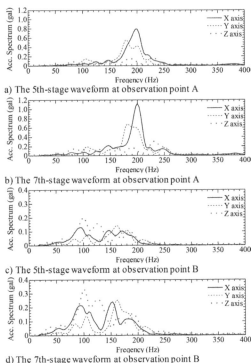

a) The 5th-stage waveform at observation point A

b) The 7th-stage waveform at observation point A

c) The 5th-stage waveform at observation point B

d) The 7th-stage waveform at observation point B

Figure 5. Frequency characteristics of the single-shot waveform

stage changed significantly de-pending on the rock conditions, such as hardness, cracking, and distance from the free surface.

2.4 Prediction method using the result of single-shot blasting measurement

Based on the results that were obtained from the consecutive blasting of cross sections in the single-shot tests, vibrations were predicted by superimposing the blasting waveforms that were measured in arbitrary time intervals over the entire course of the blasting.

For example, to predict the result of a case in which the fifth-stage blasting waveform obtained at observation point A was produced over 100 stages of blasting based on time intervals of 20ms, a time history waveform was created by overlaying the measured single-speed velocity waveforms over 100 stages with a delay of 20ms. Acceleration was further obtained by differentiating the resulting velocity waveform with respect to time, and, finally, a sensory correction based on the JIS (Japan Industrial Standard 2014) was added to obtain the predicted vibration level.

Figure 6 depicts the results of the predictive effort at each time interval. In the prediction process, the maximum value of the velocity, which is assumed to be a composite vector, and the maximum vibrational level in the vertical direction are extracted and organized over each time interval.

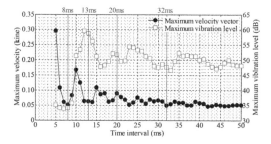

Figure 6. Prediction results for the 5th-stage waveform at observation point A

It can be observed from Fig. 6 that the maximum velocity and vibration level change significantly with the time interval and that the relations between the magnitudes of these factors are inconsistent because the frequency characteristics of the vibrational level are rectified based on the human sensory characteristics (sensory rectification).

This method can result in errors in the prediction of actual blasting results owing to the vibrational amplitude differences between the blast holes based on bedrock and other characteristics as well as the phase differences (differences between the transmission time) that are caused by the separation (propagation distance) between the working face and the maintenance target; the separation is observed to be a maximum of approximately 10m depending on the position of the blast hole. These prediction errors are described in the section related to the measurement results.

Based on the results of Fig. 6 and by focusing on the time intervals, we examined the following cases: (1) when the maximum velocity decreased over time intervals of 32ms, which was close to the time interval (30ms) of a conventional electronic detonator; (2) when the interval was as small as 8ms, which was when the maximum velocity decreased; and (3) 13ms or 20ms intervals in which the vibration level was characteristically predicted to be either small or large, respectively.

2.5 Results of test blasting

The velocity waveform that is measured at observation point A is depicted in Fig. 7. Further, Fig. 8 depicts the same result that is converted to acceleration and the undergoing frequency analysis. This figure also depicts the second-round result of the nine-stage DS detonator in the upper row and the result of the electronic detonator with an interval of 20ms in the lower row. Furthermore, the plot in Fig. 9 compares the measurement results of test blasting with the predicted results for the time interval in Section 2.4 (Fig. 6).

The blasting of a DS detonator that exhibits a high vibrating velocity in the center cut, as depicted in Fig. 7, when compared with the hole-by-hole

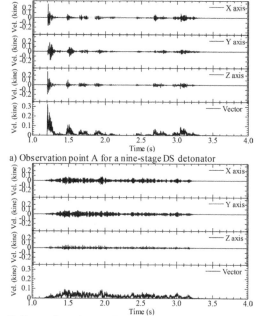

a) Observation point A for a nine-stage DS detonator

b) Observation point A at 20ms interval of electronic detonator

Figure 7. Time series of the vibration velocity waveform

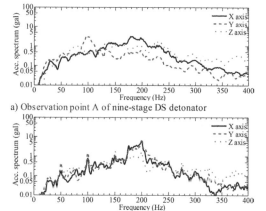

a) Observation point A of nine-stage DS detonator

b) Observation point A at 20ms interval of electronic detonator

Figure 8. Frequency characteristics

deto-nation of an electronic detonator, including the effect of the decreased minimum resistance line by the auxiliary center cut, can considerably reduce the vibration. The frequency characteristics in Fig. 8 exhibit that the maximum values are generated at n times the fundamental frequency of 50Hz (50, 100, 150Hz), corresponding to a time interval of 20ms as described in the section related to the prediction methods. Further, it also exhibits that the components possessing a frequency of less than 50Hz, which human body feels easy, considerably influence the vibration level and are small as compared to the DS detonators. Furthermore, Fig. 9 depicts

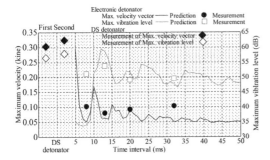

Figure 9. Comparison between the prediction and measurement for the fifth stage waveform at observation point A

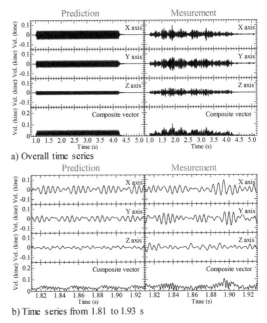

a) Overall time series

b) Time series from 1.81 to 1.93 s

Figure 10. Comparison between the prediction and measurement of electronic detonator at a 32ms interval

that although there is a prediction error, it changes by up to twice at the maximum velocity and at approximately 5dB on the vibration level depending on the time interval of the electronic detonator. This confirms that it is necessary to set a detonation time interval depending on the conditions of the site.

Here, as previously mentioned, the prediction error can be attributed to the phase difference (difference in the transmission time) between blast holes, which is caused by the diffllerence in both the vibration amplitude and the separation distance (propagation distance) between the working face and maintenance target. The observation point A is located adjacent to the working face, and the propagation distance exhibits a difference of approximately 10m, which corresponds to the width of the tunnel. Assuming that the elastic wave velocity is 3.0km/s, the transmission duration will deviate by approximately 3ms (0.003 s) when the difference in distance becomes 10m. Therefore, a phase difference of 1–3ms is always observed depending on the position of the blast hole, and the prediction error, which is dependent on the difference in the propagation distance, becomes particularly large if the detonation time interval is set to be short. To verify the influence of the phase difference, Fig. 10 compares the predicted and measured results of velocity waveforms for the detonation at 32ms intervals. From the detailed diagram depicted in Fig. 10(b), although the waveform properties are similar, it can be confirmed that the distance between the sections at which various waveforms overlap varies and that the variation in vibration amplitude leads to prediction error.

3 CONTROLLED BLASTING WITH ELECTRONIC DETONATORS

3.1 Points to be noted while setting the time interval

To optimally predict the time interval for controlled blasting and while considering the soil properties of a maintenance target, it is necessary to grasp the maintenance objective and the dominant frequency of the maintenance target or propagation path. The maintenance objectives should be controlled by speed if the maintenance target is a structure, by the vibration level if it is a resident, and by displacement if it is a factory with precision equipment, including electron microscopes; thus, different setting methods should be used to maximize the reduction effect. Furthermore, because the vibration waveform of a single-shot explosion is highly reproducible, it is possible to reduce physical quantities, such as velocity, if the detonation time interval can be set, thereby ensuring that opposite phases can be observed between the overlapping vibration waveforms. To identify the detonation time interval between the points at which opposite phases are observed, the dominant frequency of the vibration must be known. However, if the vibration level has to be managed, it should be noted that the result that is expected from a physical quantity cannot be achieved while converting the acceleration to dB owing to the interference by changes in characteristics because of the time constant and sensory correction filter.

The factor that is used for setting the optimal time interval is basically the time interval between successive detonations that result in opposite phases with respect to the dominant frequency of the maintenance target or the propagation path. Further, assuming that the vibration waveforms of a single-shot explosion are similar, we can reduce the amplitude by making the waveforms of opposite phases to overlap with each other on a 1/2-wavelength interval of the vibration waveform, as depicted in Fig. 11. The interval that

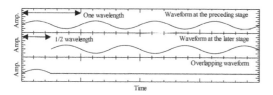

Figure 11. Conceptual diagram of the interval for creating an opposite phase

causes the opposite phase to be achieved is not only the 1/2-wavelength interval but also the 3/2-, 5/2-wavelength intervals, and so on. This indicates that the vibration waveforms are superimposed at intervals of 2/n (n = 1, 3, 5....) times the dominant frequency.

If two or more dominant frequencies are to be maintained or if the dominant frequencies are dependent on the direction, the basic approach sets a dominant frequency based on the detonation time at a non-dominant frequency by considering the frequency characteristics in each direction.

However, it is necessary to consider that the blasting effect exhibits a considerable variation if the time interval is considerably short (Hiroo 1958). For example, as the time interval decreases, the amount of energy per unit time increases and the blasting effect improves; further, cleaning may require a considerable amount of time because a significant period of time will be required for the stones to move over an extensive scattering angle range. It may also be possible that blasting will occur in the subsequent stage before the formation of a clear space by the preceding blasting and that crushing is not obtained as planned. In addition, if the time interval is considerably short, the propagation distance of an easy shot varies by approximately several meters depending on the position of the blast hole. Therefore, it is possible that the state of the non-reverse phase continues and that the reduction effect is reduced. Thus, the basic approach would be to increase the detonation time interval by as much as possible.

3.2 Applicable range of optimum-controlled blasting

Compared with the normal blasting drilling, in controlled blasting by an electric detonator, it is assumed that the propagation path to the maintenance object alters because of the progress of tunnel face drilling and that the dominant frequency of the vibration waveform also changes. In this case, because the dominant frequency is dependent on the time interval and has been well understood, we can continue to conduct blasting by monitoring the change in the

dominant frequency on the propagation route and by changing the time setting for every shift of 10 Hz.

When the dominant frequency of the maintenance target is high, there is a possibility that such a high frequency of the object cannot be detected because of the effect of the dominant frequency that originates from the time interval. In this case, it would be possible to perform four to ten stages of blasting at intervals of 300ms in the center cut (including auxiliary center cut) and the remaining easy shot at the current time interval. Further, the time interval can be reconsidered through predictive analysis by assuming one-shot waveform of the center-cut area. In addition, if we measure the vibration waveform of a single-shot detonation, we can also perform predictive analysis of approximately 10 stages of blasting using ordinary electric detonators and multistage blasting with MS + DS detonators; thus, we can proceed with construction while verifying (on an ongoing basis) the applicable range of the controlled blasting with an electronic detonator.

4 CONCLUSION

Based on the results of test blasting, we demonstrated that it was possible to effectively reduce the vibration by optimizing the detonation time interval of an electronic detonator by considering the site conditions. Further, we exhibited that it was important to confirm the dominant frequency in the maintenance target and the propagation path for obtaining the vibration-reduction effect. We also discussed a method for determining the detonation time interval and the optimal range of controlled blasting.

REFERENCES

Japan Explosives Industry Association. 2002.3. Blasting case collection, p. 80.
Yoshiharu Tanaka, Katsuzo Teramoto, Toshiyuki Ichijyo, Keiichi Sato, Masaaki Yamamoto, Kenichi Aiko. 1992.2. Discussion on blasting experiments using EDD, *Proceedings of the 24th symposium on rock dynamics*, pp. 390–394.
Keita Iwano, Toshiyuki Koshikawa, Kinya Kuriki, Takayuki Inukawa, Katsunori Fukui., 2014.12. Blasting for the reduction of environmental load under the residential area using high-precision electronic detonators, *Reports of Tunnel Engineering, Vol. 24, I-35.*
Japan Industrial Standard. JIS-C-1517 (2014). Vibration level meters: Measuring instruments used in transaction or certification.
Ito Hiroo. 1958.3. Blasting tests using electric detonators with various time intervals, *Journal of Mining Institute of Kyushu, Vol. 26, No. 3 , pp.* 133–136.

2019 Rock Dynamics Summit– Aydan et al. (eds)
© *2019 Taylor & Francis Group, London, ISBN 978-0-367-34783-3*

Estimation of impulsive forces of hydraulic breaker via transfer path analysis (TPA) method

Changheon Song, Daeji Kim, Sang-Seuk Kweon, Joo-Young Oh, Jin-Young Park,
Jong-Hyoung Kim & Jung-Woo Cho
Korea Institute of Industrial Technology, Gyeongsan-si, Republic of Korea

ABSTRACT: This study proposed a transfer path analysis (TPA) method to estimate the percussion force of hydraulic breakers. The frequency response function (FRF) obtained through modal impact experiments are applied to the TPA model. The impact stress wave generated from the piston transmitted to the housing of hydraulic breaker. The impulsive force components on the housing were measured. The force components derived from TPA model are compared with the data obtained from the standard method using strain gauges. The TPA data were in good agreement with the standard method. This new method can be used to estimate the forces and vibrations of hydraulic breakers and rock drilling machines.

1 INTRODUCTION

An excavator is often equipped with a hydraulic breaker to fracture the ground or bedrock in order to produce resource materials, aggregates, or to demolish structures (Song et al., 2017; Kwak and Chang, 2008; Giuffrida and Laforgia, 2005) (Figure 1(a)).

The hydraulic breaker comprises a directional valve, which controls the flowing direction of fluid; an accumulator for storing the hydraulic energy and for supplying necessary flow during the operation. The back head which is filled with nitrogen gas; pistons for delivering the striking energy to the chisel while performing a reciprocal motion; and a chisel which directly breaks the subject material. The hydraulic breaker is mounted onto an excavator by a coupler (Figure 1(b)).

Previous studies of hydraulic breakers include the work of Ficarella et al., (2006, 2007, 2008), who used a 1D simulation and experimental tests to investigate the performance and design of a hydraulic breaker. Several other studies have focused on improving performance, reducing the weight of the housing, and optimizing vibration and noise (Lee et al., 2003; Park et al., 2011).

In the construction equipment having percussive operations, the impact loads are continuously generated, and these are closely related to the durability of the equipment. However, few studies have sought to quantify the impact loads delivered to the housing. Because the strain gauges attached on the equipment easily break away during the percussion test, a direct measurement and data acquisition are very difficult.

This study examined the basic concept of the transfer path analysis (TPA), and the applicability of the TPA approach on mechanical equipment with impact vibrations and forces was discussed. In addition, the vibration characteristics that occurred while striking and the frequency response functions (FRF) obtained through the modal impact experiments were applied to the TPA method. As a consequence, the percussive forces, and the respective contributions to the load transfer in each component of the hydraulic breaker were evaluated. Additionally, the impulsive forces of each component derived from the TPA approach were compared with the impulsive forces that occurred in the chisel, as derived by the striking energy measured using the strain gauges, in order to verify the validity regarding the TPA method applicability to the structures.

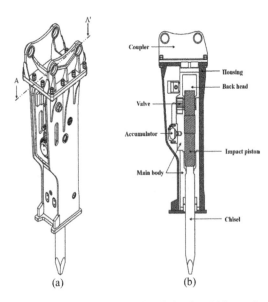

Figure 1. Schematic of the hydraulic breaker; (a) isometric view, (b) cross section A-A'.

2 VIBRATION EXPERIMENTS

2.1 Setup of vibration experiments

We estimated the impact loads delivered from the chisel to the housing. The vibrations occurring during striking were measured by attaching an acceleration sensor (Figure 2) to a position that represented the shape of the hydraulic breaker properly. The sensor was a three-axis accelerometer designed for high-impact environments. Data acquisition and analysis were managed by a SCADAS Mobile DAQ system (LMS International, 2005). Table 1 shows the conditions for the vibration experiments and the operational conditions of the excavator used in the experiments.

The hydraulic breaker operates by converting the energies of the hydraulic pressure and nitrogen gas into the kinetic energy of the striking pistons. The kinetic energy is then delivered to the rock via the chisel, thereby breaking it. The hydraulic breaker in this study strikes the steel plate (Figure 3) to create uniform impulsive forces. The setup and adjustments ensured the repeatability and reproducibility of the experiment, and minimized variations in the impulsive forces caused by the compressive strength of the rock. The setup also excluded any effects caused by the inhomogeneity and anisotropy of the rock.

Higher impulsive forces were expected during the striking of the steel plates than during actual bedrock striking. This was because the energy that would be used to break the bedrock was expected to be delivered to the housing in the form of a reaction force (in addition to the energy dissipated by vibrations and heat). Further experiments using actual bedrock are required.

Figure 4 shows the measured results of the operational vibration. The results are shown by the time trace impact signals obtained from Z-direction (Figure 2), which is the main striking direction of the hydraulic breaker. The vibration signal was obtained from sensor No. 14.

Table 1. Parameters for vibration experiments and operating conditions for the excavator.

Items	Parameters	Value
Tracking	Measurement method	Time trace
	Duration	30 s
	Increment	0.5 s
Acquisition	Bandwidth	6400 Hz
	Resolution	12.5 Hz
Excavator	Weight	36 ton
	Engine	1700 rpm
Hydraulic	Flow rate	200 lpm
	Working pressure	160 bar

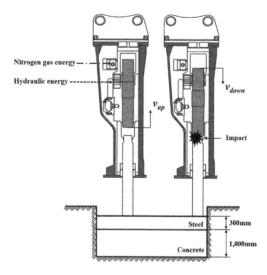

Figure 3. Mechanism of the hydraulic breaker and vibration test condition.

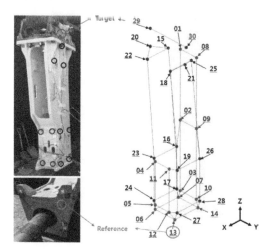

Figure 2. Installed position of tri-axial accelerometers.

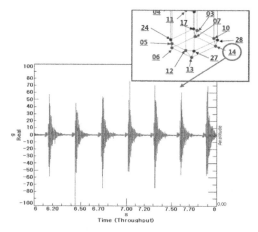

Figure 4. Impulsive signal data of acceleration measured on the operating hydraulic breaker.

720

2.2 *Modal impact experiments*

Modal impact testing can obtain the modal parameters of a structure, which are the eigenvalue (natural frequency and damping) and eigenvectors (mode shapes) of the dynamic system for the vibratory structure (Ewins, 2000; Lee and Richardson, 1992).

Figure 5 shows the results of the modal impact testing for the analysis of the hydraulic breaker's characteristics and vibration native mode. Modal impact experiments were performed to set the position of excitation for obtaining the FRFs along the y-directions on the bottom end of the hydraulic breaker (Figure 5(a)). The FRF matrix, which can be obtained through modal impact testing (Figure 5(b)), provides the mass matrix and strength matrix of the structure, but also the damping matrix, thereby enabling the derivation of characteristics that consider the dynamic system (Ewins, 2000; Lee and Richardson, 1992). Examining the major modal characteristics of the hydraulic breaker (Figure 6) showed that the bending mode often occurred in low-frequency bands. This was attributed to the breaker's scale and its overall shape.

2.3 *TPA of impulsive forces*

TPA performs calculations by using the characteristics of the impact vibrations that occur during operation and the transfer function matrix. The path is a medium that contributes to the load transfer between the subject of the striking (the main body, namely the excitation source) and the structure being struck. The calculation of the load transferred through the medium is possible only when the native characteristics of the greatest contributing medium are known (Van der Seijs et al., 2016; SIEMENS, 2014). Therefore, each path that contributes to the load transfer of the hydraulic breaker was selected (Figure 7).

The key paths of selection determined that the isolation pad is within the major transfer path. The pad was expected to show a high contribution to

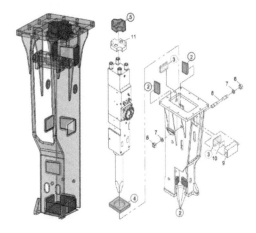

Figure 7. Positions of isolation pad in hydraulic breaker.

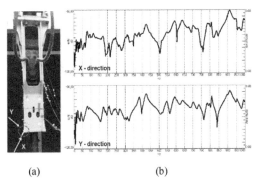

(a)   (b)

Figure 5. Modal impact experiments; (a) installation position of sensors, (b) measured frequency response function (FRF) during the tests.

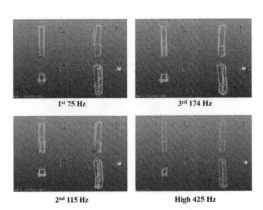

1st 75 Hz   3rd 174 Hz

2nd 115 Hz   High 425 Hz

Figure 6. Mode shapes of hydraulic breaker.

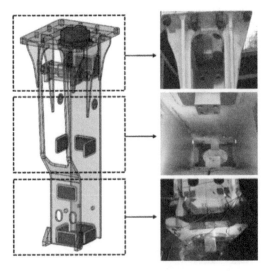

Figure 8. Tri-axial excitation jigs for FRF acquisition.

Table 2. Calculation results of impulsive forces delivered to the housing.

No.	Path	Axis	Force (kN)
1	Top	X	196.4
		Y	215.4
		Z	344.0
2	Middle	X	286.9
		Y	262.9
		Z	362.5
3	Bottom	X	381.4
		Y	410.7
		Z	399.0

the impact load transfer because, in the hydraulic breaker, percussion module and housing assembly are coupled by the pad. Therefore, the position of each isolation pad represents a position of excitation. An excitation jig identical in shape to the isolation pad was manufactured for FRF acquisition on each excitation position (Figure 8).

Additionally, Table 2 shows the results of TPA on the impulsive forces delivered to the housing via the isolation pad during the striking of hydraulic breaker.

3 VALIDATION OF TPA METHOD

3.1 *Impulsive force measurement by strain gauges*

To estimate the impulsive forces delivered to the housing through each transfer path, quantitative information of the striking force occurring at the chisel is required. The impulsive force occurring in the chisel can be utilized as the datum value for verifying the TPA calculation results. In order to calculate the impulsive force, which occurs at the chisel during the striking action of the hydraulic breaker, the impact energy measurement using the strain gauge was adopted (Park and Kim, 2006).

The impact energy of hydraulic breaker can be calculated by measuring the deformation of the strain gauge attached to the chisel. The impact energy of a single strike performed by the hydraulic breaker, was calculated by Eq. (1), which is derived from the deformation measured by the strain gauge and the elastic strain energy (Park and Kim, 2006).

$$E_i = \frac{CF^2}{A_c \cdot \sqrt{E_c \cdot \rho_c}} \int_{t_1}^{t} \varepsilon_i^2 \, dt \tag{1}$$

Where, E_i refers to the impact energy which is generated by a single strike of the chisel, A_t to the cross-sectional area of chisel, E_c is the Young's modulus of the chisel, ρ_c is the density of the chisel,

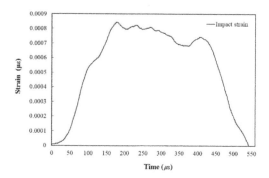

Figure 9. Shock wave of impact energy strain.

Table 3. Results of impact energy measurement.

Items	Description	Unit	Value
Chisel	Young's modulus	kgf/m²	2.10e+10
	Cross section area	m²	0.0213
Energy	Impact energy	J	4660
	Strain rate		8.42e-04
Force	Impulsive force	kN	3690

CF is the calibration factor in static calibration, and ε_i is the deformation of the chisel.

Figure 9 shows the measurement of the shock wave of the deformation caused by the striking of the hydraulic breaker. The impact energy calculated by the strain gauge, the maximum deformation, and the physical properties of the chisel are arranged in Table 3. The impulsive forces of the hydraulic breaker's chisel was calculated using Eq. (2) based on the stress-strain relation. The quantified results of the impact energy and impulsive forces, which occur during the striking action of the hydraulic breaker, are shown in Table 4.

$$F_{impulse} - \left(E_c \varepsilon_i \right) \times A_c \tag{2}$$

Here, $F_{impulse}$ refers to the impulsive force generated by the chisel, and E_c refers to the Young's modulus of the chisel.

3.2 *Results and discussions*

Through the TPA approach, the calculation of the impulsive force during the operation of hydraulic breaker was performed. Although the direction to which the main load of the hydraulic breaker is applied in the vertical direction, the impulsive forces are delivered to the housing due to the pitch, roll, and yaw motions, which are the vibration effects of the main body caused by the striking action. Therefore, all the impact loads delivered to the hydraulic breaker's housing from all directions (i.e., X, Y, and Z-direction) were considered.

The impulsive force obtained by TPA was 2850 kN, whereas that from the striking force measured by the strain gauge was 3690 kN, a difference of 22.6%. This means that the reaction force generated during the striking action of the chisel is damped by the isolation pad and is instead delivered to the housing. Therefore, the applicability of the TPA approach on a structure with impulsive forces and vibrations was positively validated.

This approach takes into account the energy loss caused by the noise, vibration and heat occurring during the chisel's strikes, and reflects the assumption that such energy loss is instead delivered to the housing in the form of a reaction force. Also, regarding the reliability of the TPA approach, the introduction of computer aided engineering (CAE) into the investigation of the load transfer through the isolation pad is necessary as well as additional physical tests on the isolation pad is also required in the future.

4 CONCLUSIONS

The TPA approach with using strain gauge was proposed for construction equipment. The results provided concentrated discussions regarding the method for quantitatively estimating the impulsive forces of the hydraulic breaker.

The applicability of TPA has been confirmed for mechanical equipment that generate shock vibration and impulsive forces.In addition, the validity of the TPA for a structure with impact vibrations and impulsive forces was verified by comparing the striking force of chisel obtained by the strain gauge method with the impulsive forces calculated by the TPA approach.

To ensure the reliability of the TPA method, comparative analyses of the impulsive forces delivered to the housing are required through conducting various experiments including those using strain gauges, load cells and actual rock. Furthermore, an additional study based on the present measurements and evaluation of the forces might lead to a new method for estimating the dynamic damping factor of various rocks.

REFERENCES

Ewins D.J. (2000), Modal Testing: Theory, Practice and Application Second Edition. RESEARCH STUDIES PRESS LTD.

Ficarella, A., Giuffrida, A., and Laforgia, D. (2006) Numerical investigations on the working cycle of a hydraulic breaker: off-design performance and influence of design parameters. International Journal of Fluid Power 7, pp.41–50.

Ficarella, A., Giuffrida, A., and Laforgia, D. (2007) Investigation on the impact energy of a hydraulic breaker. SAE Technical Paper, 01-4229.

Ficarella, A., Giuffrida, A., and Laforgia D. (2008) The effects of distributor and striking mass on the performance of a hydraulic impact machine. SAE Technical Paper, 01-2679.

Giuffrida, A., and Laforgia, D. (2005) Modelling and simulation of a hydraulic breaker. International Journal of Fluid Power 6, pp.47-56.

Kwak, K.S., Chang, H.W. (2008) Performance optimization of a fully hydraulic breaker using taguchi method. J Drive and Control 5, pp.41–48.

Lee, M., and Richardson, M. (1992), Determining the Accuracy of Modal Parameter Estimation Method. Proc. 10th International Modal Analysis Conference (IMAC X), pp.1-8.

Lee, S.H., Han, C.S., and Song, C.S. (2003) A study on the performance improvement of a high efficiency hydraulic breaker. J The Korean Soc Tribol Lubr Eng. 19(2), pp.59–64.

LMS International, (2005), SCADAS III User and Installation Manual.

Park, G.B., Park, C.H., Park, Y.S., and Choi, D.H. (2011) Optimal design for minimizing weight of housing of hydraulic breaker. Trans Korean Soc Mech Eng A 35(2),pp.207–212.

Park, J.W., and Kim, H.E. (2011), Impact energy measurement of hydraulic breaker. Key Engineering Materials: 3, pp.1669–1672.

SIEMENS (2014) Technical info issued: Advanced transfer path analysis techniques.

Song, C., Kim, D.J., Chung, J., Lee, K.W., Kweon, S.S., and Kang, Y.K. (2017), Estimation of Impact Loads in a Hydraulic Breaker by Transfer Path Analysis. Shock and Vibration, Article ID: 8564381.

Van der Seijs, M.V., Klerk, D., and Rixen, D. (2016), General framework for transfer path analysis: History, theory and classification of techniques. Mech Syst Signal Process, 68-69, pp.217–244.

2019 Rock Dynamics Summit– Aydan et al. (eds)
© 2019 Taylor & Francis Group, London, ISBN 978-0-367-34783-3

Numerical simulation of vibration in an excavated tunnel caused by blasts in adjacent tunnel

Qingbin Zhang
School of Civil Engineering, Changsha University of Science & Technology, Hunan, China

Zongxian Zhang
Oulu Mining School, University of Oulu, Finland

Congshi Wu
School of Civil Engineering, Changsha University of Science & Technology, Hunan, China

Junsheng Yang & Zhenyu Wang
School of Civil Engineering, Central South University, Hunan, China

ABSTRACT: In order to control the scale of blasting in the excavation of a tunnel and mitigate the blast-caused adverse effects, especially on its neighboring tunnel, the particle velocities of the concrete lining of an excavated tunnel caused by the blasts in its adjacent operating tunnel, were analyzed in ANSYS/LS-DYNA software. The simulated results are consistent with the in-situ measurement results, and the results show that (1) the PPV in the left part of the right tunnel close to the blast source in the left tunnel is larger than that in the right part of the right tunnel away from the blast source; (2) the ratio of the PPV in the right tunnel from the left spring line to the right spring line is in a range of 1.36-5.90, and the ratio from left sidewall to the right sidewall is 1.87-5.59; (3) the region in the right tunnel against unexcavated area in the left tunnel are affected by blasting in the left tunnel more seriously than the region in the right tunnel against excavated area in the left tunnel, and the former is 1.11-2.45 times greater than the latter, with an average value 1.64. And the left part of right tunnel closed to the blasting position in the left tunnel and situated in the negative area should be paid much attention in the actual construction of tunnels.

KEYWORDS: Numerical simulation; concrete lining; blast; adjacent tunnel; LS-DYNA

1 INTRODUCTION

Until now, rock blasting has been dominant in hard rock tunneling (Nateghi, 2012; Yi et al., 2017). Unfortunately, tunnel blasting always causes adverse impact on its lining or the lining of another tunnel located nearby. Many linings of tunnels were damaged by blasting during their construction (NAKANO et al., 1993; Wang et al., 2004; Li et al., 2014). Therefore, vibrations induced by blast must be controlled in tunneling.

Much work about the adverse effect on neighboring structures by blasting has been carried out. For example, Wu C.S. et al. (2004) deduced that the blasting vibration will not damage the concrete lining of the neighboring tunnel if the PPV (peak particle velocity) is less than 6cm/s by 2D numerical simulation and the field blasting tests. Nateghi, R (2011) deduced the empirical

relationships between scale distance and PPV in different surface and underground regions. Wang P. (2018) putted forward an innovative method to improve dynamic response behaviors of protective concrete arches subjected to blast load. Anders (2004) obtained that the shotcrete and young shotcrete without reinforcement can withstand high particle velocity vibrations without being seriously damaged by in-situ tests. Zhang et al. (2005a) simulated the effects induced by the blast on the fresh concrete in a laboratory. Xia et al. (2013) analyzed the damage of the surrounding rock and the lining system under different blast loads by field tests and numerical simulations. Lamis Ahmed et al. (2014) adapted a finite element model to analyze the young and hardening shotcrete behavior under blasting. Zhang et al. (2005b, 2016) introduced some methods that can reduce ground vibrations caused by blasts.

It is noted that most of the previous studies focused on adverse effect on the nearby structures, especially the shortcrete lining, but few are focused on the concrete lining. However, the concrete lining is important for a tunnel because it can influence the life of a tunnel. Therefore, the effect of concrete lining by blasting should be considered during the construction of neighborhood tunnels. In this paper, the vibration velocities of the concrete lining are analyzed in ANSYS/LS-DYNA software and based on the Hengshan tunnel. And, the vibration velocities and its attenuation law of the concrete lining are studied. At last, the area where the particle velocity caused by blasts is the largest and must be monitored mainly is obtained.

2 LOCATION AND GEOLOGY OF HENGSHAN TUNNEL

Hengshan tunnel consists of two adjacent parallel tunnels that are located in the south-west circle expressway of Wenzhou. The length of the left tunnel is 338 m and the length of the right tunnel is 330 m. The pillar between the two tunnels is 8 m-14 m wide. The right tunnel was excavated by full face method and the left tunnel was excavated by bench method. The right tunnel was excavated firstly and was 30-35 m ahead of the left tunnel in order to decrease the interaction of the two tunnels.

Hengshan tunnel is situated in hilly area. The lithology of upper strata is slope gravel soil and that of lower strata is felsite porphyry and sandstone, which are all hard rock. The maximum buried depth of tunnel is about 53 m. The surrounding rock in the entrances and exits of the tunnels are mainly IV and V classes, meaning that the surrounding rock is poor integrity. The two tunnels are more deeply buried in their middle parts than which in the entrances and exits of the tunnels, and the surrounding rock is class III with a good integrity. Hengshan tunnel is bidirectional with six lanes, and the designed speed is 100Km/h. A composite lining is employed in the tunnels and it consists of primary support, waterproof layer, and second lining. The detail of composite lining in class III surrounding rock is shown in Table 1.

Table 1. Support parameters of Hengshan tunnel in class III surrounding rock.

	Primary support		Second lining
Bolts(circle interval * longitudinal interval, length)	Steel mesh (type, diameter)	Shotcrete (thickness, type)	Arch ring (thickness, type)
1.2m*1.2m, 3m	Single layer, $\phi6$	18cm, C20	45cm C30 concrete

3 BLAST DESIGN

The bench blasting is used in the left tunnel and there are a total of 118 blastholes in the cross-section including 12 holes in cut blast with a V-form. All blastholes excluding cut holes are 3.5m long and 42mm in diameter, and they were fired by non-electric detonators with 7 different delay times. The explosive is No 2 rock emulsion explosive with 32 mm diameter, 300 mm length, over 3400 m/s VOD (velocity of detonation), and with a density of 1100 kg/m³.The parameters of blast design are shown in Table 2.

4 NUMERICAL ANALYSIS

According to the tunnel design, a numerical model is generated with the 3D LS-DYNA software. The distance from the tunnel to the left boundary and the right boundary are all 45 m and the distance from the tunnel to the bottom boundary is 40 m. The length of model in longitudinal is 30 m and the buried depth is 30 m. The pillar between the two tunnels is 8m. In the model, the bench method is used in the left tunnel and the tunneling face is located at the middle of the left tunnel, while the right tunnel has been built. The cut blasting is considered only in the model because the cut blasting often give rise to maximum vibration (Zhang, 2016) and the charge weight is 30 kg.

The explosive in the model is a Mat-High-Explosive-Burn material and with Jones-Wilkins-Lee (JWL) equation of state (EOS) (Lee et al., 1968).

Table 2. Parameters of blast design

No. of detonator	Delay time (ms)	Number of blast holes	Charge per blast hole(kg)	Total charge(kg)
1	0	12	2.6	31.2
3	50	8	2.6	20.8
5	110	8	2.6	20.8
7	200	13	2.6	33.8
9	310	16	2.6	41.6
11	460	11	2.6	28.6
13	650	22	Charge 2.8kg per hole for 10 holes; Charge 2.6kg per hole for 2 holes; Charge 2kg per hole for 10 holes	53.2
15	880	28	Charge 2kg per hole for 26 holes; Charge 2.6kg per hole for 2 holes	57.2
Total charge of the blast pattren (kg)				287.2

Table 3. Parameters of explosive in the model

ρ/Kg/m³	D/m/s	A/GPa	B/GPa	R_1	R_2	Ω	E_e
1000	3300	229	0.55	6.5	1.0	0.35	3.51

$$P=A(1-\frac{\omega}{R_1V})e^{-R_1V} + B(1-\frac{\omega}{R_2V})e^{-R_2V} + \frac{\omega E_e}{V} \quad (1)$$

where ω, A, B, R_1, and R_2 are user defined input parameters (Hallquist J.O, 2006). The parameters of the explosive in the model are listed in Table 3 (Doboratz B.M et al., 1985).

where ρ is the explosive density, D is the VOD (velocity of detonation) of explosive.

The Plastic-Kinematic material (Krieg R.D et al., 1976) and Johnson-Holmguist-Concrete material (Xu H. et al., 2016) are used to model the surrounding rock and the concrete lining, and the parameters are listed in Table 4 and Table 5, respectively.

where A is cohesive strength; B is pressure hardening; C is strain rate coefficient; FC is quasi-static uniaxial compress strength; N is pressure hardening exponent; T is maximum tensile hydrostatic pressure; K_1, K_2 and K_3 are pressure constants. The typical vibration waveforms are shown in Figure 1.

Table 4. Parameters of surrounding rock in the model

Elastic modulus/MPa	Density/ Kg/m³	Tangential modulus/MPa	Failure strain	Poisson
400	2700	33	1.25	0.27

Table 5. Parameters of the concrete lining in the model

ρ/Kg/m³	Shear modulus/GPa	A	B	C	N	FC/ MPa
2310	13.24	0.79	1.6	0.008	0.62	30.5
T/GPa	K1	K2	K3			
0.004	86	-173	204			

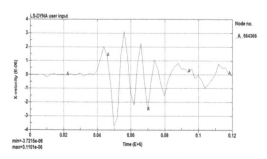

Figure 1. The particle vibration velocities of radial direction at the left sidewall.

5 DATA ANALYSIS

5.1 Compare of the in-situ measurement data and the simulation results

The vibration velocities of the concrete lining in the right tunnel which located at the same cross-section of the blast source in the left tunnel are compared between the in-situ measurement data and the simulation results, and the results are shown in Figure. 2.

All the results of simulation are less than the data of in-suit measurement because of the less charge weight in the simulation. Another, most of the ratio error of the difference to the in-suit measurement results is below 9%, which verifies the simulation results are basically coincided with the in-suit measurement value. Therefore, the simulation results are believable.

5.2 Attenuation of particle velocity

The particle velocities of the points with the nearly same distance in longitudinal direction to the blast source are analyzed. The negative area and the positive area are defined at first, depending on whether there exists an unexcavated area between the monitoring cross-section and the blast source (see Figure. 3). The vibration velocities of the simulation are listed in Table 6.

From the data in Table 6, the PPV in the positive region are larger than that in the negative region. The main reason is that the stress wave caused by the blasting in the left tunnel can arrive directly at the positive region of the right tunnel, but it cannot be transported along the straight line to the negative region of the right tunnel because of exist of the excavated area in the left tunnel. The former is 1.11~2.00 times that the latter when the monitor point is 5 m far from the blast source in the longitudinal direction, and the former is 1.23~2.45 times that the latter when the monitor point is 10 m far from the blast source in the longitudinal direction, with an average value 1.64.

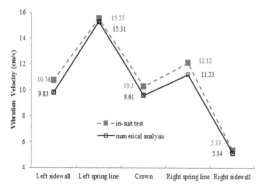

Figure 2. Compare of the data between the in-suit measurement data and the simulation results

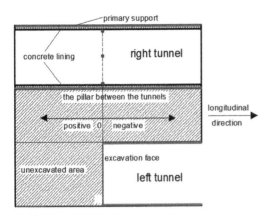

Figure 3. The defined coordinate, positive and negative areas.

Table 6. Vibration velocities of the simulation

Monitoring point	Vibration velocity at different distance from the blast source in longitudinal direction (cm/s)				
	-10 m	-5 m	0 m	5 m	10 m
Left sidewall	2.989	4.095	9.611	8.179	5.144
Left spring wall	3.250	6.162	15.310	9.613	7.948
Crown	1.549	2.198	9.834	2.931	2.480
Right spring line	1.182	1.457	11.232	1.628	1.449
Right sidewall	0.665	1.005	5.141	1.466	1.317

The vibration velocities of the left part of the right tunnel are larger than that of the right part of the right tunnel, and the reason is that the former part is closer to the blast source in the left tunnel than the latter part. The vibration velocity of left spring line in right tunnel is 1.36 ~5.90 times that of right spring line in right tunnel, and the vibration velocity of left sidewall in the right tunnel is 1.87~5.59 times that of right sidewall in the right tunnel.

6 CONCLUSIONS

The vibration velocities of the concrete lining by blasting are analyzed in LS-DYNA software based on Hengshan tunnels, and the results are listed as following.

The region in the right tunnel against the unexcavated area in the left tunnel is affected by blasting in the left tunnel more serious than the region in the right tunnel against the excavated area in the left tunnel. The particle velocity in former area is mostly 1.11~2.45 times greater than that of the latter area and with an average value 1.64.

The vibration velocity in the left part of the right tunnel is larger than that in the right part of the right tunnel. The particle velocity of the left spring line is 1.36~5.90 times that of the right spring line, and the particle velocity of the left sidewall is 1.87~5.59 times that of the right sidewall.

The left part of right tunnel closed to the blasting position in the left tunnel and situated in the negative area should be paid much attention in the actual construction of tunnels.

ACKNOWLEDGMENTS

The authors would like to acknowledge the financial support from National Nature Science Foundation of China (51508038, 51274049), the Key Disciplines Research Funds for the Changsha University of Science & Technology (13ZDXK11). The support provided by China Scholarship Council (CSC) during the visit of the first author to University of Oulu is acknowledged.

REFERENCES

Ahmed, L., & Ansell, A. 2014. Vibration vulnerability of shotcrete on tunnel walls during construction blasting. *Tunnelling and Underground Space Technology*, 42, 105-111.
Ansell, A. 2004. In situ testing of young shotcrete subjected to vibrations from blasting. *Tunnelling and Underground Space Technology*, 19(6),587-596.
Doboratz, B.M., Crawford, P. C., & Handbook, L. E. 1985. *Properties of Chemical Explosives and Explosive Simulants. UCRL-52997*, Lawerence Livermore national Laboratory, USA.
Hallquist, J. O. 2006. *LS-DYNA theory manual*. Livermore software Technology corporation, 3, 25-31.
Krieg R.D., Key S.W., 1976.Implementation of a time dependent plasticity theory into structural computer programs, *Vol. 20 of constitutive equations in viscoplasticity: Computational and Engineering Aspects* (American Society of Mechanical Engineers, New York, N.Y.,), 125–137.
Lee, E. L., Hornig, H. C., & Kury, J. W. 1968. *Adiabatic expansion of high explosive detonation products (No. UCRL-50422)*. Univ. of California Radiation Lab. at Livermore, Livermore, CA (United States).
Li, S., Li, K., Lei, G. 2014. Study of blasting vibration and deformation control for metro construction beneath existing metro tunnel in short distance. *Rock and Soil Mechanics*, 35(S2),285–289. (in Chinese)
NAKANO, K. I., OKADA, S., FURUKAWA, K., & NAKAGAWA, K. 1993. Vibration and cracking of tunnel lining due to adjacent blasting. *Doboku Gakkai Ronbunshu*, (462), 53-62.
Nateghi, R. 2011. Prediction of ground vibration level induced by blasting at different rock units. *International Journal of Rock Mechanics and Mining Sciences*, 48(6),899–908.
Nateghi, R. 2012. Evaluation of blast induced ground vibration for minimizing negative effects on surrounding structures. *Soil Dynamics and Earthquake Engineering*, 43, 133–138.

Wang M, Pan X, Zhang C. 2004. Study of blasting vibration influence on close-spaced tunnel. *Rock Soil Mech.* 25(3),412–414. (in Chinese)

Wang, P., Jiang, M., Zhou, J., Wang, B., Feng, J., Chen, H., … & Jin, F. 2018. Spalling in concrete arches subjected to shock wave and CFRP strengthening effect. *Tunnelling and Underground Space Technology*, 74, 10–19.

Wu, C.S., Li, J.X., Chen, X., & Xu, Z.P. 2004. Blasting in twin tunnels with small spacing and its vibration control. In tunneling and underground space technology. *Underground space for sustainable urban development. Proceedings of the 30th ITA-AITES world tunnel congress Singapore*, 22-27 May 2004 (Vol. 19, No. 4-5).

Xia, X., Li, H.B., Li, J.C., Liu, B., & Yu, C. 2013. A case study on rock damage prediction and control method for underground tunnels subjected to adjacent excavation blasting. *Tunnelling and Underground Space Technology*, 35, 1–7.

Xu, H., & Wen, H. M. 2016. A computational constitutive model for concrete subjected to dynamic loadings. *International Journal of Impact Engineering*, 91, 116–125.

Yi, C., Sjöberg, J., & Johansson, D. 2017. Numerical modelling for blast-induced fragmentation in sublevel caving mines. *Tunnelling and Underground Space Technology*, 68, 167–173.

Zhang, C., Hu, F., & Zou, S. 2005a. Effects of blast induced vibrations on the fresh concrete lining of a shaft. *Tunnelling and underground space technology*, 20(4),356–361.

Zhang, Z. X., & Naarttijärvi, T. 2005b. Reducing ground vibrations caused by underground blasts in LKAB Malmberget mine. *Fragblast*, 9(2),61–78.

Zhang, Z. X. 2016. *Rock fracture and blasting: theory and applications*. Butterworth- Heinemann.

2019 Rock Dynamics Summit– Aydan et al. (eds)
© 2019 Taylor & Francis Group, London, ISBN 978-0-367-34783-3

Investigation of vibration patterns occurred during TBM excavation and rock cutting tests

U. Ates, H. Copur, A.S. Mamaghani & E. Avunduk
Department of Mining Engineering, Istanbul Technical University, Istanbul, Turkey

I.S. Binen
Gulermak-Nurol-Makyol Metro Cons. J.V, Istanbul, Turkey

ABSTRACT: TBMs are widely utilized for infrastructure projects and to reach favorable excavation rates, selected TBMs should be operated to comply with the geological changes. Therefore, recognizing and understanding the geological changes during the excavation is of great importance. In that aspect, vibrations generated during the excavation can give useful information about the geological conditions. However, this topic has not been studied in detail and only limited number of researches have been conducted up to now.

This paper summarizes vibration patterns created during rock cutting by using a full scale linear rock cutting machine as well as during TBM excavation in field. Vibrations created during TBM excavation and TBM performance data recorded for 328 rings, were compared to each other and geological data. It was seen that geological conditions effects vibration pattern. In addition it was also found that broken or flat cutters have also effect on vibration. It was also seen during the laboratory tests that acceleration amplitudes were related with rock type and cutting patterns.

1 INTRODUCTION

Performance of a Tunnel Boring Machine is directly related to its specifications, geological conditions and as well as their crews' abilities to interpret these conditions and react properly, especially operators. Excavation operation generates vibration due to rock cutting process, which then is transmitted to the cutterhead and other parts of the TBM. These vibrations could help to understand geological conditions in front of the TBM without face inspections, which could save lots of time.

There are several researches made to understand forces and vibrations created during rock cutting. One of the first researches was made by Samuel and Seow (1984). Strain gauges installed to the discs' shafts of a 4.1 m diameter Robbins 146-193 model TBM and force components acting on a disc cutter was investigated. Then, frequency analyses done on force data. Another detailed research on this area was performed by Gertsch (1991, 1993) and frequency analyses was performed on the forces caused by 17" constant cross section (CCS) disc cutter during the experiments with a linear cutting machine (LCM). These researches were performed using strain-gauges instead of accelerometers to understand vibrations, which is different from the studies performed later on this area.

TBM manufacturers are also interested with measuring forces acting on cutters, as well as vibrations. To measure the amount of vibration on a cutter, wireless sensor packages were mounted to cutter holders. Changes in vibration could indicate variations in the geology, a blocky face, or a possible failure problem with the cutter, among many other things. However, as the TBMs are operated in a very dynamic environment, extensive filtering and large amounts of data are required to obtain a meaningful data set (Shanahan 2010). Understanding vibrations is also important for cutter life and design. For this purpose, Huo et al. (2014) developed a dynamic multi-degree-of-freedom model for disc cutters, which contained a cutter ring, a cutter body, two bearings and cutter shaft.

Another and more practical approach for measuring vibrations is placing accelerometers behind the cutterhead such as performed by Walter et al. (2012) and Walter (2013). TBM vibrations were correlated with geological conditions encountered along the TBM alignment and a boulder detection system were developed, Buckley (2015) continued to analyze data and improved this system later.

Different from the previous studies, this research includes vibration data both from the TBM field operations and laboratory rock cutting tests.

The vibration data related to excavation and rock cutting can be identified as random data, which occurs in contribution of many independent factors. Thus, random data analysis techniques should be used during signal analysis (Shin & Hammond 2008, Bendat & Piersol 2010, Piersol & Paez 2010, Brandt 2011).

2 METHODOLOGY

2.1 *Vibration recording equipment*

Accelerometers were becoming more affordable by the years. Micro-Electro-Mechanical Systems, in short MEMS, is a fabrication technology that have been used to manufacture accelerometers and other sensors, which has brought lower manufacturing costs and smaller sizes to accelerometers. At the beginning of the project a proof of concept was required to investigate weather vibration could be used for identifying geological changes or condition of disc cutters and MEMS type accelerometers were considered as suitable in terms of size and price.

The first set of the accelerometers used at the project, for TBM vibration recording, was DE-ACCM3D from Dimension Engineering, which had a relatively cheaper price, internal low pass filter and its reasonable sensitivity. It had 3G measurement range, which was seen to be insufficient on some occasions, however, it was successful and durable enough to finish the proof of concept part of the project. In addition to these, various types of analog accelerometers were used for vibration recordings, of which their list and specifications are given in Table 1.

As the project advanced, more robust and industrial type accelerometers were bought from PCB Electronics. Model 356A17 accelerometers were piezoelectric type and this type of accelerometers were used quite commonly for testing purposes in related industries. Their sensitivity, measurement and frequency ranges are higher than the other accelerometers used in the studied project up to now, in addition to their lower noise characteristics with special cables used for this purpose. MEMS type ADXL354BZ and ADXL356CZ from Analog Devices were also used during the laboratory tests, which have higher measurement range than the DE-ACCM3D.

For digitizing the signals coming from analog accelerometers pairs of USB-6009 and 9230 BNC model analog – digital convertors from National Instruments were used. NI USB-6009 is a 14 bit, 48 kS/s convertor with 8 analog input channels used with MEMS type accelerometers, and NI 9230 BNC is a 24 bit, 12.8 kS/s/ch convertor with 3 analog input channels. This device has IEPE support for piezoelectric accelerometers and internal low pass filter for preventing aliasing.

2.2 *Data recording and data analysis software*

A computer was dedicated for recording data in the TBM, and a custom purpose data recording software was used. The data recording software was written by using Laboratory Virtual Instrument Engineering Workbench (LabVIEW) visual programming language from National Instruments. It is capable of recording vibration data continuously. After first usage on the TBM, it was improved drastically and lots of functions were implemented for the next use including notifying by email if the acceleration levels increase for identifying geological conditions on time.

Analysis of the recorded data was performed in Matlab by using custom written scripts for this project.

2.3 *Geological and geotechnical parameters of the rocks*

Since one of the aims of the project is finding vibration parameters' relationship with excavated rock, understanding geotechnical parameters of the excavated rock is important.

The TBM vibration data was recorded during the excavation of Mecidiyekoy-Mahmutbey Metro Line in Istanbul, Turkey. The studied section was between Karadeniz Mahallesi and Tekstilkent Stations, which was composed from alteration of sandstone, claystone and siltstone with various strengths (Polat 2014, 2015). Between these two stations, rock samples were taken from the belt conveyor, Schmidt hammer tests were conducted on these samples and results were later used for analysis. Table 2 summarizes physical, mechanical and geotechnical properties of the rocks.

The laboratory cutting tests were performed using the rocks provided by Gulermak-Nurol-Makyol Metro Construction JV from the Goztepe-Umraniye-Atasehir Metro Line. The samples represent Kurtkoy and Pendik Formations. Table 3 summarizes physical, mechanical and geotechnical properties of the rocks.

Table 1. Accelerometers used in the project.

	Dimen-sion Eng.	Analog Devices	Analog Devices	PCB Elec.
Manufacturer				
Model	ACCM3D	ADX-L354BZ	ADX-L356CZ	356A17
Type	MEMS	MEMS	MEMS	Piezoelectric
Number of accelerometers	2	2	1	2
Axes	x,y,z	x,y,z	x,y,z	x,y,z
Sensitivity (mV/g)	333	200	20	500
Measurement Range (±G)	3	4	40	10
Frequency Range (Hz)	500	1500	1500	0.5-3000
Internal Analog LP Filter (Hz)	500	1500	1500	N/A

Table 2. Rock Properties of Mecidiyekoy-Mahmutbey Metro Line.

Lithology	Sandstone (w1-2)	Sandstone (w3-4)	Siltstone - Claystone
γn (g/cm³)	2.60	2.60	2.40
c (kPa)	150	110	10
$\phi°$	36	33	31
Em (MPa)	1860	1325	900
μ	0.30	0.30	0.30
σu (MPa)	21	12	6
RQD	25	13	8
GSI	50	40	23
Q	1.25	0.50	0.15
RMR	42	35	30

Table 3. Rock Properties of Goztepe-Umraniye-Atasehir Metro Line.

Lithology	Kurtkoy Formation (Siltstone)	Pendik Formation (Mudstone)
γn (g/cm³)	2.67	2.57
c (kPa)	245	305
$\phi°$	44	42
Em (MPa)	14000	4900
μ	0.31	0.11
σu (MPa)	46.5	21.5
RQD	35	20
Q	0.085	0.088
RMR	36	31

3 VIBRATION DATA RECORDED ON TBM AND ANALYSES

Combination of three EBP TBMs and NATM method were used for excavating 18.5 km Mecidiyekoy – Mahmutbey Metro Line (Ates et al. 2016, 2017, Bilgin et al. 2017, Binen et al. 2017). To investigate vibrations created during the excavation, two DE-ACCM3D model accelerometers coupled with a NI USB-6009 analog – digital convertor were mounted behind the pressure chamber of one of the TBMs and vibrations were recorded continuously for 328 rings. 6.56 m diameter EPB TBM had 8 hydraulic cutterhead motors, 46 V type disc cutters and 72 scrapers. Since there are too many moving parts involved in the excavation process a broadband noise was expected. Figure 1 shows placement of the accelerometers in the TBM as well as axes used during analyses.

During the analyses some distinct patterns were found. First of them was related to a calcite vein at one side of the excavation face. Figure 2 shows

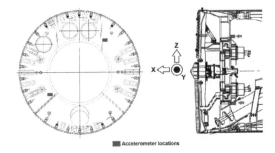

Figure 1. Accelerometer placement on TBM.

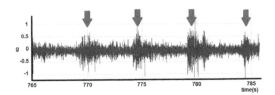

Figure 2. Cutterhead acceleration for one axis.

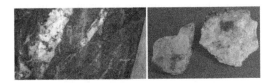

Figure 3. Calcite vein at the excavation face.

data from one channel of an accelerometer. The data was recorded when the cutterhead was turning on 3 rpm. As it can be seen in Figure 2, there are distinguishable peaks in around every 5 seconds. According to disc's placement on the cutterhead, one arm of face cutters, which were placed on + shape, passes a location in every 5 seconds when cutterhead was turning. This pattern is similar to the findings of Walter et al. (2012) and Walter's (2013).

A face inspection done after related excavation showed that there was a calcite vein present at the face that was causing acceleration peaks (Figure 3). A similar pattern was also found in other instances.

Probability analysis was also performed for the time data for each ring. Figure 4 shows cumulative probability of acceleration data which is higher than ±1G based on each excavation for 80 rings. Figure 4 shows also locations of disc cutter replacements, cutterhead speed, uniaxial compressive strength converted from Schmidt hammer results and average chamber pressure. It can be clearly seen that at the low strength zone, probability of vibrations reaching 1G is very low, moreover when the rock's strength increases vibration levels also increase. In addition to that, when the rock strength is high and the rock is not uniform, it causes high vibration

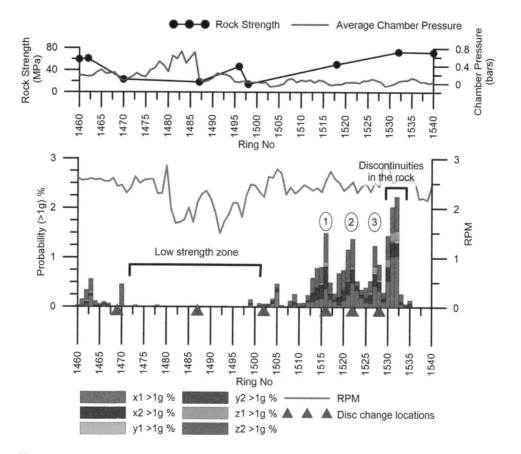

Figure 4. Cumulative probability of acceleration data.

levels, as can be seen from the figure around ring number 1530.

Disc replacement logs showed too many broken rings on this area, paired with vibration data, it is possible to say that the existence of broken disc rings change vibration patterns. At the locations were marked with 1, 2 and 3 in Figure 4, it is believed that vibrations were resulted by broken disc rings which caused scrapers to touch and scrape the rock face. After changing the broken discs with the new ones, the vibration amplitudes reduced instantly. However, as can also be seen from the figure, not all disc related problems were creating high levels of vibrations. Between the rings 1465 and 1505, 3 disc inspections and replacements took place without causing any distinct variation in vibration patterns. By taking into consideration both situations, it can be said that on high strength rocks, if multiple discs were damaged and scrapers were touching the rock face, as expected, high levels of vibrations occurs.

During the excavation of 328 rings, vibrations were only increased and showed a distinguishable change on 50% of total disc changes. In addition to that it was not possible to identify vibrations created by disc changes and geological conditions.

4 VIBRATION DATA RECORDED ON LABORATORY AND ANALYSES

Due to the complexity of TBMs, too much noise is created in addition to the relevant data. Moreover, there are more than one cutters present at the cutterhead whose vibrations are superImposed to each other and can not be recorded separately due to harsh environment of the cutterhead. As the technology improves, today it is possible to add sensors to disc housings and monitor disc rotation, temperature and other data. However, this applications are still very limited and not common.

Since understanding vibrations related with rock cutting process created during TBM excavation is quite limited, laboratory tests by full scale linear rock cutting machine was also performed to investigate vibration patterns. In a controlled environment with mostly homogenous rock surface, only one disc cutter and very little noise compared with TBMs, it was possible to see different vibration patterns for different rocks and different cutting conditions.

Two rock blocks with different geological and mechanical properties from Goztepe-Umraniye-Atasehir Metro Line were used in the full-scale linear rock cutting tests. The tests were carried out by using full-scale linear cutting machine (FLCM) at Istanbul

Technical University, Mining Engineering laboratory. Rock layers were perpendicular to the cutting direction for Pendik formation and 45° angled for Kurtkoy Formation. In addition to strain-gauges to measure the forces, 5 accelerometers were also attached to the machine. Figure 5 shows the placement of accelerometers. A 17" CCS disc was used during the tests and cutting speed was selected as 12.7 cm/s. Selected spacing and depth configurations can be seen in Table 4.

Similar to vibration data obtained from the TBM, probability density analyses were performed on the data. Figure 6 shows probability density graph of the accelerations for x channel from ADXL354BZ 4G accelerometer. As can be clearly seen from the graph, probability of acceleration amplitudes changes according to penetration, but more importantly, according to lithology. A similar pattern was seen for each channel. Since the rocks were not completely homogenous, with bedding planes, and data was obtained for a very limited length, it was not possible to compare effect of penetration; however, it will not be wrong to say that mostly, acceleration increases also with increased penetration. As the Kurtkoy Formation is stronger than Pendik Formation, probability of vibration amplitudes scattered broadly. This behavior was also seen on the TBM data, as the rock becomes stronger, vibration amplitudes increases. It was also realized that other probability distribution types could fit data better than normal distribution.

According to analyses, maximum accelerations also changed according to rock type. Maximum 16.4G acceleration was recorded during the laboratory tests for Kurtkoy Formation and 7.72G was recorded for Pendik Formation. Moreover, these acceleration numbers were recorded for the deepest penetrations, 8 mm and 10 mm, respectively.

By examining vibration data with force data and video recordings, it was also found that some vibration shocks were apparent just after a major rock chip was broken off. During this time, normal and rolling forces drooped and started to increase again. However, any relation were not found between the force and vibration amplitudes.

Figure 5. Accelerometer placement on LCM.

Table 4. Rock cutting test configuration.

Formation	Kurtkoy	Pendik
Spacing (mm)	Penetration (mm)	
70	5	5
	7	7
85	4	4
	6	6
	8	8
	-	10

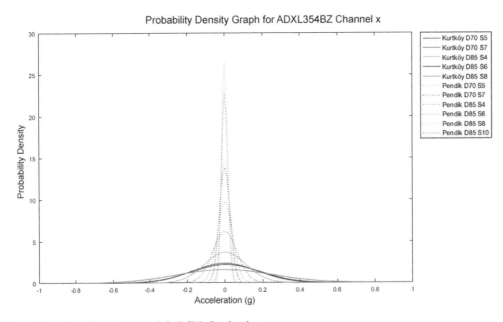

Figure 6. Probability density graph for LCM vibration data.

5 CONCLUSIONS

There are very limited researches conducted on the past regarding vibrations created on rock cutting. However, during the past few years, TBM manufacturers, as well as researchers, started to investigate vibrations.

This paper summarizes and compares vibration data recorded during TBM excavation, as well as full-scale linear rock cutting tests with rock properties. It was found that geological conditions effects vibration amplitudes during excavation and this finding was supported with laboratory tests. In addition to geological conditions, broken or worn disc cutters also created vibrations since scrapers close to these discs started to touch excavation face. The maximum acceleration recorded in the laboratory during rock cutting tests was 16.4G.

The data available is very limited and it requires extensive filtering and analyses which makes very difficult to identify rocks using vibrations, however, similar to previous studies, it is possible to say that there is relationship between rock type, accelerations and cutting conditions. For the future studies, it was planned to increase sample numbers and investigate relationship between rock properties and vibration patterns.

ACKNOWLEDGEMENTS

This work was supported by Research Fund of the Istanbul Technical University. Project Number: MDK-2017-40698.

REFERENCES

Ates, U., Binen, I.S., Acun, S., Murteza, M. & Celik, Y. 2016. EPB Tunneling Challenges in Sandy Ground. *2nd International Conference on Tunnel Boring Machines in Difficult Grounds, Istanbul, 16–18 November 2016.*

Ates, U., Binen, I.S., Acun, S. & Celik, Y. 2017. TBM Performance and Challenges Faced during Excavation of Mecidiyekoy - Mahmutbey Metro Line. *International Tunneling Symposium in Turkey: Challenges of Tunneling, Istanbul, 2–3 December 2017.*

Bendat, J.S. & Piersol, A.G. 2010. *Random Data, Analysis and Measurement Procedures.* New York: John Wiley and Sons.

Bilgin, N., Acun, S., Ates, U., Murteza, M., Celik, Y. 2017. The factors affecting the performance of three different TBMs in a complex geology in Istanbul. *Proceedings of the World Tunnel Congress 2017 – Surface challenges – Underground solutions. Bergen, Norway, 9 – 15 June 2017.*

Binen, I.S., Ates, U., Acun, S., Copur, H., Celik, Y. 2018. Effects of Cutter Types on Cutter Consumption and TBM Performance in Different and Complex Geological Conditions. *Proceedings of the World Tunnel Congress 2018, Dubai, UAE, 21-26 April 2018.*

Brandt, A. 2011. *Noise and vibration analysis: signal analysis and experimental procedures.* West Sussex: John Wiley and Sons.

Buckley, J. 2015. *Monitoring the vibration response of a tunnel boring machine: application to real time boulder detection.* M.Sc. Thesis, Colorado School of Mines

Gertsch, R. E. 1991. Disc cutter vibrations for the design and performance assesment of shaft and tunnel boring machines. *Converences on Shaft Drilling Technology Proceedings, Las Vegas, Nevada.*

Gertsch, R. E. 1993. *Tunnel boring machine disk cutter vibrations.* M.Sc. Thesis, Colorado School of Mines.

Huo, J., Sun, X., Li, G., Li, T. & Sun, W. 2015. Multi-degree-of-freedom coupling dynamic characteristic of TBM disc cutter under shock excitation. *Journal of Central South University.* 3326-3337.

Piersol A.G. & Paez T. L. (eds) 2010. *Harris' shock vibration handbook (6th Edition).* McGraw-Hill.

Polat, F. 2014. *Mecidiyekoy – Mahmutbey Metro Line, Depot, Maintenance Area and Depot Connection Lines Construction Works Geological – Geotechnical Works Report for the Section Between Yenimahalle - Mahmutbey (Km 15+535 – 22+763)* (Unpublished).

Polat, F. 2015. *Mecidiyekoy – Mahmutbey Metro Line, Depot, Maintenance Area and Depot Connection Lines Construction Works Geological – Geotechnical Works Report for the Section Between Yenimahalle – Alibeykoy River (Km 13+780 – 10+540)* (Unpublished).

Samuel, A. E., & Seow, L. P. 1984. Disc force measurements on a full-face tunnelling machine. *In International Journal of Rock Mechanics and Mining Sciences & Geomechanics Abstracts* Vol. 21, No. 2: 83–96.

Shanahan, A. 2010. Cutter Instrumentation System for Tunnel Boring Machines. *North American Tunneling 2010 Proceedings.*

Shin, K.& Hammond, J. 2008. *Fundamentals of signal processing for sound and vibration engineers,* Chichester: John Wiley and Sons.

Walter, B., Alavi Gharahbagh, E., Frank, G., DiPonio, M., & Mooney, M. 2012. Extending TBM Reliability by Detecting Boulders. *In North American Tunneling Conference, Indianapolis, IN.*

Walter, B. W. 2013. *Detecting changing geologic conditions with tunnel boring machines by using passive vibration measurements.* Doctoral dissertation, Colorado School of Mines.

2019 Rock Dynamics Summit– Aydan et al. (eds)
© *2019 Taylor & Francis Group, London, ISBN 978-0-367-34783-3*

Safe and rapid drift development in burst-prone mines through innovative rock support

M. Cai
China-Canada Centre of Deep Mining Innovation, Northeastern University, Shenyang, Liaoning China
Laurentian University, Sudbury, Ontario, Canada
Bharti School of Engineering, Laurentian University, Sudbury, Canada

ABSTRACT: Strainburst is the most frequently encountered rockburst in underground mines. Conventional method to support strainburst-prone grounds is to install rock reinforcement system using rebar and mesh first and then install yielding support system using dynamic rockbolts at a later stage. This two-stage rock support installation process is not safe and effective because it can adversely impact worker's safety and mine production schedule. Two new dynamic rockbolts, which are called Superbolts, are developed for rapid rock support in burst-prone grounds. Laboratory testing confirmed that the new rockbolts have superb capacities to achieve the goal of reinforcing and holding rock masses at the same time. Detailed test results of the MCB-Superbolts are presented in this paper. The new rockbolts are quick to install and can be used in a one-pass rock support system to facilitate safe and rapid drift development in underground mines.

1 INTRODUCTION

1.1 *Rockburst*

Failure of drifts in underground mines may be stress driven or structurally controlled. Several rock mass failure modes such as fall of ground due to wedge failure, squeezing and rockbursting have been identified in underground mine construction. In deep hard rock mines, stress-driven failure in the form of rockburst poses a great danger to workers and construction equipment. A rockburst is defined as damage to an excavation that occurs in a sudden and violent manner and is associated with a seismic event (Kaiser *et al.*, 1996). Fault-slip burst, pillar burst and strainburst are three rockburst types that occur in underground excavations and constructions. Among them, strainbursts are the mostly commonly encountered rockbursts.

1.2 *Rockburst damage*

Rockburst damages near excavations can be in the forms of static stress fracturing or strainbursting due to tangential straining, shakedown due to acceleration forces from a remote seismic event, and rock ejection by momentum or energy transfer from a remote seismic source or due to high bulking deformation rates during strainburst (Kaiser *et al.*, 1996; Cai & Kaiser, 2018). Sudden stress fracturing of rock leads to a disintegration of the rock mass and this is associated with rock mass bulking in the radial direction of tunnels. This inward movement can be described by a bulking factor (Kaiser *et al.*, 1996). Seismically induced rockfalls, which occur frequently at intersections where the span is large and the confinement of the roof rock is low, are caused by seismic waves from relatively large and remote seismic events that shake the entire volume of a potentially unstable mass of rock. Rock ejection can be caused by strainbursts, by a remote seismic event through dynamic energy or momentum transfer, or by a combination of both. Rockburst damage severity can be roughly classified into minor, moderate, and major or severe levels based on the volume of failed or displaced rock, the degree of support damage, and the violence of the energy release in terms of impact or ejection velocity. Detailed descriptions of the severity levels can be found in Kaiser *et al.* (1996).

The most immediate and severe cost of rockburst is injury and possible death of workers caught in the path of violent rock ejection and drift collapses. The second obvious cost is the production delay created by such incidents and the expense of cleaning up damaged sites. Hence, rockbursts can not only increase the safety risk but also substantially increase investment risk (Kaiser & Cai, 2012). To mitigate strainburst damage risk, effective rock support is needed.

Conventional method to support burst-prone grounds is to install rock reinforcement system using rebar and mesh first in the development cycle and then install yielding support system using dynamic rockbolts at a later stage. This two-stage rock support installation process is not safe and effective because a rockburst can occur before the rockburst support system is installed, causing fatality, injury to workers, and damage to excavations. Furthermore, it can adversely impact mine production schedule if the main production paths are not accessible due to excavation damage. In order to increase productivity without sacrificing safety, it is better to speed up rock support system installation and reduce cycle time through the application of new or emerging support technologies that are quick to install and effective in reducing rockburst damage. The goal is to combine needs for short-term (static loading) and longer-term (dynamic loading) supports and facilitate safe and rapid drift development to reduce overall development and production costs.

In the following discussion, rock support functions in highly stressed grounds are explained first. The need to reinforce rock mass and to provide energy absorption capacity using various types of rockbolts is then discussed, followed by presenting a new rock support technology that can facilitate safe and rapid drift development in strainbursting grounds.

2 ROCK SUPPORT IN HIGHLY STRESSED GROUNDS

2.1 Rock support functions

In practice, it is necessary to understand and differentiate different support functions to select an enhanced rock support system. Kaiser et al. (1996) defined three primary support functions: (1) reinforce, (2) retain, and (3) hold. A fourth function, connect, is added to the support functions as shown in Figure 1 (Cai & Kaiser, 2018). A rock support system is composed of tendon components such as friction bolt, rebar, and cablebolts and surface support components such as mesh, shotcrete, and straps. Some tendons have more capacities to reinforce rock mass while others are more suited for fulfilling the holding role. The combination of tendons and surface support components secures the rock mass in place to prevent gravity-, static stress-, and dynamic stress-driven rock mass failures.

2.2 Rock support in highly stressed grounds

It is observed from underground excavations that most stress-induced rock fractures are concentrated near the excavation boundary and are sub-parallel to the tunnel wall. 3D numerical modeling results show that the intermediate principal stress contributes to the formation of surface parallel fractures near excavation boundaries (Cai, 2008). This stress fracturing zone is often limited in extent and reinforcing this zone, in the form of installing grouted rebars or cablebolts in a grid pattern, can enhance the rock mass integrity and hence the rock mass strength.

In some cases, rock failure can be violent and releases a large amount of strain energy. Hence, rock support system must be able to yield to accommodate large rock mass bulking and dissipate dynamic energy. Cook & Ortlepp (1968) first suggested the use of yielding support in deep gold mines in South Africa. Over the years, various rock support products

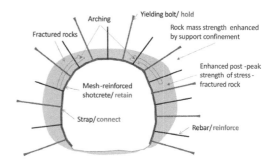

Figure 1. An illustration of two-pass rockburst support system that possesses the required support functions: reinforce, retain, hold, and connect (Cai & Kaiser, 2018).

have been developed and used in practice, such as conebolt (Jager, 1992), modified conebolt (MCB) (Simser et al., 2002), Durabar (Ortlepp et al., 2001), MCB33 (Cai & Champaigne, 2012), Garford solid bar (Varden et al., 2008), D-bolt (Li & Charette, 2010), HE-bolt (He et al., 2014), amongst others. These existing yielding bolts are intended for fulfilling the holding role (Figure 2). Reinforcement bolts such as rebars are needed in the support system to form an integrated support system. The full load and deformation capacities or the energy dissipation capacity of a support system is only achieved if all rock support elements are well integrated and connected, forming an effective rock support system (Cai & Champaigne, 2009; Cai, 2013).

Not every support component, i.e., bolt, mesh or shotcrete, serves the same rock support function. An effective rock support system is comprised of rock support components that provide all support functions (reinforce, retain, hold, and connect) required to control burst-prone ground. Rockbolts such as rebars must be used in the support system to fulfill the reinforcement role (Figure 1). Dynamic rockbolts such as MCBs and D-bolts are used in the support system to fulfill the holding role. There is no rockbolt in the existing support technology that fulfills the reinforcement and hold functions simultaneously. As stated in Cai (2013), following the rockburst support design principles we need to use rock support elements that are relatively easy to manufacture and simple as well as rapid to install during regular mine operations. Such a new rock support technology has been developed recently and the details are discussed in Section 3.

3 ROCK SUPPORT FOR SAFE AND RAPID DRIFT DEVELOPMENT IN STRAINBURST-PRONE GROUNDS

3.1 Support strategies

In underground mine drift development using the drill-and-blast excavation method, a typical work cycle includes drilling blast holes, loading and blasting,

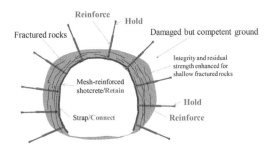

Figure 2. An illustration of one-pass, integrated rockburst support system.

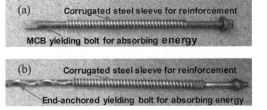

Figure 3. (a) A mini sample of patented MCB-Superbolt; (b) a mini sample of patented Versa-Superbolt.

ventilation, mucking, and rock support installation. It is estimated that about 27% of cycle time is spent on rock support installation, which is the second most time-consuming work after drilling that requires 29% of the cycle time (Kaiser *et al.*, 2003). In terms of cost, ground support constitutes about 34% of the total cost of a development cycle (Board *et al.*, 1996).

For rapid drift development, it is better to install rock support at face quickly. Therefore, an ideal rock support system for burst-prone mines is one that is fast to install, and yet immediately provides the required support capacity to deal with static and dynamic loadings and to ensure safety and excavation stability. As discussed in Section 2, an effective rock support system needs to have tendons such as rebar to reinforce the rock mass and yielding bolts such as conebolts to accommodate large wall deformation and absorb dynamic energy. One alternative is to install rebar and yielding bolts in the first pass as shown in Figure 1, in an alternate pattern. However, two types of boreholes are needed for installing reinforcing and holding rockbolts and the total time required to complete rock support installation is long, which can negatively impact the drift advance rate.

A new approach to increase drift advance rate in burst-prone mines while not sacrificing safety is to have a new rock support technology that possesses reinforcing and holding abilities and is quick to install. This concept is illustrated in Figure 2. Boreholes are drilled in the roof and walls of the drift and the composite bolts are installed with surface retaining components such as mesh and straps. The bolts are fully grouted in the borehole using resin and the red components of the bolts are intended for rock reinforcement and the blue component of the bolts are intended for energy absorption. The bolts can be installed in one pass of drift development and once installed the support system can prevent and limit rockburst damage.

3.2 *Development of new rockbolts for rock support in burst-prone grounds*

To meet the demand of safe and rapid drift development in deep underground mines, two new types of patented rockbolts, which are called Superbolts, have been developed at Laurentian University in Canada.

A Superbolt consists of a reinforcement component in the form of corrugated steel sleeve or pipe bonded to the borehole wall by grout, a yielding rockbolt passing through the reinforcement component, a nut, and a plate. The rockbolt is threaded at one end and has a yielding mechanism in the form of steel stretching or controlled anchor sliding. The inner surface of the reinforcing pipe is not bonded so that the yielding bolt can slide freely within. The outside surface of the pipe is deformed or roughened so that it can be firmly bonded to the borehole wall by resin or cement grout. The main purpose of the reinforcement component is to reinforce the rock mass near the excavation boundary where stress-induced fracturing concentrates. The reinforcement component also acts as a debonding agent for the yielding bolt. The yielding bolt can be in the form of conebolt, MCB, or end-anchored bolt with a yielding mechanism (sliding or steel stretching). The version with the inner bolt using MCB is called MCB-Superbolt (Figure 3a), and version with the inner bolt using VersaBolt is called Versa-Superbolt (Figure 3b). The outer diameters of the corrugated steel sleeve and the smooth bar of the MCB-Superbolt are 28.3 and 17.2 mm, respectively. The outer diameters of the corrugated steel sleeve and the smooth bar of the Versa-Superbolt are 32.3 and 20.6 mm, respectively. Under static and/or dynamic loading, the corrugated steel sleeve can tight the rock mass in place and increase the rock mass strength; the yielding bolt can slide or yield and thus absorb dynamic energy and accommodate large wall deformation.

3.3 *Laboratory test results*

Static and dynamic tests were conducted at CANMET's testing laboratory in Ottawa, Canada, to evaluate the load, deformation and energy absorption capacities of the Superbolts. Due to paper length limitation, only selected test results of the MCB-Superbolt are presented in the following.

The tested MCB-Superbolts, which are 2400 mm long, have a cone head in one end and a threaded section in the other end, and the bolt's shaft is covered by a 1700 mm long corrugated steel sleeve for rock reinforcement. The rockbolts are installed in 12.7 mm thick wall steel tubes to mimic boreholes in underground mines. The steel tubes used for the tests were 2194 mm in length with an internal diameter of 38 mm.

Two types of load transfer setups, direct loading and double-embedment or split-tube indirect loading, were used to test the capacities of the MCB-Superbolts. For the direct impact loading test, the rockbolts are installed in the thick-wall steel tubes and the drop weight directly impacts the plate attached at the threaded section of the rockbolts. This is referred as the continuous-tube configuration. In the split-tube test, a bolt is installed in two joining steel tubes and the drop weight impacts a steel seat welded to the lower steel tube (Figure 4). This is referred as the split-tube configuration. The direct loading test method provides the load–displacement characteristics of a rockbolt when it is loaded via the plate. For the static pull tests, only the direct loading test was conducted. For the dynamic drop tests, both direct and indirect loading tests were conducted.

Based on the load–displacement curves of a MCB-Superbolt under static loading, it is found that the MCB conebolt starts to slide when the load reaches about 100 kN. The cone starts to lock up at about 107 mm displacement and the bolt steel starts to experience yielding and strain-hardening deformation. As loading continues, the bolt load increases and the bolt fails on the smooth bar with a total displacement of 278 mm and a maximum load of 196.5 kN. The total energy absorbed by the bolt under this static loading configuration is about 37 kJ.

Dynamic drop tests were conducted using the two configurations described above. Figure 5 presents one test results of a MCB-Superbolt under direct impact loading. The input energy for each impact is 25 kJ. On the first impact, the peak load reaches 110 kN but quickly drops due to the sliding of the cone. The low load in the first 100 mm displacement is due to insufficient mixing of resin near the toe, which is a common problem in lab testing if sufficient rotation is not achieved during bolt installation. As the cone further slides, the resistance to load increases gradually. The maximum load reaches about 150 kN when the plate displacement is 438 mm. The energy absorbed by the bolt is 29.4 kJ, which is higher than the input energy due to extra bolt displacement. On the second impact, the cone head locks up at the corrugated steel sleeve and the load increases rapidly. In this case, steel stretching is the main mechanism that absorbs the dynamic energy. The peak load is 196 kN and the maximum plate displacement is 164 mm. The bolt absorbed 26.9 kJ energy in the second impact loading and it did not fail. A third impact was applied to the bolt, and the bolt absorbed 0.6 kJ energy before it failed at the threaded section. The accumulated energy absorbed by this bolt is 56.9 kJ.

Figure 6 presents one test results of a MCB-Superbolt under indirect impact loading. The input energy for each impact is 25 kJ. On the first impact, the peak load reaches 448 kN and after the corrugated steel sleeve breaks the load varies around 140 kN. The energy absorbed by the bolt is 22 kJ, which is slightly smaller than the input energy. The bolt deformed and absorbed 27.4 kJ energy on the second impact, and it failed close to the cone 120 mm from the paddle. For some reasons the cone was

Figure 4. CANMET drop test facility showing the setup for split-tube indirect loading test.

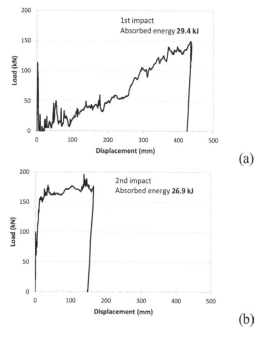

(a)

(b)

Figure 5. Load–displacement and energy–displacement curves of a MCB-Superbolt under direct impact loading: (a) first impact; (b) second impact.

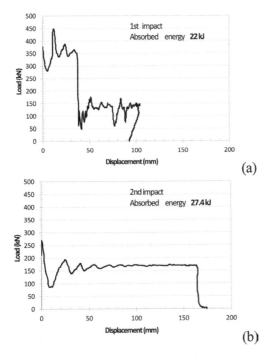

(a)

(b)

Figure 6. Load–displacement and energy–displacement curves of MCB-Superbolt under indirect impact loading: (a) first impact; (b) second impact.

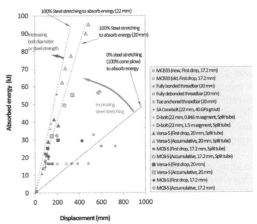

Figure 7. Comparison of energy dissipation capacities of MCB- and Versa-Superbolts (MCB-S and Versa-S) with other bolts.

locked up in the resin and there was a very small amount (14.2 mm) of cone displacement in the first impact and there was no cone displacement in the second impact. The accumulated energy absorbed by this bolt is 49.2 kJ, mostly from steel stretching. The accumulated displacement is 267 mm. The non-sliding of the bolt was probably due to the friction caused by the insertion of resin between the corrugated steel sleeve and the smooth bar during installation. If this were avoided, there would be more cone sliding to absorb more energy.

Due to time and budget constraints, only a limited number of tests were performed to test the capacities of the MCB-Superbolt. Based on the performance of the bolts, it is reckoned that the minimum energy absorption capacities of the bolt on a single impact loading under the continuous-tube and the split-tube configurations are higher than 25 kJ. If cone sliding is not prohibited, the energy absorption capacities can be higher. Note that the diameter of the bolt is only 17.2 mm and a guaranteed single energy absorption capacity of 25 kJ for two impact loadings demonstrates that the bolt can be used for dynamic rock support.

The accumulative energy absorption capacities of the MCB-Superbolts are presented in Figure 7, along with the tested energy capacities of the first impact loading. The test results of the Versa-Superbolts and other dynamic rockbolts are also shown in the figure. The superb energy absorption capacity for repeated dynamic loading of the Superbolts is obvious from the plot. It is seen from

the test results that the MCB-Superbolts can absorb about 30 kJ energy on the first impact in the direct loading configuration if there is cone sliding. The MCB-Superbolts can absorb more than 25 kJ energy on the first impact in the indirect loading configuration. In the split-tube configuration, the corrugated steel sleeve broke on the first impact loading. The test data indicate that the corrugated steel sleeve helps to absorb about 12 kJ energy. It is seen that the corrugated steel sleeve not only fulfils the role of rock reinforcement, it also helps to absorb dynamic energy and increase the effective diameter of the bolt. Compared with the MCB33 rockbolts, the MCB-Superbolts have more energy capacity at small wall displacements. In practice, it is important to design rock support that can limit the amount of allowable inward displacement to minimize rehabilitation. Cai & Kaiser (2018) suggest that the one-sided maximum practical displacement limit of drift is in the range of 200 to 300 mm. The Superbolts can provide rock support to achieve the design goal if the accumulated energy is less than 50 kJ. For the MCB-Superbolt, the sliding length (the distance between the cone head and the far end of the corrugated steel sleeve) can be configured to limit the amount of total displacement.

3.4 Discussions

The Superbolts have both rock reinforcement and holding capacities thereby can realize both support roles in one support component. Thus, it offers many advantages such as:

– One-pass rock support system to speed up drift development (Figure 2). This will translate into major cost saving compared with the conventional two-pass system (Figure 1).
– Immediate reinforcement to preserve the rock strength and provide immediate protection against

rockburst damage. The Superbolts can reinforce the rock mass and increase the rock mass strength. Using conventional two-pass system to install rockburst support can put workers at great risk because a rockburst can occur before the rockburst support is installed. Using the Superbolts, however, the problem is resolved and mining personnel are protected under enhanced support immediately after the drift is excavated and supported. This can increase workplace safety significantly.

- Reduced direct support cost because fewer labor is involved for rock support installation, fewer boreholes are drilled, and less resin is used.
- Easy installation. Installing the new rockbolt is basically the same as installing a rebar.Increased shear strength of the rockbolt.

4 CONCLUSIONS

Ground support products that can support increased stress loading, shearing and dynamic loading are invaluable assets that can increase a mine's efficiency and cost effectiveness. In response to the needs from the mining industry, two new dynamic rockbolts have been developed and their load, displacement and energy capacities tested using static pull test and dynamic drop test. The test results show that the MCB-Superbolt performs well in both direct and indirect dynamic loadings. The energy absorption capacities of the bolt on the first impact loading are at least 25 kJ in the direct and indirect dynamic loadings. The test results reveal that the accumulative energy capacities of the bolt are greater than 50 kJ.

The developed new ground support technology is most suitable for grounds with high risk of rockburst damage. The first economic benefit is a direct result of the change from a two-pass installation of ground support to a one-pass system. This will decrease the time required for drift development and the cost of labor to install rockbolts. In addition, the increased capacity (static and dynamic as well as shear strengths) of the rockbolt will decrease the likelihood of rockburst damage in underground installations. In strainbursting grounds, the reinforcement provided by the Superbolts can enhance the rock mass strength so that the rock mass will less likely to fail; even if the rock mass fails with a large energy release, the yielding of the bolt can absorb the released energy and control rock mass deformation. The Superbolts form a two-tier defence system. It is anticipated that the use of the new ground support technology can improve mine safety and reduce cost associated with dealing with rockburst risk.

ACKNOWLEDGEMENTS

The author would like to acknowledge financial and technical supports from NSERC (CRDPJ 461108-13) and Mansour Mining Technologies Inc. The author also would like to thank R. Royer and D. MacDonald of CANMET for executing the dynamic drop test.

REFERENCES

Board, M., Brummer, R., Kaiser, P.K. & Tannant, D. 1996. Kidd deep mining risk assessment. Phase 2 Report: Submitted to Falconbridge Ltd. Kidd Creek Division.

Cai, M. 2008. Influence of intermediate principal stress on rock fracturing and strength near excavation boundaries - insight from numerical modeling. *Int. J. Rock Mech. Min. Sci.*, 45(5),763–772.

Cai, M. 2013. Principles of rock support in burst-prone grounds. *Tunnelling and Underground Space Technology*, 36(6),46-56.

Cai, M. & Champaigne, D. 2009. The art of rock support in burst-prone ground. In: Tang, C.A. (eds.). *RaSiM 7: Controlling Seismic Hazard and Sustainable Development of Deep Mines*. Rinton Press. pp. 33–46.

Cai, M. & Champaigne, D. 2012. Influence of bolt-grout bonding on MCB conebolt performance. *Int. J. Rock Mech. Min. Sci.*, 49(1),165-175.

Cai, M. & Kaiser, P.K. 2018. *Rockburst support reference book. Volume I: Rockburst phenomena and support characteristics*. MIRARCO, Laurentian University, Sudbury, Ontario.

Cook, N.G.W. & Ortlepp, W.D. 1968. A yielding rockbolt. *Chamber of Mines of South Africa Research Organization Bulletin*, 14(6–8.

He, M., Gong, W., Wang, J., Qi, P., Tao, Z., Du, S. & Peng, Y. 2014. Development of a novel energy-absorbing bolt with extraordinarily large elongation and constant resistance. *Int. J. Rock Mech. Min. Sci.*, 67(April), 29-42.

Jager, A.J. 1992. Two new support units for the control of rockburst damage. In: Kaiser, P.K. & McCreath, D.R. (eds.). *Rock Support in Mining and Underground Construction* pp. 621–631.

Kaiser, P.K. & Cai, M. 2012. Design of rock support system under rockburst condition. *Journal of Rock Mechanics and Geotechnical Engineering*, 4(3),215–227.

Kaiser, P.K., Tannant, D.D. & McCreath, D.R. 1996. *Canadian Rockburst Support Handbook*. Geomechanics Research Centre, Laurentian University, Sudbury, Ontario.

Kaiser, P.K., Suorineni, F., Henning, J., Diederichs, M.S., Cai, M., Tannant, D.D. & Hajiabdolmajid, V. 2003. Guidelines on Support Selection for Safe Rapid Drift Development, Report to Canadian Mining Industry Research Organization (CAMIRO), p. 237.

Li, C. & Charette, F. 2010. Dynamic performance of the D-Bolt. In: Van Sint Jan, M. & Potvin, Y. (eds.). *Proc. 5th Int. Seminar on Deep and High Stress Mining* pp. 321–328.

Ortlepp, W.D., Bornman, J.J. & Erasmus, P.N. 2001. The Durabar - a yieldable support tendon - design rationale and laboratory results. *5th Int. Symp. on Rockburst and Seismicity in Mines* pp. 263–266.

Simser, B., Andrieus, P. & Gaudreau, D. 2002. Rockburst support at Noranda's Brunswick Mine, Bathurst, New Brunswick. In: Hammah, R., Bawden, W., Curran, J. & Telesnicki, M. (eds.). *NARMS 2002*. University of Toronto Press. 1 pp. 805–813.

Varden, R., Lachenicht, R., Player, J., Thompson, A. & Villaescusa, E. 2008. Development and implementation of the Garford dynamic bolt at the Kanowna Belle Mine. *10th Underground Operators' Conference* pp. 95–102.

2019 Rock Dynamics Summit– Aydan et al. (eds)
© 2019 Taylor & Francis Group, London, ISBN 978-0-367-34783-3

A parametric study of fault-slip in longwall mining

C. Wei, C. Zhang & I. Canbulat
School of Minerals and Energy Resources Engineering, University of New South Wales,
Sydney, Australia

ABSTRACT: Fault-slip induced by longwall mining is a significant cause of indirect coal burst. This paper summarises the results of a numerical modelling study using FLAC³ᴰ, in which a longwall face approaches a fault with different fault angles of 45°, 60°, 75° and 90°. Zero-thickness interface elements, which are capable of Coulomb sliding, were built in the numerical model to simulate the fault. The paper presents detailed parametric analysis of the influence of cover depth, friction angle of the fault plane, fault angle and approaching direction. Numerical results showed that these factors have a significant influence on shear stress drop, shear slip and the magnitude of seismic energy. In terms of fault behaviour during longwall extraction, shear stress on the fault plane above the coal seam first increased and then decreased when the longwall face was approaching the fault. The maximum magnitude of seismic events occurred when the longwall face was 15 m to 40 m away from the fault. Most seismic slip occurred on the fault plane above the coal seam with minimal movement below the coal seam.

1 INTRODUCTION

Coal burst has been a major problem in underground mining for decades (Mark, 2016). There are many factors that can result in coal burst, such as deep cover depth, massive rock layers above or below the coal seam, faults and other geological problems (Ortlepp, 2005; Iannacchione & Tadolini, 2016; Zhang et al, 2017). Of these causal factors, Ortlepp (2005) stated that fault-slip is a significant factor that can produce higher potential seismic energy than the other factors. Therefore, this paper mainly focuses on seismic behaviour along a major fault when the longwall face approaches it in various geological and geotechnical conditions. Several numerical models were built using FLAC³ᴰ to conduct a detailed parametric study to evaluate impact of the fault angle, friction angle of fault and mining depth. The magnitude of seismic events induced by fault-slip in longwall mining is evaluated.

2 NUMERICAL ANALYSIS

2.1 *Model description*

A series of global models were established using FLAC³ᴰ, as shown in Figure 1 where the green line represents a major fault and the layer in black is the coal seam. There are seven rock layers in the z-direction. The thickness of each layer is shown in Table 1. The top boundary was 150 m above the coal seam and the bottom boundary was 90 m below the coal seam. The longwall face was extracted from the starting line all the way to the fault and finally crossed the fault to stop at the end line. The starting line was 100 m to the right boundary and 480 m to

the end line. The distance between the left boundary and the end line was a minimum of 140 m, which varied depending on the fault angle in the model to ensure that the boundary effect is minimized and the mesh size is acceptable in each model.

The coal seam and the immediate roof were extracted together. The thickness of the immediate roof was four times the coal seam thickness. The abutment angle was kept constant at 20°. For the first 380 m, the excavation length was 10 m per step. It then decreased to 5 m per excavation step when the longwall face was within 100 m of the end line. An elastic goaf material with 500 MPa of Young's modulus was used to fill in the goaf area during extractions.

For boundary conditions, a vertical load determined by cover depth was put on the top boundary. The horizontal boundaries were fixed only in their normal direction. The bottom boundary was fixed in all directions. The initial horizontal stress applied in the model is given by Esterhuizen et al (2010), as shown in Equation 1:

$$\sigma_{h\max} = 1.2\sigma_v + 2.6 + 0.003E$$
$$\sigma_{h\min} = 1.2\sigma_v + 0.0015E \tag{1}$$

Where E is the elastic modulus of each rock layer and σ_v is the vertical overburden stress, $\sigma_{h\max}$ is the maximum horizontal principal stress (in x-direction) and $\sigma_{h\min}$ is the minimum horizontal principal stress (in y-direction).

The mechanical properties of the rock layers and the coal seam are summarized in Table 1. A strain softening failure criterion was applied to all rock layers. Residual tension was set at 0 MPa and the residual cohesion was set to 10% of the peak cohesion for all rock layers.

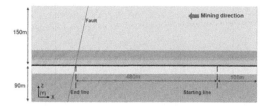

Figure 1. Schematic diagram of the basic models.

Table 1. Mechanical properties used for rock layers

Material	t	E	υ	c	φ	σ_t
	(m)	GPa		MPa	(°)	MPa
Shale	111	6	0.25	3.3	24	1
Sandstone	18	8	0.25	5.5	26	1.7
Shale	6	6	0.25	3.3	24	1
Shale	12					
Coal seam	3	3	0.25	1.2	28	1
Shale	20	6	0.25	3.3	24	1
Shale	70	7	0.25	4.5	25	1.4

t: thickness; E: Elastic modulus; υ: Poisson's ratio; c: cohesion; φ: friction angle; σ_t tensile strength

2.2 Fault properties

The FLAC^{3D} built-in zero-thickness interface elements were adopted to represent the fault. The cohesion, friction angle and tensile strength of the fault material were predefined in the model. The friction angle is one of the main factors in this parametric study. Three values adopted in the models: 15°, 25° and 35°. The cohesion and the tensile strength were set at 0 MPa. Normal and shear stiffness were set at 3.2 GPa/m.

3 PARAMETRIC STUDY AND SEISMIC SOURCE PARAMETERS

3.1 Parametric study cases

The influence of fault angle, friction angle and the cover depth on fault-slip behaviour was examined using the 10 models listed in Table 2. The fault angle was 75° in Model 5, which is the same as the fault angle in Model 2, but the approaching direction is in the reverse direction. This is the only model where the longwall face was extracted from the roof to the floor. The purpose of this model was to study the influence of the approaching direction on fault-slip behaviour.

3.2 Seismic source parameters

The following seismic parameters were used to evaluate fault-slip behaviour: seismic energy, E_S, seismic moment, M_o, and moment magnitude, M. These parameters were introduced by Aki & Richards

Table 2. Parametric study plan

Model	Fault angle (°)	Friction angle (°)	Cover depth (m)
1	90	25	800
2	75	25	800
3	60	25	800
4	45	25	800
5	75_reverse*	25	800
6	75	15	800
7	75	35	800
8	75	25	1000
9	75	25	600
10	75	25	400

*The 'reverse' means the longwall face approached the fault from hanging wall to footwall

(2002) and are widely used to study mining-induced fault-slip (Gibowicz & Kijko, 1994; Sainoki & Mitri, 2014). The parameters are:

$$E_s = 0.5\Delta\sigma \bar{D} A \tag{2}$$

$$M_0 = G\bar{D}A \tag{3}$$

$$M = \frac{2}{3}\log M_o - 6 \tag{4}$$

Where $\bar{D}$ is the average shear displacement along the fault plane, and A is the area where sliding takes place. Given the two-dimensional nature of the fault model, a circular slip area with a diameter equal to the slip length is assumed. $\Delta\sigma$ represents the stress drop defined as the average difference between the shear stress on the fault before an excavation and the shear stress after the excavation, as given by Equation 5:

$$\Delta\sigma = \frac{1}{A}\int_s \left[\sigma(t_2) - \sigma(t_1)\right] dS \tag{5}$$

4 RESULTS AND DISCUSSION

This section discusses the numerical modeling results obtained from the models listed in Table 2 and provides analysis of the parametric study. The seismic events in simulations are defined as the activities which occurred in the fault area where the fault is slipping and the shear stress of this area is decreasing (stress drop). Only seismic events with slip increase larger than 0.01 m and stress drop greater than 1 MPa were identified and analyzed.

4.1 Fault-slip behaviour in numerical simulations

A total of three parameters are defined and analysed in this section. When the longwall face was approaching the fault, the distance between them at which the fault began to slip is defined as the activation distance. The fault-slip area then increased when the longwall face was closer to the fault. The maximum slip length is defined as the maximum length of a consecutive seismic slip area along the fault plane during all excavations in each model. The seismic slip area in this section is defined as the area in which the shear stress drop is larger than 1 MPa and the shear displacement increase is greater than 0.01 m. The maximum slip is the maximum shear displacement which occurred along the fault in each model.

During longwall excavations, the fault began to slip earlier when the cover depth increased, and the friction angle and the fault angle became lower. The earliest slip occurred in Model 4, where the fault angle is 45°. In this case, the slippage of the fault plane began to occur when the longwall face was 80 m away from the fault. The smallest activation distance of 40 m was recorded in Model 10 with a cover depth of 400 m.

The friction angle has the largest influence on the maximum slip. The largest slip value was 1.53 m when the friction angle was 15° and the smallest value was 0.06 m when the friction angle was 35°. In contrast, the fault angle had limited impact on the maximum slip.

The results indicated that the friction angle has a significant impact on the maximum slip length. The maximum slip length decreased from 38 m to 5 m when the friction angle increased from 15° to 35°. In terms of the influence of fault angle, the maximum slip length remained constant at approximately 25 m when fault angle increased from 45° to 75°, but it decreased to 16 m when the fault angle was 90°.

4.2 Maximum seismic energy

The maximum seismic energy is defined as the maximum energy released by an adjoining seismic slip area during excavations in each model. Fig. 2 shows the maximum seismic energy results and the corresponding seismic magnitudes obtained from the models presented in Table 2. As evident in this figure, both the cover depth and the friction angle have a significant influence on the maximum seismic energy release. The maximum seismic energy increased exponentially when the mining depth exceeded 600 m.

The largest seismic energy events occurred in Model 6 with a magnitude of $7.31*10^8$ J, where the friction angle was 15°. The value then decreased sharply to $3.5*10^5$ J when the friction angle increased to 35°. The significant difference reflects the strong influence of the friction angle on the magnitude of released seismic energy produced by fault-slip.

In terms of the fault angle, the largest magnitude of seismic energy ($1.6*10^8$ J) was released on the fault with a 60° angle (Model 3) rather than the fault

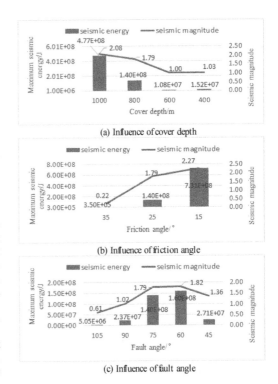

(a) Influence of cover depth

(b) Influence of friction angle

(c) Influence of fault angle

Figure 2. Maximum seismic energy and the corresponding magnitude in each model.

with a 45° angle (Model 4), although the fault began to slip 80 m away from the fault in Model 4, which is earlier than the 50 m away from the fault in Model 3.

Another important finding about the approaching direction was that the released seismic energy reduced steeply if the approaching direction was from the roof to the floor (in Model 5), which is shown in Fig. 2(c). The term '105' (180-105=75) means that the 75° fault angle is in a reverse inclined angle. In addition, the total number of seismic events in Model 5 was less than that in other models.

5 CONCLUSIONS

This paper presented a parametric study to investigate the influence of cover depth, fault angle, friction angle of the fault and approaching direction on fault-slip behaviour during a longwall extraction. FLAC^{3D} models were designed to simulate a longwall face approaching a fault with different fault angles. The numerical results show that all of these factors influence fault-slip behaviour. The released seismic energy increased with increased cover depth and decreased with decreased friction angle. The results revealed that a 60° fault angle can produce the largest seismic event, while the magnitude of seismic events sharply decreased when the approaching direction was from the roof towards the floor.

REFERENCES

Aki, K and Richards, P G, 2002. *Quantitative Seismology* (University Science Books).

Alber, M, Fritschen, R, Bischoff, M and Meier, T, 2009. Rock mechanical investigations of seismic events in a deep longwall coal mine, *International Journal of Rock Mechanics and Mining Sciences*, 46(2),408–420.

Esterhuizen, E, Mark, C and Murphy, M M, 2010. Numerical model calibration for simulating coal pillars, gob and overburden response, in *Proceedings of the 29th International Conference on Ground Control in Mining, Morgantown, WV*, pp 46–57.

Gibowicz, S J and Kijko, A, 1994. *An Introduction to Mining Seismology* (Academic Press).

Iannacchione, A and Tadolini, S C, 2016. Occurrence, prediction, and control of coal burst events in the US, *International Journal of Mining Science and Technology*, 26(1),39–46.

Itasca, 2006. FLAC3D Fast Lagrangian analysis of continua in 3D dimensions, *User's guide*.

Mark, C, 2016. Coal bursts in the deep longwall mines of the United States, *International Journal of Coal Science & Technology*, 3(1),1–9.

Ortlepp, W, 2005. RaSiM comes of age—a review of the contribution to the understanding and control of mine rockbursts, in *Proceedings of the Sixth International Symposium on Rockburst and Seismicity in Mines, Perth, Western Australia*, pp 9–11.

Sainoki, A and Mitri, H S, 2014. Dynamic behaviour of mining-induced fault slip, *International Journal of Rock Mechanics and Mining Sciences*, 66, 19–29.

Zhang, C, Canbulat, I, Hebblewhite, B and Ward, C R, 2017. Assessing coal burst phenomena in mining and insights into directions for future research, *International Journal of Coal Geology*, 179, 28–44.

T8: Nondestructive Testing Using Shock Waves

T9: Case Histories of Failure Phenomenon in Rock Engineering

2019 Rock Dynamics Summit– Aydan et al. (eds)
© 2019 Taylor & Francis Group, London, ISBN 978-0-367-34783-3

Tunnel face monitoring system for detecting and warning falling rocks

T. Tani, Y. Koga & T. Aoki
Taisei Corporation, Tokyo, Japan

T. Hayasaka & N. Honma
Nikko Denki Tsushin Ltd., Tokyo, Japan

ABSTRACT: In mountain tunneling, construction works are in the proximity of tunnel faces where falling rocks and collapse of ground are often feared. It is however difficult to improve quickly such work environments e.g., by applying remote construction and complete mechanization. In reality, an exclusive watch guard stands in place and supervises instability of the tunnel face, when charging/connecting explosives and installing tunnel supports. When the guard judges that the continuation of tunneling work is dangerous, he gives warning to the workers, so they can immediately leave from the work area. A newly-developed face monitoring system applies a high-speed camera set beside the watch guard and the image recognition technology. It promptly arrests fall of pebbles and pieces of sprayed concrete, symptoms of possible tunnel face collapse, that could lead to disaster, gives warning within a mere 0.1 seconds after the detection of movements of falling objects, and gets the construction workers to leave immediately from the dangerous spot. Effectiveness of this system in securing safety by foretelling falling rocks was verified in a mountain tunneling site.

1 INTRODUCTION

In mountain tunneling, construction workers carry out various operations near the tunnel face, depending on the stage of construction. In particular, during preparations for blasting to excavate hard rock and during installation of steel arch supports (bent H-section steel beams) to support the surrounding rock after excavation, it is necessary for the construction workers to be close to the tunnel face where the risk of injury is high. Actual operations at the tunnel face are affected by rock failure or fragility of the ground, which can cause falling rocks or fragments of sprayed concrete to fall. Of the 47 persons that were killed or injured in the 44 falling debris accidents that occurred from the year 2000 to the year 2010, 6% were killed, and 42% were out of work for one month or longer, according to a technical document (National Institute of Occupational Safety and Health, Japan, 2012).

Although development of technologies to carry out these operations automatically or by remote control of construction machinery is advancing, at present the technical level has not reached the stage where these technologies can be applied to all sites. In operations at the tunnel face, measures are therefore taken such as applying sprayed concrete to the natural ground that is exposed during excavation, or measures such as installing a rock fall protective mat above the construction workers in order to ensure safety of the workers carrying out operations close to the tunnel face. In addition, a full-time tunnel face watch guard is deployed to watch out for abnormal occurrences at the tunnel face and immediately warn the workers to evacuate, in order to avoid a serious accident (see Figures 1).

2 FALLING ROCK DETECTION AND ALARM DEVICE

2.1 *Equipment configuration and specification of the device*

Figure 2 shows a configuration of the equipment in the falling rock detection and alarm device. During operation, a camera with lighting is installed at a location (5 to 10 m) away from the tunnel face so as not to obstruct the operations. Falling objects such as falling debris

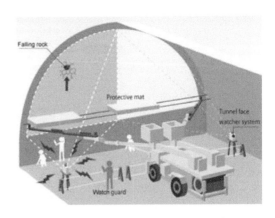

Figure 1. Workers and safety measures at tunnel face.

are detected from images taken by the camera, to form a warning system. Figure 3 shows the device installed within a tunnel and a front view of the device.

This device monitors small falling rocks of about 1 cm size, as a forewarning of larger rock falls or collapse of the tunnel face. A frame difference method is used for detection of falling rocks. In this method images taken at high speed (30 to 50 times per second) by a high-speed camera are used to generate difference images between two consecutive frames (images), to detect objects moving in the vertical direction as falling rocks. Note that the image processing and image detection technology used in this device will be described in detail in Section 3.

A Light Emitting Diode (LED) lighting is used to provide illumination over the area being monitored for falling rocks. Near infrared light is irradiated so that it does not interfere with laser pointers used by the workers or (visible) lights from the construction machinery, etc. A high-performance personal computer (PC) is included in the system for sensing and detecting falling rocks, by analyzing the images taken by the camera with an optical filter that allows only near infrared light to pass through the lens installed. This PC has necessary information

on the conditions for taking images to detect falling rocks. This includes, for example, camera angle of view, resolution, frame rate, and distance from the tunnel face to the camera. The PC is also equipped with information on the objects to be detected. This includes, for example, size of the falling rocks to be detected (minimum and maximum) and information on falling speed. The specification of the PC used is shown in Table 1.

When a falling object (falling rock or peeled sprayed concrete, etc.) is recognized, the alarm device immediately informs the workers of the danger by flashing LED warning lights and the sound of a siren.

2.2 Issues in development

– During development, the following three issues were extracted, as a result of a study of similar equipment operation.The equipment has to be able to withstand the severe environment of a tunnel construction site, such as dust, high humidity, and vibration. Also the equipment has to be concise and light so that it can be easily handled.
– The equipment has to be capable of reliably detecting small falling rocks down to about 1 cm size, regardless of the color of the stone or the background color.
– The equipment must not respond to movements other than falling rocks or to noise, and must not generate false alarms.

As for issue (1), each item of equipment was designed to have a protective enclosure (housing) with a rating of IP55 or higher in accordance with the IP standard for waterproof protective structures. Also, it was decided that the overall equipment including the PCs used for equipment control and image measurement and the LED projectors would have a structure completely free from fans.

Regarding issue (2), before the design of the device, using tunnel faces on rest days, rocks of various sizes and colors were artificially dropped

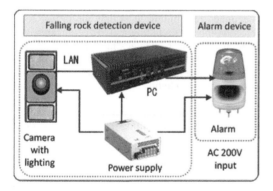

Figure 2. System configuration diagram of the face monitoring system.

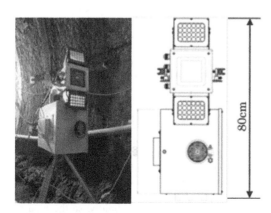

Figure 3. Exterior view of the face monitoring system.

Table 1. Specification on the PC used

Case	Box type
External dimensions	W 264.2mm × D 156.2 mm × H 66.5 mm
Expansion bay	2.5 type bay ×1 (interior 1)
Power supply voltage	DC input 9-36 V
Mass	About 3 kg
CPU	Intel Core i7-6700TE, 2.4 GHz, 8MB Cash
Memory	PC4-17000 (DDR4-2133), SO-DIMM 8 GB
Storage	SDD 2.5 inch MLC 120 GB
Usage environment	Temperature 0 to 50°C, humidity 10 to 80% (40°C)

while varying the level of illumination and the image capture speed of cameras in two different sensitivities and resolutions. This was to determine the necessary and ideal conditions for detection, which could then be reflected in the design. Note that the same tests were carried out when the sprayed concrete on the tunnel face forming the background for the detection images was both wet and dry. At present a popular 2 million pixel high-sensitivity camera is used in verification operations on site. It has been confirmed that falling rocks of size down to 1 cm can be detected under various imaging conditions.

For issue (3), there are concerns that if the device reacts to movements other than falling rocks and triggers an alarm, the workers will become used to the alarm and may not take appropriate evacuation measures when there is actually a danger. In connection with this issue it was decided that image recognition and judgment would have to be carried out using advanced image processing. These technologies are described in detail in Section 3.

3 IMPROVEMENT IN DETECTION ACCURACY BY IMAGE RECOGNITION

In this section, the result of development in image recognition technologies to prevent false alarms from being issued is described. This is the most important development key of the three main development issues.

3.1 *Method of preventing false detection*

3.2 *Overview*

Figure 4 shows a configuration of a falling rock detection algorithm. For detection of falling rocks, the images captured by the camera are input to an image acquisition unit as data in real-time. The falling rock detection is carried out based on this image data. First a falling rock detection area calculation unit determines falling rock detection exclusion areas based on difference images between frames in the past N frames (in practice N=3 to about 10). Then two falling rock detection units (individual and group falling rock detection units) detect falling rocks in the falling rock detection area, excluding those in these exclusion areas. If a falling rock is detected, data is transmitted to a falling rock detection notification unit indicating that a rock fall has occurred, and the alarm is issued.

By this method even if a worker or construction machinery is captured in the images, by setting that part as a falling rock detection exclusion area in advance, false alarms are eliminated, and only falling rocks can be accurately detected and the alarm can be correctly issued.

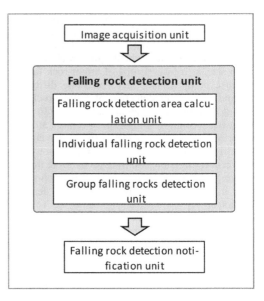

Figure 4. Schematic diagram of the falling rock detection algorithm.

3.3 *Recognition of moving bodies other than falling rocks*

The image recognition algorithm for falling rock detection includes a falling rock detection area calculation unit that recognizes moving bodies other than falling rocks such as workers or heavy machinery near the tunnel face, and identifies a falling rock detection area, excluding areas of these moving bodies; and two types of falling rock detection units that recognize moving bodies appearing in the falling rock detection area as falling rocks. In this section, the method of recognition of moving bodies and the processing method of excluding false detection are described.

The falling rock detection area calculation unit carries out processing on difference images between consecutive frames in a time series, extracts moving body areas from each of the difference images in the past N frames, and carries out a process to determine a falling rock detection area that is the area excluding the collective range of each moving body area. The method of calculation of the area for detection of falling rock candidates is explained using Figure 5. Difference images (middle row in Figure 5) are generated for two consecutive frames in the time series of the images for the last N frames (top row in Figure 5), and for each difference image a moving body area is extracted by carrying out a binarization process using a threshold value. Figure 5 shows a person raising one arm, and mainly the raised arm area is extracted as the moving body area. Note that in this case the trunk, head, the other arm, etc., apart from the raised arm have moved slightly, so these parts have also been extracted as the moving body area.

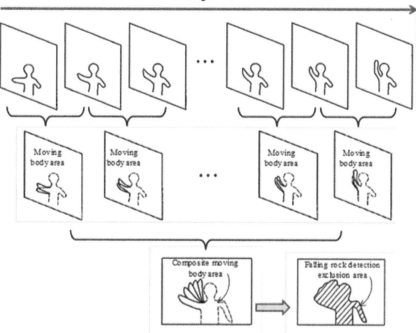

Figure 5. Recognition method for moving objects such as workers and machines.

Next, the falling rock detection area calculation unit obtains a composite moving body area (bottom row in Figure 5) by combining these moving body areas, and obtains the falling rock detection exclusion area as the composite moving body area by performing a labeling process allocating the same number to connected pixels as one group and an expansion process that is described later, so the falling rock detection area becomes the area apart from the falling rock detection exclusion area. In Figure 5, the area of the arm, the trunk, the head, the other arm, etc., that have moved in the past N frames mainly becomes the composite moving body area. Also, parts of the images that are bright and have a certain area are excluded from the falling rock detection area. This is to include reflective vests or hardhats worn by the workers in the falling rock detection exclusion area.

3.4 Improvement in falling rock detection accuracy

The falling rock detection unit includes an individual falling rock detection unit that identifies specific single moving bodies that appear in the falling rock detection area as falling rock candidates, tracks them through multiple frames, and judges whether or not they are actually falling rocks. The individual falling rock detection unit is described with reference to Figure 6. The top row in Figure 6 shows images taken in each of the three frames, and a falling rock appears in each of them. Here time passes from left to right. Also, the Y axis in Figure 6 shows the distance in the vertical downward direction as positive. As shown in Figure 6, the falling rock moves downward (natural fall) with time. The bottom row of Figure 6 shows the difference images for each frame, and in each difference images a group of falling rock candidates is included. The falling rock candidate on the upper side corresponds to the falling rock before it moves, while the falling rock candidate on the lower side corresponds to the falling rock after it has moved.

When two falling rock candidates have been detected in a group, the individual falling rock detection unit obtains the centroid calculated from the centroid positions the two falling rock candidates. It is a mathematical centroid, when the upper and lower falling rock candidates are considered in two-dimensional images. Centroid is calculated based on the shape, area, etc., of the falling rock candidate. Next, a judgment is made whether or not the position of the lower falling rock candidate in the difference image shown on the right side of the bottom row in Figure 6 comes within the natural fall area calculated from the previous difference image. The natural fall area indicates a movement range of the falling rock between frames for natural fall. For example it comes within a fixed angle and a certain fall distance relative to the previous position of the falling rock candidate. For this reason the natural fall

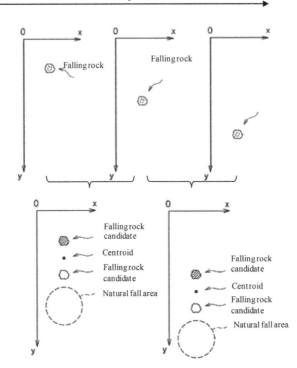

Figure 6. Detection method of individual falling rock.

area generally comes below the centroid of the falling rock candidate on the previous frame. For example, if the position of the falling rock candidate is within the natural fall area over multiple consecutive frames, the falling rock candidate is judged to be truly a falling rock.

In addition, the falling rock detection unit includes a group falling rock detection unit that detects all moving bodies appearing in the falling rock detection area as falling rock candidates, tracks all the centroids of the falling rock candidates over multiple frames, and judges whether or not they f are truly falling rocks. Consequently, it is possible to detect true falling rocks even when a multiple rock falls occur at the same time, and correctly to prevent overlooking falling rocks.

3.5 *Verification*

Verification was carried out into whether or not the moving bodies were correctly judged from the various image processing and the image recognition scheme described above and whether falling rocks only are properly notified, based on images acquired on site near the tunnel face to date and those on falling rock tests. Among 50 hours of video used in the verification, only representative verification images captured are shown here.

3.6 *Moving bodies*

Figure 7 shows a recognized image of a worker on a work platform (man cage) at the tip of a drill jumbo boom and the boom during movement. It can be seen from this figure that the boom, the worker, and randomly falling seepage water are not judged to be falling rocks. It was confirmed that image recognition could be carried out without issuing an alarm. Note that in the center of the figure, the area enclosed by the red line indicates the falling rock detection exclusion area resulting from the moving body recognition, whereas that enclosed by the green line outside the red

Figure 7. A result of moving object detection.

751

Figure 8. Confirmation of falling rock detection time in laboratory test.

line indicates the falling rock detection exclusion area after performing the expansion process. On the right-hand side of the photo, the position of water seepage is outlined in green lines. In this area, however, recognition of individual falling objects was carried out as described in Section 3.2, and this process was capable of judging that the seepage water falling continuously as multiple water drops should be included in the falling rock detection exclusion areas.

3.7 *Moving bodies*

Using the same equipment configuration, laboratory tests were carried out to confirm the time required to detect a falling rock. This was carried out by reading the position of the falling rock on the recognition image using a ruler that appeared in the same video at the time it was detected, and calculating the time until the detection of the falling rock.

The position at which detection occurred as read from the image was 5 cm (see Figures 8). According to the equation of free fall, it is equivalent to 0.1 seconds from the start of the fall to detection.

4 SUMMARY AND OUTLOOK FOR THE FUTURE

The falling rock monitoring system of the tunnel face watcher system has been completed as a technology for ensuring safety at the tunnel face, and its performance has been confirmed using test images obtained on site and laboratory tests. In particular, a method has been devised to ensure that falling rocks only are reliably detected for issuing an alarm, by using image recognition. In addition, at present the equipment size was reduced down to one third compared with the prototype. This concise equipment has been used on site since July 2018, contributing to ensuring safety of operations at the tunnel face.

For the future we are further slimming down the equipment to achieve major size reductions. In addition, it is considered necessary for specifications to be adapted to the conditions at the various sites. Verification tests will be carried out at various mountain tunneling sites with different cross-section size and geological conditions.

In January 2018 the Ministry of Health, Labour and Welfare (MHLW) has amended the "Guidelines for Falling Debris Accident Prevention Measures at Tunnel Faces in Mountain Tunneling Construction" (MHLW, Japan, 2018), and there is a perception that safety initiatives in the industry as a whole have accelerated. It is intended that devices for ensuring safety of construction and workers be actively and continuously developed in future in monitoring not only for rock falls at tunnel faces, but for slopes on working roads in mountainous areas.

REFERENCES

Ministry of Health, Labour and Welfare, 2018, Guidelines for Falling Debris Accident Prevention Measures at Tunnel Faces in Mountain Tunneling Construction (Translation from Japanese title).

National Institute of Occupational Safety and Health, Japan, 2012, Analysis of Survey of Labor Accidents due to Falling Debris at Tunnel Faces and Proposal of Prevention Measures (Translation from Japanese title), JNIOSH-TD-No. 2, pp. 5–6.

2019 Rock Dynamics Summit– Aydan et al. (eds)
© 2019 Taylor & Francis Group, London, ISBN 978-0-367-34783-3

Evaluation of the excavation disturbed zone of sedimentary rock in the Horonobe Underground Research Laboratory

K. Kubota
Central Research Institute of Electric Power Industry (CRIEPI), Chiba, Japan

K. Aoyagi & Y. Sugita
Japan Atomic Energy Agency (JAEA), Hokkaido, Japan

ABSTRACT: During the excavation of shafts and galleries in the deep subsurface for disposing of high-level radioactive waste, an excavation disturbed zone (EdZ) or excavation damaged zone (EDZ) is developed around the shafts and galleries. Such zones could influence the transfer behavior of radioactive nuclides, and it is therefore important to understand the mechanical and hydraulic behavior of the EdZ or EDZ. We performed in situ experiments such as seismic and electrical resistivity tomography and so on before, during, and after gallery excavation in galleries of 140 and 250 m in depth in an area of soft sedimentary rock in Japan. The results demonstrate that the extent of fractures induced by the gallery excavation related with EDZ was confined to about 0.45 m from the gallery wall in the 140 m gallery and to about 1 m from the gallery wall in the 250 m gallery. The extent of the unsaturated zone related with EdZ was about 1 m in the 140 m gallery, but an unsaturated zone did not appear in the 250 m gallery.

1 INTRODUCTION

During the excavation of shafts and galleries in the deep subsurface for disposing of high-level radioactive waste (HLW), an excavation disturbed zone (EdZ) or excavation damaged zone (EDZ) is developed around the shafts and galleries (Tsang et al. 2005). In such zones, several phenomena, for example the creation of fractures induced by excavation, the evolution of unsaturated zones due to the decrease in pore water pressure or degassing, and stress redistribution, will occur. These phenomena will lead to an increase in hydraulic conductivity and influence the transfer behavior of radioactive nuclides. Therefore, the development of an EdZ or EDZ may have significant impacts on the operation and long-term behavior of HLW disposal.

In this study, we performed in situ experiments such as seismic and electrical resistivity tomography and so on before, during, and after gallery excavation in an area of soft sedimentary rock in Japan in order to investigate the behavior of EdZ and EDZ.

2 OUTLINE OF THE INVESTIGATION

Our aim was to compare the characteristics of EdZ and EDZ in rocks with different geological and geo-physical properties; therefore, the investigation was conducted in galleries excavated to two different depths (140 and 250 m) located in the Horonobe

Underground Research Laboratory, Hokkaido, Japan (Figure 1). For both depths, several geological and geophysical experiments were conducted before, during, and after the excavation of horizontal galleries.

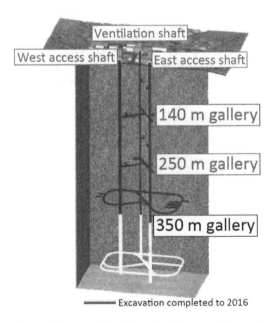

Figure 1. Layout of the Horonobe Underground Research Laboratory.

2.1 Investigation in the 140 m gallery

Figure 2 shows a schematic layout of the 140 m gallery. Two adits (adits No. 2 and No. 3) were excavated from the horizontal gallery. From these adits, horizontal boreholes were drilled before the horizontal gallery (broken lines in Figure 2; excavated from April 2009) was excavated. Measurement instruments were installed in these boreholes. We arranged the boreholes and instruments so as to avoid affecting the measurement conditions of other experiments. In this study, Borehole Television (BTV) observation, seismic tomography, electrical resistivity tomography, hydraulic conductivity analysis, measurements of water content and pore water pressure, and borehole loading tests were conducted (Kubota et al. 2017a).

2.2 Investigation in the 250 m gallery

Figure 3 shows a schematic layout of the 250 m gallery. Four boreholes (CE01, CE02, CE03, and CH01) were drilled before the horizontal gallery (red rectangles in Figure 3) was excavated. In the investigation area (blue square in Figure 3), geological observation on the wall of the gallery and boreholes, seismic tomography, electrical resistivity tomography, hydraulic conductivity analysis, and measurements of water content and pore water pressure, were conducted (Kubota et al. 2017b).

3 EXPERIMENTAL RESULTS

In this paper, we mainly demonstrate the results of the geophysical exploration.

3.1 Results in the 140 m gallery

Figure 4 shows the results of the seismic tomography and electrical resistivity tomography for the 140 m gallery. Specifically, Figure 4 shows the

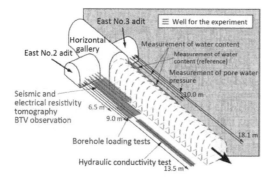

Figure 2. Schematic layout of the 140 m gallery (bird's-eye view). Solid lines show the area excavated before the experiment. Broken lines show the area excavated during and after the experiment.

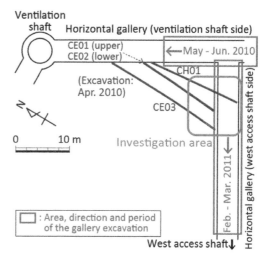

Figure 3. Schematic layout of the 250 m gallery (plane view). Red rectangles show the area excavated during and after the experiment. Well heads of boreholes CE01 and CE02 were situated in the same position but at different heights (CE01 and CE02 were 1.5 m and 0.5 m from the base plate of the gallery, respectively).

changes in P-wave velocity and electrical resistivity, relative to the values before the gallery excavation, at various times after the excavation (149 – 1759 days). The results demonstrate that P-wave velocity decreased by up to about 25 %, and electrical resistivity increased by up to about 50 %, due to the gallery excavation. The changes to these parameters were confined to within 1 m from the gallery wall. P-wave velocity and electrical resistivity changed soon after the gallery excavation started, and slightly changed in seasonably over a long period.

The water contents and pore water pressure measured at 1 and 2 m away from the gallery wall decreased after the gallery excavation. Assuming a water saturation of 100 % before the excavation, water saturation decreased to about 60 to 80 % (Yabuuchi et al. 2011; Kubota et al. 2017a). The results of the hydraulic conductivity analysis conducted at about 3 m away from the gallery wall demonstrate that hydraulic conductivity did not change due to the gallery excavation. The results of the geological observation from BTV or gallery wall demonstrate that the extent of fractures induced by the gallery excavation was confined to within 0.45 m of the gallery wall.

This suggests that the decrease in water saturation or the presence of fractures induced by the gallery excavation led to the change in P-wave velocity and electrical resistivity.

3.2 Results in the 250 m gallery

Figure 5 shows the results of the seismic tomography and electrical resistivity tomography for the

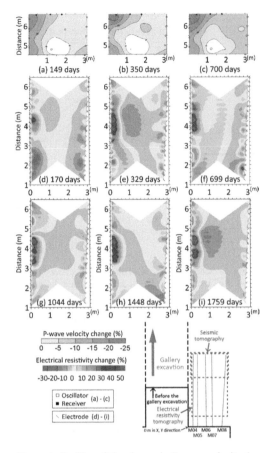

Figure 4. Profiles of the change in P-wave velocity (a - c) and electrical resistivity (d - i) for the 140 m gallery compared with before the gallery excavation. The number of days in each profile show the time of the measurement relative to the start of the gallery excavation.

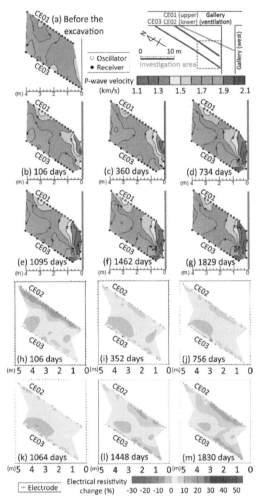

Figure 5. Profiles of P-wave velocity distribution (a - g) and electrical resistivity change (h - m) from the excavation of the 250 m gallery. The number of days in each profile shows the time of the measurement relative to the start of the gallery excavation.

250 m gallery. Specifically, Figure 5 shows changes in the profiles of P-wave velocity distribution and electrical resistivity, relative to before the gallery excavation, at various times after the excavation (106 – 1830 days). The results demonstrate that P-wave velocity decreased from about 1.6 km/s to about 1.2 – 1.4 km/s due to the gallery excavation. The P-wave velocity decrease was observed within 1 m of the gallery wall, and this extent was not expanded over the measurement period. On the other hand, the change in electrical resistivity was not due as much to the gallery excavation, but rather to seasonal change, which resulted in a change of about 10 %.

The results of the geological observation of the gallery wall demonstrate that fractures were induced by the gallery excavation. In the measurement sections within 0.5 to 1 m from the gallery wall, hydraulic conductivity (conducted in well CH01) increased by about 3 to 8 times following the gallery excavation (Kubota et al. 2017b). This

change is thought to be related to the fracturing. The decrease in P-wave velocity was related to the density of fractures induced by the gallery excavation (Aoyagi et al. 2014); accordingly, fractures were created by the gallery excavation within 1 m from the gallery wall.

At this depth, the water content was not measured in the borehole. However, we measured the water saturation of three rock samples acquired from the gallery wall in the investigation area 861 days after the start of the gallery excavation. As a result, the water saturation of all samples was more than 90% (Aoyagi et al. 2014). We also measured the electrical resistivity of rock samples from the boreholes in the investigation area of the 140 m and 250 m galleries to investigate the relationship between water saturation and electrical resistivity. The measurement results demonstrate that the change in electrical resistivity was small when water saturation was changed from 100 % to about 90 % (Figure 6). Generally, electrical resistivity was not changed much by fracture creation in the saturated condition. Accordingly, we can infer that electrical resistivity was not changed much and an unsaturated zone was not created due to the gallery excavation.

On the other hand, electrical resistivity experienced slight seasonal variation; that is, electrical resistivity in summer was lower than that in winter (Figure 7). Additionally, we measured the temperature in the borehole by putting four thermocouples in well CE02; Figure 8 shows the temporal changes of temperature in this well. The results demonstrate that temperature in summer was higher than that in winter; the difference was up to about 6 °C, and bigger near the gallery wall. Generally, electrical resistivity changes when temperature changes. When the temperature in the borehole changed from 20 to 15 °C, the electrical resistivity decreased by about 10 % (Figure 9). This suggests

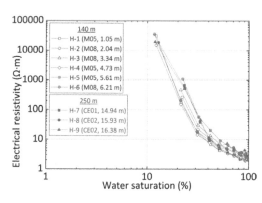

Figure 6. The relationship between water saturation and electrical resistivity of rock samples from the boreholes in the investigation areas of the 140 m and 250 m galleries.

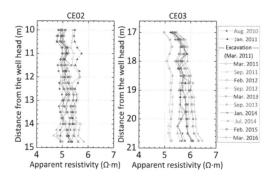

Figure 7. Apparent resistivity profiles measured in summer (red lines; July to September) and winter (blue lines; January to March) in wells CE02 and CE03.

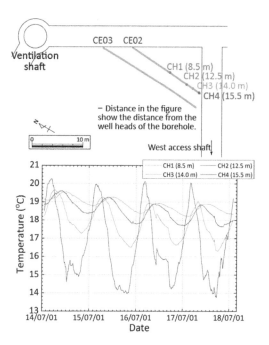

Figure 8. Temporal changes in temperature from July 2014 to August 2018 in well CE02.

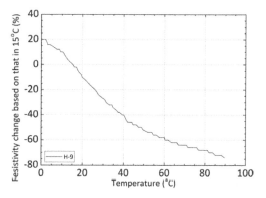

Figure 9. The relationship between the temperature and electrical resistivity of rock samples from the borehole in the investigation area of the 250 m gallery. Percentage changes in electrical resistivity are relative to that at 15 °C.

that the reason why electrical resistivity changed seasonally by about 10 % was due to temperature change in the well.

4 COMPARISON BETWEEN EXPERIMENTAL RESULTS FOR THE 140 M AND 250 M GALLERIES

We performed experiments in gallery excavation sites at two different depths, 140 and 250 m. The results for each depth showed different tenden-

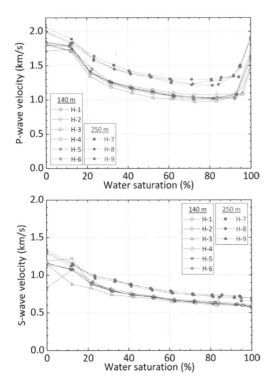

Figure 10. The relationship between water saturation and P-wave velocity (upper), S-wave velocity (lower) of rock samples from the boreholes in the investigation areas of the 140 m and 250 m galleries.

In future research, we will investigate further about mechanical and hydraulic properties at both depths (140 and 250 m) to clarify the developments and temporal changes of EdZ or EDZ.

5 CONCLUSION

In this study, we performed in situ experiments such as seismic and electrical resistivity tomography and so on before, during and after the excavation of galleries to investigate the behavior of EdZ or EDZ in the sedimentary rock. The results demonstrate that the extent of the fractures induced by the gallery excavation related with EDZ was about 0.45 m from the gallery wall in the 140 m gallery and about 1 m from the gallery wall in the 250 m gallery. The extent of the unsaturated zones related with EdZ was about 1 m from the gallery wall in the 140 m gallery, however an unsaturated zone was not observed in the 250 m gallery. One of the factors behind these differences is considered to be that the degree of diagenesis is different depending on the depth.

ACKNOWLEDGEMENTS

This study was performed as a collaborative research in the Central Research Institute of Electric Power Industry (CRIEPI) and Japan Atomic Energy Agency (JAEA).

REFERENCES

Aoyagi, K., Tsusaka, K., Kubota, K., Tokiwa, T., Kondo, K. & Inagaki, D. 2014. Investigations for a change of an excavation damaged zone with time at the 250m gallery in the Horonobe underground research laboratory, *Journal of Japan society of civil engineers, Ser. C (Geosphere Engineering)* 70: 412–423.

Kubota, K., Nakata, E., Oyama, T., Sugita, Y. & Aoyagi, K. 2017a. Evaluation of the Excavation disturbed Zone in the Horonobe Underground Research Laboratory (Part 1) - Investigation in the 140m gallery -, *CRIEPI report* N17005.

Kubota, K., Nakata, E., Suenaga, H., Nohara, S. & Aoyagi, K. 2017b. Evaluation of the Excavation disturbed Zone in the Horonobe Underground Research Laboratory (Part 2) - Investigation in the 250m gallery -, *CRIEPI report* N17006.

Tsang, C. F., Bernier, F. & Davies, C. 2005. Geohydromechanical processes in the Excavation Damaged Zone in crystalline rock, rock salt, and indurated and plastic clays - in the context of radioactive waste disposal, *International journal of rock mechanics & mining sciences* 42: 109-125.

Yabuuchi, S., Kunimaru, T., Kishi, A. & Komatsu, M. 2011. Monitoring of pore water pressure and water content around a horizontal drift through excavation – measurement at the 140m gallery in the Horonobe URL, *Journal of Japan society of civil engineers, Ser. C (Geosphere Engineering)* 67: 464–473.

cies. The distance from the gallery wall within which fractures were generated was about 0.45 m for the 140 m gallery and about 1 m for the 250 m gallery. An unsaturated zone appeared within 1 m from the gallery wall in the 140 m gallery, but no unsaturated zone was observed for the 250 m gallery. We investigated the reasons for these differences.

In both investigation areas (i.e, the 140 m and 250 m galleries), diatomaceous mudstones of the Koetoi Formation are distributed. The porosity of the rocks in the 250 m gallery (51 % to 52 %) is lower than that in the 140 m gallery (57 % to 59 %). Furthermore, the P- and S-wave velocities of the rocks in the 250 m gallery are higher than those in the 140 m gallery (Figure 10). One of the factors behind this difference is that degree of diagenesis is different at different depths; this difference in diagenesis results in a difference in strength of the rock mass, which may in turn influence the creation of the EDZ. The difference in the degree of diagenesis is also related to the difference in porosity. This porosity difference may have influenced the dissipation of pore water pressure, and the tendency of water saturation change will be different.

2019 Rock Dynamics Summit– Aydan et al. (eds)
© *2019 Taylor & Francis Group, London, ISBN 978-0-367-34783-3*

An attempt to estimation of continuous in-situ elastic modulus ahead of tunnel face in volcanic and pyroclastic rock area

K. Okazaki & T. Kurahashi
Civil Engineering Research Institute for cold region, Sapporo, Hokkaido, Japan

ABSTRACT: In this study, the elastic modulus along a borehole for the construction of a national highway tunnel in ground consisting of volcanic and pyroclastic rocks was estimated by using dynamic elastic modulus. The data used in estimating the elastic modulus were the P-wave velocity measured in the velocity logging during the boring and the P- and S-wave velocities, wet density, Poisson's ratio, and unconfined compressive strength of cores measured in the laboratory. As the results of analysis, in-situ elastic modulus along the borehole wall is used to assume the rough value of deformation modulus in the ground classification for a tunnel. And predicted tunnel deformation by in-situ strain along the borehole wall based on in-situ elastic modulus along the borehole wall is useful for setting the continuous control standard values of displacement in tunnel excavation.

1 INTRODUCTION

In conducting numerical analyses and reviewing the rock mass classification for a tunnel construction, physical property data that reflect the conditions of the site ground as much as possible and evaluation based on the mechanical strength of the site ground are necessary. Because of such necessities, a horizontal boring investigation (advanced boring) at the excavation tunnel face may be conducted during a construction. The ground conditions immediately adjacent to the excavation ahead of tunnel face are evaluated based on the P-wave velocities obtained in velocity loggings in advanced borings, deformation moduli obtained in the horizontal loading test, and the P-wave velocities obtained in the laboratory test. In the velocity logging, the S-wave velocity ahead of the tunnel face is not measured; however, the S-wave velocity ahead of the tunnel face is able to be estimated based on the P-wave velocity measured in-situ and P- and S- waves obtained in the core test. The continuous in-situ elastic modulus for the total length of the advanced borehole is able to be obtained by using the estimated S-wave velocity, wet density, and Poisson's ratio. This in-situ elastic modulus along the borehole obtained in this way enables evaluation of the ground at the construction site based on values that appropriately reflect the in-situ conditions. These elastic moduli are able to be applied to estimation of the strain in the ground around the advanced borehole and in setting the control standard displacement values that associate with the excavation of a tunnel.

In this paper, the in-situ elastic modulus along the advanced boreholes for the construction of a national highway tunnel in ground consisting of volcanic and pyroclastic rocks were obtained. The

elastic modulus was compared with the deformation modulus measured in the horizontal loading test conducted in the tunnel. We also reported on an experimental method for estimating the strain in an advanced borehole based on the elastic modulus for the wall of the advanced borehole.

2 OUTLINE OF THE ANALYSIS

2.1 *The data on the tunnel and the analysis method*

In this analysis, a total of 400 data points were used. The data set included the P-wave velocities measured in 63 advanced boreholes (Vph) of 11 national highway tunnels (total length of 15.4km) constructed by New Austrian Tunneling Method (NATM), and the P-wave velocities of bore cores (Vpc), S-wave velocities of bore cores (Vsc), wet density (ρw), and the unconfined compressive strength (σck) of the bore cores measured in the laboratory test. The Vph was measured by using the downhole logging method on the shotcreted tunnel face was applied. The Vpc and Vsc were measured in the ultrasonic propagation velocity tests. For the deformation modulus obtained from the horizontal loading test in the borehole (Ed), a total of 118 data points obtained by using the uniformly distributed loading test were used. The types of rocks of the site ground were andesite, dacite, lava, tuff breccia, and volcanic breccia. In the analysis, the elastic modulus along the advanced boreholes (Esh) were obtained using the above data points. The relationship between the Esh and the range of classified rock types of the ground, for which the Ed is able to be applied, was analyzed. Based on the relationship between the Esh and σck, the in-situ strain along the advanced borehole (εh) were estimated. The estimation accuracy was verified by comparing the estimated deformation based on the in-situ strain along the bore-hole wall with the deformation for a section for which the Vph was measured in an actual tunnel.

2.2 *In-situ elastic modulus along an advanced borehole wall*

In obtaining the Esh, the S-wave velocity along the borehole wall (Vsh) was calculated by using Equation (1).

$$Vsh = Vsc \cdot Vph / Vpc \tag{1}$$

Next, the dynamic elastic shear modulus (Gd) was calculated by using the Vsh, ρw and Equation (2).

$$Gd = \rho w \cdot Vsh^2 \tag{2}$$

Then, the Esh was obtained by using Gd, static Poisson's ratio (vs), and Equation (3).

$$Esh = 2(1+v_s) \cdot Gd \tag{3}$$

Where, Esh is the static elastic modulus obtained by using Gd and vs. After obtaining dynamic Poisson's ratio (vd) by using Vph, Vsh, and Equation (4), the value of vs necessary for obtaining Esh was determined based on the relationship between the vd and 159 data points of vs measured in three volcanic rock tunnels.

$$v_d = ((Vph / Vsh)^2 - 2) / (2((Vph / Vsh)^2 - 1)) \tag{4}$$

This value vs was input into Equation (3). The Esh is considered to be representing the condition close to the actual ground, because the ratio of Vpc to Vph, which represents the influence from the differences in the cracks in the core and those in the in-site ground, is considered.

2.3 *In-situ elastic modulus along an advanced borehole wall and the deformation of the ground from tunnel excavation*

In obtaining εh, a relational equation for Vph and Esh and that for σck were obtained. Okubo & Terasaki (1971) investigated the relationship between Vp and σck based on a total of 1,000 data points on the igneous rock, sedimentary rock, and metamorphic rock collected at dam and tunnel construction sites, and demonstrated that σck is able to be approximated as the following equation (5). In this analysis, the relationship between Vph and σck was investigated based on 400 data points on the volcanic and pyroclastic rock. After this investigation, εh was determined by using Esh and σck that corresponded to Vph.

$$\sigma ck = 10 \cdot Vp^3 \tag{5}$$

3 ANALYSIS RESULTS

3.1 *In-situ elastic modulus along the advanced borehole wall*

Figure 1 shows the relationship between the P-wave velocity (Vp) and S-wave velocity (Vs) of the core and those measured in the advanced borehole. The Vp is from 0.8 to 5.8 km/s, and the Vs is from 0.2 to 3.28 km/s. The Vp and Vs of the core and those of the borehole are found to be correlated. The Vp of the core and that of the borehole have a tendency of being in the range of about 1.6 to 1.8 times that of Vs. Therefore, it is considered to be acceptable to use Vsh, which is a calculated value, in estimating Esh.

Figure 2 shows the relationship between the v_d from velocity logging and the v_s from the unconfined compression test. v_d is from 0.25 to 0.45, and v_s is from 0.1 to 0.4. v_s is about 64 % of v_d. To obtain Esh, v_s was reduced by this percentage, and input into Equation (3).

Figure 3 shows the relationship between Vph and Ed and that between Vph and Ed. Esh is greater than Ed. The reasons for this are that, even though both are in-situ values, Ed is calculated using 0.3 for v_s, and Esh includes information from the laboratory test as Vpc, Vsc, and that the measurement methods

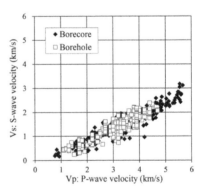

Figure 1. Vp and Vs of the core and those measured in the advanced bore-hole.

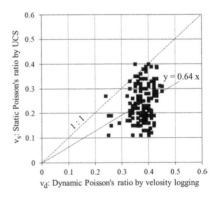

Figure 2. v_d by velocity logging and v_s by UTS.

759

and levels of strain differed in the velocity logging and horizontal loading test in the borehole.

3.2 *In-situ strain along the advanced borehole wall*

Figure 4 shows the relationship between Vph and σck. Even though these values have variations, σck was obtained in the approximation equation in the figure. The strain was determined by dividing the obtained σck by Esh.

Figure 5 shows the relationship between Vph and εh. When Vph is great, εh is small. It can be recognized that the elastic wave velocity is strain-dependent (Sugino et al, 2007).

4 DISCUSSION

4.1 *Deformation moduli in the classification of tunnel ground characteristics*

Figure 6 shows the relationship between Esh and Ed. The dotted line indicates the lower limit of Ed. The figure also shows the range for application of Ed in a ground classification (HRDB, 2018). When Esh is in the range from 5,000 to 15,000 MPa, the class of the in-site ground is determined as D2. When Esh is in the range from 15,000 to 40,000 MPa, the class of

the in-site ground is determined as D1. When Esh is greater than 40,000 MPa, the class of the in-site ground is determined as C2 or higher. By determining Esh, the rough value of Ed in the ground classification for a tunnel is able to be assumed.

4.2 *The strain and deformation along the borehole*

To predict the deformation associated with the excavation of a tunnel, deformation (Uh) based on the εh was obtained by multiplying εh by the tunnel diameter that corresponds to the class of the support pattern. Uh was verified by comparing the deformation associated with actual tunnel excavation (Ur). In this verification, Uh and Ur were compared by using a tunnel excavated in volcanic rocks as a model.

Figure 7 shows the results of the analysis for the model tunnel. When Ur is great, Uh is also shows great value except zones of 120 - 160 m, around 370 m and 670 - 950 m. Vph of 670 - 950 m correspond to the debris flow deposit, and was not measured due to geological condition. Thus, it was almost suggested that the value of Uh, which was obtained by using εh, was able to be used for rough reference

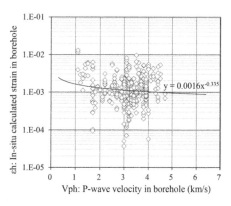

Figure 5. Vph and εh that was determined by using Esh and σck.

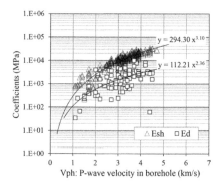

Figure 3. Vph and Ed and that between Vph and Ed.

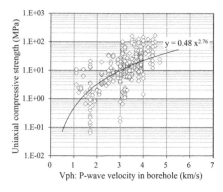

Figure 4. Vph and UTS of core (σck).

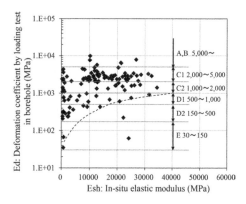

Figure 6. Esh and Ed with range for application of Ed in a ground classification.

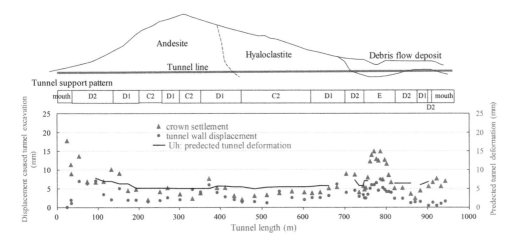

Figure 7. Results of the analysis for the model tunnel.

value of control standards in terms of displacement associated with the excavation of a tunnel.

5 SUMMARY

The followings are obtained in this analysis;The elastic moduli along an advanced borehole for the construction of a national highway tunnel excavated in ground consisting of volcanic and pyroclastic rocks were estimated. The data used in estimating the elastic modulus were the P-wave velocity measured in the velocity logging during the advanced boring and the P- and S-wave velocities, wet density, Poisson's ratio, and unconfined compressive strength of cores measured in the laboratory. These elastic moduli are considered to be values which reflect the continuous condition of the in-site ground along the advanced bore-hole well.The deformation modulus measured in the horizontal loading test conducted in the tunnel and the elastic modulus estimated in this study were compared. The result showed that it is possible to know the rough standard values of the deformation modulus, which were obtained from the horizontal loading test, in classifying the ground for the tunnel construction by using the estimated elastic modulus along the advanced borehole wall.The deformation associated with a tunnel excavation was predicted based on the strain in the borehole of the tunnel construction site estimated from the velocity logging in the borehole. The deformation occurring in the excavation of a tunnel was examined by using a tunnel constructed in ground of volcanic and pyroclastic rock as a model. The examination revealed that when the generated deformation is great, many of the data points tend to be included in the range of the warning value. It was suggested that the prediction of displacement performed in this way is able to be indicated in setting the control standard values for displacement in tunnel excavation.

ACKNOWLEDGEMENTS

We would like to express our gratitude to the departments and individuals of the Hokkaido Regional Development Bureau, Ministry of Land, Infrastructure, Transport and Tourism, Japan for providing us data for this study.

REFERENCES

HRDB; Hokkaido Regional Development Bureau, 2018. Ministry of Land, Infrastructure, Transport and Tourism: Road Design Guidelines, Vol. 4, Tunnels, 13–29. (in Japanese)

Ohkubo, T. & Terasaki, A. 1971. Physical Properties and Elastic Wave Velocity of Rocks, *Tsuchi-to-Kiso*, Geotechnical Engineering Society of Japan, Vol. 19 (7), 31–37. (in Japanese)

Sugino, Y., Kanazumi, K., Ono, K. & Fujimoto, H. 2007. Measurements of Deformation Modulus of Tertiary Mudstone with Different Levels of Strain, Geotechnology E-Forum 2007 Sapporo, No. 59. (in Japanese)

2019 Rock Dynamics Summit– Aydan et al. (eds)

Evaluation of strength of rock mass with fractures and deformation structures using homogenization theory at Tokuyama underground power house

H. Takehata, K. Nishizawa & M. Matsushita
Chubu Electric Power Co., Inc., Nagoya, Aichi, Japan

H. Kobayakawa, S. Tanaka & H. Kuno
Central Research Institute of Electric Power Industry, Abiko, Chiba, Japan

ABSTRACT: In this paper, we have demonstrated that the rock mass mechanical properties of the basalt lava was evaluated by the numerical analysis with a homogenization theory. The basalt lava around Tokuyama underground power house at the upper part of Ibi River in Gifu prefecture, Japan, included cracks and fractures with slickenside and specific deforming structures that would be formed in the process of the plate accretion and faulting. It was feared that the deformation structures would adversely affect the stability of the underground power house. Therefore, we evaluated the mechanical properties of the rock mass including the structure by a method using a numerical analysis based on homogenization theory, and then performed excavation analysis of the underground power house with the calculated mechanical properties and compared the analysis result with the observed displacement around the underground power house. As a result, it was showed that the method was practical for the evaluation of such rock-mass properties.

1 INTRODUCTION

The mechanical properties of the rock mass are affected by the geometric characteristics of discontinuous surfaces and the characteristics of the rocks that constitute them. Therefore, in order to obtain the properties, it is required to conduct in-situ tests and laboratory tests that reflect the mechanical properties for rocks and rock mass, the geometric characteristics and scales of discontinuous. However, it is not always easy to evaluate the physical and mechanical properties of the rock mass with discontinuities and fractures of spacing and persistence difficult to reflect on in-situ tests. In order to solve this problem, it has been recommended to utilize the evaluation method of strength properties using the numerical analysis effectively. According to the previous research, the evaluation method using a homogenization theory has the advantage of being able to evaluate the strength of the rock mass including the cyclical cracks and fractures from the geometric information and the mechanical properties of the constitutional components for the targeted rock mass. However, there are few cases that applied to the evaluation of rock mass to conduct the excavation analysis for a large-scale underground cavern (Kyoya et al. 1999). Therefore, this paper demonstrates that the evaluation method can be applied to the excavation analysis for the underground power

house construction at the Tokuyama hydropower plant in Gifu Prefecture, Japan.

2 GEOLOGICAL STRUCTURE AROUND THE TOKUYAMA HYDROELECTRIC POWER STATION

2.1 Project overview

The Tokuyama hydroelectric power station is in the upper stream of Ibi River which is a tributary of the Kiso River in Gifu prefecture. Chubu Electric Power Co., Inc. has participated as a power producer to the Tokuyama Dam Project and constructed the power station. This is the largest conventional hydroelectric power station except for the pumped storage power stations in the company (Noike 2010, Aoki 2015).

Table 1. the specification of the Tokuyama Hydroelectric Power Station

Item	Unit 1	unit 2
Installed capacity*	139.0 MW	24.3 MW
Max. water discharge**	82.38 m³/s	18.97 m³/s
Effective head	181.96 m	145.71 m

*The total output during simultaneous operation amounts to 161.9 MW
**The total maximum water discharge of Units 1 and 2 amount to 100.4 m³/s

2.2 Site geology

The geology at the power station site has bedrock of Paleozoic/Mesozoic Mino Zone, covered with a Quaternary layer. The bedrock consists of basalt lava, tuff, chert, and slate, and occasionally contains hundreds of meters of the rocks of them. Most of the headrace tunnel and underground power house are laid out in the bedrock of basalt lava, tuff, mélange and chert (Wakita 1988). The faults having the aspect of the boundary between lava and mélange pass through the surge tank for tailrace on the downstream side of the underground cavern for the Unit 1. Several high-angle faults run almost parallel to the boundary fault from the upstream side of the Tokuyama dam right bank to the left bank downside, and they are branching and converging (Nishizawa el al. 2013). Figure 1 shows the geological evolution of the underground cavern of the Unit 1. Some of these faults are identified as faults (F1, F2, F3) and seams (S1) crossing the underground power house.

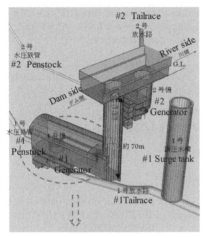

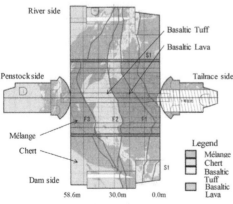

Figure 1. Upper figure showing the layout sketch of the underground structures in the Tokuyama hydropower plant. The lower one showing the developed geological map of the underground power house of the Unit.

2.3 Properties of rock

The cracks and dynamic deformation structures with slickenside have existed on the bedrock of lava and tuff around the underground power house of the Unit 1, as shown in Figure 2. The observation on the outcrop and the boring cores have revealed that the rock becomes easy to exfoliate along the cracks and deformation structures conspicuously as the rock weathers.

The tuff in this site have fluid deformation structures, which consists of laminated structures of a few mm to several tens of millimeters arranging slightly oriented like a cleavage, and in addition to the structures, the rocks have the cracks with slickenside of several 10 cm to several m intervals. Unlike the tuff, the basalt lava does not have the cracks with laminated structures but fractures with slickensides at intervals of several tens of centimeters and some directions. Also, the fractures were observed to be distributed in a mesh form around the place where the cracks and fractures were densely distributed. Since they were distributed around the E-W faults, the fractures would be formed in the process of the plate accretion and faulting.

On the other hand, the laminated deformation structure seems to be a deformed structure formed in a period considerably older than the mesh-like fractures, since its structure found in the rock mass classified as B class on the DENKEN classification system is completely adhered. It was considered that such a fluidized and mesh-like deformed structure is a structure developed in the adduct formation process.

We concerned that the behavior of these fractures with slickenside and specific deforming structures accompanying stress release of excavation of the underground power house would affect the stability of large-scale underground cavern. Especially since the basalt lava are widely distributed around the underground power house of the Unit 1, in this study we calculated its mechanical properties using the homogenization method and evaluated the method.

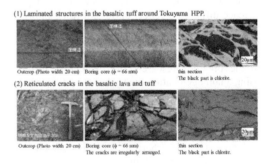

Figure 2. Photographs showing the characteristic deformation structures and cracks found in the basaltic lava and tuff around the Tokuyama hydro power plant. (1) The photographs showing the laminated structures observed in the outcrop, the boring core and the thin section in order from left side. (2) The photographs showing the reticulated cracks observed in the outcrop, the boring core and the thin section.

3 EVALUATION OF ROCK PHYSICAL PROPERTIES USING HOMOGENIZATION METHOD

3.1 Outline of analysis method using a homogenization method

The evaluation of rock physical properties by a homogenization method consists of three steps: (1) creating geometric unit cell model, (2) analyzing 3-dimensional space of the failure criteria, and (3) evaluating rock mass strength by ultimate load analysis. In this evaluation method, it has reported that the deformation characteristics obtained by (2) are accurately evaluated (Kobayakawa et al. 2007, Kobayakawa et al. 2016). However, it has concluded that the strength characteristic utilizing (3) sometimes makes it difficult to express confining stress dependency. Therefore, in this paper, the yield condition of the unit cell evaluated by the numerical calculation of (2) was used as the intensity obtained by the homogenization method. The evaluation was made by preparing a geometric model that matches the plate loading test (PLT) result of the basalt lava in C_H class and calculating the strength with respect to the geometric model.

3.2 Mechanical properties of an intact rock and a crack with slickenside

We describe various test results on the rock mass of the basaltic lava classified as C_H class, and the intact rock and the weak layer that is consisted of by the cracks with the slickenside and its periphery.

The results of the multistage direct box shear tests for the intact rock and the weak layer, the tri-

axial compression tests for intact rock and the rock including the weak layer are shown in Figure 4.

The stress-strain curve and the strain-Young's modulus relation for the results of the uniaxial compression test for the weak layer part, t are shown in Figure 5.

The average of the elastic modulus of intact rock is 44,019 MPa. Respectively in No.1 and No.2 specimen, the Young's modulus increased from 370 MPa to 900 MPa and 60 MPa to 500 MPa as the strain increased. At that time, the strain-secant modulus curve of No.1 was convex, whereas No.2 was convex downward. Also, the average value was 500 MPa.

Along with the increase in strain, the Young's modulus of the weak layer was assumed to approach that of the intact rock due to the interlocking of the rocks, but, under the stress condition of this test, that of the weak layer did not rise to that of the intact rock.

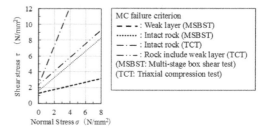

Figure 4. Normal stress vs Shear stress.

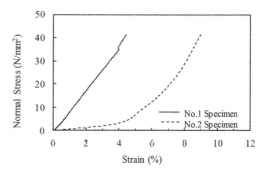

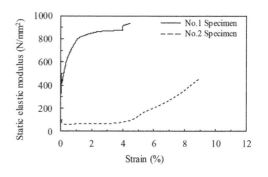

Figure 5. Result of the unconfined compression test for the weak layer. The upper graph showing Strain vs Normal stress, the lower graph showing Strain vs Static elastic modulus.

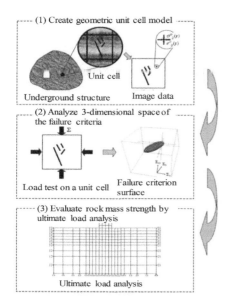

Figure 3. Outline of rock physical property evaluation based on the homogenization theory.

3.3 Modeling targeted rock mass

In a homogenization theory, a cyclic structure is assumed for the microscopic internal structure, and a unit cell is defined as the minimum structural unit. When we create a geometric model to be reflected in a unit cell, we must model the representative volume element of the targeted rock. Therefore, harmonic and cyclic cracks were modeled from the images of the place where we carried out the in-situ tests and the observation results of the surrounding boring core. The procedure to model the unit cell upon the geological survey on the lava around underground power house was shown in Figure 6.

The cracks in the bedrock of the lava around there ran mainly in E-W direction and N-S direction. The N-S cracks were high angles, while the E-W cracks were low to medium angles. From the observation of the loading face of the PLT at the underground power house, approximately two cracks in continuous north-south direction were confirmed in 30 cm square. Also, the E-W cracks, which were separated by the N-S ones, were wider than the interval of the N-S ones. Therefore, the geometric unit cell model was created as shown in Figure 7. As the two-dimensional excavation analysis to the cross-section direction of the underground power house would be carried out, the geometric model used for the analysis to a failure criterion was also prepared for the cross section in the same direction.

3.4 Evaluation of rock mass mechanical properties

3.4.1 Analysis of failure criteria

The linear elastic modulus E_w and the Poisson's ratio v_w of the weak layer are shown as follows:

$$E_w = E_R \frac{d-w}{d} + E_1 \frac{w}{d} \tag{1}$$

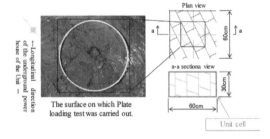

Figure 6. Procedure for creating a geometric model

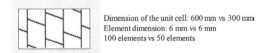

Dimension of the unit cell: 600 mm vs 300 mm
Element dimension: 6 mm vs 6 mm
100 elements vs 50 elements

Figure 7. Unit cell model in the sectional view.

$$v_w = v_j \tag{2}$$

where E_j and v_j are the Young's modulus and Poisson's ratio of the crack with slickenside, E_R is the Young's modulus of the intact rock and d is the length of an element in the unit cell. The crack width is on average 0.22 mm, while the unit cell element size is 6 mm. The equivalent thickness w is set to the crack in the element of the unit cell and the peripheral part. The equivalent thickness is obtained by performing inverse analysis based on the actual PLT results.

For the deformation of the weak layer, we set the case of nonlinear elasticity considering the crack closure. Cracks are modeled by a weak layer model. The weak layer model transposes the crack and the area surrounding it, and the model can describe the decrease in stiffness caused by the crack closure. Specifically, the increase in stiffness due to the crack closure is described by changes in the secant modulus (E^*) as

$$E^* \left(e_V^* \right) = E_0 + E_\alpha \left(1 - \exp\left[-\beta \left[e_V^* \right]^n \right] \right) \tag{3}$$

$$\left[e_V^* \right] = \frac{1}{2} \left(\left| e_V^* \right| - e_V^* \right) = \begin{cases} 0 & (e_V^* > 0) \\ e_V^* & (e_V^* \leq 0) \end{cases} \tag{4}$$

where e_V^* is the volumetric strain of the weak layer related to the crack closure, E_0, n and β are parameters determined based on experimental results.

In this study, we use the physical properties of the intact rock and the weak layers which are the N-S and the E-W cracks in section 3. as shown in Table 2.

3.4.2 Analysis results

The result of the PLTs, which was carried out in three places, and the results of analysis of the weak layer as an elastic body are shown in Table 3.

Table 2. Mechanical properties

Elastic Parameter Value	E_R(MPa)	v_R		E_j(MPa)	v_j	w(mm)
	44,019	0.3		60	0	6.0069

No-linear Parameter	E_0	E_α	n	β
Case1	9.1477	44,060.76	3.6637	63.5
Case2	9.1477	44,060.76	4.988	6,460.00

Table 3. Results on the deformation characteristics

	No.1	No.2	No.3
PLT Result [MPa]	5,265	7,600	8,743
	Vertical	Horizontal	
	E_V	E_H	
Analysis Result [MPa]	7,211	2,030	

The deformation characteristics by numerical tests could well express the increase in elastic modulus due to the increase in strain as shown in Figure 8. In this figure, the strain- indexing Young's modulus relationships for the PLT are plotted for each 0.5 N/mm² from 0 to 4.0 N/mm². As the strain increases, the analysis expresses the phenomenon in which the elastic modulus increases. The analysis expresses the phenomenon in which the elastic modulus in-creases as the strain increases.

The yield stress calculated by the numerical analysis of the unit cell evaluates the value of the element in the unit cell by changing the strain in each direction as the yield stress in that direction. Therefore, the strength characteristic of the unit cell is arranged on the plane of the first strain invariant of the stress J_1 and the second invariant of the deviation stress J_2'. The relation of J_1-$\sqrt{J_2'}$ is shown in Figure 9.

The straight line marked "** DP" in the figure is the Drucker-Prager (DP) yielding criterion calculated from the constants c, ϕ for the Mohr-Coulomb yield criterion. The yield point group of the unit cell is plotted at the position included in the curved surface of the DP criterion obtained from the result of the triaxial compression test to the intact rock of the lava. The slope of the plot obtained by least squares approximation is close to the DP criterion. In the plot of the yield point group, the distribution of the square root J_2' has a width in a place where J_1 is close to zero. Also, the horizontal strength tends to be large, but the vertical one be small. This indicates that this unit cell has anisotropy when the binding force is small. As J_1 becomes larger, the anisotropy tends to be smaller.

In the yielding condition in the analysis, it is judged that the entire unit cell yielded at the time when one element in the unit cell yielded. Therefore, it is considered that the yield point group of the unit cell indicates the value before the peak intensity of the unit cell is reached. Therefore, although ϕ seems to be close to the value of actual rock mass, it is considered that c obtained by the calculation is smaller than the value of the actual rock mass.

4 VERIFICATION BY EXCAVATION ANALYSIS

4.1 Analysis condition

We conducted the excavation analysis on the underground power house using the physical property values obtained by analysis by the homogenization method and evaluated the validity of this physical property evaluation method by comparing the result of the excavation analysis with the rock mass displacement measured during the excavation.

The excavation analysis on the underground power house was carried out by two-dimensional nonlinear elasto-plastic analysis. In the analysis, arch concrete, side wall concrete and PS anchors were modeled as shown in Figure 10. The result of physical property evaluation by homogenization method indicates that the rock which the characteristic cracks with slickenside has anisotropy in regard to deformation characteristics. Therefore, two cases were carried out in which the bedrock was regarded as isotropic materials and the horizontal and vertical elastic modulus were uniformly applied to it in the model since the anisotropic material could not be set in the code of the excavation analysis we used.

Also, the envelop type failure criterion shown in Figure 11. was set for the strength characteristic.

The initial ground pressure shown in Table 5 were used considering the results of the initial ground

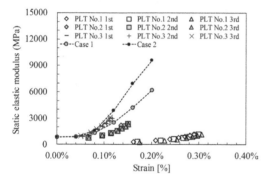

Figure 8. Elastic modulus calculated using homogenization method. Strain-elastic modulus relationship.

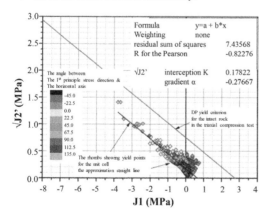

Figure 9. First strain invariant of the stress J_1 versus the second invariant of the deviation stress $\sqrt{J_2'}$.

Table 4. Bedrock physical properties used in the analysis.

Parameter	E [MPa]	c [MPa]	ϕ [°]	γ [MPa]	v
Case1	2,030	0.62	71.0	29.0	0.3
Case2	7,211	0.62	71.0	29.0	0.3

Table 5. Initial ground pressure in the analysis

Item	σ_V [MPa]	σ_H [MPa]	τ_{HV} [MPa]	σ_V/σ_H
Value	4.99	3.12	0.00	1.60

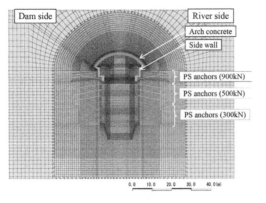

Figure 10. FEM model for the excavation analysis.

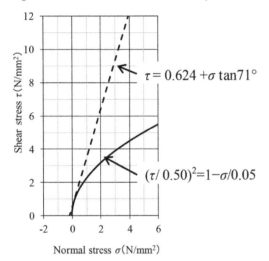

Figure 11. MC Failure criterion and envelop type failure criteria for the excavation analysis.

pressure measurement and the findings obtained from rock behavior during excavation.

4.2 Measurement of rock mass displacement

The rock mass extensometers were installed at TD 15.0 m (section A), TD 30.0 m (section B) and TD 45.0 m (section C) from the No. 1 shaft side toward the penstock side. As shown in Figure 1, the basalt lava is distributed around the power house in section A. The effect of the fault was found in the behavior of the bedrock during excavation on the river side. We compared the results of the excavation analysis and the measurement on the dam side at the section.

4.3 Analysis result and discussion

Figure 12 shows the calculation results and the measurement results for the displacement of the dam

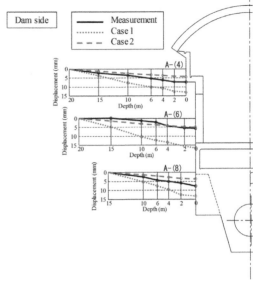

Figure 12. Measurement and analysis results at the section A.

side of the section A at the completion of excavation. The displacement indicates values with the fixed point of the anchor at the deepest part taken as the fixed point. The measurement results show almost continuous distribution from the fixed point on the dam side without the influence of the fault.

The analysis result of case 1 is larger than the measurement result. The analysis result of case 2 is almost the same as or slightly smaller than the measurement result. In other words, the measurement result is distributed between case 1 and case 2 excluding A-6.

If the analysis is performed using the code that can take account of anisotropy, it can be assumed that the result by the analysis would be distributed between the result of case 1 and case 2.

5 CONCLUSION

In this paper, we have evaluated the applicability of the evaluation method of the rock-mass physical properties based on a homogenization method to the actual rock mass.

The excavation analysis on the large-scale underground cavern was carried out using the rock strength properties calculated by the homogenization method and compared the analysis result with the measured displacement of the bedrock. As a result, it was confirmed that the rock mechanical properties obtained by the homogenization method are appropriate. Consequently, it was showed that the evaluation method was practical for the evaluation of such rock-mass properties.

REFERENCES

Aoki, T. 2015. Construction Report in Tokuyama Hydropower Plant. *Electric Power Civil Engineering* 376. 25–28 (in Japanese).

Kobayakawa, H. & Kyoya, T. 2007. Evaluation of Rock Mass Strength by a Homogenization Method with a Weak Layer Constitutive Model. *Journal of Japan Society of Civil Engineers, Ser. C (Geotechnics)* 63, No.2. 428–440 (in Japanese).

Kobayakawa, H. & Takehata, H. 2016. Elasto-plastic Analysis of Lock Shear Test. *Proceeding of 71st Annual Meeting of Japan Society of Civil engineering*, 883–884 (in Japanese)

Kyoya, T., Terada, K. and L. Oyang. 1999. Evaluation of Deformability and Strength of Rock Mass Using Properties of Intact Rocks and Digital Image of Crack Distribution. *JSCE J. Geotechnical and Geoenvironmental Engineering* 48, 131–150 (in Japanese).

Nishizawa, K., Matsushita, M. and Takehata, H. 2013. Design of Large-scale Underground Power House and Intelligence Construction for Excavation in Tokuyama Hydropower Plant. *Electric Power Civil Engineering* 365. 14–18 (in Japanese).

Noike, E. 2010. Development Planning and Technical Examination in Tokuyama Hydropower Plant. *Electric Power Civil Engineering* No.345. 39–42 (in Japanese).

Wakita, K. 1988. Origin of chaotically mixed rock bodies in the Early Jurassic to Early Cretaceous sedimentary complex of the Mino terrane, central Japan. *Bull. Geol. Surv. Japan* vol.39. 675–757 (in Japanese).

2019 Rock Dynamics Summit– Aydan et al. (eds)
© 2019 Taylor & Francis Group, London, ISBN 978-0-367-34783-3

Assessment of rock mass conditions of Ryukyu limestone formation for a rock-cut in Urasoe fault zone (Okinawa) by elastic wave velocity tomography

H.Inoue, K.Hokama

Nanjyou consultants, Haebaru-city, Okinawa, Japan

ABSTRACT: Ryukyu limestone formation is widely distributed over the Ryukyu Archipelago, and it is highly distrubed by the existence of faults resulting from the tectonic movements of the Ryukyu platelet. The faults in the Okinawa Island have generally the normal faulting characteristics. Recently a rock-cut project along the Urasoe fault zone having a NW-SE direction in the Urasoe City was undertaken and rock mass was highly fractured. The authors utilized an elastic wave tomography method to evaluate rock mass conditions besides the boring exploration. In this study, the outcomes of non-destructive elastic wave velocity tomography investigations and geological investigations are presented and they are used in the stability assessment of the rock cut.

1 INTRODUCTION

Ryukyu limestone formation is widely distributed over the Ryukyu Archipelago. However, it is highly distrubed by the existence of faults resulting from the tectonic movements of the Ryukyu platelet bounded by Phillipine Sea plate in the east and Eurosian plate in the west. The faults in the Okinawa Island have generally the normal faulting characteristics. The authors involved with a rock-cut project along the Urasoe fault zone having a NW-SE direction in the Urasoe City. The rock cut was excavated as two benches, with an overall height ranging between 7-9 m. As the site was situated within the Urasoe Fault Zone, rock mass was highly fractured with a size of gravel and several joints were recognized. The geological condition revealed that a highly fractured rock exist between two Ryukyu limestone layers. The elastic wave tomography method was utilized to evaluate rock mass conditions besides the boring exploration. The rock mass classified as highly fractured rock mass, disturbed rock mass and undisturbed rock mass zone. The authors will report the outcomes of non-destructive elastic wave velocity tomography investigations and how it is used in the stability assessment of the rock cut in this paper.

2 REGIONAL GEOLOGY AND TECTONICS

Ryukyu Islands are situated on Ryukyu arc between Kyushu Island and Taiwan. The arc is a rifting fragment of continental crust and it is roughly oriented NE-SW.

According to Kizaki (1986), tectonic evolution since the Neogene is divided into three stages. Stage 1 (late Miocene) is pre-rift sedimentation. Stage 2 (Early Pleistocene) is the initial back-arc rifting. Stage 3 (Holocene) is the back-arc rifting still in

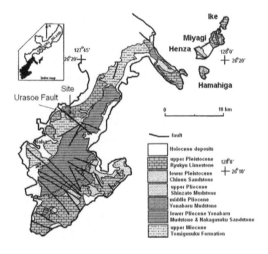

Figure 1. Geology of Southern Okinawa Island and major faults (modified from Tokashiki and Aydan 2013).

progress. The age of the basement is pre-Cenozoic and the basement rocks consist of chert and schists. Cenozoic sandstone, shale and limestone overlay the basement rocks. These rock units are followed by Pliocene Shimajiri formation and all formations are covered with Quaternary Ryukyu limestone and Holocene deposits. Figure 1 shows the geology of the southern part of Okinawa Island.

The formation of Ryukyu arc started in Miocene by rifting a detached block from Euro-Asian plate. This motion is said to be almost southward. While the Philippine Sea plate subducts beneath the rifting Ryukyu arc, the arc is bent between Taiwan and Kyushu-Palau ridge by rotation and rifting and it is fragmented into several blocks as seen in Figure 2. The geological investigations indicated that while the

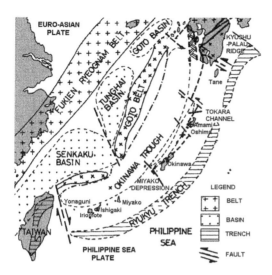

Figure 2. Tectonic features of Ryukyu Islands and their close vicinity (modified after Kizaki (1986))

southern half of the arc rotates clock-wise, its northern part rotates anti-clock-wise (Fabri and Fournier, 1999; Kizaki, 1986). As a result of rifting, rotations and bending of the arc, normal faults, dextral and sinistral faults with or without downward or upward components developed since Miocene. The normal faults are only found at the upper-most part of the crust. The faults can be broadly classified according to their strike as NW-SE and NE-SW faults.

3 LOCAL GEOLOGY

Figure 3 shows the local geology of the site. Ryukyu limestone overlays the Pliocene Shimajiri formation in the close vicinity of the site together with major

Figure 3. Local geology of the site (base map from Japan Geological Survey)

faults. The elevation difference between the south side and north side of the Urasoe fault zone, which has the normal faulting sense with a NW-SE strike, ranges between 50 and 80 meters. Furthermore, the Urasoe fault zone has a small amount of sinistral sense of offset as noted in Figure 3.

The site of rock-cut project is located along the Urasoe fault zone having in the Urasoe City. The layers dip towards SW direction with an inclination ranging between 10 to 25 degrees. The rock mass was highly fractured with a size of gravel and several joints were recognized. The geological condition revealed that a highly fractured rock exist between two Ryukyu limestone layers.

4 BORING AND ROCKMASS CONDITIONS

4.1 Boring

More than 25 boring with different depth were performed at the site. RQD values were determined and some standard penetration tests, which are quite commonly used even in the explorations of rock

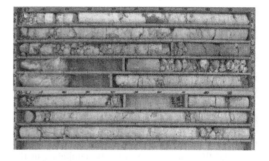

Figure 4. View of cores of a borehole BV1.

Figure 5. Outcrops and fracture in a borehole core.

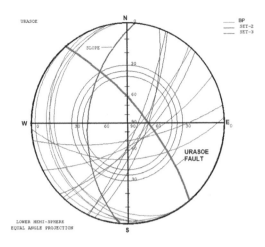

Figure 6. A stereo projection of great-circles of discontinuities.

Table 1. RMQR rating of outcrops.

Parameter	Description	Natural	Description	Disturbed
DD	Fresh	15	slight weathering	11
DSN	1-3	12-16	DSN (3sets+)	4-8
DS	0.3-1.2m	8-12	DS	1-4
DC	rough	15-22	Rough	15-22
GWSC	Wet	5	drip	3
GWAC	Highly-absorptive	5-6	H. Absorptive	4-5
RMQR		60-76	RMQR	38-53

masses. Figure 3 shows the core recovery of the boring numbered BV2 as an example. As seen from the figure, a highly fractured 2m thick zone noted. The fragments of rocks were in the size of 1-4cm. In some cores, fractures with striations and clayey material were also observed.

4.2 Outcrop observations

Discontinuities appearing at outcrops and rock cut were measured using clinometers as seen in Figure 5. The measurements were plotted using a stereographic projections as seen in Figure 6. In this figure, the Urasoe fault zone is illustrated. The dominant discontinuities seems to be closely related to the Urasoe fault zone and the fracture zone may be conjugate to the Urasoe fault zone.

4.3 Rockmass characterization

The observation of the outcrops, cores of a borehole indicated that the rock mass has more than three discontinuity sets (Figure 7). Furthermore, borings indicated some fracture zones containing clayey material. The rock mass at site may be broadly classified into three groups, namely, highly fractured zone, relatively disturbed rock mass and undisturbed rock. The rock mass was classified using the new rock classification system called "Rock Mass Quality Rating - RMQR". Table 1 gives some rating of outcrops for disturbed rock mass and undisturbed rock while Table 2 gives some ratings of highly fractured zone and gouge zone.

4.4 Rock-cut geometry

The rock-cut was designed to be with a total height of about 12m as illustrated in Figure 8. The slope angles was about 55 degrees. There are two benches with a height of 6 m.

Table 2. RMQR rating of outcrops.

Parameter	Description	Highly fractured	Description	Gouge zone
DD	Heavy degradation	3	decomposed	0-1
DSN	Four sets + random	4	crushed	0-1
DS	>0.07	1	>0.1	0
DC	slickensided	7	Thick-fill	1
GWSC	Wet	5	drip	3
GWAC	Highly-absorptive	2	Extremly absorptive	1
RMQR		22	RMQR	4-6

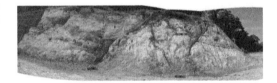

Figure 7. A view of the rock-cut

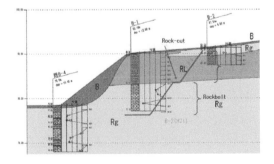

Figure 8. Rock-cut geometry

5 GEO-TOMOGRAPHIC INVESTIGATIONS

5.1 *Procedure*

The method is based on p-wave velocity. Vibration source was hammer and vibrations were measured using geophones (Figure 9). McSeis-SX was used to monitor and process the vibration data. Figure 10 illustrates one of the diagrams between distance and arrival time of p-waves for a given measurement section. Measurements were done along 5 sections.

The authors utilized Simultaneous Iterative Reconstruction Technique (SIRT) method for geo-tomographic investigations (Figure 11). This method utilizes the least square principle. It is thought to be insensitive to the errors of the measurement data. It can be used to reconstruct high quality images from even the inaccurate data with much noise. Furthermore, it is always convergent. Because of these advantages, SIRT is a good algorithm for the reconstruction of geo-tomography.

5.2 *Results*

The measurements along 5 lines were analyzed using the SIRT method. Figure 12 shows a 3D view of the contours of the p-wave velocities along each measurement lines. The velocity of elastic p-waves ranges between 0.4 to 3.4 km/s. The elastic p-wave velocity of intact Ryukyu limestone samples generally ranges between 3.5-4.9 km/s. When this fact taken into account, the condition of rock mass varies depending upon the location. In other words, the rock mass condition is highly influenced by the fracturing state due to faulting. Except the weathered surface layers, the wave velocity is expected to decrease in the fracture zones.

Figures 13 and 14 show the p-wave velocity contour on the rock-cut section. The elastic wave velocity of the highly fractured rock mass was quite low in the range of 0.4-1.8 km/s. The elastic wave velocity of roc mass along the major faults ranged between 0.8 and 1.4 km/s. The elastic wave

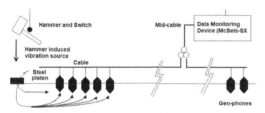

Figure 9. An illustration of the measurement technique.

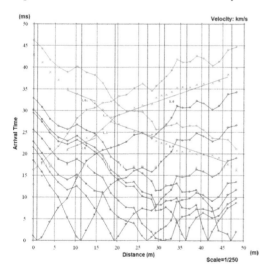

Figure 10. An example of distance - arrival time of p-waves.

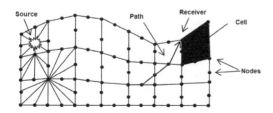

Figure 11. An illustration of the concept of SIRT method.

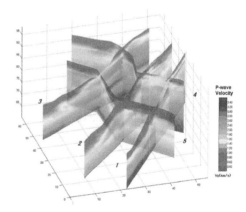

Figure 12. P-wave velocity contours along 5 measurement sections.

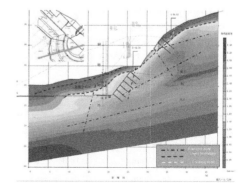

Figure 13. P-wave velocity distribution along the rock-cut section 1.

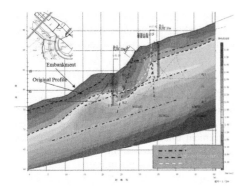

Figure 14. P-wave velocity distribution alon the rock-cut section 3.

velocity of the undisturbed zone was more than 3.2 km/s while the elastic wave.

6 THE UTILIZATION FOR THE EVALUATION ROCKMASS PARAMETERS AND STABILITY ASSESSMENT

6.1 *Interpretation of geo-tomographic investigations*

Geo-tomographic results of the investigations are used in the design of various structures and constitutes one of the fundamental parameters of rock classifications used in Japan (i.e DENKEN, KY-KOKUTETSU, NEXCO (KYU-DOROKODAN); JSEG 1992). The elastic modulus and uniaxial strength of rock mass as a fraction of those of intact rock are generally obtained from the following relations:

$$R = \frac{E_m}{E_i} = \frac{\sigma_{cm}}{\sigma_{ci}} = \left(\frac{V_{pm}}{V_{pi}}\right)^2 \qquad (1)$$

From the geo-tomographic investigations, the rock mass can be classified into four classes as given in Table 3. The reduction factor R is obtained using the average p-wave velocity of Ryukyu limestone as 4.3 km/s.

6.2 *RMQR-based evaluations*

RMQR system is used to estimate rock mass through the utilization of RMQR values presented in Subsection and intact rock properties. The properties of rock mass is obtained from the following relation (Aydan et al. 2014).

Table 3. Rock classes and their p-wave velocities.

Parameter	Vpm	R=(Vpm/Vpi)2
Undisturbed	>3.2	R >0.58
Disturbed	1.8<Vpm<3.2	0.18<R<0.58
Fracture zone	0.8<Vpm<1.8	0.026<R<0.18
Gouge	0.4<Vpm<0.8	0.007<R<0.026

Vpi=4.2

Table 3. Values of parameters used in the Eq. (2)

Property (α)	α_0	α_{100}	β
Deformation modulus	0.0	1.0	6
Poisson's ratio	2.5	1.0	0.3
Uniaxial compressive strength	0.0	1.0	6
Tensile strength	0.0	1.0	6
Cohesion	0.0	1.0	6
Friction angle	0.3	1.0	1.0

Table 4. Rock classes and their p-wave velocities.

Parameter	RMQR	R
Undisturbed	60-76	R >0.2-0.34
Disturbed	38-53	0.13<R<0.16
Fracture zone	22	R<0.045
Gouge	4-6	0.008<R<0.012

$$\alpha = \alpha_0 - (\alpha_0 - \alpha_{100})\frac{RMQR}{RMQR + \beta(100 - RMQR)} \qquad (2)$$

The property reduction factor R was calculated by using the RMQR values given in Table 4 in Eq.(1). Calculation examples are given in Table 4. As noted from the table, the material properties drastically reduced as the rock mass quality decreases.

6.3 *Comparisons and discussions*

Regarding properties of rock mass in terms of those of intact rock, we can compare the values of reduction factor R for each rock class given in Tables 3 and 4. Although there is some discrepancy for the estimations based on two entirely different approaches for undisturbed rock mass, the estimations for disturbed rock mass, fractured rock mass and gouge are quite close to each other. The utilization of s-wave velocity of rock mass, which is generally rarely used in geophysical explorations, it may yield better estimations, which may be close to those from RMQR approach.

7 STABILITY ASSESSMENT OF ROCK-CUT AND COUNTER-MEASURES

There are different techniques to evaluate the stability of the rock-cut having discontinuities. The most simple technique is known as kinematic approach. The projection of great-circles or poles on stereo-nets is often used as shown in Figure 2. In this particular site, the bedding planes may be a part of planar sliding (sliding-1) as shown in Figure 15. However, its inclination is less than the friction angle of bedding planes and it is unlikely to cause planar sliding. Another set may induce sliding (sliding-2) for slopes facing south. As the slopes of rock-cut is westward, it is unlikely to induce such planar sliding. However, there is possibility of flexural or block toppling failure due to eastward-dipping discontinuities aided by bedding planes (Figure 2 and Figure 15). They are critical for the stability of the rock-cut and possible failure modes are illustrated in Figure 16. To prevent any instability rock-cut, block formations at several sections were analyzed with the consideration of the counter-measures and material properties. Figure 17 shows an example of reinforcement suggested for Block iv as an example.

8 CONCLUSIONS

The authors involved with a rock-cut project along the Urasoe fault zone striking a NW-SE direction in the Urasoe City. The rock cut was excavated as

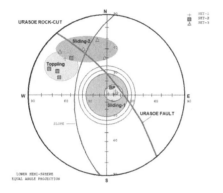

Figure 15. A stereo projection of poles of discontinuities.

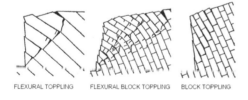

FLEXURAL TOPPLING   FLEXURAL BLOCK TOPPLING   BLOCK TOPPLING

Figure 16. Inferred possible failure modes of the rock-cut (arranged from Aydan et al. 1989; Aydan and Kawamoto 1992, Aydan 2015).

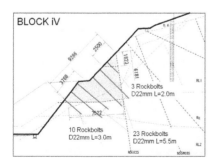

Figure 17. Reinforcement of BLOCK iv against its instability.

two benches, with an overall height ranging between 7-12 m. As the site was situated within the Urasoe Fault Zone, rock mass was highly fractured with a size of gravel and several joints were recognized.

The elastic wave tomography method was utilized to evaluate rock mass conditions besides the boring and outcrop explorations. The elastic wave velocity of the highly fractured rock mass was quite low in the range of 0.4-1.8 km/s. The elastic wave velocity of rock mass along the major faults ranged between 0.8 and 1.4 km/s. The elastic wave velocity of the undisturbed zone was more than 3.2 km/s. The stability of the rock cut was evaluated using the elastic-wave velocity profiles and orientation of major structural discontinuities. On the basis of non-destructive elastic wave velocity tomography investigations and various methods for stability assessment, the stability of rock-cut was investigated. It was found that there may be possibility of toppling failure. Some counter-measures are proposed for stabilizing rock-cut.

ACKNOWLEDGEMENTS

The authors would like to thank Prof. Ö. Aydan of the University of the Ryukyus for editorial help to prepare this paper.

REFERENCES

Aydan, Ö. (2015). Large Rock Slope Failures induced by Recent Earthquakes. Rock Mechanics and Rock Engineering, Special Issue on Deep-seated Landslides. 49(6), pp. 2503–2524.

Aydan, Ö. and Kawamoto, T (1992). The stability of slopes and underground against flexural toppling and their stabilization. Rock Mechanics and Rock Engineering, pp-143–165.

Aydan, Ö., T. Kyoya, Y. Ichikawa, T. Kawamoto and Y Shimizu (1989). A model study on failure modes and mechanisms of slopes in discontinuous rock mass. The 24th Annual Meetings of Soil Mechanics and Foundation Eng. of Japan, Miyazaki, 415, 1089–1093.

Aydan, Ö., Ulusay, R. & Tokashiki, N. 2014. A new rock mass quality rating system: Rock Mass Quality Rating (RMQR) and its application to the estimation of geomechanical characteristics of rock masses. Rock Mech Rock Eng 47: 1255–1276.

Fabbri, O. & Fournier, M. (1999): Extension in the southern Ryukyu arc (Japan): Link with oblique subduction and back arc rifting. Tectonics, 18(3),486–497.

JSEG, 1992. Rock mass classification of the Central Research Institute of Electric Power Industry (CRIEPI). In Japan Society of Engineering Geology. Rock mass classification in Japan:- Engineering Geology Special Issue, 18–22.

Kizaki, K. (1986): Geology and tectonics of the Ryukyu Islands. Tectonophysics, 125, 193–207.

Tokashiki, N. Aydan, Ö. (2013): Rock engineering issues in Ryukyu Islands. 13th Domestic Rock mechanics Symposium of Japan, Kobe.

2019 Rock Dynamics Summit– Aydan et al. (eds)
© 2019 Taylor & Francis Group, London, ISBN 978-0-367-34783-3

Method to determine coal-rock joint conditions by tomography data: Theory and lab test

L.A. Nazarova & L.A. Nazarov
Chinakal Institute of Mining of the Siberian Branch of Russian Academy of Science, Novosibirsk, Russia

P.V. Nikolenko & A.L.Karchevsky
Institute of Comprehensive Exploitation of Mineral Resources of Russian Academy of Science, Moscow, Russia

ABSTRACT: The article describes the results of uniaxial compression tests of rectangular specimens made of geomaterial with a view to investigate coal–rock contact conditions on a lab scale. Based on the acoustic sounding data, tomography was carried out to find spatial distribution of longitudinal wave velocity V in a specimen per stages of loading. Using the empirical relation $V(\sigma)$, the mean normal stress s was calculated and taken as the input data in the solution of an inverse problem for determination of boundary conditions. The model calculations show that under vertical compression, near slip sections of coal–country rock contact, areas of horizontal tension arise, which are the most probable sources of failure and outbursts.

1 INTRODUCTION

Coal cutting process is accompanied by rapid alternation of stresses in rock mass, which occasionally induces dynamic phenomena such as gas outbursts and rock bursts (Seidle 2011, Bondarenko et al. 2014, Mark & Gauna 2016, Mark 2018). With a view to controlling coal–rock mass conditions and predicting such dynamic phenomena, mines are equipped with microseismicity monitoring systems (Luxbacher et al. 2008, Zhenbi & Baiting 2012, Al Heib 2012, NIOSH 2016). At the same time, modern methods of mine seismicity data interpretation fail to detect discontinuities having linear sizes of a few centimeters (Palmer 1987, Henson & Sexton 1991, Cocker et al. 1997, Khukhuudei & Khukhuudei 2014). Such discontinuities include thin contact zones between coal and enclosing rocks, and their acoustic properties differ slightly (Mironov 1988). In the meanwhile, the presence of areas with the decreased friction factor in coal–rock contact zones can be a cause of cleavage fracture of coal face and subsequent coal and gas outbursts (Karchevsky et al. 2017). The specified areas introduce qualitative changes in stress state of coal–rock mass, which can be detected using acoustic methods of mine geophysics (Everett 2013, Telford et al. 1990).

Based on results of physical modeling and geomechanical simulation, this article justifies the method to determine coal–rock contact conditions by solving inverse boundary-value problem using seismic tomography data and an empirical relation of longitudinal wave velocity and mean normal stress established in laboratory environments.

2 EXPERIMENTAL PLANT AND LAB TESTS

The model material was epoxy resin filled with mica; the content of the latter was selected so that acoustic and deformation characteristics of the blend compound were like in coal. The specimens were made as 5 cylinders (diameter 38 mm, height 76 mm) and 8 rectangular prisms (length X=100 mm, height Y=140 mm, thickness 30 mm) with a density of 1600 kg/m³.

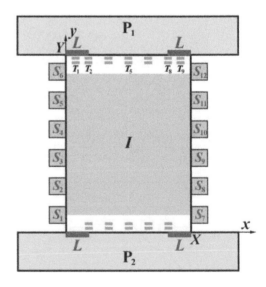

Figure 1. Experimental design and surveying system: (x,y) — Cartesian coordinates, $S_1,\ldots, S_{12}$ — piezometers, I — illuminated domain, $T_1,\ldots,T_9$ — strain sensors.

The cylindrical specimens were subjected to a standard compression test (Ulusay 2015, State Standard 1991). As a result, we determined static Young's modulus E=9.21 GPa and Poisson's ratio n=0.21, which matched up to the same indexes of medium hardness coal (Xie 2015), and calculated the Lamé parameters λ=2.76 GPa, μ=3.81 GPa.

At each step of loading, P-wave velocity V was determined by supersonic sounding. The obtained data were processed using the original method (Nazarova et al. 2016, Nazarova et al. 2017) which yielded an empirical relation:

$$V(\sigma) = A - B \exp(-\alpha\sigma / \sigma_0), \qquad (1)$$

where σ is the mean normal stress, A=3041 m/s, B=671 m/s, α=0.87 and σ_0=3 MPa.

In the main experiments, a prism was placed in the compression testing machine (Instron 300DX, maximum load 300 kN); at the ends of the specimen–plate contact (Fig. 1), special gaskets L (length l) were put to simulate low-friction zones.

On each vertical face of the plate, six piezometers $S_1,...,S_{12}$ (frequency 100 kHz) capable to operate as receivers and emitters were attached.

Figure 2 shows the assembled representation of experimental plant.

Vertical compression of the specimens was carried out at a step of 2 MPa. At each loading stage k, sounding was performed: one of the sensors S_m generated an impulse signal received by 6 sensors on the opposite face, the precision of recording was 0.2 μs. By the first arrival, P-wave travel time between S_m and S_n was determined. The set of t_{kmn} is the input data for tomography.

Two series of experiments were accomplished: travel times t_{kmn} under load W=2k MPa (k=1,2,3) were determined when:

1. the test machine plates P_1 and P_2 were rigidly clamped with the specimen (l=0);
2. the end sections of the specimen–plate contact contained low-friction areas L (Fig. 1) with a length l=25 mm.

The values of t_{kmn} were calculated as an average value of ten sounding events between S_m and S_n in both directions.

For the experimental estimation of shear stresses σ_{xy} at y=0,Y, on each of the frontal and back faces of

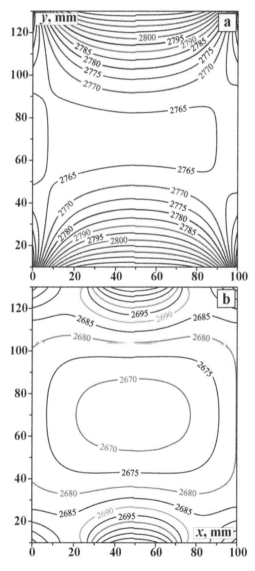

Figure 3. Isolines of P-wave velocity V (m/s) in specimens after tomography-based recovering: W=6 MPa, l=0 (a); W=4 MPa, l=25 mm (b).

Figure 2. General view of test bench.

a specimen, 9 pairs of strain sensors T (Fig. 1) were glued at a distance of 3 mm to each other and 2 mm to the horizontal boundaries. The readings of these sensors and the known value of shear modulus μ were used to calculate $\sigma_{xy}(x,Y)$ and $\sigma_{xy}(x,Y)$.

3 RECOVERING OF LONGITUDINAL WAVES VELOCITY DISTRIBUTION

Tomography of a specimen is implemented using an original iterative procedure (Nazarova et al. 2016), the stability of which is ensured by expansion of intermediate solutions in a specially constructed basis (Woodward et al., 2008). The illuminated domain I for the chosen surveying system $\{S_m\}$ is shown in the Figure 1. Figure 3 presents isolines of P-wave velocities $V(x,y)$ in I under different plate–specimen contact conditions and varied load W.

The primary cause of asymmetry in the patterns of isolines is the chaotic arrangement of filler particles in the aggregate materials of the test specimens; for this reason, for instance, P-wave travel times t_{kmn} and t_{knm} differ though the spacing of the relevant sensors is the same and the loading is symmetrical about medians $x=0.5X$ and $y=0.5Y$.

Figure 4 illustrates the aforesaid by the distribution diagrams of the mean normal stress s in the horizontal section $y=y_0$, calculated from inverse to (1) formula

$$\Sigma(x) = \sigma(x,y_0) = \frac{\sigma_0}{\alpha} \ln \frac{B}{A - V(x,y_0)}$$

by the recovered velocities V (Fig. 3) under different W and contact conditions (dashed lines: $l=0$, solid lines: $l=25$ mm).

In the experiments, the load is set so that a specimen deforms elastically, and, thus, s is proportional to W. The relations $S(x)$ are the input data for the inverse problem considered below in the article.

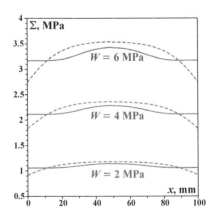

Figure 4. Distribution diagrams of the mean normal stress s in the section $y=0.2Y$ (positive values correspond to compression), dashed lines — no slip; solid lines — slip in the end sections of the contact.

4 DETERMINATION OF SHEAR STRESS AT THE SPESIMEN BOUNDARIES FROM THE SOLUTION OF INVERSE PROBLEM

Stress distribution in the experimental domain $D=\{0 \le x \le X, \, 0 \le y \le Y\}$ (Fig. 1) can be found from the solution of a system, including:
equations of equilibrium

$$\sigma_{ij,j} = 0 \qquad (2)$$

Hooke's law for plane stress state

$$\sigma_{ij} = \lambda_*(\varepsilon_{xx} + \varepsilon_{yy})\delta_{ij} + 2\mu\varepsilon_{ij} \qquad (3)$$

and the Cauchy relations

$$\varepsilon_{ij} = 0.5(u_{i,j} + u_{j,i}) \qquad (4)$$

at the boundary conditions

$$\sigma_{xx}(0,y) = \sigma_{xy}(0,y) = \sigma_{xx}(X,y) = \sigma_{xy}(X,y) = 0,$$
$$\sigma_{yy}(x,0) = P(x), \quad \sigma_{xy}(x,0) = F(x),$$
$$\sigma_{yy}(x,Y) = P(x), \quad \sigma_{xy}(x,Y) = -F(x), \qquad (5)$$

where σ_{ij} and ε_{ij} are the components of the stress and strain tensors $(i,j=x,y)$, u_i are the displacements, δ_{ij} is the Kronecker delta, summing up is carried out over repeated indexes, $\lambda_*=2\lambda\mu/(\lambda+\mu)$. System (2)–(4) is reduced to a biharmonic equation (Timoshenko & Goodier 1987)

$$\Delta\Delta\varphi = 0 \qquad (6)$$

relative to the Airy function φ in terms of which the stress tensor components are expressed

$$\sigma_{xx} = \varphi_{,yy}, \quad \sigma_{yy} = \varphi_{,xx}, \quad \sigma_{xy} = -\varphi_{,xy}. \qquad (7)$$

If the functions P and F are known, the direct problem (5)-(7) is resolvable with an accuracy to a linear function (Timoshenko & Goodier 1987). An analytical solution of the system (5)-(7) was found (Karchevsky 2016) in the form of rapidly converging series. In mining of flat coal beds, the regularities of distribution of vertical stress $P(x)$ (abutment pressure) along the strike are well known (Peng 2006, Baklashov 2004, Zhu et al. 2015).

With regard to the experimental design, we assume that at the boundaries $y=0,Y$ the constant vertical stress W is assigned

$$P(x) = W \qquad (8)$$

and from the tomography data, at the line $y=y_0$ the mean normal stress is known

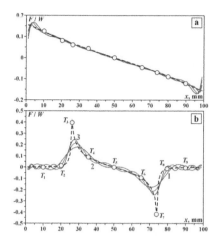

Figure 5. Recovering of the function $F(x)$ at different values of yY without (a) and with (b) low-friction sections.

$$\sigma(x, y_0) = 0.5[\sigma_{xx}(x, y_0) + \sigma_{yy}(x, y_0)] = \sum(x). \quad (9)$$

Since the shear stress (function F) at the horizontal boundaries is unknown, we formulate an inverse problem: find the function F if there is additional information (9) on direct problem solution. The inverse problem (5)–(9) was solved by minimization of residual functional (Karchevsky et al. 2017)

$$J[F] = \int_0^X [\sigma(x, y_0) - \Sigma(x)]^2 \, dx, \quad (10)$$

using the original procedure (Nazarov et al., 2013) with the modified method of conjugate gradients.

Figure 5 demonstrates the inverse problem solution: distribution of the shear stress σ_{xy} (related to W) at the horizontal boundaries without and with the low-friction sections (Figs. 5a and 5b, respectively). The circles show the measured values of σ_{xy}, the dash lines — approximation of the experimental data, curves 1, 2 and 3 are the results of recovering of the shear stress at the boundaries using the additional information (9) at y_0=118, 126 and 134 mm, respectively (Fig. 1).

Thus, it is shown that the tomography data obtained closer to the boundary produce more accurate shear stress at the boundary, and the trend is very exact. When the wanted functions jump, the values of the jumps will be recovered with an error as these values are governed by high-frequency spatial harmonics, which rapidly attenuate with distance from the boundary.

5 STRESS STATE OF SPECIMEN UNDER DIFFERENT CONTACT CONDITIONS

Let us analyze stress distribution in a specimen for different contact conditions at the horizontal boundaries. The calculations are performed in the framework of the model (2)–(5) implemented in the form of original code (Nazarova & Nazarov 2009); the

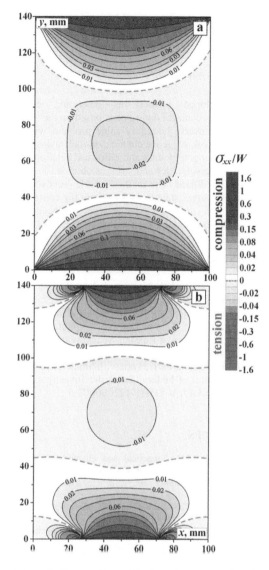

Figure 6. Contour lines of horizontal stress σ_{xx} for slip sections of different size: l=0 (a); l=25 mm (b).

functions $F(x)$ are obtained from the solution of the inverse problem by the input data (9).

Figures 6a and 6b show the contour lines of the stress σ_{xx} (related to W) for l=0 and l=25 mm, respectively. It appears that if the faces of the specimen and the test machine plates (seam and enclosing rocks *in-situ*) are in good cohesion, an extensive zone of horizontal tension exists in the center of D but the magnitude of σ_{xx} is insignificant (of the order of $0.02W$) and below the tensile strength of coal even when W=25–30 MPa (which corresponds to the lithostatic pressure at a depth of 1 km).

In case when the end sections of the horizontal boundaries are the slip zones, vertical compression generates areas of horizontal tension near them, and magnitude of stresses in these areas reaches $0.8W$. These areas are the most probable sources of failure and outbursts.

6 CONCLUSION

The developed method to determine coal–countryrock contact conditions is based on the solution of an inverse boundary-value problem within the framework of geomechanical model, using the seismic tomography data and empirical relations of elastic wave velocities and stresses obtained in laboratory conditions. The method enables recovering of variable stress state in rock mass in the course of mining, and can be included in current procedures for prediction of outburst in coal seams.

ACKNOWLEDGEMENTS

The work of P.V. Nikolenko and A.L. Karchevsky (Sections 2, 3) was supported by the Russian Science Foundation, Project No. 16-17-00029, and performed in Institute of Comprehensive Exploitation of Mineral Resources of Russian Academy of Science. The work of L.A. Nazarova and L.A. Nazarova (Sections 4, 5) was carried out with financial support from Program of Federal Scientific Investigations, Project No. AAAA-A17-117122090002-5.

REFERENCES

Al Heib, M. 2012. Numerical and Geophysical Tools Applied for the Prediction of Mine Induced Seismicity in French Coalmines. *Int. J. of Geosciences* 3(4A): 834–846.

Baklashov, I.V. 2004. *Geomechanics.* Moscow: Gornaya kniga.

Bondarenko, V., Kovalevs'ka, I., Ganushevych, K. 2014. *Progressive Technologies of Coal, Coalbed Methane, and Ores Mining.* CRC Press, Taylor & Francis Group.

Cocker, J., Urosevic, M., Evans, B.A. 1997. High Resolution Seismic Survey to Assist in Mine Planning. In A.G. Gubins (ed.) *Proceedings of Exploration 97: Fourth Decennial International Conference on Mineral Exploration.* 473–476.

Everett, M.E. 2013. *Near-Surface Applied Geophysics.* Cambridge University Press.

Henson, H., Sexton, J.L. 1991. Premine Study of Shallow Coal Seams Using High Resolution Seismic Reflection Methods. *Geophysics* 56(9): 1494–1501.

Karchevsky, A.L. 2016. Calculation of Stresses in a Coal Seam in Presence of Gas Diffusion. *J. of Applied and Industrial Mathematics* 10(4): 482–493.

Karchevsky, A.L. 2017. Determination of the Possibility of Rock Burst in a Coal Seam. *J. of Applied and Industrial Mathematics* 11(4): 527–534.

Karchevsky, A.L., Nazarova, L.A., Zakharov, V.N., Nazarov, L.A. 2017. Stress State Estimation in Coal Bed Under Random Conditions in Contact Zone with Enclosing Rocks Based on Inverse Problem Solution. *Gornyi Zhurnal* 11: 37–40.

Khukhuudei, M., Khukhuudei, U. 2014. High Resolution Seismic Reflection Survey for Coal Mine: Fault Detection. *American Geophysical Union, Fall Meeting 2014.* Abstract NS43A-3856.

Luxbacher, K.D., Westman, E.C., Swanson, P.L. Karafakis, M. 2008. Three-Dimensional Time-Lapse Velocity Tomography of an Underground Longwall Panel. *Int. J. Rock Mech. Min. Sci.* 45(4): 478–485.

Mark, C. 2018. Coal Bursts that Occur During Development: A Rock Mechanics Enigma. *Int. J. Min. Sci. Technol.* 28(1): 35–42.

Mark, C., Gauna, M. 2016. Evaluating the Risk of Coal Bursts in Underground Coal Mines. *Int. J. Min. Sci. Technol.* 26(1): 47–52.

Mironov, K.V. 1988. *Coalman-Geologist's Manual.* Moscow: Nedra.

Nazarov, L.A., Nazarova, L.A., Panov, A.V., Karchevsky, A.L., 2013. Estimation of Stresses and Deformation Properties of Rock Masses which Is Based on the Solution of an Inverse Problem from the Measurement Data of the Free Surface Displacement. *J. of Applied and Industrial Mathematics* 7(2): 234–240.

Nazarova, L.A., Nazarov, L.A. 2009. Dilatancy and the Formation and Evolution of Disintegration Zones in the Vicinity of Heterogeneities in a Rock Mass. *J. of Mining Science* 45(5): 411–419.

Nazarova, L.A., Zakharov, V.N., Shkuratnik, V.L., Nazarov, L.A., Protasov, M.I., Nikolenko, P.V. 2017. Use of Tomography in Stress-Strain Analysis of Coal-Rock Mass by Solving Boundary Inverse Problems. *Procedia Engineering* 191: 1048–1055.

Nazarova, L.A., Nazarov, L.A., Protasov, M.I. 2016. Reconstruction of 3D Stress Field in Coal–Rock Mass by Solving Inverse Problem Using Tomography Data. *J. of Mining Science* 52(4): 623–631.

NIOSH. 2016. *Seismic monitoring strategies for deep longwall coal mines.* P. Swanson, M.S. Boltz, D. Chambers. Spokane WA: U.S. Department of Health and Human Services, Centers for Disease Control and Prevention, National Institute for Occupational Safety and Health, Publication No. 2017-102, RI 9700.

Palmer, D. 1987. High resolution seismic reflection surveys for coal. *Geoexploration* 24(4-5): 397–408.

Peng, S.S. 2006. *Longwall Mining,* 2nd Edition. West Virginia University, Morgantown, W.Va.

Seidle, J. 2011. *Foundations of Coalbed Methane Reservoir Engineering.* Tulsa: PennWell Books.

State Standard of Russian Federation 28985-91. 1991. *Rocks. Method for Determination of Deformation Characteristics under Uniaxial Compression.*

Telford, W.M., Geldart, L.P., Sheriff, R.E. 1990. *Applied Geophysics.* 2nd Edition, Cambridge University Press.

Timoshenko, S.P., Goodier, J.N. 1987. *Theory of Elasticity.* 3rd Edition. New York: McGraw-Hill.

Ulusay, R. (ed.). 2015. *The ISRM Suggested Methods for Rock Characterization, Testing and Monitoring: 2007–2014.*

Woodward, M.J., Nichols, D., Zdraveva, O., Whitfield, P., Johns, T. 2008. A Decade of Tomography. *Geophysics* 73(5): VE5-VE11.

Xie, K.-C. 2015. *Structure and Reactivity of Coal.* London: Springer.

Zhenbi, L., Baiting, Zh. 2012. Microseism monitoring system for coal and gas outburst. *Int. J. Computer Sci. Issues* 9(5): 24–28.

Zhu, S., Feng, Y., Jiang, F. 2015. Determination of Abutment Pressure in Coal Mines with Extremely Thick Alluvium Stratum: A Typical Kind of Rockburst Mines in China. *Rock Mechanics and Rock Engineering* 49(5): 1943–1952.

2019 Rock Dynamics Summit– Aydan et al. (eds)
© 2019 Taylor & Francis Group, London, ISBN 978-0-367-34783-3

The utilization of drones and laser scanning technology in rock engineering

N. Okabe
Okabe Maintenance Co., Naha, Japan

K. Suzuki
Green Earth, NPO, Nishihara, Japan

ABSTRACT: Drones and laser scanning technologies are quite advanced and they are now utilized for assessing the damage and size of collapses. The authors have been developing techniques utilizing the quantification of various collapses induced by the Kumamoto earthquakes and other natural agents. The authors utilized drones and laser scanning technology for the quantifications of landslides, failure of bridge foundations induced by the Kumamoto earthquake. The authors also evaluated the size and geometrical position of tsunami boulders in Okinawa Island and Shimoji Island. In this study, the authors briefly introduce the technology utilizing drones and laser scanning technology. Then several applications of these techniques to actual events observed such as slope failures, faulting induced displacement and tsunami boulders are presented and the applicability of these technologies in Rock Engineering are discussed.

1 INTRODUCTION

The recent earthquakes such as 2016 Kumamoto earthquake, the 2011 Great East Japan earthquake, 1999 Kocaeli earthquake clearly showed that the rock slopes, cliffs bridges and their foundations and tunnels may be quite vulnerable to collapse during earthquakes. Drones and laser scanning technologies are quite advanced and they are now utilized for assessing the damage and size of collapses in many engineering projects. Furthermore, they may be utilized for long-term monitoring of rock engineering structures for maintenance projects.

The authors have been developing techniques utilizing the quantification of various collapses induced by the Kumamoto earthquakes and other natural agents. The 2016 Kumamoto earthquake caused a landslide in the vicinity of Takano village and it was one of causes of the collapse of Great Aso Bridge, the rock foundations of Choyou Bridge. Furthermore, the Tawarayama tunnel was damaged by secondary faulting and slopes failures (Aydan et al. 2018a,b). It is also known that there are many tsunami boulders in islands of Ryukyu Archipelago. The authors utilized drones and laser scanning technology for the quantifications of landslides, failure of bridge foundations induced by the Kumamoto earthquake. The authors also evaluated the size and geometrical position of tsunami boulders in Okinawa Island and Shimoji Island. Furthermore, this technology is used to evaluate some slope failures caused by heavy rainfalls and the topographical situation of cliffs in Miyako Island.

In this study, the authors briefly introduce the technology utilizing drones and laser scanning technology. Then several applications of these techniques to actual events observed such as slope failures, faulting induced displacement and tsunami boulders are presented and the applicability of these technologies in Rock Engineering are discussed.

2 PRINCIPLES OF DRONE TECHNOLOGY

2.1 Drones

Drones are essentially unmanned aerial vehicles (UAV) are equipped with high quality cameras, which can take photos at exact intervals, and gyroscopes, inertial measurement unit (IMU) and controllers to fly smoothly. For perfect fly of a drone, the Inertial Measurement Unit (IMU), gyro stabilization and flight controller technology are essential. Drones generally use three and six axis gyro stabilization technology to provide navigational information to the flight controller. An inertial measurement unit detects the current rate of acceleration using one or more accelerometers. A magnetometer may be used to assist IMU on drones against orientation drift.

Drones may be equipped with a number of sensors such as distance sensors (ultrasonic, laser, lidar), chemical sensors for digital mapping or other purposes. As Lidar, which is an acronym for *l*aser interferometry *d*etection and *r*anging, can penetrate forest canopy, they are widely used for topographical mapping.

2.2 Aerial phogrammetry

Aerial photogrammetry is used in topographical mapping using digital or digitized aerial photos of area with known control points. Aerial photographs were taken from a camera mounted on the bottom of an airplane and later were digitized. These days digital photographs are used together with record of height, position using GPS and/or other positioning sensors. The plane flies over the area to take over-lapping photographs (generally 60% overlapping) over the entire area of interest. When it is used for mapping and measuring the displacements of struc-tures following the earthquakes, three-dimensional coordinates of the common points on pre-post earth-quake photographs were determined. Hamada and Wakamatsu (1998) used this technique to determine the liquefaction induced displacements. This tech-nique is now utilized together with images from drones. However, the fundamental principles remain same.

(a) Front view

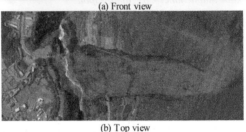

(b) Top view

Figure 2. Digital images obtained from drones using the aerial photogrammetry technique.

2.3 Laser scanning technology

The basic principle is based on the emission of a light signal (Laser) by a transmitter and receiving the return signal by a receiver. The scanner uses dif-ferent techniques for distance calculation that distin-guish the type of instrument in the receiving phase. The distance is computed from the time elapsed between the emission of the laser and the reception of the return signal or phase shift based when the computation is carried out by comparing the phases of the output and return signals. The laser scanner devices operates by rotating a pulsed laser light at high speed and measuring reflected pulses with a sensor. The scanner automatically rotates around its vertical axis and an oscillating mirror moves the beam up and down. The scanner calculate the dis-tance of a measured point together with its angular parameters. The measured points constitutes a set of points called cloud points, which are used to quan-tify the geometry of the structure or surface in 3D.

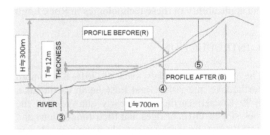

Figure 3. Comparison of longitudinal profiles before and after landslide due to the Kumamoto earthquake.

in surface ruptures due to faulting and induced high strong motions over a large area. The second earthquake was particularly destructive and caused wide-spread damage to rock engineering structures including the built-environment. The causes of the damage were high ground motions and permanent straining, which is one of well-known characteris-tics of intra-plate earthquakes associated with sur-face faulting.

The drone was used first time to estimate the geometry of the landslide body and the volume of landslide body. Fig. 2 shows the digital image of the landslide while Figure 3 compares the longitu-dinal profiles of the landslide area before and after the earthquake. As noted from the figures, it is very easy to evaluate the geometry of the slope failures for evaluating the landslide body.

3 APPLICATIONS TO SLOPES AND CLIFFS

3.1 Application of drone surveying to the slope stability problems and landslide caused by the 2016 Kumamoto earthquakes

Kumamoto prefecture suffered by two successive earthquakes occurred on April 14 and April 16, 2016 (Aydan et al. 2018a). These two earthquakes were associated with well-known faults in the region. While the first earthquake on April 14, 2016 had a moment magnitude of 6.1-6.2 (Mj 6.5), the strong motions at Mashiki town was more than 1500 cm/s². The second earthquake with a moment magnitude of 7.0 (Mj 7.3) occurred on April 16, 2016 resulted

3.2 Application of drone surveying to cliffs and steep slopes

The investigation of the possibility of the fail-ures of slopes and cliffs or the back analyses of

Figure 4. A digital image of the cliffs in the vicinity of the Gushikawa Castle remains in Itoman City.

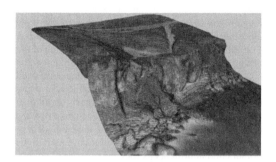

Figure 5. A digital image of the cliffs in the southern shore of Miyako Island.

the failed cliffs and slopes require exact geometry of the topography. Figs. 4 and 5 show the applications in the shore of the Gushikawa in Itoman City in the south of Okinawa island and the shore at the southern part of the Miyako Island. As noted from the figures, it is quite easy the evaluate the digitally the geometry of slopes and cliffs. Particularly the evaluation of cliffs is quite cumbersome due to overhanging of rock mass with toe erosion.

4 APPLICATIONS TO SINKHOLES

The evaluation of the geometry of sinkholes is an extremely difficult and dangerous task due to the unstable configuration and unseen cracks. A sinkhole recently occurred in a Ryukyu limestone quarry in Kumejima Island during the quarrying. The excavator fallen into the sinkhole together with its operator. Luckily no one was hurt. The authors were consulted by the owner of the quarry to carry out the evaluation of the size and geometry of the sinkhole. The drone utilizing the aerial photogrammetry technique was applied at this site and results are shown in Figs. 6 and 7. It is interesting to note that the overhanging part of the sinkholes can be accurately evaluated.

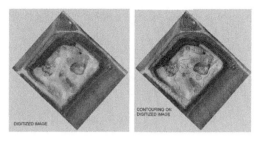

Figure 6. The evaluation of the sinkhole geometry

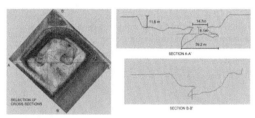

Figure 7. Selection of cross-sections and tracing

5 TSUNAMI BOULDERS

There are many tsunami boulders in major islands of Ryukyu Archipelago (Aydan and Tokashiki 2018). The largest tsunami boulders is probably the one in Shimoji Island near the Shimoji Airport. The quantification of the geometry and position of these boulders are of great importance to assess the past major earthquake and tsunami events in a given region. Both drones based on aerial photogrammetry technique and laser scanning technique were used to evaluate the tsunami boulders in Okinawa Island and Shimoji Island.

5.1 Kasakanja tsunami boulder in Okinawa Island

A drone based on the aerial photogrammetry technique was utilized to evaluate the geometry and the position of the tsunami boulder at Kasakanja of the Okinawa Island. Fig. 8 shows a view of the drone in operation near by the tsunami boulder in Kasakanja. Fig. 9 shows the topography of the investigated area together with projections on a chosen cross-section and in plan. The skill of the operator is also important when the investigations are carried out in areas where overhaning cliffs exist. As noted from Fig. 9, the geometry of the overhaning cliffs can also be acurately evaluated.

5.2 Tsunami boulder in Shimoji Island

As mentioned previously, there is Shimoji Airport with a 4 km long runway near this tsunami boulder.

Figure 8. A view of the operation the drone near to the tsunami boulder at Kasakanja.

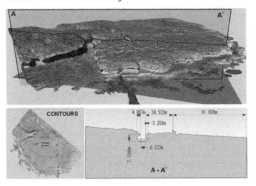

Figure 9. The processed digital topography of the Kasakanja tsunami boulder and its close vicinity.

Figure 10. A view of the tsunami boulder in Shimoji Island.

Figure 11. A digital laser scanned image of the tsunami boulder in Shimoji Island.

As drones could not fly nearby the airports due to restriction, which is automatically imposed on the drones, the laser scanning technique was used. Fig. 10 shows the actual tsunami boulder and Fig. 11 shows the laser scanned image of the boulder from the same angle. Although the laser scanning technique can evaluate the geometry of the tsunami boulder, it is somewhat affected by the existence tress and bushes. In other words, the existence of tress and bushes disturb the digital data for a proper evaluation of the geometry of the tsunami boulders.

6 COLLAPSE OF ASO GREAT BRIDGE

Aso Great Bridge collapsed due to the 2016 Kumamoto earthquake. At this site, a huge scale landslide shown in Fig. 2 occurred and hit the bridge. The authors obtained the original configuration of the bridge and topography of the site. Fig. 12 shows the area with the Aso Great Bridge was superimposed on the digital topography after the earthquake using the drone based aerial photogrammetry technique. Although the failure of the bridge was contemplated to be due to the impact forces of the landslide debris, the failure may also involve the failure of foundations of piers and abutments as rock mass made of columnar jointed andesite. This fact implies that the drone technology yields a better picture of the failure mechanism associated with the collapse of this bridge.

7 MASONRY CASTLES

There are many historical masonry structures in Okinawa Prefecture, Japan. The north-east corner of the Katsuren Castle collapsed during the 2010 Off-Okinawa Island earthquake (Fig. 13). Therefore, there is a great concern about the performance and stability of masonry structures during earthquakes and long-term in Okinawa Prefecture.

The authors have been involved with Katsuren Castle and Nakagusuku tunnel where Aydan et al. (2016) carry some long-term monitoring and strong motion observations are implemented (Figs. 14-16). The authors utilized drone-based aerial photogrammetry technique to observe the current state of Katsuren Castle and Nakagusuku Castle with a particular attention to

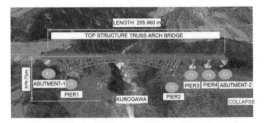

Figure 12. Re-imposed Aso Great Bridge on the digital image of the terrain after the 2016 Kumamoto earthquake.

Figure 13. Collapse and damage to the retaining walls of the Katsuren Castle during the 2010 Off-Okinawa Island earthquake (after Aydan et al. 2016, 2018b)

Figure 14. A 3D digital image of Katsuren Castle.

Figure 15. A 3D digital image of Katsuren Castle at its NE corner, where the continuous monitoring is implemented.

Figure 16. A 3D digitized image of Nakagusuku Castle.

locations where continuous measurements were undertaken by Aydan et al. (2016). These measurements are going to be repeated and compared with those form continuous monitoring results. The repetitions of the measurements using the aerial photogrammetry technique are expected to provide the overall behaviour of the castles in long term three-dimensionally. These type monitoring would be also of the first kind to utilize the drone technology in the world.

8 MAINTENANCE AND MONITORING

Japan has establised some regulations to carry out some compulsary checks on the long-term performance of infra-structures every 5 years period. For this purpose, the authorities or public and private companies and establishments owning the structures have been implementing various techniques to evaluate the state of the structures at every 5 years period. No need to say, such evaluations should be such that they should be independent of techniques employed. The techniques vary from very simple procedures to very sophisticated techniques. In this respect the utilization of the drone-based and/or laser scanning techniques could be of great use. As an application of this concept, the authors have tried to evaluate the performance of a tunnel in Okinawa prefecture. Fig. 17 shows a digital image of the tunnel during the construction phase. As the tunnels have concrete liners at the final stage of construction, it would be quite practical to evaluate the configuration of the tunnel in a 3D digital form and check its geometrical changes every five years period. This type evaluation would provide a quick evaluation of the state of the tunnel and possible locations where some degradation of support systems may occur and some unusual fracturing or deformed configurations of the liners resulting from large deformation or fracturing of the surrounding ground may be assessed. The concept described in the previous structures could be also utilize for the maintenance and long-term deformation monitoring of rock engineering structures such as tunnels, slopes, underground power houses.

Figure 17. A digital image of a tunnel under construction obtained from laser scanners.

9 MONITORING FAULTING INDUCED DEFORMATIONS

Deformation of ground may be induced due to earthquake faulting or creeping faults may also be monitored by the utilization of the drone-based and/or laser scanning techniques in a similar fashion to those described in previous structures. Fig. 18 shows an application of an application of the drone based aerial photogrammetry technique at a site during the Kumamoto earthquake. It is expected that the utilization of the the drone-based and/or laser scanning techniques would be quite usefull for the evaluations of the deformation of ground as well as structures induced by earthquake faulting or permanent movements resulting from ground liquefaction or other causes would be quite effective in years to come (Aydan et al. 2018a). These acievement may also lead to better evaluations of the effects of earthquake faulting as well as permanent ground movements on structures.

10 CONCLUSIONS

The authors explained various examples of the utilization of drones and laser scanners in the field of rock engineering in this study. The recent advances in drone and laser scanner technology together with

(a)

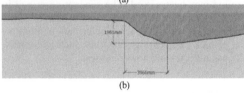

(b)

Figure 18. (a) A view of the utilization of evaluation of ground deformations induced by earthquake faulting at a site in the 2016 Kumamoto Faulting, and (b) measured subsidence.

stabilization technology to provide navigational information to the flight controller have made possible to quantify the geometry of structures and ground surfaces in a short period of time, which could not achieved several years ago. From the examples of variations, the techniques are quite effective to evaluate the effects of earthquake faulting as well as permanent ground movements on structures. Therefore, the role of drone-based and laser techniques in rock engineering would be more important and useful in years to come.

ACKNOWLEDGEMENTS

The authors would like to thank Prof. Ö. Aydan of the University of the Ryukyus for giving the chance to contribute to the 2019 Rock Dynamics Summit in Okinawa and editorial help to prepare this paper.

REFERENCES

Aydan, Ö, 2012, Ground motions and deformations associated with earthquake faulting and their effects on the safety of engineering structures. *Encyclopedia of Sustainability Science and Technology*, Springer, R. Meyers (Ed.), 3233–3253.

Aydan, Ö., 2016. Issues on Rock Dynamics and Future Directions. Keynote. ARMS2016, Bali, 20p, on USB.

Aydan, Ö. 2018. Some Thoughts on the Risk of Natural Disasters in Ryukyu Archipelago. International Journal of Environmental Science and Development, 9(10), 282–289.

Aydan, Ö., Tokashiki, N. (2018). Tsunami Boulders and Their Implications on the Mega Earthquake Potential along Ryukyu Archipelago, Japan. Bulletin of Engineering Geology and Environment. 10.1007/s10064-018-1378-3.

Aydan, Ö., Tokashiki, N., Tomiyama 2016. Development and application of multi-parameter monitoring system for historical masonary structures. 44th Japan Rock Mechanics Symposium, 56–61.

Aydan, Ö., Tomiyama, J., Matsubara, H., Tokashiki, N., Iwata, N. 2018a. Damage to rock engineering structures induced by the 2016 Kumamoto earthquakes. The 3rd Int. Symp on Rock Dynamics, RocDyn3, Trondheim, 6p, on CD.

Aydan, Ö., Nasiry, N.Z., Ohta, Y., Ulusay, R. 2018b. Effects of Earthquake Faulting on Civil Engineering Structures. Journal of Earthquake and Tsunami, 12(4), 1841007 (25 pages).

Hamada M. Wakamatsu K. 1998. A study on ground displacement caused by soil liquefaction. Geotechnical Journal JSCE; 596/III-43: 189–208.

2019 Rock Dynamics Summit– Aydan et al. (eds)
© *2019 Taylor & Francis Group, London, ISBN 978-0-367-34783-3*

Nature and distribution of cavities within the Ma on Shan Marble at Area 90 – Hong Kong

A. Indelicato
Atkins Ltd.

ABSTRACT: Marble is not very common in Hong Kong, it accounts for only 3% of the geological area. Despite this fact, marble and the cavities associated, play an important role within the building strategies of the city, being located in areas crucial for the development of new residential areas in Hong Kong. In order to better understand the issue, boreholes from Area 90 in Ma On Shan have been studied. This paper aims to study cavity distribution, size and location along with the rock profile and to provide an overview of the engineering problems associated with this geology. Ma On Shan Formation can be classified according to the current literature along with the risk that might pose to future civil engineering projects.

1 INTRODUCTION

Karst landscapes are characterized by fluted and pitted rock surfaces, shafts, sinkholes, sinking streams, springs, subsurface drainage systems and caves. The unique features and the three-dimensional nature of karst landscapes are a result of a complex interplay between geology, climate, topography, hydrology and biological factors over long time scales (Stoke T. et al. 2010).

During the Carboniferous, Guangdong Province developed into a platform-flat facies characterized by medium to thick–bedded, lenticular cross-bedded, bioclastic micro-fine crystalline limestone, argillaceous limestone, carbonaceous and calcareous shale and dolomitic limestone sequence.

Carboniferous rocks are present through the north-western part of Hong Kong. The oldest rocks do not occur at outcrop, but have been penetrated in boreholes principally in the Yuen Long and Ma On Shan areas (Sewell R.J. et al. 2000).

From the boreholes, the marble consists of bluish grey to creamy white, dolomite to calcite marble with thin (<10mm) interbeds of dark green to black meta-siltstone and is steeply dipping (70-80°) to the South East (Sewell R.J. 1996 and Sewell R.J. et al. 2000). The Ma On Shan Formation has a minimum thickness of 200m. No fossils have been recovered from the marble and contact formations are not known (Sewell R.J. et al. 2000). Joints, commonly parallel to the foliation planes, are frequently coated with chlorite. Clasts of black siltstone within the marble have been detected in some boreholes together with weak pyritization along foliation planes (Sewell R.J. 1996).

Typical features present within this type of rock are voids and enlarged discontinuities, and the extremely variable karstic surface can be considered potentially major geotechnical constrains in design

of deep pile foundation. The bedrock surface may be variable over short distances, particularly in the pure marble members, which may contain individual pinnacles of rocks up to many meters high (Irfan T.Y. 1995). Weak and compressible clayey soils may occur inside cavities which can cause large differential settlement of individual piles or group piles (Irfan T.Y. 1995). Cavities are also responsible for collapse or subsidence of the ground surface due to upward migration of voids through the soil cover by erosion which is aggravated by changes in the groundwater table due to seasonal variations or dewatering (Irfan T.Y. 1995).

With the expansion of the city and the consequently demands for new high rise residential building in those rural areas, there is an interaction between the foundations and this type of rock which has to be better understood.

2 STUDY AREA

The District is dominated by the densely vegetated, step sided slopes of Ngau Ngak Shan and Ma On Shan, this area of Hong Kong includes also a large reclaimed land known as Ma On Shan reclamation area (Sewell R.J. 1996). The Hong Kong Geological Survey mapped Ma On Shan in detail in 1990 using 28 drillholes (Sewell R.J. 1996). Compared to the Yueng Long Formation, the Ma On Shan Formation has a much better defined skarn zone and it is separated from the granite to its southwest by a fault.

Marble rock probably originated as largely sub-horizontally stratified, variably bioclastic, limestones and interbedded calcareous mudstones, siltstone, sandstone and cherts (Frost D.V. 1991). The bedding features have been commonly modified by recrystallization during metamorphism and tectonic deformation (Fletcher C.J.N. 2010).

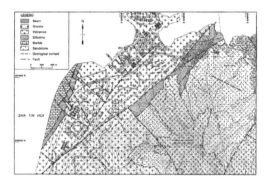

Figure 1. Simplified geology of Ma On Shan (Sewell R.J. 1996)

The Ma On Shan Formation is not exposed in the district and is encountered only in boreholes on the northern and eastern edge of the reclaimed area (Sewell R.J. 1996). It consists of bluish grey to off-white, dolomite to calcite marble with thin inter-beds of dark green to black metasiltstone. The marble displays well developed solution features and is strongly foliated with a steep angle of dip (typically 70-80°). Jointing commonly parallels the foliation planes, which are frequently coated by chlorite. Clasts of black siltstone within the marble have been also detected in some boreholes, together with weak pyritization along foliation planes (Sewell R.J. 1996).

This formation originates predominantly from carbonate and other sedimentary deposits in warm shallow seas (Campbell S.D.G., Sewell R.J. 2004). The location of a pluton on the East side, has been probably the main source of metamorphism for the marble in Ma On Shan (Figure 1).

3 CAVITIES

Marble and limestones are soluble in water with dissolved carbon dioxide as per the formula below:

$$CaCO_3 + CO_2 + H_2O \leftrightarrow Ca^{++} + 2HCO_3^-.$$

In carbonate rocks, a significant part of soil-$CO2$ is dissolved in infiltration water, produces acidic water and dissolves the carbonate rock. However, the concentration of $CO2$ is mainly controlled by plant roots and the respiration of microorganisms as well as organic matter decomposition which depends on climate, season and nature of soils and can vary within a certain range (Jeannin P.-Y. et al. 2016).The amount of water available to dissolve the calcitic rock mass can vary tremendously and it is the main controlling factor in the dissolution process (Chan Y.C. 1989). Flow rates and the water's aggressiveness (degree of chemical saturation)

mainly determine rates of cave enlargement, which originates on bedding planes and tectonic fractures (Lowe D.J. 2000).

Marble rock mass is relatively permeable (Chan Y.C. 1989). Part of the rainfall would infiltrate into rock and travel along joints. The flow volume would depend on joint width and spacing. Hence the flow is larger in fractured rocks or fault zones.

Slightly acidic water exploits any existing joints or fractures in the bedrock, gradually dissolving the bedrock and creating larger openings or conduits for the water to flow through. Over many thousands of years, this process eventually creates underground drainage systems and caves. Mechanical processes such as stream corrosion (abrasion) also come into play once subsurface conduits are of a significant size (Stoke T. et al. 2010).

The dissolution leads to the formation of cavities under the Karst surface where water is free to flow. The Cavities in the marble at Ma On Shan indicate a zone of epikarstic development similar to that at Yuen Long and the karstic top of marble is locally very uneven which has caused significant engineering problems in the development of the new town (Sewell R.J. et al. 2000).

The epikarst (Figure 2) is the zone of solutionally enlarged openings or fractures that extends from the surface down as much as 10-30m below the surface to the underlying endokarst. The epikarst zone therefore plays a critical role in the karst system, allowing water, air and other materials to be readily transferred from surface to the subsurface (Stoke T. et al. 2010).

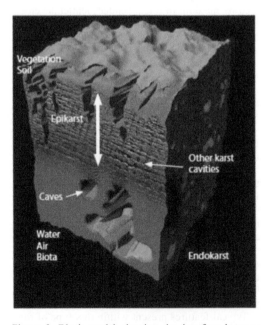

Figure 2. Block model showing the interface between Epikarst and Endokarst (Stoke T. et al. 2010)

In Ma On Shan Sewell has identified three types of cavities within marble (Sewell R.J. 1996):

a) Irregular primary solution cavities generally 1-3m high.
b) Joint related cavities sub-parallel to foliation and slip planes (<300mm wide).
c) Fault zone-related cavities (<30mm).

Joint-related cavities are more common and appear to be formed by movement of water along fracture planes or adjacent to impermeable layers within marble. All voids are interconnected because they were formed through drainage; therefore narrow fissures, wide river passages and large chambers are merely elements of a cave system (Waltham A.C. et al. 2003).

Effectively, irregular primary solution cavities could have been fault or joint related, however the original joint or fissures from which the dissolution started is erased by dissolution so one cannot be sure. Faults are usually more continuous and of wider aperture or shear zones, as such dissolution is more aggressive. This should have resulted in larger cavities, not smaller. For this reason the Sewell classification seems not appropriate for this case.

4 METHODOLOGY

The boreholes from Area 90 have been considered for this study. They have been geographically located and grouped into three areas named north, middle and south. However, only boreholes with marble and cavities have been considered for this research. The dissolution features on cores from marble area were logged with great detail. The geological issue has been approached with an engineering view in order to simplify the geology of the area.

The following assumptions have been made during the data extrapolation and analysis.

1. Marble core between cavities with a length less than 1m and described as marble with solution features, core soluted/fractured as cobble size with a little gravel. This material has been interpreted as cobble/small boulder and considered part of the filling cavities. This has been applied core with a Total core recovery (TCR) less than 85% even within competent marble.

2. Cores with a TCR<85% but still within competent rock (marble above and below with a TCR>85%) have been identified as rock.

3. Cavities less than 0.15 m within competent marble, are considered rather fault related cavities and not included into the data analysis as not part of any paleo-flow channel.

4. Engineering Rock Head (ERH) has been identified where continuous marble core has been identified where no o minimal joints or small

cavities are present and the Total Core Recovery has been higher than 85%.

The data has been extrapolated from the borehole and analyzed through an excel spreadsheet and graphically.

5 DATA ANALYSIS

Of the 77 boreholes analyzed for Area 90, 38 show one or more cavities within the core run. The total cavities analyzed exceed 100 in number. As the original size of the cavities cannot be estimated due to the continuous action of groundwater, we have grouped these features according to their actual size. However a sort of classification has been attempted in order to distinguish the type of cavities presents within the marble formation in Ma On Shan.

Figure 3 shows the distribution of the cavities present within the marble in Ma On Shan. More than 50% of them are large with a width equal or higher than 1m. The largest ones are concentrated close to the rock surface where the groundwater was more aggressive, most of the top marble cores are probably large boulders which have been transported by the action of the river. Among the largest cavities, few of them are connected, with large granitic intrusion probably connected with the faulting event which has generated the fractures within the marble where the plutonic injections passed through and then, later on, solidified. These sheared areas were probably affected by the groundwater on a later stage after the seismic and volcanic activity ceased, as water moved as concentrated flows along marble/non-marble interface as result of low permeability of the latter. Likewise, when water conducting joints meet less permeable dykes, water would collect, concentrating the flow at the dyke interface (Chan Y.C. 1989).

The study of the engineering rock head has been also conducted. The average of the engineering rock head is at -56mPD in the north and south area, in the middle area the average is around -59mPD (Figure 4). However, the amount of borehole available varies across the site, another important factor

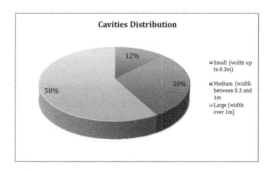

Figure 3. Cavities distribution within the Ma On Shan marble Formation

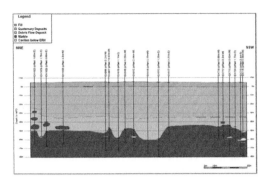

Figure 4. Geological cross-section of the Area 90

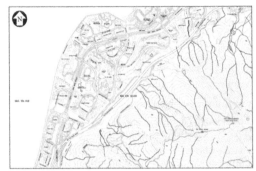

Figure 5. Topography of the area and location of main streams

is related to the standard variation values, 8-10m for the first two areas, 12 for the middle section.

The result of this study shows that whereas rock head relief may be quantified, the distribution of sinkholes and caves is so diverse, chaotic and unpredictable that a classification provides only a broad concepts of their abundance, which is in agreement with the 2003 paper published by Waltham and Fookes.

Previous studies have also highlighted the alignment of the zone of dissolution, they might give an indication of fractured zones or joint planes (Chan Y.C. et al. 1995). The presence of these areas of intersections among planes has facilitated the dissolution within the marble. From the analysis of the boreholes it is possible to identify these fractured zones which tend to dip NNE-SSW, parallel to the fault lines identified during the investigation of Ma On Shan.

6 DISCUSSION & CONCLUSION

In engineering terms, the marble formation is quite challenging and even extensive ground investigation can miss localized feature that might be discovered only during a later stage of construction. The analysis from the marble cavities in Ma On Shan has given an idea of the palaeo-channel surfaces. Considering the proximity of few streams, close to the southern mountain ranges, it is possible that in the past the same rivers that are still visible in the topographic map were cutting through the marble formation (Figure 5). The presence of a consistent thick debris flow deposit across the area might indicate that few of these rivers were flooded by channelized debris flow which occurred in the past, in turn filling the river bed and covering the marble formation.

Karst ground condition can be divided into 5 classes (Waltham A.C. et al. 2003) which provide the basis of an engineering classification that characterizes karst in terms of the complexity and difficulty to be encountered by the foundation engineer.

According to the Waltham and Fookes' classification, Area 90 marble can be interpreted as a class III which is a mature karst. The ground investigation carried out on site has not shown any pinnacles or loose pillars which are typical features of class IV, and normal in tropical regions. The lack of such features can be explained by the fact that offshore carbonate sediments normally have no karst features as most seawater is saturated with calcium carbonate. Caves typically reach width of 10m in karst of class kIV (Waltham A.C. et al. 2003), no such cave has been identified within the study area with the exception of one of 10.4m which was adjacent to a granite intrusion. Even if no typical kIV class features have been identified, those might be visible only with a more extensive ground investigation across the area. Exploration of pinnacles rockhead in a karst may demand extensive probing but there is no right answer about the number of probing needed. At many construction sites, the true ground conditions are discovered only when foundation are excavated.

Therefore the classification of the karstic profile under Area 90 is constricted to this portion of Ma On Shan only.

Civil engineering projects on marble or, more in general, karstic areas have proved to be quite challenging. A city like Hong Kong which faces a continuous expansion combined with scarcity of developing lands will experience an increase in the risks of encountering such difficult ground conditions. In an area like Ma On Shan high rise building have already been built on top of such conditions, however the costs to remedially stabilize karstic zone might become too expensive for any project. Further advancement in geological investigations and geophysical survey might be able to assist better in the identification of marble cavities and to mitigate future risks.

Although geophysical identification of ground voids has not be produced consistently reliable interpretations there is a hope that in the future the advancement on this field could provide the type of information needed before any civil engineering project could start.

ACKNOWLEDGEMENT

The Author wishes to thank Professor C.Y. Chan for his contribution and valuable assistance and Mister David Dobson for the editing work.

REFERENCES

Chan Y.C. 1989. Classification and zoning of marble site. GEO Report No. 29. Geotechnical Engineering Office, Civil Engineering Department, Hong Kong;

Chan Y.C., Pun W.K. 1995. Karst morphology for foundation design. GEO Report No. 32. Geotechnical Engineering Office, Civil Engineering Department, Hong Kong;

Culshaw M.G., Waltham A.C. 1987. Natural and Artificial cavities as ground engineering hazards in Quarterly Journal of Engineering Geology, London Vol. 20, No. 2, pp.139–150;

Fletcher C.J.N. 2010. Geology of Site Investigation Boreholes from Hong Kong. Association of Geotechnical & Geoenvironmental Specialist (Hong Kong), pp.43–44;

Frost D.V. 1991. The geology of Ma On Shan and adjacent areas. British Geological Survey Technical Report WC/91/17;

Irfan T.Y. 1995. Potential construction problems for bridge foundations in areas underlain by a complex marble formation in Hong Kong. Engineering Geology of Construction. Geological Society Engineering Geology Special Publication No.10, pp. 127–134;

Jeannin P.-Y., Hessenauer M., Malard A., Chapuis V. 2016. Impact of global change on karst groundwater mineralization in the Jura Mountains in Science of the Total Environment, 541, pp.1208–1221;

Liao Z.S., Liu X.Y. 1987. Problems of Environmental Geology in the development of Karst Water Resources in Northern China in Geological Society of Hong Kong Bulletin No.3, The Role of Geology in Urban Development, pp. 535–542;

Lowe D.J. 2000. Role of stratigraphic elements in speleogenesis: the speleoinception concept in Klimchouk A.B., Ford D.C., Palmer A.N., Dreybrodt W. (eds) Speleogenesis: Evolution of Karst Aquifers. National Speleological Society, Huntsville, pp.65–76;

Sewell R.J. 1996. Geology of Ma On Shan. Hong Kong Geological Survey Sheet Report No. 5. Geotechnical Engineering Office, Civil Engineering Department, Hong Kong;

Sewell, R.J., Campbell, S.D.G., Fletcher, C.J.N., Lai, K.W. & Kirk, P.A. 2000. The Pre-Quaternary Geology of Hong Kong. Geotechnical Engineering Office, Civil Engineering Department, The Government of the Hong Kong SAR;

Stoke T., Griffiths P., Ramsey C. 2010. Karst Geomorphology, Hydrology and Management in Pike R.G., Redding T.E., Moore R.D., Winkler R.D., Bladon K.D. (eds.) Compendium of Forest Hydrology and Geomorphology in British Columbia, Forest Science program/Forrex: 373–400,

Waltham T., Smart P.L. 1988. Civil Engineering Difficulties in the Karst of China in Quarterly Journal of Engineering Geology, London Vol. 21, No. 1, pp.2–6;

Waltham A.C., Fookes P.G. 2003. Engineering classification of karst ground conditions in Quarterly Journal of Engineering Geology and Hydrogeology, n.36, pp.101–118

2019 Rock Dynamics Summit– Aydan et al. (eds)
© 2019 Taylor & Francis Group, London, ISBN 978-0-367-34783-3

Author index

Printed and bound by CPI Group (UK) Ltd, Croydon, CR0 4YY

17/10/2024

01775694-0006